SOCIOLOGY

IN A CHANGING WORLD

SOCIOLOGY

IN A CHANGING WORLD

FIFTH EDITION

WILLIAM KORNBLUM
City University of New York, Graduate School and University Center

In collaboration with
CAROLYN D. SMITH

Harcourt College Publishers

Fort Worth Philadelphia San Diego New York Austin Orlando San Antonio
Toronto Montreal London Sydney Tokyo

Publisher	Earl McPeek
Acquisitions Editor	Lin Marshall
Market Strategist	Kathleen Sharp
Developmental Editor	Peggy Howell
Project Editor	Michele Tomiak
Art Director	David Day
Production Manager	Andrea Archer

Photo credits/Cover Photographs:
Satellite View from Earth – World Perspectives/© Tony Stone Images
Satellite View of Europe at Night – Earth Imaging/© Tony Stone Images
Football – Phil Cole/© Tony Stone Images
People Looking at Baby – ©Richard Hutching/PhotoEdit
Barn Raising – ©Robert Brenner/PhotoEdit

ISBN: 0-15-507430-X
Library of Congress Catalog Card Number: 99-64303

Address for Domestic Orders
Harcourt College Publishers, 6277 Sea Harbor Drive, Orlando, FL 32887-6777
800-782-4479

Address for International Orders
International Customer Service
Harcourt, Inc., 6277 Sea Harbor Drive, Orlando, FL 32887-6777
407-345-3800
(fax) 407-345-4060
(e-mail) hbintl@harcourt.com

Address for Editorial Correspondence
Harcourt College Publishers, 301 Commerce Street, Suite 3700, Fort Worth, TX 76102

Web Site Address
http://www.harcourtcollege.com

Harcourt College Publishers will provide complimentary supplements or supplement packages to those adopters qualified under our adoption policy. Please contact your sales representative to learn how you qualify. If as an adopter or potential user you receive supplements you do not need, please return them to your sales representative or send them to: Attn: Returns Department, Troy Warehouse, 465 South Lincoln Drive, Troy, MO 63379.

Printed in the United States of America

1 2 3 4 5 6 7 8 9 032 10 9 8 7 6 5

Harcourt College Publishers

SOCIOLOGY IN A CHANGING WORLD

A GUIDE TO LEARNING FROM YOUR TEXTBOOK

SOCIOLOGY IN A CHANGING WORLD, Fifth Edition, provides a complete educational program for students and instructors. In order to offer a thorough introduction to the principles of sociology, *Sociology in a Changing World,* Fifth Edition, has been developed around six essential principles:

- ▼ LEARNING THROUGH VISUAL PRESENTATION: An extensive visual presentation maximizes learning by reflecting contemporary cultural assumptions and tastes.
- ▼ INCORPORATING DIVERSITY: The research and theory of women and minorities is not limited to a single chapter, but is presented in appropriate contexts in all chapters.
- ▼ EXERCISING A GLOBAL PERSPECTIVE: Today, social and economic relations are increasingly global in scope, and this perspective is incorporated throughout the textbook.
- ▼ APPLYING SOCIOLOGY: By applying sociology to everyday life, students' comprehension is made more complete.
- ▼ OBSERVING SOCIAL CHANGE: The text uses visual features to develop social observation skills in studying social change as it occurs among individuals, groups, and societies.
- ▼ PRESENTING ACCESSIBLE SCHOLARSHIP: The scholarship is presented in an accessible style, promoting understanding of the material.

248 PART 2 Social Dynamics

experiences that are shared by all, especially through the mass media. For example, no one knows for sure whether the severe droughts that burned crops in the United States and Canada in 1987 and 1988 were a symptom of global warming, but scientists' warnings about this possibility, and their explanations of the "greenhouse effect" caused by high levels of carbon dioxide and other gases in the atmosphere, were made far more vivid by media coverage of the drought. Public support for environmental protection was stimulated still further by media coverage of the oil spill by the Exxon *Valdez* in Alaska in 1989.

RESEARCH ON THE CUTTING EDGE: The Celebrity Industry

Americans are hooked on fame. Famous people, or *celebrities,* are so much in demand, so much the focus of attention, so widely discussed that everywhere we turn—at the supermarket, on television, in the press—gossip about the rise and fall of celebrities crowds in on us. No surprise, then, that sociologists who study collective behavior and public opinion are seeking to better understand the origins of this explosion in the public's appetite for stories about stars and celebrities.

In a recent study, sociologist Josh Gamson (1994) shows in detail how celebrity is constructed by a thriving "fame industry." But Gamson admits that he became fascinated by fame and celebrities just the way other Americans do, by watching television, often instead of studying. "I became interested in exploring the peculiarities of celebrity culture in graduate school," he says, "through my evening encounters with *Entertainment Tonight,* a television program that can generously be called lightweight." Gamson was amazed to read that *Entertainment Tonight's* anchor, Mary Hart, "whose celebrity derives from a chipper reading of teleprompted words about entertainers," had received a star on the Hollywood Walk of Fame. But trivial as Gamson thought her fame was, he could not help watching the show and thinking about the celebrities. Trained to analyze collective behavior and fads, he realized that he needed "to tap into the weird world of what for many people in this country, including myself, had become the stuff of everyday life" (1994, p. 4).

Gamson's study examines what is done in the entertainment industry to "manufacture" celebrity, and it also looks at the audience for fame by analyzing "how people actively encounter celebrities and interpret and use celebrity images" (p. 5). Like all professional researchers, however, Gamson drew many of his research questions and investigative leads from a thorough study of the historical and sociological literature about fame.

A question that is often raised in the literature on celebrity is this: If people like Mary Hart and other television celebrities are famous, what is that fame in comparison with "the famous of the past, those names that echo in our minds—Alexander, Caesar, Cleopatra, Jesus, Mohammed, Joan of Arc, Shakespeare" (Braudy, 1997)? In the past our heroes tended to be people of great achievement with admirable qualities that set them apart from the rest of us. In premodern societies, however, word of the deeds of heroes spread relatively slowly. It could take weeks for the details of a general's actions in a great battle to spread, even through the elites of a population. With the rise of modern and more democratic societies, equipped with ever more rapid forms of communication, names and faces arise and fall with stunning rapidity. The important differences between actions and image, between truth and fiction, between deserving and undeserving, are blurred in the rush of information and the seductive power of the media.

Gamson found plenty of critical writing about fame in the modern world. Most of this literature emphasizes how visibility becomes its own reward and artificial images replace real-life ones so that people become unable to distinguish between truth and fiction and, more important, are uninterested in doing so. But Gamson found more balanced evaluations of the fame phenomenon in the work of historians and sociologists who wonder how deep the frenzy for fame actually goes in contemporary societies. From sociologist Michael Schudson, for example, Gamson learned to question how seriously consumers take the fleeting fame of many celebrities. There are probably many gradations in consumers' ability to distinguish among celebrities whose fame is purely a matter of television exposure and those, like Mark McGwire or Michael Jordan, who more clearly merit their fame because of their achievements.

These insights led Gamson to narrow the focus of his study to the entertainment industry. "Entertainment,"

New to This Edition

RESEARCH ON THE CUTTING EDGE

Research on the Cutting Edge presents award-winning sociological research that has relevance for our lives.

Features and Learning Aids

The following pages will introduce you to the many features of *Sociology in a Changing World,* Fifth Edition, and show you how to use them to enhance your study of sociology.

SOCIOLOGY VERSUS IDEOLOGY

Sociology Versus Ideology is designed to help students deal with prejudices and to help them go beyond labels and subjective judgments, to separate facts from opinion.

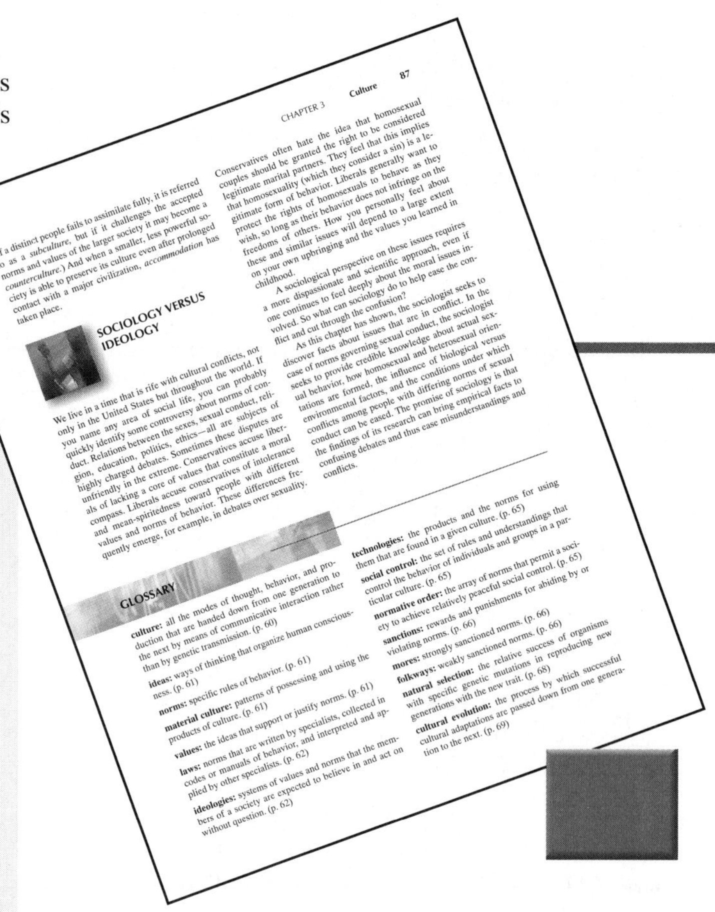

(If a distinct people fails to assimilate fully, it is referred to as a *subculture*, but if it challenges the accepted norms and values of the larger society it may become a *counterculture*.) And when a smaller, less powerful society is able to preserve its culture even after prolonged contact with a major civilization, *accommodation* has taken place.

SOCIOLOGY VERSUS IDEOLOGY

We live in a time that is rife with cultural conflicts, not only in the United States but throughout the world. If you name any area of social life, you can probably quickly identify some controversy about norms of conduct. Relations between the sexes, sexual conduct, religion, education, politics, ethics—all are subjects of highly charged debates. Sometimes these disputes are unfriendly in the extreme. Conservatives accuse liberals of lacking a core of values that constitute a moral compass. Liberals accuse conservatives of intolerance and mean-spiritedness toward people with different values and norms of behavior. These differences frequently emerge, for example, in debates over sexuality.

Conservatives often hate the idea that homosexual couples should be granted the right to be considered legitimate marital partners. They feel that this implies that homosexuality (which they consider a sin) is a legitimate form of behavior. Liberals generally want to protect the rights of homosexuals to behave as they wish, so long as their behavior does not infringe on the freedoms of others. How you personally feel about these and similar issues will depend to a large extent on your own upbringing and the values you learned in childhood.

A sociological perspective on these issues requires a more dispassionate and scientific approach, even if one continues to feel deeply about the moral issues involved. So what can sociology do to help ease the conflict and cut through the confusion?

As this chapter has shown, the sociologist seeks to discover facts about issues that are in conflict. In the case of norms governing sexual conduct, the sociologist seeks to provide credible knowledge about actual sexual behavior, how homosexual and heterosexual orientations are formed, the influence of biological versus environmental factors, and the conditions under which conflicts among people with differing norms of sexual conduct can be eased. The promise of sociology is that the findings of its research can bring empirical facts to confusing debates and thus ease misunderstandings and conflicts.

GLOSSARY

culture: all the modes of thought, behavior, and production that are handed down from one generation to the next by means of communicative interaction rather than by genetic transmission. (p. 60)

ideas: ways of thinking that organize human consciousness. (p. 61)

norms: specific rules of behavior. (p. 61)

material culture: patterns of possessing and using the products of culture. (p. 61)

values: the ideas that support or justify norms. (p. 61)

laws: norms that are written by specialists, collected in codes or manuals of behavior, and interpreted and applied by other specialists. (p. 62)

ideologies: systems of values and norms that the members of a society are expected to believe in and act on without question. (p. 62)

technologies: the products and the norms for using them that are found in a given culture. (p. 65)

social control: the set of rules and understandings that control the behavior of individuals and groups in a particular culture. (p. 65)

normative order: the array of norms that permit a society to achieve relatively peaceful social control. (p. 65)

sanctions: rewards and punishments for abiding by or violating norms. (p. 66)

mores: strongly sanctioned norms. (p. 66)

folkways: weakly sanctioned norms. (p. 66)

natural selection: the relative success of organisms with specific genetic mutations in reproducing new generations with the new trait. (p. 68)

cultural evolution: the process by which successful cultural adaptations are passed down from one generation to the next. (p. 69)

ART AND THE SOCIOLOGICAL EYE

Art and the Sociological Eye shows how the works of past and contemporary artists can be analyzed from a sociological perspective.

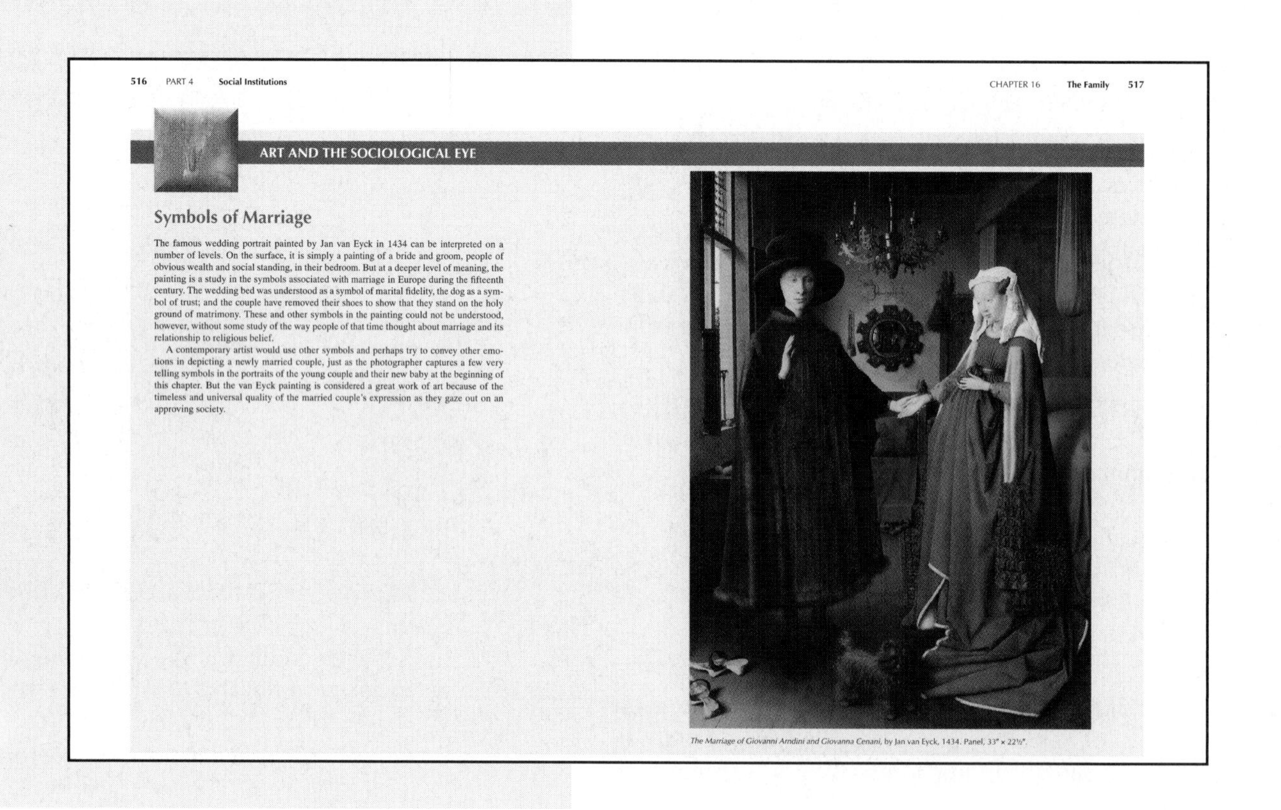

ART AND THE SOCIOLOGICAL EYE

Symbols of Marriage

The famous wedding portrait painted by Jan van Eyck in 1434 can be interpreted on a number of levels. On the surface, it is simply a painting of a bride and groom, people of obvious wealth and social standing, in their bedroom. But at a deeper level of meaning, the painting is a study in the symbols associated with marriage in Europe during the fifteenth century. The wedding bed was understood as a symbol of marital fidelity, the dog as a symbol of trust; and the couple have removed their shoes to show that they stand on the holy ground of matrimony. These and other symbols in the painting could not be understood, however, without some study of the way people of that time thought about marriage and its relationship to religious belief.

A contemporary artist would use other symbols and perhaps try to convey other emotions in depicting a newly married couple, just as the photographer captures a few very telling symbols in the portraits of the young couple and their new baby at the beginning of this chapter. But the van Eyck painting is considered a great work of art because of the timeless and universal quality of the married couple's expression as they gaze out on an approving society.

The Marriage of Giovanni Arnolini and Giovanna Cenani, by Jan van Eyck, 1434. Panel, 33" × 22½".

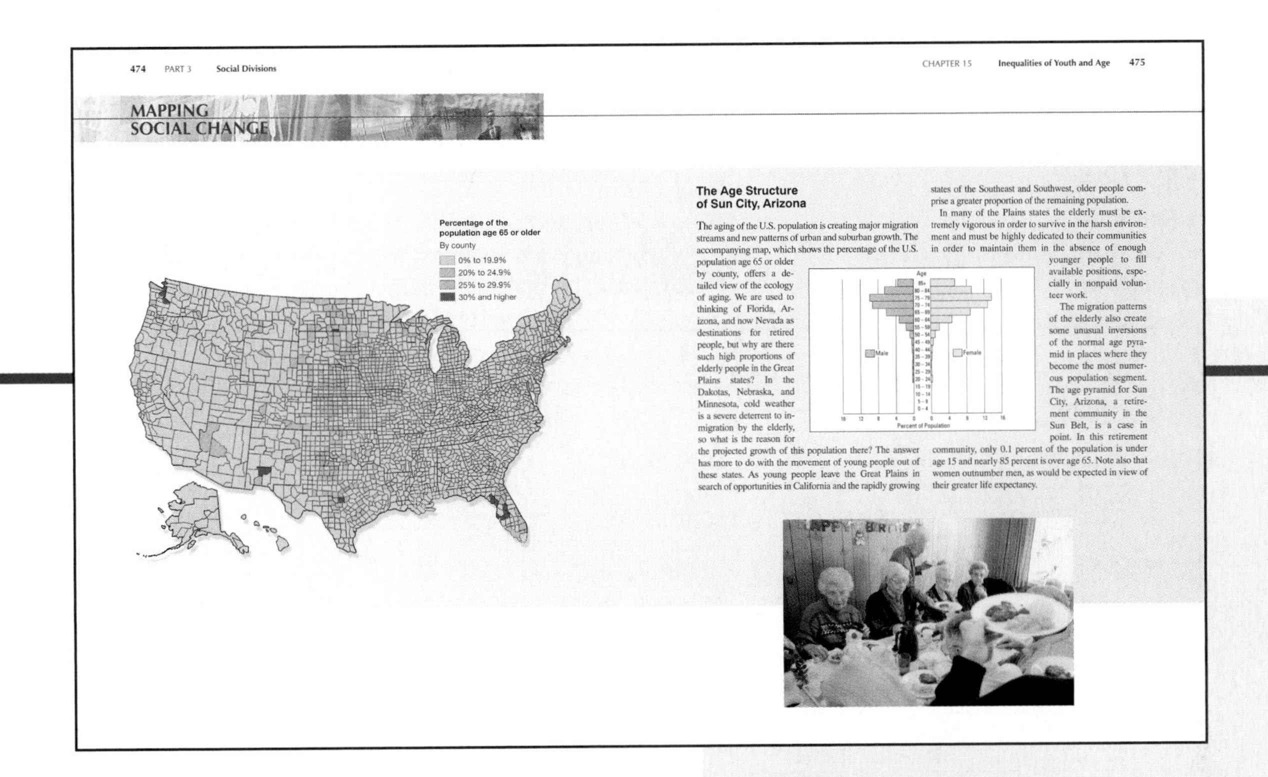

474 PART 3 Social Divisions

CHAPTER 15 Inequalities of Youth and Age 475

MAPPING SOCIAL CHANGE

Percentage of the population age 65 or older
By county
0% to 19.9%
20% to 24.9%
25% to 29.9%
30% and higher

The Age Structure of Sun City, Arizona

The aging of the U.S. population is creating major migration streams and new patterns of urban and suburban growth. The accompanying map, which shows the percentage of the U.S. population age 65 or older by county, offers a detailed view of the ecology of aging. We are used to thinking of Florida, Arizona, and now Nevada as destinations for retired people, but why are there such high proportions of elderly people in the Great Plains states? In the Dakotas, Nebraska, and Minnesota, cold weather is a severe deterrent to in-migration by the elderly, so what is the reason for the projected growth of this population there? The answer has more to do with the movement of young people out of these states. As young people leave the Great Plains in search of opportunities in California and the rapidly growing states of the Southeast and Southwest, older people comprise a greater proportion of the remaining population.

In many of the Plains states the elderly must be extremely vigorous in order to survive in the harsh environment and must be highly dedicated to their communities in order to maintain them in the absence of enough younger people to fill available positions, especially in nonpaid volunteer work.

The migration patterns of the elderly also create some unusual inversions of the normal age pyramid in places where they become the most numerous population segment. The age pyramid for Sun City, Arizona, a retirement community in the Sun Belt, is a case in point. In this retirement community, only 0.1 percent of the population is under age 15 and nearly 85 percent is over age 65. Note also that women outnumber men, as would be expected in view of their greater life expectancy.

MAPPING SOCIAL CHANGE

Mapping Social Change has been expanded to increase knowledge of world geography and understanding of the global effects of social change.

Other distinctive features that are still part of Sociology in a Changing World:

BEGINNING OF CHAPTER ELEMENTS:

Each chapter is introduced with a preview of the chapter in an organized and consistent format.

Chapter Outlines preview the topics that will be discussed in the chapter.

Opening Vignettes capture your attention with high-interest stories and introduce the subject matter of the chapter.

CHAPTER 10

GLOBAL SOCIAL CHANGE

The Nature of Social Change
Dilemmas of Social Change
Two Forces of Social Change
War and Conquest
Modernization
Social Change in Everyday Life
Gender Roles and the Family
Race and Ethnic Relations in a Postindustrial Society
Environmental Politics and Policies
Models of Change
Evolutionary Models
Cyclical Theories
Conflict Models
Functionalist Models
Applying Models of Change
Research on the Cutting Edge: The New American Reality
Visual Sociology: Social Change in Ivory Coast

One sultry African night, as the people of Blokosso were sleeping, a band of thieves crept into their houses and ransacked them. They stole radios, jewelry, and expensive articles of ceremonial clothing. But as the villagers assessed the damage early the next morning, they mourned a far greater loss. Never before in anyone's memory had a robbery like this occurred in the village. It was not the goods themselves they missed, for these could be replaced. It was the loss of a way of life, a social world, that they lamented. "We loved to sleep outside on the very hot nights," said Mr. Joseph, one of the most respected men in the village—"the women by this palm grove in the soft sand along the lagoon, the men under the lamp a bit. We like to talk and sleep with each other. We are not used to sleeping with our things."

Throughout history theirs had been a village society. Blokosso was one of several fishing villages inhabited by the Ebrié tribe. For as long as they could remember, the Ebrié people had made their living by fishing in the quiet waters of the lagoon, which was protected from the huge ocean swells by a long barrier island. But in recent decades the village had been engulfed by the rapidly growing city of Abidjan, the capital of Ivory Coast and a booming port. More and more of the tribesmen were now working in the city. The women continued to tend their garden plots, but they too were increasingly drawn into the economy of the city and felt an ever-increasing need for spending money, which they earned by renting rooms to migrant workers and selling their extra produce.

Gradually, in the time before the robbery occurred, Blokosso had ceased to be a self-sustaining village with its own economy and tribal culture. Instead, it was becoming a "village enclave," a little island of cultural homogeneity in an urban society. The robbery was a blunt statement to the villagers about how much change had occurred in a short time and how much more was likely to come.

For me, the experience of living in Blokosso at that time (during 1962 and 1963) was as much of a revelation as the robbery was to the villagers.

I had been trained as a biologist and was living in Blokosso while teaching physics and chemistry in a junior college. During the day I delighted in opening up to students the secrets of matter and energy. In the evenings, however, the villagers taught me about living in a totally different social world, one that was experiencing even more rapid social change than my own.

Mr. Joseph, my best friend in the village, was a middle-level executive in one of Abidjan's insurance agencies. He commuted across the lagoon in a motorized dugout canoe. During the day his mind was entirely devoted to business. In the evenings he dealt with the problems of a traditional African family. His three wives became increasingly jealous of one another as new wealth and their rising expectations led to new opportunities for better health and more leisure and also to new sources of conflict. I watched him and his family deal with the impact of change on their lives. I saw the villagers attempt to invoke their ancestral spirits and to use witchcraft to cope with new phenomena like robberies. The entire colony had become independent from France only 2 years before, and I saw the villagers—now citizens of a new nation called Ivory Coast—becoming interested in politics and current events.

During that time my own interests shifted from the physical to the social sciences. Nothing has ever interested me more than the villagers' questions about what social change brought to them and what it caused them to lose. When I began to study sociology, I discovered that such questions are at the core of this young science.

END OF CHAPTER ELEMENTS:

Glossary: This is an alphabetically arranged list of all boldfaced key terms that appear in the chapter, with definitions.

Chapter Summaries: For quick review, this section condenses the main points of the contents in the chapter and provides key terms in italics.

Sociology Versus Ideology: This is designed to help students deal with prejudices and go beyond labels and subjective judgments.

Where to Find It: Dr. Kornblum provides a collection of suggested readings, with a brief synopsis of each book, journal, and Internet source.

SUMMARY

A *society* is a population of people or other social animals that is organized in a cooperative manner to carry out the major functions of life. *Social structure* refers to the recurring patterns of behavior that create relationships among individuals and groups within a society. The building blocks of human societies are *groups*—collections of people who interact on the basis of shared expectations regarding one another's behavior. In every group there are socially defined positions known as *statuses*. The way a society defines how an individual is to behave in a particular status is called a *role*.

A social *institution* is a more or less stable structure of statuses and roles devoted to meeting the basic needs of people in a society. Within any given institution there are norms that specify how people in various statuses are to perform their roles. New institutions continually emerge through the process of *differentiation*.

The growth of the world's population is directly related to the evolution of human social structures, which in turn is related to changes in production technologies. The first million years of human social evolution were characterized by a hunting-and-gathering way of life. During that time the family and other kinship structures evolved, and the band became the basic territorial unit of human society. The shift to agriculture is commonly linked with the invention of the plow, but for centuries human societies had been acquiring food through the domestication of plants and animals. Some became *pastoral societies* based on the herding of animals, while others evolved into *horticultural societies* based on the raising of seed crops. However, the first large-scale agrarian societies evolved after the development of plow-and-harvest agriculture.

Agrarian societies allow people to escape from dependence on food sources over which they have no control. In such societies, people produce surpluses that can be used to feed new classes of non-food-producers such as warriors. At the same time, these societies require increasing amounts of land, and this may lead to conflicts over territory. The need to store and defend food supplies and to house non-food-producers results in the growth of villages and small cities.

The next major change in human production technologies was the industrial revolution, the shift from agriculture to trade and industry. This began in England around 1650 and spread to the United States and other nations during the next two centuries. Its impetus came not only from technological advances but also from the rise of a new social order: *capitalism*.

The shift to industrial production affects social structure in several major ways. As a consequence of the industrialization of agriculture, relatively few people live on the land, whereas increasing numbers live in cities and suburbs. Greater openness to change results in the emergence of new classes and social movements. Scientific and technical advances produce tremendous wealth, and the world "shrinks" as a result of innovations in transportation and communication.

For the individual member of a human society, adaptation to a more modern society involves a shift from *gemeinschaft* (close, personal relationships) to *gesellschaft* (well-organized but impersonal relationships). *Primary groups* like the family are supplemented, if not replaced, by *secondary groups* (organizations or associations) whose members do not have strong feelings for one another. Roles in secondary groups often conflict with roles in primary groups, a situation known as *role conflict. Role strain* occurs when a person experiences conflicting demands within a single role.

Another difference between simpler and more advanced societies is that in the former almost all statuses are *ascribed* (determined by birth or tradition), whereas in the latter there is a tendency to replace ascribed statuses with ones that are *achieved* (determined by a person's own efforts). Sometimes a particular status takes precedence over all of an individual's other statuses; such a status is referred to as a *master status*.

When they think of a society, most people in the world today think in terms of the nation-state, or nation, of which they are members. (A *state* is a society's set of political structures, and a *nation-state* is the territory within which those structures operate.) But although the members of a society often think of themselves as members of a particular nation, this is not always so, and in extreme cases the lack of a clear match between society and nation can result in a civil war.

SOCIOLOGY VERSUS IDEOLOGY

Many people mourn the loss of close-knit communities in which people were said to care for each other and help each other over difficult times. Frequently, but not always, they consider themselves conservatives and wish to recapture lost values of community solidarity. They may identify large metropolitan regions and their institutions (especially universities, major corporations, and government agencies) as the problem. With their emphasis on efficiency, profit, science, and secular values, these institutions are often blamed for their role in weakening traditional communities and the families within them. People on the more liberal side of the ideological continuum often feel that small, cohesive communities tend to stifle creativity and individual expression. They may yearn to leave the small town or community for what they perceive as the greater freedom of the city or metropolitan region. And while we may associate this familiar conflict with our own society, in fact we find similar conflicts throughout the world.

A sociological understanding of societies and social change can help resolve the conflict. Sociologists study the historical social forces that disrupt communities by changing the way people earn their livelihood or altering the way they travel or communicate. They study how the growing interdependencies of societies and communities may weaken our attachments to one another at the local level as we become more mobile. Does this sociological understanding mean that sociologists approve of all the changes that are occurring? Does it mean that they condone the weakening of local communities? Not at all. But in order to address the problems of communities and help those who wish to strengthen interpersonal attachments, sociologists first seek to know which social forces will work against these efforts and which ones will be of assistance. In fact, a sociological understanding of one's community and how it is situated in the global environment can actually help the community survive. For example, much sociological assessment goes into a community's efforts to replace declining industries like mining or heavy manufacturing by competing for new industries. In the contemporary global environment, one of the best ways to preserve communities is to strengthen the local institutions, such as community colleges, that help people adapt to change even as they assert their desire to maintain traditional values.

GLOSSARY

society: a population that is organized in a cooperative manner to carry out the major functions of life. (p. 92)

social structure: the recurring patterns of behavior that people create through their interactions, their exchange of information, and their relationships. (p. 93)

group: any collection of people who interact on the basis of shared expectations regarding one another's behavior. (p. 93)

status: a socially defined position in a group. (p. 93)

role: the way a society defines how an individual is to behave in a particular status. (p. 95)

role expectations: a society's expectations about how a role should be performed, together with the individual's perceptions of what is required in performing that role. (p. 95)

institution: a more or less stable structure of statuses and roles devoted to meeting the basic needs of people in a society. (p. 97)

differentiation: the processes whereby sets of social activities performed by one social institution are divided among different institutions. (p. 98)

pastoral society: a society whose primary means of subsistence is herding animals and moving with them over a wide expanse of grazing land. (p. 101)

horticultural society: a society whose primary means of subsistence is raising crops, which it plants ... vates, often developing an exte... ing the crops. (p. 102)

social ...

THEN AND NOW: Visual comparisons present historical perspectives on social change and encourage the student to utilize sociological concepts. A discussion of how social change has affected a particular social condition accompanies the images.

USING THE SOCIOLOGICAL IMAGINATION: This popular feature presents examples of how sociologists apply their imaginations to understanding the changes in society.

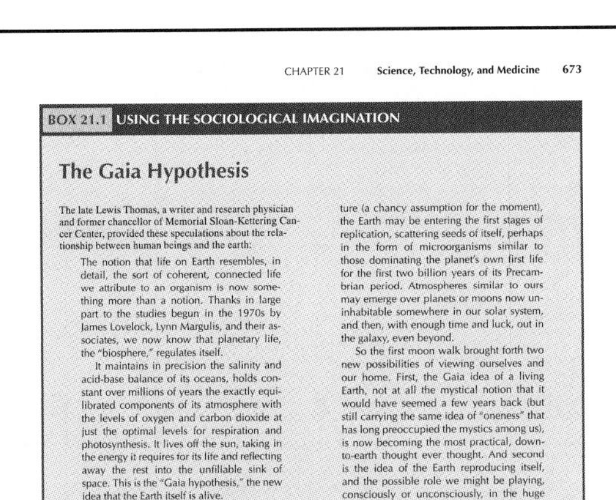

BOX 21.1 USING THE SOCIOLOGICAL IMAGINATION

The Gaia Hypothesis

The late Lewis Thomas, a writer and research physician and former chancellor of Memorial Sloan-Kettering Cancer Center, provided these speculations about the relationship between human beings and the earth:

> The notion that life on Earth resembles, in detail, the sort of coherent, connected life we attribute to an organism is now something more than a notion. Thanks in large part to the studies begun in the 1970s by James Lovelock, Lynn Margulis, and their associates, we now know that planetary life, the "biosphere," regulates itself.
>
> It maintains in precision the salinity and acid-base balance of its oceans, holds constant over millions of years the exactly equilibrated components of its atmosphere with the levels of oxygen and carbon dioxide at just the optimal levels for respiration and photosynthesis. It lives off the sun, taking in the energy it requires for its life and reflecting away the rest into the unfillable sink of space. This is the "Gaia hypothesis," the new idea that the Earth itself is alive.
>
> The one biological function the Earth does not yet perform to qualify for the formal definition of an organism is reproduction. But wait around, and keep an eye on it. In real life, this may turn out to be what it started to do twenty years ago [on the occasion of the first moon walk].
>
> Given enough time, and given our long survival as working parts of the great creature (a chancy assumption for the moment), the Earth may be entering the first stages of replication, scattering seeds of itself, perhaps in the form of microorganisms similar to those dominating the planet's own first life for the first two billion years of its Precambrian period. Atmospheres similar to ours may emerge over planets or moons now uninhabitable somewhere in our solar system, and then, with enough time and luck, out in the galaxy, even beyond.
>
> So the first moon walk brought forth two new possibilities of viewing ourselves and our home. First, the Gaia idea of a living Earth, not at all the mystical notion that it would have seemed a few years back (but still carrying the same idea of "oneness" that has long preoccupied the mystics among us), is now becoming the most practical, down-to-earth thought ever thought. And second is the idea of the Earth reproducing itself, and the possible role we might be playing, consciously or unconsciously, in the huge process.
>
> Finally, as something to think about, there is the strangest of all paradoxes: the notion that an organism so immense and complex, with so many interconnected and communicating central nervous systems at work, from crickets and fireflies to philosophers, should be itself mindless. I cannot believe it. (1989, p. 25)

in and exploit different habitats. Indeed, in the p
centuries it has changed the face of the earth (N
1996). Airports, skyscrapers, housing developme
refineries, superhighways, the Panama Canal, the
launching pad at Cape Canaveral are all prod
technology that have transformed the surround
which human beings live and work. It is this in
tionship between science and technology that is
sible for the breathtaking speed of technological
in our century. (Box 21.1 introduces the Gaia h
sis, the idea that humans and their cultures an
nologies are also part of the living earth and m
accordingly.)

The benefits of technological change are
ways distributed equally. Quite often, in fact, te
gies allow people in wealthier nations to p

GLOBAL CHANGE AND U.S. SOCIETY: Extensive examples of important worldwide changes reinforce the textbook's theme of social change and its enhanced commitment to advancing a global perspective.

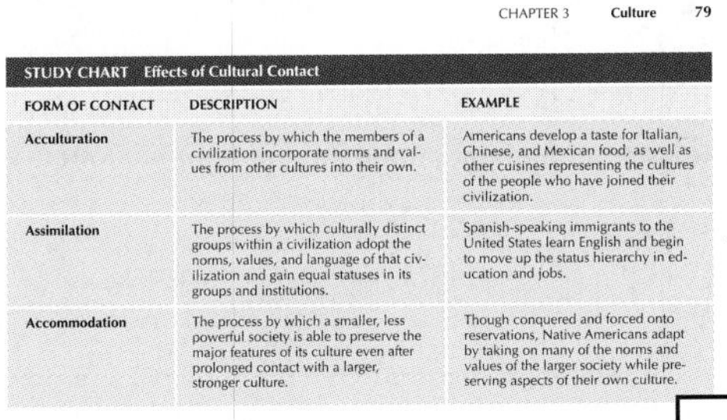

STUDY CHART Effects of Cultural Contact

FORM OF CONTACT	DESCRIPTION	EXAMPLE
Acculturation	The process by which the members of a civilization incorporate norms and values from other cultures into their own.	Americans develop a taste for Italian, Chinese, and Mexican food, as well as other cuisines representing the cultures of the people who have joined their civilization.
Assimilation	The process by which culturally distinct groups within a civilization adopt the norms, values, and language of that civilization and gain equal statuses in its groups and institutions.	Spanish-speaking immigrants to the United States learn English and begin to move up the status hierarchy in education and jobs.
Accommodation	The process by which a smaller, less powerful society is able to preserve the major features of its culture even after prolonged contact with a larger, stronger culture.	Though conquered and forced onto reservations, Native Americans adapt by taking on many of the norms and values of the larger society while preserving aspects of their own culture.

BOX 3.2 GLOBAL CHANGE AND U.S. SOCIETY

North America's African Heritage

All the important streams of migration to North America have combined to produce the culture of the United States and Canada. Pizza (Italy), hamburgers (Germany), French fries—examples of the influence of immigrant groups on American material culture abound. Of all these cultural influences, perhaps the least well understood is the legacy of the people who were brought to North America from Africa as slaves, especially during the last three decades of the eighteenth century and the first half of the nineteenth.

One of the largest groups of slaves, captured in what is now the Congo and Angola, was made up of people from many tribes who shared a common language system, Bantu. Many of these Bantu-speaking slaves were landed in the Carolinas. By the end of the Revolutionary War, the Carolinas had the largest concentration of people of African descent of any of the new states. In consequence, many of the names of towns and villages, rivers, streams, and other geographic features in those states are derived from Bantu words, often mixed with local Indian terms and transformed through long usage. Thus Pinder Town in South Carolina is derived from the Bantu word for peanut, *mpinda*. George R. Stewart, an expert on place names, concludes that hundreds of small streams, swamps, and villages in the Carolinas, Mississippi, and Georgia received their names from the African slaves (Stewart, 1968; Vass, 1979).

The strongest evidence of African influence on contemporary American culture can be found in the Sea Islands along the Atlantic coast between Charlestown, South Carolina, and Savannah, Georgia. On the islands of Port Royal, Parris (famous for its Marine training base), Ladies, Hilton Head, and St. Helena reside the Gullah people. Also known as Geechi, they are descendants of slaves whose original tribes are believed to hav
lived in Angola (from which the term *Gullah* may be de
rived) and in the areas that are now Liberia and Sierra Leone. On the plantations the Gullah slaves grew indig
and the long-fiber cotton that made the region famous i
the period before the Civil War. Some of them escaped
become leaders of slave rebellions. Others were arreste
during the Civil War for singing spirituals like "Roll, Jo
dan, Roll" to which they had added verses about their dreams of freedom. To this day the religious practices, language, and place names of the Sea Island people bea
witness to their African heritage (Creel, 1988).

A Gullah woman of the Sea Islands in a photograph taken
early in this century.

BOX 5.2 SOCIOLOGICAL METHODS

Measures of Central Tendency

The following table, taken from Lewis and Feiring's study of American families, presents information about the social network of the typical 3-year-old:

The 3-Year-Old Child's Social Network (N = 117)*

	X	Range
Number of relatives other than parents seen at least once a week	3.20	0–15
Number of child's friends seen at least once a week	4.43	0–13
Number of adults seen at least once a week	4.38	0–24

* N refers to the number of subjects in a study.

Source: Lewis & Feiring, 1982, p. 116.

To understand this table you must know something about the common measures of central tendency. The measure used in the table above is the *mean*, represented by X. To arrive at this number each child's total number of friends, relatives, and other adults seen at least once a week is added to the totals for the other children in the sample. The grand total is then divided by the number of children in the sample. The resulting number can be used to represent the entire sample.

The other common measures of central tendency are the *median* and the *mode*. The median is the number that divides a sample into two equal halves when all the numbers in the sample are arranged from lowest to highest. The mode is simply the score that occurs most frequently in the sample. Sociologists make a point of specifying which of these measures they are using. They try to avoid the term *average* in statistical tables, as it can refer to any one of these measures.

To illustrate all three measures of central tendency, let us use the following data from 10 of the children in Lewis and Feiring's sample:

Child	Number of Friends Seen at Least Once a Week
1	2
2	3
3	3
4	7
5	12
6	2
7	3
8	1
9	0
10	5
	38

In this example the mode would be 3 and the median would also be 3, but the mean would be 3.8.

degree to which this is true depends on how much social change is experienced from one generation to the next. Socialization creates the personalities and channels the behaviors of the members of a society, but that socialization is never entirely finished. Thus, in *Manchild in the Promised Land*, his masterpiece about growing up in Harlem during the period of rapid migration of blacks from the South, Claude Brown (1966) wrote that his rural-born parents did not "seem to be ready for urban life." Their values and norms of behavior made no sense to their son, who had to survive on Harlem's mean streets:

> When I was a little boy, Mama and Dad would beat me and tell me, "You better be good," but I didn't know what being good was. To me it meant that they just wanted me to sit down and fold my hands or something crazy like that. Stay in front of the house, don't go anyplace, don't get into trouble. I didn't know what it meant, and I don't think they knew what it meant, because they couldn't ever tell me what they really wanted. The way I saw it, everything I was doing was good. If I stole something and didn't get caught, I was good. If I got into a fight with somebody, I tried to be good enough to beat him. If I broke into a place, I tried to be quiet and steal as much as I could. I was always trying to be good. They kept on beating me and talking about being good. And I just kept on doing what I was doing and kept on trying to do it good. (p. 279)

Brown's story is that of a young man whose parents' experience with social change created a severe disjunction in their lives. This left them ill equipped to socialize their son for the demands of a new environment. And so he learned to survive on the streets. He became a thief and a gang fighter. But by his own account, later in his life he

SOCIOLOGICAL METHODS: Throughout the textbook, Dr. Kornblum applies research to every area of sociology. By integrating methodology, rather than confining it to a single chapter, he reinforces the importance of research.

THE PERSONAL CHALLENGE: This feature illustrates how social change is altering the lives of students and encourages them to take an active part in changing their world.

STUDY CHARTS, FIGURES, AND TABLES: These learning aids organize material and ideas for ease in studying and increase comprehension.

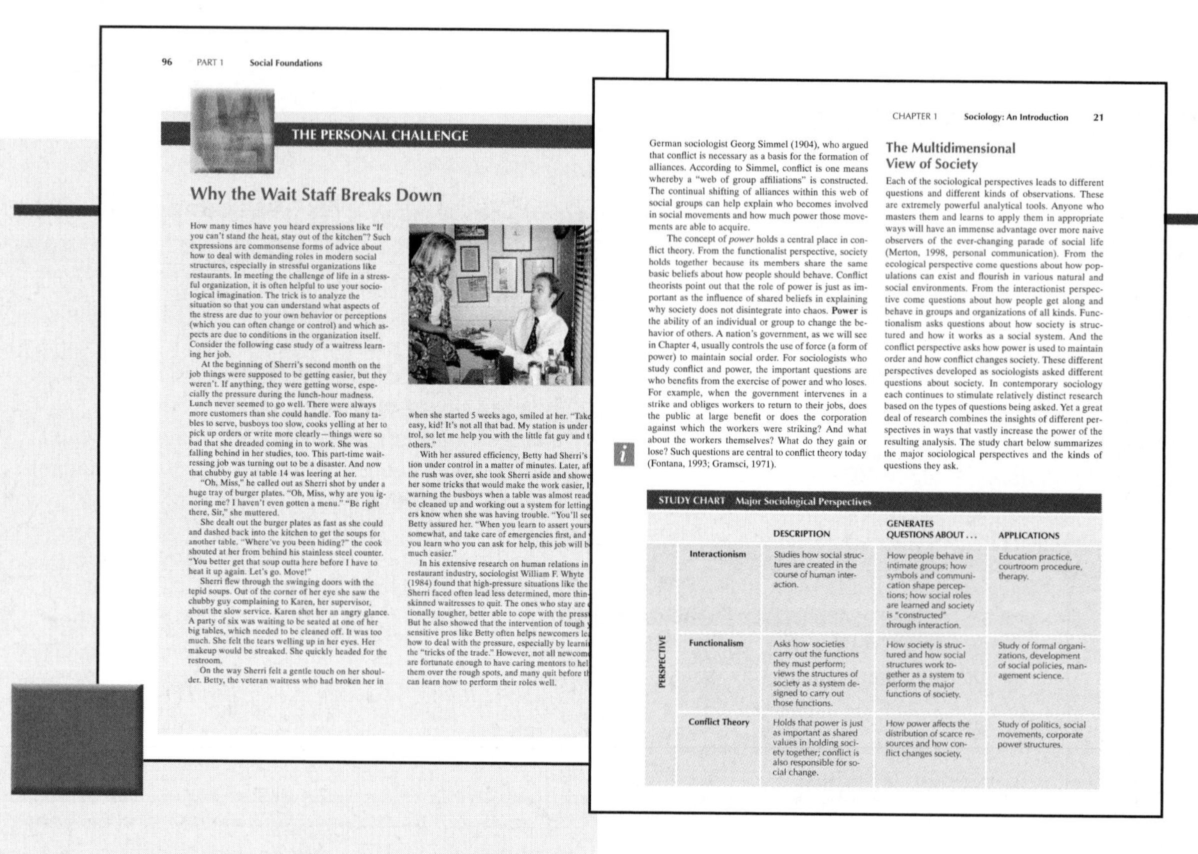

96 PART 1 Social Foundations

THE PERSONAL CHALLENGE

Why the Wait Staff Breaks Down

How many times have you heard expressions like "If you can't stand the heat, stay out of the kitchen"? Such expressions are commonsense forms of advice about how to deal with demanding roles in modern social structures, especially in stressful organizations like restaurants. In meeting the challenge of life in a stressful organization, it is often helpful to use your sociological imagination. The trick is to analyze the situation so that you can understand what aspects of the stress are due to your own behavior or perceptions (which you can often change or control) and which aspects are due to conditions in the organization itself. Consider the following case study of a waitress learning her job.

At the beginning of Sherri's second month on the job things were supposed to be getting easier, but they weren't. If anything, they were getting worse, especially the pressure during the lunch-hour madness. Lunch never seemed to go well. There were always more customers than she could handle. Too many tables to serve, busboys too slow, cooks yelling at her to pick up orders or write more clearly—things were so bad that she dreaded coming in to work. She was falling behind in her studies, too. This part-time waitressing job was turning out to be a disaster. And now that chubby guy at table 14 was leering at her.

"Oh, Miss," he called out as Sherri shot by under a huge tray of burger plates. "Oh, Miss, why are you ignoring me? I haven't even gotten a menu." "Be right there, Sir," she muttered.

She dealt out the burger plates as fast as she could and dashed back into the kitchen to get the soups for another table. "Where've you been hiding?" the cook shouted at her from behind his stainless steel counter. "You better get that soup outta here before I have to heat it up again. Let's go. Move!"

Sherri flew through the swinging doors with the tepid soups. Out of the corner of her eye she saw the chubby guy complaining to Karen, her supervisor, about the slow service. Karen shot her an angry glance. A party of six was waiting to be seated at one of her big tables, which needed to be cleaned off. It was too much. She felt the tears welling up in her eyes. Her makeup would be streaked. She quickly headed for the restroom.

On the way Sherri felt a gentle touch on her shoulder. Betty, the veteran waitress who had broken her in when she started 5 weeks ago, smiled at her. "Take it easy, kid! It's not all that bad. My station is under control, so let me help you with the little fat guy and t[he] others."

With her assured efficiency, Betty had Sherri's [sta]tion under control in a matter of minutes. Later, af[ter] the rush was over, she took Sherri aside and showe[d] her some tricks that would make the work easier, [like] warning the busboys when a table was almost read[y to] be cleaned up and working out a system for letting [the cook]ers know when she was having trouble. "You'll se[e]," Betty assured her. "When you learn to assert yours[elf] somewhat, and take care of emergencies first, and [when] you learn who you can ask for help, this job will b[e] much easier."

In his extensive research on human relations in [the] restaurant industry, sociologist William F. Whyte (1984) found that high-pressure situations like the [one] Sherri faced often lead less determined, more thin-skinned waitresses to quit. The ones who stay are [emo]tionally tougher, better able to cope with the press[ure]. But he also showed that the intervention of tough [but] sensitive pros like Betty often helps newcomers le[arn] how to deal with the pressure, especially by learni[ng] the "tricks of the trade." However, not all newcom[ers] are fortunate enough to have caring mentors to hel[p] them over the rough spots, and many quit before th[ey] can learn how to perform their roles well.

CHAPTER 1 Sociology: An Introduction 21

German sociologist Georg Simmel (1904), who argued that conflict is necessary as a basis for the formation of alliances. According to Simmel, conflict is one means whereby a "web of group affiliations" is constructed. The continual shifting of alliances within this web of social groups can help explain who becomes involved in social movements and how much power those movements are able to acquire.

The concept of *power* holds a central place in conflict theory. From the functionalist perspective, society holds together because its members share the same basic beliefs about how people should behave. Conflict theorists point out that the role of power is just as important as the influence of shared beliefs in explaining why society does not disintegrate into chaos. **Power** is the ability of an individual or group to change the behavior of others. A nation's government, as we will see in Chapter 4, usually controls the use of force (a form of power) to maintain social order. For sociologists who study conflict and power, the important questions are who benefits from the exercise of power and who loses. For example, when the government intervenes in a strike and obliges workers to return to their jobs, does the public at large benefit or does the corporation against which the workers were striking? And what about the workers themselves? What do they gain or lose? Such questions are central to conflict theory today (Fontana, 1993; Gramsci, 1971).

The Multidimensional View of Society

Each of the sociological perspectives leads to different questions and different kinds of observations. These are extremely powerful analytical tools. Anyone who masters them and learns to apply them in appropriate ways will have an immense advantage over more naive observers of the ever-changing parade of social life (Merton, 1998, personal communication). From the ecological perspective come questions about how populations can exist and flourish in various natural and social environments. From the interactionist perspective come questions about how people get along and behave in groups and organizations of all kinds. Functionalism asks questions about how society is structured and how it works as a social system. And the conflict perspective asks how power is used to maintain order and how conflict changes society. These different perspectives developed as sociologists asked different questions about society. In contemporary sociology each continues to stimulate relatively distinct research based on the types of questions being asked. Yet a great deal of research combines the insights of different perspectives in ways that vastly increase the power of the resulting analysis. The study chart below summarizes the major sociological perspectives and the kinds of questions they ask.

STUDY CHART Major Sociological Perspectives

PERSPECTIVE	DESCRIPTION	GENERATES QUESTIONS ABOUT . . .	APPLICATIONS
Interactionism	Studies how social structures are created in the course of human interaction.	How people behave in intimate groups; how symbols and communication shape perceptions; how social roles are learned and society is "constructed" through interaction.	Education practice, courtroom procedure, therapy.
Functionalism	Asks how societies carry out the functions they must perform; views the structures of society as a system designed to carry out those functions.	How society is structured and how social structures work together as a system to perform the major functions of society.	Study of formal organizations, development of social policies, management science.
Conflict Theory	Holds that power is just as important as shared values in holding society together; conflict is also responsible for social change.	How power affects the distribution of scarce resources and how conflict changes society.	Study of politics, social movements, corporate power structures.

VISUAL SOCIOLOGY: These photo essays illustrate in photographs how sociologists interpret and analyze visual information as part of their research, demonstrating how integral the visual aspect is to sociology.

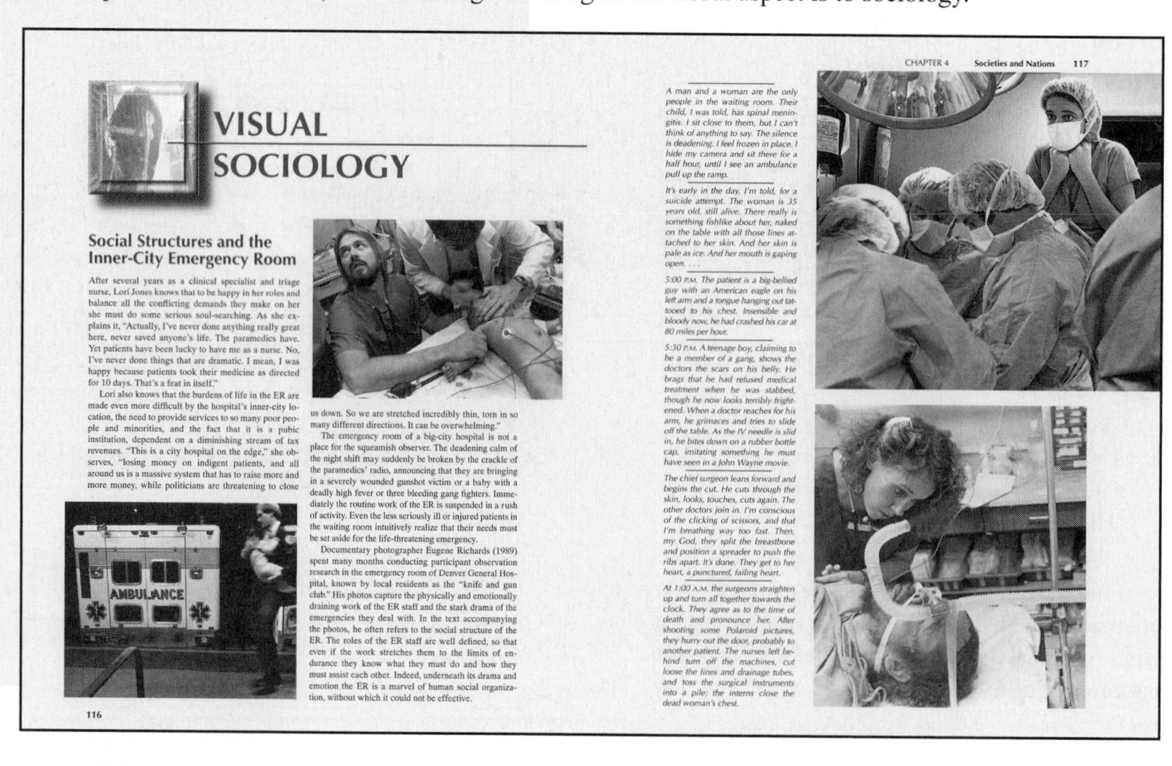

VISUAL SOCIOLOGY

Social Structures and the Inner-City Emergency Room

After several years as a clinical specialist and triage nurse, Lori Jones knows that to be happy in her roles and balance all the conflicting demands they make on her she must do some serious soul-searching. As she explains it, "Actually, I've never done anything really great here, never saved anyone's life. The paramedics have. Yet patients have been lucky to have me as a nurse. No, I've never done things that are dramatic. I mean, I was happy because patients took their medicine as directed for 10 days. That's a feat in itself."

Lori also knows that the burdens of life in the ER are made even more difficult by the hospital's inner-city location, the need to provide services to so many poor people and minorities, and the fact that it is a public institution, dependent on a diminishing stream of tax revenues. "This is a city hospital on the edge," she observes, "losing money on indigent patients, and all around us is a massive system that has to raise more and more money, while politicians are threatening to close us down. So we are stretched incredibly thin, torn in so many different directions. It can be overwhelming."

The emergency room of a big-city hospital is not a place for the squeamish observer. The deadening calm of the night shift may suddenly be broken by the crackle of the paramedics' radio, announcing that they are bringing in a severely wounded gunshot victim or a baby with a deadly high fever or three bleeding gang fighters. Immediately the routine work of the ER is suspended in a rush of activity. Even the less seriously ill or injured patients in the waiting room intuitively realize that their needs must be set aside for the life-threatening emergency.

Documentary photographer Eugene Richards (1989) spent many months conducting participant observation research in the emergency room of Denver General Hospital, known by local residents as the "knife and gun club." His photos capture the physically and emotionally draining work of the ER staff and the stark drama of the emergencies they deal with. In the text accompanying the photos, he often refers to the social structure of the ER. The roles of the ER staff are well defined, so that even if the work stretches them to the limits of endurance they know what they must do and how they must assist each other. Indeed, underneath its drama and emotion the ER is a marvel of human social organization, without which it could not be effective.

116

CHAPTER 4 Societies and Nations 117

A man and a woman are the only people in the waiting room. Their child, I was told, has spinal meningitis. I sit close to them, but I can't think of anything to say. The silence is deadening. I feel frozen in place. I hide my camera and sit there for a half hour, until I see an ambulance pull up the ramp.

It's early in the day, I'm told, for a suicide attempt. The woman is 35 years old, still alive. There really is something fishlike about her, naked on the table with all those lines attached to her skin. And her skin is pale as ice. And her mouth is gaping open. . . .

5:00 P.M. The patient is a big-bellied guy with an American eagle on his left arm and a tongue hanging out tattooed to his chest. Insensible and bloody now, he had crashed his car at 80 miles per hour.

5:30 P.M. A teenage boy, claiming to be a member of a gang, shows the doctors the scars on his belly. He brags that he had refused medical treatment when he was stabbed, though he now looks terribly frightened. When a doctor reaches for his arm, he grimaces and tries to slide off the table. As the IV needle is slid in, he bites down on a rubber bottle cap, imitating something he must have seen in a John Wayne movie.

The chief surgeon leans forward and begins the cut. He cuts through the skin, looks, touches, cuts again. The other doctors join in. I'm conscious of the clicking of scissors, and that I'm breathing way too fast. Then, my God, they split the breastbone and position a spreader to push the ribs apart. It's done. They get to her heart, a punctured, failing heart.

At 1:00 A.M. the surgeons straighten up and turn all together towards the clock. They agree as to the time of death and pronounce her. After shooting some Polaroid pictures, they hurry out the door, probably to another patient. The nurses left behind turn off the machines, cut loose the lines and drainage tubes, and toss the surgical instruments into a pile; the interns close the dead woman's chest.

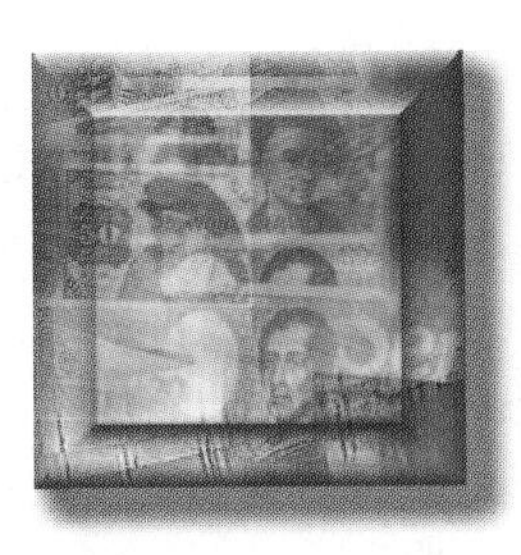

FEATURES

Art and the Sociological Eye

Global Change and U.S. Society

Mapping Social Change

Research on the Cutting Edge

Sociological Methods

The Personal Challenge

Then and Now

Using the Sociological Imagination

Visual Sociology

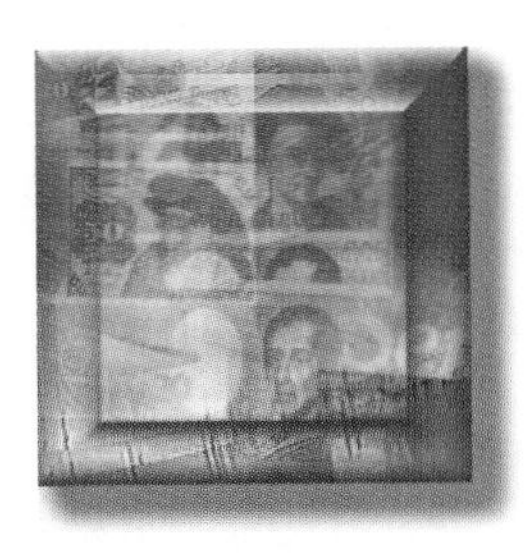

PREFACE

When I was writing the first edition of *Sociology in a Changing World,* I took the manuscript to my students at Queens College of the City University of New York. I wanted them to use the book and to criticize the manuscript as I was revising it. Some of the students were young undergraduates, tired of college textbooks whose content and design reminded them of high school books. Others were older adults going back to school after many years in the "real world"; they were wary of sociological platitudes. All wanted clear information about how sociology could matter in their professional and personal lives, and they were avid critics. Their enthusiasm for the theme of social change and the concept of the sociological imagination was evident. They found that these ideas applied in many concrete ways to their own interests and aspirations. On the other hand, they pointed out places where concepts were not presented clearly enough or where their interest flagged. They began to see the value of observing a changing world with a sociological eye.

One major way in which *Sociology in a Changing World* differs from other sociology texts is in its consistent focus on issues of social change. Moreover, it is not slanted toward one or another perspective, ideological or methodological. It presents ample evidence of the strengths and weaknesses of the interactionist, functionalist, and conflict approaches to the explanation and prediction of social change. In addition, it has a stronger grounding in demographic, ecological, and historical methods than is found in existing comprehensive texts. And throughout the text an effort is made to strike a balance between comparative and historical material and research based on American culture and the social changes that have occurred and are occurring today in the United States. Although this book is designed for the introductory course, I hope its emphasis on current sociological research and its combination of classic and new research will make it a useful reference source for students in more advanced social-science courses.

It is my hope that both instructors and students will benefit from the strength of this book and that it will convey the excitement of sociological discovery. The more I study, teach, and do research in sociology, the more I love the subject. No other area of intellectual work overlaps with so many other fields of knowledge; none straddles the sciences and the humanities so squarely or offers such varied insights into the vexing questions of our own time. The field is taking on new importance as sociological research is in increasing demand throughout the world. Major, often tumultuous changes are creating a growing demand for scientific description and analysis of human behavior. There is growing awareness that information has become a key source of wealth and power. In consequence, sociological concepts and methods are being applied with increasing frequency, and the study of sociology is attracting a far more diverse student population. This edition of *Sociology in a Changing World* is designed to address the changes in the discipline, the students, and the work of sociologists.

The fifth edition of *Sociology in a Changing World* has had the benefit of constructive criticism from both students and colleagues. I have used their comments and suggestions wherever possible and appropriate. Some readers of this text may have ideas about improvements that might be made in future editions. Please feel free to send your ideas or comments to me at the CUNY Graduate School, 565 Fifth Avenue, New York, NY 10016.

New in This Edition

In the years since the first edition was completed, I have taught the introductory course many times. Each time, I have noticed that students appreciate the comparative material in the text and especially like the emphasis on current social issues. Because of this, a new feature, *Research on the Cutting Edge,* has been added to the fifth edition. This feature, which appears in almost every chapter, showcases the research of contemporary sociologists on topics related to the content of the chapter. To the extent possible, these features draw upon award-winning studies published within the past few years.

Also new in this edition is a pedagogical element designed to stimulate critical thinking. Titled *Sociology Versus Ideology,* this element appears at the end of each chapter and shows how sociological research helps cut through the impasse created by arguments between liberals and conservatives.

The visual dimension has always been a strength of this text, and in this edition the visual features have been enhanced and expanded. Each chapter-opening vignette is now linked to the Visual Sociology feature to create a coherent "story." Many chapters also contain a new feature, titled *Art and the Sociological Eye,* that shows how fine art can be interpreted from a sociological perspective. Finally, the *Mapping Social Change* features have been expanded so that they are easier to read and interpret.

Highlights of the Fifth Edition

Sociology in a Changing World has been extensively revised to reflect continuing changes in our own and other societies. The revision process was guided by six essential principles:

- *Exercising a Global Perspective:* We live in a world in which social and economic relations are increasingly global in scope. This global perspective is incorporated throughout the textbook.

- *Observing Social Change:* A key to studying sociology is observing social change among individuals, groups, and societies around the world. The text uses visual features to help students develop their powers of social observation.

- *Applying Sociology:* By applying sociology to everyday life, students' comprehension is made more complete.

- *Learning Through Visual Presentation:* An extensive visual presentation maximizes learning by reflecting contemporary cultural assumptions and tastes.

- *Incorporating Diversity in Theory and Research:* The research of women and minorities is presented in appropriate contexts throughout the book and not limited to single chapters.

- *Presenting Accessible Scholarship:* The scholarship is presented in an accessible style, promoting comprehension of the material.

Many of the fifth edition's features are in response to these principles. For example, Visual Sociology, Mapping Social Change, Then and Now, and Art and the Sociological Eye are visually oriented presentations of sociological concepts and changing societal conditions. The Personal Challenge, which illustrates how social change is altering students' lives and encourages students to take an active part in changing their world, inspires students to observe social change, apply sociology, and exercise a global perspective.

Major content revisions have been made throughout the book in response to feedback from my colleagues. Many of the chapters—especially those on socialization, social class, education, and economic institutions—have been extensively revised. I have also added the following new discussions on important topics:

Chapter 1: greater emphasis on early contributions of women to sociology (e.g., Harriet Martineau)
Chapter 2: use of suicide in its many forms as an example throughout the chapter
Chapter 3: cultural hegemony
Chapter 5: gender and moral reasoning
Chapter 9: gender conflict in cities
Chapter 10: war and gender relations
Chapter 12: the socioeconomic status gradient in health; welfare reform and its impact
Chapter 15: interactions among age, race, and gender
Chapter 16: sociological insights into Waco and Wounded Knee; effects of immigration on religion in the United States
Chapter 17: effects of the retreat from bilingualism; class size and educational achievement; school choice
Chapter 19: gender and workplace diversity
Chapter 20: typology of political regimes

In addition, the Research on the Cutting Edge features discuss important contemporary research. Examples include global immigration to the United States (Chapter 4), the application of small-group research to economic development (Chapter 6), celebrity and the industry that has grown up around it (Chapter 8), and the relationship between poverty and place of residence (Chapter 12).

Distinctive Features

Sociology in a Changing World differs from its competitors primarily in its emphasis on social change. Throughout, the book highlights the tension between attempts to modify social institutions and efforts to maintain traditional modes of behavior. The text also tries to emphasize contemporary social-scientific research. The following are the special features that incorporate this teaching philosophy:

- *Chapter outlines* serve as advance organizers, previewing the topics that will be discussed in the chapter.

- *Opening vignettes* begin each chapter to capture the students' attention and introduce them to the subject matter. The "story" presented in each vignette is continued in the Visual Sociology section at the end of the chapter.

- *Then and Now* contrasts photos of a particular social condition in the past and today.

- *Art and the Sociological Eye* shows how fine art can be interpreted from a sociological perspective.

- *Feature boxes,* including *Sociological Methods, Global Change and U.S. Society,* and *Using the Sociological Imagination,* are found within the chapters.

- *The Personal Challenge* discusses the impact of social change on aspects of students' everyday lives.

- *Mapping Social Change* uses maps and photos to show the distribution of a condition or characteristic throughout the United States or the world.

- *Study charts* summarize material for ease of studying.

- *Research on the Cutting Edge,* the final section of each chapter, discusses current work by recognized sociological researchers in subject areas covered in the chapter, relating it to applied research and policy issues.

- *Visual Sociology,* a photo essay at the end of each chapter, shows how sociologists analyze photographs or other visual material as part of their research.

- *Chapter summaries* are found at the end of each chapter. Each summary is a thorough but concise rendering of the key concepts and relationships presented in the chapter.

- *Sociology Versus Ideology* shows how sociologists approach controversial issues and is designed to encourage critical thinking.

- *End-of-chapter glossaries* are convenient; students will not have to flip to the end of the book to review important terms and their definitions.

- *Where to Find It,* at the end of each chapter, lists a set of suggested readings that includes not only books and journal articles but also basic references and data sources and Internet resources.

Organization of the Text

In organizing the book, I resisted the temptation to radically rearrange the standard introductory sociology course. There are traditional approaches to the subject that deserve respect, such as presenting an overview of sociology and its history, introducing the basic research methods, examining the dimensions of social inequality, and including a series of chapters on major social institutions. But I have added to this solid, time-tested framework, and in some cases I have moved a chapter from its traditional position in an introductory textbook.

The first 10 chapters of the book set forth most of the concepts that any student needs to know and be able to use. This makes the book easy to adapt to courses of different length. In addition, the first three parts of the book introduce many of the basic sociological explanations of social stability and change. Topics like urbanization and social movements are therefore presented in the first half of the book rather than in later sections. Other fundamental concepts are presented in these early chapters as well. I refer repeatedly to these concepts in the second half of the book, rather than introducing them and then going on to entirely new material.

The book is divided into four parts:

Part 1, Social Foundations, introduces the "human science" of sociology: Chapter 1 traces the history of sociology and introduces the major perspectives used by sociologists, and Chapter 2 describes the methods used by sociologists in conducting their research. The next two chapters focus on some of sociology's most fundamental concepts: culture (Chapter 3) and the elements of social structure (Chapter 4).

Part 2, Social Dynamics, covers many of the processes that seem to account for social stability and change: Chapter 5 discusses socialization, and Chapter 6 examines the structure of groups and how people behave in groups of different kinds. Chapters 7, 8, and 9 move from the structure and function of groups within a society to the processes that change societies. Chapter 7 deals with deviance; Chapter 8 looks at collective behavior and social movements; and Chapter 9 explores the relationships among population growth, urbanization, and community. Chapter 10 summarizes the ways in which sociologists think about and conduct research on the causes and consequences of social change.

The five chapters in **Part 3, Social Divisions,** examine social inequality, particularly in American society. Chapter 11 introduces the concept of social stratification. Chapter 12 deals with inequalities of social class, and Chapters 13, 14, and 15 focus on social inequalities due to race, ethnicity, gender, and age.

Part 4, Social Institutions, applies the concepts and perspectives discussed earlier in the book to an analysis of several major institutions: Chapter 16 explores the changing nature of the family, Chapter 17 introduces religion, and Chapter 18 covers education and communications media. Chapter 19 deals with economic institutions, and Chapter 20 discusses political institutions. Finally, Chapter 21 analyzes the institutions of science, technology, and medicine.

Ancillary Package

Harcourt College Publishers may provide complimentary instructional aids and supplements or supplement packages to those adopters qualified under our adoption policy. Please contact your sales representative for more information. If, as an adopter or potential user, you receive textbooks or supplements you do not need, please return them to your sales representative or send them to

ATTN: Returns Department
Harcourt College Publishers
465 South Lincoln Drive
Troy, MO 63379

For the Instructor

Instructor's Manual
Written by William Kornblum and Carolyn Smith, the Instructor's Manual includes lecture outlines, instructional goals, teaching suggestions that explain the distinctive features and central concepts of each chapter, topics for discussion, and suggestions for using the charts, tables, and other visual features in the text.

Test Bank
Prepared by Carol Chenault and Cindy Wade-Harper of Calhoun Community College, Decatur, Alabama, the Test Bank includes over 3100 items. It is available in printed and software versions and the page number in the textbook is referenced for each question. The computerized test bank is available in Macintosh and Windows formats.

EXAMaster software allows you to create tests using fewer keystrokes, guiding you through the process by easy-to-follow screen prompts. *EXAMaster* has three test creation options: *EasyTest,* which lets you create a test from a single screen, *FullTest,* which gives you a larger range of options including editing of items; and *RequesTest,* a test compilation service for the instructor who has no computer. *EXAMaster* comes with EXAMRecord, a customized grade book software program.

The *Sociological Imagination* Video Segments
These 26 minute video clips are from the Dallas County Community College District telecourse in introductory sociology. These segments, highlighting relevant subject matter, are part of the *Sociological Imagination* telecourse that is coordinated with *Sociology in a Changing World.* The new telecourse study guide is also available.

Web Site
Visit the cutting edge web site for Sociology at www.harbrace.com/soc. The material related to *Sociology in a Changing World* was generated by Carol Chenault of Calhoun Community College in Decatur, Alabama. It features multiple choice quizzes for each chapter, downloadable PowerPoint slides, career resources, and 3 or 4 interactive exercises per chapter. These exercises enable students to use Internet research skills, participate in group activities centered around the Internet, and play online learning games. A special feature of this web site is "Sociology in Action," which has numerous Internet links relating to sociology in real-life.

Harcourt Video Library
A variety of updated video programs are available to enrich classroom presentations. Selections include *The Sandwich Generation: Caring for Both Children and Parents, The Good Society and Attorney General Janet Reno* among others. The videos are selected from Films for the Humanities and Sciences and from the Public Broadcasting System's CPB Video Series.

Overhead Teaching Transparencies
Classroom Lectures will be enhanced with this collection of full color transparencies that illustrate sociological concepts. Selected by Charles Faupel of Auburn University from the art programs in the Harcourt introductory sociology textbooks, they contain information to supplement material in the textbook.

PowerPoint Slides
These helpful teaching aids are available on the web site for *Sociology in a Changing World.* They can be downloaded to make a great slide presentation for the main concepts in each chapter.

For the Student

Study Guide
Written by William Kornblum and Carolyn Smith, the Study Guide provides students with a self-paced review of the text. Each chapter begins with an outline and learning objectives, followed by a fill-in-the-blank review, a matching exercise in which key terms are matched with their definitions, a self-text consisting of 15 multiple-choice and 5 true/false questions, and a short answer section that requires the student to apply concepts presented in the textbook. Answers to the matching exercises and self-tests are provided. The Study Guide also includes a section that encourages

students to think and write critically about what they have learned in each chapter of the textbook.

The *Sociological Imagination* Telecourse Study Guide

Written as a companion to the telecourse, this Study Guide has been extensively revised. Lesson guidelines include a list of goals at the beginning of every lesson, and suggested readings supplement the textbook. Each lesson provides an overview of the text and video program. The Study Guide is written by Jane Penny of Dallas County Community College District. Available also from Dallas County Community College District is the newly revised Faculty Guide and Test Bank to accompany the telecourse.

CD-ROM

The CD-ROM is an interactive student learning tool that corresponds with the fifth edition of *Sociology in a Changing World.* It provides dynamic multimedia presentations of the most complex and important concepts in sociology. Students using the CD-ROM will have personal access to resources that will enhance their learning of key concepts in sociology through contemporary multimedia articles, illustrations, software exercises, demonstrations, internet access and more.

***SimCity* Software**

This educational version of *SimCity* software is geared to sociology using environmental, economic, or geographical variables to teach the volatile nature of sociology. It allows the user to create his or her own society, an experience instructors and students will find invaluable. User's guide included.

Acknowledgments

The new edition of *Sociology in a Changing World* is the culmination of two years of steady work—weekly writing and editing for a year, plus another year of production-related tasks. The values in a book like this are the work of many minds and hands. From the contributions of the social scientists who have reviewed the manuscript and the sociology students who have commented on previous editions, to the final touches of detail before the book is printed, there are scores of people whose contributions are essential at every step in the process. It is exciting to launch a new edition of the text, just as it is gratifying to thank those who made it possible. I have had the benefit of continuity in editors and professional friends who have advised me since the first edition. But each edition also brings in new publishing professionals whose influences help keep the book fresh and up-to-date.

First among the many people who continue to influence this project is my editor and collaborator Carolyn Smith. Carolyn first encouraged me to take on this project, and when my resolve wavered, her faith in our ability to complete it sustained me. Carolyn has untangled my prose, agonized with me over reviewers' comments, and kept track of the endless details that go into creating a scholarly text. There is not one word in this book that she has not thought about; every page bears the stamp of her expertise. Guy Smith also has my gratitude for his supportiveness during the long and demanding revision process.

The visual identity of this book, its use of photographic material for sociological example and method, have always been extremely important to me and to everyone involved in its creation. Our aim has been to create a textbook that would integrate sociological ideas with the design and layout of the book. In this edition the design, created by David Day of Harcourt, and the art and photo programs continue to emphasize creative treatments of data presentation and visual sociology.

My colleagues at the City University of New York have been unstinting with their suggestions and advice. My dear friend and colleague, Vernon Boggs, was always my greatest personal ally in the making of this book. His untimely death deprives me of someone whose deep commitment to teaching and research in sociology were a sustaining example. My longtime collaborator and friend, Terry Williams, continues to influence my thinking on key issues in sociology and life. Rolf Meyersohn has been heroic in his generosity with books and articles and words of encouragement. My colleague Charles Kadushin's intimate knowledge of the publishing world has been invaluable on many occasions, as have Paul Attewell's mastery of trends in sociological research and methods, and the profound knowledge of social movements and class theory shared by Stanley Aronowitz, William DiFazio, and Bogdan Denitch. If I have managed to avoid a natural predisposition toward male bias, much of the credit must go to Susan Kornblum, Cynthia Epstein, Judith Lorber, June Nash, Barbara Katz Rothman, and many other colleagues who have shared their insights and knowledge. My colleagues Lily Hoffman, John Mollenkopf, and Charles Winick are unstinting in sharing their sociological insights and knowledge of new literatures.

The thoughtful comments we receive about the book and the encouragement of many students and colleagues are immensely rewarding. Their corrections and constructive suggestions, too, are most welcome. Many colleagues throughout the United States offer me wise counsel and scholarly advice, for which I continue to be deeply grateful. But I owe a special debt to my colleague and friend, Carol Chenault of Calhoun Community College, Decatur, Alabama, for her abiding interest in the textbook and her generosity. Professor Chenault's comparative visual sociology is featured in Chapter 14.

No single author can hope to adequately represent a field as large and complex as sociology. Over several years of textbook writing, I have learned that the comments and suggestions offered by reviewers and users of the book are essential to its success. Many dedicated Harcourt sales representatives have taken the time to introduce me to instructors who use the book. In addition, the comments of numerous reviewers have helped me to correct mistakes of fact, interpretation, and emphasis. I might disagree with some of them, but I always find them helpful.

Each edition of the book has benefited from the insights of reviewers. I gratefully acknowledge the following instructors who contributed to the fifth edition of *Sociology in a Changing World:*

Nancy Andes
University of Alaska Anchorage

Robert S. Anwyl
Miami-Dade Community College

Kathy Brace
York College of Pennsylvania

Robin Brown
Southern Union State Community College

Kim Catatt
SUNY Buffalo

Emilie Walter Cellucci
Francis Marion University M.

Carol Deming Chenault
Calhoun County Community College

B. Keith Crew
University of Northern Iowa

Lilli Downes
Hartford Community College

Darleen Garrett
Weatherford Community College

James Green
St. Thomas Aquinas College

Joanna Grey
Pikes Peak Community College

Charles F. Hanna
Duquesne University

Gary Heidinger
Roane State Community College

Randy Hodson
Ohio State University

Drew Hurley
Santa Fe Community College

Paul Kamolnick
East Tennessee State University

James R. McIntosh
Lehigh University

Carla Mueller
Lindenwood University

Priscilla Reinertsen
University of New Hampshire

Carolyn Rivera
Florence-Darlington Technical College

Craig T. Robertson
University of North Alabama

Jon Schlenker
University of Maine at Augusta

Hasan Shapari
Villanova University

Martin Simpson
Nicholls State University

Joe Charles Snell
Kirkwood College

Linda Tiemann
Blinn College

Bruce T. Wyman
Delaware Community College

George Yancey
University of Wisconsin Whitewater

The textbook continues to benefit from the helpful comments of the following instructors who reviewed for previous editions:

Barry Adams
University of Windsor

Elizabeth Almquist
University of North Texas

Rita Argiros
St. Cloud State University

Ed Armstrong
University of Wisconsin–Stout

Michele Aronica
Saint Joseph's College

Augustine Aryee
Fitchburg State College

Dorothy Balancio
Mercy College

Richard Bales
Illinois Central College

Matilda Barker
Cerritos College

Bernard Beck
Northwestern University

Scott A. Beck
East Tennessee State University

Phillip Bosserman
Salisbury State College

Cliff Broman
Michigan State University

Brent Bruton
Iowa State University

Marvin Camfield
Walters State Community College

R. E. Canjar
University of Maryland

Carole M. Carroll
Middle Tennessee State University

Chuck Carselowey
Oklahoma City Community College

Carol Deming Chenault
Calhoun Community College

Mary Beth Collins
Central Piedmont Community College

Donna K. Crossman
Greenville Technical College

Glenn Currier
El Centro College

Vasilikie Demos
University of Minnesota

Mita Dhariwal
Bakersfield College

Jack Dison
Arkansas State University

Dennis R. Dubbs
Harrisburg Area Community College

Ken Elder
Southeastern Oklahoma State University

Martin Epstein
Middlesex Community College

Douglas Farley
Niagara County Community College

Charles E. Faupel
Auburn University

Frank Forwood
Northeast Louisiana University

Michael J. Fraleigh
Bryant College

Joseph Galaskiewicz
University of Minnesota

Eric P. Godfrey
Ripon College

Kirstin Grϕnbjerg
Loyola University (Chicago)

Larry J. Halford
Washburn University

John Hall
University of California

Scarlett Hardesty
University of Arizona

Rudy Harris
Des Moines Area Community College

Denny Hill
Georgia Southern College

Lily M. Hoffman
City College in New York

Drew Hurley
Santa Fe Community College

Mike Hurtado
Citrus College

Benton Johnson
University of Oregon

Mark Kassop
Bergen Community College

Laura Kramer
Montclair State College

Kathryn Kuhn
University of Texas

Martin L. Levin
Emory University

Ruby C. Lewis
Dekalb College–Central Campus

Theodore Long
Merrimack College

Mits Maeda
Cypress College

Joseph Marolla
Virginia Commonwealth University

George Martin
Montclair State College

Allan McCutcheon
University of Delaware

Virginia McKeefery-Reynolds
Northern Illinois University

Anthony J. Mendonca
Community College of Allegheny County–Allegheny Campus

Jerry L. L. Miller
University of Arizona

Ephraim Mizruchi
Syracuse University

Karen Mundy
Lee College

Ann S. Oakes
Idaho State University

James D. Orcutt
Florida State University

Larry M. Perkins
Oklahoma State University

Novella Perrin
Central Missouri State University

Dick Phillips
Kalamazoo Valley Community College

Judy Porta
Las Positas College

David L. Preston
San Diego State University

Leroy Reed
Central Florida Community College

Joseph Ribal
El Camino College

J. D. Robson
University of Arkansas

John Roman
University of Maine

Jefferry P. Rosenfeld
Nassau Community College

Ellen Rosengarten
Sinclair Community College

Wesley Shrum
Louisiana State University

Richard L. Simpson
University of North Carolina

Joel C. Snell
Kirkwood Community College

Gary Spencer
Syracuse University

Larry D. Stokes
University of Tennessee–Chattanooga

Paul Stowell
Wenatchee Valley College

Lorene Taylor
Valencia Community College

Doren Thompson
Washtenaw Community College

Marcella Thompson
University of Arkansas–Fayetteville

Kathleen Tiemann
Mercer University

Jacquelyn A. Troup
Cerritos College

William Trush
Kalamazoo Valley Community College

Tim Tuinstra
Kalamazoo Valley Community College

David Van Merlo
St. Charles County Community College

Steven L. Vasser
Mankato State University

C. Edwin Vaughan
University of Missouri

Theodore C. Wagenaar
Miami University

Eric Wagner
Ohio University

Joe Walsh
Bucks County Community College

Chaim I. Waxman
Rutgers, the State University of New Jersey

Jack Weller
University of Kansas

Timothy Wickham-Crowley
Georgetown University

Kenneth Wilson
East Carolina University

Steven R. Wilson
Temple University

Leslie E. Wooten
Cleveland State Community College

Thomas J. Yacovone
Los Angeles Valley College

Stacey G. H. Yap
Plymouth State College

Richard Yinger
Palm Beach Community College

Wayne Zatapek
Tarrant County Junior College–Northeast

Herbert Zeigler
Chesapeake College

William W. Zellner
Doane College

In preparing this edition of the textbook, I needed the help of a team of professional editors and publishers who would recognize the strengths of the book and work with me to make improvements while keeping the basic elements of the original content and design. I found such people in Lin Marshall, Peggy Howell, Kathleen Sharp, Earl McPeek, Chris Klein, and Ted Buchholz of Harcourt. They all encouraged me whenever I had good ideas, and were sensitive in their criticism. My thanks also go to David Day for his text design and supervision of the art program, to Sandra Lord for the development of the photo program, to Michele Gitlin for her insightful copyediting, and to Andrea Archer and Michele Tomiak for their skillful production management and editorial coordination.

Thanks also go to my students at the City University of New York, both graduates and undergraduates, for keeping me "turned on" to sociology.

William Kornblum

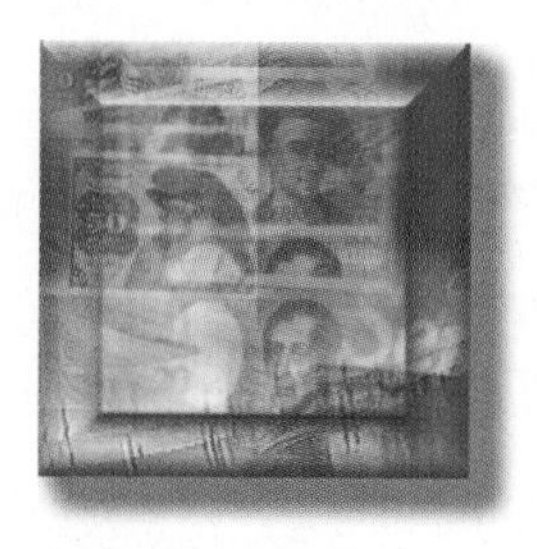

About the Author

William Kornblum is a professor of sociology at the Graduate School of the City University of New York, where he helps train future instructors and researchers in the social sciences. He also teaches undergraduates at various campuses of the City University, including Queens College, Hunter College, and City College.

A specialist in urban and community studies, Kornblum began his teaching career with the Peace Corps in the early 1960s, when he taught physics and chemistry in French-speaking West Africa (see the introduction to Chapter 10). He received his doctorate in sociology from the University of Chicago in 1971. He has also taught at the University of Washington at Seattle and worked as a research sociologist for the U.S. Department of the Interior. At the CUNY Graduate School, he directs research on youth and employment and on urban policy. With his longtime research partner, Terry Williams, he recently coauthored *The Uptown Kids,* a sociological portrait of teenagers and young adults growing up in high-rise public housing projects. He is also the principal investigator of Project TELL, a longitudinal study of the ways in which home computers can improve the life chances of young people at risk of dropping out of school.

The author's other publications include *Blue Collar Community,* a study of the steel-making community of South Chicago; *Growing Up Poor* and *The Uptown Kids* (with Terry Williams), studies of teenagers growing up in different low-income communities in the United States; and *Social Problems,* a comprehensive textbook about social problems and social policies in the United States.

To Susan, my partner in the Great Adventure, and Morris Janowitz, my mentor

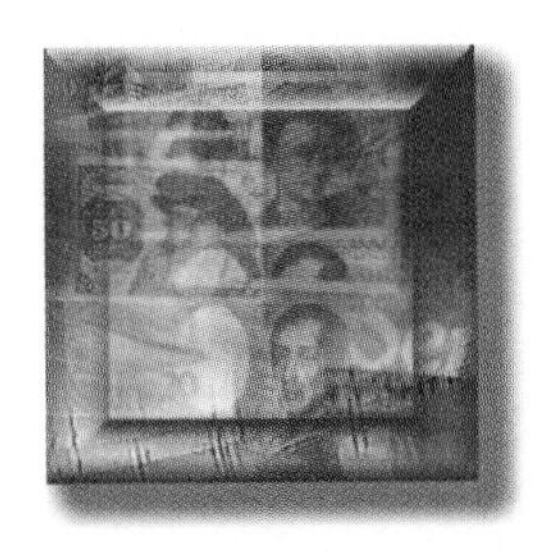

Brief Contents

PART 1
Social Foundations

PART 2
Social Dynamics

PART 3
Social Divisions

PART 4
Social Institutions

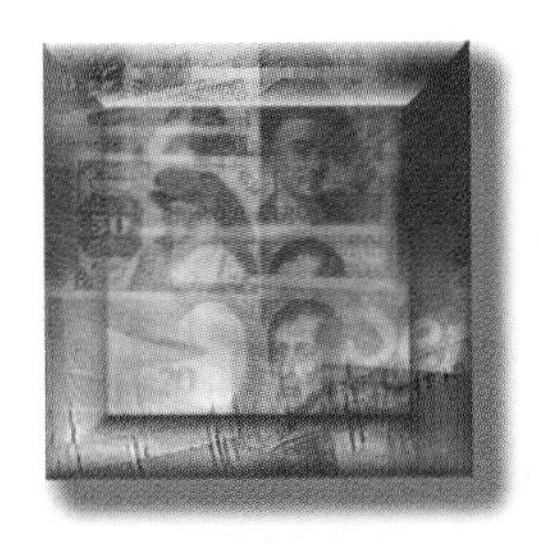

Contents

PART 3 Social Divisions 325

Chapter 11

Stratification and Social Mobility 326

Chapter 12

Inequalities of Social Class 358

Chapter 13

Inequalities of Race and Ethnicity 396

Chapter 20

Politics and Political Institutions 640

Chapter 21

Science, Technology, and Medicine 670

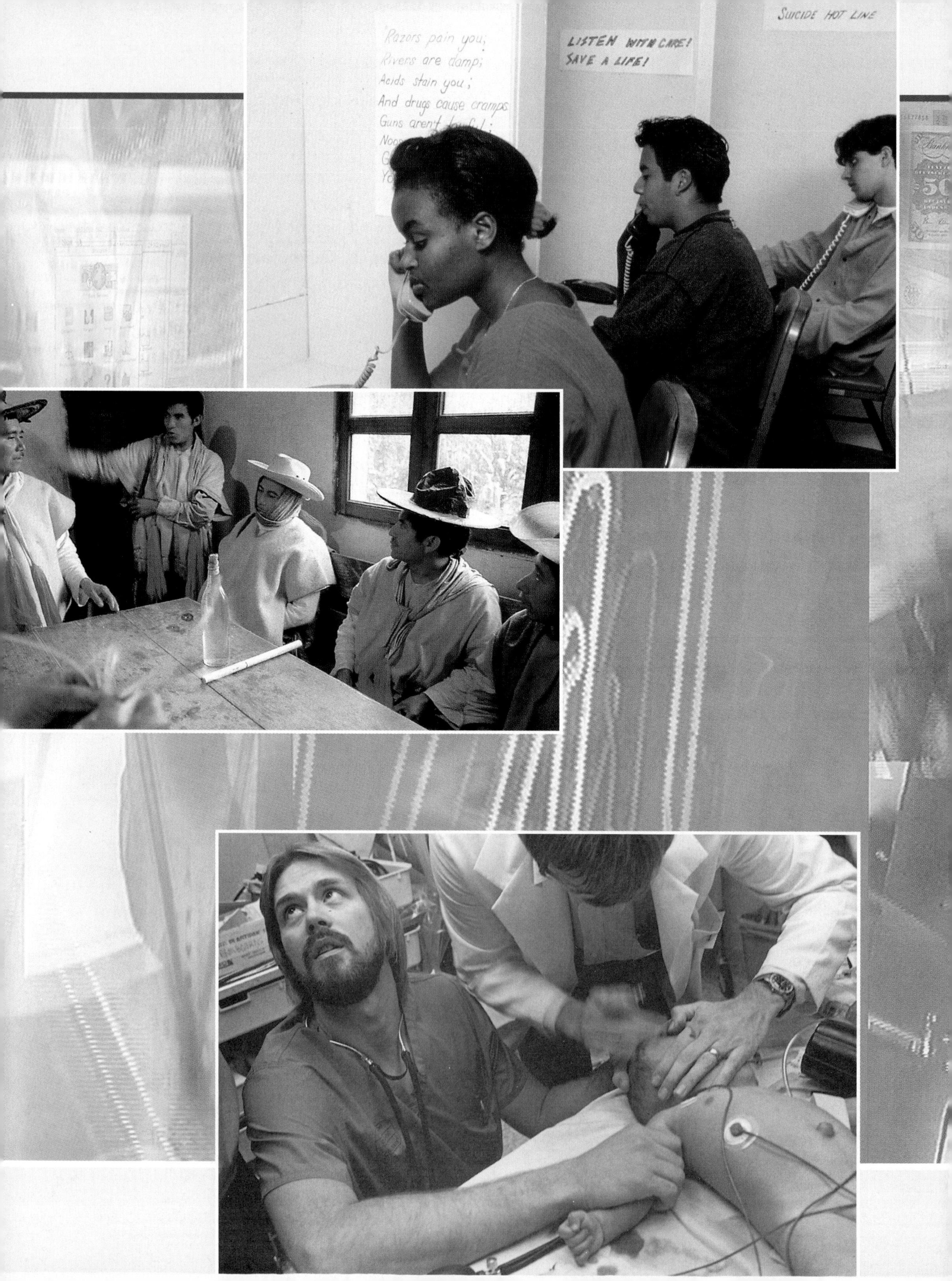
Razors pain you;
Rivers are damp;
Acids stain you;
And drugs cause cramps
LISTEN WITH CARE!
SAVE A LIFE!
SUICIDE HOT LINE

PART 1

SOCIAL FOUNDATIONS

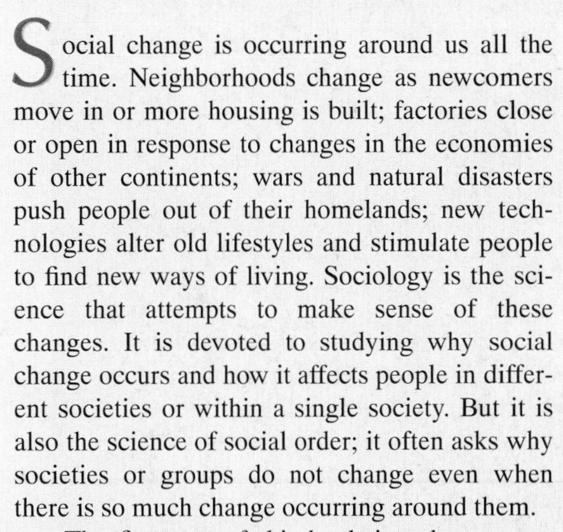

Social change is occurring around us all the time. Neighborhoods change as newcomers move in or more housing is built; factories close or open in response to changes in the economies of other continents; wars and natural disasters push people out of their homelands; new technologies alter old lifestyles and stimulate people to find new ways of living. Sociology is the science that attempts to make sense of these changes. It is devoted to studying why social change occurs and how it affects people in different societies or within a single society. But it is also the science of social order; it often asks why societies or groups do not change even when there is so much change occurring around them.

The first part of this book introduces some basic sociological concepts. Chapter 1 shows how sociology developed into a science that can help us understand the world around us and how it changes. Chapter 2 describes the methods sociologists use in conducting their research. When we talk about culture in Chapter 3, we are referring to the making of human consciousness—how we think and communicate and how these processes make society itself possible. Chapter 4 introduces the elements of social structure and explains how various kinds of societies have developed over time.

CHAPTER 1

SOCIOLOGY: AN INTRODUCTION

Everybody loves a parade. Throughout the world significant social occasions are marked by parades. The people walking or marching in the parade enjoy dressing in the appropriate uniforms or costumes and showing off before an admiring public. Spectators enjoy the pageantry, the excitement, the bands, the costumes, and much more. To sociologists and other keen observers, parades are an occasion to observe people displaying many of the attributes of their societies that they value and enjoy.

Mardi Gras parades are among the world's most joyous and exciting public displays. In New Orleans, where Mardi Gras parades are a famous feature of the city's culture, the colorful floats and often raucous masked "krews" dancing through different parts of the city and its suburbs are a major tourist attraction. Social scientists observe that Mardi Gras, or Carnival as it is often called, "expresses the society of New Orleans" (Trillin, 1998). An insightful and knowledgeable observer can watch how people from different social backgrounds—rich, poor, black, white, women, men, and so on—interact in the parades. And the way the parades and other Mardi Gras celebrations, especially the famous masked balls, have changed over the last few decades tells us a great deal about how New Orleans society is changing. The city's population is now more than 60 percent African American, so it is no wonder that elected officials have recently passed ordinances denying parade permits to krews like Comus and Rex, two of the oldest and most elite clubs, that exclude people of color. Nor is it surprising that as visitors and tourists from all over the United States flock to New Orleans for Mardi Gras, some of the parades include krews and floats representing other communities.

These larger forces of social change are visibly transforming the nature of the Mardi Gras parades. A sociologist can use observations of Mardi Gras to study the way the local society of New Orleans functions—for example, who presents a float and which members of the city's elite still participate in the parade. Such observations can also reveal the way conflict is affecting New Orleans society, perhaps resulting in more integrated parades in the center of the city and more segregated ones in the suburbs. The observant sociologist can also watch how interactions among parade participants and bystanders reflect issues of gender and sexuality. Some masked revelers cross-dress in order to try on different identities, and by watching them a skillful observer can learn a lot about how being a man or a woman is related to forms of communication and interaction as much as it is to biology. All of these subjects will come up in later chapters of this book. The point here is that sociologists love parades for what they can reveal about underlying social relations.

THE SOCIOLOGICAL IMAGINATION

It may take some thought to see why various kinds of parades tell us a great deal about our own and other societies. After all, one might argue that a parade is a special event. It is not meant to be "read" for its deeper social meanings. And no doubt most of the people who watch Mardi Gras parades are caught up in the fun of the event—by the creativity of the costumes, the flirtations that go on, the music and dance, and perhaps the drinking. They might be offended if someone asked them to look more analytically at the larger significance of their behavior. But a sophisticated understanding of social life requires some imagination. While some participants in an event like a parade may be entirely caught up in the moment, others may also have fun while simultaneously thinking about what is going on at a deeper level.

One of the main goals of sociology courses is to help you develop this ability to both participate in social life and step back and analyze the broader meanings of what is going on. This ability is often called the **sociological imagination.** In this book we hope to help you develop this special insight, which will equip you to use sociological knowledge in your daily life. Most of all, we hope to enable you to use your sociological imagination to gain wisdom about the society in which we all participate and for whose future we are all responsible.

Most people need some help in developing a sociological imagination. This is especially true when it comes to understanding their own place in what might be thought of as the "parade" of social life. People with a limited sociological imagination often fail to distinguish between social forces and personal troubles. If they are excluded from the "parade" because they are unemployed, they blame themselves for failing to do better; if they divorce, they blame each other. When they see crime, they blame "human nature"; when they see success, they praise individual achievement. But this tendency to think of life as a series of individual mistakes or successes blinds them to the fact that social conditions also shape individual lives, often in ways for which individuals can hardly be held accountable. And the habit of seeing events mainly in terms of how they affect individuals blinds people to the possibility of improving the way their society is organized.

Sociologists are concerned with how **social conditions** influence our lives as individuals. Social conditions are the realities of the life we create together as social beings. Conditions such as poverty or wealth or crime and drug use, for example, differ from biological facts (facts concerning our behavior and needs as animals) and psychological facts (facts about our patterns of behavior as individuals). Sociologists do not deny that psychological facts are important. Differences among individuals help some people cope with stress better than others, seize opportunities that others allow to slip by, or fail where others succeed. But before saying that the success or failure of an individual or group is due to psychological causes, the sociologist tries to look at how social conditions such as poverty or wealth, war, or changes in the availability of jobs affect the individual's chances of success.

According to sociologist C. Wright Mills, who made famous the term *sociological imagination,* people often believe that their private lives can be explained mainly in terms of their personal successes and failures. They fail to see the links between their own individual biographies and the course of human history. Often they blame themselves for their troubles without grasping the effects of social change on their lives. "The facts of contemporary history," Mills cautions, "are also facts about the success and the failure of individual men and women."

> When a society is industrialized, a peasant becomes a worker; a feudal lord is liquidated or becomes a businessman. When classes rise or fall, a man is employed or unemployed; when the rate of investment goes up or down, a man takes new heart or goes broke. When wars happen, an insurance salesman becomes a rocket launcher; a store clerk, a radar man; a wife lives alone; a child grows up without a father. (1959, p. 3)

According to Mills, neither a person's biography nor the history of a society can be understood unless we take into account the influence of each on the other. The social forces of history—war, depression or recession, increases in population, changes in production and consumption, and many other social conditions—become the forces that influence individuals to behave in new ways. But those new ways of behavior themselves become social forces and, in turn, shape history.

To take just one example, the not-so-distant ancestors of many African Americans were brought to the Western Hemisphere in slave ships. African Americans have experienced slavery, war, emancipation, segregation, and rural and urban poverty. In reaction to the historical forces that deprived them of full citizenship in the United States, African Americans developed a variety of behaviors, from the spirituals that expressed their deep feelings of religious faith and protest, to boycotts and demonstrations against segregation. These protests and demonstrations, which often take the form of a special kind of parade, themselves became powerful social forces that continue to shape the history of the American people.

By applying the sociological imagination to events such as the rapes of defenseless women and girls in Kosovo or the bombing of the federal office building in Oklahoma City, one can begin to "grasp history and biography and the relations between the two within society" (Mills, 1959, p. 6). One of the main objectives of this book is to help you apply your sociological imagination to an understanding of the social forces that are shaping your own place in the parade—that is, the forces that are shaping your biography and those of the people you care about. The sociological imagination can help us avoid needlessly blaming ourselves for the troubles we encounter in life. It can help us understand, for example, why some people are rich and powerful but many others are not; why the benefits of good health care or enriching education are available to some but not to others; or why women may find themselves resenting the men in their lives. The sociological imagination helps us sort out which facts about ourselves are explained by our place in society and which ones are a result of our own actions. Above all, the sociological imagination can suggest ways in which we can realistically bring about change in our lives and in society itself.

Sociology: The Human Science

Sociology is the scientific study of human societies and human behavior in the many groups that make up a society. Sociologists must ask difficult, sometimes embarrassing, questions about human life in order to explore the consequences of cataclysmic events like those that shut down factories or enslave an entire people. To understand the possible futures of people who confront such drastic changes, sociologists are continually seeking knowledge about what holds societies together and what makes them bend under the impact of major forces such as war and migration.

The Social Environment

The knowledge sociologists gather covers a vast range. Sociologists study religious behavior; conduct in the military; the behavior of workers and managers in industry; the activities of voluntary associations like parent-teacher groups and political parties; the changing relationships between men and women or between aging individuals and their own elderly parents; the behavior of groups in cities and neighborhoods; the activities of gangs, criminals, and judges; differences in the behaviors of entire social classes—the rich, the middle classes, the poor, the down-and-out; the way cities grow and change; the fate of entire societies during and after revolutions; and a host of other subjects. But how to make sure the information gathered is reliable and precise, how to use it to build theories of social cohesion and social change—that is the challenge faced by the young science of sociology.

As in any science, there are many debates in sociology about the appropriate ways to study social life and about which theories or types of theories best explain social phenomena. Most sociologists, however, would agree with the following position:

> Human actions are limited or determined by "environment." Human beings become what they are at any given moment not by their own free decisions, taken rationally and in full knowledge of the conditions, but under the pressure of circumstances which delimit their range of choice and which also fix their objectives and the standards by which they make choices. (Shils, 1985, p. 805)

This statement expresses a core idea of sociology: Individual choice is never entirely free but is always determined to some extent by a person's environment. In sociology, *environment* refers to all the expectations and incentives established by other people in a person's social world. For the sociologist, therefore, the environment within which an individual's biography unfolds is a set of people and groups and organizations, all with their own ways of thinking and acting. Certainly each individual has unique choices to make in life, but the social world into which that person was born—be it an Indian reservation, an urban ghetto, a comfortable suburb, or an immigrant enclave in a strange city—determines to varying degrees what those choices will be.

Levels of Social Reality

In their studies of social environments, sociologists look at behaviors ranging from the intimate glances of lovers to the complex coordination of a space shuttle launch. Thus for purposes of analysis we often speak of social behavior as occurring at three different levels of complexity: micro, middle, and macro.

The **micro level** of sociological observation is concerned with the behaviors of the individual and his or her immediate others—that is, with patterns of interaction among a few people. One example is Erving Goffman's studies of the routine behaviors of everyday life. Goffman's research showed how seemingly insignificant ways of acting in public actually carry significant meanings. Thus, in a study titled "Territories of the Self" (1972), Goffman categorized some of the ways in which we use objects as "markers" to claim a personal space:

> Markers are of various kinds. There are "central markers," being objects that announce a territorial claim, the territory radiating outward from it, as when sunglasses and lotion claim a beach chair, or a purse a seat in an airliner, or a drink on a bar the stool in front of it. . . . There are "boundary markers," objects that mark the line between two adjacent territories. The bar used in supermarket checkout counters to separate one customer's batch of articles from the next is an example. (pp. 41–42)

The last time you placed your sweater or book on the empty seat next to you on a bus, you told yourself that when someone came for the seat you would take up your things. But you hoped that the stranger who was coming along the aisle would get your message and choose another seat; you would claim your extra space as long as possible. You communicated all this by the manner in which you placed your marker and the persistence with which, by your body language, you defended "your" space.

Some sociologists deal almost exclusively with a much larger scale, or **macro level,** of analysis. The macro level of social life refers to whole societies and the ways in which they are changing—that is, to revolutions, wars, major changes in the production of goods and services, and similar social phenomena that involve very large numbers of people. One example of macrosociological analysis is the study of how the shift from heavy manufacturing to high-tech industries has affected the way workers earn their livings. Another is the study of how the invasion and settlement of the American West in the nineteenth and early twentieth centuries gave rise to the beliefs and actions that drove Native Americans onto reservations.

Middle-level social phenomena are those that occur in communities or in organizations such as businesses and voluntary associations. Middle-level social forms are smaller than entire societies but larger than the micro-level social forms in which everyone involved knows everyone else or is in close proximity to the others (as on a bus or in a classroom). The drama that surrounds the firing of a coach on a sports team, and the reorganization of personnel that often follows, is an example of social change at the middle level of social analysis.

Each of these passengers in a railroad station waiting room has claimed a personal territory. Each has taken a separate bench and has chosen to sit at one end of the bench. If another person came to that bench, he or she would probably choose to sit as far away as possible.

STUDY CHART Levels of Sociological Analysis

ANALYTICAL LEVEL		SOCIAL BEHAVIORS STUDIED	TYPICAL QUESTIONS
	Macro	Revolutions; intercontinental migrations; emergence of new institutions.	How are entire societies or institutions changing?
	Middle	Relations in bureaucracies; social movements; participation in communities, organizations, tribes.	How does bureaucracy affect personality? Do all social movements go through similar stages?
	Micro	Interaction in small groups; self-image; enactment of roles.	How do people create and take roles in groups? How are group structures created?

These three levels of sociological analysis can be helpful in understanding the experience of the immigrants from Asia, Africa, or Latin America who may be appearing in your town or community, or one nearby. Macro-level social forces like war or overpopulation may account for the influx of immigrants. Middle-level social forces such as the availability of work at low wage and skill levels, or the presence of earlier arrivals from other nations, may help explain why certain immigrant groups become concentrated in particular communities within the United States. And at the micro level of analysis there will be, especially at first, important differences in the way immigrant and native-born people interact on a daily basis.

The study chart above presents the three basic levels of sociological analysis, as well as some examples of the types of studies conducted at each level. Throughout this book we show how the sociological imagination can be applied at different levels of society. In this chapter we set the stage by describing the basic perspectives from which modern sociologists approach the study of social conditions. In the next chapter we outline the procedures and methods used by sociologists in conducting their research. We begin here with a brief description of the origins and development of the science of sociology.

FROM SOCIAL THOUGHT TO SOCIAL SCIENCE

Like all the sciences, sociology developed out of prescientific longings to understand and predict. The central questions of sociology have been pondered by the world's great thinkers since the earliest periods of recorded history. The ancient Greek philosophers believed that human societies inevitably arose, flourished, and declined. They tended to perceive the past as better than the present, looking back to a "golden age" in which social conditions were presumed to have been better than those of the degraded present. Before the scientific revolution of the seventeenth century, the theologians and philosophers of medieval Europe and the Islamic world also believed that human misery and strife were inevitable. As the Bible put it, "The poor always ye have with you." Mere mortals could do little to correct social conditions, which were viewed as the work of divine Providence.

The Age of Enlightenment

The roots of modern sociology can be found in the work of the philosophers and scientists of the Great Enlightenment, which had its origins in the scientific discoveries of the seventeenth century. That pivotal century began with Galileo's "heretical" proof that the earth was not the center of the universe; it ended with the publication of Isaac Newton's *Principia Mathematica.* Newton is often credited with the founding of modern science. He not only discovered the laws of gravity and motion but, in developing the calculus, also provided later generations with the mathematical tools whereby further discoveries in all the sciences could be made.

Hard on the heels of this unprecedented progress in science and mathematics came a theory of human progress that paved the way for a "science of humanity." Francis Bacon in England, René Descartes and Blaise Pascal in France, and Gottfried Wilhelm Leibniz in Germany were among the philosophers who recognized the social importance of scientific discoveries. Their writings emphasized the idea of progress guided by human reason and opposed the dominant notion that the human condition was ordained by God and could not be improved through human actions (Bury, 1932; Nisbet, 1969).

Today we are used to inventions crowding one upon another. Between the childhood of our grandparents and our own adulthood, society has undergone some major transformations: from agrarian to industrial production;

from rural settlements and small towns to large cities and expanding metropolitan regions; from reliance on wood and coal as energy sources to dependence on electricity and nuclear power; from typewriters to computers. But in the seventeenth century people were used to far more stability. Ways of life that had existed since the Middle Ages were not expected to change in a generation.

The rise of science transformed the social order. As was often said at the time, science "broke the cake of custom." New methods of navigation made it possible to explore and chart the world's oceans and continents. Applied to warfare, scientific knowledge enabled Europeans to conquer the peoples of Africa, Asia, and the Western Hemisphere. In Europe, those conquests opened up new markets and stimulated new patterns of trade that hastened the growth of some regions and cities and the decline of others. The entire human world had entered a period of rapid social change that continues today and shows no signs of ending.

The Age of Revolution

The vehicle of social change was not science itself, since relatively few people at any level of society were practicing scientists. Rather, the modern era of rapid social change is a product of the many new social ideas that captured people's imagination during the eighteenth century. The series of revolutions that took place in the American colonies, in France, and in England all resulted in part from social movements unleashed by the triumphs of science and reason. The ideas of human rights (that is, the rights of all humans, not just the elite), of democracy versus rule by an absolute monarch, of self-government for colonial peoples, and of applying reason and science to human affairs in general—all are currents of thought that arose during this period.

The revolutions of the eighteenth century loosed a torrent of questions that could not even have been imagined before. The old order of society was breaking down as secular (i.e., nonreligious) knowledge replaced sacred traditions. The study of laws and lawmaking and debates about justice in society began to replace the idea that kings and other leaders had a "divine right" to rule. Communities were breaking apart; courts and palaces and great estates were crumbling as people struggled to be free. What would replace them? Would the rule of the mob replace the rule of the monarch? Would greed and envy replace piety and faith? Would there be enough opportunities in the New World for all the people who were being driven off the land in the Old World? Would the factory system become the new order of society, and if so, what did that imply for the future of society?

No longer could the Scriptures or the classics of ancient Greece and Rome be consulted for easy answers to such questions. Rather, it was becoming evident that new answers could be discovered through the **scientific method:** repeated observation, careful description, the formulation of theories based on possible explanations, and the gathering of additional data about the questions that followed from those theories. Why not use the same methods to create a science of human society? This ambitious idea led to the birth of sociology. It is little wonder that the French philosopher Auguste Comte thought of sociology, even in its infancy, as the "queen of sciences," one that would soon take its rightful place beside the reigning science of physics. It was he who coined the term *sociology* to designate the scientific study of society. Comte believed that the study of social stability and social change was the most important subject for sociology to tackle. He made some of the earliest attempts to apply scientific methods to the study of social life.

The Great European Sociologists

In the nineteenth century an increasing number of philosophers and historians began to see themselves as specializing in the study of social conditions and social change. They attempted to develop global theories of social change based on the essential qualities of societies at different stages of human history, and they devoted much of their attention to comparing existing societies and civilizations, both past and present. As Comte put it, the age of discovery had revealed such an array of societies that "from the wretched inhabitants of Tierra del Fuego to the most advanced nations of Western Europe" there is such a great diversity of societies that comparisons among them will yield much insight into why they differ and how they change (1971/1854, p. 48).

The early sociologists tended to think in macrosociological terms. Their writing dealt with whole societies and how their special characteristics influence human behavior and social change. Karl Marx, for example, was both highly appreciative and extremely critical of the societies of his day. His analysis of those societies led him to predict major upheavals arising from conflicts between the owners of wealth and the impoverished workers. The French sociologist Émile Durkheim did not agree with Marx's prediction that violent revolutions would transform society. Durkheim's theories explain social change as resulting from population growth and from changes in the ways in which work and community life are organized. Still another pioneering European theorist, Max Weber, was the first to understand the overwhelming importance of bureaucratic forms of social organization in modern societies and to point out their increasing dominance in the lives of individuals. (See Box 1.1.)

Sociology: A Global Science

As it originated in the work of Auguste Comte, represented here by his statue in the Place de La Sorbonne in Paris, sociology was based on the study of history and philosophy. In the nineteenth century it was dominated by university scholars with a deep understanding of Western social thought. There were a few exceptions, such as W. E. B. Du Bois and the brilliant Muslim social thinker Ibn Khaldun, but for the most part the founders of sociology were European men.

Women of the nineteenth century were often barred from the professions, including academic professions, and had difficulty conducting research or finding other outlets for their ideas. A notable exception was Harriet Martineau (1802–1876), who is thought to have been the first woman to contribute to the emerging field of sociology. Martineau brought Comte's works to English readers and wrote important sociological interpretations of the early phases of capitalism and modernity. She was also a social reformer whose writings on the plight of childen and women in British factories were highly influential (Hoecker-Drysdale, 1992).

Today sociology is far more diverse in every aspect, and women throughout the world are well represented among the field's most innovative minds. The photo of Manjula Giri shows a modern sociologist from a developing nation, Nepal, who is both a university-trained scholar and a dedicated social activist. Giri has written about the failure of revolution in Nepal to improve the conditions of rural women. In her own village she is using her sociological skills to help local women to develop a women's farming cooperative. She has also started literacy classes and is planning to expand her work to other villages in the region. Giri is an example of the scholar-activist, a dual role that is increasingly common throughout the poorer regions of the world.

Manjula Giri

Auguste Comte

Harriet Martineau

BOX 1.1 USING THE SOCIOLOGICAL IMAGINATION

Sociological Pioneers of the Nineteenth Century

Most sociologists would agree that the nineteenth-century social theorists who had the greatest and most lasting influence on the field were Karl Marx, Émile Durkheim, and Max Weber. All three applied the new concepts of sociology to gain an understanding of the immense changes occurring around them.

German-born Karl Marx (1818–1883) became a radical philosopher as a young man and was embroiled in numerous insurrections and attempts at revolution in Germany and France. Forced to flee Germany after the abortive European revolution of 1848, he lived and worked in England for the rest of his life. Often penniless, Marx worked for hours on end in the library of the British Museum, where he developed the social and economic theories that would have a major influence on sociological thought. His famous treatise *Capital* is a detailed study of the rise of capitalism as a dominant system of production. In this work and elsewhere, Marx set forth an extremely powerful theory to explain the transformations taking place as societies became more industrialized and urbanized. He believed that those transformations would inevitably end in a revolution in which the workers would overthrow capitalism, but he also felt that revolution could be hastened through political action.

Karl Marx

Émile Durkheim (1858–1917) was the founder of scientific sociology in France. His books, among which the best known are *The Division of Labor in Society, Rules of the Sociological Method,* and *Suicide,* were pioneering examples of the use of comparative data to assess the directions and consequences of social change. The first university professor with a chair in the "social sciences," Durkheim was soon surrounded by a brilliant group of academic disciples who were deeply interested in understanding the vast changes that occurred in societies as they became more populous, more urbanized, and more technologically complex. In 1898 Durkheim and his colleagues established the first scientific journal in sociology, *L'Année sociologique* (*The Sociological Year*). This journal and much of Durkheim's own writing were among the first examples of the application of statistics to social issues. (See the discussion of suicide in Chapter 2.)

Max Weber (1864–1920) was a German historian, economist, and sociologist. Weber's life, like Durkheim's, spanned much of the second half of the nineteenth century and the early decades of the twentieth. Like the other early sociologists, therefore, Weber witnessed the tumultuous changes that were bringing down the old order. He saw monarchies tottering in the face of demands for democratic rule. He observed new industries

Each of these three sociological pioneers based his theories on detailed reviews of the history of entire societies. Marx wrote about the origins of free enterprise and capitalism in the Western nations. Durkheim wrote about the transformations that occur as societies evolve from hunting and gathering to industrial production. Weber produced volumes about the organizations found in precapitalist and capitalist societies. But to become a science rather than merely a branch of philosophy, sociology had to build on the research of its founders. The twentieth century brought changes of such magnitude at every level of society that sociologists were in increasing demand. Their mission was to gain new information about the scope and meaning of social change.

The Rise of Modern Sociology

We credit the European social thinkers and philosophers with creating sociology, but nowhere did the new science find more fertile ground for development than in North America. By the beginning of the twentieth century, sociology was rapidly acquiring new adherents in the United States and Canada, partly owing to the influence of European sociologists like Marx and Durkheim, but even more because of the rapid social changes occurring in North America at the time. Waves of immigrants to cities and towns, the explosive growth of population and industry in the cities, race riots, strikes and labor strife, moral crusades against crime

i

Émile Durkheim

and markets spanning the globe and linking formerly isolated peoples. He saw and described the rise of modern science and jurisprudence and modern ways of doing business. The growing tendency to apply rational decision-making procedures, rather than merely relying on tradition, was for Weber a dramatic departure from the older ways of feudal societies and mercantile aristocracies. Weber compared many different societies to show how new forms of government and administration were evolving.

Max Weber

All three of these pioneers in sociology were scholars of great genius. They were also political activists. Marx, of course, was the most revolutionary of the three and devoted much of his energy to the international socialist movement. Durkheim was a lifelong socialist but was more moderate than Marx. While he took stands on many political issues, he did not devote himself to political activities. As a young man Weber had been involved in the movement to create a unified German nation, but as a mature scholar he developed a belief in "value-free" social science. A social scientist might draw research questions from personal political beliefs, but the research itself must apply scientific methods. This view, which Durkheim also shared, did much to advance sociology to the level of a social science rather than just a branch of philosophy.

and vice and alcohol, the demand for woman suffrage—these and many other changes caused American sociology to take a new turn. There was increasing demand for knowledge about exactly what changes were occurring and who was affected by them. In North America, therefore, sociologists began to emphasize the quest for facts about changing social conditions—that is, the empirical investigation of social issues.

Empirical information refers to carefully gathered, unbiased data regarding social conditions and behavior. In general, modern sociology is distinguished by its relentless and systematic search for empirical data to answer questions about society. Journalists, for example, also seek the facts about social conditions, but they must cover many different events and situations and present them as "stories" that will attract their readers' interest. Because they usually cannot dwell on one subject very long, journalists frequently must content themselves with citing examples and quoting experts whose opinions may or may not be based on empirical evidence. In contrast, sociologists study a situation or phenomenon in more depth, and when they do not have enough facts they are likely to say, "That is an empirical question. Let's see what the research tells us, and if the answers are inconclusive we will do more research." Evidence based on measurable effects and outcomes is required before one can make an informed decision about an issue.

To use the sociological imagination to ask relevant questions and to seek answers to those questions backed by evidence that can be verified by others are among the chief goals of modern sociology. Anyone can make assertions about society or about why people behave the way they do. "I think it's human nature to act selfishly, no matter what kind of education people have" is a common assertion that is not backed up by any solid evidence. As you read this book you will learn that to strengthen your sociological imagination you must learn how to apply evidence to your views and admit that your opinions can be modified by that evidence.

The Social Surveys The empirical focus of American sociology began largely as an outgrowth of the reform movements of the late nineteenth and early twentieth centuries. During this period the nation had not yet recovered from the havoc created by the Civil War. Southern blacks were migrating to the more industrialized North in ever-increasing numbers, and at the same time millions of European immigrants were finding ill-paid jobs in the larger cities, where cheap labor was in great demand. By the turn of the century, therefore, the nation's cities were crowded with poor families for whom the promise of "gold in America" had become a tarnished dream. In this time of rapid social change, Americans continually debated the merits of social reform and proposed new solutions for pressing social issues. Some called for socialism, others for a return to the free market, or a ban on labor organizations, or an end to immigration, or the removal of black Americans to Africa. But where would the facts to be used in judging those ideas come from?

In order to gain empirical information about social conditions, dedicated individuals undertook numerous "social surveys." Jacob Riis's (1890) account of life on New York's Lower East Side; W. E. B. Du Bois's (1967/1899) survey of Philadelphia blacks; Emily Balch's (1910) depiction of living conditions among Slavic miners and steelworkers in the Pittsburgh area; and Jane Addams's famous *Hull House Maps and Papers* (1895), which described the lives of her neighbors in Chicago's West Side slum area—these and other carefully documented surveys of the living conditions of people experiencing the effects of rapid industrialization and urbanization left an enduring mark on American sociology. Box 1.2 presents an excerpt from the landmark survey conducted by Du Bois, the first black sociologist to gain worldwide recognition. Du Bois helped direct sociological research to racial and social issues in minority communities, using empirical data to provide an objective account of the dismal social conditions of northern blacks at the turn of the century.

The Chicago School and Human Ecology By the late 1920s the United States had become the world leader in sociology. The two great centers of American sociological research were the University of Chicago and Columbia University. At these universities and others influenced by them, two distinct approaches to the study of society evolved. The *Chicago school* emphasized the relationship between the individual and society, whereas the major East Coast universities, which were more strongly influenced by European sociology, tended toward macro-level analyses of social structure and change.

The sociology department at the University of Chicago (the oldest in the nation) extended its influence to many other universities, especially in the Midwest, South, and West. At that time the department was under the leadership of Robert Park and his younger colleague Ernest Burgess. Park in particular is associated with the Chicago school. His main contribution was to develop an agenda for sociological research that used the city as a "social laboratory." Park favored an approach in which facts concerning what was actually occurring among people in their local communities (at the micro and middle levels) would be collected within a broader theoretical framework. That framework attempted to link macro-level changes in society, such as industrialization and the growth of urban populations, to patterns of settlement in cities and to how people actually lived in cities.

In one of his essays on this subject, Park began with the idea that industrialization causes the breakdown of traditional "primary-group" attachments (those of family members, age-mates, or clans). After stating the probable relationship between the effects of industrialization and high rates of crime, Park asked several specific questions:

> What is the effect of ownership of property . . . on truancy, on divorce, and on crime?
>
> In what regions and classes are certain kinds of crime endemic? In what classes does divorce occur most frequently? What is the difference in this respect between farmers and, say, actors?
>
> To what extent in any given [ethnic] group . . . do parents and children live in the same world, speak the same language, and share the same ideas, and how far do the conditions found account for juvenile delinquency in that particular group? (1967/1925a, p. 22)

This set of research questions, of which those quoted here are a small sample, inspired and shaped the work of hundreds of sociologists who were influenced by the Chicago school. To this day Chicago remains the most systematically studied city in the United States, although similar research has been carried out in other

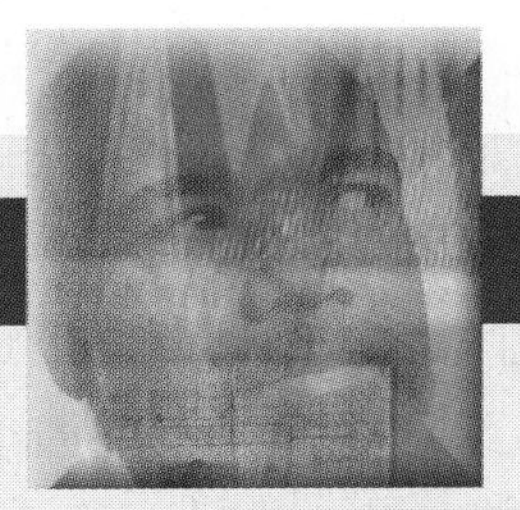

THE PERSONAL CHALLENGE

Social Change and Sociological Careers

It is reasonable to wonder, as you work your way through this course, what the study of sociology can do for you. All introductory college courses are designed to acquaint students with fields that they may want to delve into further in the future. But perhaps more than other introductory courses, sociology can help you understand how society and social change are affecting your own life and the lives of people around you. Sociology is not unique in that sense, but it offers some powerful concepts and methods for understanding how the world is changing and with what consequences, personal and otherwise.

President Bill Clinton reads sociology. So do thousands of leaders in business and government who want to get a handle on how current social changes may be altering the way they go about their professional and personal lives. Sociology is a growing field throughout the world. As the pace of global social change quickens—with new cities, new technologies, and new forms of cooperation and conflict appearing all the time—the contributions of sociological theory and practice become increasingly important.

If you should decide to major in sociology and pursue a career in this exciting field, you will join approximately 25,000 professional sociologists currently at work in North America. Of this number, the majority are employed in college teaching. Approximately one third work in careers in applied fields outside colleges and universities (Donow, 1990; Lyson & Squires, 1993). (For more information on careers in sociology, see the Appendix at the end of the book.) Of course, you need not become a professional sociologist to find that knowledge of the field gives you great advantages in understanding society and social change.

Sociology is an excellent background for careers in law, business, public administration, health, and law enforcement, to name just some of the major professional areas that are voracious consumers of sociological information. Market research, which seeks to predict and shape consumer behavior, is a form of applied sociology. The allied field of opinion research, which charts changes in beliefs and attitudes, is also based almost entirely on sociological principles and methods. Professionals in the health care industry need to know about changes in diet and other health-related behaviors (smoking, exercise, and so on) among different population groups, and people with backgrounds in sociology are often asked to fill these information needs. Indeed, wherever major changes are occurring you will find sociologists at work along with other professionals, charting the changes and attempting to predict their consequences.

The study of sociology beyond the introductory level can lead to advanced skills in computation, data analysis, and statistics. Research experience in the field can greatly increase your ability to understand social situations and analyze the functioning or malfunctioning of all types of organizations. This can be a valuable skill in business or social work. People with sociological training are also better able to get along with people from different cultural backgrounds. As the world continues to shrink and people from different cultures and societies come into greater contact with one another, sociological skills become increasingly valuable.

large cities throughout the nation. From studies of the linguistic diversity of African peoples to attempts to understand the subtle negotiations by which youth gangs divide up an urban "turf," the insights of the Chicago school remain a vital aspect of contemporary sociology (Glausiusz, 1997).

As you can see from the types of questions Park asked, the distinctive orientation of the Chicago school was its emphasis on the relationships among social order, social disorganization, and the distribution of populations in space and time. Park and Burgess called this approach **human ecology.** With many modifications, it remains an important, though not dominant, perspective in contemporary sociology.

Human ecology, as Park and others defined it, is the branch of sociology that is concerned with population growth and change. In particular, it seeks to discover how populations organize themselves to survive and prosper. Human ecologists are interested in how groups that are organized in different ways compete and cooperate. They also look for forms of social organization that may emerge as a group adjusts to life in new surroundings.

A key concept for human ecologists is *community.* There are many ways of defining this term, just as there are many ways of defining most of the central concepts of sociology. From the ecological perspective, however, the term *community* usually refers to a population that

BOX 1.2 SOCIOLOGICAL METHODS

The Social Survey

W. E. B. Du Bois was one of the first American sociologists to publish highly factual and objective descriptions of life in American cities. An African American, Du Bois earned his doctorate in philosophy at Harvard before the turn of the twentieth century, when sociology was still regarded as a subfield of that discipline. But, as can be seen in this excerpt from his account of black life in Philadelphia about a century ago, Du Bois was able to sharpen his arguments about the effects of racial discrimination with simple but telling statistics.

W. E. B. Du Bois

> For a group of freedmen the question of economic survival is the most pressing of all questions; the problem as to how, under the circumstances of modern life, any group of people can earn a decent living, so as to maintain their standard of life, is not always easy to answer. But when the question is complicated by the fact that the group has a low degree of efficiency on account of previous training; is in competition with well-trained, eager and often ruthless competitors; is more or less handicapped by . . . discrimination; and finally, is seeking not merely to maintain a standard of living but steadily to raise it to a higher plane—such a situation presents baffling problems to the sociologist. . . .
>
> And yet this is the situation of the Negro in Philadelphia; he is trying to better his condition; is seeking to rise; for this end his first need is work of a character to engage his best talents, and remunerative enough for him to support a home and train up his children well. The competition in a large city is fierce, and it is difficult for any poor people to succeed. The Negro, however, has two especial difficulties: his training as a slave and freedman has not been such as make the average of the race as efficient and reliable workmen as the average [native-born] American or as many foreign immigrants. The Negro is, as a rule, willing, honest and good-natured; but he is also, as a rule, careless, unreliable and unsteady. This is without doubt to be expected in a people who for generations have been trained to shirk work; but an historical excuse counts for little in the whirl and battle of breadwinning. Of course, there are large exceptions to this average rule; there are many Negroes who are as bright, talented and reliable as any class of workmen, and who in untrammeled competition would soon rise high in the economic scale, and thus by the law of the survival of the fittest we should soon

carries out major life functions (e.g., birth, marriage, death) within a particular territory. Human ecology does not assume that there will ever be a "steady state" or an end to the process of change in human communities. Instead, it attempts to trace the change and document its consequences for the social environment. What happens when newcomers "invade" a community? In what way are local gangs a response to recent changes in population or in the ability of members of a community to compete for jobs? Not only do populations change, but people's preferences and behaviors also continually change. So do the technologies for producing the goods and services we want. As a result, our ways of getting a living, our modes of transportation, and our choices of leisure activities create constant change, not just in communities but in entire societies.

The Chicago school became known for this "ecological" approach, the idea that the study of human society have left at the bottom those inefficient and lazy drones who did not deserve a better fate. However, in the realm of social phenomena the law of survival is greatly modified by human choice, wish, whim and prejudice. And consequently one never knows when one sees a social outcast how far this failure to survive is due to the deficiencies of the individual, and how far to the accidents or injustice of his environment. This is especially the case with the Negro. Every one knows that in a city like Philadelphia a Negro does not have the same chance to exercise his ability or secure work according to his talents as a white man. Just how far this is so we shall discuss later; now it is sufficient to say in general that the sorts of work open to Negroes are not only restricted by their own lack of training but also by discrimination against them on account of their race; that their economic rise is not only hindered by their present poverty, but also by a widespread inclination to shut against them many doors of advancement. . . .

What has thus far been the result of this complicated situation? What do the mass of the Negroes of the city at present do for a living, and how successful are they in those lines? And in so far as they are successful, what have they accomplished, and where they are inefficient in their present sphere of work, what is the cause and remedy? These are the questions before us, and we proceed to answer the first in this chapter, taking the occupations of the Negroes of the Seventh Ward first. . . .

Of the 257 boys between the ages of ten and twenty, who were regularly at work in 1896, 39 percent were porters and errand boys; 25.5 percent were servants; 16 percent were common laborers, and 19.5 percent had miscellaneous employment. The occupations in detail are as follows:

Total population, males 10 to 20		651	
Engaged in gainful occupations		257	
Porters and errand boys		100	39.0%
Servants		66	25.5%
Common laborers		40	16.0%
Miscellaneous employment			
Teamsters	7		
Apprentices	6		
Bootblacks	6		
Drivers	5		
Newsboys	5		
Peddlers	4		
Typesetters	3		
Actors	2		
Bricklayers	2	51	19.5%
Hostlers	2		
Typists	2		
Barber	1		
Bartender	1		
Bookbinder	1		
Factory hand	1		
Rubber worker	1		
Sailor	1		
Shoemaker	1		
		257	100.0%

Note: This simple table includes much useful information, but modern tables present data more efficiently. See Chapter 2 for a detailed discussion of the presentation of data in tables.

Source: Du Bois, 1967/1899, pp. 97–99.

should begin with empirical questions about population size, the distribution of populations over territories, and the like. The human ecologists recognized that there are many other processes by which society is shaped, but their most important contribution to the discipline of sociology was to include the processes by which populations change and communities are formed.

Modern ecological theories also consider the relations between humans and their natural environment. We will see that the way people earn their livelihood, the resources they use, the energy they consume, and their efforts to control pollution all have far-reaching consequences not only for their own lives but also for the society in which they live. These patterns of use and consumption also have an increasing impact on the entire planet—so much so that ecological problems are becoming an ever more important area of sociological research. (See the Mapping Social Change feature on pages 16 and 17.)

MAPPING SOCIAL CHANGE

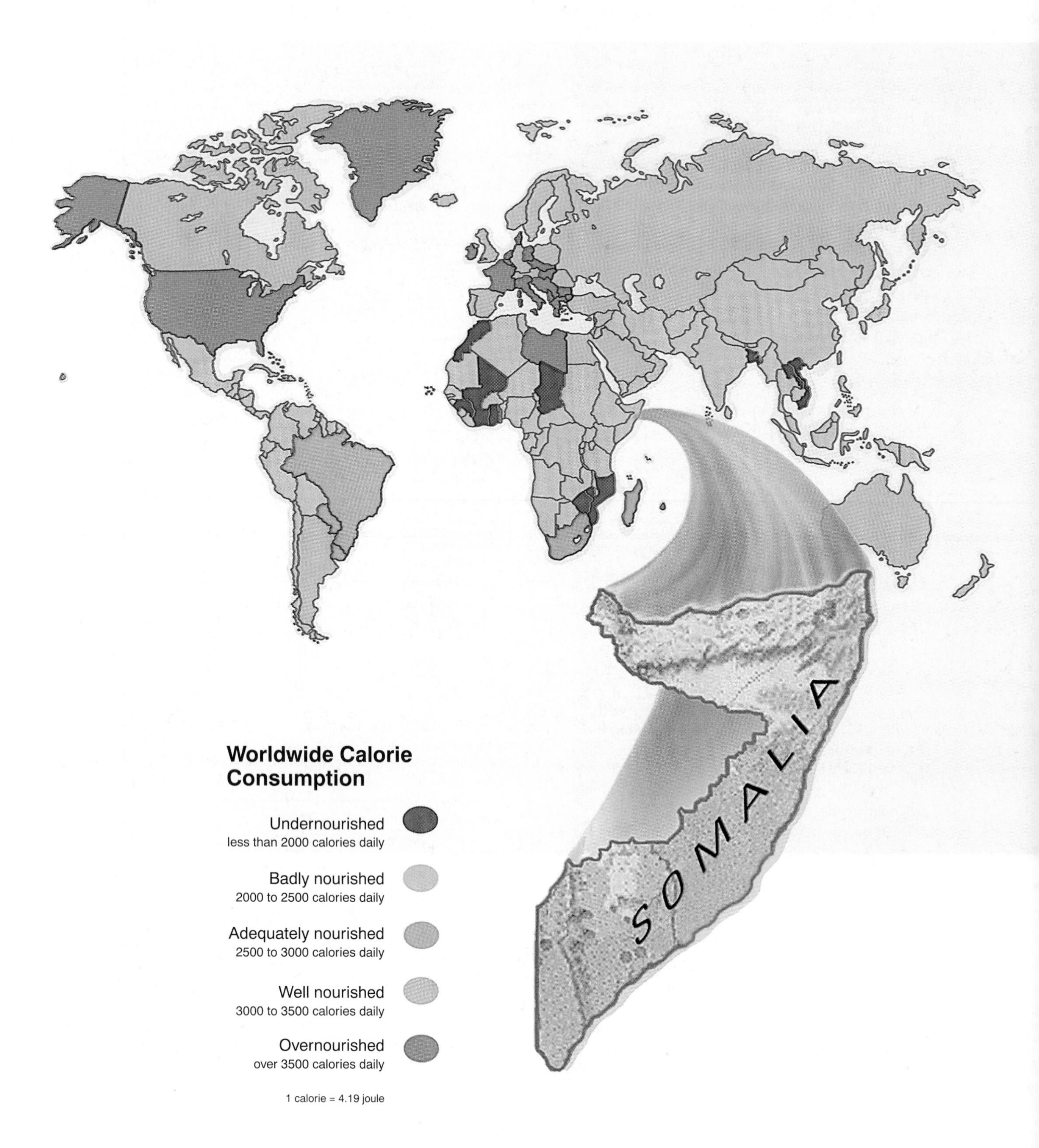
SOMALIA
Worldwide Calorie Consumption
Undernourished
less than 2000 calories daily
Badly nourished
2000 to 2500 calories daily
Adequately nourished
2500 to 3000 calories daily
Well nourished
3000 to 3500 calories daily
Overnourished
over 3500 calories daily
1 calorie = 4.19 joule

Worldwide Calorie Consumption and Areas of Famine

From a biological perspective, the success of any species is measured by how well it meets the broad requirements of population growth and maintenance. Every day the world's 5 billion people seek and obtain enough food to convert into bodily energy, a minimum of perhaps 1,500 calories a day for survival at starvation levels (although this varies greatly with climate and other factors). More than 70 percent of the world's population is inadequately nourished (that is, obtains fewer than 2,500 calories a day), often while engaging in hard physical labor, whereas a smaller proportion, including most (but by no means all) of the North American population, lives comfortably well above the daily minimum.

Ecological theories help explain malnutrition and starvation in some areas of the world. In Somalia, for example, the combination of a desert environment and warring political factions produces social instability that leads to persistent famine. The problem is not so much that food is unavailable as that supplies cannot be delivered or distributed effectively.

MAJOR SOCIOLOGICAL PERSPECTIVES

Sociological *perspectives* are sets of ideas and theories that sociologists use in attempting to understand various problems of human society, such as the problem of population size, the problem of conflict between populations, the problem of how people become part of a society, and other issues that we will encounter throughout this book. Although human ecology remains an important sociological perspective, it is by no means the only one employed by modern sociologists. Other perspectives, to which we now turn, guide empirical description and help explain social stability and social change.

Interactionism

Interactionism is the sociological perspective that views social order and social change as resulting from all the immense variety of repeated interactions among individuals and groups. Families, committees, corporations, armies, entire societies—indeed, all the social forms we can think of—are a result of interpersonal behavior in which people communicate, give and take, share, compete, and so on. If there were no exchange of goods, information, love, and all the rest—that is, if there were no interaction among people—obviously there could be no social life at all.

The interactionist perspective usually generates analyses of social life at the level of interpersonal relationships, but it does not limit itself to the micro level of social reality. It also looks at how middle- and macro-level phenomena result from micro-level behaviors or, conversely, how middle- and macro-level influences shape the interactions among individuals. From the interactionist perspective, for example, a family is a product of interactions among a set of individuals who define themselves as family members. But each person's understanding of how a family ought to behave is a product of middle- and macro-level forces: religious teachings about family life, laws dealing with education or child support, and so on. And these are always changing. You may have experienced the consequences of changing values that cause older and younger family members to feel differently about such issues as whether a couple should live together before marrying. In sum, the interactionist perspective insists that we look carefully at how individuals interact, how they interpret their own and other people's actions, and the consequences of those actions for the larger social group (Blumer, 1969b; Frank, 1988).

The general framework of interactionism contains at least two major and quite different sets of issues. One set has to do with the problems of exchange and choice: How can social order exist and groups or societies maintain stability when people have selfish motives for being in groups—that is, when they are seeking to gain as much personal advantage as they can? The second set of issues involves how people actually manage to communicate their values and how they arrive at mutual understandings. Research and explanations of the first problem fall under the heading of "rational choice" (or exchange theory), while the second issue is addressed by the study of "symbolic interaction." In recent years these two areas of inquiry have emerged as quite different yet increasingly related aspects of the study of interaction.

Rational Choice—The Sociological View Adam Smith, whose famous work *The Wealth of Nations* (1910/1776) became the basis for most subsequent economic thought, believed that individuals always seek to maximize their pleasure and minimize their pain. If over time they are allowed to make the best possible choices for themselves, they will also produce an affluent and just society. They will serve others, even when they are unaware that they are doing so, in order to increase their own benefit. They will choose a constitution and a government that protects their property and their right to engage in trade. They will seek the government's protection against those who would infringe on their rights or attempt to dominate them, but the government need do little more than protect them and allow them to make choices based on their own reasoning.

You may have already encountered this theory, known as *utilitarianism,* in an economics or political science course. In sociology it is applied to a variety of issues. Often this rational-choice view of interaction is referred to as *exchange theory* because it focuses on what people seem to be getting out of their interactions and what they in turn are contributing to the relationship or to the larger group. In every interaction something is being exchanged. It may be time or attention, friendship, material values (e.g., wages or possessions), or less easily calculated values like esteem or allegiance. The larger the number of interacting members, the more complex the types of exchanges that occur among them. When people perceive an interaction as being one-way, they begin to feel that they are being exploited or treated unfairly and will usually leave the relationship or quit the group (Homans, 1961). In industry, for example, if workers feel that they are not being paid enough for their work they may form a union, bargain collectively with the bosses, or even go on strike. But in doing so each worker will weigh potential benefits against potential losses—losses in pay, in esteem, in friendship, and so forth. The choices are not always easy, nor are the motivations always obvious. When many values are involved, the rational calculation of benefits and costs becomes even more difficult.

Rational-choice models of behavior prompt us to look at *patterns* of behavior to see how they conform to and depart from normal expectations of personal profit and loss. But those models do not always identify the underlying values. How we learn what to value in the first place, how we communicate our choices and intentions, how we learn new values through interaction—all are subjects that require other concepts besides those found in rational-choice theories of behavior. Such questions lead us toward research on how human interaction is actually carried out and understood by people in their daily lives.

Symbolic Interactionism When we make choices about our interactions with other people, we may be said to be acting rationally. But there are likely to be other forces shaping our behavior as well. For example, you may select a particular course because the instructor is rumored to be good. But what does "good" mean—clear and well organized? an easy grader? friendly? humorous? The dimensions of a choice can be complicated, and we may not be aware of everything that goes into our decisions. For example, you may, without realizing it, choose a course as much to be with certain other people as to be in that particular course. Our choices tell other people about us: what we like, what we want to become, and so on. Indeed, the way people dress, the way they carry themselves (body language), the way they speak to each other, and the gestures they make convey a great deal of information that is not always intentional or expressed in speech. There are levels of communication that give information without speaking it, or speak one thing and mean another. But words are of great importance too, and the content of communication is made explicit in words and sentences. Sociologists refer to all these aspects of behavior as *symbolic interaction.* From the symbolic-interactionist perspective, "society itself tends to be seen as a mosaic of little scenes and dramas in which people make indications to themselves and others, respond to those indications, align their actions, and so build identities and social structures" (Rock, 1985, p. 844; see also Hughes, 1958; Hutchinson, 1985; Goffman, 1959).

Symbolic interactionists call attention to how social life is "constructed" through the mundane acts of social communication. For example, in all the choices students make—their joining of friendship groups, their learning of the informal rules of the school, their challenging and breaking of those rules—the social order of student society, or "college culture," is actually "constructed." Erving Goffman, whose work was mentioned earlier, is known for his research on these processes. Goffman applied the symbolic-interactionist perspective to the study of everyday interactions like rituals of greeting and departure, of daily life in asylums and gambling houses, and of behavior in streets and public places. His work examines how people behave in social situations and how their "performances" are rated by others.

The power of symbolic interactionism lies in its ability to generate theories about how people learn to play certain roles and how those roles are used in the social construction of groups and organizations. However, if we want to think sociologically about more complex phenomena, such as the rise of bureaucratic organizations or the reasons that some societies experience revolutions, we also need the concepts developed by two other perspectives: functionalism and conflict theory.

i

Functionalism

Is society simply the sum total of countless micro-level exchanges and communicative interactions, or do the organizations within a society have properties independent of the actions of individuals? When we speak of the family, the army, the corporation, or the laboratory, we generally have in mind an entity marked by certain specific functions, tasks, and types of behavior. The army requires that its members learn to engage in armed combat, even if that is not what they will be doing most of the time. The family requires that its members behave in nurturant ways toward one another. The farm requires that those who run it know how to plant and harvest. Individual interactions may determine how well a given person performs these various tasks, but it is the larger organization—the army, the family, the farm—that establishes specific ways of behaving, of doing the work of that organization, which the individual must master. In this sense the organization, which exists longer than any of its members, has its own existence.

The **functionalist** perspective in sociology asks how society manages to carry out the functions it must perform in order to maintain social order, feed large masses of people each day, defend itself against attackers, produce the next generation, and so on. From this perspective the many groups and organizations that make up a society form the structure of human society. This social structure is a complex system designed to carry out the essential functions of human life. The function of the family, for example, is to raise and train a new generation to replace the old; the function of the military is to defend the society; the function of schools is to teach the next generation the beliefs and skills they will need to maintain the society in the future; and a major function of religion is to develop shared ideas of morality.

When a society is functioning well, all its major parts are said to be "well integrated" and in equilibrium. But periods of rapid social change can throw social structures out of equilibrium. Entire ways of life can lose their purpose or function. When that happens, the various structures of society can become poorly integrated, and what were formerly useful functions can become "dysfunctional."

Consider an example. In agrarian societies, in which most people work the land, families typically include three generations, with many members in each generation all living close to one another. Labor is in great demand; many hands are needed where there are no machines to perform work in fields and barnyards and granaries. The emphasis on early marriage and large numbers of children found in the agrarian family is highly functional for such a society. But when the society industrializes and its agriculture becomes mechanized, families may continue to produce large numbers of children even though the demand for farmhands has decreased. When they grow up, those children may migrate to towns and cities. The migrants are likely to continue to value large families and to have numerous children, but if there are a limited number of jobs in the cities they may join the ranks of the unemployed and their children may grow up in poverty. In this situation the family can be said to be poorly integrated with the needs of the society; the value of large family size has become dysfunctional: It no longer contributes to the well-being of groups or individuals.

Conflict Theory

A major flaw in the functionalist perspective is the fact that we have rarely seen anything approaching equilibrium in human societies. Conflict and strife appear to be as basic to society as harmony, integration, and smooth functioning. In the twentieth century alone, two world wars and many civil wars disrupted the lives of millions of people. Almost as devastating was the Great Depression of the 1930s, the most severe economic slump in modern history. Worst of all were the nightmares of the Nazi Holocaust and the purges of Stalinist Russia, in which more than 20 million people were exterminated.

In Sarajevo the physical and emotional scars of war will last long after the cessation of hostilities. As wars become more destructive, the study of social conflict and its solutions takes on even greater urgency.

The world wars, the Depression, and the Holocaust shocked and demoralized the entire world. They also called into question the optimism of the nineteenth- and early twentieth-century social philosophers, many of whom believed in the promise of progress through modern science and technology. Between 1914 and the end of World War II, modern ideas and technologies were used for horrible purposes often enough to disillusion all but the most ardent optimists. Bewildered intellectuals and political leaders turned to sociology to find some explanation for those horrors.

One explanation was provided by Marxian theory. According to Marx, the cause of conflict in modern times could be found in the rise of capitalism. Under capitalism, forms of exploitation and domination spread. For example, in the early period of industrial capitalism workers were forced to work 12 hours a day, 6 days a week; in less developed areas of the world large populations were virtually enslaved by the new colonial powers.

At the heart of capitalism, for Marx, is conflict among people in different economic classes, especially between those who control wealth and power and those who do not. Marx argued that the division of people in a society into different classes, defined by how they make a living, always produces conflict. Under capitalism this conflict occurs between the owners of factories and the workers. Marx believed that class conflict would eventually destroy or at least vastly modify capitalism. His theory is at the heart of what has come to be known as conflict theory or the **conflict perspective.**

In the 1960s, when protests against racism and segregation, the Vietnam War, pollution of the environment, and discrimination against women each became the focus of a major social movement, the conflict perspective became more prominent. It clearly was not possible to explain the rapid appearance of major social movements with theories that emphasized how the social system would function if it were in a state of equilibrium. Even Marxian theory did not do a very good job of predicting the protest movements of the 1960s or their effects on American society. The environmental movement and the women's movement, for example, were not based on economic inequalities alone, nor were the people who joined them necessarily exploited workers. Sociologists studying the role of conflict in social change therefore had to go beyond the Marxian view. Many turned to the writings of the

German sociologist Georg Simmel (1904), who argued that conflict is necessary as a basis for the formation of alliances. According to Simmel, conflict is one means whereby a "web of group affiliations" is constructed. The continual shifting of alliances within this web of social groups can help explain who becomes involved in social movements and how much power those movements are able to acquire.

The concept of *power* holds a central place in conflict theory. From the functionalist perspective, society holds together because its members share the same basic beliefs about how people should behave. Conflict theorists point out that the role of power is just as important as the influence of shared beliefs in explaining why society does not disintegrate into chaos. **Power** is the ability of an individual or group to change the behavior of others. A nation's government, as we will see in Chapter 4, usually controls the use of force (a form of power) to maintain social order. For sociologists who study conflict and power, the important questions are who benefits from the exercise of power and who loses. For example, when the government intervenes in a strike and obliges workers to return to their jobs, does the public at large benefit or does the corporation against which the workers were striking? And what about the workers themselves? What do they gain or lose? Such questions are central to conflict theory today (Fontana, 1993; Gramsci, 1971).

The Multidimensional View of Society

Each of the sociological perspectives leads to different questions and different kinds of observations. These are extremely powerful analytical tools. Anyone who masters them and learns to apply them in appropriate ways will have an immense advantage over more naive observers of the ever-changing parade of social life (Merton, 1998). From the ecological perspective come questions about how populations can exist and flourish in various natural and social environments. From the interactionist perspective come questions about how people get along and behave in groups and organizations of all kinds. Functionalism asks questions about how society is structured and how it works as a social system. And the conflict perspective asks how power is used to maintain order and how conflict changes society. These different perspectives developed as sociologists asked different questions about society. In contemporary sociology each continues to stimulate relatively distinct research based on the types of questions being asked. Yet a great deal of research combines the insights of different perspectives in ways that vastly increase the power of the resulting analysis. The study chart below summarizes the major sociological perspectives and the kinds of questions they ask.

STUDY CHART Major Sociological Perspectives

PERSPECTIVE	DESCRIPTION	GENERATES QUESTIONS ABOUT . . .	APPLICATIONS
Interactionism	Studies how social structures are created in the course of human interaction.	How people behave in intimate groups; how symbols and communication shape perceptions; how social roles are learned and society is "constructed" through interaction.	Education practice, courtroom procedure, therapy.
Functionalism	Asks how societies carry out the functions they must perform; views the structures of society as a system designed to carry out those functions.	How society is structured and how social structures work together as a system to perform the major functions of society.	Study of formal organizations, development of social policies, management science.
Conflict Theory	Holds that power is just as important as shared values in holding society together; conflict is also responsible for social change.	How power affects the distribution of scarce resources and how conflict changes society.	Study of politics, social movements, corporate power structures.

VISUAL SOCIOLOGY

All photography in this book is "visual sociology" in the sense that we have chosen pictures that highlight social situations, social relationships, and social institutions. But in assembling a group of photos we can communicate more subtle cues about what it feels like to be in a particular situation or relationship. Through photographs we can convey a great deal about the culture, social structure, and patterns of interaction that characterize a particular social setting, at the same time that we leave the reader free to make his or her own judgments. Sociologists routinely use photography to focus on types of behavior, on the way people sort themselves out in space and time, on patterns of inequality, and much more. These are all aspects of visual sociology that appear in this volume.

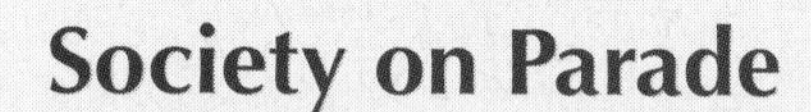

Society on Parade

Everybody loves a parade, and no one more than the sociologist. We see parades as events that intentionally display the people and the values we cherish. Parades also yield endless sociological insights into other people's social worlds. The way particular sociologists analyze a parade may tell us something about their concerns and values as well. Take the example of the Rose Bowl parade.

The Rose Bowl parade takes place in Pasadena, California, on New Year's Day. It is one of the most celebrated annual parades in the Unites States. It is both a local and a national event, and it attracts an immense television audience (which is also attracted by the promise of a major university football game). It is a

A saint's day parade in a devout Roman Catholic village of Central or South America offers another glimpse into the interactions that mark a special event in a village society. Who carries the village's patron saint? By what rules of social standing are people arranged in the parade? Simply by describing who's who in the parade, a local informant could offer many insights into the way the village is organized and how its people think and feel.

An ethnic parade like those that honor Saint Patrick, the patron saint of Ireland, reveals how an ethnic group views itself. Its rituals, its food, and all the conflicts that threaten its solidarity can be observed from the sidelines. The spectator can also gain insights into how the group deals with relations between children and adults, men and women, adults and the elderly. By contrast, a small and somber parade of mourners can offer a version of the celebration of death that may jar observers into not taking for granted the meanings of their own death rituals.

A military parade may be a joyous or extremely ominous event, depending on whether one is on good terms with the generals or not. From a functionalist perspective, this is a display of force. Official communications may portray it as a way of demonstrating that the nation has the capacity to defend itself against other nations. But in many parts of the world military parades are also meant to let citizens know in no uncertain terms who controls the use of deadly force. From a conflict perspective, they reveal who has power in the society and how that power is used to favor some groups, who are usually featured in the parade, over others, who often are not included among the marchers.

celebration for the community and an opportunity for it to be in the national spotlight. A functionalist sociologist would argue that the parade serves to affirm the society's values of competition and success and, in the process, helps the merchants of Pasadena. A sociologist with a more critical eye might call attention to the way the parade affirms values that perpetuate inequality and social injustices. The Rose Bowl Queen may be beautiful and talented, but a critical interpretation might emphasize how she is set on a pedestal and brought along to grace the male field of combat. From this perspective the Queen symbolizes women as beautiful objects to be possessed by men with power and money. An interactionist analysis might see the parade and its viewers as a form of public theatre in which the symbols of American culture are displayed in endless variations as the different floats pass by.

SUMMARY

Sociology is the scientific study of human societies and of human behavior in the groups that make up a society. It is concerned with how *social conditions* influence our lives as individuals. The ability to see the world from this point of view has been described as the *sociological imagination.*

Sociologists study social behavior at three levels of complexity. *Micro-level* sociology deals with behaviors that occur at the level of the individual and immediate others. The *middle level* of sociological observation is concerned with how the social structures in which people participate actually shape their lives. *Macro-level* studies attempt to explain the social processes that influence populations, social classes, and entire societies.

The scientific discoveries of the seventeenth century led to the rise of the idea of progress, as opposed to the notion of human helplessness in the face of divine Providence. In the eighteenth century, revolutions in Europe and North America completely changed the social order and gave rise to new perspectives on human social life. Out of this period of social and intellectual ferment came the idea of creating a science of human society. Sociology, as the new science was called, developed in Europe in the nineteenth century. During that formative period a number of outstanding sociologists shaped and refined the new discipline. Among them were Karl Marx, Émile Durkheim, and Max Weber.

In the twentieth century sociology developed most rapidly in North America, spurred by the need for empirical information concerning social conditions. Numerous "social surveys" were conducted around the turn of the century, and by the late 1920s two distinct approaches to the study of society had evolved at American universities. The "Chicago school" focused on the relationship between the individual and society, while the major East Coast universities leaned toward macro-level analysis.

Under the leadership of Robert Park and Ernest Burgess, the Chicago school developed the approach known as *human ecology.* This perspective emphasizes the relationships among social order, social disorganization, and the distribution of populations in space and time. Fundamental to this approach is the concept of community, meaning a population that carries out major functions within a particular territory.

Modern sociologists employ other basic perspectives besides human ecology. *Interactionism* is a perspective that views social order and social change as resulting from all the repeated interactions among individuals and groups. One version of this approach is rational-choice or exchange theory, which focuses on what people seem to be getting out of their interactions and what they contribute to them. Another is the symbolic-interactionist perspective, which studies how social structures are actually created in the course of human interaction.

Functionalism, in contrast, is concerned primarily with the large-scale structures of society; it asks how those structures enable society to carry out its basic functions. In the decades since World War II this perspective has been strongly challenged by the *conflict perspective,* which emphasizes the role of conflict and *power* in explaining not only why societies change but also why they hold together.

None of the major sociological perspectives is fully independent of the others. Each emphasizes different questions and different observations about social life. Used in combination, they greatly increase our ability to understand and explain almost any aspect of human society.

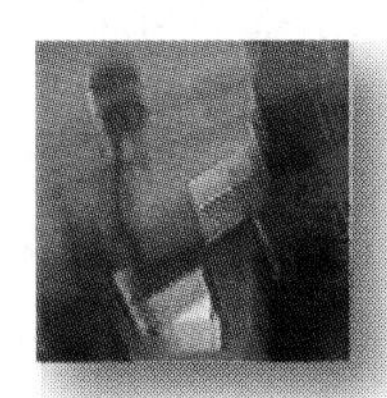

SOCIOLOGY VERSUS IDEOLOGY

Dr. Paul Lazarsfeld, a highly influential sociologist who pioneered the application of scientific methods to social issues, liked to tell a parable about a sociologist who interviews a centipede (this requires a leap of imagination) in order to find out how the centipede manages to coordinate the movements of all those legs. In thinking about the question, the centipede becomes confused and loses track of how her legs actually work. She becomes all tangled up in her legs because for the first time someone asked her to think about something that she had always taken for granted. Sociology can disturb people in a similar manner by making them think about behavior and beliefs that they take for granted. Does that make it a subversive science?

Some people believe that any discipline that asks searching questions and does not simply accept answers like "Well, that is how we do things around here" is threatening. Others equate sociology with "social engineering"; that is, they fear that once a sociologist identifies a problem in social relations, the next move is to recommend some kind of intervention—a social program or other device for changing the way things are. As with all stereotypes, there is some truth in these fears. At times the simple asking of penetrating questions makes us uneasy, and we may begin to question behaviors or opinions that we hold dear. At other times the facts that sociologists reveal about injustices, or about social conditions like crime, lead us to want to make intentional changes.

Does that make sociologists social engineers? No more so than in many other professions. We live in a world where people of all descriptions seek to improve their societies, their businesses, their communities, their schools, and almost every other aspect of their social world. In this book you will find many examples of situations in which sociologists who have studied disturbing trends, such as divorce or family stress, have also suggested ways to ameliorate the conditions that cause suffering. But they have little capacity to put their ideas into practice. That requires political processes and a great deal of public debate.

GLOSSARY

sociological imagination: according to C. Wright Mills, the ability to see how social conditions affect our lives. (p. 4)

social conditions: the realities of the life we create together as social beings. (p. 4)

sociology: the scientific study of human societies and human behavior in the groups that make up a society. (p. 5)

micro-level sociology: an approach to the study of society that focuses on patterns of social interaction at the individual level. (p. 6)

macro-level sociology: an approach to the study of society that focuses on the major structures and institutions of society. (p. 6)

middle-level sociology: an approach to the study of society that focuses on relationships between social structures and the individual. (p. 6)

scientific method: the process by which theories and explanations are constructed through repeated observation and careful description. (p. 8)

human ecology: a sociological perspective that emphasizes the relationships among social order, social disorganization, and the distribution of populations in space and time. (p. 13)

interactionism: a sociological perspective that views social order and social change as resulting from all the repeated interactions among individuals and groups. (p. 18)

functionalism: a sociological perspective that focuses on the ways in which a complex pattern of social structures and arrangements contributes to social order. (p. 19)

conflict perspective: a sociological perspective that emphasizes the role of conflict and power in society. (p. 20)

power: the ability to control the behavior of others, even against their will. (p. 21)

WHERE TO FIND IT

BOOKS

Critical Social Theories (Ben Agger; Westview, 1998). A useful review of critical sociological theories, with up-to-date examples of critical sociology in the contemporary world.

Modern Sociological Theory: Key Debates and New Directions (Derek Layder; UCL Press, 1997). A valuable introduction to the basic questions and debates in sociology.

Great Jobs for Sociology Majors: A Career Guide (Stephen E. Lambert; VGM Career Publications, 1997). A good survey of the kinds of work sociologists do and the kinds of jobs people with various levels of sociological training can aspire to.

Harriet Martineau: First Woman Sociologist (Susan Hoecker-Drysdale; Berg Publishers, 1992). A thorough review of the work, life, and sociological contributions of one of the most influential sociological minds of the nineteenth century.

The Sociological Imagination (C. Wright Mills; Oxford University Press, 1959). Still the best and most passionate statement of what the sociological imagination is, by the person who coined the phrase.

Invitation to Sociology: A Humanist Perspective (Peter Berger; Doubleday Anchor, 1963). A lively look at the science and craft of sociology by an author with a predominantly interactionist perspective.

Sociology for Pleasure (Marcello Truzzi; Prentice Hall, 1974). A collection of studies of offbeat social groups and situations—gypsies, cults, nudist-camp visitors, and many others—that reminds the reader that sociology can illuminate little-known social worlds.

Masters of Sociological Thought (Lewis Coser; Harcourt Brace, 1977). An indispensable source for anyone interested in in-depth treatments of the contributions made by the founders of sociology.

JOURNALS

Contemporary Sociology. The official journal of book reviews in sociology.

American Sociological Review. The journal of the American Sociological Association. Articles in this journal are often quite technical and may be somewhat advanced for the beginning student, but they offer a good perspective on current research.

American Journal of Sociology. The oldest journal in sociology; a treasure trove of articles going back to the early decades of the century. Consult the index for earlier papers, and recent issues for excellent new research.

OTHER SOURCES

Encyclopedia of Sociology (Edgar F. Borgatta, ed.; Macmillan, 1992). A comprehensive collection of essays that define and discuss the concepts and methods of sociology and social science.

Sociological Abstracts. A set of reviews of existing literature on a variety of social-scientific subjects. The abstracts are organized by topic and offer brief overviews of original research papers and other articles.

INTERNET RESOURCES

American Sociological Association (www.asanet.org). The official page of the American Sociological Association aims to serve sociologists in their work, advance sociology as a science and profession, and promote the use of sociology in society.

The Dead Sociologists Society (diogenes.baylor.edu/WWWproviders/Larry_Ridener/dss/deadsoc.html). A page with a good sociological sense of humor. Provides links to pages devoted to some of the founders of sociology, such as Émile Durkheim, Karl Marx, Max Weber, and many others.

Sociological Timeline, University of Missouri (www.missouri.edu/~socbrent/timeline.htm). Provides a fascinating look at the development of sociology and its contributions to social change.

CHAPTER 2

THE TOOLS OF SOCIOLOGY

Have you or any of your friends ever seriously considered committing suicide? This may not be the kind of question most people want to discuss openly, yet suicide is a serious social issue in the United States and many other nations. Perhaps the most tragic aspect of the phenomenon is the number of suicides by teenagers and young adults who have their entire adult lives ahead of them. What can a person do to counteract persistent thoughts about suicide? To whom can one turn when the urge to kill oneself becomes overwhelming? And what are the causes of suicide? These are all important sociological questions.

Most people who have not studied sociology assume that suicide is closely related to an individual's mental state. They will cite cases of people who ended their lives because of intense feelings of depression, or hopelessness, or shame due to crime, debt, loss of a career, or similar misfortunes. In other words, for the most part people who lack a sociological imagination will attribute suicide to personal problems that overwhelm the individual and make life seem intolerable. In some cases they may also cite examples of groups of people who have committed suicide together, as members of the Heaven's Gate and People's Temple cults did. But these examples of mass suicide tend to be viewed as exceptions to the notion that suicide is an individual act based on an intense desire to "end it all."

Sociologists do not deny that people who kill themselves are suffering from intensely negative feelings. But they do not believe that all suicide is caused by intense mental states. There is too much evidence suggesting that other forces are operating on individuals who commit suicide. For example, in recent research on suicide in different nations, sociologists have found that the highest numbers of suicides occur in Russia, China, and the nations of the former Soviet Union. Among the Western democracies, numbers of suicides vary considerably. Finland, Austria, Denmark, and France have the highest levels of suicide. The United States is in the middle of the range, and Italy and Greece are at the bottom.

What explains these marked differences in how many people in a given society kill themselves? When sociologists ask a question like that, they are asking not why an individual commits suicide but why an entire society has a high level of suicides. This is an important question, because if you live in a society where suicide is more common, it may be more likely that you will consider this fatal option at some point in your life.

The study of suicide, like the study of any form of human behavior, raises many issues of scientific observation and interpretation. Consider this question: If in a given year 1,000 people kill themselves in a small nation and 1,000 people kill themselves in another nation with a far larger population, can we say that both nations have the same amount of suicide? Clearly we need to consider the difference in population size when comparing the levels of suicide in these nations. Measurement of social phenomena like suicide is a vital aspect of understanding the behavior and possibly preventing it. This chapter, therefore, is devoted to the basic scientific methods that sociologists employ when trying to understand complex human behaviors.

APPLYING THE SOCIOLOGICAL IMAGINATION

As citizens of a democratic nation, we are encouraged to form opinions about the social issues of our time. Among the most troubling issues we face are many that involve life and death. Capital punishment, living wills, abortion, fertility drugs, suicide, assisted suicide—all pose extremely difficult ethical problems. These problems also indicate the rapid pace of social change in our lives. For the most part, our grandparents did not have to face many of these issues.

As in all democracies, candidates seeking our votes are eager to tell us what to believe about social issues and what society should do about them. But often the claims and recommendations we hear are contradictory. The explanations given for such problems as suicide, for example, may suggest very different solutions. How is an informed citizen to make decisions about the causes of social conditions or what should be done to remedy them? This is where sociology comes in. Sociology brings scientific methods to bear on these debates.

Most people associate social-scientific research with questionnaires, opinion surveys, and statistical reports. This is unfortunate. True, sociologists throughout the world use these research techniques, but they also use many other methods to explore social conditions. In this chapter we review many of the most common research tools used by sociologists. We will show how sociological research conforms to the rules of the scientific method and how the findings lead to changes in theories and to new research.

Anyone who embarks on scientific research, no matter what the subject or the method used, will go through many intellectual and emotional ups and downs before the work is done. Very often it will appear that the research is easy to define and conduct; soon, however, it will seem far more complicated, and the research questions themselves may appear less well defined than they did at the outset. This can be just as true in biology or chemistry as it is in sociology or another social science. Nevertheless, in all research projects certain basic steps must be completed, although not always in the order listed here:

1. *Deciding on the problem.* At first this may be simply a subject or topic of interest. Eventually, however, it must be worded in the form of a specific research question or questions in order to provide a focus for the rest of the work.
2. *Reviewing the literature.* Usually others have conducted research on the same topic. Find their reports and use them to determine what you can accomplish through your own research.
3. *Formulating research questions.* The work of others points to questions that have not been answered. Your own interest, time, resources, and available methods help determine what specific questions within your broad topic you can actually tackle.
4. *Selecting a method.* Different questions require different types of data—which in turn suggest different methods of data collection and analysis. Some of those methods are described in detail in this chapter.
5. *Analyzing the data.* Data are analyzed at each stage of the research, not just when preparing the final report.

In this chapter we act as if this step-by-step procedure is always followed by researchers in the social sciences. But do not assume that this procedure can be applied to every research question. One does not always progress easily from one step to another, and often one must go back and repeat earlier steps. In addition, some research questions require that we devise new ways of conducting research or new combinations of existing research methods.

Formulating Research Questions

A good deal of information is required even to know what situations or events one should study in conducting research on a particular social issue. General questions about societies or social behavior have to be translated into specific questions that can be studied using observations and measures of all kinds. Émile Durkheim's study of suicide (1951/1897) provides a good example of the process by which a sociologist converts a broad question about social change into the specific questions to be addressed in an empirical study—one that gathers evidence to describe behavior and to prove or disprove explanations of why that behavior occurs. These explanations are often, but not always, stated in the form of a **hypothesis,** a statement that expresses an informed (or "educated") guess regarding the possible relationship between two or more phenomena.

In his study of suicide Durkheim challenged the intellectuals and scientists of his day. Through the presentation of verifiable statistical evidence on suicide rates in different societies, he demonstrated that he could predict where and when suicides would be more numerous. Psychological reasons might account for why a particular individual committed suicide, but Durkheim showed that "social variables" such as religion or fluctuations in economic conditions could explain differences in the number of suicides from one society or region to another.

The question that eventually led Durkheim to his empirical study of suicide actually had nothing to do with suicide. He began by thinking about the consequences of large-scale social change in Western nations. In particular, he believed that industrialization and the rapid growth of cities weakened people's attachment to their families and communities. As they became increasingly anonymous and isolated, they were more likely to engage in a variety of self-destructive acts, the most extreme being suicide. In Durkheim's view, the act of suicide could be explained as much by social variables like rates of marriage and divorce as by individual psychological variables like depression or despair. Thus for Durkheim the study of suicide was a way of exploring the larger concept of integration or lack of integration into society: He sought to discover whether people who were less well integrated into society (i.e., more isolated from other people) were more likely to commit suicide.

If this view was correct, Durkheim reasoned, the rates of suicide among various populations should vary along with measures of social integration. He therefore formulated these hypotheses, among others:

- Suicide rates should be higher for unmarried people than for married people.
- Suicide rates should be higher for people without children than for people with children.
- Suicide rates should be higher for people with higher levels of education (education emphasizes individual achievement, which weakens group ties).
- Suicide rates should be higher in Protestant than in Catholic communities (Protestantism places more stress on individual achievement than Catholicism does, and this in turn weakens group ties).

Each of these hypotheses specifies a relationship between two variables that can be tested—that is, proved true or false—through empirical observation. In sociology, **variables** are characteristics of individuals, groups, or entire societies that can vary from one case to another. In the hypotheses just presented, the suicide rate is a social variable. Religion, education, marital status, and number of children are other variables in these hypotheses. The techniques Durkheim used to establish a set of hypotheses became a model for modern social-scientific research.

Let us examine one of Durkheim's classic hypotheses about suicide to see what it can teach us about applying the scientific method to social issues. In the hypothesis stating that suicide rates should be higher for unmarried people than for married people, Durkheim was proposing that there is a relationship between two variables: incidence of suicide and incidence of marriage. The suicide *rate* is a measure of suicide that takes into account the size of the population. So is the rate of marriage, as we will see shortly.

In this hypothesis the suicide rate is referred to as the **dependent variable;** it is the one we are trying to explain in terms of another variable, marriage rates. Marriage rates are referred to as the **independent variable** in this hypothesis. An independent variable is a factor that the researcher believes causes changes in the dependent variable. It is necessary to be extremely careful in making statements about causality, however. All social phenomena are caused by interactions among a number of variables rather than by only one or two. Thus we may say that lower marriage rates are one cause of suicide in a society. We know that there are other causes as well. (We discuss the issue of causality more fully later in the chapter.)

But, clearly, the story is more complicated than that. Some people commit suicide despite the fact that they are married and loved. Today some people think about suicide as an alternative to a lingering hospital death. (See Box 2.1.) As Durkheim first showed a century ago, suicide can take many forms and have many causes in different societies. Since then, sociologists have argued that before anyone can develop a set of testable hypotheses about the causes of suicide (or any other complex social situation or behavior), it is necessary to know a great deal about the actual experiences of the people involved. Thus even a sociologist who has never contemplated suicide or been part of a suicidal cult can attempt to see the world from the viewpoint of people in suicidal situations to find out what their daily lives are like. If they exist, such studies are usually the first ones consulted when the sociologist begins research on a particular problem.

Reviewing the Literature

Perhaps the insights one would need to understand the issues surrounding suicide are already available in "the

Then  and Now

Suicide in the Public Spotlight

The stock market crash of 1929 led to the suicides of people who had lost fortunes and faced immediate bankruptcy. More recently, the popular rock star Kurt Cobain killed himself. Both examples illustrate the key findings of Durkheim's pioneering studies of suicide. The suicide of an individual stockbroker or celebrity can be explained by the individual's state of mind at the time. However, the tendency for people to kill themselves when their fortunes change suddenly, or for celebrities to live out-of-control lives that often end in suicide, can be explained by social influences that cause people to feel lost or adrift. Durkheim labeled this feeling *anomie* and traced it to a lack of integration of the individual into social groups and communities.

literature"—in existing books or journal articles, published statistics, photos, and other materials. There is no need to conduct new research if the answers sought are already available. Most sociological research therefore begins in the library with a "review of the literature."

To stimulate our sociological imagination about suicide or any other social phenomenon, we need to look at a variety of studies that deal with this issue. But it takes some imagination just to think of the kinds of studies to look for. The various sociological perspectives described

BOX 2.1 GLOBAL CHANGE AND U.S. SOCIETY

"Designer Death"

A key feature of social change throughout the world is that it tends to make choices available that people in traditional societies never dreamed of. In Holland and Israel, panels of physicians, social scientists, and laypersons are asked to review citizens' requests for cosmetic surgery. They often debate whether a person actually needs to have breast implants or a facelift; if they approve the procedure, the central insurance system will pay for the operation. In most modern nations, pension plans and social security programs allow people to embark on new activities like photography or ballroom dancing at an age when their parents would more likely have been rocking quietly on the porch. Sexuality, political beliefs, styles of dress, and tastes in entertainment all present more choices and possibilities than ever before. In fact, the array of choices can become confusing and controversial as traditional norms are challenged by new possibilities.

One of the most controversial of all the new choices people now face is the decision to terminate one's life voluntarily. In the United States this area of medical ethics has been brought forcefully to our attention by Dr. Jack Kevorkian, the notorious "death doctor" whose "assisted suicides" led to his conviction for murder in Michigan. From a sociological viewpoint, Kevorkian's grim example and all the other new choices we face in dealing with death represent yet another set of customs (what sociologists call *norms,* as we will see in the next chapter) that are being challenged, for better or worse, in the rapidly changing modern world. The sociologist's role is to chart these changes and try to understand how people are coping with the confusion and the possibilities they present.

Not only in the United States but also in the Netherlands and the Scandinavian nations, where assisted suicides have become more frequent, many people are taking a sociological view of the issue. A sociological perspective on voluntary or assisted suicide can provide the facts and interpretations that help cut through the many moral debates that surround such a momentous issue. The literature will offer many comparisons among societies that show how suicide has been understood and dealt with throughout history and in numerous societies. One of the best places to start such a literature search, as we have already seen, would be the work of Émile Durkheim.

In his classic writings on suicide and the law, Durkheim pointed out that in ancient Greece and Rome suicide was considered a crime against society and the victim was denied the traditional rites of burial. "Also," Durkheim noted, "his hand was cut from his body and buried separately" (quoted in Lukes & Scull, 1983, p. 136). In other societies suicide has been or still is considered a sin against God. In early Islamic societies suicide was considered a major sin and crime because, according to Mohammed, "A man dies only by the will of God according to the book which fixes the term of his life." Throughout Christendom in the feudal period, suicide was also considered a grave crime; the victim's property was confiscated and the body was often publicly defiled.

But if it has been so severely punished, how is it that in our own time suicide has become so much more a matter of individual choice? Here is Durkheim's answer:

> Originally society is everything, the individual nothing. Consequently, the strongest social feelings are those connecting the individual with the collectivity; society is its own aim. Man is considered only an instrument in its hands; he seems to draw all his rights from it and has no counter-prerogative, because nothing higher than it exists. But gradually things change. As societies become greater in volume and density, they increase in complexity, work is divided, individual differences multiply, and the moment approaches when the only remaining bond among the members of a single human group will be that they are all men. (quoted in Lukes & Scull, p. 143)

Developments in law and medicine in the United States tend to create powerful examples elsewhere in the world. This fact—combined with the severe conflict that many people, and many religious groups, feel about any form of chosen death—suggests that the controversy over assisted suicide is unlikely to be resolved in the near future.

Dr. Jack Kevorkian.

in Chapter 1 can help organize the search, especially when the perspectives are framed as questions.

Who? How Many? Where? Who are we talking about and where are they found? This is a way of phrasing the ecological perspective, which suggests why that perspective is helpful in beginning research on an issue. In researching almost any subject involving human behavior, it is helpful to ask who is involved and in what numbers, and where the behavior in question occurs. The ecological perspective gives rise to two types of studies: community studies and demographic studies.

Community studies are among the richest research traditions in sociology. They portray the typical day-to-day life of a particular population. An example of a community study of a suicidal group is John R. Hall's *Gone From the Promised Land: Jonestown in American Cultural History* (1987). Hall was not a member of the infamous People's Temple, but he had lived in a number of communes and religious communities, where he collected firsthand accounts of what it is like to be part of an intense community such as the one led by Jim Jones. If a student wanted to explore what led members of the Heaven's Gate community to commit mass suicide in March 1997, Hall's study would be an excellent resource for ideas and comparisons. A search of the literature on the Heaven's Gate suicide would also turn up many firsthand accounts about what it is like to live among people who are obsessed with extraterrestrials and UFOs.

Demographic studies are also useful in studying the magnitude of a major social phenomenon like suicide. Such studies provide counts of people in various relevant population categories—in this case, people who are isolated in one way or another, or who have actually attempted or succeeded in committing suicide, or who may be considering assisted suicide because of advanced illness. Demographic studies of suicide offer a wealth of information about this drastic behavior from nations throughout the world. In a 1998 study, for example, the French National Institute for Demographic Studies reported that "France has the sad privilege to be classed among the nations of the world with the highest rates of suicide. If one eliminates the states of the old communist bloc and China, France, with a suicide rate in the neighborhood of 20 per 100,000, is in the fourth position behind Finland, Denmark, and Austria. All the other nations have a lower suicide rate: Japan (15 per 100,000), Sweden (15), Germany (14), Norway (13), Canada (13), the United States (12), the United Kingdom (7), Italy (7), Greece (3)" (Kremer, 1998).

How Is the Situation Defined? Interactionist approaches to extreme phenomena like suicide tend to look at how people who decide to take their own lives actually perceive their situation and how their interactions with others influence their behavior. In the case of suicide such studies are not common, but sociological research can often draw upon firsthand accounts. This is especially true of sensational suicides like the Heaven's Gate case: Since the 1997 mass suicide there have been numerous testimonials by former cult members describing how followers surrender their personal autonomy to the will of the leader and accept the leader's extreme definition of the situation (Bearak, 1997; Brooke, 1997). For the Heaven's Gate cult, the definition of the situation was that all who killed themselves would be immediately transported to a better existence among the superior beings accompanying the Hale-Bopp comet. In fact, it was possible to follow the development of the deadly cult's beliefs on the Internet, which was also used to recruit new cult members (Levy, 1997).

What Groups or Organizations Are Involved? This question stems from the functionalist perspective. It asks how society is organized to deal with a social issue or problem. Functionalist sociologists are concerned with how social policies actually function, as opposed to how they are supposed to function. Studies of social interventions to prevent suicide, for example, would describe how community crisis hotlines are established and who operates them, and ideally would include some measures of how effective they are. These measures might include how many calls are received and the nature of the calls (e.g., actual suicide threats, people in need of counseling, parents or guardians of possible teenage suicides, and so on). There might also be some description of how many times the potential suicide victim was saved through the actual intervention of crisis center personnel.

What Difference Does Power Make? This is a key question arising from the conflict perspective. Studies that take a conflict approach to suicide would be likely to compare suicide rates for men and women around the world. The differences would be striking. In many nations women are far more likely to kill themselves than men are. Why this is so has a lot to do with men's power and women's lack of power in different societies. In many parts of India, for example, a widowed woman is expected either to kill herself upon her husband's death or to submit herself entirely to the will of her husband's family, in which case she often serves essentially as a house slave, a condition that itself can drive a person to suicide. In some African and Latin American societies where there are arranged marriages or where women are routinely sexually exploited, we also find higher suicide rates for women than for men. In consequence, many sociologists argue that changes in the relative power of women, resulting from greater literacy, access to occupations, and basic protections under the law, would improve the balance of power between the sexes and lead to more equal rates of suicide, in addition to many other benefits.

The foregoing are only a few of the studies one would find in a review of the sociological literature on suicide. They are not cataloged according to the basic sociological perspectives. Rather, one needs to know what questions each of the sociological perspectives raises about an issue. For any research subject, if you ask questions about who and where and how people interact, what organizations and policies guide their actions, and who has the power, you will be well on your way toward a thorough review of what is known about the issue you plan to study.

THE BASIC METHODS

Once the researcher has specified a question, developed hypotheses, and reviewed the literature, the next step is to decide on the method or methods to be used in conducting the actual research. Sociological research methods are the techniques an investigator uses to systematically gather information, or *data,* to help answer a question about some aspect of society. The variety of methods that may be used is vast, with the choice of a method depending largely on the type of question being asked. The most frequently used methods are observation, experiments, and surveys. These are summarized in the study chart below.

Observation

Participant Observation Much sociological research requires direct observation of the people being studied. Community studies like Hall's work on the People's Temple are based on lengthy periods spent observing a particular group. This research method is rarely successful unless the sociologist also participates in the daily life of the people he or she is observing—that is, becomes their friend and a member of their social groups. Therefore, the central method of community studies is known as **participant observation.** The sociologist attempts to be both an objective observer of events and an

STUDY CHART The Basic Methods

METHOD	DESCRIPTION	EXAMPLE
Participant Observation	A form of field observation in which the observer participates to some degree in the lives of the people being observed.	Martin Sanchez-Jankowski spends many months with youth gangs, gaining their trust and learning gang norms regarding violence.
Unobtrusive Measures	Observational techniques that measure behavior but intrude as little as possible into actual social settings.	Media researchers find out what radio stations different types of cars are tuned to in order to learn about the listening tastes of different segments of the public.
Controlled Experiment	An experimental situation in which the researcher manipulates an independent variable in order to observe and measure changes in a dependent variable.	Judith Gerone of the Manpower Research and Development Corporation conducts controlled experiments to find out what types of training and job programs are most effective in reducing state welfare rolls.
Field Experiment	An experimental situation in which the researcher observes and studies subjects in their natural setting.	Milton Rokeach meets three delusional mental patients, each of whom believes he is Jesus Christ. He introduces them to each other to learn how they incorporate conflicting information into their self-presentations.
Survey	Research in which a sample of respondents drawn from a specific population respond to questions either in an interview or on a questionnaire.	A team of National Opinion Research Corporation sociologists survey the public about their sexual behavior and attitudes.

actual participant in the social milieu under study—not an easy task for even the most experienced researcher. In such situations the observer faithfully records his or her observations and interactions in *field notes,* which supply the descriptive data that will be used in the analysis and writing phases of the study.

An excellent example of a study based on participant observation is Donna Gaines's study *Teenage Wasteland* (1998). Gaines spent many months "hanging out" and speaking with teenagers in a relatively affluent suburban community adjacent to a large central city. Many of the teenagers lived lives devoted to school, family, and part-time jobs. But others became what Gaines called "suburbia's dead-end kids." They experimented with drugs, were attracted to satanism and heavy metal rock music, and flirted with violence and suicide, sometimes with tragic consequences. These young people, who often thought of themselves as "burnouts," described their lives to Gaines because she listened to them and showed that she could be trusted. In the beginning they tested her by giving her information and telling her not to share it with other kids. Once they saw that she did not divulge secrets, she was able to learn much more about their behavior. And by listening carefully she also learned how some kids who seem to be at a dead end are actually struggling to create a better world for themselves, often against great odds.

This kind of research describes the quality of life of the people involved, and for that reason it is often called *qualitative* research in order to distinguish it from the *quantitative* research methods we will consider shortly. James Coleman observes that in qualitative research "we report a stream of action in which the interlinking of events suggests how the [social] system functions" (1964, p. 222).

Another excellent example of a study based on participant observation is Douglas S. Harper's classic, *Good Company* (1982). Harper spent many months riding on freight trains and living in "jungles" with hoboes. His goal was to describe how hoboes, or tramps or bums as they are often called, actually live and how they learn to trust or distrust one another in a world considered deviant by members of "respectable" society. Harper had to learn the nomadic and often dangerous life of the hobo, and this meant that he needed to establish a relationship with someone in that life who could serve as his mentor. The successful participant observer usually finds at least one such person. In Harper's case, a hobo named Carl made it possible for him to participate in, and therefore to understand and write about, the hobo's world.

Qualitative research carried out "in the field" where behavior is actually occurring is the best method for analyzing the processes of human interaction. If Donna Gaines had not actually spent a good deal of time with suburban teenagers, she would not have obtained nearly as many insights into their behavior. A shortcoming of this approach, however, is that it is usually based on a single community, group, or social system, which makes it difficult to generalize the findings to other social settings. Thus community studies and other types of qualitative studies are most often used for exploratory research, and the findings serve as a basis for generating hypotheses for further research.

Unobtrusive Measures Observation can employ numerous other techniques besides direct participation. Among these are **unobtrusive measures**—that is, observational techniques that measure the effects of behavior but intrude as little as possible into actual social settings. Here are some examples:

> The floor tiles around the hatching-chick exhibit at Chicago's Museum of Science and Industry must be replaced every 6 weeks. Tiles in other parts of the museum need not be replaced for years. The selective erosion of tiles, indexed by the replacement rate, is a measure of the relative popularity of exhibits.
>
> Chinese jade dealers have used the pupil dilation of their customers as a measure of the client's interest in particular stones. (Webb et al., 1966, p. 2)
>
> To study changes in the diet of Americans, investigators sifted through more than 2,000 household trash collections to measure the amount of fat being trimmed from meat packages. (Rathje, 1993)

Observations like these can be transformed into useful measures of the variables under study. Their nature is limited only by the researcher's creativity, and they intrude far less into people's lives than do interviews or participant observation.

Visual Sociology An increasingly popular set of observational techniques involves the use of photography and videotape, or visual sociology. These techniques can be just as obtrusive as interviewing, if not more so, but they can also be used in unobtrusive ways, depending on the questions being asked. A classic example of the use of photographic data is the important work of the French anthropologist Claude Levi-Strauss. His extensive explorations in Brazil in the late 1930s provided priceless images of what life was like for the indigenous tribes of the Amazonian region before the most powerful effects of conquest and forced modernization began to be felt. Levi-Strauss created a photographic record of his interactions with the tribespeople; the results offer some rare glimpses of what social-scientific field research in simpler societies was like during the early part of the twentieth century. Many of his photos

also show how rapid social change was altering the life of the Indians (Levi-Strauss, 1995).

Another powerful example of visual sociology is found in the work of Letizia Battaglia and Franco Zecchin (1989), whose photographic studies of violence and Mafia influence in the Sicilian town of Palermo offer unique insights into the way organized crime pervades the life of a city. Their photos, such as the one shown here of women mourning for a loved but criminal family member, reveal how life is lived among people for whom organized crime and murder are a fact of daily existence in a world of poverty and relative lack of opportunity.

Applying your sociological imagination to visual material can be both interesting and informative. Levi-Strauss's work and the photos of Battaglia and Zecchin are only two examples of the rich vein of visual sociology that is captured in photographs. Similar examples will appear throughout later chapters of this book. In addition to photographs, works of fine art can provide insight into the societies within which they were created. Some examples are presented in the Art and the Sociological Eye feature on pages 38 and 39.

Experiments

Although for both moral and practical reasons sociologists do not have many opportunities to perform experiments, there is a large body of literature in the social sciences, especially social psychology (an interdisciplinary science that draws ideas and researchers from both sociology and psychology), that is based on experiments. There are two experimental models that social scientists use frequently. The first and most rigorous is the controlled experiment conducted in a laboratory. The second is the field experiment, which is often used to test public policies that are applied to some groups but not to others.

Controlled Experiments The **controlled experiment** allows the researcher to manipulate an independent variable so as to observe and measure changes in a dependent variable. The experimenter forms an **experimental group** that will experience a change in the independent variable (the "treatment") and a **control group** that will not experience the treatment but whose behavior will be compared with that of the experimental group. (The control group is similar to the experimental group in every other way.) This type of experiment is especially characteristic of studies at the micro level of sociological research.

Consider an example. Which line in Figure 2.1(b) appears to match the line in Figure 2.1(a) most closely? Could anything persuade you that a line other than the one you have selected is the correct choice? This simple diagram formed the basis of a famous series of experiments conducted by Solomon Asch in the early 1950s. They showed that the opinion of the majority can have an extremely powerful influence on that of an individual.

Asch's control group consisted of subjects[1] who were seated together in a room but were allowed to make their judgments independently. When they looked at sets of lines like those in Figure 2.1, the subjects in this group invariably matched the correct lines, just as

FIGURE 2.1 Lines Used in the Asch Experiment on Conformity

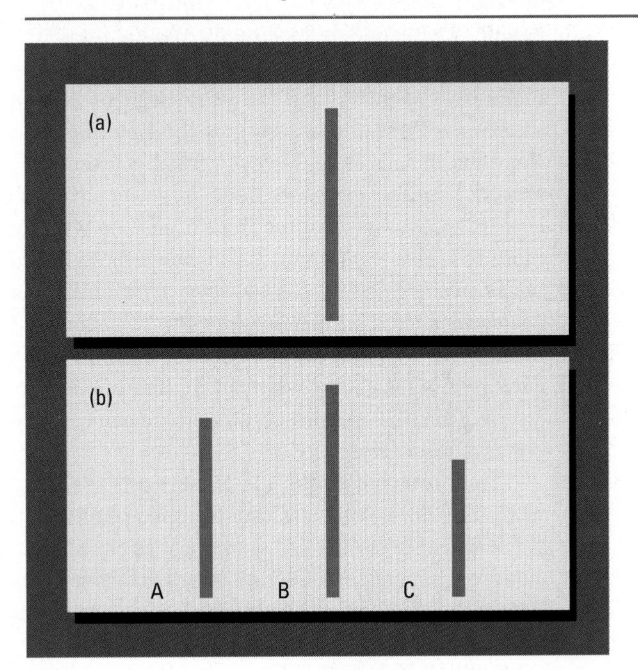

Cards like these were used in the Asch experiment. Subjects were asked to judge the lengths of various lines by comparing them with the three lines on the bottom card. The line on the top card quite obviously matches line B on the bottom card; all of the judgments were this simple.

[1] The term *subject* refers to a person who participates in a controlled experiment.

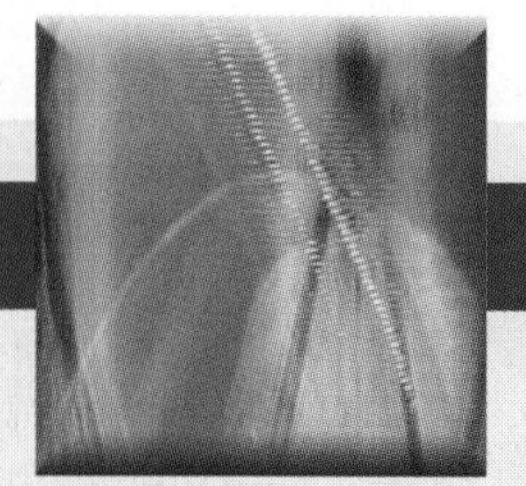

ART AND THE SOCIOLOGICAL EYE

"Reading" Art for Social Meaning

Drawing and painting give us our earliest evidence of human beings' desire to understand social life and to communicate that understanding to others. For instance, although we cannot know in much detail how the preliterate cave dwellers of southern Europe and northern Africa lived during the late Ice Age more than 10,000 years ago, cave paintings like the hunting scene shown here can give us glimpses into the way prehistoric humans felt about their lives. Note how carefully the artist rendered the hunting scene, with special attention to the placement of the hunters before the onrushing antelope. In the very choice of the hunting subject, the ancient artist is communicating a message about what was uppermost in the minds of our early ancestors.

Cave painting from Los Caballos, Spain.

The art of any of the world's peoples, past or present, can likewise be "read" for evidence of how people organize their society, for the nature of interpersonal relationships, for what they value as sacred or simply enjoyable, and for what they despise. Thus the painted limestone carving of the Egyptian prince Rahotep and his wife Nofret, probably executed by an official artist of the pharaoh in about 2610 B.C., reveals many details of the lives of people who lived more than three millennia ago. Look at how the couple's dress reveals differences in what must have been considered proper appearance for men and women of their social standing. Her breasts are covered, but his chest is bare; his only ornaments are a thin necklace and an equally thin mustache, whereas her hair and jewelry are far more elaborate. The sculptures also suggest that the princely rulers of one of the world's first civilizations could have had either dark or light skin.

But let us not confuse sociology and fine art. These artists were not sociologists, nor were they necessarily seeking to make points about their society. Their goals were to use the forms of everyday life to create something that would please them and their patrons, to capture the nuances of shape and form, to convey the emotional quality of a scene. The task of the sociologist is to read their art for its social meaning, without detracting from the magic that makes art beautiful and timeless.

Prince Rahotep and His Wife Nofret, artist unknown, about 2610 B.C. Painted limestone, height 47″.

you no doubt have. But in the experimental group a different result was produced by the introduction of an independent variable: group pressure.

Asch's experimental group consisted of subjects who were asked to announce their decisions aloud in the group setting. Each subject was brought into a room with eight people who posed as other subjects but were actually confederates of the experimenter. When the lines were flashed on a screen, those "subjects" all chose a line that was not the matching one. When it was the real subject's turn to choose, he or she was faced with the unanimous opinion of a majority of "subjects" who had picked the wrong line. Thirty-two percent of the real subjects went along with the majority and chose the wrong line as well. And even among the subjects Asch called "independent" (the 68 percent of the real subjects who gave the correct response despite the pressure of the majority) there was a great deal of variation. Some gave the correct response at all times, whereas others gave it only part of the time. This conforming response was less likely when there was at least one other person in the group who also went against the majority. By varying the number of people who said that the shorter line was longer, Asch was varying the degree of group pressure experienced by the subject. Higher levels of the independent variable (group pressure) thus produced more "errors," or choices of the wrong line (Asch, 1966).

Field Experiments **Field experiments** are used extensively in evaluating public programs that address specific social problems. In these experiments there is usually a "treatment group" of people who participate in the program and a control group of people who do not. In one example of this type of experiment, Angelo Atondo, Mauro Chavez, and Richard Regua of California's Evergreen Valley College attempted to test the theory that if students at risk of failure are linked up with adult mentors from their own communities, their chances of success will improve (Bashi, 1991). This popular theory stems in part from an influential article in the *Harvard Business Review* titled "Everyone Who Makes It Has a Mentor" (Collins & Scott, 1978). Many successful people from all walks of life, especially people who have escaped from impoverished backgrounds, say that they benefited greatly from the guidance of a mentor who steered them toward constructive goals. Most mentors do not actually tutor the individuals they spend time with; instead, they teach through conversation and example.

Researchers have found little evidence to prove that mentors actually make a difference. Unless one can compare students who had mentors with similar students who did not, it is impossible to say with any scientific certainty—that is, beyond individual stories of success or failure—whether the mentors made a difference. The Evergreen Valley College researchers therefore assigned 115 entering Latino students with low English proficiency to Latino mentors with excellent English skills. Their control group was composed of a comparable number of Latino students with similar characteristics; these students were not assigned to mentors. Both groups attended the same classes and took the same exams. At the end of the semester 89 percent of the students with mentors passed the freshman English course, but only 46 percent of the students in the control group passed the same course. Later, more of the students with mentors went on to four-year colleges. While much more controlled research of this nature needs to be done before we can fully understand how mentors help students overcome their educational problems, this example shows how a field experiment can advance our understanding of a social issue like success and failure in school (Bashi, 1991).

Sociologists can conduct "natural experiments" when two similar groups receive differing treatments and the results can be measured. An example of such an experiment deals with problems of teenagers, including early pregnancy, school failure, and suicide. Gregory J. Duncan and Jeanne Brooks-Gunn (1997) studied the lives of teenagers growing up in low-income neighborhoods. By comparing children from families with very similar social characteristics (income, race, ethnic background, education, etc.), they could identify the effects of certain characteristics of their surroundings (e.g., rates of poverty, crime, suicide, and drug abuse) on their behavior.

The Hawthorne Effect A common problem of experimental studies is that just by paying attention to people in an experimental group the researcher may be introducing additional variables. This problem was recognized for the first time in the late 1930s, when a team of researchers led by Elton Mayo conducted a famous series of experiments at Western Electric's Hawthorne plant. The purpose of the study was to determine the effects of various environmental and social conditions on the workers' productivity. One hypothesis was that improvements in the physical features of the workplace (such as better lighting) and improved social conditions (coffee breaks, different methods of payment, and so on) would result in greater productivity (the dependent variable). But it appeared that *no matter what the experimenters did,* the workers' productivity increased. When the experimenters improved the lighting, the workers' productivity increased over that of the control group. When they dimmed the lights, productivity went up again. This was true even when the workers were subjected to somewhat worse conditions or were returned to their original conditions.

At first this series of experiments was considered a failure, and the researchers concluded that the variables they were introducing had little to do with the changes in productivity that resulted. But on further reflection they realized that the real independent variable was not better working conditions but, simply, *attention.* The workers liked the attention they were getting; it made them feel special, so they worked harder. This realization led the researchers to design experiments dealing with the effects of attention from supervisors and better communication between workers and their managers; those experiments led to a new philosophy of worker-management relations, described in Chapter 18. Today the term **Hawthorne effect** is used to refer to any unintended effect that results from the attention given to subjects in an experiment.

The Hawthorne effect occurs in a variety of experimental situations. For example, in the Evergreen Valley College experiment described earlier, it is possible that the students with mentors were more successful than the students in the control group because they were getting extra attention from adults. Over time the effects of that attention may disappear and the experimenters may conclude that once the novelty of the extra attention wears off, students with mentors do not do appreciably better than others. Sophisticated evaluations of educational change try to anticipate the Hawthorne effect by making sure that the experimental group does not get more attention than the control group.

Survey Research

A sociological survey asks people to give precise information about their behavior, their attitudes, and at times the behavior and attitudes of others (e.g., other members of their households). There is a world of difference between the modern sociological survey and the "social surveys" conducted around the turn of the twentieth century (see Chapter 1). Those surveys attempted to present an unbiased, factual account of the social conditions of a specific community; their findings could not be applied to other groups. In contrast, today's survey techniques make it possible to *generalize* from a small sample of respondents to an entire population. An example is election polls, in which a small sample of 1,000 or 2,000 respondents can be used to predict how the entire electorate will vote in a presidential election. Done properly, the modern survey is one of the most powerful tools available to social scientists.

Surveys are the central method used in election polling, market research, opinion polling, television ratings, and a host of other applications. The most ambitious and most heavily used sociological surveys are national censuses. A *national census* is a full enumeration of every member of the society—a regular system for counting its people and determining where and under what conditions they live, how they work and gain income, their patterns of family composition, their age distribution, their levels of educational attainment, and related data. Without these measures, a nation cannot plan intelligently for the needs of its people. The United States conducts a national census every 10 years. The census for the year 2000, the twenty-second in the nation's history, will be the largest and most complex ever undertaken; it will enumerate about 265 million people in about 115 million housing units.

Once a national census has gathered the basic demographic and ecological facts about a nation's people, it is possible to add more information through the use of smaller and far less costly **sample surveys.** In the United States, this is done regularly through the Census Bureau's Current Population Survey (CPS). From the CPS we get monthly estimates of employment and unemployment, poverty, births, deaths, marriages, divorces, social insurance and welfare, and many other indicators of the social well-being and problems of the American people.

Opinion Polls Another type of survey research, pioneered by sociologists earlier in this century, is the *opinion poll.* Today opinion polling is a $2.5 billion industry (Rothenberg, 1990). Opinion polls are used by marketing firms to help corporations make decisions about their products. Political candidates and their staffs use them to measure the progress of their campaigns. And elected officials use polls to monitor public opinion on key political and economic issues.

A good example of an opinion poll is the General Social Survey (GSS) conducted by the National Opinion Research Center (NORC). This poll uses a relatively small sample of respondents, selected at random, whose opinions about a wide variety of issues can be generalized to the entire population of American adults.

To get a better idea of how opinion surveys are carried out, let us look at a series of results from the GSS on environmental issues. Figure 2.2 on page 43 shows how Americans responded to questions about how they feel about levels of government spending for environmental preservation and environmental regulations. Since the questions were asked almost every year during the last two and a half decades, the series of responses provides a strong indication of trends in public opinion on this important issue. Clearly, a majority of people in the United States favor environmental preservation (and spending for such programs). In recent years that support, while still strong, has decreased. Similar data for other social issues, such as health care and crime, would show that the public is increasingly wary of government spending in these areas. One can also see, by looking at

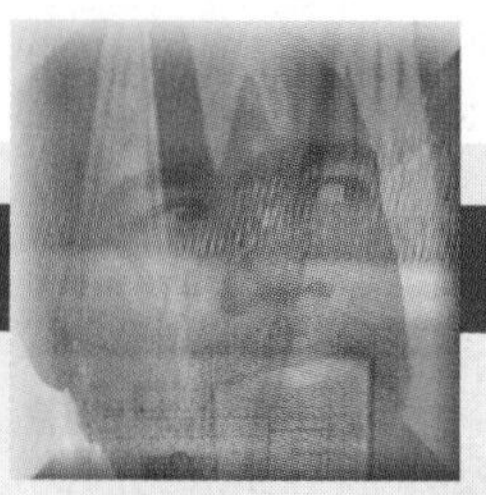

THE PERSONAL CHALLENGE

The Census: Science or Politics?

A debate is raging over how to conduct the U.S. census for the year 2000, a debate that hinges directly on sociological methods. The way the debate is resolved may also determine many other aspects of social policy in the coming decade.

The Bureau of the Census, a branch of the federal Department of Commerce, has the constitutional responsibility to provide a decennial count or enumeration of the resident population of the United States for the purpose of establishing proportional representation of the states in Congress. In the many decades since the original census, the population count has been put to many other uses in government and business. Companies use the census to target their marketing campaigns. Construction firms and urban planners use the census to determine how much and what kind of housing their regions will need. Medical experts and executives rely on the census for information about the health needs of the populations they serve. Census data thus are vital to many areas of life in a modern society. But why is the method used to count the population a subject of controversy?

In 1998 the director of the Census Bureau, demographer Martha Farnesworth Riche, resigned in protest over Congress's failure to approve her agency's plans to conduct the 2000 census. She and her staff advocated the use of standard sampling techniques for estimating segments of the population that are usually undercounted. These include poor Americans of all races, African Americans, Latinos, people in isolated rural areas, and Native Americans. Demographers argue that these groups have been undercounted by traditional enumeration methods (Preston, Elo, & Foster, 1998). To make matters worse, when their numbers are not adequately represented in official census figures, these groups are not adequately represented by elected officials, nor do the regions in which they live receive the share of medical and other benefits for which they would be eligible if the census counts reflected their actual numbers (Choldin, 1998). When census takers go from door to door to enumerate residents of households, they tend to miss poor people and minorities because their neighborhoods often are more difficult to work in, and because these groups are likely to be more mobile than people in better housing and often distrust enumerators and hence fail to respond.

For many decades demographers have been urging the Census Bureau to continue enumerating in its traditional fashion but to also conduct sample surveys, which would provide additional estimates of population size and characteristics in problematic urban and rural areas. The data from the sample surveys would then be used to correct undercount errors. But some elected officials fear that this method could change election districts in such a way that poor and minority people would have more representation than they now have in Congress and state legislatures—which could alter the balance of political power in the nation, since poor and minority people are more likely to vote for Democratic candidates than for Republicans. For their part, critics of the sampling plan argue that the Constitution specifies an *enumeration,* which implies house-to-house counting, and that if people are not counted, that is their problem (Safire, 1998).

What are your thoughts on this issue? As you think about the problem, you might want to consider the fact that these days almost all elected officials use samples of voters, drawn by methods similar to those that would be used to address the census undercount problem, to determine citizens' opinions on issues and how they may vote. Sampling and estimates are widely used. Is the problem one of science, or is it one of politics?

Martha Farnesworth Riche.

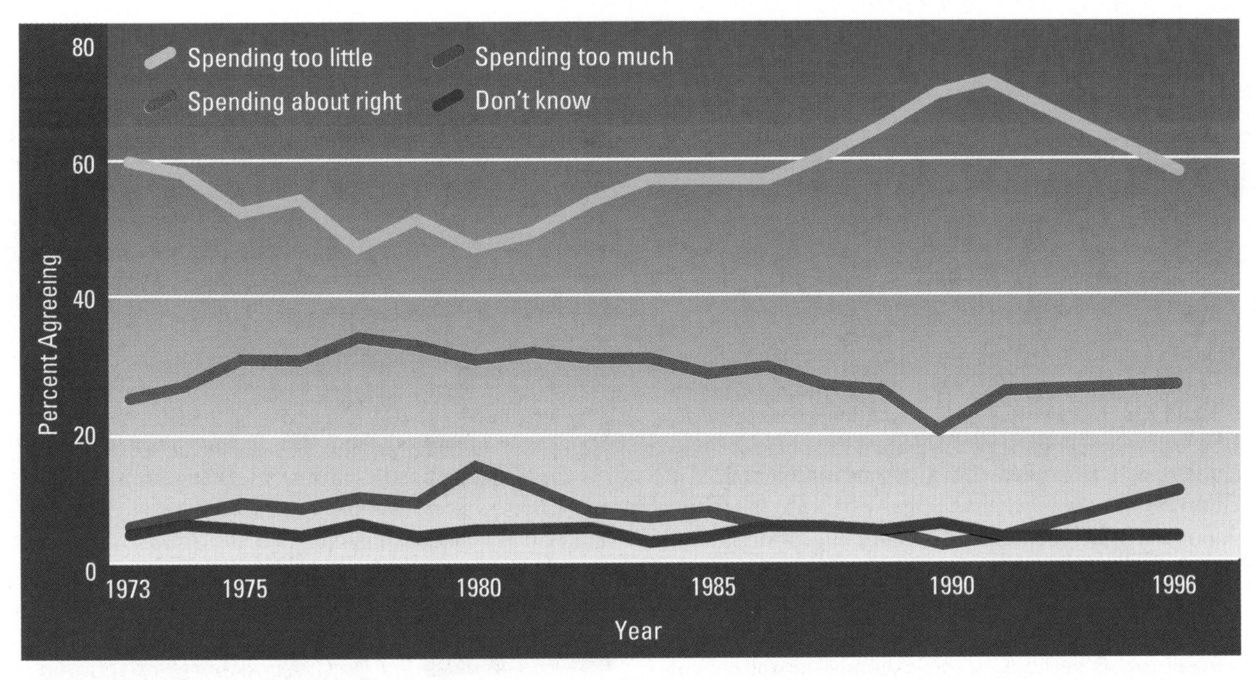

Source: National Opinion Research Center, 1996.

FIGURE 2.2 **Public Support for Environmental Action, 1973–1996**

the "Don't know" response, that most of the Americans polled had an opinion on the issue.

Since poll takers often wish to confront respondents with the possible consequences of their opinions, questions about spending of tax money on a social issue like the environment are quite helpful. In essence, people are being asked not only about their support of a policy but also about their willingness to see their tax dollars spent for that purpose. The results in Figure 2.2 show that an increasing majority of Americans feel that spending for the environment should be increased. This sentiment became less prevalent during the energy crisis of the late 1970s but increased dramatically in recent years as concern about such problems as acid rain and global warming has outstripped spending on these problems by the federal government.

The NORC General Social Survey is conducted each year, if possible, among a nationwide random sample of at least 1,500 adults. The term **sample** refers to a set of respondents selected from a specific population. The first step in selecting a sample is to define the population to be sampled; in our example the population consists of adult American citizens. The next step is to establish rules for the *random* selection of respondents. The goal of this procedure is to ensure that within the specified population everyone has an equal chance, or *probability,* of being selected to answer the survey questions. A sample in which potential respondents do not all have the same probability of being included is called a *biased sample.* To avoid bias, respondents must be selected by some process of random sampling.[2] In other words, any form of "volunteering" to be interviewed must be eliminated. (For example, the researcher cannot intentionally select his or her friends to be part of the sample.) Only a random sample can be considered truly representative of the target population.

Surveys that are not based on random samples can be very misleading. In a famous example from the early days of surveys, researchers predicted that candidate Alf Landon would defeat the incumbent president, Franklin D. Roosevelt, in the 1936 election. They based their prediction on data from a telephone survey, but at the time many low-income households did not have telephones. The survey therefore overrepresented the opinions of those with higher incomes, who were less likely to vote for Roosevelt. Roosevelt won by a landslide, and since then political pollsters have been much more alert to the problem of sample bias. But many contemporary surveys also produce biased results. Most mail surveys and many telephone surveys that claim to be conducting research are actually

[2] Random sampling is accomplished by a variety of statistical techniques that are discussed in more advanced courses.

marketing surveys designed to tap the opinions of particular subgroups rather than a representative sample of the entire population.

Before leaving the subject of opinion polling, we should mention one additional point. Polls are subject to *sampling error.* A sampling error of plus or minus 3 percentage points, for example, would be normal in a national sample of 1,500 to 2,000 respondents. This is important information for anyone who reads and interprets poll data. It means that a difference of 3 percent or less between the percentages of American adults expressing certain opinions could be due to chance rather than to a real difference in the distribution of opinions. In other words, it is possible that just by chance a higher percentage of people with a certain opinion was included in the sample than is actually the case in the total population. The possibility of sampling error is more critical when there is a very small difference between two sets of responses—for example, when 51 percent of the sample favors one presidential candidate and 49 percent favors another. In this example the survey analysts would have to conclude that the election is "too close to call."

Questionnaire Design Just as important as careful selection of the sample to be interviewed is the design of the *instrument,* or questionnaire, to be used in the survey. Questionnaire design is both a science and an art. Questions must be worded precisely yet be easily understood. Above all, the questions must be worded so as to avoid biasing the answers. For example, the General Social Survey asked whether the United States was spending too much, too little, or about the right amount on "assistance to the poor." About 66 percent of the respondents answered "too little." But when the word "welfare" was substituted for "assistance to the poor," the findings were almost reversed, with slightly less than 50 percent responding that too much was being spent on welfare (Kagay & Elder, 1992). (See Box 2.2.)

One of the first decisions to make in designing survey questions is whether they should be open or closed. **Closed questions** require respondents to select from a

BOX 2.2 SOCIOLOGICAL METHODS

Avoiding Bias in Survey Questions

In *Asking Questions: A Practical Guide to Questionnaire Design* (1982), survey researchers Seymour Sudman and Norman Bradburn of the National Opinion Research Center describe a common abuse of survey methods: the use of biased questions. In a questionnaire mailed to them by a lobbying group the authors noticed questions like these:

1. Do you feel there is too much power concentrated in the hands of labor union officials?

   yes ____ no ____

2. Are you in favor of forcing state, county, and municipal employees to pay union dues to hold their government jobs?

   yes ____ no ____

3. Are you in favor of allowing construction union czars the power to shut down an entire construction site because of a dispute with a single contractor, thus forcing even more workers to knuckle under to union agencies?

   yes ____ no ____

These questions violate even the simplest rules of objectivity in questionnaire design. As the authors point out, they are "loaded with nonneutral words: 'forcing,' 'union czars,' 'knuckle under.'" Such questions are "deceptive and unethical, but they are not illegal" (p. 3).

To eliminate bias, one would have to begin by rephrasing the questions. For example, the first question could read something like this:

1. Which answer best sums up your feeling about the amount of power held by labor union officials today?

   a. too much power
   b. about the right amount of power
   c. too little power
   d. don't know

Fair questions avoid nonneutral words and give the respondent an opportunity to answer on either side of the issue or to say that he or she does not know. Whenever you read the results of a survey, ask yourself whether the questions are free of obvious bias, if they are specific or general, and if they are threatening or nonthreatening to the respondent. Such an evaluation can make a great difference in how you view the answers.

set of answers, whereas **open questions** allow them to say whatever comes to mind. The items in Box 2.2 are examples of closed questions. An example of an open question would be "Please tell me how you feel about your neighborhood." In answering this type of question the respondent can say that he or she does not like the neighborhood at all or give one or more reasons for liking it. The interviewer attempts to write down the answers in the respondent's own words.

When researchers wish to include both forced-choice categories and freely given opinions in a single survey, the questionnaire usually begins with open questions and then shifts to closed questions on the same subject. Survey instruments that rely on open questions are often called *interview guides*. They guide the researcher's questions in certain directions and emphasize certain categories of information, but they also allow the interviewer to follow up respondents' comments with further questions.

Research Ethics and the Rights of Respondents

In asking any type of question in a survey or other type of sociological study, the researcher must be aware of the rights of respondents. Much sociological research deals with the personal lives and inner thoughts of real human beings. Although most of that research seems relatively innocent, there are many times when the questions asked or the behaviors witnessed may be embarrassing or even more damaging. In one famous example, Laud Humphreys (1970, 1975) studied interactions between men seeking casual sexual encounters in public restrooms. Many of his colleagues attacked him for invading the men's privacy. Others defended him for daring to investigate what had until then been a taboo subject (homosexuality and bisexuality). After all, they reasoned, Humphreys was careful to keep the men's identities secret. But he also followed some of the men home and conducted interviews with them there. They were not aware that he knew about their homosexual activities, and upon reflection Humphreys admitted that he had deceived his subjects.

Sociologists continue to debate the ethical dilemmas raised by Humphreys's research and by studies like the Asch experiment (in which the subject is duped and may feel embarrassed). As a result of such controversies, the federal government now requires that research involving human subjects be monitored by "human subjects review panels" at all research institutions that receive federal funding. A research review must be undertaken before a study can be funded or approved for degree credit. To pass the review, the researcher must provide proof that he or she has taken precautions to protect the fundamental rights of human subjects. Those rights include privacy, informed consent, and confidentiality.

The right of **privacy** can be defined as "the right of the individual to define for himself, with only extraordinary exceptions in the interest of society, when and on what terms his acts should be revealed to the general public" (Westin, 1967, p. 373). **Confidentiality** is closely related to privacy. When a respondent is told that information will remain confidential, the researcher may not pass it on to anyone else in a form that can be traced to that respondent. The information can be pooled with information provided by other respondents, but extreme care must be taken to ensure that none of the responses can be traced to a particular individual.

Informed consent refers to statements made to respondents (usually before any questions are asked) about what they are being asked and how the information they supply will be used. It includes the assurance that their participation is voluntary—that is, that there is no compulsion to answer questions or give information. The respondent should be allowed to judge the degree of personal risk involved in answering questions even when an assurance of confidentiality has been given.

The ethics of social research require careful attention. Before undertaking research in which you plan to ask people questions of any kind, be sure you have thought out the ethical issues and checked with your instructor about the ethics of your methods of collecting data and presenting them to the public.

i

ANALYZING THE DATA

Survey research normally is designed to generate numerical data regarding how certain variables are distributed in the population under study. To understand how such data are presented and analyzed, refer to Tables 2.1, 2.2, and 2.3 on the next page. We will use these tables as a framework for a discussion of some of the basic techniques of quantitative data presentation and analysis. Other types of sociological data and analytical techniques are explained at appropriate points in later chapters.

In approaching a statistical table, the first step is to read the title carefully. The title should state exactly what information is presented in the table, including the *units of analysis*—that is, the entity (e.g., individual, family, group) to which a measure applies. In Tables 2.1 and 2.2, the units of analysis are households; in Table 2.3, they are married-couple families.

TABLE 2.1 Types of Households in the United States, 1970 and 1997 (in thousands)

Type of Household	1970	1997
All households	**63,401**	**101,018**
Nonfamily households	**11,945**	**30,777**
Family households	**51,456**	**70,241**
No own children under 18	22,725	35,575
With own children under 18	28,731	34,665
Married-couple family	44,728	53,604
No own children under 18	19,196	28,521
With own children under 18	25,532	25,083
Male householder	1,228	3,847
No own children under 18	887	2,138
With own children under 18	341	1,709
Female householder	5,500	12,790
No own children under 18	2,642	4,916
With own children under 18	2,858	7,874

Source: Adapted from *Statistical Abstract,* 1998.

TABLE 2.2 Types of Households in the United States, 1970 and 1997 (as percentage of total households)

Type of Household	1970	1997
All households	**100.0**	**100.0**
Nonfamily households	**18.8**	**30.5**
Family households	**81.2**	**69.5**
No own children under 18	35.8	35.2
With own children under 18	45.3	34.3
Married-couple family	70.5	53.1
No own children under 18	30.3	28.2
With own children under 18	40.3	24.8
Male householder	1.9	3.8
No own children under 18	1.4	2.1
With own children under 18	0.5	1.7
Female householder	8.7	12.7
No own children under 18	4.2	4.9
With own children under 18	4.5	7.8

TABLE 2.3 Married-Couple Families, by Number of Own Children Under 18, 1970 and 1997 (as percentage of total households)

	1970	1997
Married-couple families	**100.0**	**100.0**
No own children under 18	42.9	53.2
With own children under 18	57.1	46.8

Also check the source of the information presented in the table. This is usually given in a source note at the end of the table. The source indicates the quality of the data (census data are considered to be of very high quality) and tells the reader where those data can be verified or where further information can be obtained.

Frequency Distributions

As you begin to study the table, make sure you understand what kind of information is being presented. The numbers in Table 2.1 are **frequency distributions.** For each year (1970 and 1997), they show how various types of households were distributed in the U.S. population. Frequency distributions indicate how many observations fall within each category of a variable. Thus, within the category of "Married-couple family" for the variable "Type of Household," Table 2.1 indicates that there were 63,401,000 households in the United States in 1970 and that the total increased to 101,018,000 by 1997.

Table 2.1 can tell us a great deal about social change in the United States. For example, social scientists use the term *household* to designate all the people residing at a given address, provided that the address is not a hospital, school, jail, army barracks, or other "residential institution." A household is not necessarily a married-couple family, as is evident from the categories listed in Table 2.1. There are also many households in which a male, with or without his own children, or a female, with or without her own children, is the primary reporting adult. There are also millions of nonfamily households, in which the adults reporting to the census takers are not related. These include people who are roommates, people who are cohabiting, people who are sharing a house or apartment but do not consider themselves roommates, and many other possibilities.

In earlier times it was considered unfortunate and a bit odd for a woman who was not a widow to head a household with children. It was assumed that a true family was composed of a male head, his wife, their children, and anyone related to them by blood who lived on their premises. The numbers in Table 2.1 show, however, that the number of female-headed households more than doubled between 1970 and 1997. The data also show that there was an increase in the number of family households (from 51.5 million in 1970 to 70.2 million in 1997), and almost as great a jump in the number of nonfamily households (from 11.9 million in 1970 to 30.8 million in 1997).

Percent Analysis

Comparing the numbers for 1970 and 1997 in Table 2.1 can be misleading. Remember that these are absolute numbers. They show that the number of nonfamily households has increased and that the number of female-headed and male-headed households has also increased. However, so has the number of married-couple families. And so has the total number of households. How, then, can we evaluate the importance of these various changes?

To compare categories from one period to another, we need a way of taking into account changes in the overall size of the population. This can be done through **percent analysis.** Here is an example:

Table 2.1 shows that between 1970 and 1997 the number of male-headed households with one or more children rose from 341,000 to 1,709,000—a fivefold increase. But how important is this increase in view of the increase in the total population between 1970 and 1997? By using the total number of households in each year as the base and calculating the percentage of each household type for that year, we can "hold constant" the effect of the increase in the total number of households. In the case of male-headed households with one or more children, we find that the increase as a percentage of all households (from 0.54 percent in 1970 to 1.69 percent in 1997) is not as important as the increase in absolute numbers would suggest. The calculation is as follows:

$$341{,}000/63{,}401{,}000 \times 100 = 0.54$$
$$1{,}709{,}000/101{,}018{,}000 \times 100 = 1.69$$

Table 2.2 presents each household type as a percentage of total households in that year. In this way it eliminates the effect of the increase in overall population size between 1970 and 1997. The percentages in Table 2.2 reveal some significant changes that would not be evident from a comparison of absolute numbers. For example, although Table 2.1 seems to indicate that there has been an increase in married-couple families, Table 2.2 shows that when we control for the overall increase in households, married couples actually declined as a proportion of the total, from 70.5 percent in 1970 to 53.1 percent in 1997. The proportion of female-headed families, especially those with children, increased during the period. So did the proportion of nonfamily households.

But we cannot yet view these findings as definite. We must examine them carefully to be sure we are making the right kinds of comparisons. Look again at Table 2.2. In the category of "Married-couple family," those with no children of their own decreased as a proportion of total households—from 30.3 percent in 1970 to 28.2 percent in 1997. But this apparent decrease is due to the overall increase in other types of households. It is still not clear whether, among married-couple families, those without children decreased in relation to those with children. Indeed, if we take only the category of "Married-couple families" and compare subcategories within it, as shown in Table 2.3, we see that as a *proportion of married-couple households,* those without children actually increased from 42.9 percent to 53.2 percent.

This example should convince you to pay close attention to the comparisons made in numerical tables. In this example we see that married couples with children accounted for a smaller percentage of all married-couple families in 1997 than in 1970, but we would not have seen this without making the additional comparison presented in Table 2.3.

Correlations

The term **correlation** refers to a specific relationship between two variables: As one varies in some way, so does the other. Figure 2.3 shows a correlation that people in Europe are particularly concerned about. The figure suggests that there is a correlation between the rate of suicide among men and the rate of unemployment among young people. We can observe from the graph that the two variables (rate of suicide and rate of unemployment) vary together: As one increases, so does the other. This correlation is another confirmation of Durkheim's original hypothesis that the rate of suicide in a society is related (correlated) to major changes in how people in the society behave (i.e., whether they marry, find jobs, have children, and so on). No wonder, then, that according to the French sociologists who gathered these data, the correlation between unemployment and suicide shows that suicide "is an expression of social distress, not only a matter of personal relationships" (quoted in Kremer, 1998, p. 8).

The search for correlations is a common strategy in many kinds of research. In market research, for example, the investigator seeks correlations between the use of certain products and other social variables. (Thus the consumption of beer correlates with the proportion of male consumers in a target population such as baseball fans; the consumption of light beer correlates with the proportion of older, more weight-conscious male consumers; and so on.)

If you take other courses in sociology and statistics, you will learn techniques for calculating the strength of correlations among variables and for sorting out the effects of different variables on the dependent variable you are studying. Although the calculation of

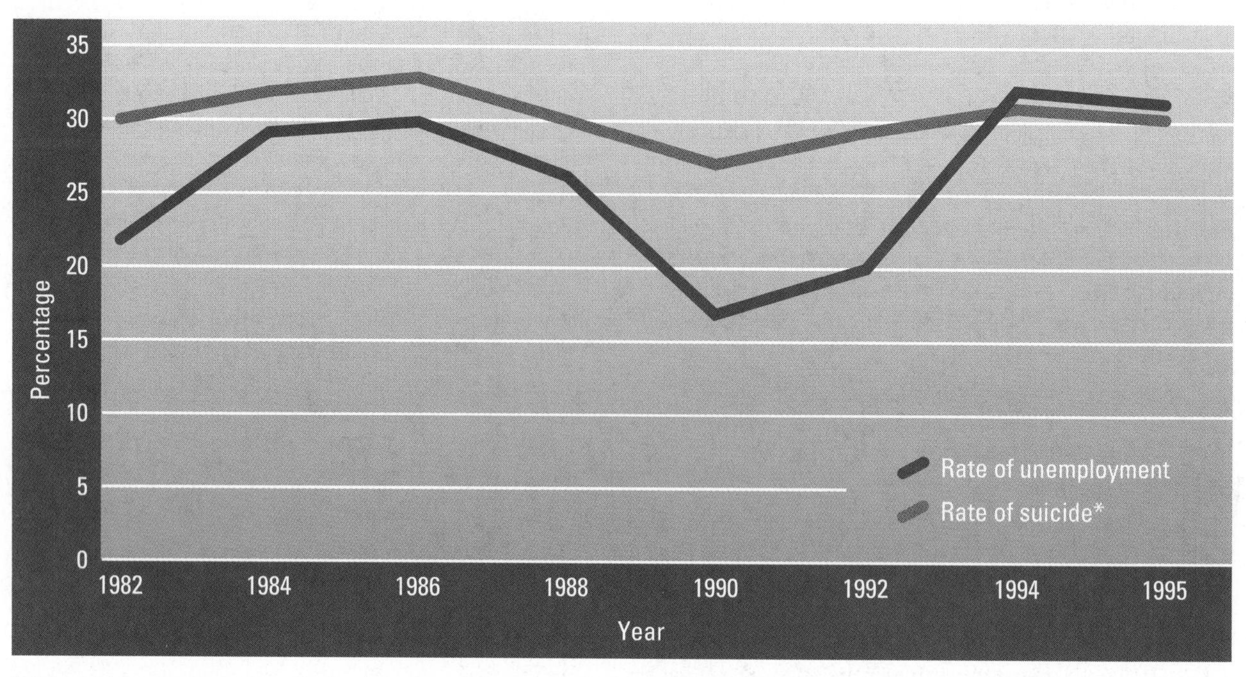

*Male suicides per 100,000; male youth 2 years or less out of school, in percent.

FIGURE 2.3 **Rates of Youth Unemployment and Suicide in France, 1982–1995**

correlation is covered in statistics courses, for our purposes it will be helpful to know that the measure of correlation between two variables, termed the *correlation coefficient,* can vary between +1.0 and –1.0, with 0.0 representing no measured correlation at all. A correlation coefficient of +1 would indicate that the variables are positively and perfectly related—a change in one variable produces an equivalent change in the same direction (increase or decrease) in the other variable. A correlation coefficient of –1 would mean that the variables are perfectly inversely related—a change in one produces an equivalent change *in the opposite direction* in the other. In reality most variables are not perfectly correlated, and correlation coefficients usually fall somewhere between these two extremes.

Correlation and Causation Although correlation can be very useful in the analysis of relationships among variables, it must not be confused with causation. For example, the French sociologists who studied the correlations presented in Figure 2.3 do not claim that unemployment causes suicide; they only point out that the two variables are closely related. They are cautious about making claims about causality because often a strong correlation that seems to indicate causality is in fact a spurious or misleading relationship. Take the example of storks and babies. The fable that storks bring babies seemed to be based on a statistical reality: In rural Holland until recent decades there was a correlation between storks nesting in chimneys and the presence of babies in those households. The more storks, the more babies. In fact, however, the presence of babies in the home meant more fires in the fireplace and more heat going up the chimney to attract storks to nest there. Storks do not bring babies; babies, in effect, bring storks. But the real causal variable is heat, something that was not suggested in the original commonsense correlation. Correlations may suggest the possibility of causality, but one must be extremely careful in making the leap from correlation (association) to causation.

Mapping Social Data

Data about individuals or households can be mapped in precise ways, providing a powerful tool for analyzing sociological data (Monnier, 1993). In a sense, the mapping of data is a form of correlation. The variable under study—it could be poverty, crime, educational achievement, an illness like AIDS, or any other variable—is mapped over a given geographic area, thereby correlating the variable with location.

Figure 2.4 shows an early example of the mapping of social data. Walter Reckless, a sociologist working in Chicago during the early decades of the twentieth century, was interested in finding out where vice crimes like prostitution were occurring. First he counted and noted the locations of all the brothels known from police records. Then he created a map that showed each location as a dot. The map revealed that many brothels were located in poor neighborhoods inhabited mainly by recent immigrants from European nations, and by African Americans who had migrated from the South. The map led Reckless to ask some additional questions: Were brothels located in the immigrant and African American communities because the residents used the services of prostitutes? Or was there another explanation? To find the answer, Reckless had to find out about the clientele served by the brothels. His observations revealed that relatively few immigrants or African Americans patronized them. The majority of the clients were well-off white males. Once he had gained the trust of the madams and the gangsters who sponsored them, he learned that "respectable" men preferred to visit brothels in neighborhoods far away from their own and that the police were far more willing to "look the other way" when prostitution and other vices were concentrated in immigrant and racial ghettos than when the same kinds of businesses were located in more affluent neighborhoods (Reckless, 1933).

Contemporary sociologists are making increasing use of sophisticated mapping techniques. An example is presented in Figure 2.5, which maps the spatial distribution of white-collar workers in San Antonio, Texas. This is an example of research by sociologist Michael White (1987), who studies the characteristics of neighborhoods in U.S. cities and towns. It shows that households with white-collar workers tend to be concentrated on the northern edges of the central city, extending outward into the northern suburbs. Of course, different cities will show different patterns of concentration. Such maps provide a useful way of showing how people's choices about where to live can shape an entire metropolitan region.

In the San Antonio map, the units of area categorized by percentage of white-collar workers living there are examples of *census tracts.* A census tract is a small geographic area within a metropolitan region defined by the U.S. Bureau of the Census. All maps of sociological data use some kind of areal unit, such as counties or census tracts. In later chapters we will present additional examples of mapping of sociological data.

FIGURE 2.4 **Vice Areas in Local Communities of Chicago, 1930**

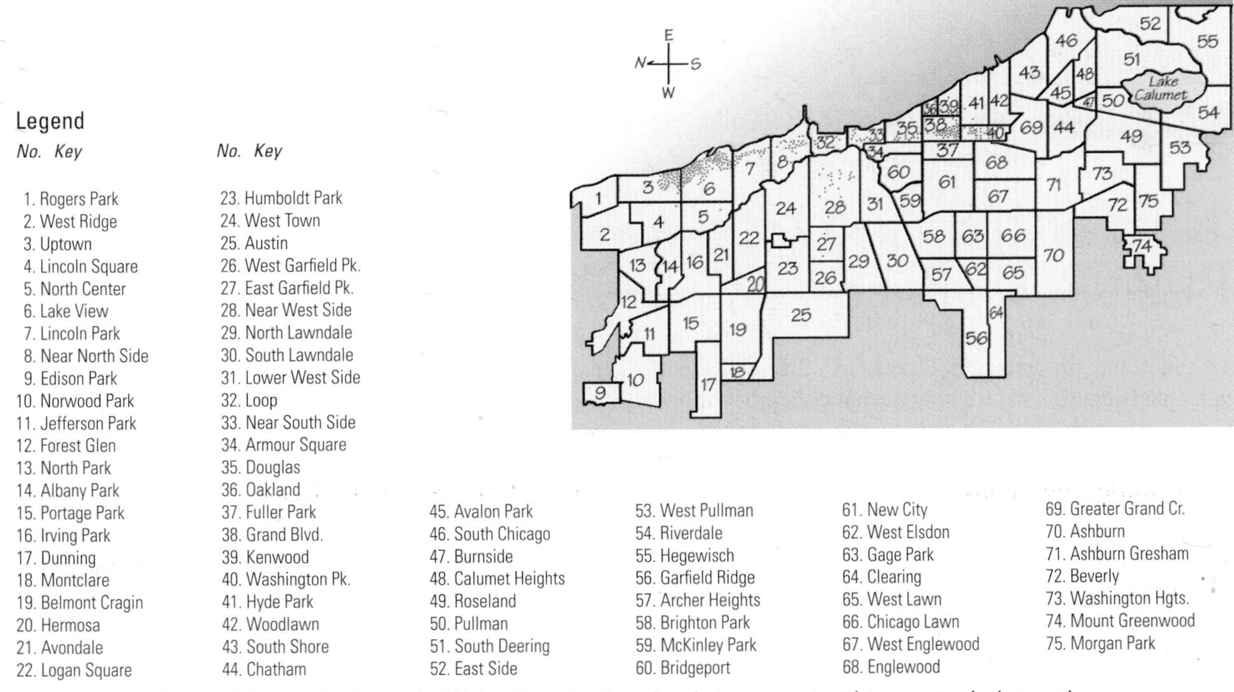

This is a reworking of the original map in Walter Reckless' study of vice areas in Chicago. Each dot on the map represents a house of prostitution known to the Chicago police at the time of the study.

Source: Reckless, 1933.

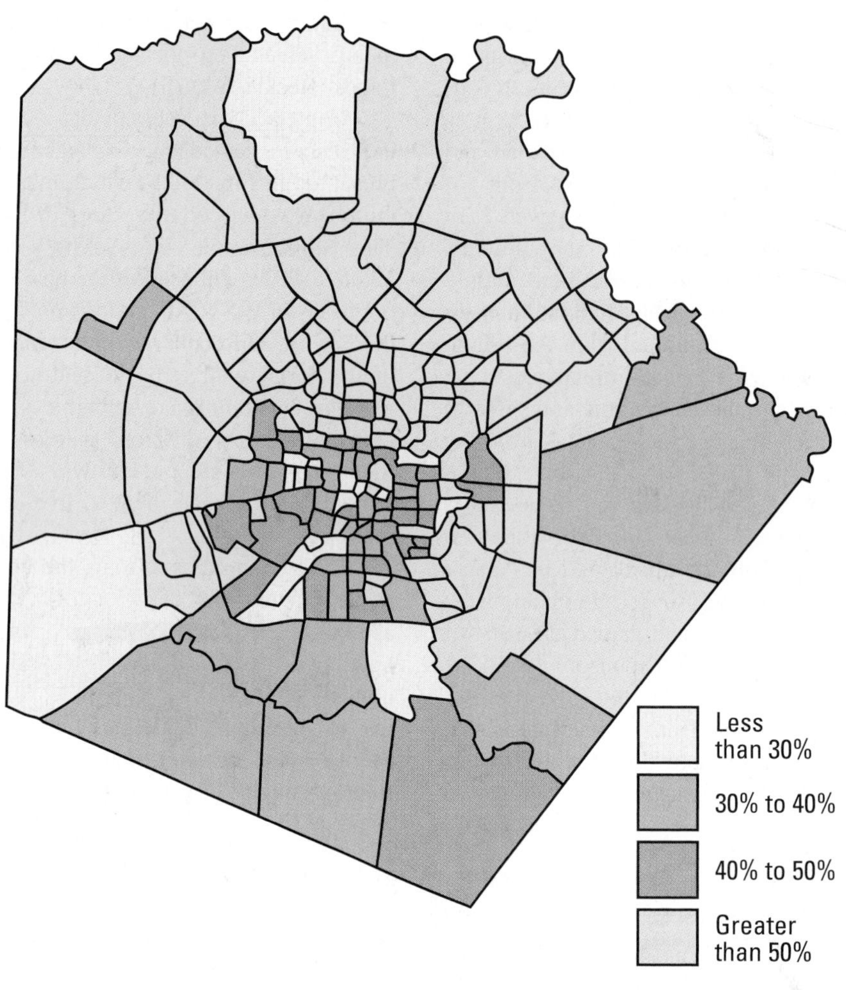

Source: White, 1987.

FIGURE 2.5 **Percentage of White-Collar Workers in San Antonio, by Census Tract**

THEORIES AND PERSPECTIVES

The material presented in Tables 2.1, 2.2, and 2.3 is valuable because it shows how the nation's most important survey, the census, reveals fundamental changes in the structure of the population. But how do these facts about household composition relate to larger trends and issues? The census reveals trends, but it does not explain those trends unless an investigator armed with a set of hypotheses goes to work on the facts to make them prove or disprove a theoretical point.

A **theory** is a set of interrelated concepts that seeks to explain the causes of an observable phenomenon. Some theories attempt to explain an extremely wide range of phenomena, while others limit their explanations to a narrower range. In physics, for example, Newton's theory of gravitation related the force of gravity to the mass and distance between objects, but it did not try to explain why the force of gravity existed in the first place or how it was related to other forces, such as electromagnetism, that could be observed in nature. Einstein's theory of relativity, on the other hand, attempted to explain the relationships among all natural forces, and it predicted forces that had not yet been observed, such as those that are released by nuclear fission.

It is often said that sociology lacks highly predictive theories like those of the physical sciences, but this is a debatable point. If theories are judged by their ability to explain observable phenomena and predict future events, then sociology has some powerful theories. At the beginning of the twentieth century Émile Durkheim used his theory of social integration to predict the social conditions that would increase rates of suicide. The same theory of social integration predicted the conditions that would lead to the rise of totalitarian regimes like that of the Nazis in Germany. Max Weber's theory of bureaucracy is valuable in explaining the experiences we are likely to have in organizations of all kinds. And Karl Marx's theory of class conflict is still the dominant explanation of the revolutions that occurred in feudal and early capitalist societies. No single sociological theory can explain all the complexities of human social life and social change, but in this respect sociology is not very different from other social and physical sciences. A variety of economic theories compete for the attention of policy makers, and the theory of relativity has not fully explained the physical forces that produced the universe.

To cope with the many levels of social explanation, sociologists come to their work armed with **theoretical perspectives:** sets of interrelated theories that offer explanations for important aspects of social behavior. Like the methods of observation and analysis discussed in this chapter, theoretical perspectives are tools of sociological research. They provide a framework of ideas and explanations that helps us make sense of the data we gather. The basic theoretical perspectives are the ones discussed in Chapter 1: interactionism, functionalism, and conflict theory. We also rely on the ecological perspective for descriptive data regarding communities and populations. At times these perspectives offer competing explanations of social life; at other times they seek to explain different aspects of society; and at still other times they are combined in various ways.

Our knowledge about societies and how they change is advanced through scientific competition among sociologists who approach their subjects from different theoretical perspectives. Toward the end of most chapters of this book you will find a feature that illustrates how practicing sociologists use questions and hypotheses derived from these perspectives in their research. To the extent possible, this feature will present "cutting-edge," award-winning sociological research.

VISUAL SOCIOLOGY

Suicide as a Form of Protest

Suicide as a form of protest is one of the most profound and extreme acts imaginable. Any form of public death—be it an execution, a disaster, a brutal murder, or a suicide—sends shock waves through society and stimulates a great deal of emotion, especially among the witnesses. This has probably been true throughout history, but in our age of widespread mass media, sensational events are instantly visible to far greater numbers of people than ever before. A consequence of this change is that suicide deaths may have an immense impact on public feelings and opinions.

The suicide of a Buddhist monk by self-immolation during the Vietnam War was instantly broadcast around the world, not only as a photograph but as a televised event. This act instantly conveyed the message that an important element of the South Vietnamese public opposed the actions of the U.S.-supported government then in power. For antiwar protesters in Europe and the United States, the event sent a clear moral message: Others who opposed the war must redouble their efforts to swing public opinion against the war.

Not all suicides motivated by protest have the desired effect. At times they may have completely unintended consequences. The public suicide by a distraught man, Daniel Jones, in April 1998 had almost no effect on the bureaucracy he was protesting against. Instead, it aroused a storm of protest against the television stations that interrupted their daytime programming, some of which was directed at children, to show the standoff between the man and the authorities, which ended when he shot himself with a revolver.

This protest suicide contributed to an ongoing debate about the propriety of showing such events on national television. Parents often argue that they do not have an opportunity to make decisions about what their children should be allowed to watch when a program is interrupted to present coverage of a sensational event like a public suicide. The news media, on the other hand, argue that they are merely bringing the news to a public that is hungry for knowledge about major events. Media critics point out, however, that this incident was not a major news event in that the issue the man was protesting was largely personal—in contrast to the Vietnam War, which was a subject of intense media coverage.

The suicide, also in 1998, of a prominent Catholic bishop in Pakistan is a far more important example of the use of suicide as a form of political and moral protest. Bishop John Joseph killed himself in protest over the execution of a young Catholic villager named Ayub Masih, who was found guilty of blasphemy after praising

Salman Rushdie's *Satanic Verses.* Here the issues involve not only the problems of the individual but also the rights of religious minorities (i.e., Catholics) in a society with one major religion (Islam). In this example, the bishop was distraught over his inability to influence events in Islamic Pakistan or to bring the problem to the attention of world leaders. His protest suicide did have the effect he desired; it embarrassed the Pakistani government. In the long run, however, more sustained protests by masses of people are usually necessary to bring about lasting political or social change.

From a sociological viewpoint, the foregoing examples represent a special type of suicidal behavior. These individuals, like all people who choose to end their lives, are no doubt motivated by many different feelings, notably suffering and despair. But their decision to link their death to a social cause raises many other sociological issues. Is the protest justified? Should the event be widely publicized? Is the person a hero for having made the supreme sacrifice for a cause? These are all examples of the moral and sociological issues raised by protest suicides.

SUMMARY

In all scientific research certain basic steps must be completed: deciding on the problem, reviewing the literature, formulating research questions, selecting a method, and analyzing the data. Not all researchers follow these steps in the order given, and often one or more steps must be repeated.

The first step in designing social research is formulating the question—that is, asking a question about a social situation that can be answered through the systematic collection and analysis of data. Often the research question is expressed in the form of a hypothesis. A *hypothesis* states a relationship between two *variables* that can be tested through empirical observation. The variable that is to be explained is the *dependent variable.* The other variable, a factor that the researcher believes causes changes in the dependent variable, is the *independent variable.*

Before collecting new data, a professional researcher reviews as much existing research and other data sources as possible. This "review of the literature" sometimes supplies all the data necessary for a particular study.

The most frequently used research methods in sociology are observation, experiments, and surveys. Observation may take the form of *participant observation,* in which the researcher participates to some degree in the life of the people being observed. It may also take the form of *unobtrusive measures,* or observational techniques that measure behavior but intrude as little as possible into actual social settings. Visual sociology involves the use of photography and videotape to observe people in a variety of settings.

Sociological experiments can take one of two basic forms: the controlled experiment and the field experiment. In a *controlled experiment* the researcher establishes an *experimental group,* which will experience the "treatment" (a change in the independent variable), and a *control group,* which will not experience the treatment. The effect of the treatment on the dependent variable can be measured by comparing the two groups. *Field experiments* take place outside the laboratory and are often used in evaluating public programs designed to remedy specific social problems. The treatment group consists of people who experience a particular social program, and the control group consists of comparable people who do not experience the program. A common problem of experimental studies is the *Hawthorne effect,* which refers to any unintended effect resulting from the attention given to subjects in an experiment.

The third basic method of sociological research, the survey, asks people to give precise information about their behavior and attitudes. The most ambitious surveys are national censuses; the data obtained in a census can be supplemented by smaller, less costly *sample surveys.* A *sample* is a selection of respondents drawn from a specific population. To ensure that everyone in the specified population has an equal chance of being selected, the respondents must be selected by some process of random sampling.

Questionnaire design is an important aspect of survey research. Questions must be precisely worded, easy to understand, and free of bias. *Closed questions* require the respondent to select from a set of answers, whereas *open questions* allow respondents to say whatever comes to mind.

Sociological researchers must always consider the rights of human subjects. *Privacy* is the right to decide the terms on which one's acts may be revealed to the public. *Confidentiality* means that the researcher cannot use responses in such a way that they can be traced to a particular respondent. In order for respondents to give *informed consent,* they must be told how the information they supply will be used and they must be allowed to judge the degree of personal risk involved in answering questions.

The data gathered in a survey are usually presented in the form of statistical tables. In reading a table it is important to know what the units of analysis are and what kinds of data are being presented. *Frequency distributions,* given in absolute numbers, reveal the actual size of each category of a variable, but to compare the numbers for different years it is necessary to use *percent analysis.* Analyzing data often leads to the discovery of *correlations,* or specific relationships between two variables. Correlation should not be confused with causation. Mapping of data about individuals or households also provides a powerful tool for analyzing sociological data.

Once data have been presented and analyzed, they can be used to generate new hypotheses. The types of hypotheses that might be developed depend on the researcher's *theoretical perspective*—a set of interrelated theories that offer explanations for important aspects of social behavior. The functionalist, interactionist, and conflict perspectives give rise to quite different hypotheses.

SOCIOLOGY VERSUS IDEOLOGY

Durkheim's study of suicide helped establish sociology as a scientific discipline. Suicide has also been the subject of numerous philosophical, theological, and literary debates. Hamlet's soliloquy—"To be or not to be, that is the question," among the most famous passages in all of literature—comes to mind. And as we have seen, social change throughout the world is confronting a growing number of people with this life-or-death choice. No wonder there is so much controversy over the concept of physician-assisted suicide.

Most conservatives feel that suicide must be strictly banned. Some liberals feel that individuals should be free to choose to escape from suffering. Some people believe that government should almost never dictate what individuals can and cannot do, so long as there is no harm to others. Those with this belief may call themselves libertarians. They are opposed by many people who, on religious grounds, feel strongly that taking any life, including one's own, is a sin. What can knowledge of sociology, and sociological methods in particular, do to cut through these many conflicting values and beliefs?

Sociological research methods cannot provide moral guidance, but they can point to empirical evidence that can help people better understand these opposing views. After all, Durkheim wrote extensively on how different societies have regarded suicide through history. The facts he gathered showed that, almost without exception, individual suicide (as opposed to group suicides or the suicide of a widow following the death of her husband) has been viewed as behavior that must be suppressed for the good of the larger society. This important fact helps us understand why so many people in societies throughout the world strongly oppose the idea of assisted suicide. Just because it has been socially condemned in the past does not automatically justify continuing to ban it, but the findings of sociological research show us that assisted suicide is an issue for society as a whole, not merely a matter of individual choice.

GLOSSARY

hypothesis: a statement that specifies a relationship between two or more variables that can be tested through empirical observation. (p. 31)

variable: a characteristic of an individual, group, or society that can vary from one case to another. (p. 31)

dependent variable: the variable that a hypothesis seeks to explain. (p. 31)

independent variable: a variable that the researcher believes causes a change in another variable (i.e., the dependent variable). (p. 31)

participant observation: a form of observation in which the researcher participates to some degree in the lives of the people being observed. (p. 35)

unobtrusive measures: observational techniques that measure behavior but intrude as little as possible into actual social settings. (p. 36)

controlled experiment: an experimental situation in which the researcher manipulates an independent variable in order to observe and measure changes in a dependent variable. (p. 37)

experimental group: in an experiment, the subjects who are exposed to a change in the independent variable. (p. 37)

control group: in an experiment, the subjects who do not experience a change in the independent variable. (p. 37)

field experiment: an experimental situation in which the researcher observes and studies subjects in their natural setting. (p. 40)

Hawthorne effect: the unintended effect that results from the attention given to subjects in an experimental situation. (p. 41)

sample survey: a survey administered to a selection of respondents drawn from a specific population. (p. 41)

sample: a set of respondents selected from a specific population. (p. 43)

closed question: a question that requires the respondent to choose among a predetermined set of answers. (p. 44)

open question: a question that does not require the respondent to choose from a predetermined set of answers; instead, the respondent may answer in his or her own words. (p. 45)

privacy: the right of a respondent to define when and on what terms his or her actions may be revealed to the general public. (p. 45)

confidentiality: the promise that the information provided to a researcher by a respondent will not appear in any way that can be traced to that respondent. (p. 45)

informed consent: the right of respondents to be informed of the purpose for which the information they supply will be used and to judge the degree of personal risk involved in answering questions, even when an assurance of confidentiality has been given. (p. 45)

frequency distribution: a classification of data that describes how many observations fall within each category of a variable. (p. 46)

percent analysis: a mathematical operation that transforms an absolute number into a proportion as a part of 100. (p. 47)

correlation: a specific relationship between two variables. (p. 47)

theory: a set of interrelated concepts that seeks to explain the causes of an observable phenomenon. (p. 50)

theoretical perspective: a set of interrelated theories that offer explanations for important aspects of social behavior. (p. 51)

WHERE TO FIND IT

BOOKS

Handbook of Survey Research (Peter Rossi et al.; Basic Books, 1983). Provides comprehensive, though rather technical, coverage of issues in the design, administration, and analysis of sociological survey research.

The Practice of Social Research (Earl Bobbie; Wadsworth, 1992); and ***Social Research Methods: Qualitative and Quantitative Applications,*** 3rd ed. (William Lawrence Neuman; Allyn & Bacon, 1997). Two practical texts that introduce the ways in which sociologists actually formulate their research questions and conduct their research.

Unobtrusive Measures: Nonreactive Research in the Social Sciences (Eugene J. Webb et al.; Rand McNally, 1966). A pathbreaking treatise on how to develop scientific techniques for observing social life in nondisruptive ways.

In the Field: Readings on the Field Research Experience, 2nd ed. (Carolyn D. Smith and William Kornblum, eds.; Praeger, 1996). A selection of personal accounts by a group of noted ethnographic researchers, designed to give students a sense of what it is actually like to conduct ethnographic research, especially participant observation.

How to Lie With Statistics (Darrell Huff; Norton, 1954). A charming and valuable book about the uses and misuses of quantitative data. To know how to lie with statistics is also to know how to unmask those who would lie and to credit those who use numbers and statistics properly.

Participatory Action Research (William F. Whyte, ed.; Sage, 1991). A collection of essays by social scientists who are engaged in action projects in the United States and Third World nations. The projects are designed to create social changes that will improve food supplies, agricultural practices, or industrial production. The book highlights ways in which the researcher can be an active participant in social change without losing scientific credibility.

OTHER SOURCES

Statistical Abstract of the United States (U.S. Bureau of the Census). The first source consulted for quantitative facts about the population of the United States. Published annually.

Census of the United States. An invaluable tool for all the social sciences, but one that requires a patient user. There are volumes for all states and cities in the United States. The tables show population totals and subtotals by sex, age, race, and occupation. Published every 10 years on the basis of analysis of the data gathered in the national census.

The American Statistics Index (Congressional Information Service Inc.). The best single source for statistics collected by agencies of the U.S. government. Published monthly and collected into an annual volume, this index offers the most complete list of public statistics available in the United States.

American Demographics. A magazine devoted to applications of census and opinion survey data to marketing and other studies of consumer behavior.

INTERNET RESOURCES

Sociology Department, Princeton University (www.princeton.edu/~sociolog/). A good example of how modern sociology depeartments are using the Internet. Offers links with other sociology departments on each continent, domestic and international research institutes, data archives, and Web pages for academic journals.

Society for Applied Sociology (www.indiana.edu/~appsoc/). Provides many links to other Web sites and addresses the three goals of the society: to provide a forum for sociologists and others interested in applying sociological knowledge, to enhance understanding of the interrelationship between sociological knowledge and sociological practice, and to increase the effectiveness of applied sociological research and training.

CHAPTER 3

CULTURE

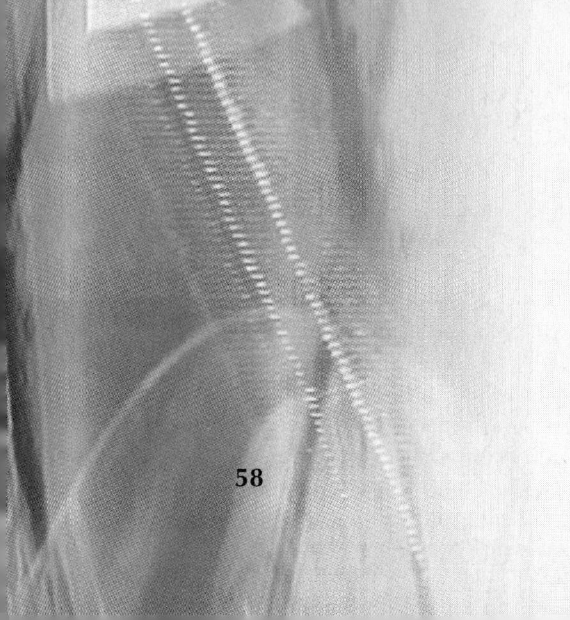

For Native Americans, the coming of the Europeans to the New World marked the beginning of a long, drawn-out disaster. This is how sociologist Russell Thornton, himself a Native American, describes it:

> In the centuries after Columbus these "Indians" suffered a demographic collapse. Numbers declined sharply; entire tribes, often quickly, were "wiped from the face of the earth." This is certainly true of the American Indians on the land that was to become the United States of America. For them the arrival of the Europeans marked the beginning of a long holocaust, although it came not in ovens, as it did for the Jews. The fires that consumed North American Indians were the fevers brought on by newly encountered diseases, the flashes of settlers' and soldiers' guns, the ravages of "firewater," the flames of villages and fields burned by the scorched-earth policy of vengeful Euro-Americans. The effects of this holocaust of North American Indians, like that of the Jews, was millions of deaths. In fact, the holocaust of the North American tribes was, in a way, even more destructive than that of the Jews, since many American Indian peoples became extinct. (Thornton, 1987, pp. xv–xvi)

Others have shown that even when Indians and settlers tried to live as neighbors it was very difficult for them to do so (Cronon, 1983; Warhus, 1997). Seemingly simple differences could result in major conflicts. One major source of conflict could be found in the different ways in which Indians and settlers related to the land. The settlers tended to think of the bounty offered by their new environment as inexhaustible. They were not nearly as aware of the rhythms of nature as the Indians were, nor did they know how to live in harmony with nature without destroying its resources. So when they saw the Indians moving from one hunting or fishing ground to another as the seasons changed, they concluded that the Indians were not "improving the land" as proper farmers did. Over and over again, Europeans justified the taking of the Indians' forests and meadows by claiming that the Indians had no sense of ownership of the land because they did not clear it and farm it.

It would be difficult to overestimate the shock and confusion the Indians must have felt as they were continually driven from their traditional hunting and fishing grounds by the encroaching Europeans. The resulting conflicts were often bloody, with injuries and deaths on both sides, but the effects of the Europeans' technological superiority were especially devastating. The Europeans' cannon and rifles gave them the ultimate power to inflict their will on the Indians. No wonder that wherever they encountered the

Europeans, the Indians began to covet the powerful new killing tools made of iron and steel. No wonder that the Europeans never doubted, even as they learned from the Indians how to survive in their new environment, that their own way of life was the only "true" civilization. Indeed, so powerful did the notion of European superiority become that even today we celebrate the "discovery" of the New World by European explorers. Too often we forget that what happened in 1492 was not the discovery of a new world but the establishment of contact between two worlds, both already old (Jennings, 1975; Marks, 1998).

THE MEANING OF CULTURE

Was the European way of life really superior? This question is a subject of stormy controversies throughout the Western Hemisphere. Although conquest by European invaders and settlers spelled the doom of many Native American societies, some of those societies managed to survive and are struggling to maintain their way of life within larger societies such as those of Canada, the United States, Mexico, Bolivia, Brazil, and Peru. An example is provided by the Mayan Indians of the Mexican highlands, as illustrated in the Visual Sociology feature near the end of the chapter. And even when cultural contact occurs on a more equal basis, as it does when people from different modern cultures meet, it is difficult to achieve understanding and arrive at mutually beneficial forms of cooperation. Why is this so? Many of the answers may be found in the study of human culture.

In everyday speech the word *culture* refers to pursuits like literature and music. But these forms of expression are only part of the definition of culture. To the social scientist, "a humble cooking pot is as much a cultural product as is a Beethoven sonata" (Kluckhohn, 1949, quoted in Ross, 1963, p. 96). Culture is a basic concept in sociology because it is what makes humans unique in the animal kingdom. All the familiar forms of social organization, from the simplest family to the most complex corporation, depend on culture for their existence. But although societies cannot exist without cultures, the two are not the same thing. *Societies* are populations that are organized to carry out the major functions of life. A society's *culture* consists of all the ways in which its members think about their society and communicate about it among themselves.

We can define **culture** as all the modes of thought, behavior, and production that are handed down from one generation to the next by means of communicative interaction—speech, gestures, writing, building, and all other communication among humans—rather than by genetic transmission, or heredity. This definition encompasses a vast array of behaviors, technologies, religions, and so on—in other words, just about everything made or thought by humans. Among all the elements of culture that could be studied, social scientists are interested primarily in aspects that explain social organization and behavior. Thus, although they may analyze trends in movies and popular music (which significantly affect behavior in the modern world), they are even more likely to study aspects of culture that account for phenomena like the behavior of people in corporations or the conduct of scientific research.

Quilt making is a highly social folk art form. This quilt, created by University of Mississippi graduate student Patti B. Melton, is actually a sociology paper about quilt making reproduced on an original quilt. It blends ideas with material culture.

Dimensions of Culture

The culture of any people on earth, no matter how simple it may seem to us, is a complex set of behaviors and artifacts. A useful framework for thinking about culture was suggested by Robert Bierstedt (1963). Bierstedt views culture as having three major dimensions: **ideas,** or ways of thinking that organize human consciousness; **norms,** or accepted ways of doing or carrying out ideas; and **material culture,** or patterns of possessing and using the products of culture. Figure 3.1 presents some examples of these three dimensions, along with two aspects of culture—*ideologies* and *technologies*—that combine elements from more than one dimension.

Ideas As indicated in Figure 3.1, theories about how the physical world operates (scientific knowledge), strongly held notions about what is right and wrong (values), and traditional beliefs, legends, and customs (folklore) are among the most important types of ideas found in any given culture. Of these, values are especially important because people feel so strongly about them and because they often undergo changes that result in social conflict (as we can see in contemporary debates about "a woman's right to control her reproductive destiny" versus "a fetus's right to life").

Values are socially shared ideas about what is right. Thus for most people in North America education is a value; that is, they conceive of it as a proper and good way to achieve social standing. Loyalty to friends and loved ones, patriotism, the importance of religion, the significance of material possessions—these and other values are commonly found in our culture and many others, but of course there are wide differences in how people interpret these values and in the extent to which they adhere to them. In small tribal and peasant societies, there tends to be far greater consensus on values than in large, complex industrial societies. In a large and diverse society like ours, there is bound to be a good deal of conflict over values. Some people are satisfied with the way wealth and power are distributed, for example, while others are less satisfied with the status quo. Some people feel that it is desirable to attempt to improve one's own well-being and that society as a whole gains when everyone strives to be well-off. Others assert that the value of individual gain conflicts with the values of community and social cohesion; too great a gap between the well-off and the not-well-off creates suffering, envy, crime, and other problems.

In societies undergoing rapid and far-reaching social change, such as the United States, there is bound to be conflict over values. Controversies over "family values," the sanctity of marriage, obligations to children, and similar issues tend to be particularly intense because people's individual experiences differ greatly. The more diversity there is among the people who share a basic culture, the more conflict there is likely to be over certain aspects of that culture. Such disputes often turn on ideas about how to behave—that is, on the norms governing behavior in a society.

FIGURE 3.1 **Dimensions of Culture**

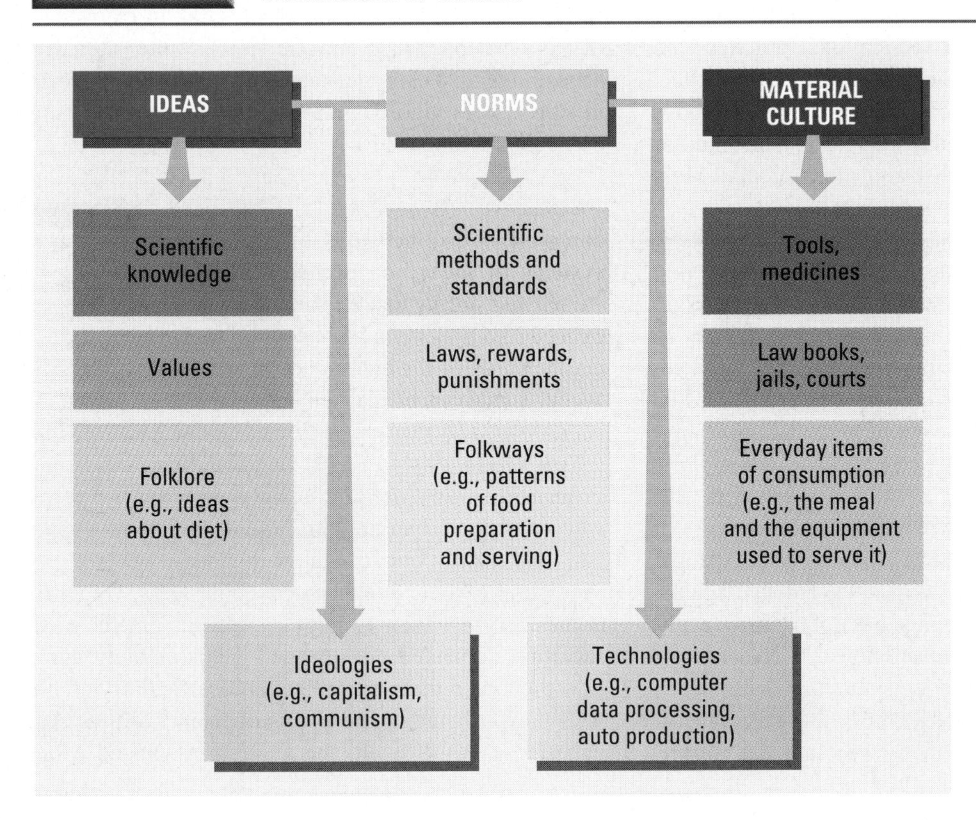

Norms From people's beliefs in what is right and good—that is, from their values—are derived the norms, or rules of behavior, of a society. Values are more abstract than norms; they are the ideas that support or justify norms. Norms are more specific. They are "the adjustments which human beings make to the surrounding environment. We may think of them as solutions to recurring problems or situations" (Nisbet, 1970, p. 225). But norms involve more than behavior. Any given norm is supported by the idea that a particular behavior is correct and proper or incorrect and improper. "The moral order of society is a kind of tissue of 'oughts': negative ones which forbid certain actions and positive ones which [require certain] actions" (Nisbet, 1970, p. 226).

If we think about a complex aspect of everyday life like driving a car, it is evident that without norms life would be far more chaotic and dangerous than it already is. When we drive we keep to the right, obey traffic lights and speed limits, and avoid reckless behavior that could cause accidents. These are among the many norms that allow the automobile to be such an essential article of North American culture.

Examples of norms are easy to find. Take the college or high school classroom. The classroom is organized according to norms of educational practice: There should be a textbook or books; there should be class discussion; there should be assignments, exams, grades. Note that these "shoulds" correspond to actual behavior. Norms usually refer to behavior that we either approve or disapprove of. However, members of a culture often disagree about how a particular norm operates. Thus in the case of classroom organization there is considerable debate about what educational practices are most effective. In Israel, for example, there is debate among orthodox and liberal Jews over whether boys and girls should be separated in classrooms, and similar debates divide fundamentalist and more secular Muslims in other Middle Eastern societies.

In societies where values are in conflict or where changes are occurring rapidly, norms are also subject to change. This can readily be seen in the case of smoking. The norms governing smoking are changing rapidly as a result of changing values regarding health. Until recent years the norms governing smoking in the United States were very liberal. One could light up almost anywhere, and nonsmokers were expected not to complain. Recently, however, greater emphasis on fitness and health, as well as new knowledge about the dangers of secondhand smoke, have increased the value people place on clean air and have tipped the balance against the earlier norms. But the change has not occurred without conflict between smokers and nonsmokers and their representatives in courts and legislatures.

Laws are norms that are included in a society's official written codes of behavior. Laws are often developed by a specialized occupational group, such as priests in the ancient world and legislators, judges, and lawyers in the modern world. Some of the oldest examples of laws are the Code of Hammurabi, the ancient Babylonian code that specified punishments equal to the gravity of the crime (e.g., an eye for an eye, a tooth for a tooth); the Ten Commandments, written by God, according to the Old Testament, on stone tablets at least 1,600 years before the birth of Christ; and the codes of Confucius, written and interpreted by the royal scribes of the ancient Chinese empire. In our time, laws are the special province of professional lawmakers, but all citizens participate in a network of laws and other written regulations that govern daily life. Basic behaviors like driving, going to school, marrying, and investing are greatly influenced by laws governing the conduct of these activities.

At times we may grumble or protest against laws that we find overly restrictive. And there is always some group planning to protest a law that is under consideration in a legislative body like Congress. But when we regard the world beyond our own society we see that the rule of law could save millions of lives if only it could be achieved in places like Guatemala, Rwanda, Bosnia, Yugoslavia, or many of the nations of the former Soviet Union. On the other hand, even in nations where the rule of law is generally accepted, social scientists often find that new laws have unintended consequences. For example, when extremely strict laws against adult drug-dealing were passed, the dealers began recruiting juveniles to sell drugs. The lawmakers never intended the new laws to encourage teenagers to become dealers, but in many states that is what happened (Bayer, 1991). Quite often laws have unintended consequences because they were created in response to vigorous efforts by people who espouse particular ideologies.

Ideologies As indicated in Figure 3.1, **ideologies** comprise two dimensions of culture; they are sets or systems of ideas and norms. Ideologies combine the values and norms that all the members of a society are expected to believe in and act upon without question. A classic study of the emergence of an ideology was Max Weber's analysis of the link between Protestantism and capitalism, *The Protestant Ethic and the Spirit of Capitalism* (1974/1904). Weber noticed that the rise of Protestantism in Europe coincided with the rise of private enterprise, banking, and other aspects of capitalism. He also noticed that a majority of the most successful early capitalists were Protestants. Weber hypothesized that their religious values taught them that salvation depended not on good deeds or piety but on how they lived their entire lives and particularly on how well they adhered to the norms of their "callings" (occupations). As a result, the Protestants placed a high

Then and Now

The Once Glamorous Habit

How quickly norms can change, and how much confusion those changes can bring. Smoking is a fine example. In less than four decades this behavior changed from a glamorous activity to an act that many people consider deviant.

For most of the twentieth century smoking was extremely popular. The majority of Americans smoked, and smoking was even more popular in England and on the European continent. Humphrey Bogart and Lauren Bacall smoked, as did most other movie stars, as well as many athletes and famous journalists. Smoking reached its peak of popularity during and just after World War II. Children often began smoking steadily by age 14, encouraged to do so by ubiquitous advertising that associated smoking with success, sensuality, and, in the case of the famous Marlboro cowboy, masculinity.

Today much has changed. Smoking has been shown to be extremely dangerous to health—not only the health of the smoker but also the health of people exposed to secondhand tobacco smoke. In consequence, the norms and values associated with smoking are being challenged throughout most Western nations. New norms restricting smoking in public places and in most offices are creating a deviant group of smokers who must linger outside their offices to satisfy their craving and enjoy a cigarette.

value on frugality and abstinence. To prove that they were worthy of salvation, they devoted themselves tirelessly to commerce and plowed their profits back into their firms. But Catholics, who did not share these values, were less single-mindedly dedicated to their business ventures. They often spent their profits on good deeds rather than investing them in their businesses. Weber attempted to show how a set of religious values and norms combined with economic norms to create the ideology of capitalism.

A contemporary example of an ideology may be found among religious fundamentalists. Christian fundamentalists in the United States generally share a set of ideas and norms of behavior that include the value of prayer, the value of family and children, the negative value of abortion and secular humanism (an ethical system based on scientific knowledge rather than on religious teachings), the belief in salvation and redemption for one's sins, and other values.

Values and Ideologies in Conflict Most societies experience conflict among people with differing values and ideologies. At times this conflict can have serious outcomes, such as widespread violence among antagonistic factions or parties, as occurs in Northern Ireland between Protestants and Catholics, in the Middle East between Jews and Arabs, and in the United States between supporters and opponents of abortion rights. We return continually to the subject of conflicting values and resultant conflict among social groups because the subject is so important in understanding social life and social change. In discussing values and ideologies, however, it is important to point out that in almost every known society there are inherent conflicts in the values the society claims to support. In most democratic societies, for example, the values of liberty (freedom) and equality are cherished. But there is inherent conflict in these two central values. As Isaiah Berlin, one of the most influential social thinkers of the twentieth century, explained:

> Perfect liberty is not compatible with perfect equality. If man is free to do anything he chooses, then the strong will crush the weak, the wolves will eat the sheep, and this puts an end to equality. If perfect equality is to be attained, then men must be prevented from outdistancing each other, whether in material or in intellectual or in spiritual achievement, otherwise inequalities will result. (1998, p. 60)

We will see in later chapters that the desire of some groups to achieve total equality or freedom continues to produce ideologies in which the goals of action—that is, equality or freedom—lead people to assert that "the ends justify the means" used to achieve those ends, even if they include violating norms like "Thou shalt not steal" or "Thou shalt not kill." Such ideologies led to some of the twentieth century's most atrocious episodes of mass violence and destruction. But the inherent conflict among values can also be a creative aspect of society. As we seek to decrease the gap between the haves and the have-nots in the United States and elsewhere while at the same time attempting to preserve as much individual freedom as possible, we realize that a just society is one in which people are free to seek the best peaceful means of balancing their conflicting interests and values.

Material Culture The third dimension of culture identified by Bierstedt is material culture. Material culture consists of all the things that a society produces. Mundane things like pots and pans or the wooden eating bowls of nonindustrial societies; immensely complex systems of things, such as the space shuttle; cherished items of religious worship like rosary beads or Indian fetish necklaces—all take their shape and purpose from the ideas of the culture that produces them. Members of societies that place a high value on science and efficiency are used to seeing these values expressed in material objects. For example, we design our houses to conserve energy and create desirable combinations of view and privacy. We may take older forms like the ranch house or the Cape Cod bungalow and modify them to suit modern purposes; in this way our houses combine tradition with usefulness. The particular form that appeals to us is usually a result of many different ideas, including what we know from our own upbringing, what we can afford, how much space we require, and the environmental conditions we anticipate.

The same is true throughout the world. If one travels in Africa, for example, one will find large areas in which people build square houses with thatched roofs or roofs of corrugated metal. Suddenly one will pass into a region where the people build round houses with steeply pitched roofs. In another region people may live in sprawling apartment-like complexes built of sand and mortar. When asked why they prefer a particular kind of dwelling, people usually say, "Because we have always built our houses this way." In other words, their answers are based on ideas of what is right for them. But on further discussion a traveler with a sociological imagination will discover that the different building forms also express functional ideas. The mud-walled "apartments" of the desert village seem cramped and dark to Western eyes, but the people of the village will explain that they often sleep outdoors and the dark, cool rooms are most useful on very hot days. On the other hand, the tribespeople in a society that builds round houses may give a spiritual explanation for the shape of their dwellings; they may believe that corners are places where evil spirits can lurk. In sum, different cultures may give different reasons for the form and function of a material element such as a house, but in all cultures one will find a material culture that stems from the society's most important ideas (Kidder, 1985; Rybczynski, 1987).

Technologies Technologies are another aspect of culture that spans two of Bierstedt's major dimensions. (Refer again to Figure 3.1.) **Technologies** are the things (material culture) and the norms for using them that are found in a given culture (Bierstedt, 1963; Ellul, 1964). Without the norms that govern their use, things are at best confusing and at worst useless or dangerous. In the United States, for example, new telecommunications technologies based on computers, modems, and networks like the Internet are a powerful means of communicating information across long distances. But norms for using these new technologies—such as norms about privacy, freedom of speech, and personal accountability—are only now developing. There is serious conflict, for example, about whether pornographic messages directed at juveniles or e-mail messages among colleagues in an office are protected under the norms of free speech, or whether they are subject to public scrutiny and perhaps social sanction. Because the Internet is a relatively new form of communication, the norms (i.e., laws) that apply to print and TV communication have not yet been fully worked out and applied to this new communications medium.

Guns present another case in which norms governing use of the things themselves are very much in dispute. Most Americans agree that people should be free to own and use hunting weapons—but what about handguns, assault rifles, machine guns, and rocket launchers? What, if any, restrictions should there be on the distribution and use of such weapons? As we will see in later chapters, the debate about appropriate norms for controlling the distribution and possession of deadly firearms that are clearly not designed for hunting is an indication of a deeper cultural conflict over the right to bear arms versus the state's monopoly over the use of force.

Norms and Social Control

Norms are what permit life in society to proceed in an orderly fashion without violence and chaos. Shared norms contribute enormously to a society's ability to regulate itself without constantly resorting to the coercive force of armies and police. The term **social control** refers to the set of rules and understandings that control the behavior of individuals and groups in a culture. Park and Burgess (1921) noted that throughout the world there are certain basic norms that contribute to social control: "All groups have such 'commandments' as 'Honor thy father and mother,' 'Thou shalt not kill,' 'Thou shalt not steal'" (p. 787).

The wide array of norms that permit a society to achieve relatively peaceful social control is called its **normative order.** It is the normative order that creates what Morris Janowitz has termed "the capacity of a social group, including a whole society, to regulate itself" (1978, p. 3). This self-regulation requires a set of "higher moral principles" and the norms that express those principles. The norms define what makes a person a "good" member of the culture and society.

The most important norms in a culture are often taught as absolutes. The Ten Commandments, for example, are absolutes: "Thou shalt not kill," "Thou shalt not steal," and so on. Unfortunately, people do not always extend those norms to members of another culture. For example, the same explorers who swore to bring the values of Western civilization (including the Ten Commandments) to the "savage" Indians of the New World thought nothing of taking the Indians' land by force. Queen Elizabeth I of England could authorize agents like Sir Walter Raleigh to seize remote "heathen and barbarous" lands without viewing this act as a violation of the strongest norms of her own society (Jennings, 1975; Snipp, 1991). Protests by the Indians often resulted in violent death, but the murder of Indians and the theft of

This nineteenth-century painting shows a woman depositing her newborn infant in a "receptacle for foundlings" probably associated with a church or cathedral. Women who violated the norms of legitimate fertility by bearing children out of wedlock were often forced to give up their infants.

Young people who engage in radical piercing are implicitly challenging conventional notions about personal adornment. Their behavior does not violate the mores of their culture, but the new folkways they are choosing may be disapproved of in some parts of society.

their land was rationalized by the notion that the Indians were an inferior people who would ultimately benefit from European influence. In the ideology of conquest and colonial rule, the Ten Commandments did not apply.

In the social sciences, punishments and rewards for adhering to or violating norms are known as **sanctions.** Cultural norms vary according to the degree of sanction associated with them. Rewards can range from a smile to the Nobel Prize; punishments may vary in strength from a raised eyebrow to the electric chair. During the period of European colonial expansion, the murder of a native was far less strongly sanctioned than was the murder of another European.

The most strongly sanctioned norms are called **mores.**[1] They are norms that people consider vital to the continuation of human groups and societies, and therefore they figure prominently in a culture's sense of morality. **Folkways,** on the other hand, are far less strongly sanctioned. People often cannot explain why these norms exist, nor do they feel that they are essential to the continuation of the group or the society. Both terms were first used by William Graham Sumner in his study *Folkways* (1940/1907). Sumner also pointed out that laws are norms that have been enacted through the formal procedures of government. Laws often formalize the mores of a society by putting them into written form and interpreting them. But laws can also formalize folkways, as can be seen, for example, in laws governing the wearing of clothing in public places.

Sumner also pointed out that a person who violates mores is subject to severe moral indignation whereas one who violates folkways is not. People who violate folkways such as table manners or dress codes may be thought of as idiosyncratic or "flaky," but those who violate mores are branded as morally reprehensible. Thus among prisoners there are norms for the treatment of fellow prisoners, based on the types of crimes they committed "outside." Rapists and child molesters are moral outcasts; their offenses are the most reprehensible. Such offenders are often beaten and tormented by other prisoners.

Box 3.1 presents a typology of norms according to their degree of sanction and their mode of development—that is, whether they are formal or informal. Laws and other norms, such as company regulations or the rules of games and sports, are known as *formal norms*. They differ from *informal norms,* which grow out of everyday behavior and do not usually take the form of written rules, even though they too regulate our behavior. For example, when waiting to enter a movie theater, it is usually permissible to have one member of a

[1] This term, pronounced "morays," is the plural of the Latin word *mos,* meaning "custom."

BOX 3.1 SOCIOLOGICAL METHODS

Constructing a Typology of Norms

Typologies are ways of grouping observable phenomena into categories in order to identify regularities in what may appear to be a great variety of observations. For example, there are so many social norms that the average person has no hope of ever sorting them out without some kind of system for organizing them. In the chart presented here, the subject is norms of various types. The sociologist constructs the types by comparing various dimensions along which norms may differ. The norms listed in the Ten Commandments, for example, differ in many ways. The norm "Thou shalt not kill" not only is generally believed and passed along from one generation to the next but also is codified in laws. So is the commandment "Thou shalt not steal," which is an important part of the written laws of our society. But the commandment "Remember the Sabbath day, to keep it holy" is not a written law in the United States, at least not in the federal statutes.

In some states and communities there are laws specifying that businesses must close on Sunday, but not all religious groups recognize Sunday as the Sabbath. In addition, ideas about what behaviors are appropriate on the Sabbath are changing, and laws governing those behaviors are being challenged. These differences indicate that norms may differ according to whether they are informally taught to new generations or whether they are formal, written "laws of the land."

Another dimension along which norms may differ is the degree to which they are sanctioned—that is, the degree to which adherence is rewarded and violation is punished. The norm that men do not wear hats indoors is relatively weak. On the other hand, the norm that men and women do not casually display (or "flash") their genitals is strongly sanctioned.

Using these two comparative dimensions—mode of development (formal vs. informal) and degree of sanction (weak vs. strong)—we can create four categories: (1) norms that are informal and are weakly sanctioned (e.g., table manners, dress fashions), (2) norms that are informal but are strongly sanctioned (e.g., adultery), (3) norms that are part of the formal legal code but are weakly sanctioned (parking regulations, antismoking laws, etc.), and (4) norms that are formal laws and are strongly sanctioned (e.g., capital offenses like murder of a police officer or treason in wartime).

By juxtaposing the two dimensions, each with its two categories, we arrive at a fourfold classification of norms (see chart). But as Max Weber (1947) observed, such a classification is an "ideal-typical" arrangement of observations in a form that accentuates some aspects and neglects others. These ideal types (folkways, mores and taboos, misdemeanor laws, and felonies or capital offenses) are useful because they establish standards against which to compare the norms of other cultures. For example, not all cultures treat the norm about the Sabbath the way Americans do. In parts of Israel or the Islamic lands one could be arrested for violating the Sabbath laws. Typologies like this one also help identify areas of social change, as when one compares the way the Sabbath was treated in American laws early in the twentieth century, when stores were obliged to close on Sunday, with the way it is treated today, when Sunday is often viewed as another shopping day.

But the aspects of society described by ideal types are rarely so uncomplicated in real life. Even the norms that seem most formal and unambiguous, such as the prohibition against murder, become murky under some conditions, as in cases of self-defense or in war, or in arguments about the death penalty and abortion. Studies of how actual behavior departs from the ideal-typical version invariably offer insights into how cultures and social structures are changing in the course of daily life.

Mode of Development	Degree of Sanction: Relatively Weak	Degree of Sanction: Relatively Strong
Informal	Folkways, fashions	Taboos, mores
Formal	Misdemeanor laws, some rules, guidelines, civil rights laws	Capital-offense laws, felony laws

small group save a place in line for the others, who may arrive later. And in a "pickup" basketball game a player can call a foul and the opposing player usually cannot contest the call. Of course, there are times when such norms are disputed, depending on how the people involved define the situation. In the case of the basketball game, when the player on whom the foul is called disagrees with the call—and the score is extremely close—different definitions of the situation can lead to conflict.

Culture and Social Change

Cultures in all societies change, and much of social science is devoted to trying to understand and predict those changes. Norms and values that once were thought to be odd or criminal may come to be shared by the majority of the society's members, and the opposite can occur as well. In the Olympic Games of the 1990s, for example, American women dominated many events in which they had previously been weak performers. In the past, American women were often reluctant to lift weights and improve their strength for fear of being considered unfeminine. As the level of world competition increased, Americans found ways to overcome the older values that said women had to be soft in order to be feminine. Star female athletes continue to demonstrate to the American public that a woman can be muscular and athletic and still project a "feminine" identity.

Figure 3.1 and Box 3.1 present some key aspects of human culture, but they leave out all the social processes whereby elements of culture are produced and changed and diffused from one society to another. An unsuspecting reader might even conclude that a society's norms and values are shared equally by all of its members, which is hardly an accurate view of culture. Any given norm is likely to be obeyed to varying degrees by different people within a culture. Consider the norms about marriage, fertility, and parenthood in the United States. All may agree that young couples should not have children out of wedlock. They may also agree that abortion is not a desirable means of birth control. Unmarried couples who are expecting a child resolve the conflict between these norms in different ways: Some obtain an abortion; others get married; still others decide to have the baby out of wedlock, perhaps giving it up for adoption. Some unmarried mothers decide to raise their children alone. One way or another, such events in people's intimate family lives require difficult choices among often contradictory norms.

When large numbers of people reject the normative order for private gain, as appears to be occurring in some regions of the former Soviet Union, the result may be higher rates of crime and civil strife. Similarly, as populations grow and societies become larger and more complex through immigration and urban growth, conflicts over values and norms are likely to become more severe. In times of great social change and conflict, cultures tend to become more punitive and coercive in attempting to maintain order, often without success. Even where there is much consensus about norms, however, new ideas, styles, and technologies contribute to changes in norms and behaviors.

Another important issue in understanding culture and social change is the question of the extent to which human culture is determined by biological factors, if at all. Are there any features that all human cultures share—any norms, for example, that are found in all known societies? And what happens when two cultures exist within the same society or nation? The remainder of this chapter explores these questions, beginning with one that has long been a subject of lively debate: the connection between culture and biology.

CULTURE, EVOLUTION, AND HUMAN BEHAVIOR

Of all the species of living creatures on this planet, human beings are the most widely distributed. The early European explorers—Columbus, Magellan, Cook, da Gama, and many others—marveled at the discovery of human life thriving, more or less, in some of the earth's most inhospitable environments. Nevertheless, they were convinced that the "savages" they encountered were inferior to the more powerful Europeans; in fact, they had difficulty accepting the idea that the native peoples were fully human. In other words, they considered them biologically as well as culturally inferior.

Groups that are in conflict often accuse their enemies of being biologically inferior. The Nazis claimed that the Jews were a biologically inferior people. Serbian Orthodox Christians think of the Muslim Kosovars as innately inferior, just as some people in the United States believe that African Americans and other groups, such as Puerto Ricans, are inferior. These and other examples of racial and ethnic prejudice will be discussed more fully in later chapters. Here we need to emphasize that differences in culture frequently are wrongly viewed as innate biological differences between distinct peoples.

Students of sociology must see beyond this tendency to confuse cultural and biological differences. To do so, it is first necessary to understand how the famous naturalist Charles Darwin revolutionized modern thinking on these issues. Darwin's theory of natural selection is a starting point for scientific explanations of the differences among the animal species, including the human species (Eisley, 1961; Gould, 1996).

Darwin's Theory of Natural Selection

Darwin's theory of **natural selection** is the central explanation of how living species evolve to adapt to their changing environments. It is based on the observation that unexpected physical changes, or *mutations,* in organisms occur more or less randomly from one generation to the next. When those mutations improve an individual organism's ability to survive in its environment, the new traits are "selected for"; that is, an individual that possesses those traits is more likely to survive, and hence more likely to reproduce and pass on its traits to the next generation, than individuals that lack the new traits. Over a few generations such mutations can become so extensive that two species are created where formerly there was only one. This process of natural selection accounts for the great diversity of animal and plant life on the earth and for the ability of animals and plants to adapt to new environments.

Darwin's theory was based on his empirical observations of the natural world, especially those he had made as a scientist aboard the HMS *Beagle* during an extensive voyage of exploration in the early 1830s. On that voyage Darwin had observed, in both fossils and living specimens, that some species of birds, turtles, and other animals had modified their physical form in ways that seemed to "fit" or "adapt" them to their environment. It took him almost two decades of study and reflection to make sense of his observations. After all, much of what he had observed directly challenged the fundamental beliefs of most of the religious and scientific leaders of his day. For when Darwin considered all the information, and especially his observations of similar species on different islands, it became clear to him that God had not created all the living things on earth at once—instead, they had been evolving through natural selection over many millions of years. And this theory could be applied to all other species, including humans.

The Social Darwinists

The theory of natural selection had a dramatic impact on biological science but an even greater effect on the prevailing views of human society. Among the social thinkers who were most profoundly influenced by Darwin was Herbert Spencer of England, whose writings in sociology and philosophy were to dominate intellectual life in much of the Western world from 1870 to 1890.

According to Spencer, the fact that humans, unlike other species, have remained similar even on different continents must be explained by the fact that we adapt to changes in our environment through the use of culture rather than through biological adaptation. This process, which can be termed **cultural evolution,** parallels biological evolution in that the most successful adaptations are handed down to the next generation (Geertz, 1973). Spencer had this process in mind when he coined the phrase "survival of the fittest." He meant that the people who are most successful at adapting to the environment in which they find themselves—that is, the better-educated, wealthier, more powerful people—are most likely to survive and to have children who will also be successful.

Spencer concluded that it is impossible, by means of intentional action, to improve on the course of cultural evolution. The task of sociology, as he saw it, is to discover that course through empirical observation and analysis. Sociologists should not engage in efforts to reform society—to aid the poor, for example. To do so would be futile, and it could have the damaging effect of violating the principle of survival of the fittest in favor of the "artificial preservation of those least able to take care of themselves" (Spencer, 1874, p. 343).

Spencer's view, which came to be known as **social Darwinism,** claimed to explain why some people prospered during the industrial revolution while others barely scraped by. The people who were being pushed off the land and into the factories and slums of the cities were less well equipped culturally to succeed in an urban environment than people who could innovate and invent. There was not much for government to do, according to Spencer's theory, besides keep the peace and let the most competitive groups in society flourish. In so doing, those groups would give less competitive groups a chance to survive, if not to thrive.

Darwin's theory of evolution thus spawned a revolution in social thought. It also produced North America's first prominent sociologist and social philosopher, William Graham Sumner. Sumner (1963/1911) developed a theory of society based on humans' need to adapt to an environment of scarcity. According to his theory, when resources, especially land, were plentiful, there would be peace, and institutions based on democratic forms of government could thrive. But eventually, when population growth made it necessary to distribute the same amounts of resources among more people, societies would lean toward oligarchy (rule by the most powerful) and greater use of coercion and force.

By the end of the nineteenth century, efforts to apply the theory of natural selection to human societies had reached their logical extreme. Largely as a result of the work of Spencer, Sumner, and their followers, early sociologists in the United States and other Western societies favored the view that Western culture, with its emphasis on competition within the capitalist system, was clearly superior to all others and that the people who were most successful at competing within that system were to be considered superior human beings.

The next generation of social scientists, who came to prominence around the turn of the twentieth century, rejected this theory. In Chicago and other U.S. cities where the industrial revolution was still occurring, they saw mounting evidence that wealth itself brings privilege to the children of the wealthy, regardless of whatever innate or learned traits they may possess. Cultural evolution, they argued, is a result of the development of more effective institutions and is not related to the innate qualities of individuals. Successful business firms, for example, are able to prosper in a highly competitive environment because of their superior organization. That is, their success can be explained without recourse to arguments about the genetic fitness of their leaders (Swedberg, 1994).

Sociobiology

The tendency to explain social phenomena in terms of biological causes such as physiology or genes is known as *biological reductionism.* For example, theories of crime proposing that there are genes that produce criminal behavior reduce the explanation of crime to biological causes. Some form of biological reductionism has emerged in every generation since Darwin's time.

The most recent version of biological reductionism is **sociobiology.** This term, coined by the Harvard biologist Edward O. Wilson (1975, 1998), refers to efforts to link genetic factors with the social behavior of animals. When applied to human societies, sociobiology has drawn severe criticism from both social scientists and biologists. Nevertheless, some sociologists support the sociobiological hypothesis that genes can explain certain aspects of human society and behavior (Caplan, 1978; Herlihy, 1995; Mascie-Taylor, 1990; Maxwell, 1991; Quadagno, 1979; Sahlins, 1976). Because this hypothesis is so controversial it deserves a closer look.

Let us take as an example the incest taboo, one of the strongest and most widespread norms in human life. The social scientist tends to explain the incest taboo as a cultural norm that is necessary for the existence of the family as a social institution. The family is an organized group with a need for well-defined statuses and roles. Should the different statuses within the family become confused, as would undoubtedly happen if sexual intimacy were permitted between children and their parents or between brothers and sisters, it would be difficult to maintain the family as a stable institution (Davis, 1939; Malinowski, 1927).

Sociobiology takes a different view. For the sociobiologist, the incest taboo evolved in response to biological conditions:

> . . . a deeper, more urgent cause, the heavy physiological penalty imposed by inbreeding. Several studies by human geneticists have demonstrated that even a moderate amount of inbreeding results in children who are diminished in overall body size, muscular coordination, and academic performance. More than 100 recessive genes have been discovered that cause hereditary disease . . . a condition vastly enhanced by inbreeding. (Wilson, 1979, p. 38)

Throughout most of the history of human evolution, sociobiologists point out, humans did not have any knowledge of genetics. Thus "the 'gut feeling' that promotes . . . sanctions against incest is largely unconscious" (Wilson, 1979, p. 40). Individuals with a strong predisposition to avoid incest passed on more of their genes to the next generation because their children were less likely to suffer from the illnesses that result from inbreeding. And over many centuries of natural selection of individuals who did not inbreed, humans developed "an instinct [to avoid inbreeding] which is based on genes" (Wilson, 1979, p. 40).

This leap, from the observation of a strong and persistent norm like the incest taboo to the belief that certain human behaviors are genetically programmed, is an example of the sociobiological hypothesis regarding human nature. Sociobiologists have proposed a hypothesis in which not only the incest taboo but also aggression, homosexuality, and religious feelings are genetically programmed, and they believe that future discoveries by geneticists will prove their hypothesis correct (Cecco & Parker, 1995).

Although it is true that genes set limits on human abilities and can be shown to influence many aspects of brain functioning, the hypothesis that genetic programming establishes complex forms of normative behavior is not supported by direct evidence: There is as yet no proof that such genes or sets of genes actually exist (Lewontin, 1982, 1995). Nevertheless, the rules of science require that we not reject the sociobiological hypothesis and that it remain an open area of investigation.

The counterargument to sociobiology is that the human brain and other physical attributes are products of interaction between cultural and biological evolution and that in the past hundred thousand years of human evolution there has been relatively little organic change in our species. Instead, the important developments in human life have occurred as a result of social and cultural change. This widely accepted view of culture denies that humans have innate instincts such as an instinct to avoid incest. It argues instead that the great advantage of culture in human evolution is its creation of a basis for natural selection that would not be dependent on genes but would allow humans to adapt relatively quickly to any physical or social environment (Geertz, 1973; MacWhinney, 1998).

Archaeological evidence shows that humans were using tools for primitive agriculture and making jewelry and personal adornments more than 30,000 years ago (Stevens, 1988). No doubt they began using fire even earlier. Once humans could use fire and hand tools and simple weapons, the ability to make and use these items increased the survival chances of those who possessed these skills. This would create conditions for natural selection in which the traits being "selected for" were those that had to do with the manipulation of cultural and social aspects of life. These abilities (dexterity, linguistic ability, leadership, social skills, and so on) would in turn influence the further development of the human brain, again through the process of natural selection.

This view is supported by research on primate behavior showing that the higher primates use simple tools and that they teach this cultural technique to their young (Schaller, 1964). Jane Goodall (1968, 1994), for example, described how chimpanzees use sticks to probe for termites and chew leaves to produce a pulp to be used as a sponge to draw water out of tree stumps. These and many other instances of rudimentary culture among animal species demonstrate that culture is not unique to humans; that is, we are not alone in using culture to aid in adaptation. What happened in the case of humans that did not happen (or at least has not happened yet) in any other species is that at a certain stage in human prehistory our ability to alter our cultures in response to changing conditions developed so quickly that the human species entered a new realm of social life. In other words, culture became self-generating. And once it had been freed from genetic constraints, culture had no limits.

LANGUAGE AND CULTURE

Perhaps the most significant of the inventions made possible by culture is language. In fact, the learning of culture takes place through language. From our enormous capacity to use language is derived our collective memory (myths, fables, proverbs, ballads, and the like), as well as writing, art, and all the other media that shape human consciousness and store and transmit knowledge. Note that although the capacity to learn language appears to be innate (Chomsky, 1965; MacWhinney, 1998), language does not occur outside a cultural setting and indeed is the most universal dimension of human cultures.

To return to the example at the beginning of this chapter, as savage as they may have appeared to Europeans, the Native Americans had language. They could learn Spanish, just as the explorers could have learned their language if they had made an effort to do so. But without a common language that they could use to communicate with each other, the Indians and the explorers misunderstood each other far more than they would have if they had possessed this powerful communication tool. Without language, neither side could explain to the other its strange behaviors and different ways of dealing with the physical world.

Research With Other Primates

What is unique about human language? Primatologists have shown that our closest evolutionary kin, the great apes (especially chimpanzees and pygmy chimpanzees), can learn language to some extent. Although their throats are not capable of producing the sounds that humans mold into language, apes do have the capacity to use language; that is, they can grasp the meanings of words as symbols for things and relationships. Through the use of sign language, or special languages using typewriters and other devices, apes can be taught a limited vocabulary.

To the surprise of many skeptics, Francine Patterson taught Koko, a female gorilla, American Sign Language. Koko learned more than 300 words. She also disproved the theory that apes can learn words but cannot invent new concepts. Koko invented sign words for *ring* ("finger bracelet") and *mask* ("eye hat") and was even able to talk in sign language about her feelings of fear, happiness, and shame (Hill, 1978). Subsequent research with a pygmy chimpanzee named Kanzi revealed that apes may actually be able to learn language through observation and imitation, the way a child does, rather than through long and difficult training (Eckholm, 1985). Sue Savage-Rumbaugh demonstrated that Kanzi could recognize English syntax, the patterns in the ordering of signs or words that give sentences different meanings. For example, Kanzi could distinguish between "Throw the potato at the ball" and "Throw the ball at the potato." Savage-Rumbaugh also showed that Kanzi could acquire new words simply by listening and watching as other chimps learn (Savage-Rumbaugh & Shanker, 1998). This kind of imitative language learning was formerly thought to be beyond the ape's mental capacities.

Fascinating as these experiments are, they only confirm the immense difference in communicative ability between humans and the apes. After months of training, an adult ape can use language with no more skill than an average 2½-year-old human infant (Dreifus, 1998; Harris, 1983). On the basis of the research conducted so far, primate researchers conclude that the apes' innate ability to learn language is severely restricted. No amount of training can produce in apes the more advanced uses of language, including complex sentences containing abstract concepts, that are found in all normal humans regardless of their culture (Eibl-Eibesfeldt, 1989). For example, every human language allows its speakers to express an infinite number of thoughts and ideas that can persist even after their originators are gone. This property of human language, which is not shared by any other known species (Eisley, 1970; Rymer, 1992a, 1992b), allows human groups to transmit elements of their culture from one generation to the next.

Does Language Determine Thought?

So complete is the human reliance on language that it often seems as if language actually determines the possibilities for thought and action in any given culture. Perhaps we are actually unable to perceive phenomena for which we have no nouns or to engage in actions for which we have no verbs. This idea is expressed in the **linguistic-relativity hypothesis.** As developed by the American linguists Edward Sapir and Benjamin Whorf in the 1930s, this hypothesis asserts that "a person's thoughts are controlled by inexorable laws or patterns of which he is unconscious. . . . His thinking itself is in a language—in English, in Sanskrit, in Chinese. And every language is a vast pattern-system, different from others" (Whorf, 1961, p. 135).

This observation was based on evidence from the social sciences, especially anthropology. For example, Margaret Mead's field research among the Arapesh of New Guinea had revealed that the Arapesh had no developed system of numbers. Theirs was a technologically simple society, and therefore complex numbering systems were not much use to them. They counted only "one, two, and one and two, and dog" (dog being the equivalent of four and probably based on the dog's four legs). To count seven objects, the Arapesh would say "dog and one and two." Eight would be "two dog," and twenty-four would come out as "two dog, two dog, two dog." It is easy to see that in this small society one would quickly become tired of attempting to count much beyond twenty-four and would simply say "many" (Mead, 1971).

Other cultures have been found to have only a limited number of words for colors, and as a result they do not make some of the fine distinctions between colors that we do. And in a famous example Whorf argued that many languages have ways of referring to time that are very different from those found in English and other Indo-European languages. In English we have verb tenses, which lead us to make sharp distinctions among past, present, and future time. In Vietnamese, in contrast, there are no separate forms of verbs to indicate different times; the phrase *tôi đi về*, for example, could mean "I'm going home" or "I went home" or "I will go home." The language of the Hopi Indians also lacks clear tenses, and it seemed to Whorf that this made it unlikely that the Hopi culture could develop the systems of timekeeping that are essential to modern science and technology.

Thus in its most radical form the linguistic-relativity hypothesis asserts that language actually determines the possibilities for a culture's norms, beliefs, and values. But there is little justification for this extreme version of the hypothesis. The Arapesh did not have a developed number system, but they easily learned to count using the Western base-10 system. Once they were exposed to the money economy of the modern world, most isolated cultures formed words for the base-10 number system or else incorporated foreign words into their own vocabularies. They had no difficulty understanding the use of money, and this too became incorporated into so-called primitive cultures.

A more acceptable version of the linguistic-relativity hypothesis recognizes the mutual influences of thought and language. One does not determine the other. For example, someone living in Canada or the northern parts of the United States is likely to have a much larger vocabulary for talking about snow (*loose powder, packed powder, corn snow, slush,* and so on) than a person from an area where snow is rare. A person who loves to watch birds will have a much larger vocabulary about bird habitats and bird names than one who cares little about bird life. In each case we realize that in order to share the world of bird-watchers or of winter sports fans, we will need to learn new ways of seeing and of talking about what we see. So, although the extreme version of the linguistic-relativity hypothesis is incorrect, it has been a valuable stimulus toward the development of a less biased view of other cultures. We now understand that a culture's language expresses how the people of that culture perceive and understand the world and, at the same time, influences their perceptions and understandings.

CROSSING CULTURAL LINES

Global transactions of all kinds are a feature of the contemporary world. Success in such transactions often hinges on a person's ability to move comfortably among different cultures and subcultures. But unless we can see ourselves as others see us, we take for granted that our own cultural traits are natural and proper and that traits that differ from ours are unnatural and somehow wrong. Our ways of behaving in public, our food and dress and sports—all of our cultural traits have become "internalized," so they seem almost instinctive. But once we understand another culture and how its members think and feel, we can look at our own traits from the perspective of that culture. This *cross-cultural perspective* has become an integral part of sociological analysis.

Ethnocentrism and Cultural Relativity

The ability to think in cross-cultural terms allows people to avoid the common tendency to disparage other cultures simply because they are different. However, most people live out their lives in a single culture. Indeed, they may go so far as to consider that culture superior to any other, anywhere in the world. Social scientists term this attitude **ethnocentrism.** Ethnocentrism refers to the tendency to judge other cultures as inferior in terms of one's own norms and values. The other culture is weighed against standards derived from the culture with which one is most familiar. The explorers' assumption that the Native Americans could benefit from the adoption of European cultural traits is an example of ethnocentric

THE PERSONAL CHALLENGE

Social Change and Intercultural Relations

It can take some courage to cross cultural boundaries. People who travel for any length of time in a new culture, especially one whose language and norms are quite different from their own, often experience a sense of disorientation and confusion known as "culture shock." Sophisticated travelers learn how to cope with the feeling of being foreign. Businesspeople who work in global markets, for example, learn to feel at ease in strange cultures and not assume that they will be able to find a corner of home wherever they are. However, students who wish to become more culturally sophisticated may not need to travel very far. The resources for a cross-cultural experience could be available close to home.

More than 700,000 people from different cultures and societies settle in the United States each year. While most of them are concentrated in a few "gateway" cities like Miami, Los Angeles, and New York, increasing numbers are moving throughout the United States. It is likely that people from Asia, Latin America, Africa, or the Caribbean are moving into your own community. If so, it is also likely that their children are encountering various forms of social conflict and prejudice brought on by their foreignness. Adolescents and young adults are especially likely to express prejudice against those who are different. And the differences may not always be due to language or foreign forms of dress and diet. Minority or gay students may also feel like outsiders.

Reaching out to newcomers or those who are different can bring immense rewards. Cross-cultural friendships can change one's life in very positive ways. Mundane aspects of one's own life that are taken for granted can become subjects of animated discussion. By crossing cultural boundaries one can learn new ways of cooking, new ways of thinking about beauty, and much more. Often it takes some determination and a sense of security to resist the pressure to conform to the dominant group. But there is a wide and changing world of people who are different from us and from each other. By reaching out to those who are different, you can begin to develop the ability to move among them in a global environment.

behavior. But such obvious ethnocentrism is not limited to historical examples. We encounter it every day—for example, in our use of the term *American* to refer to citizens of the United States and not to those of Canada and the South American nations. Another example is our tendency to judge other cultures by how well they supply their people with the consumer goods we prize rather than by how well they adhere to their own values.

To get along well in other parts of the world, a businessperson or politician or scientist must be able to suspend judgment about other cultures, an approach that is termed **cultural relativity.** Cultural relativity entails the recognition that all cultures develop their own ways of dealing with the specific demands of their environments. This kind of understanding does not come automatically through the experience of living among members of other cultures. It is an acquired skill.

There are limits, however, to the value of cultural relativity. It is an essential attitude to adopt in understanding another culture, but it does not require that one avoid moral judgment entirely. We can, for example, attempt to understand the values and ideologies of the citizens who supported the Nazi regime, even though we abhor what that regime stood for. And we can suspend our outrage at racism in our own society long enough to understand the culture that produced racial hatred and fear. But as social scientists it is also our task to evaluate the moral implications of a culture's norms and values and to condemn them when we see that they produce cruelty and suffering. To a large extent, worldwide outrage at South Africa's policies of apartheid (racial separation and enforced inferiority for black citizens) helped bring about the demise of those policies in the early 1990s.

MAPPING SOCIAL CHANGE

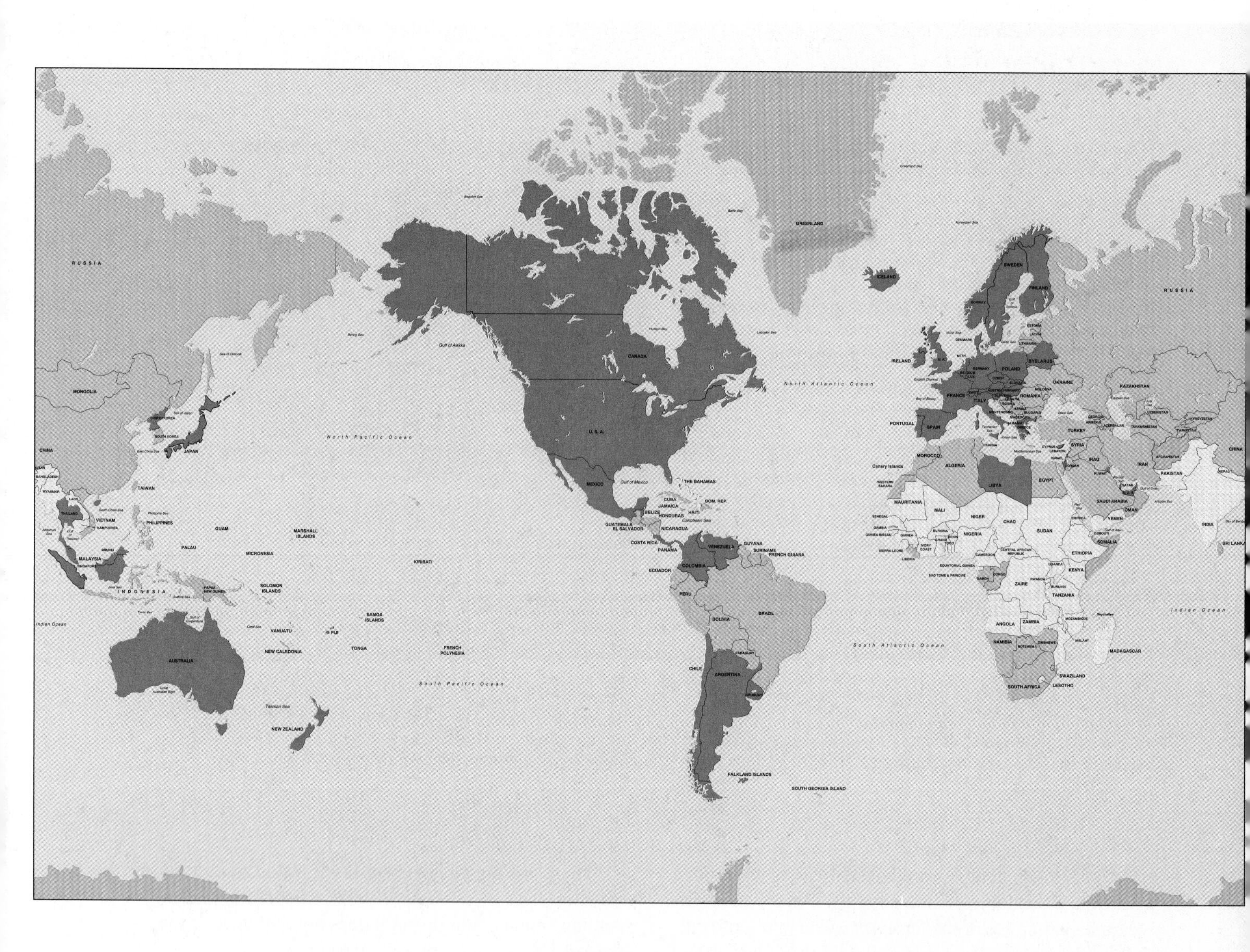

Women and Cultural Change

Throughout the world poverty is regarded as a measure of a society or a culture's development. The higher the proportion of a population living in poverty (as measured by standard U.S. definitions), the less developed that population is considered to be. Female poverty and deprivation in particular are widely regarded as among the foremost indicators of development. The relative poverty of women, known as *gender poverty*, is closely related to other indicators of development, such as infant mortality or exclusion of women from the paid labor force.

Culture plays a significant role in gender poverty, since one of the primary indicators of how well a nation is doing in reducing gender poverty is the female literacy rate. The accompanying map shows the proportion of girls enrolled in secondary schools (high schools) in nations classified at various levels of human development. This is a good measure of a nation's commitment to give women the cultural skills (language, math, science, and so on) they need if they are to participate in its economic and social life. The female secondary enrollment ratio is determined by dividing the number of girls in secondary schools by the number of boys in those schools. More developed societies often have more female than male secondary school students, so their ratios may exceed 100, whereas the ratios for less developed nations like Bangladesh are as low as 54 or even lower.

Level of Human Development

High Human Development

Medium Human Development

Low Human Development

Cultural Hegemony: Myth or Reality?

When one culture's values, norms, and products become dominant and diminish the strength of another existing culture, the stronger culture is said to exert hegemony over the weaker one. The term **hegemony** refers to dominance or undue power or influence. A hegemonic culture is one that dominates other cultures, just as a hegemonic society is one that exerts undue power over other societies (Gramsci, 1992/1965,1995). In Europe and parts of Asia there is a lively debate about whether the culture of the United States endangers the continued existence of local cultures. English is the international language of business and science. American movies and popular music, fast food, and fashions are coveted around the world. Nations with less widespread languages and less powerful cultural institutions (e.g., media industries, universities, fashion industries) fear that their young people will lose interest in their own cultures and embrace the values, language, and norms of Americans. Discussions of the global influence of McDonald's or the Disney company appear frequently in social-scientific essays and cultural critiques in nations where American culture is popular.

Within the United States there are also sociologists and others who worry that the powerful commercial cultural institutions that make certain types of food, music, movies, fashions, and magazines popular throughout the nation endanger the existence of unique local subcultures. These critics fear that regional differences are decreasing and being replaced by a homogeneous culture that lacks diversity (Lieske, 1993; Oldenburg, 1997).

How realistic are these fears? Like all generalizations about social life, they contain some truth and much myth. Further research and more factual information are needed to understand whether American culture is indeed hegemonic, as its critics claim. But it is clear that in many regions of the world there are societies with extremely strong and distinct cultures that are able to produce their own adaptations of American commercial culture. India, for example, is the second leading producer of commercial movies in the world. There is a huge audience for Indian films, which combine the technology of Western films with the music, language, norms, and values of India's many subcultures. And as Europe becomes more economically unified it is likely that the strength of European cultural institutions will improve, despite the linguistic divisions that make some Europeans feel vulnerable to the influence of American culture.

At the same time, there can be no doubt that on the world stage the United States is more than just another culture. It represents a powerful civilization whose cultural features have diffused across national boundaries to exert a lasting influence over the cultures of other societies. Does this mean that the United States exerts hegemony over other cultures? This question is a subject of continuing research and debate. Meanwhile, it is worth noting that much of what we think of as distinctive about American civilization originated in European, African, Asian, and Hispanic cultures. In consequence, we devote the rest of this chapter to a discussion of civilizations and cultural change.

CIVILIZATIONS AND CULTURAL CHANGE

Civilizations are advanced cultures. They usually have forms of expression in writing and the arts, powerful economic and political institutions, and innovative technologies, all of which strongly influence other cultures with which they come into contact. Some civilizations, like those of ancient Egypt and Rome, declined thousands of years ago and exist today mainly in museums and in the consciousness of scholars, artists, and scientists. Others are living civilizations with long histories, like those of China and India, which were conquered by other civilizations and are rising again in forms that combine the old with the new. Islamic civilization is an example of such a civilization; much of the unrest in the Islamic world is due to conflicts that pit orthodox leaders like the late Ayatollah Khomeini against Western-influenced leaders like the late King Hussein of Jordan. Then there are the dominant civilizations of North America, Europe, the former Soviet Union, and Japan. They are dominant because they compete on a world scale to "export" their ideas and their technology—in fact, their entire culture or "blueprint for living."

Like most of the principal concepts in the social sciences, the concept of civilization can be elusive. It is used in many different contexts, in popular language as well as in social-scientific usage. In popular speech the word *civilization* is often used to make negative comparisons between people who adhere to the norms of polite conduct, and are therefore said to be "civilized," and those who are "uncouth" and act like "barbarians" or "savages." This is how the word was understood by the explorers of Columbus's time, and colonial conquest was justified in part as an effort to civilize the barbarians.

In his effort to trace the origins of Western notions of what is civilized and what is barbaric behavior, the German sociologist Norbert Elias (1978/1939) showed that much of what we call "civilized" behavior is derived from the norms of the courts of medieval Europe.

Elias based his study on accounts by medieval writers of the spread of "courtesy" (the manners of the court) to other levels of society. The following are some examples from a thirteenth-century poem on courtly table manners:

> A man of refinement should not slurp his spoon when in company; that is the way people at court behave who often indulge in unrefined conduct.
>
> It is not polite to drink from the dish, although some who approve of this rude habit insolently pick up the dish and pour it down as if they were mad.
>
> A number of people gnaw a bone and then put it back in the dish—this is a serious offense. (p. 85)

Through his analysis of writings on manners, Elias demonstrated that Western norms concerning the control of bodily functions, from eating and sleeping to blowing one's nose, arose in the Middle Ages as the courts consolidated their power over feudal societies and exported the standards of courtly behavior to the countryside. Such behavior became a sign that a person was a member of the upper classes and not a serf or a savage.

In the social sciences, the most common use of the term *civilization* stems from the study of changes in human society at the macro level, which often requires comparisons among major cultures. In this context, a **civilization** is "a cultural complex formed by the identical major cultural features of a number of particular societies. We might, for example, describe Western capitalism as a civilization, in which specific forms of science, technology, religion, art, and so on, are to be found in a number of distinct societies" (Bottomore, 1973, p. 130). Thus Italy, France, Germany, the United States, Sweden, and many other nations that have made great contributions to Western civilization all have private corporations, and their normative orders, laws, and judicial systems are quite similar (Gramsci, 1995). Even though each may have a different language and each differs in the way it organizes some aspects of social life (European and North American universities define academic degrees differently, for example), they share similar norms and values and can all be said to represent Western civilization.

Effects of Cultural Contact

Although sociologists do not distinguish among cultures in terms of how "civilized" they are, this standard is often applied by members of different cultures when they come into contact with one another. The contact between Christopher Columbus and his crew and the natives of the West Indies is typical of such episodes. The explorers represented the relatively advanced civilization of Europe, while the natives whom the Europeans first encountered represented a much simpler culture, one that seemed totally "uncivilized" to the explorers. Throughout the period of world exploration, from the late fifteenth to the early nineteenth centuries, explorers and traders brought back reports of "savages" living in every part of the globe. These episodes of cultural contact (and often conflict as well) shaped the history of the next two centuries and continue to influence human existence.

Through such processes as exploration and conquest, civilizations invariably spread beyond their original boundaries. Figure 3.2 shows how through much of the nineteenth century England spread its version of Western civilization throughout the world as it conquered tribal peoples and established colonies in Africa and Asia. Colonial rule brought *cultural imperialism*, the imposition of a new culture on the conquered peoples. This meant that colonial peoples had to learn the languages of their conquerors, especially English, Spanish, French, Portuguese, and Dutch (or Afrikaans). Along with language came the imposition of ideologies like Christianity in place of older beliefs and religions.

According to the historian Fernand Braudel, "The mark of a living civilization is that it is capable of exporting itself, of spreading its culture to distant places. It is impossible to imagine a true civilization which does not export its people, its ways of thinking and living" (1976/1949, p. 763). In his research on the contacts and clashes between the great civilizations surrounding the Mediterranean Sea during the 1500s, Braudel uses three important sociological concepts to explain the spread of civilizations around the world: acculturation, assimilation, and accommodation. (See the study chart on page 79.)

Acculturaltion People from one civilization incorporate norms and values from other cultures into their own through a process called **acculturation.** Most acculturation occurs through intercultural contact and the borrowing or imitation of cultural norms. But there have been many instances of acculturation through cultural imperialism, in which one culture has been forced to adopt the language or other traits of a more dominant one. Thus people in societies that were colonized in the nineteenth century were forced to learn the language of the conquering nation.

Aspects of our culture that we take for granted usually can be shown to have traveled a complicated route through other cultures to become part of our way of life. Braudel's study of the Mediterranean civilizations shows that many of the plants and foods that are commonly found around the Mediterranean Sea, and later were

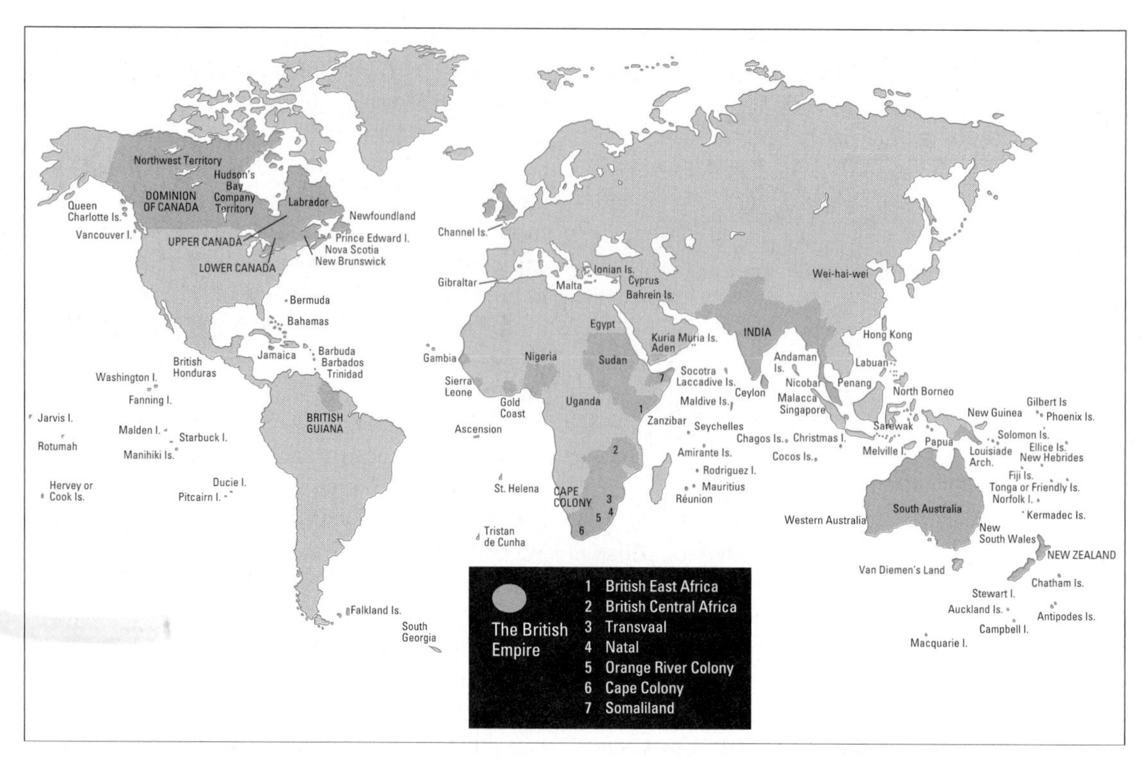

This map shows the extent of the British Empire in 1901. At the peak of its power, the empire had spread English civilization throughout the world. But its influence was strongest in peripheral, less industrialized lands. In a mere half-century most of the empire would crumble, but the effects of English civilization on language and social institutions throughout the world would endure.

Source: Eldridge, 1978.

FIGURE 3.2 **The British Empire in 1901**

imported to the New World, were themselves borrowed from other cultures and incorporated into those of the Mediterranean societies. The process of acculturation can be traced in many other aspects of everyday life. As Ralph Linton has written:

> Our solid American citizen awakens in a bed built on a pattern which originated in the Near East. . . . He throws back covers made from cotton, domesticated in India. . . . He slips into his moccasins, invented by the Indians of the Eastern woodlands. . . . He takes off his pajamas, a garment invented in India, and washes with soap invented by the ancient Gauls. . . . He puts on garments whose form originally derived from the skin clothing of the nomads of the Asiatic steppes . . . and ties around his neck a strip of bright-colored cloth which is a vestigial survival of the shoulder shawls worn by the seventeenth-century Croatians. Before going out for breakfast he glances through the window, made of glass invented in Egypt, and if it is raining puts on overshoes made of rubber discovered by the Central American Indians and takes an umbrella, invented in southeastern Asia. (1936, p. 326)

The concept of acculturation can also be applied to how a newcomer people adopts the cultural ways of the host society. But acculturation is rarely a one-way process: At the same time that they are becoming more like their hosts in values and behavior, newcomers teach members of the host society to use and appreciate aspects of their own culture. Because of the cultural diversity of its population, examples of acculturation are especially prevalent in the United States. For example, even small towns in the interior of the country have at least one Chinese restaurant, but often the food served in those restaurants is a highly acculturated form of the cuisine of China. Indeed, most of the things we think of as part of the American way of life—from hamburgers, pizza, and baseball to democracy and free enterprise—originally were aspects of other cultures. That they have become incorporated into American culture through acculturation, and in the process have become changed from their original forms, does not deny the fact of their "foreign" origin (see Box 3.2).

Assimilation and Subcultures When culturally distinct groups within a larger civilization adopt the

STUDY CHART Effects of Cultural Contact

FORM OF CONTACT	DESCRIPTION	EXAMPLE
Acculturation	The process by which the members of a civilization incorporate norms and values from other cultures into their own.	Americans develop a taste for Italian, Chinese, and Mexican food, as well as other cuisines representing the cultures of the people who have joined their civilization.
Assimilation	The process by which culturally distinct groups within a civilization adopt the norms, values, and language of that civilization and gain equal statuses in its groups and institutions.	Spanish-speaking immigrants to the United States learn English and begin to move up the status hierarchy in education and jobs.
Accommodation	The process by which a smaller, less powerful society is able to preserve the major features of its culture even after prolonged contact with a larger, stronger culture.	Though conquered and forced onto reservations, Native Americans adapt by taking on many of the norms and values of the larger society while preserving aspects of their own culture.

BOX 3.2 GLOBAL CHANGE AND U.S. SOCIETY

North America's African Heritage

All the important streams of migration to North America have combined to produce the culture of the United States and Canada. Pizza (Italy), hamburgers (Germany), French fries—examples of the influence of immigrant groups on American material culture abound. Of all these cultural influences, perhaps the least well understood is the legacy of the people who were brought to North America from Africa as slaves, especially during the last three decades of the eighteenth century and the first half of the nineteenth.

One of the largest groups of slaves, captured in what is now the Congo and Angola, was made up of people from many tribes who shared a common language system, Bantu. Many of these Bantu-speaking slaves were landed in the Carolinas. By the end of the Revolutionary War, the Carolinas had the largest concentration of people of African descent of any of the new states. In consequence, many of the names of towns and villages, rivers, streams, and other geographic features in those states are derived from Bantu words, often mixed with local Indian terms and transformed through long usage. Thus Pinder Town in South Carolina is derived from the Bantu word for peanut, *mpinda.* George R. Stewart, an expert on place names, concludes that hundreds of small streams, swamps, and villages in the Carolinas, Mississippi, and Georgia received their names from the African slaves (Stewart, 1968; Vass, 1979).

The strongest evidence of African influence on contemporary American culture can be found in the Sea Islands along the Atlantic coast between Charlestown, South Carolina, and Savannah, Georgia. On the islands of Port Royal, Parris (famous for its Marine training base), Ladies, Hilton Head, and St. Helena reside the Gullah people. Also known as Geechi, they are descendants of slaves whose original tribes are believed to have lived in Angola (from which the term *Gullah* may be derived) and in the areas that are now Liberia and Sierra Leone. On the plantations the Gullah slaves grew indigo and the long-fiber cotton that made the region famous in the period before the Civil War. Some of them escaped to become leaders of slave rebellions. Others were arrested during the Civil War for singing spirituals like "Roll, Jordan, Roll" to which they had added verses about their dreams of freedom. To this day the religious practices, language, and place names of the Sea Island people bear witness to their African heritage (Creel, 1988).

A Gullah woman of the Sea Islands in a photograph taken early in this century.

language, values, and norms of the host civilization and their acculturation enables them to assume equal statuses in the social groups and institutions of that civilization, we refer to that process as **assimilation.** When groups become assimilated into American society, for example, people often say that they have been "Americanized." Assimilation has been a major issue for immigrant groups in North America, as it still is for immigrants all over the world. It is no surprise, then, that we continually see articles in the press that ask such questions as: Will the various Hispanic peoples in America give up their language over time? Will American Jews marry members of other groups and lose their distinct identity? Will Italian Americans gradually forget their cultural heritage and come to think of themselves as "100 percent Americans" (Brimelow, 1995)? These are the kinds of questions that form the subject matter of racial and ethnic relations in pluralistic societies like those of the United States and the commonwealth of former Soviet republics. Both of these societies (which are also civilizations) are composed of a multitude of peoples, each of which once had its own culture but is under pressure to become assimilated into the dominant civilization.

When a culturally distinct people within a larger culture fails to assimilate fully or has not yet become fully assimilated, we say that it is a **subculture** within the larger culture. (The term is also applied to groups that have had significantly different experiences from those of most members of the society.) People who maintain their own subculture generally share many of the values and norms of the larger culture, but they retain certain rituals, values, traditions, and norms—and in some cases their own language—that set them apart. Thus we speak of African American, Latino, Native American, and a host of other subcultures in the United States. As explained in Chapter 13, these are also known as *ethnic groups,* since their members have a sense of shared descent, a feeling of being "a people" with a history and a way of life that exists within a larger and more culturally diverse society.

i

Ethnic subcultures are created out of the experience of migration or invasion and subsequent adaptation to a host culture. But subcultures are also created out of the experience of people in complex societies who actively seek to create and maintain a way of life distinct from that of other members of their society (Fischer, 1987; Gans, 1976). For example, there are subcultures in large cities composed of artists and other people whose livelihood depends on the arts: theater people, rock musicians and record producers, visual artists, gallery owners, art critics, curators. In many U.S. cities, and in some in Western Europe, there are quite distinct gay and lesbian subcultures. Because people who form homosexual relationships often experience hostility from heterosexuals, they tend to develop their own particular norms of communication and social behavior. When a subculture that challenges the accepted norms and values of the larger society establishes an alternative lifestyle, we call it a **counterculture.** The hippies of the 1960s, along with members of New Left political groups, activists in the women's movement, and environmental activists, formed a counterculture that had a significant influence on American politics and foreign policy during the Vietnam War years (Roszak, 1969).

Frequently subcultures are under intense pressure to conform to the society's most widely accepted norms and values or to adapt to new ways, new technologies, or the like. Social scientists often investigate the changing norms and values of such subcultures—the gay subcultures, the subculture of rock musicians and other celebrities, or the subcultures of occupational groups like Appalachian miners, Wall Street lawyers, doctors, and construction workers, among others (Backer, 1982; Bosk, 1979; Eisenberg, 1998; Erikson, 1976; Millman, 1976; Smigel, 1964). Frequently this research is valuable in predicting trends in such areas as drug use, popular music, or patterns of labor–management conflict and cooperation (Boggs & Meyersohn, 1988; Flores, 1988; Nyden, 1984).

Accommodation and Resistance Throughout history many societies have withstood tremendous pressure to become assimilated into larger civilizations. But even greater numbers have been either wiped out or fully assimilated. Only a century ago, for example, one could still find many hundreds of hunter-gatherer societies throughout the world. Today there are probably fewer than a hundred, and these live in the most isolated regions of the earth.

Larger and smaller societies do not usually develop ways of living together without the smaller one becoming extinct or totally assimilated into the larger one. But when the smaller, less powerful society is able to preserve the major features of its culture even after prolonged contact, **accommodation** is said to have occurred. For example, in the Islamic civilization of the Middle East before the creation of Israel in 1948, Jews and other non-Muslims usually found it rather easy to maintain their cultures within the larger Arab societies. Compare this pattern of accommodation with the experience of the Jews in Spain, who were forced to leave in 1492 in one of the largest mass expulsions in history.

Accommodation requires that each side tolerate the existence of the other and even share territory and social institutions. The history of relations between Native Americans and European settlers in the Western

Hemisphere is a complex story of resistance and accommodation. Throughout the period of conquest, expansion, and settlement by the Europeans there was continual resistance by the native peoples. This resistance took many forms, including refusal to adopt Christianity, to speak English or Spanish, to sell goods and services to the settlers, and to fight in the Europeans' wars. Resistance did not save Native Americans from death by disease, military conquest, or famine, but it did allow them to maintain their cultures and to borrow from the settlers the cultural customs that were most advantageous to them. For example, the Plains Indians adopted horses from the Spanish explorers, which completely changed their culture, and much later they borrowed trucks from American culture, which helped them adapt to modern ranching.

Most cultures in the world today have had to confront the influence of at least one expanding civilization. In some instances the result has been prolonged conflict and the eventual annihilation of the smaller society. However, throughout the world it is as easy to find examples of failures in accommodation as it is to find examples of sharing and cooperation among cultures. Many of the conflicts that lead to severe social unrest, such as the conflict between Islamic and Eastern Orthodox Bosnians, or among warring tribal peoples in Rwanda, are based on cultural differences. Such conflicts may also occur *within* a culture. In the United States, for example, cultural differences are becoming more visible, and so are cultural conflicts. In no area of contemporary life is this more true than in the realm of intimacy and sexual behavior.

RESEARCH ON THE CUTTING EDGE: Changing Norms of Intimacy

Many critics of contemporary American society blame the rising divorce rate and the increase in single-parent families on changes in the norms of intimacy that occurred during the social movements and counterculture of the 1960s. They seek a return to "traditional" norms of intimacy, including premarital celibacy, monogamy, and heterosexuality. These and other critics of diversity in sexual norms view the 1950s as a "golden age" of traditional norms regarding the conduct of intimate relations.

Sociologists who study trends in American culture point out that the 1950s were by no means a golden age of traditional behavior. Changes in American culture brought on by mass advertising, television, the growing equality of men and women, and the increasingly widespread desire for individual happiness had been developing throughout the century. These longer-term changes, they reason, should be reflected in statistics on divorce rates. On the other hand, if changes in sexual norms were a result of the social movements of the 1960s, there should be evidence of such a connection in statistics on divorce rates over time.

Figure 3.3 shows that there is some justification for each of these positions, but far more empirical support for the view that rising divorce rates are not a product of the tumultuous 1960s. Clearly, divorces have been increasing in frequency over a period going well back into the nineteenth century. These data call into question any "golden age" theory of stability in American couples.

A survey by the National Opinion Research Center explored Americans' sexual attitudes and behaviors in great detail, using extremely careful interviewing methods (Laumann et al., 1994). This study found that there have been vast changes, as well as a good deal of continuity, in the norms of American culture regarding sexual behavior. Figure 3.4 shows a major shift in norms regarding premarital sex with the person one eventually marries. Only about a third of the cohort of people born between 1933 and 1942, who came of age sexually during the 1950s, admit that they had sex with their spouse before marriage. For the cohort born between 1963 and 1974, who reached the age of sexual activity during the 1980s, the situation was almost completely reversed: The vast majority of men and a lesser majority of women report having had sex with their spouse before marriage.

These findings support the theory that there have been major changes in the norms of sexual intimacy since the 1950s. They do not, however, shed light on the relationship between what people *think* the norms are and how they actually *behave.* Many Americans still believe in the norm of premarital celibacy but find that it conflicts with a newer norm, the idea that couples should find compatibility and pleasure in their sexual relationship. The

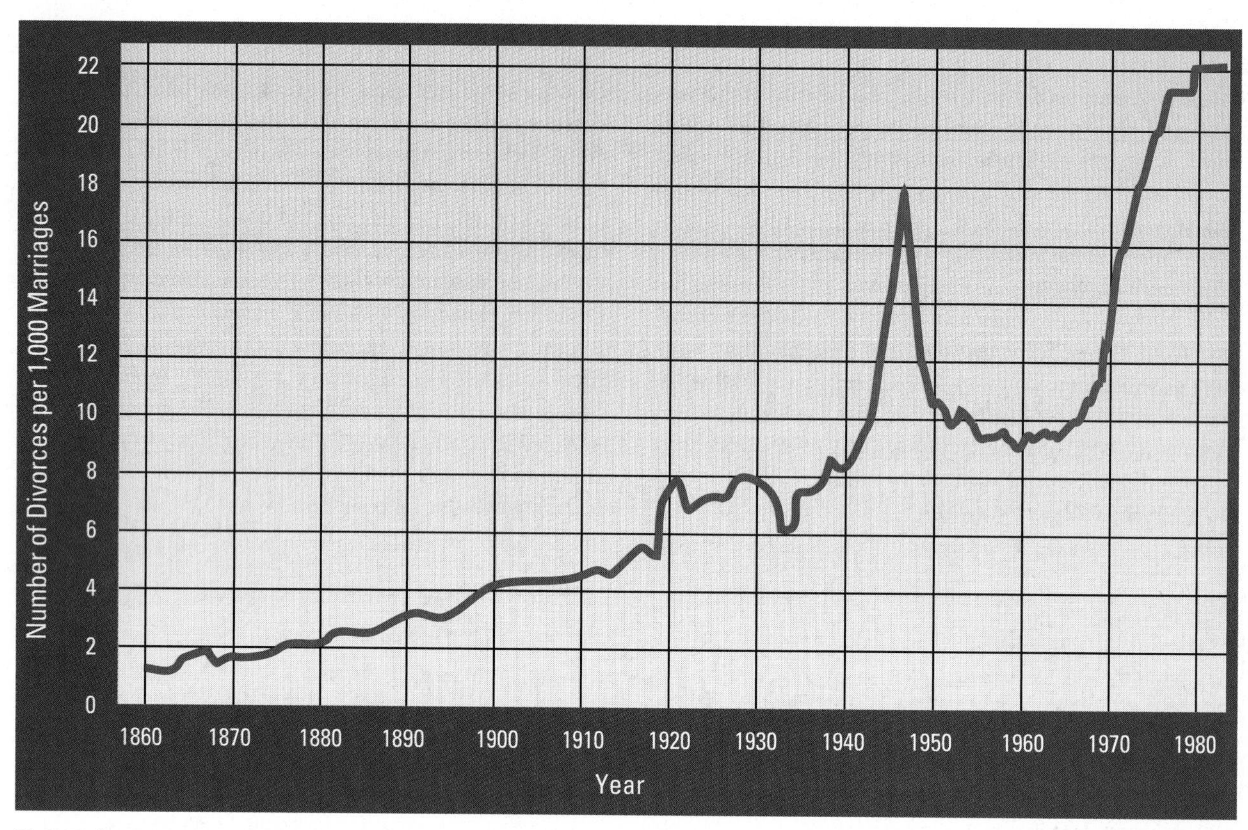

Figures for 1860–1920 represent the number of divorces per 1,000 existing marriages; figures for 1921–1980 represent the number of divorces per 1,000 married women age 15 and over.

Source: Cherlin, 1981.

FIGURE 3.3 **Divorce Rate, United States, 1860–1980**

changes in behavior shown in Figure 3.4 indicate that sanctions against premarital sexual experimentation have weakened, while the norm that couples should be happy in their sex lives has become stronger.

But there is never a complete substitution of new norms for old ones in a large and diverse population like that of North America. In consequence, we will always see differences in how norms are understood and in how they guide actual behavior. And these differences, especially when they concern major areas of moral conduct, will continue to produce conflict over norms and values.

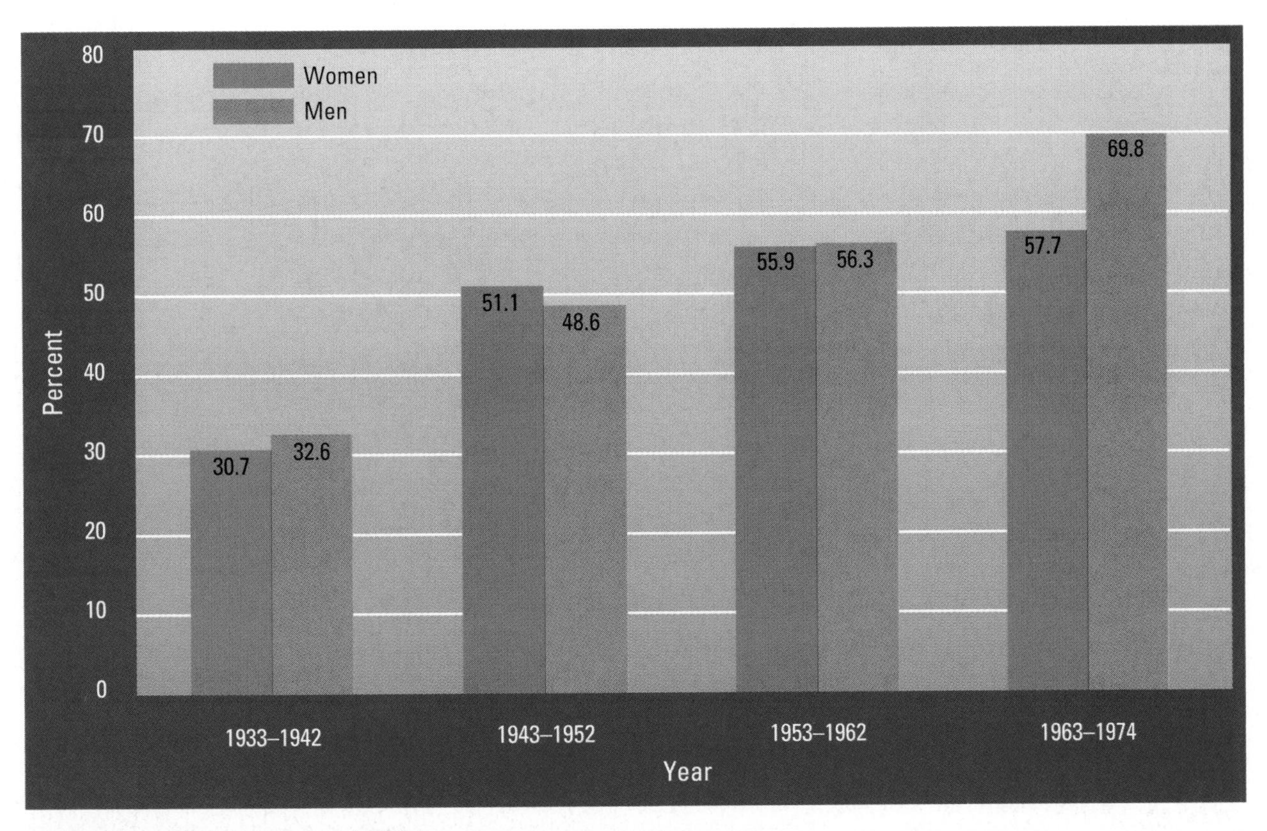

Source: Adapted from Laumann et al., 1994.

FIGURE 3.4 **Premarital Sex With Eventual Spouse, by Birth Cohort, United States**

VISUAL SOCIOLOGY

Living Maya

The people in these photos are Mayan Indians. They live in the mountains and jungles of Chiapas in southern Mexico and in part of Guatamala. They are descendants of the ancient Mayans, whose civilization flourished in Central America before 900 A.D. The Mayan nobility was literate and, like the Egyptians, intensely interested in astronomy and the construction of calendars. The ancient noble palace and observatory of Palenque, deep in the Lacondon rainforest of Chiapas, was one of the most outstanding archaeological achievements of Mayan civilization. But the people in these photos are our contemporaries. They live in villages, each of which is clustered around a market and a small but much loved Catholic church. The men in one of the photos are holding a session of the village court. In the foreground a man occupies the time by plaiting palm fronds to make a hat. The standing man is one of the parties to the dispute, as is the one arguing in the background. The seated men are civil officials.

Men's and women's lives are separated for much of the time. In each village the women engage in specialized crafts in addition to their other chores. In this highland region most women are potters or weavers. Their crafts are an example of material culture passed from one generation to the next for centuries. The woven patterns and symbols on the cloth they are wearing can be seen in the clothing of ancient Mayans depicted on pottery and wall paintings.

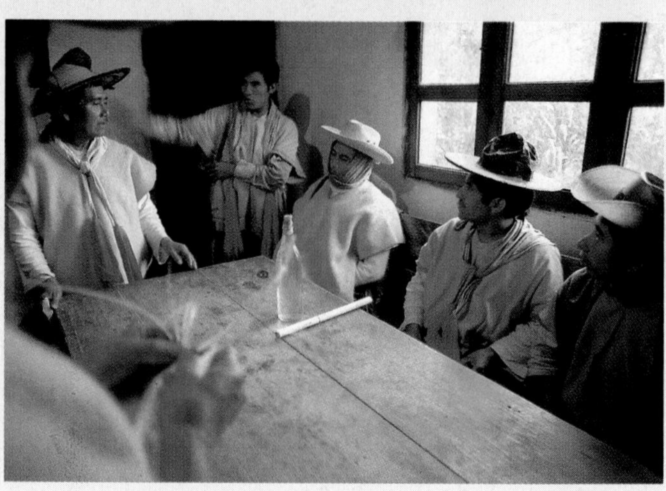

These highlanders also believe in traditional spirits and powers. The shaman with his back to the camera is searching the forest for flowers and branches that will become herbal cures and potions. Most Mayans can identify highland plants in stunning detail, but the shaman has experienced mystical dreams that have given him special knowledge of the magical powers of forest life. The botanical lore of the Mayans is legendary. In fact, the active ingredient in birth-control pills was discovered after study of an herb well known to Mayan women.

Contemporary Mayans are extremely poor. Their subsistence agricultural and crafts economy is adequate to allow them to preserve and reproduce their rich culture. But they crave access to better land. There are huge gaps in power and privilege between the Indians and the wealthy, often absentee, landowners.

Chiapas is also a place of rapid change and revolution. The region is home to the Zapatista guerillas. Occupation by government troops and the threat of violence from either side pose a growing danger to the Mayan villagers and their culture.

SUMMARY

In the social sciences, *culture* refers to all the modes of thought, behavior, and production that are handed down from one generation to the next by means of communicative interaction. Sociologists are concerned primarily with aspects of culture that help explain social organization and behavior.

Culture can be viewed as consisting of three major dimensions: ideas, norms, and material culture. *Ideas* are the ways of thinking that organize human consciousness. Among the most important ideas are *values,* socially shared ideas about what is right. *Norms* are specific rules of behavior that are supported or justified by values; *laws* are norms that are included in a society's official written codes of behavior. *Ideologies* combine ideas and norms; they are systems of values and norms that the members of a society are expected to believe in and act on without question. A society's *material culture* consists of all the things it produces. *Technologies* combine norms and material culture; they are the things and the norms for using them that are found in a given culture.

Social control refers to the set of rules and understandings that control the behavior of individuals and groups in a culture. The wide array of norms that permit a society to achieve relatively peaceful social control is called its *normative order. Sanctions* are rewards and punishments for adhering to or violating norms. Strongly sanctioned norms are called *mores;* more weakly sanctioned norms are known as *folkways.*

One of the most hotly debated questions in the social sciences is how much, if at all, human culture is determined by biological factors. According to Darwin's theory of *natural selection,* mutations in organisms occur more or less randomly. Mutations that enable an individual organism to survive and reproduce are passed on to the next generation. It is this process that permits animals and plants to adapt to new environments.

Herbert Spencer and other social thinkers, who came to be known as *social Darwinists,* attempted to apply Darwin's theory to humans' ability to adapt to social environments. Spencer used the phrase "survival of the fittest" to describe this ability. People who were able to survive in the urban environment created by the industrial revolution were viewed as superior human beings.

A more recent attempt to attribute social phenomena to biological processes is *sociobiology,* which refers to efforts to link genetic factors with the social behavior of animals. According to the sociobiologists, such behaviors as incest avoidance, aggression, and homosexuality may be genetically programmed in human beings. As yet there is no evidence that such genes or sets of genes actually exist.

A more widely accepted view of culture denies that humans have innate instincts and states that at a certain stage in prehistoric times human culture became self-generating. Thus human evolution is not dependent on genes; instead, cultural techniques allow humans to adapt to any physical or social environment.

The learning of culture is made possible by language. Although apes have been taught to use language to some extent, human language is unique in that it allows its speakers to express an infinite number of thoughts and ideas that can persist even after their originators are gone. According to the *linguistic-relativity hypothesis,* language also determines the possibilities for a culture's norms, beliefs, and values. A less extreme form of that hypothesis recognizes the mutual influences of culture and language.

The notion that one's own culture is superior to any other is called *ethnocentrism.* To understand other cultures it is necessary to suspend judgment about those cultures, an approach termed *cultural relativity.* Recently there has been much debate over whether the culture of the United States is exercising undue dominance or *hegemony* over other cultures throughout the world. While more research is needed to answer this question, it is clear that in many regions of the world there remain extremely strong and diverse cultures that are able to maintain their own values and traditions.

Similarities among cultures have resulted from the processes by which cultures spread across national boundaries and become part of a larger, more advanced culture called a civilization. A *civilization* may be defined as a cultural complex formed by the identical major cultural features of a number of particular societies.

A key feature of civilizations is that they invariably expand beyond their original boundaries. The spread of civilizations can be explained by three processes: acculturation, assimilation, and accommodation. When people from one civilization incorporate norms and values from other cultures into their own, *acculturation* is said to occur. The process by which culturally distinct groups within a larger civilization adopt the language, values, and norms of the host civilization and gain equal statuses in its institutions is termed *assimilation.*

(If a distinct people fails to assimilate fully, it is referred to as a *subculture,* but if it challenges the accepted norms and values of the larger society it may become a *counterculture.*) And when a smaller, less powerful society is able to preserve its culture even after prolonged contact with a major civilization, *accommodation* has taken place.

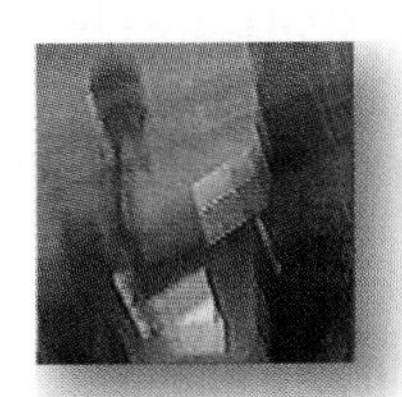

SOCIOLOGY VERSUS IDEOLOGY

We live in a time that is rife with cultural conflicts, not only in the United States but throughout the world. If you name any area of social life, you can probably quickly identify some controversy about norms of conduct. Relations between the sexes, sexual conduct, religion, education, politics, ethics—all are subjects of highly charged debates. Sometimes these disputes are unfriendly in the extreme. Conservatives accuse liberals of lacking a core of values that constitute a moral compass. Liberals accuse conservatives of intolerance and mean-spiritedness toward people with different values and norms of behavior. These differences frequently emerge, for example, in debates over sexuality. Conservatives often hate the idea that homosexual couples should be granted the right to be considered legitimate marital partners. They feel that this implies that homosexuality (which they consider a sin) is a legitimate form of behavior. Liberals generally want to protect the rights of homosexuals to behave as they wish, so long as their behavior does not infringe on the freedoms of others. How you personally feel about these and similar issues will depend to a large extent on your own upbringing and the values you learned in childhood.

A sociological perspective on these issues requires a more dispassionate and scientific approach, even if one continues to feel deeply about the moral issues involved. So what can sociology do to help ease the conflict and cut through the confusion?

As this chapter has shown, the sociologist seeks to discover facts about issues that are in conflict. In the case of norms governing sexual conduct, the sociologist seeks to provide credible knowledge about actual sexual behavior, how homosexual and heterosexual orientations are formed, the influence of biological versus environmental factors, and the conditions under which conflicts among people with differing norms of sexual conduct can be eased. The promise of sociology is that the findings of its research can bring empirical facts to confusing debates and thus ease misunderstandings and conflicts.

GLOSSARY

culture: all the modes of thought, behavior, and production that are handed down from one generation to the next by means of communicative interaction rather than by genetic transmission. (p. 60)

ideas: ways of thinking that organize human consciousness. (p. 61)

norms: specific rules of behavior. (p. 61)

material culture: patterns of possessing and using the products of culture. (p. 61)

values: the ideas that support or justify norms. (p. 61)

laws: norms that are written by specialists, collected in codes or manuals of behavior, and interpreted and applied by other specialists. (p. 62)

ideologies: systems of values and norms that the members of a society are expected to believe in and act on without question. (p. 62)

technologies: the products and the norms for using them that are found in a given culture. (p. 65)

social control: the set of rules and understandings that control the behavior of individuals and groups in a particular culture. (p. 65)

normative order: the array of norms that permit a society to achieve relatively peaceful social control. (p. 65)

sanctions: rewards and punishments for abiding by or violating norms. (p. 66)

mores: strongly sanctioned norms. (p. 66)

folkways: weakly sanctioned norms. (p. 66)

natural selection: the relative success of organisms with specific genetic mutations in reproducing new generations with the new trait. (p. 68)

cultural evolution: the process by which successful cultural adaptations are passed down from one generation to the next. (p. 69)

social Darwinism: the notion that people who are more successful at adapting to the environment in which they find themselves are more likely to survive and to have children who will also be successful. (p. 69)

sociobiology: the hypothesis that all human behavior is determined by genetic factors. (p. 70)

linguistic-relativity hypothesis: the belief that language determines the possibilities for thought and action in any given culture. (p. 72)

ethnocentrism: the tendency to judge other cultures as inferior to one's own. (p. 72)

cultural relativity: the recognition that all cultures develop their own ways of dealing with the specific demands of their environments. (p. 73)

hegemony: undue power or influence. (p. 76)

civilization: a cultural complex formed by the identical major cultural features of a number of societies. (p. 77)

acculturation: the process by which the members of a civilization incorporate norms and values from other cultures into their own. (p. 77)

assimilation: the process by which culturally distinct groups in a larger civilization adopt the norms, values, and language of the host civilization and are able to gain equal statuses in its groups and institutions. (p. 80)

subculture: a group of people who hold many of the values and norms of the larger culture but also hold certain beliefs, values, or norms that set them apart from that culture. (p. 80)

counterculture: a subculture that challenges the accepted norms and values of the larger society and establishes an alternative lifestyle. (p. 80)

accommodation: the process by which a smaller, less powerful society is able to preserve the major features of its culture even after prolonged contact with a larger, stronger culture. (p. 80)

WHERE TO FIND IT

BOOKS

Cultural Encounters and Conflicts (Charles Philip Issawi; Oxford University Press, 1998). A perceptive study of the difficulties and rewards of interactions across cultures, with emphasis on the troubled relations between Westerners and Middle Easterners.

Beyond Cultural Imperialism: Globalization, Communications and the New International Order (Peter Golding and Phil Harris, eds.; Sage, 1997). A critical sociological study of cultural hegemony and global communications.

Cultivating Differences: Symbolic Boundaries and the Making of Inequality (Michele Lamont and Marcel Fournier, eds.; University of Chicago Press, 1992). A collection of essays about how human cultures create and destroy boundaries among groups and societies.

The Interpretation of Cultures (Clifford Geertz; Basic Books, 1973). A set of essays that explore the many meanings of human culture, including a fine essay on the relationship between culture and human social and physical evolution.

Cultural Anthropology, 4th ed. (Marvin Harris; HarperCollins, 1995). A basic textbook that covers in detail the cultures of nonindustrial peoples.

JOURNALS

Technology and Culture. A quarterly journal devoted to studies of the interrelationships between cultural and technological change.

Signs. A quarterly journal that supplies valuable data on women in various societies and cultures.

Daedalus. The journal of the American Academy of Arts and Sciences. It devotes entire issues to subjects of great importance in the world—such as AIDS, computers, violence, and nationalism—and contains some of the best writing on social and cultural change.

INTERNET RESOURCES

American Anthropology Association (www.ameranthassn.org/). The home page for professional anthropologists. An excellent place to look for information about the social-scientific study of cultures, especially those in other parts of the world.

The United Nations (www.un.org/ and www.unsystem.org/). The UN's home page and index page, presenting information about cultures and societies around the world.

Amnesty International (www.Amnesty.org/). A good place to start if you are looking for information about human rights and culture conflicts throughout the world. Information about the clashes between ethnic groups in Africa, the problems of leaders of opposition movements, and the like can be found here or through links from this site.

The Southern Institute at Tulane University (www.tulane.edu/~so-inst/index.html). Provides access to discussion groups, databases, and research publications, especially on issues of culture and race in the United States and elsewhere in the world.

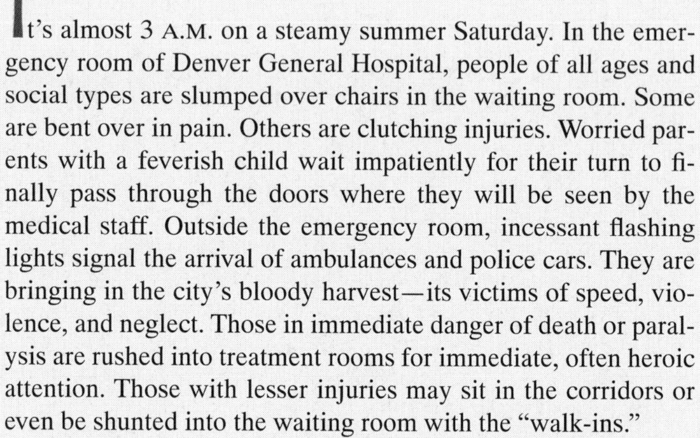

CHAPTER 4

SOCIETIES AND NATIONS

It's almost 3 A.M. on a steamy summer Saturday. In the emergency room of Denver General Hospital, people of all ages and social types are slumped over chairs in the waiting room. Some are bent over in pain. Others are clutching injuries. Worried parents with a feverish child wait impatiently for their turn to finally pass through the doors where they will be seen by the medical staff. Outside the emergency room, incessant flashing lights signal the arrival of ambulances and police cars. They are bringing in the city's bloody harvest—its victims of speed, violence, and neglect. Those in immediate danger of death or paralysis are rushed into treatment rooms for immediate, often heroic attention. Those with lesser injuries may sit in the corridors or even be shunted into the waiting room with the "walk-ins."

Clinical specialist Lori Jones often serves as the ER's triage nurse. Her job is to assess the gravity of the symptoms patients present. She must choose those who need to be seen immediately by doctors and place those with less serious conditions on the waiting list. "You cannot refuse care," she explains. "When you set a little sign outside with a light on it that says, 'Twenty-four-hour emergency room' you cannot refuse to see anybody. Then we turn around and teach everybody that in this day of limited resources and overcrowding, you can and must refuse care to those you don't believe need it" (Richards, 1989, p.72). Jones has the unenviable job of deciding who gets care immediately, who gets it later, and who waits untreated through the long night. "The psych patient and the homeless person wants to stay in the ER to get seen immediately. He doesn't want to get put out in the waiting room or the walk-in clinic, where the wait could be two, four, five hours. And he knows from experience that the likelihood of getting a real bed upstairs is much greater if he comes in through the emergency room than the walk-in clinic" (p.72). Most of the homeless street people who use the emergency room, Jones continues to explain, "aren't bag people, don't have backpacks. Many of them don't push carts around and they don't carry big plastic garbage sacks of possessions around. They have only the clothes on their bodies. So if they can get into Denver General, it may be loud, but at least it's warm. At least we can't rob them. They will get a meal. Nobody will kick them or smash their head" (p.73).

Inside the ER the staff faces one immediate crisis after another. Death lurks around the hospital gurneys, as many of us know from watching *ER,* one of the most popular programs in the history of American television. But Dr. Jim Rappaport has just experienced the death of an elderly man with cardiac failure who he thought he might be able to save. "Death is a very nebulous thing sometimes. . . . At what point does he die? I don't know. The heart can stop working, yet each cell is still alive." Dr. Rappaport and his team fought to revive this patient, but at some point they knew it was hopeless. Yet they continued their efforts and finally said, "Okay, he's dead. . . . But at some time, at some point, maybe even hours ago when he was being carried over here in the ambulance, his soul left him" (p.94).

For Dr. Paul Thurman, director of emergency medical services at Denver General, the threat to the quality of emergency care, and to the ultimate survival of the ER, is "without a doubt, selfishness. The reality is that about 20 percent of our society cannot or will not care for themselves. So the rest of society has got to make a decision about whether or not they're going to pick up the tab for that 20 percent." But as Dr. Thurman sees it, the society he lives in is increasingly indifferent to the needs of the poor and those who are otherwise incapable of caring for themselves. "So we have to decide what kind of society we want to live in. Do we want to live in a Calcutta where people die in the streets because there's nobody to help them and care for them? Or do we want to live someplace like Denver that has one of the finest indigent-care and trauma emergency systems in the country?" (p.163).

The director and his staff of doctors, nurses, paramedics, ambulance drivers, and administrative personnel are struggling to provide care in the face of dwindling resources and budget cuts. The frustrations of trying to maintain an effective health care delivery organization in this negative environment weigh heavily on him and his entire staff. As he says, "We're fighting desperately to keep alive, but I'm tired of working at a resource-poor hospital where we get punished for doing a good job. I'm tired of living in a society where city councilmen can stand up and say, 'It's okay to take care of patients, but don't do too good a job of it'" (p.163).

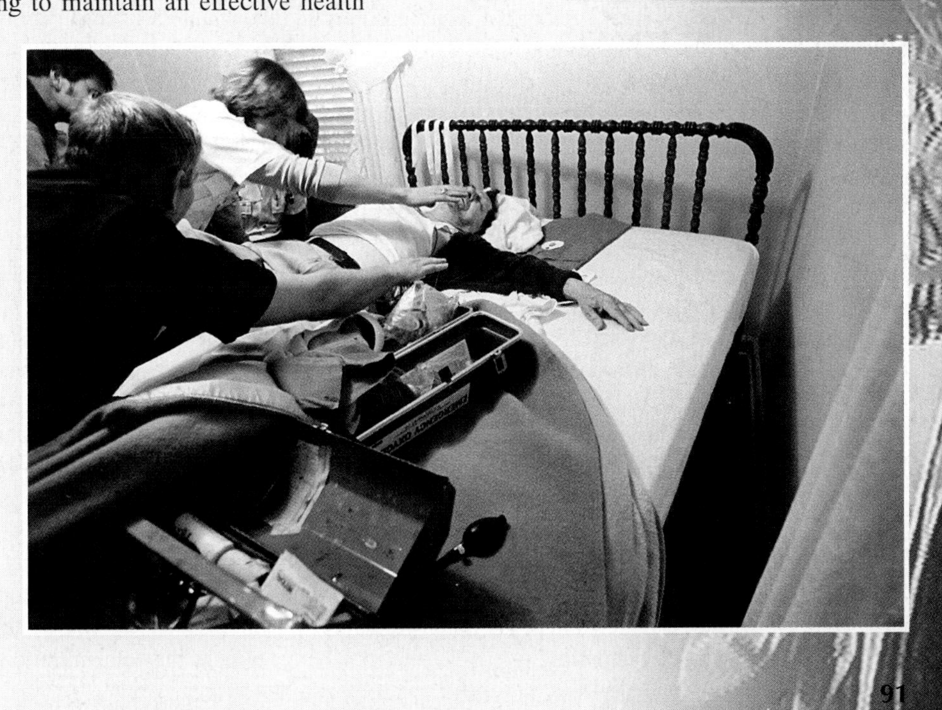

THE SOCIAL ORDER

Amid the chaos of incoming emergencies and frightened people, the emergency room staff maintain a definite social order. Despite the confusion and stress, all the professionals go about doing their jobs. As we see in the preceding vignette, and in the Visual Sociology feature at the end of the chapter, these jobs are not always routine or easy. The triage nurse, for example, must watch as children and families wait anxiously for treatment while less deserving but more serious medical cases are treated before them. The chief of medical services worries that even an effective social organization like that of the ER at Denver General is threatened by forces in the larger social environment within which the ER exists. Is the ER unique in facing these challenges? By no means. The members of every human social structure (organization, group, family, congregation, etc.) experience different kinds of stress as they seek to carry out their functions. The larger social environment is always changing and exerting new forces that shape and alter the social structure, whatever its nature. One of the main goals of this chapter is to indicate the common features of the many different types of human social structures and show how they influence individual behavior.

Of course, humans are not the only animals capable of social organization. We can learn a lot about life in human societies by comparing ourselves with other social animals. If you look closely at an ant colony or a beehive, for example, you can see a remarkable amount of organization. The nest or hive is a complete social world with workers, warriors, queen, drones, and so on. Each individual ant or bee has something to do and seems to do it quite well. But the differences between human societies and those of social animals like bees are even more important than the similarities. Unlike a human actor, an individual social animal is able to perform only a certain number of innate (inborn) tasks. The human can learn an infinite number of tasks. The individual bee is born a worker or a queen. Lori Jones, the clinical specialist in the ER, can be a mother, a voter, a taxpayer, a congregation member, a driver, a singer, and pretty much whatever else she sets her heart on becoming.

We often assume that we know how society works and how to steer our way through it. But there will be times—especially when, like the medical director, we are learning to adapt to changing social environments—when we will be unsure of what is expected of us and how we should perform. Worse still, there may be times when it seems that society itself is threatened, that its continued existence as we know it is endangered. We catch glimpses of the breakdown of society during riots or wars or severe economic recessions.

At earlier times in human history, plagues and famines were frequent reminders that people had little control over their own destinies. Today plagues and famines still occur, but we are more often faced with real or potential crises of our own making: war, genocide, environmental disasters, drug addiction, criminal violence. Thus, if we are to continue to exist and thrive as a species, it is vital that we study societies and social structures—how they hold together, how they change, and why they sometimes seem to fall apart.

As in any science, we begin with some basic terms and definitions. The next few pages introduce the principal elements of social structure. (See the study chart on page 93.) Later sections of the chapter apply these and related concepts to an analysis of how societies have developed since the beginning of human history and how differences among societies affect the lives of individuals. We also discuss the important distinction between societies and nation-states.

Society and Social Structure

The term **society** refers to a population of people (or other social animals) that is organized in a cooperative manner to carry out the major functions of life, including reproduction, sustenance, shelter, and defense. This definition distinguishes between societies and populations. The notion of a population implies nothing about social organization, but the idea of a society stresses the *interrelationships* among the members of the population. In other words, a population can be any set of individuals that we decide to count or otherwise consider, such as the total number of people living between the

STUDY CHART Elements of Social Structure

STRUCTURAL ELEMENT	DESCRIPTION	EXAMPLE
Group	Any collection of people who interact on the basis of shared expectations regarding one another's behavior.	A discussion group; a Bible study class; a local union.
Status	A socially defined position in a group.	Orderly, practical nurse, registered nurse, resident, chief resident (all statuses in a hospital ward).
Role	The way a society defines how an individual is to behave in a particular status.	The doctor diagnoses and treats illnesses; the nurse provides care to patients under the doctor's supervision.
Role Expectations	A society's expectations about how a role should be performed, together with the individual's perceptions of what is required in performing that role.	A major league center fielder is expected to have a batting average over .300, drive in more than 75 runs, and cover the field with a minimum of errors.
Institution	A more or less stable structure of statuses and roles devoted to meeting the basic needs of people in a society.	The military is the primary institution devoted to providing national defense.

Rio Grande and the Arctic Circle, whereas a society is a population that is organized in some way, such as the population of the United States or Canada or the Amish people of Pennsylvania. In the modern world most societies are also (but not always) nation-states.

Social structure refers to the recurring patterns of behavior that people create through their interactions, their exchange of information, and their relationships (Mark, 1998). We say, for example, that the family has a structure in which there are parents and children and other relatives who interact in specific ways on a regular basis. The larger society usually requires that family members assume certain obligations toward each other. Parents are required to educate their children or send them to schools; children are required to obey their parents until they have reached an age at which they are no longer considered dependent. These requirements contribute to the structure of relationships that is characteristic of the family. (The social scientist's method of diagramming family relationships, shown in Box 4.1, creates a graphic depiction of family social structure.)

Throughout life individuals maintain relationships in an enormous range of social structures, of which families are just one. There are many others. People may be members of relatively small groups like the friendship or peer group and the work group. They may also be members of larger structures like churches, business organizations, or public agencies. And they may participate in even more broadly based structures, such as political groups and party organizations, or interest groups like the National Rifle Association or Planned Parenthood. All of these social structures are composed of groups with different degrees of complexity and quite different patterns of interaction. A military brigade, for example, is far more complicated than a barbershop quartet, and people behave quite differently in each. But both are social structures. In such structures our time, our activities, and even our thoughts may be "structured" according to the needs and activities of the group.

Elements of Social Structure

Groups The "building blocks" of societies are social groups. A **group** is any collection of people who interact on the basis of shared expectations regarding one another's behavior. One's immediate family is a group; so are a softball team, a seminar, a caucus, and so on. But a collection of people on a busy street—a crowd—is not a group unless for some reason its members begin to interact in a regular fashion. Usually a crowd is composed of many different kinds of groups—couples, families, groups of friends, and so on. They may be molded into a single group in response to an event that affects them all, such as a fire that creates the need for the orderly evacuation of a building.

Statuses In every group there are socially defined positions known as **statuses.** Father, mother, son, daughter, teacher, student, and principal are examples of

BOX 4.1 SOCIOLOGICAL METHODS

Using Kinship Diagrams to Portray Social Structure

Kinship diagrams like the one shown here provide a visual model of one type of social structure, that associated with family statuses extending over more than one generation. Such diagrams are used in both anthropology and sociology to denote lines of descent among people who are related by blood (children and their parents and siblings) and by marriage. To understand the chart one must know the meanings of the symbols used; these are explained in the key to the chart. The chart applies these symbols to show how cross-cousin marriages between members of different bands link the bands together into a larger structure of kinship networks.

The chart illustrates the social structure of a hunter-gatherer group. One individual, Ego, is used as a point of reference, and kinship links are traced from Ego's offspring, parents, grandparents, and more distant relatives.

In most hunter-gatherer societies it is not possible for the entire society to travel and hunt within the same territory because of scarcities of game and edible vegetation. Instead, the society is divided into natal bands—in this example, those formed by the male line of descent (grandfather, father, Ego, Ego's son).

The striped coloring within the kinship symbols refers to another characteristic of hunter-gatherer societies: Marriages within the band cannot occur because of the incest taboo, and there are specific rules governing marriage outside the band. Note that Ego has married his father's sister's daughter, or his first cousin on his father's side. Ego's wife's brother has married a woman from outside their band. This couple has come to live in the husband's natal band. Their daughter has married Ego's son, another cross-cousin marriage. Ego's daughter, on the other hand, will eventually marry someone from another band and live outside her natal band. By means of this custom the balance of women coming into and going out of the band is preserved.

Cross-cousin marriages, which are taboo in most Western societies, permit the hunter-gatherer band to develop a strong network of interfamily and interband kinship ties. This network widens over the generations and extends the ties of kinship throughout these small, mobile societies.

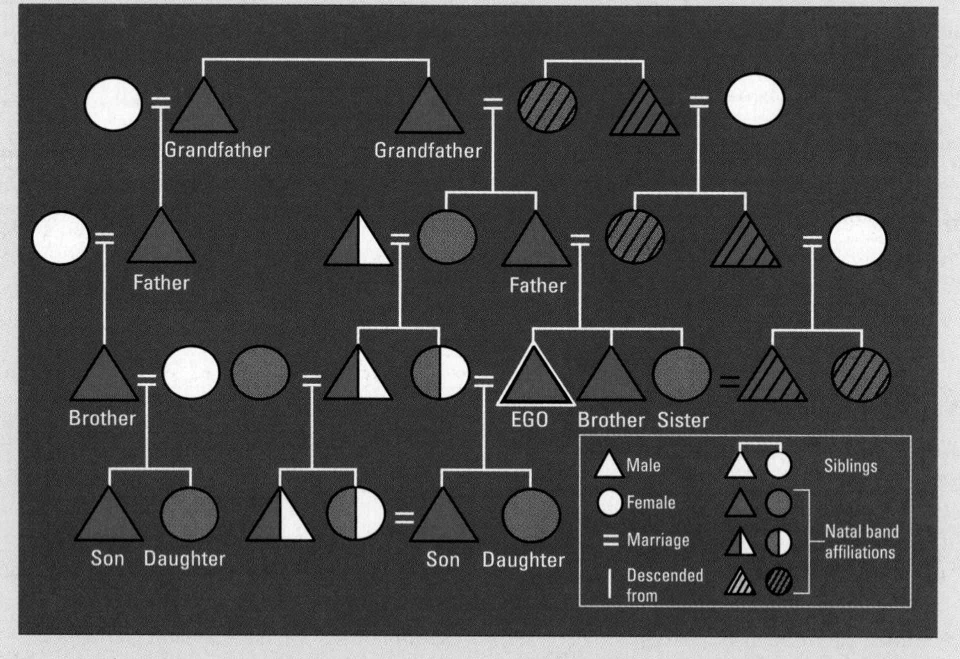

Source: Adapted from "The Origin of Society," by Marshall D. Sahlins. Copyright © September 1960 by Scientific American, Inc. All rights reserved.

familiar statuses in the family and the school. There are an infinite number of statuses in human societies. In a corporation, for example, the statuses range from president and chief executive officer to elevator operator and janitor. Between these two extremes there could be thousands of other statuses.[1] Moreover, the corporation can always create new statuses if the need arises. Thus, in the 1970s, when American society agreed to combat racism

[1] Like many other sociological terms, *status* has more than one meaning. It can refer to a person's rank in a social system and also to a person's prestige—that is, the esteem with which others regard him or her. The fundamental meaning is the one we use here.

and sexism in business and government, many corporations invented the status of affirmative action director to provide equal opportunities for workers of both sexes and all racial and ethnic groups. In the 1990s, as the larger society became more deeply concerned about environmental pollution and drug abuse, corporations began to create new statuses like pollution control manager, drug counselor, or ethics officer (Kelley, 1998). Each of these statuses then became part of the "corporate structure."

Human societies rely heavily on the creation of statuses to adapt to new conditions like environmental pollution and drug use, and one can often observe this adaptation occurring in daily life. In the family, for example, it is increasingly common (though not universally condoned) for young adults to cohabit before marriage or after divorce. As a large-scale phenomenon, this is a relatively recent trend in American society, so new that we have not defined very well the new status of the person who participates in such a relationship. Do we say *boyfriend* or *girlfriend, lover, mate, significant other,* or what? The awkwardness of these terms is due to the fact that this is an emerging status that our society has not yet fully accepted or defined.

This example highlights an essential point about human social structure: It is never fixed or perfectly formed but instead is always changing and adapting to new conditions. Often the process of change involves much conflict and uncertainty, and often there is little consensus about how one should perform in a given status. Should the president of a corporation be an aloof, aggressive leader who directs subordinates with little regard for their feelings? Or should the president show concern for employees' feelings and personal needs and perhaps thereby gain greater loyalty and motivation? This is just one of thousands of dilemmas arising from questions about how we should act in a given social status. To clarify our thinking about statuses in groups and the behaviors associated with those statuses, sociologists make a distinction between a *status* and a *role.*

Roles and Role Expectations The way a society defines how an individual is to behave in a particular status is referred to as a **role.** In the ER, for example, Lori Jones has a number of distinct roles to play. She specializes in giving care to homeless and psychiatric patients, but she also often serves as triage nurse, and then her role involves deciding which patients must wait while others receive care for more serious conditions. It bothers her that the ER offers care to all but has to turn some away, yet she manages to perform her role to the satisfaction of her superiors in the organization. Clearly, the ways in which people actually perform a role may vary widely. They are the product of **role expectations,** the society's expectations about how a role should be performed, together with the individual's perceptions of what is required in performing that role.

To appreciate the importance of role expectations you need only think of the mothers and fathers of your close friends: All hold the same statuses, but how different their behavior is! Part of that difference is due to personality—to psychological variables—but another part is due to how the individual mothers and fathers perceive what is expected of them in the statuses they hold. One mother and father may have been raised to believe that children should work to support the family. They will insist that their children get early job experience. Another couple may have been taught that childhood is too short and should be prolonged if possible. Other things being equal (e.g., adequate family income), that couple will not encourage their children to find jobs before they are more or less the same age the parents were when they went to work.

Social change makes for even more debate and anxiety about role expectations, a subject we discuss in more detail later in the chapter. For now, consider the example of a mother who is also an attorney (something that was rather rare before women gained greater access to professional training). Because of the demands of her profession, she may be unable to take on a leadership position in the school PTA. However, being active in the PTA may have been one of her role expectations for motherhood before the opportunity for a professional career (and the income it provides) became attractive. Now she may feel harried by the pressure of her conflicting statuses of mother and attorney—to say nothing of her other possible statuses, such as daughter, citizen, consumer, committee member, and so on. She may demand that her spouse perform tasks not traditionally associated with the status of husband, and he may or may not modify his original role expectations about that status; in either case, there is likely to be some conflict in the family as it adjusts to these changes.

Another family, in which the mother is also in the labor force but there is no father in the home, has even more adjusting to do. The older children may take on parental roles far sooner than they might have in a two-parent family, but they may also resent this added responsibility and take out their anger on themselves and their siblings. In still another family, one that conforms to the tradition in which the mother is a homemaker and the father works outside the home, the pressures and conflict created by multiple role expectations may not occur in the early years. But what happens when the father retires and gives up his lifelong status of breadwinner? This is a time when traditional families frequently experience strain.

In sum, sociologists are interested in the way roles and statuses affect individual behavior. They also study how the lack of roles—for example, in the form of jobs—can affect people's lives. They do not deny the importance of personality, but they place greater emphasis on the influence of social structure as an explanation of individual behavior.

THE PERSONAL CHALLENGE

Why the Wait Staff Breaks Down

How many times have you heard expressions like "If you can't stand the heat, stay out of the kitchen"? Such expressions are commonsense forms of advice about how to deal with demanding roles in modern social structures, especially in stressful organizations like restaurants. In meeting the challenge of life in a stressful organization, it is often helpful to use your sociological imagination. The trick is to analyze the situation so that you can understand what aspects of the stress are due to your own behavior or perceptions (which you can often change or control) and which aspects are due to conditions in the organization itself. Consider the following case study of a waitress learning her job.

At the beginning of Sherri's second month on the job things were supposed to be getting easier, but they weren't. If anything, they were getting worse, especially the pressure during the lunch-hour madness. Lunch never seemed to go well. There were always more customers than she could handle. Too many tables to serve, busboys too slow, cooks yelling at her to pick up orders or write more clearly—things were so bad that she dreaded coming in to work. She was falling behind in her studies, too. This part-time waitressing job was turning out to be a disaster. And now that chubby guy at table 14 was leering at her.

"Oh, Miss," he called out as Sherri shot by under a huge tray of burger plates. "Oh, Miss, why are you ignoring me? I haven't even gotten a menu." "Be right there, Sir," she muttered.

She dealt out the burger plates as fast as she could and dashed back into the kitchen to get the soups for another table. "Where've you been hiding?" the cook shouted at her from behind his stainless steel counter. "You better get that soup outta here before I have to heat it up again. Let's go. Move!"

Sherri flew through the swinging doors with the tepid soups. Out of the corner of her eye she saw the chubby guy complaining to Karen, her supervisor, about the slow service. Karen shot her an angry glance. A party of six was waiting to be seated at one of her big tables, which needed to be cleaned off. It was too much. She felt the tears welling up in her eyes. Her makeup would be streaked. She quickly headed for the restroom.

On the way Sherri felt a gentle touch on her shoulder. Betty, the veteran waitress who had broken her in when she started 5 weeks ago, smiled at her. "Take it easy, kid! It's not all that bad. My station is under control, so let me help you with the little fat guy and the others."

With her assured efficiency, Betty had Sherri's station under control in a matter of minutes. Later, after the rush was over, she took Sherri aside and showed her some tricks that would make the work easier, like warning the busboys when a table was almost ready to be cleaned up and working out a system for letting others know when she was having trouble. "You'll see," Betty assured her. "When you learn to assert yourself somewhat, and take care of emergencies first, and when you learn who you can ask for help, this job will be much easier."

In his extensive research on human relations in the restaurant industry, sociologist William F. Whyte (1984) found that high-pressure situations like the one Sherri faced often lead less determined, more thin-skinned waitresses to quit. The ones who stay are emotionally tougher, better able to cope with the pressure. But he also showed that the intervention of tough yet sensitive pros like Betty often helps newcomers learn how to deal with the pressure, especially by learning the "tricks of the trade." However, not all newcomers are fortunate enough to have caring mentors to help them over the rough spots, and many quit before they can learn how to perform their roles well.

Organization in Groups Groups vary greatly in the extent to which the statuses of their members are well or poorly defined. The family is an example of a group in which statuses are well defined. Although parents carry out their roles in different ways, the laws of their society define many of the obligations of parenthood, placing certain limits on what they can and cannot do. Other groups are much more informal. We may form groups for brief periods in buses, hallways, or doctors' offices. There are roles and statuses in these groups also, but they are variable and ill defined. All of the people riding on a bus are passengers, but they do not interact the way the members of a family do. At most, we expect civility and "small talk" from other passengers on a bus, but we demand affection and support from other members of our family.

Social change is constantly making the structure of groups more varied and complex. To continue with our earlier example, the family may seem to have well-defined statuses, but high rates of divorce and remarriage have increased the proportion of families in which one parent is a stepparent (see Chapter 16). Consider the difference in role expectations for "mother" or "father" in a situation in which each spouse must interact with a new spouse, a new set of children, and a new set of in-laws as well as an ex-spouse and children in the first family, plus the old in-laws, plus his or her own parents. Balancing role expectations among the often competing demands of these groups can be a daunting task, one that was far less common when the norms of society made it difficult to divorce.

Groups also vary greatly in the ways in which they are connected with other groups into a larger structure known as an organization. An army platoon, for example, includes the well-defined statuses of private, corporal, sergeant, and lieutenant, each with specific roles to play in training and combat. Platoons are grouped together under the leadership of higher officers to form companies; this pattern is repeated at higher levels to create the battalion and the brigade.

Figure 4.1 shows the formal structure of a typical army combat brigade. It does not show the brigade's informal organization, which consists of the ways in which its members actually relate to one another on the basis of friendships or animosities or mutual obligations of various kinds. (The importance of informal organization within larger groups, such as bureaucracies, is discussed more fully in Chapter 6.)

Social Institutions

The structure of most of the important groups in society is determined by shared definitions of the statuses and roles of their members. When such statuses and roles are designed to perform major social functions, they are termed **institutions.** In popular usage the word *institution* generally refers to a large bureaucratic organization like a university or hospital or prison (usually with cafeterias that serve "institutional food"). But although the word has this meaning in everyday language, the sociological use of the term should not be confused with this meaning.

In sociology, an institution is a more or less stable structure of statuses and roles devoted to meeting the basic needs of people in a society. The family is an institution that controls reproduction and the training of

FIGURE 4.1 The Structure of a Typical Combat Brigade

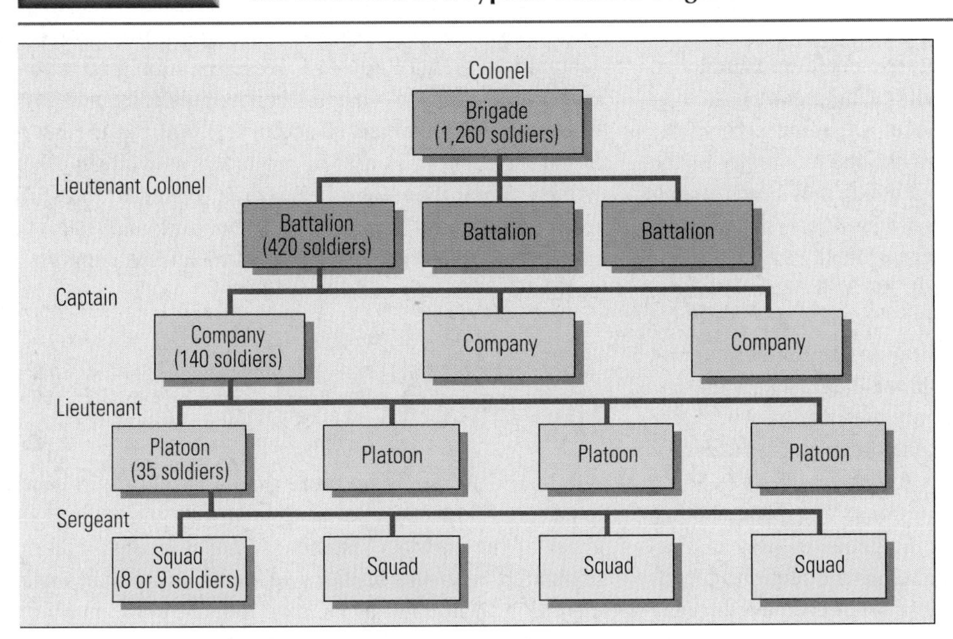

new generations. The market is an institution that regulates the production and exchange of goods and services. The military is an institution that defends a society or expands its territory through conquest. Any particular family, corporation, or military unit is a group or organization within one of these institutions.

Within any given institution there are norms that specify how people in various statuses are to perform their roles. Thus to be a general or a new recruit or a supplier of military hardware is to have a definite status in a military institution. But to carry out one's role in that status is to behave in accordance with a normative system—a set of mores and folkways that distinguish a particular institution from others. For example, in a classic research study about the process of becoming a military recruit, Sanford Dornbusch (1955) described how all the signs of a person's status in civilian life, from fashions in dress to the use of free time, are erased in boot camp. The new status of recruit must be earned by adhering to all the norms of military life. So it is with every social institution: Each has a specific set of norms to govern the behavior of people within that institution.

Institutions of religion, education, politics, and economics tend to become more complex over time as the society changes. It is helpful, therefore, to think of large-scale societies like the United States as having a number of institutional *sectors* (sets of closely related institutions). For example, the economy is an institutional sector that includes markets, corporations, and other economic institutions. Politics is an institutional sector containing legal, executive, legislative, and other political institutions. Each of these institutions, in turn, is composed of numerous groups and organizations. The Chicago Mercantile Exchange is a market organization; the House of Representatives is a legislative organization. In this book, as in much social-scientific writing, the term *institution* is used to refer to institutional sectors (e.g., religion) as well as to the institutions included within a sector (e.g., Buddhism or Christianity).

The history of human societies is marked by the emergence of new institutions like the university or the laboratory. In fact, a dominant feature of human societies is the continual creation of new social institutions, a feature that the social theorist Talcott Parsons (1951, 1966) labeled **differentiation.** By this term, Parsons meant the processes whereby sets of social activities performed by one social institution are divided among different institutions. In small-scale agrarian societies, for example, the family not only performs reproductive and training functions but is also the primary economic institution. Over time the family gradually loses some of its functions to new institutions. The processes of differentiation result in the emergence of new institutions designed to manage economic production (corporations), train new generations (schools), develop new technologies (science), or perform other important social activities. (Many of those institutions are discussed in detail in Part 4.) Why does this differentiation take place? We will see in the next section that new institutions are particularly likely to emerge when populations grow, the need for coordination of their activities increases, and new demands are placed on older institutions that they cannot entirely fulfill.

POPULATIONS AND SOCIETIES

The remarkable growth of the world's human population is the great biological success story of the last million years. Throughout most of the evolution of the human species, the world's population remained relatively small and constant. At the end of the Neolithic period, about 8000 B.C., there were an estimated 5 million to 10 million humans, concentrated mainly in the Middle East, East Africa, southern Europe, and a few fertile river basins in India, China, Latin America, and Central America. The overall range of human existence was much wider—populations were moving north as the last Ice Age receded about 12,000 years ago—but the human population was nowhere near as widely distributed as it is today. By the time of Jesus there were an estimated 200 million people on the earth, and by the start of the industrial age, about the year 1650, there were an estimated 500 million. By 1945, however, the population had reached about 2.3 billion, and it is now close to 6 billion. This trend is illustrated in Figure 4.2.

What explains the shape of this population curve, the gradual rise in the world's population during the late Stone Age, the increasing rate of growth in the early millennia of recorded history, the explosive growth after 1650? The answers have to do with the changing production technologies of societies and their ever-increasing ability to sustain their populations, not only those directly involved in acquiring food but the ever-larger numbers of non-food-producers as well. In other words, human population growth is related to the shift from hunting and gathering to agriculture and, later, to industrial production as the chief means of supplying people with the necessities of life.

The First Million Years: Hunting and Gathering

A human lifetime is no more than a twinkling in time. What are 70 or 80 years compared with the billions of years of the earth's existence? What is one generation compared with the millions of years of human social evolution? Yet many of us hope to leave some mark on society, perhaps to change it for the better, to ease some

i

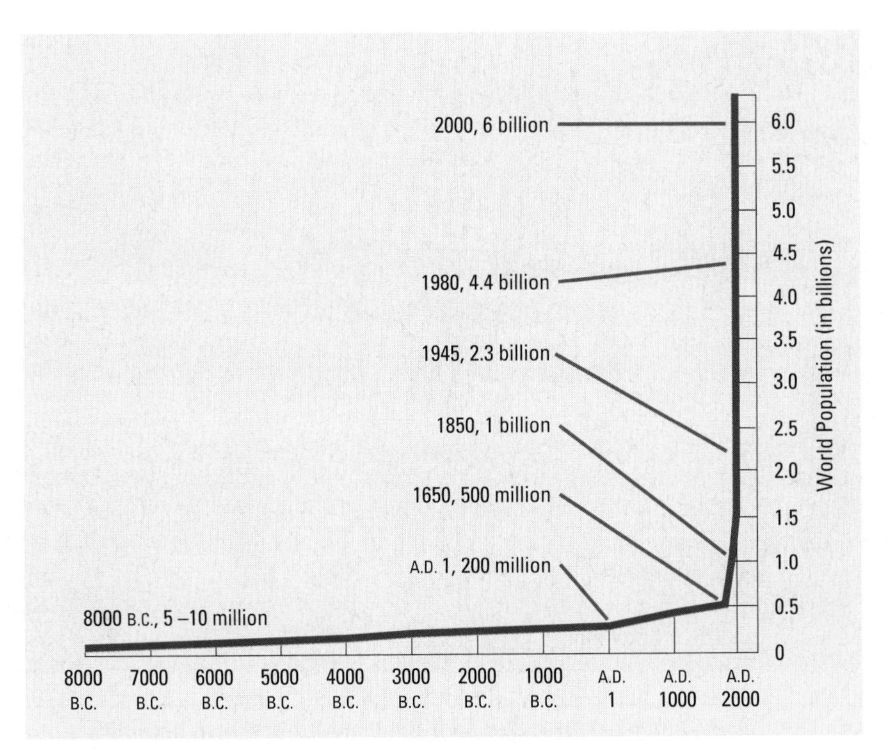

This chart shows world population growth since 8000 B.C. Note that the first, more gradual increase in world population is associated with the shift to agriculture, but that its effects are dwarfed by the impact of the industrial revolution, which began in the eighteenth century.

Source: Data from Office of Technology Assessment.

FIGURE 4.2 **World Population Growth from 8000 B.C. to A.D. 2000**

suffering, to increase productivity, to fight racism and ignorance . . . what vaulting ambitions! And what a radical change from the worldview of our ancestors! The idea that people can shape their society—or even enjoy adequate shelter and ample meals—is widely accepted today. But for most of human history mere survival was the primary motivator of human action, and thus a fatalistic acceptance of human frailty in the face of overwhelming natural forces was the dominant worldview.

For most of the first million years of human evolution, human societies were developing from those of primates. Populations were small because humans, like other primates, lived on wild animals and plants. These sources of food are easily used up and their supply fluctuates greatly, and as a result periods of starvation or gnawing hunger might alternate with bouts of gorging on sudden windfalls of game or berries. Thus the hunting-and-gathering life that characterized the earliest human populations could support only extremely small societies; most human societies therefore had no more than about 60 members.

Recent archaeological evidence indicates that some hunting-and-gathering societies began to develop permanent settlements long before the advent of agriculture. The emergence of farming was one of the changes that accelerated human social evolution, but the new evidence shows that in parts of what is now Europe and the Middle East there were stable settlements of hunter-gatherers as early as 30,000 to 20,000 years ago. These rather large and complex societies were most firmly established in the area that is now Israel at the end of the last Ice Age, some 13,000 to 12,000 years ago (Henry, 1989; Stevens, 1988).

Despite the slow pace of human evolution until about 35,000 years ago, some astonishing physical and social changes occurred during that long period, changes that enabled human life to take the forms it does today. Among them were the following:

- The development of an upright posture (which freed the hands for eventual use of tools) and an enlarged cerebral cortex, making possible vastly increased cognitive abilities and the development of language.
- Social control of sexuality through the development of the family and other kinship structures and the enforcement of the incest taboo.
- The establishment of the band of hunter-gatherers as the basic territorial unit of human society, coupled with

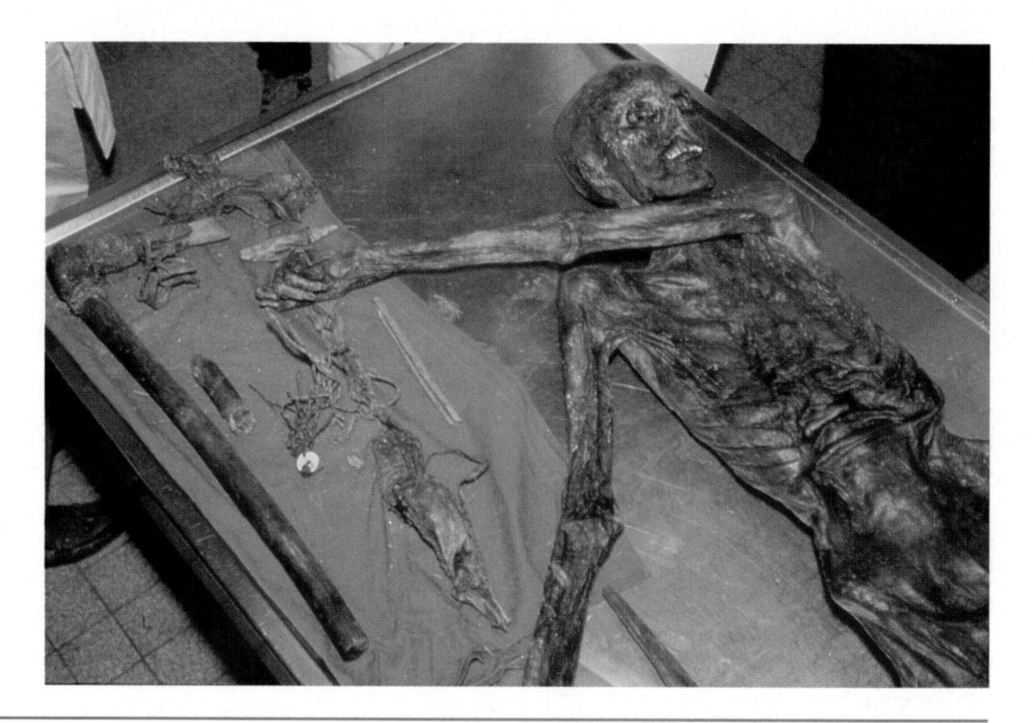

The "Ice Man," discovered in 1991 by hikers in the Italian Alps, is shown here with his hatchet and other items he was carrying when he died. These remains, among the oldest samples of human flesh and organs ever discovered, are providing a wealth of knowledge about daily life in hunting-and-gathering societies 5,000 years ago.

the development of kinship structures that linked bands together into tribes. Within the band, the family became the primary economic unit, organizing the production and distribution of food and other necessities.

The chart in Box 4.1 illustrates how kinship ties link hunting-and-gathering bands together into a larger society. Anthropologists and sociologists often represent social structures with diagrams of this type. They show how people are related to one another and what their status is in relation to any other person in the social structure.

By the end of the last Ice Age, many aspects of this evolutionary process were more or less complete. Human societies had fully developed languages and a social structure based on the family and the band. To be cast out of the band for some wrongdoing—that is, to be considered a deviant person (see Chapter 7)—usually meant total banishment from the society and eventual death, either by starvation or as a result of aggression by members of another society (Salisbury, 1962). But warfare and violence were not typical of early human societies, as the social anthropologist Marshall Sahlins has pointed out:

> Warfare is limited among hunters and gatherers. Indeed, many are reported to find the idea of war incomprehensible. A massive military effort would be difficult to sustain for technical and logistic reasons. But war is even further inhibited by the spread of a social relation—kinship—which in primitive society is often a synonym for "peace." Thomas Hobbes' famous fantasy of a war of "all against all" in the natural state could not be further from the truth. War increases in intensity, bloodiness, duration, and significance for social survival through the evolution of culture, reaching its culmination in modern civilization. (1960, p. 82)

On the other hand, one must not romanticize the hunter-gatherers. Their lives were far more subject to the pressures of adaptation to the natural environment than has been true in any subsequent form of society. Individual survival was usually subordinated to that of the group. If there were too many children to feed, some were killed or left to die; when the old became infirm or weak, they often chose death so as not to diminish the chances of the others. Thus the frail Eskimo grandfather or grandmother wandered off into the snowy night to "meet the polar bear and the Great Spirit." Additional thousands of years of social evolution would pass before the idea that every person could and should survive, prosper, and die with dignity would even occur to our ancestors.

The Transition to Agriculture

For some time before the advent of plow-and-harvest agriculture in the Middle East and the Far East, hunting-and-gathering societies were supplementing their diets with foods acquired through the domestication of plants and animals. In this way they were able to avoid the alternation of periods of feast and famine caused by reliance on animal prey. Karl Marx was the first social theorist to observe that social revolutions like the shift to agriculture or industrial production are never merely the result of technological innovations such as the plow or the steam engine. The origins of new forms of society are to be found within the old ones. New social orders do not simply burst upon the scene but are created out of the problems faced by the old order. Thus, as they experimented with domestication of animals and planting of crops, some hunting-and-gathering societies were evolving into nomadic shepherding or **pastoral societies** in which bands followed flocks of animals. Others were developing into **horticultural societies** in which the women raised seed crops and the men combed the territory for game and fish (Wilford, 1997).

Regarding this momentous change in the material basis of human survival, the historian William McNeill made the following observation:

> The seed-bearing grasses ancestral to modern cultivated grains probably grew wild eight or nine thousand years ago in the hill country between Anatolia and the Zagros Mountains, as varieties of wheat and barley continue to do today. If so, we can imagine that from time immemorial the women of those regions searched out patches of wheat and barley grasses when the seeds were ripe and gathered the wild harvest by hand or with the help of simple cutting tools. Such women may gradually have discovered methods for assisting the growth of grain, e.g., by pulling out competing plants; and it is likely that primitive sickles were invented to speed the harvest long before agriculture in the stricter sense came to be practiced. (1963, p. 12)

As a result of these and other innovations, agriculture became the productive basis of human societies. Pastoral societies spread quickly throughout the uplands and grasslands of Africa, northern Asia, Europe, and the Western Hemisphere, and grain-producing societies arose in the fertile river valleys of Mesopotamia, India, China, and—somewhat later—Central and South America. (See Figure 4.3.) Mixed societies of shepherds and marginal farmers wandered over the lands between the upland pastures and the lowland farms.

The First Large-Scale Societies We have reached the beginning of recorded history (around 4000 B.C.), which was marked by the rise of the ancient civilizations of

FIGURE 4.3 **Agricultural Origins**

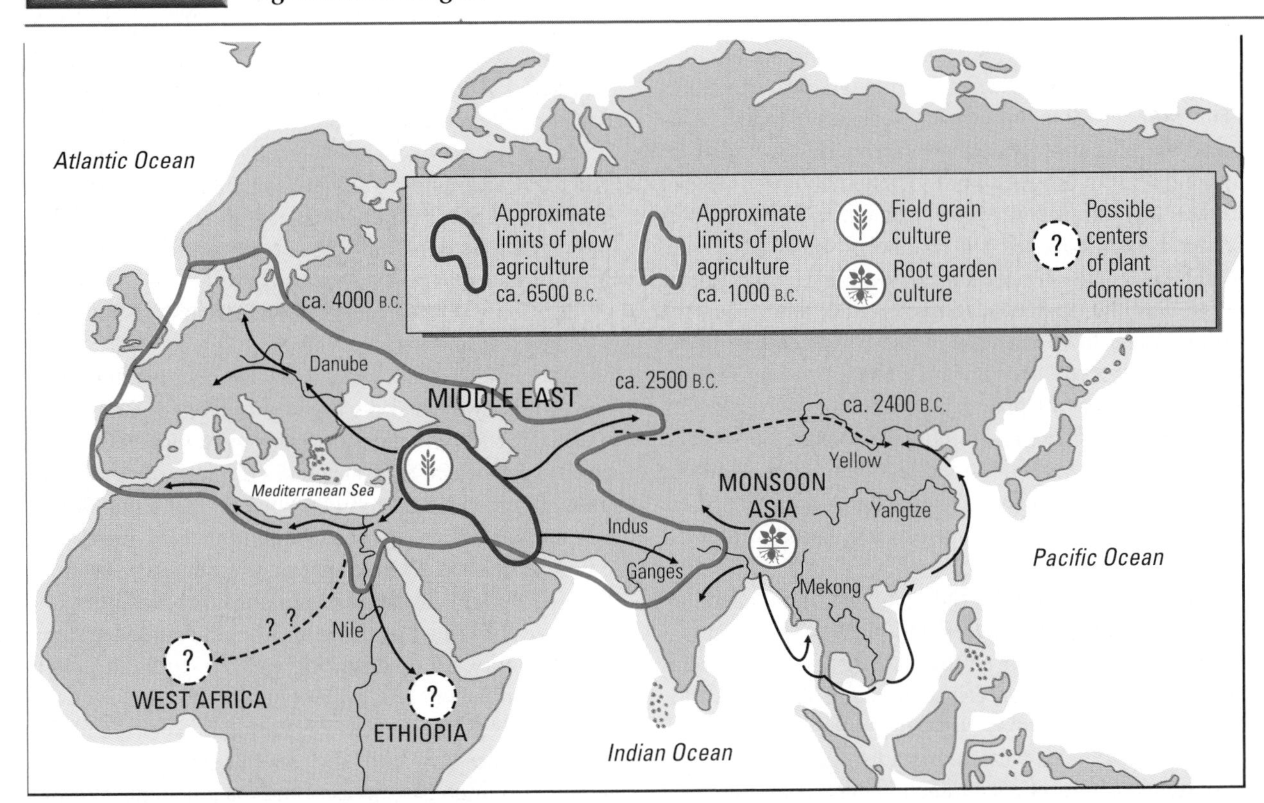

Source: McNeill, 1963.

Sumer, Babylonia, China and Japan, Benin, and the Maya, Incas, and Aztecs. The detailed study of these societies is the province of archaeology, history, and classics. But sociologists need to know as much as possible about the earliest large-scale societies because many contemporary social institutions (e.g., government and religion) and most areas of severe social conflict (e.g., class and ethnic conflict) developed sometime in the agrarian epoch—between 3000 B.C. and A.D. 1600.

From the standpoint of social evolution, the following dimensions of agrarian societies are the most salient (Braidwood, 1967):

- Agriculture allows humans to escape from dependence on food sources over which they have no control. Agrarian societies produce surpluses that do not merely *permit* but *require* new classes of non-food-producers to exist. An example is the class of warriors, who defend the surplus or add to it through plunder.

- Agriculture requires an ever-larger supply of land, resulting in conflicts over territory and in wars with other agricultural or pastoral societies.

- The need to store and defend food surpluses and to house the nonagrarian classes results in new territorial units: villages and small cities.

New Social Structures Freed from direct dependence on undomesticated species of plants and animals, agrarian societies developed far more complex social structures than were possible in simpler societies. Hunting-and-gathering societies divided labor primarily according to age and sex, but in agricultural societies labor was divided in new ways to perform more specialized tasks. Peoples who had been conquered in war might be enslaved and assigned the most difficult or least desirable work. Priests controlled the society's religious life, and from the priestly class there emerged a class of hereditary rulers who, as the society became larger and more complex, assumed the status of pharaoh or emperor. Artisans with special skills—in the making of armaments or buildings, for example—usually formed another class. And far more numerous than any of these classes were the tillers of the soil, the "common" agricultural workers and their dependents.

The process whereby the members of a society are sorted into different statuses and classes based on differences in wealth, power, and prestige is called **social stratification** and is described in detail in Chapter 11. In general, societies may be **open,** so that a person who was not born into a particular status may gain entry to that status, or **closed,** with each status accessible only by birth. A key characteristic of agrarian societies is that their stratification systems became extremely closed and rigid. Because the majority of the people were needed in the fields, there were few opportunities for people to move from one level of society to another. These therefore were closed societies.

The emergence of agrarian societies was based largely on the development of new, more efficient production technologies. Of these, the plow and irrigation were among the most important. The ancient agrarian empires of Egypt, Rome, and China are examples of societies in which irrigation made possible the production of large food surpluses, which in turn permitted the emergence of central governments led by pharaohs and emperors, priests and soldiers. Indeed, some sociologists have argued that because large-scale irrigation systems required a great deal of coordination, their development led to the evolution of imperial courts and such institutions as slavery, which coerced large numbers of agrarian workers into forced labor (Wittfogel, 1957).

By the fifteenth century the world's population was organized into a wide variety of societies with many different types of cultures and civilizations. Figure 4.4 depicts the spread of human life over more than 150 million square kilometers of land in the year 1500, before the age of exploration and the industrial revolution. The 76 cultures and civilizations shown include hundreds of different societies, ranging from simple hunting-and-gathering societies to more advanced agrarian ones to civilizations that were already expanding and exporting their culture through conquest and trade. The areas numbered 1 to 27 on the map were inhabited by simple societies of hunters, gatherers, and fishers; those numbered 28 to 44 were inhabited by nomads and stockbreeders; areas 45 to 60 were inhabited by agricultural peoples, primarily peasants using hoes rather than plows; and the remaining areas were more advanced civilizations. With the exception of the Mayan and Incan civilizations, all of the latter were found in Asia, the Middle East, and parts of Europe.

The areas where the earliest civilizations emerged are to this day the most densely inhabited parts of the world. In 1500 these civilizations were marked by their ability to produce food surpluses through the use of the plow, domesticated animals, wheels, and carts. Above all, they are marked by the importance of towns and cities as centers of administration and religious practice. Although the Mayan and Incan civilizations did not have the wheel or the plow, their advanced systems of astronomy, art, and writing make a strong argument for their inclusion as world civilizations.

The explorations led by Portugal, Spain, and England in the sixteenth and seventeenth centuries resulted in conquest and European settlement in many

Then and Now

The Valley of Mexico

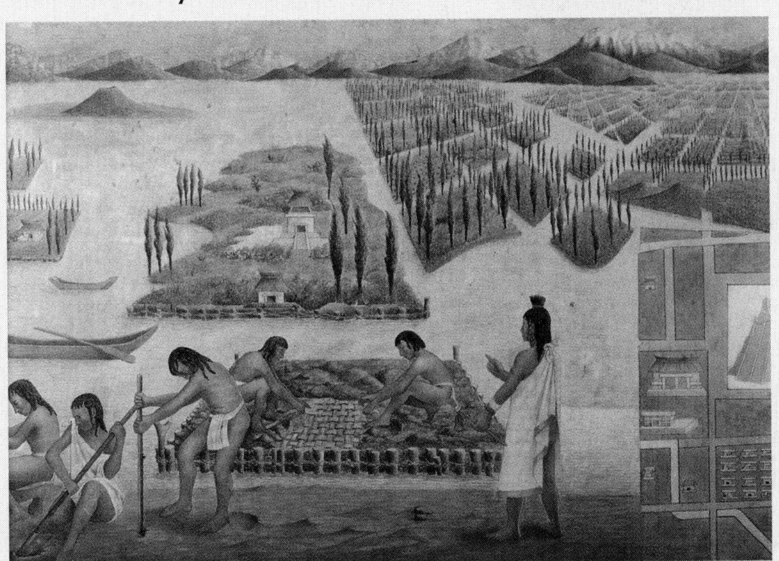

This detail from a famous mural in Mexico's Palacio Nacional shows what the Valley of Mexico looked like in 1519, when conquistador Hernán Cortés and a small band of Spanish soldiers began their conquest of the strife-weakened Aztec empire. Under the emperor Montezuma, the Aztec capital, Tenochtitlan, was a marvel of social structure, the product of a highly developed and complex social system governed primarily by a priestly caste. As the mural suggests, Tenochtitlan was carefully planned to allow its inhabitants to enjoy the temperate climate of the high valley and was served by a sophisticated irrigation system (Iglesia, 1990).

Today, less than 500 years after the conquest, the Valley of Mexico is blanketed by Mexico City and the many urban communities surrounding it. The social structure of Mexico City is extremely complex, because the city is both Mexico's largest commercial center and its national capital. The city is surrounded by ancient volcanoes, and the valley floor is prone to earthquakes. Despite the problems created by rapid growth in an environmentally sensitive area, Mexico City continues to grow and is now one of the two or three largest metropolitan centers in the world.

parts of the Western Hemisphere and Africa. Conquest caused radical disruptions in the internal development of the world's cultures and civilizations. But even more drastic would be the changes caused by the industrial revolution of the eighteenth and nineteenth centuries.

The Industrial Revolution

In 1650, when Holland, Spain, and England were the world's principal trading nations (Block, 1990), the population of England was approximately 10 million,

FIGURE 4.4 Civilizations, "Cultures," and Primitive Peoples, ca. 1500

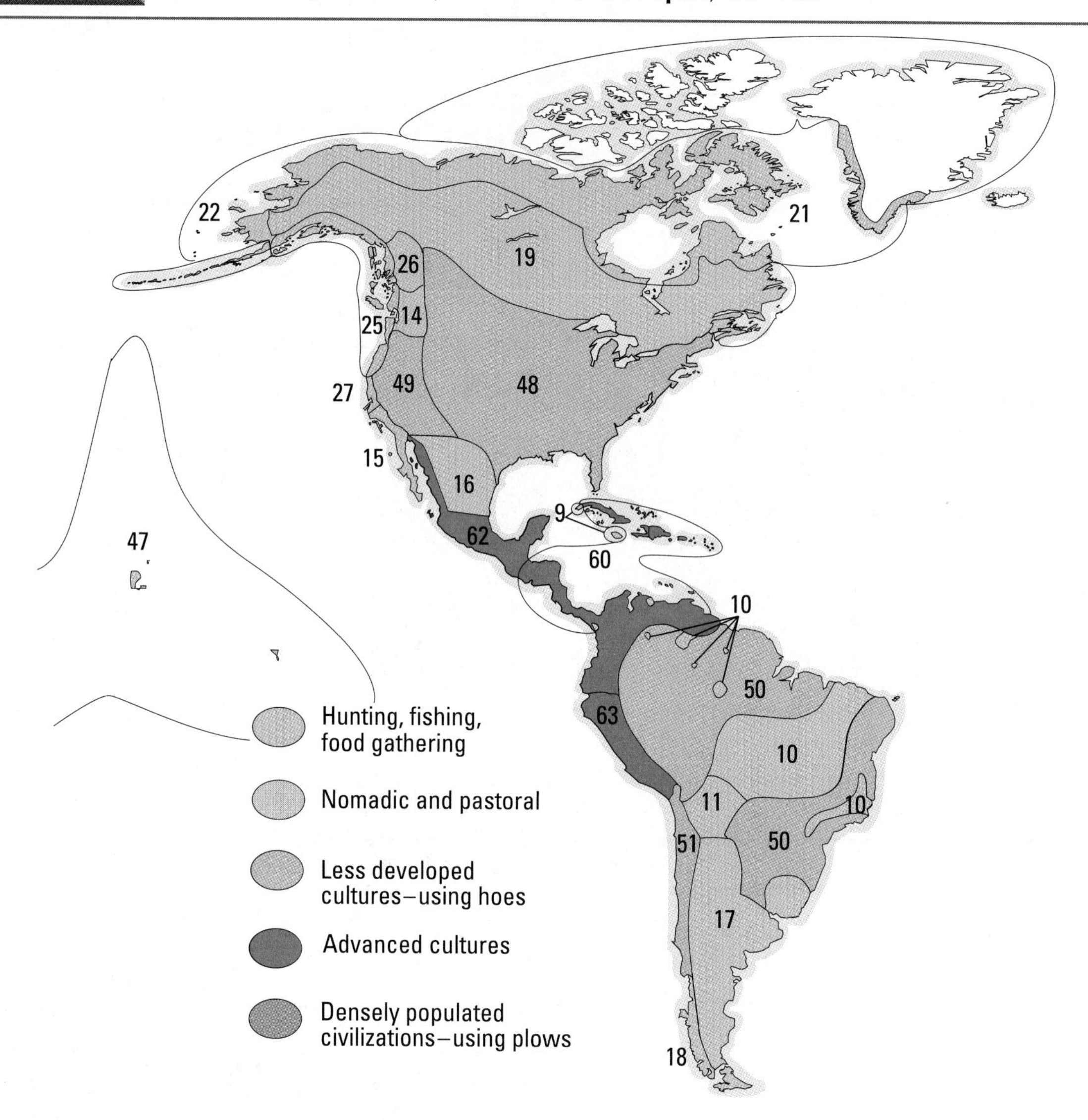

1. Tasmanians
2. Congo Pygmies
3. The Vedda (Ceylon)
4. Andamanese
5. Sakai and Semang
6. Kubu
7. Punan (Borneo)
8. Negritos of the Philippines
9. Ciboneys (Antilles)
10. Gê-Botocudos
11. Gran Chaco Indians
12. Bushmen
13. Australians
14. Great Basin
15. Lower California
16. Texas and northeastern Mexico
17. Patagonia
18. Indians of the southern coast of Chile
19. Athabascans and Algonkin (northern Canada)
20. Yukaghir
21. Eastern and central Eskimos
22. Western Eskimos
23. Kamchadal, Koryak, Chukchi
24. Ainu, Gilyak, Gol'dy
25. Northwest coast Indians (United States and Canada)
26. Columbia Plateau
27. Central California
28. Reindeer-herding peoples
29. Canary Islands
30. Sahara nomads
31. Arabian nomads
32. Pastoral mountain peoples in the Near East
33. Pastoral peoples of the Pamir region and Hindu Kush
34. Kazakh-Kirghiz
35. Mongols
36. Pastoral Tibetans
37. Settled Tibetans
38. Western Sudanese
39. Eastern Sudanese
40. Somali and Galla of northeastern Africa
41. Nilotic tribes
42. East African stock-rearing peoples

Sources: Adapted from Hewes, 1954; and Braudel, 1981.

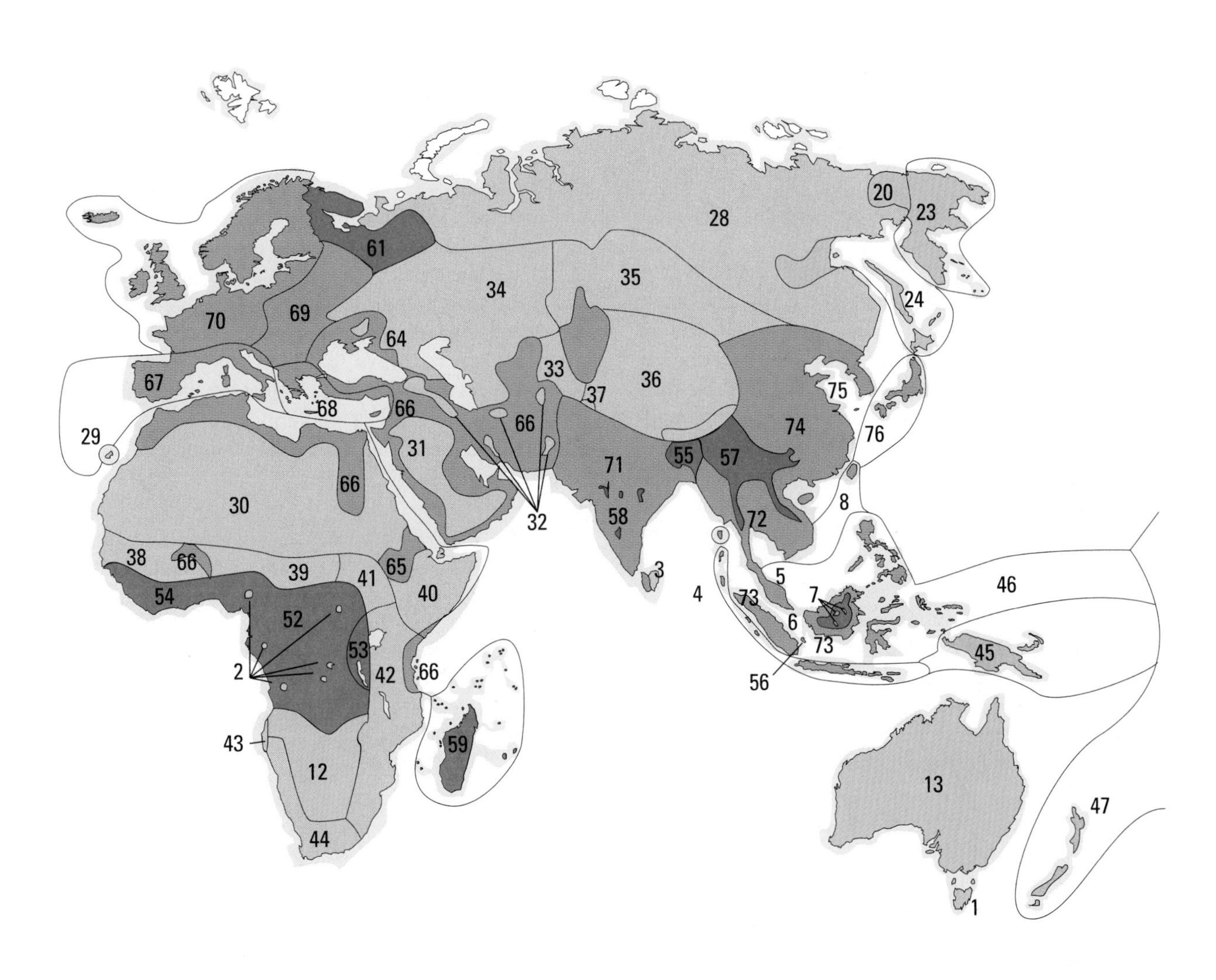

43. Western Bantu
44. Hottentots
45. Melanesian Papuans
46. Micronesians
47. Polynesians
48. American Indians (eastern United States)
49. American Indians (western United States)
50. Brazilian Indians
51. Chilean Indians
52. Congolese peoples
53. Lake-dwellers of East Africa
54. Guinea coasts
55. Tribes of the Assam and the Burmese highlands
56. Tribes of the Indonesian highlands
57. Highland peoples of Indochina
58. Mountain and forest tribes of central India
59. Malagasy
60. Caribbean peoples
61. Finns
62. Mexicans, Maya
63. Peru and the Andes
64. Caucasians
65. Ethiopia
66. Settled Muslims
67. Southwestern Europe
68. Eastern Mediterranean
69. Eastern Europe
70. Northwestern Europe
71. India (this map does not differentiate between Muslims and Hindus)
72. Lowlands of Southeast Asia
73. Indonesian lowlands
74. Chinese
75. Koreans
76. Japanese

of which about 90 percent earned a livelihood through farming of one kind or another. Just 200 years later, in 1850, the English population had soared to more than 30 million, with less than 20 percent at work in fields, barns, and granaries. England had become the world's first industrial society and the center of an empire that spanned the world.

A similar transformation occurred in the United States. In 1860, on the eve of the Civil War, there were about 30 million people in the United States. Ninety percent of that population, a people considered quite backward by the rapidly industrializing English, were farmers or people who worked in occupations directly related to farming. A mere 100 years later, in 1960, only about 8 percent of Americans were farmers or agricultural workers, yet they were able to produce enough to feed a population exceeding 200 million. Today less than 3 percent of Americans work on farms and ranches. These dramatic changes in England and America—and in other nations as well—occurred as a result of the industrial revolution.

The industrial revolution is often associated with innovations in energy production, especially the steam engine. But the shift from an agrarian to an industrial society did not happen simply as a result of technological advances. Rather, the industrial revolution was made possible by the rise of a new social order known as **capitalism.** This new way of organizing production originated in nation-states, which engaged in international trade, exploration, and warfare. Above all, the industrial revolution depended on the development of *markets,* social structures that would function to regulate the supply of and demand for goods and services throughout the world (Polanyi, 1944).

The transition from an agrarian to an industrial society affects every aspect of social life. It changes the structure of society in several ways, of which the following are among the most significant:

- The industrialization of agriculture allows many more people to be supported by each agrarian worker than ever before. Only a relatively small number of people live on the land; increasing numbers live in towns, cities, and suburbs.

- Industrial societies are generally far more receptive to social change than agrarian societies. One result is the emergence of new classes like industrial workers and scientific professionals (engineers, technicians) and new social movements like the women's movement and the movement for racial equality.

- Scientific, technical, and productive institutions produce both unparalleled wealth and unparalleled destructive capacities.

- The world "shrinks" as a result of innovations in transportation and communication. A "global society" develops, but at the same time the unevenness of industrialization leads to conflicts that threaten world peace.

These are only some of the important features of industrial societies. There are many others, which we analyze throughout this book. We will present research that shows how the types of jobs people find and the ways in which they spend their leisure time change as technology advances and their expectations change. We will also look into the ways in which other major institutions—especially the family—adapt to the changes created by the industrial revolution. In fact, we will continually return to the industrial revolution as a major force in our times.

The Theory of Postindustrial Society The United States and most other societies that experienced industrialization in earlier periods are now considered by sociologists to be "postindustrial societies." Even though these societies continue to be important producers of industrial goods like steel, glass, aluminum, and many other products that are essential to industrial manufacturing, their industrial plants are highly automated and employment in manufacturing is no longer the largest and fastest-growing sector of the economy. As sociologist Fred Block (1990) observes:

> [Ours is] a strange period in the history of the United States because people lack a shared understanding of the kind of society in which they live. For generations, the United States was considered as an industrial society, but that definition is no longer compelling. Yet no convincing alternative has emerged in its absence. (p. 1)

In this book we continue to group the United States and the European nations among the world's industrial societies because they are major producers of manufactured goods. However, it should be noted that the businesses that produce those goods are dwindling in number owing to the effects of the industrial revolution in other areas of the world. Also, for the first time in U.S. history the number of jobs in services exceeds that of jobs in goods-producing industries. Far more jobs are created in restaurants and hotels each year than in steel mills and other kinds of factories (Kasarda, 1989). In addition, there is an ever-increasing demand for highly educated and trained workers in new industries that deal with high-speed transmission of information. Computers and telecommunications technologies are surely produced by "industries," but they are not the type of mass-employment "smokestack" industries that were the hallmark of earlier stages of the industrial revolution.

The theory of postindustrial society explores why these fundamental changes have come about and their impact, not only on postindustrial societies themselves but also on the cultures of other societies. There are also many questions about what will happen as the world's societies become ever more interdependent—and decisions made in metropolitan centers like Tokyo, Los Angeles, and London affect the lives of villagers in the most remote parts of the world (Giddens, 1984; Sassen, 1994). We will return to the issues raised by theories of industrialization and the emergence of postindustrial society throughout this book. For it is clear that while the most advanced and powerful nations are undergoing an economic transformation, in many other parts of the world the industrial revolution is in full swing, and in still other, more remote areas older forms of agrarianism and tribalism continue to thrive. (See Box 4.2.)

SOCIETY AND THE INDIVIDUAL

From Gemeinschaft to Gesellschaft

Imagine that you grew up in an agrarian village or a town in a slowly industrializing society and then came to live in the United States, as millions of people do each decade. A difficult part of that experience would be getting used to the impersonality of modern American society when compared with the close relationships you had with people you had known all your life in your native society. American society would seem to be composed of masses of strangers organized into highly impersonal categories. You would have to get used to being a shopper, an applicant, a depositor, a sports fan, a commuter, and so on, and it would be necessary to shift from one to another of these roles several times in a day or even within a half hour.

Sociologists often describe this experience as a change from **gemeinschaft** (the close, personal relationships of small groups and communities) to **gesellschaft** (the well-organized but impersonal relationships of modern societies). These are German terms taken from the writings of the social theorist Ferdinand Tönnies (1957/1887). Complex industrial societies, Tönnies argued, have developed gesellschaft social structures like factories and office bureaucracies to such a degree that they tend to dominate day-to-day life in the modern world.

The American sociologist Charles H. Cooley applied this distinction to the micro level of society in his discussion of primary versus secondary groups. "By **primary groups,**" Cooley wrote,

> I mean those characterized by intimate face-to-face association and cooperation. They are primary in several senses, but chiefly in that they are fundamental in forming the social nature and ideals of the individual. [Such a group] involves the sort of sympathy and mutual identification for which "we" is the natural expression. (1909, p. 23)

Secondary groups, in contrast, are groups in which we participate for instrumental reasons—that is, in order to accomplish some task or set of tasks. Examples are school classrooms, town committees, and political party organizations. We do not have intimate relationships with other members of secondary groups; at least, we do not normally expect to do so and are not concerned about the fact that our relationships in such groups involve only a limited range of emotions.

Of course, when we speak of intimacy in primary groups we do not always refer to love and warm feelings. The intimacy of the members of a family refers to a highly charged set of emotions that can range from love to hate but rarely includes indifference. But in a secondary group, indifference, or lack of personal involvement, is the norm. We therefore do not usually develop strong feelings toward all the other members of the organizations for which we work or the associations to which we belong, even though we may have one or more close friends among our workmates. The idea that people do not have to form primary attachments in work organizations also protects people with less power from those with more. Women in subordinate positions, for example, often need protection from men in more powerful positions who make sexual advances. Modern laws on sexual conduct in the workplace classify such behavior as illegal harassment. But the fact that we need such laws points up the degree of conflict we often experience in performing our roles.

Role and Status in Modern Societies

Roles in Conflict Gesellschaft forms of social organization are more complex than gemeinschaft forms in terms of the number of statuses people hold and, thus, the number of roles they must perform. One result of this greater complexity is that roles in secondary groups or associations often conflict with roles in primary groups like the family. Much of the stress of life in modern societies is due to the anxiety we experience as we attempt to balance the demands of our various roles. This anxiety is captured in the terms *role conflict* and *role strain.*

Role conflict occurs when, in order to perform one role well, a person must violate another important role. Parents who are also employees may experience this kind of conflict when their supervisors ask them to put in extra time, which cuts into the time they are able to spend with their children. The late Felicia Schwartz, a leader in

BOX 4.2 GLOBAL CHANGE AND U.S. SOCIETY

Social Structures in a Changing World

Why all the noise about globalization, global influences on social change, and so on? Are these merely fashionable buzz words, or are they central issues in our own lives and those of our neighbors and friends? To answer these questions we need only think about some of the most important changes occurring in the social groups and structures around us and try to see what, if anything, these changes have to do with major social forces acting on a global, rather than merely national, level. Consider the following major areas of change, just for starters:

- *The revolution in information technologies.* Personal computers linked to worldwide information networks did not exist when your parents were children. They are making it possible for manufacturers to locate production facilities wherever in the world they can find low labor costs and a favorable business environment. This accelerates the trend whereby other parts of the world become specialized in manufacturing while in the United States there is a great deal of advanced technological production and many people work in service industries.
- *The rapid flow of information.* Through television, the Internet, modern long-distance phone systems, and jet travel, the globe "shrinks." Events that occur in once remote places like Somalia or Kosovo become subjects of dinner-table discussion in middle America. People in India and China read about American society and long for the kinds of opportunities Americans enjoy.

We hear so much about trends like these that there is a risk of becoming bored by them. Yet sociologists recognize these trends as the underlying causes of a great many fascinating and often heart-wrenching social phenomena. For

example, if we look at two extremes of the recent wave of immigration to the United States we see evidence of the powerful effects of these global social changes. At one extreme is the illegal immigration of the very poorest and most exploited workers. Labor shortages at the lowest levels of the pay scale make it attractive for unscrupulous entrepreneurs to "import" what is essentially slave labor. Sociologist Peter Kwong (1998) has studied this trend as it is experienced by unskilled Chinese workers. The most notorious example of the illegal importation of indentured or quasi-slave labor, according to Kwong, was the work of professional immigration criminals, known as "Dragons," who attempted to land a freighter, the *Golden Venture,* in New York harbor in 1994. Instead, the smugglers' ship grounded on the ocean shore, and many of the hapless

research on women, work, and social policy, argued that many corporations informally classify their female employees into two categories: those who seem to make their career their primary life concern, and those who seek to balance career and family obligations. Often they reward the former and subtly discriminate against the latter by placing them on what some have called the "mommy track," in which the woman is allowed more flexibility to perform family roles at the cost of promotion to more demanding positions within the company. The 1993 Family and Medical Leave Act attempts to resolve this conflict between work and family roles by requiring corporations to provide options for family leave without penalties; nevertheless, role conflict stemming from the competing demands of family and work remains widespread, especially for working women but for men as well (Cramer & Boyd, 1995; Siegel, 1992).

Research on how people behave in disasters also illustrates some of the effects of role conflict. For example, a police officer or firefighter who is on duty during a disaster that affects his or her own family may leave an assigned post to see to the safety of loved ones, thereby violating one role in order to fulfill another (Killian, 1952). Throughout the world efforts are being made to train women in disaster-related skills so that some of the conflicts that male disaster workers experience may be eased through greater sharing of roles (Gramling & Krogman, 1997; Sapir, 1997).

Role strain occurs when people experience conflicting demands in an existing role or cannot meet the expectations of a new one. For example, when she performs the duties of a triage nurse, Lori Jones experiences severe role strain. She has been taught that care must be provided to those in need, but she must balance

friends who have already immigrated to the United States, and to learn from them about areas of expanding educational and business opportunity (Mitra, 1997). For many years these nations have been exporting a significant proportion of their most highly educated and technically trained people to the United States, but this trend is changing in some important ways. There is still a steady stream of Chinese and Indian university graduates who immigrate to obtain further training in U.S. universities. Many of these skilled immigrants seek careers in the U.S. telecommunications and computer industries. But this immigration, often known as a "brain drain," is no longer a one-way process. In fact, many skilled Chinese and Indian computer programmers and computer scientists are returning to their own nations to begin startup firms of their own (Das & Srinivasan, 1997). No wonder, then, that there is a regular circulation of Asian technicians and businesspeople between Silicon Valley in California and the rapidly expanding information technology centers in Delhi, Hong Kong, and elsewhere in China and India (Carnoy, Castells, & Benner, 1997).

immigrants were drowned as they attempted to swim through the surf to freedom. If they had been successfully smuggled into the city, the Dragons would have kept them in low-wage jobs in restaurants and illegal manufacturing concerns (sweatshops) for many years while they paid off the cost of their transport at exorbitant rates.

At the other extreme of responses to the forces of globalization is immigration by the most educated and highly trained elites of Third World nations. In India and China, for example, people with access to computers and the Internet are able to stay in touch with their kin and

that role expectation against the fact that care is a scarce resource in a busy hospital and must be rationed (Chambliss, 1996; Goodman-Draper, 1995).

Role strain in the form of anxiety over poor performance is at least as common as role strain caused by conflicting expectations. For example, the unemployed head of a family feels severe stress as a result of inability to provide for the family's needs (Bakke, 1933; Jahoda, 1982; Jahoda, Lazarsfeld, & Zeisel, 1971). The mother of a newborn baby often feels intense anxiety over how well she can care for a helpless infant, a feeling that is heightened if she herself is young and dependent on others (Mayfield, 1984).

Ascribed Versus Achieved Statuses Role conflicts may occur in simpler societies, but they are far more common in societies undergoing rapid change. One reason for the relative lack of role conflict in simpler, more stable societies is that in such societies a person's statuses are likely to be determined by birth or tradition rather than by anything the person achieves through his or her own efforts. These **ascribed statuses** (peasant, aristocrat, slave, and so forth) usually cannot be changed and hence are less likely to be subject to different role expectations. Such statuses are found in industrial societies too (statuses based on race or sex are examples), but they tend to become less important in modern institutions than **achieved statuses** like editor, professor, or Nobel Prize winner.

We expect to be able to achieve our occupational status, our marital and family statuses, and other statuses in the community and the larger society. Nevertheless, there is some tension in modern societies between the persistence of ascribed statuses and the ideal of achieved statuses. The empirical study of that tension and of efforts to replace ascribed with achieved statuses (e.g., through equality of educational opportunity) is another major area of sociological research. We will show in later chapters that many of the social movements studied by sociologists arose as different groups in a society (e.g., African Americans, wage workers, immigrants, and the poor) organized to press for greater access to economic and social institutions like corporations and universities.

Master Statuses One reason so many groups have had to organize to obtain social justice is related to the way statuses operate in many societies. Although any person may fill a variety of statuses, many people find that one of their statuses is more important than all the others. Such a status, which is termed a **master status,** can have very damaging effects (Hughes, 1945). A black man, for example, may be a doctor, a father, and a leader in his church, but he may find that his status as a black American takes precedence over all of those other statuses. The same is often true for women. A woman may be a brilliant scientist and a leader in her community, and fill other statuses as well, but she may find that when she deals with men her status as a woman is more important than any of the others.

The effects of a master status are also felt by people who have been in prison, by members of various racial and ethnic groups, by Americans overseas (Smith, 1994, 1996), and by members of many other types of groups. For example, recent studies have shown that people who are noticeably overweight or obese often find that their status as a "fat person" is a master status that denies them opportunities to be appreciated for their performance in other statuses, such as student or worker.

Patterns of discrimination and prejudice stemming from the problem of master statuses are not easily corrected. We will see at many points in this book that a major source of social change is the efforts of some groups to eliminate those patterns. But when changes actually occur in a society they often take the form of new national laws, such as laws against racial segregation or gender discrimination. This raises the question of the distinction between societies and nations.

SOCIETIES AND NATION-STATES

For most people in the world today, the social entity that represents society itself is the nation-state. The assumed correspondence between society and nation can be seen in the fact that expressions like "the United States" and "American society" are often used interchangeably. Moreover, most people think of their society in terms of national boundaries; thus, if you were asked to name the society of which you are a member, you would be more likely to say "the United States" than "California" or "Chicago" or "the University of Texas." Yet as we will see shortly, societies and nations are by no means the same thing.

The State

To understand the distinction between society and nation, we need to begin with the concept of the state. In a lecture at Munich University, Max Weber described a state as

> a human community that (successfully) claims *the monopoly of the legitimate use of physical force* within a given territory. . . . The right to use physical force is ascribed to other institutions or to individuals only to the extent to which the state permits it. The state is considered the sole source of the "right" to use violence. (quoted in Gerth & Mills, 1958, p. 78; emphasis in original)

The **state** thus may be defined as a society's set of political institutions—that is, the groups and organizations that deal with questions of "who gets what, when, and how" (Lasswell, 1936). The **nation-state** is the largest territory within which those institutions can operate without having to face challenges to their sovereignty (their right to govern). Weber was careful to note that the state has a monopoly over the use of force, which under certain circumstances it grants to other agencies (e.g., state and municipal governments). But the state gains this right—the source of its power to influence the behavior of citizens—from the people themselves, from their belief that it is *legitimate* for the state to have this power. As we will see in later chapters, the concepts of power and legitimacy are essential to understanding the workings of the modern state; indeed, power is a basic concept at all levels of human social behavior. Here, however, we must examine the idea of a state as it operates in the concept of nation-state or, simply, nation.

The Nation-State

"One nation, under God, indivisible, with liberty and justice for all." We can say the words in our sleep. We do not usually give much thought to the significance of what we are doing when we repeat the Pledge of Allegiance. Yet repeating the pledge is a highly significant action: It enhances the legitimacy of the state and thus helps create the nation—in this case, the nation known as the United States of America.

But do all the inhabitants of the United States of America think of themselves as members of one nation? To a large extent they do, and this is one of the greatest strengths of that nation. Yet at various times in American history certain groups—African Americans, Native Americans, and the Amish, for example—have thought of themselves as separate peoples. And the idea of an "indivisible" nation was fought out in one of the bloodiest wars the world has ever known, the American Civil War of 1861–1865.

In the century since the Civil War, the United States has not experienced any real tests of its national solidarity. Countries like Canada, Lebanon, Zimbabwe, South Africa, Iran, Northern Ireland, and the nations that were formerly part of Yugoslavia—to name only a few—have been far less fortunate. For them, the issue of creating a national identity that can unite peoples who think of themselves as members of different societies remains a burning question.

Between 1945 and 1970, as peoples all over the world adjusted to the breakup of the European colonial empires, more than 100 new states were created, many of them in the poorer regions of Africa and Asia. In these new nations, as in many of the nation-states that existed before the 1950s, the correspondence between national identity and society is often problematic, a situation that frequently produces social upheavals. In Nigeria, for example, the stability—indeed, the very existence—of the nation-state is endangered by animosities among the various societies included within its boundaries. Although the nation has rich oil resources and is the largest of the sub-Saharan nations, its per capita income fell to $250 in 1993 from $1,000 in 1980. This decline is due largely to conflicts among the major tribal groups and the failure of the nation-state's leaders to overcome those conflicts. Although there are scores of tribal and ethnic groups in the nation's population, the largest are the Hausa (21 percent), the Yoruba (20 percent), the Ibo (17 percent), and the Fulani (9 percent), each of which is afraid that the others might gain power at their expense (Olojede, 1995). The political instability of the nation is worsened by the ever-present threat of military dictatorship, which is to some degree a consequence of these ethnic divisions. (See the Mapping Social Change feature on pp. 112–113.)

South Africa provides another example of differing levels of national and social identity. In 1994 the minority white government ended its policy of apartheid (racial separation) and yielded political power to the black majority under the leadership of Nelson Mandela and the formerly banned African National Congress. But black South Africans themselves often experience conflicts derived from differing tribal identities. The larger tribal populations, such as the Zulus, have their own kings and political leaders. Some are afraid that the incorporation of their people into a modern nation-state will cause them to lose their identity and rights as Zulus. As a result, there have been numerous episodes of conflict and violence between Zulu separatists and the members of the African National Congress.

The lack of a clear match between society and nation can be seen in the case of entire groups who think of themselves as "a people" (e.g., the Ibo, the Jews, Native Americans, the Kosovars) as well as in smaller groups. You and your friends are part of "American society," by which we mean the populations and social structures found within the territory claimed by the nation-state known as the United States of America. But you are also part of a local community with smaller structures and more gemeinschaft or primary relationships. Indeed, this community level of society may have greater meaning for you in your daily life than society at the national level.

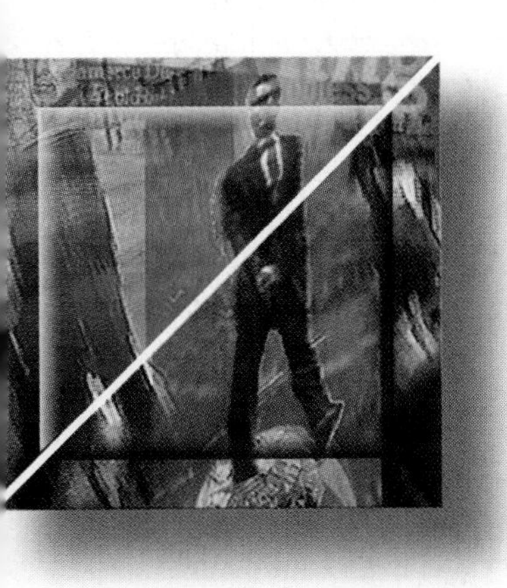

RESEARCH ON THE CUTTING EDGE: Global Immigration to the United States

Throughout the world, one of the most difficult problems people face in their communities and societies is that of accommodating cultural diversity. The outbreaks of racism and ethnic hatred that occur in communities in the United States are similar in their origins, if not in their intensity, to those occurring in places like Kosovo. In each case religious, racial, or ethnic divisions in the population of the nation-state create animosities that sometimes escalate into acts of cruelty and outbreaks of violence. In the United States, the origins of cultural diversity lie in the fact that throughout its history the nation has attracted immigrants from many other countries, making it a "nation of immigrants." The

MAPPING SOCIAL CHANGE

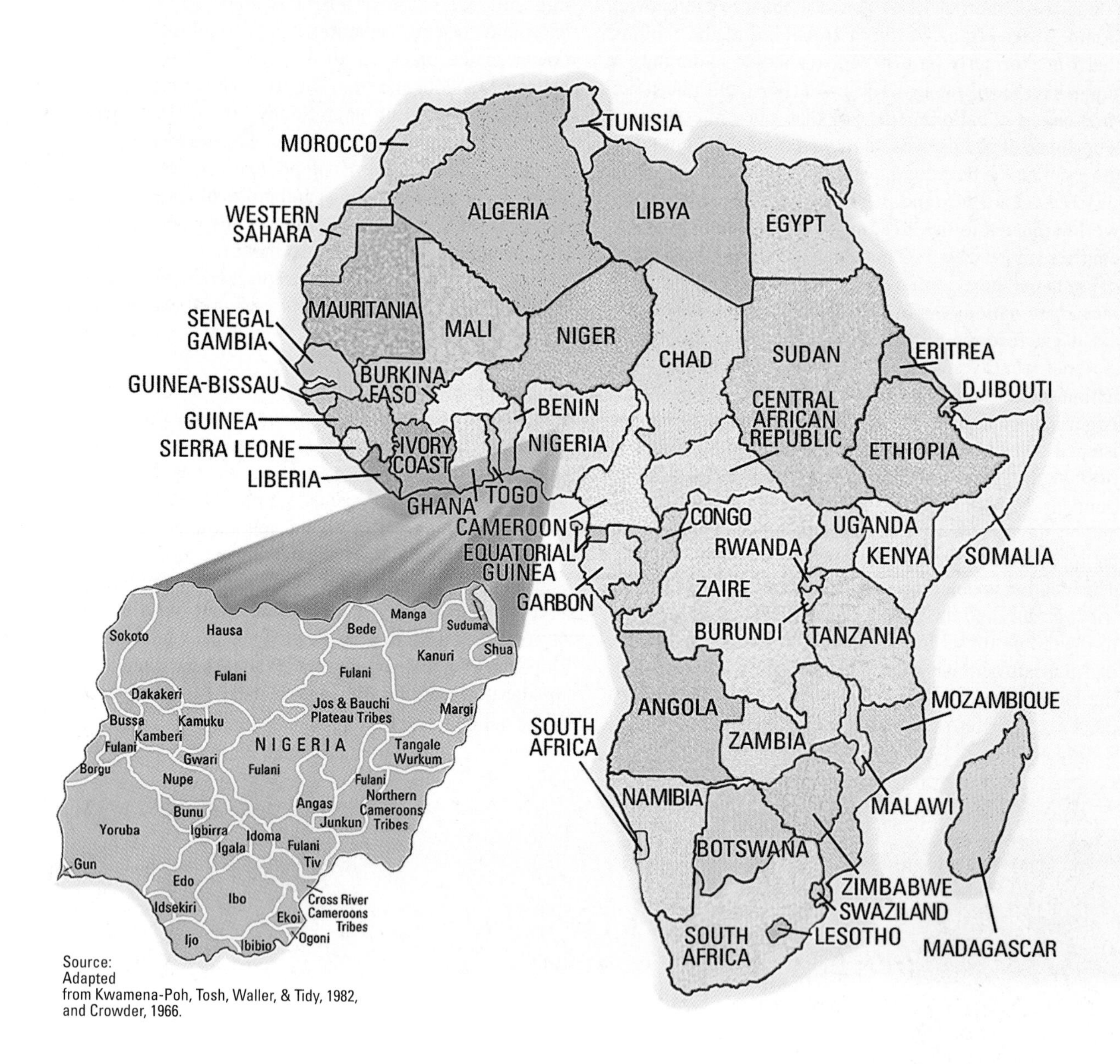

Source: Adapted from Kwamena-Poh, Tosh, Waller, & Tidy, 1982, and Crowder, 1966.

Tribal Societies Within the Modern State of Nigeria

Nigeria is the largest nation of Africa south of the Sahara desert. As the detailed map shows, it is composed of many societies with their own cultures and languages. Some, like the Hausa, are spread out over vast territories in the northern savannas; others, like the Tiv and Ibo, are clustered in the more densely populated coastal plains and on the delta of the mighty Niger River. Nigeria was once considered a leader in the development of democratic political institutions. It had a lively cultural and intellectual life and produced some of Africa's best novelists and playwrights. In 1998, after many years of dictatorship, the Nigerian military began to relinquish its harsh rule over the nation, but at this writing the persistence of regional and ethnic animosities continue to hamper the emergence of national leadership. Conflict between its many societies, often over the disposal of oil revenues from the Niger delta region, and rampant political and economic corruption continually threaten the viability of this nation-state's democratic institutions.

Ken Saro-Wiwa, shown in the photo, was one of the leaders of the effort to create a modern Nigerian nation even as he fought for the well-being of his own region. He is now considered to be a martyr of Nigeria's democratic hopes. Despite protests throughout the world, this celebrated writer, environmental activist, and political dissident was executed by the Nigerian dictatorship late in 1995. Saro-Wiwa was the founder and leader of the movement for the survival of the Ogoni People in Nigeria's oil-rich Niger delta region. Although the government and the Dutch Shell Corporation had taken billions of dollars' worth of oil from their lands, the Ogoni, a small tribal society in the Cross River region, still live in mud huts and dig for yams with bamboo sticks. Acid rain and oil pollution have destroyed many productive farms. Saro-Wiwa was attempting to organize a peaceful social movement to protest these conditions. His arrest and execution is an example of how environmental issues increasingly become confounded with intergroup politics in this land of tribal conflict.

waves of new immigrants often had difficulty adjusting to U.S. society, and their more established neighbors tended to resent the newcomers and to react with hostility.

Given its global importance, it is no wonder that ever since the founding of sociology the study of immigration and its consequences has held a central place in sociological research. This has been particularly true in the United States. Many of the research questions explored by the Chicago school sociologists had to do with the fate of different immigrant and migrant groups in the cities. Would the newcomers get decent jobs? Would they be able to retain their native languages while they learned English? Were they more prone to criminal behavior and mental illness than more established, native-born populations? Would they compete for jobs with other less advantaged groups, such as African Americans and native-born Hispanics? Would they form segregated ethnic enclaves with others from the same background? Would they eventually vote as an ethnic bloc? These and many other questions about the newcomers were at the core of empirical research in sociology during the early decades of this century. Today, because of the rapid pace of new immigration, these issues are once again on the frontier of sociological research.

One of the foremost students of immigration and its impact on American society is Alejandro Portes of Johns Hopkins University. In *Immigrant America* (Portes & Rumbaut, 1996) Portes shows that, for the first time since before World War I, the United States "has become anew a nation of immigrants." Each year during the 1980s an average of about 600,000 immigrants and refugees were legally admitted into this country, and "a sizable if uncertain number of others enter and remain without legal status, clandestinely crossing the borders or overstaying their visas" (p. xvii). Unlike earlier flows of immigrants, the vast majority of today's newcomers are from Asia and Latin America.

About 40 percent of the 6 million newcomers who came to the United States during the 1980s and 1990s settled in the six cities shown in Figure 4.5. This figure shows data for a single year, but if each of the numbers given is multiplied by 10, the result is a rough estimate of the number of people from the major recent immigrant groups who came to these cities during the decade. In fact, as Portes and his coauthor Rubén Rumbaut observe, people from more than 100 different nations arrived in the United States during this period—although the largest groups are those shown in the figure, and the preferred places for first settlement by these immigrants are the cities shown. Figure 4.6 shows the principal destinations of the eight largest immigrant groups.

Research by Portes and his colleagues shows that in Miami and other major U.S. cities the new immigrants are revitalizing old urban neighborhoods. Although concern over their growing numbers and

FIGURE 4.5 Composition of Immigrant Flows to Six Major Metropolitan Destinations in a Typical Year

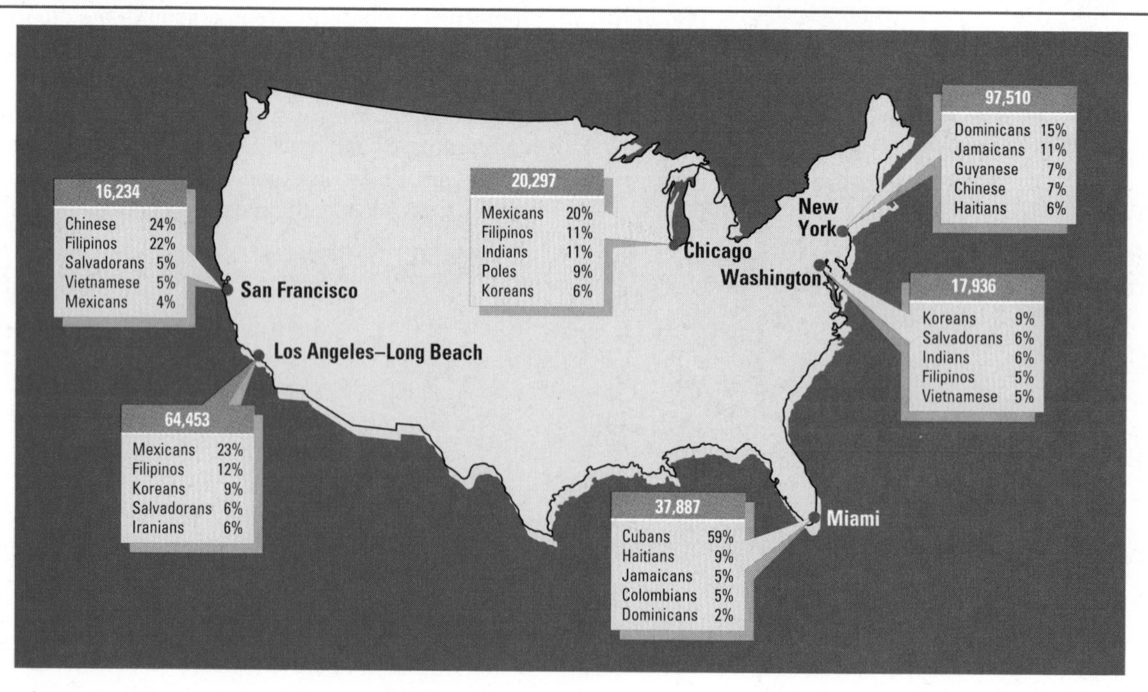

Note: Chinese include immigrants from mainland China only.

Source: Adapted from Portes & Rumbaut, 1996.

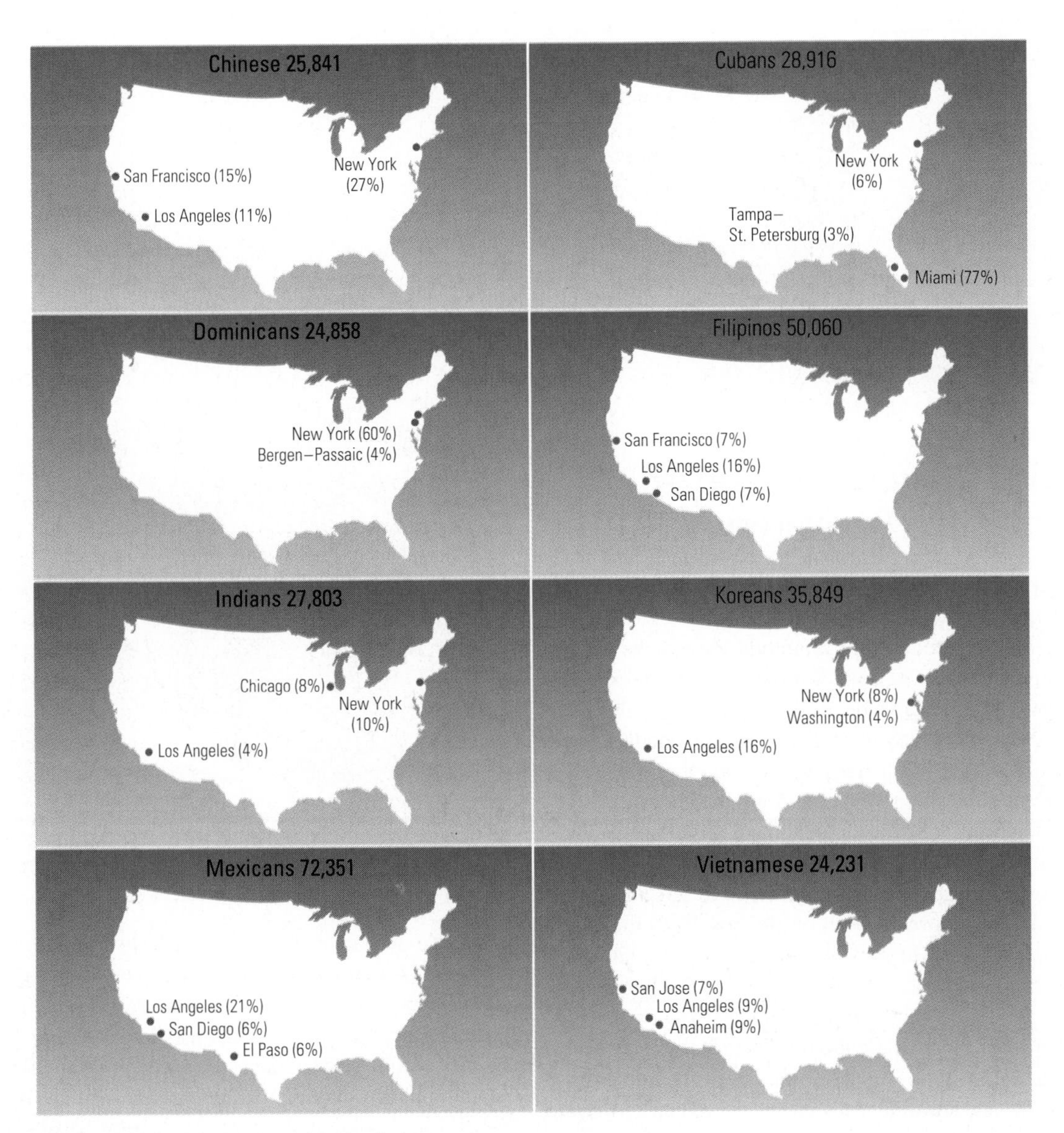

Note: Chinese include immigrants from mainland China only.

Source: Adapted from Portes & Rumbaut, 1996.

FIGURE 4.6 **Principal Destinations of the Eight Largest Immigrant Groups in a Typical Year**

economic influence has caused anti-immigration sentiment and demands for lower immigration quotas, sociological research shows that once they achieve legal status and stable living conditions, the vast majority of immigrants become highly productive citizens.

Throughout the world, between the beginning of World War I in 1914 and the end of the 1970s, an estimated 120 million people were forced to move from one nation to another because of either political violence or severe economic disruption (Kincaid & Portes, 1994; Zolberg, 1981). Being uprooted in this way means leaving behind old statuses and roles, giving up a way of life, and attempting to re-create these statuses and roles in a new society. It is little wonder, then, that patterns of migration and immigration, and their social consequences, have figured so centrally in the history of sociology.

VISUAL SOCIOLOGY

Social Structures and the Inner-City Emergency Room

After several years as a clinical specialist and triage nurse, Lori Jones knows that to be happy in her roles and balance all the conflicting demands they make on her she must do some serious soul-searching. As she explains it, "Actually, I've never done anything really great here, never saved anyone's life. The paramedics have. Yet patients have been lucky to have me as a nurse. No, I've never done things that are dramatic. I mean, I was happy because patients took their medicine as directed for 10 days. That's a feat in itself."

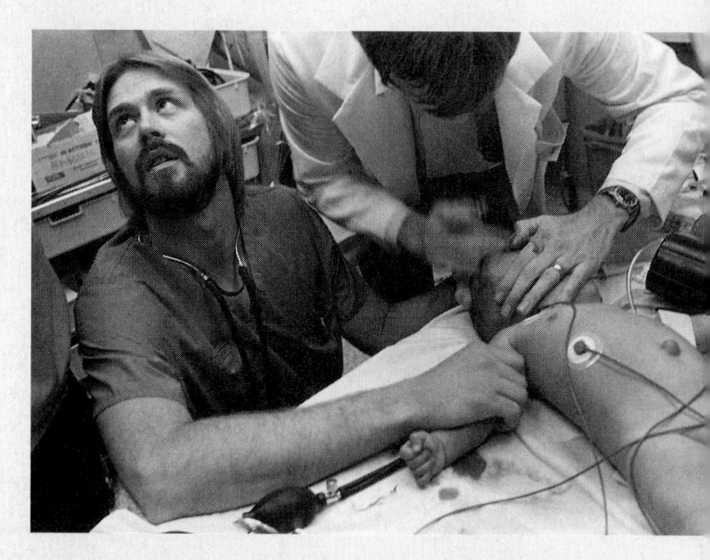

Jones also knows that the burdens of life in the ER are made even more difficult by the hospital's inner-city location, the need to provide services to so many poor people and minorities, and the fact that it is a pubic institution, dependent on a diminishing stream of tax revenues. "This is a city hospital on the edge," she observes, "losing money on indigent patients, and all around us is a massive system that has to raise more and more money, while politicians are threatening to close us down. So we are stretched incredibly thin, torn in so many different directions. It can be overwhelming."

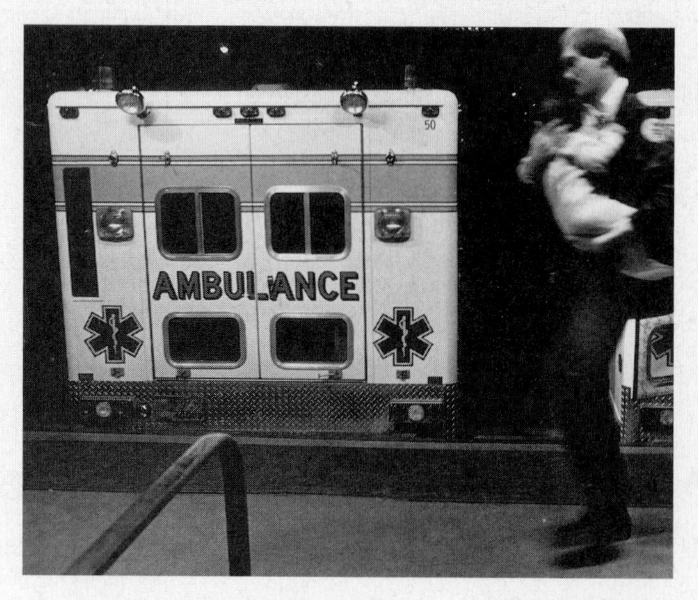

The emergency room of a big-city hospital is not a place for the squeamish observer. The deadening calm of the night shift may suddenly be broken by the crackle of the paramedics' radio, announcing that they are bringing in a severely wounded gunshot victim or a baby with a deadly high fever or three bleeding gang fighters. Immediately the routine work of the ER is suspended in a rush of activity. Even the less seriously ill or injured patients in the waiting room intuitively realize that their needs must be set aside for the life-threatening emergency.

Documentary photographer Eugene Richards (1989) spent many months conducting participant observation research in the emergency room of Denver General Hospital, known by local residents as the "knife and gun club." His photos capture the physically and emotionally draining work of the ER staff and the stark drama of the emergencies they deal with. In the text accompanying the photos, he often refers to the social structure of the ER. The roles of the ER staff are well defined, so that even if the work stretches them to the limits of endurance they know what they must do and how they must assist each other. Indeed, underneath its drama and emotion the ER is a marvel of human social organization, without which it could not be effective.

A man and a woman are the only people in the waiting room. Their child, I was told, has spinal meningitis. I sit close to them, but I can't think of anything to say. The silence is deadening. I feel frozen in place. I hide my camera and sit there for a half hour, until I see an ambulance pull up the ramp.

It's early in the day, I'm told, for a suicide attempt. The woman is 35 years old, still alive. There really is something fishlike about her, naked on the table with all those lines attached to her skin. And her skin is pale as ice. And her mouth is gaping open. . . .

5:00 P.M. The patient is a big-bellied guy with an American eagle on his left arm and a tongue hanging out tattooed to his chest. Insensible and bloody now, he had crashed his car at 80 miles per hour.

5:30 P.M. A teenage boy, claiming to be a member of a gang, shows the doctors the scars on his belly. He brags that he had refused medical treatment when he was stabbed, though he now looks terribly frightened. When a doctor reaches for his arm, he grimaces and tries to slide off the table. As the IV needle is slid in, he bites down on a rubber bottle cap, imitating something he must have seen in a John Wayne movie.

The chief surgeon leans forward and begins the cut. He cuts through the skin, looks, touches, cuts again. The other doctors join in. I'm conscious of the clicking of scissors, and that I'm breathing way too fast. Then, my God, they split the breastbone and position a spreader to push the ribs apart. It's done. They get to her heart, a punctured, failing heart.

At 1:00 A.M. the surgeons straighten up and turn all together towards the clock. They agree as to the time of death and pronounce her. After shooting some Polaroid pictures, they hurry out the door, probably to another patient. The nurses left behind turn off the machines, cut loose the lines and drainage tubes, and toss the surgical instruments into a pile; the interns close the dead woman's chest.

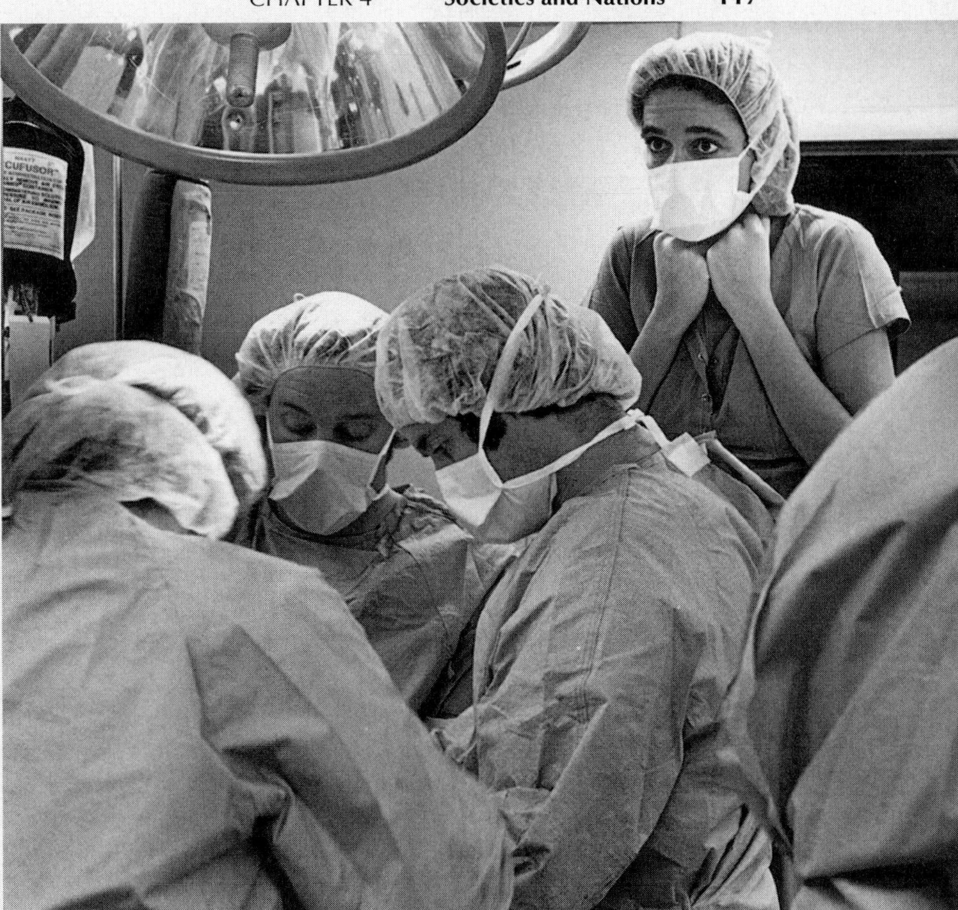

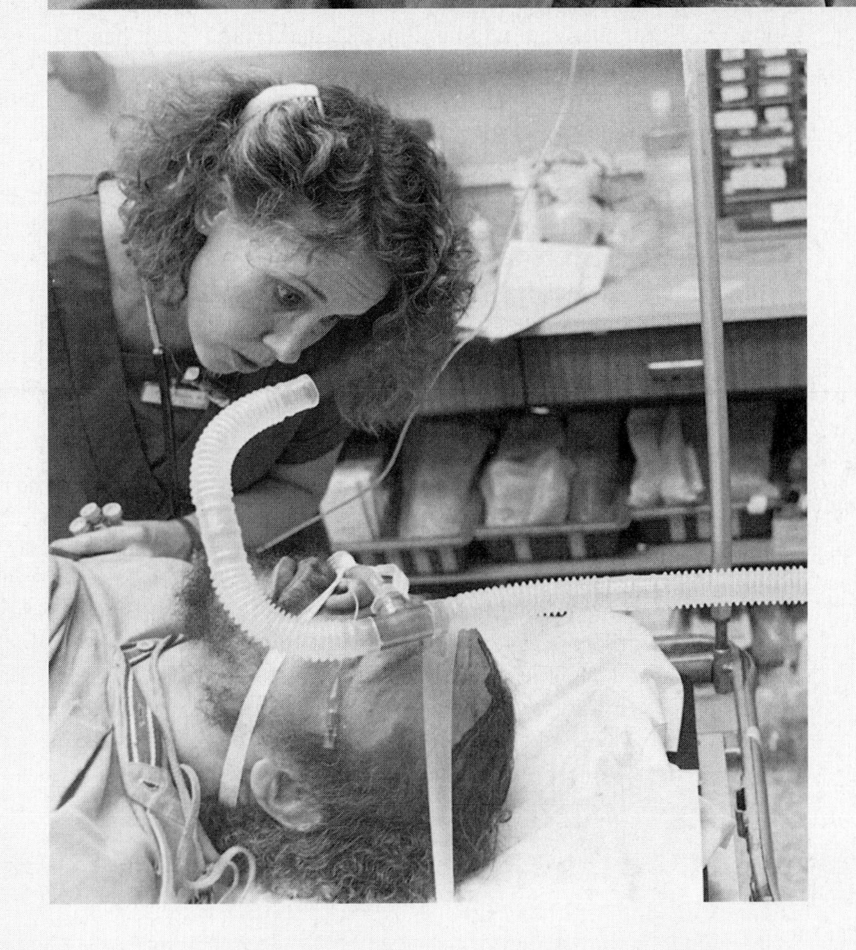

SUMMARY

A *society* is a population of people or other social animals that is organized in a cooperative manner to carry out the major functions of life. *Social structure* refers to the recurring patterns of behavior that create relationships among individuals and groups within a society. The building blocks of human societies are *groups*—collections of people who interact on the basis of shared expectations regarding one another's behavior. In every group there are socially defined positions known as *statuses*. The way a society defines how an individual is to behave in a particular status is called a *role*.

A social *institution* is a more or less stable structure of statuses and roles devoted to meeting the basic needs of people in a society. Within any given institution there are norms that specify how people in various statuses are to perform their roles. New institutions continually emerge through the process of *differentiation*.

The growth of the world's population is directly related to the evolution of human social structures, which in turn is related to changes in production technologies. The first million years of human social evolution were characterized by a hunting-and-gathering way of life. During that time the family and other kinship structures evolved, and the band became the basic territorial unit of human society. The shift to agriculture is commonly linked with the invention of the plow, but for centuries human societies had been acquiring food through the domestication of plants and animals. Some became *pastoral societies* based on the herding of animals, while others evolved into *horticultural societies* based on the raising of seed crops. However, the first large-scale agrarian societies evolved after the development of plow-and-harvest agriculture.

Agrarian societies allow people to escape from dependence on food sources over which they have no control. In such societies, people produce surpluses that can be used to feed new classes of non-food-producers such as warriors. At the same time, these societies require increasing amounts of land, and this may lead to conflicts over territory. The need to store and defend food supplies and to house non-food-producers results in the growth of villages and small cities.

The next major change in human production technologies was the industrial revolution, the shift from agriculture to trade and industry. This began in England around 1650 and spread to the United States and other nations during the next two centuries. Its impetus came not only from technological advances but also from the rise of a new social order: *capitalism*.

The shift to industrial production affects social structure in several major ways. As a consequence of the industrialization of agriculture, relatively few people live on the land, whereas increasing numbers live in cities and suburbs. Greater openness to change results in the emergence of new classes and social movements. Scientific and technical advances produce tremendous wealth, and the world "shrinks" as a result of innovations in transportation and communication.

For the individual member of a human society, adaptation to a more modern society involves a shift from *gemeinschaft* (close, personal relationships) to *gesellschaft* (well-organized but impersonal relationships). *Primary groups* like the family are supplemented, if not replaced, by *secondary groups* (organizations or associations) whose members do not have strong feelings for one another. Roles in secondary groups often conflict with roles in primary groups, a situation known as *role conflict. Role strain* occurs when a person experiences conflicting demands within a single role.

Another difference between simpler and more advanced societies is that in the former almost all statuses are *ascribed* (determined by birth or tradition), whereas in the latter there is a tendency to replace ascribed statuses with ones that are *achieved* (determined by a person's own efforts). Sometimes a particular status takes precedence over all of an individual's other statuses; such a status is referred to as a *master status*.

When they think of a society, most people in the world today think in terms of the nation-state, or nation, of which they are members. (A *state* is a society's set of political structures, and a *nation-state* is the territory within which those structures operate.) But although the members of a society often think of themselves as members of a particular nation, this is not always so, and in extreme cases the lack of a clear match between society and nation can result in a civil war.

SOCIOLOGY VERSUS IDEOLOGY

Many people mourn the loss of close-knit communities in which people were said to care for each other and help each other over difficult times. Frequently, but not always, they consider themselves conservatives and wish to recapture lost values of community solidarity. They may identify large metropolitan regions and their institutions (especially universities, major corporations, and government agencies) as the problem. With their emphasis on efficiency, profit, science, and secular values, these institutions are often blamed for their role in weakening traditional communities and the families within them. People on the more liberal side of the ideological continuum often feel that small, cohesive communities tend to stifle creativity and individual expression. They may yearn to leave the small town or community for what they perceive as the greater freedom of the city or metropolitan region. And while we may associate this familiar conflict with our own society, in fact we find similar conflicts throughout the world.

A sociological understanding of societies and social change can help resolve the conflict. Sociologists study the historical social forces that disrupt communities by changing the way people earn their livelihood or altering the way they travel or communicate. They study how the growing interdependencies of societies and communities may weaken our attachments to one another at the local level as we become more mobile. Does this sociological understanding mean that sociologists approve of all the changes that are occurring? Does it mean that they condone the weakening of local communities? Not at all. But in order to address the problems of communities and help those who wish to strengthen interpersonal attachments, sociologists first seek to know which social forces will work against these efforts and which ones will be of assistance. In fact, a sociological understanding of one's community and how it is situated in the global environment can actually help the community survive. For example, much sociological assessment goes into a community's efforts to replace declining industries like mining or heavy manufacturing by competing for new industries. In the contemporary global environment, one of the best ways to preserve communities is to strengthen the local institutions, such as community colleges, that help people adapt to change even as they assert their desire to maintain traditional values.

GLOSSARY

society: a population that is organized in a cooperative manner to carry out the major functions of life. (p. 92)

social structure: the recurring patterns of behavior that people create through their interactions, their exchange of information, and their relationships. (p. 93)

group: any collection of people who interact on the basis of shared expectations regarding one another's behavior. (p. 93)

status: a socially defined position in a group. (p. 93)

role: the way a society defines how an individual is to behave in a particular status. (p. 95)

role expectations: a society's expectations about how a role should be performed, together with the individual's perceptions of what is required in performing that role. (p. 95)

institution: a more or less stable structure of statuses and roles devoted to meeting the basic needs of people in a society. (p. 97)

differentiation: the processes whereby sets of social activities performed by one social institution are divided among different institutions. (p. 98)

pastoral society: a society whose primary means of subsistence is herding animals and moving with them over a wide expanse of grazing land. (p. 101)

horticultural society: a society whose primary means of subsistence is raising crops, which it plants and cultivates, often developing an extensive system for watering the crops. (p. 102)

social stratification: the process whereby the members of a society are sorted into different statuses. (p. 102)

open society: a society in which social mobility is possible for everyone. (p. 102)

closed society: a society in which social mobility does not exist. (p. 102)

capitalism: a system for organizing the production of goods and services that is based on markets, private property, and the business firm or company. (p. 106)

gemeinschaft: a term used to refer to the close, personal relationships of small groups and communities. (p. 107)

gesellschaft: a term used to refer to the well-organized but impersonal relationships among the members of modern societies. (p. 107)

primary group: a small group characterized by intimate, face-to-face associations. (p. 107)

secondary group: a social group whose members have a shared goal or purpose but are not bound together by strong emotional ties. (p. 107)

role conflict: conflict that occurs when in order to perform one role well a person must violate the expectations associated with another role. (p. 107)

role strain: conflict that occurs when the expectations associated with a single role are contradictory. (p. 108)

ascribed status: a position or rank that is assigned to an individual at birth and cannot be changed. (p. 109)

achieved status: a position or rank that is earned through the efforts of the individual. (p. 109)

master status: a status that takes precedence over all of an individual's other statuses. (p. 110)

state: a society's set of political structures. (p. 110)

nation-state: the largest territory within which a society's political institutions can operate without having to face challenges to their sovereignty. (p. 110)

WHERE TO FIND IT

BOOKS

Comparative National Development (Douglas Kincaid and Alejandro Portes; University of North Carolina Press, 1994). A comprehensive review of the changing social structures of the world's nations, with a balanced approach to economic, demographic, and cultural variables.

The Consequences of Modernity (Anthony Giddens; Stanford University Press, 1990). A thorough treatment of the elusive but centrally important concept of social structure.

Population and Food: Global Trends and Future Prospects (Tim Dyson; Routledge, 1996). An excellent overview of recent trends and research on the world's population and its changing capability to obtain adequate diets. Overall, the conclusions point to general improvement in the world's food situation, but there remain areas of grave concern owing to political instability, environmental degradation, and persistent malnutrition.

The Rise of the West (William K. McNeill; University of Chicago Press, 1970). Not as slanted toward Western civilizations as its title may suggest, this masterful history of the world's major civilizations is a valuable source of material on the social structures and ideas that shaped the modern world.

The Great Transformation: The Political and Economic Origins of Our Time (Karl Polanyi; Beacon, 1944). Essential reading on the sociological implications of the shift to a market economy during the eighteenth and nineteenth centuries.

An Embarrassment of Riches (Simon Schama; University of California Press, 1988). A brilliant cultural history of Dutch civilization with many implications for present-day sociology. Schama deals with the problem of how a society that achieved wealth rather early in its history also sought to share that wealth and to prevent its people from becoming profligate.

The Division of Labor in Society (Émile Durkheim; Free Press, 1964). Durkheim's classic treatment of the impact of the rise of industrial urban societies. The famous preface to the second edition presents a succinct statement of how social change forever alters older forms of community.

OTHER SOURCES

Historical Statistics of the U.S. From Colonial Times to 1970 (U.S. Bureau of the Census). An invaluable source of tables and charts showing changes in population, national origins, work, place of residence, and many other vital indicators of social change in the United States.

Statistical Yearbook (United Nations Educational, Scientific, and Cultural Organization [UNESCO]). Published annually, this is one of the best sources of comparative information about the world's nations and the trends occurring in them.

INTERNET RESOURCES

University of Michigan (www.umich.eduasq). Offers many resources and links to other sites dealing with issues of social structure, bureaucracy, and administration.

The History Chart (www.hyperhistory.com/online n2/History n2/a.html). Offers a good schematic rendering of the history of the world from the perspective of human experience.

PART 2

SOCIAL DYNAMICS

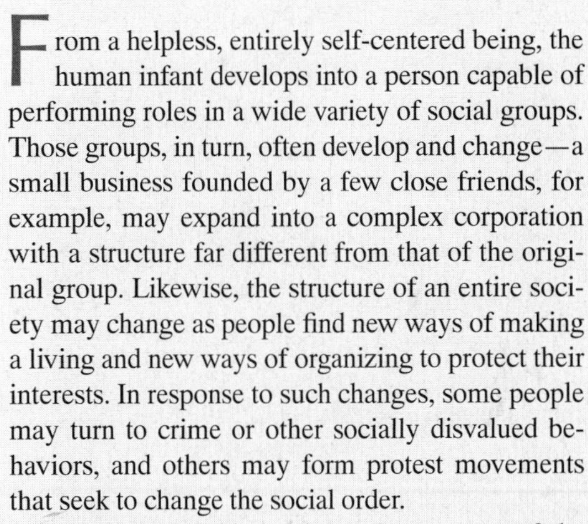

From a helpless, entirely self-centered being, the human infant develops into a person capable of performing roles in a wide variety of social groups. Those groups, in turn, often develop and change—a small business founded by a few close friends, for example, may expand into a complex corporation with a structure far different from that of the original group. Likewise, the structure of an entire society may change as people find new ways of making a living and new ways of organizing to protect their interests. In response to such changes, some people may turn to crime or other socially disvalued behaviors, and others may form protest movements that seek to change the social order.

In Part 2 of this book we cover many of the processes that seem to account for social stability and change. In Chapter 5 we discuss socialization, the process by which individuals become fully participating members of their society. One major outcome of socialization is the ability to function in groups, the subject of Chapter 6. Chapter 7 discusses how and why people deviate from the norms of their culture, and Chapter 8 looks at collective behavior and social movements, phenomena that involve large numbers of people in a society and often bring about major social change. Chapter 9 examines the changes in the nature of human settlements that have accompanied the growth of the world's population. Finally, Chapter 10 summarizes the ways in which sociologists think about and conduct research on the changes that affect the daily lives of people in societies throughout the world.

CHAPTER 5

SOCIALIZATION

His parents were professional educators who knew where to seek help. They had done everything they could to get their violent son on the right track. Yet despite their best efforts, Kip Kinkel of Springfield, Oregon, was suspended from school on a spring day in 1998 for hiding a gun in his locker. The next day the boy, age 15, walked into his school's cafeteria and began shooting at fellow pupils with a .22-caliber semiautomatic rifle. Two students died; many others were wounded. Once in custody, Kip attempted to stab a police officer with a knife he had concealed in his sock. Kip's rampage of violence, one of a rash of school shootings that have occurred since 1997, came shortly after the even more ghastly, and coldly premeditated, stalking and shooting of schoolchildren in Jonesboro, Arkansas, by two boys, one 11, the other 13 (Brandon, 1998).

What do these tragedies tell us about human behavior and social life? With the nation transfixed in horror and grief in the aftermath of these almost unimaginable murders, the broadcast media seek the opinions of experts, who are prompted to come up with explanations in the form of sound bites. Newspaper readers see headlines like this one from the *Chicago Tribune:* BIOLOGICAL MAKEUP CITED ON QUESTION OF WHY KIDS GO ON KILLING SPREES (McCall, 1998). Another headline, from the *St. Louis Post-Dispatch,* offers this alternative explanation: EXPERTS BLAME YOUTH VIOLENCE ON THE CULTURE, ACCESS TO

GUNS (Autman, 1998). The public is bombarded with articles about the perpetrators' family lives, seeking evidence of abuse, family violence, the effects of divorce or neglect, or any other experiences that might help us understand how children so young could act so brutally. There is also speculation about whether some of these juveniles displayed early signs of an innate propensity toward evil and violence: Did they torture animals? Were they reputed to be cruel to other children? These behaviors are often viewed as possible evidence of some kind of genetic predisposition to commit the unthinkable acts we seek to explain.

The school murders raise many questions beyond why some children fail to develop a moral sense or social conscience. All of us feel violent impulses on occasion. What keeps us from acting out our violent fantasies, especially when we are so often exposed to violence in our culture (through films, video, and music, for example)? What are the influences of family, schools, peers, and other social groups on these behaviors? And what can be done in families and in schools to identify children who are prone to violence, and to intervene before tragedy strikes?

Kip Kinkel, 15, was charged with four counts of aggravated murder in connection with the shooting deaths of two high school students and the deaths of his parents. Here he is being led to his arraignment in Eugene, Oregon.

We should remember, however, that these episodes of deadly violence are only part of a much larger problem of interpersonal violence among peers that can be seen throughout the world. In England, recent studies have exposed shocking levels of bullying of schoolchildren whom their peers suspect of homosexual feelings or propensities (Chaudhary, 1998). In Japan, a rash of child suicides due to school bullying has shocked that society just as the epidemic of school murders has stunned people in the United States (Fredman, 1995). But the school killings cannot help but make us wonder whether some individuals are simply "born evil." For social scientists these questions are of central importance. We understand, however, that there are no simple explanations. All investigations of extremely bad children call attention to the question of why and how the vast majority of us—even those of us who have grown up in quite stressful or deprived social environments—learn to behave as responsible members of society. (The Visual Sociology feature at the end of this chapter provides an inspiring look at how a group of homeless children have learned to capture their world, with its joys and hardships, in photos that also convey a great deal of sociological insight.)

BECOMING A SOCIAL BEING

Violence in schools is particularly disturbing to us because we think of schools as places we can trust to be safe. Families send their children to school so that they can be taught the rules of behavior in their culture. They hope that the school will help their children prepare for and even rehearse the roles they will eventually perform in their society, even if they also learn to think critically and question their parents' norms and values. This process, which occurs in families, schools, churches, neighborhoods, and elsewhere in society, and through which the culture of a society is passed on from one generation to the next, is the subject of this chapter.

Socialization is the term sociologists use to describe the ways in which people learn to conform to their society's norms, values, and roles. People develop their own unique personalities as a result of the learning they gain from parents, siblings, relatives, peers, teachers, mentors, and all the other people who influence them throughout their lives (Corsaro, 1997; Elkin & Handel, 1989). From the viewpoint of society as a whole, however, what is important about the process of socialization is that people learn to behave according to the norms of their culture. How people learn to behave according to cultural norms—that is, the way they learn their culture—makes possible the transmission of culture from one generation to the next. In this way the culture is "reproduced" in the next generation (Gonzalez-Mena, 1998; Parsons & Bales, 1955).

Socialization occurs throughout life as the individual learns new norms in new groups and situations. However, for purposes of analysis socialization can be divided into three major phases. The first is *primary socialization.* It refers to all the ways in which the newborn individual is molded into a social being—that is, into a growing person who can interact with others according to the expectations of society. Primary socialization occurs within the family and other intimate groups in the child's social environment. *Secondary socialization* occurs in later childhood and adolescence, when the child enters school and comes under the influence of adults and peers outside the household and immediate family. *Adult socialization* is a third stage, when the person learns the norms associated with new statuses such as wife, husband, journalist, programmer, grandparent, or nursing home patient (Elkin & Handel, 1989).

There are a number of unresolved and highly controversial issues in the study of socialization, and we will explore several of them in this chapter.

Nature and nurture. What is the relative strength of biological (i.e., genetic) versus social influences on the individual? This issue, often referred to as the nature-nurture problem, is raised most strikingly by the actions of Kip Kinkel and the other juvenile murderers. The "nature" side of the nature-nurture controversy asserts that there are genetic or other innate factors that explain these extreme failures of socialization. The "nurture" side of the debate seeks explanations in the children's social environments—their families, their peer groups, their schools, their involvement in television and other media. Social scientists tend to lean toward explanations based on nurture, but as we will see in this chapter, biology plays a significant role in behavior and in forming an individual's personality. Neither biological nor sociological factors alone explain the complex behaviors involved in successful socialization (Steen, 1996).

The social construction of the self. A second controversy in the study of socialization is the question of how a person's sense of self becomes established. We all learn to play many different roles—but how do the influences of others in our social world affect our role playing, and how do those experiences help form our sense of ourselves? How do people learn to conform to society's norms and to take the roles that society makes available to them?

Influences on socialization. How do different social environments, such as the affluent suburban school or the slum neighborhood or the military boot camp, influence socialization? In other words, how do different social environments produce different kinds of people? What are the influences throughout life of different agencies of socialization and different experiences with other people?

Gender socialization and sexual identity. Gender socialization refers to the ways in which we become the girls and boys and, gradually, the men and women of our society and culture. All the controversies over whether behavior is innate or learned are intensified when we consider the differences and similarities in the socialization of males and females. This is especially true in reference to the ways in which we are socialized to acquire a sexual identity.

In this chapter we explore each of these questions in detail. We will be concerned primarily with the socialization of "normal" members of society—people who are able to perform roles, to feel empathy for others, to express emotions and yet control feelings that are antisocial, to nurture others and raise children who will also be able to nurture, and to take on new roles as they grow older. But the failures of socialization can also tell us a great deal about what is involved in creating the social being. For instance, consider the case of Charles Manson, who was convicted for his part in the 1969 Tate–La Bianca murders. Not long before the murders, Manson had been released after serving a 10-year prison term. In fact, Manson, who was born in a state penitentiary, spent 17 of his first 35 years in more than a dozen penal institutions. His history is one of complete neglect. He was beaten with a heavy paddle, and he beat others. He was sodomized, and he sodomized others at knifepoint. In fact, he was so undersocialized for life outside prison that when he was due to be released he pleaded to be allowed to stay there.

The issue in cases like Manson's is not whether we should excuse what such individuals have done but whether we can learn from their tragedies so as to prevent others. And scientists, biological and social, still have a good deal to learn. Socialization is an extremely complex process. Some people who were abused and neglected can nonetheless become good parents, despite the odds. Others who seem to have experienced all the right influences can end up doing evil things, again despite the odds (Corsaro, 1997; Wrong, 1961). Even if we could trace all the social influences on a person's development, there would remain many unanswered questions about the combined influences of the person's genetic potential and his or her social experiences—that is, the relative importance of nature and nurture.

NATURE AND NURTURE

Throughout recorded history there have been intense debates over what aspects of behavior are "human nature" and what aspects can be intentionally shaped through nurture or socialization. During all the centuries of prescientific thought, the human body was thought to be influenced by the planets, the moon, and the sun, or by various forces originating within the body, especially in the brain, heart, and liver. In the ancient world and continuing into the Middle Ages, blood, bile, phlegm, and other bodily fluids were thought to control people's moods and affect their personalities. Human behavior and health could also be affected by evil spirits or witches. While they are no longer dominant explanations of human behavior, none of these ideas has ever entirely disappeared.

In the eighteenth century a new and quite radical idea of the "natural man" emerged. The idea that humans inherently possess qualities such as wisdom and rationality, which are damaged in the process of socialization, took hold in the imaginations of many educated people. Social philosophers like Jean-Jacques Rousseau believed that if only human society could be improved, people would emerge with fewer emotional scars and limitations of spirit. This belief was based on the enthusiasm created by scientific discoveries, which would free humans from ignorance and superstition, and new forms of social organization like democracy and capitalism, which would unleash new social forces that would produce wealth and destroy obsolete social forms like aristocracy.

In the United States, Thomas Jefferson applied these ideas in developing the educational and governmental institutions of the new nation. Much later, in the nineteenth century, Karl Marx and other social theorists applied the same basic idea of human perfectibility in their criticisms of capitalism. A revolutionary new society, they predicted, would overturn the worst effects of capitalism and finally realize Rousseau's promise that a superior society could produce superior people. As these examples indicate, many of the most renowned thinkers of the last two centuries have rejected the belief that nature places strict limits on what humans can achieve.

The Freudian Revolution

Sigmund Freud was the first social scientist to develop a theory that addressed both the "nature" and "nurture" aspects of human existence (Nagel, 1994; Robinson, 1994). For Freud, the social self develops primarily in the family, wherein the infant is gradually forced to control its biological functions and needs: sucking, eating, defecation, genital stimulation, warmth, sleep, and so on. Freud shocked the straitlaced intellectuals of his day by arguing that infants have sexual urges and by showing that these aspects of the self are the primary targets of early socialization—that the infant is taught in many ways to delay physical gratification and to channel its biological urges into socially accepted forms of behavior.

Freud's model of the personality is derived from his view of the socialization process. Freud divided the personality into three functional areas, or interrelated parts, that permit the self to function well in society.[1] The part from which the infant's unsocialized drives arise is termed the **id.** The moral codes of adults, especially parents,

[1] Freud never expected that actual physical parts of the brain that correspond to the id, ego, and superego would be discovered. Instead, he was referring to aspects of the functioning personality that are observed in the individual.

become incorporated into the part of the personality that Freud called the **superego.** Freud thought of this part of the personality as consisting of all the internalized norms, values, and feelings that are taught in the socialization process. In addition to the id and the superego, the personality as Freud described it has a third vital element, the **ego.** The ego is our conception of ourself in relation to others, in contrast with the id, which represents self-centeredness in its purest form. To have a "strong ego" is to be self-confident in dealing with others and to be able to accept criticism. To have a "weak ego" is to need continual approval from others. (The popular expression that someone "has a big ego" and demands constant attention actually signals a lack of ego strength in the Freudian sense.)

In the growth of the personality, according to Freud, the formation of the ego or social self is critical, but it does not occur without a great deal of conflict. The conflict between the infant's basic biological urges and society's need for a socialized person becomes evident very early. Freud believed that the individual's major personality traits (security or insecurity, fears and longings, ways of interacting with others) are formed in the conflict that occurs as the parents insist that the infant control its biological urges. This conflict, Freud believed, is most severe between the child and the same-sex parent. The infant wishes to receive pleasure, especially sexual stimulation, from the opposite-sex parent and therefore is competing with the same-sex parent. To become more attractive to the opposite-sex parent, the infant attempts to imitate the same-sex parent. Thus for Freud the same-sex parent is the most powerful socializing influence on the growing child, for reasons related to the biological differences and attractions between the child and the opposite-sex parent.

Contemporary sociologists who are influenced by Freud's biologically and socially based theories have used his concepts of same-sex attraction and imitation of the same-sex parent's behavior to explain differences between men and women. Alice Rossi (1977), for instance, argues that women's shared experience of menstruation and childbearing creates a strong bond between mothers and daughters. Nancy Chodorow (1978) claims that women's earliest experiences with their mothers tend to convince them that a woman is fulfilled by becoming a mother in her turn; thus women are socialized from a very early age to "reproduce motherhood." Research on socialization has shown that men also are strongly influenced by the same-sex parent. Fathers often serve as models of behavior whom boys will emulate throughout their lives.

Freud's theory includes the idea that the conflicts of childhood reappear throughout life in ways that the individual cannot predict. The demands of the superego ("conscience") and the "childish" desires of the id are always threatening to disrupt the functioning of the ego, especially in families in which normal levels of conflict are either exaggerated or suppressed. Note, however, that Freud focused on the traditional family, consisting of mother, father, and children. The more families depart from this conventional form, the more we need to question the adequacy of Freudian socialization theory.

Behaviorism

In the early decades of the twentieth century Freud's theory was challenged by a different branch of social-scientific thought, known as **behaviorism.** In contrast to Freud and others who saw many human qualities as innate or biologically determined (nature), behaviorists saw the individual as a *tabula rasa,* or blank slate, that could be "written upon" through socialization. In other words, individual behavior could be determined entirely through social processes (nurture). Behaviorism asserts that individual behavior is not determined by instincts or any other "hardware" in the individual's brain or glands. Rather, all behavior is learned.

Behaviorism traces its origins to the work of the Russian psychologist Ivan Pavlov (1927). Pavlov's experiments with dogs and humans revealed that behavior that had been thought to be entirely instinctual could in fact be shaped or **conditioned** by learning situations. Pavlov's dog, one of the most famous subjects in the history of psychology, was conditioned to salivate at the sound of a bell. The dog would normally salivate whenever food was presented to it. In his experiment, Pavlov rang a bell whenever the dog was fed. Soon the dog would salivate at the sound of the bell alone, thereby showing that salivation, which had always seemed to be a purely biological reflex, could be a conditioned, or learned, response as well.

The American psychologist John B. Watson carried on Pavlov's work with an equally famous series of experiments on "Little Albert," an 11-month-old boy. Watson conditioned Albert to fear baby toys that were thought to be inherently cute and cuddly, such as stuffed white rabbits. By presenting these objects to Albert at the same time that he frightened him with a loud noise (i.e., a negative stimulus), Watson showed that the baby could be conditioned to fear any fuzzy white object, including Santa Claus's beard. He also showed that through the systematic presentation of white objects accompanied by positive stimuli he could extinguish Albert's fear and cause him to like white objects again.

On the basis of his findings, Watson wrote:

> Give me a dozen healthy infants, well-formed, and my own specified world to bring them up in and I'll guarantee to take any one at random and train him to become any type of specialist I might select—doctor, lawyer, artist, merchant-chief and, yes, even beggar-man and thief, regardless of his talents, penchants, tendencies, abilities, vocations, and race of his ancestors. (1930, p. 104)

For the behaviorist, in other words, nature is irrelevant and nurture all-important.

Behaviorists who followed Watson—the most famous being B. F. Skinner—developed even more effective ways of shaping individual behavior. Skinner and his followers reasoned that in order to avoid failures in socialization it is necessary to completely control all the learning that goes on in the child's social environment (Skinner, 1976). Sociologists are critical of the notion that it is possible to control the world of the developing person. They argue that while the behaviorists may show us how some types of social learning take place, psychological research often does not deal with real social environments. It has very little to say about what is actually learned in different social contexts, how it is learned (or not learned), and the influences of different social situations on the individual throughout life (Elkin & Handel, 1989). One type of situation that has been of interest to sociologists studying socialization processes is that of the child reared in extreme isolation.

Isolated Children

The idea that children might be raised apart from society, or that they could be reared by wolves or chimpanzees or some other social animal, has fascinated people since ancient times. Romulus and Remus, the legendary founders of Rome, were said to have been raised by a wolf. The story of Tarzan, a boy of noble birth who was abandoned in Africa and raised by apes, became a worldwide bestseller early in this century and has intrigued readers and movie audiences ever since. However, modern studies of children who have experienced extreme isolation cast doubt on the possibility that a truly unsocialized person can exist.

The discovery of a **feral** ("untamed") **child** always seems to promise new insights into the relationship between biological capabilities and socialization, or nature and nurture (Davis, 1947). Each case of a child raised in extreme isolation is looked upon as a natural experiment that might reveal the effects of lack of socialization on child development. Once the child has been brought under proper care, studies are undertaken to determine how well he or she functions. Invariably those studies show that victims of severe isolation are able to learn, but that they do so far more slowly than children who have not been isolated (Malson, 1972).

In the late 1930s a noted sociologist, Kingsley Davis, was called to investigate the case of Anna, an 8-year-old girl who had been born out of wedlock to an impoverished and mentally impaired mother. For years Anna's mother had kept her locked up in an attic room to avoid her own father's anger. Aside from brief visits to deliver her food, Anna had almost no human contact. When Davis examined her she was unable to speak or walk, yet in her smile and obvious pleasure at human contact she seemed to show the potential to learn language and become socialized in other ways.

Anna did respond to Davis and others who tried to toilet train her and teach her to speak. But her progress was slow, and unfortunately she died at the age of 10 from a disease that may have been caused by her previous isolation. The researchers were unable to determine whether her learning deficiencies were due to innate mental retardation or whether her retardation was a result of her cruel isolation in infancy.

A more recent case of social isolation in childhood is that of Genie, a girl born to a psychotic father and a blind and highly dependent mother. For the first 11 years of her life Genie was strapped to a potty chair in an isolated room of the couple's suburban Los Angeles home. From birth she had almost no contact with other people. She was not toilet trained, and food was pushed toward her through a slit in the door of her room. When she was discovered by child welfare authorities after the mother told neighbors about the child's existence, the father committed suicide. Genie was placed in the custody of a team of medical personnel and child development researchers.

In the first few weeks after she was discovered, everyone who observed Genie was shocked. At first glance she looked like a normal child, with dark hair and pink cheeks and a placid demeanor. Very quickly, however, it became clear that she was severely impaired. She walked awkwardly and was unable to dress herself. She had virtually no language ability—at most, she knew a few words, which she pronounced in an incomprehensible babble. She spit continuously and masturbated with no sense of social propriety. In short, she was a clear case of a child who had been deprived of social learning and in consequence was severely retarded in her individual and social development. She was alive, but she was not a social being in any real sense of the term. She had not developed a sense of self, nor had she the basic ability to communicate that comes with language learning.

For years researcher Susan Curtiss (1977) studied Genie's slow progress toward language learning. Curtiss showed that Genie could learn many more words than a mentally retarded person would be expected to learn, but that she had great difficulty with the more complex rules of grammar that come naturally to a child who learns language in a social world. Genie's language remained in the shortened form characteristic of people who learn a language late in life. Most significant, Genie never mastered the language of social interaction. She had great difficulty with words such as hello and thank you, although she could make her wants and feelings known with nonverbal cues.

Extensive tests showed that in many ways Genie was highly intelligent. But her language abilities never advanced beyond those of a third-grader. Genie gradually

Then  and Now

The Indomitable Human Spirit

Blind and deaf as a result of a serious illness when she was only 19 months old, Helen Keller appeared incapable of learning. In the late nineteenth and early twentieth centuries people like Helen were relegated to back wards of custodial care institutions or to back bedrooms in affluent homes. However, through the patient instruction of a gifted teacher, Helen learned to understand the connection between objects and the words for them. Suddenly this profoundly handicapped girl had a future in the social world. Helen Keller became a world-famous symbol of the unconquerable human spirit, blazing a trail that other people with disabilities and special needs have followed ever since.

In a more recent example of the triumph of will over biological destiny, 19-year-old Craig Ludin is shown at his high school graduation. Craig was the first student with Down syndrome to graduate from a regular high school in his school district.

Genie drew this picture in 1977. At first she drew the picture of her mother and labeled it "I miss Mama." Then she drew more. The moment she finished, she took the researcher's hand and placed it next to what she had just drawn. She motioned for the researcher to write and said, "Baby Genie." Then she pointed under her drawing and said, "Mama hand." The researcher wrote the corresponding letters. Satisfied, Genie sat back and stared at the picture. There she was, a baby in her mother's arms.

learned to adhere to social norms (e.g., she stopped spitting), but she never became a truly social being. Eventually the scientists who worked with her came to the conclusion that the most severe deprivation, the one that was the primary cause of her inability to become fully social and to master language, was her lack of emotional learning and especially her feelings of loss and lack of love. Genie herself expressed her inner longings in the picture above, which she drew at the age of about 16. Never fully capable of independent living, Genie has spent her life in a home for developmentally disabled adults (Rymer, 1992a, 1992b).

The Need for Love

All studies of isolated children point to the undeniable need for nurturance in early childhood. They all show that extreme isolation is associated with profound retardation in the acquisition of language and social skills. However, they cannot establish causality, since it is always possible that the child may have been retarded at birth. Despite their lack of firm conclusions, studies of children reared in extreme isolation have pointed researchers in an important direction: They suggest that lack of parental attention can result in retardation and early death. This conclusion receives further support from studies of children reared in orphanages and other residential care facilities, which have shown that such children are more likely to develop emotional problems and to be retarded in their language development than comparable children reared by their parents (Koler & Freeman, 1994; Spitz, 1945).

In a series of studies that have become classics in the field of socialization and child development, the primate psychologist Harry Harlow showed that infant monkeys reared apart from other monkeys never learned how to interact with other monkeys (Harlow, 1986; Harlow & Harlow, 1962); they could not refrain from aggressive behavior when they were brought into group situations. When females who had been reared apart from their mothers became mothers themselves, they tended to act in what Harlow could only describe as a "ghastly fashion" toward their young. In some cases they even crushed their babies' heads with their teeth before handlers could intervene. Although it is risky to generalize from primate behavior to that of humans, these studies of the effects of lack of nurturance bear a striking resemblance to studies of child abuse in humans. This research generally shows that one of the best predictors of abuse is whether the parent was also abused as a child (Keniston, 1977; Ornish, 1998; Polansky et al., 1981; Talbot, 1998).

These findings confirm our intuitive knowledge that nurturance and parental love play a profound, though still incompletely understood, role in the development of the individual as a social being. These and related findings offer support for social policies that seek to enrich the socialization process with nurturance from other caring adults—for example, in early-education programs. Yet many researchers and policy makers remain convinced that biological traits place limits on what individuals can achieve, regardless of the kind of nurturance they receive. In recent years proponents of this view have focused on the controversial question of how intelligence affects achievement (see Box 5.1).

A Sociological Summary

Neither extreme of the nature-nurture debate presents a complete picture of socialization. Nature may endow individuals with greater or lesser innate abilities, yet despite those differences most people learn to function as social beings. And while humans have an infinite capacity to

BOX 5.1 USING THE SOCIOLOGICAL IMAGINATION

Is IQ Destiny?

In a study titled *The Bell Curve,* Richard Herrnstein and Charles Murray (1994) argue that inequality is increasing in the United States and other industrialized nations because of biological differences among various population groups in these societies. They claim that the growth in occupations requiring advanced education and technical skills is creating a demanding new economic environment in which many people are doomed to failure. They do not believe that efforts to address inequalities through preschool programs, improvements in public education, and the like will make a difference. The real obstacle, they argue, is intelligence, or rather the lack of it in vast numbers of people. Since by their reasoning intelligence is an innate trait that is distributed unequally among various subgroups of the population, social intervention cannot do much to equalize the economic effects of differences in intelligence.

Like many traits that vary from one person to another, such as height and weight, scores on intelligence tests are distributed in the shape of a "normal" or bell curve. Most people's IQ scores fall near the center of the distribution, while some are at either extreme. About these facts there is no argument. Murray and Herrnstein further assert, however, that scores on the bell curve of intelligence are creating what they call "cognitive classes"—that is, categories of people whom they label "very bright," "bright," "normal," "dull," and "very dull" (see graph). They note that 5 percent of the U.S. population falls within the left and right extremes of the curve, and another 20 percent are in Class II and Class IV. By this reasoning, approximately 50 million residents of the United States are classified in the lower cognitive classes.

Herrnstein and Murray also discuss the implications of the fact that the mean scores of some black and Hispanic population groups fall below those of whites and Asians. The authors attribute these differences to genetic factors rather than to the long-term effects of economic and social deprivation. We will return to some of the controversies and misunderstandings created by this research in Chapter 13. Here it will suffice to note that Herrnstein and Murray demonstrate a correlation between what they term cognitive classes (based on IQ scores) and poverty due to inability to succeed in a demanding economic environment (see table).

Most social scientists who have commented on Herrnstein and Murray's research are strongly opposed to the conclusions they reach, for several reasons. First, there has been much criticism of IQ as a single measure of intelligence. Many experts believe that intelligence is far too complex to be represented by a single measure like IQ (Gardner, 1983). Second, there is evidence of cultural and middle-class biases in the questions used to test IQ; examination of the test items reveals that they would be far more familiar to middle-class test takers than to those from a disadvantaged background. Third, the authors of *The Bell Curve* have been criticized for asserting that correlation is the same as causality. Just because IQ and poverty are correlated does not mean that IQ causes poverty. In Northern Ireland, for example, Catholic Irish individuals score lower on IQ tests than do Protestant Irish, but these differences in IQ are not found in the United States, where there are no differences in wealth or advantage between the two groups as there are in Northern Ireland. In other words, socialization in an impoverished environment could cause low IQ scores, rather than the other way around.

White Poverty, by Cognitive Class

Cognitive Class	Percentage in Poverty
I Very Bright	2
II Bright	3
III Normal	6
IV Dull	16
V Very Dull	30
Overall average	7

Source: Herrnstein & Murray, 1994.

Defining the Cognitive Classes

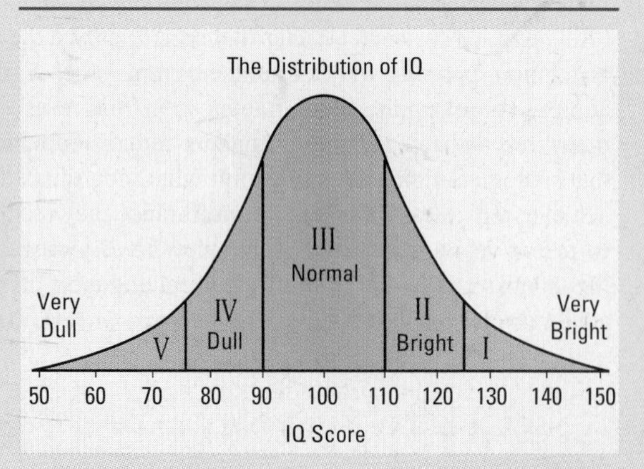

Caution: The labels imposed on this IQ curve and the scores used as boundaries between "cognitive classes" are those of Herrnstein and Murray and do not represent the thinking of many other social scientists.

Source: Herrnstein & Murray, 1994.

learn behaviors of all kinds, through socialization they learn the particular behaviors required to function within their own culture and society. Nevertheless, the nature-nurture debate will endure and will continue to stimulate new research on the biological and psychological bases of behavior. In the meantime, most sociological research will focus on the following hypotheses:

1. *The social environment can unleash or stifle human potential.* Genetic and other biological traits establish broad boundaries for individual achievement, but the environment in which a person is raised can cause his or her potential to be realized more or less fully within those boundaries (Bronfenbrenner & Ceci, 1994). (This insight is found in numerous folktales and jokes. "Who was the greatest artist who ever lived?" a newly defunct art critic asked Saint Peter on his way into heaven. "Why, it was Lucius Cadmore," answered Saint Peter without a moment's hesitation. "But I never heard of him," said the sorely perplexed critic. "That's because he never had a chance to paint anything," Saint Peter explained.)

2. *The social environment presents an ever-changing array of roles and expectations.* Through socialization people learn to perform roles as members of a particular culture and society. As the society and its culture change, so do definitions of what makes a person well socialized. These changing definitions create endless possibilities for misunderstandings and conflict within the groups and institutions in which socialization takes place.

THE SOCIAL CONSTRUCTION OF THE SELF

The self is the outcome of socialization; it may be defined as "the capacity to represent to oneself what one wishes to communicate to others" (Elkin & Handel, 1989, p. 47). Genie, an isolated, unsocialized child, did not develop a self in early childhood. She did not learn to formulate in her own mind the words that would express her feelings to others. Through socialization, most children learn to convert cries of discomfort, hunger, or fear into socially understandable symbols like "Want bottle" or "Go out now." These utterances show that the young child has learned to recognize his or her inner states and communicate them to others. "The child who can do this is on his or her way to becoming human, that is, to being simultaneously self-regulating and socially responsive" (Elkin & Handel, 1989, p. 47).

In sociology the self is viewed as a social construct: It is produced or "constructed" through interaction with other people over a lifetime. Studies of how the self emerges therefore usually take an interactionist perspective.

Interactionist Models of the Self

The early American sociologist Charles Horton Cooley (1956/1902) was an astute observer of human interactions. As he watched young children interact with adults and other children, he saw that they continually pay attention to how others respond to their behavior. He realized that they are seeking cues in others' behavior that reflect their own—that tell them how they look, how well they express themselves, how well they are performing a task, and so on. Cooley reasoned that as we mature, the overall pattern of these reflections becomes a dominant aspect of our identity—that is, of how we conceive of ourselves. He proposed that through these processes we actually become the kind of person we believe others think we are. He called this identity the "looking glass self." The looking glass self is the reflection of our self that we think we see in the behaviors of others around us.

Language, Culture, and the Self Cooley's insight into the role of others in defining the self was the foundation for the view of the self proposed by George Herbert Mead. With Cooley, Mead believed strongly that the self is a social product. We are not born with selves that are "brought out" by socialization. Instead, we acquire a self by observing and assimilating the identities of others (Grodin & Lindlof, 1996; Nisbet, 1970). The vehicle for this identification and assimilation is language. As Mead wrote, "There neither can be nor could have been any mind or thought without language; and the early stages of the development of language must have been prior to the development of mind or thought" (quoted in Truzzi, 1971, p. 272).

This view places culture at the center of the formation of the self. The kind of person we become is in large part a result of the cultural influences that surround us during socialization. Through interaction with people who are Catholic, for example, one takes on the language, the jokes, the style of a person of that religion. If the father is a firefighter and the mother a nurse, certain attitudes about service to society and about illness and danger will carry over to the child. If the same child plays on sports teams with children in the neighborhood, the norms and values of those children and their parents will become part of the child's experience and will be incorporated into his or her personality. Another child, growing up on a Sioux Indian reservation, would learn some of the same values—such as fair play, reward for achievement, and good citizenship—but would also learn the norms and values of the Sioux (e.g., reverence for one's ancestors and for the natural environment).

As each person learns the norms of his or her culture and its various ways of communicating—whether through language, dress, or gestures—and as each experiences the influences of a particular family and peer group, a unique self is formed. The self, thus, is a product of many influences and experiences; every person emerges with a personality of his or her own, and each has incorporated to varying degrees the values of the larger society and of a particular subculture.

Role Taking: The Significant Other and the Generalized Other For Mead (1971/1934), two of the most important activities of childhood are play and games. In play, the child practices taking the roles of others. If you watch preadolescent children play, you will see them continually "trying on" roles: "You be the mommy and I'll be the teacher, and you'll come to school to find out why . . ." or "Pretend I'm Michael Jordan and you're Reggie Miller and the score is tied in the fourth quarter." They are reenacting the dramas of winning and losing, or calling into question the behaviors of the schoolroom, or trying to understand sickness and death.

This idea of **role taking** is central to the interactionist view of socialization. It refers to the way we try to look at social situations from the standpoint of another person from whom we seek a response. Mead believed that children develop this ability in three stages, during which they gain their sense of self and learn to act as persons in society. He labeled them the preparatory, play, and game stages. (See the study chart below.)

In the *preparatory stage* the child attempts to mimic the behavior of people who are significant in his or her life. **Significant others** are people who loom large in our lives, people who appear to be directly involved in winning and losing, achieving and failing. They tend to be people after whom we model our behavior—or whose behavior we seek to avoid. In the preparatory stage the child's significant others are those who respond to calls for help and shape social behavior like language.

In Mead's second stage, the *play stage,* children play at being others who are significant in their lives:

> They play at being the others who are significant to them. They want to push the broom, carry the umbrella, put on the hat, and do all the other things they see their parents do, including saying what their parents say. The story is told of the four-year-old playing "daddy" who put on his hat and his coat, said "good bye," and walked out the front door, only to return a few minutes later because he didn't know what to do next. He had taken as much of his father's work role as he could see and hear—the ritualized morning departure. (Elkin & Handel, 1989, p. 49)

As children grow in age and experience, they enter what Mead called the *game stage.* To take part in a game, a child must have already learned to become, in a symbolic sense, all the other participants in the game. Mead called this the ability to "take the role of the generalized other." Thus, in a baseball game:

> [The child] must know what everyone else is going to do in order to carry out his own play. He has to take all of these roles. They do not have to be present in consciousness at the same time, but at some moments he has to have three or four individuals present in his own attitude, such as the one who is going to throw the ball, the one who is going to catch it, and so on. (Mead, 1971/1934, p. 151)

When we are able to take the role of the generalized other, we know that rules apply to us no matter who we are. We know, for example, that rules about not smoking in the school building apply equally to students, parents, and teachers and that those who violate the rules will not be excused because of their status (Mead, 1971/1934).

The **generalized other** is a composite of all the roles of all the participants in the game. A person who participates in a game like baseball, for example, has developed the capacity for role taking and now (again using Mead's phrase) "takes the role of the generalized other." When little children play team games, they often have a very hard time taking specific roles. Watch them play basketball or soccer, for example, and you will often find them clumping together in an effort to get the

STUDY CHART Stages in the Development of Role Taking

STAGE	DESCRIPTION	EXAMPLE
Preparatory	The child attempts to mimic the behavior of people who are significant in his or her life.	A toddler tries to walk in her mother's shoes.
Play	Children play at being people who are significant in their lives, such as parents.	Three children play house: "You be the mommy, I'll be the daddy, and Joey can be the baby."
Game	Each child is able to become, in a symbolic sense, all the other participants in the game.	A group of neighborhood kids gather at a playground to play baseball.

ball. As children mature, their growing sense of the generalized other makes them better able to understand all the roles on a team and learn their own specific roles and positions in games and team sports.

The generalized other represents the voice of society, which is internalized as "conscience." For some people, the generalized other demands perfection and strict adherence to every rule. For others, the generalized other may be extremely demanding where sports and other games are concerned but much more relaxed about achievement in school or adherence to the norms of property. For still others, the generalized other may insist on amassing large amounts of money as the primary indicator of success, or it may require community service and not value financial success at all. Within any given culture, such variations will be wide but will tend to follow certain easily recognized patterns.

Role Playing and "Face Work"

Why do Americans and Asians often have so much difficulty understanding each other even when they are playing the same role (e.g., student) and speaking the same language? Why are adolescents so "hung up" about their dealings with other kids? And why is it that in some communities seemingly small insults are treated as major signs of disrespect? These are all examples of the kinds of problems that arise when people actually play the roles for which they have been socialized. They can be analyzed according to the rules of interaction known as *face work.*

As Erving Goffman (1965) defines it, *face* is the positive social value a person claims for herself or himself by acting out a specific set of socially approved attributes (politeness, humor, strength, cuteness, sensitivity, etc.). Through his close observations of seemingly routine greetings, formulas of politeness, and the give-and-take of small talk, Goffman identified the rules of interaction whereby people seek to present a positive image of themselves, their "face." Once they have established a specific image, they seek to defend it against any possible threat that might cause them to "lose face." The concept of saving or losing face is found in cultures throughout the world. Indeed, in most Asian cultures the rules of face work are even more elaborate than in the West. (The importance of such unwritten rules of interaction is explored more fully in Chapter 6.)

Most of us take it for granted that we want to maintain our self-respect and not lose face or be embarrassed in social situations. But we do not give much thought to the actual interactions that serve to maintain face.

> Just as the member of any group is expected to have self-respect, so also he is expected to sustain a standard of considerateness; he is expected to go to certain lengths to save the feelings and the face of others present, and he is expected to do this willingly and spontaneously because of emotional identification with the others and with their feelings. . . . The person who can witness another's humiliation and unfeelingly retain a cool countenance himself is said in our society to be "heartless," just as he who can unfeelingly participate in his own defacement is thought to be "shameless." (Goffman, 1965, p. 10)

The way we apply these rules of face work in performing our roles differs according to the prestige of the people involved. When firing a junior clerk, for example, the boss does not go to nearly the same lengths to take into account the employee's feelings as he does when firing a vice president. Recent studies of role playing among adolescents in inner-city communities draw on these insights. Where there is very little prestige of any kind for people to share, "fronting"—or pretending to play roles that one cannot really perform (e.g., great ballplayer, ladies' man)—is extremely common. So, therefore, are potential threats to one's face. Face work in such communities—among gang members, for example—can become a deadly business, especially in situations where one senses disrespect (i.e., feels that one is being "dissed") (MacLeod, 1995; Williams & Kornblum, 1994). Emotions play a strong part in these interactions. When people believe that they have seriously lost face, their feelings can run extremely high, especially if their peers do not hurry in to defend them.

Rapid social change tends to heighten the difficulties people have in playing the roles they have been socialized to perform. As Goffman points out:

> A person's performance of face-work, extended by his tacit agreement to help others perform theirs, represents his willingness to abide by the ground rules of social interaction. Here is the hallmark of his socialization as an interactant. If he and the others were not socialized in this way, interaction in most societies and most situations would be a much more hazardous thing for feelings and faces. (1965, p. 31)

As populations become more diverse through the processes of social change (e.g., immigration and migration), the possibilities for cultural confusion over the rules of successful role playing in school or on the street can multiply dramatically.

This interactionist perspective on socialization emphasizes how people become social actors and how they intuitively adopt the rules and rituals of interaction, such as face work, that exist in their cultures. But it does not address how people acquire their notions of morality. We know that people generally learn their values and ideas of morality as young children in the family, but research on child development shows that the acquisition of morality is not a simple matter.

Theories of Moral Development

A friend asks for help on an exam. Suddenly rules about cheating come into conflict with the norm that friends help each other. We face such moral conflicts all the time, and often they involve higher stakes or greater risk. Throughout life we face a wide variety of moral dilemmas, and these have a significant effect on our personalities. In consequence, social scientists have devoted considerable study to the processes through which people develop concepts of morality. Among the best-known students of moral development are the Swiss child psychologist Jean Piaget and the American social psychologists Lawrence Kohlberg and Carol Gilligan.

Piaget stands with Freud as one of the most important and original researchers and writers on child development. In the 1920s he became concerned with how children understand their environment, how they view their world, and how they develop their own personal philosophies. To discover the mental processes unique to children he used what was then an equally unique method: He spent long hours with a small number of children simply having conversations with them. These open-ended discussions were devoted to getting at how children actually think. In this way Piaget discovered evidence for the existence of ideas that are quite foreign to the adult mind (Berman, 1997; Elkind, 1970). For example, the child gives inanimate objects human motives and tends to see everything as existing for human purposes. In this phase of his research Piaget also described the egocentric aspect of the child's mental world, which is illustrated by the tendency to invent words and expect others to understand them.

In the later phases of his research and writing, Piaget devoted his efforts to questions about children's moral reasoning—the way children interpret the rules of games and judge the consequences of their actions. He observed that children form absolute notions of right and wrong very early in life, but that they often cannot understand the ambiguities of adult roles until they approach adolescence. This line of investigation was continued by the American social psychologist Lawrence Kohlberg, whose theory incorporates Piaget's views on the development of children's notions of morality.

Kohlberg's theory of moral development emphasizes the cognitive aspects of moral behavior. (By *cognitive* we mean aspects of behavior that one thinks about and makes conscious choices about, rather than those one engages in as a result of feelings or purely intuitive reactions.) In a study of 57 Chicago children that began in 1957 and continued until the children were young adults, Kohlberg presented the children with moral dilemmas like the following:

> A husband is told that his wife needs a special kind of medicine if she is to survive a severe form of cancer. The medication is extremely expensive, and the husband can raise only half the needed funds. When he begs the inventor of the drug for a reduced price, he is rebuffed because the inventor wants to make a lot of money on his invention. The husband then considers stealing the medicine, and the child is asked whether the man should steal in order to save his wife. (Kohlberg & Gilligan, 1971)

On the basis of children's answers to such dilemmas at different ages, Kohlberg proposed a theory of moral development consisting of three stages: (1) *preconventional,* in which the child acts out of the desire for reward and the fear of punishment; (2) *conventional,* in which the child's decisions are based on an understanding of right and wrong as embodied in social rules or laws; and (3) *postconventional,* in which the individual develops a sense of relativity and can distinguish between social laws and moral principles. Very often subjects in the preconventional and conventional stages will immediately assume that stealing is wrong in the situation Kohlberg has posed, but postconventional thinking in older children will cause them to debate the fairness of rules against stealing, in view of the larger moral dilemma involved.

Gender and Moral Reasoning Kohlberg's studies have been criticized for focusing too heavily on the behavior of boys and men from secure American families and not exploring possible alternative lines of moral reasoning that may prevail among females or people from differing cultural and racial backgrounds (Garcia, 1996; Gilligan et al., 1988; Wren, 1997). Pioneering work by social psychologist Carol Gilligan, an early collaborator of Kohlberg, has produced an impressive body of evidence that demonstrates the propensity of females to make moral choices on the basis of a somewhat different line of reasoning from that generally followed by males. Gilligan's research, and that of others who have followed her lead, shows that females are more likely than males to base moral judgments on considerations of caring as well as justice or law. More than their male counterparts, females tend to look for solutions to moral dilemmas that also serve to maintain relationships. Caring solutions that consider the needs of both sides are therefore more often invoked by females.

A good example of this difference appears in the work of D. Kay Johnston (1988), in which adolescent boys and girls were presented with dilemmas taken from Aesop's fables. The young people were read a fable that presents a moral dilemma and then asked what they understood the problem to be and how they would solve it. In the fable of the dog in the manger (see Figure 5.1), the problem is clearly that the dog has taken sleeping space from the deserving ox. Some adolescents judged the situation purely in terms of which animal had the right to the space, and made statements like "It's [the ox's] ownership and nobody else had the right to it." Others sought a caring solution that would take into consideration both animals' needs, and made statements like "If there's enough hay, well, this is one way, split it. Like, if they could cooperate."

The table in Figure 5.1 shows that boys were more likely than girls to give solutions based on rights, while girls were more likely than boys to choose solutions that emphasized caring. Some chose solutions that combined the two approaches. As Gilligan notes, "An innovative aspect of Johnston's design lay in the fact that after the children had stated and solved the fable problems, she asked, 'Is there another way to think about this problem?' About half of the children . . . spontaneously switched orientation and solved the problem in the other mode" (from Gilligan et al., 1988, p. xxi). On the basis of this and much subsequent research, Gilligan concludes that by age 11 most children can solve moral problems both in terms of rights (a justice approach) and in terms of response (a caring approach). The fact that a person adopts one approach in solving a problem does not mean that he or she does not know or appreciate others.

Gilligan and others who study moral development and gender point out that adolescence is a critical time in the development of morality and identity. However, in schools and elsewhere in society the message that comes across is that norms, values, and the most highly esteemed roles require that there be a "right way" to feel and think. Most often this right way is associated with the justice focus—and the caring focus is silenced, along with the voices of girls and others to whom it appears to be a valuable alternative mode of moral reasoning (Wren, 1997). In adolescent girls and many minority students, this form of silencing can be detrimental to the development of the self in social situations, a subject of immense importance to which we will return at many points in later chapters (Taylor, Gilligan, & Sullivan, 1995).

FIGURE 5.1 **The Dog in the Manger**

A dog, looking for a comfortable place to nap, came upon the empty stall of an ox. There it was quiet and cool and the hay was soft. The dog, who was very tired, curled up on the hay and was soon fast asleep.

A few hours later the ox lumbered in from the fields. He had worked hard and was looking forward to his dinner of hay. His heavy steps woke the dog, who jumped up in a great temper. As the ox came near the stall the dog snapped angrily, as if to bite him. Again and again the ox tried to reach his food, but each time he tried the dog stopped him.

Moral Orientation of Spontaneous Solution for the Dog in the Manger Fable, by Gender

Orientation	Female	Male
Rights (justice)	12	22
Response (caring)	15	5
Both	3	1

Source: Adapted from Johnston, 1988, p. 57.

SOCIAL ENVIRONMENTS AND EARLY SOCIALIZATION

None of the theories of development we have reviewed touches on the many social and natural environments in which human beings are socialized. Suppose, for example, that Kip Kinkel, whom we met at the beginning of the chapter, had been raised in a group home for adolescents whose parents were in prison or had been killed. His murderous behavior would very likely have been attributed to lack of adult love or involvement, rather than to possible genetic or other biological factors. Yet we also know that the quality of care and nurturing found in group homes can vary quite widely. How socialization actually occurs in any social environment is an empirical question and a subject of social-scientific research.

Socialization occurs in many different settings, some of which are more desirable than others. Children are born into relatively affluent homes in American suburbs, into the squalid slums of Calcutta, the icy wastes of Siberia, the extreme poverty of the Somalian desert, and myriad other environments—yet in every case those who survive infancy are socialized to live and perhaps flourish in these diverse settings. Cross-cultural studies attempt to show how socialization processes vary in different societies and how they are affected by demographic trends (trends in population size and composition) within societies. This comparative view of socialization attempts to bring together the often competing viewpoints of biological, psychological, and sociological explanations of developmental patterns.

Within the United States, there are important differences from one region or state to another in the degree of stress children experience in their home environment and in the "success" of their socialization. A crude but

In the early decades of this century, when children routinely worked in textile mills and coal mines, the environment in which they were socialized forced them to take on adult roles at an early age. For some children growing up in America this is still true, for different reasons.

extremely useful measure of stress in children's socialization environment is the percentage of children living in households with incomes below the official poverty line. Similarly, measures of the percentage of children living in single-parent households—which are more likely than other households to have below-average income and in which the parent may have less opportunity to interact with the child—are useful indicators of the outcomes of socialization. As shown in Figures 5.2 and 5.3, these indicators have risen in recent decades. The circumstances reflected in measures like these have many causes and significant consequences for social change. For example, because teenage mothers are more likely than nonteenagers to be single mothers and to be living in poverty—two negative stresses on a child's socialization environment—teenage pregnancy becomes an important issue. A major consequence of the high rate of teenage fertility in the United States is that proportionately more babies are born into homes in which they will not receive the attention and material benefits that will help them realize their full genetic potential (Furstenberg, 1976). In other words, these children start off at a disadvantage in that their mothers have not yet reached maturity themselves and are likely to be single parents who may be unable to provide their children with economic security, love, and discipline.

FIGURE 5.2 **Percent of Children in Poverty, 1975–1996**

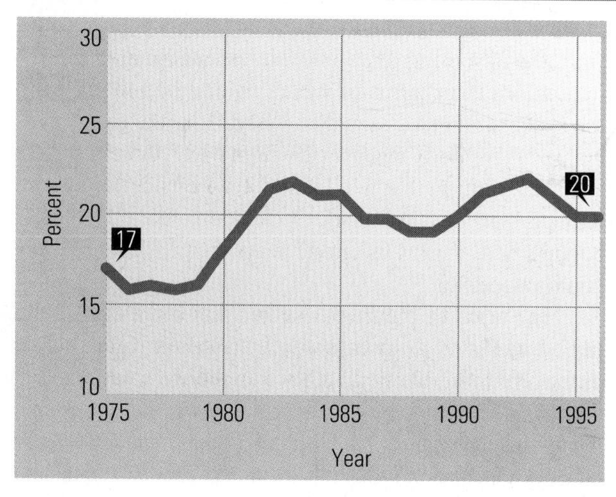

Source: Annie E. Casey Foundation, 1998.

FIGURE 5.3 **Percent Families With Children Under 18 Headed by a Single Parent, 1975–1997**

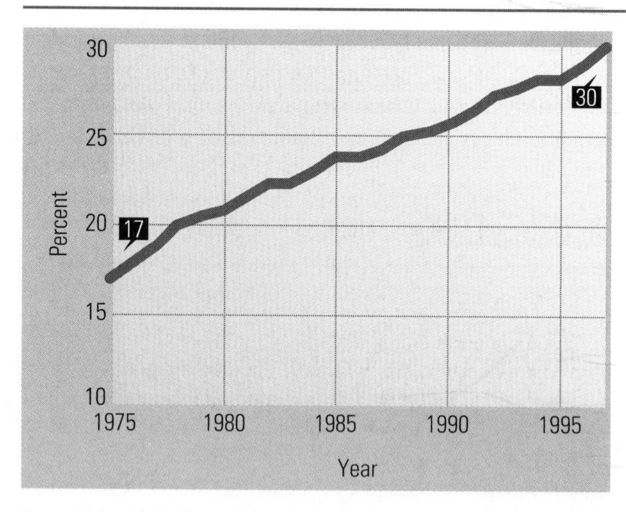

Source: Annie E. Casey Foundation, 1998.

These indicators are published annually by the Annie E. Casey Foundation of Baltimore, which is devoted to improving the quality of children's lives in the United States.

Urie Bronfenbrenner, one of the nation's most respected and original researchers in the field of childhood socialization, summarizes the conclusions of numerous studies with these two propositions:

> Proposition 1. In order to develop normally, a child needs the enduring . . . involvement of one or more adults in care of and joint activity with the child.
>
> Proposition 2. The involvement of one or more adults in care of and joint activity with the child requires public policies that provide opportunity, status, resources, encouragement, example, and, above all, time for parenthood, primarily by parents, but also by other adults in the child's environment, both within and outside the home. (1981, p. 39)

No society can pass legislation requiring that a growing child experience love in the family or household. It can, however, formulate public policies that support the efforts of families to raise healthy children and socialize them well. It can do this, in part, by helping to provide children with the attention of caring adults outside the family, especially in agencies of socialization such as schools.

Agencies of Socialization

Agencies of socialization are the groups of people, along with the interactions that occur within those groups, that influence a person's social development throughout his or her lifetime (Elkin & Handel, 1989; Gonzalez-Mena, 1998); **agents of socialization** are individuals, such as parents and teachers, who socialize others. The most familiar agencies of socialization are the family, schools, socializing agencies in the community, the peer group, and the mass media. Later in life adults may experience further socialization in the workplace, in universities, or in the military, to cite a few common examples. And people who wish to change self-destructive behaviors like drug addiction or alcoholism may be resocialized in "12-step" programs such as Alcoholics Anonymous. (Resocialization is discussed further in the next section.)

Within all agencies of socialization a number of processes are continually occurring that shape the individual's development. Among these are direct instruction, imitation or modeling of behavior, and reinforcement of particular behaviors (e.g., through rewards or punishments). Many of these processes appear in what sociologists call **anticipatory socialization.** Whenever an individual plays at a role he or she is likely to assume later in life, anticipatory socialization is taking place. By dressing up in adult clothing, for example, or by playing house, a child imitates the behavior of adults. The adolescent who attends a senior prom is being socialized in anticipation of a time when she or he will be expected to participate in formal social events. And a worker in a corporate setting may rehearse before coworkers a presentation she or he will later make to a potential client.

Another important aspect of agencies of socialization is that they must continually deal with social change. For example, as guns have become more readily available in large cities, peer groups have had to cope with the ever-present danger of violent death. As divorce and remarriage rates have risen, many families have had to deal with new structures that add much more complexity to family socialization patterns. As a result, the influence of social change on socialization patterns has become an important area of sociological research.

In the following pages we briefly consider several agencies of socialization: the family, the schools, the community, the peer group, and the mass media. We will encounter many of these agencies in later chapters as well.

The Family The family is the primary agency of socialization. It is the environment into which the child is born and in which his or her earliest experiences with other people occur—experiences that have a lasting influence on the personality. Family environments vary greatly, not only in terms of such key variables as parents' income and education but also in terms of living arrangements, urban versus rural residence, number of children, relations with kin, and so on. Much contemporary research centers on the effects of different family environments on the child's development (Berman, 1997; Gonzalez-Mena, 1998). For example, Michael Lewis and Candice Feiring's intriguing study of 117 families, "Some American Families at Dinner" (1982), shows that a typical 3-year-old child interacts regularly with a network of kin, friends, and other adults who may play a significant role in the child's early socialization. (See Box 5.2.)

One of the most significant changes occurring in American society today is the increase in the proportion of children growing up in poor families (20 percent of all children under age 18) or in no family at all. In the early 1990s, nearly 10 percent of all children in the United States were growing up in households with no parent present. Most of these children (76 percent) were being socialized in the homes of grandparents and relatives; the remainder were living in group homes and with nonrelatives (Gross, 1992). There has also been a steady increase in the number of single-parent families, the vast majority of which are headed by women.

But changes in the way families are organized or how they cope with changing social conditions are not a new phenomenon. The values and ways of parents are never entirely valid for their children, although the

BOX 5.2 SOCIOLOGICAL METHODS

Measures of Central Tendency

The following table, taken from Lewis and Feiring's study of American families, presents information about the social network of the typical 3-year-old:

The 3-Year-Old Child's Social Network (N = 117)*

	X	Range
Number of relatives other than parents seen at least once a week	3.20	0–15
Number of child's friends seen at least once a week	4.43	0–13
Number of adults seen at least once a week	4.38	0–24

* *N* refers to the number of subjects in a study.

Source: Lewis & Feiring, 1982, p. 116.

To understand this table you must know something about the common measures of central tendency. The measure used in the table above is the *mean*, represented by X. To arrive at this number each child's total number of friends, relatives, and other adults seen at least once a week is added to the totals for the other children in the sample. The grand total is then divided by the number of children in the sample. The resulting number can be used to represent the entire sample.

The other common measures of central tendency are the *median* and the *mode*. The median is the number that divides a sample into two equal halves when all the numbers in the sample are arranged from lowest to highest. The mode is simply the score that occurs most frequently in the sample. Sociologists make a point of specifying which of these measures they are using. They try to avoid the term *average* in statistical tables, as it can refer to any one of these measures.

To illustrate all three measures of central tendency, let us use the following data from 10 of the children in Lewis and Feiring's sample:

Child	Number of Friends Seen at Least Once a Week
1	2
2	3
3	3
4	7
5	12
6	2
7	3
8	1
9	0
10	5
	38

In this example the mode would be 3 and the median would also be 3, but the mean would be 3.8.

degree to which this is true depends on how much social change is experienced from one generation to the next. Socialization creates the personalities and channels the behaviors of the members of a society, but that socialization is never entirely finished. Thus, in *Manchild in the Promised Land,* his masterpiece about growing up in Harlem during the period of rapid migration of blacks from the South, Claude Brown (1966) wrote that his rural-born parents did not "seem to be ready for urban life." Their values and norms of behavior made no sense to their son, who had to survive on Harlem's mean streets:

> When I was a little boy, Mama and Dad would beat me and tell me, "You better be good," but I didn't know what being good was. To me it meant that they just wanted me to sit down and fold my hands or something crazy like that. Stay in front of the house, don't go anyplace, don't get into trouble. I didn't know what it meant, and I don't think they knew what it meant, because they couldn't ever tell me what they really wanted. The way I saw it, everything I was doing was good. If I stole something and didn't get caught, I was good. If I got into a fight with somebody, I tried to be good enough to beat him. If I broke into a place, I tried to be quiet and steal as much as I could. I was always trying to be good. They kept on beating me and talking about being good. And I just kept on doing what I was doing and kept on trying to do it good. (p. 279)

Brown's story is that of a young man whose parents' experience with social change created a severe disjunction in their lives. This left them ill equipped to socialize their son for the demands of a new environment. And so he learned to survive on the streets. He became a thief and a gang fighter. But by his own account, later in his life he was greatly influenced by people who had studied the social sciences and created well-functioning institutions, special schools in particular, that could bring out his talents and socialize him for a more satisfying and constructive life than he had led as a child.

i

The Schools For most of us, regardless of what our home life is like, teachers are generally the first agents of socialization we encounter who are not kin. In some cases children are also influenced by agents of socialization in the church (e.g., priests or rabbis), but for most children the school is the most important agency of socialization after the family. Children experience many opportunities to perform new roles in school (e.g., as student, teammate, etc.). No wonder, then, that in these first "public" appearances they tend to be highly sensitive to taunts or teasing by other children. Even very young children can become distraught when they feel that they are not wearing the "right" clothes.

Schools are institutions where differences between the values of the family and those of the larger society come into sharp focus. In some families, for example, there may be great concern about any form of sex education in school. Indeed, much of the conflict over educational norms—what ought to be taught, whether there should be prayer in school, and the like—stems from differences between the values of some families and the values that many educators wish to teach. Such conflicts point to the exceedingly difficult situation of schools in American society. Research on school–family and school–community relations has shown that the schools are expected to conserve the society's values (by teaching ideals of citizenship, morality, family values, and the like) and, at the same time, to play a major role in dealing with innovation and change (by expanding the curriculum to include new knowledge, coping with children's perceptions of current events, addressing past patterns of discrimination, and the like) (Goslin, 1965; Meier, 1995).

In Chapter 18 we will return to questions of whether the schools help reproduce the status quo and maintain inequalities in society, and whether they can actually affect the lives of disadvantaged students. Although many individuals credit schools (and even particular teachers) with bringing about important improvements in their lives, there is much debate on these questions in the social sciences.

The Community Schools may be the most important agency of socialization outside the family, but there are a number of other significant agencies of socialization in most communities—including day care centers, scout troops, churches, recreation centers, and leagues of all kinds. These agencies engage in many and varied forms of socialization. Parades on Memorial Day and the Fourth of July, for example, reinforce values of citizenship and patriotism. Participation in team sports instills values of fair play, teamwork, and competitive spirit. The uniforms and equipment required for these activities stimulate shopping trips to suburban malls and downtown department stores, and those trips prepare children and adolescents for the time when they will be consumers (McKendrick, Brewer, & Plump, 1982; Newman, 1988).

Of all these agencies of socialization, the day care center is perhaps the most controversial. Polls often show that Americans have doubts about the effects of day care on very young children, and scandals involving charges of child abuse by day care workers intensify those fears. But when both parents work outside the home, or when single parents must work to support their children, day care centers may play a critical role in socialization. Many studies have shown that good-quality centers are not harmful to children and in some cases may be beneficial, but the norm requiring that mothers stay home to care for young children remains strong in many communities (Kammerman, 1995).

The central question addressed by these studies is the quality of the care received. A recent longitudinal study of 1,153 children at 10 different child care centers revealed that in the better centers (as measured by indicators like the qualifications of the caregivers and the amount of individual attention received by the children) children actually made gains in language and thinking abilities that were superior to those made by children from similar backgrounds who were not enrolled in high-quality day care. This effect was especially strong for children from lower-income households (Garrett, 1997). On the negative side, however, the researchers found that the more time young children spent in centers with lower-quality care, the more likely they were to experience problems in forming close attachments with their mothers.

Conservative critics of day care often argue that mothers of young children should stay home and care for them. But about three of every five mothers with children under age 3 are in the labor force. They and their working spouses are dependent on various forms of child care. Whatever the reasons for this dependence, it is a reality of contemporary life. High-quality child care is becoming an increasingly pressing social need as a consequence of long-term changes in the economy that have resulted in large numbers of dual-earner families. At present there are far too many low-quality day care facilities—as many as 40 percent of the total, according to one recent estimate—providing care to children in the United States (Population Reference Bureau, 1998).

The Peer Group In the United States the peer group tends to be the dominant agency of socialization in middle and late childhood. **Peer groups** are interacting groups of

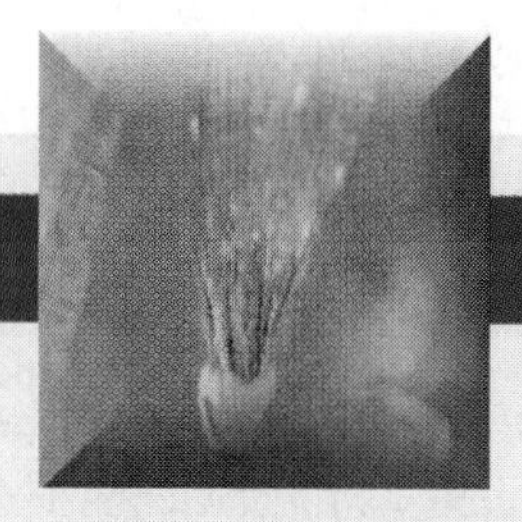

ART AND THE SOCIOLOGICAL EYE

Graffiti—Art or Vandalism?

The teenagers who do it call it different things in different cities—writing, tagging, bombing, and even art. Most adults think of it as graffiti and consider it either a nuisance or, if it defaces their property, outright vandalism. Sociologists who study teenage behavior often explain that graffiti can send many kinds of signals. Gangs often use initials or symbols to mark their turf in an urban community. Aspiring young artists, influenced by hip-hop styles of dress and music, may become interested in graffiti as an art form. Sometimes this art is executed as a protest against the larger society, as when large graffiti murals are drawn at night on buses or trains or on public buildings. Part of the creative effort involves the daring exploit of producing art that is against the law. In some cities, youth who produce this underground or protest art are known as "bombers" because they make sudden, stealthy strikes on a building or a train. Bombers look down on mere "taggers," who scrawl their highly stylized initials ("tags") everywhere they go, even on the far more artistic bombers' work.

No matter how much we might understand and appreciate well-executed graffiti art, however, from the viewpoint of the larger society graffiti art is a defacement of public property. Its removal is costly; its effects can be demoralizing; and it can be read as a sign that a community's young people, or at least some of them, are alienated and angry. Realizing this, many youth workers and community leaders have found ways to reach out to graffiti artists and guide their energy and creativity into socially constructive or at least more acceptable channels. In New Orleans, for example, the Ya Ya Collective has made graffiti art into products that young people can reproduce and sell. This builds their self-esteem and provides funds for other constructive youth activities. In New York City, where graffiti art is extremely popular, a sympathetic industrialist has hired graffiti artists to paint enormous murals on the walls of an old factory building in an industrial section of the city. In Philadelphia, Chicago, Cleveland, and many other U.S. cities, graffiti artists and their friends are applying their skills to the creation of public murals commemorating the victims of gang violence or the drug trade, and in so doing they are learning to use their art in positive ways.

people of about the same age. Among adolescents, peer groups exert a strong influence on their members' attitudes and values. Studies confirm the high degree of importance adolescents and adults alike attach to their friendship groups. Adolescents typically acquire much of their identity from their peers and consequently find it difficult to deviate from the norms of behavior that their peer group establishes (Gans, 1962, 1984; Homans, 1950; McAndrew, 1985; W. F. Whyte, 1943). In fact, the peer group may become even more important than the family in the development of the individual's identity.

i

There is often a rather high level of conflict within families over the extent to which the peer group influences the adolescent to behave in ways that are not approved of by the family. Even where conflict is limited, the peer group usually provides the child's first experience with close friendships outside the family. The peer group becomes the child's age-specific subculture—that is, a circle of close friends of roughly the same age, often with shifting loyalties.

The peer group usually engages in a set of activities that are not related to adult society. The peer subculture may, for example, include games that adults no longer play and may have forgotten. British researchers Iona and Peter Opie (1969) conducted a classic study of the games children play. They identified hundreds of games that were known by children between the ages of 6 and 10 all over England. They also identified the common elements in those games, such as chasing, pretending, or seeking, and found that in many of the games the

central problem is to choose an "It," a seeker or chaser. The children's efforts to avoid being "It" seemed to express their desire to avoid roles that make them different from the others.

When children's peer groups are faced with conflict and social change—for example, in communities where there are high levels of poverty and demographic change, with groups of culturally distinct people moving in and out—the group often organizes itself in gangs for self-defense and aggression. There are many such gangs in large American cities characterized by rapid immigration, widespread drug use, and the ready availability of cheap handguns. Under these conditions both male and female peer groups may become extremely dangerous environments for socialization, yet they exert a strong attraction on adolescents who seek protection and companionship in what is perceived as a hostile world (Sanchez-Jankowski, 1991; Williams, 1989).

The Mass Media The most controversial agency of socialization in American society is the mass media. In debates about the effects of the media on socialization, television comes under the greatest scrutiny because of the number of hours children spend in front of the "electronic baby-sitter." Estimates of how much television children watch differ, depending on the methodology used to conduct the study (especially since simply having the set turned on does not mean that the children are actually watching). Many studies have shown that the amount of television viewed varies, depending on whether the child comes from a poor home with few alternatives or a more affluent one where other activities are available. Children from poor homes in urban communities often have the set on for seven or more hours a day, as opposed to children from more affluent homes, where the set is turned on for three or four hours daily (Elkin & Handel, 1989; Wellins, 1990).

The effects of all this television viewing on children and adolescents are a subject of intensive research. In particular, the effects of seeing violent acts on television or listening to violent music like gangsta rap are hotly debated. There is no doubt that the amount of violence shown on TV is immense. George Gerbner, one of the nation's foremost researchers on TV violence, has been monitoring the number of violent acts shown on TV. His data indicate that the average child will have seen about 100,000 acts of violence on TV before graduating from elementary school—and the number is far higher for children from poor neighborhoods (Gerbner, 1990).

We will return to the subject of the media and their effects in Chapter 18. It should be noted here, however, that the effects of television on socialization are by no means all negative. Television, radio, and the movies provide windows onto social worlds that most people cannot otherwise enter. Children, adolescents, and adults in modern American society learn far more about current events, social issues, and the arts than people did before the advent of television.

SOCIALIZATION THROUGH THE LIFE COURSE

Childhood socialization is the primary influence on the individual. Whatever happens to a person later in life, the early childhood experiences that shaped the social self will continue to influence that person's attitudes and behavior. But people are also affected by their *life course:* the set of roles they play over a lifetime and the way those roles change as a consequence of social change. The life course of a person who came of age just before World War II, for example, would typically have involved service in the armed forces for men (perhaps including combat duty) and work in a factory (or noncombatant military service) for women. The life course of children born in the 1960s or 1970s is far less likely to include a long or intense experience of warfare. Natural disasters, major changes in the educational system or in the political stability of their society—these and other possible changes can influence the roles people play during their life course.

Changes in the culture of a society can also significantly alter the roles people play during their lives. In the first half of the twentieth century, for example, the role of wife was less likely to include periods of single parenthood than it is now. Fifty years ago the role of an elderly person was far more likely to involve relative poverty and dependence on younger family members than it does now. The need to adjust to changes in society and culture, and the fact that so many people are able to make such adjustments (though often with great difficulty) are evidence that socialization is not finished in childhood but continues throughout the life course (Riley, Foner, & Waring, 1988; Sorenson, Weinert, & Sherrod, 1986).

Adult Socialization

Two of the influences that produce socialization after childhood are significant others (especially new friends) and occupational mobility (especially new jobs). When one moves into a new neighborhood, for example, one often makes new friends, and their influence may lead to new activities. A person who has always been uninvolved in politics may, through the influence of new friends, become committed to a social movement such as the women's movement. New friends can also introduce a person to dangerous or unlawful activities such as drug use. One's *core identity*—the part of the self formed in early childhood that does not change easily—may prevent one from being overly influenced by a given peer group later in life, or it may actually cause that influence to be stronger (Elkin & Handel, 1989).

Such factors as stress and changing physical health may also have an influence. In a study of how people change their patterns of behavior later in life, Marjorie Fiske and David A. Chiriboga (1990) found that everyday hassles and the boredom of a routine, predictable existence have a greater impact on the changes adults seek in their lives than was indicated by earlier research.

In any new activity the newcomer or recruit must learn a new set of norms associated with the roles of the organization. Socialization associated with a new job, for example, often requires that a person learn new words and technical terms. The individual also needs to interact with a new peer group, usually composed of people whose work brings them together for long hours. And in most jobs the person will eventually be faced with choices between loyalty to peers and loyalty to those higher in the organizational hierarchy who can confer benefits like raises and promotions. In these and most other adult socialization experiences, the choices people make will be highly influenced by the individual's core identity, which continually shapes his or her responses to new situations and challenges (Rosow, 1965).

Resocialization

Many adults and even adolescents experience the need to correct certain patterns of prior social learning that they and others find detrimental. This need usually cannot be met through the normal processes of lifelong socialization, especially since the individual's negative behaviors may be leading toward a personal crisis and causing immense pain for family members and friends. Such a condition often responds to specific efforts at resocialization.

Resocialization is a process whereby individuals undergo intense, deliberate socialization designed to change major beliefs and behaviors. It is often aimed at changing behaviors like excessive drinking, drug use, and overeating—particularly common in affluent societies, where individuals are exposed to a great deal of choice and many pleasurable stimuli. Perhaps most widespread and pernicious among the addictions, alcoholism affects millions of people in North America and is involved in much of the marital strife that leads to family breakup. Alcoholics Anonymous (AA) is a voluntary group program that uses techniques of resocialization and peer example to help alcoholics reject old behaviors and learn new, more positive ones that do not involve drinking. The AA approach to voluntary resocialization, which emphasizes personal change and positive growth as the individual passes through 12 steps of recovery, has been successfully applied to problems of obesity and overeating (Overeaters Anonymous) and narcotics use (Narcotics Anonymous). Many other self-help support groups assist people in dealing with almost any kind of problem that they previously struggled with in isolation (Makela, 1996).

Much successful resocialization takes place in what are often called **total institutions** (Goffman, 1961). These are settings where people undergoing resocialization are isolated from the larger society under the control of a specialized staff whose members themselves may have experienced the same process of resocialization. All aspects of the inmates' daily lives are controlled, and supervision is constant. Drug treatment centers, for example, are often set up as total institutions for the resocialization of addicts. Even the smallest rewards, like extra time alone or more freedom to walk around the grounds, are controlled by the staff and bestowed only on inmates who have made progress toward the resocialization goals.

In the resocialization process, the staff first attempts to tear down the individual's former sense of self. This stage may include various forms of degradation and abasement in which the individual is forced to reject the undesired thoughts and behaviors. In the next stage the staff rewards the individual's attempts to build a new sense of self that conforms to the goals of the resocialization process (e.g., a sober person who accepts responsibility for previous wrongs and achieves new interests and new awareness of personal strengths). However, while total institutions can be extremely powerful environments for resocialization, they sometimes run the risk of creating people who need to remain in their controlling environment—that is, people who have developed a need to be "institutionalized."

STUDY CHART Theories of Socialization

THEORIST	DESCRIPTION OF THEORY
Sigmund Freud	Through socialization the infant is gradually forced to channel its biological urges into socially acceptable forms of behavior. Major personality traits are formed in the conflict between the child and its parents, especially the same-sex parent.
George Herbert Mead	Socialization is a process in which the self emerges out of interaction with others, not only in early infancy but throughout life. Role taking is central to socialization; role-taking ability develops through interaction during childhood in the preparatory, play, and game stages.
Jean Piaget	Children develop definite awareness of moral issues at an early age but cannot deal with moral ambiguities until they mature further. This insight is incorporated in Kohlberg's preconventional, conventional, and postconventional stages of moral development.
Erik Erikson	Throughout the life course the individual must resolve a series of conflicts that shape the person's sense of self and ability to perform social roles successfully. Central to this process is identification, in which the individual chooses other people as role models.
Carol Gilligan	Children tend to develop different ways of resolving moral dilemmas. Some (more often male) tend to rely on strict rules of right and wrong, while others (most often but not always female) tend to make judgments based on notions of fairness and cooperation. In societies where male voices are dominant, issues of cooperation and fairness are passed over, to the detriment of female socialization for leadership and achievement.

Erikson on Lifelong Socialization

The theories of social psychologist Erik Erikson are especially relevant to the study of adult socialization and resocialization. Erikson agreed with Freud that a person's sense of identity is shaped by early childhood experiences. But Erikson also focused on the many changes occurring throughout life that can shape a person's sense of self and ability to perform social roles successfully. He demonstrated, for example, that combat experiences can produce damaged identities because soldiers often feel guilty about not having done enough for their fallen comrades (this is sometimes called "survivor guilt"). For Freud, in contrast, war-produced mental illness was always related to problems that the soldier had experienced in early childhood.

In *Childhood and Society* (1963), Erikson's central work on the formation of the self, the concept of identification takes center stage. **Identification** is the social process whereby the individual chooses other people as models and attempts to imitate their behavior in particular roles. Erikson noted that identification with these role models occurs throughout the life course. He pointed out that even older people seek role models who can help them through difficult life transitions.

For Erikson, every phase of life requires additional socialization to resolve the new conflicts that inevitably present themselves. Table 5.1 presents the basic conflicts that each person must resolve throughout life, and indicates how a positive role model can help in resolving those conflicts. (The major theories of socialization are summarized in the study chart on page 145.)

Resocialization at any age usually requires a strong relationship between the individual and the role model. In the military or in a law enforcement agency, for example, the new recruit is usually paired with a more experienced person who offers advice and can show the recruit the essential "tricks" of the new role—"how it's really done." Thus, in Alcoholics Anonymous, once the self-confessed "drunk" faces up to that negative identity, he or she is taught new ways of thinking, feeling, and acting that do not involve drinking. An important step in this process is the development of a relationship with a *sponsor,* a person who has been through the same experiences and has made a successful recovery. In fact, one of AA's aims is to enable the resocialized alcoholic to eventually perform as a sponsor—that is, as a role model—for someone else who is going through the same process.

GENDER SOCIALIZATION

No discussion of socialization would be complete without some mention of gender socialization. **Gender socialization** refers to the ways in which we learn our gender identity and develop according to cultural norms of "masculinity" or "femininity." Gender is not synonymous with biological sex; rather, "scholars use the word *sex* to refer to attributes of men and women created by their biological characteristics and *gender* to refer to the distinctive qualities of men and women (or masculinity and femininity) that are culturally created" (Epstein, 1988, p. 6). By *gender identity* we mean "an individual's own feeling of whether she or he is a woman or a man, or a girl or a boy" (Kessler & McKenna, 1978, p. 10).

TABLE 5.1 Erikson's View of Lifelong Socialization

Stage of Life	Conflict	Successful Resolution in Old Age*
Infancy	Trust vs. mistrust	Appreciation of interdependence and relatedness.
Early Childhood	Autonomy vs. shame in the development of the will to be a social actor	Acceptance of the cycle of life, from integration to disintegration.
Play Age	Initiative vs. guilt in the development of a sense of purpose	Humor; empathy; resilience.
School Age	Industry vs. inferiority in the quest for competence	Humility; acceptance of the course of one's life and unfulfilled hopes.
Adolescence	Identity vs. confusion; struggles over fidelity to parents or friends	A sense of the complexity of life; merger of sensory, logical, and aesthetic perception.
Early Adulthood	Intimacy vs. isolation in the quest for love	A sense of the complexity of relationships; value of tenderness and loving freely.
Adulthood	Generativity vs. stagnation in interpersonal relationships	*Caritas* (caring for others) and *agape* (empathy and concern).
Old Age	Integrity vs. despair	Wisdom and a sense of integrity strong enough to withstand physical disintegration.

* Erikson did not believe that elderly people necessarily resolve these conflicts entirely, but when they do, the results are as shown here.

Our ideas about what it means to be a man or a woman are different from our ideas about the anatomical definitions of *male* and *female.* These ideas are shaped by the values and socialization practices of our culture, but once they have become part of the self, they are usually extremely strong.

The strength and pervasiveness of gender socialization are illustrated by the case of Ron Kovic, the author of *Born on the Fourth of July.* Kovic saw himself as a red-blooded American male, a typical product of socialization for boys raised in his community. He was taught to be patriotic, to want to defend his country, to want to prove his manhood in combat and sexual conquest. When he became paralyzed in Vietnam and lost his manhood (in the narrow, conventional sense of that term), he began to doubt everything—including himself, his country, and his friends. His story raises some important issues, not least of which is how we form our ideas of manliness and womanliness and whether those ideas are helpful in meeting the demands of a changing world.

The story of Ron Kovic is a striking example of how powerful gender socialization is and the effects it can have on a person's life. Elsewhere in this book (especially in Chapter 14) we present other examples of the importance of gender in explaining life chances. But here let us take a more mundane example of the results of gender socialization: the way men and women view their own bodies. Men tend to think of their own bodies as "just about perfect," but women tend to think of themselves as overweight, at least when they compare themselves with a mental image of an attractive woman. These are the findings of a survey of 500 college-age men and women conducted by April Fallon and Paul Rozin (cited in Goleman, 1985a).

A striking indication of the role of agencies of socialization in the social construction of gender identity is the influence of the mass media in promoting ideal body types for men and women. These influences have a lot to do with why men tend to be happy with their bodies while many girls and women are extremely unhappy with theirs. Men dominate the media—radio, television, newspapers, and magazines—that communicate American culture to mass audiences. Girls and women are continually exposed to images of women created by men, and those images more often than not portray women as sex objects, to be judged by the shape and size of their breasts, the length of their legs, and the tightness of their skin (Chernin, 1981; Lurie, 1981). It is little wonder, then, that women feel worse about themselves than do men—about whom it has been said that it is sexy to be a sagging warrior or a somewhat wrinkled philosopher.

Among the many momentous changes that are occurring in the world today, those affecting gender socialization are among the most profound. Research on socialization in Asia and Africa, for example, shows how norms that favor boys over girls act to the detriment of female development. Studies from Bangladesh, India, Nepal, and Pakistan reveal that boys receive more food than girls, even though girls are often required to expend more calories on field and household work. In many poor nations it is common for boys to be fed before girls and for the higher-quality nutrients, especially proteins, to be consumed before the girls are allowed to eat. So strong are gender norms in those countries that most mothers are convinced that this practice is necessary when food is scarce (Jacobson, 1992). As a result of this and other socialization practices that place females at a disadvantage, the health and learning capacities of women are imperiled from an early age in many parts of the world.

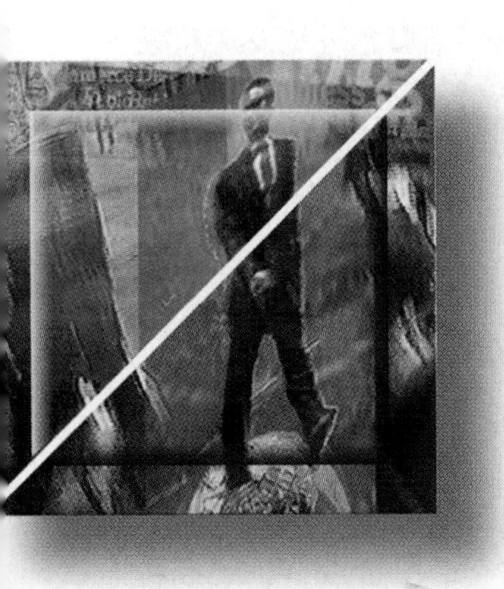

RESEARCH ON THE CUTTING EDGE: Socialization and the "Time Bind"

Sociologist Arlie Russell Hochschild never ceases to bring unexpected new empirical findings to bear on important issues of socilization. Her previous book was a study of the nature of American men and women's participation in the work of keeping up the home and taking care of the family. Its title was *The Second Shift: Working Parents and the Revolution at Home* (1989). The "second shift," as Hochschild uses the term, refers to the work that parents do at home after their day's work in the labor force. Not surprisingly, she found that women who work full-time continue to bear more than their share of responsibility for the home and children. Men may be devoting more time to child care and domestic chores than they did a generation or two ago, but women still bear the brunt of responsibility and as a result have far less disposable time than their husbands or male partners.

More recently, Hochschild has looked into what is going on among men and women at work, and at the implications of changes in male and female employment for

Americans report feeling more stress in coping with family and work responsibilities. No wonder women like the executive mother shown here are often referred to as "jugglers."

children in dual-earner households. As she puts it, before she began her new study she kept thinking about a "nagging question":

> Given longer workdays—and more of them—how could parents balance jobs with family life? Or, to put the matter another way, was life at work winning out over life at home? If so, was there not some way to organize work to avoid penalizing employees, male and female, for having lives outside of work and to ease the burden on their children? (1997, p.7)

Hochschild also knew that "in 1952, 12.6 percent of married mothers with children under age 17 worked for pay; by 1994, 69 percent did so, and 58.8 percent of wives with children age one or younger were in the workforce. Many of these wives also had a hand in caring for elderly relatives" (p. 6). And studies also show that instead of cutting back at work, more women were working at full-time rather than part-time jobs and men with young children were not working fewer hours to make up the difference but were actually working longer hours. Why was this happening, and with what consequences for children?

To explore her question through an empirical study of actual behavior at home and at work, Hochschild spent over a year observing and interviewing workers and managers at the Americo Corporation, a large manufacturing firm located in a small midwestern city (the names of both are disguised to protect the identity of her informants). This was a "family-friendly" corporation with extremely advanced family leave policies and its own high-quality corporate child care center. In order to avoid losing highly trained female workers and managers, the company instituted a number of measures to restore the "work–family balance." These included options for part-time work, job sharing, and flextime. The company's top officials were interested in knowing if these measures were helping; they gave Hochschild free run of the facilities and access to anyone she wanted to speak with. She interviewed 130 women and men who were employees at all levels of the corporation; followed six families through an entire day of their home and work lives; made detailed observations of employees' comings and goings at work; and closely observed the routines at the child care center.

The most startling finding emerging from Hochschild's interviews and other data is that relatively few men or women took advantage of the family options offered to them at work. Children placed in the day care center were spending an average of eight hours per day there. By the end of their long days they were clearly anxious to be with their parents. But the parents were reluctant to take advantage of opportunities offered to them by the company to spend more time with their children and somewhat less time at work. In fact, many of the parents, both women and men, said that they did not want to give up the socializing they did at work, and that their relationships with peers at work were as important to them as the money they earned. Others distrusted the company's efforts to provide family care options. Both men and women feared that going on flextime might indicate to their supervisors that they were not devoting their full energy to the company and its goals. The company was doing everything it could to encourage workers to spend more time with their families, yet its own culture stressed values like intense commitment to the work group, which worked against its stated family-friendly policies.

Before the end of Hochschild's study, Americo's management and labor force underwent a wrenching restructuring involving job combinations and reassignments. These changes were the result of the company's stated need to meet the challenge of global competition from nations with no such family leave policies. For the Americo employees at every level, however, the restructuring confirmed their earlier suspicions that to show anything but complete commitment to the workplace and its groups was to take a serious personal risk.

It is dangerous to generalize from Hochschild's study of employees in one large corporation to the entire society. But we do know that Americans are spending more time at work (Robinson & Godbey, 1997), that they perceive themselves to be short of time, and that they feel stress in juggling home and family responsibilities (Schor, 1992). Hochschild suggests some Americans are spending more time at work because they want to rather than because they need to and that this trend can have immense implications for the way they raise their children. In addition to the need for far more high-quality day care, Hochschild advocates a "time movement" in which American women and men recognize the need to devote more time to their children and families and demand greater support for their family time needs.

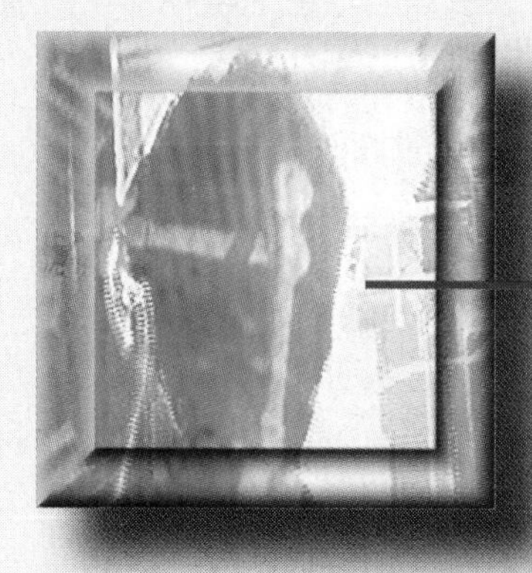

VISUAL SOCIOLOGY

Shooting Back

The noted social scientist Robert Coles tells the story of a 10-year-old homeless girl he encountered in a shelter in Boston. The girl had drawn a picture of herself, but suddenly she picked up a black crayon and drew an X over the piece of paper she had been using. Saddened, Coles asked her why she had destroyed the picture. "That," she said, "is what people think of us, but they are wrong, and I know they are, and if I could, I'd tell them!"

A project undertaken by UPI photographer Jim Hubbard and other news photographers gave homeless children an opportunity to "tell them" through the medium of photography. The photographers lent the children cameras and accompanied them as they learned how to use the equipment and began taking pictures of their friends and the environment in which they lived.

That environment was grim indeed. In Hubbard's words:

> Shelter life is a journey into despair. It is life on the edge. Many of the shelters I visited were the scenes of round-the-clock violence, drug dealing, abuse and cases of parental neglect, and widespread chaos. They were places not fit for a child. . . . Between 1987 and 1989, five children died [at one of the shelters]. Two were stabbed to death by their overburdened father. One was hit by a train behind the shelter while playing the children's favorite game of tag with the train. There wasn't even a playground for these children except on worn-out mattresses that they pulled from the trash. (Hubbard, 1991, p. 4)

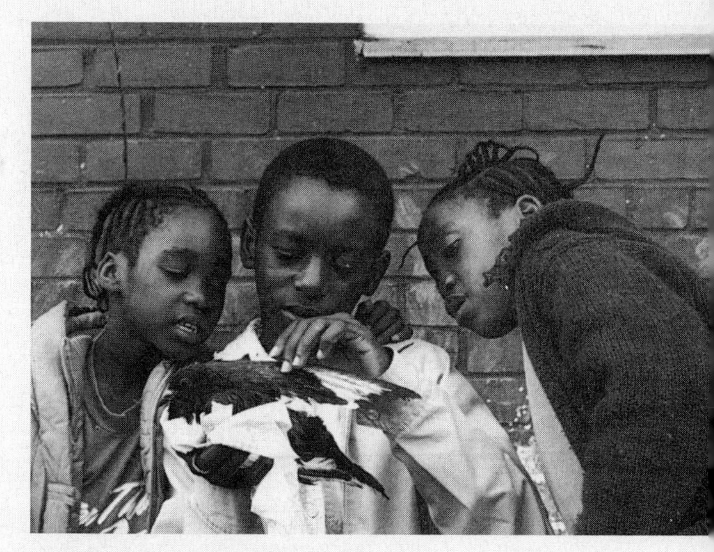

Charlene's comment—*My favorite picture that I took: that's me and my brother and my sister with a pigeon. My brother had found a pigeon, and the pigeon had been shot in the wing. My brother was trying to fix it, so I just told my brother to look at it, and I just put the timer on, and focused it, and I ran over there to get in the picture. I want to be a photographer.*

Comment to Charlene from schoolchild—*I saw your picture,* Birds, *in the exhibiit in New York. I saw it from way across the room and I felt drawn to it. . . . The girl on the left, her teeth show, like she understood pain and in that instant, experienced it for the bird.*

Chris's comment—*This is the kids on the train tracks, jumping. They are at the shelter. The best part of the shelter is the train track.*

The children's photographs provide an eloquent statement about their world. In the midst of despair,

Shawn talking with Jim—*A lot of the pictures that I had—I had sold them. Those pictures you all had blown up for me last time.* You sold them? *Yeah—to the people I took the pictures of.* The pictures you took of people here you sold to the people? Yeah. You are becoming a major businessman. I mean, I never heard of such a great thing. You were formerly selling crack. Now you are selling photographs. You may not make quite as much money, you probably won't get shot, you probably won't go to prison for selling photographs, plus it's a creative thing, follow? I'm going to help you on this, if you let me.

they found occasional moments of enjoyment and hope. They played, fought, and took care of younger siblings. Their pictures contain ample evidence of agents of socialization at work—parents, peers, older brothers and sisters, and the news photographers who taught them the basics of photography. When Hubbard asked the young photographers to comment on their pictures (a technique of visual sociology known as *photo elicitation*), they responded with warmth and intelligence. Some of their photographs, along with their comments, are reproduced here.

In 1989 Hubbard and his colleagues established a nonprofit center in Washington, D.C., to teach photography and darkroom skills, painting, drawing, and creative writing. In 1990 a selection of the children's photographs were exhibited at a gallery in Washington, and in 1991 many of them were published in a book titled *Shooting Back.* As Coles explains in his preface to the book, the children "shot back" with their eyes, mind, and spirit; their pictures are visual statements about the world in which they are growing up and a moral statement about the society that allows such a world to exist.

Patricia Williams, with daughter Charlene, and Vanessa Johnson, with son Dion (another *Shooting Back* photographer), before they were interviewed on the *CBS Morning Show* in 1990.

What is notable about the Shooting Back project in the context of debates about juvenile murderers like those described at the beginning of the chapter is that although shelter kids often experience the kinds of negative influences that are said to cause juvenile violence, they often become extremely creative in finding ways to survive and grow into productive adults. But their success often depends on a helping hand from caring adults like Jim Hubbard.

Comment from Calvin—*When I go back into the shelters, you know, I understand what they're feeling. I don't talk to them a lot about the environment because I know it's a low-life situation. They become your friends, so they open up to you more if you talk to them with the sense, like, I've been there before.*

SUMMARY

Socialization refers to the ways in which people learn to conform to their society's norms, values, and roles. Primary socialization consists of the ways in which the newborn individual is molded into a person who can interact with others according to the expectations of society. Secondary socialization occurs in childhood and adolescence, primarily through schooling; and adult socialization refers to the ways in which a person learns the norms associated with new statuses.

Among the most basic questions in the study of human socialization is that of "nature" versus "nurture": To what extent does the development of the person depend on genetic factors, and to what extent does it depend on learning? The first social scientist to develop a theory that addressed this issue was Sigmund Freud. Freud believed that the personality develops out of the processes of socialization through which the infant is gradually forced to control its biological urges. He divided the personality into three functional areas: the *id,* from which unsocialized drives arise; the *superego,* which incorporates the moral codes of elders; and the *ego,* or one's conception of oneself in relation to others.

In the growth of the personality, the formation of the ego or social self is critical. According to Freud, this takes place in a series of stages in which conflict between the demands of the superego and those of the id is always threatening to disrupt the functioning of the ego.

Behaviorism asserts that all behavior is learned. It originated in the work of Ivan Pavlov, who showed that behavior thought to be instinctual could in fact be shaped or *conditioned* by learning situations. This line of research was continued by John B. Watson, whose experiments revealed the ability of conditioning to shape behavior in almost any direction.

Studies of *feral children,* who have experienced extreme isolation or have been reared outside human society, show that such children are able to learn—but do so far more slowly than children who have not been isolated in early childhood. Other studies have found that normal development requires not only the presence of other humans but also the attention and love of adults. Children raised in orphanages and other nonfamily settings are more likely to develop emotional problems and to be retarded in their language development than comparable children reared by their parents.

Interactionist models of socialization stress the development of the social self through interaction with others. One of the earliest interactionist theories was Charles Horton Cooley's concept of the "looking glass self," the reflection of our self that we think we see in the behaviors of other people toward us. This concept was carried further by George Herbert Mead, who emphasized the importance of culture in the formation of the self. Mead believed that when children play, they practice *role taking,* or trying to look at social situations from the standpoint of another person. This ability develops through three stages. During the preparatory stage, children mimic the behavior of the *significant others* in their social environment. During the play stage, they play at being others who are significant in their lives. During the third stage, the game stage, they develop the ability to take the role of the *generalized other*—that is, to shape their participation according to the roles of the other participants.

In playing the roles for which they have been socialized, people adhere to rules of interaction known as "face work." They seek to present a positive image of themselves, their "face," and to avoid being embarrassed or "losing face."

Lawrence Kohlberg proposed a three-stage sequence of moral development in which the child's moral reasoning evolves from emphasis on reward and punishment to the ability to distinguish between social laws and moral principles. Other sociologists, especially Carol Gilligan, have challenged Kohlberg's theory on the ground that it does not distinguish between moral reasoning based on rules and justice (most common in males) and moral reasoning based on fairness and cooperation (most common in females).

Studies of the environments in which socialization occurs have found that normal development requires the involvement of one or more adults in the care of the child, as well as public policies that promote such involvement.

Agencies of socialization are the groups of people, along with the interactions that occur within those groups, that influence a person's social development. Within all agencies of socialization one finds a great deal of *anticipatory socialization,* in which the individual plays at a role that he or she is likely to assume later in life. After the family, the most important agencies of socialization are the schools. Other socializing agencies include day care centers, churches, leagues, and other associations. The dominant agency of socialization outside the family is the *peer group,* an interacting group of people of about the same age. Peer groups exert a significant influence on the individual from adolescence on. The mass media, especially television, are another significant agency of socialization in American society.

The roles a person plays over a lifetime are influenced by social change and by changes in the culture of his or her society. Socialization after childhood often occurs as a result of the influence of significant others and occupational mobility. A person's core identity shapes that individual's responses to new situations and challenges. *Resocialization* may occur at any time during adulthood. Sometimes people undergo resocialization to correct patterns of social learning that they and others find detrimental. Erik Erikson focused on *identification,* the social process whereby the individual chooses adults as role models and attempts to imitate their behavior.

An important aspect of socialization is *gender socialization,* or the ways in which we learn our gender identity and develop according to cultural norms of masculinity and femininity. Gender identity is an individual's own feeling of whether he or she is a male or a female.

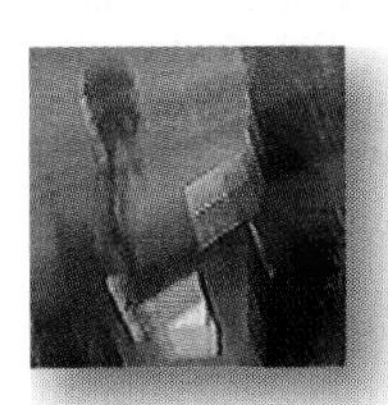

SOCIOLOGY VERSUS IDEOLOGY

Whom can we blame when children act in extremely antisocial ways? Some people immediately seek to place blame on the parents. They search for clues in the parents' behavior. Did they drink? Were they separated or divorced? Was there abuse in the home? Did the parents themselves engage in antisocial behavior? Other people seek explanations in the children's biological nature. Perhaps the children are suffering from imbalances in brain chemistry or unfortunate combinations of genetic influences that cause them to behave in antisocial ways. These arguments hardly exhaust the possibilities for debate, but they are heard so frequently that they may become a substitute for a careful search for the true causes of the behavior in question.

Sociologists cut through these debates by recognizing that we cannot yet fully explain episodes like the epidemic of juvenile murders in the late 1990s. We can gather facts and weigh the influences that help explain the behavior, but we need always to be aware that neither a biological nor a purely social explanation is sufficient. At this stage of our effort to understand such behavior, we need to explore how both biological and social factors influence an individual's actions.

GLOSSARY

socialization: the processes whereby we learn to behave according to the norms of our culture. (p. 126)

id: according to Freud, the part of the human personality from which all innate drives arise. (p. 127)

superego: according to Freud, the part of the human personality that internalizes the moral codes of adults. (p. 128)

ego: according to Freud, the part of the human personality that is the individual's conception of himself or herself in relation to others. (p. 128)

behaviorism: a theory stating that all behavior is learned and that this learning occurs through the process known as conditioning. (p. 128)

conditioning: the shaping of behavior through reward and punishment. (p. 128)

feral child: a child reared outside human society. (p. 129)

role taking: trying to look at social situations from the standpoint of another person from whom one seeks a response. (p. 134)

significant other: any person who is important to an individual. (p. 134)

generalized other: a person's internalized conception of the expectations and attitudes held by society. (p. 134)

agencies of socialization: the groups of people, along with the interactions that occur within those groups, that influence a person's social development. (p. 139)

agents of socialization: individuals who socialize others. (p. 139)

anticipatory socialization: socialization that prepares an individual for a role that he or she is likely to assume later in life. (p. 139)

peer group: an interacting group of people of about the same age that has a significant influence on the norms and values of its members. (p. 141)

resocialization: intense, deliberate socialization designed to change major beliefs and behaviors. (p. 145)

total institution: a setting in which people undergoing resocialization are isolated from the larger society under the control of a specialized staff. (p. 145)

identification: the social process whereby an individual chooses role models and attempts to imitate their behavior. (p. 146)

gender socialization: the ways in which we learn our gender identity and develop according to cultural norms of masculinity and femininity. (p. 146)

WHERE TO FIND IT

BOOKS

The Child in the Family and the Community, 2nd ed. (Janet Gonzalez-Mena; Merrill, 1998). A valuable review of research and theory about child socialization in families, with a strong emphasis on cross-cultural differences and multicultural societies.

Child, Family, Community: Socialization and Support, 3rd ed. (Roberta Berns; Harcourt Brace, 1993). A basic text in the field of socialization, with a good review of relevant sociological research.

The Child and Society, 5th ed. (Frederick Elkin and Gerald Handel; Random House, 1989). A basic text that offers in-depth coverage of most aspects of childhood socialization.

Not in Our Genes (R. C. Lewontin, Steven Rose, and Leon Kamin; Pantheon, 1984). A strong but fair critique of biological thinking in the field of human development.

In a Different Voice (Carol Gilligan; Harvard University Press, 1982); ***Mapping the Moral Domain*** (Carol Gilligan et al.; Harvard University Press, 1988); and ***Between Voice and Silence: Women and Girls, Race and Relationship*** (Jill McLean Taylor, Carol Gilligan, and Amy M. Sullivan; Harvard University Press, 1995). In these important volumes of her research, both solo and collaborative, the pioneering social psychologist Carol Gilligan presents theory and data to support her argument that classic research and theory on moral development are biased toward male development and hence neglect important aspects of socialization and personality development.

JOURNALS

Social Psychology Quarterly. A journal of the American Sociological Association that is devoted to recent research on "the processes and products of social interaction."

Sociological Studies of Child Development. An annual review of sociological research on socialization and child development. Begun in 1986, it represents many of the newest trends in this active area of sociological research.

Journal of Health and Social Behavior. A journal of the American Sociological Association that specializes in "sociological approaches to the definition and analysis of problems bearing on human health and illness" and often contains important articles on child welfare, socialization, and problems of health and development through the life cycle.

OTHER SOURCES

The Kids Count Data Book. (Annie E. Casey Foundation). An annual statistical review of the conditions under which children in the United States are being socialized; uses the best available data to measure the educational, social, economic, and physical well-being of children.

Basic Handbook of Child Psychiatry. (Joseph D. Noshpitz, ed.; Basic Books, 1979). A two-volume work that contains many excellent articles about human development. The focus is interdisciplinary, although there is also an emphasis on abnormal socialization and treatment.

Statistical Yearbook. Published annually by the United Nations. This is a valuable source of comparative data on infant mortality and other indicators of the health of children, as well as statistical material on education, literacy, and other measures of socialization.

INTERNET RESOURCES

Foundation for Children (www.unicef.org/). Offers excellent resources and links to other Web sites dealing with child development throughout the world, including problems of child labor, slavery, and abuse.

Kaiser Family Foundation (www.kff.org). Contains features on many aspects of socialization, sexuality, health, and agencies of socialization, with links to other important sites.

Foster Parent Home Page (http://fostercare.org/FPHP/). Includes debates on orphanages, surveys on child abuse and neglect, and analyses of media coverage of foster care.

Annie E. Casey Foundation (www.aecf.org). Publishes *Kids Count,* an annual review of the situation in the United States and elsewhere in the world. A valuable starting point for research on the problems of socialization.

Gender and Society, Trinity College (www.Trinity. edu/~mkearl/gender.html). Features much current thinking on gender socialization and related topics, with links to fine sites with even more resources on gender socialization.

American Psychological Association (www.apa.org/pubinfo). Provides links to topics like work and family as they relate to recent study and research.

CHAPTER 6

INTERACTION IN GROUPS

"I used to fight a lot when I was young. Every afternoon I think I was the main event. Every afternoon. There was times when I was scared to fight, but I used to fight, because if I didn't fight and I didn't win my sister would beat the shit out of me when I got home, or my cousins they would hit me, you know" (Moore, 1991, p. 62).

The speaker is a member of a Los Angeles gang; she is being interviewed by sociologist Joan Moore. An expert on gangs, Moore has spoken with generations of gang members about changes in gang activity and behavior. She is especially interested in finding out why the level of violence in gangs has been increasing in recent years. In this interview a young Mexican American woman describes the intense pressure she felt from her peers in the gang, and from members of her family as well, to be tough and to fight. When Moore asks her about losing fights, she reacts with great emotion: "No, I couldn't lose. They didn't care how punched up I came home. I had to be a winner. That was the motto

we had, you know, and if they beat you up you don't cry, you don't say nothing, you don't snivel" (Moore, 1991, p. 62).

The gang members Moore interviews—many of whom she has known for more than 20 years—always mention the need not to back down, not to "be a punk." This emphasis on gaining and maintaining the respect of fellow gang members is an often-recurring theme in the study of gangs. Moore's research also shows that when good jobs and legitimate careers are hard to find—as is often the case in poor minority neighborhoods—the level of violence attached to maintaining respect rises. And as more gang members have served time in prison, their options for life outside the confined world of the neighborhood and its gang have diminished. In consequence, the gang peer group may have a stronger and more lasting influence on this young woman than would have been true in the past. (Of course, gangs are only one of many types of friendship groups. For insights into other types of friendship groups, and glimpses of the strong emotions friends often express within them, see the Visual Sociology feature at the end of the chapter.)

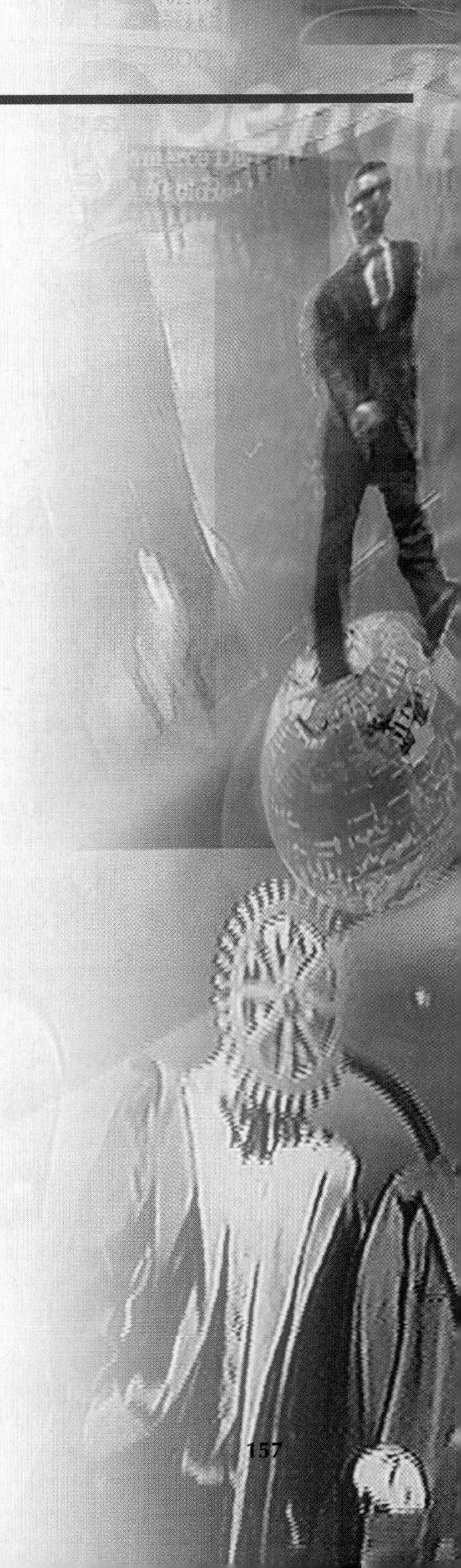

THE IMPORTANCE OF GROUPS

Moore's study of Los Angeles gangs is rich in examples of how people interact in groups. It also shows how frequently the strong norms or values of a peer group lead gang members to behave violently in particular situations. Other sociologists confirm Moore's findings about the importance of "respect" or honor among gang members (Horowitz, 1985; Sanchez-Jankowski, 1991). Martin Sanchez-Jankowski, who has conducted extensive comparative research on gangs throughout the United States, concludes that, contrary to what people often believe, most gang members do not enjoy fighting. Indeed, those who are not among the gang's leaders fear violence. However, as he reports, "Fear of physical harm . . . is not sufficient to completely deter gang violence, because most gang members . . . believe that if you do not attack, you will be attacked" (1991, p. 177).

The social fabric of modern societies is composed of millions of groups. Gangs are only one example. Some groups are as intimate as a pair of lovers. Others, like the modern corporation or university, are extremely large and are composed of many interrelated subgroups. We need to perform well in all of these groups, and to have this ability is to be considered successful by others. But the knowledge needed for success in a group with a formal structure of roles and statuses is quite different from the "smarts" needed for success in intimate groups.

In this chapter we analyze some important aspects of the way people behave in groups. We begin by describing the most common types of groups and showing how they can be linked together by overlapping memberships. This first portion of the chapter illustrates the functionalist view of group interaction. It looks at the structure of groups and the way they function in society. It also provides typologies of groups and insights into how they are organized. However, simply describing categories of groups and their structure does not explain how groups maintain the commitment of their members or why they change. In the middle portion of the chapter we examine these questions, focusing on the give-and-take that occurs in groups of different types and on how these interactions produce group norms. This section illustrates interactionist research on behavior in groups. In the final section of the chapter we show how formal organization and bureaucracy operate to control conflict and cope with change in groups. The concepts introduced here will be used over and over again in other chapters, for, as we saw in Chapter 4, groups of all kinds are the building blocks of society.

Characteristics of Social Groups

We sometimes group people together artificially in the process of analyzing society (e.g., all the teenagers in a particular community). This results in a **social category,** a collection of individuals who are grouped together because they share a trait deemed by the observer to be socially relevant (e.g., sex, race, or age). But there are also groups in which we actually participate and in which we have distinct roles and statuses. These are **social groups.** A social group is a set of two or more individuals who share a sense of common identity and belonging and who interact on a regular basis. Group members are recruited according to specific criteria of membership and are bound together by a set of membership rights and mutual obligations. Throughout this chapter, unless indicated otherwise, when we refer to groups we mean social groups.

Group identity, the sense of belonging to a group, means that members are aware of their participation in the group and know the identities of other members of the group. This also implies that group members have a sense of the boundaries of their group—that is, of who belongs and who does not belong. Groups themselves have a social structure that arises from repeated interaction among their members. In those interactions the members form ideas about the status of each and the role each can play in the group (Hare et al., 1994; Holy, 1985; Homans, 1950). A skilled member of a work group, for example, may become the leader while an unskilled member may become recognized as the person others instruct and send on errands.

Through their interactions, group members develop feelings of attachment to each other. Groups whose members have strong positive attachments to each other are said to be highly cohesive, while those whose members are not very strongly attached to each other are said to lack cohesion. Groups also develop norms governing behavior in the group, and they generally have goals such as performing a task, playing a game, or making public policy. Finally, because groups are composed of interacting human beings we must also recognize that all groups have the potential for conflict among their members; the resolution of such conflicts may be vital to continued group cohesion.

Membership criteria, awareness and boundaries, a clear social structure, cohesion, norms and goals, and the possibility of conflict are among the dimensions that define different groups. If we briefly compare two types of groups—say, a computer bulletin board and a married couple—the importance of these dimensions becomes apparent. Membership in the couple is exclusive and cohesion is based on intimacy and trust. If he comes home and says, "Honey, I brought a new woman into our relationship—I think you'll like her," she is likely to throw him out—unless they happen to come from a culture that practices polygyny, and even then the sudden introduction of a new woman into the household will disrupt the cohesion of the original marital group. Among members of the computer bulletin board, in contrast, the only criterion for membership may be ownership of the proper equipment and payment of dues. Not all the members will know each other, and new members can be added without any sense of reduced cohesion as would occur in the case of the married couple.

Beaches are excellent places to observe the way people group themselves in public. In this photo, note the way people cluster in primary groups and tend to leave as much space as possible between themselves and other groups.

Group Size and Structure

If one walks along a beach on a nice summer day, the variety of groups one sees is representative of the range of social groups in the larger society. Many of the groups on the beach are composed of family members. Others are friendship groups, usually consisting of young men and women of about the same age. Some of these are same-sex groups; others include both males and females. Then there are groups that form around activities like volleyball. Still other groups may be based on occupational ties; the lifeguards and their supervisors or the food service workers are examples of such groups.

Primary and Secondary Groups The distinction we made in Chapter 4 between primary and secondary groups helps clarify some of the differences among the groups one encounters on a sunny beach or anywhere else in society. Charles Horton Cooley defined **primary groups** as "those characterized by intimate face-to-face association and cooperation. They are primary in several senses, but chiefly in that they are fundamental in forming the social nature and ideals of the individual" (1909, p. 25). Out on the beach, the family groups are, of course, examples of primary groups. So are the groups of friends one observes lying near each other on the sand. And among the food service workers or the lifeguards there are probably some smaller friendship groups or cliques within the larger occupational group.

The occupational groups serving the beachgoers are examples of **secondary groups** (a concept we also introduced in Chapter 4 but need to use again here). The concept of a secondary group follows from Cooley's definition of primary groups, although he did not actually use the term (Dewey, 1948). Secondary groups are characterized by relationships that involve few aspects of their members' personalities. In such groups the members' reasons for participation are usually limited to a small number of goals. The bonds of association in secondary groups are usually based on some form of contract, a written or unwritten agreement that specifies the scope of interaction within the group. All organizations and associations, including companies with employers and employees, are secondary groups.

Within most secondary groups one can usually find a number of primary groups based on regular interaction and friendship. Scottish clans and Native American tribes, for example, are secondary associations that link individuals together through a network of kin relations. But within the clan or tribe there are many primary groups, often composed of men or women of roughly the same age.

Group Size and Relationships As the number of people in a group increases, the number of possible relationships among the group's members increases at a greater rate. When there are two people in the group, there is just one relationship. But when a third person is added there are three possible relationships (A–B, B–C, A–C). When a fourth person joins the group there are six possible relationships; a group of only six people includes fourteen possible relationships. Box 6.1 on pages 162–163 describes this phenomenon in more detail.

When a group has more than six members, the relationships among the members become so complex that the group is likely to break up into smaller groups, at least temporarily. One sees this quite frequently at parties or other social gatherings. Five or six people can carry on a conversation in which all participate. But as more people arrive, the group tends to break up into several smaller groups of two, three, or four people. The difficulty of maintaining relationships as group size increases becomes greater when there are limited resources, such as seats in cars or at restaurants, and the group is forced to define the smaller groups that will interact more intimately. Intimacy and feelings of closeness to others decrease with increased group size.

The more possible relationships there are, the more likely it is that some of those relationships will be troubled by conflict and jealousy. In consequence, the larger the group, the more unstable it tends to be and the more likely it is to break up into smaller groups in which there are fewer possible relationships and a greater likelihood of positive bonds among all members. If a larger group devoted to specific goals is to maintain its cohesion in the face of the pressures created by primary attachments among its members, it will need a formal, agreed-upon structure of leaders and rules. Thus a group of friends who decide to meet regularly to play a favorite game (it could be golf, basketball, or any other activity) will find that as their numbers grow they need a formal structure—an elected or appointed leader, a calendar of events, a membership list, and written rules of conduct. In short, it will be transformed from a primary group into a secondary group or, more specifically, a formal organization.

Dyads and Triads The basic principles of group structure and size were first described by the pioneering German sociologist Georg Simmel (1858–1918). Simmel perceived the need to study behavior in small and larger groups because groups are the basic units of life in all societies. Among his other contributions, he pointed out the significance of what he termed **dyads,** groups composed of only two people, and **triads,** groups composed of three people. He recognized that the strongest social bonds are formed between two people, be they best friends, lovers, or married couples. But he saw that dyads are also quite vulnerable to breakup, since if the single relationship on which it is based ends, the dyad ends as well.

When the dyadic bond is strong (and the two who share it often jealously guard their intimacy), the introduction of a new person into the group frequently causes problems. Conventional wisdom has much to say about this matter ("Two's a couple, three's a crowd"), and if one observes children's play groups it soon becomes clear that much of the conflict they experience has to do with the desire to have an exclusive relationship with a best friend. In families, too, the shift from a dyad to a triad often creates problems. A couple who are experiencing a high level of conflict may believe that a baby will offer them a challenge they can meet together, thereby renewing their love for each other. At first this may seem to be the case, but as the infant makes more demands on the couple's energy and time, the father may feel that the baby is depriving him of attention from his wife, and he may become jealous of the newcomer. For some couples, the addition of a child can actually increase the chance of divorce—resulting in a stable mother–child dyad and a single male.

The instability of triads can also be seen among siblings. The third child often has a rough time of it. Not only is the third child "the baby" who imagines it impossible ever to perform as well as the older siblings, but he or she has committed the unintentional sin of breaking the bond between the two older children. Whenever one of the latter becomes intimate and loving toward "the baby," the other will demand attention and remind the third child that he or she is the newcomer. This point should not be exaggerated; sibling triads can be stable and happy. In general, however, the addition of a third child increases the possibility of conflict and coalitions within the family (Caplow, 1969; Deaux & Wrightsman, 1988).

Communities

At a level of social organization between the primary group and the larger institutions of the nation-state are communities of all descriptions. **Communities** are sets of primary and secondary groups in which the individual carries out important life functions such as raising a family, earning a living, finding shelter, and the like. Communities may be either territorial or nonterritorial. Both include primary and secondary groups. In general, **territorial communities** are contained within geographic boundaries, whereas **nonterritorial communities** are networks of associations that form around shared goals. When people speak of a "professional community," such as the medical or legal community, they are referring to a nonterritorial community.

Territorial communities are populations that function within a particular geographic area, and this is by far the more common meaning of the term as it is used both in everyday speech and in social-scientific writing (Lowry, 1993; Suttles, 1972). Territorial communities are usually composed of one or more *neighborhoods.* The neighborhood level of group contact includes primary groups (particularly families and peer groups) that form attachments on the basis of proximity—that is, as a result of living near one another.

In studies of how people form friendship groups in suburban neighborhoods, both Herbert Gans (1976) and Bennett Berger (1961) found that proximity tended to explain patterns of primary-group formation better than any other variable except social class. Thus families that move into new suburban communities tend to find friends among others of the same social class who live near enough to allow ease of interaction and casual visiting; the same is true of students living in dorms on college campuses. On the international level, observers of the ethnic violence in Rwanda have reported that where members of the Hutu and Tutsi tribes live close to each other and cooperate in bringing in the harvest, they are far less likely to bear grudges and resort to violence than members of the same tribes who do not interact in their daily lives (Bonner, 1994).

That physical propinquity and social-class similarities explain how friendships form may seem self-evident to us. Moving into new neighborhoods or into dorms is a constant feature of life in a large urban society. But the idea that mere proximity, rather than kinship or tribal membership, can explain patterns of association would shock many people in simpler agrarian societies, where mobility is not a fact of everyday life. As Robert Park (1967/1926) noted, it is only under conditions of rapid and persistent social change that proximity helps explain why people form groups.

Networks

In modern societies people form their deepest friendships in face-to-face groups, but these groups are, in turn, integrated into larger and more impersonal secondary-group structures that may extend well beyond the bounds of territorial communities. William F. Whyte opened *Street Corner Society* (1943), his study of street-corner peer groups, with this observation:

> The Nortons were Doc's gang. The group was brought together primarily by Doc, and it was built around Doc. When Doc was growing up, there was a kid's gang on Norton Street for every significant difference in age. There was a gang that averaged about three years older than Doc; there was Doc's gang, which included Nutsy, Danny, and a number of others; there was a group about three years younger, which included Joe Dodge and Frank Bonilli; and there was still a younger group, to which Carl and Tommy belonged. (p. 3)

Like the gangs Joan Moore studies in contemporary Los Angeles, the Nortons were active in a network of neighborhood-level peer groups. They would compete against other groups of boys of roughly the same age in a yearly round of baseball, bowling, and occasional interneighborhood fights (the latter occurring mainly in their younger teenage years). But the Nortons were also integrated into the community through some of its secondary associations. For example, at election time they were recruited by the local political party organization, and some were recruited by the racketeers who controlled their low-income neighborhood. Party organizations and organized crime groups are examples of secondary associations that extend outside territorial communities.

In-Groups and Out-Groups The "corner boys" studied by Whyte tended to be hostile toward similar groups from other street corners and even toward certain boys from their own territory. These distinctions between groups are common and are referred to as *in-group–out-group distinctions.* The **in-group** consists of one's own peers, whereas **out-groups** are those whom one considers to be outside the bounds of intimacy. Simmel observed that in-group–out-group distinctions can form around almost any quality, even one that many people would not consider meaningful at all. Thus, in a study of juvenile groups in a Chicago housing project, Gerald Suttles (1972) found that distinctions were made between boys who lived in lighter or darker brick buildings. Similarly, American children who have lived in foreign countries for several years may find themselves relegated to out-groups when they return to school in the United States (Smith, 1994, 1996).

In-group–out-group distinctions are usually based on such qualities as income, race, and religion. In "Cornerville," the community Whyte studied, group boundaries were based on educational background,

BOX 6.1 SOCIOLOGICAL METHODS

Diagramming Group Structure

A variety of techniques, collectively known as *sociometry,* are used to study the structure of groups. The three most frequently used sociometric methods are basic group diagrams, diagrams that indicate the valence of group bonds, and sociograms, which chart individuals' preferences in groups.

At their simplest, group diagrams show the number of members in a group. Each member is represented by the same symbol, and the presence of a relationship between two members is shown by a line. The group diagrams in the figure labeled A show how the number of possible relationships increases geometrically as the number of members in the group increases arithmetically.

In a slightly more sophisticated type of diagram, a bond between group members is shown only if it actually exists in the group's interactions. Where there is no relationship between two members, no line is drawn between them. In a peer group it is usual for all the members to have a relationship, even if they do not all share the same strong feelings of friendship. But in work groups it is not uncommon for people to have relationships that are based on cooperation in carrying out specific tasks. If their work does not bring them together, they may not have any relationship at all.

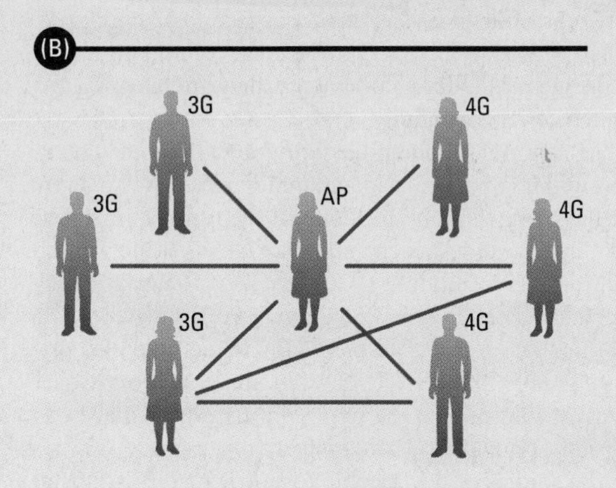

Diagram B represents a group in a school where the third grade (3G) and fourth grade (4G) teachers who teach language arts are members of a committee convened by an assistant principal (AP). The diagram shows

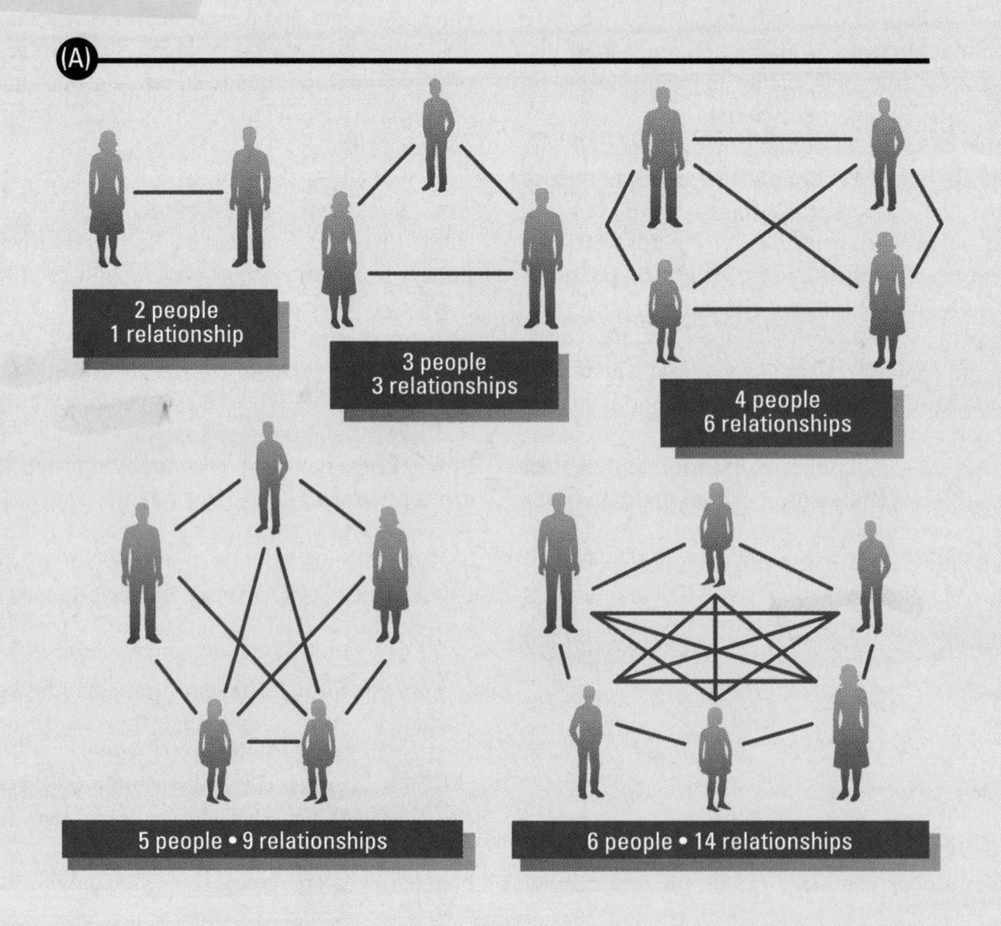

that three of the teachers share relationships (either through cooperation in projects at work or through friendship), while the other teachers are linked to one another only through their work with the assistant principal.

Valence is the feeling that exists between any two people in a group. It refers to the value they place on their relationship. The most basic valence in a relationship is positive or negative—that is, like or dislike. In most friendship groups there is a balance of valences; otherwise the group will very likely break up. In diagram C, for example, Joe likes Dave and Marty and they reciprocate that liking, but Dave and Marty do not like each other. Joe must either convince the other two to like each other or choose between them.

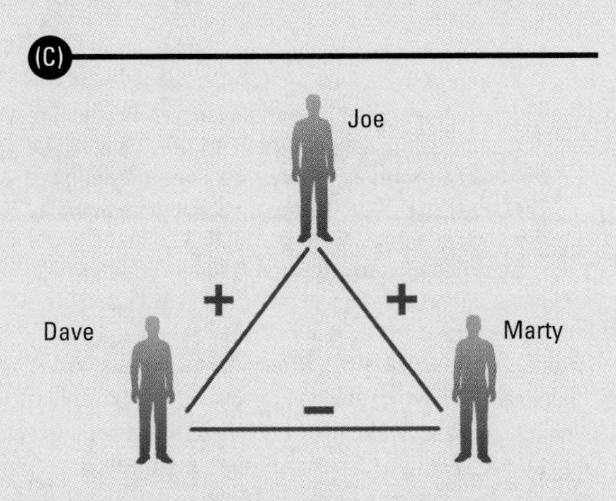

A difference in valence between two people can be indicated with a double line and two different signs. In diagram D, we see that Dave now likes Marty but Marty does not reciprocate. Perhaps he would like to have an exclusive relationship with Joe and views Marty as a threat. But now the pressure is all on Marty. If Joe can show him that Dave likes him and if Dave continues to try to get along with him, the group is very likely to form a balanced triad, as shown in diagram E.

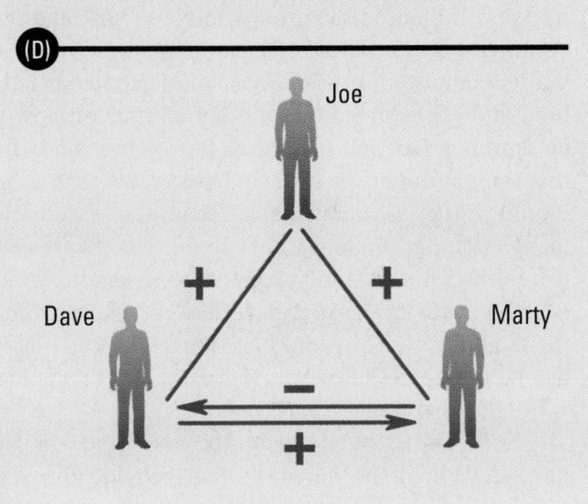

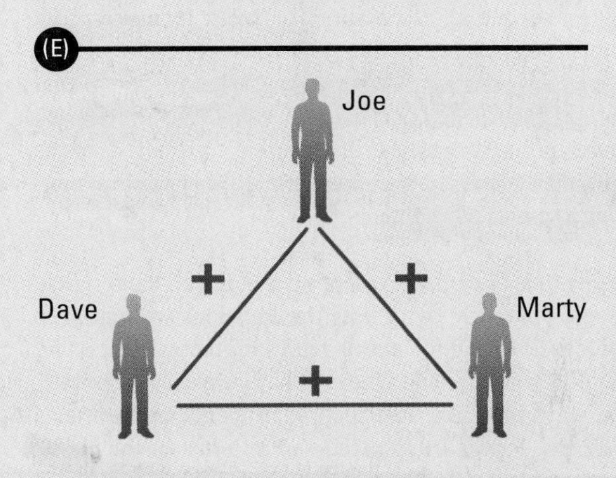

In a larger group, such as a school classroom or a club, the members express their preferences for each other in many ways. They call each other to talk about the events of the day or week; they gossip about each other; they give each other advice or criticism. These preferences result in a structure of cliques that exists within the larger group. When social scientists wish to chart such a structure, they simply ask all the members of the group to list the three or four individuals in the group with whom they would most like to spend time. By indicating each choice with a line, the researchers create a diagram known as a sociogram. Diagram F is a simple sociogram showing preferences among nine boys and girls. It reveals, among other things, that there is a clique centering on Dave and another centering on Latasha.

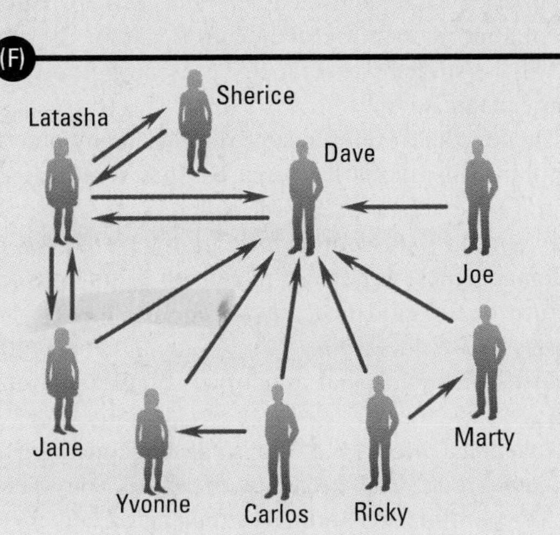

particularly college versus noncollege education. One of Whyte's informants explained this distinction:

> In Cornerville the noncollege man has an inferiority complex. He hasn't had much education, and he has that feeling of inferiority. . . . Now the college man felt that way before he went to college, but when he is in college, he tries to throw it off. . . . Naturally, the noncollege man resents that. (Whyte, 1943, p. 79)

In-group–out-group distinctions often make it difficult for secondary associations to attract members from both groups. For example, when two ethnic or racial groups in a community make in-group–out-group distinctions, they often find themselves drifting toward different political parties or forming distinct factions within the same party rather than uniting to find solutions to common problems.

Reference Groups An important area of sociological research is the study of how the decisions we make are influenced by significant or relevant others—that is, by people we admire and whose behavior we try to imitate. The ways in which other people influence our attitudes and behavior are often discussed in terms of the **reference group,** a group that the individual uses as a frame of reference for self-evaluation and attitude formation (Deaux & Wrightsman, 1988; Merton, 1968). Some reference groups set and enforce certain values and norms, while others serve as a standard for comparison (Hare, 1992).

An example of the influence of reference groups can be seen in a study of students at Bennington College conducted by Theodore Newcomb (1958). Newcomb found that conservative students often joined groups of like-minded peers who would affirm the values they had brought with them from home. They also tended to go home for visits more often than liberal students did. By their senior year, however, they were likely to have shifted their attitudes to match those of a more liberal reference group, since the majority of the students and faculty at Bennington were liberal or radical.

The concept of reference groups has many practical applications. In market research, for example, it is common practice to determine where potential customers tend to seek help in forming their attitudes—which magazines, writers, or television commentators they turn to for guidance. These *opinion leaders,* as they are sometimes called, are the customer's reference group in a particular area of consumption (Katz, 1957).

Note that a reference group is not the same thing as a primary group. The friends with whom you spend much of your time are a primary group and are likely to act as a reference group as well. But we can have numerous reference groups, each of which influences us in different areas. As you choose a profession or career, you will be influenced by the values and norms of reference groups in that field. As you gain new statuses, such as that of parent, voter, or supervisor, you will look to other reference groups: child care experts, political commentators, and the like. A reference group thus can be quite limited in its relationship to you as an individual—its influence may act on what you are to become rather than on what you are now.

Social Network Analysis The study of whom people associate with, how those choices are made, and the effects of those choices on social structure and individual personality is known as *social network analysis.* The British social scientist Elizabeth Bott, one of the founders of this branch of sociology, describes a social network as follows:

> Each person is, as it were, in touch with a number of people, some of whom are directly in touch with each other and some of whom are not. . . . I find it convenient to talk of a social field of this kind as a network. The image I have is of a set of points, some of which are joined by lines. The points of the image are people, or sometimes groups, and the lines indicate which people interact with each other. (1977, p. 256)

The diagrams in Box 6.1 illustrate social networks.

Social network analysis grew out of studies of the choices people make in becoming members of various social groups. Today network analysis extends beyond this rather narrow subject to studies illustrating how the interconnectedness of certain members of a society can produce interaction patterns that may have a lasting influence on the lives of people both within and outside the network (Cochran et al., 1990). Sociologist Aubrey Bonnett (1981), for example, studied *susus,* the rotating credit associations formed by West Indian ethnic groups in New York and other American cities. These associations are actually networks composed of numerous small groups of neighbors and kin who join with other similar groups, pay an entrance fee, and take turns borrowing funds from the association at far lower interest rates than a bank would charge. Bonnett showed that these associations bring together members of immigrant groups with different histories and values. As one *susu* organizer told him, "In my *susu* I have had members from all the islands. . . . You would be amazed to know how it has helped to get ourselves together, yes, this is really a West Indian thing" (p. 67).

Sociologists who study the rich and powerful have described the networks that provide the social

Then and Now

From Haymarket Square to Waco

Although they occurred almost a century apart, the bombing in Chicago's Haymarket Square in 1886 and the tragic assault and fire at Waco, Texas, in 1993 reveal many similarities in the way radical groups and agencies of the state become enmeshed in situations that lead to tragedy. The Haymarket bombing took place during a demonstration calling for an eight-hour workday held by a small, intensely cohesive group of committed anarchists, most of whom were immigrants from Europe, especially Germany. The anarchists were devoted to the destruction of the capitalist system of production and private property in the United States and elsewhere. Their leaders, including Albert Parsons, were speaking at the rally when Chicago police began forcing them to disperse. A bomb was thrown into the crowd, killing seven police officers and three bystanders and wounding at least 100 others. Authorities immediately blamed the anarchists and their followers for the explosion, and their leaders were jailed. Four were hanged; one committed suicide. But their guilt was never well established, and after intense lobbying by labor groups and civil rights activists, Illinois Governor John Peter Altgeld pardoned the remaining anarchists in 1893.

The followers of David Koresh, known as the Branch Davidians, were a dedicated religious group, with little concern for destroying capitalism but very firm ideas about the need to live a highly spiritual life apart from what they perceived as a corrupt society. They believed that they needed to arm themselves against a hostile world. Accusations of child abuse and felonies by their leader only confirmed the group's fears and desire to protect themselves. A long and sensational armed standoff between federal law enforcement agencies and the Branch Davidians ended in gunfire and the deadly fire pictured here. The Waco tragedy, like the Haymarket bombing, became symbolic both of the excesses of ideologically motivated and isolated extremist groups and of the blundering authoritarianism of law enforcement agencies (Niebuhr, 1995).

contacts essential for success in entering and climbing upward in the corporate hierarchy (Cookson, 1997). In their book *Preparing for Power,* for example, Peter Cookson and Caroline Persell (1985) describe how young men from elite social backgrounds develop strong attachments with others like them at prestigious prep schools, primarily (but not entirely) located in New England or Virginia, such as Phillips Academy (founded in 1778) and Woodbury Forest School in Virginia (founded in 1889). Through attendance at these schools, the children of the wealthy acquire a great deal of "cultural capital." All students gain some forms of cultural capital by mastering a range of academic and social skills, but prep school students acquire a special form of cultural capital that prepares them for life among the elite. They learn how to speak; how to behave at formal social gatherings; how to behave as a guest at a country home; what colleges, clubs, companies, and communities one should seek to enter; and what kind of spouse one should marry. In short, prep schools "integrate new brains with old wealth to revitalize the upper classes" (Cookson & Persell, 1985, p. 187).

Prep school students' social networks come into play when they apply to elite colleges, particularly Ivy League institutions like Harvard, Yale, and Princeton. Many of the administrators and admissions officers at the most prestigious colleges were former prep school students themselves and tend to show preference for applicants from those schools. Figure 6.1 shows that when graduates of elite prep schools apply to the most competitive colleges, even those with low SAT scores are highly likely to gain admission. Later in life they will continue to benefit from the social bonds formed at prep school. As can be seen in Figure 6.2, business managers from upper-class backgrounds (those who are most likely to have attended elite prep schools) are more likely to become chief executive officers or to hold seats on the boards of major corporations than those from other social-class backgrounds.

In other research on social networks, sociologists have demonstrated how the networks of individuals and groups within which people interact provide social support in times of stress or illness. Interestingly, this research often finds that people's social-support networks shrink as they become elderly and infirm and lose the ability to offer support or other resources in exchange (Fischer, 1982; Peek, Zsembik, & Cloward, 1997).

FIGURE 6.1 **Acceptance at Highly Selective Colleges, by Prep School Status and SAT Scores**

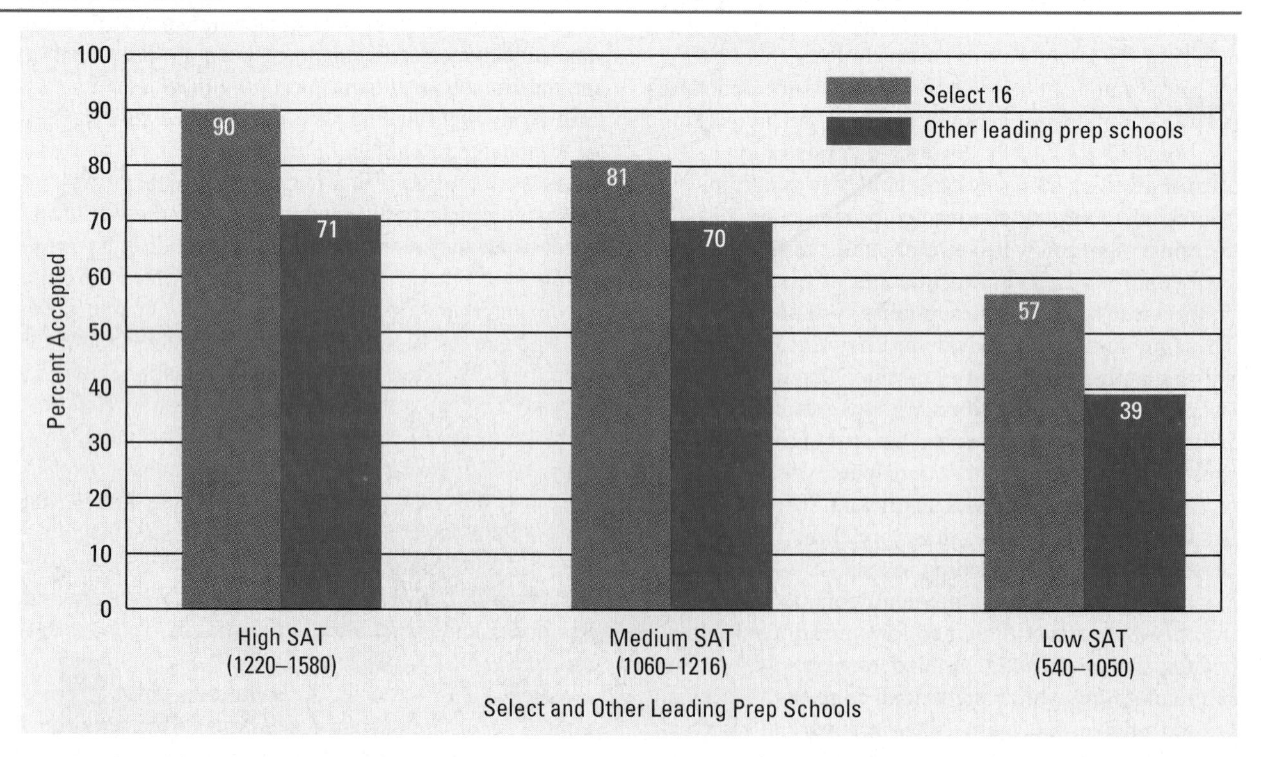

Source: Adapted from Cookson & Persell, 1985.

FIGURE 6.2 **Achievement of High Corporate Position, by Manager's Class Background***

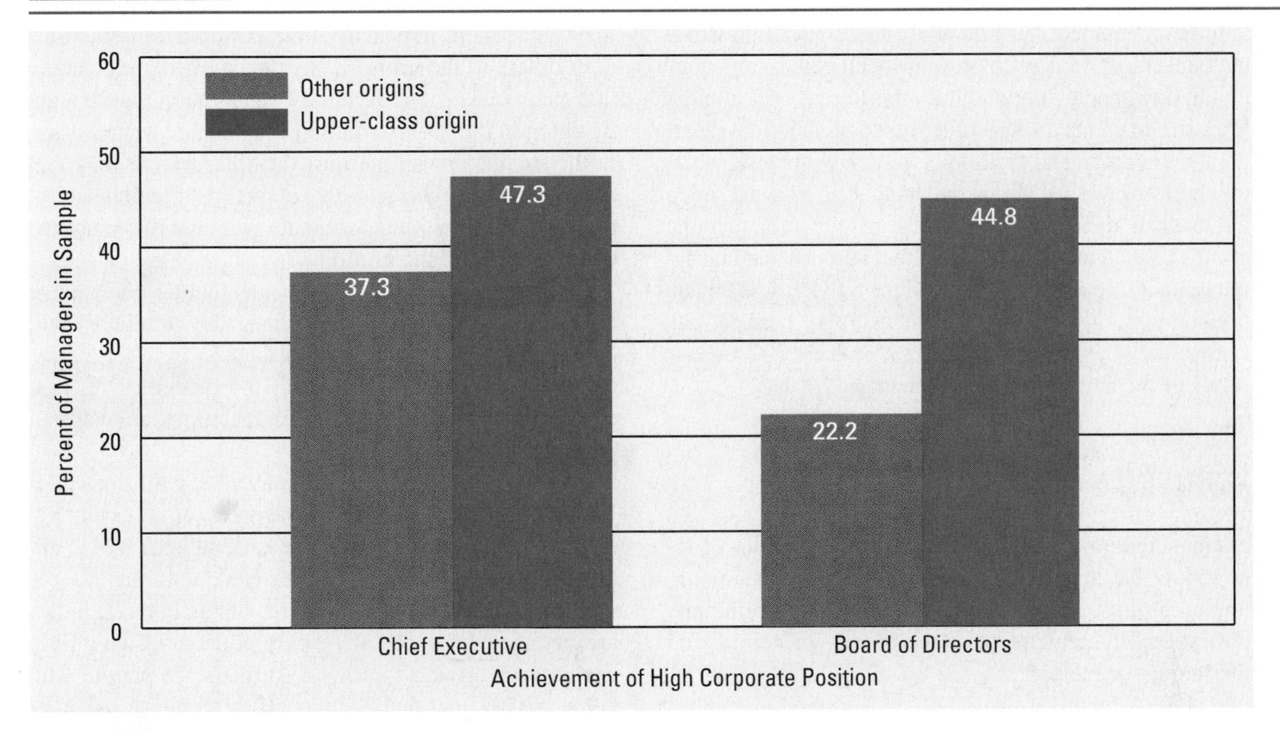

*Includes all college backgrounds.

Source: Adapted from Cookson & Persell, 1985.

INTERACTION IN GROUPS

In the preceding section we discussed the basic building blocks of social structure: the primary group and the secondary group. We saw that these groups are embedded in more complex structures of groups, such as communities and social networks. We also saw how reference groups act to confirm an individual's values and behavior, and how the distinctions we make between in-groups and out-groups help us maintain the boundaries of the groups in which we participate. But these descriptions do not tell us what principles of interaction operate to produce cohesion or instability in groups. In this section, therefore, we take a closer look at interaction in groups.

Simply to observe that a society is a fabric of groups is to neglect two important questions: How can groups stay together in the first place, and how can a fabric of groups produce a cohesive society? Indeed, all around us we see evidence that groups constantly fall apart. Romantic ardor cools; marriages fail; friends argue and seek new friends; successful rock bands split up; businesses go bankrupt; entire nations are fragmented by social and economic upheavals. Certainly new groups are forming all the time, but others are continually breaking up.

When a group disintegrates we often hear explanations like "I felt that I was only giving and not getting"; "We started to grow in separate directions"; "I no longer shared their values"; or "He [she, they] wanted too much of the [credit, money, power]." In short, when one listens to explanations of why groups fail to stay together, it sounds as if people are continually seeking more rewards for themselves or blaming others for taking too much of whatever rewards are available—too much attention, affection, approval, prestige, money, power, or whatever. If this is the case, it does not seem likely that groups could have much cohesiveness or stability.

One can imagine even more serious effects of human greed and self-serving behavior. The seventeenth-century English philosopher Thomas Hobbes was deeply concerned about the relationship between the individual and society. How is society, or any group within it, possible if every individual puts his or her own interests above those of everyone else? Hobbes warned that the sum total of everyone's selfish acts would be a "war of all against all." In *Leviathan,* his monumental treatise on this problem, he concluded that humanity needs an all-powerful but benevolent dictator to prevent social chaos.

Today people throughout the world reject this notion. We strive to create a social order based on individual freedom, democracy, and the rule of law. Nevertheless, wherever we look we see major threats to such a social order. Often such threats emerge when people feel that the groups they count on, such as political parties and legislatures, are not acting fairly. This brings us back to the question of how, given individual selfishness, humans can achieve order, cooperation, and trust in their associations rather than coercion, terror, and chaos.

Principles of Interaction

Social scientists have identified certain principles of interaction that help explain both stability and change in human groups. Among them are the pleasure principle, the rationality principle, the reciprocity principle, and the fairness principle.

The Pleasure Principle People seek pleasure and avoid pain—although this basic principle implies nothing about what any particular person considers pleasurable or painful. As we will see shortly, this is a very personal matter that depends on the norms that have been internalized by the individual. In fact, many people seem to find pain or discomfort desirable; examples include athletes and dieters, who seek pain or deprive themselves of pleasure in order to experience the thrill of athletic victory or the satisfaction of wearing a smaller size.

The pleasure principle does not address the question of what people value; all it states is that once one knows what kind of pleasure one wants, one will seek more of it and less of other values that yield less personal pleasure (Bentham, 1789; Homans, 1961). When applied to behavior in groups, the pleasure principle simply means that over time people will continue to interact in a group in which they experience the pleasure of reward but will withdraw from groups in which the pain they experience outweighs their pleasure.

The Rationality Principle In social life, a gain in pleasure may require some pain as well. For example, when two members of a group of friends become very close and find pleasure in their intimacy, the others may feel left out, and this causes some pain both to the two intimate friends and to those who feel left out. In larger groups, when a rule such as a ban on smoking is made, some members will derive more pleasure than others from the cleaner air. Those whose smoking pleasure is denied will experience an increase in pain. Consciously or not, people in groups calculate whether they think they will benefit personally from continued interactions with others in the group. Thus the smokers may decide that their loss of pleasure outweighs the gains they get by being in the group, and they may decide to withdraw. In the small friendship group, the other friends may feel that the pain of their jealousy outweighs the pleasure of their friendship with the intimate pair, and they may exclude them from the group.

The rationality principle means that people change their behavior according to whether they think they will be worse or better off as a consequence. In other words, in making decisions about their interactions with others, people tend to make rough calculations of costs and benefits.

Does the idea that people seek a net gain from their interactions mean that they are greedy or self-centered or lacking in motives other than materialism? Not at all. Of course, there are extremes: Hedonists are people who can enjoy only the pleasures of the body. Egotists are people who especially enjoy praise and have little capacity for giving to others. Altruists are people who enjoy giving to others, often at the expense of other pleasures. But most people strive to behave in ways that give them physical pleasure and ego gratification ("boosts" or "strokes") and also give them the satisfaction of helping others. How far they go in either direction depends on what other members of the group are doing (how giving the others are, how selfish, how self-sacrificing, etc.)—which brings us to the principle of reciprocity in interaction.

The Reciprocity Principle In group interactions people usually adhere to a norm stating that what others do for you, you should try to do for others. This is one of the world's most commonly followed rules of interaction; it is known as the norm of reciprocity (Homans, 1950). The Roman statesman Cicero recognized this principle when he wrote, "There is no duty more indispensable than that of returning a kindness. All men distrust one forgetful of a benefit." But it is not necessary to go back to ancient Rome for evidence that reciprocity is vital to continuing group interaction. Suppose that you and two classmates meet at a local pub. You pick up the check, and the other two agree to do so on future occasions. The next time, one of them pays the check. The third time, the third friend pleads a lack of cash and you each end up paying your own share. You and your second friend will feel that this is an embarrassing lapse on the part of your third friend. If such behavior continues, you will feel that the third friend takes more than he or she gives or thinks his or her company is worth the extra money paid by you. It is unlikely that this threesome will stay together very long if such imbalances continue.

Under what conditions could such a group continue in spite of the third friend's violation of the norm of reciprocity? One condition involves a loss of prestige for the third friend. You might ask him or her to make up the debt by doing something else for you. In that case the third friend takes a subordinate position in the group, doing your bidding to make up for the inability to reciprocate on an equal basis (Homans, 1961; Mauss, 1966/1925). On the other hand, suppose that the nonreciprocator is a movie star or a powerful politician. Then the other two members of the threesome might continue to pay in the hope that the third member would continue to spend his or her valuable time with them and that there would be an even greater reward for them in the future. This kind of relationship is quite common in social life. Social climbers seek ways to flatter and please someone who can confer prestige on them. Power seekers do the same in order to gain positions in which they will have power over others. A suitor may forgo reciprocity for quite a while in the hope of eventual conquest. In these situations and others like them, the participants in the interaction have an intuitive sense of what they are giving and what they are receiving or hope to receive in the future.

In the Bible the norm of reciprocity is stated in the form of the Golden Rule. "Do unto others as you would have others do unto you" in effect urges individuals to go beyond expectations of immediate reciprocity and set examples of higher moral behavior in the hope that others will follow them. In real life, however, people tend to prefer that their gains be as immediate as possible.

The Fairness Principle We see ample evidence that people tend to expect certain kinds of treatment from others and that they tend to become angry when they do not receive it, especially when they themselves feel that they have done what is expected of them in the situation. We want the rules to apply equally to everyone in the game, be it a friendly game of pool or the more complex game of corporate strategy. When we are not rewarded in the same ways as others, we say that we are being treated unfairly.

The unfair condition in which a person or group has come to expect certain rewards for certain efforts, yet does not get them while others do, is called *relative deprivation.* In times of economic depression, for example, all groups in society must make do with less, and all feel deprived in comparison to their previous condition. If the economy begins to improve and some groups begin to get higher wages or more profits while others experience only limited improvement, members of the less successful group will feel deprived relative to others and will very likely become angry even though their own plight has actually improved somewhat (Merton & Kitt, 1950). The concept of relative deprivation is useful in explaining some aspects of group behavior and appears again in Chapter 8 when we discuss how major social changes like revolutions occur.

People's ideas about what is fair in their interactions with others often conflict with simple calculations of gain and loss. The fact that they might come out ahead in an interaction does not guarantee that they will feel good about it and continue the interaction. This was demonstrated in a series of studies by Daniel Kahneman and his associates (1986). A large number of respondents were asked to judge the fairness of this situation:

> A landlord rents out a small house. When the lease is due for renewal, the landlord learns that the tenant has taken a job very close to the house and is therefore unlikely to move. The landlord raises the rent $40 more than he was planning to.

An economist might argue that the tenant should think of the situation in terms of whether the rent increase is offset by the savings in the cost of travel to work (in terms of money, time, convenience, etc.). If the tenant still comes out ahead, it will make sense to sign the new lease (the rationality principle). But more than 90 percent of the respondents in Kahneman's survey said that the landlord was being unfair. In real life, when people feel that a transaction like this is unfair they often end the interaction, even at some cost to themselves. In this case people intuitively felt that the landlord was unfairly taking advantage of a gain made by the tenant (the new job) without adding anything to the property to earn the right to raise the rent.

The pure-rationality model of behavior predicts that people will act in their own interests so as to maximize their profit. If we were simple profit takers, however, our feelings about fairness would not play such a strong part in explaining group interaction (Frank, 1988). But in fact these feelings are very powerful. Kahneman and his associates examined the strength of these feelings in an experiment in which they asked subjects to divide $20 with another player whom they could not see but were told was in the room. Only two choices were given: to give $10 to each player or to keep $18 and give the other player $2. Out of 161 subjects, 122—76 percent—offered the even split. From this the experimenters concluded that most people are motivated by their own ideas of fairness, presumably as a result of socialization over many years.

In real life, there are many examples of situations in which notions of fairness outweigh the principle of rationality. Barbers and beauticians do not charge more for haircuts on Saturday, nor do ski resorts usually charge more for lift tickets on holidays. Although they might like to profit from the higher demand at those times, they are afraid that people would regard the higher prices as unfair (Frank, 1988). On the other hand, there are many instances in which services are offered at higher rates during peak seasons or bargain offers such as early-bird specials or off-peak fares are used to encourage people to accept services at times other than those they might prefer.

Applications of the fairness principle can be seen in a great many group situations. In the classic study conducted by Elton Mayo and his associates at Western Electric's Hawthorne plant (see Chapter 2), the observers often noticed that workers used various forms of joking and sarcastic teasing to enforce the group's norms. They had a strong sense of what amount of work they should turn out, both individually and as a group, in order to merit their pay: a fair day's work for a fair day's pay. In the plant's Bank Wiring Room, where workers hand-wired electrical circuits, it was common to see the members of a work group gang up and in a joking way hit a fellow worker on the shoulder when he produced more than the other members of the group. This practice was called "binging." A worker was binged by his coworkers if he produced too much in a day or if he did not produce enough, according to their definition of a fair day's work.

The Hawthorne researchers quickly realized that binging was part of the work group's culture. That culture—the group's norms of conduct and the ways in which its members interpreted those norms—was unique to that group and applied only when the group was at work. The culture of this work group had evolved over many years, its norms functioning to control the workers' responses to the demands of the company's managers. Thus, in a later analysis of the Bank Wiring Room observations, George Homans (1951) made the following comments:

> [The workers] shared a common body of sentiments. A person should not turn out too much work. If he did, he was a "rate-buster." The theory was that if an excessive amount of work was turned out, the management would lower the piecework rate so that the employees would be in the position of doing more work for approximately the same pay. On the other hand, a person should not turn out too little work. If he did he was a "chiseler"; that is, he was getting paid for work he did not do. (p. 235)

The Economic Person Versus the Social Person

The four basic principles of interaction in groups—pleasure, rationality, reciprocity, and fairness—contain some obvious (and fortunate) contradictions. People calculate their individual gain and loss in interactions; they act according to their own interests in many situations. Does this mean that people will always act selfishly? Not always, for we have seen that people also have deeply ingrained notions of reciprocity and fairness, which often prompt them to act in favor of the needs of the group or of society as a whole. Indeed, many situations that arise in the social world demonstrate the "war of all against all" that would be created in a society whose members always acted in what they perceive as their self-interest.

Consider this example. It's the Friday evening rush hour. The streets downtown are crowded with cars, their drivers eager to get home. Although the light is about to turn red, some drivers move their cars into the intersection. They expect the traffic ahead of them to move forward so that they will be clear of oncoming traffic from the cross street. But alas, the traffic ahead does not move and the intersection is blocked. Now the cars on the cross street have a green light, but they cannot move. Fearing that they will miss their first opportunity to move, and amid much horn blowing and gnashing of teeth, some of these drivers also squeeze into the intersection. This chaotic and frustrating situation is known as gridlock. It is the result of a number of drivers acting in what they perceive as their own interests and not thinking about the needs of all the drivers as a group. The gridlock example is one among a great many that could be cited in which it is actually advantageous for individuals to behave in a prosocial manner rather than to maximize what they perceive as their immediate advantage (Hechter & Kanazawa, 1997; Shotland, 1985).

In many situations, however, there is serious conflict over what the best use of resources might be (Sen, 1997). Questions about who should have access to good medical care, quality education, or a safe and healthy environment involve considerations of individual gain versus the public good. It is not only because of the human tendency toward selfishness that such conflict arises. Often there are disagreements about what different groups or classes of individuals *deserve.* We will come back often to this issue—when we discuss social inequalities, for example, and when we look at the problems involved in increasing equality of opportunity in modern societies.

Communication and Behavior in Groups

"Human interaction," Herbert Blumer observed, "is mediated by the use of symbols, by interpretations, or by ascertaining the meaning of one another's actions" (1969b, pp. 179–180). By "mediated" Blumer meant that we do not normally respond directly to the actions of another person. Instead, we react to our own interpretations of those actions. When we see other drivers moving into an intersection against the light, for example, we begin to feel that everyone is out for themselves and we would be fools to hold back. These interpretations are made in the interval between the stimulus (the other person's action) and the response (our own action). Thus, in the account at the beginning of the chapter, the gang member must frequently decide whether her honor is being attacked or, perhaps more important, whether her gang sisters and family members will demand that she defend her honor. Her decision to fight or not to fight will be guided by her interpretation and by her need to gain standing in the eyes of her peers.

What factors explain how people decide how to act and whether they should act for their own benefit or according to some notion of what is good for society? This question has been the subject of much research in the social sciences. Let us therefore take a brief look at some research on how people define their needs in social situations.

Definitions of the Situation "Situations we define as real, are real in their consequences" stated W. I. Thomas (1971/1921), a pioneer in the study of social interaction. By this he meant that our understandings or definitions of what is occurring around us—whether they are correct or not—guide our subsequent actions. Thus, in a study of how dying patients are treated by medical groups in hospital emergency rooms, David Sudnow (1967) found that when a patient was brought in with no heartbeat, the person's age had a great deal to do with what happened next. The arrival of a younger patient who seemed to be dying would produce a frenzy of attempts to restart the heart. At times the entire group of medical personnel would become involved. But an old person with no heartbeat was far more likely to be pronounced dead on arrival, with little or no mobilization of the medical group. The patient's aged appearance caused the members of the group to define the situation as one in which urgent and heroic efforts were not required.

In another study of medical groups, Charles Bosk (1979) found that surgeons developed subtle techniques for covering up the mistakes of residents in training, depending on how a particular situation was defined. They would invoke those techniques when they believed a mistake was due to lack of knowledge. But mistakes they believed to be caused by carelessness would often be exposed. The guilty resident would be held up to public scorn, often at a high cost to the new surgeon's status among other doctors.

Both of the foregoing studies began by questioning how definitions of the situation account for the ways in which people interact. The emergency room teams that Sudnow observed were performing an unofficial cost–benefit analysis in deciding when to apply heroic measures to heart attack victims. The surgeons in Bosk's study were resolving a conflict between their teaching and medical functions: Young surgeons need to learn and practice, but patients need to receive the best possible care. The surgeon in training needs to maintain a good reputation among other doctors and nurses, even if he or she makes mistakes. Such conflicts are resolved by norms that define the legitimacy of mistakes in different situations. The subtle cues that define each situation are communicated through phrases, gestures, and other symbolic behavior, as well as through explicit evaluations of what was good and bad about a particular operation.

Ethnomethodology Studies based on this interactionist perspective use a variety of techniques. One of these is **ethnomethodology,** the study of the underlying rules of behavior that guide group interaction. This approach was used by Harold Garfinkel (1967) in a classic series of studies of how we use verbal formulas to create a flow of communication that we feel is normal and that we can understand. Garfinkel asked his students to engage in conversations with friends and family members in which they violated some of the simple norms that most people follow in carrying on a conversation. Here is an example:

VICTIM: [*waving cheerily*] How are you?

STUDENT: How am I in regard to what? My health, my finances, my schoolwork, my peace of mind, my . . . ?

VICTIM: [*red in the face and suddenly out of control*] Look! I was just trying to be polite. Frankly, I don't give a damn how you are. (Garfinkel, 1967, p. 44)

As this brief example shows, when the norms of spoken interaction that "entitle" us to continue or maintain a conversation are violated, the conversation often cannot continue because there is no longer a mutual area of interaction within which both parties tacitly agree to make sense.

The Dramaturgical Approach Another technique used in research on interaction in groups is the **dramaturgical approach.** This approach is based on the recognition that much social interaction depends on how we wish to impress those who may be watching us. For example, Erving Goffman (1965) observed that people change their facial expressions just before entering a room in which they expect to find others who will greet them. Couples who are fighting when they are "backstage" often present themselves as models of friendship when they are "frontstage"—that is, when they are in the presence of other people. And many social environments, such as hotels, restaurants, and funeral parlors, are explicitly set up with a front and a back "stage" so that the public is spared the noisy and sometimes conflicted interaction occurring "behind the scenes." Using these and other strategies to "set a stage" for our own purposes is known as **impression management.**

In a well-known study that applied this "dramaturgical" view of group interaction, James Henslin and Mae Briggs (1971) drew upon Briggs's extensive experience as a gynecological nurse. Their research showed how doctors and nurses play roles that define the situation of a pelvic examination as unembarrassing and attempt to save each participant's "face" or self-image. According to Henslin and Briggs, the pelvic examination is carried out in a series of scenes. First the doctor and the patient discuss the patient's condition. If the doctor decides that the patient needs a pelvic examination, the doctor will leave the room; this is the end of scene 1. In scene 2 the nurse enters. Her role is to work with the patient to create a situation in which the body of the patient is hidden behind sheets and her depersonalized pelvic area is exposed for clinical inspection. Through these preparations the patient becomes symbolically distanced from the doctor, allowing the next scene, the pelvic examination itself, to be desexualized. The props, the stage setting (the examining room), and the language used help define the situation as a nonsexual encounter, thereby saving everyone involved from embarrassment.

Altruism and the Bystander Effect Intentional efforts to set a stage for purposes of impression management are common in hospitals, restaurants, and many other social settings where patterns of behavior are predictable and regular. But in other public settings, especially streets, parks, and public transportation, people often try to hide behind their anonymity. They shy away from behavior that will be noticed, as illustrated in the following example of the *bystander effect.*

Late one night in 1964, Kitty Genovese was returning home from a social gathering when she was attacked by a man who had been following her. He caught up with her in front of a bookstore in the shadow of her apartment building. As she screamed for help, he stabbed her in the chest. Lights went on in some of the apartments, but no one intervened or called the police. One neighbor shouted from a window: "Let that girl alone." The stalker retreated and waited, and when the neighbor no longer seemed to be watching he resumed his attack. He stabbed and raped the young woman in a stairwell of her apartment house, but the police did not receive a call for help from any of her neighbors until thirty minutes after she had first cried out. Interviewed later, the neighbors said things like "I was tired" or "I assumed others were helping" or "Frankly, we were afraid."

This incident is often cited to illustrate the callousness and indifference of urban dwellers (Frank, 1988). It might seem from this and similar incidents that when faced with danger and other costs of helping (e.g., time in court, dealing with the police, loss of time from work) people make fairly selfish calculations of gain and loss. But what about all the unselfish acts of heroism one reads about? Victims are dragged from fires and submerged cars and broken ice by self-sacrificing civilians, many of whom die in their efforts to help others. Did Kitty Genovese simply have a bad group of neighbors? Empirical evidence suggests that the situation is more complicated than that (Batson, Batson, & Todd, 1995).

Studies by Bibb Latané and John Darley (1970) have shown under many experimental conditions that bystanders are likely to offer help to victims. But their research also shows that the presence of other people who do not help increases the chances that no one will help. In one simple experiment, a graduate student begins interviewing the subject. The interviewer excuses himself or herself for a moment and enters an adjacent room. Suddenly the waiting subject hears cries for help: "Oh my God, my foot . . . I . . . I can't get this thing off!" In about 70 percent of the trials, the subject rushes into the room and offers help. But when there is another person in the first room who just sits there and seems to be associated with the experimenter in some way, only 7 percent of the subjects intervene to help the "victim."

In the experiment just described, the helper can see that no one else has offered help. In the Kitty Genovese case, on the other hand, people could conveniently assume that others must be helping. The point is that when responsibility seems to be diffused rather than falling on a particular individual, people are more likely to avoid helping others. But when people cannot avoid defining the situation as involving them, they are likely to throw the rationality principle to the winds and intervene altruistically despite the potential costs to themselves (Deaux & Wrightsman, 1988; McIntyre, 1995).

Interaction and Group Structure

The examples just discussed involve decisions and strategies that individuals use in deciding whether to behave more selfishly or more altruistically in a given situation. But as members of groups we often explain our actions on the basis of our status in the group rather than on the basis of our individual expectations. "I must do this because I'm expected to lead," we might say, or "I can't decide that without talking to the others and especially to [the group's leader]." These explanations suggest that the structure of groups is a major factor in explaining behavior. But the basic principles of interaction—pleasure seeking, rationality, reciprocity, and fairness—can be used to explain the emergence of group structure in the first place (Beebe, 1997; Pillari, 1998). In an early and now famous series of experiments, sociologist George Homans and others provided scientifically verifiable data on the emergence of group structure through interactions.

The procedure is as follows. Six college students of the same sex are hired to participate in a small-group experiment. All are unknown to one another before the experiment. The specific task they are asked to perform is of little importance except that it must involve all the members of the group in a cooperative effort and must not be so difficult that they become frustrated or so easy that they become bored.

Initially the group has no structure. All the subjects are peers, and all have come for the same reason: to earn some money for participating in the experiment. There are no predetermined statuses or roles. The students listen to a description of what they are to do and then are left alone to get the work under way. Actually they are being observed through a one-way mirror, and all their interactions are being recorded and counted.

Table 6.1 summarizes (or *aggregates*) the data obtained by counting all the interactions—including small utterances like "Oh" and "I see"—occurring in a group of six male subjects meeting for 18 one-hour sessions. This table shows who initiated each interaction and to whom it was directed, including interactions directed toward the group as a whole. The participants are ranked from high to low on this variable (i.e., initiating an interaction). Thus subject 1 spoke 1,238 times to subject 2, spoke 961 times to subject 3, and so on (Homans, 1961). Subject 1 initiated 3,506 utterances to specific individuals (e.g., "Why don't you . . . ?") and 5,661 utterances to the group as a whole (e.g., "Why don't we . . . ?"), for a total of 9,167 utterances over all the sessions—more than any of the other participants. You can see that subject 6 spoke only about one ninth as often as subject 1 and that most of what he said was directed to subject 1 (470 utterances). In fact, subject 1 received the most interaction from all the other participants, illustrating that people who initiate interactions are more likely to receive them. In later interviews the experimenters also found that subject 1 was the most respected member of the group.

What makes the person who initiates and receives the most communications the most respected member of the group? The reasons vary with the specific tasks the group is performing, but if we assume competence at those tasks, usually the person at the center of the communication is spending a lot of time helping others as well as helping the group accomplish its goals. That person not only is concerned with his or her own performance (the rationality principle) but also takes pleasure in helping the others—giving approval, voicing criticism, making suggestions, and so on (the pleasure principle). In exchange for this help, the other members of the group give approval, respect, and allegiance to the central individual (the reciprocity principle).

Is the person who initiates the most interactions respected simply because he or she talks a lot? Common

TABLE 6.1 Aggregate Matrix for 18 Sessions of Six-Man Groups

From Individuals	Initiated to Individuals: 1	2	3	4	5	6	Total to Individuals	To Group as Whole	Total Initiated
1		1,238	961	545	445	317	3,506	5,661	9,167
2	1,748		443	310	175	102	2,778	1,211	3,989
3	1,371	415		305	125	69	2,285	742	3,027
4	952	310	282		83	49	1,676	676	2,352
5	662	224	144	83		28	1,141	443	1,584
6	470	126	114	65	44		819	373	1,192
Total Received	5,203	2,313	1,944	1,306	872	565	12,205	9,106	21,311

Note: Differences between the total number of interactions initiated and the number received are due to the fact that some utterances are answered indirectly by means of utterances directed at the entire group. The diagonal cells are empty because the experiments did not count utterances by a subject to himself.

Source: *Social Behavior,* by George Caspar Homans. Copyright © 1961 by Harcourt Brace Jovanovich, Inc. Reprinted by permission of the publisher.

sense tells us that this is far from true. People who talk a lot but have little to contribute soon find that no one is paying attention to them. Their rate of interaction then drops sharply. Table 6.1 cannot show such changes in interaction patterns, but they are frequently observed in small-group studies. Other patterns have also been identified. Robert F. Bales and Philip Slater (1955), for example, observed that the person who initiated the most interactions—both in getting tasks done and in supporting the suggestions of others—often came to be thought of as a leader. The group began to orient itself toward that person and to expect leadership from him or her. A second member of the group, usually the one who initiated the second-highest number of interactions, was often the best-liked person in the group.

Bales and his colleagues concluded that groups often develop both a "task leader" (or instrumental leader) and another leader, whom they called a "socioemotional leader" (or expressive leader). The former tends to adhere quite strictly to group norms and to take the lead in urging everyone to get the work done. But this often leaves other members with ruffled feelings. The socioemotional leader is the person who eases the group over rough spots with jokes, encouragement, and attention to the group's emotional climate. Thus most classes have a class clown whose antics help release the tension of test situations, and most teams have a respected member who can also joke around and get people to relax under pressure. At times these roles are performed by the same person, but generally as the group settles into a given task there is an informal division of labor between the two kinds of leaders (Beebe, 1997).

Recall the Bank Wiring Room observations mentioned earlier. The workers at the equipment wiring tables are an example of small-group structure. Their leader was the person who could turn out the most work in the shortest time and could also help the others do their work. The workers with higher status in the group were those who showed greater competence at their jobs and stronger allegiance to the norms of the group. Newer workers with lesser skills and less shared experience with the others were spoken to less and were less central to the group's interactions. Over time, as these workers were "binged" for producing too little or too much, the other workers would see how they handled the criticism and would form opinions of them accordingly.

One of the most striking aspects of the Hawthorne studies is that they revealed how norms about fairness in the work group are enforced through a particular kind of joking ritual (binging). But why are workers binged when they produce too much? That behavior points to the conflict between the norms of the small work group (which are based on fairness) and those of the larger corporation in which the group is embedded (which are based on rationality). The company would like to obtain greater productivity from the workers, but the workers have their own ideas about what is fair and their own ways of enforcing the group norm.

This example brings us to the principle that as group size increases, or as the number of groups in an organization increases, the leaders of the larger organization must invest in ways of controlling the behavior of groups within it (Hechter, 1987). Frequently this investment takes the form of agents of control such as foremen, accountants, inspectors, and the like. These roles become part of the formal structure known as bureaucracy.

FORMAL ORGANIZATIONS AND BUREAUCRACY

Informal organizations are groups whose norms and statuses are generally agreed upon but are not set down in writing. The Bank Wiring Room observations are a classic example of the dynamics of informal organizations. The members of the work groups in the wiring room followed group norms that limited their output. Usually such groups have leaders who help create and enforce the group's norms but have no formal leadership position in the company.

Formal organizations have explicit (often written) sets of norms, statuses, and roles that specify each member's relationships to the others and the conditions under which those relationships hold. Organization charts and job descriptions are typical of such organizations. Formal organizations take a wide variety of forms. For example, the New England town meeting is composed of residents of a town who gather to debate and discuss any issues the members wish to raise. The tenants' association of an urban apartment building is also composed of people who reside in a specific place, but its scope of action is usually limited to housing issues. Both of these are formal organizations, since there are rules defining who may participate and the scope and manner of that participation. As is true in many formal organizations, the members of town meetings and tenants' associations try to arrive at decisions through some form of democratic process—that is, by adhering to norms that allow the majority to run the organization but not to infringe on the rights of the minority.

A familiar type of formal organization is the **voluntary association.** People join groups like the PTA or the Rotary Club to pursue interests that they share with other members of the group. Voluntary associations are usually democratically run, at least in principle, and have rules and regulations and an administrative staff. Churches, fraternal organizations, political clubs, and neighborhood improvement groups are examples of voluntary associations often found in American communities. Sociologists study these associations in order to understand how well or poorly people are integrated into their society.

Bureaucracies are another common type of formal organization. A **bureaucracy** is a specific structure of statuses and roles in which the power to influence the actions of others increases as one nears the top of the organization; this is in marked contrast to the democratic procedures used in other kinds of organizations. Voluntary associations, for example, usually have some of the elements of bureaucracies, but they are run as democratic structures in which power is based on majority rule rather than on executive orders as is the case in pure bureaucracies like General Motors or the U.S. Army.

We owe much of our understanding of bureaucracies to the work of Max Weber, who identified the following typical aspects of most bureaucratic organizations:

1. *Positions with clearly defined responsibilities:* "The regular activities required for the purposes of the organization are distributed in a fixed way as official duties."
2. *Positions ordered in a hierarchy:* The organization of offices "follows the principle of hierarchy; that is, each lower office is under the control and supervision of a higher one."
3. *Rules and precedents:* The functioning of the bureaucracy is governed "by a consistent system of abstract rules" and the "application of these rules to specific cases."
4. *Impersonality and impartiality:* "The ideal official conducts his office . . . in a spirit of formalistic impersonality . . . without hatred or passion, and hence without affection or enthusiasm."
5. *A career ladder:* Work in a bureaucracy "constitutes a career. There is a system of 'promotions' according to seniority, or to achievement, or both."
6. *The norm of efficiency:* "The purely bureaucratic type of administrative organization . . . is from a purely technical point of view, capable of attaining the highest degree of efficiency" (Weber, 1958/1922). (See the study chart below.)

Weber believed that bureaucracy made human social life more "rational" than it had ever been in the past. Rules, impersonality, and the norm of efficiency are some of the ways in which bureaucracies "rationalize" human societies. By this Weber meant that society becomes dominated by groups organized so that the interactions of their members will maximize the group's efficiency. Once the group's goals have been set, the officials in a bureaucracy can seek the most efficient means of reaching those goals. All the less rational behaviors of human groups, such as magic and ritual, are avoided by groups organized as bureaucracies. But the people in a bureaucracy may not take full responsibility for their actions. For this and other reasons, therefore, Weber had some misgivings about the consequences of the increasing dominance of bureaucratic groups in modern societies.

STUDY CHART Characteristics of Bureaucracy

CHARACTERISTIC	EXAMPLE
Positions Clearly Defined	Job or position descriptions detail the responsibilities of each job, avoiding confusion about the duties of each jobholder.
Positions Ordered in a Hierarchy	Positions in the company or agency are ranked from top to bottom so that each position reports to another and supervisory responsibilities are clear.
Rules and Precedents	"The way we do things" in an organization is often written down, becoming "the book" of rules to be followed by its members.
Impersonality and Impartiality	At least in principle, actual performance on the job, rather than personal likes or favors, determines each individual's performance rating.
Career Ladder	New jobs are posted so that employees have a chance to move up in the organization on the basis of achievement, seniority, or both.
Norm of Efficiency	The company continually seeks to increase its efficiency by increasing productivity per employee; this can lead to layoffs or downsizing.

TABLE 6.2 Maximum Shocks Administered in Four Varients of Milgram's Experiment

Shock Level	Verbal Designation and Voltage Level	1 Remote ($n = 40$)	2 Voice Feedback ($n = 40$)	3 Proximity ($n = 40$)	4 Touch Proximity ($n = 40$)
	Slight Shock				
1	15				
2	30				
3	45				
4	60				
	Moderate Shock				
5	75				
6	90				
7	105			1	
8	120				
	Strong Shock				
9	135		1		1
10	150		5	10	16
11	165		1		
12	180		1	2	3
	Very Strong Shock				
13	195				
14	210				1
15	225			1	1
16	240				
	Intense Shock				
17	255				1
18	270			1	
19	285		1		1
20	300	5*	1	5	1
	Extreme Intensity Shock				
21	315	4	3	3	2
22	330	2			
23	345	1	1		1
24	360	1	1		
	Danger: Severe Shock				
25	375	1		1	
26	390				
27	405				
28	420				
	XXX				
29	435				
30	450	26	25	16	12
	Mean maximum shock level	27.0	24.53	20.80	17.88
	Obedient subjects (%)	65.0	62.5	40.0	30.0

Note: The four variants of Milgram's experiment differed in terms of the degree of proximity between the "teacher" and the "learner." The variants were: (1) *remote* (different rooms, in which the learner could be seen but not heard); (2) *voice feedback* (different rooms, in which the learner could be seen and heard); (3) *proximity* (both in the same room); and (4) *touch proximity* (both in the same room, in which the teacher was told to force the learner's hand onto the electrode).

* Indicates that in experiment 1, five subjects administered shock of 300 volts.

Source: *Obedience to Authority,* by Stanley Milgram. Copyright © 1974 by Stanley Milgram. Reprinted by permission of Harper & Row Publishers, Inc.

Bureaucracy and Obedience to Authority

In Chapter 2 we discussed the experiments on conformity conducted by Solomon Asch. Asch's demonstration of the power of group pressure raised serious questions about most people's ability to resist such pressure. It also led to many other studies of conformity. None of those studies is more powerful in its implications and disturbing in its methods than the series of experiments on obedience to authority conducted by Stanley Milgram.

Milgram's study was designed to "take a close look at the act of obeying." As Milgram described it:

> Two people come to a psychology laboratory to take part in a study of memory and learning. One of them is designated as a "teacher" and the other as a "learner."
>
> The experimenter explains that the study is concerned with the effects of punishment on learning. The learner is conducted into a room, seated in a chair, his arms strapped to prevent excessive movement, and an electrode is attached to his wrist. He is told that he is to learn a list of word pairs; whenever he makes an error, he will receive electric shocks of increasing intensity.
>
> The real focus of the experiment is the teacher. After watching the learner being strapped into place, he is taken into the main experimental room and seated before an impressive shock generator. Its main feature is a horizontal line of thirty switches, ranging from 15 volts to 450 volts, in 15-volt increments. There are also verbal designations which range from SLIGHT SHOCK to DANGER—SEVERE SHOCK. The teacher is told that he is to administer the learning test to the man in the other room. When the learner responds correctly, the teacher moves on to the next item; when the other man gives an incorrect answer, the teacher is to give him an electric shock. He is to start at the lowest shock level (15 volts) and to increase the level each time the man makes an error, going through 30 volts, 45 volts, and so on. (1974, pp. 3–4)

The "learner" is an actor who pretends to suffer pain but receives no actual shock. The subject ("teacher") is a businessperson or an industrial worker or a student, someone who has been recruited by a classified ad offering payment for spare-time work in a university laboratory.

Milgram was dismayed to discover that very large proportions of his subjects were willing to obey any order given by the experimenter. In the basic version of the experiment, in which the "learner" is in one room and the "teacher" in another from which the "learner" is visible but cannot be heard, 65 percent of the subjects administered the highest levels of shock, while the other 35 percent were obedient well into the "intense shock" levels.

Milgram used a functionalist argument to explain the high levels of obedience revealed in his experiment: In bureaucratic organizations people seek approval by adhering to the rules, which often absolve them of moral responsibility for their actions. But he also explored the conditions under which conflict will take place—that is, the conditions under which the subject will rebel against the experimenter. As can be seen in Table 6.2, in situations in which subjects are forced to confront the consequences of their behavior their ability to rely on "duty" to justify that behavior seems to be diminished. This effect is presented in graphic form in Figure 6.3.

FIGURE 6.3 Mean Maximum Shocks in Four Variants of Milgram's Experiment

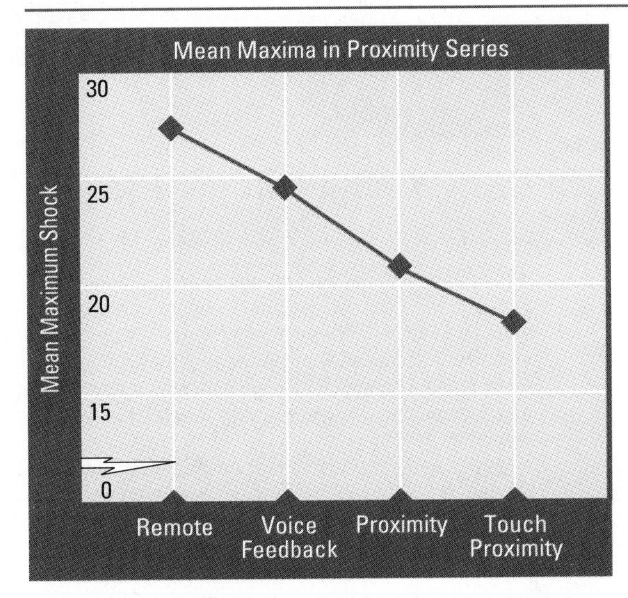

This graph illustrates the relation between the maximum shocks administered and the four variants of proximity between the "teacher" and the "learner" in Milgram's experiment.

Source: *Obedience to Authority,* by Stanley Milgram. Copyright © 1974 by Stanley Milgram. Reprinted by permission of Harper & Row Publishers, Inc.

When Milgram published the results of these experiments, he was criticized for deceiving his subjects and, in some cases, causing them undue stress. The controversy created by this study was one of the factors leading to the establishment of rules for the protection of human subjects in social-scientific research (see Chapter 2).

Commitment to Bureaucratic Groups

Another question related to the impact of bureaucracy on individuals is how bureaucratic groups—which are based on unemotional and "rational" systems of recruitment, decision making, and reward—can sometimes attract strong commitment from their members. One possible answer is ideology: People believe in the goals and methods of the bureaucracy. Another explanation is that people within a bureaucracy form primary groups that function to maintain their commitment to the larger organization. There is considerable research evidence showing that both explanations are valid—both ideology and primary-group ties may operate to reinforce

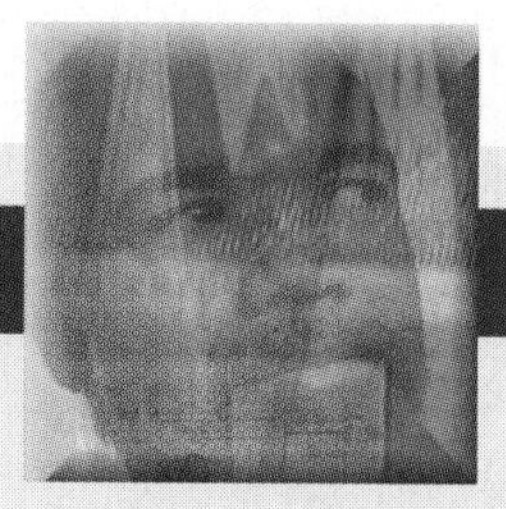

THE PERSONAL CHALLENGE

Social Change and Business Scams

As corporations and smaller companies seek the efficiency and profits that are vital to their continued existence, they tend to eliminate jobs. People who had come to expect a secure income and familiar workmates are often faced with the challenge of forging a new way of life. Increasingly they are turning to self-employment (Siwolop, 1995).

Millions of people dream of owning their own business. But the rise in the number of individuals who want to build their own business organizations has also spawned an increase in the number of unscrupulous or shady characters hoping to profit from the mistakes of the unsuspecting or naive entrepreneur. Some insights into the way scam groups operate and a good measure of sociological skepticism can go a long way toward avoiding being taken in.

The National Consumers League, a nonprofit organization devoted to helping people avoid being swindled, receives more than 25 complaints a day from people all over the United States who have been defrauded in their efforts to start businesses. Viki Blum of Seattle, Washington, is one example. After losing a job with a major corporation, she invested $30,000 of her savings, including severance pay, in electronic game machines that she hoped to place in stores in her community. Her "territory" would yield a steady income. But to her dismay she discovered that the out-of-state company from which she bought the machines had sold her defective equipment. It had also sold similar machines to other entrepreneurs in the area, so that even if she got the equipment fixed she would have nothing like an exclusive territory to work in. When she sought to get her money back, she found that the company had moved and could not be located.

A sociologist in Viki's network of friends could probably have helped her avoid losing her hard-earned savings by asking these questions:

- *What commitments does the vendor have to the investor's community?* If the vendor is located elsewhere, especially if it is an out-of-state company, it may not need to worry about its reputation in the investor's community.
- *What do specialists in the investor's social network have to say about the offering?* We live in a world of divided labor and increasing specialization. This means that there are people in our own social networks, and in the various groups we belong to, who have expertise that they may be willing to share. Had Viki Blum sought the advice of a lawyer friend, she might have decided to be more cautious before investing.
- *Whom can the potential investor trust?* Scam groups often offer enticements like "profits with only part-time work," or "exclusive territory," or "no experience necessary." These are very attractive but should arouse suspicion. To counter suspicion, the scam outfit may use "singers," people who serve as references to sing the praises of the business scheme. The singers may be receiving special rewards for promoting the venture to new investors. We tend to want to trust others, especially when they are saying what we want to hear. But trust must be earned. The potential investor should meet with the vendor face-to-face and have an opportunity to ask questions and examine financial statements covering at least 3 years.

people's commitment to the organization. A number of studies have addressed this question.

Primary Groups in Bureaucracies In one of those studies, Morris Janowitz and Edward Shils were assigned by the U.S. Army to study the attitudes of German soldiers captured during World War II. They were attempting to discover what had made the German army so effective and committed that isolated units continued to fight even in the face of certain defeat. As Shils explained it, he and Janowitz "discovered the influence of small, closely knit groups on the conduct of their members in the performance of tasks set them from the outside" (Janowitz & Shils, 1948, p. 48).

The soldiers of the German army behaved in this way not because of their ideological commitment to the Nazi cause but because of their loyalty to small combat units whose members had become so devoted to one another that to continue to fight even in the face of defeat, so as not to be dishonored as a group, seemed the only possible course of action. The close primary-group structure of the German army was one of the secrets of its success. The other was an extremely efficient organization. Ideology figured very little in its effectiveness.

Ideological Primary Groups Primary groups would seem to have no place in a "pure" bureaucracy, yet the study just cited, together with the Hawthorne experiments and other empirical studies, shows that there is no bureaucracy that completely eliminates primary groups. In the case of the German army, it appears that the development of primary groups among combat troops increased the effectiveness of the organization as a whole. Another important study of this subject was Philip Selznick's (1952) analysis of what made the Bolsheviks so effective that they won out in the fierce competition for political dominance in Russia. Selznick's data show that the ideologically based primary group was the key element in the Bolsheviks' organization. Small, secret "cells" of devoted communists were organized in neighborhoods, factories, army units, farms, and universities. Through years of tense and often dangerous political activity, the members became extremely devoted to one another and to their revolutionary cause. According to Selznick, this doubling of ideological and primary-group cohesion made the Bolsheviks themselves an "organizational weapon."

In the contemporary world we see many examples of ideological groups with extremely high levels of cohesion. Very often these groups are devoted to the leadership and teachings of a forceful and highly attractive leader whose ideas exert a powerful influence over the behavior of the group's members. We often focus on the extremes of such behavior, such as the mass suicide by the members of Heaven's Gate, but many ideological groups can have beneficial effects on their members. This is especially true of groups devoted to personal or spiritual growth and positive behavior change. Indeed, the ability of people to form groups freely and to leave them if they so desire is one of the distinguishing features of societies in which the rule of law protects the individual's right of association. A problem of some ideological groups, however, is that they may not grant their members this freedom (Elshtain, 1997).

GROUPS IN COMPLEX SOCIETIES

As we saw in Chapter 4, in simpler tribal or peasant societies all social structures are primary groups. The tribal village is usually so small that everyone knows everyone else. The family, the extended family, the age- and sex-segregated peer groups, and even the village itself all allow for everyone to know everyone else and to interact on an almost daily basis. Typically these societies lack a secondary-group level of life. In traditional peasant societies, for example, there often were no associations that brought together soldiers or farmers or artisans in groups that extended beyond the limited bounds of the village. And yet that level of social life is what makes contemporary societies so efficient.

The differences between simpler and more complex societies can also make interactions among members of those societies confusing and frustrating (see Box 6.2). For example, when they were first assigned to peasant villages in Africa and Asia, American Peace Corps volunteers were unprepared to cope with situations in which villagers came to them with requests for money or help with marital and family problems. For the young Americans, such requests were appropriate only among members of primary groups—and the villagers did not know how to engage in interactions that are typical of secondary groups (Luce, 1964).

In his most famous theoretical work, *The Division of Labor in Society* (1964/1893), Émile Durkheim reasoned that as a society becomes larger and more complex there is a vast increase in interdependence among its members, as the labor needed to feed, house, educate, communicate with, transport, care for, and defend them becomes more complex. Faced with more choices regarding what work to prepare for, whom to interact with, and how to live life, the individual gains much more freedom than is available to a member of a simpler society, who (for better or worse) is confined to a narrow set of village primary groups. But Durkheim also saw that the modern person could be overwhelmed by the choices open to him or her and might find it difficult to decide what groups to join and what norms to conform to.

Max Weber shared Durkheim's concern about the prospects for human happiness in the modern world. According to Weber (1958/1922), the fundamental paradox of modern society was the rise of bureaucracies, which greatly expanded the efficiency and rationality of society but "disenchanted the world." He meant that the modern formal organization, which is so effective in science, the military, the state (government), economic production, and almost every other aspect of modern life, at the same time deprives human life of a sense of magic and spirituality. The rise of fundamentalist movements in the Islamic world, for example, is in part a reaction against the influence of bureaucratic rationality in the institutions of secular states such as Egypt and Algeria.

Even in Western societies, where the process of bureaucratization is extremely advanced, people may feel overwhelmed by bureaucracy. The individual often feels like a tiny cog in a huge set of interlocking organizations. He or she fills specific statuses and enacts clearly defined roles yet feels more and more oppressed. Weber deplored the type of person that he felt was produced by the organizations of modern society: "a petty routine creature, lacking in heroism, human spontaneity, and inventiveness" (Weber, 1922, in Gerth & Mills, 1958, p. 50). In the United States, however, democracy was always linked with citizens' freedom of association and their tendency to join groups that could protect them against the excesses of large bureaucracies, on the one hand, and offer them the intimacy of small-group sociability, on the other.

BOX 6.2 GLOBAL CHANGE AND U.S. SOCIETY

Communication and Interaction

Global social change continually produces waves of refugees and immigrants. For example, the new freedom to travel in the former Soviet satellite nations has produced thousands of ambitious immigrants. Civil strife in Central and Latin America, repression in some Asian societies (e.g., China and North and South Korea), growing poverty in the Philippines and Albania, and continuing turmoil in Cambodia and the Middle East have pushed families and individuals out of their homes and into the growing tide of worldwide migration. All of these changes are having profound effects on the linguistic diversity of North America and on how people in the Americas communicate and interact.

The United States continues to hold a global reputation as a place where immigrants can work to achieve success—and since it remains the nation with the most liberal immigration policies, it is no surprise that well over 500,000 legal immigrants and refugees (and an untold number of illegal entrants) seek to establish themselves in the "golden land" each year. In California alone, during the 1980s about 3.6 million immigrants brought their languages and cultural norms to the nation's most diverse state. But no part of the nation missed the increases in cultural and linguistic diversity that result from immigration. Between 1980 and 1990, for example, the number of foreign-language speakers in the state of Georgia (not known as a primary destination of immigrants) increased from 134,000 to 285,000. All told, by the beginning of the 1990s more than 32 million residents of the United States did not speak English in their homes (U.S. Bureau of the Census, 1990).

The need for English as a Second Language (ESL) courses in schools and community centers throughout the nation is only one indication of the kinds of demands increased linguistic diversity is placing on communities where immigration rates are high. In time most Americans come to appreciate the diversity of languages spoken in the United States as a great cultural resource. But the welter of different languages spoken in some communities can also be a source of conflict. The extreme rapidity of the change has given rise to numerous controversies over how, and more specifically in what language, people should be expected to interact. For example, in 1989 a federal judge ruled that the Los Angeles suburb of Pomona had violated the civil rights of Korean and Chinese businesspeople by requiring them to place English signs on their shops. On the other hand, in the same community Filipino nurses lost a suit against a hospital that would not allow them to speak their native language, Tagalog, among themselves at work.

Elsewhere in the nation there is an active movement to make English the "official" language of the United States. This is accompanied by a crusade against the offering of bilingual education for children from other language groups. The problems of interaction across lingusitic divides are greatest in states with large proportions of foreign-language speakers, such as New Mexico (33.5 percent according to the last census), California (31.5 percent), Texas (25.4 percent), Hawaii (24.8 percent), and New York (23.3 percent). Conversely, these states benefit the most from the wealth of new phrases and styles of communication stemming from their growing linguistic diversity.

A Nation of Joiners

Early in the nineteenth century the French aristocrat and social theorist Alexis de Tocqueville visited the United States. Like other European intellectuals, he considered the new nation a bold and controversial social experiment. He had doubts about the capacity of common people to govern themselves without the guiding hand of a strong aristocracy. But as he traveled through the former colonies he was pleasantly surprised by what he observed. He was particularly impressed by the number of associations and small groups to which people belonged. Americans, he wrote, are "a nation of joiners" (1956/1840).

Tocqueville's phrase is often quoted in descriptions of democratic societies. Democratic theory states that when people are given the freedom to associate with whomever they choose, they will form a complex civil society—that is, a society characterized by a never-ending and always changing array of freely formed nongovernmental groups (see Chapter 8). In fact, as we will see in later chapters, dictators who wish to curtail individual freedom invariably attempt to control the groups people can belong to.

The number of voluntary associations in the United States and other democratic nations has increased steadily since Tocqueville's day. But what kinds of groups are people forming? What needs are they seeking to satisfy through their group interactions? And how does social change affect the types of groups they form?

A study by sociologist Robert Wuthnow directly addresses these questions. Wuthnow's findings, described in his book *Sharing the Journey* (1994), are based on a large survey carried out by the Gallup Poll in which respondents were asked about their group affiliations and the benefits they derived from them. Table 6.3 presents Wuthnow's estimates of the major types of groups to which people in the United States belong. Religious groups engaging in Bible study and Sunday school classes are by far the largest category; self-help groups in which people attempt to resocialize themselves,

TABLE 6.3 Groups in the United States

	Estimated Numbers	
	Members	Groups
Sunday school classes	18–22 million	800,000
Bible study groups	15–20 million	900,000
Self-help groups	8–10 million	500,000
Special interest groups		
Political/current events	5–10 million	250,000
Book/discussion groups	5–10 million	250,000
Sports/hobby groups	5–10 million	250,000

Source: Wuthnow, 1994.

to replace old and often destructive patterns of behavior with new and more desirable ones, are the second largest category.

Sociologist Frank Reissman (1984) has been tracking the enormous growth of self-help groups over the past 30 years. His inventory of such groups continues to grow rapidly. Part of the reason for this growth, he believes, lies in the increasing need people feel to overcome compulsive behaviors or behavior that is harming their social attachments. There are groups for people recovering from addictions to gambling, smoking, marijuana use, and shoplifting; there are support groups for parents, grandparents, widows, and divorcees; there are groups for homosexuals, former fundamentalists, people with eating disorders, and families of sex offenders (Wuthnow, 1994).

As Table 6.4 indicates, the appeal of these "people-changing" groups is quite wide. The table also shows that there are some differences in who joins small groups—more women than men, more middle-aged people than young ones, more people in the West than in the South or the Midwest, more members among the college educated. But the differences may be less important than the similarities in terms of race, urban or rural residence, and income.

TABLE 6.4 Small-Group Membership in the United States (percentage of each category that is currently in a small group that meets regularly and provides caring and support for its members)

National	40
Women	44
Men	36
Age 18–34	35
Age 35–49	42
Age 50 and over	45
High school or less	37
Some college	43
College graduates	48
Income less than $20,000 per year	39
Income $20,000–$39,000 per year	42
Income $40,000 or more per year	43
White Anglo	40
Black	41
Hispanic	46
South	39
Midwest	39
Northeast	41
West	45
In large cities	41
In medium cities	42
In towns or rural areas	40

Source: Wuthnow, 1994.

Wuthnow's study revealed that people who join small groups are owning up to a deeply felt need to be part of a face-to-face primary group in which they attain a sense of community with others. Table 6.5 shows that

TABLE 6.5 Meeting the Need for Community (percentage of group members who have (A) felt each need and (B) met each need fully)

	A	B
Having neighbors with whom you can interact freely and comfortably	93	43
Being able to share deepest feelings with someone	94	57
Having friends who value the same things in life you do	98	58
Having people in your life who give you deep emotional support	98	66
Being in a group where you can discuss your most basic beliefs and values	90	50
Having friends you can always count on when you're in a jam	97	64
Having people in your life who are never critical of you	83	29
Being part of a group that helps you grow spiritually	90	53
Having cooperation rather than competition with people at work	85	31
Having people you can turn to when you feel depressed or lonely	96	62
Knowing more people in your community	95	32

Source: Wuthnow, 1994.

there are often wide gaps between people's needs and the ability of group interactions to actually satisfy those needs. But these gaps do not seem to produce widespread disillusionment. In their search for a feeling of belonging and community, people continue to join groups of all kinds.

RESEARCH ON THE CUTTING EDGE: Small Groups, Small Loans, Big Change

Professor Muhammed Yunus's dream is that "Maybe our great-grandchildren will go to museums to see what poverty was." If the professor's dream is realized, he may in fact be one of the most heroic figures in the museum. His invention, the Grameen Bank of Bangladesh, has become a model for establishing systems of "microenterprise credit," in which the owners of extremely small businesses, who are essentially destitute, are loaned very small amounts of money and required to join a small group of others like them who have also received small loans. The group meets regularly in their village, discusses ways to improve their business practices, and supports each other's efforts to repay the loans. Yunus, who is trained in economics and sociology, has one of the finest sociological imaginations possible. His vision of bringing financial and political power to the poorest of the poor, most of whom are women in extremely impoverished villages, has become a model for similar microenterprise credit systems throughout the world (Cabral,1998).

Muhammed Yunus (right).

Yunus based his innovative banking system on the findings of research on the sociology and economics of poverty in the villages surrounding his university in Bangladesh. His students were sent out to speak with the landless "hard core" poor who make up about half of the nation's population. They asked simple questions: "How many families had food stocks for a year, for six months, for one month? How many lived day-to-day, what sort of work did they do and how much did they earn?" (Bornstein, 1996, p. 38). The students learned that for the most part the poorest of the poor were either rickshaw pullers, men who pull or pedal a two-wheeled cart with a passenger, or bamboo weavers, women who use bamboo strands to weave mats or baskets of bamboo. In the case of the male rickshaw drivers, they discovered that none owned their own rickshaw. Each day they paid rent to its owner, after which they had almost nothing left for themselves and their families. For the women, the story was similar: They typically needed to borrow money for materials from a trader, who then purchased their final product, leaving the women with only pennies at the end of a long day of labor. "It was a form of bonded labor, of slavery," Yunus recalled (Bornstein, 1996, p.39). Yunus and his students calculated that it would take a mere $26, distributed among the 42 men and women they had spoken with, to enable each man to buy his own rickshaw and each woman to purchase enough materials to enable her to work independently of the traders. Yunus was extremely disturbed by these findings. He felt disgust toward the social sciences and toward "a society that spent a great deal of time and money teaching fancy development theories in the classroom while permitting this sort of exploitation to go on down the road. . . . I felt extremely ashamed of myself being part of a society which could not provide $26 to forty-two able, skilled human beings who were trying to make a living" (quoted in Bornstein, 1996, p. 39).

In the weeks and months that followed, Yunus learned that bankers would not make loans to the poor people in the villages, even at his urging, because the villagers had no capital to put down as collateral to secure the loans. So he borrowed $300 and began making small loans himself. He needed the help of a dedicated woman, Priti Rani Barua, who could speak to the women in the villages, something that as a man he was not allowed to do. He insisted that the loans would not be charity. They would be repaid on time. Only landless villagers would be eligible. And finally, to the extent possible, they would try to lend to women, who were most likely to be denied commercial credit.

Next, Yunus and Barua convened groups of villagers and asked them how the loans should be administered. From these discussions came the idea of forming groups of 5 to 10 villagers. There would be rickshaw groups, milk cow groups, bamboo furniture weaving groups, and chili pepper trading groups. Each group would meet weekly to make payments on their loans and discuss ways of improving their business practices. The groups would be self-supervising. To receive loans, they would agree to follow certain simple rules, the most important of which was that each member was jointly responsible for everyone else's loan. The loans would be staggered, with only two or three members of a group receiving funds at any one time, so that if these recipients failed to repay on time, others in the group could not receive their loans. This rule applies peer pressure and the reciprocity principle to the system and is at the heart of its success.

In addition, Yunus made another rule: All participants were required to open an extremely small savings account, even if they deposited only a penny a week. Finally, Yunus required his students, and later his employees, to attend the group meetings so that all rules would be followed religiously. (See Figure 6.4.)

After many ups and downs and years of improving their microenterprise lending and collecting system, Yunus's groups, now organized into what is known as the Grameen Bank, number well over 35,000. More than 2 million extremely poor people in Bangladesh are participating in the microenterprise credit groups, and the Bank has branched out in many directions to assist villagers in their efforts to improve agriculture, create midsize businesses, secure education, and much more. Monsoons and floods periodically threaten a group's ability to make repayments, and there have been numerous crises to overcome, but still the bank continues to grow, partly because it has become a large and innovative corporation with more than 14,000 dedicated managers who share Yunus's vision (Holcombe, 1995).

But the secret of the bank's success lies in the power of small groups to motivate, pressure, and support their members. Grameen is not a panacea for poverty or a replacement for other strategies, but it has demonstrated what people with a sociological imagination and detailed knowledge of local conditions can achieve with extremely modest resources.

FIGURE 6.4 **The Sixteen Decisions**

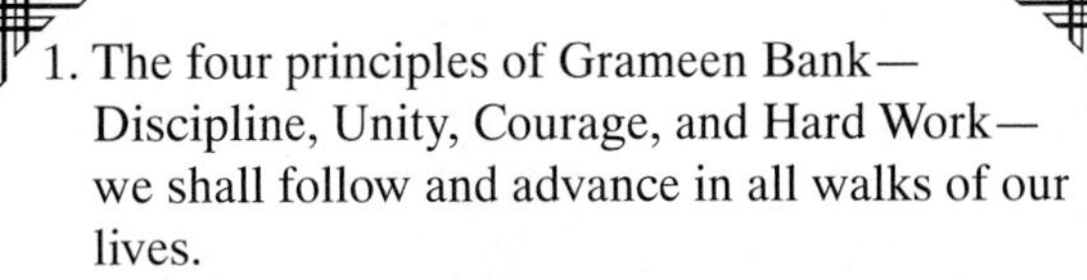

1. The four principles of Grameen Bank—Discipline, Unity, Courage, and Hard Work—we shall follow and advance in all walks of our lives.
2. Prosperity we shall bring to our families.
3. We shall not live in dilapidated houses. We shall repair our houses and work toward constructing new houses at the earliest.
4. We shall grow vegetables all the year round. We shall eat plenty of them and sell the surplus.
5. During the plantation seasons, we shall plant as many seedlings as possible.
6. We shall plan to keep our families small. We shall minimize our expenditures. We shall look after our health.
7. We shall educate our children and ensure that they can earn to pay for their education.
8. We shall always keep our children and the environment clean.
9. We shall build and use pit-latrines.
10. We shall drink tubewell water. If it is not available, we shall boil water or use alum.
11. We shall not take any dowry in our sons' weddings, neither shall we give any dowry in our daughters' weddings. We shall keep the center free from the curse of dowry. We shall not practice child marriage.
12. We shall not inflict any injustice on anyone, neither shall we allow anyone to do so.
13. For higher income we shall collectively undertake bigger investments.
14. We shall always be ready to help each other. If anyone is in difficulty, we shall all help him.
15. If we come to know of any breach of discipline in any center, we shall all go there and help restore discipline.
16. We shall introduce physical exercise in all our centers. We shall take part in all social activities collectively.

Source: Bornstein, 1996.

VISUAL SOCIOLOGY

Friendship Groups

These photos capture some aspects of the universal experience known as friendship. They show friendship groups in societies throughout the world—from the young men and women in an urban street gang, to a group of wealthy English boarding school students, to scenes from African villages. They convey the emotional quality of interaction among friends, but to the sociological observer they also reveal some characteristic features of friendship groups.

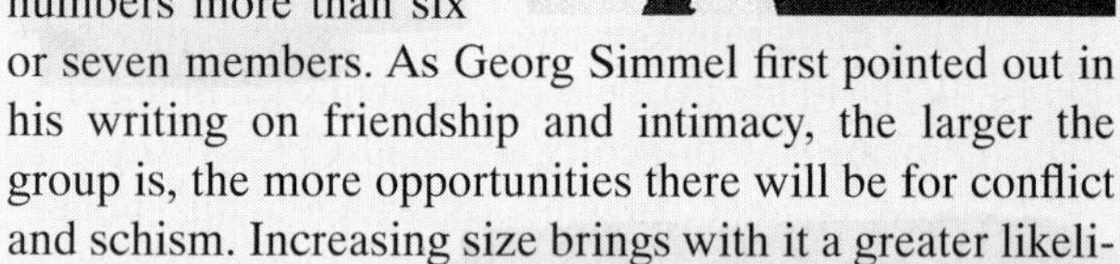

First, note the size of the groups. None numbers more than six or seven members. As Georg Simmel first pointed out in his writing on friendship and intimacy, the larger the group is, the more opportunities there will be for conflict and schism. Increasing size brings with it a greater likelihood of jealousy, competition for attention, and conflicting values, and these tensions tend to break larger friendship groups into smaller, more intimate ones like those shown here.

A second observation that we can draw from these photos has to do with gender differences in friendship groups. Children's friendship groups are involved in anticipatory socialization, in taking roles that will later be incorporated

into adult statuses. Thus the girls are seen holding dolls and practicing the interactions that occupy so much of their mothers' time. And unlike the typical male friendship group, female groups often have the responsibility of caring for younger siblings, as we see in at least one of these photos. Males more typically engage in activities that involve physical prowess and changes in the environment. We also see in these photos that gender segregation is normally the rule in friendship groups, although there may be many exceptions, such as when games are played in school.

Finally, these photos remind us of the emotional complexity of friendship. Friends can fight and make up, but people who are less than friends often cannot fight without ending the relationship. And consider the photo of the men hugging each other. Is this scene likely to occur in American culture? Perhaps today it is, but until quite recently such a show of emotion between men, even between a father and his son, was discouraged in our culture. Today this norm is changing, although in many other cultures it is still far more acceptable for men to hug and kiss each other than it is in North America.

SUMMARY

The social fabric of modern societies is composed of millions of groups of many types and sizes. Unlike a *social category,* a collection of individuals who are grouped together because they share a particular trait, a *social group* is a set of two or more individuals who share a sense of common identity and belonging and who interact on a regular basis. Those interactions create a social structure composed of specific statuses and roles.

A *primary group* is characterized by intimate, often face-to-face, association and cooperation. *Secondary groups* are characterized by relationships that involve few aspects of the personality; the members' reasons for participation are usually limited to a small number of goals. As the number of people in a group increases, the number of possible relationships among the group's members increases at a faster rate. A group composed of only two people is known as a *dyad.* The addition of a third person to form a *triad* reduces the stability of the group.

At the level of social organization between the primary group and the institutions of the nation-state are *communities. Territorial communities* are contained within geographic boundaries; *nonterritorial communities* are networks of associations formed around shared goals. Territorial communities are usually composed of one or more neighborhoods in which people form attachments based on proximity.

Groups formed at the neighborhood level are integrated into networks that may extend beyond geographic boundaries. Key factors in the formation of networks are *in-group–out-group* distinctions. Such distinctions can form around almost any quality but are usually based on such qualities as income, race, or religion. Another type of group is the *reference group,* a group the individual uses as a frame of reference for self-evaluation and attitude formation. The study of whom people associate with, how those choices are made, and the effects of those choices is known as social network analysis.

Social scientists have identified certain principles of interaction that help explain both stability and change in human groups. Among them are the pleasure principle, the rationality principle, the reciprocity principle, and the fairness principle. The balance among these principles varies from one situation to another, with economic motives dominating in some instances and social needs winning out in others. From an interactionist perspective, an important factor determining how people behave in a given instance is their definition of the situation.

Studies of interaction in groups use a variety of techniques. *Ethnomethodology* is the study of the underlying rules of behavior that guide group interaction. The *dramaturgical approach* regards interaction as though it were taking place on a stage and unfolding in scenes. The strategies that people use to set a stage for their own purposes are known as *impression management.* Research on the bystander effect has shown that when responsibility seems to be diffused, people are more likely to avoid helping others. But when people define the situation as involving them, they will intervene despite the potential costs to themselves.

Research on small groups has shown that they tend to develop two kinds of leaders, a "task leader" who keeps the group focused on its goals and a "socioemotional leader" who creates a positive emotional climate.

Informal organizations are groups with generally agreed-upon but unwritten norms and statuses, whereas *formal organizations* have explicit, often written, sets of norms, statuses, and roles that specify each member's relationships to the others and the conditions under which those relationships hold. A *voluntary association* is a formal organization whose members pursue shared interests and arrive at decisions through some sort of democratic process. A *bureaucracy* is a formal organization characterized by positions with clearly defined responsibilities, the ordering of positions in a hierarchy, governance by rules and precedents, impersonality and impartiality, a career ladder, and efficiency as a basic norm.

One effect of the increasing dominance of bureaucracies in modern societies is the possibility that individuals will not take full responsibility for their actions. A study of obedience to authority conducted by Stanley Milgram raised serious questions about people's ability to resist pressure to carry out orders for which they are not personally responsible. Milgram also found, however, that rebellion against authority is more likely when individuals who rebel have the support of others. Other studies have found that commitment to bureaucratic organizations is greatest when it is supported by ideology or by strong primary-group attachments.

As a society becomes larger and more complex, it tends increasingly to be characterized by secondary groups and organizations. These make the society more efficient, but they can also cause confusion and unhappiness. Durkheim pointed out that members of complex societies have greater freedom to choose what groups to join and what norms to conform to but that they can be overwhelmed by the choices open to them.

Weber examined the effects of the rise of bureaucracies in modern societies, finding that the individual often comes to feel like a tiny cog in a huge set of interlocking organizations.

Democratic theory states that when people are given freedom of association they will form a complex civil society. This theory is supported by research on group membership in the United States. Religious and self-help groups are two types of groups that Americans join in large numbers.

SOCIOLOGY VERSUS IDEOLOGY

People who dislike government and its influence on our lives often argue that whenever we try to intervene in the workings of the market we become mired in bureaucratic rules and regulations that make our lives miserable. We may want to preserve the environment, prevent child abuse, or protect consumers against unscrupulous businesses, but each time we pass laws designed to achieve these goals we also encourage the development of a bureaucracy that limits our freedom and presents obstacles to entrepreneurial innovation.

Those who believe that it is essential for a nation to protect its citizens in various ways often reluctantly defend the need for regulations and the bureaucracy to enforce them. How, they ask, can we protect people from fraudulent business practices unless we monitor the quality of products offered for sale? And if it takes a bureaucracy with officials and rules to do this important work, that is the cost of vigilance.

The sociologist is not content with this ideological stalemate between conservative critics of the bureaucratic "nanny state" and its liberal defenders. Both sides generally concede that bureaucracy is inevitable if laws are to be enforced. But sociologists who study bureaucracies often learn that there are many possibilities for limiting their size without throwing away their goals. And there are many ways of allowing citizens to have a greater say in the means of enforcement and of streamlining regulations and compliance requirements so that they do not stifle innovation. From the sociologist's perspective, people locked in ideological debates fail to consider the fact that, when asked, citizens of a democratic state can often come up with creative solutions to the problem of bureaucratic growth—solutions that do not require a complete return to the unregulated past.

GLOSSARY

social category: a collection of individuals who are grouped together because they share a trait deemed by the observer to be socially relevant. (p. 158)

social group: a set of two or more individuals who share a sense of common identity and belonging and who interact on a regular basis. (p. 158)

primary group: a social group characterized by intimate, face-to-face associations. (p. 159)

secondary group: a social group whose members have a shared goal or purpose but are not bound together by strong emotional ties. (p. 159)

dyad: a group consisting of two people. (p. 160)

triad: a group consisting of three people. (p. 160)

community: a set of primary and secondary groups in which the individual carries out important life functions. (p. 160)

territorial community: a population that functions within a particular geographic area. (p. 161)

nonterritorial community: a network of relationships formed around shared goals. (p. 161)

in-group: a social group to which an individual has a feeling of allegiance; usually, but not always, a primary group. (p. 161)

out-group: any social group to which an individual does not have a feeling of allegiance; may be in competition or conflict with the in-group. (p. 161)

reference group: a group that an individual uses as a frame of reference for self-evaluation and attitude formation. (p. 164)

ethnomethodology: the study of the underlying rules of behavior that guide group interaction. (p. 171)

dramaturgical approach: an approach to research on interaction in groups that is based on the recognition that much social interaction depends on the desire to impress those who may be watching. (p. 171)

impression management: the strategies one uses to "set a stage" for one's own purposes. (p. 172)

informal organization: a group whose norms and statuses are generally agreed upon but are not set down in writing. (p. 174)

formal organization: a group that has an explicit, often written, set of norms, statuses, and roles that specify each member's relationships to the others and the conditions under which those relationships hold. (p. 174)

voluntary association: a formal organization whose members pursue shared interests and arrive at decisions through some sort of democratic process. (p. 174)

bureaucracy: a formal organization characterized by a clearly defined hierarchy with a commitment to rules, efficiency, and impersonality. (p. 175)

WHERE TO FIND IT

BOOKS

Small Group Research: A Handbook (A. Paul Hare, Herbert H. Blumberg, Martin F. Davis, and Valerie Kent; Ablex, 1994). A recent and comprehensive survey of classic and contemporary research on human behavior in small groups.

Communicating in Small Groups (Stephen Beebe; Longman, 1997). A good source of practical advice about how to interact successfully in groups, especially in business and educational settings.

Sharing the Journey: Support Groups and America's New Quest for Community (Robert Wuthnow; Free Press, 1994). A good empirical study with excellent historical and theoretical material on the rise of support groups and their purpose in people's lives.

The Presentation of Self in Everyday Life (Erving Goffman; Doubleday, 1959). One of the classics in the sociology of interaction at the micro level. A must-read for anyone wanting to gain greater insight into the dramaturgical approach to the analysis of social life.

Studies in Ethnomethodology (Harold Garfinkel; Prentice-Hall, 1967). A brilliant and controversial collection of original papers that illustrates what scientific concern for the most micro and subjective aspects of interaction can yield in the way of fresh insights into how people act in society.

Passions Within Reason: The Strategic Role of the Emotions (R. H. Frank; Norton, 1988). A highly readable treatment of a complex and often difficult subject: the relationships between emotion and social behavior. The emphasis is on rational behavior and its limits in the real world.

Extraordinary Groups: An Examination of Unconventional Life-Styles, 4th ed. (William M. Kephart; St. Martin's Press, 1991). A valuable collection of essays about unconventional groups that manage to maintain their solidarity despite their deviance from the mainstream. Includes material on ethnic groups, self-help groups like Alcoholics Anonymous, and unusual occupations.

JOURNALS

Administrative Science Quarterly. One of the principal journals in the field of organizations and behavior. A valuable source for theoretical and empirical studies of bureaucratic organizations like corporations, government agencies, and universities.

Human Organization. Another good source of research on the problems and possibilities of efforts to improve human organizations. More diverse in its scope than *Administrative Science Quarterly* but also oriented toward economic organizations like firms, labor unions, and regulatory agencies.

Urban Life and Culture. A journal specializing in articles about human interaction in urban settings. A source of new empirical studies from the interactionist perspective.

INTERNET RESOURCES

Society for the Study of Symbolic Interaction (www.sun.soci.niu.edu). A home page about symbolic interaction and the sociologists who use this method in their work, with links to related sites.

Ethnomethodology (www.bekkoame.or.ji). A highly creative site designed by an unofficial representative of the world of ethnomethodology, with many links to other fascinating sites.

Bureau of Labor Statistics (http://stats.bls.gov/blshome.html). Displays several pages and features on work and organizations in the U.S. economy, with links to international organizations like the International Labor Organization. Also provides links to many research reports and publications.

U.S. Department of Labor (www.dol.gov). Offers a great deal of information and data about people in organizations, especially companies and corporations, as well as a link to the Economics and Statistics Administration, the source of much of the statistical, economic, and demographic information collected by the federal government.

CHAPTER 7

DEVIANCE AND SOCIAL CONTROL

I see them lining up for breakfast well before seven in the morning. They look exhausted and threadbare. Some have slept all night in chairs in the "day room." Others are drifting in from the street, where they have a secret hiding place to sleep—perhaps over a grate that in winter will blow warm air on them and keep them from freezing. This morning, for example, I see Willie W. and his partner Samantha. They are crack addicts who roam the streets at night in search of soda and beer cans, which they will sell to other street folk for half of the 5-cent deposit price because they need immediate cash for a "hit" of cocaine. This morning I also spot the loan shark who waits outside the Drop-in Center to see if anyone who owes him money is around so he can collect on debts or threaten those who are delinquent in their loan payments.

As I climb the stairs to the Center, I am greeted by several staff members, all of whom were formerly homeless but are now proudly on their way to work in the Center's kitchen or in its various recycling and building maintenance programs. They greet me as "The Professor." The reason I appear so frequently in their midst is that I am the chair of the Center's board of directors. Its founder and executive director is a former sociology doctoral student at my university. Two of my closest friends are also deeply involved in its programs. Although I have no shortage of responsibilities to occupy my time, I could hardly refuse when I was asked to serve on the board. It made sense for me to take my turn leading the board and participating in the Center's activities. Nor am I at all unusual in this regard. Most professional sociologists find that their research and writing leads to invitations to take active roles in organizations that seek to put into practice what they have recommended in their research. I chose to make a commitment of time and energy to the Grand Central Neighborhood Social Services Agency because I have been involved in research with the city's street population for almost three decades.

I believe that the problems of homeless adults in urban centers raise fascinating sociological issues about how a good society might deal with people who deviate from its norms and are "down and out." I find it sad that in one of the world's richest cities there are people living on the streets. Often they are mentally ill or addicted, or both. Equally often they are younger people who have been pushed out of crowded apartments after the breakup of a family. On the streets they find many opportunities to make some money through various forms of hustling.

Race, ethnicity, and gender also play a part in the story. Two generations ago the city's homeless were primarily white males. Today the majority of street people are black and Hispanic males from extremely poor backgrounds. An increasing number of street people are women. About 20 percent of those who come to the Center are female. All of the people who spend time at the Center deviate from the norms of success that Americans value so highly. Many have served time in prison—but in fact they frequently come to the Center to escape from predatory felons on the streets. At the Center we are there to offer a helping hand, but we insist that everyone take responsibility for their behavior. Those who can do so often begin the long and difficult process of rehabilitation.

Like many other volunteers, I spend time at the Center because it gives me a sense of gratification to share my skills with those in need. As a professional sociologist, I also serve because it is a way to gain knowledge about society's outcasts and the approaches the city is developing to address their needs.

WHAT IS DEVIANCE?

Almost every American political campaign will touch on issues of crime and deviance and what should be done about offenders, be they juveniles, elderly, or hardened criminals. In this chapter we will look at some current trends in deviance and crime and also examine issues of social control—that is, what societies do to attempt to curb deviance. Research on these and related issues is a major field within sociology. No one can offer definitive answers, but sociologists are frequently looked to for research on trends in deviance and criminal behavior and for explanations of these trends.

Deviance, broadly defined, is behavior that violates the norms of a particular society. But because all of us violate norms to some degree at some time or other, we must distinguish between *deviance* and *deviants.* Deviance can be something as simple as dyeing one's hair purple or wearing outrageous clothing or becoming tipsy at a stuffy party. Or it may be behavior over which the individual has little control, such as being homeless and living on the street, or it may consist of more strongly sanctioned departures from the society's norms—such acts as rape, mugging, and murder. Not all deviance is considered socially wrong, yet it can have negative effects for the individual. For example, "whistle-blowers" who publicize illegal or harmful actions by their employers deviate from the norms of bureaucratic organizations and are often threatened with the loss of their jobs. Yet at the same time they benefit the public by calling attention to dangerous or illegal activities. In recent years, therefore, Congress and the courts have tried to encourage whistle-blowers and prevent reprisals against them. In 1992, for example, a federal judge awarded $7.5 million to Christopher Urda as a share of the $55.6 million the U.S. government saved as a result of his testimony against his employer (Stevenson, 1992b).

A **deviant person,** by contrast, is someone who violates or opposes a society's most valued norms, especially those that are valued by elite groups. Through such behavior deviant individuals become disvalued people, and their disvalued behavior provokes hostile reactions (Davis, 1975; Goffman, 1963; Sagarin, 1975; Schur, 1984). *Deviant* may be a label attached to a person or group. Or the word may refer to behavior that brings punishment to a person under certain conditions.

Cases of influence peddling or extramarital sexual liaisons by high government officials raise an important question: What are the conditions under which violations of norms are punished? Here is an area of conduct in which there is often some uncertainty about what is legal and what is merely sleazy. Such questions reveal that deviance is not absolute. As sociologist Kai Erikson explains it, deviance "is not a property inherent in certain forms of behavior; it is a property conferred upon these forms by the audience which directly or indirectly witnesses them" (1962, p. 307). Some of us may believe that influence peddling is deviant, while others may believe that it is acceptable. Which of our views become the norm—and which are to be enforced through rewards and punishments—is just as important as the behavior itself. This point is illustrated in Erikson's classic study of deviance among the New England Puritans, discussed in Box 7.1.

The study of deviance is central to the science of sociology, not only because deviance results in major social problems like crime, but also because it can bring about social change. Indeed, the Puritans were a deviant group in England at the time of their emigration to the New World. They challenged the authority of the king and many of the central norms that upheld the stratification system of feudal England. In the Massachusetts Bay Colony they created their own society, one in which they were no longer deviant. But, as Erikson showed, they also created their own forms of deviance, which reflected their unique problems as a society on a rapidly changing social frontier.

In fact, if you think about it, every aspect of daily life—driving, walking, going to a show, shopping, church attendance, political behavior, or sports—is controlled by the norms of a given culture. And thousands of deviant acts, such as cutting in front of someone in a line, are frowned upon by people asserting their normative expectations ("Excuse me, I was ahead of you here"). All these norms and assertions constitute a system of informal social control without which society

BOX 7.1 USING THE SOCIOLOGICAL IMAGINATION

The Salem Witch-Hunt

> No one knows how the witchcraft hysteria began, but it originated in the home of the Reverend Samuel Parris, minister of the local church. In early 1692, several girls from the neighborhood began to spend their afternoons in the Parris kitchen with a slave named Tituba, and it was not long before a mysterious sorority of girls, aged between nine and twenty, became regular visitors to the parsonage. We can only speculate about what went on behind the kitchen door, but we know that Tituba had been brought to Massachusetts from Barbados and enjoyed a reputation in the neighborhood for her skills in the magic arts. As the girls grew closer together, a remarkable change seemed to come over them: perhaps it is not true, as someone later reported, that they went out into the forest to celebrate their own version of a black mass, but it is apparent that they began to live in a state of high tension and shared secrets with one another which were hardly becoming to quiet Puritan maidens. (Erikson, 1966, p. 141)

The girls quickly drew the concerned attention of Salem's ministers and its doctor. Unable to understand much about their hysterical state or to deny their claim that they were possessed by the devil, the doctor pronounced the girls bewitched. This gave them the freedom to make accusations regarding the cause of their unfortunate condition.

Tituba was the first to be accused and jailed. She was followed by scores of others as the fear of witches swept through the community. Soon women with too many warts or annoying tics were accused, tried, and jailed for their sins. Then the executions began. In the first and worst of the waves of executions, in August and September of 1692, at least 20 people were killed, including one man who was pressed to death under piled rocks for "standing mute at his trial."

For the sociologist, the Salem witch-hunt of 300 years ago has meaning for today's world. In an award-winning study, Kai T. Erikson (1966) showed that the crime of witchcraft can be understood as a sign of "social disruption and change." In Europe only a half-century earlier, thousands of accused witches had been burned and hanged during the turbulent period when the European states were emerging from feudalism. The witch craze in backwoods New England also came during a time of great change. In 1692 the orthodox Puritan way of life was ending: Puritanism was being watered down by rough wilderness ways and urbane city values. The settlements were no longer hemmed in by the wilderness, and the Indians had been pushed away.

In his study of the Salem witch craze, Erikson showed that the punishment of suspected witches served as a defense against the weakening of Puritan society. By casting out the "witches," the Puritans were reaffirming their community values: strict adherence to religious devotion, fear of God, abstinence from the pleasures of secular society (drink, sex, music, dance), and the like. The trial and punishment of the so-called witches illustrates Émile Durkheim's earlier discovery that every society creates its own forms of deviance and in fact needs those deviant acts. The punishment of deviant acts reaffirms the commitment of a society's members to its norms and values and thereby reinforces social solidarity. On the surface, Durkheim argued, deviant acts may seem to be harmful to group life, but in fact the punishment of those who commit such acts makes it clear to all exactly what deviations are most intolerable. By Durkheim's reasoning, the stark images of punishment—the guillotine, the electric chair, the syringe, the wretched life behind bars—become opportunities to let the population know that those who threaten the social order will be severely judged.

The Trial of George Jacobs, August 5, 1692. Oil on canvas by T. H. Matteson, 1855. Jacobs was accused of witchcraft.

would dissolve into chaos, confusion, and hostility (Gibbs, 1994).

The ways in which a society prevents deviance and punishes deviants are known as **social control.** As we saw in Chapter 3, the norms of a culture, the means by which they are instilled in us through socialization, and the ways in which they are enforced in social institutions—the family, the schools, government agencies—establish a society's system of social control. In fact, social control can be thought of as all the ways in which a society establishes and enforces its cultural norms. It is "the capacity of a social group, including a whole society, to regulate itself" (Janowitz, 1978, p. 3).

The means used to prevent deviance and punish deviants are one dimension of social control. They include the police, prisons, mental hospitals, and other institutions responsible for applying social control, keeping order, and enforcing major norms. But if we had to rely entirely on official institutions to enforce norms, social order would probably be impossible to achieve (Chwast, 1965). In fact, the official institutions of social control deal mainly with the deviant individuals and groups that a society fears most. Less threatening forms of deviance are controlled through the everyday interactions of individuals, as when parents attempt to prevent their children from wearing their hair in the style of reggae musicians or piercing their tongues.

In this chapter we look first at the dimensions of deviance and at how its meanings emerge and change as a society's values change. Then we explore the major sociological perspectives on deviance and social control. The final section deals with the ecology of deviance—its distribution among people in different subcultures and income categories—and ends with an introduction to the organization and functioning of institutions of social control such as prisons.

DIMENSIONS OF DEVIANCE

Deviance is an especially controversial topic: There is usually much disagreement not only about which behaviors are deviant and which are not but also about which behaviors should be strongly punished and which should be condoned or punished only mildly. The debate over whether abortion should be legal is a good example of such disagreement. As Kai Erikson noted, "Behavior which qualifies one man for prison may qualify another for sainthood since the quality of the act itself depends so much on the circumstances under which it was performed and the temper of the audience which witnessed it" (1966, pp. 5–6).

Consider South African president Nelson Mandela. Mandela was released from a maximum-security prison in 1990 after serving almost 30 years of a life sentence for his leadership of the movement to end apartheid, South Africa's racial caste system. The dominant white minority in South Africa regarded Mandela and other opponents of apartheid as criminals. But the black majority viewed him as a hero and a martyr; indeed, black South Africans revere Mandela the way people in the United States revere George Washington. Until the black majority began to mobilize world opinion and other nations began to enforce negative sanctions against South Africa, the white minority was able to define Mandela's opposition to apartheid as a criminal activity.

The power to define which acts are legal and which are illegal is an important dimension of deviance. In the United States, the power of some groups to define certain acts as deviant helps explain why influence peddling is not a crime under most circumstances while cultivating marijuana is. In this connection it is interesting to note that during the 1960s, when the children of powerful members of society began smoking marijuana, legislators found themselves under pressure to relax the enforcement of marijuana laws.

The fact that people in power can define what behavior is deviant and determine who is punished fails to explain differences in definitions of deviance in societies with different cultures. Behavior regarded as normal in one society may be considered highly deviant in another. For example, in the United States the drinking of alcoholic beverages is considered normal, but in orthodox Islamic culture it is forbidden. Even within a society, members of certain social groups may behave in ways that are considered deviant by others. Thus the official norms of the Catholic church do not permit the use of artificial methods of birth control, yet large numbers of Catholics use condoms, birth control pills, and other contraceptives. Similarly, the norms of Judaism require that the Sabbath be set aside for religious observance, yet many Jews work on the Sabbath and do not attend synagogue; in parts of Jerusalem they have occasionally been attacked by orthodox Jews for violating that norm. Clearly, differences in values are another important source of definitions of deviance and of disagreements about those definitions.

Ascribed Statuses

Another dimension of deviance has to do with attributes that are ascribed unavoidably at birth (e.g., race or physical appearance) or that a person cannot control (e.g., having a convict as a parent), in contrast to statuses that are achieved through actual behavior, which is usually voluntary (see the discussion of ascribed versus achieved statuses in Chapter 4). A criminal is deviant in ways that a mentally ill person is not, and it is criminal behavior that is most costly to society. Yet in many situations a mentally ill person is labeled as deviant, and this label may actually drive him or her toward criminality. A related issue is how people who deviate from generally accepted norms manage to survive in societies where they are considered outsiders. In fact, deviant people often form their own communities with their own norms and values, and these deviant subcultures sustain them in their conflicted relations with "normal" members of society.

An important point is that deviant subcultures, which engage in prostitution, gambling, drug use, and other deviant behaviors, could not exist were they not performing services and supplying products that people in the larger society secretly demand. It would be wise, therefore, not to draw the distinction between deviant and normal people too sharply. Many people deviate from the norm, and their deviations create opportunities for others whose identities and occupations are deviant.

In sum, three dimensions of social life—power, culture, and voluntary versus involuntary behavior—give rise to the major forces operating in any society to produce the forms of deviance that are typical of that society.

Deviance and Stigma

To narrow the range of phenomena we must deal with in discussing deviance, let us keep in mind Erving Goffman's (1963) distinction between stigma and deviance. "The term **stigma,**" Goffman stated, "refers to an attribute that is deeply discrediting" and that reduces the person "from a whole and usual person to a tainted and discounted one" (p. 3). People may be stigmatized because of ascribed statuses like mental illness, eccentricity, membership in a disvalued racial or nationality group, and so on. In some instances the stigma is visible and obvious to all, as in the case of a disfigured person like John Merrick, commonly known as the Elephant Man. Suffering from a disease that grossly distorted his face, Merrick was rejected by society even though he was a highly intelligent person. In other cases stigma is revealed only with growing acquaintance, as in the stigma attached to the children of convicts. For the stigmatized person, the disqualifying trait defines the person's master status (Becker, 1963; Scull, 1988) (see Chapter 4). A blind person, for example, may be an excellent musician and a caring parent, but the fact that he or she is blind will outweigh these achieved statuses except in unusual cases like that of Stevie Wonder.

John Hurt, starring as the pathetically disfigured sideshow attraction in the film *The Elephant Man,* manages to escape his sadistic master through the help of the other freaks in the circus.

Stigmatized people deviate from some norm of "respectable" society, but they are not necessarily social deviants. The term *deviant,* Goffman argued, should be reserved for people "who are seen as declining voluntarily and openly to accept the social place accorded them, and who act irregularly and somewhat rebelliously in connection with our basic institutions" (1963, p. 143). Among the people Goffman classified as social deviants are "prostitutes, drug addicts, delinquents, criminals, jazz musicians, bohemians, gypsies, carnival workers, hoboes, winos, show people, full time gamblers, beach dwellers, homosexuals, and the urban unrepentant poor." Goffman's list is somewhat whimsical and leaves out more recent groups that are considered socially deviant, such as Internet pornographers; his point, however, is that these are examples of social groups that "are considered to be engaged in some kind of collective denial of the social order. They are perceived as failing to use available opportunity for advancement in the various approved runways of society" (1963, p. 144).

According to the definition of stigma, the population of social deviants is smaller than that of stigmatized individuals; only some stigmatized behaviors are socially deviant. Deviant behaviors are characterized by denial of the social order through violation of the norms of permissible conduct. This point should be kept in mind as we continue our discussion of criminals and other people who are considered social deviants.

Deviance and Crime

Much of the study of social deviance focuses on crime. **Crime** is usually defined as an act, or the omission of an act, for which the state can apply sanctions. Those sanctions are part of the criminal law, a set of written rules that prohibit certain acts and prescribe punishments to be meted out to violators (Kornblum & Julian, 1998). But the questions of which specific behaviors constitute crime and how the state should deal with them are often controversial.

In every society there are some behaviors that almost everyone will agree are criminal and should be punished, and other behaviors that some consider criminal but others do not. All societies punish murder and theft, for example, but there is far more variation in the treatment of adultery, prostitution, and pornography. Indeed, the largest number of "crimes" committed in the United States each year are so-called public order crimes such as public drunkenness, vagrancy, disorderly conduct, prostitution, gambling, drug addiction, and certain homosexual interactions. Many sociologists claim that these are "victimless crimes" because they generally cause no physical harm to anyone but the offenders themselves (Silberman, 1980; Thio, 1998). Not all social scientists agree with this view, however. Some point out that crimes like prostitution actually inflict damage on society because they are usually linked with an underworld that engages in far more serious and costly criminal activities (Wilson, 1977).

In the United States, crime is considered one of the nation's most serious social problems. According to the Federal Bureau of Investigation (FBI), one violent crime occurs every 17 seconds and one property crime every 3 seconds. The most serious, most frequently occurring, and most likely to be reported crimes are called *index crimes* by the FBI because they are included in its Crime Index, a commonly used measure of crime rates. These categories of crime, along with crime rates, are shown in Table 7.1.

For many decades criminologists and other social scientists have criticized the FBI's crime statistics on the ground that they do not reflect differences in effectiveness among crime-reporting agencies. For example, a more professional law enforcement agency in one community could show higher crime rates than an agency in another community that has done a haphazard job of collecting the statistics. Another serious problem is that there is no way of knowing the crime rates for the total population (rather than just for those who actually reported crimes). Since 1973, therefore, the Census Bureau and the Law Enforcement Assistance Administration have conducted semiannual surveys in which respondents are asked whether they or their businesses have been victims of robbery, rape, assault, burglary, or other forms of theft. In general, these *victimization surveys* show that the overall rate of serious crime is actually between two and three times higher than the reported Crime Index.

TABLE 7.1 The Crime Rate,* 1972, 1982, 1996

	1972	1982	1996
Crime Index, total	3,961	5,604	5,079
Violent crime	401	571	634
Property crime	3,560	5,033	4,445
Murder	9	9	7
Forcible rape	23	34	36
Robbery	181	239	202
Aggravated assault	189	289	388
Burglary	1,141	1,489	943
Larceny-theft	1,994	3,085	2,976
Motor vehicle theft	426	459	526

* Rate per 100,000 inhabitants, calculated on number of offenses before rounding.

Source: *Statistical Abstract,* 1998.

DEGREE OF CONSENSUS	STRENGTH OF SANCTION: Weak	STRENGTH OF SANCTION: Strong
Weak	• Recreational drugs • Homosexuality • Abortion	• Sale of whiskey during Prohibition • Prostitution • Abortion before 1973 Supreme Court ruling
Strong	• Schizophrenia • Driving while intoxicated • Public drunkenness • Corporate crime • Wife or child beating	• Major crimes (felonies) • Treason

FIGURE 7.1 A Typology of Deviance in the United States

Deviance and Changing Values

One of the factors that make it especially difficult to gather and interpret statistics on crime is that definitions of crime and deviance are constantly changing as the society's values change. Almost every week, for example, there are reports in the media of men who have battered and even killed their wives, often after serving brief prison sentences for previous attacks on their spouses. Why, many ask, are law enforcement agencies unable to prevent and punish spouse abuse? Advocates of new laws and greater investment in prevention point out that until fairly recently a married woman was viewed as the husband's household possession. Although both husband and wife had sworn to "love, honor, and cherish" their partner, these norms were applied far more forcefully to women than to men. This double standard is changing as men and women become more equal before the law, but in the minds of many judges and juries this change is not yet complete, and men who abuse their spouses often receive relatively light sentences (Heitzeg, 1996).

In any society, agreement on particular aspects of crime and deviance can range from weak (in cases in which there is much controversy) to strong (in cases in which there is little disagreement). Negative sanctions, or punishments, can also range from very weak to very strong. Capital punishment is the strongest sanction in the United States, followed by life imprisonment. Minor fines or the suggestion that a person undergo treatment for a behavior that is viewed as deviant are relatively weak sanctions. Nor are all sanctions formal punishments meted out according to law. Some deviance can be punished by means of shunning or the "silent treatment," and milder infractions can be controlled by simply poking fun at the person in an attempt to change his or her behavior.

Figure 7.1 uses these distinctions to construct a typology of deviance as it is generally viewed in the United States today. Figure 7.2 suggests what such a typology would have looked like before the Civil War. Together they highlight some of the continuities and changes in patterns of deviance in the United States

FIGURE 7.2 Deviance in the United States Before the Civil War

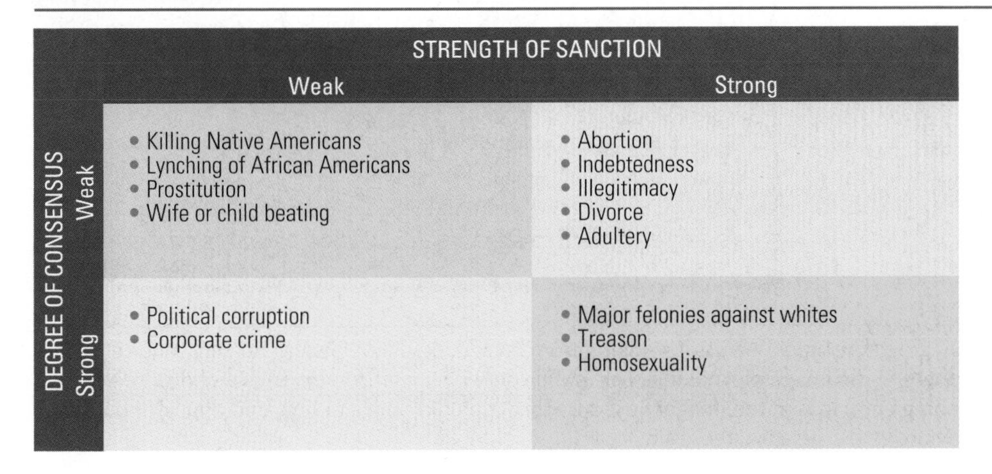

over the last century and a half. You will probably find points in them to argue with—which is further evidence of the difficulty of classifying deviant behaviors in a rapidly changing society.

Driving while intoxicated (DWI) is classified as a deviant behavior, but only in the last 10 years have DWI laws been enforced systematically. Why? We all fear the drunk driver and agree that driving while intoxicated is dangerous, but because so many people are social drinkers this norm is frequently violated. Consequently, we have placed relatively mild sanctions on DWI except when the outcome is a fatal accident. In recent decades, however, organized groups opposed to drunk driving have helped change public attitudes, and the sanctions against DWI have become stronger. In a study of this issue, Joseph Gusfield (1981) found that the highway death rate had steadily decreased after 1945, but that the public's perception of the relationship between highway deaths and drunk driving had sharpened as a result of media coverage and lobbying by citizens' groups.

Our perception and treatment of homosexuality is another example of significant change in Americans' attitudes toward a deviant behavior and the strength of the sanctions invoked. A century ago homosexuality was widely regarded as a serious form of deviance. Homosexuals were persecuted whenever they were exposed. Today there is far less agreement on this subject (Stolberg, 1997). To be sure, homosexuality continues to be viewed as deviant by the majority of nonhomosexuals, but it is tolerated to the extent that gay men and women are able to create their own communities and openly fight against such sanctions as discrimination in the labor market. Homosexuals have had some success in gaining passage of laws that would grant same-sex couples the legal standing of de facto married couples. The decision by the Walt Disney Company to extend health and other insurance benefits to same-sex domestic partners of employees is a highly visible example of the increasing legitimacy of homosexual households, although the company has come under intense fire from groups that oppose homosexuality on religious grounds.

At the same time, however, the perception of AIDS as a "gay disease" has added to the stigma attached to the homosexual identity (Altman, 1987; Bayer, 1988). This can be seen in the strenuous attack by Senator

Changing Americans' views of smoking from an accepted to a deviant activity requires changes in the image of smoking. This public-service advertisement uses a child's prize-winning entry to turn the imagery used by tobacco companies to link smoking with outdoor activities into a negative representation.

Jesse Helms of South Carolina on federal funding for AIDS research. The senator argues that because AIDS is a consequence of "disgusting and immoral behavior," it should not be singled out for special federal funding. Public health officials point out that AIDS is the leading cause of death among men and women aged 25 to 45 and that it is inaccurate to view it as a "gay disease," since it is spreading most rapidly among heterosexuals. Nevertheless, attacks on AIDS funding by Helms and other conservatives are an indication of the extent of the conflict over what constitutes deviance in sexual behavior.

Elsewhere in the world, the growing emphasis on religious orthodoxy in Islamic nations has led to attacks on writers and intellectuals who question Muslim norms. The Egyptian Nobel Prize winner Naguib Mahfouz, the Iranian novelist Salman Rushdie, and Bangladeshi physician, poet, and essayist Taslima Nasrin are among those who have experienced such attacks (Weaver, 1994). In the case of Nasrin, because she speaks and writes against the traditional norms that require women to remain in the home, to be bought as brides, and otherwise to be treated as male possessions, fundamentalist religious leaders have offered a reward (about $2,500) to anyone who succeeds in killing her. Nasrin is now in hiding in Europe. "I know if I ever go back I'll have to keep silent, stay inside my house," she says. "I'll never lead a normal life in my country, until my death" (quoted in Weaver, 1994, p. 60).

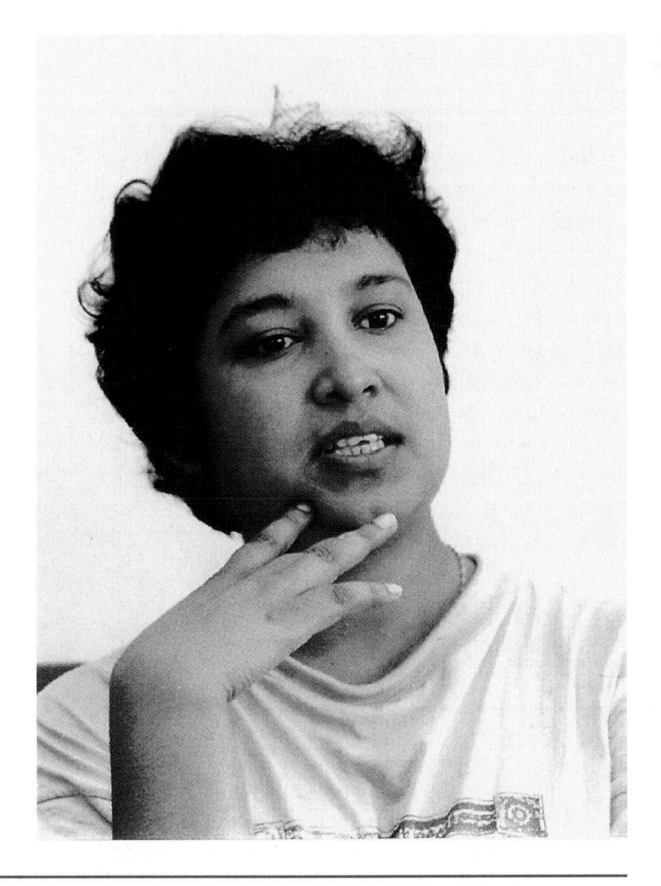

Taslima Nasrin, a world-renowned feminist author and poet from Bangladesh, is in hiding to avoid arrest or mob violence because of her criticism of Islamic gender norms. She is considered a courageous intellectual by many in the West but a deviant criminal by key political and religious authorities in her own country.

Deviant Subcultures

Even when the majority of the population can be said to support a particular set of values, it will contain many subcultures whose lifestyles are labeled deviant by the larger population (Rubington & Weinberg, 1996). A deviant subculture includes a system of values, attitudes, behaviors, and lifestyles that are opposed to the dominant culture of the society in which it is found (Brake, 1980). The members of the subculture are also members of the larger society; they have families and friends outside the subculture with whom they share many values and norms. But within the subculture they pursue values that are opposed to those of the larger or "mainstream" culture. Subcultures evolve their own rather insulated social worlds or communities, with local myths ("X is a slick dealer who once beat the tables in Vegas and was asked to leave the casino"), ways of measuring people's reputations ("he's a punk," "he's cool"), rituals and social routines, particular language or slang, formal and informal uniforms (e.g., the attire worn by bikers), and symbols of belonging such as tattoos (Simmons, 1985).

Subcultures of drug users and dealers provide many illustrations of the ways in which people who engage in deviant activities often develop their own vocabulary and norms. In his fascinating accounts of the subculture of crack dealers and crack-cocaine addicts, urban sociologist Terry Williams (1989, 1992) shows how everyone in the crack subculture—addicts, dealers, and hangers-on—uses the language of the famous TV series *Star Trek* in referring to the drug and its use. The drug itself is often called "Scotty," and smoking the drug is referred to as "beaming up." As addicts and dealers become more deeply enmeshed in the frantic world of crack-cocaine use, they spend increasing amounts of time in the company of people who speak their language and see the world as they do.

Similar examples of the power of the deviant subculture abound in the work of another sociologist, Janet Foster. Foster (1990) observed the lives of young criminal men in an English slum. In the following account of the feelings of a criminal's wife toward the police, Foster shows how the norms and attitudes of the criminals become assimilated into the thinking of women who

Then and Now

Tattoos in Ritual and Fashion

Throughout history, as people from different cultures have come into contact with each other, norms that are part of the core culture of one society have been adopted by deviant subcultures in other societies. In time these deviant symbols tend to become less deviant, mildly acceptable, and even fashionable. A fascinating example of this process is the spread of tattooing into the cultures of Europe and the United States, where at present tattoos are a major fashion statement.

European and American sailors who traveled through the fabled Pacific isles and Oceania in the eighteenth century were impressed by the *tattaw,* or tattoos, proudly worn by people of high status like the Maori chief shown here. The sailors adopted tattoos as a subcultural symbol of their "exotic" experience.

Later, criminals and prostitutes in European cities took to tattoos as a way of signifying their allegiance to their trades. With almost endless time on their hands, prisoners made amateur tattoos (requiring hours of painful jabs with an inked pen) into symbols of their opposition to the "screws" and "pigs"—their jailors and tormentors. Soldiers of the colonial powers, such as the French legionnaire shown here, also adopted tattoos as a way of boasting of their various campaigns.

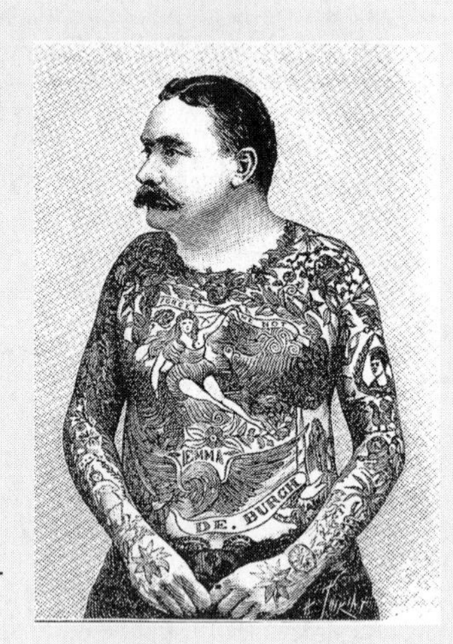

have relationships with the men but are not themselves criminals:

> The two coppers came up that had nicked [arrested] him, and they said, "You all right, mate?" I thought how can you stand there and be so friendly? One of them looked at me and said, "Why don't you take a seat with the baby, love," and all this. I thought what are you being so nice for, you could be depriving my kid of a father here. . . . I know it's not their fault and they're for doing what they're paid to do, and it was more fool Jim for doin' it, you know. It was just bad luck that he got caught. That must be the first time I felt any real resentment against the police. (quoted in Foster, 1990, p. 104)

Professional criminals—including major drug dealers, bookmakers and gamblers, hired killers, and loan sharks—who think of their illegal activities as occupations,

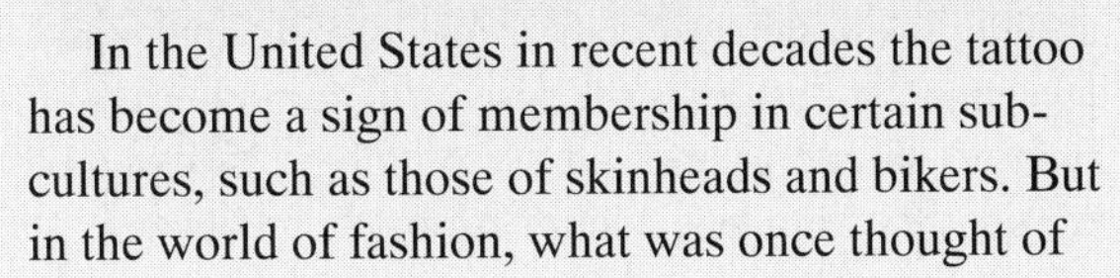

In the United States in recent decades the tattoo has become a sign of membership in certain subcultures, such as those of skinheads and bikers. But in the world of fashion, what was once thought of as deviant becomes daring and may eventually be fashionable, as seen in the contemporary use of the tattoo as a "fashion statement" among young people at almost all levels of society.

often become part of secret crime organizations. The best known of these (but hardly the only one) is the Mafia. Originally a Sicilian crime organization, the Mafia spread to the United States via Sicilian immigration in the late nineteenth century. Today families associated with the Mafia are active in most large cities in North America and elsewhere and often extract millions of dollars from legitimate businesses in exchange for "protection." In recent years there have been some successful crusades against the Mafia, but competing crime organizations are always ready to engage in lucrative criminal activities as the Mafia is weakened (Ianni, 1998).

The subculture of organized crime is not unique to North America. In Japan, rates of homicide and mugging are small fractions of those in U.S. cities, yet organized crime thrives there (Buruma, 1984; Kaplan, 1998). The Yakuza, as the crime gangs are called, control many illegal businesses, just as they do in other countries, and they often resort to violence to discipline their members. As in the United States, organized crime in Japan is most directly associated with behaviors that are considered deviant, such as prostitution, gambling, and drug use, and a distinct subculture has emerged among those who engage in such behaviors.

In sum, many deviant subcultures flourish because they provide opportunities to engage in behavior that is pleasurable to some people but is considered deviant in "respectable" (conventional) society. Clearly, therefore, the line between what is normal and what is deviant is not nearly as distinct as one might believe from official descriptions of the norms of good conduct. On the other hand, one must recognize that membership in a deviant subculture, especially for those without money and power, often leads to exploitation and early death. In view of these problematic aspects of deviant subcultures, sociological and other explanations of deviant behavior take on special importance.

THEORETICAL PERSPECTIVES ON SOCIAL DEVIANCE

Biological Explanations of Crime

Early in the twentieth century the Italian criminologist Cesare Lombroso (1911) claimed to have proved that criminals were throwbacks to primitive, aggressive human types who could be recognized by physical features like prominent foreheads, shifty eyes, and red hair. Although Lombroso's theory has been thoroughly refuted by modern researchers, efforts to link body type with crime seem destined to reappear from time to time. In the 1940s, for example, the psychologist and physician William Sheldon announced that body type was correlated with crime. He believed that human beings could be classified into three types: *ectomorphs,* or thin people; *endomorphs,* or soft, fat people; and *mesomorphs,* or people with firm, well-defined muscles. The latter were most prone to crime, according to Sheldon, but he neglected to account for the possibility that mesomorphs simply might have led harder lives than less muscular individuals and were better equipped to commit crimes that require strength (Glueck & Glueck, 1950).

The more modern biological view is represented in the influential work of social scientists Richard J. Herrnstein and James Q. Wilson. Their research has led them to conclude that both biology and social environment play important roles in producing criminals. In commenting on a recent study by researchers at the U.S. Department of Justice reporting that more than one third of adult criminals in U.S. jails have at least one parent who is or was a criminal, Herrnstein argues that whatever determines criminality "is transmitted both genetically and environmentally. So kids brought up in criminal families get a double exposure" (quoted in Butterfield, 1992b, p. A16).

A study by Deborah W. Denno offers support for this view. In her research on 1,000 African American boys from poor neighborhoods of Philadelphia, Denno (1990) found that a disproportionate number of individuals who had become criminals had histories of childhood hyperactivity. This genetically transmitted condition makes it difficult for children to concentrate and succeed in school. It also can make it hard for them to hold jobs. Thus Denno sees the causal chain as one of biological hyperactivity leading to social failure and then to crime.

But to many social scientists such studies are not convincing. Marvin E. Wolfgang, one of the nation's leading experts on criminality, points out that all the studies deal with people from poor neighborhoods, where a high proportion of people will be sent to jail whether they are related or not and whether they were hyperactive as children or not. Hyperactivity is also found in children from wealthier families but is not so often associated with criminal careers because the families can afford special schools and programs to deal with the condition (Butterfield, 1992b).

Social-Scientific Explanations of Deviance and Crime

Early sociological explanations of social deviance were also influenced by the biological concept of disease (Thio, 1998). The earliest of these explanations viewed crime as a form of social pathology, or social disease, that could be attributed to the evils of city life. Corruption, criminality, and depravity were thought to be bred in slums and to infect innocent residents of those grim communities, just as typhoid spread in unclean surroundings and was passed by contact from one victim to another (Bulmer, 1986). Thus many early juvenile training schools were built in rural areas where they would be isolated from the "corrupting" influence of cities (Platt, 1977).

The early sociological view of deviance and crime as resulting from "social pathologies" tended to rely on subjective terms and circular reasoning. To refer to crime or other forms of deviance as pathologies, for example, does not provide a theory of causality. It merely attaches a scientific-sounding term to behavior that the larger society finds wrong or offensive. It does not offer any insight into why the behavior occurs, what benefits it might provide for those who engage in it, and how it evolves from other behaviors or social conditions. In consequence, the view of deviance as a kind of disease was replaced by more objective and verifiable theories derived from the functionalist, interactionist, and conflict perspectives of modern sociology.

The Functionalist Perspective

The functionalist theorist Robert K. Merton (1938) developed a useful typology of deviance based on how people adapt to the demands of their society. Merton's aim was to discover "how some social structures *exert a definite pressure upon certain persons in society to engage in nonconformist rather than conformist conduct*" (p. 672). "Among the elements of social and cultural structure," Merton continued, "two are important for our purposes. . . . The first consists of culturally defined goals, purposes, and interests. . . . The second . . . defines, regulates, and controls the acceptable modes of achieving these goals" (pp. 672–673).

Merton's explanation of deviance is based on the concept of **anomie,** or normlessness. In his view, anomie results from the frustration and confusion people feel when what they have been taught to desire cannot be achieved by the legitimate means available to them. Merton believes that people in North America and other modern societies exhibit high levels of anomie because they are socialized to desire success in the form of material well-being and social prestige. For many, however, the means of attaining these culturally defined ends, such as hard work and saving, seem to be out of reach. This is especially true for people who experience rapid social changes, such as the closing of factories, and find themselves deprived of opportunities to attain what they have come to expect from life.

Consider some examples. Possession of money is a culturally defined goal. Work is a socially acceptable mode of achieving that goal; theft is not. Mating is also a culturally defined goal. Courtship and seduction are acceptable means of achieving it; kidnap and rape are not. But if theft and rape are unacceptable means of achieving culturally defined goals, why do they exist? According to Merton, the gap between culturally defined goals and acceptable means of achieving them causes feelings of anomie, which in turn make some people more likely to choose deviant strategies of various kinds.

Through socialization we learn the goals and acceptable means of our society. Most of us would love to be rich or powerful or famous. We accept these goals of our culture. We also accept the legitimate means of achieving them: education, work, the electoral process, plastic surgery, acting school, and so on. We are conformists. But not everyone accepts either the cultural goals or the accepted means of achieving them. Some become "innovators" in that they explore (and often step over) the frontiers of acceptable goal-seeking behavior; others (e.g., hoboes) retreat into a life that rejects both the goals and the accepted means; some rebel and seek to change the goals and the institutions that support them; and still others reject the quest for these precious values while carrying out the rituals of social institutions. (In many bureaucracies, for example, one can find ritualists who have given up the quest for promotion yet insist on receiving deference from people below them in the hierarchy.) Merton's typology thus is based on whether people accept either the cultural goals of their society, the acceptable means of achieving them, or neither. Figure 7.3 presents this typology and

FIGURE 7.3 **The Merton Typology—Modes of Adaptation by Individuals Within the Society or Group**

MODE OF ADAPTATION	CULTURAL GOALS	INSTITUTIONAL MEANS
Conformity	+	+
Innovation	+	–
Ritualism	–	+
Retreatism	–	–
Rebellion	+/–	+/–

Merton explained that "+" signifies acceptance, "–" signifies elimination, and "+/–" signifies rejection and substitution of new goals and standards. The line separating "Rebellion" from the other roles signifies that the individual no longer accepts the society's culture and structure; other individuals, although they may deviate, continue to accept the society's culture and structure.

MODE OF ADAPTATION	THE POOR	THE MIDDLE CLASS	THE RICH
Conformity	The working poor	The suburban family	The wealthy civic leader
Innovation	The mugger	The embezzler	The stock manipulator
Ritualism	The chronic welfare recipient	The resigned bureaucrat	The hedonist
Retreatism	The wino or junkie	The skidding alcoholic	The bohemian
Rebellion	The bandit	The anarchist	The fascist

These examples are meant to illustrate possible outcomes, not stereotypes. Perceptions of approved goals and means may vary. In some cases, for example, the very rich, who feel that they could be even richer were it not for legal obstacles in their path, may choose to bend the rules. Also, a number of cultural values besides wealth shape the likelihood that a member of a given class will behave as he or she does.

FIGURE 7.4 Examples of Social Roles Based on the Merton Typology

Figure 7.4 gives examples of all the deviant types in Merton's framework.

You may be wondering what there is in Merton's theory that explains why some poor people resort to crime while so many others do not. The answer lies in the way people who experience anomie gravitate toward criminal subcultures. For example, adolescents who choose to steal to obtain things that their parents cannot afford must learn new norms—they must learn how to steal successfully and must receive some approval from peers for their conduct (Cloward & Ohlin, 1960; McNamara, 1994). As a result, they drift toward deviant peer groups. But this explanation presents a problem for the functionalist perspective: If there are deviant subcultures, the idea that there is a single culture whose goals and means are shared by all members of society is called into question. The presence of different subcultures within a society suggests another possible cause of deviant behaviors: conflict between groups.

Conflict Perspectives

Cultural Conflict In an influential essay, "Crime as an American Way of Life," Daniel Bell (1962) observed that at the turn of the twentieth century cultural conflict developed between the "Big City and the small-town conscience. Crime as a growing business was fed by the revenues from prostitution, liquor, and gambling that a wide-open urban society encouraged and that a middle-class Protestant ethos tried to suppress with a ferocity unmatched in any other civilized society" (p. 128). This example of conflict between the official morality of the dominant American culture and the norms of subcultures that do not include strictures against gambling, drinking, prostitution, and the like encouraged the growth of criminal organizations. In the United States as in many other societies, such organizations thrive by supplying the needs of millions of people in cities, towns, and rural areas who appear to be law-abiding citizens yet engage in certain deviant behaviors.

The prohibition of alcoholic beverages in the United States from 1919 to 1933 is an example of how cultural conflict can lead to situations that encourage criminal activity. Prohibition has been interpreted as an effort by the nation's largely Protestant lawmakers to impose their version of morality on immigrant groups for whom consumption of alcohol was an important part of social life. Once they were passed, however, laws against the production and sale of alcoholic beverages created opportunities for illegal production, bootlegging (smuggling), and illegal drinking establishments (speakeasies). These in turn supported the rise of organized crime syndicates. The current laws against drugs like marijuana and cocaine have similar effects in that they lead to the clandestine production and supply of these illegal but widely used substances. Similar opportunities to make illegal profits arise when states pass laws raising the drinking age (Gusfield, 1966, 1996; Kornblum & Julian, 1998).

To recognize these effects of cultural conflict is not to condone the sale and use of illegal drugs. Rather, it is to be aware that whenever laws promoted by the powerful impose a set of moral standards on a minority, illegal markets are created that tend to be supplied by criminal organizations. Such organizations often act ruthlessly to control their markets, as can be seen in the murder of Mexican border guards who surprise drug smugglers, the killing of cocaine traffickers by rival dealers in Miami or Los Angeles, or the periodic "wars" that break out between organized crime families in many large cities.

Marxian Conflict Theory For Marxian students of social deviance, the cultural conflict explanation is inadequate because it does not take into account the effects of power and class conflict. Marxian sociologists believe that situations like Prohibition do not occur just because of cultural conflict (Felson, 1998). They happen because the powerful classes in society (i.e., those who own and control the means of production) wish to control the working class and the poor so that they will produce more. From the Marxian perspective, as criminologist Richard Quinney (1980) points out, "crime is to be understood in terms of the development of capitalism" (p. 41), as it was in Marx's original analysis. From this perspective, most crime is essentially a form of class conflict—either the have-nots taking what they can from the ruling class, or the rich and their agents somehow taking what they can from the poor.

The economic "robber barons"—John Jacob Astor, John D. Rockefeller, J. P. Morgan, Leland Stanford, Andrew Carnegie, and many others—amassed huge fortunes in the period of booming industrial growth following the Civil War. But they often resorted to illegal means in pursuing the culturally approved goal of great wealth. Among other tactics, they used violence to drive settlers off land they had purchased or to break strikes by workers, and they were not above manipulating prices in order to drive out competitors and monopolize the markets for steel, oil, coal, precious metals, food products, and numerous other goods (Chernow, 1998). In Merton's typology of deviance, their actions would classify them as "innovators" (see Figure 7.4), but from a Marxian viewpoint they were merely carrying out "the logic of capitalism," which was based on the exploitation of the poor by the rich and powerful.

Marxian students of deviance point out that legal definitions of deviant behavior usually depend on the ability of the more powerful members of society to impose their will on the government and to protect their actions from legal sanctions. Thus the crimes of the robber barons almost always went unpunished. Definitions of what is criminal and who should be punished are generally applied more forcefully to the poor and the working class (Balbus, 1978; Quinney, 1978; Turk, 1978).

Marx and his collaborator Friedrich Engels recognized that the working class (or *proletariat,* as they called it) would resort to individual crimes like robbery when driven to do so by unemployment and poverty, but they believed that the workers would be more likely to form associations aimed at destroying capitalism. The chronic poor, on the other hand, would form a class that Marx called the *lumpenproletariat,* people who were unable to get jobs in the capitalist system or were cast off for not working hard enough or for being injured or sick. Marx did not believe that members of this class would join forces with the proletariat. Instead, they would act as spies, informers, and thugs whose services could be purchased by the rich to be used against the workers. Marx agreed with other thinkers of his time who called the lumpenproletariat the "dangerous class" created by capitalism; from its ranks came thieves, prostitutes, gamblers, pickpockets, con artists, and contract murderers.

It can be helpful to think of crime and deviance as symptoms of the class struggles that occur in any society and to show that laws that define and punish criminal behaviors are often imposed by the powerful on those with less power. But to attribute crime as we know it to the workings of capitalism is to suggest that if capitalism were abolished, crime would vanish. This clearly is not the case. In China, for example, there are thousands of prisoners in a vast chain of prison camps. Many of those prisoners were convicted of the deviant act of opposing the dominant regime (Ning, 1995; Wu & Wakeman, 1995). And in Cuba homosexuality is considered a serious crime. It is severely punished because the society's leaders think of it as an offense against masculinity and a symbol of a "decadent" subculture. As these examples indicate, societies with Marxian ideologies are hardly free from their own forms of deviance and crime.

The Interactionist Perspective

Functionalist theories explain deviance as a reaction to social dysfunctions; conflict theories explain it as a product of deviant subcultures or of the type of class struggle that occurs in a society in a particular historical period. Neither of these approaches accounts very well for the issues of recruitment and production (Rubington & Weinberg, 1996). *Recruitment,* in this context, refers to the question of why some people become deviants while others in the same social situation do not. *Production* refers to the creation of new categories of deviance in a society.

Recruitment Through Differential Association In 1940 the sociologist and criminologist Edwin H. Sutherland published a paper, "White-Collar Criminality," in which he argued that official crime statistics do not measure the many forms of crime that are not correlated with poverty. Outstanding among these are *white-collar crimes*—that is, the criminal behavior of people in business and professional positions. In Sutherland's words:

> White-collar criminality in business is expressed most frequently in the form of misrepresentation in financial statements of corporations, manipulation in stock exchange, commercial bribery, bribery of public officials directly or indirectly in order to secure favorable contracts and legislation, misrepresentation in advertising and salesmanship, embezzlement and misapplication of funds, short weights and measures and misgrading of commodities, tax frauds, misapplication of funds in receiverships and bankruptcies. These are what Al Capone called "the legitimate rackets." These and many others are found in abundance in the business world. (1940, pp. 2–3)

Sutherland was pointing out that an accurate statistical comparison of the crimes committed by the rich and the poor was not available. But his paper on white-collar crime also set forth an interactionist theory of crime and deviance:

> White-collar criminality, just as other systematic criminality . . . is learned in direct or indirect association with those who already practice the behavior; and . . . those who learn this criminal behavior are segregated from frequent and intimate contacts with law-abiding behavior. Whether a person becomes a criminal or not is determined largely by the comparative frequency and intimacy of his contacts with the two types of behavior. This may be called the process of differential association. (pp. 10–11)

The concept of **differential association** offered an answer to some of the weaknesses of functionalist and conflict theories. Not only did it account for the prevalence of deviance in all social classes but it also provided clues to how crime is learned in groups that are culturally distinct from the dominant society. For example, in the 1920s sociologist Clifford Shaw (1929) observed that some Chicago neighborhoods had consistently higher rates of juvenile delinquency than others. These were immigrant neighborhoods, but their high rates of delinquency persisted regardless of which immigrant groups lived there at any given time. Sutherland's theory explained this pattern by calling attention to the culture of deviance that had become part of the way of life of teenagers in those neighborhoods. According to Sutherland, the teenagers became delinquent because they interacted in groups whose culture legitimated crime. It was not a matter of teenage delinquents deviating from conventional norms because the approved means of achieving approved goals were closed to them. Rather, they acted as they did because the culture of their peer group made crime an acceptable means of achieving desired goals.

In an empirical study that tested Sutherland's theory, Walter Miller (1958) found that delinquency in areas with high rates of juvenile crime was in fact supported by the norms of lower-class teenage peer groups. In 3 years of careful observation Miller found that delinquent groups had a set of well-defined values: *trouble, toughness, smartness, excitement, fate,* and *autonomy.* Whereas other groups felt that it was important to stay out of trouble, the delinquent groups viewed trouble—meaning fighting, drinking, and sexual adventures—as something to brag about, as long as they didn't get caught. Toughness as shown by physical prowess or fearlessness; smartness as evidenced by the ability to con or outsmart gullible "marks"; the excitement to be found in risking danger successfully; one's fate as demonstrated by luck or good fortune in avoiding capture; and the autonomy that crime seemed to provide in the form of independence from authorities—all were values of delinquent groups that distinguished them from nondelinquent groups in the same neighborhoods.

A study by criminologist Donald R. Cressey (1971/1953) supported the theory of differential association with the finding that embezzlers generally had to learn how to commit their crime by associating with people who could teach them how to commit it with the greatest likelihood of avoiding suspicion. Furthermore, Cressey also found that people who became embezzlers often had serious personal problems (marital difficulties, gambling, and the like) that directed them toward people who would influence them to commit this form of white-collar crime. When bank officials had problems of this sort, they were vulnerable to suggestions of deviant means of gaining money and solving personal problems while maintaining the outward signs of respectability.

Not all deviants are people whose means of achieving success have been blocked or who are acting out some form of class struggle or have associated with a deviant group. For example, many alcoholics and drug users are not thought of as deviant, either because their behavior is not considered serious or because it is not witnessed by other people. From an interactionist perspective, the key question about such people is how their behavior is understood by others (Rubington & Weinberg, 1996). The central concepts that attempt to answer this question are *labeling* and the idea of the *deviant career.*

Labeling According to symbolic-interactionist theory, deviance is produced by a process known as **labeling,** meaning a societal reaction to certain behaviors that labels the offender as a deviant. Most often, labeling is done by official agents of social control like the police, the courts, mental institutions, and schools, which distribute labels like "troublemaker," "hustler," "kook," or "blockhead" that stick, often for life (Becker, 1963; Erikson, 1962; Gusfield, 1981; Lerman, 1996). Here is what occurs in the labeling process, according to Howard Becker:

> Social groups create deviance by making the rules whose infraction constitutes deviance and by applying those rules to particular people and labeling them as outsiders. From this point of view, deviance is not a quality of the act the person commits, but rather a consequence of the application by others of rules and sanctions to an "offender." The deviant is one to whom that label has been successfully applied; deviant behavior is behavior that people so label. (1963, p. 9)

In a famous experiment that tested the effects of labeling, David Rosenhan (1973) and eight other researchers were admitted to a mental hospital after pretending to "hear voices." Each of these pseudopatients was diagnosed as schizophrenic. Before long many of the patients with whom the pseudopatients associated recognized them as normal, but the doctors who had made the diagnosis continued to think of them as schizophrenic. As the pseudopatients waited in the lunch line, for example, they were said to be exhibiting "oral-acquisitive behavior." Gradually the pseudopatients were released with the diagnosis of "schizophrenia in remission," but none was ever thought to be cured.

Rosenhan and his researchers also observed that not only did the diagnosis of schizophrenia label the patient for life but the label itself became a justification for other forms of mistreatment. The doctors and hospital staff disregarded the patients' opinions, treated them as incompetent, and often punished them for infractions of minor rules. The hospital's social atmosphere was based on the powerlessness of the people who were labeled mentally ill.

Rosenhan's study accelerated the movement to reform mental institutions and to deinstitutionalize as many mental patients as possible. But studies of deinstitutionalized mental patients (many of whom are homeless), together with research on the problems of released convicts, have shown that the labels attached to people who have deviated become incorporated into their definitions of themselves as deviant. In this way labeling at some stage of a person's development tends to steer that person into a community of other deviants, where he or she may become trapped in a *deviant career* (Bassuk, 1984).

Deviant Careers "There is no reason to assume," Becker (1963) points out, "that only those who finally commit a deviant act actually have the impulse to do so. It is much more likely that most people experience deviant impulses frequently. At least in fantasy, people are much more deviant than they appear" (p. 26). Becker suggests that the proper sociological question is not why some people do things that are disapproved of but, rather, why "conventional people do not follow through on the deviant impulses they have" (p. 26). According to Becker and other interactionists, the answers are to be found in the individual's commitment to conventional institutions and behaviors.

Commitment means adherence to and dependence on the norms of a given social institution. The middle-class youth's commitment to school, which has developed over many years of socialization in the family and the community, often prevents him or her from giving in to the impulse to play hooky. "In fact," Becker asserts, "the normal development of people in our society (and probably in any society) can be seen as a series of progressively increasing commitments to conventional norms and institutions" (p. 27).

Travis Hirschi studied how people become committed to conventional norms. Such commitment, he found, emerges out of the interactions that create our social bonds to others. When we are closely tied to people who adhere to conventional norms, we have little chance to deviate. And as we grow older, our investment in upholding conventional norms increases because we feel that we have more to protect. This process of lifetime socialization in groups produces a system of social control that is internalized in most members of the society (Hirschi & Gottfredson, 1980, 1984). (See Chapter 5 for more on socialization over the life course.)

By the same token, a person who once gives in to the impulse to commit a deviant act and is caught, or who becomes a member of a deviant group because it has recruited him or her, gradually develops a commitment to that group and its deviant culture. The reasons for taking the first step toward deviance may be many and varied, but from the interactionist perspective "one of the most crucial steps in the process of building a stable pattern of deviant behavior is likely to be the experience of being caught and publicly labeled as deviant" (Becker, 1963, p. 31).

In a well-known empirical study of youth gangs, William J. Chambliss (1973) applied both the conflict and interactionist perspectives. For 2 years Chambliss observed the Saints, a gang of boys from rather wealthy families, and the Roughnecks, a gang of poor boys from the same community. Both gangs engaged in car theft and joyriding, vandalism, dangerous practical jokes, and fighting—and in fact the Saints were involved in a larger overall number of incidents. But the Roughnecks were more frequently caught by authorities, described as "tough young criminals headed for trouble," and sent to reform school. The wealthier youths were rarely caught and were never labeled delinquent.

STUDY CHART Theoretical Perspectives on Deviance

PERSPECTIVE	DESCRIPTION	CRITIQUE
BIOLOGICAL THEORIES	Crime and other forms of deviance are genetically determined.	There is evidence of effects of socialization.
SOCIOLOGICAL THEORIES		
Social Pathology	The deviant or criminal person is a product of a "social sickness" or social disintegration.	A popular notion, but discredited among professionals because it suggests no real theory of causality.
Functionalism	Deviance and crime result from the failure of social structures to function properly. Every society produces its own forms of deviance.	Does not explain why some people drift into deviant subcultures.
Conflict Theories		
Cultural conflict	Cultural conflict creates opportunities for deviance and criminal gain in deviant subcultures (e.g., prohibitions create opportunities for organized crime).	Explains a narrow range of phenomena.
Marxian theory	Capitalism produces poor and powerless masses who may resort to crime to survive. The rich employ their own agents to break laws and enhance their power and wealth.	Crime exists in societies that have sought to eliminate capitalism.
Interactionist Theories		
Differential association	Criminal careers result from recruitment into crime groups based on association and interaction with criminals.	Excellent explanation of recruitment but not as effective in explaining deviant careers.
Labeling	Deviance is created by groups that have the power to attach labels to others, marking particular people as outsiders. It is extremely difficult to shed a label once it has been acquired, and the labeled person tends to behave in the expected manner.	Not always supported by empirical evidence. People can also use labels (e.g., drunk) to change their behavior.

Chambliss observed that the Saints had access to cars and could commit their misdeeds in other communities, where they were not known. The Roughnecks hung out in their own community and performed many of their antisocial acts there. A far more important explanation, according to Chambliss, pertained to the relative influence of the boys' parents. The parents of the Saints argued that the boys' activities were normal youthful behavior; the boys were just "sowing their wild oats." Their social position enabled them to influence the way their children's behavior was perceived, an influence the parents of the Roughnecks did not have. Thus the Roughnecks were caught and labeled and became increasingly committed to deviant careers, but the Saints escaped without serious sanction.

In commenting on Chambliss' study, sociologists often note that members of both gangs engaged in deviant acts, referred to as **primary deviance,** but that only the Roughnecks were labeled delinquent by the police and the juvenile courts. As a result of that labeling, many of the Roughnecks went on to commit acts that sociologists call **secondary deviance**—that is, behaviors appropriate to someone who has already been labeled delinquent. (For example, in juvenile detention centers teenage offenders often learn deviant skills such as how to deal in drugs.) The distinction between primary and secondary deviance is useful because it emphasizes that most of us deviate from cultural norms in many ways; but once we are *labeled* deviant we tend to commit additional deviant acts in order to fulfill the negative definitions society has attached to us (Goode, 1994; Schur, 1971).

Reasonable as the labeling perspective appears, it has not always been borne out by empirical research. Some studies have found that people who have been labeled delinquent after being caught and convicted of serious offenses go on to commit other deviant acts, but other studies have found that labeling can lead to decreased deviance and a lower probability of further offenses (Rubington & Weinberg, 1996). (The various theoretical approaches to crime that we have discussed are summarized in the study chart opposite.)

CRIME AND SOCIAL CONTROL

We have seen that crime as defined by a society's laws is only one aspect of the range of behaviors included in the study of deviance. But the general public and policy makers in government are most concerned with crime and its control. In this section, therefore, we touch on some of the most hotly debated issues in the control of crime and the treatment of criminals.

At the beginning of the chapter we defined social control in broad terms as all the ways in which a society establishes and enforces its cultural norms. Certainly it is true that without socialization and the controlling actions of social groups like the family, schools, the military, and corporations there would be much more anomie, crime, and violence. But in considering a society's means of controlling crime, sociologists most often study what might be called "government social control"—that is, the society's legal codes; the operation of its judicial, police, penal, and rehabilitative institutions; and the ways in which its most powerful members promote their views of crime, deviance, and social control (Black, 1984; Scull, 1988). The rates of crime and the ways in which crime affects communities in different parts of the society establish the need for these institutions of social control, as we will see in the following analysis of the ecology of crime in American society.

Ecological Dimensions of Crime

Table 7.2 shows that the incidence of different types of crime varies from one region of the country to another, with homicides highest in the southern states, and rape and assault lowest in the northeastern states. Ecological data like these are an essential starting point in research on the incidence of crime and provide some insights into why such regional differences exist. In the South, for example, high rates of gun ownership contribute to higher homicide rates. The higher average age of the population in the Northeast contributes to lower rates of rape and assault in that region, while greater disparities between rich and poor lead to a higher rate of robbery.

Another contribution of ecological analysis can be seen in data on the relationship between crime and the age composition of a population. Some political leaders and social scientists favor larger police forces and tougher law enforcement, and they argue that these policies are responsible for the decrease in the U.S.

TABLE 7.2 The Crime Rate,* by Region of the United States, 1995

	Northeast	Midwest	South	West
Crime Index, total	4,180	4,751	5,742	6,083
Violent crime, total	611	588	738	770
Murder	6.2	6.9	9.8	9.0
Forcible rape	24.9	40.1	40.9	38.6
Robbery	260	182	212	242
Aggravated assault	319	359	474	481
Property crime, total	3,570	4,164	5,004	5,312
Burglary	758	741	1,137	1,121
Larceny-theft	2,296	2,821	3,337	3,435
Motor vehicle theft	516	451	530	766

* Offenses known to the police, per 100,000 population.

Source: FBI, 1997.

crime rate over the past few years. However, it must not be forgotten that the age composition of the U.S. population is changing rather quickly. Among other things, the baby boom generation has moved out of the young-adult years, those in which people are most likely to commit crimes. (About half of all reported violent crimes are committed by males between the ages of 15 and 24.) Thus, before taking credit for reducing crime rates, law enforcement agencies and advocates of tough anticrime measures must make certain that lower crime rates are not simply a result of a reduction in the most crime-prone portion of the population.

Still another variable that lends itself to ecological analysis is the effect of possession of firearms on homicide rates. The United States far outpaces other industrialized nations in number of homicides. No other nation even comes close. This fact is often thought to reflect a greater overall level of violence in the United States than in other comparable nations. However, in a study comparing rates of violent crime in Denmark and northeastern Ohio, investigators found that rates of assault did not differ in the two areas. What did differ were rates of homicide using firearms; such crimes were much more frequent in Ohio. In Denmark the possession of handguns is banned, whereas in the United States "50 percent of all households have guns and one in five has a handgun" (Mawson, 1989, p. 239).

As mentioned earlier in the chapter, studies of criminal victimization, which ask people about their actual experiences with crime, have provided a major breakthrough in our understanding of crime in complex societies. Of every 1,000 crimes, slightly more than half are reported to the police. Of those, about 15 percent result in arrests. Twenty years ago less than half of all arrests led to convictions, and less than half of those convicted were sentenced to custody. Today, with more forceful law enforcement strategies, including tougher sentencing guidelines and mandatory jailing of third-time offenders in California and other states, these figures have improved. Nevertheless, the probability that a criminal who is arrested will be sentenced to prison remains low. On the other hand, while the probability of arrest for a single property crime, such as theft of an automobile, is low, criminals who commit many crimes do tend to be arrested eventually; this is especially true for those who commit violent crimes (Butterfield, 1998b).

Most sociological studies of **recidivism** (the probability that a person who has served a jail term will commit additional crimes and be jailed again) indicate that there is an overall probability of 50 percent that recidivism will occur. Recidivism is most frequent among young minority males from poor backgrounds who are drug and/or alcohol addicts and have been imprisoned on numerous previous occasions for crimes against property (Land, 1989). In recent years there have been forceful efforts to reduce the types of recidivism that pose the greatest threat to society. People who have been convicted of violent offenses and people who have served prison sentences for sex crimes are especially likely to be refused parole. In the case of sex offenders, community notification policies, in which neighbors are informed that a former sex offender has moved into their community, are being implemented in many states (Zonana, 1997). The high crime rates of recent decades, together with public support for longer sentences and less lenient parole policies, have resulted in record high rates of imprisonment, as we will see shortly. Despite the decline in crime rates in the last few years, states are finding it necessary to build new prisons even as corrections professionals search for alternatives to imprisonment, especially for nonviolent offenders.

Institutions of Coercive Control: Courts and Prisons

When the United States consisted mainly of small agrarian communities, there were jails and courts, but a great deal of social control of deviance and crime was carried out by the local institutions of the family or the church. Parents, for example, were expected to control their children; if they did not, they would lose the respect of other members of the community. But as the size, complexity, and diversity of societies increase, the ability of local institutions to control all of the society's members is diminished (Wirth, 1968/1938). Societies therefore develop specialized, more or less coercive institutions to deal with deviants. Courts, prisons, police forces, and social-welfare agencies grow as the influence of the community on the behavior of its members declines.

Capital Punishment—Cruel and Unusual? Capital punishment provides an illustration of the issues raised by the use of coercive forms of social control. Only recently (during the twentieth century in some nations) has there been ongoing debate about the morality of capital punishment (Phillips, 1998). In simpler societies and earlier civilizations, execution was more than a penalty carried out on an individual; it was also an occasion for a public ceremony. People attended beheadings and hangings, and hawkers sold food and favors as the crowd waited for the bloody pageant. The villain's death reaffirmed their common values, their solidarity as a people who could purge evil elements from their midst.

Today, by contrast, the value of capital punishment is a matter of heated debate, as we have seen recently in the dramatic sentencing of Timothy McVeigh for what is clearly "the U.S. crime of the century." Although many people believe that the death sentence is necessary as a means of discouraging some individuals from committing terrible crimes, others are convinced that there is no justification for putting another person to death, the more so because there is always the possibility that the person did not commit the crime for which he or she was executed. A recent study of this grim possibility provides evidence of as many as 18 wrongful executions in the United States since 1900 (Radelet & Bedau, 1992). In one of a series of studies of the equity of capital punishment, criminologists Marvin Wolfgang and Marc Riedel (1973) found that blacks were more likely than whites to be executed for the same crimes and that people who could afford good lawyers were far more likely to escape execution than those who lacked the means to hire the best legal defenders.

In 1972, swayed by evidence like that gathered by Wolfgang and Riedel and by arguments that capital punishment could be considered "cruel and unusual punishment" according to the Constitution (most other Western nations had already banned the death penalty), the Supreme Court ruled in a 5-to-4 decision that in the absence of clear specifications for when it might be used, the death sentence violated the Eighth Amendment to the Constitution. But Congress has since passed new legislation that legalized the death sentence, and the Court has not overruled it. As a result, capital punishment has been reinstated in many states; there are now between 15 and 30 executions a year in the United States, with Texas, Florida, Mississippi, and Arizona leading in numbers of executions carried out. Indeed, in many states the debate about capital punishment has shifted away from its morality to questions about technology, such as the use of lethal injections versus the electric chair.

But does capital punishment have the effect of deterring people from committing murder, as its advocates claim? Much of the evidence on this subject is negative. As Figure 7.5 shows quite clearly, little encouragement is found in comparisons among states that have the death penalty, states that have had it with interruptions, and states that have never had it. Yet in interviews with criminals charged with robbery, James Q. Wilson (1977) found evidence that fear of the death penalty discouraged them from carrying guns. There is still considerable opposition to the death penalty among people who feel that it represents cruel and unusual punishment, but surveys indicate that the tide of public opinion has turned toward support for it. They also show, for better or worse, that the public's increased concern about crime has led to greater emphasis on use of the death sentence as retribution for murders that have already been committed and less emphasis on the possible deterrent effect of capital punishment.

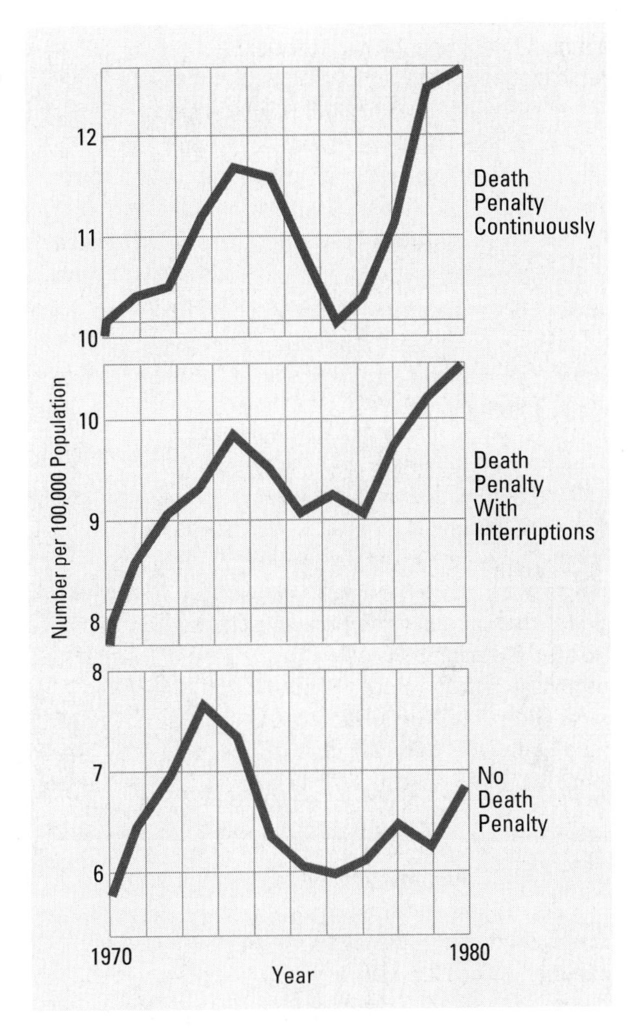

States that had the death penalty continuously: Arizona, Arkansas, Connecticut, Florida, Georgia, Montana, Nebraska, Nevada, Oklahoma, Texas, Utah, Virginia. States without the death penalty: Alaska, District of Columbia, Hawaii, Iowa, Maine, Michigan, Minnesota, New Jersey, North Dakota, West Virginia, Wisconsin. All other states had the death penalty with interruptions.

Source: Zeisel, 1982.

FIGURE 7.5 **Homicide Rates in States With and Without the Dealth Penalty, 1970–1980**

Plea Bargaining—A Revolving Door? Capital punishment addresses a small but sensational part of the crime problem. Many more criminals will commit other types of crimes, and for them too the issue of punishment and rehabilitation is controversial. At present, judges and police officers feel hampered by the relatively limited array of treatments for criminal offenders, as well as the backlogs in the courts and the overcrowding of prisons. One means of reducing the pressure on the judicial system is **plea bargaining,** in which a person who is charged with a crime agrees to plead

(continued on page 216)

MAPPING SOCIAL CHANGE

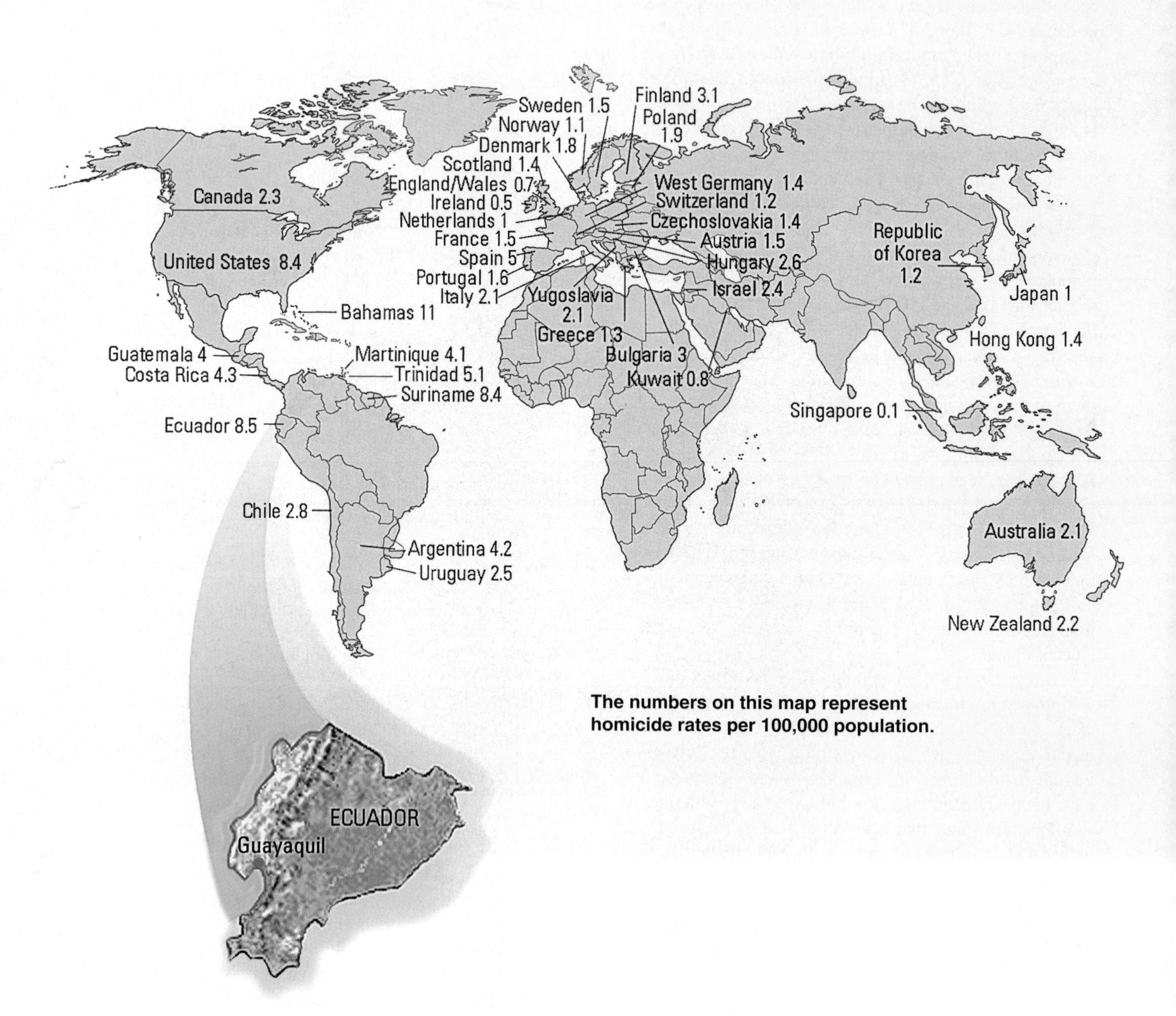
Finland 3.1
Sweden 1.5
Poland 1.9
Norway 1.1
Denmark 1.8
Scotland 1.4
England/Wales 0.7
Ireland 0.5
Netherlands 1
France 1.5
Spain 5
Portugal 1.6
Italy 2.1
West Germany 1.4
Switzerland 1.2
Czechoslovakia 1.4
Austria 1.5
Hungary 2.6
Israel 2.4
Yugoslavia 2.1
Greece 1.3
Bulgaria 3
Kuwait 0.8
Canada 2.3
United States 8.4
Bahamas 11
Guatemala 4
Costa Rica 4.3
Martinique 4.1
Trinidad 5.1
Suriname 8.4
Ecuador 8.5
Chile 2.8
Argentina 4.2
Uruguay 2.5
Republic of Korea 1.2
Japan 1
Hong Kong 1.4
Singapore 0.1
Australia 2.1
New Zealand 2.2
ECUADOR
Guayaquil
The numbers on this map represent homicide rates per 100,000 population.

Worldwide Homicide Rates

Although homicide rates in the United States have declined in the past 4 years, there is little likelihood of a decline to levels characteristic of the European democracies. There are nations with higher homicide rates, such as Ecuador, the Bahamas, and perhaps Russia. All are undergoing rapid social change, but none enjoy the degree of affluence found in the United States. In Ecuador, high homicide rates are associated with rapid urbanization in cities like Guayaquil, where murders occur at extremely high rates in sprawling slums. In the Bahamas, the international drug markets—which often include stops in this island nation—are an important influence. In Russia, the abrupt changes accompanying the transition from communism to capitalism have produced an increase in homicide rates.

The United States shares certain characteristics with each of these nations. Like Ecuador and other Latin American nations, it has a frontier tradition that fosters and still maintains norms of rugged individualism, including the use of guns. It also has a widening gap between rich and poor, with the poor often concentrated in ghetto slums. Drug activity contributes significantly to homicide rates, as in the Bahamas. And as in Russia, much of the deadly violence occurring in the United States stems from the activities of criminals in organized gangs.

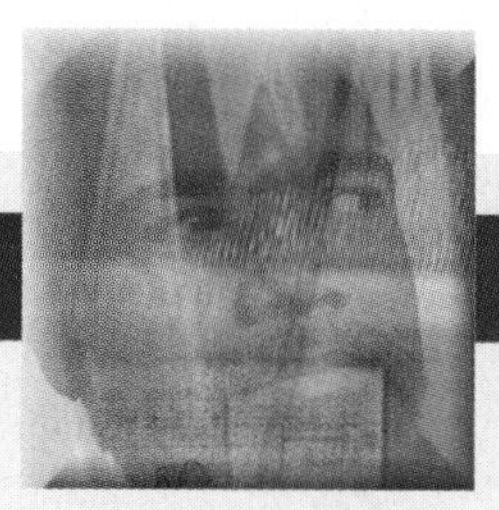

THE PERSONAL CHALLENGE

Oklahoma City and the Death Penalty

Sometimes it hurts to debate the issue of capital punishment. Feelings about this subject usually run strong and deep. But because it is a social policy that can be changed, the death penalty is not an issue that an informed citizen can simply "duck" while others with deep convictions seek to sway public opinion. Two people whose lives were shattered by Timothy McVeigh, the convicted bomber of the federal office building in Oklahoma City, offer differing guidance on how to meet the personal challenge of capital punishment.

> *Donetta Apple:* On the day of the bombing my brother Tom Hawthorne's company was on strike, and he went into Oklahoma City to do some Social Security paperwork for an acquaintance. He was at the Murrah building purely by chance; he could have done his errands any other day. On that day Tim McVeigh took something very precious from me—my brother. But he also took something from the whole country—our sense of security. And he did it deliberately. That's why I think he should die.
>
> Before the blast, I just couldn't support the death penalty because its application seemed so arbitrary. One jury would hand down a death sentence in an ambiguous case when the criminal didn't seem cold or calculating, and then another jury would send a murderer who committed terrible, terrible atrocities off to prison for life. I figured that maybe only a perfect society could really pass judgment on another human being, and we weren't there yet.
>
> But there is nothing arbitrary or ambiguous about McVeigh's conscious decision to do what he did. He chose to park that truck, put in his earplugs, and walk off. When he did that, he took away the rights of 168 people to ever make decisions of their own again. My brother and the others can't elect to work, or play, or spend time with their families. So I don't want McVeigh to have the freedom to even get a drink of water in his cell. If those 168 victims can't make the most basic of choices, why should he? McVeigh has to pay for the choice he made on April 19, 1995—and he has to pay with his life.
>
> *Bud Welch:* Most people believe Timothy McVeigh should be put to death. I certainly understand their anger: My daughter Julie, a Spanish-language translator who worked for the Social Security Administration at the Murrah building, lost her life in the building. I am filled with rage at McVeigh. But I don't think he should be executed.
>
> I'm not a minister or philosopher—and I'm not an anti-death-penalty crusader. But executing a murderer is just another kind of murder. When Julie was at Marquette University, we'd debate politics during the long drive between Oklahoma City and Milwaukee. One time she said something I've recalled a lot recently: "Dad, the death penalty doesn't teach us anything but hate."
>
> McVeigh shouldn't get off easy. Lock him up for good, with no chance to get out. Is that punishment enough? The part of me that still screams "kill him" doesn't think so. But my Catholicism teaches that even he has a soul, and we must at least try to save him—and even try to forgive him. I'm still too angry to deal with that now. But I'll have to be forgiving if I am to have peace. That would be harder if he is executed. I don't want McVeigh's death on my head. A lady from Texas called me and said her husband had been murdered. After his killer was executed, she began to feel guilty. She thought knowing the murderer was dead would help ease her grief, but it didn't—and I don't think it would help me, either.
>
> I'm not trying to win converts. I just want people to think hard about the costs of the death penalty. Killing McVeigh won't bring my daughter back. The only way I can go on is to continue to believe in the sanctity of life—even a mass murderer's. (Quoted in *Newsweek,* June 16, 1997, p. 30.)

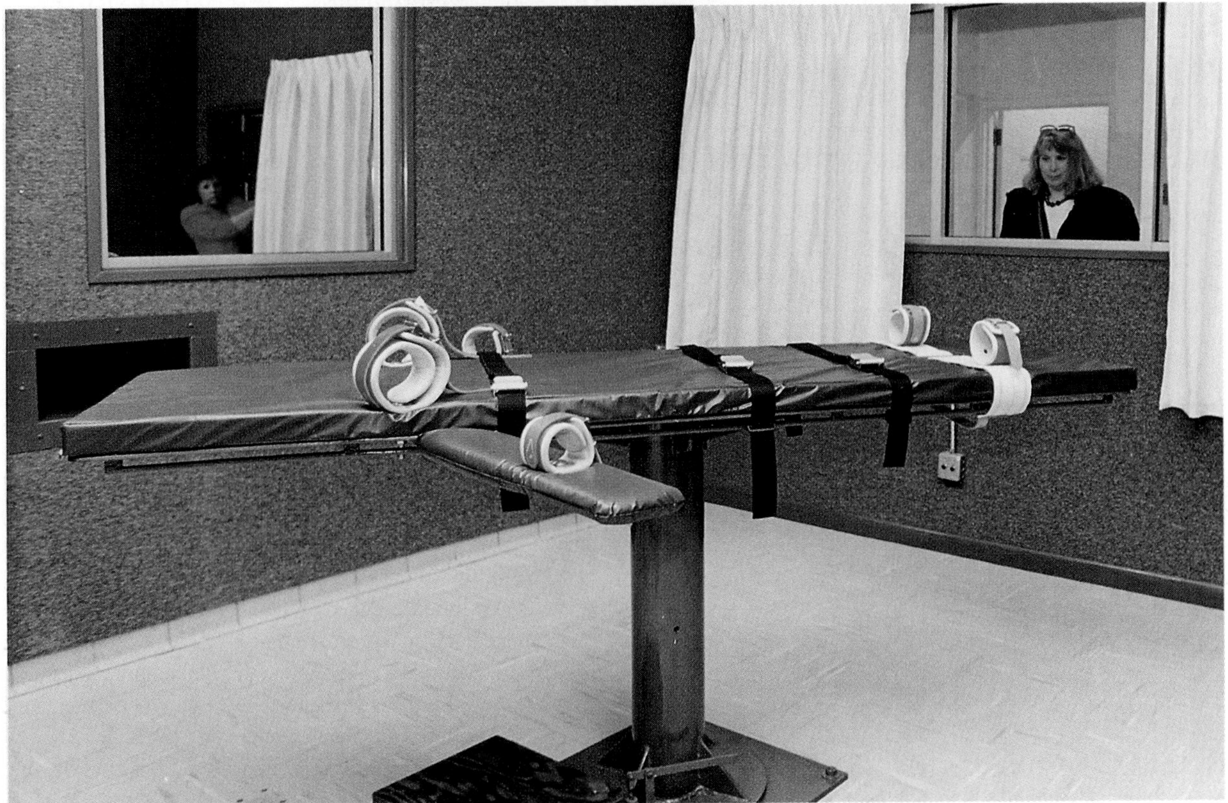

In earlier periods of history, executions were public spectacles and often considered a form of entertainment. In modern societies that still practice execution, there is much less emphasis on public spectacle and much more on the technology employed to perform the execution.

guilty to a lesser charge and thereby free the courts from having to conduct a lengthy and costly jury trial. The shorter sentences that result from this system diminish somewhat the size of prison populations. Plea bargaining has been criticized as "revolving-door justice"; however, criminologists estimate that an increase of only 20 percent in the number of offenders who are tried and imprisoned would place an intolerable burden on the correctional system (Reid, 1993).

Prisons—Schools for Crime? Despite the small proportion of crimes resulting in arrest and imprisonment, prisons remain the social institution with the primary responsibility for dealing with criminals. From 1980 to 1997 the total number of Americans in prisons of all kinds more than tripled (see Figure 7.6). In addition to the prison population, there are more than 3.5 million people on probation or parole. It is estimated that within a decade more than 7.5 million people will be under some form of law enforcement surveillance in the United States (Butterfield, 1998a).

FIGURE 7.6 **Inmates in Federal, State, and Local Prisons, 1980–1997**

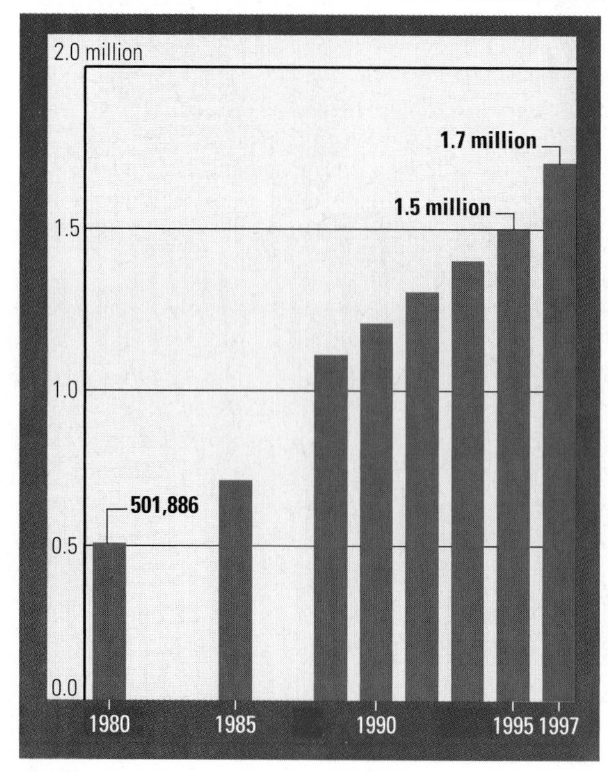

Source: Bureau of Justice Statistics.

The largest proportion of the new prisoners were arrested for the sale or possession of felonious quantities of narcotics and were imprisoned under mandatory sentencing laws that had been passed as part of the national war on drugs. But the arrests were by no means equally distributed among drug users and dealers. According to researchers with the Federal Sentencing Project, for young Americans from ghetto communities and many other impoverished areas "going to prison has become as inevitable . . . as going to college is for middle-class kids" (quoted in Butterfield, 1992a, p. E4). While there is widespread agreement on the need to punish drug dealers and users, there is growing debate over the desirability of sending them to prison, since it is highly likely that a young person who is imprisoned will learn more diverse and devious ways of breaking the law.

The arguments used to support reliance on prison as the central institution of punishment and reform are mainly functionalist in nature. The functions of prisons are said to be deterrence, rehabilitation, and retribution (i.e., punishment) (Goode, 1994; Hawkins, 1976). As Bruce Jackson, a well-known student of prison life, explains it, a prison is supposed to deter criminals from committing crimes ("its presence is supposed to keep those among us of weak moral strength from actions we might otherwise commit"); it is supposed to rehabilitate those who do commit crimes ("within its walls those who have, for whatever reason, transgressed society's norms are presumably shown the error of their ways and retooled so they can live outside in a more acceptable and satisfactory fashion"); and it is supposed to punish criminals—the only function that it clearly fulfills (1972, p. 248).

Sociologists who study prisons note their similarity to other kinds of total institutions—social environments in which every aspect of the individual's life is subject to the control of authorities. In these institutions inmates are deprived of their former statuses through haircuts, uniforms, and the like (Goffman, 1961). (See Chapter 5.) The goal of this process is to socialize the inmates to behave in ways that suit the organization's needs. But in most such organizations there is also a strong inmate subculture consisting of norms that specify ways of resisting the officials in favor of such values as mutual aid and loyalty among the inmates (Rubington & Weinberg, 1996). Thus, in Bruce Jackson's (1972) collection of case histories of criminals and prison inmates, a seasoned convict summed up the world of the total institution as follows: "A penitentiary is like a prisoner of war camp. The officials are the enemy and the inmates are captured. They're on one side and we're on the other" (p. 253). But the same prisoner also described another aspect of life in total institutions. Often, as the following passage reveals, the inmates develop "institutional personalities." As they try to conform to the norms of the organization, they become dependent on the routines and constraints of minimal freedom and, hence, minimal responsibility:

> I like things to be orderly. You might say I'm a conservative, I like the status quo. If something's going smooth I like that. . . . Some people . . . kind of like it. Get institutionalized. You know, there's a lot of security in a place like this for a person. . . . They'll tell you when to get up, they'll tell you when to go to bed, when to eat, when to work. They'll do your laundry for you. You don't have to worry about anything else. (p. 253)

Only relatively recently, beginning mainly in the second half of the twentieth century, has the goal of rehabilitation—the effort to return the criminal to society as a law-abiding citizen—been taken seriously as a function of prisons. Critics of the prison system, many of whom argue from a conflict perspective, often claim that far from rehabilitating their inmates, prisons in fact function as "schools for crime" (Califano, 1998). As early as 1789 the English philosopher and economist Jeremy Bentham wrote that "prisons, with the exception of a small number, include every imaginable means of infecting both body and mind. . . . An ordinary prison is a school in which wickedness is taught by surer means than can ever be employed for the inculcation of virtue" (pp. 351–352). In the intervening years many studies have attempted to show under what conditions this is true and what can be done to increase the deterrent effects of prison and prevent it from becoming a community that socializes criminals.

Yet there is much sociological research that defends prisons as a means of deterrence and a necessary form of retribution (Allen, 1997). James Q. Wilson (1977) holds to the functionalist position: Societies need the firm moral authority they gain from stigmatizing and punishing crime. He believes that prisoners should receive better forms of rehabilitation in prison and be guaranteed their rights as citizens once they are outside again. But, he states, "to destigmatize crime would be to lift from it the weight of moral judgment and to make crime simply a particular occupation or avocation which society has chosen to reward less (or perhaps more) than other pursuits" (p. 230). Some states have taken the desire to get tough on crime to the extreme of reinstating chain gangs.

Whatever their sociological perspective, all students of the American prison system agree that by far the least successful aspect of prison life is rehabilitation. Research conducted in the 1970s consistently found that the only effective rehabilitation programs in prisons, measured by lower rates of recidivism, involve job training and education (Martinson, 1972). Indeed, during the 1980s, as prison populations more than doubled, sociologists and criminologists found that men from very poor and minority backgrounds were beginning to regard prison as the only way they could obtain job training, health care, and other social services. Prisons, after all, provide more adequate housing than the large urban shelters for the homeless, and in prison one gets three meals a day and, in some cases, a chance to improve one's skills and self-confidence.

The conservative mood of the United States in the 1990s had a profound impact on rehabilitation programs in federal and state prisons. Throughout the United States, legislative and administrative actions reduced educational opportunities for offenders, especially by limiting federal and state funding for prison post-secondary education programs (Tewksbury & Taylor, 1996). Some legislators, showing far more interest in punishment than in rehabilitation, supported a proposal that would void all court-ordered efforts to correct inhumane conditions in prisons (Kunen, 1995). As one of the nation's leading researchers on crime and prisons has observed, in our society "we have an exaggerated belief in the efficacy of imprisonment. The real problem is that we make life really terrible for some people and then blame them when they become dangerous" (Norval Morris, quoted in Butterfield, 1992a).

RESEARCH ON THE CUTTING EDGE: Living With Deviant Labels

In 1994 sociologist James W. Trent, Jr., won a prestigious award from the American Association on Mental Retardation for his research on the effects of labeling people with mental limitations. Trent's research directly challenges theories such as those of Robert Herrnstein and Charles Murray (1994), which defend labels like "normal" and "below normal" and claim that there are strict biological limits on what people with low intelligence can achieve. (See Box 5.1 for a discussion of this issue.) One of the people described in Trent's research is Emma Jane Nixon, a white woman from a well-to-do family in North Carolina who had given birth to a "Negro" child almost 60 years earlier, during the heyday of segregation. In the community where she lived, and in the eyes of her prominent family, the illegitimate birth was an intolerable embarrassment. Although Nixon, then in her twenties, had done reasonably well in school, her family and local social-welfare officials concluded that her behavior proved that she was mentally retarded. She was placed in a state home for the "feebleminded" and lived in such institutions for the rest of her life. Intelligence tests given to her in 1975 showed that her IQ was 85, about average for people of her age.

Before her death in her mid-eighties, Nixon experienced all the theories and deviant labels that could be applied to people with limited intelligence or mental retardation. As Trent (1995) explains it, from the time she was first institutionalized as "feebleminded" until her death,

> The only official changes in Nixon's circumstances involved the labels used to describe her: "Moron" (a subset of "feebleminded") gave way to "mental defective," then to "mentally retarded" (still the most commonly used term) and, finally, "developmentally disabled." Every few decades, it seemed, her condition earned a few new syllables. (p. 18)

Just as the labels changed, so did theories about how to deal with people of supposed very low intelligence.

The 1840s may have been the best time for people who were then referred to as "idiots." The best treatment centers were run like small families in which everyone shared responsibilities and all assumed that the patients could make progress toward becoming independent and productive. Often they were trained to work on farms, which always needed more workers. Later in the nineteenth century, dramatic social changes such as rapid urbanization, immigration, and "boom and bust" economic cycles led to new labels and new theories about people of low intelligence. The more neutral-sounding term "feebleminded" began to replace the term "idiot," but more important, the professionals assigning the labels began to believe that urban vice, crime, vagrancy, illegitimacy, prostitution, and the decline of neighborhood cohesion made such people easy prey for shrewd and criminal characters. It would be best, they asserted, to institutionalize these people for their own protection.

During good economic times, the institutions tended to have enough resources to provide pleasant and positive environments, but each successive economic downturn produced reductions in public funding. In time, thousands of people labeled as feebleminded—or more recently as mentally retarded, learning disabled, profoundly retarded, or developmentally disabled—were living in squalid conditions in prisonlike state institutions.

In the 1970s an alliance of civil liberties advocates concerned about inhumane conditions, along with state officials who wanted to shift the cost of care to the federal government, succeeded in influencing Congress to include "mental retardation" under Medicare and the Supplemental Social Security program. This allowed states to deinstitutionalize persons labeled as mentally retarded and move them into federally subsidized community nursing homes and group homes. But this "community mental health" approach had limited success in addressing the problems of labeling.

The popular film *Forrest Gump* presented the idea that "a person ceases to be mentally retarded . . . the moment people quit defining him as such" (Trent, 1995, p. 18). Because his mother and other people in his life refused to accept the limiting label, Gump could go on to become a star football player, a war hero, a ping-pong champion, a millionaire, and a spiritual leader. In 1992, acting on the same theory, the American Association on Mental Retardation introduced a new policy stating that people are never mentally retarded in absolute terms; instead, they have differing "intensities of functional needs." By restructuring themselves to meet those needs, communities can essentially "cure" mental retardation by providing environments in which everyone can function to the best of his or her ability.

As Trent points out, this position converges in ironic ways with the ideas of those who see mental retardation as an immutable biological condition. Herrnstein and Murray (1994) argue for a return to old-fashioned caring neighborhoods that can offer mentally retarded citizens "a valued niche in society where they can live valued lives and help out in doing the mundane chores of daily life"—a position very close to that of the community mental health approach but without any suggestion of public funding.

According to Trent, this idea is wishful thinking. "Ridding people of labels is certainly better than declaring them permanently disabled," he argues, "but in America today neither approach is adequate" (1995, p. 22). Massive cuts in social spending make it highly unlikely that neighborhoods will have the resources to provide caring environments in which labels need not apply. Trent continues:

> When the economy is unstable, as the past 150 years have shown, mentally retarded people land in institutions or jail or on the streets more often than they land in caring communities. . . . Without jobs and humane social programs, the

> most open-minded society offers its mentally retarded citizens little more than the illusion of integration and equality. (1995, p. 22)

The same reasoning could be applied to many other deviant populations in need of attention. In an era of shrinking public budgets, sociologists like James Trent will be carefully monitoring the actual experiences of people who must struggle to live with deviant labels.

VISUAL SOCIOLOGY

The Tunnel People

You get to meet some weird people coming through the tunnel. We met young graffiti artists. In fact we met one of the best graffiti artists—his name was Freedom-Chris. He just did a mural down in our place. It must be something like sixty-foot by thirty-foot. Then we had a fellow by the name of Sane. He died recently—fell off a bridge. And he did a piece right on our wall about the mind, intellectuality, that man is screwing up the whole world. And it really looked nice.

During the first few decades of the twentieth century homeless people, especially homeless men, were shunted into the "skid row" areas of cities in the United States. Skid row was, and in some cities still is, a downtown area where there were cheap hotels and cafeterias and religious missions—all of which served the needs of homeless men, day laborers, migrant workers, chronic alcoholics, and other destitute people who sought to survive in the city while maintaining socially disapproved lifestyles. But skid row areas have been no more immune to major social change than other parts of cities. Changes in the need for cheap labor in seaports and regional shipment centers have decreased the demand for day-rate workers. The population of hoboes has decreased as well, and so has the propensity for some men to go on drunken binges for weeks at a time, drifting into skid row to enjoy its cheap bars and anonymous hotels. Most recently, the rising value of downtown real estate has driven out the older skid row institutions. So where can homeless, alcoholic, drug-addicted, mentally ill, or destitute men and women live in the contemporary city?

Very often these people still live in the downtown area, under boxes, or on the periphery of the city, in vacant lots or alleys. In the largest cities, however, these people are often driven from these more visible spots. Some find shelter underground, in railroad tunnels and abandoned water tunnels. Photographer Margaret Morton and sociologist Terry Williams have conducted extensive studies of the lives of homeless people underground. These photos are from Morton's photo essay about the tunnel people of midtown Manhattan. They show how desperately people strive to create the semblance of a "normal" life under the most discouraging conditions.

Bob's last job was as a cook in a hospital, but *Coke and amphetamines started coming back to me because I had an unlimited supply. . . . And I gave up everything, gave up everything that I had. And I moved into a place called the Amtrak Tunnel.*

I take medication for asthma and epilepsy. It doesn't matter where you live, what you're doing. If it comes, it comes. There's nothing you can do about it. I go out every day. I have friends. I go to the bar to see my friends and stuff like that. I don't hide. But everybody doesn't know everything about me either because it's none of their business. I don't ask them about them, and they don't ask me about me. So, that's the way it is. Because some of them are my friends, they know where I live. But then there are others who are just acquaintances that I see when I see, and that's it. And they don't know nothing. They don't have to.

I have always done most of the cooking. In most cases, people will heat up stuff down there. But cooking stews, soups—I'm an excellent cook. I love to cook. People say, "That was amazing!" I've never cooked professionally, I've never had a job as a cook. They thought I was in the military or something, which I wasn't.

I got my little family [eighteen cats], and that's enough. I don't need any more. They are all different, and if you're feeling bad they make you feel better. They're not like people, they're not two-faced. They don't have one side and then have another the next time. So that's why I love my animals. And they know they can depend on me.

SUMMARY

Deviance, broadly defined, is behavior that violates the norms of a particular society. The deviant label is attached to a person who violates or opposes a society's most valued norms. The ways in which a society encourages conformity to its norms and prevents deviance are known as *social control.*

There is usually much disagreement not only about which behaviors are deviant, but also about which ones should be condoned or punished only mildly. An important dimension of deviance is the power of some groups in society to define which acts are legal and which are illegal. Another dimension has to do with attributes that a person cannot control, in contrast to actual behavior. In addition, some people are thought of as deviant because of their membership in a group that deviates from the norms of the larger society. To the rest of society they are deviants, but within the group they are not deviant but are conforming to the group's norms.

Deviance should be distinguished from *stigma.* A stigmatized person has some attribute that is deeply discrediting, such as a disfiguring disease, but is not necessarily a social deviant. The study of deviance is concerned with social deviants—that is, people who voluntarily violate the norms of their society. In particular, the study of deviance focuses on criminal deviance: acts or omissions of acts for which the state can apply sanctions.

As a culture's values and norms change, so do its notions of what kinds of behavior are deviant and how they should be sanctioned. The extent to which the members of a society agree on whether or not a particular behavior is deviant can range from weak (in cases in which there is much controversy) to strong (in cases in which there is little disagreement). Negative sanctions, or punishments, can also range from very weak to very strong.

A deviant subculture includes a system of values, attitudes, behaviors, and lifestyles that are opposed to the dominant culture of the society in which it is found. Many deviant subcultures are harmful to society because they sustain criminal occupations. Others provide opportunities to engage in behavior that is pleasurable to many people but is considered deviant in "respectable" society. The boundaries between what is normal and what is deviant are not distinct.

Biological explanations of deviance relate criminality to physical features or body type. These explanations influenced the earliest sociological theories of deviance, which viewed crime and other forms of social deviance as varieties of "social pathology" that could be attributed to the evils of city life. This view has been replaced by more objective and verifiable theories drawn from the basic perspectives of modern sociology.

Functionalist theories of deviance include Robert Merton's typology based on how people adapt to the demands of their society. In this view, through socialization people learn what goals are approved of in their society and the approved means of achieving those goals. Individuals who do not accept the approved goals and/or the legitimate means of achieving them are likely to engage in deviant behaviors.

Functionalist theories have been criticized for assuming that there is a single set of values shared by all the members of a society. Conflict theorists stress the relationship between cultural diversity and deviance. The two main types of conflict theories are cultural conflict theories and Marxian theories. Cultural conflict theories concentrate on the ways in which conflicting sets of norms result in situations that encourage criminal activity. Marxian theories place more emphasis on class conflict, explaining various types of crime in terms of the social-class position of those who commit them.

Interactionist theories of deviance focus on the issues of recruitment (the question of why some people become deviant whereas others do not) and production (the creation of new categories of deviance in a society). Edwin H. Sutherland's theory of *differential association* holds that whether or not a person becomes deviant is determined by the extent of his or her association with criminal subcultures. Interactionists believe that deviance is produced by the process of *labeling,* in which the society's reaction to certain behaviors is to brand or label the offender as a deviant. Once acquired, such a label is likely to become incorporated into the person's self-image and to increase the likelihood that he or she will become committed to a "deviant career."

Studies of the incidence of crime often start with ecological data on crime rates. Studies of criminal victimization, which ask people about their actual experiences with crime, have made a major contribution to the understanding of crime in complex societies.

The methods used to control crime change as societies grow and become more complex. In larger, more diverse societies the ability of local institutions to control all of the society's members is diminished. Such societies tend to develop standardized, more or less coercive institutions to deal with deviants. Among the most prominent institutions of social control in modern societies are courts and prisons.

The primary functions of prisons are said to be deterrence, rehabilitation, and punishment. However, prisons do not seem to deter crime, and only recently has the goal of rehabilitation been taken seriously. Numerous studies have found that prisons are not successful in rehabilitating their inmates and in fact often serve as "schools for crime." The only rehabilitation programs that appear to be effective are those that give inmates job training and work experience.

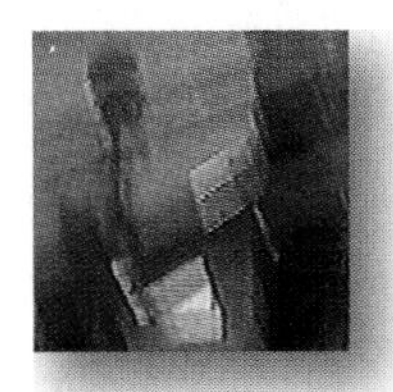

SOCIOLOGY VERSUS IDEOLOGY

Issues of crime and punishment are certain to divide people in a democratic society into at least two camps. Some will dwell on society's need to set examples or isolate evildoers from innocent people (deterrence) and to punish offenders (retribution), while others, who do not necessarily deny the need for deterrence and retribution, will also plead for rehabilitation. Are we doomed to be forever torn between "hard-liners" and people who are said to be "soft on criminals" because they seek rehabilitation of those who can be helped? What does a sociological perspective on these issues add to the debate?

Sociologists can help resolve these conflicts in at least two important ways: (1) by focusing on the legitimate emotions people express in their views and (2) by seeking to bring new facts to the debate. Whenever people debate issues of crime and punishment, strong feelings are involved. People who take a hard line on punishment often dwell on the outrage they feel about heinous crimes like murder and rape. People who argue for the humanity of the criminals too often fail to own up to the rage they would feel if they or a loved one was victimized. It will do no good to continue the debate before people have had a chance to voice their feelings on these issues. But once they have done so, the sociologist can enlighten the argument with useful facts. Does capital punishment deter homicide, for example? The facts do not support its deterrent value, so we probably need to understand that we apply the death penalty as the supreme form of retribution or revenge. If that is what the majority of us want in a democracy, that will be our policy. It is also of vital importance that the policy be administered fairly, and if we can show that it is not, there is a good argument for revising the policy.

Our reasoning must proceed in similar ways for retribution and rehabilitation issues involving other crimes. What do the facts show us? Who can benefit from rehabilitation and release into society, and who poses too great a threat to society to be permitted to be free again? These are empirical questions. Research can help us answer them, and thus arrive at wiser social control policies, but only if we have sorted through our emotions beforehand.

GLOSSARY

deviance: behavior that violates the norms of a particular society. (p. 192)

social control: the ways in which a society encourages conformity to its norms and prevents deviance. (p. 194)

stigma: an attribute or quality of an individual that is deeply discrediting. (p. 195)

crime: an act or omission of an act that is prohibited by law. (p. 196)

anomie: a state of normlessness. (p. 203)

differential association: a theory that explains deviance as a learned behavior determined by the extent of a person's association with individuals who engage in such behavior. (p. 206)

labeling: a theory that explains deviance as a societal reaction that brands or labels as deviant people who engage in certain behaviors. (p. 206)

primary deviance: an act that results in the labeling of the offender as deviant. (p. 208)

secondary deviance: behavior that is engaged in as a reaction to the experience of being labeled as deviant. (p. 208)

recidivism: the probability that a person who has served a jail term will commit additional crimes and be jailed again. (p. 210)

plea bargaining: a process in which a person charged with a crime agrees to plead guilty to a lesser charge. (p. 211)

WHERE TO FIND IT

BOOKS

Street Women (Eleanor Miller; Temple, 1986). An empirical study of how women become involved in crime and organize their lives as street hustlers.

The Outsiders (Howard Becker; Free Press, 1963). An original application of the interactionist and labeling perspectives to the study of deviance and crime, using empirical examples.

Capital Punishment and the American Agenda (Franklin E. Zimring; Cambridge University Press, 1989). An analysis of why we have returned to earlier policies regarding capital punishment, by one of the nation's leading criminologists and proponents of rehabilitation.

Thinking About Crime, rev. ed. (James Q. Wilson; Vintage, 1985); and ***Crime*** (James Q. Wilson; ICS Press, 1995). These works present a more conservative analysis of crime and crime control than that presented by Zimring and others. Wilson is a major proponent of swifter punishment and more effective enforcement of criminal laws.

Stigma: Notes on the Management of Spoiled Identity (Erving Goffman; Prentice Hall, 1982). A classic study of how people who differ from the norm are often labeled as deviant; explores the consequences of such labeling.

Constructions of Deviance: Social Power, Context, and Interaction (Patricia A. Adler and Peter Adler, eds.; Wadsworth, 1994). A collection of contemporary articles on theories and research about many aspects of deviant behavior.

OTHER SOURCES

Crime in the United States (Uniform Crime Reports). An annual report on criminal offenses, arrests, and law enforcement employment. Data are reported by local police departments to the Federal Bureau of Investigation and published by the U.S. Department of Justice.

National Crime Survey. A continuing series of reports on the incidence of crime throughout the United States; published by the U.S. Department of Justice.

Sourcebook of Criminal Justice Statistics. A comprehensive annual compilation of statistics on criminal and related matters; issued by the Bureau of Justice Statistics.

Understanding and Preventing Violence (Albert J. Reiss and Jerry A. Roth, eds.; National Academy Press, 1993). A summary volume of a six-volume report on violence in the United States. Contains excellent reviews of the literature on gun control, interpersonal violence, and the biological and social antecedents of violent behavior.

INTERNET RESOURCES

Federal Bureau of Investigation (www.fbi.gov/). Provides several sources of information about current and historical investigations; provides links to many other information sources.

U.S. Department of Justice (www.usdoj.gov/). A regularly updated home page with links to various agencies and projects.

Violence Against Women Office (www.usdoj.gov.vawo/). Offers information on the National Domestic Violence Hotline, copies of federal legislation and regulations, ongoing research reports and studies, and a Domestic Violence Awareness Manual.

Federal Bureau of Prisons (www.bop.gov/). Provides statistics on the federal prison population.

The National Gay and Lesbian Task Force (www.ngltf.org/gi.html). A "progressive civil rights organization" with a home page that tracks political issues of concern to lesbians and gays and suggests activities for those who want to advance their civil rights.

CHAPTER 8

COLLECTIVE BEHAVIOR, SOCIAL MOVEMENTS, AND MASS PUBLICS

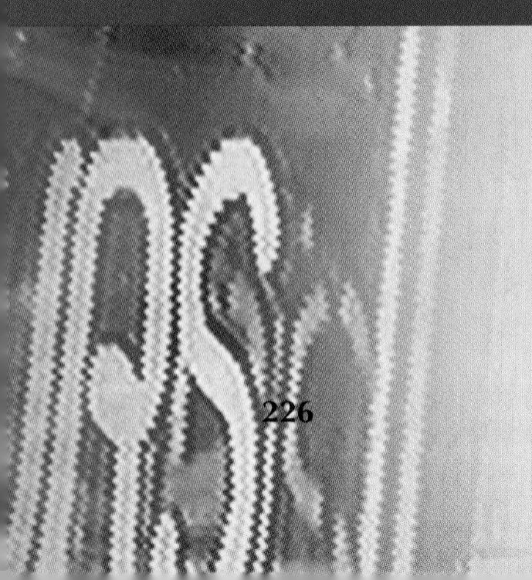

Victor Ayala did not intend to become one of the leaders of the AIDS awareness movement in his community, but it gradually became clear to him that he must do so. His story is about how a person with a strong conscience becomes involved in movements that bring about social change. In the early 1980s, as a college student, he volunteered to spend some time each week working in a gay men's crisis center. There he was trained in how to answer basic questions about sexually transmitted diseases and safe sex. Over time he also learned how to counsel people whose loved ones were desperately ill. Victor's involvement in the gay rights movement and his work at the crisis center were part of what convinced him to pursue a graduate degree in sociology. He wanted to keep improving his ability to help, and for him this meant learning more about social research.

Throughout the 1980s the AIDS epidemic worsened. People close to Victor became ill and died after much suffering and anguish. Most of them were still relatively young, their talents not fully realized.

As Victor pursued his doctorate in sociology and the time came for him to choose a dissertation topic, he sought a way to combine his social activism with his sociological research. AIDS had already taken a great emotional toll on him through the loss of loved ones, and now he felt a need to use his research to help combat the disease. He was aware that the insidious illness was spreading to more and more poor people from minority backgrounds, and to more women, through unsafe sex, prostitution, and the sharing of needles for intravenous drug use. Victor could see evidence of these trends in the inner-city neighborhoods surrounding the community college where he worked as a counselor. He therefore decided to volunteer in the AIDS ward of a nearby hospital. He was assigned to the counseling staff and rapidly became expert at providing guidance and assistance to the indigent and often homeless AIDS patients who were crowding the ward in ever-larger numbers.

Victor realized that if he kept careful notes of his experiences in the hospital and conducted systematic interviews with the patients, he would be able to gather data for a dissertation on AIDS and homeless people. At the same time, through his work as an advocate for patients' rights and his contacts with people in social service agencies outside the hospital, he became known as someone who could speak with authority about the impact of AIDS on indigent populations. He was often asked to speak at conferences and public meetings where other AIDS activists sought to convince officials to increase education and intervention efforts among homeless people, not just among more affluent urban residents. At night Victor worked with AIDS patients in the hospital, and during the day he continued to teach and do counseling at the community college. Whenever possible he juggled his busy schedule to make time for speaking engagements at the political events that were so important in changing the public's understanding of the disease. "I didn't want to be so busy," Victor remembers, "and it took a lot out of me, but someone had to speak from my viewpoint as a Latino working with poor people. I felt responsible."

Victor wrote his dissertation and published it as a book, *Falling Through the Cracks: AIDS and the Urban Poor* (Ayala, 1996). His was the first book to describe how AIDS affects the lives of very poor people, and it contains valuable insights about how to reach this "at risk" population with information about AIDS. It is being read by many people in the movement for AIDS awareness. Now Victor is busier than ever, frequently speaking about AIDS in his own community and elsewhere in the nation. "You know, Bill," he explains, "I never intended to become so involved in politics and rallies and all that, but it came with my sociological training and with my life, so I guess I should learn to accept it as part of how we deal with the disease."

Victor Ayala.

THE MEANINGS OF DISORDER

In the early years of the AIDS epidemic, many people panicked. They were afraid that they could contract the disease by kissing or even by being in the same room with a person infected with HIV. Victor Ayala saw many innocent people suffer from the fearful reactions of friends and colleagues. His decision to work for the gay men's health center was his way of helping to ease the fear and panic created by the AIDS crisis in his community.

In this chapter we look closely at various types of collective behavior, many of which are far more violent than the AIDS panic, others far more benign and even playful. Momentous episodes, like the riots in Los Angeles after the trial of the police officers accused of beating Rodney King, are the most severe form of collective behavior. The race riots that occurred during the civil rights movement in the United States, the huge rallies and demonstrations that ushered in the Nazi regime in pre–World War II Germany, the strikes and picketing that spurred the labor movement, the demonstrations that enlivened the women's movement in the 1970s, the riots and demonstrations in Romania and other Eastern European dictatorships in 1989, and the rioting in the streets of Jakarta that deposed the Suharto regime in 1998—these are all examples of episodes of collective behavior that have shaped history in our own time.

But there are less extensive and less dangerous incidents of collective behavior in modern societies that are nonetheless important. Think of the panic that occurs when a local bank or savings and loan association is rumored to be short of funds, or the joyful anticipation of the crowds that form outside ticket offices when a popular rock star announces a concert. These events also have economic and social significance: Banks fail when panicked depositors suddenly withdraw their savings; rock promoters wait in gleeful anticipation of full houses as the lines of ticket buyers form.

Many sectors of the American economy profit from the behavior of large numbers of people. The fast-food industry, for example, depends on the behavior of the millions of Americans who stop for burgers and fries as they stream home from beaches or concerts or movies. Every year the toy industry produces gimmicks like Beanie Babies, and fortunes are made and lost on the accuracy of predictions about whether the public will embrace such notions. The computer industry also has profited from fads and crazes: In the early 1990s people rushed to buy more powerful personal computers with CD drives and modems in order to join the "information society," only to realize that this new information channel was fraught with problems of its own, such as the possibility that children using the computer would be exposed to pornography and hate speech.

The study of mass or collective behavior encompasses a wide range of phenomena and presents many problems of classification and explanation. Yet this subject deserves our attention because of its influence on social change. The student occupation of Beijing's Tiananmen Square; the strikes and protests in Poland, Czechoslovakia, Hungary, and Romania that toppled Communist dictatorships; the antitax movements occurring throughout the United States—all attest to the immense power of people who are moved to engage in collective action for social change.

In our attempt to understand the nature and importance of these behaviors, we turn first to the concepts that define their similarities and differences. Then we turn to theories regarding why these episodes occur, how social movements mobilize, and with what consequences. Finally, we discuss the concept of mass publics, especially the leisure and consumption behavior of large populations.

The Nature of Collective Behavior

The term **collective behavior** refers to a continuum of unusual or nonroutine behaviors that are engaged in by large numbers of people. At one extreme of this continuum is the spontaneous behavior of people who react to situations they perceive as uncertain, threatening, or extremely attractive. The violence that occurred in the recent race riots in Los Angeles and Miami is one example of spontaneous collective behavior. Another is the sudden action of coal miners who decide that conditions in the mines are unsafe and walk off their jobs in a sudden and unplanned wildcat strike. Such behaviors

Then  and Now

Continuity and Change in Racial Violence

Racial rioting and violence has erupted periodically in American history. These two photos, separated by more than 70 years, have a chilling similarity. In the first, white men caught up in the frenzy of violence that accompanied the infamous Chicago riot of 1919 are stoning a black man. The second photo shows the beating of a white man during the Los Angeles riots in 1992. Both events are generally credited with having brought problems of American race relations to the center of public attention.

are not governed by the routine norms that control behavior at the beach or on the job (Smelser, 1962).

At the other extreme of the continuum of collective behaviors are rallies, demonstrations, marches, protest meetings, festivals, and similar events. These involve large numbers of people in nonroutine behaviors, but they are organized by leaders and have specific goals. When workers in a union plan a strike, for example, their picketing and rallies are forms of organized collective behavior whose purpose is to demonstrate their solidarity and their determination to obtain their demands. When blacks marched in commemoration of those who died in the 1919 race riot in Chicago, the event was organized to build solidarity and publicize a new determination to resist racism. In such cases the organization that plans the event and uses collective behavior to make its feelings or demands known is a social movement.

Social movements are intentional efforts by groups in a society to create new institutions or reform existing ones. Such movements often grow out of more spontaneous episodes of collective behavior; once they are organized, they continue to plan collective events to promote their cause (Blumer, 1978; Genevie, 1978). Some movements, like the antiabortion movement, resist change; others, like the labor movement, have brought about far-reaching changes in social institutions. Still others, like the gay and lesbian movement, seek to gain the rights other groups have won. In this sense they are movements for full inclusion in society by people who feel that they are discriminated against. We will see in later chapters that many social movements in a multiethnic and culturally diverse society challenge discrimination and deviant labels (Tarrow, 1994).

The American labor movement provides many examples of the ways in which spontaneous episodes of collective behavior and the social movements associated with them can change the course of a society's development. During the nation's stormy transition from an agrarian to an industrial society, workers fought against the traditional right of employers to establish individual wage rates and to hire and fire workers as they pleased. They demanded instead the right to organize unions that could negotiate a collective wage rate (collective bargaining) for each category of workers. They also demanded better working conditions and benefits. Mass picketing, sit-down strikes, and pitched battles between workers and company agents or the police were everyday events in the 1890s.

Often the workers joined in walkouts, rioting, or other spontaneous collective actions when faced with mine disasters or intolerable working conditions. From these episodes of bitter conflict emerged the modern labor movement. In the ensuing decades, culminating in the period of New Deal legislation in the 1930s, the workers' demand for collective bargaining was institutionalized. By this we mean that the right to join unions and to bargain collectively was incorporated into the nation's laws. Labor unions thereby became recognized organizations and collective bargaining the recognized means of settling labor disputes. The unions, in turn, had to discipline their workers by specifying in labor agreements the conditions under which collective behavior such as strikes and other actions could be used.

By the end of World War II, labor unions and collective bargaining had become legitimate social institutions. In recent decades, however, drastic declines in industrial employment in North America and Europe, combined with renewed attacks on labor unions by employers, have led to a decrease in the ranks of organized labor. But major social movements that affect the lives of millions of people often go through cycles of growth and decline followed by a rebirth of activity. It remains to be seen whether this will be true of the labor movement (Geoghegan, 1991).

Spontaneous collective behavior, from which social movements often arise, can take many forms. Such behaviors range from the demonstrations and riots that mark major revolutions to the fads, fashions, crazes, and rumors that sweep through modern societies with such rapidity that what is new and shocking one day will be a subject of nostalgia the next. Some of these types of collective behavior become associated with social movements, but many do not. For example, the sensational and tragic death of England's Princess Diana in 1997 resulted in the greatest spontaneous outpouring of feelings that observers had ever seen in a culture known for keeping emotions suppressed (the meaning of the phrase "keeping a stiff upper lip"). So many mourners came to Buckingham Palace and to Diana's home, Kensington Palace, leaving bouquets in Diana's memory, that Queen Elizabeth was compelled by public opinion to give a speech on the subject of the grief felt by the royal family. However, while displays of popular feeling for Diana and her memory have continued since her death, and there is an endless stream of articles about every aspect of her life in the print and electronic media, there is no actual social movement associated with these events.

A Typology of Spontaneous Collective Behaviors

Sociologists who study collective behavior try to discover the conditions under which different types of spontaneous collective behavior occur and why these events do or do not become linked to social movements. Typologies that classify these phenomena are an important first step toward understanding why collective behavior assumes particular forms and develops through recognizable stages.

Crowds and Masses Students of spontaneous forms of collective behavior often begin by examining the structure of such behavior—that is, by discovering whether the people who engage in a particular kind of behavior are in close proximity to one another or whether they are connected in a more indirect way. In this regard it is useful to distinguish between crowds and masses. A **crowd** is a large number of people who are gathered together in close proximity to one another (e.g., at a demonstration or a football game). A **mass** is more diffuse; it does not occur in a physical setting. A mass is a large number of people who are all oriented toward a set of shared symbols or social objects (Lofland, 1981); an example is the audience for a particular television program. Collective behavior can occur in crowds, in masses, or in both at once.

Motivating Emotions The actual behavior a crowd or mass generates depends largely on the emotions the people involved feel are appropriate to express in a particular situation. They may be motivated by many desires—for excitement, for a change, for material gain, and so on—but outwardly most will express an emotion that the norms of the event suggest is appropriate. In the witch-hunt described in Chapter 7, the people involved expressed fear and hysteria (although, as is true of all such events, some were "just going along with the crowd"). Hostility aroused by anger, desire for revenge, or enraged hatred is another common motivating emotion in episodes of collective behavior. Lynch mobs are one kind of hostile crowd. The crowd of angry strikers that violently opposes "scabs" trying to cross a picket line is another. At the mass level, the outbreak of animosity toward Islamic people immediately after the bombing of the federal office building in Oklahoma City in 1995 is an example of collective behavior provoked by hostility.

Joy is a third important emotion that motivates crowds and masses. The Mardi Gras celebrations in Latin America and New Orleans involve large, joyful crowds that create a "moral holiday" in which behaviors are tolerated that would not be acceptable in public at other times. At the mass level, the immense interest and joy that spread through the U.S. population during the 1998 baseball season, when Mark McGwire and Sammy Sosa were locked in a battle to break the single-season home run record, raised the sport to new levels of popularity and stimulated many examples of joyful collective behavior, especially in St. Louis and Chicago.

By cross-classifying the most significant emotions that motivate collective behavior (fear, hostility, and joy) with the structural dimensions of crowd and mass, sociologist John Lofland created a typology that includes a wide range of spontaneous collective behaviors. This typology is presented in Figure 8.1. Can you think of examples of your own for the various types?

FIGURE 8.1 **Lofland's Typology of Spontaneous Collective Behaviors**

BASIC EMOTION	SETTING: Crowd	SETTING: Mass
Fear	• Panic exodus from burning theater • Hostages taken by terrorist groups aboard an airliner	• Natural calamities • Red scares • Crime waves • Three Mile Island • Salem witch-hunts
Hostility	• Political rallies, marches • Lynch mobs • Race riots	• Scapegoating of public figures • Waves of cross burnings
Joy	• Revival meetings • Rallies in Nazi Nuremberg • Mardi Gras carnival • Rock concerts • Sports events	• Gold rushes • Punk fashion • Jogging • Disco • *Star Wars, Star Trek* • Pet rocks

Source: Adapted from *Social Psychology: Sociological Perspectives,* edited by Morris Rosenberg and Ralph H. Turner. Copyright © 1981 by The American Sociological Association. Reprinted by permission of Basic Books, Inc., Publishers.

Where, for example, would you place the Boston Tea Party, the riots that often occur during spring break in Florida, or the crowds of fans who came out to pay last respects to singer Jerry Garcia in 1995?

The Lofland typology is useful as a means of classifying some types of collective behavior, but remember that often there are interactions among these categories: Mass behavior may turn into crowd behavior; spontaneous collective behavior may generate a social movement. For example, when the stock market crashed in 1929 at the beginning of the Great Depression, crowds of panic-stricken investors spilled onto the streets of the financial districts of New York, Chicago, and San Francisco. There, within sight of one another, people found their fears magnified. And those fears quickly spread through the mass of Americans. The panic started among people who had made investments on credit, but it spread to people who merely had savings in local banks. Crowds of terrified savers descended on the banks, which were unable to handle the sudden demand for withdrawals because they themselves had been investing in stocks that were suddenly worthless. As a result, millions of Americans lost all their savings.

In the aftermath of the stock market crash and the resulting mass panic, social movements arose to seek reforms like investment insurance that would protect investors against similar episodes in the future. During the depression that followed the crash, citizens' fears about their lack of retirement or pension savings led to the development of new institutions like Social Security and deposit insurance. Today the American Association of Retired Persons, an organization that represents the interests of older people who once protested as an unorganized mass, is the largest single lobbying group in American politics.

Dimensions of Social Movements

There are as many types of social movements as there are varieties of spontaneous collective behavior. Just think of how many "causes" can motivate people to take organized action. We have a social-welfare movement, an antitax movement, a civil rights movement, a women's movement, an environmental protection movement, a consumer movement, an animal rights movement, and on and on. Within each of these movements numerous organizations attempt to speak for everyone who supports the movement's goals. Actually, within any large social movement there are likely to be a number of different social movement organizations (sometimes referred to as SMOs) that may have different, and at times competing, ideas about how to achieve the movement's goals (Goode, 1992). As we will see shortly, the way such organizations operate and the goals they seek depend to a large extent on the characteristics of the social movement itself.

Classifying Social Movements Social movements can be classified into four categories based on the goals they seek to achieve, as follows:

1. *Revolutionary movements* seek to overthrow existing stratification systems and institutions and replace them with new ones (McAdam, McCarthy, & Zald, 1996). The Russian Bolsheviks who founded the Communist party were a revolutionary social movement that sought to eliminate the class structure of Russian society. They also believed that the institutions of capitalism—the market and private property—must be replaced with democratic worker groups or *soviets* whose efforts would be directed by a central committee. Meanwhile certain other existing institutions, especially religion and the family, had to be either eliminated or drastically altered.
2. *Reformist movements* seek partial changes in some institutions and values, usually on behalf of some segments of society rather than all. The labor movement is basically reformist. It seeks to alter the institutions of private property by requiring the owners of businesses to bargain collectively with workers concerning wages and working conditions and to reach an agreement that will apply to all the workers in the firm. Collective bargaining replaces the old system of individual contracts between workers and employers, but it does not totally destroy the institution of private property.
3. *Conservative movements* seek to uphold the values and institutions of society and generally resist attempts to change them, unless their goal is to undo undesired changes that have already occurred. The conservative movement in the United States seeks to reinforce the values and functions of capitalist institutions and to support the traditional values of such institutions as the family and the church.
4. *Reactionary movements* seek to return to the institutions and values of the past and, therefore, to do away with some or all existing social institutions and cultural values (Cameron, 1966; McAdam, McCarthy, & Zald, 1996). The Ku Klux Klan is part of a reactionary social movement that seeks a return to the racial caste system that was supported by American legal institutions (laws, courts, and the police) until the 1954 Supreme Court decision that declared the "separate but equal" doctrine unconstitutional.

Other Types of Movements Although they cover some of the movements that have had the greatest impact on modern societies, the preceding four categories do not include all the possible types of movements. Herbert Blumer (1969a), for example, identified a fifth category: *expressive social movements,* or movements devoted to the expression of personal beliefs and feelings. Those beliefs and feelings may be religious or ethical or may involve an entire lifestyle, as in the case of the punk movement of the 1980s. An important aspect of expressive social movements is that their members typically reject the idea that their efforts ought to be directed at changing society or the behavior of people who do not belong to their movement. Their quest is for personal expression, and if others choose to "see the light," that is welcome.

Some of the various New Age groups and organizations now popular in the United States fit quite well into the expressive social movement category. Non–Native Americans who appropriate some aspects of Native American spirituality (including the use of sweat lodges or the practice of vision quests), but do not proselytize, are examples of people involved in an expressive social movement. So are men who follow the teachings of poet Robert Bly, practicing collective drumming to bond together and express suppressed emotions. But, like the followers of any social movement, there can be those in expressive social movements whose extreme beliefs lead to tragedy, as in the 1996 case of the 7-year-old pilot whose parents encouraged her to follow her vision at all costs. Such advice resulted in her death, as well as the deaths of her father and her flight instructor.

Other social scientists who study social movements identify still more categories or develop subcategories that combine elements of those just presented. (See the study chart below.) For example, in his classic study *The Pursuit of the Millennium* (1961), Norman Cohen analyzed what are known as *messianic* or *millenarian movements,* which are both revolutionary and expressive. A millenarian movement promises its followers total social change by miraculous means. These movements envision a perfect society of the future, "a new Paradise on earth, a world purged of suffering and sin, a Kingdom of the Saints" (p. xiii). Most millenarian movements are secular, but the term is derived from the Christian concept of the millennium, the return of Christ as the messiah who would save the world after a thousand years. Often millenarian movements begin as small movements with a charismatic leader (someone who seems to possess special powers) and a few devoted followers.

An example of a millenarian movement is the People's Temple, whose 900 members committed suicide by drinking cyanide-laced Kool-Aid in Jonestown, Guyana, on November 18, 1978. In his insightful study of the events leading up to the tragedy, sociologist John Hall (1987) demonstrates that popular interpretations of this shocking story are inadequate. The conclusion reached in most previous accounts is that the movement's leader, Jim Jones, was a deranged personality who acted as the Antichrist and led his misguided followers to their deaths. Such accounts, according to Hall, may be comforting expressions of moral outrage, but by exaggerating the influence of an individual personality they lead to wrong ideas about how such tragedies can occur.

In his analysis of the Jonestown tragedy, Hall shows that the members of the People's Temple were part of a cohesive social movement. Many were of modest means, some quite poor; in addition, most were black. They felt alienated from the larger society and saw the Temple as their community and often as their family. It is only by understanding the strength of their

STUDY CHART Types of Social Movements

TYPE	DESCRIPTION	EXAMPLE
Revolutionary	Seeks to overthrow and replace existing stratification systems and institutions.	Bolsheviks
Reformist	Seeks partial changes in some institutions and values, usually on behalf of some segments of society rather than all.	Labor movement
Conservative	Seeks to uphold the values and institutions of society and resists attempts to change them.	Conservative movements
Reactionary	Seeks to return to the institutions and values of the past and do away with existing institutions and values.	Ku Klux Klan
Expressive	Devoted to the expression of personal beliefs and feelings.	Punk movement
Millenarian	Combines elements of revolutionary and expressive movements.	People's Temple

attachments to one another and to their belief in Jones's vision of a better world to come that one can make sense of what appears to be a senseless, insane episode of collective behavior. Faced with what they perceived as imminent attacks on the religious community they were building in the jungle, some Temple members worked themselves into a frenzy in which they became convinced that by drinking the poison they were "stepping over" into a better world. Others may or may not have felt these emotions; we cannot ever know. Probably some people thought the occasion was a practice suicide. In any case, Hall's study shows that Jones planned the act of "collective martyrdom" but that the vast majority of the Temple members followed to the end because of their unshaken belief in the goals of their social movement.

SOCIAL MOVEMENT THEORY

Early theories of collective behavior were derived from the notion that hysteria or contagious feelings like hatred or fear could spread through masses of people. One of the first social thinkers to develop a theory of collective behavior was Gustave LeBon. An aristocratic critic of the emerging industrial democracies, LeBon believed that those societies were producing "an era of crowds" in which agitators and despots are heroes, "the populace is sovereign, and the tide of barbarism mounts" (1947/1896, pp. 14, 207). In its usual definition, he pointed out, "the word 'crowd' means a gathering of individuals of whatever nationality, profession, or sex, and whatever be the chances that have brought them together." But from a psychological viewpoint a crowd can create conditions in which "the sentiments and ideas of all the persons in the gathering take one and the same direction, and their conscious personality vanishes" (quoted in Genevie, 1978, p. 9).

LeBon attributed the strikes and riots that are common in rapidly urbanizing societies to the fact that people are cut off from traditional village social institutions and jammed into cities, creating a mass of strangers. This view of early industrial societies was clearly exaggerated; research has shown that urban newcomers seek to form new social attachments, often drawing upon networks of people who came to the city from the same rural area or small town. But more important, research on crowds and riots has shown that despite the appearance of a "mob mentality" and psychological "contagion," people who participate in riots, demonstrations, and other forms of collective behavior actually have many different motivating emotions and may only *appear* to be following along blindly as if they were part of a herd. Also, the contagion theories of LeBon and others are not useful in understanding the full range of collective behaviors and social movements that arise from the strains and conflicts characteristic of rapidly changing industrial societies (Rochon, 1998).

A Continuum of Collective Behavior

Collective behaviors and social movements can be placed along a continuum according to how much they change the societies in which they occur. At one end of the continuum are revolutions; at the other are fads or crazes. In the twentieth century there were three revolutions that had a major impact on world history: the communist revolutions in Russia and China and the fascist revolution that brought Hitler to power in 1933. These can be placed at the high end of the continuum. At the low end are fads and crazes that may excite us for a while but usually do not bring about lasting change in the major structures of society (Turner, 1974). Some people become so completely involved in a fad or craze (as has occurred, for example, among collectors of Beanie Babies) that they devote all their energies to the new activity.

Between the two extremes of revolutions and fads are social movements that capture the attention of masses of people and have varying effects on the societies in which they occur. The women's movement, the civil rights movement, and the environmental protection movement are examples of movements with broad membership in many nations and the continuing power to bring about social change. The gay liberation movement, the antiabortion movement, the evangelical movement, and the hospice movement (see Chapter 15) are examples of more narrowly focused movements that are concerned with a single issue or set of issues but also exert powerful pressure for change in the United States and other societies.

There are also movements that attract large numbers of people for a brief time but then begin to falter and have a limited impact. An example is the men's spiritual and social renewal movement, which has found expression in the Million Man March by African Americans and the mass rallies held by the Promise Keepers. The latter is a social movement organization led by a former University of Colorado football coach. It calls upon Christian men to become "promise keepers, not promise breakers" and promotes racial reconciliation, family values, and male leadership within the family. After much initial success in attracting large crowds of men to stadiums throughout the United States, in 1998 the group faltered as its leaders tried to find other ways of raising money besides charging high admission prices for its rallies. Unable to cover expenses, the organization was forced to fire at least two thirds of its staff as its leaders struggled to attract donations. At this writing it appears that significant donations are being made and the Promise Keepers will continue to function, but it is far from certain whether they will have a lasting impact on society (Sahagun & Stammer, 1998; Wheeler, 1998).

Theories of collective behavior and social movements often seek to explain the origins and effects of revolutionary and reform movements (Kimmel, 1990; McAdam, McCarthy, & Zald, 1988). In the remainder of this section we discuss several such theories. We begin by examining the nature of revolutions and revolutionary movements. Then we consider theories that attempt to explain why major revolutions and revolutionary social movements have arisen and what factors have enabled them to succeed. Note that these are macro-level theories: They explain revolutions and the associated social movements as symptoms of even larger-scale social change.

Theories of Revolution

Sociologists often distinguish between "long revolutions," or large-scale changes in the ecological relationships of humans to the earth and to one another (e.g., the rise of capitalism and the industrial revolution), and revolutions that are primarily social or political in nature (Braudel, 1984; Wolf, 1984a, 1984b). The course of world history is shaped by long revolutions, whereas individual societies are transformed by social revolutions like those that occurred in the United States, France, and Russia (Bobrick, 1997).

Political and Social Revolutions Sociologist Theda Skocpol (1979) makes a further distinction between political and social revolutions. **Political revolutions** are transformations in the political structures and leadership of a society that are not accompanied by a full-scale rearrangement of the society's productive capacities, culture, and stratification system. **Social revolutions**, on the other hand, sweep away the old order. They not only change the institutions of government but also bring about basic changes in social stratification. Both political and social revolutions are brought about by revolutionary social movements as well as by external forces like colonialism.

The American Revolution, according to Skocpol's theory, was a political revolution. The minutemen who fought against British troops at the Battle of Concord were a political revolutionary group that did not at first seek to change American society in radical ways. On the other hand, the revolutions that destroyed the existing social order in Russia and China in the twentieth century were social revolutions. Their goal was to transform the class structure and institutions of their societies.

Skocpol's analysis further shows that the actual outcomes of social revolutions depend on the kinds of coalitions formed among revolutionary parties and larger social classes. The communist revolutionaries in Russia, for example, did not join forces with the peasants as the Chinese communist revolutionaries did. One consequence was that in the former Soviet Union the peasants were persecuted during the revolution; their land was taken from them and many were forced to live on collective farms. These disruptions of the society's agrarian base severely weakened the economy. In China, however, while there were many disruptions of rural life after the communists took power, food production continued to to be carried out in villages and small landholdings. Peasant communist leaders replaced feudal landowners, but agricultural production did not suffer as it did in the former Soviet Union.

Why do revolutionary social movements arise, and what makes some of them content mainly with seizing power while others call for a complete reorganization of society? For most of this century the answer has been some version of Marxian conflict theory. This theory uses detailed knowledge of existing societies to predict the shape of future ones. According to Marx, the world would become capitalist; then capitalist markets would come under the control of monopolies; impoverished workers and colonial peoples would rebel in a generation of mass social movements and revolutions; and a new, classless society would be created in which the workers would own the means of production. (Marx's theory of class conflict leading to revolution is discussed in Chapters 7 and 11.)

Relative Deprivation The idea that the increasing misery of the working class would lead workers to join revolutionary social movements was not Marx's only explanation of the causes of revolution. He also believed that under some conditions, "although the enjoyments of the workers have risen," their level of dissatisfaction could rise even faster owing to the much greater increase in the "enjoyments of the capitalists, which are inaccessible to the worker" (Marx & Engels, 1955, Vol. 1, p. 94). This is a version of the theory known as **relative deprivation** (Stouffer et al., 1949). According to this theory, the presence of deprivation (i.e., poverty or misery) does not by itself explain why people join revolutionary social movements. Instead, it is the feeling of deprivation relative to others. We tend to measure our own well-being against that of others, and even if we are doing fairly well, if they are doing better we are likely to feel a sense of injustice and, sometimes, extreme anger. It is this feeling of deprivation relative to others that gives rise to revolutionary social movements.

Alexis de Tocqueville came to the same conclusion in his study of the causes and results of the French Revolution. He was struck by the fact that the revolution did not occur in the seventeenth century, when the economic conditions of the French people were in severe decline. Instead, it occurred in the eighteenth century, a period of rapid economic growth. Tocqueville concluded that "revolutions are not always brought about by a gradual decline from bad to worse. Nations that have endured patiently and almost unconsciously the most overwhelming oppression often burst into rebellion against the yoke the moment it begins to grow lighter" (1955/1856, p. 214). The Zapatista rebellion in the Mexican state of Chiapas is an example of a revolutionary movement that began among very poor Indian peasants but became a broader movement during a time of relative economic prosperity (Preston, 1998; Vilas, 1993).

Valuable as they are as broad theories of revolution, neither Marxian theory nor Tocqueville's analysis can fully explain the rise of revolutionary movements throughout the contemporary world, nor do they help much in understanding the changes major social movements go through as their goals and organizations evolve over time. These and related questions are analyzed more directly in recent theories of protest cycles and changes within social movements, to which we now turn.

Protest Cycles, Action Frames, and Charisma

In analyzing the revolutions that have occurred during the past 500 years, sociologist Charles Tilly concludes that there are no "neat formulations of standard, recurrent conditions for forcible transfers of state power" (1993, p. 237). However, he agrees with other social scientists who have pointed out that revolutionary social movements and other forms of protest often occur in cycles or waves of unrest. He cites the waves of protest that swept through Europe and North America in the late 1960s and the current wave of fundamentalist movements in much of the contemporary world. During such waves, he observes:

> One set of demands seems to incite another, social movement organizations compete with each other for support, demands become more extreme for a while before subsiding. As this happens, activists often experiment with new ways of organizing, framing their demands, combating their enemies and holding on to what they have. (1993, p. 13)

Waves of Protest No theory adequately explains why particular waves of protest sweep through societies. But Sidney Tarrow (1994), another astute student of protest cycles, argues that waves of protest often result from major social shocks such as an international economic recession, the end of a period of warfare, or hasty actions by one or more governments. Such an episode occurred in Poland in the early 1980s when the communist regime announced large increases in the price of bread. The movement that ensued, known as Solidarity, demonstrated the weakness of that regime.

During cycles of protest, there is a dramatic increase in social conflicts of all kinds, often resulting in violent episodes of collective behavior—riots, demonstrations, and counterdemonstrations. The heightened sense that change is in the air, often brought on by the activities of a few "early risers" among protest leaders (e.g., Lech Walesa and Mahatma Gandhi), triggers a variety of processes, including imitations of the protest elsewhere (e.g., Martin Luther King, Jr., and Nelson Mandela, who followed the example set by Gandhi). The wave of protest may also build momentum through diffusion, often spreading from larger cities to smaller towns and rural areas. New ideologies arise to justify collective action and help the movement mobilize a following (Tarrow, 1994).

Collective Action Frames The old saying that "the pen is mightier than the sword" refers to the power of ideas to shape our view of society and why it needs to be changed. In this century we have seen the revolutionary power of ideas embodied in *The Communist Manifesto* of Karl Marx and Friedrich Engels, Hitler's *Mein Kampf* ("My Struggle"), and the "Little Red Book" containing the sayings of Chairman Mao Ze-dong. In our own time such books as Newt Gingrich's *To Renew America* and William Bennett's *Book of Virtues* have presented ideas that motivate various conservative movements in the United States. Millions of people latch on to these concepts, not because they are incapable of forming their own ideas but because all of us are continually trying out ideas that seem to explain our times and help us make sense of our experiences in society. When faced with quandaries about what to do, what actions we should take or support, we often seek out concepts that are phrased as "collective action frames," sets of beliefs and interpretations of events that inspire and justify social movements (Snow & Benford, 1992). Collective action frames are described in more detail in Box 8.1.

Charisma Leadership plays an extremely important part in the success or failure of revolutionary social movements. The leaders of such movements are often said to have almost supernatural powers to inspire and motivate masses of followers. Max Weber (1968) called this ability **charisma.** A charismatic leader appears to

BOX 8.1 USING THE SOCIOLOGICAL IMAGINATION

Collective Action Frames

Sociologist William A. Gamson (1992) is one of the leading students of political activism in the United States. His recent research on social movements focuses on the mental frameworks people develop in their thinking about social issues. He notes that social movements strive to offer one or more *collective action frames.* These are ways of thinking that justify the need for and desirability of a particular course of action. After studying the writings and speeches of activists in many different social movements, Gamson concluded that collective action frames have three important components:

1. *An injustice component,* which includes a sense of moral indignation or outrage against a perceived injustice and the people who are said to be responsible for the condition.
2. *An agency component,* which embodies the idea that not only is there injustice, but together "we" can do something about the condition.
3. *An identity component,* which defines who "we" are, usually in opposition to a "they" with different tastes and values.

Gamson studies collective action frames by recording conversations in which people discuss social issues. In exploring the injustice component, for example, he listened carefully for statements like "That is just wrong," "That's unfair," "That really burns me up," or "That pisses me off." Such statements are marked by "explicit moral condemnation, unqualified by offsetting arguments and unchallenged by other group members," as in this example:

CHARACTERS:
Marjorie, a waitress, in her forties.
Judy, a data entry clerk, in her thirties.
Several others who don't speak in this scene.

(The group is near the end of a discussion of nuclear power.)

MARJORIE: They shouldn't have taken all that money to nuclear power and everything—and you've got kids starving in America.
JUDY: Yeah.
MARJORIE: You've got homeless people. Where's your values? [*pause*] They suck. They really do suck.
JUDY: You're on tape.
[*laughter*]
MARJORIE: I don't care what I'm on. Still—it's obvious—when you pay millions and millions of dollars in nuclear plants when people in America are starving—
JUDY: Right.
MARJORIE: And you've got homeless people, no matter what they are—whether they're drunks or they're—whatever they are. Mentally ill people and you've got them living on a street. And you've got a family of five people—I worked for Legal Services of Greater Boston and I had people in the Milner Hotel, mothers with five kids in one room. Living. And we don't have places for them, but we have places to build nuclear plants. That's garbage. That's garbage! (Gamson, 1992, pp. 48–49)

Sociologists listen carefully to the way people like Marjorie frame issues. In this case the frame is injustice and the point being made is that too much money is being spent on nuclear plants while other, more pressing human needs go unmet. In this dialogue Marjorie is outspoken and not afraid of having her opinions recorded, whereas Judy is far more passive and concerned about voicing her opinions. The sociologist would probably conclude that within her circle of friends and coworkers Marjorie is a leader who often frames issues for others. As you listen to conversations like this one, observe how the leaders frame issues for the other participants.

possess extraordinary "gifts of the body and spirit" that mark him or her as specially chosen to lead. The voices that Joan of Arc heard, which convinced her to take up arms to save France, gave her the inspired commitment to her cause that is a central feature of charisma. Mahatma Gandhi, probably this century's greatest example of a charismatic leader, swayed the Indian masses with his almost supernatural spirituality and courage. Similarly, the extraordinary gifts of oratory, faith, and energy displayed by Martin Luther King, Jr., allowed him to emerge as the most influential leader of the American civil rights movement.

Often the movement's followers themselves attribute special powers to their leader. This process may produce additional myths about the leader's powers and add to whatever personal magnetism the leader originally possessed. For example, Hitler was no doubt a fiery orator, but his followers' frenzied reception of many of his ideas contributed to the perception that he possessed exceptional charisma.

The charisma of the leader may eventually pose problems for the movement. Every social movement must incorporate the goals and gifts of its leaders into the structure of the movement and eventually into the

Mahatma Gandhi, probably the most revered and charismatic political figure of the twentieth century, is shown here during a light moment with his granddaughters. Gandhi's techniques of nonviolence and passive resistance inspired Martin Luther King, Jr., and other leaders of political protest movements. But his followers often endured great personal sacrifice and physical danger, for his political strategies called for nonviolent civil disobedience in the name of higher moral values.

institutions of society. But how can this be done without losing track of the movement's original purpose and values? Scholars who follow Weber's lead in this area of research have termed this problem the *institutionalization of charisma* (Shils, 1970). Weber recognized that the more successful a movement becomes in taking power and assuming authority, the more difficult it is to retain the zeal and motivation of its charismatic founders. Once the leaders have power and can obtain special privileges for themselves, they tend to resist continued efforts to do away with inequalities of wealth and power. Perhaps the most instructive example of this problem is the institutionalization of the ideals of Marx and Lenin in the organization of the Bolshevik party in Russia after the revolution of 1917.

"Workers of the world, unite," Marx urged in *The Communist Manifesto;* "you have nothing to lose but your chains" (Marx & Engels, 1969/1848). The vision of a society in which poverty, social injustice, repression, war, and all the evils of capitalism would be eliminated formed the core of Marxian socialism. The founders of the Bolshevik party realized, however, that revolutions are not made with slogans and visions alone (Schumpeter, 1950). The unification of the workers required an organization that not only would carry forward the revolution but also would take the lead in mobilizing the Russian masses to work for the ideals of socialism. The charismatic leader who built this organization was Vladimir Ilyich Lenin.

Under the leadership of Lenin (and far more ruthlessly under his successor Josef Stalin), any Bolshevik who challenged party discipline was exiled or killed. Because the goals of the Bolsheviks were so radical and the need for reform so pressing, Lenin and Stalin argued that the party had to exert total control over its members in order to prevent spontaneous protests that might threaten the party's goals. By the same logic, other parties had to be eliminated and dissenters imprisoned. Such efforts by an elite to exert control over all forms of organizational life in a society are known as *totalitarianism.* All social movements that believe the ends they seek justify the means used to achieve them risk becoming totalitarian because they sacrifice the rule of law and the ideal of democratic process to gain power (Harrington, 1987; Howe, 1983).

Co-optation Weberian theory predicts that social movements will institutionalize their ideals in a bureaucratic structure. And the more successful they are in gaining the power to change society, the more they will be forced to control the activities of their members. In fact, once a movement is institutionalized, with an

This famous photo from the Holocaust shows Jewish families being rounded up for transport to concentration camps. Under Hitler, the National Socialists (Nazis) often resorted to public acts of terror to establish their absolute rule over German society. Such totalitarian regimes often use terror to destroy mutual trust among citizens.

administration and bureaucratic rules, its leaders tend to influence new charismatic leaders to become part of their bureaucracy, a process known as *co-optation.* In the labor movement, for example, maverick leaders who demand reform and instigate protests in their plants are often given inducements such as jobs in the union administration to keep them under control (Geschwender, 1977; Kornhauser, 1952). In politics, local party leaders may identify and co-opt a local leader who they think might pose a threat in the future, offering inducements such as support in an election campaign in return for loyalty to the party. In this way an outspoken environmentalist and history professor named Newt Gingrich was co-opted by the Republican leaders of Carrollton, Georgia, early in his political career.

SOCIAL MOVEMENTS AND CIVIL SOCIETY

Most of us will probably never belong to a revolutionary social movement, at least not one devoted to the violent overthrow of a government or society. Nor will many of us experience the deadening force of totalitarian rule. Most of us therefore will not find ourselves helping to rebuild the voluntary associations and communities, the leagues, the hobby groups, the unions and auxiliaries of our societies, as the people of many Eastern European nations are doing today. But many readers of this book will become members of organizations and groups of all kinds, which in turn may become affiliated with broader social movements.

In contemporary societies an ever-growing and changing array of groups and organizations are formed in the public sphere of social activity. Known as **civil society,** this sphere of nongovernmental, nonbusiness social activity is composed of millions of church congregations, sports leagues, amateur arts groups, charitable associations, ethnic associations, and much more. The Mexican sociologist Carlos M. Vilas explains it as follows:

> Civil society refers to a sphere of collective actions distinct from both the market and "political society"—parties, legislatures, courts, state agencies. Civil society is not independent of politics, but clearly, when people identify themselves as "civil society," they are seeking to carve out a relatively autonomous sphere for organization and action. (1993, p. 38)

In parts of the world where the gap between rich and poor is widening, people who participate in civil society are often recruited into movements calling for social and economic justice. Vilas observes, for example, that in Latin America activity in social movements "creates a confluence of the poor and middle classes . . . to confront the traditional alliance of the rich and powerful with the state" (1993, p. 41).

In more affluent regions of the world, civil society also flourishes and provides a rich source of recruits for social movements. As noted in Chapter 6, increasing membership in voluntary associations appears to be a long-term trend in the United States. National surveys conducted in the 1950s found that about one third of adult citizens were members of voluntary associations; today more than two thirds of American adults say that they belong to a voluntary association (Caplow et al., 1991). Since 1970 the number of registered nonprofit associations of all kinds has increased from somewhat over 10,000 to almost 23,000 (*Statistical Abstract,* 1998). While a significant minority of Americans are not members of any voluntary organizations or congregations, the majority are active members. From the standpoint of social movements, the importance of these facts is that many of the activists in social movements are recruited from the ranks of voluntary associations (e.g., through mailing lists or membership meetings).

Even people who never join social movements will surely find that their lives are changed by them. The social critic H. L. Mencken once said that if three Americans are in a room together for more than an instant, two of them will try to change the morals of the third. Campaigns against smoking, drinking, rock lyrics, violence on television, wearing of furs, abortion, and many other behaviors would seem to confirm this notion. So too would patterns in the growth of different types of voluntary associations. Since 1980 the number of nonprofit voluntary associations devoted to educational and cultural issues has increased by about 36 percent while the number devoted to public affairs has almost doubled (*Statistical Abstract,* 1998).

Resource Mobilization and Free Riders

What is the relationship between civil society and social movements? In sociology this is known as the *resource mobilization* question, referring to the fact that social movements need to mobilize existing leaders and organizations rather than simply relying on the participation of people who happen to be moved to action (Klandermans, 1997; Zurcher & Snow, 1981). A second, related issue is known as the *free rider problem.* This refers to the tendency of many people not to lend their support and resources—time, money, and

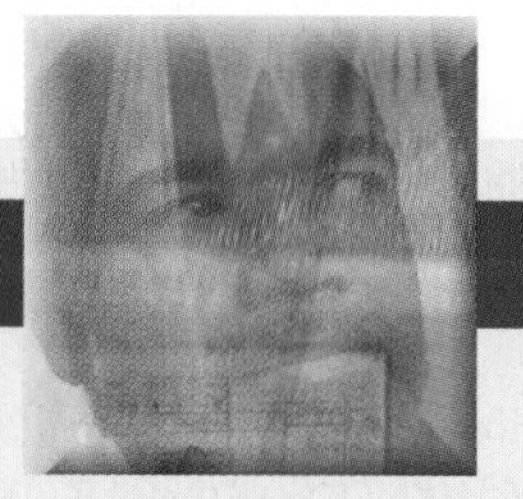

THE PERSONAL CHALLENGE

Social Change and Social Movements

i

Wherever it occurs, social change results in an explosive growth of social movements. As societies around the world become enmeshed in global interrelationships, as old ways of making a living decline and new ones appear, as worldwide currents of thought such as nationalism, fundamentalism, or movements for equality sweep through their societies, people try to join forces and create social movements to cope with those changes. Every social movement has active members who seek to attract new recruits and will appeal to you for your attention and allegiance. The choices you make can help shape the future of your community and society. They can also change your life in both positive and, sometimes, negative ways.

As this chapter shows, millions of Americans are part of one or more social movements. However, there may be important differences between a social movement and a particular organization seeking to advance that movement's goals. For example, you may oppose abortion and consider yourself part of the pro-life movement, but you may not actually be a member of a pro-life organization within that movement. Similarly, you may favor better working conditions for workers and consider yourself an ally of the labor movement, but you may not actually be a member of a specific labor union within that broader social movement. If you continue to support the movement, attend information meetings, take part in demonstrations, and so on, before long you will be recruited for actual membership in an organization.

What questions can one ask about a social movement organization in deciding whether or not to join it? Here are a few especially important ones:

1. *Does the organization demand total allegiance to its leaders and goals?* How much will you have to give up in order to become active in the organization? Many organizations require total commitment. Some demand that their members cut themselves off from previous friends and even from family members. This is especially true of cultlike political and religious organizations, which often make extreme demands on their members. New recruits are often brought along slowly so that at first the demands are not too severe. So look closely at what is expected of those who are in the organization's higher ranks and how much personal autonomy they must sacrifice. Beware of any organization that requires you to break other social ties so that you can devote yourself fully to its aims and activities.

2. *How does the organization arrive at strategies for achieving its goals?* All social movements must deal with the problem of how to make their demands and aims known to the public. Does the organization you are thinking of joining use democratic processes in deciding on its strategies? Do the members actually have a say in those decisions, or do they merely put a rubber stamp on policies that come "from above"? If the organization claims to be run democratically, you should be able to come up with real evidence that this is so. For example, does it give a full accounting of its finances to members? Does it have a history of corruption? If so, is it open about what it has done to replace corrupt leaders?

3. *What is the organization's stance on civil disobedience and the rule of law?* Many social movement organizations are formed to protest laws that are believed to be unjust. Their members challenge those laws by publicly

leadership—to social movements but to reap the benefits anyway (Marwell & Ames, 1985; Olson, 1965).

Are these issues important? One need only think of all the American workers who have benefited from the gains won by the labor movement without themselves making the effort to form unions. Or consider the rapid growth of the environmental protection movement in the late 1960s and early 1970s. What explains the speed with which that movement captured a central place in public policy debates? In studying the growth of the movement for environmental quality, sociologist Carol Kronus (1977) found that out of a sample of 209 existing organizations in a midwestern city, more than half gave material and moral support to the new movement. And she found that the groups that did link up with the movement were ones whose goals were in agreement with those of the movement and whose members believed in taking action to improve the quality of life in their community. Free rider groups, on the other hand, agreed that "something should be done about pollution" but felt that other groups should take action before they were able to do so.

The free rider hypothesis predicts that some proportion of the potential members of a social movement

violating them. Those who do so risk fines and even imprisonment. Civil disobedience has a long and honorable history in democratic societies. The abolition of slavery in the United States and of apartheid in South Africa, for example, probably would never have happened without civil disobedience. But how far should one go in challenging laws, even when they are believed to be unjust? There are many forms of civil disobedience, from sitting down and blocking normal traffic to engaging in acts of violence. Beware of any organization that prides itself on its ability to resort to violence. Be extremely wary of any group that makes quick and undemocratic decisions that lead to civil disobedience.

Many organizations that espouse goals you agree with will argue that the ends justify the means, that the social crisis is so grave or the movement's goals so important that the rights and safety of group members and the public must be sacrificed. Think long and hard about the implications of this stance—consider, for example, the tragic history of the Soviet communists or the harm done by the Oklahoma City bombers—before you decide to join this kind of organization. Never hesitate to be an active citizen and a member of social movement organizations, but make sure you ask critical questions about any organization that seeks your allegiance.

will not lend their energies to the movement but will nonetheless hope that the movement will provide them with the benefits they desire. But how extensive will the free rider problem be? How many people will act out of self-interest and try to reap the rewards of the movement's activities without making any personal sacrifices? These are empirical questions on which there is a growing body of research findings.

One study of the free rider problem is Edward J. Walsh and Rex H. Warland's (1983) study of the social movement that emerged after the near-meltdown of a nuclear reactor on Three Mile Island (TMI) near Middletown, Pennsylvania, in 1979. After the accident, antinuclear organizations grew in strength throughout the TMI area. Many of them opposed efforts to start up a second reactor at TMI, which had not been damaged. Walsh and Warland found that of those who opposed the start-up of the undamaged reactor, 87 percent were free riders in that they had not contributed in any way to the organizations opposing the start-up. And of those in favor of the start-up, 98 percent were free riders. Only a very small percentage—6 percent of those in opposition and 1 percent of those in favor—actually participated. Although this is a small proportion, it does show that

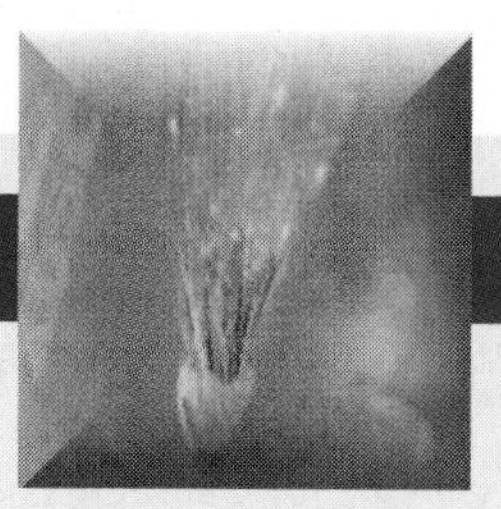

ART AND THE SOCIOLOGICAL EYE

Images of Collective Behavior

Artists often capture for future generations the feelings and images associated with collective behavior and social movements. We see many examples of this in famous paintings in museums throughout the world. People's daily lives are being severely disrupted, for better or worse, by great social movements or far-reaching changes in the fate of entire social classes. There is a dynamism in these paintings; the people they portray are being swept up in the great events of their time.

Simone Martini's depiction of *The Road to Calvary*, opposite, portrays a scene in which Christ, the Virgin and saints, and the Roman soldiers and commoners all are caught up in a chaotic and violent event. The artist sets the scene against the backdrop of a fortified town, suggesting the strains in the medieval social order of Europe that were occurring as a result of the accelerating renaissance in art, science, and commerce. The Renaissance threatened the long-standing dominance of the church and the feudal state. Thus the painting makes a strong connection between the changes that occurred in Jesus' time and those that were occurring in Europe during the fourteenth century.

Revolutions, civil wars, and invasions of the weak by the powerful are often either consequences or causes of lasting social change. It is no wonder, then, that scenes of war are so often portrayed in paintings. Francisco de Goya's *The Executions of the Third of May*, below, is a fierce rendering of the brutality of the Napoleonic conquest of Spain. Firing squad executions like these took place in Madrid after the weak Spanish monarchy had refused to resist the invasion. But it is not necessary to know the details of this moment in

The Executions of the Third of May, by Francisco de Goya, 1808. Oil on canvas, 8′ 8¾″ × 11′ 3⅞″.

The Road to Calvary, by Simone Martini, about 1340. Panel, 10″ × 6″.

history. Goya's painting gains its impact from the way the artist portrays the behavior of individuals under extreme conditions. The soldiers' faces are hidden. The soldiers are presented as automatons following orders rather than as humans. The dying civilians, in contrast, are all too human in their reactions. Some stare defiantly at their executioners while others cower with their faces covered.

there are people who are ready to be mobilized as activists on an important social issue. Not everyone is an apathetic or alienated free rider.

The TMI research suggests that a large majority of people will be free riders and will not join a social movement even when they strongly desire it to succeed and to win victories that will benefit them directly. Other research, however, indicates that the tendency to be a free rider is far less prevalent than the TMI study suggests (Garner & Tenuto, 1997). Data show that more than 85 percent of American households make charitable contributions of some kind. And public television and radio stations do succeed, after much persuasion, in raising the funds they need to continue operating, even though many people who enjoy listening or viewing would just as soon be free riders. Findings from thousands of laboratory experiments conducted since the early 1970s also contradict the free rider hypothesis, showing instead that many people make commitments and sacrifice individual gain in order to advance a cause. Still, the relatively low proportion of people who join a cause, vote in elections, or voluntarily support institutions like National Public Radio promises to make the free rider problem a subject of continuing social-scientific research (Frank, 1988; Hechter, 1987).

In predicting the success of any given social movement, resource mobilization theory points to the need to consider how well a movement enlists or mobilizes the resources available to it—for example, how well it deals with gender differences in recruiting members or how well it keeps recruits actively committed. Research by Doug McAdam (1992) on the experience of men and women in the civil rights movement during the early 1960s shows that organizations in the movement were constantly struggling with the problem of gender. White women who volunteered as civil rights workers often had more experience in organizations and civil rights activities than white male applicants but were less likely to be selected to participate. Those who were selected were often assigned jobs that were considered "women's work" unless they protested against such treatment. Indeed, McAdam found that many of the women who participated in the civil rights movement later became active in the women's movement, in part because of the discrimination they had encountered in the earlier movement. Studies of participation in other social movements during the 1960s have yielded similar findings: Women are available to be mobilized by social movements, but once they become involved they often find themselves faced with issues of gender inequality.

Mobilization of the law is another aspect of resource mobilization theory that has been studied extensively in recent years (Hoffman, 1989). For example, in his research on how activists attempt to use laws and the courts to advance their cause, sociologist Robert Burstein (1991) shows that activists seeking to end discrimination against minorities in the labor force are increasingly petitioning the courts under the Equal Employment Opportunity (EEO) laws. Burstein and other researchers have found that social movements are more likely to use the tactics of protest and demonstration when they are outside the framework of the law, but when laws have been passed in response to their demands, they become more likely to initiate lawsuits to achieve their goals. Burstein shows that the more a group has mobilized its resources as a protest movement, the more it will be able to mobilize the law in its favor as well. Thus racial minorities and women's groups, two well-mobilized civil rights groups, do much better at winning enforcement of EEO laws than religious or ethnic minorities, which have less experience organizing and mobilizing members and money for their causes.

MASS PUBLICS AND PUBLIC OPINION

Most of the examples of social movements discussed in this chapter have had a lasting impact on the societies in which they occurred. In this section we apply the insights gained from earlier sections to some of the less world-shaking, yet still exciting, aspects of collective behavior in American life.

Over the past century North America has been transformed from a continent of wide-open spaces and agrarian settlements to the earth's technologically most advanced and socially most mobile region. Indeed, for the rest of the world North America is synonymous with both the best and the worst aspects of what has come to be known as "modern" society.

According to Reinhard Bendix (1969), **modernization** consists of all the political and economic changes that accompanied industrialization. Among the indicators of modernity are urbanization, the shift from agricultural to industrial occupations, and increasing literacy as well as the demand for greater political participation by the masses. Bendix is quick to point out that his definition does not necessarily apply everywhere in the world. Modernization may not always mean what it has meant in Western societies. But it is probable that, despite setbacks in some nations, modernization will always involve the extension of rights, values, and opportunities from the elites to the masses in a society.

None of these changes in the organization of societies came about without collective protests and social movements. For every radical movement there was a reactionary one, for every revolution a counterrevolution or at least an attempt at one. And social movements continue to exert pressure for change in our norms and institutions. The extension of voting rights to women and blacks had to be won through political struggle. Nor did the benefits of new technologies come automatically to American workers; they had to be won by the labor movement. Likewise, the rapidly expanding cities of North America became scenes of collective protest as new populations of immigrants and new occupational groups fought for "a piece of the pie" (Lieberson, 1980). It is hardly surprising, therefore, that so much sociological research is devoted to studying how modern society came about, discovering how much further some groups have to go to gain their share of the benefits of modernity, and analyzing the role of collective behavior in these major social changes.

Consider just one important consequence of the labor movement of the late nineteenth and early twentieth centuries: the change in the length of the workweek. "Our lives shall not be sweated, from day until night closes," went the famous refrain of the Industrial Workers of the World (known as the Wobblies). "Hearts starve as well as bodies; give us bread but give us roses." Figure 8.2 shows how the demand for enough leisure to enjoy the finer things of life produced dramatic changes in the hours of work. Today we take the 8-hour day and the 5-day week as the standard for full-time employment, but in 1919 American steelworkers shut down the entire industry in protest against the 12-hour day and the 6-day week (Brody, 1960). The concessions won by them and other workers were an essential part of the extension of the benefits of modernity to working people (McAdam & Snow, 1997). The increase in leisure created by these changes in hours and days of work, along with greater social mobility owing to increased affluence, contributed to the development of mass publics (see Box 8.2).

FIGURE 8.2 **Average Weekly Work Hours, 1850–1963**

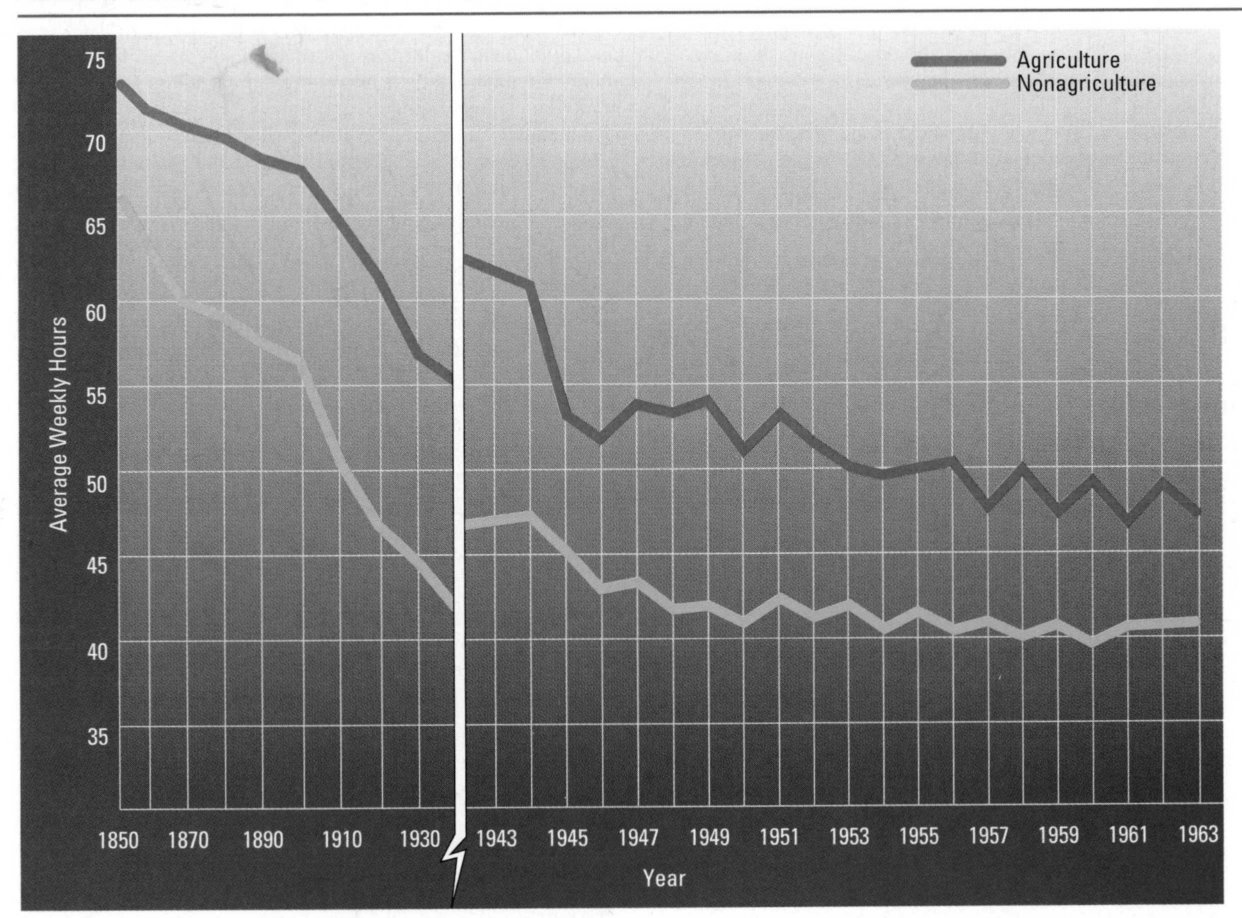

There was a brief increase in average hours of work between 1940 and 1943 (not shown on the chart), due to the labor shortages that occurred during World War II.

Source: Data from Bureau of Labor Statistics; adapted from Dankert, Mann, & Northrup, 1965.

BOX 8.2 USING THE SOCIOLOGICAL IMAGINATION

Amusing the Millions

The idea that the United States is becoming "a nation of spectators" was a popular sociological notion in the 1950s. A contemporary version of this "spectatoritis" thesis is the claim that greater affluence and increased access to television and other media have transformed Americans into a nation of overweight "couch potatoes." A more sophisticated analysis of these issues is provided by the study of leisure in the United States and other advanced industrial nations. Indeed, the study of leisure is an important frontier of sociological research.

Since the 1940s the pursuit of leisure activities of all kinds has become a passion for those who can afford the time. This is true throughout most of the industrial world, not just in the United States and Canada (Machlis & Tichnell, 1985). People are spending more money on leisure pursuits and engaging in more diverse pleasures than ever before. In ancient Rome and feudal Europe only the nobility had leisure; everyone else toiled ceaselessly in the fields or at their crafts. The limited leisure available to the masses took the form of religious festivals and games. The notion that common people could enjoy leisure in specified blocks of time—weekends, holidays, vacations, retirement—would have been viewed as revolutionary.

The rise of modern social movements changed the expectations of "common folk" about many values, including leisure. For example, at the turn of the twentieth century the labor movement called for an end to the 12-hour workday. At around the same time, the conservation movement was pressing for preservation of the nation's virgin forests and the creation of national parks. The woman suffrage movement marched for women's right to vote, which would eventually lead to demands by women for more leisure time.

One consequence of the increase in leisure time in North America has been the steady growth of institutions devoted to leisure and recreation. It is estimated that more than 350 million visits are paid to national parks in the United States each year and that approximately 700 million visits are made to state parks. More than 45 million tickets to major league baseball games are sold each year, and about 15 million people attend NFL football games. About 415 million rounds of golf are played each year, more than 60 million people go bowling, and at least 9 million people own an outboard motor. There are some 1,100 opera companies in the nation; they perform for at least 14 million people a year. Try maneuvering through the traffic jam caused by a college football game when 60,000 fans are trying to get home for dinner. Imagine the same scene repeated simultaneously at thousands of other locations, some smaller and some larger, and you can begin to grasp the impact of leisure pursuits.

Data like those just presented tell only part of the story of leisure's impact on a modern society. In the United States, people's leisure preferences increasingly shape the landscape itself. Highways extend to beaches and mountain recreation sites. Large portions of states like Florida, Montana, Wyoming, Arizona, and California are managed as leisure and recreation areas by powerful state and national land management agencies. Private vendors of recreation services (motels, tennis courts, restaurants, shops, etc.) crowd the tourist gateways to major attractions like Yellowstone Park or the Great Smokies. In major cities, leisure institutions like theaters, symphony orchestras, museums, sports complexes, and parks are vital to the local economy and culture. Without them, cities would become dark ghost towns after business hours. Urban planners and city leaders therefore compete to attract talent and develop such institutions into tourist and cultural centers. Leisure for the millions translates into jobs for hundreds of thousands of people in every region of the country.

Mass Publics

Mass publics are large populations (regional or national) of potential spectators or participants who engage in collective behavior of all kinds. Typically this behavior consists of the formation of crowds, audiences, or streams of buyers and voters, but it can also include crazes, panics, and the spreading of rumors. Thus a massive traffic jam can turn into a dangerous panic; a joyful victory celebration can become a violent riot like the one that occurred after the Chicago Bulls won the National Basketball Association championship in 1992. The panic selling that occurred on Wall Street in 1987 and again in 1998, resulting in a massive loss of investors' capital, is a reminder that mass publics continue to exert important effects even when supposed safeguards are in place. For this reason alone, social scientists, urban planners, and governmental leaders increasingly recognize the need to cooperate in anticipating the behaviors of mass publics.

This recognition has come after some painful experiences. A famous case of failure to anticipate the possible reaction of mass publics occurred with the 1938 radio dramatization of H. G. Wells's novel *The War of the Worlds.* This broadcast, which vividly described an invasion of the earth by Martians, was presented to the radio audience in documentary fashion, beginning with a fictitious news flash and continuing with reports of a spacecraft landing in a New Jersey field and descriptions of the invading Martian army. Not realizing that the broadcast was only a dramatic presentation, hundreds of thousands of Americans began gathering in panicky crowds, while thousands of others jammed telephone lines in efforts to reach loved ones.

A few years later, when the dramatization was broadcast by Radio Quito using Ecuadorian place names, "the initial reaction was the same as in the United States. Multitudes poured out of the city in all directions, running and driving as far and as fast as they could" (Klass, 1988, p. 48). When listeners found out that the program had been a dramatization, they were so angry at the radio station that they burned it to the ground. As a result of experiences like these, broadcasters became far more careful in presenting fictitious accounts on the radio (Cantril, 1982/1940).

The availability of more time for leisure pursuits, the development of a mass market for automobiles, and the technological revolution in communications and the mass media have all exerted an immense influence on the lifestyles of mass publics, who in turn shape the society in which they live. For example, through their demands for roads, leisure facilities, and services that cater to a highly mobile lifestyle, Americans have transformed the physical landscape. Rural scenes of farms and small towns still exist, but throughout the nation they are being enveloped by networks of suburbs and shopping malls and pleasure grounds (stadiums, amusement parks, etc.) linked together by a labyrinth of highways (Carter, 1975; Flink, 1975; Thomas, 1956).

Mass publics have also created the conditions that make possible whole new industries. The hot dog, for example, was an innovation that allowed large numbers of people to eat while strolling along the boardwalk at Coney Island. But fast food soon became an industry and even, as McDonald's founder Ray Kroc (1977) described it, an art form:

> Consider, for example, the hamburger bun. It requires a certain kind of mind to see beauty in a hamburger bun. Yet, is it any more unusual to find grace in the texture and softly curved silhouette of a bun than to reflect lovingly on the hackles of a favorite fishing fly? Or the arrangement of textures and colors in a butterfly wing? . . . Not if you regard the bun as an essential material in the art of serving a great many meals fast. [p. 99]

The thousands of teenagers and young adults who work in the fast-food industry may not sing rhapsodies to buns and burgers; nevertheless, their industry owes its existence to the behavior of hungry, mobile multitudes—that is, mass publics.

Public Opinion

The presence of mass publics sets the stage for the emergence of public opinion as a powerful force in modern societies. **Public opinion** refers to the values and attitudes of mass publics. The kinds of behavior that develop out of public opinion include fads, fashions, demands for particular goods and services, voting behavior, and much more.

Public opinion is shaped in part by collective behavior, especially social movements. An example is the rise of the conservative movement in the United States in the 1980s. The impact of this movement can be seen in the contrast between the results of the 1964 and 1984 presidential elections. In 1964 Barry Goldwater, who espoused many of the ideals of the current conservative movement, was overwhelmingly defeated by Lyndon Johnson, an outspoken proponent of liberal social-welfare policies. Twenty years later the pendulum of public opinion had swung to the other extreme, and the staunchly conservative Ronald Reagan defeated a liberal candidate, Walter Mondale, by an even greater percentage of the electoral vote than Johnson had won in 1964.

During the 1992 presidential election campaign there seemed to be a backlash against extreme conservatism, reflected in the voters' rejection of Republican emphasis on "family values" as code words for disapproval of homosexuals, single parents, and other people with unconventional lifestyles. But then, Republican victories in the 1994 congressional elections showed that conservative values were at least temporarily in ascendance once again. Supported by active social movements, particularly the antiabortion forces and those promoting Christian conservative causes, the Republicans faced the problem of satisfying the demands of those movements while persuading others that the party could accommodate their often conflicting needs. Democrat Bill Clinton's reelection in 1996 and the small but surprising Democratic gains in the 1998 off-year elections indicate how difficult it can be to predict public opinion and to satisfy mass publics.

Public opinion is also shaped by more fleeting shifts in the national mood as citizens respond to experiences that are shared by all, especially through the mass media. For example, no one knows for sure whether the severe droughts that burned crops in the United States and Canada in 1987 and 1988 were a symptom of global warming, but scientists' warnings about this possibility, and their explanations of the "greenhouse effect" caused by high levels of carbon dioxide and other gases in the atmosphere, were made far more vivid by media coverage of the drought. Public support for environmental protection was stimulated still further by media coverage of the oil spill by the Exxon *Valdez* in Alaska in 1989.

RESEARCH ON THE CUTTING EDGE: The Celebrity Industry

Americans are hooked on fame. Famous people, or *celebrities,* are so much in demand, so much the focus of attention, so widely discussed that everywhere we turn—at the supermarket, on television, in the press—gossip about the rise and fall of celebrities crowds in on us. No surprise, then, that sociologists who study collective behavior and public opinion are seeking to better understand the origins of this explosion in the public's appetite for stories about stars and celebrities.

In a recent study, sociologist Josh Gamson (1994) shows in detail how celebrity is constructed by a thriving "fame industry." But Gamson admits that he became fascinated by fame and celebrities just the way other Americans do, by watching television, often instead of studying. "I became interested in exploring the peculiarities of celebrity culture in graduate school," he says, "through my evening encounters with *Entertainment Tonight,* a television program that can generously be called lightweight." Gamson was amazed to read that *Entertainment Tonight*'s anchor, Mary Hart, "whose celebrity derives from a chipper reading of teleprompted words about entertainers," had received a star on the Hollywood Walk of Fame. But trivial as Gamson thought her fame was, he could not help watching the show and thinking about the celebrities. Trained to analyze collective behavior and fads, he realized that he needed "to tap into the weird world of what for many people in this country, including myself, had become the stuff of everyday life" (1994, p. 4).

Gamson's study examines what is done in the entertainment industry to "manufacture" celebrity, and it also looks at the audience for fame by analyzing "how people actively encounter celebrities and interpret and use celebrity images" (p. 5). Like all professional researchers, however, Gamson drew many of his research questions and investigative leads from a thorough study of the historical and sociological literature about fame.

A question that is often raised in the literature on celebrity is this: If people like Mary Hart and other television celebrities are famous, what is that fame in comparison with "the famous of the past, those names that echo in our minds—Alexander, Caesar, Cleopatra, Jesus, Mohammed, Joan of Arc, Shakespeare" (Braudy, 1997)? In the past our heroes tended to be people of great achievement with admirable qualities that set them apart from the rest of us. In premodern societies, however, word of the deeds of heroes spread relatively slowly. It could take weeks for the details of a general's actions in a great battle to spread, even through the elites of a population. With the rise of modern and more democratic societies, equipped with ever more rapid forms of communication, names and faces arise and fall with stunning rapidity. The important differences between actions and image, between truth and fiction, between deserving and undeserving, are blurred in the rush of information and the seductive power of the media.

Gamson found plenty of critical writing about fame in the modern world. Most of this literature emphasizes how visibility becomes its own reward and artificial images replace real-life ones so that people become unable to distinguish between truth and fiction and, more important, are uninterested in doing so. But Gamson found more balanced evaluations of the fame phenomenon in the work of historians and sociologists who wonder how deep the frenzy for fame actually goes in contemporary societies. From sociologist Michael Schudson, for example, Gamson learned to question how seriously consumers take the fleeting fame of many celebrities. There are probably many gradations in consumers' ability to distinguish among celebrities whose fame is purely a matter of television exposure and those, like Mark McGwire or Michael Jordan, who more clearly merit their fame because of their achievements.

These insights led Gamson to narrow the focus of his study to the entertainment industry. "Entertainment," he argues, "is clearly the dominant celebrity realm in this century; it is also the most fully rationalized and industrialized. It is therefore typically used as a model for the development of celebrity in other realms (politics, for example). . . . Understanding entertainment celebrity promises to help us comprehend celebrity as a general cultural phenomenon: its particular dynamics, its place in everyday lives, its broader implications" (1994, p. 5).

Gamson sees in the rise of mass entertainment the technological possibilities for more instant renown. Today's celebrities are the "products" of an efficient fame industry. No doubt many of the older generation of movie stars, like Elizabeth Taylor and John Wayne, actually developed their fame over long careers of outstanding acting. They and hundreds like them seem different from television personalities whose fame appears to be based on mere exposure in the mass media. But Gamson's research shows that the rise of the mass entertainment industries earlier in the twentieth century also spurred the rise of the modern fame industry. The need for movie studios to maintain the fame of their actors helped account for the emergence of occupations designed to keep important names and faces before the public: the agent, the publicist, the celebrity photographer. As more media channels have developed, the pressure to attract viewers by showing them people who are famous has also increased, and the meaning of fame has been watered down.

Through extensive interviews and observations in the Los Angeles and New York celebrity industries, Gamson shows just how fame is manufactured. The entertainment industry is "much like other commodity-production systems." People working in it "speak primarily in the language of commerce and machinery," and "marketing plays a key role in marching products to distributors and consumers, depending especially on strategies of product transformation and the building of consumer loyalty" (1994, p. 58). Over and over again, those whose job is to make and evaluate, buy and sell famous people described to Gamson the ins and outs of creating and, sometimes more difficult, maintaining celebrity. The photo of Cher, which Gamson includes in his study, is one of many he analyzes to reveal the processes of "celebrity making." In this photo, which is designed to show the often frenetic Cher in a serene mood, we get a glimpse of the artificiality of the situation she was in at the time. And in the photo of "Two Actors" Gamson points out that humorous commentary on celebrity itself can be used in the "hype" of stardom.

VISUAL SOCIOLOGY

The NAMES Project Quilt

Victor Ayala, the AIDS activist we met at the beginning of the chapter, did not need to have his consciousness raised by other activists before he became involved in fighting the disease and the fear and scorn its victims often suffer. He had firsthand experience with the disease early in its history. But many other people have come to understand the personal meaning of the AIDS epidemic through an extraordinary social movement: the NAMES Project, and the famous quilt it has produced. This movement uses visual sociology in the most creative way one can imagine. As AIDS activist Elizabeth Taylor writes:

> The quilt is a moving depiction of stories of the loved ones of human beings who have died of AIDS. It reflects the true spirit of America. In their tragedy and grief over loss immeasurable, contributors to the quilt have used art and love to keep the spirit of their loved ones alive. (quoted in Ruskin, 1988, p. 7)

The NAMES Project and the quilt are the inspiration of Cleve Jones. When the actor Marvin Feldman, his dear friend of 14 years, died of AIDS at the age of 33, Jones went to his backyard with paint, stencils, and a sheet to create a memorial. He remembers spending the entire day thinking about Marvin: "I thought about why we were best friends and why I loved him so much. By the time I finished the piece, my grief had been replaced by a sense of resolution and completion." Jones realized that if people got together to make quilts as memorials to lost loved ones, the activity could be a redeeming and mobilizing way of confronting the AIDS epidemic.

A Quaker with years of political activism in his background, Jones became a gay activist when he was 18. In 1986, he appeared on *60 Minutes* and discussed his reactions to the discovery that he was HIV positive. Shortly thereafter, in Sacramento, he was attacked by two men who stalked him, calling him "faggot," and stabbed him in the back. During his convalescence, he thought more about the idea of the NAMES Project and the quilt, and after his recovery he joined with AIDS activists throughout the world to advertise the project. Before long, hundreds of groups of friends and family members were producing quilt panels to commemorate the loss of their loved ones.

On October 11, 1987, the quilt was displayed on the Mall in Washington, D.C.; it covered an area larger than two football fields. It was subsequently exhibited throughout the United States, focusing attention on the need for increased funding for research and the allocation of additional medical resources to the AIDS epidemic.

"There is nothing beautiful about AIDS. It is a hideous disease. . . . With the Quilt, we're able to touch people in a new way and open their hearts so that they no longer turn away from it but rather understand the value of all those lost lives."

WASH. D.C.
AUG. 2,1983
"I CAME HERE TODAY TO ASK THAT THIS NATION WITH ALL ITS RESOURCES AND COMPASSION NOT LET MY EPITAPH READ HE DIED OF RED TAPE."
Roger Gail Lyon

"The saddest thing, I believe, for Roger Lyon," says his devoted friend, Cindy McMullin, "was how unfair it was to have to die without knowing how it—AIDS—is going to end."

Sixty-year-old homemaker Marian Mosner of Lincoln, Nebraska, made this panel for a man who worked with her son in Denver. She did not know her son's friend well but says, "I thought that there would be no recognition from his family. I felt bad about that. I feel bad about all the people who die of AIDS that nobody knows."

SUMMARY

The term *collective behavior* is used to refer to a continuum of unusual or nonroutine behaviors that are engaged in by large numbers of people. At one end of the continuum is the spontaneous behavior of people reacting to situations they perceive as uncertain, threatening, or extremely attractive. At the other end are events that involve large numbers of people in nonroutine behaviors but are organized by leaders and have specific goals. The set of organizations that plan such events is a *social movement.*

The study of spontaneous forms of collective behavior often begins by distinguishing between crowds and masses. A *crowd* is a large number of people who are gathered together in close proximity to one another. A *mass* is a large number of people, not necessarily in close proximity, who are all oriented toward a set of shared symbols or social objects. Collective behavior can occur in crowds, in masses, or in both at once. The actual behavior that a crowd or mass generates depends largely on the emotions people involved in the events feel are appropriate to express in those situations. The most significant categories of emotions that motivate collective behavior are fear, hostility, and joy.

Social movements have been classified into four types based on the goals they seek to achieve. Revolutionary movements aim to overthrow existing stratification systems and social institutions; reformist movements seek partial changes in some institutions and values; conservative movements attempt to uphold the existing values and institutions of society; and reactionary movements seek to return to the institutions and values of the past. In addition, there are expressive social movements, or movements devoted to the expression of personal beliefs and feelings, and millenarian movements, which are both revolutionary and expressive. Within any large social movement there are likely to be a number of different social movement organizations, or SMOs.

Early theories of collective behavior were based on the notion that hysteria or contagious feelings like hatred or fear could spread through masses of people. Gustave LeBon attributed the strikes and riots that are common in rapidly urbanizing societies to a mob mentality created by the presence of large numbers of strangers in crowded cities.

Sociologists often distinguish between "long revolutions," or large-scale changes in the ecological relationships of humans to the earth and to one another, and revolutions that are primarily social or political. *Political revolutions* are transformations in the political structures and leadership of a society that are not accompanied by a full-scale rearrangement of the society's productive capacities, culture, and stratification system. *Social revolutions* not only change the institutions of government but also bring about basic changes in social stratification.

According to Marx, revolutions would occur as a result of the spread of capitalism: Impoverished workers and colonial peoples would rebel against the capitalists and create a new classless society. Also, Marx and Tocqueville pointed to the role of *relative deprivation,* noting that the feeling of deprivation relative to others—not the presence of deprivation itself—may give rise to revolutionary social movements.

More recent analyses have shown that revolutionary social movements often occur in cycles or waves. Waves of protest may result from major social shocks. During such periods there is a dramatic increase in social conflicts, often motivated by collective action frames, or sets of beliefs and interpretations of events.

Successful leaders of social movements are often said to have almost supernatural powers to inspire and motivate their followers. Max Weber called this ability *charisma.* Over time, however, those leaders' goals must be incorporated into the structure of the movement, a process that is referred to as the institutionalization of charisma. However, the more successful the movement, the more difficult it is to maintain the zeal of its founders. In extreme cases the process can end in totalitarianism, or efforts by an elite to control all forms of organizational life in a society.

The sphere of public nongovernmental, nonbusiness social activity is termed *civil society.* People who participate in civil society are often recruited into social movements. The resource mobilization question refers to the need to mobilize existing leaders and organizations to achieve a movement's goals. A related issue is the free rider problem, the tendency of many people not to lend their support and resources to social movements but to reap the benefits anyway.

In recent years activist groups have made increasing use of laws and the courts. Social movements are more likely to use the tactics of protest when they are outside the framework of the law; when laws have been passed in response to their demands, they become more likely to initiate lawsuits to achieve their goals.

Mass publics are large populations of potential spectators or participants who engage in collective behavior. Such factors as increased leisure time, the almost universal use of automobiles, and the technological revolution in communications and the mass media have had an immense influence on the lifestyles of mass publics, which in turn shape the society in which they live.

The presence of mass publics makes possible the emergence of *public opinion,* or the values and attitudes of mass publics. The behavior that develops out of public opinion can take a variety of forms, including fads, fashions, and demands for particular goods and services. Public opinion is shaped in part by collective behavior, especially social movements. It is also affected by experiences that are shared by all the members of society through the mass media.

SOCIOLOGY VERSUS IDEOLOGY

Activists in social movements sometimes face extremely difficult moral choices. Should they risk their careers by devoting so much of their time to a cause? Under what circumstances will they break the law in order to make their point to the authorities and the public? And the more deeply held their ideals are, the more often they face serious and potentially life-altering, or even life-threatening, dilemmas. Many social movements include extremists who believe that in order to create change it is necessary to take drastic action. We see evidence of this in bombings of abortion clinics, standoffs between armed militias and federal agents, violence associated with certain labor union conflicts, and similar situations. In the twentieth century, in fact, many of the worst crimes against humanity were perpetrated by leaders who claimed that "the ends justify the means," meaning that the goals of the movement are so important that any methods deemed necessary to attain them are justified, including assassination, violent protests, attacks on opponents or the authorities, and even bombings. Considering the social cataclysm that resulted from these tactics when they were used by the Nazis in Germany or the Stalinists in the Soviet Union, it might be safe to assume that ends never justify means that violate human freedom or due process under the law.

But then what about the abolitionists, or the civil rights marchers, or the suffragettes, or any of hundreds of other examples of social movement activists who on occasion have felt it necessary to break laws and even go to jail to call attention to their causes? If we examine their law-breaking behavior, we usually find that they stopped short of proclaiming that their goals justified any and all means, no matter how violent. They often had to repudiate the hotheads among them who were ready to take the most extreme actions. It is not always easy to be a voice of moderation in a social movement, however, and there may be times when protests against laws that the group considers immoral seem to justify violation of those laws. But if there is one outstanding sociological and moral lesson of the bloody twentieth century, it is that as a blanket rule the ends do not justify the means. That idea opens the way to the most extreme anti-democratic, anti-human behaviors, which often subvert the central ideals of the movement itself.

GLOSSARY

collective behavior: nonroutine behavior that is engaged in by large numbers of people responding to a common stimulus. (p. 228)

social movement: organized collective behavior aimed at changing or reforming social institutions or the social order itself. (p. 230)

crowd: a large number of people who are gathered together in close proximity to one another. (p. 231)

mass: a large number of people who are all oriented toward a set of shared symbols or social objects. (p. 231)

political revolution: a set of changes in the political structures and leadership of a society. (p. 235)

social revolution: a complete transformation of the social order, including the institutions of government and the system of stratification. (p. 235)

relative deprivation: deprivation as determined by comparison with others rather than by some objective measure. (p. 235)

charisma: a special quality or "gift" that motivates people to follow a particular leader. (p. 236)

civil society: the sphere of nongovernmental, nonbusiness social activity carried out by voluntary associations, congregations, and the like. (p. 239)

modernization: a term used to describe the changes that societies and individuals experience as a result of industrialization, urbanization, and the development of nation-states. (p. 244)

mass public: a large population of potential spectators or participants who engage in collective behavior. (p. 246)

public opinion: the values and attitudes held by mass publics. (p. 247)

WHERE TO FIND IT

BOOKS

Social Movements: Readings on Their Emergence, Mobilization, and Dynamics (Doug McAdam & David S. Snow; Roxbury, 1997). An excellent collection of recent studies and theoretical analysis of social movements, with especially good articles on how movements emerge, become organized, and mobilize supporters.

Claims to Fame: Celebrity in America (Josh Gamson; University of California Press, 1994). A brilliant study of the celebrity industry and the audiences who make celebrity one of the fastest growing and most lucrative markets in the United States.

Frontiers in Social Movement Theory (Aldon Morris & Carol McClurg Mueller, eds.; Yale University Press, 1992). A collection of recent studies of social movements that includes examples from most new areas of research in this rapidly changing field.

The Crowd in History (George Rudé; Wiley, 1964). A fascinating account of the influence of riots, demonstrations, and protests on the history of the industrializing Western world.

Social Movements in an Organizational Society (Meyer N. Zald & John McCarthy, eds.; Transaction, 1987). A thorough analysis of competing explanations of the evolution of social movements.

A History of Recreation: America Learns to Play, 2nd ed. (Foster Rea Dulles; Appleton, 1965). A history of recreation in America that covers the transformation of the nation from one in which religion dominated all leisure to one characterized by differentiated leisure institutions such as theater, popular music, and sports of all kinds.

Collective Behavior (Ralph H. Turner & Lewis M. Killian; Prentice Hall, 1987). Widely regarded as the best text on collective behavior and social movements; includes excellent discussions of the major concepts.

The Logic of Collective Action (Mancur Olson; Harvard University Press, 1965). A classic treatise on the costs and benefits of individual choices to engage in collective action; includes a good analysis of free riders.

Women's Movements in America: Their Successes, Disappointments, and Aspirations (Rita J. Simon & Gloria Danziger, eds.; Praeger, 1991). Assesses the relative success of a major social movement in securing positive change for women in the United States; contains excellent bibliographical material and references to other works.

OTHER SOURCES

The Gallup Poll. An annual compilation of polls taken throughout the year by Gallup, Inc.; available in the reference section of most college and university libraries.

The Utne Reader. A compilation of articles, many written by sociologically informed journalists, about current American trends, with much material on major social movements.

General Social Survey: Cumulative Codebook. Published by the National Opinion Research Center (NORC), University of Chicago. An annual survey that provides measures of the demographic characteristics and behaviors of a large sample of Americans. It is available through the data archives of most colleges and universities, along with other research reports published by the NORC.

INTERNET RESOURCES

Project on Social Research for Social Change, Cornell University (www.cornell.edu). Provides information about grassroots social movements, with links to many related sites.

The AFL-CIO Home Page (www.alfcio.org). An example of an older social movement organization that is adapting to new conditions and technologies.

U.S. Census Bureau: The Official Statistics (www.census.gov). The official site for the U.S. Census Bureau continually updates labor and social statistics. The current U.S. population count and current economic indicators are popular features of the site.

CHAPTER 9

POPULATION, URBANIZATION, AND COMMUNITY

Holly Sidford is part of a small band of activists who are changing the face of American cities. Sidford is responsible for the Urban Parks Initiative at the Lila Wallace Reader's Digest Foundation. Grants from the foundation are helping community leaders in cities throughout the United States do a better job of restoring and taking care of their parks and open spaces. Many cities are also making plans to expand their parklands, often after years of frustrating efforts. The urban parks activists are committed to making local and regional parks more accessible to people who, for one reason or another, have not used parks much but would like to if some changes were made. Sidford and her band of park activists are making these changes happen.

Sidford began her career as a guide in historic parks. She noticed that most visitors to those parks were people with high levels of education and the means to pay for relatively expensive vacation travel. Since then she has devoted herself to making cultural institutions like museums and historic places—and now urban parks—more inviting and accessible to people with less education and income. As a foundation executive who can disburse funds to worthy projects, she is in a position to help make positive changes occur. But money alone is not the answer. She knows from long experience that all the money in the world can be wasted quite quickly on ill-conceived projects or even on worthy ones that do not have the right combination of talented people to carry them out. A large part of her job and that of her staff at the Reader's Digest Foundation involves finding people who are already working on projects related to urban parks—projects that could be even more successful with some help from a major funder.

Why this emphasis on parks? With all the serious problems of cities—crowding, crime, poverty, illness, lack of adequate shelter, and environmental pollution, to name only a few—how do Sidford and others justify spending a large portion of the foundation's disposable funds on improving urban parks?

The Visual Sociology feature at the end of this chapter offers one answer to this question. Bruce Davidson's lyrical vision of people enjoying New York's Central Park almost makes one forget that a decade ago the city's leaders worried that the park was becoming deteriorated and dangerous. If a city's leaders do not see to it that there are green spaces and parks where people from different neighborhoods can relax outdoors or enjoy themselves at games of all kinds, the American urban environment becomes, as economist and former ambassador John K. Galbraith warned, a place of "private affluence and public squalor." Throughout the nineteenth century, inspiring thinkers like Henry David Thoreau, the author of *Walden,* Frederick Law Olmsted, the great American landscape architect who designed parks all over the United States, and Jane Addams, a founder of the field of social work, all realized that land needed to be set aside and used for public recreation and relaxation. Rich people could always afford to get away to the country, often to opulent estates with horses and private lakes. But the families of working people could afford no such luxuries. Unless some choice properties were taken off the market and set aside for use by all the city's residents, there would be no public green space at all, no place for less fortunate Americans in the rapidly growing cities to breathe cleaner air and picnic with their families on a grassy lawn.

The parks and recreation movement is still very much alive and is largely responsible for the fact that national, state, and municipal parks exist at all. But like any public good that is used by everyone, parks and open space are subject to a variety of threats. Municipal budget cuts may encourage those who wish to use urban green spaces for private, profit-making purposes or for improved transportation. A few sensational crimes in public spaces may discourage efforts to improve or restore them. Rapid changes in a city's population may remove old community leaders from the scene while new ones have yet to emerge on the public stage. These and many other obstacles to the maintenance and wise use of precious urban parks and open spaces are the problems toward which Sidford's Urban Parks Initiative is directed. Thanks to the work of local community activists, her grants are helping to restore and revitalize parks in Houston, New Orleans, Oakland, Chicago, New York, Cleveland, Portland, Boston, Buffalo, St. Louis, Louisville, and many other cities where parks and green spaces are essential to the quality of urban community life.

Sidford knows that it helps immensely to have money to give to worthy causes, but she also realizes the importance of people with energy and the sociological imagination necessary to understand that without public parks our quality of life would be greatly diminished. As we will see in this chapter, the rapid growth of the world's population, especially its urban populations, is raising questions about how best to meet the needs of billions of people throughout the world. In this context, efforts to ensure that urban residents have access to parks and other public places take on even greater significance.

PEOPLE, CITIES, AND URBAN GROWTH

The human population continues to spread throughout the world. Although the vast majority of the world's people live in rural villages, in modern societies they typically live in cities or metropolitan areas rather than on farms or in small towns. In fact, 90 percent of Americans live within 25 miles of a city center (Anderton, 1997; Berry, 1978). This means that they live in the human-built environment of cities and suburbs; relatively few live in the "country" environment of rural areas. It also means, among many other things, that settlement occurs in unstable places that invite disaster.

The study of population growth and the growth of cities as centers of demographic and social change has long been a central field in sociology. In this chapter, therefore, we will see how increasing population size and other important changes in populations are connected with the rise of cities and metropolitan areas throughout the world. At the end of the chapter we will return to the study of natural and social disasters, which are an increasingly frequent consequence of life on an urbanizing planet.

The Population Explosion

Before we can study the phenomenon of urbanization, we must understand the impact of population growth on the formation and expansion of cities. In Chapter 4 we described the extremely rapid growth of the world's population in the past three centuries. If you refer back to Figure 4.2, you can readily see that the *rate* of population growth increased dramatically in the twentieth century, giving rise to the often-used term *population explosion.* The world's population is currently estimated at 5.88 billion (McFalls, 1998), and according to recent United Nations forecasts it may increase to 8.7 billion by 2025 (Haub, 1997). This would represent growth over a 30-year period that is roughly equal to the entire world population in 1965 (Mitchell, 1998).

At present about 90 million new people are added to the world's population each year. Most forecasts predict that this rapid growth will begin to decline by mid-century, but by then there could be as many as 14 billion people on a very crowded planet (American Assembly, 1990). Demographers point out, however, that crop production, famine and war, contraception, economic development, and many other variables affect population. The potential effects of these variables make it extremely difficult to predict trends in population growth. It is certain, though, that rapid population growth, especially in the poorer regions of the world, will continue in coming decades and will pose severe challenges to human survival and well-being (Sanderson & Scherbov, 1997). The pressures of population growth on resources of food, space, and water have produced changes in economic and social arrangements throughout history, but never before have those changes occurred as rapidly or on as great a scale as they are occurring today (Cohen, 1998).

Malthusian Population Theory Is there a danger that the earth will become overpopulated? The debate over population growth is not new. For almost two centuries social scientists have been seeking to determine whether human populations will grow beyond the earth's capacity to support them. The earliest and most forceful theory of overpopulation appeared in Thomas Malthus's *An Essay on Population* (1927–1928/1798). Malthus attempted to show that population size normally increases far more rapidly than the food and energy resources needed to keep people alive. Couples will have as many children as they can afford to feed, and their children will do the same. This will cause populations to grow *geometrically* (2, 4, 8, 16, 32, . . .). Meanwhile, available food supplies will increase *arithmetically* (2, 3, 4, 5, 6, . . .) as farms are expanded and crop yields increased. As a result, population growth will always threaten to outstrip food supplies. The resulting poverty, famine, disease, war, and mass migrations will act as natural checks on rapid population growth.

History has proved Malthus wrong on at least two counts. To begin with, we are not biologically driven to multiply beyond the capacity of the environment to support our offspring; Malthus himself recognized that people could limit their reproduction through delay of marriage or celibacy. The second fault of Malthusian theory is its failure to recognize that technological and institutional change could expand available resources rapidly enough to keep up with population growth. This has occurred in the more affluent regions of the world, where improvements in the quality of life have tended to outstrip population growth. Improvements in agricultural technology have also increased the yield of crops in some of the less developed parts of the world, such as India. But rates of population growth and exhaustion of environmental resources (firewood, water, grazing land) are highest in the poorest nations (Furedi, 1997). Will reductions in population growth rates and increases in available resources also occur there? Or will the Malthusian theory prove correct in the long run? The theory known as the *demographic transition* provides a framework for studying this question, but before we explore this theory we need to know more about measuring population change.

Rates of Population Change Populations change as a consequence of births, deaths, out-migration, and in-migration. The relationships of these variables to a society's total population are expressed in the *basic demographic equation:*

$$P_t = P_0 + (B - D) + (M_i - M_o)$$

where

P_0 = the census count for an earlier period

P_t = the census count for the later period

B = total births between P_0 and P_t

D = total deaths between P_0 and P_t

M_o = out-migration between P_0 and P_t

M_i = in-migration between P_0 and P_t

Once they know the absolute values of the terms in this equation, demographers usually convert them into percentages in order to compare populations of different sizes. (We discussed the usefulness of such conversions in the discussion of percent analysis in Chapter 2.)

The most basic measures of population change are *crude rates,* or the number of events of a given type (e.g., births or deaths) that occur in a year divided by the midyear population (Bogue, 1969). Thus the **crude birthrate** (CBR) is the number of births occurring during a year in a given population divided by the midyear population, and the **crude death rate** (CDR) is the number of deaths occurring during a year divided by the midyear population. These fractions are usually expressed as a rate per thousand persons. They are "crude" because they compare the total number of births or deaths with the total midyear population, when in fact not all members of the population are equally likely to give birth or to die.

The **rate of reproductive change** is the difference between the CBR and the CDR for a given population. It is a measure of the *natural increase* of the population; that is, it measures increases due to the excess of births over deaths and disregards in- and out-migration. At present there are several nations in which the rate of reproductive change is zero or less, meaning that there is no natural population growth. Germany, for example, had a CBR of 10 and a CDR of 10 in 1998, for an annual rate of increase of zero. In Austria, the CBR was 11 and the CDR 10, for an annual rate of increase of 0.1 percent. In the United States, the rate of population growth is about 0.6 percent, representing an increase of about 1.6 million people per year. These rates are in dramatic contrast with the annual growth rates of countries like Liberia and Guatemala, which are above 3 percent. Rates of population growth in selected countries are listed in Table 9.1.

We can easily see from Table 9.2 that an annual rate of population growth of only 1 percent will lead to an increase of almost 270 percent in a century. Since World War II the world's population has been increasing at a rate of more than 1.5 percent—with projections surpassing 6 billion by the year 2000. The processes of population growth and the effects of its control are summarized in the phenomenon known as the demographic transition.

The Demographic Transition The **demographic transition** is a set of major changes in birth and death rates that has occurred most completely in urban industrial nations in the last 200 years. We saw in Chapter 4 that the rapid increase in the world population in the past century and a half was due in large part to rapid declines in death rates. Beginning in the second half of the eighteenth century and continuing until the first half of the twentieth, there was a marked decline in death rates in the countries of northern and western Europe. Improvements in public health practices were one factor in that decline. Even more important were higher agricultural yields owing to technological changes in farming methods, as well as improvements in the distribution of food as a result of better transportation, which made cheaper food available to more people (Bacci, 1997;

TABLE 9.1 Population Growth and Doubling Time, Selected Countries

Country	Population (millions)	Annual Growth Rate* (percent)	Doubling Time** (years)
India	988.7	1.9	37
China	1,242.5	1.0	69
Brazil	162.1	1.4	48
Bangladesh	123.4	1.8	38
Nigeria	121.8	3.0	23
Pakistan	141.9	2.8	25
Indonesia	207.4	1.5	45
Russia	146.9	–0.5	—
Mexico	97.5	2.2	32
United States	270.2	0.6	116

* Annual rate of natural increase.
** Number of years in which the population will double at current rate of increase.

Source: Population Reference Bureau, 1998.

TABLE 9.2 Relationship of Population Growth per Year and per Century

Population Growth per Year (percent)	Population Growth per Century (percent)
1	270
2	724
3	1,922

Source: Data from Worldwatch Institute.

Vining, 1985). At the same time, however, birthrates in those countries remained high. The resulting gap between birth and death rates produced huge increases in population. It appeared that the gloomy predictions of Malthus and others would be borne out.

In the second half of the nineteenth century, birthrates began to decline as couples delayed marriage and childbearing. As a result of lower birthrates, the gap between birth and death rates narrowed and population growth slowed. This occurred at different times in different countries (Figure 9.1 graphs the demographic transition in Sweden), but the general pattern was the same in each case: a stage of high birthrates and death rates (the *high growth potential* stage) followed by a stage of declining death rates (the *transitional growth* stage) and, eventually, by a stage of declining birthrates (called the stage of *incipient decline* because it is possible for population growth rates to decrease at this point). These three stages are summarized in the study chart on the following page.

In the second, or transitional growth, stage of the demographic transition, the population not only grows rapidly but undergoes changes in its age composition.

FIGURE 9.1 The Demographic Transition: Sweden, 1691–1963

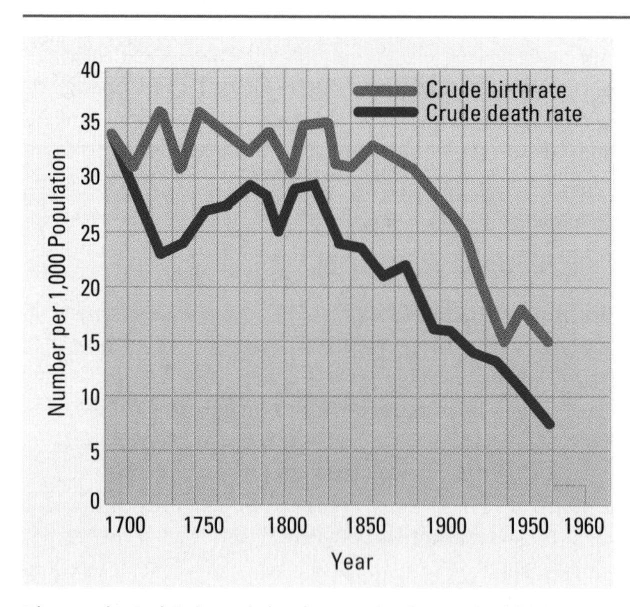

The peaks in birth and death rates in the early 1800s are due to social unrest and war. The drop in deaths and simultaneous rise in births in the early 1700s was a result of peace, good crops, and the absence of plagues.

Source: Judah Matras, *Population and Societies*, Englewood Cliffs, N.J., © Prentice-Hall, 1973. Permission of Armand Colin Éditeur.

Because people now live longer, there is a slight increase in the proportion of elderly people. There is also a marked increase in the proportion of people under age 20 as a result of significant decreases in infant and child mortality. This is the stage in which many less developed countries find themselves today: Death rates

STUDY CHART The Demographic Transition

STAGE OF POPULATION GROWTH	TYPE OF SOCIETY	MAIN DEMOGRAPHIC FEATURES
High Growth Potential	Most types of preindustrial societies.	High death rates due to infant mortality and low life expectancy; high rates of fertility; relatively low rates of increase.
Transitional Growth Stage	Most societies in early stages of urban and industrial development where basic public health measures are being introduced (e.g., safe drinking water systems).	Decreasing mortality rates; continuing high rates of fertility.
Incipient Decline	Societies in more advanced stages of urban and industrial development where people are delaying marriage and are more likely to use birth control.	Decreasing rates of mortality and fertility, with low or even negative rates of increase.

dropped in the twentieth century because of improved medical care and public health measures, as well as increased agricultural production. Yet in these societies birthrates have remained high, causing phenomenal increases in population, especially in the younger, more dependent age groups.

No population has entered the third stage of the demographic transition without limiting its birthrate. This can be achieved by encouraging couples to marry later and postpone childbearing or by preventing pregnancies or births through various birth control techniques. In advanced industrial societies, many couples use both approaches, making their own decisions about whether and when to have children. In other societies, such as China and India, the state has attempted to limit population growth by promoting birth control through educational programs or, in the case of China, by imposing penalties on couples who have more than a prescribed number of children. Such measures have had some transitory success, but often at a high political cost to national leaders. In China, for example, the strict one-child rule is not being enforced rigorously in areas undergoing rapid industrial and commercial development, where labor is in demand (Faison, 1997).

Note that economic and social development is essential if the demographic transition is to occur. Death rates cannot decrease, or food supplies increase, without progress in social institutions like public health, medicine, and transportation. People in more developed societies tend to limit their family size because they seek economic advancement and wish to delay marriage and childbearing until they can support a family.

In many highly industrialized nations, on the other hand, population growth rates have fallen below the rate necessary to maintain the population at the existing level. If a fertility rate of two children or less per couple became the norm for an entire population, the growth rate would slow to zero or even a negative rate. The United States, New Zealand, Japan, Australia, and Canada all have total fertility rates of 2.0 or less (Population Reference Bureau, 1998). Similar low rates are appearing in many European nations. This is a highly significant development; if it continues over a generation or more, it could result in negative rates of natural increase and, possibly, other consequences such as slower economic development and labor shortages. Continued low fertility could also result in increased immigration from countries with high birthrates to countries with lower birthrates.

A full exposition of the processes and politics of population control is not possible here. Suffice it to say that most sociologists would agree with the demographer Philip Hauser (1957) that low population growth rates are due primarily to a combination of delay of marriage, celibacy, and use of modern birth control techniques. This trend is beginning to occur in many less developed nations (Furedi, 1997; Ofosu-Amaah, 1998).

Figure 9.2 illustrates the demographic transition in Singapore, a nation in Southeast Asia that has undergone this important population change more recently. Note that the decline in deaths, which took almost a century and a half in Western industrial nations, took less than half a century in Singapore. The decline in births began in the late 1950s, spurred by rapid economic development, the new desire for smaller families, and widespread availability of contraceptives. Whereas in Western nations like Sweden this phase of

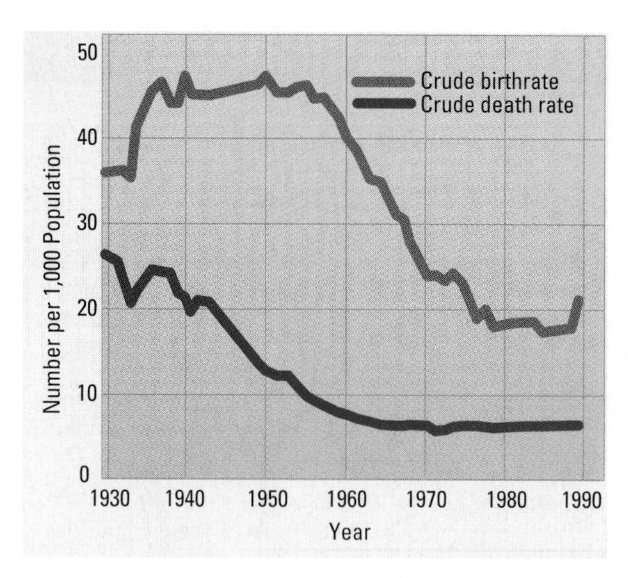

Source: Davis, 1991.

FIGURE 9.2 **Crude Birth and Death Rates: Singapore, 1930–1988**

declining births unfolded over more than a century, in Singapore this phase occurred in only about 20 years (between about 1955 and 1975). Although world population growth remains a serious problem, the demographic transition in formerly poor nations like this one shows that population control efforts along with economic development can have dramatic effects (Davis, 1991).

Population Growth and Urbanization Studies of the demographic transition in Europe (Bacci, 1997; Laslett, 1972, 1983) have concluded that its specific course in any given society depends on a complex combination of factors: higher age at marriage and fewer couples marrying, use of birth control techniques, increased education, migration to other countries, and rural–urban migration. Some of these factors are discussed in Box 9.1. In each case, however, the bulk of the population growth was absorbed by the cities. Rapid urbanization thus to some extent is an outgrowth of the demographic transition.

Urbanization refers to the proportion of the total population that is concentrated in urban settlements. Although urbanization and the growth of cities have occurred together, it is important to distinguish between the two. Cities can continue to grow even after the majority of the population is urbanized and the society's dominant institutions (e.g., government agencies, major markets, newspapers, television networks) are located in its urban centers.

Urbanization contributes to the lower birthrates that are characteristic of the third stage of the demographic transition. As Louis Wirth pointed out, "The decline in the birthrate generally may be regarded as one of the most significant signs of the urbanization of the Western world" (1968/1938, p. 59). In the city a variety of factors lead to the postponement of marriage and childbearing. For one thing, living space is limited. For another, newcomers to the city must find jobs before they can even think about marrying, and often they lack the ties to family and kin groups that might encourage them to marry and have children.

Studies of declining birthrates in non-Western countries find that—contrary to the experience of cities in the West, where urbanization has brought declining birthrates—the populations of some cities in Asia and Africa are increasing faster than those of rural areas. The reason seems to be that rapid change in rural areas, especially the mechanization of agriculture, pushes rural people to the cities, where they often live in shantytowns and villagelike settlements. In those settlements the rural tradition of large families is not quickly altered. In addition, because they have greater access to health care during pregnancy and childbirth, the infant death rate may be lower among the migrants than in rural villages. However, when these populations become part of the urban economy, they too begin to limit their family size (Gugler, 1997; Lowey, 1990).

In sum, urbanization is closely linked with rapid increases in population, yet the nature of life in cities tends to limit the size of urban families. Cities grow primarily as a result of migration, but new migrants do not find it easy to form families. Thus, in his research on the changing populations of Western industrial cities, Hauser (1957) showed that birthrates were lowest in the areas of the city that had the highest proportions of new migrants. There is also mounting evidence from archaeology that child sacrifice in ancient cities may have helped control urban populations (Browne, 1987). The eventual result of large-scale migration to cities may be a slowdown in population growth: As an increasing proportion of a society's population lives in cities, the rate of growth of the population as a whole tends to drop.

The Growth of Cities

In Chapter 4 we noted that cities became possible when agricultural populations began to produce enough extra food to support people who were not directly engaged in agriculture, such as priests, warriors, and artisans. Changes in the technology of food production made it possible for increasingly larger populations to be supported by the same number of agricultural workers. This has been a central factor in the evolution of cities, but as we will see shortly, the most dramatic increases in urban populations have occurred only in the past 150 years.

BOX 9.1 GLOBAL CHANGE AND U.S. SOCIETY

The Power of Rural–Urban Migration

Immigration and migration are social forces that continually bring about change of all kinds. In Eastern Europe, for example, the animosities among nationality groups that threaten to cause war and violence can often be traced to earlier periods of immigration and resettlement of populations due to war and conquest. In the United States as well, immigration continually produces social change—especially in the major cities that serve as the gateways for newcomers.

Migration, which includes both immigration from a foreign nation and movement within national boundaries, is, along with birth and death rates, a key factor affecting population size. People have migrated from one region or country to another throughout history, compelled to seek new homes because they have exhausted their food supplies or been driven out by invaders or because they believe they will find a better life in the new land. Although people have migrated from rural areas to the city throughout history, rural–urban migration has become especially pronounced in the last two centuries.

The forces that impel residents of rural areas to migrate to cities are not fully understood. There is continual debate over whether rural people are "pushed" into cities by conditions beyond their control or "pulled" to cities by the attractions of city life. Of course, both push and pull factors affect rural–urban migration, but it is not clear exactly how they interact. It appears, however, that "it is the push of existing rural circumstances which suggests to the rural resident that things might be better in the urban area" (Breese, 1966, p. 80).

Several factors may be responsible for pushing people out of rural areas. Chief among these is overpopulation, which reduces the amount of food and work available per rural resident. Others include lack of opportunities to obtain farmland (e.g., as a result of *primogeniture,* in which only the oldest son inherits land) and the seasonal nature of employment in agriculture. When rural people who are experiencing these conditions become aware of higher living standards in urban places, they may develop a sense of relative deprivation (see Chapter 8). "They view with great interest the reported higher income, access to education, and other rumored facilities of the urban area" (Breese, 1966, pp. 80–81). The feeling of relative deprivation is intensified by improvements in communication and transportation that provide rural people with more feedback about the advantages of life in the city.

A significant pull factor is the presence of relatives and friends in the city. These individuals can be called on for help when the rural migrant arrives in search of a new home and a job. This factor produces what is known as *chain migration,* a pattern in which a network of friends and relatives is transferred from the village to the city over time. The result is the formation of a small, homogeneous community within the city.

Chain migration operates across national boundaries as well as between rural and urban areas in the same country. For example, a large proportion of the immigrants to American cities in the twentieth century were rural people from European and Asian countries. Once a few hardy souls establish a foothold, their families and friends can join them. In this way small ethnic communities are formed within the city—like the Chinatowns of New York and San Francisco, the Slovenian community in Cleveland, or the Mexican community in Chicago. Recently this pattern has been established in suburban portions of metropolitan areas as well as in central cities. Thus the suburbs of metropolitan areas in the South and West have become home to large numbers of immigrants from Vietnam (Portes & Rumbaut, 1996).

These Mexican immigrants are crossing the Rio Grande in hope of finding work in the United States.

The Urban Revolution The increasing tendency of people throughout the world to live in cities has been referred to as the *urban revolution.* The extent of this "revolution" can be grasped by comparing a few figures. In 1800, only 3 percent of the world's people lived in cities with populations above 5,000, and of this proportion, a mere 2.4 percent lived in cities with populations above 20,000. Between 1800 and 1970, a period during which the world's population increased fourfold, the percentage of people living in cities with 5,000 or more inhabitants increased elevenfold, whereas that of people living in cities with 100,000 or more inhabitants increased almost fourteenfold. By 1970, fully one third of the world's population lived in cities (Gugler, 1997; Vining, 1985).

These data indicate not only that increasing percentages of the world's population are living in cities but also that the cities themselves are larger than ever before. The growth of cities in this century has given rise to the concept of the **metropolitan area,** in which a central city is surrounded by a number of smaller cities and suburbs that are closely related to it both socially and economically. Most people in the United States live in large metropolitan areas, as Box 9.2 demonstrates.

Large-scale urbanization is a relatively recent development in human history. As Kingsley Davis (1955) pointed out:

> Compared to most other aspects of society—such as language, religion, stratification, or the family—cities appeared only yesterday, and urbanization, meaning that a sizable proportion of the population lives in cities, has developed only in the last few moments of man's existence. (p. 429)

Although there were a few cities as early as 4000 B.C., they were very small and were supported by large rural populations. The famous cities of ancient times were minuscule by modern standards: Babylon covered roughly 3.2 square miles, Ur some 220 acres (Davis, 1955).

Preindustrial cities like Ur and Nineveh, early Athens and Rome, and the ancient Mayan cities were vastly different from the cities we know today. They did not grow around a core of office buildings and retail outlets the way industrial cities do. Instead, they were built around temples or other ceremonial buildings (e.g., Notre Dame in Paris). Close by the temple one could find the palaces of the rulers and the courtyards of the royal families. Parade grounds and public shrines made the core of the ancient city a spacious place where the city's population could meet on special occasions. On the outskirts of these ancient cities one found not the rich, as in contemporary cities with well-developed transportation networks, but the poor, who lived in hovels and were often pushed from one location to another according to the whims of those with more wealth and power (Gugler, 1997; Sjoberg, 1968).

A variety of factors limited the size of cities. Among them were farming methods that did not produce enough surplus food to feed many city dwellers, the lack of efficient means of transporting goods over long distances, inadequate technology for transporting water in great quantities, and the lack of scientific medicine (which made urban living, as Davis put it, "deadly"). Not until about 1800 did large-scale urbanization become possible.

The speed with which urbanization has changed the size and layout of cities is remarkable: "Before 1850 no society could be described as predominantly urbanized, and by 1900 only one—Great Britain—could be so regarded. Today . . . all industrial nations are highly urbanized, and in the world as a whole the process of urbanization is accelerating rapidly" (Davis, 1968, p. 33). By 1990, over 45 percent of the world population lived in urban areas, and by the year 2000 this figure should be nearly half—40 percent of the population in developing countries and 78 percent in developed countries (*Statistical Abstract,* 1997). Africa now has the fastest rate of urbanization; and as the twenty-first century begins, about 70 percent of the population of Latin America will live in urban areas—a rate of urbanization comparable to that of North America and Europe.

As Table 9.3 (p. 266) shows, many metropolitan regions in older advanced nations like the United States (especially Greater New York), Great Britain (Greater London), and Germany (Rhein–Ruhr) reached their peak of growth decades ago and are now growing slowly if at all. In contrast, in many parts of Latin America and Asia metropolitan regions are experiencing explosive growth and attracting vast populations of urban newcomers, many of whom are living as squatters on vacant land.

Rapid urbanization occurring throughout the world brings together diverse groups of people in cities that often are not prepared to absorb them. The problems caused by such urbanization are immense. They include housing, educating, and caring for the health of newcomers; preventing gang violence and intergroup hatred; and many other difficult tasks. Moreover, as the world becomes ever more urbanized, populations become increasingly interdependent. Urban populations are supported, for example, by worldwide agricultural production, not just by the produce grown in the surrounding countryside. In the same way, the problems of one major city or one large urbanizing region can no longer be thought of as isolated from the problems of the older, more affluent urbanized regions.

Urban Societies Urbanization produces *urban societies.* By this we mean not only that cities are the

BOX 9.2 SOCIOLOGICAL METHODS

Defining Metropolitan Areas

Formal definitions of urban places changed considerably in the twentieth century. Traditionally, an urban place had been defined as any incorporated area with 2,500 or more inhabitants. But this definition became highly unsatisfactory as cities expanded and their surrounding areas became increasingly urbanized, especially after World War II. As early as 1910 the U.S. Bureau of the Census introduced the concept of the *metropolitan district,* or metropolitan area. This concept has been refined quite often as urban growth has engulfed the rural areas adjacent to cities. Moreover, as cities and their suburbs have begun to run into each other, it has become necessary to distinguish between central cities and the surrounding metropolitan areas.

In 1984 the Census Bureau established new definitions of urban areas that remain in effect today. The basic types of urban areas are the metropolitan statistical area, the primary metropolitan statistical area, and the consolidated metropolitan statistical area. A *metropolitan area* is a geographic area with a population of at least 100,000 consisting of a nucleus with a population of at least 50,000 together with adjacent communities that have a high degree of economic and social integration with that nucleus. The term *statistical areas* refers to the fact that these definitions are used to organize numerical data about urban regions. Federal, state, and local government agencies use these geographic and statistical definitions to monitor social change and carry out essential planning functions.

A *metropolitan statistical area (MSA)* is a metropolitan area that is not closely associated with other metropolitan areas and typically is surrounded by nonmetropolitan counties. A *primary metropolitan statistical area (PMSA)* is defined as a large urbanized county or cluster of economically interdependent counties with a population of at least 100,000 that is part of a metropolitan area with more than 1 million inhabitants. A metropolitan area that contains two or more PMSAs is a *consolidated metropolitan statistical area (CMSA).* A CMSA therefore is equivalent to what sociologists often call a "megalopolitan region," such as the Dallas–Fort Worth area or the Los Angeles area.

Central cities are the officially designated largest and most densely populated places in metropolitan areas. There are central cities that are not included in a metropolitan area.

The accompanying graph shows the proportions of the U.S. population living in rural, urban, metropolitan, and nonmetropolitan areas. The organization of data shown in the graph is useful because the curve above the area for each category shows its relative trend in population growth. In order to find out how many millions of people lived in a metropolitan area in 1980, for example, subtract the reading at the bottom of the area for that year from the reading at the top of the area for the same year. For 1980 the line at the bottom of the area falls at about 50 million and the line at the top at about 225 million; thus there were about 175 million residents of metropolitan areas in the United States in that year.

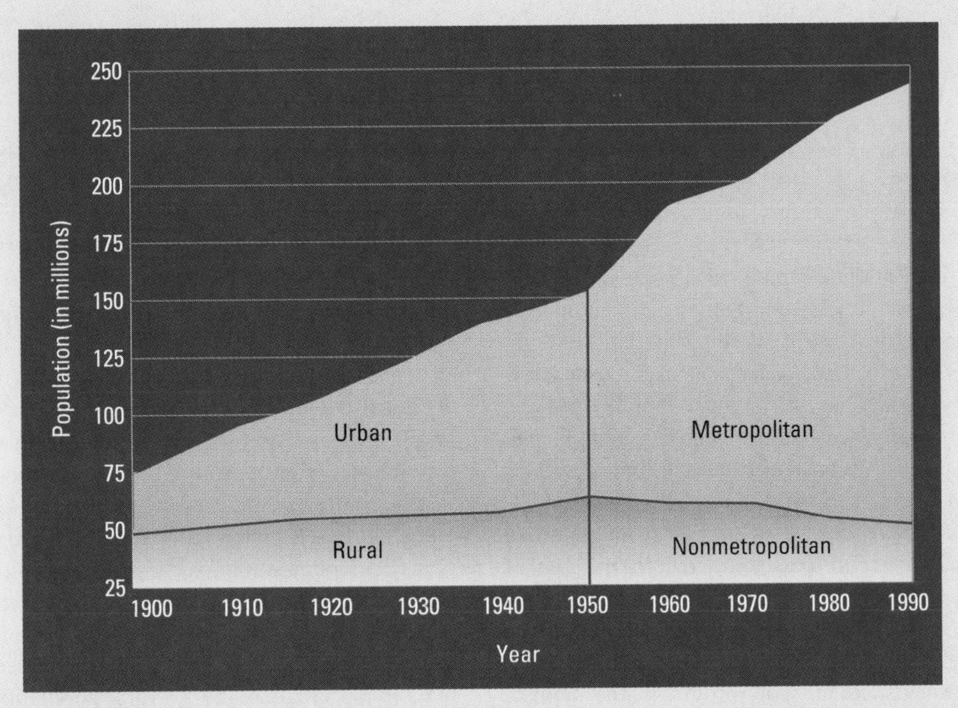

Source: Council on Environmental Quality, Census Bureau.

TABLE 9.3 Populations of the World's 30 Largest Urbanized Areas, 1950–2000

Rank in Year 2000	Country	Urbanized Area	Thousands of Inhabitants 1950	1975	2000
1	Mexico	Mexico City	3,050	11,610	25,820
2	Japan	Tokyo–Yokohama	6,736	17,668	24,172
3	Brazil	São Paulo	2,760	10,290	23,970
4	India	Calcutta	4,520	8,250	16,530
5	India	Bombay	2,950	7,170	16,000
6	United States	NewYork–N.E. New Jersey	12,410	15,940	15,780
7	Rep. of Korea	Seoul	1,113	6,950	13,770
8	Brazil	Rio de Janeiro	3,480	8,150	13,260
9	China	Shanghai	10,420	11,590	13,260
10	Indonesia	Jakarta	1,820	5,530	13,250
11	India	Delhi	1,410	4,630	13,240
12	Argentina	Buenos Aires	5,251	9,290	13,180
13	Pakistan	Karachi	1,040	4,030	12,000
14	Iran	Teheran	1,126	4,267	11,329
15	Bangladesh	Dhaka	430	2,350	11,160
16	Egypt	Cairo–Ginza–Imbaba	3,500	6,250	11,130
17	Iraq	Baghdad	579	3,830	11,125
18	Japan	Osaka–Kobe	3,828	8,649	11,109
19	Philippines	Manila	1,570	5,040	11,070
20	United States	Los Angeles–Long Beach	4,070	8,960	10,990
21	Thailand	Bangkok–Thonburi	1,440	4,050	10,710
22	United Kingdom	London	10,369	10,310	10,510
23	USSR	Moscow	4,841	7,600	10,400
24	China	Beijing	6,740	8,910	10,360
25	Germany	Rhein–Ruhr	6,853	9,311	9,151
26	Peru	Lima–Callao	1,050	3,700	9,140
27	France	Paris	5,525	8,620	8,720
28	Nigeria	Lagos	360	2,100	8,340
29	Italy	Milan	3,637	6,150	8,150
30	India	Madras	1,420	3,770	8,150

Note: The urban places listed above are areas of dense contiguous settlement whose boundaries do not necessarily coincide with administrative boundaries; they usually include one or more cities and an urbanized fringe. The list includes all such places projected to contain at least 8 million inhabitants in the year 2000.

Source: Lowey, 1993.

cultural and institutional centers of a society but also that urban life has a pervasive influence on the entire society (Durkheim, 1964/1893; Weber, 1962/1921; Wirth, 1968/1938). Today the United States is spanned by interstate highways that link the nation's rapidly growing urban and suburban places and carry traffic through rural areas at high speeds. Waterways, forests, hills, and valleys are channeled and cut and bulldozed to make way for expanding settlements. Once considered far from the urban scene, national parks and forests now receive millions of visitors from the metropolitan centers. And in an urban society more and more people, even those living in isolated rural communities, share in the mass culture of that society—the television and radio programs, the movies, the books and magazines, all of which stress themes that appeal to people who are familiar with metropolitan living. In an urban society not everyone lives in the cities, but no one can escape the pervasive influence of urban centers.

The concept of an urban society may become clearer if we look at a society that has not become fully urbanized, such as India. Until the early twentieth century the rate of city growth in India was relatively slow. But beginning in the 1920s the populations of India's cities increased dramatically, and by 1960 seven Indian cities had populations surpassing a million. Yet India has not become an urban society. It remains, in the words of Noel P. Gist (1968), "a land of villagers." Because of persistently high birthrates and declining death rates in the villages, India's rural population is growing almost as fast as its urban population. And although the growth of the cities has affected village life, especially

Urban sociologists have always looked at the relationship between the planned and unplanned aspects of urban growth and the social groups that represent each aspect. Usually the poor and the immigrants represent unplanned additions to the city, while the rich and the powerful guide the processes of urban planning and organization to suit their interests.

through the efforts of city-trained teachers, health officials, administrators, and storekeepers, it has had little effect on the social structure of rural India. Thus, as a whole, Indian society exhibits the extremes of rural isolation and urban dynamism, with all the chaos and poverty that are characteristic of societies undergoing major social change (World Bank, 1997).

Throughout the twentieth century sociologists devoted considerable study to urbanization and the changes that accompany it. The ways in which the growth of cities and metropolitan regions alters the surface of the planet are part of the sociological study of urbanization. So are questions about how cities change our experience of community and our relationships with others. Finally, sociologists continue to focus on the changing patterns of inequality and conflict that occur in metropolitan regions. The remainder of this chapter explores these issues in greater detail.

THE URBAN LANDSCAPE

Urban Expansion

The effects of urbanization began to be felt in American society in the mid-nineteenth century. As growing numbers of people settled in the West, waves of immigrants from Ireland, Germany, Italy, and many other parts of central and southern Europe streamed into the cities of the East. In the 1840s and 1850s, for example, approximately 1.35 million Irish immigrants arrived in the United States, and in just 12 years, from 1880 to 1892, more than 1.7 million Germans arrived (Bogue, 1985). In other chapters we have more to say about the impact of the great migrations from China and Korea and Latin America, the importation of slaves from Africa, and the large numbers of people of all races and ethnic groups who continue to arrive in U.S. cities. The point here is that for a century and a half, North American cities have been a preferred destination of people from all over the world; as a result, they have received numerous waves of newcomers since their period of explosive growth in the nineteenth century.

The science of sociology found early supporters in the United States and Canada partly because the cities in those nations were growing so rapidly. It often appeared that North American cities would be unable to absorb all the newcomers who were arriving in such large numbers. Presociological thinkers like Frederick Law Olmsted, the founder of the movement to build parks and recreation areas in cities, and Jacob Riis, an advocate of slum reform, urged the nation's leaders to invest in improving the urban environment, building parks and beaches, and making better housing available to all (Cranz, 1982; Kornblum & Lawler, 1999). As we saw in Chapter 1, these reform efforts were greatly aided by sociologists who conducted empirical research on the social conditions in cities. In the early twentieth century many sociologists lived in cities like Chicago that were characterized by rapid population growth and serious social problems. It seemed logical to use empirical research to construct theories about how cities grow and change in response to major social forces as well as more controlled urban planning.

The founders of the Chicago school of sociology, Robert Park and Ernest Burgess, attempted to develop a dynamic model of the city, one that would account not only for the expansion of cities in terms of population and territory but also for the patterns of settlement and land use within cities. They identified several factors that influence the physical form of cities. Among them are "transportation and communication, tramways and

Then and Now

Street Urchins and Trailer Kids

A century ago, as millions of newcomers from throughout the world crowded into U.S. cities in search of opportunity, rapid social change produced some of the world's most notorious urban slums. Jacob Riis, a photojournalist and amateur sociologist, brought the harsh living conditions in those slums to the attention of the world in his famous photos of "how the other half lives." Children who spent much of their time on the streets and survived by their wits were called "street urchins" (or, as Riis called them, "Street Arabs").

In today's world the central city is no longer viewed as a place to find new opportunities. Mobile families and individuals travel long distances, seeking manufacturing and service jobs outside major city centers. They often find what they hope will be temporary lodging in trailer parks. Poor and lacking health insurance and other forms of social security, residents of trailer parks tend to be stigmatized by those who are more fortunate. They and their children are sometimes called "trailer trash," a term that reveals the deep prejudices that continue to divide Americans, even in a society that is far more affluent than the one described by Riis.

telephones, newspapers and advertising, steel construction and elevators—all things, in fact, which tend to bring about at once a greater mobility and a greater concentration of the urban populations" (Park, 1967/1925a, p. 2). The important role of transportation is described in one of Park's essays:

> The extent to which . . . an increase of population in one part of the city is reflected in every other depends very largely upon the character of the local transportation system. Every extension and multiplication of the means of transportation connecting the periphery of the city with the center tends to bring more people to the central business district, and to bring them there oftener. This increases the congestion at the center; it increases, eventually, the height of office buildings and the values of the land on which those buildings stand. The influence of land values at the business center radiates from that point to every part of the city. (1967/1926, pp. 57–58)

The Concentric-Zone Model Park and Burgess based their model of urban growth on the concept of *natural areas*—that is, areas in which the population is relatively homogeneous and land is used in similar ways without deliberate planning. In Park's words:

> Every great city has its racial colonies, like the Chinatowns of San Francisco and New York, the Little Sicily of Chicago. . . . Most cities have their segregated vice districts . . . their rendezvous for criminals of various sorts. Every large city has its occupational suburbs, like the Stockyards in Chicago, and its residential enclaves, like Brookline in Boston. (1967/1925a, p. 10)

Park and Burgess saw urban expansion as occurring through a series of "invasions" of successive zones or areas surrounding the center of the city. For example, migrants from rural areas and other societies "invaded" areas where housing was cheap. Those areas tended to be close to the places where they worked. In turn, people who could afford better housing and the cost of commuting "invaded" areas farther from the business district, and these became the Brooklines, Gold Coasts, and Greenwich Villages of their respective cities.

Park and Burgess's model, which has come to be known as the *concentric-zone model,* is portrayed in Figure 9.3. (Figure 9.4 applies the model to Chicago.) Because the model was originally based on studies of Chicago, its center is labeled "Loop," the term that is commonly applied to that city's central commercial zone. Surrounding the central zone is a "zone in transition," an area that is being invaded by business and light manufacturing. The third zone is inhabited by workers who do not want to live in the factory or business district but at the same time need to live reasonably close to where they work. The fourth or residential zone consists of higher-class apartment buildings and single-family homes. And the outermost ring, outside the city limits, is the suburban or commuters' zone; its residents live within a 30- to 60-minute ride of the central business district (Burgess, 1925).

Studies by Park, Burgess, and other Chicago school sociologists showed how new groups of immigrants tended to become concentrated in segregated areas within inner-city zones, where they encountered suspicion, discrimination, and hostility from ethnic groups that had arrived earlier. Over time, however, each group was able to adjust to life in the city and to find a place for itself in the urban economy. Eventually many of the immigrants were assimilated into the institutions of American society and moved to desegregated areas in outer zones; the ghettos they left behind were promptly occupied by new waves of immigrants (Kasarda, 1989).

Note that each zone is continually expanding outward. Thus, Burgess wrote, "If this chart is applied to Chicago, all four of these zones were in its early history included in the circumference of the inner zone, the present business district. The present boundaries of the [zone in transition] were not many years ago those of the zone now inhabited by independent wage-earners" (1925, p. 50). Burgess also pointed out that "neither Chicago nor any other city fits perfectly into this [model]. Complications are introduced by the lake front, the Chicago River, railroad lines, historical factors in the location of industry, the relative degree of the resistance of communities to invasion, etc." (pp. 50–51).

The Park and Burgess model of growth in zones and natural areas of the city can still be used to describe patterns of growth in cities that were built around a central business district and that continue to attract large numbers of immigrants. But this model is biased toward the commercial and industrial cities of North America, which have tended to form around business centers rather than around palaces or cathedrals, as is the case in so many other parts of the world. Moreover, it fails to account for other patterns of urbanization, such as the rise of satellite cities and the rapid urbanization that occurs along commercial transportation corridors.

Satellite Cities Outside the city of Detroit lies the town of River Rouge, long famous as a center of steel and automobile production. Outside the city of Toronto lies Hamilton, also a smoky manufacturing town where

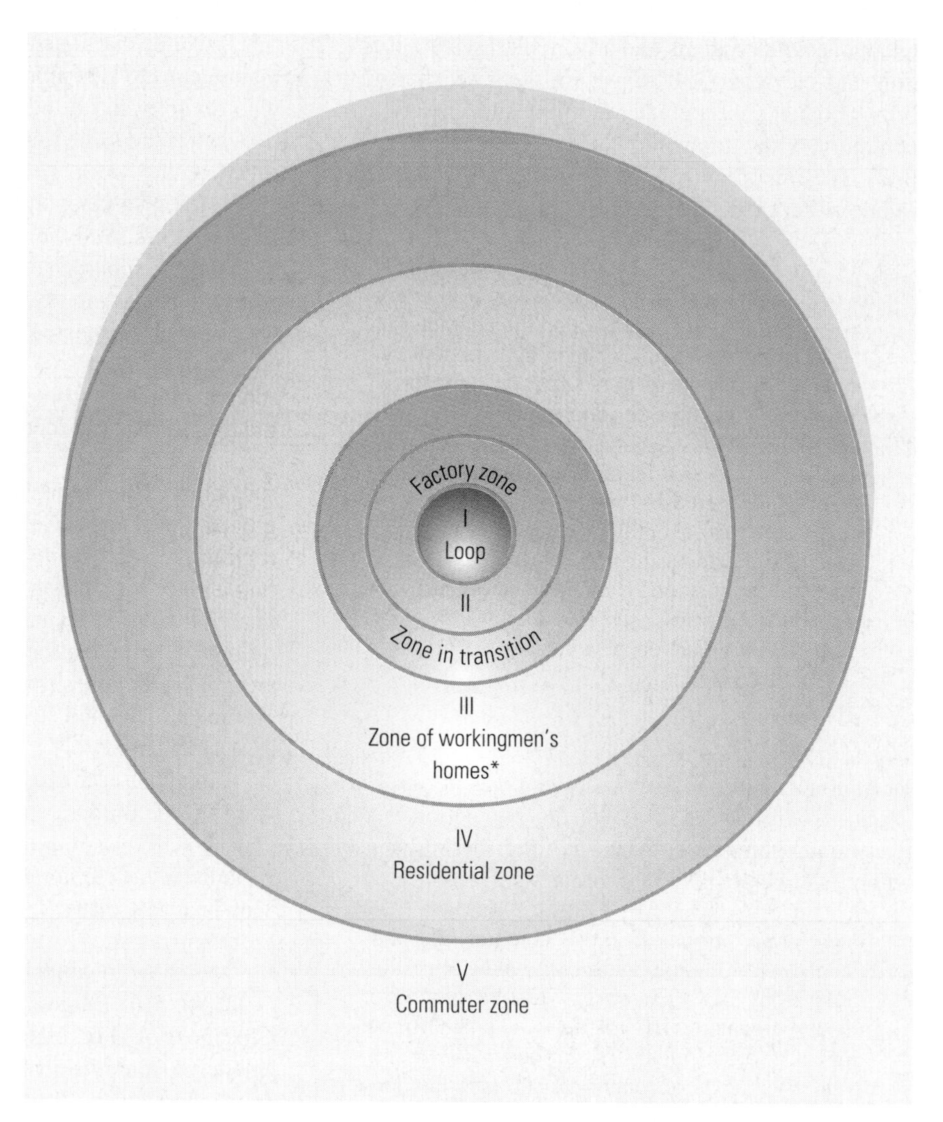

*Workers' homes ("workingmen" was the accepted terminology of the time).

Source: Park & Burgess, 1967/1925.

FIGURE 9.3 **The Concentric-Zone Model**

much of Canada's steel is produced. And outside Chicago is Gary, Indiana, another major center of heavy industrial production. Outside New York City there are many other *satellite cities,* some devoted to heavy industries whose needs for space, rail and water service, and energy make it impossible to locate them in the central business districts.

Other satellite cities are devoted to less environmentally stressful industries. Fort Worth, Texas, for example, is becoming a major center for white-collar industries, such as publishing, that are leaving older central-city locations in search of cheaper space and a well-educated labor force. Still others, like Tysons Corner, Virginia, near Washington, D.C., have grown around large shopping centers and mall complexes. In the second half of the twentieth century the growth of these and other satellite cities was accelerated by public investment in the interstate and metropolitan highway systems, a point to which we will return shortly in discussing the rise of metropolitan urban systems (Baldassare, 1986; Clay, 1994; Fishman, 1987).

Strip Development The growth of satellite cities specializing in the production of a particular commodity or product is typical of the period of rapid industrial growth that occurred before World War II. A more current model of urbanization is known as *strip develop-*

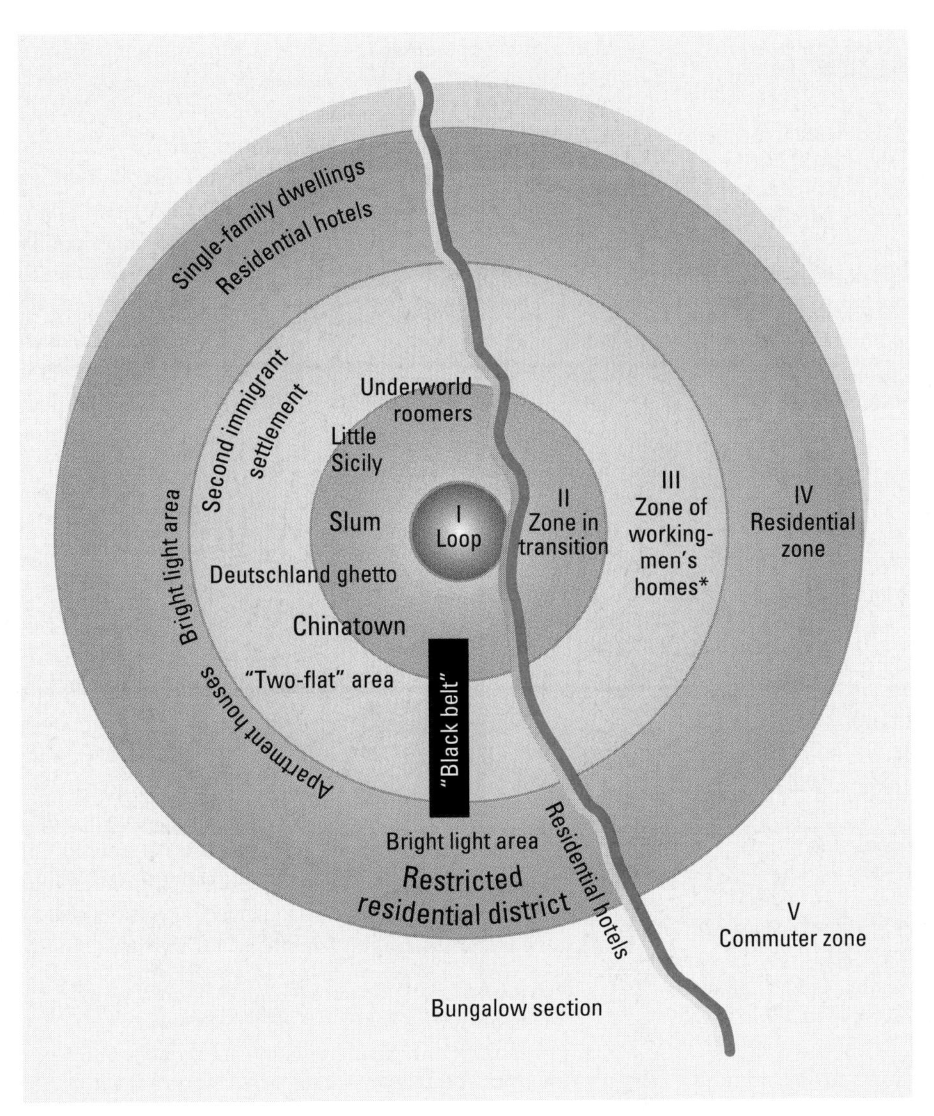

*Workers' homes ("workingmen" was the accepted terminology of the time).

Source: Park & Burgess, 1967/1925.

FIGURE 9.4 **The Concentric-Zone Model Applied to Chicago**

ment; it is shown in schematic form in Figure 9.5. Strip I represents a typical nineteenth- or early twentieth-century farming area bisected by a road and a stream. In time the bridge over the stream and the path along the stream became a road that intersected the original road (Strip II), creating an intersection that stimulated the growth of a village with a mill; families living in the village depended on wages from the mill. As the village grew into a town, population growth and increasing automobile traffic created the need for wider roads and bypass roads, as shown in Strips III and IV. Specialization as an "automobile convenient" strip, with the development of more motels, a drive-in theater, and drive-in businesses, can be seen in Strip V. At this time the residential functions along the strip began to disappear and the area assumed an increasingly commercial character.

Strip VI shows the strip's present stage of development. A spur of the limited-access highway system creates new growth around a cloverleaf (including a new community college situated between the artificial lake and the shopping center). Note that along the original main road are office buildings and a modern industrial park rather than the smaller retail businesses that once defined the town center. Indeed, as the strip has developed and surrounding roads have been enlarged to form part of the metropolitan highway system, the town center itself has largely disappeared (Clay, 1980). The strip development

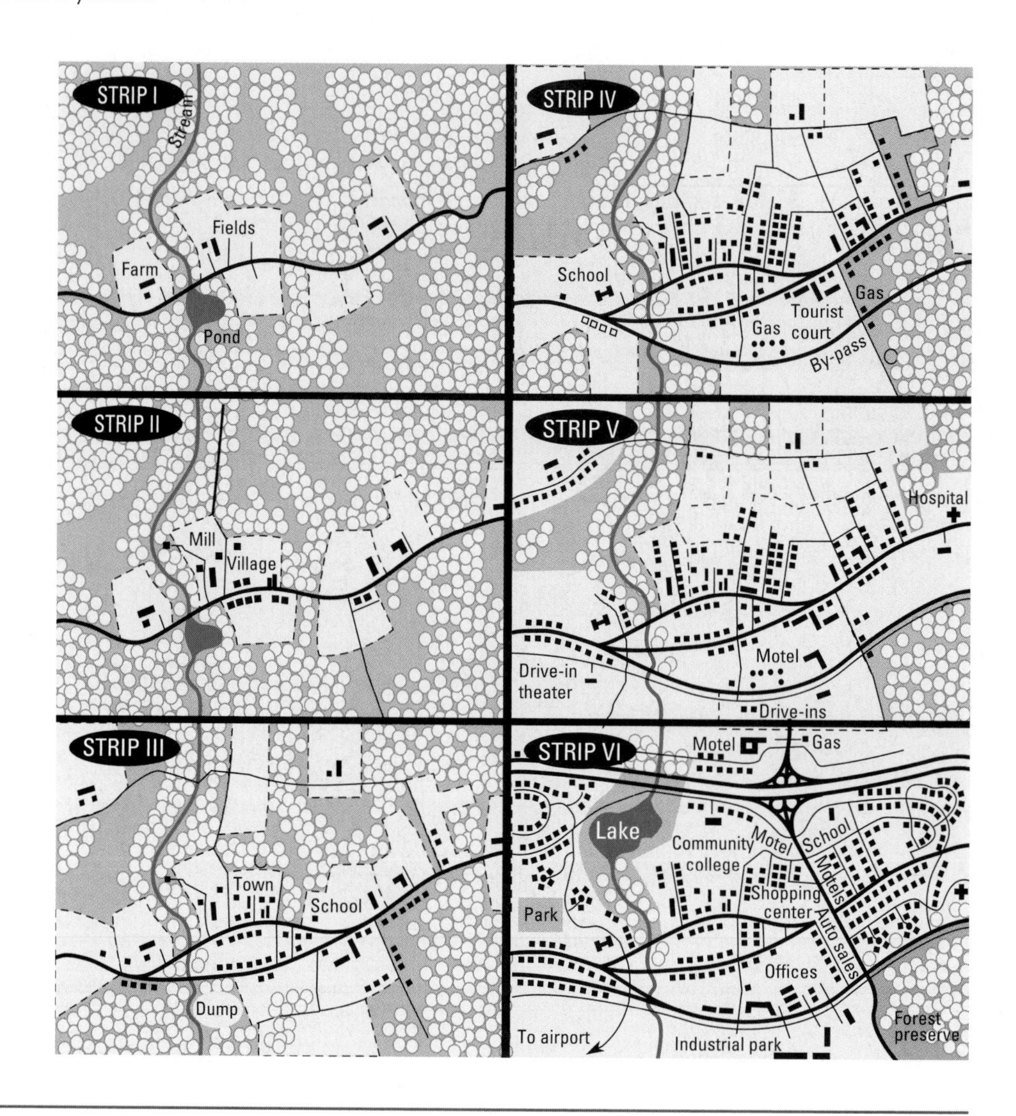

Source: Clay, 1980.

FIGURE 9.5 **Strip Development**

model thus describes the incorporation of smaller communities and towns into a larger metropolitan area.

Metropolitan Areas

Megalopolis After 1920 new metropolitan areas developed largely as a result of the increasing use of automobiles and the construction of a network of highways covering the entire nation (Flink, 1988). The shift to automobile travel brought former satellite cities within commuting distance of the major industrial centers, thereby adding to the size of those metropolitan areas. In the South and Southwest, new metropolitan areas developed. In fact, in recent years these have become the fastest growing urban areas in the nation.

Since World War II sociologists have been studying an increasingly important urban phenomenon: the emergence of large multinuclear urban systems. The term **megalopolis** is used to describe these vast complexes, whose total population exceeds 25 million. Jean Gottmann (1978) pointed out that a megalopolis is not "simply an overgrown metropolitan area"; rather, it is a system of cities distributed along "a major axis of traffic and communication" (p. 56). Gottmann identified six megalopolises: the American Northeastern megalopolis, the Great Lakes megalopolis, the Tokaido megalopolis in Japan, the megalopolis in England (the London area), the megalopolis of northwestern Europe (extending from Amsterdam to the Ruhr), and the Urban Constellation in China (centered on Shanghai). Four others are developing rapidly: the Rio de Janeiro–São Paulo complex in Brazil, the Milan–Turin–Genoa triangle in northern Italy, the Valley of Mexico, and the urban swath extending from San Diego to the San Francisco Bay area.

A megalopolis is characterized by an "intertwined web of relationships between a variety of distinct urban centers . . . expressed partly in a physical infrastructure consisting of highways, railways, waterways, telephone lines, pipelines, water supply and sewage systems crisscrossing the whole area, and partly in more fluid networks, such as the flows of traffic, the movement of people and goods, the flows of telephone calls [and] of mail" (Gottmann, 1978, p. 57). Despite their interdependence, however, "the sizes and specializations of the various . . . components [of a megalopolis] are extremely varied, as demonstrated by the diverse characteristics of the cities, towns, villages, suburban and rural areas that form the vast system" (p. 57). Therefore a megalopolis can best be described as a huge social and economic mosaic.

The recent history of Los Angeles provides a good example of the development of a megalopolis (see Figure 9.6). Between 1960 and 1970 the population of Los Angeles increased by more than 2 million, double the growth of Chicago and more than that of New York and San Francisco put together (Smith, 1968). Today the Los Angeles metropolitan area continues to grow, although at a somewhat slower rate. As a result of its extraordinary growth, the region must struggle to control the effects of air pollution from more than 3 million automobiles. It must also struggle to furnish adequate supplies of water for its residents. In a review of conflicts between Angeleno leaders and the residents of smaller towns near water sources, sociologist John Walton (1992) has shown that control over water and other natural resources, especially by powerful groups that dominate the regional real estate markets, is a key to understanding the history not only of the Los Angeles metropolitan region but of the entire desert West.

Another problem affecting many of the world's megalopolitan regions is the possibility of major disasters. Many are located in disaster-prone areas—either on coastal shorelines, as in the case of the Boston–Washington corridor, or worse, along major fault lines of earthquake activity, as in the case of Los Angeles (Davis, 1998).

Decentralization One effect of the growth of megalopolitan areas is *decentralization,* in which outlying areas become more important at the expense of the central city. This trend is not new. In the 1960s and 1970s large numbers of middle-income city dwellers moved to

FIGURE 9.6 **The Los Angeles–Anaheim–Riverside Megalopolis**

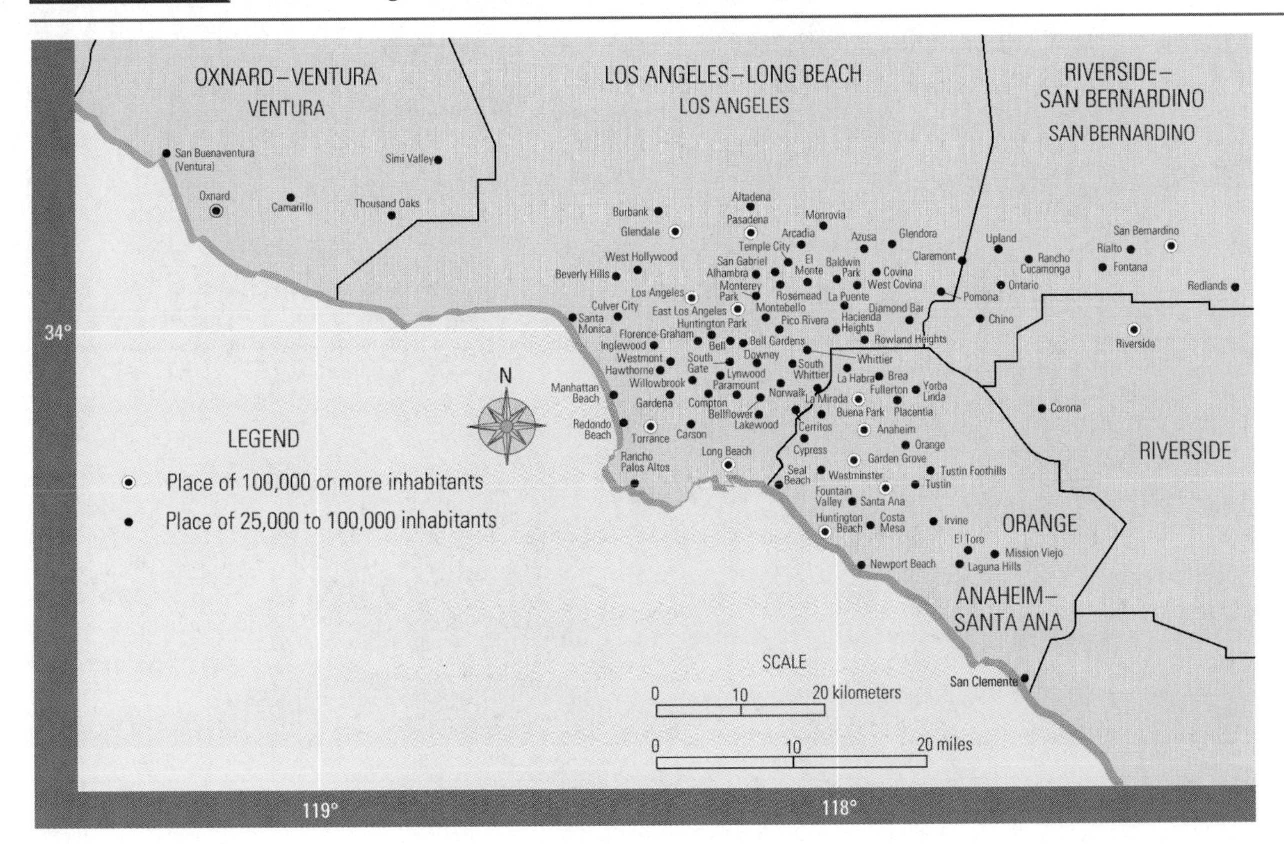

Source: Census Bureau.

suburban areas while the poor remained in the central cities. Business and industry also moved to the suburbs, giving rise to widespread speculation that vibrant central cities would become a thing of the past. Recently, however, the central cities of some metropolitan areas—such as New York, Philadelphia, Chicago, and Baltimore—have shown renewed vitality. Far from decaying, they have become major financial and cultural centers serving the needs of huge populations. On the other hand, medium-sized cities like Gary, Indiana, and Paterson, New Jersey, have suffered, because their central business districts have little to offer suburban dwellers in the way of financial services like banking and insurance or cultural attractions like theaters and symphony orchestras (Kasarda, 1988).

An important feature of megalopolitan areas—especially as they become more decentralized—is their diversity. These huge urban regions include many different kinds of communities: ethnic communities in both central cities and suburbs, middle-class "bedroom" suburbs, industrial towns, areas devoted to truck gardening or dairy farming, "second home" communities (e.g., beachfront areas), and so on. Each meets the economic and cultural needs of a specific urban population. Sociologists have devoted considerable study to these urban communities, and we discuss their findings in the next section.

URBAN COMMUNITIES

"The city," wrote Robert Park, is more than a set of "social conveniences—streets, buildings, electric lights, tramways, and telephones, etc.; something more, also, than a mere constellation of . . . courts, hospitals, schools, police, and civil functionaries of various sorts. The city is, rather, a state of mind, a body of customs and traditions. . . . It is a product of nature, and particularly of human nature" (1967/1925a, p. 1).

The connection between the city and human nature has been a recurrent theme in literature throughout history. Many literary images of the city are negative. In the Bible, for example, the cities of Sodom and Gomorrah are symbols of the worst aspects of human nature. The poet Juvenal complained that Rome produced ulcers and insomnia and subjected its residents to burglars and dishonest landlords. American literature also contains many negative images of the city. Thomas Jefferson, for instance, wrote, "I view great American cities as pestilential to the morals, health, and the liberties of man." And Henry David Thoreau escaped from the city to Walden Pond in an effort to "rediscover his soul" (Fischer, 1976).

Eastern European sociologists note that cities like Sarajevo were cosmopolitan centers marked by higher levels of ethnic and religious intermarriage than the surrounding areas, which were more homogeneous. They attribute some of the bitterness and hostility of the Bosnian war to antiurban sentiment in the mountain villages and towns.

Social scientists who study cities have devoted a great deal of attention to the tension between "community" and "individualism" as it relates to life in cities (Lofland, 1998). Country dwellers have been thought of as "happily ensconced in warm, humanly rich and supportive social relationships: the family, neighborhood, town," whereas city dwellers are "strangers to all, including themselves. They are lonely, not emotionally touching or being touched by others, and consequently set psychically adrift" (Fischer, 1976, p. 19). On the other hand, country dwellers are sometimes viewed as "stifled by conventionality, repressed by the intrusion and social control of narrow-minded kin, neighbors, and townsmen," whereas city dwellers are "free to develop individual abilities, express personal styles, and satisfy private needs" (p. 20). These views of the city are obviously contradictory, and much research has been devoted to the question of how urban life affects individuals and communities. In this section we look at some of the findings of that research and the theories of urban life that have been proposed on the basis of those findings.

The Decline-of-Community Thesis

Early studies of the nature and effects of urban life were dominated by efforts to evaluate the differences between rural and urban societies. They tended to reach rather gloomy conclusions. We have already noted (see Chapter 4) that Ferdinand Tönnies described the process of urbanization as a shift from gemeinschaft (a community based on kinship ties) to gesellschaft (a society based on common interests). Émile Durkheim reached a similar conclusion: Small rural communities are held together by ties based on shared ideas and common experiences, whereas urban societies are held together by ties based on the interdependence of people who perform specialized tasks. Both Tönnies and Durkheim believed that urban life weakens kinship ties and produces impersonal social relationships.

In a 1905 essay, "The Metropolis and Mental Life," Georg Simmel focused on the effects of urban life on the minds and personalities of individuals. According to Simmel, cities bombard their residents with sensory stimuli: "Horns blare, signs flash, solicitors tug at coattails, polltakers telephone, newspaper headlines try to catch the eye, strange-looking and strange-behaving persons distract attention" (quoted in Fischer, 1976, p. 30). The urban dweller is forced to adapt to this profusion of stimuli, which Stanley Milgram (1970) termed *psychic overload.* The usual way of adapting is to become calculating and emotionally distant. Hence the image of the city dweller as aloof, brusque, and impersonal in his or her dealings with others.

This view of the effects of urban life found further expression in the work of Louis Wirth, especially in his essay "Urbanism as a Way of Life" (1968/1938). Wirth began by defining the city as "a relatively large, dense, and permanent settlement of socially heterogeneous individuals" (p. 28). He then attempted to show how these characteristics of cities produce psychological stress and social disorganization. The primary psychological effect of urban life, according to Wirth, is a weakening of the individual's bonds to other people. Without such bonds, the individual must deal with the crises of life alone; often the result is mental illness. In other cases the city dweller, again because of the absence of close ties to friends or kin, lacks the restraints that might prevent him or her from engaging in antisocial behaviors.

Wirth linked social disorganization to the diversity that is characteristic of cities. Unlike rural residents, city dwellers work in one place, live in another, and relax in yet another. They divide their social lives among coworkers, neighbors, friends, and kin. Their jobs, lifestyles, and interests are extremely varied. As a result, no single group, be it the family, the friendship group, or the neighborhood, controls their lives. In Wirth's view, this absence of social control produces anomie or normlessness (see Chapter 7). Urban dwellers frequently do not agree on the norms that should govern their lives, and hence they are likely to either challenge existing norms or ignore them. Consequently, instead of being controlled by the norms of primary groups, the lives of city dwellers are controlled by impersonal agencies like banks and police forces.

One consequence of the impersonality of urban life, some argue, is greater callousness among city dwellers. An often-cited example is the case of Kitty Genovese, discussed in Chapter 6. After that episode, in which a young woman was murdered while 38 residents of nearby apartments—who heard her cries—did nothing, many commentators called attention to the callous character of city dwellers. Subsequent research on bystander apathy revealed that the presence of many other people tends to diffuse the sense of responsibility. We are less likely to take action if we have reason to believe that someone else will do so (Hunt, 1985; Latané & Darley, 1970). In this sense, the bigger the city and the more one is surrounded by strangers, the more likely it is that such behavior will occur—but this does not mean that city dwellers are alienated from one another when they are among people they know, or that when alone they would not help someone in trouble.

On the basis of many years of research on urban social interaction, together with an extensive review of the literature, urban sociologist Lyn Lofland (1998) concludes that city dwellers adhere to a set of norms that govern behavior among strangers in the city. These norms stress civility along with inattention, concern along with avoidance of any appearance of prying. To some observers the urban dweller's behavior might seem hostile or unconcerned, but according to Lofland

In urban societies, where there are large numbers of people who are strangers to one another, sociologists have noted a propensity to wear identifying "uniforms" that indicate the kind of person one wishes to know and interact with.

this is largely a misunderstanding. And although urbanites may feel less responsibility for others when surrounded by strangers, this condition of city life is hardly evidence for the decline-of-community thesis.

The Persistence of Urban Communities

The idea that urbanization leads to the decline of community has been criticized on a number of grounds. Rural life is nowhere near as pleasant as some urban sociologists have assumed it to be; evidence of this is the almost magical attraction that cities often have for rural people. On the other hand, urban social disorganization is not as extensive as the early urban sociologists believed. Many city dwellers maintain stable, intimate relationships with kin, neighbors, and coworkers. Moreover, urban life is not necessarily stressful or anomic.

Subcultural Theory A more recent view of urban life sees the city as a "mosaic of social worlds" or intimate social groups. Numerous studies have shown that the typical urban dweller does not resemble the isolated, anomic individual portrayed by Simmel and Wirth. In fact, communities of all kinds can be found in cities. Many urban dwellers, for example, are members of ethnic communities that have not become fully assimilated into the melting pot of American society and are unlikely to do so. They may be the children or grandchildren of immigrants who formed ethnic enclaves within large cities in the late nineteenth and early twentieth centuries, or they may be recent immigrants themselves, trying to build a new life in a strange culture.

But group ties among urban dwellers are not based solely on ethnicity. They may be based on kinship, occupation, lifestyle, or similar personal attributes (Fischer, 1976; Hummin, 1990; Lofland, 1998). Thus many cities contain communities of college students, elderly people, homosexuals, artists and musicians, wealthy socialites, and so on. Although the members of any given group do not always live in the same neighborhood, they are in close touch with one another much of the time. Their sense of community is based not so much on place of residence as on the ability to come together by telephone, in special meeting places like churches or synagogues, or even in restaurants and bars (Duneier, 1992; Fischer, 1976; Oldenburg, 1997).

An example of this point of view, known as *subcultural theory,* is Illsoo Kim's (1981) detailed study of the Korean community in New York City. This growing ethnic community has developed since the passage of the Immigration Act of 1965, which eliminated the nationality quotas established earlier in the twentieth century. Largely because of population pressure in South Korea (which has led to overcrowded cities and high rates of unemployment), about 750,000 Koreans have immigrated to the United States since 1970, with more than 100,000 finding new homes in the city of New York.

Like all immigrants, the Koreans have had to create a new way of life. They have had to find new ways of making a living, and they have had to adapt to a new culture. The first problem has been solved mainly by opening small businesses, the second by establishing neighborhoods where the immigrants can maintain their own culture while they and their children learn the values and norms of Western culture. Kim explains the Koreans' inclination to open small businesses (mainly grocery stores) in this passage:

> Old-timers frequently tell newcomers that "running a *jangsa* (commercial business) is the fastest way to get ahead in America." The language barrier partly explains this inclination. . . . This is an insuperable barrier to most Korean immigrants; it deprives them of many opportunities. . . . This fact, combined with differences in both the skills demanded and the system of rewards in the United States, means that occupational status cannot be transferred from the homeland to the new land. A high proportion of Korean immigrants were thus forced to turn to small retail businesses. (1981, pp. 102–103)

Many of the businesses established by Korean immigrants are geared to Korean ethnic tastes and cultural needs. For example, Korean immigrants are often unwilling to change their diet and will buy most of their food from Korean-owned food stores. Other shops import books, gifts, magazines, and other items from South Korea. Numerous Korean travel agencies satisfy the immigrants' strong desire to visit their homeland. But Korean small businesses are not limited to serving the needs of the immigrant population. A growing number cater to all racial and ethnic groups and supply typically American goods and services.

In summing up the situation of Korean immigrants in New York, Kim points out that "until they completely master the American language, education, and culture, Koreans will be forced to rely on one another" (p. 319). For this reason they are likely to maintain their own community within the city for at least two generations. However, in time their commitment to education and a better life will cause them to become more fully assimilated into American society, and their ties to the Korean community will be weakened.

The case of the Koreans echoes that of many other immigrant groups, both past and present, as well as that of other groups, such as homosexuals and artists. The details may differ somewhat, but the tendency to establish ethnic or cultural enclaves in large cities is universal. In a well-known study titled *The Urban Villagers,* Herbert Gans (1984) described an Italian neighborhood in Boston. This community was a long-established one with many of the characteristics that are typical of ethnic communities everywhere. But at the time that Gans did his research, the neighborhood was being destroyed by an urban renewal project. Its residents were forced to find new homes in other parts of the city, to join new churches, attend new schools, seek out new places to buy food and clothes. For some, the upheaval was so great that it produced depression and anomie (Fried, 1963).

It seems reasonable to conclude that the effects of urban life on communities and individuals are more complex than Durkheim, Simmel, and Wirth suggested. Certainly, social disorganization occurs in cities, but so does social *re*organization. Communities that are uprooted by urban renewal eventually may be re-formed in other parts of the city or its suburbs. The new community may be less homogeneous than the old one, but it is a community nonetheless.

The Suburbs Like city dwellers, suburban dwellers have been said to lack the close attachments that are thought to characterize rural communities. In fact, before the late 1960s most social scientists had a rather dismal view of suburban life. Suburbs had grown rapidly after World War II as large numbers of middle-class Americans left the central cities in search of a less crowded, more pleasant lifestyle. Many suburban dwellers were corporate employees, and in the 1950s and early 1960s they became the subject of a widely held stereotype. The suburbs, it was said, were "breeding a new set of Americans, as mass-produced as the houses they lived in, driven into a never-ending round of group activity ruled by the strictest conformity. Suburbanites were incapable of real friendships; they were bored and lonely, alienated, atomized, and depersonalized" (Gans, 1967, pp. xxvii–xxviii).

In *The Levittowners,* a classic study of a new suburban community in New Jersey, Gans (1967, 1976) challenged this stereotype. He found that the residents of Levittown lost no time in forming attachments to one another. At first they associated only with their neighbors, but before long they formed more extensive associations based on shared interests and concerns. Moreover, far from being bored and isolated, a majority of the Levittowners were satisfied with their lives and felt that Levittown was a good place to live. Gans concluded that "new towns are merely old social structures on new land" (1967, p. vii). In other words, the contrasts between central-city and suburban life are generally exaggerated. Other researchers have reached similar conclusions. They note that as suburban communities age and newcomers buy homes from people whose children are grown, these neighborhoods go through ethnic, racial, and generational changes that are not so different from the patterns occurring inside the cities (Jackson, 1985).

Some comparative research has focused on women born in suburban communities after the 1950s who never lived in other types of communities (Stimson, 1980; Wekerle, 1980). For example, Sylvia Fava (1985) found that younger women born in suburban communities do not think of suburban life as stifling or as limiting their opportunities. They may prefer to live in older, more diverse suburban communities within easy commuting distance of the jobs and nightlife of the central cities, but they normally do not feel that their suburban lifestyle is unsatisfactory. Indeed, as conditions continue to deteriorate in some inner-city communities—with increasing homelessness, exposure to drug markets, and violence—both older and newer suburban areas become more attractive to people seeking to escape the dangers they perceive in urban life (Lofland, 1998).

For the past few decades the phenomenon of "white flight" from older inner-city communities to newer and presumably safer suburban communities has been a subject of concern among urban sociologists and policy makers (Wilson, 1987). Recent data show, however, that African Americans and other minority groups are also fleeing the central cities at an accelerating rate. Demographer William Frey notes that "minority suburbanization took off in the 1980s both as the black middle class came into its own and as more assimilated Latinos and Asians translated their moves up the socioeconomic ladder into a suburban life style" (quoted in De Witt, 1994, p. A1). This outward migration threatens to leave central-city communities with even greater concentrations of poor people. In cities with large numbers of immigrants, this discouraging trend is countered by increasing racial, ethnic, and class diversity (Huang, 1994), but in other cities the trend toward greater concentration of poor people inside the city and more affluent people in the suburbs is a growing problem. Recent research by Paul Jargowsky (1997) on ghettos and barrios (predominantly Latino neighborhoods) shows that middle-class migration, combined with the flight of industries from central cities, has produced dramatic increases in the number of people living in city neighborhoods with household poverty rates of 40 percent or more. Figure 9.7 demonstrates this disturbing trend, whose implications will become evident in later chapters on social inequality.

Private Communities The growth of private communities throughout the United States is another indication of the widespread fear of urban life (Garreau, 1991; McKenzie, 1994). Recent data indicate that about 4 million Americans live in private communities

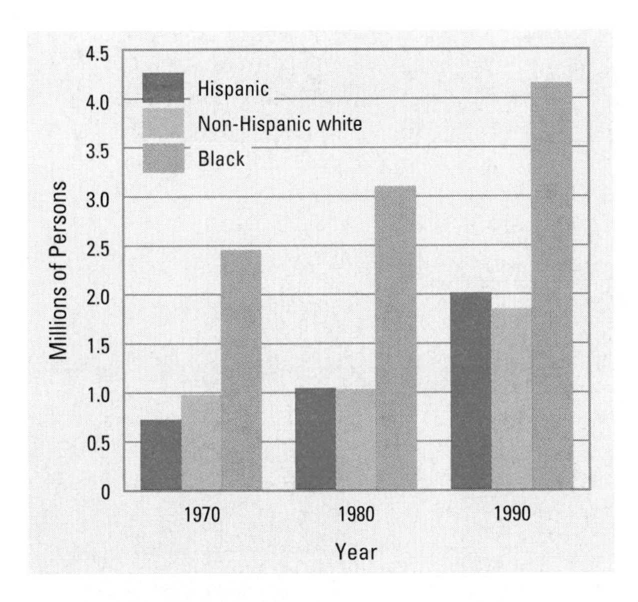

Source: Jargowsky, 1997.

FIGURE 9.7 **Residents of High-Poverty Neighborhoods, by Race/Ethnicity, 1970–1990**

where public through-traffic is prohibited. For those who can afford to live there, these heavily patrolled and regulated communities represent a decision to place individual comfort and family safety above more social efforts to create communities. An even larger number of Americans, about 28 million, live in communities represented by private community associations. These include condominiums and cooperatives that require all residents to be members of the association, obey its rules, and pay fees to support its operations. Although they are less restrictive than private communities, these community associations also seek to control public access.

These private and quasi-private communities are the fastest growing types of communities in the United States, and there is some concern about what this growth implies. Supporters of such communities dwell on their safety and their ability to maintain property values. Critics like Gerald Frug, an expert on local government, believe that people are trying to capture the feeling of life in a village without many other qualities of village life. "The village was open to the public," Frug observes. "The village did not have these kinds of restrictions. The village had poor people, retarded people. Somebody could hand you a leaflet. These private communities are totally devoid of random encounters" (quoted in Egan, 1995, p. A22). Frug and other critics of the trend toward private communities fear that as people choose to live only with others like themselves, "they will become less likely to support schools, parks or roads for everyone else" (p. A22).

CITIES AND SOCIAL CHANGE

In the preceding section we stressed the presence of a wide variety of communities and subcultures within cities. We noted that contemporary social scientists see the city as a place where many different communities coexist and thrive rather than as a place where people are isolated and do not have a feeling of belonging to a particular social group or community. At the same time, however, there is no escaping the fact that various urban communities are often in conflict. For example, ethnic and racial communities may clash in violent confrontations over such issues as the busing of children to achieve school integration. In this section we examine the origins and implications of such conflicts.

Social change in an urban society is likely to be felt most deeply by city dwellers, for it is in the cities that people are most densely congregated. In recent decades this has been the case in North America as manufacturing jobs have been "exported" to Asia and Latin America. As a result of this shift, American cities are undergoing massive social change. "America's major cities are different places today from what they were in the 1960s," concludes John Kasarda (1989, p. 28), one of the nation's foremost urban sociologists. New modes of transportation, new communication technologies such as satellites and computers, and new industrial technologies (e.g., automation and recycling) are transforming our cities from production and distribution centers to administrative, financial, and information centers. "In the process," writes Kasarda, "many blue-collar jobs that once constituted the economic backbone of cities and provided employment opportunities for their poorly educated residents have either vanished or moved." Many of those jobs have been replaced by "knowledge-intensive white-collar jobs with educational requirements that exclude many with substandard education" (p. 28).

Inequality and Conflict

As noted earlier, American cities and metropolitan regions have always attracted streams of migrants and immigrants. Migrants arrive from other parts of the nation—blacks from the rural South, Chicanos from the Southwest. Immigrants come from foreign lands—Haiti, Poland, China, Korea, Italy, and elsewhere. Until the 1970s, except during economic depressions or

MAPPING SOCIAL CHANGE

Urban Locations and Environmental Risk

Millions of people live in highly unstable physical environments where there is a high risk of natural disasters like earthquakes, hurricanes, or floods. For instance, the map shows the regions of the world where earthquake activity is greatest. Along the West Coast of the United States and Latin America, in the Valley of Mexico, in much of Japan, and on the Pacific Rim, where rates of urban growth are among the highest in the world, the risk of devastating earthquakes is an undeniable fact of life. Or is it?

Our knowledge of where and how earthquakes occur has improved considerably in the last 30 years, but the cities in question began growing rapidly long before the risks were well understood. People continue to migrate to these risky environments, often denying that the risks actually apply to them. But denial

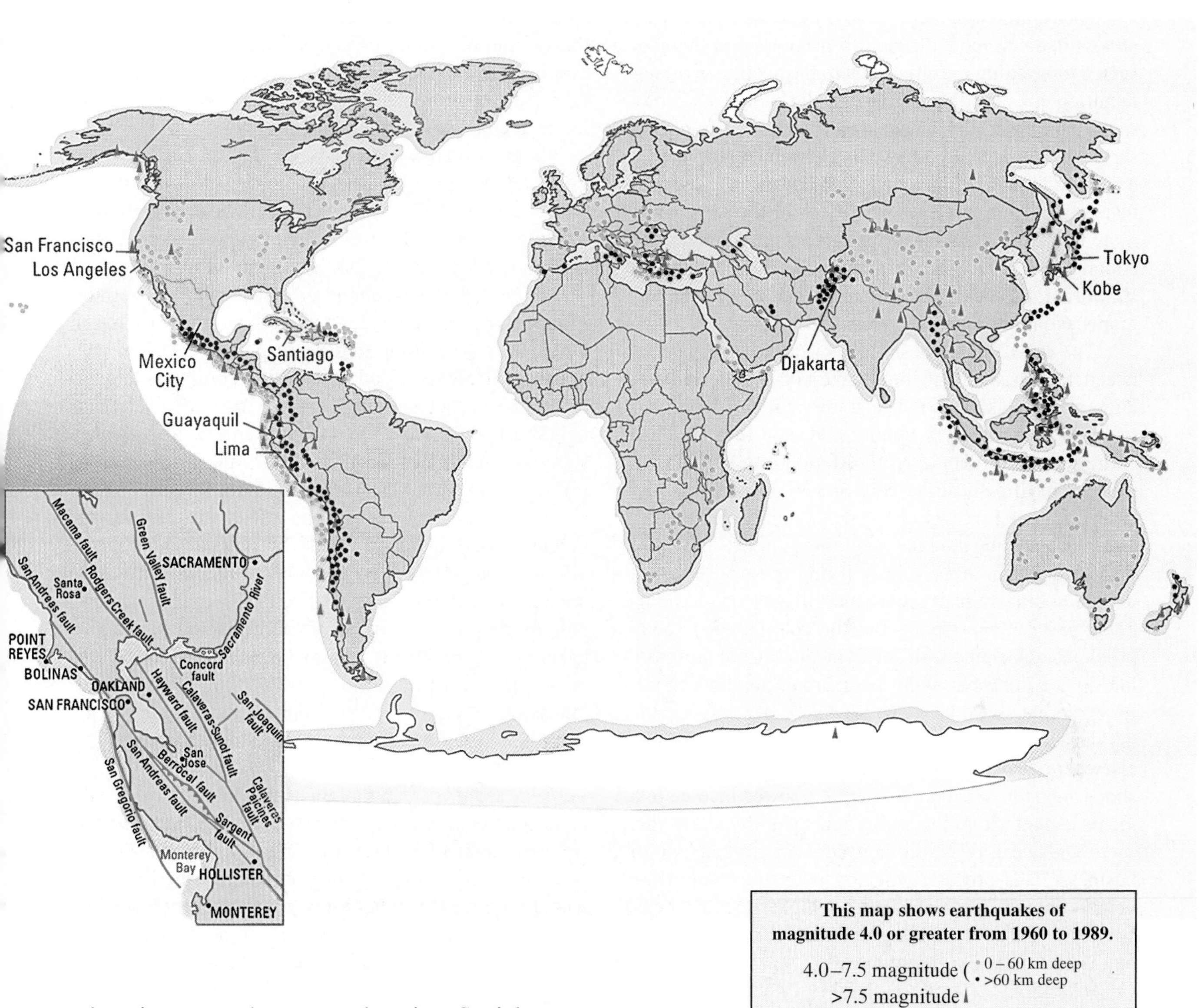

This map shows earthquakes of magnitude 4.0 or greater from 1960 to 1989.

4.0–7.5 magnitude (· 0 – 60 km deep • >60 km deep

>7.5 magnitude

alone is not an adequate explanation. Social policies that encourage the growth of such cities—through favorable mortgage and insurance terms, for example—also play a significant role. Social policies that require better preparation for earthquakes, better design and construction of buildings and freeways, and advance planning for disaster relief are among the consequences of recent disasters. But the desirable coastal locations of vulnerable cities will result in continued population growth and an increasing need for effective planningto avert disasters.

recessions, the new arrivals had no difficulty finding work in the mills and factories that produced textiles, clothing, steel, rubber, glass, cars, and trucks. And much of their education was gained while working. Formal schooling was not as crucial to job success as it is today.

Today, although there are still many manufacturing jobs, it is much more difficult to make a good living at such a job. Within the cities, where the decline in manufacturing jobs has been greatest, the number of low-status jobs, especially in restaurants and other low-paying services, has increased. So has the number of highly paid jobs in management and the professions, positions that require high levels of education. At the same time, more than 20 million legal immigrants have entered the country since 1970, usually settling in the cities. The result, often, has been fierce competition between immigrants and more established residents.

The net effect of these trends is a growing gap between the "haves" and the "have-nots," especially in large cities (Silk, 1989). The study of *social stratification*—of patterns of inequality and how they produce entire classes of people with differing opportunities to succeed in life—is at the core of sociology and is discussed in detail in later chapters. Certainly the gap between those who enjoy the good things in life and those who spend their lives simply trying to survive is not unique to cities. Small towns and villages also have affluent and poor residents. But the conflicts and social problems associated with social inequality are most visible in the cities. In many poor urban neighborhoods, for example, wealthier city dwellers are buying the buildings in which the poor live, forcing them to move elsewhere. In neighborhoods where this occurs, the shops and stores that once catered to lower-income residents cannot afford the higher rents paid by stores that serve the wealthy newcomers and are forced out of business. This process, whereby poor and dilapidated neighborhoods are renovated by higher-income newcomers while poor residents and merchants are pushed out, is known as *gentrification.*

Status Conflict In the last section we noted that an urban renewal project had dislocated large numbers of Italian Americans from their Boston neighborhood. Such projects have sometimes been labeled "urban removal" because they remove poor people from decaying neighborhoods and force them to find homes elsewhere. Recently urban renewal has fallen into disfavor and *redevelopment* has become popular. Redevelopment does not involve the destruction of entire neighborhoods, but it may have similar effects. For example, the redevelopment of New York City's Times Square by the Walt Disney Company and other powerful groups, both public and private, is restoring a sleazy area to its former glory as an entertainment center. The explicit goal is to make the area attractive to "decent" people and to eliminate the prostitutes, drug and pornography dealers, and petty criminals that have made Times Square a symbol of vice throughout the world (Kornblum, 1995; McNamara, 1995; Mollenkopf, 1985). This is a typical instance of *status conflict,* in which different groups (for example, theater and restaurant owners versus owners of "adult" establishments) vie for territory, occupational advantages, and other benefits that will enhance their prestige in the neighborhood or community.

Some urban sociologists see the city as divided into "defended neighborhoods" or territories (Lofland, 1998; Suttles, 1972). This concept was originally developed by Park and Burgess, who viewed such neighborhoods as a type of "natural area" within the city. The defended neighborhood is a territory that a certain group of people consider to be their "turf" or base, which they are willing to defend against "invasion" by outsiders. Neighborhood defense is a common element of urban conflict, although the means used to defend neighborhoods may differ. For example, in wealthy suburban neighborhoods it may take the form of zoning regulations that establish minimum lot sizes of an acre or more. Such regulations effectively defend the neighborhood against invasion by people who cannot afford the large lots or by developers who would like to erect apartment buildings (Perin, 1977). In less affluent neighborhoods, defense is often conducted by neighborhood improvement groups, which sometimes engage in vigilante action when they fear racial "invasion" (Hamilton, 1969). In poor neighborhoods, defense is very often the province of street-corner gangs.

Gender Conflict In the nineteenth century many social commentators were dismayed at the increasingly frequent sight of women walking alone or with other women in city streets. Women unaccompanied by men were thought to be either immoral or innocently risking their honor or their lives in a world of predatory males. As Lyn Lofland observes, "The increasing presence of 'respectable' women in urban public space served as especially vivid and provocative evidence of women's unnatural and undesirable abandonment of hearth and home" (1998, p.128). Of course, Lofland realizes that "poor women had rarely, in the earliest cities, been barred from the streets," whereas the "spatial freedom of privileged women has been culturally and historically quite variable." What moralists were protesting against was the presence of middle-class or "respectable" women in public places.

Today, too, one can find many examples of the conflicts experienced by women in urban public places. On their lunch breaks, for example, women are routinely harassed by ogling men and then accused of inviting such behavior by wearing provocative clothing. In Islamic cities, especially in more orthodox nations like Pakistan or Afghanistan, women are prohibited from appearing in public at all; if they must be out on the streets, they are required to hide their faces and completely cover their bodies. Conflicts that result from efforts by men to "keep women in their place" thus take many forms and are a widespread form of status conflict, one that is especially visible in cities. It may be true that racial conflict is more strident and violent, but for a person with a sociological imagination, gender conflicts are easy to spot.

Racial Conflict Perhaps the most common source of intergroup conflict in American cities has been racial tension. In the belief that the presence of black residents reduces the prestige of a neighborhood and lowers the value of real estate there, whites have resisted efforts by blacks to obtain housing in "their" neighborhoods. This resistance has taken a wide variety of forms, ranging from discrimination by real estate agents to outright violence against black families moving into white neighborhoods. When blacks have succeeded in gaining a foothold in previously all-white neighborhoods, many white families have sold their homes and moved to the suburbs (Massey & Denton, 1993).

For racial and ethnic minorities, the desire to move out of segregated ghettos is based on the need to establish or maintain a residential community while pursuing opportunities to build a better life. This is among the key findings of studies of immigrant groups in South Florida. In looking at the difficulties that Haitian and Cuban refugees encounter in finding jobs in the Miami area, Alejandro Portes and Alex Stepick (1985) found that Cubans could draw upon kin and ethnic networks in Miami neighborhoods in looking for jobs but that the Haitians had no such networks to help them adapt to their new environment.

The riots that took place in 1992 in Los Angeles, Atlanta, and other major U.S. cities were a tragic example of the relationship between social change and racial violence in urban centers. Although the conditions of life had been deteriorating in central-city ghettos throughout the 1980s, it was not until televised bloodshed brought this situation to the attention of the majority of Americans that significant numbers of people began to demand that the federal government take action. But this is by no means a new phenomenon in American social life. As we saw in Chapter 8, conflict between racial and other groups often raises public consciousness and contributes to major social change (Rothenberg, 1992).

i

RESEARCH ON THE CUTTING EDGE: Coping With Disasters in Urban Communities

Riots, epidemics, earthquakes, terrorist bombings, wars, floods, tornadoes, and hurricanes are among the many kinds of disasters that may strike densely populated human settlements. Of course there are important differences in the causes and outcomes of purely social disasters like riots and purely natural ones like earthquakes. But from a sociological vantage point social and natural disasters have a great deal in common. And because we can expect both natural and social disasters to occur with increasing frequency on our urbanizing planet, the study of disasters is a thriving area of sociological theory and research.

Recent sociological research on disasters has focused on the longer-term consequences for the victims. In this area of research, the work of sociologist Kai Erikson is particularly noteworthy. Erikson's powerful descriptions of responses to disasters and his development of the sociological theory of trauma are having significant and positive impacts on victims' lives. In a number of precedent-setting cases Erikson has convinced juries and judges that victims have suffered sociological damage, especially in the form of loss of community.

Erikson's recent work on social disasters focuses on the consequences of what he calls "a new species of trouble" that is particularly common in urban industrial societies. The trouble may take the form of toxic spills, nuclear accidents, mercury poisoning, or any of a large number of other disasters attributable to human negligence, greed, or combinations of both. The result is often an abrupt and shattering loss of community.

Contamination and pollution of the earth and water beneath these houses, caused by the criminal dumping of toxic wastes, forced the residents to abandon their homes. Sociologist Kai Erikson has identified such disasters as a "new species of trouble" that has profound consequences for its victims.

In one such case, Erikson interviewed the residents of East Swallow Drive in Fort Collins, Colorado. This middle-income neighborhood of private homes with middle-aged or elderly residents was the site of a gasoline spill with devastating consequences. The local service station had been experiencing "inventory losses" over a number of years and had finally replaced three "incontinent" fuel tanks with new ones. Not long afterward, nearby residents began smelling gasoline fumes in their homes. An investigation revealed that there were large pools of gasoline in the ground under the homes along East Swallow Drive. One house near the service station was condemned, and in the months that followed, as Erikson describes it, "waves of concern began to radiate in slow-motion ripples down East Swallow Drive as people began to understand that a plume of petroleum was moving eastward with the sureness of a glacier" (1994a, pp. 99–100).

A group of homeowners brought suit against the companies involved in the spill. As is often the case, however, the companies were able to hire teams of lawyers who could delay the suit for months. Eventually lawyers for the residents asked Erikson to prepare testimony on the sociological damage caused by the spill. He and an assistant conducted observations in the neighborhood and interviewed the residents. Later Erikson shared transcripts of the interviews with the residents and asked them for additional comments. This method, it turned out, had powerful effects. As one resident noted, "I was greatly relieved to see what other people said. . . . Seeing this report gives me a little reassurance that I wasn't feeling something that other people were not feeling" (quoted in Erikson, 1994a, p. 138).

In his report Erikson argued that the personal distress the residents were experiencing—in the form of anxiety, sleeplessness, anger, and other emotions—was not only a sign of psychological suffering but also a symptom of sociological trauma due to the loss of their attachments to the neighborhood. As one resident said, "Well, we feel that the value of our property has taken a nose dive. It is our only security. We don't feel that it's salable. If we can't sell it, we can't relocate, we can't move, we can't expand or change. We feel that we're sort of stuck" (quoted in Erikson, 1994a, p. 232).

Erikson convinced the court not only that the residents had lost property value in terms of dollars, but also that they had suffered a greater loss in a sociological sense. He noted that they had invested a great deal more than money in their homes:

> It is well understood by social and behavioral scientists who are familiar with accidents of this

> kind that the loss of a home can be profoundly traumatic. I wrote in my report on the Buffalo Creek flood of 1972 [in West Virginia], for example, that a home in which a person has made a real personal investment "is not simply an expression of one's taste; it is the outer edge of personality, a part of the self itself." (1994a, p. 233)

Similarly, a well-known study of the impact of a tornado in Texas reported that people who had lost their homes felt as if they had lost part of themselves (Moore, 1958). And an equally well-known study of an urban relocation project in Boston's East End was titled simply "Grieving for a Lost Home" (Fried, 1963). In the East Swallow Drive case, the residents benefited from Erikson's sociological insight: Eventually the corporations responsible for the damage had to settle with the residents and pay not only for the damage to their property but for their loss of community as well.

Erikson and other sociologists who study disasters hope that as a consequence of their research—and their successes in court—there will be more safeguards against negligence and more emphasis on anticipating the risks of natural disasters. In the near term, however, Erikson fears that the balance of power is shifting toward corporations and other powerful actors, to the disadvantage of the victims or potential victims of community disasters.

VISUAL SOCIOLOGY

Frederick Law Olmsted's Vision of Urban Parks

Every city in the world faces the dilemma of how—in a crowded urban environment where land is at a premium—to provide parks and open space for its residents. Paris, London, Beijing, Moscow, New York, Tokyo, Rio, and other great cities have set aside places where people can congregate to stroll, enjoy their leisure, and get some exercise. But accepting the idea that parks and open spaces are part of what makes a city great required many generations of struggle and many contributions by people with unusual talent and personal commitment. Holly Sidford, whom we met at the beginning of the chapter, is attempting to carry on the work started by pioneers in the field of parks and recreation. Among those pioneers, perhaps none is as important as Frederick Law Olmsted.

Olmsted, shown here in a photo from the mid-nineteenth century, is known as one of the originators of the field of landscape architecture. He is even more famous as the creator of great urban parks, including Prospect Park in Brooklyn, Central Park in Manhattan, Golden Gate Park in San Francisco, and major parks in

Chicago, Buffalo, Louisville, and many other cities. As a result of a voyage of exploration to the West, particularly the Yosemite area, Olmsted wrote a report on the need to preserve the valley and other places of similar beauty. He knew that cities and towns would develop in rural areas throughout the nation, and that there would be a need to bring nature into the city. The national park system in the United States was inspired by the work of Olmsted and a few other dedicated individuals.

How much we owe to Olmsted and the other creators of the parks and recreation movement is evident in Bruce Davidson's photographic study of life in Central Park. Davidson's photos capture much of what makes parks so essential to urban life. We see the park as a place where people come at all times of the year for the exercise that we now view as vital to health throughout the life span. We see the park as a place where rich and poor share space and experiences. The park is a place where love blooms and where the passages from one stage of life to another are commemorated in weddings, communion parties, and the like. And, as Olmsted hoped, we see that the park remains a place where children and their parents can find peaceful nooks and crannies within which they can explore nature even in the heart of one of the world's most congested urban environments.

From time to time there is violence in Central Park, as there is in all urban public spaces, and people who do not experience the park on a regular basis may become afraid to visit it. In fact, studies conducted over the past 20 years have shown that the park is perceived as safer and more enjoyable than ever before—another tribute to the pioneering vision of Olmsted and the continuing work of dedicated professionals like Holly Sidford (Kornblum & Lawler, 1999).

SUMMARY

Populations change as a consequence of births, deaths, out-migration, and in-migration. The *crude birthrate* is the number of births occurring during a year in a given population divided by the midyear population. The *crude death rate* is the number of deaths occurring during a year divided by the midyear population. The *rate of reproductive change* is the difference between the crude birthrate and the crude death rate for a given population. Since World War II the world population has been increasing at a rate of more than 1.5 percent, with projections surpassing 6 billion by the year 2000.

The *demographic transition* is a set of major changes in birth and death rates that has occurred most completely in urban industrial nations in the past 200 years. It takes place in three stages: (1) high birth and death rates, (2) declining death rates, and (3) declining birthrates. These stages are accompanied by changes in the age composition of the population.

Urbanization is closely linked with rapid increases in population, but at the same time the nature of life in cities tends to limit the size of urban families. Cities grow primarily as a result of migration (which is often caused by population increases in rural areas), but new migrants do not find it easy to form families.

The increasing tendency of people throughout the world to live in cities has been referred to as the urban revolution. Not only are increasing proportions of the world's population living in cities, but the cities themselves are larger than ever before. The growth of cities in the twentieth century gave rise to the concept of the *metropolitan area,* in which a central city is surrounded by a number of smaller cities and suburbs that are closely related to it both socially and economically.

The growth of cities should be distinguished from *urbanization,* which refers to the proportion of the total population that is concentrated in urban settlements. The end result of urbanization is an "urban society." Not only do cities serve as the cultural and institutional centers of such societies, but urban life has a pervasive influence on the entire society.

Sociologists have devoted a great deal of study to the processes by which cities expand and to patterns of settlement within cities. An early model of urban expansion was the concentric-zone model developed by Park and Burgess. In this model a central business district is surrounded by successive zones or rings devoted to light manufacturing, workers' homes, higher-class apartment buildings and single-family homes, and a commuters' zone. This model is limited to commercial and industrial cities that formed around business centers and does not account for the rise of satellite cities and the rapid urbanization that occurs along commercial transportation corridors. The growth of satellite cities was especially rapid before World War II. A more current model of urbanization is known as strip development and describes the incorporation of smaller communities and towns into a larger metropolitan area.

Metropolitan areas have expanded greatly since the mid-twentieth century, largely as a result of the increasing use of automobiles and the construction of a network of highways covering the entire nation. In some areas this growth has created large multinuclear urban systems that are described by the term *megalopolis.* One effect of the development of such areas is decentralization, in which outlying areas become more important at the expense of the central city.

Social scientists who have studied the effects of urban life have been particularly concerned with the tension between community and individualism as it relates to life in cities. Early studies of urban life tended to conclude that it weakens kinship ties and produces impersonal social relationships. Urban life was also thought to produce "psychic overload" and anomie. More recently these conclusions have been criticized by researchers who have found that many city dwellers maintain stable, intimate relationships with kin, neighbors, and coworkers and that urban life is not necessarily stressful or anomic.

Subcultural theory sees the city as a mosaic of social worlds or intimate social groups. Communities of all kinds can be found in cities. Those communities may be based on ethnicity, kinship, occupation, lifestyle, or similar personal attributes. Suburban dwellers also have been found to be far less bored and isolated than was once supposed.

Occasionally various communities within cities come into conflict. Such conflict may arise out of different class interests or the conflicting goals of different groups within the city. Some urban sociologists see the city as divided into "defended neighborhoods" or territories whose residents attempt to protect them from "invasion" by outsiders.

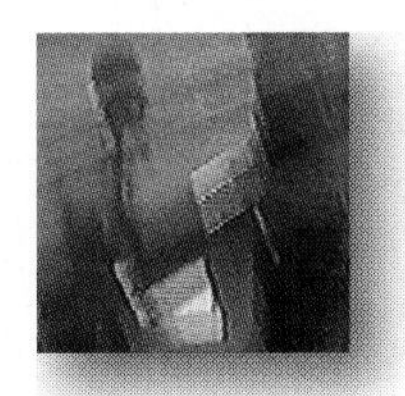

SOCIOLOGY VERSUS IDEOLOGY

In Robert Frost's famous poem, "Mending Wall," two neighbors cooperate in the work of repairing a stone wall that separates their properties. The poet wonders why his neighbor keeps saying, "Good fences make good neighbors." Why, the poet wonders, do they make good neighbors?

> Isn't it where there are cows?
> But here there are no cows.
> Before I built a wall I'd ask to know
> What I was walling in or walling out,
> And to whom I was like to give offense.

But the poet's neighbor is determined that the wall shall stand:

> He will not go behind his father's saying,
> And he likes having thought of it so well
> He says again, "Good fences make good neighbors."

In our social lives, and especially in our communities, we often see that fences and walls denote private property. They also mark off important boundaries in public spaces or bar access to dangerous places. But the fences around gated or private communities offend some people who are walled out. They feel excluded and not neighborly toward those inside. From their perspective, good fences prevent neighborliness. The people inside the walls clearly feel that the neighbors they choose to associate with form a community. They feel that they have every right to live where and how they please as long as it does no harm to others.

From a sociological viewpoint, both attitudes are understandable. The sociologist, however, considers the larger implications of these ideologies. She or he is unlikely to want to live in a society where private communities are banned and, at the same time, unlikely to want to see such communities become the norm. The sociologist is armed with facts that show the benefits of life in gated communities for those who live in them, as well as the liabilities for the larger society when the privileged isolate themselves from the rest of the population.

GLOSSARY

crude birthrate: the number of births occurring during a year in a given population, divided by the midyear population. (p. 259)

crude death rate: the number of deaths occurring during a year in a given population, divided by the midyear population. (p. 259)

rate of reproductive change: the difference between the crude birthrate and the crude death rate for a given population. (p. 259)

demographic transition: a set of major changes in birth and death rates that has occurred most completely in urban industrial nations in the past 200 years. (p. 259)

urbanization: a process in which an increasing proportion of a total population becomes concentrated in urban settlements. (p. 262)

metropolitan area: a central city surrounded by a number of smaller cities and suburbs that are closely related to it both socially and economically. (p. 264)

megalopolis: a complex of cities distributed along a major axis of traffic and communication, with a total population exceeding 25 million. (p. 272)

WHERE TO FIND IT

BOOKS

Western Times and Water Wars: State, Culture, and Rebellion in California (John Walton; University of California Press, 1992). A seminal work on the relationship of protest movements and collective behavior to metropolitan growth; describes the struggle to control vital natural resources in rapidly growing but arid regions.

Recent Social Trends in the United States, 1960–1990 (Theodore Caplow, Howard M. Bahr, John Modell, and Bruce A. Chadwick; McGill-Queens University Press, 1991). A compilation of statistical tables that document trends in population, urbanization, and other significant aspects of social change, based on data from the U.S. Census Bureau and other government agencies, and from major polling organizations, including the General Social Survey and the Gallup Poll.

The Apple Sliced (Vernon Boggs, Gerald Handel, and Sylvia F. Fava; Bergin and Garvey, 1984). A fascinating collection of sociological studies of New York City by field researchers who know the city well and are able to describe its institutions in vivid terms.

The Public Realm: Exploring the City's Quintessential Social Territory (Lyn Lofland; Aldine, 1998). An inquiry into the social life of urban public space, with far-reaching implications for the study of urban communities.

Relations in Public (Erving Goffman; Harper, 1971). Goffman's brilliance in analyzing social interaction is applied to the study of street life in urban environments.

Close-Up: How to Read the American City, 2nd ed. (Grady Clay; University of Chicago Press, 1980). A brilliant, humorous, and extremely well-illustrated guide to the physical ecology of American urbanization in the second half of the twentieth century.

The Urbanization of the Third World (Joseph Gugler, ed.; Oxford University Press, 1988). A collection of original essays by leading urban researchers, tracing most of the outstanding trends and problems of urbanization in the developing areas of the world.

The Population of the United States, 3rd ed. (Douglas L. Anderton; Free Press, 1997). A comprehensive sourcebook on historical and contemporary trends in the U.S. population.

Streetwise: Race, Class, and Change in an Urban Community (Elijah Anderson; University of Chicago Press, 1990). An insightful and original study of life in two adjacent neighborhoods, one populated primarily by lower-income minority households, the other mixed in terms of class and race but undergoing gentrification. Through detailed ethnographic research the author documents changes in public behavior, norms of sexual conduct, drug use and control, and family life.

JOURNALS

Population and Development Review. Presents international research on population change and the effects of population variables on other aspects of social and economic development. A valuable source of comparative data and studies of population problems in the Third World.

Urban Affairs Quarterly. A leading journal of original research on urban social change.

OTHER SOURCES

Urban Affairs Annual Reviews (Sage Publications). Annual volumes on social change in urban regions, focusing on the impact of technologies and changing economic institutions on urban social classes and city governments.

Demographic Yearbook. Published annually by the United Nations, Department of International Economic and Social Affairs. Presents statistics on population, birth and death rates, marriage and divorce, and economic characteristics.

INTERNET RESOURCES

The Population Council (www.popcouncil.org/). One of the best places to begin looking at the wealth of resources on the Internet that deal with population growth, population control, and studies of population processes.

The Urban Institute (www.urban.org/). One of the nation's premier think tanks concerned with urban problems. Its Web site contains many of its publications.

The Latinos Web (www.catalog.com/favision/resource.htm). Offers a wide selection of research data on migrations and population changes among Latinos, with world trends and links as well.

CHAPTER 10

GLOBAL SOCIAL CHANGE

One sultry African night, as the people of Blokosso were sleeping, a band of thieves crept into their houses and ransacked them. They stole radios, jewelry, and expensive articles of ceremonial clothing. But as the villagers assessed the damage early the next morning, they mourned a far greater loss. Never before in anyone's memory had a robbery like this occurred in the village. It was not the goods themselves they missed, for these could be replaced. It was the loss of a way of life, a social world, that they lamented. "We loved to sleep outside on the very hot nights," said Mr. Joseph, one of the most respected men in the village—"the women by this palm grove in the soft sand along the lagoon, the men under the lamp a bit. We like to talk and sleep with each other. We are not used to sleeping with our things."

Throughout history theirs had been a village society. Blokosso was one of several fishing villages inhabited by the Ebrié tribe. For as long as they could remember, the Ebrié people had made their living by fishing in the quiet waters of the

lagoon, which was protected from the huge ocean swells by a long barrier island. But in recent decades the village had been engulfed by the rapidly growing city of Abidjan, the capital of Ivory Coast and a booming port. More and more of the tribesmen were now working in the city. The women continued to tend their garden plots, but they too were increasingly drawn into the economy of the city and felt an ever-increasing need for spending money, which they earned by renting rooms to migrant workers and selling their extra produce.

Gradually, in the time before the robbery occurred, Blokosso had ceased to be a self-sustaining village with its own economy and tribal culture. Instead, it was becoming a "village enclave," a little island of cultural homogeneity in an urban society. The robbery was a blunt statement to the villagers about how much change had occurred in a short time and how much more was likely to come.

For me, the experience of living in Blokosso at that time (during 1962 and 1963) was as much of a revelation as the robbery was to the villagers.

I had been trained as a biologist and was living in Blokosso while teaching physics and chemistry in a junior college. During the day I delighted in opening up to students the secrets of matter and energy. In the evenings, however, the villagers taught me about living in a totally different social world, one that was experiencing even more rapid social change than my own.

Mr. Joseph, my best friend in the village, was a middle-level executive in one of Abidjan's insurance agencies. He commuted across the lagoon in a motorized dugout canoe. During the day his mind was entirely devoted to business. In the evenings he dealt with the problems of a traditional African family. His three wives became increasingly jealous of one another as new wealth and their rising expectations led to new opportunities for better health and more leisure and also to new sources of conflict. I watched him and his family deal with the impact of change on their lives. I saw the villagers attempt to invoke their ancestral spirits and to use witchcraft to cope with new phenomena like robberies. The entire colony had become independent from France only 2 years before, and I saw the villagers—now citizens of a new nation called Ivory Coast—becoming interested in politics and current events.

During that time my own interests shifted from the physical to the social sciences. Nothing has ever interested me more than the villagers' questions about what social change brought to them and what it caused them to lose. When I began to study sociology, I discovered that such questions are at the core of this young science.

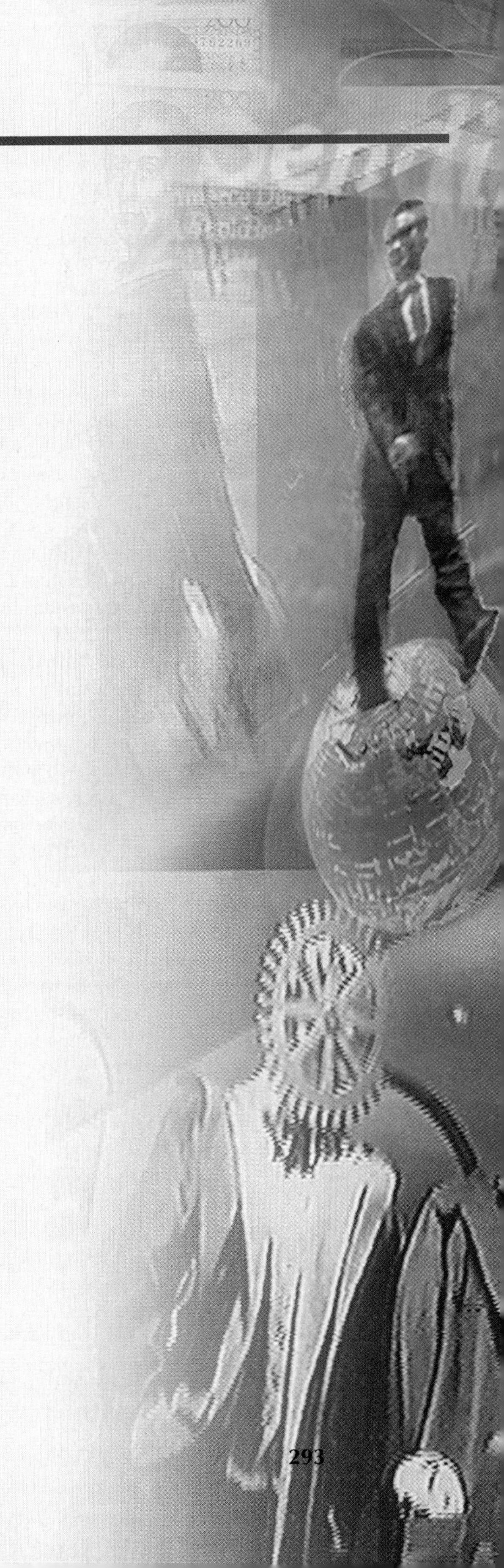

THE NATURE OF SOCIAL CHANGE

Throughout this book we deal with social changes of all kinds. The term **social change** refers to variations over time in the ecological ordering of populations and communities, in patterns of roles and social interactions, in the structure and functioning of institutions, and in the cultures of societies. We have seen that such changes can result from social forces building within societies (**endogenous forces**) as well as from forces for change exerted from the outside (**exogenous forces**).

Ebrié is not a written language. The villagers speak their history to each other and to their children as they gather in the evenings around their home fires. When I was living with them they often told stories about how, in the distant past before the whites came, a tribal village sometimes grew too large. When that happened, severe conflict would result. Factions would form and there would be fighting over land or other resources. Soon there would be a major split: The stronger faction would stay and keep the existing village while the less powerful faction would establish a new village on unused land along the lagoon. This is an example of people adjusting to endogenous change caused by population increase, a very common cause of social change.

The Ebrié also told stories about recent changes in their lives, changes brought on mainly by colonial conquest and, more recently, by their own efforts to create a new nation and maintain its social and economic growth. Colonial rule and rebellion against it are examples of social change produced by exogenous forces, particularly the influence of a powerful conquering society attempting to impose its culture on conquered peoples.

Social change affects people's lives all over the world, sometimes for the better, sometimes for the worse. As I sat under the palm trees with my Ebrié friends, we endlessly debated issues of social change. In many ways our debates were about the difference between change and progress. "Before the white man came," my friends told me, "we had time but no watches. Now we have watches but no time." The village primary school was known in Ebrié as the "paper house" where children struggled to master the three R's in order to be able to earn money, or "white man's paper." But there was also general agreement that many of the technologies introduced by the Europeans were of immense value. Health care innovations that could prevent debilitating diseases like malaria or schistosomiasis were considered a great blessing. Yet the most prestigious Western invention from the villagers' viewpoint was the camera, for with it, they exclaimed, "our children can see their ancestors."

The people of Ivory Coast and other societies that are undergoing rapid social change do not see the future as a matter of becoming Westernized. Their struggle to maintain aspects of their traditional way of life as they build a modern national state is documented in the Visual Sociology section at the end of the chapter. They are aware of a decline in their sense of community and mourn the weakening of their culture, but they actively embrace the aspects of modernity that will permit more of them to lead longer and perhaps more enjoyable lives. For my Ebrié friends there is no question of returning to an earlier state. Change is inevitable. They want to help make the change, however, and to do this they understand that they must compete in modern social institutions like schools and businesses while at the same time they attempt to preserve their village life as best they can.

Dilemmas of Social Change

Social change is a dominant theme of this book because it exerts such powerful influences on every aspect of our lives. Macro-level changes in the way the entire society is organized—such as the organization of metropolitan regions around automobile transportation—are relatively rapid. Our great-grandparents traveled in railroad sleeping cars and never imagined vast shopping malls with acres of automobile parking. Global social change accounts for the decline of whole sectors of the U.S. economy and the rise of new ones. At the middle level of social organization—in our communities and workplaces—social change can have drastic and immediate impacts. At this writing, for example, there is a trend toward reducing the size and cost of public institutions. Governmental agencies of all kinds are under attack, and their employees are experiencing the stress of cutbacks and layoffs.

At the micro level, social change also has dramatic impacts. Changing norms of sexuality, for example, may give rise to new and unexpected situations. As gay people realize their need to express their sexuality, they also wish to "come out" and speak frankly about their lifestyle with family members. Issues like abortion, interracial marriage, divorce, and drug abuse create similar conflicts and pressures. When we argue about norms within our intimate circle of family or friends, the result may be even greater confusion. Today no aspect of life seems secure and free from conflict. (More comparative examples of social change at different levels of the social order are presented in Table 10.1.) This raises the question of whether social change is always occurring everywhere—that is, whether social change is universal.

Is Social Change Universal? It would seem that ever since the industrial revolution about three centuries ago social change has been occurring throughout the world. For thousands of years people had lived in simpler societies that might or might not have been experiencing social change as a consequence of such processes as population growth, factional conflict, war, and disease. But with the advent of industrialization, urbanization, colonial conquest, and global warfare, all the societies of the earth were increasingly brought into what sociologist Immanuel Wallerstein (1974) has called the capitalist world system. Trade and markets, often dominated by the more powerful nations, enmesh all the world's people in a network of relationships and interdependencies that make self-sufficiency and social equilibrium ever more difficult to attain. As we have noted in earlier chapters, perhaps no change in human history has been as far-reaching and universal in its consequences as the transition from an agrarian to an urban, industrial way of life brought about by the industrial revolution. In sum, social change may be more or less rapid in different societies and in different parts of the world, but the

TABLE 10.1 Examples of Effects of Social Change

		Tribal Society (Ebrié)	Metropolitan Society (United States)
Macro-Level Changes	Population growth	Creates new markets for Ebrié real estate; adds newcomers to villages.	Causes new conflicts and need for social control in increasingly congested regions.
	Globalization: economic and cultural	Gives tribe members jobs at branches of multinational companies; threatens forests and waters if environmental controls are not established.	Encourages manufacturing concerns to move their production overseas; forces people to find more jobs in services and high-tech manufacturing.
Middle-Level Changes	Urbanization	Surrounds villages with new neighborhoods and non-tribal strangers.	Enmeshes most residents in metropolitan regions, where their community boundaries are blurred and diversity becomes an issue.
	Increased crime	Causes villagers to distrust strangers.	Engenders efforts to control crime and encourages privatization.
Micro-Level Changes	Rising material expectations	Leads village women to rebel against polygyny.	Pushes more women into the labor force, creating the need for child care.
	Changing norms of conduct	Creates a sense of anomie, confusion over expectations.	Increases the level of conflict in communities over sexuality, public conduct, and the like.

forces of global social change are felt everywhere. Thus, at least in our own era, it can be said that social change is universal.

Is Social Change Progress? Those who promote social change of one kind or another often equate change with progress, but in fact much social change is extremely difficult and disruptive for the people who live through it. "Progress" is defined differently, depending on the values of those who are trying to assess whether a particular change constitutes progress (Harper, 1993). In some instances social change has positive consequences for many people, and perhaps even for an entire society. Fifty years ago, for example, the United States was a highly segregated society, unable to claim that it was realizing the democratic vision of its founders. Today, after decades of struggle, many aspects of segregation remain, but many of the worst aspects of racism have been reduced or eliminated through protest, compromise, new laws, and other forms of social change. However, for every instance of positive social change, there will be others that people do not agree on or that are seen as negative or too complicated to assess. When factories close and communities lose vital jobs, that is a form of social change that most people in the affected communities do not think of as progress. Yet when the same jobs become available in Mexico, people there rejoice at their good fortune. Similarly, when political leaders persuade local police forces to protect the rights of homosexuals, gay people may feel that progress has been made. Yet in the same communities there will be conservative residents who wish to return to a past in which homosexuals were considered sinners who did not deserve equal rights. On the world scene, the vast majority of people in the former Soviet empire regard the fall of Stalinist communism as progressive social change. But the people of Bosnia and Kosovo, caught in a bloody conflict that erupted in the power vacuum left by the end of communist rule, are not at all sure that they are witnessing progress.

i

Can We Control Social Change? The examples of the Bosnian and Albanian conflicts and changing race relations in the United States raise a fundamental question about social change: Is it something we can control intentionally, or is it a result of powerful social forces that are beyond our ability to shape or modify? The answer is that some social change is intentional and some is far beyond our power to control. In a society governed by laws and democratic processes, it appears possible to reduce injustice and even to reduce some of the worst forms of inequality. But global forces of social and economic change are extremely difficult to shape even if we can learn to accommodate and adjust to them (Etzkowitz & Glassman, 1991).

Even when we engage in intentional social change, the results are often surprising and unanticipated. Before the Social Security and Medicare systems were created, the elderly experienced extremely high rates of poverty and untreated illness. These intentional and hotly debated programs changed that situation dramatically. However, it is impossible to control the rapid rise in the number of elderly people who will be covered by these programs in coming decades. This threat to the financial stability of the system may result in a weakening of the scope of Social Security and Medicare coverage in the future.

Can Social Change Be Predicted? Social science has had mixed success in predicting social change. The master trends of our era, such as economic globalization, population growth, urbanization, technological change, and the rising expectations of oppressed people, are well known. Their consequences can be understood and predicted, to a degree. Demographers can predict with reasonable certainty the major population trends that will affect societies over a 20- or 30-year period. Survey researchers can predict the outcomes of election campaigns or referenda on major social issues like immigration or crime. But we have limited ability to predict technological changes or to anticipate the outbreak of wars or the appearance of new social movements. Yet these are some of the most important sources of social change. In fact, one difference between the "pop" sociology books sold in supermarkets and the sociological research conducted at universities is the caution with which the university researchers discuss social change (Chirot, 1994a).

Sociological researchers usually make projections based on current trends, with carefully stated assumptions about what will happen if those trends continue into the future. For example, we know that as members of the huge baby boom cohort of the U.S. population grow older and face death, the media will pay more attention to issues involving aging, death, and dying, just as they paid more attention to issues such as day care when the baby boomers were forming families. But other aspects of social change, especially those due to technological innovations, are far more difficult to predict. We have difficulty predicting their appearance and limited success predicting their consequences.

Consider an example. For decades sociological writers who produce forecasts for popular media have been predicting that computers and telecommunications networks would spell the end of the central city as a place of work and entertainment for masses of people (Naisbitt & Aburdene, 1990; Toffler, 1970). However, research exploring this prediction shows that the new technologies have very mixed consequences for urban downtown areas. In large cities like Los Angeles,

Chicago, New York, and Baltimore, there is an unpredicted renaissance of central-city business and entertainment and an explosion of new small businesses, often based on applications of computer and communications technologies. But it is also true that the new communications technologies make it easier for employers in the information industries to distribute work among employees at many different locations. Hence, the future of central cities as employment-generating environments is extremely difficult to predict.

In sum, there are major trends that we can study and use to predict some important aspects of continuing social change, but there are others that are far beyond the power of any social scientist to actually predict. The breakup of the Soviet empire in 1989, the outbreak of war in the Balkans, the dramatic changes in U.S. social policies after the conservative sweep of the 1994 congressional elections, the sudden collapse of many Asian economies in the late 1990s—none of these events was actually predicted, even though there were some indications that they might occur.

TWO FORCES OF SOCIAL CHANGE

War and modernization are among the most powerful and pervasive forces that produce social change at every level of social life. Of course, as we have noted, social change is caused by many other social forces, especially technological innovation and population growth or mobility—but we will see in this section that war and modernization also stimulate these other forces of change. Many of the most significant technological innovations of the twentieth century were developed in response to the crisis of war. And many of the greatest movements of people over the planet were first set in motion by the disruptions caused by warfare or modernization (Chirot, 1986, 1994a, 1994b; Janowitz, 1978).

War and Conquest

War is among the greatest and certainly the most violent of the forces that produce social change. Ironically, the deadliest forms of warfare are associated with the rise of modern civilizations. As societies have become more advanced in their command of technology and their social organization, the devastation caused by war has increased. The wars fought by so-called primitive societies were frequently ritual affairs; the combatants often withdrew from the field after a single skirmish. Although not all preindustrial warfare was so ritualistic, relatively few combatants were killed because the technologies for killing were so limited compared with those available today.

In the Middle Ages, often viewed as a warlike time owing to the influence of knights and Crusaders, the rate of fatalities among warriors in battle was about 2 percent. In contrast, in World War I the proportion was 40 percent. Modern warfare is increasingly dangerous not only for the combatants but for civilians as well. In the Vietnam War more than 75 percent of those who died were civilians (Galtung, 1985; McNeill, 1982).

Any evaluation of the place of war in social change must consider four broad questions. First, what are the ecological effects of war on human populations? Second, how do wars help shape the consciousness and culture of a people? Third, how does warfare affect the roles of women in society? And fourth, how does war change the institutions of societies?

The Ecological Impact of War Casualties and conquest are the major ecological effects of war. People all over the world were shocked by televised scenes of oil fires and polluted coastlines during the 1991 Persian Gulf War, but the most significant ecological impacts of the Gulf War were felt by human populations. Like an epidemic of cholera or bubonic plague, war accounts for extraordinary and rapid declines in population. Pitirim Sorokin (1937) estimated that between A.D. 1100 and 1925 about 35.5 million people died in European wars alone. World War I claimed the lives of about 8.4 million soldiers and about 5 million civilians, and in World War II about 17 million military personnel and about 34 million civilians died. It is estimated that the Soviet Union lost about 15 million people during World War II and that in China about 22 million perished. Germany lost 3.7 million, Japan about 2.2 million, and the United States slightly under 300,000 (Beer, 1981).

When millions of soldiers are killed, entire populations are unbalanced for more than a generation. Many women remain single or become widows and either do not have children or raise children alone. The effect may be to reduce population pressure on food and other resources, but at the same time there are labor shortages and economic disarray due to the loss of so many skilled workers.

War also results in large-scale shifts in population and rapid acceleration of economic change. For example, the western parts of the United States and Canada experienced their most rapid growth as a result of mobilization for war during the first half of the twentieth century. New dams, new electric power plants, new factories to produce all kinds of goods were built. San Diego, Los Angeles, San Francisco, Portland, Seattle, and Vancouver all experienced massive population growth, as did many inland centers of industry and

Early in 1941, this was unused land at the tip of Los Angeles Harbor. As a result of full-scale mobilization for World War II production, entire metropolitan regions were transformed into massive industrial production centers like the one shown here.

agriculture. The end of the war saw continued growth in the western states as young families who had come west during the war decided to settle there permanently.

For the losers in war, the ecological consequences of defeat are usually far more dramatic. Population loss, economic subjugation, the imposition of a foreign language and culture, and forced movement to new towns and industrial areas are common. During more than a century of genocidal wars waged by whites against Native Americans in the nineteenth and early twentieth centuries, the consequences for the losers were death, expulsion, and banishment to reservations. For tribal peoples of Africa, invasion, war, and conquest led to colonial rule and rapid social change, often imposed through taxation, labor gangs, military draft, and similar means. In areas of the world that are torn by war today—such as the former states of Yugoslavia (especially Bosnia and Kosovo) and African countries like Rwanda, Burundi, Somalia, and Liberia—war has brought genocidal ethnic conflicts, mass expulsions ("ethnic cleansing"), and years of abject poverty in squalid refugee camps.

The Cultural Impact of War War changes a society's culture by stamping the memories of chaos and cruelty, heroism and camaraderie on entire generations. Years after a major war its effects on values and norms continue to be felt (Schuman & Scott, 1989). One need only think of the impact of the American Civil War on the former Confederate states, an impact that remains strong today and can be seen in Civil War memorials, rebel yells, the conduct of interracial relations, and North–South animosities. Recent sociological research shows that even for generations that did not experience war firsthand, the memory or threat of warfare is perceived as extremely important. Table 10.2 shows that people will cite specific wars first among "national events or changes that seem especially important to you"—even if they themselves were not alive during the wars they mention.

The impact of a major war can also be seen in the damage done to the minds and bodies of the survivors. In addition to the thousands who are maimed and mutilated, thousands more suffer from *post-traumatic stress disorder,* in which the shock of war continues to haunt the victim; others experience *survivor guilt,* the feeling of shame that many survivors feel because they escaped the fate of their comrades. These effects, which occur at the micro level of interpersonal relations, have been described by the Italian social observer Primo Levi, who was imprisoned at Auschwitz and liberated by Russian soldiers. When the soldiers encountered the piles of

During World War II, American GIs brought their culture to places throughout the world where they lived and fought. Here the orchestra of the First Cavalry Division takes time out for a jam session on Los Negros Island.

dead and the groans of the dying, Levi reports, they felt shame: "They did not greet us, nor smile; they seemed oppressed. . . . It was the shame which the just man experiences when confronted by a crime committed by another, and he feels remorse because of its existence" (1989, p. 72).

This kind of shame and guilt pervades a culture that has been torn by war. People feel that those who died were the best, the most valorous of society's members. Levi describes the shame he felt at the deaths of Chaim, a watchmaker who tried to teach him how to survive in the camp, and Szabo, a tall, silent Hungarian

TABLE 10.2 U.S. Respondents' Rank Ordering of Important Events as They Perceive Them

	Age					
Event/Change	18–29	30–39	40–49	50–59	60–69	70 Plus
World War II (265)	14	16	24	29	30	23
Vietnam (144)	18	18	13	2	4	1
Space exploration (93)	8	6	8	10	6	8
Kennedy assassination (62)	3	8	10	3	1	1
Civil rights (77)	7	7	5	7	6	3
Nuclear war, threat of (55)	6	5	6	4	2	3
Communication/transportation (46)	1	4	4	5	3	9
Depression (43)	3	3	2	5	7	13
Computers (23)	2	1	2	3	2	0
Terrorism (43)	4	2	0	1	1	0
Moral decline (28)	2	2	2	2	4	1
Women's rights (20)	1	2	3	0	2	1
Other event/change (357)	30	26	22	29	33	37
	100	100	100	100	100	100
	(289)	(312)	(200)	(167)	(165)	(110)

Source: Schuman & Scott, 1989.

peasant who needed more food than others yet never failed to help his weaker companions. He tells of his guilt over the fact that Robert, a professor at the Sorbonne, died even though he "spread courage and trust all around him," and that Baruch, a longshoreman from Livorno, died on the first day because he hit back when the guards beat him. "These, and innumerable others, died not despite their valor but because of it" (p. 83).

War and Gender Relations Throughout history women and girls have suffered particular cruelties during wars and conquest. As refugees from warfare, or as civilians in the midst of deadly battle, women are often abused, degraded, raped, and killed, sometimes along with their children. At this writing thousands of Albanian women have fled to the mountains of Kosovo, knowing that to stay in the path of their Serbian attackers is to risk rape, torture, and the possible violent death of their children. In Afghanistan, the victory of the Taliban fundamentalist army has brought untold suffering for women. They are subjected to the strictest possible religious rules, which deny them even the most basic opportunities for education and health care. In these examples, and many others that could be cited, women have been caught in the middle of vicious local conflicts that have developed largely as a consequence of the end of the cold war and its uneasy equilibrium of competing superpowers. But all wars in the modern age have taken an extremely high toll on women, as thousands of war widows can attest.

Women are by no means merely passive victims of warfare. Sociologist Suzanne Staggenborg, a researcher on gender, family, and social movements, notes that "Women's peace movements had existed in many countries since the early twentieth century and included international organizations such as the Women's International League for Peace and Freedom, which was founded in 1915" (1998, p. 45). Women's opposition to war is nothing new, however; the ancient Greek playwright Aristophanes dealt with the subject in *Lysistrata,* a semiserious comedy in which the women of Athens go on strike in an effort to avert war with Sparta. During the twentieth century the massive mobilizations of the two world wars—in which the vast majority of women and men had their lives disrupted and their careers dramatically altered by wartime industrial mobilization, military service, civilian defense, and the untimely death of loved ones who were also parents and breadwinners—had lasting effects on gender relations. These experiences increased the desire of women to enjoy the same rights and responsibilities as men, including the right to serve in the military. We return to this major aspect of social change later in this chapter and in Chapter 14.

War's Impact on Social Institutions The structure of a society, especially its major social institutions, may be drastically changed by war and preparation for war. The mobilization of large numbers of people and the marshaling of new technologies for military purposes have a centralizing effect on social institutions. In the United States, for example, the growth of large research universities in the 1960s was accelerated by huge investments in applied science and technology after the Soviet Union became the first nation to launch a space satellite. Universities that were capable of developing new science programs grew rapidly, and their administrations gained greater power. The power and influence of the national government has also grown, often at the expense of local governmental institutions, as a consequence of the two world wars and the arms race. Providing for national defense is extremely expensive and requires that the central government be granted increased taxing powers.

The French sociologist Raymond Aron (1955) has called the twentieth century "the century of total war" because of the capacity of warfare to shape the destiny of entire regions and because of the unprecedented power of nuclear and other weapons. Aron's phrase also captures the transforming power of modern war, its ability to alter societies. In the United States, for example, there is no doubt that the mobilization necessary to fight two world wars, major regional wars in Korea and Vietnam, and innumerable smaller skirmishes in Africa, Latin America, and Asia contributed to the controversial growth of the federal government during the twentieth century. The impetus for creating many of the institutions of the welfare state, such as Aid to Families with Dependent Children (the former national welfare program), was the need to care for war widows and their children. The massive growth of the federal budget deficit during the latter part of the century is attributable in large measure to the expenditure of billions of dollars on military institutions (Galbraith, 1995).

Elsewhere in the world, the dominance of military institutions and the fragility of legal and governmental institutions produce "garrison states" in which economic growth and the rule of law are subordinated to the needs of the military (Lasswell, 1941). The end of the cold war has led to political instability in middle Europe, parts of Africa, and many areas once dominated by the former Soviet Union. It has also caused unprecedented economic and social changes in Western Europe. Political instability in many regions of the world is used to justify continued military spending; in the United States, military spending still exceeds $250 billion annually. Environmentalists and other opponents of high levels of military expenditures argue that individuals cannot comprehend the magnitude of military spending unless it is presented in terms of trade-offs in-

Then and Now

Two Images of the Warrior

When the Plains Indians and the U.S. Cavalry fought bloody battles during the period of westward expansion in the nineteenth century, warfare was brutal and deadly, especially for the less well-equipped Indian warriors. But as deadly as it was, war was not nearly as devastating for noncombatant populations as it is today. A century ago warfare required guns and bayonets, cannons, and a great deal of organization and logistical support, but it did not involve sophisticated technologies like those represented by the modern fighter or bomber plane. As combat has become less a matter of brute force and more dependent on brains and technology, the possibility that well-trained women can compete as warriors has become a reality. This is a change in the conduct of war that would have been inconceivable to General Custer or Sitting Bull.

dicating what alternative social choices would cost. Table 10.3 presents some examples of such trade-offs.

Modernization

A second major source of social change is the set of trends that are collectively known as **modernization.** This term encompasses all the changes that societies and individuals experience as a result of industrialization, urbanization, and the development of nation-states. These processes occurred during a period of two or more centuries in the Western nations and Japan, but they are taking place at a far more rapid rate in the former colonial societies that are today's new nations.

The term *modernization* should be used cautiously, in its sociological sense rather than as a value judgment about different societies. It does not mean that we can judge life in modern societies as better or more satisfactory or more humane than life was for people in societies like the one the Ebrié once knew. As noted earlier, modern societies have developed the capacity to cause more destruction and human suffering than any simpler society could possibly have caused. And we have seen

TABLE 10.3 Trade-Offs Between Military and Social or Environmental Priorities

Military Priority	Cost	Social/Environmental Priority
Trident II submarine and F-18 jet fighter programs	$100,000,000,000	Estimated cost of cleaning up the 10,000 worst hazardous waste dumps in the United States.
Stealth bomber program	$68,000,000,000	Two thirds of estimated costs to meet U.S. clean water goals by the year 2000.
2 weeks of world military expenditure	$30,000,000,000	Annual cost of the proposed UN Water and Sanitation Decade.
3 days of global military spending	$6,500,000,000	Cost to fund Tropical Forest Action Plan over 5 years.
Development cost for Midgetman ICBM	$6,000,000,000	Annual cost to cut sulfur dioxide emissions by 8–12 million tons/year in the United States to combat acid rain.
2 days of global military spending	$4,800,000,000	Annual cost of proposed UN Action Plan to halt Third World desertification over 20 years.
10 days of European Economic Community military spending	$2,000,000,000	Annual cost to clean up hazardous waste sites in ten European Economic Community countries by the year 2000.
1 Trident submarine	$1,400,000,000	Global 5-year child immunization program against six deadly diseases, preventing 1 million deaths a year.
2 months of Ethiopian military spending	$50,000,000	Annual cost of proposed UN Anti-Desertification Plan for Ethiopia.
1 nuclear weapon test	$12,000,000	Cost for installation of 80,000 hand pumps to give Third World villages access to safe water.
1-hour operating cost, B-1B bomber	$21,000	Cost of community-based maternal health care in ten African villages for 10 years.

Source: Adapted from Renner, 1989.

in other chapters that many of the advantages enjoyed by the most modern nations have come at a high cost to simpler, less modern societies.

The concept of modernization does not assume that change is irreversible. For example, the rise of nation-states throughout the world during the past few centuries does not imply the end of loyalties based on a sense of "peoplehood" that conflict with the sense of shared citizenship in a nation (see Chapter 4). The bloodshed in the former Yugoslavia and in many places on the African continent bears heartrending witness to the strength of feelings about peoplehood as opposed to loyalty to the modern nation-state with its laws and governments.

Nevertheless, the term *modernization* summarizes most of the major changes, for better or worse, that societies throughout the world are experiencing, albeit at differing rates and with varying amounts of social disruption (Chirot, 1994a). Neil Smelser (1966) associates modernization with the following set of changes:

1. *In the realm of technology,* a developing society is changing from simple and traditionalized techniques toward the application of scientific knowledge.
2. *In agriculture,* the developing society evolves from subsistence farming toward the commercial production of agricultural goods. This means specialization in cash crops, purchase of nonagricultural products in the market, and often agricultural wage labor.
3. *In industry,* the developing society undergoes a transition from the use of human and animal power toward industrialization proper, or men working for wages at power-driven machines, which produce commodities marketed outside the community of production.
4. *In ecological arrangements,* the developing society moves from the farm and village toward urban concentrations. (pp. 110–111; emphasis added)

These processes can take place simultaneously, but this is not always the case. Many societies mechanize their agriculture and begin to produce cash crops for foreign markets before their cities and urban forms of employment have begun to grow rapidly. This was the case, for example, in Sri Lanka (formerly Ceylon), Indonesia, and many of the newer African nations.

Smelser and others who study modernization have shown that "technical, economic, and ecological changes ramify through the whole social and cultural fabric" (1966, p. 111). In the political sphere of life, we see the authority systems of the village giving way to domination by the institutions of nation-states. In the area of education, as societies attempt to produce workers who can meet the needs of new industries, new educational institutions are established. In the area of religion, there is a decrease in the strength of organized religions. Families change as traditional extended families adapt to new economic institutions that demand greater mobility.

Patterns of inequality in societies also change. Older patterns of gender inequality are modified (and often replaced by new forms of inequality) as women are in greater demand to fill positions in new economic institutions. And the emergence of a new class, the wage workers, increases the power of the common people, usually adding to their determination to become better educated and to participate more fully in political life. None of these changes is inevitable or irreversible; workers, for example, may see their unions "busted" in times of recession or economic change. But in the long run all of these trends are likely to appear in a modernizing society.

Modernity, Technology, and "Cultural Lag" The central part that technology plays in modern societies has led some sociologists to view it as a basic principle of social change. The classic statement of this view is that of William Fielding Ogburn (1942). Ogburn hypothesized that inventions affect the size of populations, which in turn influences the course of history. (For example, overpopulation often leads to wars and migrations.) Some inventions affect population directly: Improvements in sanitation, the development of cures for fatal diseases, and more effective contraceptive techniques are examples. But inventions can also have indirect effects on population. For example, techniques that improve crop yields or permit long-term storage of food surpluses make it possible to support a larger population with a given amount of farmland. And improvements in military technology (e.g., the use of horses in warfare, the invention of gunpowder, and the development of the armored tank) have had dramatic effects on the conduct of war and hence on population size.

Ogburn also proposed the theory known as **cultural lag.** In his words, "A cultural lag occurs when one of two parts of culture which are correlated changes before or in greater degree than the other part does, thereby causing less adjustment between the two parts than existed previously" (1957, p. 167). This theory is most often applied to the adaptation of social institutions to changing technologies. For example, the industrial revolution gave birth to many kinds of machines, often with moving parts that made them dangerous to use. The rates of injury and death resulting from industrial accidents climbed rapidly in the decades following the introduction of the new machines to the United States around 1870. Such accidents spelled disaster for workers and their families, since it was hard to prove that the employer was responsible for the accident. It was not until around 1910 that the concepts of employer liability and workers' compensation were adopted, a lag of about 40 years.

One problem with the cultural lag theory is that it fails to account for the effects of social power. For example, workers who sought compensation for the costs of industrial accidents did not have nearly as much power as the owners of the machines. When this power imbalance changed as a result of the labor movement, it became possible to enact legislation that would protect the workers.

The Postmodernist Critique For more than two centuries the forces of modernity have been changing societies throughout the world. In fact, all species of life on the earth are affected by these broad-ranging changes in the way humans exist. It is not surprising, then, that the forces of modernity also stimulate a wide variety of countertrends and reactions. The desire for gender and racial equality that accompanies modernity is countered by antifeminist and racial supremacy movements. The increasing influence of scientific thought is countered by revivals of fundamentalist religious beliefs and opposition to scientific education. The demand by some Christian conservative groups that public schools teach an anti-Darwinian view of creation, sometimes called "creation science," is an example of this trend in the United States.

Many other social movements are direct attacks on modernity itself or on specific aspects of modern thought. One of those aspects is *rationality,* which Max Weber viewed as a central principle of modernity (see Chapter 6). When applied to economic and governmental affairs, rationality fosters the development of bureaucracies, which, at least in principle, apply rational codes and regulations to the conduct of their affairs. The modern emphasis on rationality in scientific thought has been accompanied by an increase in *secularism*—separation of church and state, abolishment of state religions, and protection of free speech (including agnosticism and atheism). However, many critics point out that along with the bounty provided by scientific rationality have come the destructiveness of the nuclear age and the erosion of religious and ethical values.

In recent decades the critique of modernity in the older developed nations has given rise to a diffuse school of thought known as *postmodernism* (Foucault, 1973, 1984; Lash, 1992). Postmodernism can be defined as a critique of modern societies and cultures. It argues that science, rationality, and all the "isms" of the modern world—capitalism, socialism, behaviorism, to name only a few—deprive human lives of spirituality, mystery, myth, and diversity of expression. Modern economic institutions transform popular forms of music and art into mass-produced products. Postmodernist critics also believe that the major institutions of modern societies, including markets, laboratories, clinics, the military, and other bureaucratically controlled administrations, have become instruments of social control by powerful elites (Harper, 1993). According to postmodernist theories (which in some ways resemble New Age ideas), this control is weakening as people feel increasingly estranged from modern institutions and turn to new expressions of fantasy, myth, sexuality, and styles of dress and architecture. In health care, for example, disillusionment with some aspects of modern medicine leads people to try a wide range of alternatives, such as folk healing, holistic medical practices, and remedies that have not been scientifically tested (Deierlein, 1994).

Postmodernism remains influential as a critique of modernity and an explanation for the proliferation of often zany styles and behaviors in today's world. It is especially popular in academic circles, but aside from stimulating the development of some new cultural products (e.g., styles of architecture that refer to many different periods) it has not had any perceptible influence on modern corporations or other bureaucracies. The postmodernist critique is an important reminder that modernization has many negative effects, but it has not stimulated social movements that seriously challenge the growing influence of modern social institutions such as global corporations. The same cannot be said for other antimodernist cultural movements, such as fundamentalism, as we will see shortly.

According to postmodernist theories, styles of dress and grooming and preferences for particular lifestyles and tastes in music will become ever more diverse. At the same time, there will be less and less consensus on what the norms of society should be.

Modernization in Developing Nations Social scientists often use the term *Third World* to refer to nations that have won independence from colonial dominance in the decades since World War II. If the "first world" is that of the capitalist nation-states and the "second world" that of the former communist nations, the Third World nations are those that are not aligned with either of these "worlds" but are united in their need to survive in an environment dominated by more politically or economically powerful nations.

But the term *Third World* can be misleading, since in the past 30 years many of these nations have made strides toward modernity. Moreover, trends in global social and economic change have produced impoverished areas—sometimes called the "fourth world"—in affluent nations where illegal immigrants or stigmatized minorities work under conditions similar to those found elsewhere in the Third World (Sassen, 1991, 1998). For this reason, we prefer to use the term *developing nations* or *modernizing nations.* A **developing nation** is one that is undergoing a set of transformations whose effect is to increase the productivity of its people, their health, their literacy, and their ability to participate in political decision making. These transformations occur at different rates in different nations. Such differences are evident when we compare nations like Mexico and Brazil, which are industrializing rapidly, with nations like Chad and Mali in the interior of Africa, which are having far more trouble achieving the hallmarks of modernization (see Table 10.4).

The theory of modernization as we have described it implies that modernization will occur in a similar fashion in every society. But the differing experiences of the developing nations call this view into question. Not only do we often see the industrialization of agriculture (i.e., the growth of huge mechanized farms) without the rise of industrial cities, or the growth of cities without a decline in the strength of organized religions or the emergence of modern educational institutions, but we also see the rise of antimodernist social movements in some of these nations (Randall, 1998). Events in the Islamic world are a case in point. In Algeria, Pakistan, Iran, Saudi Arabia, Libya, Sudan, and other Islamic nations, a fundamentalist religious movement has been gaining strength in the past quarter century. This resurgence of traditional Islamic beliefs and

TABLE 10.4 Indicators of Social Change in Selected Nations

	Bolivia	Chile	Cuba	Egypt	El Salvador	India	Mexico	Pakistan	Turkey	United States
Percentage of population living in urban areas	57.7	85.8	72.8	44.8	50.4	26.8	71.3	31.5	68.8	76.2
Percentage of population literate	79.5	81.8	95.7	51.4	74.1	52.0	85.9	35.0	82.3	95.5
Daily per capita caloric intake	2,094	2,582	2,833	3,335	2,663	2,395	3,146	2,315	3,429	3,732
Daily newspaper circulation per 1,000 population	52	63	122	44	90	21	142	6	71	238
Energy consumption per capita (mil. Btu)	—	55	37	27	—	12	58	13	43	352
Military expenditure as percentage of GDP	2.4	1.9	1.6	4.1	1.2	2.9	0.6	6.0	4.1	4.3
Percentage of population under age 15	40.6	30.5	22.8	38.0	38.7	35.2	35.9	46.3	32.9	22.0
Life expectancy at birth (years), male	60.9	71.5	73.9	65.4	65.0	58.7	69.0	62.0	69.1	73.4
Life expectancy at birth (years), female	65.9	77.4	77.6	69.5	72.1	59.8	75.0	64.0	74.0	79.6

Source: Data from *Britannica Book of the Year,* 1997; *Statistical Abstract,* 1998.

practices denies that modernization must be accompanied by the rejection of religious faith, by the separation of religion and government, or by more democratic political participation. These and other aspects of the Western version of modernity are being strongly challenged by the Islamic fundamentalist movement (Ashmawi, 1998; Rejwan, 1998; Naipaul, 1998).

Studies of these antimodernist movements are being conducted throughout the world. For example, in her studies of Algerian women confronting the antimodernist movement sweeping through the Arab nations, the Algerian American sociologist Marnia Lazreg (1994) found that these movements are misnamed. In reality, she concludes, they are actually radical attempts to seize political power in which the movement's leaders use the religious fervor of the masses to achieve their goal.

The rise of antimodernist movements is not limited to the Islamic nations. Similar movements can be seen in the United States. Conservative groups plead for a return to more traditional values, and some radical groups advocate a return to self-sufficient communities that would engage in farming on a small scale. The effects of such movements on a society's institutions show that modernization does not necessarily follow a single direction or imply a single set of changes (e.g., the decline of religious faith, the rise of science, or the growth of industrial cities).

Another challenge to this view of modernization is posed by the fact that the world's resources of raw materials, water, and energy are far less plentiful than they once were. Today there are serious doubts about whether those resources are adequate to permit the poor nations to become developed to anywhere near the extent that the Western nations have, or whether the rich nations can continue to grow as they have in the past.

Modernization and Dependency Some sociologists argue that the development of the more advanced modern nations actually impedes development in the newer nations, or at least channels it in directions that are not always beneficial. In a general statement of this theory, André Gunder Frank (1966) questioned the idea that the less developed societies are merely at an earlier stage of modernization than the advanced nations. He cited the development of one-crop economies in many parts of Latin America as evidence of how social forces in the developed nations actually transform the tropical countryside. According to Frank, when peasants give up subsistence agriculture and trading in local markets because their land has been absorbed into huge banana or coffee plantations, the result is a form of underdevelopment that did not exist before, one in which the peasantry is transformed into a class of landless rural laborers.

Immanuel Wallerstein (1974; Hopkins & Wallerstein, 1996) has proposed a more general theory that he calls *world system theory.* In this theory he divides the world into core states, semiperipheral areas, and peripheral areas. The **core states** include the United States, England, France, Russia, Germany, and Japan—which are the most technologically advanced nations and which dominate the banking and financial functions of the world economy. The **semiperipheral areas** are places like Spain and Portugal, the oil-producing nations of the Middle East, and Brazil and Mexico. In these areas industry and financial institutions are developed to some extent, but they remain dependent on capital and technology provided by the core states. The **peripheral areas** include much of Africa, Asia, and Latin America. They supply basic resources and labor power to the core states and the semiperipheral areas. This world system, Wallerstein asserts, is based on various forms of economic domination and does not require political repression.

Wallerstein's theory has the drawback of suggesting that the so-called core states do not include any areas of production that may resemble the peripheral, dependent regions more than they do the fully modernized nations. There is evidence that even the most modern nations contain such regions. For example, in a study of sharecropping in California, Miriam Wells (1996) showed that large California berry growers have been dividing their land into small plots and renting them to low-income farm laborers. Wells contends that the growers have adopted this strategy as a means of avoiding the higher costs of unionized farm labor. "The sharecropper is responsible for maintaining the plots, for harvesting and packing the fruit, and for hiring and paying whatever labor is necessary to accomplish these tasks" (p. 17). In return, the sharecropper receives from 50 to 55 percent of the proceeds minus the costs of handling, loading, hauling, and marketing the crop. Wells points out that modernization theory views sharecropping as an obsolete form of production, yet it can reappear even in the most advanced societies under certain conditions, such as lack of machinery for harvesting. Once again, therefore, we see that modernization is not a unilinear process with inevitable outcomes for every society or nation. People in all societies, including our own, experience social change in an infinite number of ways, as we will see in the next section.

SOCIAL CHANGE IN EVERYDAY LIFE

People often experience social change as highly problematic, and they often blame themselves for not coping with it more effectively. But as we saw in Chapter 1, the sociological imagination requires that we ask how our own troubles are related to larger social forces. And we have seen that modernity brings with it many contradictions. In seeking new opportunities in education, leisure pursuits, intimate relationships, and political participation, we also create new problems for ourselves. Thus, according to social theorist Ralf Dahrendorf (1990), anomie has become an element of the lives of many people, especially those who are still on the way to attaining full membership in their societies. The Ebrié felt such anomie when their peaceful habit of sleeping outside their houses was shattered by a robbery. But in many parts of the world the experience of anomie is even stronger and has stimulated some creative responses, as can be seen in Box 10.1.

Like most social theorists, Dahrendorf does not believe that we can or should attempt to reverse the course of modernization. Instead, he suggests, we need flexible institutions that can be adjusted without disruption. We are far from achieving such flexibility, however. When we look at the range of problems we encounter in our daily lives, we can readily see how much more change is needed. To appreciate this point more fully, it will be helpful to review some of the dilemmas of modern life that result from rapid social change. These and related topics will be discussed in more detail in later chapters.

Gender Roles and the Family

Clearly, one of the most significant social changes in Western societies in the past quarter century has been the changing definition of women's roles. The entry of women into the labor force is only one indicator of this aspect of social change. Today over 70 percent of all married women with children are at work in offices, factories, and other workplaces, compared with only about 4 percent in 1890 (*Statistical Abstract,* 1998). But it should be noted that there have always been subgroups in the population, especially African Americans and immigrants, for whom women's wages were necessary to the family's survival. Even among women who "kept house" a century ago, some 20 percent took in lodgers and earned cash in this fashion (Aldous, 1982; Modell & Hareven, 1973).

BOX 10.1 GLOBAL CHANGE AND U.S. SOCIETY

Singing Out Against Anomie

Bruce Springsteen sings about the feelings of loss that people experience when their factories are torn down. The murdered singer Selena captured the hearts of millions by singing popular Tejano songs that described her people's trials in coping with new social and economic conditions along the rapidly industrializing Texas-Mexico border. And in 1994 Susan Aglukark, a popular Eskimo singer from the frozen shores of Hudson Bay, became the first of her people ever to sign a major recording contract. EMI Music Canada signed her to produce more of her haunting songs about the struggle of aboriginal people in the Eskimo-controlled region of Nunavut, which extends well above the Arctic Circle in the Northwest Territories.

Aglukark's Inuit or Eskimo name is Uuliniq, which means "scarred from burns." Although she is not literally scarred—the name is purely symbolic—the same symbolic name could easily be applied to all of Canada's 35,000 Eskimo people. Like people living at the edges of industrialization and urbanization throughout the world, they have experienced severe disruptions in their traditional way of life. They can no longer maintain their existence as a hunting-and-gathering people, not only because of depleted supplies of game but also owing to the far-reaching effects of changing lifestyles and technologies. For example, their need for gasoline to power generators and snowmobiles and for televisions and prepared foods like coffee and sugar makes it impossible for them to maintain their traditional culture.

As a result of these social changes, many of Aglukark's people are experiencing anomie, or normlessness. Caught between the modern and the traditional, not knowing how to find work or where to go for advice, they often turn to alcohol and drugs, become angry and abusive, or even commit suicide. Aglukark's music has elements of rock and country gospel, but the words deal with subjects such as a girlhood friend who has taken her own life. She hopes that her songs will help young people everywhere deal with the feelings that lead them to despair and sometimes to suicide. Since this is a particular problem in aboriginal communities, she is concerned with showing her people that "life is the most beautiful thing."

Susan Aglukark is proud to be identified with the newly formed Eskimo region of Nunavut. She notes that the effort to cope with change has challenged her people for many generations: "The trouble was in the past, it was forced on us. At least now we may have some control over our future" (quoted in Farnsworth, 1994, p. A4).

For both women and men, these changes often produce feelings of guilt and stress. One writer on the subject (Fallows, 1985) urged young mothers to reject career goals so that they could raise their children themselves. On the basis of observations of day care centers, she had concluded that day care is a poorly developed institution that cannot substitute for maternal care. Arguments of this nature are common. But other studies show that high-quality day care does not impede a child's development. Moreover, advocates of gender equality question the assumption that child rearing should remain the primary role of women.

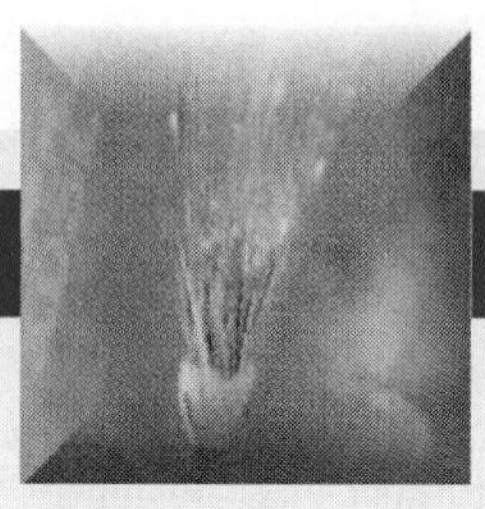

ART AND THE SOCIOLOGICAL EYE

Impressions of Social Change

Pierre-Auguste Renoir's painting of a gala scene in a nineteenth-century Parisian cafe is considered a masterpiece of impressionist modern art. But it also affords a glimpse into the immense changes that were occurring among Europe's middle classes late in the nineteenth century.

The impressionists, among whom Renoir was a pioneer, were rebelling against the strict rendering of pompous historical subjects that constituted "establishment art." They wanted to bring art outdoors, into the shimmering light of the everyday world. The impressionists were fascinated by the play of light on their subjects. They reveled in the freedom they felt to paint personal impressions of a scene rather than rendering it as if the canvas were a photographic plate. The scenes they captured, like this one in a cafe, were indicative of the social changes occurring around them. Here were women and men dancing in public. The women, especially, were exhibiting their dress and their emotions freely and in doing so were asserting their right to be full citizens of the modern city. No wonder the impressionists' art was considered daring and even dangerous by the cultural conservatives of their day.

Bal au Moulin de la Galette, by Pierre-Auguste Renoir, 1876. Oil on canvas, $51\frac{1}{2}'' \times 69''$.

This assumption has the effect of depriving women of opportunities to contribute their skills and talents to society. If day care is inadequate, a more practical response would be to improve it through public funding, education of child care workers, and the like, rather than forcing women to return to their traditional role as homemakers (Bergmann & Hartmann, 1995; Wrigley, 1995).

These arguments call attention to the need for more research on the future of families and other intimate groups in which children are reared. Some sociologists attempt to show how day care can be improved. Others conduct research on how people actually cope with changes in the household division of labor. Are we moving toward a "symmetrical society" in which men and women share equally in household and occupational pursuits? If so, the evidence suggests that we still have a long way to go. In an analysis of data from 555 couples, Carmi Schooler, Joanne Miller, and their associates (1985) found that typically the husband's sphere of household work "tends to be limited to household repairs, whereas wives are responsible for and actually do a vastly wider range of the household tasks" (p. 112).

Research on time use conducted by John Robinson confirms these findings. Robinson's analysis of detailed diaries of how men and women use their time each day shows that, on the average, women spend about thirty hours a week on child care, housework, and family care whereas men spend less than ten hours a week on such activities. While men often assume other responsibilities for the family, including putting in extra time at work, the data show that they are not yet close to sharing equally in family chores (Robinson, 1988; Robinson & Godbey, 1997). Women who have young children and work full-time are the most harried segment of the population; they have only an average of thirty-one hours a week of free time, as opposed to thirty-six hours a week for women who stay home with their children.

Another major issue related to changes in the family is the requirement that welfare recipients work at paid jobs in order to remain eligible for welfare benefits. This policy also involves the matter of day care, since single parents must have adequate care for their children while they gain the skills and experience needed to enter the labor force. Are we willing, as a society, to invest in social-welfare institutions like day care? The answer awaits further debate and political conflict. So too does the question of what, if anything, we can do about problems like teenage pregnancy, sexually transmitted diseases, or high rates of divorce. Changes in values give us more choices, more opportunities to realize our potential as individuals, but at what cost to society (Rochon, 1998)?

Race and Ethnic Relations in a Postindustrial Society

i

Another significant area of social change is race relations. When we look at the bitter ethnic strife occurring in Rwanda or Kosovo, we often congratulate ourselves on the progress our society has made toward racial and ethnic equality. Racial discrimination in the United States, which until fairly recently was supported by laws in many parts of the nation and by informal norms elsewhere, has decreased a great deal as a result of the civil rights movement of the 1960s. Since that stormy decade, blacks have made gains in all of our society's major institutions. Voting rights, greater access to education and jobs, achievement in sports, and full civil rights for blacks are often taken for granted. Yet as the 1992 Los Angeles riots clearly demonstrated, much needs to be done to address racial inequality. In the words of philosopher and sociologist Cornel West,

> The verdict that sparked the incidents in Los Angeles was perceived to be wrong by the vast majority of Americans. But whites have often failed to acknowledge the widespread mistreatment of black people, especially black men, by law-enforcement agencies, which helped ignite the spark. The Rodney King verdict was merely the occasion for deep-seated rage to come to the surface. This rage is fed by the "silent" depression ravaging the country—in which real weekly wages of all American workers since 1973 declined nearly 20 percent, while at the same time wealth has been upwardly distributed.
>
> The exodus of stable industrial jobs from urban centers to cheaper labor markets here and abroad, housing policies that have created "chocolate cities and vanilla suburbs," white fear of black crime and the urban influx of poor Spanish-speaking and Asian immigrants—all have helped erode the tax-base of American cities just as the federal government has cut its supports and programs. The result is unemployment, hunger, homelessness, and sickness for millions. (1992, p. 24)

In the chapters that follow we delve far more deeply into the causes and consequences of inequalities of income, race, gender, and age, as well as their importance for social change. It is worth noting here, however, that West's critical observations signal some of the ways in which inequalities of wealth and opportunity overlap with inequalities of race and tend to exacerbate the problem of race relations in American society.

Changes in the economic structure of society supply the main explanation for the growing divergence between the haves and the have-nots. Manufacturing jobs and many kinds of blue-collar service jobs are becoming far less numerous, while the number of jobs in white-collar service industries like finance, insurance, and banking is growing rapidly. Despite high overall levels of employment during the 1990s, higher levels of production of high-technology items like computers have not yet made up for the loss of well-paid jobs in heavy manufacturing industries like steel, autos, glass, and rubber. Blacks and Hispanics are more severely affected by these changes because they have long depended on heavy industry as a source of jobs.

Environmental Politics and Policies

A third area in which social change touches the individual is public policy. *Public policies* are laws and administrative regulations that are formulated by governments to control, regulate, or guide behavior. What public policies are likely to emerge as the United States is transformed into a postindustrial society and as other regions of the world industrialize? Clearly, one area of policy that will gain in urgency involves measures to safeguard the earth's environment. The contemporary exploration of space, applying the most advanced technologies, has allowed us to see the planet as a whole, which in turn has stimulated the development of environmental thinking and efforts to develop policies that will reduce problems such as global warming. However, to the extent that environmental policies impinge on people's livelihoods and on their reproductive behavior, these policies are bound to be extremely controversial and difficult to formulate (Perutz, 1992).

Environmental issues like the exhaustion of food resources, the spread of deserts, the destruction of forests by acid rain, and the denuding of large tracts of land for fuelwood are fast gaining a high place on the agenda of world politics. Figure 10.1 shows that population growth in excess of food-producing capacity is occurring in the Middle East and many parts of Africa; that fuelwood is scarce in India, Africa, and eastern Brazil; and that deserts are expanding on all the continents. Victims of "food insecurity" (people who lack sufficient food for normal health and physical activity) now total more than 100 million (Brown, Flavin, & French, 1997).

What are the political implications of these physical and social conditions? Gro Harlem Brundtland,

FIGURE 10.1 **Worldwide Patterns of Environmental Stress**

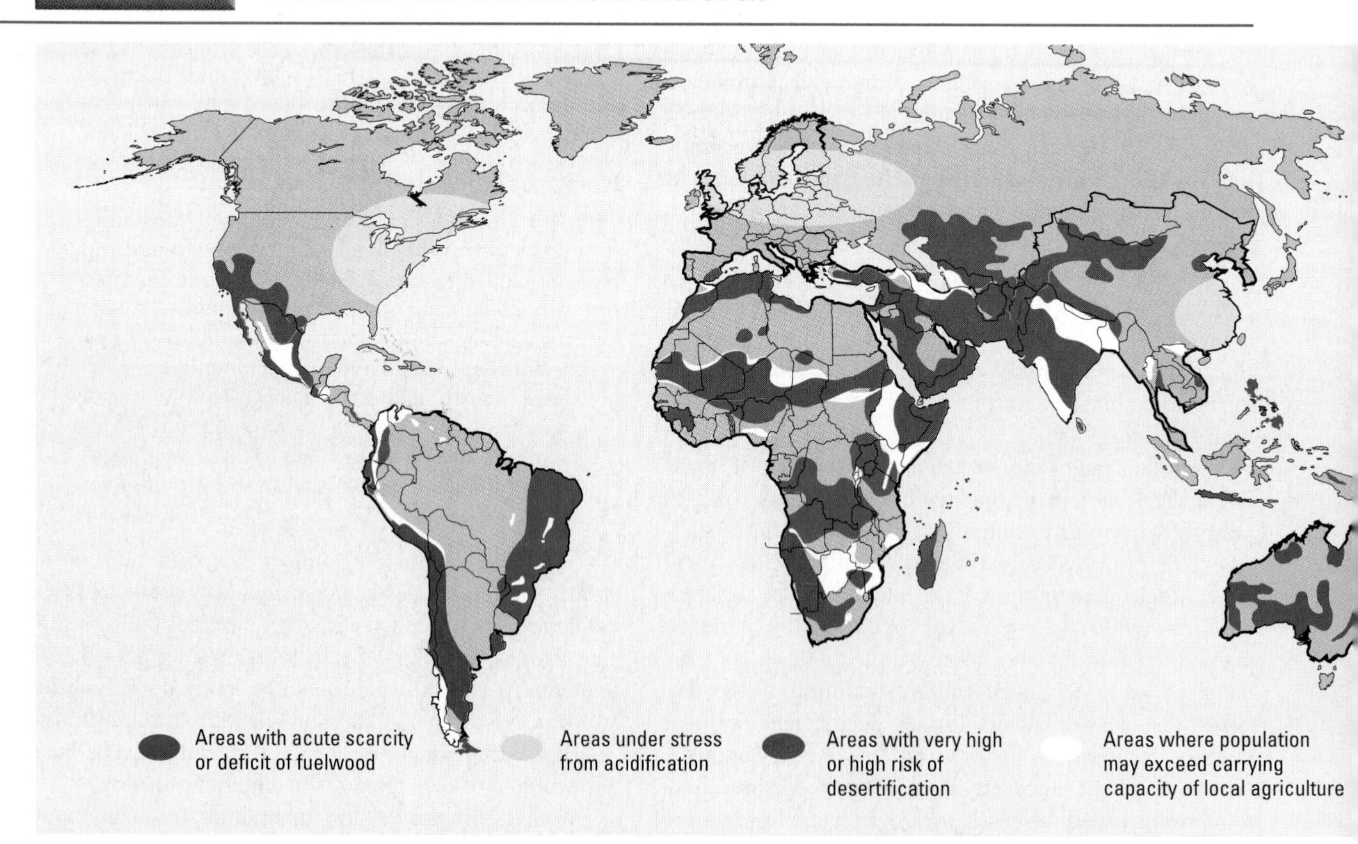

Land degradation results from a variety of human activities. Shown are regions threatened by overharvesting of firewood, acid rain, desertification, and stress induced by efforts to feed more people than the land is actually able to support.

former chair of the World Commission on Environment and Development, answers that "to secure our common future, we need a new international ethic based on the realization that the issues with which we wrestle are globally interconnected." The only way different nations can pursue their own self-interest, she adds, is for the United Nations or a similar world political body to become the institution in which environmental policy is made. And that policy cannot "insult the poor and tell them that they must remain in poverty to 'protect the environment'" (1989, p. 190). Clearly, the wealthy nations must make equal sacrifices to achieve a stable environment.

As the worldwide environmental crisis worsens, citizens of the United States and Canada will increasingly be faced with the need to regulate their economies and to end practices that contribute to pollution. This will be a costly and politically wrenching process, as evidenced by the difficulties encountered in developing policies for solid-waste disposal or the abatement of acid rain. A divided society, in which many people wish to be responsive to the environmental crisis but many others do not wish to think about it, is certain to produce conflict for years to come—unless a crisis such as flooding due to global warming creates such an obvious threat that unity of purpose is achieved overnight. Thus we face a dilemma brought on by great changes in the environment coupled with slower changes in the ability of political institutions to deal with them. This situation points to the need to look more closely at the models that sociologists use in assessing social change and predicting its course.

MODELS OF CHANGE

"We are such stuff as dreams are made on, and our little life is rounded with a sleep." These words from Shakespeare's *The Tempest* capture one of the basic difficulties of studying social change. In one short lifetime we catch glimpses of our ability to create a better world, but we can never realize all our ambitions, nor can we know what will become of our achievements or the problems we leave to later generations. Prices may shoot up again if inflation rates rise; long lines may form at gasoline pumps during a new oil shortage; or we may be faced with new medical problems like AIDS or dengue fever that affect the lives of large numbers of people. To cope with these changes, we may join new social movements and attempt to build new institutions or work to improve the ones we already have. We may also record our desire for a better society in cultural products of all kinds—in poetry and plays and novels, in film and music, and perhaps in social-scientific studies. Yet we know that social change will continue after we are gone, and we wonder whether it is possible to foresee what will happen to our society and civilization in the distant future.

Sociologists have often attempted to develop models of social change that span many generations and predict the future of whole societies or civilizations. Of course, none of these theories can be tested using data from actual experience. As Robert Nisbet has observed, "None of us has ever seen a civilization die, and it is unimaginable, short of cosmic disaster or thermonuclear holocaust, that anyone ever will" (1969, p. 3). Instead, he continues,

> We see migrations and wars, dynasties toppled, governments overthrown, economic systems made affluent or poor; revolutions in power, privilege, and wealth. We see human beings born, mating, child-rearing, working, worshiping, playing, educating, writing, philosophizing, governing. We see generation succeeding generation, each new one accepting, modifying, rejecting in different proportions the works of preceding generations. We see, depending upon our moral or esthetic disposition, good and evil, greatness and meanness, tragedy, comedy, and bathos, nobility and baseness, success and failure. (p. 3)

Nisbet's point is that we can trace trends in all of these areas, but it is extremely difficult to develop theories that can explain them all and, more important, predict the decline of existing societies and civilizations or the rise of new ones. Even when we believe we are witnessing the birth of a new society out of the chaos of revolution or war, it takes many generations to distinguish what is truly new, in terms of culture and social structure, from what has been carried over from the past. Despite these difficulties, models of change that seek to predict the future of entire societies or civilizations can be helpful. They allow us at least to compare new ideas about social change with those that have been in use for many decades. In this section, therefore, we review the most significant models of large-scale social change that have been proposed by sociologists in the past two centuries.

Evolutionary Models

Many of the founders of sociology were strongly influenced by evolutionary views of social change even before Darwin's theory seemed to offer an analogy between biological and social evolution (Chirot, 1994a; Nisbet, 1969). The main components of the nineteenth-century evolutionary model are as follows:

1. *Social change is natural and constant.* Social order exists even while change occurs; social change, on the other hand, is the means of attaining higher levels of social order.

2. *Social evolution has a direction.* Societies become increasingly complex. Émile Durkheim, for example, viewed societies as evolving from simpler forms based on similar segments like villages into more complex forms based on division of labor and the resultant interdependence among ever-larger numbers of people.
3. *Social evolution is continuous.* Change occurs as a result of social forces acting within a society, even without exogenous influences like colonialism. This happens through a steady series of stages. Many, but not all, evolutionary theories interpret social change as progress toward better conditions of life.
4. *Change is necessary and proceeds from uniform causes.* Because social change occurs naturally, continuously, and from within, it must be necessary. In other words, whether or not we want change, it will occur because of the logic of social evolution. And social evolution will be similar in all societies because all societies are similar in their ways of dealing with the dilemmas of human existence.

Two assumptions implicit in the nineteenth-century model of social evolution have been strongly criticized by twentieth-century social scientists. The first is that all of the world's societies would eventually resemble those of Western Europe in their institutions and even in their cultural values and ideologies. The second is that social evolution represents progress.

Modern evolutionary theorists refer to earlier models as *unilinear* because they predict that all societies will inevitably come to resemble Western societies. A less ethnocentric version of evolutionary theory is found in *multilinear* models of social change. These models do not assume that large-scale change in a society represents progress, and they attempt to account for the values that are lost as well as for those that are gained.

Multilinear models also emphasize that one must study each society separately in order to discover the evolutionary stages that are unique to a particular society as well as those that have been experienced by other societies (Lenski & Lenski, 1982; Sahlins, 1960; Steward, 1955). Thus societies like Ivory Coast, where the Ebrié live, have not developed much heavy industry and may never do so. On the other hand, Ivory Coast has a relatively advanced agricultural base that provides the surpluses needed for modernization to take place in other areas of social life.

Multilinear models can be useful in accounting for the erratic course of modernization in the Islamic world, or in helping the Ebrié in their attempts to understand what aspects of their village culture might remain viable even as their tribe becomes incorporated into a modern African state. But such models do not tell us why societies like Great Britain have declined from the heights of imperial power or why the ancient civilizations of Egypt and Rome flourished, declined, and are now known primarily through courses in ancient history and classics. An alternative viewpoint is found in cyclical theories of social change, which attempt to explain such phenomena by placing the possibility of decline at the same level as that of growth and "progress."

Cyclical Theories

In 1918, when Europe had been devastated by World War I, Oswald Spengler, a German schoolteacher turned historian, published a controversial book titled *The Decline of the West.* Spengler's gloomy thesis was that all societies pass through stages that are roughly equivalent to the life stages of human beings: infancy, youth, adulthood, and old age. The West, he argued, had passed through its maturity in the eighteenth century and was now experiencing a long period of decline. This process was inevitable and irreversible. There was nothing anyone could do to change its course.

A similar but more positive "rise and fall" theory of social change was developed by the British historian Arnold Toynbee (1972). In his "challenge and response" model, Toynbee suggested that all societies grow and decline as they respond to the challenges posed by their physical and social environments. For example, England needed to solve basic ecological problems, such as the fact that it is a small island nation with a limited supply of farmland. It responded by emphasizing foreign trade and using its superior naval power to protect its trade routes. When its naval power declined, it had to find new ways of meeting its challenges or face further decline. Thus for Toynbee the rise and fall of a society is accounted for by continual innovation in response to changes in its environment. However, although this theory is quite convincing as an explanation of history, it says little about what any particular society may expect in the way of challenges in the future.

Pitirim Sorokin (1937), a Russian immigrant who greatly influenced American social theory in the early twentieth century, also developed a cyclical theory of social change. His theory attempted to account for why a society or civilization might change in a particular way. All societies, he wrote, are continually experiencing social change, and such change originates in their cultures. This is because cultures are not unified but are marked by opposing sets of values, norms, and lifestyles. At one extreme of a society's cultural system is its "ideational culture," so named because it stresses spiritual values, hard work, self-denial, and a strong moral code. At the other extreme is its "sensate culture." This set of cultural traits encourages sensory experiences, self-expression, and gratification of individual desires. Neither of these extremes can produce a stable society. Ideational culture results in benefits that are enjoyed by pleasure seekers in later generations, and the society will decline until this trend is reversed and ideational traits are emphasized once again. At some point in its history a society may combine these two cultural tendencies. Then, according to Sorokin, it has reached its "idealistic point." But such a golden age is not likely to last forever.

Conflict Models

Cyclical theories of social change, with their cycles of rise and fall and their brief golden ages, may seem to confirm the common notion that history repeats itself (Chirot, 1994a). This notion is erroneous, however. Societies may experience similar events, such as wars or revolutions, at different times in their history, but the actual populations and issues involved are never exactly the same. More important, cyclical theories fail to deal with changes in social institutions or class structures. This is where conflict models of social change are most useful. They argue that conflict among groups with different amounts of power produces social change, which leads to a new system of social stratification, which in turn leads to further conflict and further change.

"The history of all hitherto existing societies," wrote Karl Marx and Friedrich Engels in *The Communist Manifesto* (1969/1848), "is the history of class struggles." As we will see in the next chapter, social classes are defined by how people make their living or gain their wealth. Marx argued that struggles between classes (e.g., workers vs. managers and company owners) are the source of social change in every period of history. In any society the main conflicting classes will be the exploiters and the exploited: those who control the means of production and those whose labor power is necessary to make those means of production actually produce. The exploited workers could become a revolutionary class—that is, one that could bring about an entirely new social order. But this can occur only when changes in the means of production—new technologies like the factory system, for example—make older classes obsolete.

We know that revolutionary class conflict has not occurred in many capitalist societies, nor did the former communist societies of Eastern Europe or the Soviet Union succeed in eliminating worker exploitation. Yet as we look around the world at the struggles between the haves and the have-nots, the rulers and the ruled, the rich nations and the poorer nations, we cannot help but apply Marxian categories and test Marx's theory of social change over and over again.

Some modern conflict theorists depart from the Marxian view of social change, finding conflict among many different kinds of groups and in every social institution. For Ralf Dahrendorf (1990, 1997), this conflict produces social change at all times, but the change is not always revolutionary. We cannot change our laws, our bureaucracies, or even our families, for example, without first experiencing conflicts among various group and individual interests. (An example of such a conflict is the half-humorous, half-serious debate that occurred in 1992 between Vice President Dan Quayle and television character Murphy Brown over what is meant by "family values.") In most cases, however, it is only when the deprivation experienced by whole classes or status groups is extreme that conflict is likely to be violent and to produce the unrest that could end in revolutionary social change.

Functionalist Models

From a functionalist perspective, social change occurs as a result of population growth, changes in technology, social inequality, and efforts by different groups to meet their needs in a world of scarce resources. There is no prediction of rise and fall or unilinear changes like those we find in early evolutionary theory or even in classical Marxian theory. Instead, the functionalist model sees change as occurring on so many fronts that it seems incredible that society can exist at all.

One of the dominant figures in functionalist social theory, Talcott Parsons (1960), developed a *homeostatic* model of society. As change occurs, said Parsons, a society's institutions attempt to restore it to something approaching equilibrium. Conflict is minimized through the emergence of legitimate governing institutions; decisions are made about who governs and with what form and degree of authority. Adjustments are also made in economic institutions: New occupational roles develop; old roles decline; wages and status rankings such as occupational prestige explain who gets what rewards. Cultural institutions, schools, the arts, the media, and religious institutions maintain the shared values that support our feeling that our government is legitimate, that a certain amount of inequality is required to maintain individual initiative, and that opportunities are distributed as well as can be expected.

Parsons and other functionalist theorists do not contend that efforts to adapt to change, to create an integrated, well-functioning social system, always work. The integrated functioning of social institutions can be disrupted, sometimes quite severely, when some institutions experience rapid change while others are slow to adapt. Technological innovations in health care, for instance, make possible longer lives for some individuals at great expense, but other economic and political institutions have been slow to develop norms for distributing

STUDY CHART Models of Change

MODEL	DESCRIPTION
Evolutionary	
Unilinear	Social change is natural and constant, has a direction, and is continuous. Change is necessary and proceeds from uniform causes.
Multilinear	Large-scale change in a society does not represent progress. Each society must be studied separately to discover the evolutionary stages unique to it.
Cyclical	
Oswald Spengler	All societies pass through stages like the life stages of humans, and eventually decline.
Arnold Toynbee	Societies grow and decline as they respond to the challenges posed by their physical and social environments.
Pitirim Sorokin	Social change originates in a society's culture, which alternates between "ideational" and "sensate."
Conflict	
Karl Marx	Social change results from conflict between social classes, which are defined by how people make their living or gain their wealth.
Ralf Dahrendorf	Social change results from conflicts among many different kinds of groups and in every social institution.
Functionalist	
Talcott Parsons	As change occurs, a society's institutions attempt to restore equilibrium.

these costly benefits among the members of society and finding ways of paying for them. (All the models and theories of social change discussed in this section are summarized in the study chart above.)

Applying Models of Change

The last two models we described—the conflict and functionalist models—clearly reflect two of the theoretical perspectives that are among the basic conceptual tools of sociology. They are most useful at the global or macro level of analysis. But they can also be used, along with the interactionist perspective, to explain social change at the micro and middle levels of social analysis. We can best illustrate this point with a closing example.

How can we apply the basic sociological perspectives to understand what is happening to the Ebrié as they become part of a bustling West African city? Starting from an ecological viewpoint, we must ask how the Ebrié are managing to make the transition from rural to urban life. We see them becoming more dependent on a worldwide economy. Changes in the value of their nation's currency, for example, affect them now just as much as the failure of a yam crop might have in an earlier time. The interactionist perspective helps us listen as they speak about how these changes affect them. It lets us see how they are adapting to a life led largely among strangers. Functionalism helps us understand how some of their tribal institutions can survive in a modern society. We see new roles emerging, and at the same time we see old ones declining in the face of new values like those regarding the place of women in society. Finally, the conflict perspective tells us that a good part of the answer to how much of Ebrié culture will persist and how much will be lost depends on the extent to which the Ebrié themselves resist assimilation into the national culture. Their struggle to maintain their language, their forms of worship, and their worldview will continue, even if it produces some conflict among themselves.

RESEARCH ON THE CUTTING EDGE: The New American Reality

Reynolds Farley is one of the leading demographers and students of social change in the United States. He is a master at using census data to chart the course of major social changes in U.S. society and culture. For Farley, the censuses conducted every 10 years "reveal social and economic trends in a uniquely rich and detailed matter . . . they reveal how people are now adapting—some with great success and others not well at all—to the massive social and economic trends that will make the United States of 2000 extremely different from what it was in generations past" (Farley, 1996, p. 1). Farley also uses large national surveys and opinion polls to supplement his demographic analyses of social trends. His analyses show how different generations of Americans have experienced the major events that have shaped their actions and values, and it is this aspect of his work that is highlighted here.

For social scientists who look at the effects of long-term and short-term trends, "social change occurs largely—but not entirely—on a cohort basis. The word **cohort** refers to all persons born in a specified time span" (Farley, 1996, p. 24). Generations differ most according to the social forces and cultural values that influence their life decisions in their late adolescent and young adult years. People who were young adults at the outbreak of World War II, for example, were likely to have had their adult lives disrupted by military service or early widowhood; on the other hand, they benefited from the array of programs created to help veterans and their families after the war, including education scholarships (the GI bill) and subsidized housing loans. The early baby boomers, who became adults during the late 1960s, experienced major upheavals in sexual mores and attitudes about military service and gender and race relations. "Generation X" (as it is often called) includes some of the readers of this text, or their older siblings. This cohort "grew up after social values had changed, but they entered a competitive job market, one in which those with specialized training would do well, but those lacking it would have a much tougher time finding a high paying job than did their parents. Their standard of living will, quite likely, fall below that of their parents—the early baby boomers" (Farley, 1996, p. 24). Perhaps the generation of Americans who are now in their late teens and early twenties will be thought of as the "millennium generation." They will share many of the same opportunities and concerns as Generation X, but perhaps they will be more open to new political and cultural ideas, or, depending on the economic conditions they experience or their exposure to war and civil strife, their values may be channeled in directions that we cannot anticipate. In consequence, the cohort analysis of social change ends, for now, with Generation X (see Table 10.5).

For Reynolds Farley and other students of social change, cohorts are often identified by decade. The adult behavior of these cohorts has been influenced by many different events and changes, but few are more vivid than the civil rights movement and the gender revolution. The two oldest twentieth-century cohorts, the grandparents and parents of the baby boom generation (refer again to Table 10.5), are now in their advanced years. In their youth and early adulthood the United States remained a racially segregated society—and thus it is no surprise that earlier in the century, when these older cohorts were coming of age, it was far more common than it is today for people to approve of segregation in jobs, transportation, neighborhoods, and education. Later cohorts, born after the civil rights era of the 1960s, are far more likely to express the antisegregationist opinions of their time (see Figure 10.2).

With regard to gender relations, women in the pre–baby boom cohorts were expected to become homemakers while their husbands worked outside the home. Even though millions of families did not conform to this cultural ideal, it remained the ideal, especially among more affluent Americans. In consequence, women were not expected to aspire to the same levels of social, economic, intellectual, or artistic achievement as men. Table 10.6 and Figure 10.3 show the dramatic changes in these norms that occurred during the twentieth century, especially since the 1970s. We see significant change beginning with the World War II cohort, and the rate of change accelerates for the two baby boom cohorts and levels off in Generation X, in which there is a discernable return to more traditional attitudes among men. A comparison of Figures 10.2 and 10.3 also shows that there is far more consensus among white respondents on attitudes about race than on attitudes about gender.

No doubt cultural change along the critical dimensions of race and gender will continue. But what will the direction and magnitude of these changes be in your generation and those to come? Because of the uncertainties of global political and economic conditions, we cannot predict the answers to these questions. However, thanks to analyses of census and survey data by Farley and others, we have a good understanding of the behavior of earlier cohorts, making possible intelligent comparisons in the future.

TABLE 10.5 Twentieth-Century Birth Cohorts

Years of Birth	Name of Cohort	Became Young Adults	Key Events at That Time	Ages in Census of 1960	1970	1980	1990
1966–1975	Generation X	Mid-1980s through 1990s	Era of economic polarization			5–14	15–24
1956–1965	Late baby boom	Mid-1970s through 1980s	Era of employment restructuring		5–14	15–24	25–34
1946–1955	Early baby boom	Mid-1960s through 1970s	Era of civil rights and sexual revolutions	5–14	15–24	25–34	35–44
1936–1945	World War II	Mid-1950s through 1960s	Post–World War II boom	15–24	25–34	35–44	45–54
1926–1935	Parents of the baby boom	Mid-1940s through 1950s	Post–World War II boom	25–34	35–44	45–54	55–64
1916–1925	Parents of the baby boom	Mid-1930s through 1940s	World War II	35–44	45–54	55–64	65–74
1906–1915	Grandparents of the baby boom	Mid-1920s through 1930s	Depression	45–54	55–64	65–74	75–84

Source: Farley, 1996.

FIGURE 10.2 Percentage of Whites Giving the Equal Opportunity Response to Questions About Principles of Racial Equity, 1940–1995

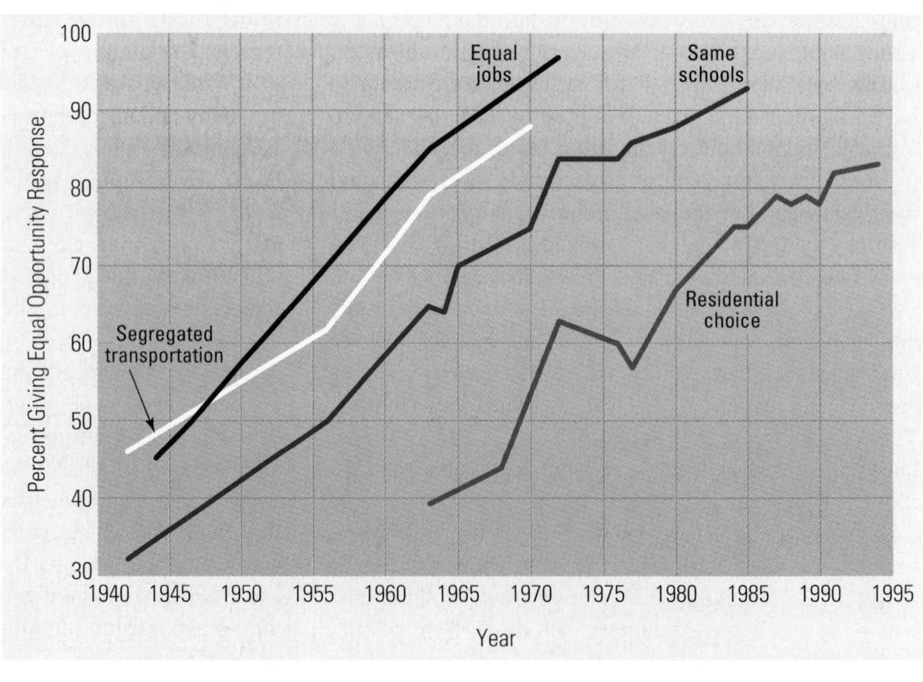

Source: Farley, 1996.

TABLE 10.6 Attitudes About Roles of Employed Women (by birth cohort)*

	Generation X Less Than 25	Late Baby Boom 25–34	Early Baby Boom 35–44	World War II 45–54	Parents of Baby Boom 55 and Over
	It is more important for a wife to help her husband's career than to have one herself (% agree)				
Women	13	13	16	24	51
Men	20	17	17	29	47
	It is better for everyone involved if the man is the achiever outside the home and the woman takes care of the home and family (% agree)				
Women	18	24	23	36	63
Men	29	28	32	42	67
	A preschool child is likely to suffer if his or her mother works (% agree)				
Women	27	30	34	41	56
Men	47	42	46	55	69
	A working mother can establish just as warm and secure a relationship with her children as a mother who does not work (% agree)				
Women	80	77	79	79	53
Men	69	64	61	55	43

*Column heads reflect cohort members' age in 1990.

Source: Farley, 1996.

FIGURE 10.3 **Percentage of National Samples Giving Liberal Responses to Questions About Women Combining Home and Labor Market Activities, 1975–1995**

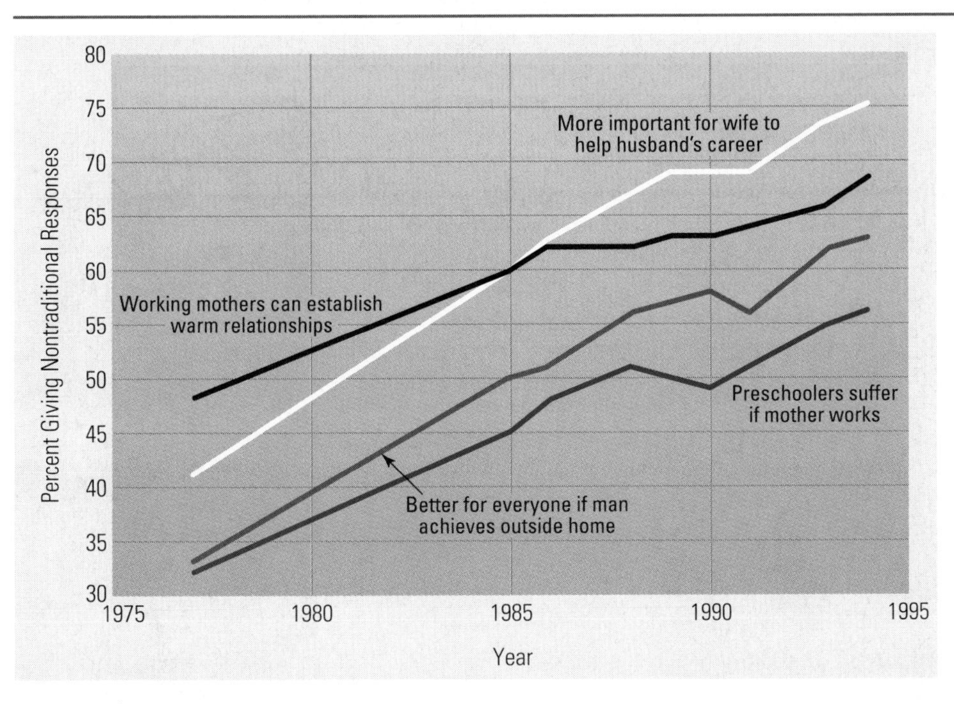

Source: Farley, 1996.

VISUAL SOCIOLOGY

Social Change in Ivory Coast

The Ebrié people with whom I lived and worked many years ago are now almost fully incorporated into the sprawling modern metropolis of Abidjan, the capital of Ivory Coast. But as these photos show, the Ebrié continue to maintain many aspects of their traditional way of life even as they continually adapt to the demands of an urban environment. Men and women who have become educated often commute across the lagoon from Blokosso to the commercial center of the city, where they work as managerial and professional employees. But other Ebrié, like the young women shown walking past a modern building, commute from the village to engage in commerce in one of the city's many open air markets. These women are likely to wear the traditional sarong or pagne and gracefully carry their wares balanced on their heads.

Blokosso men still ply some of their traditional trades; outstanding among them is goldsmithing. Like many of the ethnic groups of this region of West Africa, including the Ashanti people of neighboring Ghana, the Ebrié culture has a rich tradition of jewelry making. Men and women proudly wear gold ornaments on feast days, and many of these highly ornate gold pieces are cast in the Ebrié villages of the Abidjan region, including Blokosso. When I left the village one of my friends, Mr. Robert, gave me a gold elephant as a symbol of our years spent together.

The Ebrié people are mainly Christian, although many Muslims live in their villages as well. The village of Blokosso is home to one of the region's finest Harrist churches, shown here. Harris was an African missionary who preached an African form of Protestantism that became extremely popular earlier in the twentieth century among the lagoon peoples of Liberia, Ivory Coast, and Ghana. Harrist church ceremonies are lively and colorful, with many aspects of traditional African music and dance incorporated into the Protestant service.

Over decades, as cities have developed around them, many of the Ebrié villages have become hemmed in by the crush of poorly housed urban migrants. Adjame, shown here, was once a peaceful Ebrié village but has become an urban slum. Does the same fate await Blokosso? Perhaps. But that village is fortunate to be located along the lagoon, along with luxury hotels and villas. It is unlikely that migrants to the region will crowd this part of the city. Thus the people of Blokosso are able to maintain their delicate balance between traditional and modern African ways of life, at least for the time being.

This lighthearted sculpture by a traditional African artist shows a missionary or colonial administrator being ferried around in a dugout canoe. Colonial regimes were a major exogenous source of social change.

SUMMARY

Social change refers to variations over time in the ecological ordering of populations and communities, in patterns of roles and social interactions, in the structure and functioning of institutions, and in the cultures of societies. Such changes can result from forces building within societies (*endogenous forces*) as well as from forces exerted from the outside (*exogenous forces*). Changes at the micro, middle, and macro levels of social life usually are interrelated.

No social change in human history has been as far-reaching and universal in its consequences as the transition to an urban, industrial way of life. However, not all people who are experiencing such changes think of them as progress. Nor can all forms of social change be controlled, although some changes are intentional. Similarly, while social scientists can understand and predict the master trends of their era, they have limited ability to anticipate major forms of change such as wars, social movements, or technological changes.

One of the major forces that produce social change is war. The primary ecological effects of war are casualties and conquest. War also results in large-scale shifts in population and rapid acceleration of economic change. It can affect a society's culture in a variety of ways, and it may drastically change the structure of a society, especially its major social institutions.

A second major source of social change is *modernization*—the changes that have taken place in societies throughout the world as a result of industrialization, urbanization, and the development of nation-states. These changes include a shift from simple techniques toward the application of scientific knowledge; an evolution from subsistence farming toward the commercial production of agricultural goods; a transition from the use of human and animal power toward the use of power-driven machines; and a movement from the farm and village toward urban concentrations. These processes may or may not take place simultaneously.

Some sociologists view technology as a basic principle of social change. They recognize, however, that social institutions are often slow to adapt to changing technologies. This recognition forms the basis of the theory known as *cultural lag*. Postmodernist critics argue that technology and rationality deprive human lives of spirituality; they also believe that major institutions of modern societies have become instruments of social control by powerful elites.

The so-called Third World nations are those that have won independence from colonial dominance in the decades since World War II. Such nations are also called *developing nations* or modernizing nations. They are undergoing a set of transformations whose effect is to increase the productivity of their people, their health, their literacy, and their ability to participate in political decision making. Wallerstein's world system theory divides the world into *core states, semiperipheral areas,* and *peripheral areas.*

People often experience social change as highly problematic. In pursuing new opportunities in education, leisure activities, intimate relationships, and political participation, they may find themselves without a clear set of norms to guide their lives and hence may experience anomie. The entry of large numbers of women into the labor force, for example, has upset the traditional norms of family life. Similarly, the civil rights movement has greatly reduced racial discrimination in the United States, but for a large proportion of the black population these gains have been offset by changes in the structure of the economy.

A third area in which social change touches the individual is public policy, which may involve trade-offs between conflicting goals such as eliminating poverty and protecting the environment.

Sociologists have often attempted to develop models of social change that can be used to predict the future of whole societies or civilizations. Many of the founders of sociology favored an evolutionary model in which social change is seen as natural and constant; all societies inevitably become increasingly complex through a steady series of stages. Modern evolutionary theorists refer to such models as unilinear because they predict that all societies will undergo the same process of change. Multilinear models emphasize that one must study each society separately to discover the evolutionary stages that are unique to a particular society.

A variety of theories have taken a cyclical view of social change, in which civilizations rise and fall, respond to a series of challenges, or alternate between two opposing sets of cultural values. Conflict theorists argue that conflict among groups with different amounts of power produces social change, which leads to a new system of social stratification, which in turn leads to further conflict and further change. From a functionalist perspective, social change occurs as a result of population growth, changes in technology, social inequality, and efforts by different groups to meet their needs in a world of scarce resources. The latter two perspectives can be applied to change at the micro and middle levels of social life as well as to macro level changes.

SOCIOLOGY VERSUS IDEOLOGY

Do you agree with this statement: *Conservatives oppose social change while liberals favor it.* If you agree, you had better get your sociological imagination in gear. In fact, there are social ideas or policies that many conservatives favor, such as drastic deregulation of business activity, that could lead to radical social change—just as there are beliefs held by many liberals, such as opposition to school vouchers and privatization of public schools, that support the status quo and oppose social change. In reality, the terms *conservative* and *liberal* are inadequate to describe the great diversity of opinions among so-called liberals and conservatives. For example, the Harvard sociologist Daniel Bell, who developed the theory of postindustrial society, thinks of himself as a cultural conservative because he is opposed to the influences of mass media, but also as a political liberal because he supports public education and various forms of government regulation of business. Conservatives who place individual liberties above other values (known as libertarians) find themselves arguing in favor of legalization of marijuana use. Liberals who think of themselves as social democrats support labor unions and fight against policies that would weaken existing labor laws in favor of a freer labor market. As sociological thinkers, therefore, we need to analyze each aspect of social change in terms of who benefits and who loses in society and what the change implies for existing social institutions rather than whether a certain policy seems to be, on the surface, conservative or liberal.

GLOSSARY

social change: variations over time in the ecological ordering of populations and communities, in patterns of roles and social interactions, in the structure and functioning of institutions, and in the cultures of societies. (p. 294)

endogenous force: pressure for social change that builds within a society. (p. 294)

exogenous force: pressure for social change that is exerted from outside a society. (p. 294)

modernization: a term used to describe the changes experienced by societies and individuals as a result of industrialization, urbanization, and the development of nation-states. (p. 301)

cultural lag: the time required for social institutions to adapt to a major technological change. (p. 303)

developing nation: a nation that is undergoing a set of transformations whose effect is to increase the productivity of its people, their health, their literacy, and their ability to participate in political decision making. (p. 304)

core state: a technologically advanced nation that has a dominant position in the world economy. (p. 306)

semiperipheral area: a state or region in which industry and financial institutions are developed to some extent but that remains dependent on capital and technology provided by other states. (p. 306)

peripheral area: a region that supplies basic resources and labor power to more advanced states. (p. 306)

cohort: all persons born in a specified time span. (p. 315)

WHERE TO FIND IT

BOOKS

How Societies Change (Daniel Chirot; Pine Forge, 1994). A review of theories of change by one of the leading sociologists of social change; reflects on the many meanings of the failure of Soviet-style regimes.

The New American Reality (Reynolds Farley; Russell Sage, 1996). A thorough empirical study of the major demographic and cultural trends that shaped the lives of Americans during the twentieth century.

The Wretched of the Earth (Frantz Fanon; Grove, 1968). A classic and burning analysis of the psychology of colonial peoples and their yearning for independence even in the face of violence and bloodshed.

Social Change and History (Robert A. Nisbet; Oxford University Press, 1969). A gracefully written history of ideas of social change.

Technology, the Economy, and Society: The American Experience (Joel Colton and Stuart Bruchey, eds.; Columbia University Press, 1987). A collection of 11 articles assessing the role of technology in transforming everything from agriculture, industry, and the structure of corporations to politics, law, the military, education, and religion.

JOURNALS

Politics and Society. A critical journal devoted to studies of social change and theories of social change.

Economic Development and Cultural Change. A journal of economic and sociological writing about the interrelationships between culture and economics, and their influence on social change in developing nations.

OTHER SOURCES

Human Development Report. Published by the United Nations Development Programme and Oxford University Press. An essential primary source for anyone interested in studying trends in global social change; emphasizes such factors as literacy, poverty, political participation, gender equality, the status of children, and much more.

Statistical Yearbook. Published annually by the United Nations, Department of International Economic and Social Affairs. Presents general socioeconomic statistics.

INTERNET RESOURCES

National Association of Community Action Agencies (www.nacaa.org/). A good place to look for what people at the grassroots level are doing to bring about positive social change.

Social Science Research Council (www.ssrc.org/). A professional social science organization that has sponsored research on social change for many decades.

Sage Publications (www.sagepub.com/). An online catalog of books and research monographs dealing with all aspects of social change, including works by sociologists in many nations.

PART 3

SOCIAL DIVISIONS

An American who was born in the second half of the twentieth century has lived through events that dramatically changed the fortunes of many groups in the U.S. population. The economic growth and prosperity of the 1950s brought the promise of relative affluence for all, or almost all, Americans. Blue-collar workers and their families were encouraged to buy houses and cars, to take trips on the new interstate highway system, to send their children to college. Blacks and other minority groups won major advances in civil rights and economic security. Women, too, increasingly rejected their status of cultural and economic inferiority. We will see in this part of the book, however, that these and other victories are incomplete. Social divisions remain, even though in some instances they are becoming less severe.

The next five chapters look in more detail at social inequality, particularly in American society. Chapter 11 shows how the members of a society are sorted into social layers or strata and how a society's stratification system changes as its economic and political structure changes. Chapter 12 deals with inequalities of social class. After a discussion of how inequalities are defined and measured, the chapter moves on to an analysis of how social inequalities due to lack of wealth, education, prestige, and other rewards affect people's chances of achieving success and "the good life." Chapter 13 examines how present patterns of racial and ethnic inequality have evolved from historical patterns of intergroup hostility, conflict, and cooperation. In Chapters 14 and 15 we will see how the roles a culture defines as appropriate for women and men of different ages are related to patterns of inequality in that society.

CHAPTER 11

STRATIFICATION AND SOCIAL MOBILITY

Sebastião Salgado, a Brazilian with training in the social sciences, is widely recognized as one of the world's foremost photographers. He travels throughout the world taking powerful photographs of people who are experiencing the wrenching consequences of changes in their traditional way of life. He has received awards for his photographic coverage of droughts in Africa and poverty throughout the world. But he may be best known for his global photo studies of the industrial revolution. Indeed, he terms his photos of women and men at work throughout the world "an archaeology of the industrial age." In Salgado's words, "Concepts of production and efficiency are changing, and, with them, the nature of work. The highly industrialized world is racing ahead and stumbling over the future. In reality, this telescoping of time is the result of the work of people throughout the world, although in practice it may benefit few" (1993, p. 7).

The developed world, of which the United States is the recognized superpower, contains a fifth of the world's population but enjoys a far greater proportion of the world's wealth. As Salgado sees it, "The remaining four-fifths, who could theoretically benefit from surplus production, have no way of becoming consumers. They have transferred so many of their resources and wealth to the prosperous world that they have no way of achieving equality." Thus Salgado sees the world divided in many ways, especially on the basis of unequal access to wealth and power: "the first world in a crisis of excess, the third world in a crisis of need, and, at the end of the century, the second world—that built on socialism—in ruins" (1993, p. 7).

Although the affluent nations are undergoing rapid social change due to automation, the technological revolution in information processing, and the export of industrial jobs overseas, many of the world's poorer regions are still experiencing the first shocks of the industrial revolution. People who are desperate for income are forced out of their villages and into the industrial labor market. In Gujurat, India, for example, Salgado finds a dam being built that will irrigate an area of 185 million acres: "And in place of a desert, cotton and grain will grow." A large number of women are working on the dam. The men do the hardest physical work as "human bulldozers, tearing at the earth with the shoulders, legs, and hands. But the women carry the earth away" (1993, p.19). They do this six days a week; "on the seventh day they stay home preparing food for the week ahead." Nevertheless, these workers, like so many of the others whom Salgado photographs, maintain their inner strength and their dignity.

This dignity and grace under severe conditions is the subject of some of Salgado's best photography, as illustrated in the Visual Sociology feature later in this chapter. Coal miners in India, working in temperatures of 120 degrees, descend daily into the mines while their wives, mothers, and daughters toil equally long hours under the hot sun. In Bangladesh the industrial revolution centers on textile production, and Salgado finds scenes reminiscent of the "satanic" mills of the early nineteenth century that so shocked British reformers and socialists like Karl Marx and Friedrich Engels. In Spain, Salgado spends time photographing women working in the canning factories. Usually they are the wives of the fishermen he has also photographed as they tend nets in the face of storms and declining catches. "The fish-freezing plants," Salgado writes, "are putting the canning industry out of business. . . . Fishing is no longer what it was, the old canning industry is dying, ancient factories are in a state of abandonment" (1993, p. 8).

Salgado succeeds in overcoming the "compassion fatigue" that many people in affluent nations experience when confronted by scenes of harsh working conditions, exploitation, and the faces of the "working poor." He does this by spending so much time with his subjects that his photographs transcend their woes and capture their dignity and indomitable spirit.

THE MEANING OF STRATIFICATION

Numerous social-scientific studies have demonstrated that all human societies produce some form of inequality (Bendix & Lipset, 1966; Harris, 1980; Murdock, 1949; Sen, 1992). In the simplest societies this inequality may be due to the fact that one family's fields produce more than another's do, that one family has accumulated a greater herd than others have, or that one family has produced a larger number of brave warriors and thus has received more esteem from the other families in the tribe. In other societies, periods of scarcity due to droughts, famines, or changes in the migration patterns of game animals may allow some clans or large families who were already relatively well-off to become wealthier and more powerful. These advantages may then be passed on to the next generation and become part of the society's social structure (Nussbaum & Sen, 1993; Temkin, 1993). But as societies become more and more complex, encompassing ever-larger populations and more elaborate divisions of labor, these simple forms of inequality give way to more clearly defined systems for distributing rewards among members of the society. And those systems result in the classification of families and other social groups into rather well-defined layers, or *strata.* In each society the various strata are defined by how much wealth people have, the kinds of work they do, whom they marry, and many other aspects of life.

In this chapter we will see how inequality in societies results in systems of social stratification. **Social stratification** is a society's system for ranking people hierarchically (i.e., from high to low) according to various attributes such as income, wealth, power, prestige, age, sex, ethnicity, and religion. In this chapter we will deal primarily with stratification by wealth, power, and prestige.

Caste, Class, and Social Mobility

In every society people are grouped into different categories according to how they earn their living. This produces an imaginary set of horizontal social layers, or strata, that are more or less closed to entry by people from outside any given layer. Societies that maintain rigid boundaries between social strata are said to have **closed stratification systems;** societies in which the boundaries are easily crossed are said to have **open stratification systems.**

In open societies it is possible for some individuals and their families, and even entire communities, to move from one stratum to another; such movement is termed **social mobility.** A couple whose parents were unskilled workers may become educated, learn advanced job skills, and be able to afford a private house instead of renting a modest apartment as their parents did. Such a couple is said to experience **upward mobility.** If they have enough wealth to make their parents comfortable and to help other family members, the entire family may enjoy upward social mobility. If everyone with the same education and skills and the same occupation experiences greater prosperity and prestige, the entire occupational community is said to be upwardly mobile. But in an open society fortunes can also decline. People with advanced skills—in engineering or higher education, for example—may find that there are too many of them around. They may not be able to afford the kind of housing, medical care, education for their children, and other benefits that they have come to expect. When this occurs they are said to experience **downward mobility.**

The best examples of closed societies are found in caste societies. **Castes** are social strata into which people are born and in which they remain for life. Membership in a caste is an **ascribed status** (a status acquired at birth) rather than an **achieved status** (one based on the efforts of the individual). Members of a particular caste cannot hope to leave that caste. Slaves and plantation owners formed a caste society in the United States before the Civil War. The slaves were captives; runaway slaves were pursued and returned to their masters. Their children were born into slavery. Plantation owners, on the other hand, had amassed great wealth, especially in the form of land, and this wealth was passed on to their children. On occasion a plantation family might lose its wealth, or another family might acquire a plantation and the wealth and prestige that went with it, but this form of social mobility was rather infrequent and did not alter the caste nature of the plantation system.

Today much of modern India remains influenced by caste-based inequalities. (See Box 11.1.) South Africa, on the other hand, is an example of a society that is moving away from a rigid caste system. Under apartheid, blacks in South Africa were a racially defined caste that was kept at the bottommost rungs of the stratification order by violent repression and laws mandating racial segregation. Although this situation has been changing since the remarkable transition to majority African rule, the extreme poverty of the black population indicates that most aspects of the caste system are far from ended.

During the Persian Gulf War in 1991 many Americans discovered that Kuwait and other nations of the Middle East have strong ruling castes. In Kuwait the ruling caste is composed mainly of members of a single large, immensely wealthy family—the Al-Sabahs—many of whom live on the Italian and French rivieras for much of the year. The Al-Sabahs rule Kuwait as a closed leadership caste, and after the Gulf War they were careful to reimpose their rule and forestall any dem-ocratic movement that could challenge their authority.

Classes, like castes, are social strata, but they are based primarily on economic criteria such as occupation, income, and wealth. England is famous for its social classes and for the extent to which social class defines how people are thought of and how they think of themselves. George Orwell, the author of *Animal Farm* and *1984,* also wrote about the English class system earlier in the twentieth century, when he lived among the homeless and the migrant workers at the very bottom of the class system, and among miners whose lives were sacrificed to the dangers and diseases of the coal mines. In the following passage, he described how class affected his own life:

> All my notions—notions of good or evil, of pleasant and unpleasant, of funny and serious, of ugly and beautiful are essentially *middle class* notions; my taste in books and food and clothes, my sense of honour, my table manners, my turns of speech, my accent, even the characteristic movements of my body, are the products of a special kind of upbringing and a special niche half way up the social hierarchy. (quoted in Campbell, 1984)

Classes are generally open to entry by newcomers, at least to some extent, and in modern societies there tends to be a good deal of mobility between classes. Moreover, the classes of modern societies are not homogeneous—their members do not all share the same social rank. There are variations in people's material well-being and in how much prestige they are accorded by others. Within any given class, these variations produce groups, known as **status groups,** that are defined by how much honor or prestige they receive from the society in general.

The concept of status groups is illustrated by "high society." The nobility of England is one of the world's most prestigious status groups, despite the well-publicized marital troubles of the royal family. In the United States, people with names like Rockefeller, du Pont, Lowell, Roosevelt, Harriman, and many others, who are of Western European Protestant descent, often have more prestige than people with just as much wealth who are of Italian or Jewish or African American descent. In cities throughout the United States these very rich and prestigious families, who form a status group as defined by wealth and reputation, also tend to form groups that interact among themselves and play significant roles in their communities. The society pages of metropolitan newspapers devote most of their gossip to the philanthropic activities and private affairs of these families. Yet most of the old wealthy families in North America came from quite modest origins, as revealed by the underlying meanings of their names. Rockefeller, for example, means "a dweller in rye fields"; du Pont, "one who lives near the bridge"; Harriman, "a manservant"; Roosevelt, "one living near a rose field." Moreover, people with prestigious family names typically enjoy fortunes gained from activities that were once considered too lowly to permit entry into polite society. Thus the Fords were looked down on by high society because their fortune was linked to "smelly gasoline"; the Whitneys, whose fortune a century ago spread over five states and 36,000 acres of palatial homes, were looked down on by older millionaire families because their funds were derived from 5-cent trolley fares rather than from railroad freight charges. In sum, both money and family prestige—gained by living expensively and engaging in public philanthropy—are required for entry into the highest levels of upper-class society (Domhoff, 1983; Hacker, 1997).

Life Chances

Rankings from high to low are only one aspect of social stratification. The way people live (often referred to as their *lifestyle*), the work they do, the quality of their food and housing, the education they can provide for their children, and the way they use their leisure time are all shaped by their place in the stratification system.

The way people are grouped with respect to access to scarce resources determines their **life chances**—that

BOX 11.1 USING THE SOCIOLOGICAL IMAGINATION

The Caste System in India*

His name was Natwar Singh, and he was the very image of what you could and could not achieve if you were well-born in Chirora. Natwar . . . was a Thakur, the principal landlord caste in central Uttar Pradesh, and he owned 35 *bighas,* or 17½ acres—a very big farm by local standards. The average farmer owned about 2 acres. After much prodding, he agreed that his annual profit might go as high as $1,500.

Natwar was a passionate adherent of the tenet that physical labor is beneath the dignity of a Thakur, a belief that idled a good many people in the village. I once asked him what he did with himself all day long. He was stumped. Finally, his friend Tripathy, the village merry-andrew, said, "Checking."

Always at the periphery of Natwar's large, cheerful entourage was a character named Guria. He was gloomy and taciturn, with rudely chopped hair and dark, moony eyes. Always he wore the shapeless, grimy pale shirt and baggy white pajamas common throughout northern India.

Guria was a member of one of the "backward" castes, a term used in contrast to the "forward" castes, the Brahmins and Thakurs. He was a Mallah, meaning fisherman—this because his father had been a fisherman, though Guria himself had turned to farming. Practically everything in the village turned on caste, but not quite in the way I had expected. I had assumed, for example, that the Scheduled Caste** farmers were denied sharecropping work out of age-old prejudice, but it turned out to be something far more modern. Landowners, according to Indrapal Singh, were "afraid that the cultivators will claim the farmer's land, and since the Government always comes down on the side of the Scheduled Castes, they may get it." It hadn't happened yet, Indrapal conceded, but you never know.

In Uttar Pradesh, I was told, almost 45 percent of government jobs are reserved for these minorities, and there are preferential loans and scholarships set aside for them. In a magnificent display of ire, Natwar Singh even composed an entire English sentence for me: "All facilities Scheduled Caste, no facilities other caste."

"But what about the Scheduled Caste families in Chirora?" I objected. "They're miserably poor."

"You're right," said Tripathy with a grin. "But they used to be poorer."

The old caste harmony, based on an unquestioned tyranny, is giving way to a modest level of rivalry. The Scheduled Castes have made a tiny bid for power; the upper castes, so accustomed to unquestioned authority, are wildly overreacting. Caste relations have come to resemble class relations as the ritual element of caste has begun to dwindle. Elsewhere in India, especially in areas where powerful landlords exercise feudal control over entire villages, the competition between the landless and the landowning class has led to riots and horrible violence. In Chirora, where—more typically—no one owns more than 20 acres, it's just raised the temperature a few degrees.

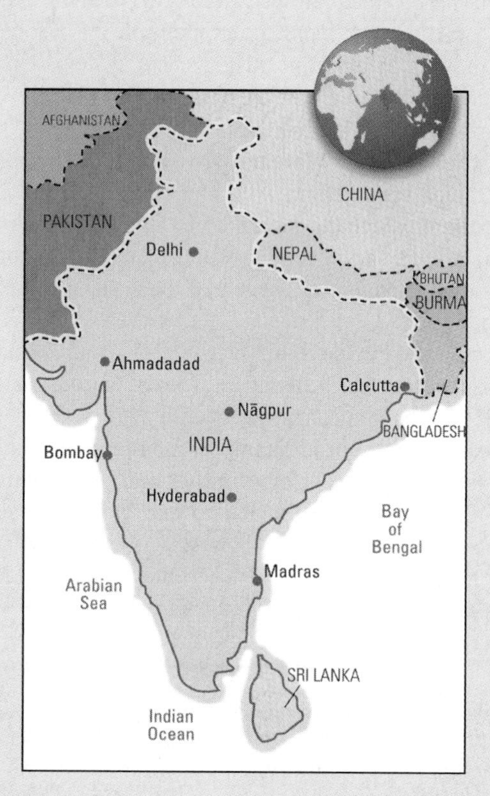

Perhaps one can gauge the power of new ideas by the impression they have made on Chirora. No one, for example, was embarrassed at the degraded status of their wives and daughters. India's extremely tentative feminist movement had made no inroads at all. Women, especially married women, were invisible. They stayed behind brick walls in their darkened homes, kneading dough, chopping vegetables, nursing fires, spoiling babies. At times as I walked down a dusty lane, I could see clusters of women circling their saris over their heads, alarmed lest a stranger see them and even look them in the eye. The rule of purdah, which Hindu India inherited from its Muslim occupiers, was complete in Chirora.

Certain women did not qualify for modesty, and thus could be seen outside the home—unmarried girls, widows, and Scheduled Caste women, who had to work with men in the fields. An unmarried girl is an insignificant thing, a pair of hands and a mouth, if also a giggle. A villager with three boys and two girls will answer, "Three," when asked how many children he has. Nowadays many of the girls graduate from eighth grade at the local school, but with few exceptions they don't go any further. And once a girl marries at 14 or 15 her life is fixed; her future is her mother's past.

* Source: Traub, 1984. Copyright © 1984 by The New York Times Company. Reprinted by permission.

** Formerly known as Untouchables but now called "Scheduled Castes" because they are scheduled to receive affirmative-action benefits.

is, the opportunities they will have or be denied throughout life: the kind of education and health care they will receive, the occupations that will be open to them, how they will spend their retirement years—even where they will be buried. The place in a society's stratification system into which a person is born (be it a comfortable home with access to good schools, doctors, and places to relax or a home that suffers from the grinding stress of poverty) has an enormous impact on what he or she does and becomes throughout life. A poor child may overcome poverty and succeed, but the experience of struggling out of poverty will leave a permanent mark on his or her personality. And most people who are born poor will not attain affluence and leisure even in the most open society.

Inequality, Poverty, and Health

Differences in income and wealth can sometimes mean the difference between life and death. In some developing nations health conditions are changing for the better as public health measures such as improved water supply and sewage systems take effect. Nevertheless, the correlation between poverty and death remains extremely high in many nations, and in the United States it has increased in recent years. These trends are so important that social scientists and international planning experts often use health statistics as a barometer of how well a nation is doing in improving the gap between the haves and the have-nots in its population.

Three indicators—mortality of children under age 5, the adult illiteracy rate, and the incidence of income poverty—are often used as primary measures of social change when comparing different nations (UNDP, 1997). We see in Figure 11.1 that some nations and regions have made progress toward reducing these indicators of severe ill health, illiteracy, and poverty. But the needs remain immense. An estimated 1.3 billion people must survive on the equivalent of less than $1 a day. Nearly 1 billion people, of whom the majority are women, are illiterate, and well over 1 billion face severe health threats due to lack of access to safe water (UNDP, 1997). And as we see in Figure 11.2, under-5 mortality rates are as much as three times higher in families in which the mother has no education than in families in which the mother has 7 or more years of education.

Figure 11.3, on page 333, compares mortality rates for developed and less developed nations, and also indicates how much greater risk males face (compared with females) of dying before age 60.

The single most important factor in determining survival, however, is income. At the extreme of poverty, in Africa, with annual per capita income of US$800, life expectancy is 52 years. At the other end of the income spectrum, in the United States, with annual per capita income of US$20,000, average life expectancy is 75 years. There are, however, low-income regions that have invested in the health and safety of their populations (e.g., some states of India), and there are affluent regions that still rank relatively low on measures of health and survival (Mosley & Cowley, 1991; Sen, 1993).

FIGURE 11.1 **Indicators of Social Change**

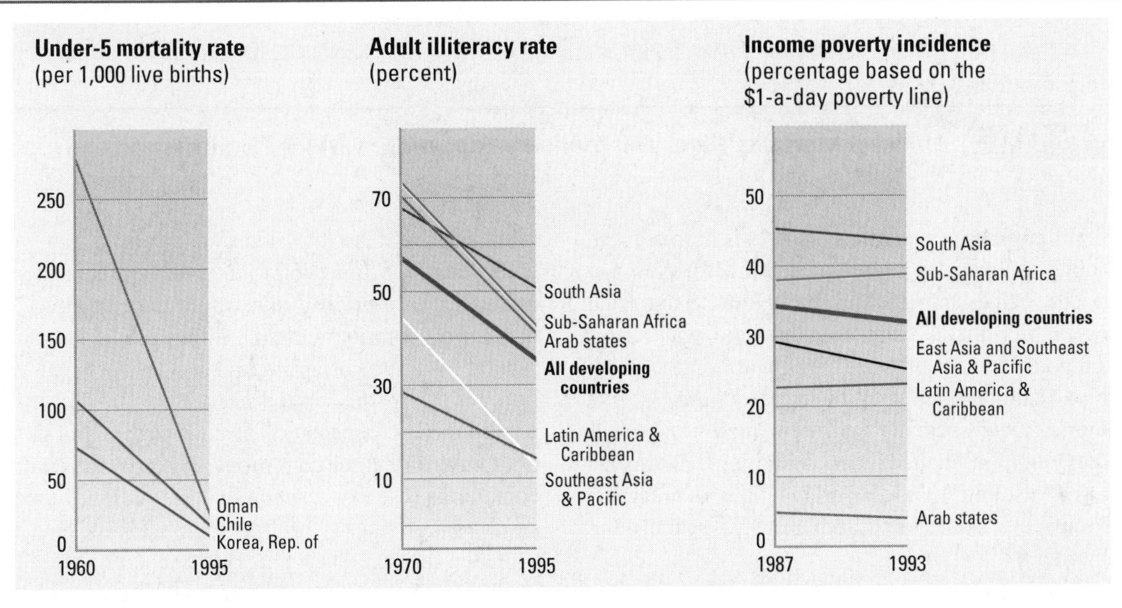

Source: UNDP, 1997.

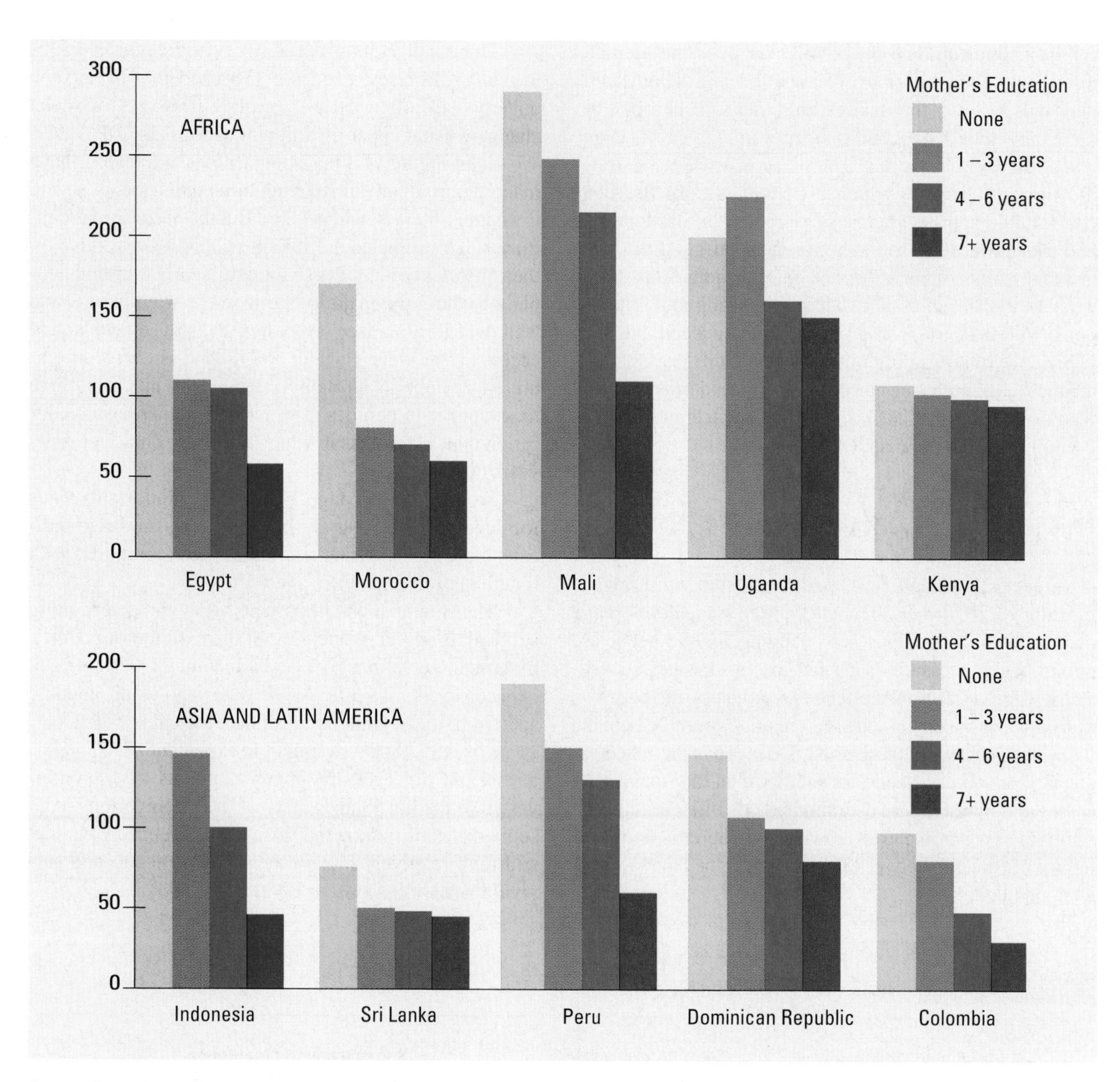

Source: Rutstein, 1991.

FIGURE 11.2 **Under-5 Mortality Rates and Mother's Education, Various Countries**

In impoverished nations like Ethiopia and Bangladesh, the primary causes of death before old age are diseases such as amoebic dysentery (severe diarrheal diseases), which result from contaminated water supplies and poor sewage systems. In more affluent nations, the rate of death before age 60 increases for adolescents and young adults, especially males, owing to the greater risks they take in their leisure activities (dangerous sports, drug use, etc.) and the higher rates of interpersonal violence in some of these nations (especially the United States).

Studies of health and inequality in the United States show that over the past four decades the health and mortality gap between the rich and the poor has widened considerably. From 1960 to the mid-1980s the degree of inequality in mortality rates more than doubled. Racial inequality compounds this difference. Medicaid, the federal insurance program that serves more than 30 million people in the United States, has significantly improved the life expectancy of poor people, but poor diet, poor environmental conditions, high risk of contracting contagious diseases (especially new strains of tuberculosis), and high rates of interpersonal violence have helped to widen the mortality gap. Because similar trends have been occurring in affluent nations of Europe, where universal health insurance is the norm, medical sociologists

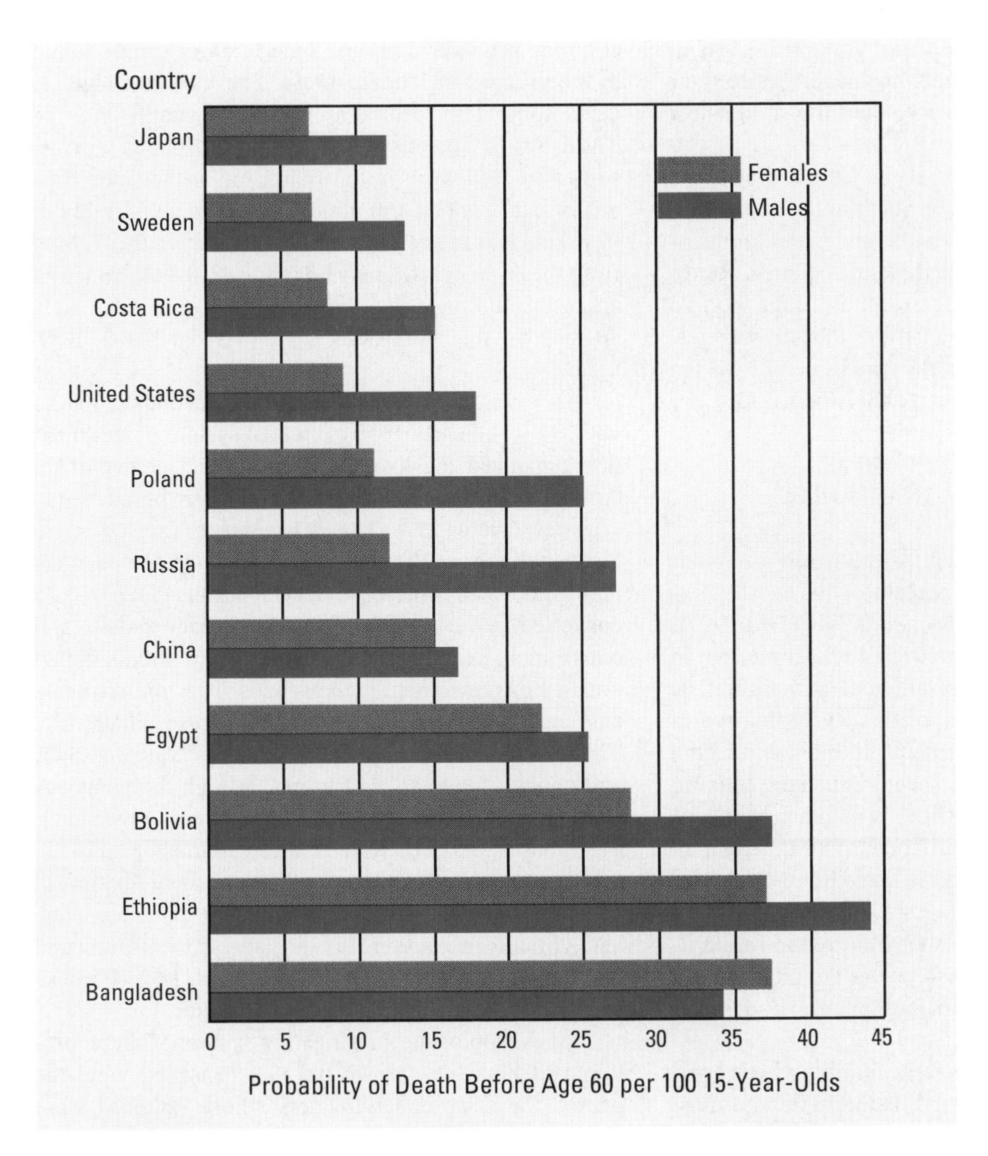

Source: Feachem et al., 1993.

FIGURE 11.3 **Risk of Dying Before Age 60 for 15-Year-Old Males and Females, Selected Countries**

warn that better insurance systems for the United States will not automatically narrow the gap in the absence of improvements in eating and smoking habits and other lifestyle changes among the poor (Pear, 1993).

STRATIFICATION AND THE MEANS OF EXISTENCE

The principal forces that produce stratification are related to the ways in which people earn their living. In the nonindustrial world the majority of the people are small farmers or peasants. When they look up from their toil in the fields, they see members of higher social strata—the landlords, the moneylenders, the military chiefs, the religious leaders. These groups control the peasants' means of existence: the land and the resources needed to make it produce. Even when farmers or peasants are citizens of a modern nation-state with the right to vote and to receive education, health care, and other benefits, when harvests are poor the landowners still take their full share while the peasants must make do with less. The system of stratification that determines their fate depends on how much the land can yield. In contrast, in the stratification systems of industrial societies, whether capitalist or socialist, most people are urban wage workers whose fates are determined by the managers of public or private firms. If the firms are no longer productive or consumers no longer desire their products, urban workers may lose their jobs and suffer economic hardship.

The stratification systems of the United States and Canada, where less than one twentieth of the population works the land, are most relevant to understanding people's life chances in modern industrial societies. But to understand the conditions of life for most of the world's population we must study inequalities in rural villages, where well over two thirds of the earth's people till the soil and fish the rivers and oceans. (See Figure 11.4.)

Stratification in Rural Villages

In 1997 the poorest 20 percent of the world's population earned a miserable 1.1 percent of the world's income, down from 1.4 percent in 1991 (UNDP, 1997). The vast majority of these impoverished people live in rural villages. Of course, not all rural agriculturalists are impoverished, but in most of the developing world their situation is extremely difficult. In India and China more than 1.2 billion people spend their lives coaxing an existence from the soil. Millions of other rural villagers squeeze a modest livelihood from the land in Central and South America, Africa, and Southeast Asia. Social divisions in these villages are based largely on land ownership and agrarian labor. Yet even in rural villages inequalities of wealth and power are increasingly affected by world markets for agricultural goods and services, as shown in Box 11.2.

In peasant societies the farm family, which typically works a small plot of land, is the basic and most common productive group. Such families can be found in the villages of modern India. The Indian village reveals some important dimensions of stratification in Third World societies. For example, women are assigned to hard work in the fields and at the same time are expected to perform almost all of the household duties. This is true even in families that are well-off. Men from the higher castes may be innovators, but women of all castes and male members of the "backward" castes do most of the productive work (Myrdal, 1965; Redfield, 1947).

In China before the communist revolution of the late 1940s and early 1950s, a feudal system of stratification organized the lives of peasants and gentry alike. Among the peasants there were four broad strata (Tawney, 1966/1932). The rich peasants (analogous to Natwar Singh in Box 11.1) had enough land to meet their basic needs and to produce a surplus that could be converted into cash at local markets. They usually had one or more draft animals, and they often hired less fortunate villagers to help in their fields. The "middle peasants" were the second stratum of Chinese village life. They had a small plot of land, barely adequate shelter, and enough food and fuel to get through the winter. A small surplus in good years allowed them to own their own houses and even to have a few animals to provide meat on feast days. Only a very few of these "luxuries" were available to the third stratum, the poor peasants, and virtually none were to be had by the fourth and worst off, the tenants and hired laborers. Most Chinese peasants were in this impoverished stratum.

At the top of the stratification system of prerevolutionary China (i.e., above the rich peasants) were the gentry, the class of landowners whose holdings were

FIGURE 11.4 **Percent of Workforce Engaged in Agriculture, Various Countries**

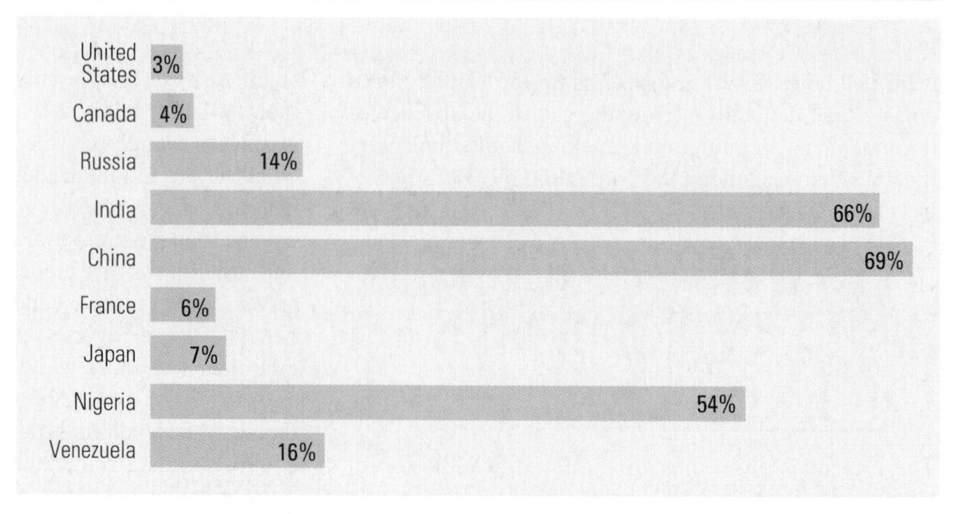

Sources: Lye, 1995; Reddy, 1994.

BOX 11.2 GLOBAL CHANGE AND U.S. SOCIETY

Sweetness and Power

Sugar, tobacco, tea, coffee, cocoa, pepper—these and other substances derived from plants that thrive in tropical climates were once unknown in the colder regions of the earth. Of them all, sugar has had perhaps the greatest influence. Sugar can be combined with cocoa to produce chocolate, fermented to produce rum, refined and sold in its pure form, or baked in candy and confections of all kinds. An exotic crop known mainly to European royalty in the fifteenth century, sugar had become an addiction of the masses by the eighteenth century. The craving for sugar and rum stimulated the growth of slavery in the New World and the transformation of entire Caribbean islands into sugar plantations (Mintz, 1985).

Most people do not realize that even today the production of sugar requires extremely dangerous and debilitating labor. This is especially true in Florida, where soft soils prohibit mechanized harvesting of the cane. Let a Jamaican sugar harvester who was brought to Florida as a contract worker continue the description:

> In America, we work the roughest way to make a living. Coming over, they ask if you're willing to work seven days a week, willing to wash your own clothes, willing to eat poor, willing to obey, work in all style of weather, eat rice seven days. . . .
>
> A lot of accident happen in the cane. When you working, your hand is sweaty, maybe the bill [cane knife] slip and fly up, maybe take your hand or your foot. Maybe take your partner. . . . Your hand become like part of a machine. Look at my palm, the bill rest here—it make a channel for its shape. When you go home, that one hand you have been cutting with you can't use. It's no good for anything else; you got to use the other one until it heal. . . .
>
> We got to watch dangerous things . . . snake, bobcat, ants climbing up your pants leg to sting you, the cane stalk strike you in the eye, pierce your eardrum. Sometime to make money we got to eat no lunch, we got no time for it, sick and still working. . . . You got to be wise and understand yourself and be quiet, otherwise they send you home. It is very disappointing, the way we are treated. We are slaves. They pay us money, but really they buy us. (quoted in Wilkinson, 1989, pp. 56–57)

Investigations of working conditions for contract labor in Florida and other cane-producing areas of the United States have spurred sporadic efforts to improve conditions on the plantations. But little has changed. In fact, the officials of the cane-producing companies make free use of arbitrary firings and deportations, and they prefer foreign workers who can be easily controlled and summarily dismissed and sent home. A section of a report published by the Florida Fruit and Vegetable Growers Association, "Why Foreign Workers Are More Productive," explains how raw power maintains the flow of sugar:

> The "unique and awesome form of management power" that sugar cane growers exert over their foreign workers provides a supermotivated workforce. As a Vice President of U.S. Sugar once said, "If I had a remedy comparable to breaching"—that is, firing and deporting—"an unsatisfactory worker which I could apply to the American worker, they'd work harder too." (quoted in Wilkinson, 1989, p. 64)

large enough to allow them to live in relative comfort and freedom from labor:

> What made the lives of the gentry so enviable to the working peasants was the security they enjoyed from hunger and cold. They at least had a roof over their heads. They had warm clothes to wear. They had some silk finery for feast days, wedding celebrations, and funerals. . . . The true landlords among them did no manual labor either in the field or in the home. Hired laborers or tenants tilled the fields. Servant girls and domestic slaves cooked the meals, sewed, washed, and swept up. (Hinton, 1966, p. 37)

As our examples from contemporary India and prerevolutionary China show, in rural villages the facts of daily life are determined largely by one's place in the local system of agricultural production. The poor peasant family, with little or no land, hovers on the edge of economic disaster. Work is endless for adults and children alike. Meals are meager; shelter is skimpy; there is not much time for play. Among the gentry, who have large land holdings and hired help to ease the burdens of work, there are "the finer things of life": education, ample food and shelter, music and games to pass the time. Of course, wealth brings additional responsibilities. Participation in village or regional politics takes time away from pleasure, at least for some. So does charity work. But the power that comes with such activities provides opportunities to amass still more wealth. Thus it is that, with few exceptions, the poor remain poor while the rich and powerful usually become richer and more powerful. And between the rich and the poor there are other strata—the middle peasants, the middle castes with skills to sell—whose members look longingly at the pleasures of the rich and console themselves with the fact that at least they are not as unfortunate as the humble poor in the strata below them.

How does this scheme of social stratification compare with that found in industrial societies? We will see that there is more mobility in industrial societies but that, just as in rural societies, to be born into the lower strata is to be disadvantaged compared with people who are born into higher strata.

Stratification in Industrial Societies

The industrial revolution profoundly altered the stratification systems of rural societies. The mechanization of agriculture greatly decreased the number of people needed to work on the land, thereby largely eliminating the classes of peasants and farm laborers in some societies. This dimension of social change is often called **structural mobility:** An entire class is eliminated as a result of changes in the means of existence. As noted in Chapter 4, the industrial revolution transformed the United States from a nation in which almost 90 percent of the people worked in farming and related occupations into one in which less than 10 percent did so. Similar changes took place in England and most of the European nations and are now taking place in many other parts of the world.

But structural mobility did not end with the industrial revolution. Today automation, foreign competition, and technological advances are creating new patterns of structural mobility. Older smokestack industries like steel and rubber have been steadily losing factories and jobs; newer industries based on information and communication technologies have been creating plants and jobs, but the people who have become "superfluous" as a result of the closing of their plants are not always willing to move to the new jobs, nor are they often trained

These men and women from the Carpenters Union are marching to demonstrate their solidarity with workers in other parts of the world. Labor unions like this one are organized to represent the social-class interests of workers.

to meet the demands of such jobs. Thus structural mobility often leads to demands for social policies like job training programs and unemployment insurance (Bluestone, 1992).

A second major change brought on by the industrial revolution was a tremendous increase in **spatial mobility** (or geographic mobility). This term refers to the movement of individuals, families, and larger groups from one location or community to another. The increase in spatial mobility resulted from the declining importance of the rural village and the increase in the importance of city-centered institutions like markets, corporations, and governments. Increasingly one's place of work became separate from one's place of residence; people's allegiance to local communities was weakened by their need to move, both within the city and to other parts of the nation; and as a result social strata began to span entire nations. Working-class people created similar communities everywhere, as did the middle classes and the rich (Janowitz, 1978; Seligson & Passe-Smith, 1993).

Despite these immense changes, our relationship to the means of existence is still the main factor determining our position in our society's stratification system. We continue to define ourselves to one another first and foremost in terms of how we make a living: "I am a professor; she is a doctor; he is a steelworker." Once we have dealt with the essentials of our existence—essentials that say a great deal about the nature and the quality of our daily lives—we go on to talk about the things we like to do with our lives after work or after educating ourselves for future work.

STRATIFICATION AND CULTURE

Why do people accept their "place" in a stratification system, especially when they are at or near the bottom? One answer is that they have no choice; they lack not only wealth and opportunities but also the power to change their situation. But lack of power does not prevent people from rebelling against inequality. Many also believe that their inferior place in the system is justified by their own failures or by the accident of their birth. If people who have good cause to rebel do not do so and instead support the existing stratification system, those who do wish to rebel may feel that their efforts will be fruitless.

Another reason people accept their place in a stratification system is that the system itself is part of their culture. Through socialization we learn the cultural norms that justify our society's system of stratification. The rich learn how to act like rich people; the poor learn how to survive, and in so doing they tacitly accept being poor. Women and men learn to accept the places assigned to them, and so do the young and the old. Yet despite the powerful influence of socialization, at times large numbers of people rebel against their cultural conditioning. To understand their reasons for doing so, we need to examine the cultural foundations of stratification systems.

The Role of Ideology

You will recall from Chapter 3 that an ideology is a system of ideas and norms that all the members of a society are expected to believe in and act on without question. Every society appears to have ideologies that justify stratification and are used to socialize new generations to believe that existing patterns of inequality are legitimate. In the United States, for example, most people embrace the ideology of the American dream, the idea that in America anyone who works hard can achieve success and wealth. At the same time, we know that the odds of achieving great wealth are very low. This is one reason people love to hear stories about poor or hardworking people of modest means whose lives are transformed by sudden lottery winnings. In the nineteenth and early twentieth centuries, the "rags-to-riches" stories of Horatio Alger were extremely popular. In Alger's first novel, *Ragged Dick,* the central character is a poor but honest boy who comes to the city looking for work. As he walks the streets he sees a runaway carriage; he leaps onto the horses and stops them. In the carriage is a beautiful young woman who turns out to have a rich father. The father takes Dick into his business, where he proves his great motivation and becomes highly successful.

The people of prerevolutionary China were also guided by ideology; they believed in the teachings of Confucius (551–479 B.C.), which emphasized the need to accept one's place in a well-ordered, highly stratified society (McNeill, 1963). Similarly, the castes of Hindu India are supported by religious ideology. The *Rig-Veda* taught that Hindu society was, by divine will, divided into four castes, of which the Brahmins were the highest because they were responsible for religious ceremonies and sacrifices (Majundar, 1951; McNeill, 1963). Over time, other castes with other tasks were added to the system as the division of labor progressed and new occupations developed. Still another powerful ideology had its origins in Europe before the spread of Christianity. Tribal peoples in what is now France and Germany associated their kings with gods, and that association became stronger in the feudal era (Dodgson, 1987).

Religious teachings often serve as the ideologies of civilizations, explaining and justifying the stratification systems associated with them. But this relationship has not held in every historical period nor for every religious movement. Originally, for example, the teachings of Jesus opposed the stratification systems of both the Roman Empire and the Jewish people. Jesus preached a gospel of love and claimed that "The last shall be first" and "It is easier for a camel to pass through the eye of a needle than for a rich man to enter heaven." He showed sympathy for prostitutes and outcasts such as moneylenders. No wonder his teachings appealed to the poor and downtrodden and enraged the wealthy and powerful. But over many centuries Christ's teachings were incorporated into church doctrine and organization, and by the Middle Ages Christianity was the ideology underlying the stratification system of kings, lords, merchants, and peasants. The vicars of the church upheld that system by affirming its legitimacy in coronations and royal weddings. They also presided over the execution of heretics who challenged the system, which was viewed as divinely ordained.

In our own era the civil rights movement in the United States, the movement to end apartheid in South Africa, the struggle of the Northern Irish Catholics for independence from Britain, and other social movements often invoke the ideology of radical Christianity. "We Shall Overcome," the theme song of the civil rights movement, was borrowed from an African American Baptist spiritual, "I Shall Overcome," and transformed into a moving song of hope and protest with religious overtones.

Stratification at the Micro Level

These relationships between religious ideologies and the stratification systems of civilizations are macro-level examples of how culture maintains stratification systems from one generation to the next. But we can also see the connection between culture and stratification in the micro-level interactions of daily life. The way we dress—whether we wear expensive designer clothes, off-the-rack apparel, or secondhand clothing from the Salvation Army—says a great deal about our place in the stratification system. So does the way we speak, as anyone knows who has been told to get rid of a southern or Brooklyn accent in order to "get ahead." Our efforts to possess and display **status symbols**—material objects or behaviors that convey prestige—are encouraged by the billion-dollar advertising industry. Many other examples could be given, but here we will concentrate on two sets of norms that reinforce stratification at the micro level: the norms of *deference* and *demeanor* (Goldhamer & Shils, 1939).

Deference By **deference** we mean the "appreciation an individual shows of another to that other" (Goffman, 1958, pp. 488–489). In popular speech the word *deference* is often used to indicate how one person should behave in the presence of another who is of higher status. Formulas for showing deference differ greatly within different subcultures. Among teenage gang members, for example, there are ways of showing deference through greeting rituals. Not performing those rituals properly may cause a gang member to feel "dissed" (disrespected), which can lead to violence. (Recall the discussion of "face work" in Chapter 5.) These formulas for showing deference illustrate how our society's stratification system is experienced in everyday life. In the United States, for example, we learn to address judges as "Your Honor" and feel embarrassed for the plaintiff who begins a sentence with "Excuse me, Judge." In most European countries, with their histories of more rigid stratification, people who want to show deference go further and address the judge as "Your Excellence."

But deference is not a one-way process. Erving Goffman has pointed out that the act of paying deference to someone in a higher status frequently obligates that other person to pay some form of deference in return: "High priests all over the world seem obliged to respond to offerings [of deference] with an equivalent of 'Bless you, my son'" (Goffman, 1958, p. 489). The point here is that deference is often

In this display of deference, then Prime Minister Margaret Thatcher makes her famous "deep curtsy" to Queen Elizabeth II.

symmetrical, in that both participants defer according to their place in the stratification system. Through deferent behavior and the appropriate response, both parties affirm their acceptance of the stratification system itself. Intuitively we all know this. When we are stopped by a police officer we may become deferential, using the most polite forms of address ("Yes, sir," "No, sir," and the like) in order to avoid embarrassment. The officer, in turn, may attempt to find out our place in the stratification system and use the appropriate forms of address in speaking to us.

Demeanor Our **demeanor** is the way we present ourselves—our body language, dress, speech, and manners. It conveys to others how much deference or respect we believe is due us. Here again the interaction can be symmetrical. The professor must make the first move toward informality in relations with students, for instance, but among professors of equal rank there is far more symmetry. The move toward informal demeanor, such as the use of first names, can be initiated by whoever feels most comfortable in his or her status. On the other hand, asymmetry in the use of names can be used to reinforce stratification, as in an office where the secretaries are addressed by their first names but address their supervisor by his or her last name plus a title such as Mr., Dr., or Professor.

Stratification and Social Interactions These largely taken-for-granted aspects of how we carry out or "construct" social stratification in our social interactions can have far-reaching effects. Maya Angelou, Richard Wright, and other African American writers describe vividly the extreme shuffling and obsequious deference demanded of their parents in the era of segregation. Even to lift one's eyes to admire a white woman could cause trouble for a black man in the Jim Crow regions of the United States. And as we saw in Chapter 7, failure to carry out the rules of demeanor—to twitch, to have a runny nose, to encroach on another person's space—can cause a person to be labeled deviant and to be cast out of the "acceptable" strata of society.

Deference, demeanor, and other ways in which we behave according to the micro-level norms of stratification further reinforce our sense of the correctness of those norms. Thus in times of rebellion against a society's stratification system those norms are explicitly rejected. During the 1960s, for example, students from wealthy families often wore jeans and tie-dyed shirts in social situations that would normally call for suits or dresses. Long hair and beards, refusal to wear a bra, and other symbolic acts in violation of generally accepted norms of demeanor were intended to communicate rejection of the society's stratification system.

POWER, AUTHORITY, AND STRATIFICATION

When the macro dimensions of social stratification change, the changes may be reflected in the behavior of people at the micro level, and those changes, in turn, accelerate change throughout the entire society. Thus, for example, the civil rights and women's movements have altered the norms of demeanor for blacks and women. Blacks insist on being referred to as blacks or African Americans if their ancestors were of African origin, rather than as Negroes. Women refuse to be called "girls" by men and especially by their supervisors, although some women continue to use the term among themselves. These may seem to be trivial matters, but when people demand respect in everyday interactions, they are also demonstrating their determination to create social change at a more macro level—in other words, to bring about a realignment of social power.

Max Weber defined **power** as "the probability that one actor within a social relationship will be in a position to carry out his own will despite resistance" (1947, p. 152). This is a very general definition; it applies

This engraving from prerevolutionary France depicts the bourgeoisie, or Third Estate, as a woman being forced to support the burden of the clergy and the nobility, the dominant "estates" before the revolution.

BOX 11.3 USING THE SOCIOLOGICAL IMAGINATION

Feudal Relationships

Imagine a peasant in a feudal society whose daughter wishes to marry a young man from another village. She cannot do so without the permission of the lord of the manor, who controls the destiny of everyone dwelling on his land. The peasant must wait for the day when the lord holds court, settling disputes and hearing petitions. Finally the day arrives. The peasant waits nervously until the reeve signals him to come forward. He approaches timidly, his wife following close behind. With the utmost humility he bows, pressing the knuckles of both hands against his forehead, while his wife makes the deepest possible curtsy. He makes his request as briefly as possible and silently awaits the lord's reply.

Feudalism is a form of social organization that has appeared throughout the world as societies have experienced the so-called agrarian revolution. It took somewhat different forms in China, Africa, Japan, and Europe, but it has always been characterized by a fundamental principle: the subordination of one person to another. As social historian Marc Bloch (1964) wrote, the essence of feudalism is "to be the 'man' of another man. . . . The Count was the 'man' of the king, as the serf was the 'man' of his manorial lord" (p. 145). In this system, known as *vassalage,* it was understood that women shared the loyalties of their male masters.

The swearing of *homage* signified and cemented the relationship of vassalage. In Bloch's words:

> Imagine two men face to face; one wishing to serve, the other willing or anxious to be served. The former puts his hands together and places them, thus joined, between the hands of the other man—a plain symbol of submission, the significance of which was sometimes further emphasized by a kneeling posture. At the same time, the person proffering his hands utters a few words—a very short declaration—by which he acknowledges himself to be the "man" of the person facing him. The chief and subordinate kiss each other on the mouth, symbolizing accord and friendship. Such were the gestures—very simple ones, eminently fitted to make an impression on minds so sensitive to visible things—which served to cement one of the strongest bonds known in the feudal era. (1964, pp. 145–146)

The basic relationship of feudalism, signified in the act of homage, is shown in the diagram on page 341. In essence, the vassal exchanges independence for protection, but this relationship can also entail more material exchanges: The lord can give his vassal a manor with the associated land and villagers, and the vassal can provide wealth and soldiers when the lord requires them.

Sociologists study feudalism and its relationships not only because some institutions derived from feudalism still exist today but also because the exchange of independence for protection continues to appear in

equally to a mugger with a gun and to an executive vice president ordering a secretary to get coffee for a visitor. But there is a big difference between the types of power used in these examples. In the first, illegitimate power is asserted through physical coercion. In the second, the secretary may not want to obey the vice president's orders yet recognizes that they are legitimate; that is, such orders are understood by everyone in the company to be within the vice president's power. This kind of power is called **authority.**

Even when power has been translated into authority, there remains the question of how authority originates and is maintained. This is a basic question in the study of stratification. As we saw earlier, the fact that people in lower strata accept their place in society requires that we examine not only the processes of socialization but also how power and authority are used to maintain existing relations among castes or classes.

A brief review of the causes of the French Revolution of 1789 provides a case study in the relationship between power and authority on the one hand and stratification on the other: Why did the French people accept the rule of absolute monarchs for so long before they finally ended that rule in a bloody revolution? In answering this question we can apply all the aspects of stratification discussed so far. We will begin with the feudal stratification system.

The major strata of French society, called *estates,* were the nobility, the clergy, the peasantry, and the bourgeoisie (merchants, shopkeepers, and artisans). Each estate had its own institutions and culture, causing its members to feel that they were part of a unified community with its own norms and values. The estates were linked together by the time-honored norms of feudalism: vassalage and fealty. A *vassal* was someone who received a grant of land from a lord. In return he swore

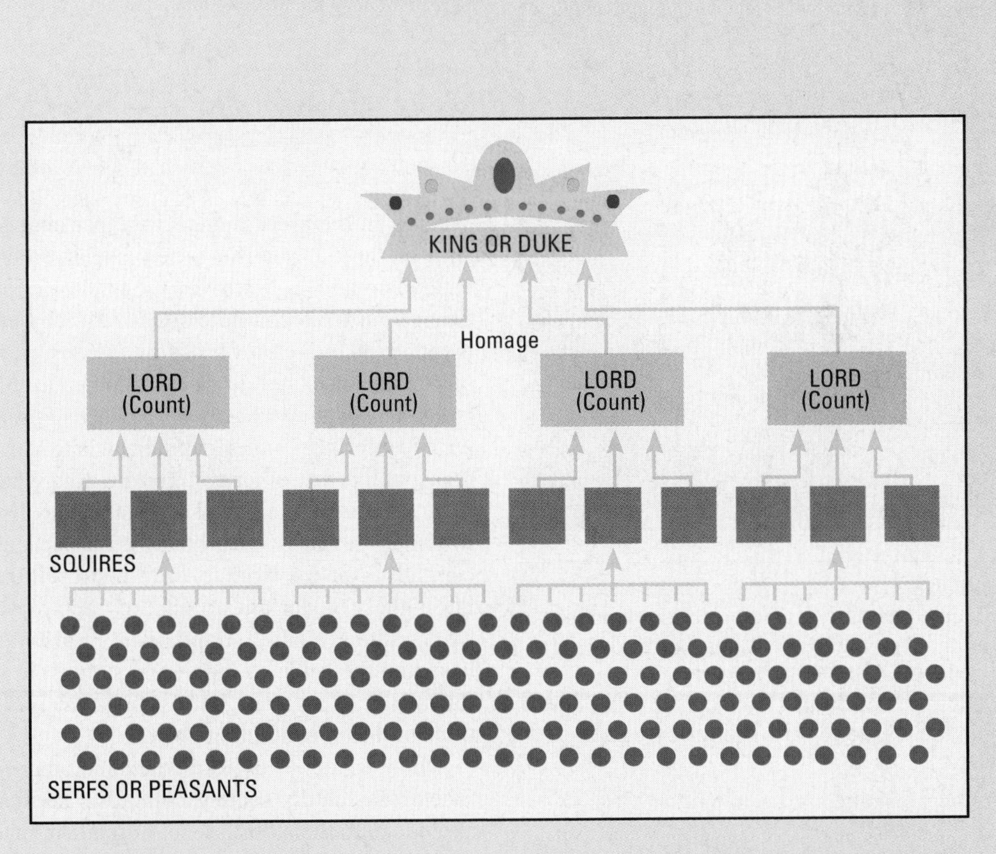

modern social institutions. For example, organized crime "families" often resemble feudal systems, as do many political party organizations and some business corporations. As people in modern societies fight for greater equality in economic and legal institutions, they often demand an end to what they describe as "paternalistic, feudal relations" in those institutions.

fealty, an oath of service and loyalty, to that lord. The lord, in turn, swore fealty to a more powerful lord, until one reached the highest level of society, the king. (The feudal stratification system is described in detail in Box 11.3.) In such a well-ordered and legitimate system, where were the seeds of revolution? The answer lies in how power, authority, and changing modes of production combined to shake the foundations of feudal France.

In any system of stratification there are likely to be conflicts. Those conflicts may be caused by the ambitions of a particular leader or group, but sometimes the characteristics of the system, and especially of the way its powerful members try to cope with change, give rise to conflict. Both types of conflict existed in prerevolutionary France. In order to compete with the other major European powers, the king had to raise armies and send fleets abroad, both of which cost huge sums of money and required thousands of men. But the vow of fealty extended only from the king to the highest level of the nobility. The nobles, in turn, were responsible for seeing that their vassals provided more money and men. This system placed severe constraints on the king, who could not easily make direct demands on all his subjects. The king was dependent on the nobles and, through them, on the lesser lords. In return, the lesser lords often demanded more power, thereby challenging the king's authority. In order to reduce the power of the lords, the king went to the clergy and the bourgeoisie for assistance.

As the cities grew and the trading and manufacturing capacities of the bourgeoisie expanded, the king was increasingly successful at drawing on the wealth of the bourgeoisie and undermining the power of the feudal lords. The court flourished; in fact, it seemed that the king's power had become absolute. But beneath the

Although feudalism is no longer dominant, its former strength is reflected in the villages of southern France that date from that era. In those villages, perched amid the foothills of the Maritime Alps, the houses of commoners were clustered around a large chateau that dominated not only the village but also the surrounding agricultural fields.

pomp and display of the court the feudal stratification system was in ruins. The weakened nobility could no longer hold the fealty of the peasants. Throughout the countryside the peasants groaned under the heavy debts imposed by their feudal masters, and they began to question the right of the nobility to tax them.

Meanwhile, in the cities and towns the rapidly growing bourgeoisie, along with the new social stratum of urban workers, clearly understood that the king's power derived from their productive activities. They felt that they were not adequately rewarded for their support. At the same time, another new stratum, the intellectuals—people who had been educated outside the church and the court—was creating a new ideology and promoting it through new institutions like the press. This new ideology—"liberty, equality, fraternity"—helped spur revolutionary social movements, but the underlying cause of the revolution was a drastic change in the stratification system, in which the new middle class of wealthy business owners (the bourgeoisie) eventually challenged the ancient aristocracy.

In our own era, revolutions, social upheavals, and the wars often associated with them continue to occur in response to extremes of social stratification. In the early 1990s, for example, peasants in the impoverished province of Chiapas, Mexico, staged an uprising that shook the political and economic institutions of Mexican society. The Chiapas peasants and their rebel leaders continue to press for reforms in the regional economy and government.

STRATIFICATION IN THE MODERN ERA

We saw in the preceding section that under the leadership of the bourgeoisie a new social order known as capitalism largely destroyed the feudal system. Capitalism is a form of economic organization based on private ownership and control of the means of production (land, machines, buildings, etc.). Whereas under feudalism land was viewed as belonging to a family as a matter of birthright, under capitalism land can be bought and sold like any other commodity. And whereas feudal labor systems were based on institutions like serfdom, in which peasants were bound to a particular lord, in capitalist systems workers are free to sell their labor to the highest bidder and to endure unemployment when work is unavailable. Capitalism is also an ideology based on the value of individual rights; the ideology of feudalism, in contrast, stressed the reciprocal obligations of different strata of society.

In descriptions by Karl Marx and other observers of nineteenth-century society, capitalism seemed almost to explode onto the world scene. Capitalism made possible the dramatic change in production methods that we call the industrial revolution, which transformed England and Europe from a world of towns and villages, courts and cathedrals to a world centered on markets, factories, and crowded cities. A few places where the physical structures of feudalism remain, such as Mont-St.-Michel in France or Dubrovnik in Croatia, continue to illustrate the contrast between the old order and the new.

The industrial revolution began in England and other parts of Europe in the late seventeenth century. Today it is still occurring in other parts of the world, including China, India, and Africa. The former colonial outposts of England, France, and Germany are only now undergoing the transformation from rural agrarian societies to urban industrial states in which an ever-decreasing portion of the population is engaged in farming. Although not all of these industrializing societies have capitalist economies, they all are developing a class of managers, entrepreneurs, and political power brokers who resemble the bourgeoisie of early capitalism.

The Great Transformation

It is not enough, however, to describe the industrial revolution in terms of the spread of industrial technology and urban settlement. Technological innovations like the steam engine, the railroad, the mechanization of the textile industry, and new processes for making steel and mining coal were accompanied by equally important innovations in social institutions.

In the words of sociologist and historian Karl Polanyi (1944), the industrial revolution was a "Great Transformation." For the first time in human history, the market became the dominant institution of society. By *market* we do not refer only to the places where villagers sold produce or traded handicrafts. Rather, the market created by the industrial revolution was a social network that gradually extended over the entire world and linked buyers and sellers in a system that governed the distribution of goods of every imaginable type, services of all kinds, human labor power, and new forms of energy like coal and fuel oil.

Among other key elements of the Great Transformation were these:

- Goods, land, and labor were transformed into commodities whose value could be calculated and translated into a specific amount of gold or its equivalent—that is, *money* (Marx, 1962/1867; Schumpeter, 1950; Weber, 1958/1922).

- Relationships that had been based on ascribed statuses were replaced with relationships based on *contracts.* A producer hired laborers, for example, rather than relying on kinship obligations or village loyalties to supply workers (Polanyi, 1944; Smith, 1910/1776).

- The business firm or *corporation* replaced the family, the manor, and the guild as the dominant economic institution (Weber, 1958/1922).

- Rural people, displaced from the land, began selling their labor for *wages* in factories and commercial firms in the cities (Davis, 1955).

- In the new industrial order, demands for *full political rights* and *equality of opportunity,* which originated with the bourgeoisie, slowly spread to the new class of wage workers, to the poor, and to women, especially in societies in which revolutions created more open stratification systems (Bendix, 1969; de Tocqueville, 1980/1835; Mannheim, 1941).

The last point deserves further comment. The demand for full political rights originated with the bourgeoisie, which was to become the capitalist class in industrial societies—the owners of the factories, heavy equipment, and other means of producing wealth. To enhance their ability to do business and make profits, members of the bourgeoisie were interested in removing various feudal barriers to economic activity. They therefore called for an end to feudal restrictions on the sale of land, greater freedom for workers to move from one employer to another, the elimination of aristocratic privileges like the right to charge tolls for the use of roads and waterways that happened to cross a lord's property, and an end to voting rights based solely on aristocratic birth. These demands for economic and political rights then spread to the new class of wage workers and to other groups in society. Although the bourgeoisie originally sought rights for itself, it had to form alliances with workers and professionals in order to topple the feudal aristocracy. (Note that these demands originally applied only to men. As we will see in Chapter 14, similar movements for equal participation by women have taken many more generations to develop and gain influence.)

Class Consciousness and Class Conflict

The Great Transformation had a profound impact on the stratification systems of modern societies. It produced new and powerful social classes and thereby changed the way people thought about their life chances and the legitimacy of their society's institutions. As Marx wrote in the last chapter of *Capital,* "The owners merely of labor-power, owners of capital, and landowners, whose respective sources of income are wages, profit, and ground rent . . . constitute the three big classes of modern society based upon the capitalist mode of production" (1962/1867, pp. 862–863). Here as elsewhere Marx defined classes in terms of the "modes of production" that are characteristic of a society in a given historical period. The workers, by far the largest class in modern societies, must sell their labor to a capitalist or landowner in return for wages. The capitalists and landowners are far less numerous than the workers, but because they own and control the means of production, they command more of everything of value than the workers do.

Of course, Marx recognized that in societies that had not yet undergone the Great Transformation there were classes that were very different from those of capitalist societies. The peasants of Russia, the slaves of America, and the exploited Indians of Latin America were "precapitalist" classes. Marx predicted that these strata would eventually be transformed into what he termed the *proletariat,* workers who did not own the means of production and had to survive by selling their labor.

Marx believed that the class of wage workers created by capitalism would inevitably rise up against the capitalist class and create a classless socialist society. As noted earlier, this prediction was proved wrong: Proletarian revolutions never occurred in the most industrialized nations. Yet Marx's description of how members of the working

Then and Now

Bolsheviks and Capitalists

Much of the social thought and ideology of the twentieth century was shaped by the Russian Revolution. The Bolsheviks, shown here storming the Winter Palace, put an end to the reign of the Russian czars and kindled hopes for a classless society not only among their own followers but among impoverished and politically repressed people throughout the world. However, the theories and ideals of socialism, when put into practice by dictatorial leaders like Lenin and Stalin, soon became corrupted. Although many impoverished Russians benefited from the revolution, the classless society never emerged. Instead, a new Soviet class system emerged, with the Communist party elite at the top. Today, as Russia transforms itself into a capitalist nation, the old party elite and the newly rich class of entrepreneurs are combining to form a powerful new class at the top of Russian society. Other classes—especially blue-collar workers who were protected under Soviet socialism—are experiencing downward mobility, at least during the painful transition to a new social order.

class become conscious of their situation as a class remains an important subject in the study of stratification.

Sources of Class Conflict "The history of all hitherto existing societies is the history of class struggles," wrote Marx and his collaborator Friedrich Engels in *The Communist Manifesto* (1969/1848, p. 11). In the modern era "society as a whole is more and more splitting up into two great hostile camps, into two great classes directly facing each other, bourgeoisie and proletariat" (p. 11). Why did Marx and other observers of the capitalist system believe that conflict between the bourgeoisie and the proletariat was inevitable? The answer can be found by taking a closer look at Marx's analysis of the evolution of capitalism.

To begin with, Marx observed the misery of industrial workers (many of whom were children) in the smoky factories of industrial England. He saw that however wretched the workers were, there would always be what he called a "reserve army" of unemployed people who were willing to work for lower wages than those who already had jobs. He noticed that the capitalists (and the intellectuals they paid to argue in their defense) always blamed the workers themselves for the miserable conditions in which they were forced to exist. If they were hungry, the capitalists claimed, it was because they did not work hard enough or because they spent their pay on too much alcohol or because they could not curb their sexual passions and bore too many children. Thus the capitalists refused to accept blame for the misery of the working class under capitalism.

Marx also argued that business competition would eliminate less successful firms and result in monopolies, which would control prices and wages and thereby contribute still more to the impoverishment of the workers. Moreover, the capitalists had the power to determine who ran the government and who controlled the police and the army. If the workers were to rebel, the armed forces and police would intervene as agents of the capitalists. Through these means the workers and the unemployed would be forced to remain a huge, helpless population that could be manipulated by the capitalists. Over time, according to Marx and his followers, these masses of people would become increasingly conscious of their plight and would unite in a revolution that would destroy the power of the capitalists and their allies.

But why would workers become more conscious of their situation as a class, rather than merely remaining miserable in their impoverished families and communities? This problem of class consciousness became a central issue in Marxian thought. The study of changing patterns of class consciousness remains a major subject in sociology to this day.

Objective and Subjective Classes In thinking about how a social class might be able to take collective action, revolutionary or otherwise, Marx distinguished between *objective* and *subjective* classes. All capitalist societies, according to Marx, have **objective classes**—social classes that an observer can identify by simply looking at people's visible, specific relationships to the means of production in their society. The workers, for example, are easily identified as an objective class that does not own the means of production and sells its labor power in return for wages. The capitalists are the objective class that owns the means of production—the machines and property and railroads—and buys the labor power of the working class.

These facts would be visible to an observer equipped with an understanding of what makes a social class, but they might not always be evident to the members of those classes themselves. Thus Marx identified **subjective class** as the extent to which the people in a given stratum of society actually perceive their situation as a class. For example, if the workers in the chicken processing industry in the American South are not aware that their low wages and miserable working conditions are similar to the conditions faced by millions of workers in other industries elsewhere in the world, and if they do not understand that they can improve their fortunes only by taking some of the power of another class (the capitalists), they are not yet a subjective class. Without this awareness of their situation, the workers are said to lack **class consciousness.** And without class consciousness they cannot form the political associations that will allow them to fight effectively against the capitalists.

In an often quoted passage, Marx described the peasants of France as an objective class because of their shared experience as agriculturalists with small landholdings, but he was doubtful about their ability to form a subjective class:

> A small holding, a peasant and his family; alongside them another small holding, another peasant and another family. A few score of these make up a village, and a few score of villages make up a Department. In this way, the great mass of the French nation is formed by simple addition of homologous magnitudes, much as potatoes in a sack form a sack of potatoes. . . . In so far as there is merely a local interconnection among these small holding peasants, and the identity of their interests begets no community, no national bond and no political organization among them, they do not form a class. (1963/1869, p. 124)

The working class, Marx believed, would be different from the peasantry because its members would become conscious of their shared interests as a class.

The Classless Society Marx and other observers of early capitalism believed that the growing conflict between the working class or proletariat and the capitalist class or bourgeoisie would produce revolutions. In those revolutions the proletariat and its allies would depose the bourgeoisie and establish a new social order known as socialism. Under socialism the key institutions of capitalism—private ownership of the means of production, the market as the dominant economic institution, and the nation-state controlled by the bourgeoisie—would be abolished. The new society would be classless because the economic institutions that produced classes would have been eliminated and all the members of society would collectively own the means of production.

In fact, Soviet-style socialism as it developed under the Bolshevik dictatorship of Lenin and then Joseph Stalin did not produce a classless society. After the revolution of 1917 a new class of Communist party officials, military generals, and government bureaucrats, known as the *nomenklatura,* replaced the older aristocratic ruling class.

As the communist elite used its power to secure privileges that were not granted to those in the lower classes, the Marxian dream of a classless society gave way to cynicism. And in the decade since the demise of the Soviet Union the extremely rapid conversion of the economy from socialism to capitalism has given rise to an entirely new class of self-made millionaires. Elsewhere in the world, however, in societies that did not experience the abuses and rigidity of Soviet communism, the Marxian view of class struggle remains one of the powerful forces giving rise to social movements to reduce inequality.

Social Mobility in Modern Societies: The Weberian View

Well before the failure of Soviet communism, Max Weber took issue with Marx's view of social stratification. Marx had defined social class in economic terms; classes are based on people's relationship to the means of production. Weber challenged this definition of social class. People are stratified, Weber reasoned, not only by how they earn their living but also by how much honor or prestige they receive from others and how much power they command. A person could be a poor European aristocrat whose lands had been taken away during a revolution, yet his prestige could be such that he would be invited to the homes of wealthy families seeking to use his social status to raise their own. Another person could have little money compared with the wealthy capitalists, and little prestige compared with the European aristocracy, yet could command immense power. The late Mayor Richard J. Daley of Chicago (the father of the present mayor) was such a person. He was born into a working-class Irish American family, and although he was rather well-off he was not rich. Yet his positions as mayor and chairman of the Cook County Democratic party organization made him a powerful man. Indeed, Mayor Daley's fame as a politician who led a powerful urban party organization but did not enrich himself in doing so helped his son continue the family tradition of political leadership: In 1989 Richard M. Daley was elected mayor of Chicago.

For Weber and many other sociologists, therefore, wealth or economic position is only one of at least three dimensions of stratification that need to be considered in defining social class. Prestige (or social status) and power are the others. We need to think of modern stratification systems as ranking people on all of these dimensions. A high ranking in terms of wealth does not always guarantee a high ranking in terms of prestige or power, although they often go together.

Other challenges to Marx's view of stratification focus on social mobility in industrial societies. Contrary to Marx's prediction, modern societies have not become polarized into two great classes, the rich and the poor. Instead, there is a large middle class of people who are neither industrial workers nor capitalists (Wright, 1989, 1997), and there is considerable social mobility, or movement from one class to another. (See the study chart below.)

Forms of Social Mobility Social mobility can be measured either within or between generations. These two kinds of mobility are termed *intragenerational* and *intergenerational* mobility. **Intragenerational mobility** refers to one's chances of rising to or falling from one social class to another within one's own lifetime. **Intergenerational mobility** is usually measured by comparing the social-class position of children with that of their parents. If there is a great deal of stability from one generation to the next, one can conclude that the stratification system is relatively rigid.

STUDY CHART Types of Social Mobility

TYPE	DESCRIPTION	EXAMPLE
Structural	Entire occupations emerge and grow rapidly while others decline because of technological or other major changes.	Workers in high-tech industries (e.g., computers); farmers.
Individual or Family		
Intergenerational	Children experience mobility in either direction in comparison with their parents.	Upward—father a blue-collar worker, son a lawyer; daughter a professor, mother a factory worker. Downward—parents professionals, children waiters and aspiring actors.
Intragenerational	A person experiences an upward or downward shift in economic and social status.	A steelworker becomes a janitor, earning less; his wife goes to work to help make ends meet.

Unfortunately, not all mobility is upward. Downward mobility involves loss of economic and social standing. It is a problem for families and individuals at all levels of the class structure in the United States and elsewhere. In its broadest sense, downward mobility can be defined as "losing one's place in society" (Newman, 1988, p. 7). In fact, the term encompasses many different kinds of experiences. The married woman who works part-time and then is divorced, loses the family house, and must move with her children to a small apartment and work full-time is experiencing downward mobility. The couple who have been living with the wife's mother in public housing and are forced out during a check of official rosters and must seek refuge in a homeless shelter also experience downward mobility. The affluent young couple living in a downtown condominium experience downward mobility when he loses his job at a brokerage firm and she must go on maternity leave while he is looking for a new job. Often entire groups of people experience downward mobility, as happened to U.S. citizens of Japanese descent when their property was confiscated during World War II, and to people of Indian descent when they were expelled from Kenya and Uganda in the 1970s.

The General Social Survey, conducted annually by the National Opinion Research Center (NORC), charts changes in people's opinions about their prospects for mobility in either direction, using questions like these:

> Compared to your parents when they were the age you are now, do you think your own standard of living now is much better, somewhat better, about the same, somewhat worse, or much worse than theirs was?
>
> When your children are at the age you are now, do you think their standard of living will be much better, somewhat better, about the same, somewhat worse, or much worse than yours is now? (NORC, 1994, p. 199)

Fear of downward mobility weighs heavily on the minds of many Americans. A central feature of the American dream is that our standard of living, and that of our children, will continue to improve. But that vision is challenged by the realities of global competition and the "export" of jobs to lower-wage regions of the world (Block, 1990; Harrison, Tilly, & Bluestone, 1986). Indeed, the findings of the General Social Survey show that more respondents claim to have experienced an improvement in their standard of living compared to their parents than believe their children will also enjoy such mobility. These results are presented in Figure 11.5. We

FIGURE 11.5 **Perceptions of Mobility Opportunities (respondent's own and children's)**

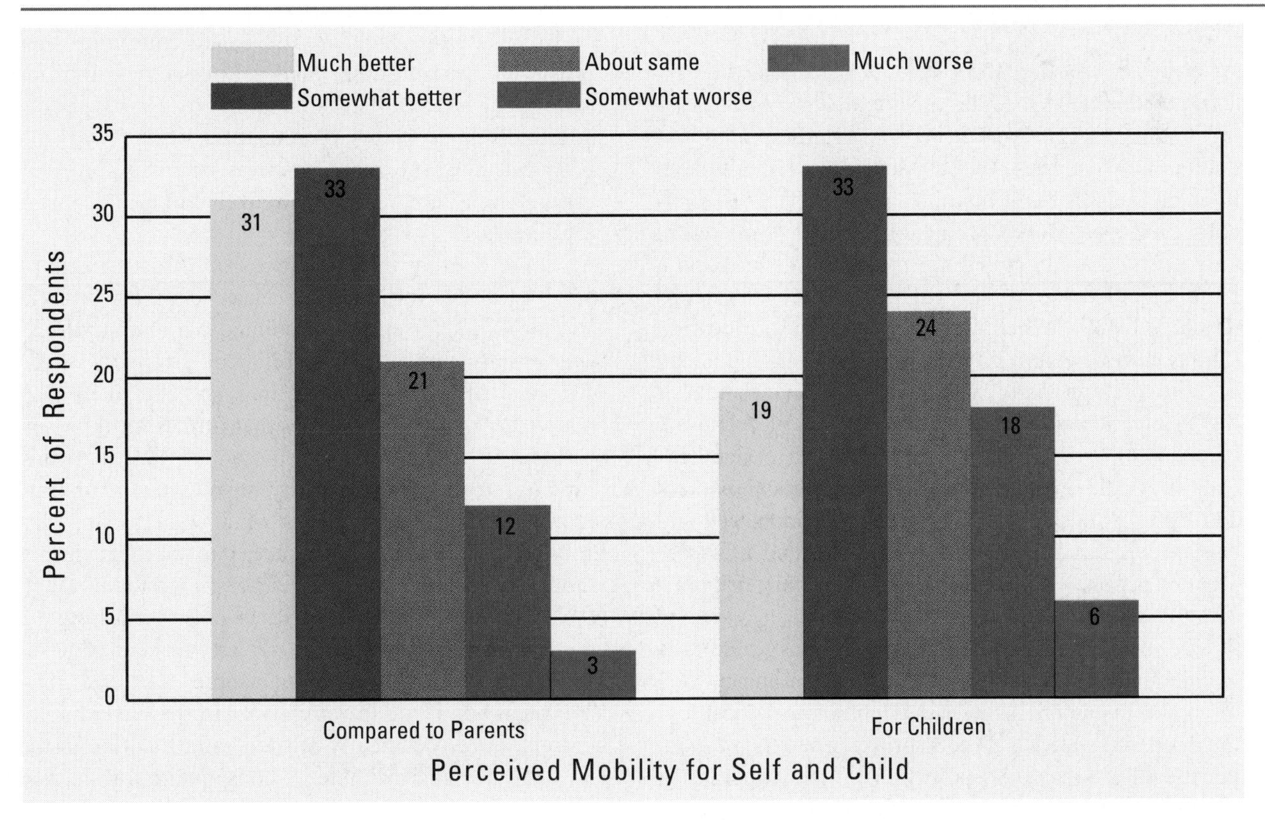

Source: NORC, 1994.

will see in later chapters that this doubt about future opportunities for one's children is an extremely powerful force in contemporary politics.

THEORIES OF STRATIFICATION

"The first lesson of modern sociology," wrote C. Wright Mills in his study of white-collar workers, "is that the individual cannot understand his own experience or gauge his own fate without locating himself within the trends of his epoch and the life-chances of all the individuals of his social layer" (1951, p. xx). We have examined the concepts of stratification and life chances and have seen that they are related to the way people earn their living in different societies. But what about the origins of "social layering" or stratification? That subject will be discussed in the next chapter with specific reference to the social-class system of the United States and other nations. In this section we examine some of the basic theories of social stratification. Each of the major sociological perspectives continues to generate important insights into how stratification and inequality emerge and operate in different societies.

Conflict Theories

As mentioned earlier, Marx's theory of stratification asserts that capitalist societies are divided into two opposing classes, wage workers and capitalists, and that conflict between these two classes will eventually lead to revolutions that will establish classless socialist societies. However, Marx's prediction was not borne out in any society that attempted to implement his ideas on a large scale. Deny it as they might, all of those societies developed well-defined systems of social stratification (Djilas, 1982; Szelenyi, 1983; Temkin, 1993). Each had an elite of high party officials; an upper stratum of higher professionals, scientists, managers of economic enterprises, local party officials, and high police officials; a middle level of well-educated technical workers and lower professionals; a proletariat of industrial and clerical workers and military personnel; and a bottom layer of people who were disabled, criminals, or political outcasts.

No persuasive evidence shows that class conflict is heightening the division between workers and owners of capital in capitalist societies. Conflict does exist, but the industrial working class is shrinking and the new occupational groups do not always share the concerns of the industrial workers. Moreover, reforms of capitalist institutions have greatly improved the workers' situation, thereby reducing the likelihood that the revolution predicted by Marx will occur.

Modern conflict theorists agree with Marx's claim that class conflict is a primary cause of social change, but they frequently debate both the nature of the class structure and the forms taken by class conflict. Thus Erik Olin Wright (1979, 1997) notes that Marxian theorists agree that workers who are directly engaged in the production of goods are part of the working class. However, he makes the following observation:

> There is no such agreement about any other category of wage-earners. Some Marxists have argued that only productive manual workers should be considered part of the proletariat. Others have argued that the working class includes low-level, routinized white-collar workers as well. Still others have argued that virtually all wage-laborers should be considered part of the working class. (1979, p. 31)

This disagreement stems, of course, from the fact that there is far greater diversity within all the classes of modern societies than Marx or his contemporaries imagined there would be. In addition to the bourgeoisie (the owners of large amounts of capital) and the petit bourgeoisie (the owners of small firms and stores), there is a constantly growing professional class, a class of top managers and engineers, another class of lower-level managers, and a class of employees with special skills (e.g., computer specialists, nurses and other medical personnel, and operating engineers). Perhaps many of these people should think of themselves as part of the working class, but they normally do not because they are earning enough to enable them to live in middle-class communities.

Some conflict theorists focus on other aspects of social stratification besides class conflict. Melvin Tumin (1967), for example, pointed out that stratification systems "limit the possibility of discovery of the full range of talent available in a society"; that they create unfavorable self-images that further limit the expression of people's creative potential; and that they "function to encourage hostility, suspicion, and distrust among the various segments of a society" (p. 58). Problems like wasted talent and poor self-image are among what Richard Sennett and Jonathan Cobb (1972) term "the hidden injuries of class," meaning the ways in which a childhood of poverty or economic insecurity can leave its mark on people even after they have risen out of the lower classes. In the next chapter we look more specifically at social stratification in the United States and examine its consequences in human terms.

The Functionalist View

The functionalist view of stratification was originally stated by Talcott Parsons (1937, 1940) and Kingsley Davis and Wilbert Moore (1945). This theory holds that social classes emerge because an unequal distribution of rewards is essential in complex societies. Such societies need to reward talented people and channel them into roles that require advanced training, personal sacrifice, and extreme stress. Thus the unequal distribution of rewards, which allows some people to accumulate wealth and deprives others of that chance, is necessary if the society is to match the most talented individuals with the most challenging positions.

Research by sociologists in the former Soviet Union often provided support for the functionalist idea that extreme "leveling" deprives people of the motivation to achieve more than a minimum of skills. On the basis of extensive surveys of workers in Russia and other former Soviet republics, sociologist Tatyana Zaslavskaya (cited in Aganbegyan, 1989) argues that unless there are incentives in the form of high wages and other advantages (such as better housing in areas where good housing is in short supply), engineers and scientists will resent their situation and will not work hard. In the United States the same arguments are used to justify higher salaries for doctors, lawyers, and other professionals. Too much equality, it is said, reduces the incentive to master difficult skills, and as a result the entire society may suffer from a lack of professional expertise.

Critics of the functionalist view of inequality and stratification point to many situations in which people in positions of power or leadership receive what appear to be excessive benefits. In the early 1990s, during the most prolonged economic recession the United States had experienced since the Great Depression of the 1930s, the multimillion-dollar salaries of many corporate executives were frequently criticized in business magazines and news analyses. For example, the late Roberto Goizueta, former chief executive officer of the Coca-Cola Company, was criticized for taking home more than $10 million in salary and bonuses. Many other executives were earning equally high salaries even when their companies were not doing as well as they might in international competition. With the annual pay of top Japanese executives often running 5 to 10 times less than that of their American counterparts, it became difficult for the American corporate elite to argue that these enormous rates of executive compensation are "functional."

From the point of view of those who are critical of social inequality, the large sums paid to a few people seem wrong, especially when so many others are struggling just to survive. Indeed, the heads of large corporations in the United States often earn more than 50 times as much as the average employee of those corporations. Is such great disparity "functional" when it produces enormous gaps between the very rich and the working classes? Functionalist theory claims that it is, for in a capitalist system of free enterprise top executives will seek the firms that are most willing to reward them for their talents. Those firms will benefit, and so will their workers.

The Interactionist Perspective

Conflict theory explains stratification primarily in economic terms. So does functionalist theory. Both trace the existence of certain classes to the central position of occupation, income, and wealth in modern life. But neither goes very far toward explaining the prestige stratification that occurs *within* social classes. Among the very rich in America, for example, people who have stables on their property tend to look down on people with somewhat smaller lots on which there is only a swimming pool. And rich families who own sailing yachts look down on equally rich people who own expensive but noisy power boats. The point is that within economic classes people form status groups whose prestige or honor is measured not according to what they produce or how much wealth they own but according to what they buy and what they communicate about themselves through their purchases. Rolex watches and BMW cars are symbols of membership in the youthful upper class. Four-wheel-drive vehicles equipped with gun racks and fishing rods are symbols of the rugged and successful middle- or working-class male. Armani suits are symbols of urbane professionalism; tweed suits and silk blouses are signs that a woman is a member of the "country club set." All of these symbols of prestige and group membership change as groups with less prestige mimic them, spurring a search for new and less "common" signs of belonging (Dowd, 1985).

Our tendency to divide ourselves up into social categories and then assert claims of greater prestige for one group or another is of major significance in our lives. The interactionist perspective on stratification therefore may not be very useful in explaining the emergence of economic classes, but it is essential to understanding the behaviors of the status groups that form within a given class. Those behaviors, in turn, often define or reinforce or challenge class divisions. The stratification system, in this view, is not a fixed system but is created over and over again through the everyday behaviors of millions of people.

RESEARCH ON THE CUTTING EDGE: Stratification and Human Development

Today it is no longer possible to think of a society's stratification system separately from those of other societies. More than ever before, the life chances of people in the United States and other highly developed nations are influenced by changes in the structure of industrial employment and the worldwide distribution of jobs. In 1998, for example, many nations of Southeast Asia were struck by economic crises. Indonesia experienced widespread social unrest in the wake of financial instability. Japan's economy, one of the strongest in the world, faltered badly. These national crises worsened the plight of millions of impoverished people in developing nations throughout the world.

These and other global trends are carefully measured and documented by the United Nations Development Programme (UNDP) in its annual *Human Development Report.* A team of social scientists from around the world contributes to the analysis of trends in inequality and the assessment of the progress many nations have made in addressing the gap between the haves and have-nots in their countries. A set of social indicators of inequality are at the heart of the report and are what make it so useful in charting trends over time.

The Human Poverty Index

Because income varies so widely across nations, it is extremely difficult to compare national levels of poverty. An income of less than $18,000 a year for a family of four in the United States is considered below the poverty level, but the same income in India or China would qualify a family as comfortable. In consequence, the researchers developed the Human Poverty Index, an indicator of poverty based on measures of "deprivation in three essential elements of human life—longevity, knowledge, and decent living standard" (UNDP, 1997, p.18). The first dimension, longevity, a measure of vulnerability to death, is measured by the percentage of people expected to die before age 40. The second dimension, knowledge, is measured by the percentage of adults who are illiterate. The third dimension relates to a decent standard of living and is represented by a composite of three variables: the percentage of people with access to health services, the percentage with access to safe water, and the percentage of malnourished children under age 5. The calculation of the Human Poverty Index from these measures is a complex process that yields the percentage of a nation's population that is experiencing the multiple effects of poverty on their daily lives and on the development of their children.

Among the developing countries, those with the lowest Human Poverty Indexes are Trinidad and Tobago (4.1%), Cuba (5.1%), Chile (5.4%), Singapore (6.6%), and Costa Rica (6.6%). The five highest poverty measures are found in the African nations of Mali (54.7%), Ethiopia (56.2%), Burkina Faso (58.3%), Sierra Leone (59.2%), and Niger (66.0%). Haiti (46.2%) has the highest index among nations of the Western Hemisphere. Honduras (22.0% in 1997), one of the poorer nations of Central America, experienced the devastation of Hurricane Mitch in 1998, and as a result much of its population has been plunged into dire poverty. In India (36.7%), Pakistan (46.8%), and Bangladesh (48.3%), long-term inequalities of social caste continue to produce persistently high rates of poverty.

High poverty index percentages for an entire nation can be somewhat misleading. In India, for example, the state of Kerala, with a firm commitment to decreasing inequality, the Human Poverty Index is only 1.5%, compared to 36.7% for India as a whole.

Gender Disparities

No contemporary society treats its women as well as its men. UNDP's *Human Development Report* summarizes the situation as follows:

> In developing countries there are still 60% more women than men among illiterate adults, female enrollment even at the primary level is 13% lower than male enrollment, and female wages are only three-fourths of male wages. In industrial countries unemployment is higher among women than men, and women constitute three-fourths of the unpaid family workers. (1997, p. 39)

In the world's poorest nations women face double deprivation: Human development problems such as lack of medical care, lack of clean water, extreme poverty, and unemployment affect all but the comfortable elite, but women in these countries still face even higher levels of risk and deprivation. On the other hand, some developing nations outperform much richer nations in achieving gender equality in political, economic, and professional activities: "Barbados is ahead of Belgium and Italy, Trinidad and Tobago outranks Portugal, and the Bahamas leads the United Kingdom. France lags behind Suriname, Colombia, and Botswana and Japan behind China, Guatemala, and Mexico" (UNDP, 1997, p.41).

Achieving Greater Equality

The *Human Development Report* points out that in most societies some members of the upper classes have a vested interest in perpetuating poverty. The poor form "a mobile pool of low-paid and unorganized workers" who are useful for doing the dirty, dangerous, and difficult work that others often refuse to do. In industrial nations like the United States, many such jobs are taken by people from the poorest backgrounds and often by immigrants, legal and illegal. In many nations—including Russia and Nigeria, to mention two notable examples—patterns of corruption among the elites make it even more difficult to narrow the gap between rich and poor.

Reports about worldwide inequality and poverty often seem to suggest that the problems are so great that improvements are unlikely. But despite the magnitude of the challenge, skepticism and cynicism are unwarranted. After all, the report notes that since the 1990 World Summit on Children, the first-ever summit on human issues, rates of child mortality have declined in all regions of the world. Seven million more children's lives are being saved each year than in 1980. Immunization against major diseases like polio now cover over 80 percent of the world's children. On other fronts, such as reducing poverty among the aged or ensuring greater access to medical care, efforts to improve conditions are having demonstrable success. While it is true that AIDS and other major diseases continue to wreak havoc in many regions, the record shows that when the world's nations make concerted efforts to address social ills related to poverty and disease, they can have major successes.

What is the price tag for poverty eradication? The *Human Development Report* estimates the cost of achieving basic social services for all in the developing countries at about $40 billion a year over the 10-year period between 1995 and 2005 (see Table 11.1). However, as the report concludes: "Lack of political commitment, not financial resources, is the real obstacle to poverty eradication. Eradicating absolute poverty is eminently affordable" (1997, p. 112).

TABLE 11.1 Cost of Achieving Universal Access to Basic Social Services

Need	Annual Cost (in U.S. dollars)
Basic education for all	$6 billion
Basic health and nutrition	$13 billion
Reproductive health and family planning	$12 billion
Low-cost water supply and sanitation	$9 billion
Total for basic social services	$40 billion

Source: UNDP, 1997.

VISUAL SOCIOLOGY

Extreme Poverty

Sebastiao Salgado spends long periods living with the people he photographs and using his camera to document their work and their efforts to overcome extreme poverty. "Salgado photographs from the inside," claims the Uruguayan writer Eduardo Galeano. "He remained in the Sahel desert for fifteen months when he went there to photograph hunger. He traveled in Latin America for seven years to garner a handful of photographs" (Salgado, 1990, p. 11).

The photos included in this sample of Salgado's "concerned photography" document the agony of extreme poverty. The great pit shown in one of the photos is a gold mine in the province of Paras, Brazil, where more than 50,000 mud-covered men carry heavy bags of dirt up slippery ladders in the hope that now and then one of the bags will yield a nugget of gold.

These haunting photographs capture two very different poverty-producing conditions. The plight of the miners is a result of the exploitation of cheap human labor in a jungle gold rush. The photos of starvation in the Sahel document the conditions of severe drought compounded by extreme political anarchy. Thousands of people have died in Ethiopia, Mali, Somalia, and other drought-stricken areas—not because food was unavailable but because violence and war kept supplies from those without guns and power.

Children's ward in the Korem refugee camp. (Ethiopia, 1984.)

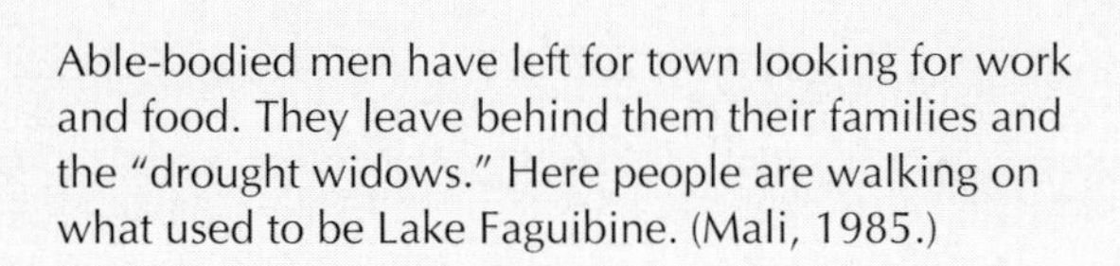

Able-bodied men have left for town looking for work and food. They leave behind them their families and the "drought widows." Here people are walking on what used to be Lake Faguibine. (Mali, 1985.)

Overall view of the Sierra Pelada gold mine, where more than 50,000 workers were employed. (Brazil, 1986.)

Women carry ground rock for road repair near Jaipur, India.

Tens of thousands of refugees from the great drought in Brazil's northeast gather near cities. Often the only way to survive is to fight with the vultures for the spoiled food in garbage dumps. (Fortaleza, Brazil, 1983.)

SUMMARY

Social stratification refers to a society's system for ranking people hierarchically according to various attributes such as wealth, power, and prestige. Societies in which there are rigid boundaries between social strata are said to have *closed stratification systems*, whereas those in which the boundaries are easily crossed are said to have *open stratification systems*. Movement from one stratum to another is known as *social mobility*.

Most closed stratification systems are characterized by *castes*, or social strata into which people are born and in which they remain for life. Membership in a caste is an *ascribed status* (a status acquired at birth), as opposed to an *achieved status* (one based on the efforts of the individual).

Open societies are characterized by *classes*, which are social strata based primarily on economic criteria. The classes of modern societies are not homogeneous; within any given class there are different groups defined by how much honor or prestige they receive from the society in general. Such groups are sometimes referred to as *status groups*. The way people are grouped with respect to their access to scarce resources determines their *life chances*—the opportunities they will have or be denied throughout life.

The principal forces leading to social stratification are created by the means of existence in a given society. Hence, for small farmers or peasants (the majority of the world's population) social strata are based on land ownership and agrarian labor, with the members of the lowest strata doing the hardest work and those at the top of the stratification system living in relative comfort. Modern industrial societies are characterized by *structural mobility* (the elimination of entire classes as a result of changes in the means of existence) and *spatial mobility* (the movement of individuals and groups from one location to another).

People accept their place in a stratification system because the system itself is part of their society's culture. The facets of culture that justify the stratification system are learned through the processes of socialization. The system is often justified by an ideology. At the micro level, the norms of everyday interactions, especially *deference* and *demeanor*, serve to reinforce the society's stratification system.

Changes in stratification systems may have as much to do with realignments of social power as with economic or cultural changes. *Power* has been defined as "the probability that one actor within a social relationship will be in a position to carry out his own will despite resistance." Legitimate power is called *authority* and is a major factor in maintaining existing relationships among castes or classes.

The rise of industrial capitalism had far-reaching effects on stratification systems. According to Karl Marx, capitalism divided societies into classes based on ownership of the means of production. The largest of these classes, the workers, must sell their labor to capitalists or landowners in return for wages. In time, the workers would become conscious of their shared interests as a class and would rebel against the capitalist class. The outcome of the revolution would be a classless society.

Marx defined social class in economic terms. Max Weber took issue with this definition and pointed out that people are stratified not only by wealth but also by how much honor or prestige they receive from others and how much power they command. Marx's view of stratification is also challenged by studies of social mobility in industrial societies, which have shown that there is considerable movement between classes.

Modern conflict theorists, like Marx, believe that class conflict is a primary cause of social change. They disagree, however, on the nature of the class structure of capitalist societies. Functionalist theorists believe that classes emerge because an unequal distribution of rewards is necessary in order to channel talented people into important roles in society. This view has been criticized because it fails to account for the fact that social rewards in one generation tend to improve the life chances of the next generation; nor does it explain why talented people from lower-class families often are unable to obtain highly rewarded positions. From the interactionist perspective, the stratification system is not a fixed system but, rather, one that is created out of everyday behaviors.

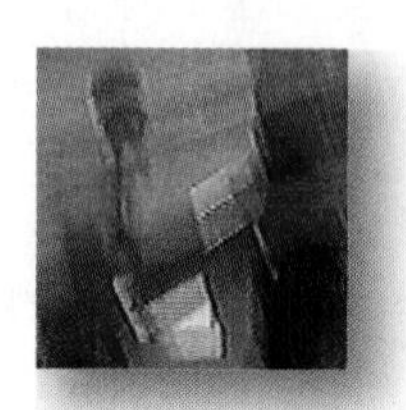

SOCIOLOGY VERSUS IDEOLOGY

Many people argue that it is useless for governments to make efforts to address inequalities of social class.

The poor will always be poor, they argue, because they are responsible for their own plight. And besides, they may add, any major effort to address existing inequalities will require that money and other resources be taken away from the wealthy and more productive classes in the form of higher taxes or other forms of confiscation. On the liberal side of the ideological divide, people with strong ideological views often argue that the state (government) needs to make far more aggressive efforts to reduce inequalities of class. The rich and the well-off, they say, are the obstacle. Their wealth and power must be reduced so that people in less advantaged strata of societies may benefit.

Sociologists often hold strong personal views on these issues. But when they examine the empirical evidence, they find that neither approach provides an adequate model for positive social change. Doing nothing to address poverty has often resulted only in increasing misery and eventual social unrest and bloody revolution. Supporting revolution or confiscation of wealth in the name of equality, on the other hand, has most often resulted in greater suffering and new forms of inequality. What has worked, though imperfectly, sociologists point out, are policies that "level the playing field" by offering improved educational and health care opportunities. And sociological research confirms that social security policies protect all of a society's members, but especially the poor, the elderly, and children, from the consequences of severe inequalities. Inequalities of social class remain problematic in these societies, but they do not pose a serious threat to social peace and individual achievement.

GLOSSARY

social stratification: a society's system for ranking people hierarchically according to such attributes as wealth, power, and prestige. (p. 328)

closed stratification system: a stratification system in which there are rigid boundaries between social strata. (p. 328)

open stratification system: a stratification system in which the boundaries between social strata are easily crossed. (p. 328)

social mobility: movement by an individual or group from one social stratum to another. (p. 328)

upward mobility: movement by an individual or group to a higher social stratum. (p. 328)

downward mobility: movement by an individual or group to a lower social stratum. (p. 328)

caste: a social stratum into which people are born and in which they remain for life. (p. 328)

ascribed status: a position or rank that is assigned to an individual at birth and cannot be changed. (p. 328)

achieved status: a position or rank that is earned through the efforts of the individual. (p. 328)

class: a social stratum that is defined primarily by economic criteria such as occupation, income, and wealth. (p. 329)

status group: a category of people within a social class, defined by how much honor or prestige they receive from the society in general. (p. 329)

life chances: the opportunities an individual will have or be denied throughout life as a result of his or her social-class position. (p. 329)

structural mobility: movement of an individual or group from one social stratum to another caused by the elimination of an entire class as a result of changes in the means of existence. (p. 336)

spatial mobility: movement of an individual or group from one location or community to another. (p. 337)

status symbols: material objects or behaviors that indicate social status or prestige. (p. 338)

deference: the respect and esteem shown to an individual. (p. 338)

demeanor: the ways in which individuals present themselves to others through body language, dress, speech, and manners. (p. 339)

power: the ability to control the behavior of others, even against their will. (p. 339)

authority: power that is considered legitimate both by those who exercise it and by those who are affected by it. (p. 340)

objective class: in Marxian theory, a social class that has a visible, specific relationship to the means of production. (p. 345)

subjective class: in Marxian theory, the way members of a given social class perceive their situation as a class. (p. 345)

class consciousness: a group's shared subjective awareness of its objective situation as a class. (p. 345)

intragenerational mobility: a change in the social class of an individual within his or her own lifetime. (p. 346)

intergenerational mobility: a change in the social class of family members from one generation to the next. (p. 346)

WHERE TO FIND IT

BOOKS

The Quality of Life (Martha Nussbaum and Amartya Sen, eds.; Oxford University Press, 1993). A collection of essays on the relationships between inequalities of caste and class and their impact on the quality of life.

Bringing Class Back in Contemporary and Historical Perspectives (Scott C. McNall, Rhonda F. Levine, and Rick Fantasia, eds.; Westview, 1991). A collection of essays dealing with theories and empirical studies of stratification in comparative and historical perspective.

The Structures of Everyday Life: Civilization and Capitalism, 15th–18th Century (vol. 1), ***The Wheels of Commerce*** (vol. 2), ***The Perspective of the World*** (vol. 3) (Fernand Braudel; Harper, 1981). A work of historical and sociological scholarship that is only beginning to influence the social sciences. The great French social historian shows in marvelous detail how social stratification has been largely explained by the way people earn their daily bread.

The Modern World System (Immanuel Wallerstein; Academic Press, 1974). A sociologist's account of the transformations in social structure and inequality that were brought about by the emergence of two capitalist institutions: the market and the firm.

The Communist Manifesto (Karl Marx and Friedrich Engels; Penguin, 1969/1848). Marx and Engels's classic description of the role of class conflict in history, with a lucid analysis of the consequences of the rise of industrial capitalism and the inevitability of a proletarian revolution. Eclipsed by events in the modern world, *The Communist Manifesto* continues to inspire readers who are critical of human exploitation and imperialism.

The American Perception of Class (Reeve Vanneman and Lynn Weber Cannon; Temple University Press, 1987). Emphasizes the extent to which there is class consciousness even in periods that are relatively lacking in class conflict, and shows how the perception of one's class membership is directly related to one's actual place in the structure of inequality.

OTHER SOURCES

World Development Report (United Nations Development Programme). An invaluable annual summary of world trends in social class and of efforts to address the worst consequences of global inequality.

The Europa Yearbook: A World Summary (Europa Publishers). An annual reference work that provides much detailed information on political, economic, and commercial institutions throughout the world.

INTERNET RESOURCES

International Sociological Association (www.ucm.es/OTROS/isa). Provides a fine introduction to global sociological work by professional researchers.

Population Reference Bureau (www.prb.org/prb/). Provides demographic data and links to other sources of useful deographic statistics.

The World Bank (www.worldbank.org/). An excellent source of comparative global statistics on indicators of social stratification in different nations and regions of the world, with particular emphasis on the developing nations.

Rural Sociological Society (www.lapop.lsu.edu/rss). Balances the usual urban and industrial emphasis in sociology with access to research on rural topics and themes.

CHAPTER 12

INEQUALITIES OF SOCIAL CLASS

"What I always wonder," he says with a slight shrug, "is how the young people are going to find a way to get ahead now that the mills are gone." Ed Sadlowski is standing in the economically depressed streets of South Chicago, the neighborhood where he was born, where he raised his five children, and where most of them are now living and raising their own children.

"You know, Bill," he continues as we walk slowly past the deserted mill gates, "the United States makes as much steel as we did back when these plants had thousands of workers in them. It's just that we make it so much more efficiently. The productivity of the individual workers is so much greater because of the new technologies, computer control of machines, robotization, and all the rest. So the old system of mass employment in industries like steel is never going to return. The industrial proletariat that Marx wrote about is gone, at least in this country, but not elsewhere in the world where industry is newer."

Edward Sadlowski is among the nation's leading experts in industrial history and the sociology of labor–management relations.

We met many years ago when I was working in a local steel mill while gathering material for a book about this famous industrial community. Sadlowski's grandfather came to Chicago as an immigrant from Poland, drawn like millions of others by the chance to use his strong back and hands in the search for a new and better life in America. His father worked in the mills too. He was one of the union's early organizers. From an early age Ed was taught that the unions in steel and railroads and mining were responsible for bringing workers a fairer share of the profits of their industries. To do so they had to constantly struggle against the mill owners, who often seemed ready to kill them rather than relinquish their control or share the wealth.

When Sadlowski was 18 he too entered the mills, but he was able to join an apprenticeship program in which he learned a skilled trade as a machinist. He immediately became active in the union and was elected to represent his workmates as their grievance officer. This meant that he was a shop floor lawyer who assisted workers when they were disciplined or fired. Their union contract afforded the workers a common pay scale and scheduled increases in their hourly wages, as well as health and other benefits that they had not had before the unions gained legal recognition in the 1930s. Overtime pay and the right to a semblance of due process in disciplinary matters were also among the union's achievements.

Eventually Sadlowski was elected president of the union at U.S. Steel's South Works, Chicago's largest steel mill until it was torn down in the late 1980s. Later in his career he ran unsuccessfully for president of the U.S. and Canadian steelworkers' union, but for his entire working life he has remained a staunch union supporter.

As we walk past acre after acre of abandoned railroad tracks and ore docks on the shore of Lake Michigan, Sadlowski wonders about the future of manufacturing in the United States. There are so many groups whose mobility depended until very recently on well-paid jobs in plants with union contracts. In Chicago there are thousands of African Americans and Chicanos whose progress toward greater economic security has been stymied by plant closings. And from his travels throughout industrial America, Sadlowski understands only too well the consequences of rapid deindustrialization—consequences that are reflected in the harsh realities faced by the workers who appear in the Visual Sociology section of this chapter. "Too many people are looking for too few good jobs," he comments. "You know, the kind that pay a decent wage you can raise a family on, and that pay for adequate health care."

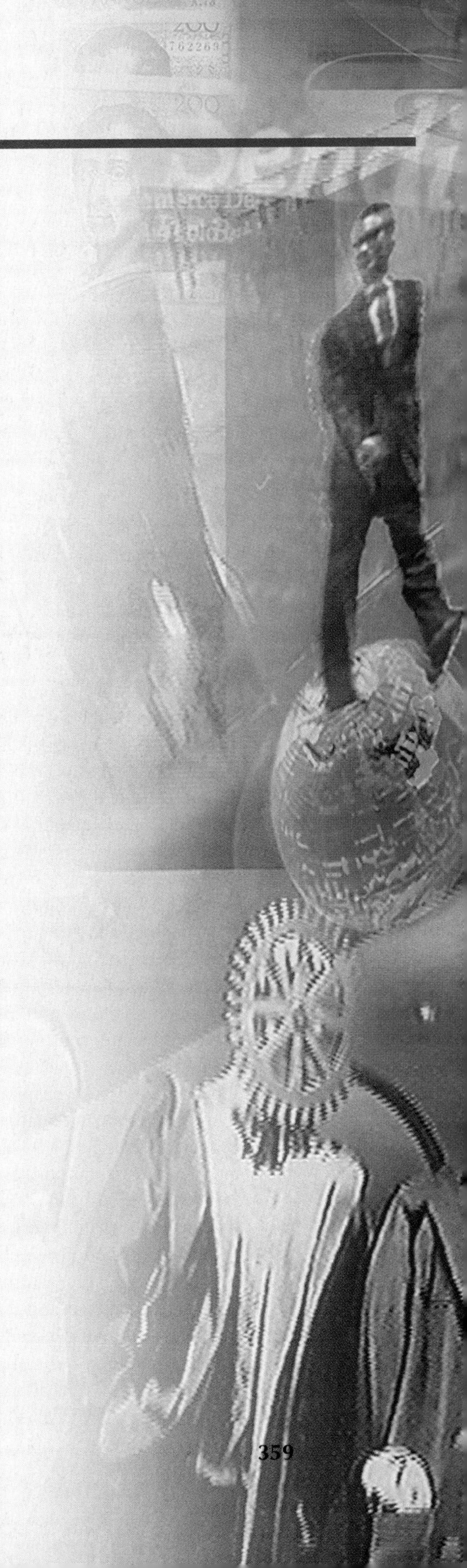

DIMENSIONS OF SOCIAL INEQUALITY IN AMERICA

The Sadlowskis and other families in South Chicago are struggling to adapt to a set of social changes that are global in scope. People in different parts of the world are experiencing these changes in differing ways. South of the U.S. border, for example, parents in hundreds of communities miss their children, who have migrated to manufacturing towns along the border. And in the industrial centers of North America people mourn the loss of jobs to countries or regions where labor costs are lower. These changes are forcing many people to go back to school and gain new skills and credentials so that they can be more competitive in changing labor markets. Union members in manufacturing industries are being forced to end strikes and go back to work to avoid losing their jobs to "replacement workers." But these are only some of the realities of life in a stratified and rapidly changing industrial society. The U.S. economy also generates immense wealth and a growing number of jobs. The society boasts an enormous middle class, many of whose members are far more affluent than their parents and grandparents ever dreamed of becoming. Thus social mobility and class conflict, the subjects of this chapter, remain important aspects of life in the United States and elsewhere in the world. No wonder they are central issues in contemporary sociology.

In the United States, despite high rates of employment in recent years, inequalities of class and status have worsened (Krugman, 1992, 1995; Wolff, 1995). In the nation's poorest communities—in urban ghettos and on Indian reservations and in the hill country of the South—many people doubt whether the conditions of poverty and depression in which they live can ever be improved. Sociological research has shown that the rich are doing better than ever and that the gap between the haves and the have-nots is widening. In different regions of the nation the details of social inequality differ, but the basic patterns of wealth and prestige and power are the same (Gans, 1995; Wilson, 1996).

This chapter further develops many of the terms and concepts defined in the preceding chapter and applies them to the study of inequality and stratification in the United States. First we discuss the ways in which sociologists measure inequality and their changing views of how inequalities result in the formation of social classes. Then we examine social classes and life chances in the United States today and analyze the influence of social-class position on opportunities for upward mobility. We will see that there is nothing static about social inequality in the United States; changes in the class structure of North America are occurring even as this chapter is being written.

Measures of Social Inequality

The basic and most readily available measures of inequality in any society are income, wealth, occupational prestige, and educational attainment. Of these measures, income is the one that is most often used to give an initial view of social-class inequality. A common method of comparing incomes in a large population like that of the United States is to divide the entire population into equal fifths, or *quintiles,* on the basis of personal or household income. The researcher starts at the lowest level of reported income and continues upward until one fifth of the population has been accounted for. This establishes the upper and lower income boundaries for that quintile. The same procedure is used to identify each of the other four quintiles; the average income for each quintile can then be compared with the average incomes for the other four quintiles. In Table 12.1, therefore, we see that in 1996 the average household income for the top quintile was $115,514, whereas for the bottom fifth it was $8,596.

By comparing the average incomes of quintiles at different points in time, social scientists and policy makers can see how changes in income over time are distributed among different strata of the population. For example, Table 12.1 allows us to compare incomes between 1976 and 1996. A comparison of household

TABLE 12.1 How Households Divided the Nation's Income: 1976 and 1996 (in 1996 dollars)

1976			1996	
Share of All Income	**Segment Average**	**Household Segments**	**Segment Average**	**Share of All Income**
43.3%	$85,335	Top 20%	$115,514	49.0%
24.8	48,876	Second 20%	54,922	23.2
17.1	33,701	Middle 20%	35,486	15.1
10.4	20,496	Fourth 20%	21,097	9.0
4.4	8,672	Bottom 20%	8,596	3.7
100.0%	**$39,416**	**All households**	**$47,123**	**100.0%**
16.0%	**$126,131**	**Richest 5%**	**$201,684**	**21.4%**

Source: Hacker, 1997.

incomes for the quintiles shows that in 1996 the richest 20 percent took in 49 percent of all income while the poorest 20 percent earned only 3.7 percent of total income. These figures represent important changes during the 20-year period. The richest quintile increased its income share by 5.7 percent (from 43.3 to 49.0 percent), whereas all the other quintiles experienced a decline in their share of total income. The table also shows that the very richest 5 percent earned over 21 percent of all income in 1996, an increase from 16 percent in 1976. These changes point to the widening gap between rich and poor in the United States; they also show that the condition of the middle classes, as measured by income, is barely improving. The average income earned by households in the middle 20 percent increased by less than $2,000 during the 20-year period ($33,701 in 1976; $35,486 in 1996)—even though, as we will see, there were more workers per household earning income.

Critics point out that these figures do not include medical and other noncash benefits received by many

For the upper classes, food becomes a symbol of status; its cost and the artistry of its presentation assume an importance well beyond mere nutrition.

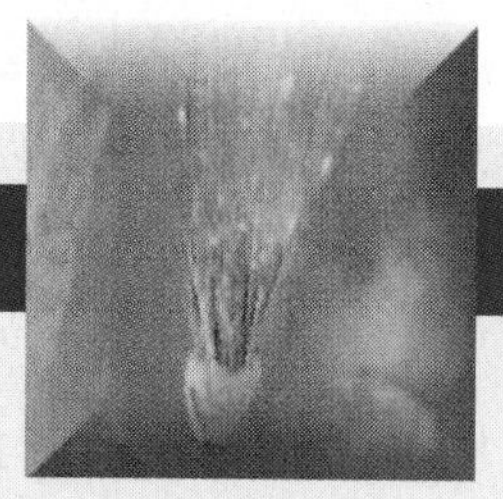

ART AND THE SOCIOLOGICAL EYE

Picturing the Consequences of Social-Class Inequality

In two famous paintings from the mid-twentieth century, we see artists offering their own personal insights into the consequences of class inequalities in the United States. Reginald Marsh's view of Depression-era Chicago, *Tattoo and Haircut,* depicts the life of single men living on Skid Row, the term still used for inner-city areas that cater to the needs of migrant laborers, alcoholics, and people who are "down on their luck." Marsh conveys the sense of boredom and resignation that is common among residents of Skid Row.

Ben Shahn's painting, *Miners' Wives,* depicts the anxiety and suffering of women in a poor working-class community. Their grief is made more vivid by the distance between the doorway of their home and the mine and its owners, who appear in the background. What explains the grim face of the woman in the foreground and her tightly clenched hands? Without being explicit about it, Shahn may be suggesting that the women are awaiting news of the fates of their absent husbands. Mining is one of the world's most dangerous occupations. Too often the wives of miners find themselves holding anxious vigils in the hope that their husbands and sons will return after an explosion or sudden cave-in deep underground.

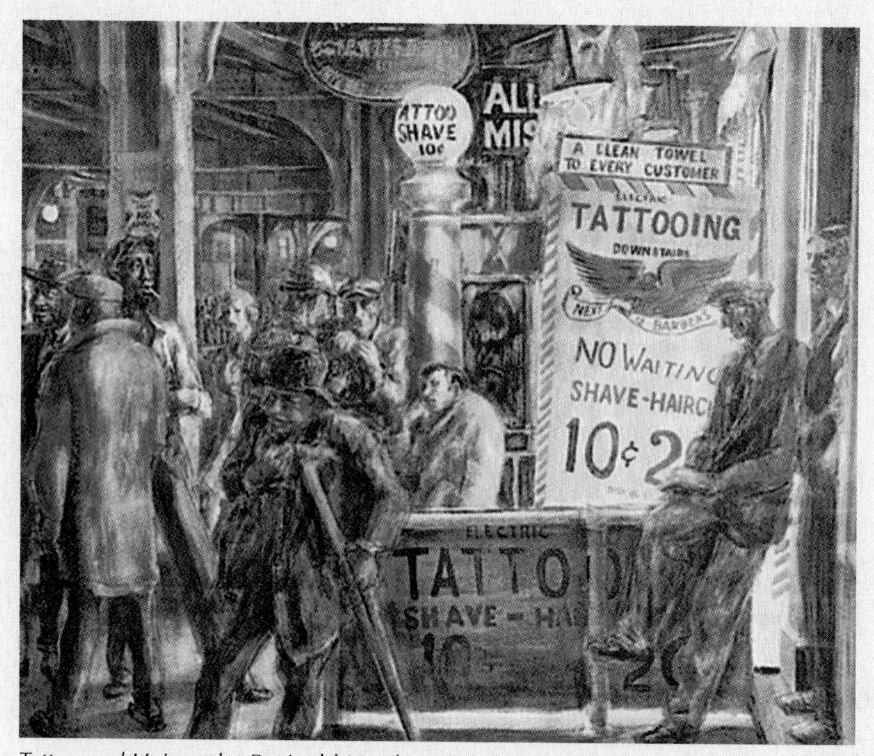

Tattoo and Haircut, by Reginald Marsh, 1932. Tempera on masonite, 46½″ × 48″.

Miners' Wives, by Ben Shahn, 1948. Egg tempera on board, 48″ × 36″.

poorer Americans. They also note that these are pretax figures; taxes have the effect of reducing income disparities. There is some merit to these criticisms. Nevertheless, when public benefits and the effects of taxes are accounted for, the proportional results shown in Table 12.1 still hold. (Gramlich et al., 1993; Hacker, 1997). Moreover, if one studies the distribution of wealth rather than of income alone, the picture is even more skewed in favor of the rich. If we measure wealth in terms of the net financial worth of households—that is, their total assets—only about 6 percent of American households have a net worth of $250,000 or more, whereas over 25 percent have a net worth of less than $5,000.

Recent studies show that the United States has replaced Great Britain and Ireland as the Western industrial nation with the largest gap between rich and poor (Hacker, 1997; Wolff, 1995). The percentage of total private wealth owned by the richest 1 percent of the U.S. population is rising quickly, while the opposite is true in Great Britain—which, unlike the United States, has never been thought of as an egalitarian society.

Figure 12.1 shows how net worth is distributed among some of the most significant aspects of household wealth. The richest 10 percent of families own homes valued at about 35 percent of the total worth of American homes. The same 10 percent, however, own almost 80 percent of other real estate, more than 90 percent of securities (stocks and bonds), and about 60 percent of the money in bank accounts (Hacker, 1997). Remember, too, that inequality expressed in dollars of income or possession of property and stocks translates into large differences in what people can spend on health care, shelter, clothing, education, books, movies, trips to parks and museums, vacations, and other necessities and comforts.

The distribution of educational attainment and occupational prestige is more nearly equal than the distribution of wealth and income. **Educational attainment,** or number of years of school completed, has become more equal among major population groups in the United States, but important inequalities remain. The educational attainment of the nation's black population, for example, has risen dramatically in the past generation. In 1960 only 20 percent of African Americans (who were age 25 or older at the time) completed 4 years of high school or more, while the comparable figure for whites was 43 percent. Today about 75 percent of African Americans have completed 4 years of high school or more, compared to 83 percent of whites. However, only about 14 percent of African Americans have completed 4 years or more of college, compared to almost 25 percent of whites. And for Latinos the gap remains far wider. Approximately 55 percent of Latinos in the U.S. population (age 25 or older) have completed 4 years of high school or more, and about 10 percent had completed 4 years of college or more in 1997 (*Statistical Abstract,* 1998).

Women, on the other hand, have nearly attained parity with men in terms of educational attainment. Eighty-two percent of adult women and 82 percent of men in the U.S. population had completed high school or beyond in 1997. Twenty-six percent of men and 22 percent of women had completed 4 years of higher education or more. We see in Table 12.2, however, that educational attainment does not yield the same benefits for women as for men.

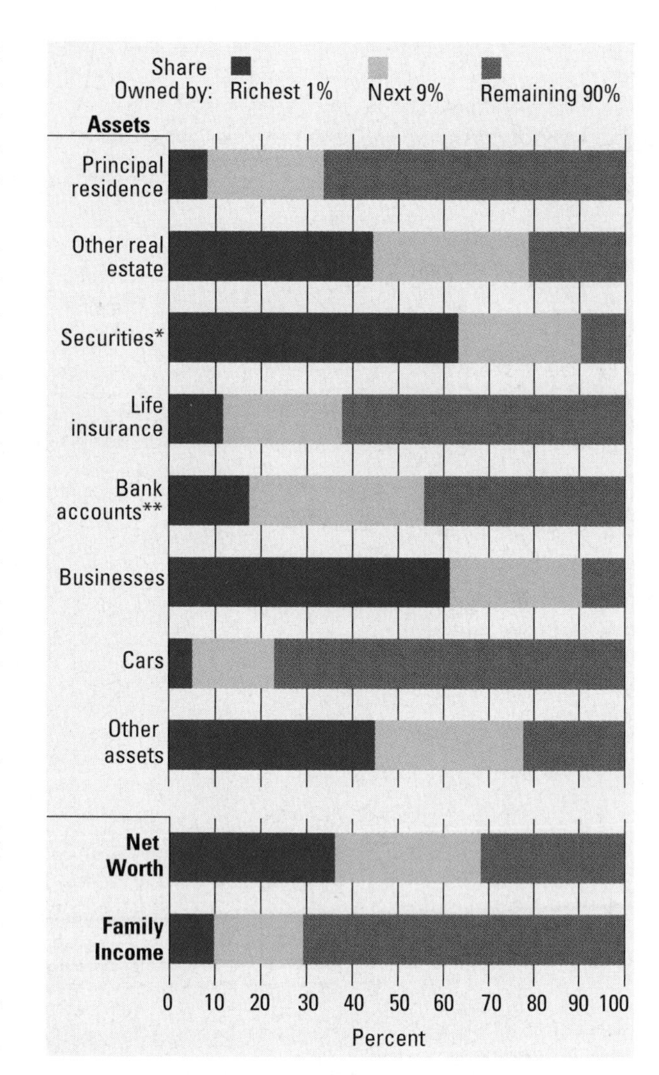

*Includes stocks, bonds, and trusts.
**Includes all deposits in checking, savings, and other accounts.

Source: Federal Reserve, 1996.

FIGURE 12.1 **Proportion of American Families Owning Various Forms of Wealth**

TABLE 12.2 Payment by Degrees: 1975 vs. 1995 *(average earnings of full-time workers, age 35 to 44 in the years' actual dollars)*

1975	Men	Women
High school diploma only	$14,007	$ 7,774
Bachelor's degree only	21,152	10,560
Bachelor's per $1,000 high school earnings	1,510	1,358
1995	**Men**	**Women**
High school diploma only	$32,689	$22,257
Bachelor's degree only	57,196	36,901
Bachelor's per $1,000 high school earnings	1,750	1,658
Advantage for bachelor's degree (change from 1975 to 1995)	+15.8%	+22.1%

Source: Hacker, 1997.

The past two decades have seen an important increase in the relationship between educational attainment and income. The advantage enjoyed by individuals with college educations over those with only high school diplomas has always been significant, but as shown in Table 12.2, it increased by almost 16 percent for men and by over 22 percent for women between 1975 and 1995. In 1975 a man with a college diploma could expect to earn $1,510 for every $1,000 earned by a male worker with only a high school diploma. These averages were somewhat lower for women. In 1995, by contrast, a female worker with a college degree earned, on the average, $1,658 for every $1,000 earned by a woman with only a high school diploma, and the figures are even higher for males. We will explore the reasons for these differences later in this chapter and in Chapter 14.

In a rapidly changing economy like that of the United States, educated people are in increasing demand and education thus becomes a more common route to upward mobility. Improvements in **educational achievement**—in basic reading, writing, and computational skills—are increasingly vital to individual careers. For the entire society, improvements in educational achievement represent improvements in the nation's "human capital," the wealth-producing capacity of its people (Harbison, 1973). But educational attainment and educational achievement are not always correlated. When students are moved through the school system without meeting certain minimum achievement criteria, attainment figures mask significant gaps in achievement. As a result, large segments of the population are unable to achieve nearly as much as they might, either for themselves or for their society.

Another aspect of social inequality, **occupational prestige,** is measured using surveys of how people throughout a society rate different jobs. Such surveys find that the prestige people attach to an occupation is heavily influenced by the education required for the job or the authority it offers its holder, as well as by income (Blau & Duncan 1967; Bose & Rossi, 1983; NORC, 1998). Thus many occupations that do not pay extremely high salaries are highly rated nonetheless. Table 12.3 shows how selected occupations are ranked according to their prestige.

These measures of wealth, income, education, and occupational prestige indicate that not all Americans share equally in "the good life." But what are the larger consequences of these patterns of inequality, both for individuals and for society? Do they combine to form an identifiable system of social-class stratification in the United States? Do marked patterns of inequality contradict the American dream, the vision of the United States as a place where hard work and sacrifice will lead to success and material comfort? In the rest of this chapter we explore these questions, but before doing so it is helpful to look at how views of inequality have changed over the past century and a half, as the United States was transformed from an agrarian society to an urban industrial society and, finally, to a postindustrial society.

Changing Views of Social Inequality

The dominant view of American society before the industrial revolution was the Jeffersonian view, which envisioned a society in which the majority of families lived on their own farms or ran small commercial or manufacturing enterprises (Lipset, 1979; Pyle, 1996; Wright, 1997). This view emphasized the value of economic self-sufficiency through hard work—a value that has persisted to the present time. Yet even in Jefferson's time the pattern of agrarian and small-town independence was not universal. In the early nineteenth century American society was already developing a rather complex and diverse set of stratification systems. In New England and the Middle Atlantic states, the Jeffersonian ideal of rural farmers and town-dwelling tradesmen, all with similar degrees of power and prestige, was reflected in reality. But in the larger cities a landless wage-earning class was emerging, as was a class of factory owners, bankers, and entrepreneurs. Throughout most of the southern states, in contrast, the major classes were the plantation owners and their families and staffs, the town merchants and tradesmen, and the marginal "poor white" farmers; in addition to these classes there was a caste: the slaves.

TABLE 12.3 Prestige Scores of Selected Occupations

Occupation	Prestige	Occupation	Prestige
Physician	95.8	Private secretary	60.9
Mayor	92.2	Floor supervisor in a hospital	60.3
Lawyer	90.1	Supervisor of telephone operators	60.3
College professor	90.1	Plumber	58.7
Architect	88.8	Police officer	58.3
City superintendent of schools	87.8	Manager of a supermarket	57.1
Owner of a factory employing 2,000 people	81.7	Car dealer	57.1
Stockbroker	81.7	Practical nurse	56.4
Advertising executive	80.8	Dental assistant	54.8
Electrical engineer	79.5	Warehouse supervisor	54.2
Building construction contractor	78.9	Assembly-line supervisor in a manufacturing plant	53.8
Chiropractor	75.3	Carpenter	53.5
Registered nurse	75.0	Locomotive engineer	52.9
Sociologist	74.7	Stenographer	52.6
Accountant	71.2	Office secretary	51.3
High school teacher	70.2	Inspector in a manufacturing plant	51.3
Manager of a factory employing 2,000 people	69.2	Housewife	51.0
Office manager	68.3	Bookkeeper	50.0
Administrative assistant	67.8	Florist	49.7
Grade school teacher	65.4	Tool machinist	48.4
Power house engineer	64.5	Welder	46.8
Hotel manager	64.1	Wholesale salesperson	46.2
Circulation director of a newspaper	63.5	Telephone operator	46.2
Social worker	63.2	Auto mechanic	44.9
Hospital lab technician	63.1	Typist	44.9
Artist	62.8	Typesetter	42.6
Electrician	62.5	Postal clerk	42.3
Insurance agent	62.5		

(Continued)

Native Americans were not included in any of these stratification systems. By the 1830s they were already a *pariah group*—excluded from the stratification system and considered too different or inferior to be eligible for citizenship. As a consequence of their pariah status in many parts of the nation, together with the settlers' desire to inhabit their lands, most Native Americans were steadily pushed west of the Mississippi.

The Impact of the Depression A century later, in the mid-1930s, a quite different view of inequality had emerged. In the midst of the most severe economic and social depression in world history, President Franklin D. Roosevelt claimed that "one third of the nation [was] ill-housed, ill-clad, and ill-nourished." Faced with mass poverty and unemployment, with able-bodied workers selling apples on street corners, the grandparents of many of today's college students questioned the reality of the American dream. Many observers saw heightened conflict between workers and the owners of capital. To many intellectuals it seemed that the dire predictions of Karl Marx were coming true and that only a communist revolution could save millions of people from starvation. Largely as a result of the Depression and fear of social disorder, legislators in the United States passed reform measures like the Social Security Act to alleviate the worst effects of poverty.

Muncie, Indiana, made famous by Helen and Robert Lynd in their study *Middletown* (1929), provided numerous examples of the gap in income and lifestyle between the working class and the owners or managers of capital. In the decades since 1880, industrialization had transformed Muncie from an agricultural trade center with 6,000 inhabitants into a bustling city of more than 36,000. According to the Lynds (1937), Muncie's businessmen "lived in a culture built around competition, the private acquisition of property, and the necessity for eternal vigilance in holding on to what one has" (p. 25). Below them was a middle class of managers, teachers, and clergymen who believed in the right of the business class to control wealth and power. In contrast, "across the railroad tracks from this world of businessmen [was] the other world of wage earners—constituting a majority of the city's population" (p. 25).

TABLE 12.3 *(Continued)*

Occupation	Prestige	Occupation	Prestige
Beautician	42.1	Fruit harvester, working for own family	26.0
Piano tuner	41.0	Blacksmith	26.0
Landscape gardener	40.5	Housekeeper	25.3
Truck driver	40.1	Flour miller	25.0
House painter	39.7	Stock clerk	24.4
Hairdresser	39.4	Coal miner	24.0
Pastry chef in a restaurant	39.4	Boardinghouse keeper	23.7
Butcher in a shop	38.8	Warehouse clerk	22.4
Washing machine repairman	38.8	Waitress/waiter	22.1
Automobile refinisher	36.9	Short-order cook	21.5
Someone who sells shoes in a store	35.9	Baby-sitter	18.3
Cashier	35.6	Rubber mixer	18.1
File clerk	34.0	Feed grinder	17.8
Dress cutter	33.3	Garbage collector	16.3
Cattledriver, working for own family	33.0	Box packer	15.1
Cotton farmer	32.4	Laundry worker	14.7
Metal-container maker	31.1	Househusband	14.5
Hospital aide	29.5	Salad maker in a hotel	13.8
Fireman in a boiler room	29.2	Janitor	12.5
Floor finisher	28.8	Yarn washer	11.8
Assembly-line worker	28.3	Maid (F)/household day worker (M)	11.5
Book binder	28.2	Bellhop	10.6
Textile machine operator	27.9	Hotel chambermaid (F)/hotel bedmaker (M)	10.3
Electric-wire winder	27.6	Carhop	8.3
Vegetable grader	27.4	Person living on welfare	8.2
Delivery truck driver	26.9	Parking lot attendant	8.0
Shirt maker in a manufacturing plant	26.6	Rag picker	4.6
Person who repairs shoes	26.0		

Source: Adapted from Bose & Rossi, 1983; NORC, 1998.

As a result of the Depression, there was more hostility between economic classes in Muncie in the late 1930s than there had been when the Lynds first studied Muncie's class system a few years before the Depression. The workers and their families were far less likely to feel that the business class had a right to its wealth, and they were increasingly drawn to social movements, especially the labor movement, that challenged the power of the business class. The business class itself was far less unified because the managers of the large corporations moving into the city did not feel obligated to join the clubs and churches of the established elite. The Lynds' study is described in more detail in Box 12.1.

Yankee City and the Black Metropolis In the mid-twentieth century a number of studies of inequality in American communities refined our knowledge of how income and prestige combine to produce stratification in modern communities. The research of William Lloyd Warner on social class in a New England town and Drake and Cayton's classic study of the caste system in an urban ghetto are important examples because they describe class relations more clearly than earlier studies did.

Warner's study was conducted in the seaside town of Newburyport, Massachusetts. Once famous for its whaling fleet and Yankee sea captains, Newburyport had become a small city with 17,000 inhabitants. It had a number of textile and shoe factories and had managed to hang on to quite a few of its Yankee families, descendants of the hardy seafarers of the nineteenth century. For this reason, Warner dubbed the community "Yankee City."

Here is how Warner began his description of the class structure of Yankee City:

> Studies of communities in New England clearly demonstrate the presence of a well-defined social-class system. At the top is an aristocracy of birth and wealth. This is the so-called "old family" class. The people of Yankee City say the families who belong to it have been in the community for a long time—for at least three generations and preferably many generations

BOX 12.1 USING THE SOCIOLOGICAL IMAGINATION

Social Classes in Middletown: 1925–1985

Muncie, Indiana, or "Middletown," is the most famous community in American sociology. In 1924 and 1925, and again in 1935, Helen and Robert Lynd studied this small midwestern industrial city. They selected Muncie because it "was not extraordinary in any way and so could be taken as a good specimen of American culture, at least of its midwestern variant" (Caplow et al., 1983, p. 3). Their books, *Middletown* (1929) and *Middletown in Transition* (1937), were "the first to describe the total culture of an American community with scientific detachment. They were also the first to replicate a community study in order to trace the velocity and direction of social change" (Caplow et al., 1983, p. vii).

When the Lynds first studied Middletown, they became convinced that social-class divisions, especially those between the working class and the owners of local firms, were reaching a state of crisis. When they returned during the Great Depression, they continued to find evidence that the business class dominated the city's social institutions. In one famous passage, a local factory worker described the influence of the "X family": "If I'm out of work I go to the X plant; if I need money I go to the X bank, and if they don't like me I don't get it; my children go to the X college; when I get sick I go to the X hospital" (Lynd & Lynd, 1937, p. 74).

The X family (actually the Ball family) owned the large glassworks in Middletown. The Ball brothers had built a fortune making glass jars and then had expanded into real estate, railways, banking, and retail stores. They worked so hard that they had little time to enjoy the pleasures their wealth could obtain. But their children appeared to be creating a distinctive upper class with exclusive country clubs, farms where their horses could be maintained, and private planes.

In the late 1970s Theodore Caplow, Howard Bahr, Bruce Chadwick, and their collaborators returned to Middletown to see how the community had changed over more than 40 years. They found that the city's stratification system had changed a great deal since the Depression. The population had increased from about 40,000 to more than 80,000. This size increase brought new patterns of mobility:

> The local dominance of a handful of rich families that looked so threatening in 1935 quietly faded away during the decades of prosperity that followed World War II. Hundreds of fortunes were made in the old ways and new—building subdivisions and shopping centers; trading in real estate; selling insurance, advertising, farm machinery, building materials, fuel oil, trucks and automobiles, furniture. . . . Middletown's new rich . . . lived much less ostentatiously than their industrial predecessors, and much of their money was spent away from Middletown (for yachts in Florida, condominiums in Colorado, boarding schools for their children, and luxury tours to everywhere for themselves). . . . The handful of families whose wealth antedated World War II adopted the same style. The imitation castles of the X, Y, and Z families were torn down or converted for institutional uses. (1983, p. 12)

The distinctive upper class that the Lynds saw emerging in Middletown in the 1930s had vanished by the 1970s:

> Meanwhile, at the lower end of the socioeconomic scale, life-styles were becoming more homogeneous. The residential building boom that began after World War II continued, year after year, to submerge the flat, rich farmlands at the edge of town under curved subdivision streets bordered by neat subdivision houses with various exteriors but nearly identical interiors. They all had central heating, indoor plumbing, telephones, automatic stoves, refrigerators, and washing machines. (pp. 12–13)

By the 1970s the factory workers of Middletown were much better off than they had been in 1935. They enjoyed job security, health insurance, and paid vacations, and their average incomes were higher than those of many white-collar workers. These changes had come about largely as a result of the activities of labor unions, which had been excluded from Middletown's factories in 1935 but were accepted soon afterward (Caplow et al., 1983). But in the 1980s the tide turned again, and Middletown, like many other industrial cities, began losing manufacturing jobs and gaining more low-paying service jobs.

Today the community of Middletown is far less self-contained than it was in the early decades of the twentieth century. It is subject to the influence of outside forces such as the shift of manufacturing jobs to other regions and even to other countries. Members of all social classes are more dependent on impersonal institutions such as corporations, national labor unions, and international markets. As a result, the old upper class has less influence, and the class structure is less cohesive and much less clear-cut than it was in the 1930s.

more than three. "Old family" means not only old to the community but old to the class. (Warner, Meeker, & Calls, 1949, p. 12)

The new families of Yankee City's upper class, who possessed wealth but had not gained "old family" status, "came up through the new industries—shoes, textiles, silverware—and finance" (p. 12). Below this upper class, people in Yankee City identified an upper-middle class of highly respectable people who "may be property owners, such as storekeepers, or highly educated professionals, but they do not have the wealth or family status to be included in the upper class" (p. 13). Below these classes Warner's respondents identified three "common man" levels. These made up the lower-middle class, which was composed primarily of clerks, skilled tradesmen, other white-collar workers, and some skilled manual workers. Directly below this class, and most difficult for the respondents to separate from those above and below it, was the "upper-lower class," or the people Yankee City residents identified as "poor but honest workers, who more often than not are only semi-skilled or unskilled" (p. 14).

In Yankee City the lowest class, which Warner called the "lower-lower class," was described as the "low-down Yankees [and immigrants] who live in the clam flats" and have a bad reputation among the classes above them. "They are thought to be improvident and unwilling or unable to save money . . . and therefore often dependent on the philanthropy of the private or public agency and on poor relief" (p. 14). The six "prestige classes" that Warner identified on the basis of the Yankee City interviews are listed in Table 12.4, along with the percentage of the city's population in each class and the proportion of income spent on necessities by members of each class. It is interesting that Warner's terms, such as *upper-middle class* and *lower-middle class,* have become part of our everyday vocabulary.

TABLE 12.4 Prestige Classes in "Yankee City" and Proportion of Income Spent on Necessities of Life, 1941*

Prestige Class	Percent of Total Population	Percent Spent on Necessities
Upper-upper	1.4	33
Lower-upper	1.6	35
Upper-middle	10.2	51
Lower-middle	28.1	59
Upper-lower	32.6	66
Lower-lower	25.2	75

*"Necessities of life" include food, shelter, and clothing.

Source: *The Social Life of a Modern Community,* by W. Lloyd Warner and Paul S. Lunt, copyright © 1941. Reprinted by permission of Yale University Press.

At about the same time that Warner was conducting his study, other social scientists were attempting to describe stratification in larger, more complex communities. Among those efforts was the first full-scale study of Chicago's African American community. It was carried out by St. Clair Drake and Horace Cayton in the 1940s and published under the title *Black Metropolis.*

Studies of inequality in southern communities had already documented the existence of a racially based caste system in the South (Dollard, 1937). But did this racial caste system also operate in northern cities like Chicago? At the time of Drake and Cayton's study, Chicago's black population was almost entirely concentrated in a single urban ghetto. A **ghetto** is a section of a city that is segregated either racially or culturally (e.g., by race, religion, or ethnicity). Large numbers of blacks worked outside the ghetto, but even those who could afford housing elsewhere were excluded from other neighborhoods by long-established patterns of residential segregation. In addition, despite the integration of workers in industrial occupations, especially steel and meatpacking, Drake and Cayton (1970/1945) showed that Chicago's blacks were barred from more comfortable, better-paying jobs in other fields. They called this pattern of occupational segregation, in which blacks could rise only so far, the *job ceiling.* In their view the job ceiling, together with the segregation of blacks in the ghetto, constituted a northern form of the racial caste system (see Figure 12.2). Many of the persistent racial inequalities in contemporary U.S. society have their origins in this urban job ceiling and the ways in which blacks and other people of color adapted to this form of stratification earlier in the century. We return to this subject in the next chapter.

Social Classes in Postindustrial Society In the decades since these studies were conducted, stratification in the United States has been strongly affected by technological and social changes. As we saw in Chapter 10, the study of these changes, which are thought to be producing a "postindustrial" society, is a lively area of sociological theory and research. The black workers whose struggles for dignity and economic security are described in *Black Metropolis,* the auto workers whom the Lynds encountered in Middletown, the members of Ed Sadlowski's family who worked in the steel mills—all were part of an earlier form of industrial society in which masses of workers were employed in large-scale manufacturing enterprises. By the 1990s, however, three quarters of all workers were engaged in providing services rather than producing goods of any kind (*Statistical Abstract,* 1990). According to Daniel Bell (1973), the principal theorist of "postindustrial society," this

FIGURE 12.2 The Job Ceiling in Government and Private Industry Prior to the Civil Rights Movement

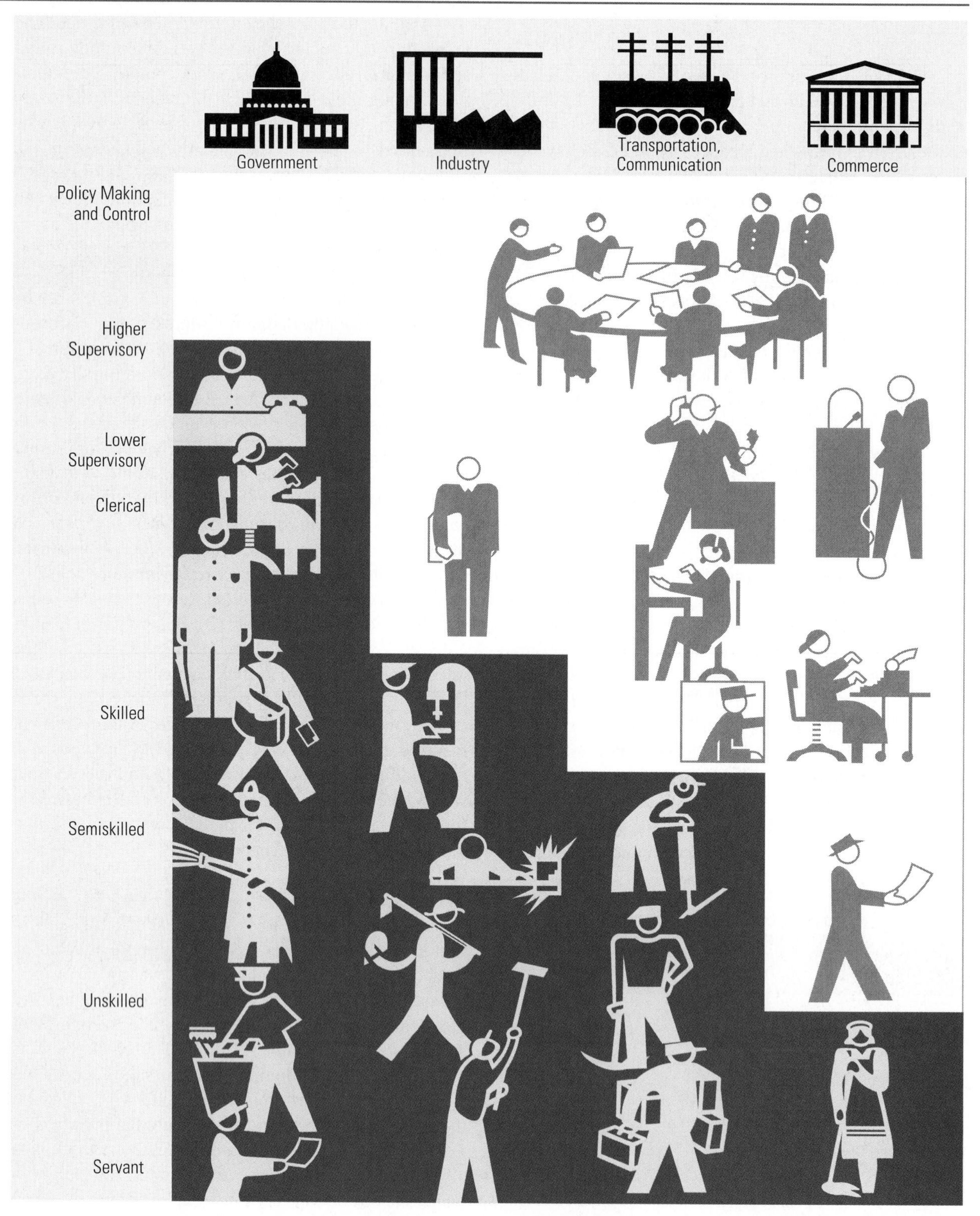

Source: *Black Metropolis,* copyright © 1945 by St. Clair Drake and Horace R. Cayton; renewed by Susan C. Woodson. Reproduced by permission of Harcourt Brace.

change represents a "revolution taking place in the structure of occupations and, to the extent that occupation determines other modes of behavior . . . it is a revolution in the class structure of society as well" (pp. 125–126). Bell and other sociologists believe that we now live in a postindustrial society in which what counts is not "raw muscle power" but scientific, technical, and financial information. The dominant person in this new order of society is the professional, who is equipped by education and training to supply the skills and information that are in most demand.

Although Marx argued that social-class divisions in industrial societies would become sharper and that class conflict would become more bitter, many American sociologists find evidence to the contrary. Their research shows that changes in the structure of the American economy, and especially in the types of jobs it produces, have resulted in a blurring of class lines and an easing of class conflict between industrial workers and the owners and managers of the means of production. More people work in offices, fewer in factories. More people live in metropolitan suburbs, fewer in industrial communities surrounding large factories. The union electrician or plumber wears a work shirt but maintains a home in a suburb with neighbors who work in offices. The computer operator wears a white shirt and has graduate training in computer science but may receive an hourly wage. The social-class membership of these employees and many others appears to be more than a simple matter of worker versus owner. Some sociologists argue, therefore, that the various dimensions of stratification—educational attainment, occupational prestige, income, wealth, and family status—overlap in complicated ways that make it difficult for people to form well-defined ideas about social-class membership (Hodge & Treiman, 1968; Wolf, 1998). Moreover, such factors as race, ethnicity (national origin), religion, and lifestyle crisscross economic class divisions and further blur what may once have been clearly defined class boundaries (Jackman & Jackman, 1983; NORC, 1998).

However, other sociologists are convinced that inequalities in income, wealth, and family status still produce clear social-class divisions in the American population (Piven, 1997). Although it is true that industrial workers make up a declining proportion of the labor force, they are still a large segment of the population, as are the poor below them. And these populations continue to demand social-welfare policies like unemployment insurance and medical benefits, in keeping with their interests as less-advantaged classes. Members of the more affluent middle and upper classes also recognize the interests that they share; they may demand such benefits as tax reductions or aid to college students (Hacker, 1997; NORC, 1998). Thus some sociologists contend that the American public does recognize class divisions and that class inequalities do play an important role in determining life chances. Although this debate continues, the empirical evidence shows that people in the United States have a good idea of the social class to which they belong.

Class Awareness in America Today

As mentioned in the preceding chapter, Marx reasoned that social class has both subjective and objective dimensions. In his view, objective social class depends on a person's relationship to the means of production (whether the person owns capital or land, sells his or her labor power, engages in a profession, etc.). Subjective social class is determined by how a person thinks about his or her membership in a social class. In Marx's view, many people who objectively are members of the working class because they sell their labor power and do not own capital in fact identify with the concerns of the owners of capital because they help run the businesses that employ them; in other words, they think of themselves as members of a higher class (subjective class) but their occupation places them in a lower class (objective class).

Research on class awareness distinguishes between these two views of social class. The *subjective method* of measuring social-class membership uses interviews in which respondents give their opinions about their class rankings. This method of assigning people to classes is based on personal opinions. The *objective method* uses indicators of rank such as occupational prestige (see Table 12.3), place of residence and type of home, level of education, and income, which are combined to form a composite index. These measures are not affected by the respondents' opinions. Researchers have found high correlations between the data generated by both methods (Hollingshead, 1949; Kahl, 1965). However, there is a tendency for people who work in blue-collar occupations, and therefore are in the working class by objective measures, to identify themselves as members of the middle class (NORC, 1998; Vanneman & Cannon, 1987).

Sociologists often use opinion surveys when they are studying subjective perceptions of class membership, but they use the objective method when they are interested in the effects of class membership on other aspects of a person's life, such as political views or access to health care.

Over the last 25 years the NORC's General Social Survey has found that an average of about 5 percent of Americans perceive themselves as being in the lower class (NORC, 1998). In contrast, sociologists Robert and Mary Jackman found that a higher proportion of respondents perceive themselves as poor. This difference probably occurs because the term *lower class* has a negative connotation for many people.

The National Opinion Research Center measures subjective social class by asking thousands of respondents this question: "People talk about social classes such as the lower class, the working class, the middle class, and the upper class. Which of these classes would you say you belong to?" Most people in the survey—at least 97 percent—assign themselves to a social class.

TABLE 12.5 Social Class, by Sex and Race

Social Class by Sex	Lower Class	Working Class	Middle Class	Upper Class	No Class	Total
Male						
Percent	4.5	47.4	45.0	3.1	0	100
N	607	6,423	6,087	425	0	13,542
Female						
Percent	5.7	45.0	46.2	3.1	0.01	100
N	977	7,681	7,899	524	1	17,082
Social Class by Race	**Lower Class**	**Working Class**	**Middle Class**	**Upper Class**	**No Class**	**Total**
White						
Percent	4.1	44.2	48.7	3.0	0	100
N	1,049	11,380	12,537	783	1	25,750
Black						
Percent	11.6	58.6	28.2	3.6	0	100
N	482	2,350	1,170	150	0	4,152
Other						
Percent	7.3	51.8	38.6	2.2	0	100
N	53	374	279	16	0	722
Total number of cases		30,624				

Source: NORC, 1998.

The data in Table 12.5 show that almost 45 percent of men and 46 percent of women classify themselves as members of the middle class, while far fewer black Americans (28 percent) do so. And even though the category "lower class" has a negative connotation for many people, we see in the table that almost 6 percent of women rank themselves in this social class and almost 12 percent of African Americans do so. These and many similar survey studies led the Jackmans to conclude that "at their basis, classes take shape in the public awareness as clusters of people with similar socioeconomic standing" (1983, p. 217).

The term *socioeconomic standing* (usually referred to as **socioeconomic status** or **SES**) requires further definition. When people think of social-class divisions in American society, they first assign various occupations to broad class ranks. When there is confusion about how an occupation is ranked, people tend to think of other aspects of social class, such as family prestige, education, and earned income. And as they think about these factors they tend to reach a consensus about what social class they belong to and what classes others should be assigned to.

But that consensus is by no means perfect. Some blue-collar occupations—for example, many skilled trades, such as plumber and electrician—are relatively well paid; in fact, they are better paid than many office jobs. This leads people to classify the holders of such jobs as members of the middle class, while at the same time they may assign holders of jobs in offices and stores to the working class. Remember also that the distinction between the lower-middle and upper-middle classes was the most difficult one for Warner's respondents in Yankee City and the other communities he studied.

In addition to this ambiguity in how people distinguish among social classes, there is the problem of how classes are perceived in rural, as opposed to urban, areas. Most studies of social inequality in America exclude the farm population, which is relatively small—about 6 million people, or less than 2.4 percent of the total population (Smiley, 1992; *Statistical Abstract,* 1998). But since the early years of the nation's history farmers have occupied a special place in American culture and the U.S. economy. The problems faced by farmers in a society that is steadily eliminating family farms have caused them to develop a strong class identity despite differences in their actual wealth. Hence, in the next section, which describes America's major social classes, we will include farmers and people living in rural areas dominated by agriculture, forestry, and mining.

SOCIAL CLASS AND LIFE CHANCES IN THE UNITED STATES

The rich have far more money than the poor, and they tend to have more education and a great deal more wealth, as measured by the value of homes, cars, investments, and much else. These facts are clear from the tables presented earlier in the chapter. But what differences do these inequalities make in people's lives? Social scientists often answer this question by analyzing the life chances of people born into different social classes. By *life chances* they mean the relative likelihood that individuals will have access to the opportunities and benefits that the society values. Compare, for example, the life chances of a child born into a family in which the mother and father earn slightly more than the minimum wage by working in restaurants and supermarkets with the life chances of a child born into the home of a police officer and a teacher, and compare both of these with the life chances of a child born into the home of a successful banker. Will these differences in the circumstances of birth affect the quality of education each child is likely to receive? Will the children's access to quality health care differ? Will the differences in the social class of their families influence who they are likely to marry? Will it make a difference in the likely length of their lives? The answer to all of these questions is, emphatically, yes. But being born into a given social class does not determine everything about an individual's life chances. The more a society attempts to equalize differences in life chances—by improving health care for the poor, for example, or by creating high-quality institutions of public education—the more it reduces the impact of social class on life chances.

Numerous social-scientific studies have shown that one's social class tells a great deal about how one will behave and the kind of life one is likely to have. Later in this section we will discuss specific social classes, but first we present some typical examples of the relationship between social class and daily life. These examples apply primarily to American society, but many of the same conditions can be found in other societies as well.

Class and Health A child born into a rich upper-class family or a comfortable middle-class family is far less likely to be premature or have a low birth weight than one born into a working-class or poor family. And a baby born into a family in which the parents are working at steady jobs is far less likely to be born with a drug addiction or AIDS or fetal alcohol syndrome than a child born to parents who are unemployed and homeless. These and many other disparities contribute to what is often called "the socioeconomic status (SES) gradient in health" (Sapolsky, 1998).

The SES gradient in health exists in all of the world's nations and is based on a complex combination of social class and culture. But the gradient is particularly marked in the United States:

> For example, in the United States the poorer you are, the more likely you are to contract and succumb to heart disease, respiratory disorders, ulcers, rheumatoid disorders, psychiatric diseases, or a number of types of cancer. And this is a whopper of an effect: In some cases disease or mortality risk increases more than fivefold as you go from the wealthiest to the poorest segments of our society, with things worsening each step of the way. (Sapolsky, 1998, p. 46)

Among adults, a salaried member of the upper class who directs the activities of other employees is less likely to be exposed to toxic chemicals or to experience occupational stress and peptic ulcers than wage workers at their machines and computer terminals. Those workers, in turn, are more likely to have adequate health insurance and medical care than the working poor—dishwashers, migrant laborers, temporary help, low-paid workers, and others whose wages for full-time work do not elevate them above the poverty level (Ellwood, 1988). The working poor are the largest category of poor Americans, and like those who lack steady jobs they often depend on local emergency rooms for medical care and report that they have neither family doctors nor health insurance. The same poor and working-class population is also more likely to smoke, consume alcohol, and be exposed to homicide and accidents—while receiving less police protection—than members of the classes above them.

Education Children of upper-class families are more likely to be educated in private schools than children from the middle or working classes. Sociologists have shown that education in elite private schools is a means of socializing the rich. As we saw in Chapter 5, Cookson and Persell's (1985) study of socialization in elite American prep schools found that "preppies" develop close ties to their classmates, ties that often last throughout life and become part of a network they can draw upon as they rise to positions of power and wealth. The segregation of upper-class adolescents in prep schools also serves the purpose of limiting dating and marriage opportunities to members of the same class.

Although middle-class parents are more likely than rich parents to send their children to public schools, they tend to select suburban communities where the schools are known to produce successful college applicants. The public schools that serve the middle classes spend more per pupil than the schools attended by working-class and poor children, and they offer a wider array of special services in such areas as music, sports, and extracurricular activities. Children in the middle and upper classes also tend to have parents who insist that they perform well in school and who can help them with their schoolwork. Moreover, children from working-class and poor families are more likely to drop out of school than children from upper-class families.

Politics The poor and members of the working class are more likely to vote for Democratic party candidates,

Then and Now

From Food Baskets to Food Stamps

The late nineteenth-century drawing and the modern photograph shown here represent two quite different approaches to aiding the poor. The drawing depicts a well-off family bringing a basket of food to the poor. This form of individual charity, which the rich often thought of as their responsibility and obligation, frequently engendered hostility rather than gratitude. The modern welfare system replaces the individual gift with "entitlements"—that is, payments and other services to poor people who qualify for them. But the modern system is inadequate to provide a dignified quality of life. And as evidenced by the antiwelfare sentiment prevailing in the United States in the1990s, the notion of entitlement to welfare is an easy political target, since many middle- and working-class voters are persuaded that the welfare system unfairly takes money from them to sustain "undeserving" poor people.

whereas upper- and middle-class voters are more likely to choose Republican candidates. Throughout the industrialized world, voters with less wealth, prestige, and power tend to vote for candidates who promise to reduce inequalities, while voters with higher socioeconomic status tend to choose candidates who support the status quo. Thus in their studies of social-class identification the Jackmans (1983) found that 48.5 percent of poor respondents and 43 percent of respondents who assign themselves to the working class believe the federal government should be doing more to achieve full employment and job guarantees, as opposed to only 24 percent of upper-middle-class respondents. They also found that about 48 percent of working-class respondents believe that "some difference" in levels of income (though less than currently exists) is desirable to sustain people's motives to achieve, whereas almost 50 percent of upper-middle-class respondents feel that a "great difference" is desirable. The most recent survey data (NORC, 1998) confirm the Jackmans' findings. Members of all classes tend to agree that whatever differences there are ought to be based on individual achievement rather than on advantages inherited at birth. Members of the working class and the poor, however, are more likely than members of other classes to vote for candidates who propose measures that would increase equality of opportunity.

Many other examples could be presented to illustrate the influence of social class on individuals in American society. But social-class divisions also affect the society as a whole. Let us turn, therefore, to an examination of some key characteristics of the major social classes in the United States, and to some of the consequences of the different life chances experienced by these classes.

Photographer Barbara Norfleet was studying the lifestyles of the very wealthy in the United States when she caught this patrician gentleman stepping off his yacht on a summer day in Massachusetts.

The Upper Classes

For the past 10 years the General Social Survey has found that approximately 4 percent of Americans classify themselves as members of the upper class (NORC, 1998). And as we saw at the beginning of the chapter, the richest members of the upper class, about 1 percent of the total population, control almost 40 percent of all personal wealth in the United States. However, sociologists regard the upper class as divided roughly into two subgroups: the richest and most prestigious families, who constitute the elite or "high society" and tend to be white Anglo-Saxon Protestants, and the "newly rich" families, who may be extremely wealthy but have not attained sufficient prestige to be included in the communities and associations of high society.

At the beginning of the twentieth century, families whose wealth came from railroads and banking looked down on "upstarts" like Rockefeller and Ford, whose money came from oil and automobiles. More recently, the great manufacturing families—the du Ponts (chemicals), the Rockefellers (oil), the Carnegies and Mellons (steel and coal)—have questioned the upper-class status of families like the Kennedys, whose wealth originally came from merchandising and whose Irish descent marked them off from the rich Protestant families.

Throughout much of the twentieth century the number of family-owned fortunes declined as large corporations like General Motors, Boeing, and IBM gained greater economic power. But as we see in Table 12.6, the major family fortunes still represent billions of dollars of capital controlled by the members of immensely wealthy families, many of whom are not descended from earlier Yankee families. (Other familiar names, such as Bill Gates of Microsoft or Michael Dell of Dell Computers or even Oprah Winfrey, certainly one of the best-known and richest women in the United States, do not appear on the list because this compilation of family fortunes includes only those that have been active for more than one generation.)

TABLE 12.6 The 20 Largest Family Fortunes

Family	Fortune Founded	1996 Worth (in billions)
Walton (retailing)	1962	$24.8
du Pont (chemicals)	1802	13.9
Mars (candy)	1911	12.0
Rockefeller (oil)	1870	9.9
Newhouse (publishing)	1922	9.0
Haas (jeans)	1873	8.6
Bass (oil)	1930s	8.1
Cox (newspapers)	1898	8.0
Cargill (grain)	1865	7.9
Dorrance (soup)	1876	7.7
Pritzker (finance)	1902	6.0
Mellon (banking)	1869	5.8
Lauder (cosmetics)	1946	4.1
Scripps (newspapers)	1870s	3.6
Upjohn (pharmaceuticals)	1885	3.2
Ziff (publishing)	1927	3.0
Smith (machinery)	1889	2.8
Davis (groceries)	1925	2.3
Chandler (newspapers)	1894	2.1
Gund (food, banking)	1919	2.1

Source: Hacker, 1997.

Members of the upper class tend to create special places in which to live and relax. They often maintain apartments in exclusive city neighborhoods as well as country estates in secluded communities. They send their children to the most expensive private schools and universities, maintain memberships in the most exclusive social clubs, and fly throughout the world to leisure resorts frequented by members of their own class. Examples of these resorts and enclaves of great wealth include Vail and Aspen in Colorado, Palm Springs in California, Palm Beach in Florida, and large stretches of the New England coastline.

A question that sociologists continually debate is whether the upper class in America is also the society's ruling class. The hit film *Titanic* presents a vision of the social-class system in which the wealthy are indeed a ruling class—but in reality the question of whether great wealth also confers the ability to rule by controlling political actors is open to empirical research and debate. Elite theorists like C. Wright Mills and William Domhoff argue that the upper class not only holds a controlling share of wealth and prestige, which it can pass along to its children, but also maintains a virtual monopoly over power in the United States (Zweigenhaft & Domhoff, 1998). A ruling class, Domhoff states, "is socially cohesive, has its basis in the large corporations and banks, plays a major role in shaping the social and political climate, and dominates the federal government through a variety of organizations and methods" (1983, p. 1). Mills (1956) attempted to show that this ruling class produces a *power elite* composed of its most politically active members plus high-level employees in the government and the military. The power elite, these sociologists claim, is the leadership arm of the ruling class.

Other sociologists and political scientists contend that there is no single, cohesive ruling class with an identifiable power elite that carries out its bidding. This is known as the *pluralist* concept (Polsby, 1980; Sullivan, 1998). These researchers agree that there is a readily identifiable upper class in American society. However, its power is not exerted in a unified fashion because there are competing centers of wealth and power within the upper class; moreover, its members' viewpoints on social policy are often opposed to one another. (We return to the debate between the power elite and pluralist theories in Chapter 20.)

The Middle Classes

Unlike the upper classes, in which family status establishes who is included in the highest stratum and who is not, the middle class includes far more varied combinations of wealth and prestige. As we saw in Table 12.5, subjective measures of social class show that about 45 percent of Americans classify themselves in the middle class, although this figure is lower for African Americans (28%). But this large proportion of Americans who identify themselves as members of the middle class is by no means a homogeneous population. The middle class is often referred to as the "middle classes" to emphasize the existence of strata with differing income, education, and access to wealth.

One group of Americans who usually identify themselves as members of the middle class are highly educated professionals. Most often they have attended graduate schools and built successful careers as engineers, lawyers, doctors, dentists, stockbrokers, or corporate managers. But their comfortable incomes, often well above $100,000 a year, may also be derived from family-owned businesses. Members of this class typically live in the suburbs. They join expensive country clubs, are active in community affairs, and spend a good deal of time transporting their children to "enriching" activities.

Another segment of the middle class is composed of people whose income is derived from small businesses, especially stores and community-oriented enterprises (in contrast with large corporations that have offices at many locations). Often referred to as the *petit bourgeoisie* or independent small-business owners, the members of this class may be found among the leaders of local chambers of commerce and other business associations that advocate local economic growth. Erik Olin Wright and his colleagues (1982, 1997) found that of the 9 percent of Americans who employ other workers, the majority are members of the petit bourgeoisie and employ fewer than 10 people.

The American concept of the middle class is based far more on patterns of consumption and the American dream of shared affluence than on the kinds of economic realities that Marx used to describe social classes. Americans have always believed that they could achieve individual affluence and become part of a great middle class. Thus in subjective terms the middle class is the largest single class in American society, but in cultural terms it is highly diverse because so many different lifestyles are represented within it.

The dominant image of the middle class from the end of World War II until the mid-1970s was of a suburban population living in relatively new private homes (Scott, 1998). The culture of the suburban middle class was thought to be shaped by the experience of frequent changes of residence and long-distance commuting, together with status symbols like "the ranch house with its two car garage, lawn and barbecue, and the nearby church and shopping center" (Schwartz, 1976, p. 327). The suburban middle class was said to be oriented toward family life and suspicious of the offbeat (Fava, 1956; Riesman, 1957). However, empirical research by a number of sociologists, including Bennett Berger (1961), Herbert Gans (1967, 1976), and M. P. Baumgartner (1988), has shown that suburban communities are far from homogeneous, that many people who think of themselves as members of the working class can be found in them, and that there is no easily identified middle-class suburban culture.

Another change in the nature of the middle class has to do with education. In the 1950s and 1960s a college diploma was often viewed as a passport to the American middle-class way of life. But we will see later that, although education remains the primary route to upward mobility for people without inherited wealth, the college diploma has lost some of its power to open the door to middle-class prestige.

The most general points that sociologists can make about the middle class are that its members tend to be employed in nonmanual occupations and that they usually have to work hard to afford the material things that are more easily acquired by the upper-middle class. Identification with the middle class is likely to be highest among teachers, middle- and lower-level office managers and clerical employees, government bureaucrats, and workers in the uniformed services.

Effects of Economic Insecurity Political candidates attempt to win the support of the middle class because they recognize that large numbers of voters think of themselves as members of that class, rather than as rich or poor. But as we have seen, the idea of an all-inclusive American middle class is more a myth than a reality for increasing numbers of working Americans. Between the extremes of poverty and wealth are, instead of a unified middle class, two increasingly distinct groups that think of themselves as middle class but face far different life chances and economic realities.

In the more secure segment of the middle class are highly skilled, highly educated professionals and managerial women and men who are doing rather well, although they increasingly experience income stagnation and must find ways to supplement their incomes. But below this segment there is a larger stratum of less skilled and less educated people who are experiencing ever-greater economic insecurity due to stagnant incomes and declining living standards. Appealing to this growing insecure middle-class stratum is a central goal of political leaders.

Politics aside, it is doubtful that the problems of the middle class can be solved through social policy. Recent figures released by the Department of Labor show that people in the middle income quintile are now spending almost half their incomes on three basic necessities of life: housing (approximately 32 percent), utilities (approximately 8 percent), and health care (approximately 7 percent). A generation ago these expenditures accounted for only one third of a typical middle-class income. (Spending on food has remained fairly constant.) Also, during the period from 1975 to 1995 the average price of a house rose by nearly 330 percent while the consumer price index increased by 220 percent. As a result, fewer young people can afford to buy a private home, perhaps the central symbol of membership in the American middle class (Cassidy, 1995). The difficulties Americans are experiencing in realizing the dream of owning their own home are clearly shown in Table 12.7. The proportion of Americans who live in detached single-family homes declined from 1970 to 1990, while the proportion living in mobile homes and in apartments increased. These indicators point to some of the most far-reaching aspects of social change in our nation and others like it, and in later chapters we will have many reasons to examine their consequences for adults and children in all social classes.

TABLE 12.7 Where Americans Live

	1970	1990
Detached single-unit homes	66.3%	59.0%
Trailers or mobile homes	3.1	7.2
Apartments and other housing	30.6	33.8

Source: Hacker, 1997.

The Working Class

Forty-four percent of white Americans and almost 57 percent of African Americans identify themselves as members of the working class (NORC, 1998). This is the class that is undergoing the most rapid and difficult changes in America today. Indeed, as with the middle class, those who classify themselves as members of the working class are extremely diverse. There are skilled craftspeople, plumbers, carpenters, machinists, and others who earn wages protected by strong unions, and factory workers whose class interests may or may not be protected by labor unions and whose plants may be threatened with closings or layoffs due to automation and the globalization of manufacturing. And there are people in the working class who work long hours yet cannot earn enough income to lift them very far above the poverty threshold. These struggling working people and their families are sometimes referred to as the "working poor."

The most important characteristic of the working class is employment in skilled, semiskilled, or unskilled manual occupations. Another distinguishing feature of this class is union membership; in fact, labor unions refer to their members as "working people." However, the proportion of workers who are union members is declining. In the mid-1970s approximately 23 percent of the American labor force belonged to labor unions. Today that proportion has declined to less than 16 percent, largely as a result of the decline in factory employment. One result of this trend is that many workers have less protection against arbitrary changes in their working conditions and earnings (Aronowitz, 1998). Another is that the wages of working people are decreasing—especially in semiskilled and unskilled blue-collar jobs. This controversial issue has been thoroughly documented in the research of Richard Freeman (1994), who has shown that the decreases in union membership and hourly wages account for as much as half of the increase in income inequality referred to in Table 12.1.

There are two major divisions within the American working class. The first is often known as the industrial working class and consists of blue-collar workers who work in large manufacturing industries and are members of industrial unions. The United Automobile Workers, the United Steelworkers, the International Ladies' Garment Workers' Union, the United Mine Workers, and the International Brotherhood of Teamsters are examples of such unions (Aronowitz, 1998; Aronowitz & DiFazio, 1994; Burawoy, 1980). The second major division of the American working class is composed of workers employed in skilled crafts, especially in the construction trades (Halle, 1984; LeMasters, 1975).

Sociologists have found that the industrial working class is more conscious of its class identity and more prone to believe that its fortunes as a class depend on its

These hotel workers, proudly showing off the luxurious accommodations they help maintain, are part of the working class. In fact, far more people are now employed in restaurants and hotels than in metal, automobile, and textile production combined.

ability to win in labor–management conflicts. Members of the skilled trades, by contrast, are more likely to be uncertain about their class identity or to think of themselves as members of the middle class. For both segments of the working class, financial worries are a routine fact of life (Halle, 1984). The situation of women in this class is especially problematic, since they are often the first to suffer the effects of layoffs and plant closings (Rosen, 1987), an issue that is discussed further in Chapter 14.

There is more racial and ethnic diversity in the working class than in other classes. White workers in the working class are far more likely to work alongside black and Hispanic workers, for example, than are white members of other classes. This situation often brings workers of different races and ethnic backgrounds into competition when jobs are in short supply or an industry is shrinking as a result of automation. The animosity produced by such competition is due largely to the fear of losing one's job and skidding downward into the ranks of the poor.

TABLE 12.8 Which Americans Are Poor?

Number and Percentage in Each Group Who Are Poor

Group	Number (in millions)	Percentage
In total population	**36.1**	**13.8**
Children under 18	14.7	20.8
Men 18–64	7.5	9.5
Women 18–64	10.9	13.3
Men 65 and over	0.8	6.2
Women 65 and over	2.5	13.6
White	16.3	8.5
Black	9.9	29.3
Hispanic	8.6	30.3
Asian	1.4	14.6

Americans Below the Poverty Line, 1960–1995

Year	All Families	Persons Over 65	Children Under 18
1960	18.1%	35.2%	26.9%
1970	10.1	24.6	15.1
1980	10.3	15.7	18.3
1990	10.7	12.2	20.6
1995	10.8	10.5	20.8

Source: Hacker, 1997.

The Poor

Studies of subjective class membership usually underestimate the poor population because many people in this class think of themselves as working people who are underpaid or "down on their luck." This may be particularly true of the NORC measures of subjective social class (see Table 12.5), which use the term "lower class" instead of "poor." In an earlier study by Mary and Robert Jackman (1983), almost 8 percent of Americans classified themselves as poor. In the NORC study, about 5 percent of the sample said that they were in the "lower class." Whichever term one uses, however, these measures greatly underestimate the number of poor people in the United States. Official census statistics show that approximately 14 percent of Americans—about 36.5 million people—are living in poverty (*Statistical Abstract,* 1998). These figures count only people whose household incomes are below the poverty threshold income of about $250 a week for a family of four. As shown in Table 12.8, African Americans, Hispanics, and children (the majority of whom live in female-headed families) are disproportionately represented among the poor (O'Hare, 1996).

Poverty: A Relative Concept Social scientists often point out that poverty is not an absolute concept but must be measured relative to the standards of well-being in particular societies. A family of four in the United States with an annual income of $16,036 is poor by official standards, yet compared with equivalent families in India or Africa it is doing quite well. The U.S. family probably has permanent shelter, even if it is in a slum or a run-down neighborhood. It probably receives some form of government income supplement, if only in the form of food stamps and federal medical insurance (Medicaid). The poor family in Africa or India, by contrast, may be living without permanent shelter, forced to beg and sift through scraps to obtain enough food to survive. Although there are poor families and individuals in the United States who exist under similar conditions, most of the U.S. poor are not yet in immediate danger of starvation.

Does this mean that the poor in the United States are better off? In absolute terms (that is, in comparison with the poor in very poor nations), they are not deprived of the necessities of life. In relative terms (that is, in comparison with the more affluent majority in the United States), they are severely deprived. This distinction between absolute and relative deprivation is important in explaining how people feel about being poor. It is no comfort to a child in a poor family in the United States whose parents cannot afford fancy sneakers or a color TV that there are people starving in Africa. In short, people feel poor in comparison with others around them who have more.

In the United States official poverty measures have always taken into consideration the fact that poverty should be measured in relation to the condition of

BOX 12.2 SOCIOLOGICAL METHODS

Measuring Poverty in the United States

The official method for measuring poverty in the United States was established in the mid-1960s by Social Security Administration statistician Mollie Orshansky. Orshansky rejected the practice of choosing an arbitrary level of income, such as $3,000 a year (a figure that was commonly selected at the time), to calculate how many Americans were poor. Instead, she argued, one should calculate simple but nutritionally adequate food budgets for families of different sizes. She assumed that food accounts for about one third of total family expenditures, basing this assumption on a food consumption survey done in 1955 (Orshansky, 1965). By this reasoning, "any family whose income was less than three times the cost of the minimum food budgets of the Department of Agriculture was classified as poor" (Ruggles, 1992).

This measure is still used today with minor changes, such as multiplying by 4 instead of 3 on the basis of data showing that food accounts for about one fourth of average family income. The measure is also adjusted for inflation. The calculations allow government agencies, especially the Social Security Administration and the Department of Agriculture, to establish poverty thresholds for households of different sizes; the official poverty threshold for a family of four in 1996 was $16,036. (Thresholds are somewhat lower for farm families because they can grow some of their own food.) Note that the official thresholds do not include noncash forms of income like food stamps and Medicaid.

Patricia Ruggles, one of the nation's leading experts on the measurement of poverty, claims that current poverty thresholds, based on the Orshansky formula corrected for inflation, underestimate the number of people living in poverty in the United States. Today, she argues, the average family spends only one sixth of its annual income on food. This means that the cost of a family's basic "market basket" should be multiplied by 6 rather than by 4 to establish a poverty threshold. Ruggles also observes that it is necessary to continually recalculate the basic food requirements of an average family because as new foods come to the market, there are changes in the public's definitions of what is required for an adequate diet. According to Ruggles's revised formula, the poverty threshold for a family of four would be about $25,220 instead of approximately $16,000. This would place almost 30 percent of the U.S. population below the poverty threshold.

Ruggles realizes that her estimates are higher than those most politicians and policy makers would care to accept. She believes that relative measures that include items in the "market basket" that people consider essential are subject to criticism by policy makers who wish to minimize the official poverty rate. "Not all needs or desires are generally considered equal in judging whether or not some should be counted as poor," she notes. "The need to eat regularly and to have someplace warm and dry to sleep is widely recognized; the need to own a particular brand of sneakers or jeans, while deeply felt by many teenagers, is rarely considered of equal importance by policy makers" (Ruggles, 1990, p. 9).

Ruggles's updated estimates modify the original Orshansky calculations to account for the lower share of an average family's budget spent on food. Her calculations do not actually consider such items as decent clothing and sneakers. Were they to do so, her estimates of the proportion of the population living in poverty would be even higher than the estimates given here.

others who have more of the "essentials" of life. Box 12.2 describes how the poverty threshold is defined.

The Working Poor One of the chief causes of poverty is that people who are working full-time are not being paid wages that give them enough income to raise them above the poverty line. Two thirds of American children living in poverty have a parent who works full-time for wages that are too low to lift the household out of poverty (Tucker, 1998). Of course, many of these households are one-parent families, but in a study of two-parent families living in poverty, David Ellwood found that 44 percent had at least one member who was working full-time. Low wages, Ellwood states, are a major cause of poverty:

> Work does not always guarantee a route out of poverty. A full-time minimum-wage job (which pays $5.15 per hour) does not even come close to supporting a family of three at the poverty line. Even one full-time job and one half-time job at the minimum wage will not bring a family of four up to the poverty line. (1988, p. 88)

Since so many of the jobs created during the economic expansion of the 1980s were low-wage jobs, rates of employment and poverty have risen simultaneously.

Another reason for the increase in poverty is the increase in the number of single-parent, female-headed families. The breakup of marriages or long-term relationships leaves women alone with the responsibility for raising small children and earning the income to do so. Such families often become poor because it is more difficult for a woman to support a family alone than it is for a man. As we will see in Chapter 15, one result of this trend toward the "feminization" of poverty is that a rising proportion of the nation's children are growing up in poor female-headed families and are deprived of the advantages enjoyed by children from more affluent homes.

A common stereotype of the poor in the United States is that they are heavily concentrated in inner-city minority ghettos. Ellwood's study of poverty shows this notion to be false, as can be seen in Figure 12.3. The poor in large central cities (those living in moderate- or high-poverty neighborhoods) account for just 19 percent of the total poor population, with only about 7 percent concentrated in high-poverty neighborhoods. Fully 29 percent of the poor reside in rural and small-town communities, and 19 percent live in the affluent suburbs of large cities. In short, the poor live everywhere and are not concentrated in a single type of community.

One reason for this pattern is that the poor are an extremely diverse population. A large portion of the poor are aged people living on fixed incomes (Social Security or very modest savings or pensions). Other categories of poor people include marginally employed rural workers and part-time miners in communities from Appalachia to Alaska; migrant farm workers in agricultural areas throughout the United States; chronically unemployed manual workers in the industrial cities; disabled workers and their families; and people who have been displaced by catastrophes like hurricanes, drought, or arson. For all of these groups, poverty brings enormous problems of insecurity and instability. Many poor people do not even know where they will be living next month or next year. This problem is especially acute for farm families who have been driven from their land.

Farmers and Farm Families

In analyzing the situation of the poor in America, it is helpful to look at what is occurring in rural areas, where a large proportion of the poor live. Although many rural communities are experiencing renewed economic growth as factories relocate to areas where the costs of energy, space, training, and wages are lower, in communities based on agricultural production poverty rates continue to increase. A survey of 1,700 American farmers (Robbins, 1985) indicated that approximately 36 percent owed more than 40 percent of their assets (land and equipment, livestock, and crops in the ground) to lending institutions. Within this category of heavily indebted farmers, 15 percent (or more than 5 percent of all farmers) owed 70 percent or more of their total assets. The consequences of these debt pressures can be seen in the changing demographics of farm families. Figure 12.4 shows that the proportion of young farmers is decreasing while that of farmers over age 65 is increasing; overall, the number of farmers is decreasing steadily each year.

FIGURE 12.3 **Geographic Distribution of the Poor (percent of total below poverty line)**

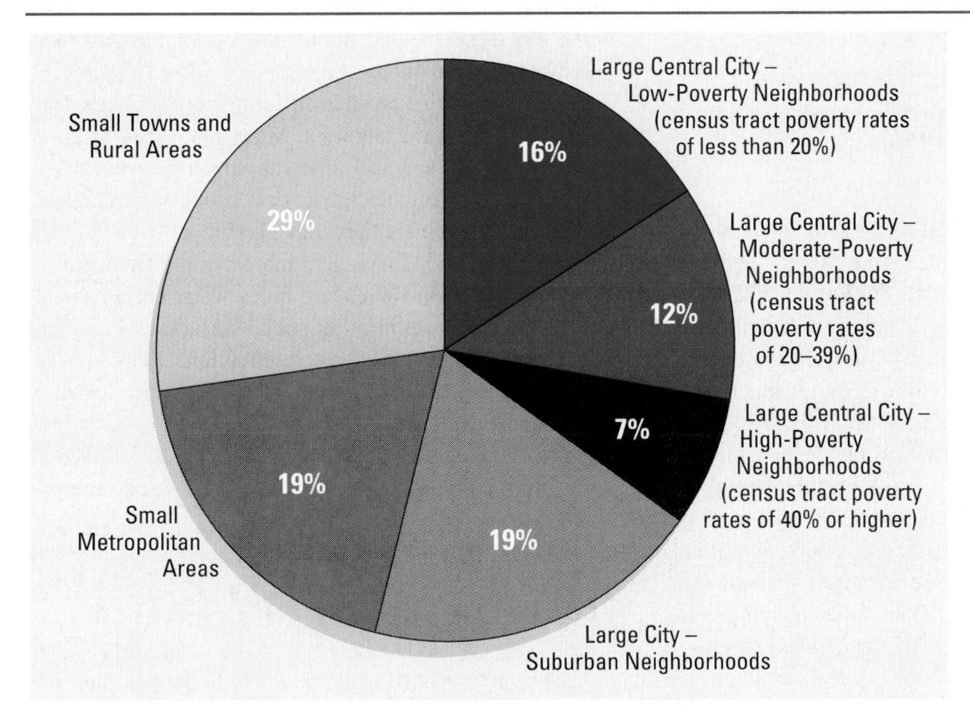

The percentages in this chart have changed somewhat since Ellwood conducted his study, but the basic distribution remains the same (the percentages do not add up to 100 because of rounding).

Source: Ellwood, 1988.

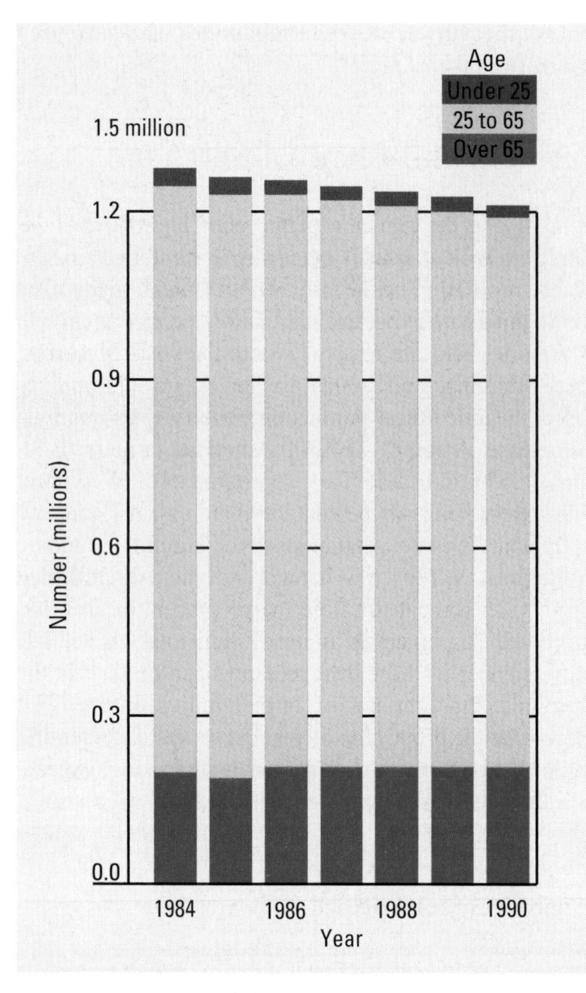

Source: Bureau of Labor Statistics.

FIGURE 12.4 **Number of Farmers, by Age, 1984–1990**

This situation repeats a condition that has been part of American social life for more than a century as the nation has been transformed from a rural to an urban society (Dyer, 1996). Although the dream of owning a family farm, or keeping one in the family, has always been an important part of American culture, the rising cost of farm technology, together with the need for ever larger amounts of land to make farms profitable, has steadily pushed farm families off the land. Very large farms devoted to the production of a narrow range of crops and farm products now dominate North American agriculture. In the United States, more than half of total farm production comes from only about 4 percent of the nation's farms. In consequence, more farmers are employed by large agribusiness operations, and each decade a smaller percentage own family farms (Albrecht, 1997; Strange, 1988).

Larger farms are typically owned by corporations that may also can, freeze, package, or otherwise prepare the crop for sale in supermarkets. These agribusinesses may be highly efficient producers, but they have some negative social consequences for rural counties. Studies by rural sociologists have shown that communities dominated by large farms are characterized by a declining population, a decreasing number of local businesses and civic associations, and an increase in poverty. One study done for the U.S. Congress in 200 of the richest agricultural counties in the United States found that the more farm size increases, the more the poverty rate in the county also increases as people are displaced from the land and must search for marginal employment (Office of Technology Assessment, 1986). In 1998, as a consequence of low grain prices and extreme heat in the Great Plains, family farmers in some Plains states were experiencing the kind of severe crisis that frequently makes their existence precarious. In North Dakota, for example, it is estimated that some 2,000 family farmers (out of about 30,000) will be forced out of business (Clairborne, 1998).

As a social class, farmers still take pride in the fact that the United States has the highest level of agricultural production in the world, but this does little to alleviate their difficulties. Moreover, as a class American farmers are continually at the mercy of a fickle world economy, a condition that they increasingly share with farmers in Europe and other industrialized regions.

One method farmers have traditionally used to hold on to the land they love is for some family members to find work in a nearby town while others continue to work the land. In fact, about half of all farmers farm part-time and work at wage-paying jobs full-time. A growing number of women in farm families work full- or part-time in rural factories. Many also work at home producing knitwear and other apparel on a piecework basis or performing electronic data processing via computer networks. Still others have service jobs.

The strategy of trying to maintain the household's economic position by adding more wage earners is increasingly common in other social classes as well. This is a trend with far-reaching implications, not only for the way Americans think about social class but for all of the society's institutions. For farmers, however, the numbers are discouraging. As an economic class they are diminishing steadily, especially as young people leave farming for other occupations.

MORE EQUALITY?

A large proportion of the world's people are extremely poor. India, China, Bangladesh, the Philippines, Malaysia, most of sub-Saharan Africa, the mountain regions of Latin America, and many other parts of the world are inhabited by large numbers of people with drastically limited incomes and only the most rudimentary health care, education, and other resources. If poverty continues to increase in those areas and the hopes of impoverished millions for a better life are frustrated, it will become increasingly difficult to maintain peace, achieve slower population growth, and prevent massive environmental damage.

Poverty is reduced through better health care, more schools, more food and shelter, and more jobs, all of which require economic and social development. The experience of the more affluent nations indicates that such development is likely to produce larger working and middle classes and reduce the rate of poverty. But the persistence of poverty and near-poverty in the United States and other industrialized nations shows how difficult it is to promote greater equality elsewhere in the world.

The presence of millions of poor people in a country as affluent as the United States is a major public policy issue as well as a subject of extensive social research. However, policy debates on this issue are often clouded by problems of definition. Most Americans will say that they believe in equality, citing the claim of the nation's founders that "all men are created equal." But when pressed to define equality they often fail to distinguish between **equality of opportunity** (equal opportunity to achieve material well-being and prestige) and **equality of result** (actual equality in levels of material well-being and prestige). Americans may believe that opportunities to succeed should be distributed equally and that the rules that determine who succeeds and who fails should be fair. But their commitment to the ideal of equality falters in the face of their belief that hard work and competence should be rewarded and laziness and incompetence punished. Thus many Americans believe that poverty is proof of personal failure. They do not stop to ask whether poor Americans have ever been given equality of opportunity, nor do they stop to notice how hard many poor people work (Jencks, 1994; Liebow, 1967).

Sociologists who study inequality in modern societies usually ask how equality of opportunity can be increased and the gaps between the haves and the have-nots decreased. Yet they are also highly aware of how difficult it is to narrow that gap in the United States or any other society. In response to those who believe that government can do nothing effective in its efforts to reduce poverty, sociologists point to the kinds of hard facts presented in Table 12.8. The table clearly shows that poverty among people over age 65 has been reduced dramatically, from 35.2 percent in 1960 to 10.5 percent in 1995, largely as a consequence of Social Security legislation and Medicare. Poverty among children, on the other hand, was greatly reduced between 1960 and 1980 but has been increasing rapidly since the 1980s.

The role the government should play in alleviating poverty is a major issue in American society today. It is frequently asserted, both in government and in the social sciences, that government has no business engaging in large-scale redistribution of income in the interests of the lower classes. This viewpoint, which is represented in the writings of conservative social scientists like Milton Friedman (1962), George Gilder (1982), and Charles Murray (1984, 1988), as well as in current social policies, agrees that public funds must be spent to provide the poor with a "safety net" of programs to prevent suffering, but it also holds that programs to redistribute wealth infringe on individual freedom and the right of private property.

Welfare Reform

The extent to which declines in the income and well-being of poor children in the United States are a consequence of changes in the welfare system is an extremely controversial subject. In 1996 Congress enacted historic changes in the nation's welfare laws, which now require each state to ensure that a rising percentage of its adult aid recipients engage in approved work. The head of each family on government financial assistance ("welfare") is required to find a job within 2 years after assistance payments begin. The great majority of families may receive benefits for no longer than 5 years, and states may impose even shorter time limits if they choose to do so. From 1994 to March 1998 the number of women on welfare—either the old Aid to Families with Dependent Children (AFDC) or the newer Temporary Assistance to Needy Families—fell by 1.8 million cases (36 percent). But one may ask whether this new policy actually improves the well-being of children. That question, in turn, leads to many different empirical questions, especially about the actual job experience of former welfare recipients (Burtless, 1998).

Some of the best data on these issues come from the state of Wisconsin, an early pioneer in welfare reform. Data for more than 25,000 former welfare recipients in Milwaukee who were required to participate in the new "welfare-to-work" initiative show that over a one-year period 18,000, or 72 percent, found at least one job that paid unemployment insurance. Despite this success, the researchers also found that "the jobs carry high unemployment risks and offer lousy pay. . . . One third of the parents who entered employment in the first quarter of 1996, for example, had no recorded earnings in the first quarter of 1997, and about one-quarter of the remaining parents earned less than $500 in the same quarter" (Burtless, 1998, p. 6). Job turnover was also high. The 18,000 people who found work held a total of more than 42,000 jobs in a 15-month period. Over half of these jobs were obtained from temporary help agencies.

Another disturbing fact is that welfare reforms have been implemented during a period of sustained economic growth and record high demand for entry-level workers. A downturn in the U.S. economy would likely make it even more difficult for former welfare recipients, the majority of whom have limited educational credentials, to find work that will sustain their families or offer reasonable income stability.

Recent studies also point out that the children of former welfare recipients who find jobs are more likely to receive inadequate supervision at home, which also increases the likelihood that these vulnerable workers will temporarily leave the labor force. Figure 12.5 illustrates the "multiple jeopardy" faced by single mothers: It compares those who never received welfare benefits with those who received benefits for 1 to 24 months and those who were longer-term welfare recipients (more than 60 months). Workers in the latter group have fewer resources at home and on the job that would enable them to work through children's illnesses or other household difficulties. These and many other indicators point to the need for the United States to make greater investments in quality child care institutions and early-childhood education programs, a subject to which we return in later chapters.

FIGURE 12.5 "Multiple Jeopardy" Among Working Mothers in Milwaukee

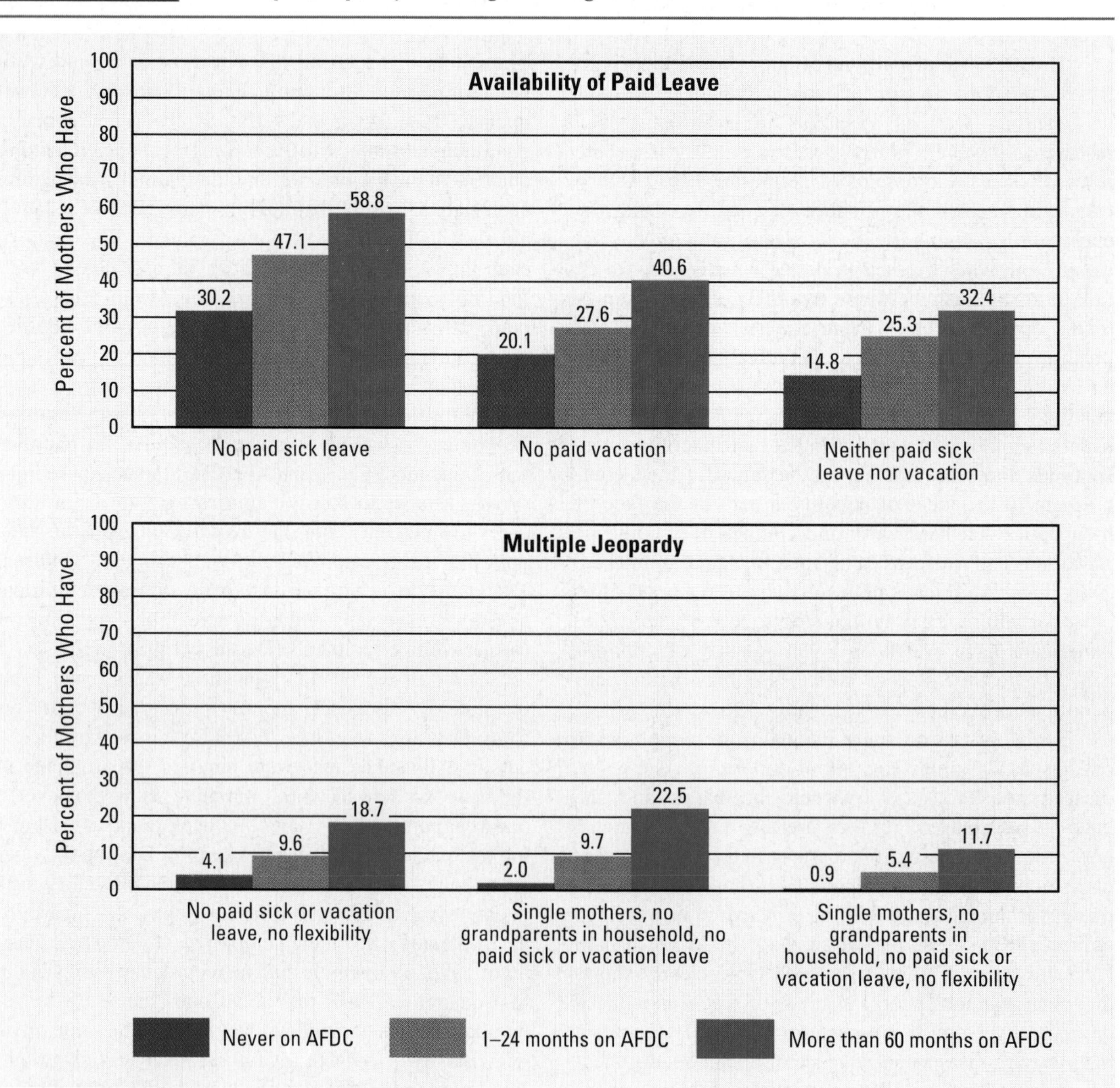

Source: Heyman & Earle, 1997.

Is There a War Against the Poor?

Many sociologists who study welfare reform are highly critical of the new policies—not because they require people to work rather than receive benefits, but because they seem to increase the number of children living in poverty. Some also argue that the new policies are essentially designed to promote competition for low-wage jobs, which will ensure that the working poor remain poor. Under both past and present policies, sociologist Herbert Gans observes, poverty is so persistent that one can well ask whether it might have positive social functions. With large doses of irony he goes on to list why he believes that legislators are waging a "war against the poor." He describes a number of "positive" functions of poverty:

- The existence of poverty ensures that society's dirty work will be done. . . . Society can fill these jobs by paying higher wages than for "clean" work, or it can force people who have no other choice to do the dirty work.

- Because the poor are required to work at low wages, they subsidize a variety of economic activities that benefit the affluent. For example, domestics subsidize the upper middle and upper classes, making life easier for their employers.

- Poverty creates jobs for a number of occupations and professions that serve or "service" the poor, or protect the rest of society from them. . . . Penology would be minuscule without the poor, as would the police.

- The poor can be identified and punished as alleged or real deviants in order to uphold the legitimacy of conventional norms. (1985, pp. 155–161)

These and other hidden functions of poverty, Gans argues, are far outweighed by its dysfunctions—suffering, violence, and waste of human and material resources. And they could be eliminated by "functional alternatives" such as higher pay for doing dirty work, or by the intentional redistribution of income. He believes welfare payments and tax credits for the poor should be increased so that they can qualify for the housing and other subsidies available to better-off Americans (Gans, 1995).

Other sociologists offer proposals that focus on education or local and regional economic development. We saw earlier that success in the job market is directly related to educational attainment. In fact, in today's economy the lack of a high school diploma puts a job seeker at a tremendous disadvantage. And when the job seeker has limited education and is also African American or Hispanic and a young mother, the disadvantages she faces can be overwhelming. (See Chapters 13 and 14 for more on this subject.) But sociologists like William J. Wilson point out that many poor people are poor because they happened to be born into households in very poor communities. If public policies were directed toward improving job prospects in these communities, they argue, the gap between the haves and the have-nots would be narrowed as well. In the final section of the chapter, therefore, we turn to issues of poverty and place.

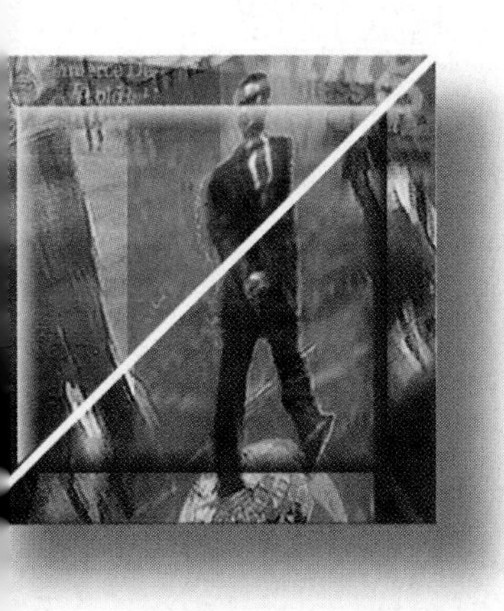

RESEARCH ON THE CUTTING EDGE: Poverty and Place

To be born into a household in the Delta region of rural Mississippi, or the arid and marginal farmlands of the arid Southwest, or an Indian reservation in the Southwest or the mountain states almost ensures that a child will experience material deprivation. The family may be as supportive and caring as any affluent family, but poverty always exacts a toll on the child's development. Poor places typically offer poorer schooling, less adequate health care, skimpier recreational opportunities, limited entry-level job prospects, higher crime rates, and higher rates of substance abuse. Rural poverty affects fewer people than poverty in urban or suburban areas, but it covers far larger segments of the American landscape, as is demonstrated in the Mapping Social Change feature on pages 386–387.

i

MAPPING SOCIAL CHANGE

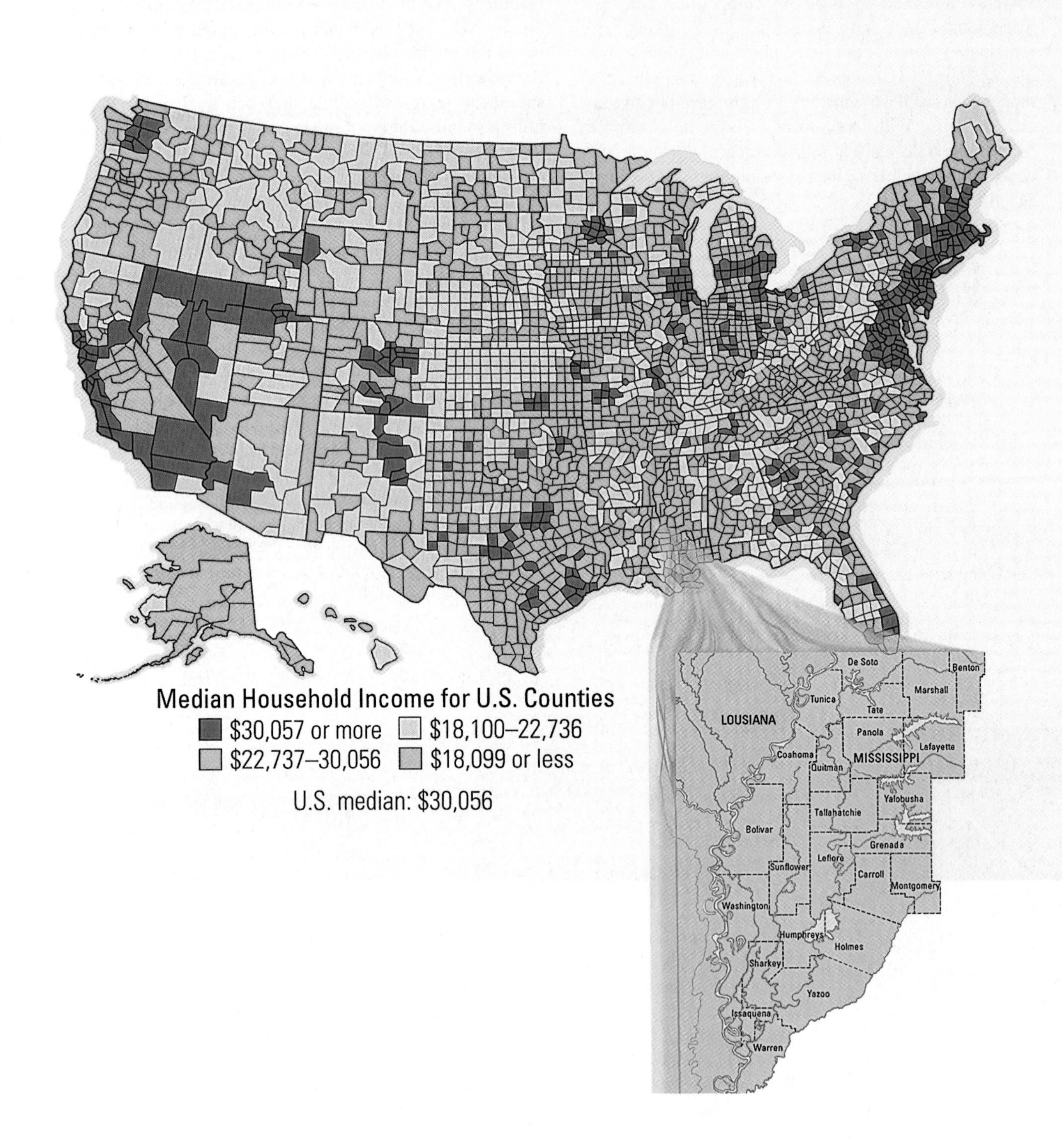

Poor Counties in the United States

This map shows all of the counties of the Untied States, grouped by average income. The extremes of wealth and poverty in U.S. society are apparent. Or are they? Actually, such a map is more useful in revealing areas of poverty outside central cities than within them. This is because impoverished inner-city communities usually comprise only a portion of a densely urban county that may also include affluent and middle-class communities. Their incomes raise the average so that the pockets of inner-city poverty do not appear on the map.

This map still reveals a great deal about the places where poor people live. For example, along the Mississippi river, especially in Arkansas and Mississippi, we see groups of counties whose residentsí average income is in the lowest range. The same is true in parts of the arid Southwest. Why are there such high concentrations of poor residents in these counties?

The Mississippi delta region has rich agricultural land, but most of it is owned by a few white families and agricultural corporations. The land is farmed by African American and poor white workers who earn extremely little. Thus, the delta region has a pattern of agrarian stratification that results in high rates of poverty. In the impoverished counties of the Southwest, the land often contains mineral and water resources that are controlled by a small number of wealthy individuals and corporations; Chicanos, Native Americans, and poor whites do not share in the wealth created by these resources, and the counties in which they live have low average incomes.

The isolated poor counties of the mountain states in the West often suffer from a newer form of poverty. In these areas economic growth is due to tourism, second-home developments, skiing, and other pursuits of the affluent upper and upper-middle classes. Farms, small-town businesses, and the trailer parks where the resort workers live may be located in a nearby valley, which may be part of a separate county with a lower average income.

Classification of counties by median income masks the existence of many high-poverty communities and neighborhoods in urban areas. The red zones on the map, which denote the counties with the highest median incomes, do not reveal the high rates of poverty in neighborhoods in and around New York City, Philadelphia, and Baltimore, or Chicago and Detroit, or the Los Angeles–San Diego corridor, to name only a few of the most outstanding areas in which poverty is concentrated. As sociologist Paul Jargowsky notes, regardless of the ethnic or racial makeup of these neighborhoods, "The concentration of poverty indicates the percentage of the poor who not only have to cope with their poverty but also that of those around them" (1997, p. 21).

Jargowsky's recent research on the complex relationships between poverty and place draws our attention to places known in popular speech as slums, ghettos, or low-income barrios. Each of these terms is commonly used to talk about an urban neighborhood that seems to be composed primarily either of poor white households or of poor black or Latino households. Jargowsky's goal was to better understand the population composition of such urban places and to use that knowledge to determine what can be done to improve the lives of their residents. To do so he needed to go beyond stereotypical ideas of what makes a slum or a ghetto by gathering quantitative evidence about who is poor in such places and why.

Jargowsky uses the census tract as his basic unit of analysis. Recall from Chapter 2 that census tracts are relatively small places in cities and towns, usually from one to four square blocks, within which the U.S. census enumerates the population and describes its characteristics (total population, education, income, race, ethnic descent, employment, and so forth). To the degree possible, the Census Bureau does not change the boundaries of census tracts from one 10-year period to another; thus changes in population characteristics can be measured in a set of small units of area that are relatively unchanging. These census tracts do not always correspond to people's perceptions of their immediate neighborhood, but they are small enough so that they can be thought of as neighborhood-like areas.

Jargowsky calculated the poverty rate of each census tract by dividing the number of people living below the official poverty threshold by the total population of the tract. When he looked for tracts with poverty rates of 90 percent or more, he found only 67 (out of the total of 60,000) in the entire nation. If he included tracts with poverty rates of 60 percent or more, he found that they had a total of only 1.4 million residents, or considerably less than 1 percent of the U.S. population. By choosing a poverty rate of 40 percent or more as the definition of a high-poverty neighborhood, he located 3,417 tracts in the most recent census. Of these, about 85 percent were located in metropolitan areas. There were an average of 630 people living in each of these tracts. When he went to visit a smaller sample of these tracts, he and his colleagues made the following observation:

> Such neighborhoods were predominantly minority. They tended to have a threatening appearance, marked by dilapidated housing, vacant units with broken or boarded-up windows, abandoned and burned-out cars, and men "hanging out" on street corners. In Philadelphia and Chicago, the housing stock consisted of high-rise public housing projects or multistory tenement buildings built right to the sidewalks. In Detroit, the high-poverty areas consisted of single-family homes built decades ago for auto workers earning the unheard of wage of $5 a day; now old and undermaintained, some had simply fallen down and only rubble remained. In Jackson, Mississippi, such neighborhoods had tin-roofed shacks. In Pine Bluff, Arkansas, trailers sat rusting alongside tiny Depression-era houses. (Jargowsky, 1997, p. 11)

The appearance of neighborhoods with poverty rates of only 20 to 40 percent was different. There were fewer housing units in obvious disrepair, fewer signs of unemployment and street gangs, fewer burned-out buildings and cars. But fewer does not mean none. There were still plenty of signs of economic and social stress. The point here, Jargowsky notes, is that there was no neat demarcation between "nice" neighborhoods and "bad" ones. All distinctions were a matter of degree.

Jargowsky's research is important because it shows that poor neighborhoods are extremely diverse in terms of race and ethnicity. What are often called "white slums" usually are also home to some people of color. What are termed African American ghettos are rarely composed entirely of African Americans, and poor neighborhoods inhabited largely by Latinos, which are often called barrios, are rarely homogeneous either. However, as Figure 12.6 shows, African Americans living in high-poverty neighborhoods constitute the largest proportion of all residents of high-poverty areas, and although this proportion decreased somewhat from 1970 to 1990 (the latest census period for which these data are available), the declines have been modest at best.

Jargowsky explains that the concentration of poor African American households in high-poverty areas is a consequence of continuing patterns of residential segregation. In addition, the flight of manufacturing jobs from inner-city industrial communities and the exodus of middle-class black and white residents from the inner-city areas relegates more of the nation's poor black population to high-poverty neighborhoods, where their children face the further obstacles of high crime rates and poor schooling.

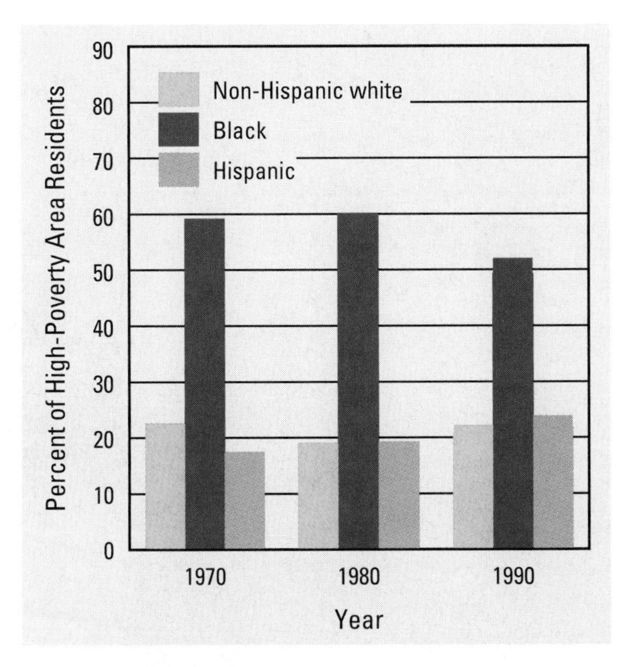

Source: Jargowsky, 1997.

FIGURE 12.6 **Racial Composition of High-Poverty Areas, 1970–1990**

In other research Jargowsky shows that some cities, especially in the South, have succeeded in reducing the numbers of people living in high-poverty areas. This will be difficult to accomplish in the older manufacturing cities, he warns, but it can be done through a combination of economic incentives, housing policies, and investments in the infrastructure of inner-city communities. If this effort is not made, he warns, the gap between the haves and the have-nots is likely to widen. We must find the political will "and the means to ensure that millions of our fellow citizens need not live in economically devastated and socially isolated neighborhoods. Finally, we must find a new and viable structure for metropolitan areas in the twenty-first century, a larger urban *community* rather than an agglomeration of separate and antagonistic *places*" (1997, p. 213).

VISUAL SOCIOLOGY

The Changing Landscape of Labor

Michael Jacobson-Hardy's photographs contrast the work environment in traditional, basic industries such as paper mills, foundries, textile mills, and shipyards with that in newer, high-technology industries such as computer manufacturing and aircraft production. In addition to examining the physical environment in which industrial production occurs, his images give visibility and voice to the workers themselves—the men and women whose lives and livelihoods have been affected most directly by the social and economic transformations now under way. The photographer explains his approach as follows:

> In 1989 I decided to seek out the effects of capitalism on workers and to record their stories through photography. The division of labor in a factory mirrors society as a whole. I sought out the impact of class oppression on different ethnic, racial, and gender groups. I became enamored with creating dignified portraits of entry-level workers. Images of the rich and famous are plentiful. It is the factory worker who needs visibility and a voice. One worker I met told me, "I've been working in this factory for forty years and nobody's ever asked to take my picture." (Jacobson-Hardy, 1995, p. 43)

It's a very dirty job [extracting oil from metal shavings]. When you walk downstairs, it's a . . . big cloudy mist of oil. Sometimes you come home and you're oil from head to toe.

When I started in the paper industry, it was still . . . a trade. You had to fight to get to be a machine tender. . . . It would take you sometimes 15 or 20 years to get to that position. . . . You had to wait. You had to grow into your job.

The plan is to do away with manufacturing. A lot of good people have been laid off. Both of my parents once worked here and lost their jobs. We can't compete. I like it here. I like the people I work with. Three and a half years ago we had about 32,000 employees—now, there are about 6,000. Every 3 or 4 months, more people get laid off. It's a slow process.

When the mills shut down some workers never even bothered to return to pick up their pay. . . . We just lost track of them. . . . What people got, if they were old enough, wouldn't be enough money to last for a week. . . . They had the worst pension plans you ever saw in your life. It wouldn't amount to a pot of beans.

The photos shown here are part of an exhibition titled "The Changing Landscape of Labor: Workers and Workplace." Although the focus of the exhibition is New England, the issues addressed are relevant to the United States as a whole, as can be seen in the accompanying comments by workers.

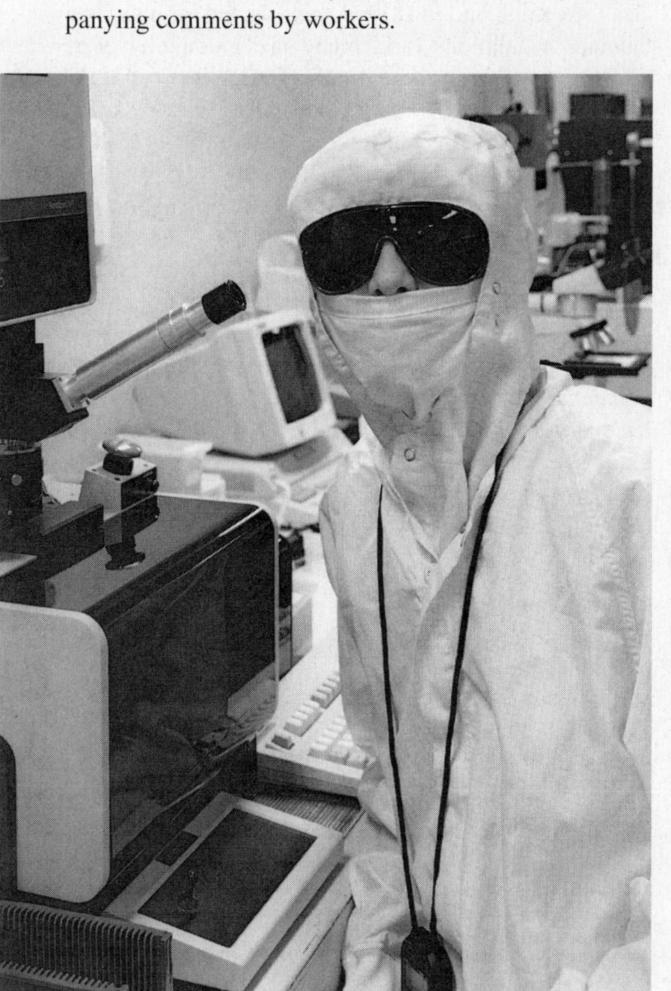

I'm tired. I want to quit. . . . I have to work until I'm 65. Maybe I'll never make it.

SUMMARY

The basic measures of inequality in any society are income, wealth, *occupational prestige*, and *educational attainment.* In American society the distribution of educational attainment and occupational prestige is more nearly equal than the distribution of wealth and income.

Sociological views of inequality in America have changed as the nation has been transformed from an agrarian society to an urban industrial society and then to a postindustrial society. The Jeffersonian view of America envisioned a society in which the majority of families lived on their own farms or ran small commercial or manufacturing enterprises. However, this view did not apply to the larger cities, the southern states, or Native Americans. During the Great Depression, the effects of industrialization tended to increase hostility between workers and the owners of businesses. In the mid-twentieth century, several important studies of inequality in American communities revealed the existence of a complex social-class system as well as a racial caste system.

The shift from an economy based on manufacturing to one based on services has resulted in a blurring of class lines and an easing of class conflict between industrial workers and the owners and managers of the means of production. Nevertheless, some sociologists argue that Americans continue to recognize social-class divisions. When people are asked what social class they belong to, the largest proportion say that they are members of the middle class. They base their class assignments on *socioeconomic status,* which is derived primarily from occupation but also takes into account family status, education, and earned income.

Social-class position has important consequences for the daily life of individuals and households. Members of the upper classes tend to have better health and more adequate health care than people in the lower classes. They are also likely to receive more and better education. In politics, the poor and members of the working class generally support the Democratic party while those in the middle and upper classes support the Republican party.

The upper class is estimated at about 1 percent of the U.S. population but controls 40 percent of all personal wealth in the United States. This class may be divided into the wealthiest and most prestigious families, who make up the elite or "high society," and families that have acquired their money more recently. Sociologists continually debate whether the upper class in America is also the society's ruling class.

The middle class, the largest class in American society, is culturally extremely diverse and hence is often referred to as the "middle classes." In the past it was thought to be associated with a family-oriented, conservative, suburban lifestyle, but recent studies have shown that there is no easily identified middle-class suburban culture.

The working class, which accounts for 44 percent of white Americans and almost 57 percent of African Americans, is undergoing rapid and difficult changes as production technologies change and industrialization spreads throughout the world. Members of this class are employed in skilled, semiskilled, or unskilled manual occupations, and many are union members. The working class can be divided into industrial workers and those employed in skilled crafts. There is more racial and ethnic diversity in the working class than in other classes.

Estimates of the proportion of the population living in poverty vary widely, depending on the standard used to define poverty. According to official statistics, about 15 percent of Americans are living in poverty. A significant proportion of the poor have jobs that do not pay enough to support their families. Another large percentage of poor families are single-parent families headed by women. Other categories of poor people include aged people with fixed incomes, marginally employed rural workers and part-time miners, chronically unemployed manual workers, and disabled workers and their families. Another group of people in danger of becoming poor are farmers.

Policy debates on the issue of poverty are often clouded by problems of definition. Although many Americans believe in *equality of opportunity,* they are less committed to the ideal of *equality of result.* Most sociologists agree that it is impossible to achieve a completely egalitarian society; instead, they concentrate on how much present levels of inequality can and should be reduced. In 1996 Congress enacted major reforms in the welfare laws, which now require that adult aid recipients engage in approved work. Data on the effects of the reforms indicate that they have had mixed results and may actually have worsened conditions for poor children.

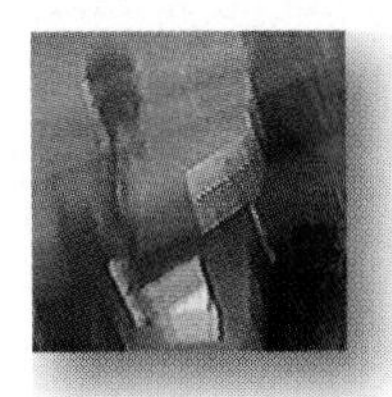

SOCIOLOGY VERSUS IDEOLOGY

It is often asserted that the United States does not have a class system like that of England or other European nations. In the United States, the argument goes, a child born into a poor family who is willing to work hard can get an education and find a good job, and then the sky is the limit. Critics often argue, on the other hand, that the American dream of success regardless of one's origins might be a myth. They point to poor schools and lack of opportunity in low-income communities as clear examples of the existence of classes in American society.

Sociologists bring facts and concepts to bear on this argument. To those who argue against the existence of classes, they can demonstrate (as we have in this chapter) major differences in the material well-being and life chances of different socioeconomic strata in U.S. society. But they can also show that many people continue to believe in the possibility of success through individual effort, and that as a result Americans are less likely to make the same sharp class distinctions that Europeans might make. Thus the sociological facts of class inequality are complex. The United States does have a class system that includes extremes of wealth and poverty that are more severe than those found in many other nations, and it is true that the class one is born into has a significant effect on one's life chances. However, there is more than enough mobility in U.S. society to maintain a strong belief in the American dream.

GLOSSARY

educational attainment: the number of years of school an individual has completed. (p. 364)

educational achievement: mastery of basic reading, writing, and computational skills. (p. 365)

occupational prestige: the honor or prestige attributed to specific occupations by adults in a society. (p. 365)

ghetto: a section of a city that is segregated either racially or culturally. (p. 369)

socioeconomic status (SES): a broad social-class ranking based on occupational status, family prestige, educational attainment, and earned income. (p. 372)

equality of opportunity: equal opportunity to achieve desired levels of material well-being and prestige. (p. 383)

equality of result: equality in the actual outcomes of people's attempts to improve their material well-being and prestige. (p. 383)

WHERE TO FIND IT

BOOKS

Poverty and Place: Ghettos, Barrios, and the American City (Paul Jargowsky; Russell Sage, 1997). A valuable new study with an excellent foreword by William J. Wilson. Presents recent findings and analysis of the relationships among social class, poverty, and place or residence.

Class Counts: Comparative Studies in Class Analysis (Erik Olin Wright; Cambridge University Press, 1997). A study of social class in the United States and elsewhere that bases measures of objective social class on the extent to which one has authority in the workplace. Presents good details on the relationship between race and gender and class differences in the United States and comparable nations.

Money: Who Has How Much and Why (Andrew Hacker; Simon & Schuster, 1997). A fact-filled account of trends in inequality and social class in the United States, with as much detail about the upper class as about the poor.

Class in Twentieth-Century American Sociology: An Analysis of Theories and Measurement Strategies (Michael D. Grimes; Praeger, 1991). A comprehensive review of the way social class has been studied in American sociology, with emphasis on why recent decades have brought renewed emphasis on the presence of classes and class conflict, in opposition to the functionalist theories and studies that were prevalent before the 1970s.

The War Against the Poor (Herbert Gans; Basic Books, 1995). A timely and thorough review of antipoverty policies and debates by one of the world's foremost experts on the subject.

JOURNALS

Monthly Review. A monthly journal of critical social-scientific writing and research on class and inequality. Most articles are Marxian in orientation.

The Public Interest. A valuable interdisciplinary social science journal with highly readable articles on issues of class, race, poverty, and social policy, with special emphasis on policy debates.

Focus. The official publication of the Institute on Poverty of the University of Wisconsin. An invaluable source for current sociological and economic research on poverty and social policy (welfare reform, family policies, early education, and much more).

OTHER SOURCES

Current Population Reports. Published monthly by the Bureau of the Census, U.S. Department of Commerce. Presents data on income and employment patterns for American households.

Handbook of Economic Statistics. Published annually by the U.S. Department of Labor. Compares economic statistics for selected nations, permitting measures of relative inequality and stratification.

Handbook of Labor Statistics. A compilation of statistical data published annually by the U.S. Department of Labor. Presents material on labor conditions and labor-force characteristics for the United States and selected foreign countries.

INTERNET RESOURCES

U.S. Census Bureau (www.census.gov/). Includes useful demographic data. Click on "Subjects A–Z" and then, for example, on "P" to look up Poverty or on "W" to look up Wealth.

The National Center for Children in Poverty (http:/cpmcnet.columbia.edu/dept/nccp/). Has updated information and ongoing research on the effects of changes in welfare policy on the lives of poor children and their communities.

Institute for Research on Poverty (www.ssc.wisc.edu/irp). Offers access to the most recent research on the causes and consequences of poverty and social inequality in the United States.

Cato Institute (www.cato.org/). A conservative research institution devoted to criticism of the welfare state. Advocates market alternatives rather than government-driven policies to address problems of social stratification and inequality.

CHAPTER 13

INEQUALITIES OF RACE AND ETHNICITY

When she was only 20 years old and still an undergraduate architecture student at Yale University, Maya Lin had an astounding experience. The young Asian American's design for a monument to commemorate the 58,156 Americans killed in the Vietnam War was chosen from more than 1,500 entries in an international competition. Earlier in the semester she had submitted the design as a project in her architecture course, but the professor did not like the concept very much. However, Lin had a strong vision of how the polished black granite monument would offer people a chance to reflect on the meaning of the war and remember those who died in it. She believed in her work.

After spending many weeks revising a one-page statement describing the intent of her design, Lin submitted the drawings to the competition's jury, never expecting to win. The rest is history. The Vietnam Veterans Memorial has become the most popular and moving piece of commemorative sculpture in the nation. Visitors run their fingers along the wall and often place bouquets, photographs, poems, or other items below the spot where their loved one's name is inscribed. "I knew people would have a personal experience as they searched for the names of those they knew," Lin (1995) explains, "but I never expected that they would leave things there. That was a surprise to me."

Although the monument commemorates the Vietnam War and does not directly address racism or ethnic conflict, Lin had to deal with a fair amount of racial abuse as the monument was being constructed. Mingled with the praise were letters that said, among other hateful things, that "a gook should not have been chosen to design a Vietnam War memorial." Even at the dedication, which Lin attended with her family, there were racial slights that wounded her deeply. But the popularity of the monument, and the way it brings together Americans of all races to contemplate death and heroism, more than compensate for the hurt. And as described in the Visual Sociology section of the chapter, the universal appeal of her vision has resulted in other commissions.

THE MEANING OF RACE AND ETHNICITY

How can we explain prejudice and racial discrimination and the inequalities they engender? And how can we measure their effects on individuals and social institutions? Many social scientists focus their research on why some groups in a society have been subordinated and the consequences of that subordination for them and their children. The effects of various forms of subordination, such as slavery, expulsion, and discrimination, on a society's patterns of inequality are at the center of the study of race and ethnicity, not only in the United States but throughout the world.

The United States is not unique in the extent to which inequality and hostility among its ethnic and racial groups result in severe social problems. In Canada, conflict between Anglo and French Canadians in Quebec periodically threatens national unity. In Germany there have been severe riots as native-born Germans have pressed for limits on the nation's acceptance of refugees from Eastern European nations that were formerly dominated by the Soviet Union. In what used to be Yugoslavia, hostility among Croats, Serbs, Bosnians, and Albanians has produced the bloodiest ethnic wars in recent history. In the former Soviet Union there is great fear that ethnic nationalism unleashed by the demise of the communist dictatorship will produce even more terrible wars because there are nuclear arms in the former Soviet republics.

Given its importance throughout the world, it is no wonder that the study of racial and ethnic relations has always been a major subfield of sociology. Most societies include minority groups, people who are defined as different according to the majority's perceptions of racial or cultural differences. And in many societies, as the ironic song from the musical *South Pacific* goes, "You've got to be taught to be afraid/of people whose eyes are differently made/or people whose skin is a different shade./You've got to be carefully taught." Sociologists try to get at the origins of these fears and groundless distinctions that categorize people as different and influence their life chances, often in dramatic ways.

In this chapter we first look at how race and ethnicity lead to the formation of groups that have a sense of themselves as different from the dominant group. In the second section of the chapter we analyze patterns of inequality and intergroup relations, while the third section discusses cultural consequences, particularly in American society. This is followed by a presentation of social-scientific theories that seek to explain the phenomena of intergroup conflict and accommodation. Rather than discussing each of the major racial and ethnic groups in American society in turn, throughout the chapter we present examples from the experience of blacks, Hispanic Americans, Native Americans, and other ethnic groups that have played an important part in American history.

Race: A Social Concept

Of the millions of species of animals on earth, ours, *Homo sapiens,* is the most widespread. For the past 10 millennia we have been spreading northward and southward and across the oceans to every corner of the globe. But we have not done so as a single people; rather, throughout our history we have been divided into innumerable societies, each of which maintains its own culture, thinks of itself as "we," and looks upon all others as "they." Through all those millennia of warfare, migration, and population growth, we have been colliding and competing and learning to cooperate. The realization that we are one great people despite our immense diversity has been slow to evolve. We persist in creating arbitrary divisions based on physical differences that are summed up in the term *race.*

In biology, **race** refers to an inbreeding population that develops distinctive physical characteristics that are hereditary. Such a population therefore has a shared genetic heritage (Coon, 1962; Marks, 1994). But the choice of which physical characteristics to use in classifying people into races is arbitrary. Skin color, hair form, blood type, and facial features such as nose shape and eyefolds have been used by biologists in such efforts. In fact, however, there is a great deal of overlap among the so-called races in the distribution of these traits. Human groups have exchanged their genes through mating to such an extent that any attempt to identify "pure" races is bound to be fruitless (Alland, 1973; Dobzhansky, 1962; Gould, 1981).

Yet doesn't common sense tell us that there are different races? Can't we see that there is a Negro, or "black," race of people with dark skin, tightly curled hair, and broad facial features; a Caucasian, or "white," race of people with pale skin and ample body hair; and a Mongoloid, or "Oriental," race of people with yellowish or reddish skin and deep eyefolds that give their eyes a slanted look? Of course these races exist. But they are not a set of distinct populations based on biological differences. The definitions of race used in different societies emerged from the interaction of various populations over long periods of human history. The specific physical characteristics that we use to assign people to different races are arbitrary and meaningless—people from the Indian subcontinent tend to have dark skin and straight hair; Africans from Ethiopia have dark skin and narrow facial features; American blacks have skin colors ranging from extremely dark to extremely light; and whites have facial features and hair forms that include those of all the other supposed races. There is no scientifically valid typology of human races; what counts is what people in a society define as meaningful. In short, race is a social concept that varies from one society to another, depending on how the people of that society feel about the importance of certain physical differences among human beings. In reality, as Edward O. Wilson (1979) has written, human beings are "one great breeding system through which the genes flow and mix in each generation. Because of that flux, mankind viewed over many generations shares a single human nature within which relatively minor hereditary influences recycle through ever-changing patterns, between the sexes and across families and entire populations" (p. 52).

i Racism

Throughout human history, many individuals and groups have rejected the idea of a single human people. Tragic mistakes and incalculable suffering have been caused by the application of erroneous ideas about race and racial purity. They are among the most extreme consequences of the attitude known as racism.

Racism is an ideology based on the belief that an observable, supposedly inherited trait, such as skin color, is a mark of inferiority that justifies discriminatory treatment of people with that trait. In their classic text on racial and cultural minorities, Simpson and Yinger (1953) highlighted several beliefs that are at the heart of racism. The most common of these is the "doctrine of biologically superior and inferior races" (p. 55). Before World War I, for example, many of the foremost social thinkers in the Western world firmly believed that whites were genetically superior to blacks in intelligence. However, when the U.S. Army administered an IQ test to its recruits, the results showed that performance on such tests was linked to social-class background rather than to race. And when investigators controlled for differences in social class among the test takers, the racial differences in IQ disappeared (Kleinberg, 1935).

Since that time there have been other attempts to demonstrate innate differences in intelligence among people of different races. The most recent of these efforts is Herrnstein and Murray's controversial study, *The Bell Curve* (1994), which we discussed in Chapter 5. This study argues that because scores on intelligence tests are distributed along a "normal" bell curve, there will necessarily be a large number of people whose scores fall well below the mean and who are not capable of performing well in situations requiring reasoning ability or other academic skills. Because the bell curves for African Americans and whites are different, with that of blacks peaking at a somewhat lower mean score, the authors argue that group differences in IQ establish biological limits on ability. This conclusion, in turn, leads them to argue that programs intended to correct differences in educational opportunities, such as Head Start, are doomed to failure.

Critics of these conclusions argue that studies like *The Bell Curve* merely dredge up old academic justifications for the status quo of racial and class inequality (Gould, 1995). There is a great deal of evidence that IQ tests are biased against members of minority groups and that something as complex and elusive as intelligence cannot be summarized by a single score on a test. The history of efforts to address inequalities in education shows that when they have access to high-quality educational programs, minority students quickly begin to achieve at the same levels as white students.

The notions that members of different races have different personalities, that there are identifiable "racial cultures," and that ethical standards differ from one race to another are among the other racist doctrines that have been debunked by social-scientific studies over the past 50 years. But even though these doctrines have been discredited, we will see shortly that they continue to play a major role in intergroup relations in many nations. Racist beliefs in the innate inferiority of populations that are erroneously thought of as separate races remain one

Then  and Now

Reaching Higher Ground

"Do you know the hymn that goes, 'The great trees are bending/The Good Lord is sending/His people up to higher ground'?" The speaker is Polly Heidelberg, a founder of the civil rights movement in Meridian, Mississippi. Among the events that have burned themselves into her memory are the murders of three civil rights workers—James Chaney, Michael Schwerner, and Andrew Goodman—who worked together in Mississippi during the "Freedom Summer" of 1964. Chaney was an African American resident of Meridian; Schwerner and Goodman were volunteers from northern colleges. And Polly Heidelberg was one of the last people the young men saw before they set out for the little village of Philadelphia, Mississippi, where they were murdered and hastily buried in an earthen dam. (Their murder, one of the pivotal moments of the civil rights movement, is the subject of the 1988 film *Mississippi Burning.*) "Miz" Heidelberg, as she is known in Meridian, was Chaney's neighbor, and she feels responsible for "bringing him into the movement." She knew all three men, and even now when she thinks of them her eyes cloud over and she prays for their souls.

As she speaks, Heidelberg and I are sitting in a white-owned coffee shop in a white neighborhood of Meridian. One of her granddaughters, a child of about 4, has finished eating and is playing quietly on the floor. I am scribbling Heidelberg's words on

of the major social problems of the modern world. And this tendency to denigrate socially defined racial groups extends to members of particular ethnic groups as well.

Ethnic Groups and Minorities

Ethnic groups are populations that have a sense of group identity based on a distinctive cultural pattern and, usually, shared ancestry, whether actual or assumed. Ethnic groups often have a sense of "peoplehood" that is maintained within a larger society (Kornblum, 1974; Teitelbaum & Weiner, 1995). Their members usually have migrated to a new nation or been conquered by an invading population. In the United States and Canada, a large proportion of the population consists of people who either immigrated themselves or are descended from people who immigrated to the New World or were brought there as slaves. If one traces history back far enough, it appears that everyone living in the Western Hemisphere can trace his or her origins to other continents. Even the Native Americans are bea notepad while attempting to eat an ample southern breakfast. An interview with Polly Heidelberg is fun. She often slips riddles into her speech to gently tease the listener. She is also a deeply religious and patriotic person, and her speech is filled with allusions to God and country.

Heidelberg tells me about the movement to integrate the armed forces during World War II, about the bitterness of racial strife during the 1950s, and about the exhausting, exhilarating years of the 1960s. There is no end to the movement as she sees it. Civil rights was only one aspect of the struggle for true equality of opportunity. Now there must be more progress toward economic equality.

We have finished our breakfast and the child is becoming restless. Heidelberg asks me, "That line in the hymn that mentions higher ground, what do you think higher ground is?"

Stammering a bit, I begin by saying something about "loftier values, greater spirituality, . . ." but Miz Heidelberg gently interrupts me.

"Sure, you could think about it that way. But to me 'higher ground' is right where we are now. It's you and me and the child having breakfast in this coffee shop here in Meridian, Mississippi."

lieved to have crossed the Bering Strait as migrant peoples between 14,000 and 20,000 years ago.

European conquests of the Americas in the sixteenth and seventeenth centuries resulted in people of Iberian (Spanish and Portuguese) origins becoming the dominant population group in Central and South America. People of English, German, Dutch, and French origins became the dominant population groups throughout much of North America. By the late 1770s almost 80 percent of the population of the 13 American colonies had been born in England or Ireland or were the children of people who had come from those countries. A century later, owing to the influence of slavery and large-scale immigration, the populations of the United States and Canada were far more diverse in terms of national origins, but people from England were still dominant.

According to sociological definitions of the term *dominance,* a dominant group is not necessarily numerically larger than other population groups in a nation. More important in establishing dominance is control or ownership of wealth (farms, banks, manufacturing concerns, and private property of all kinds) and political

BOX 13.1 GLOBAL CHANGE AND U.S. SOCIETY

Periods of Migration and Settlement in the United States

In 1790, when the republic was newly formed, the vast majority of U.S. citizens and resident foreigners (77 percent) had come from Great Britain and the counties that are now included in Northern Ireland. People from Germany (7.4 percent), Ireland (4.4 percent), the Netherlands (3.3 percent), France (2 percent), Mexico (1 percent), and other countries were present in the population, but with the exception of the Irish they were also primarily white Protestants. Blacks—involuntary immigrants—and their children actually accounted for 19 percent of the total population, but neither they nor Native Americans were citizens.

1820–1885: The "Old" Northwest European and Asian Migration

By about 1830 the mass migration of different groups into the new nation was under way. The largest flows of immigrants came from Ireland and Germany, with well over a million people from each of these countries coming to the United States to escape political or economic troubles. Smaller waves of immigrants originated in the northern European nations of Sweden, Norway, and Denmark.

Although people from northern Europe made up the most important immigrant groups, far exceeding the continuing influx from Great Britain, 1885 marked the beginning of a flow of immigrants from China that continued at high rates for the next 30 years. Chinese immigrants settled primarily in the West and contributed immensely to the development of the western states, although they met with sporadic and often violent hostility.

1885–1940: The "Intermediate" Migration From Southern and Eastern Europe and the Beginning of Heavy Immigration From Mexico

From 1880 to 1900 the flow of immigrants from Italy, Poland, Russia, the Baltic states, and southern Europe (Serbs, Croats, Slovenes, Romanians, Bulgarians, Greeks, etc.) exceeded the flow of people from older immigrant groups by about 4 to 1. In 1907 alone more than 250,000 Italians and 338,000 Poles and other central Europeans were admitted through New York's Ellis Island. Many of these immigrants were Jewish, Roman Catholic, or Orthodox Christian. The newcomers' tendency to live in neighborhoods known as "Little Italy" or "Little Poland" convinced more established Americans that the immigrants would never learn English or become fully American. The mass arrival of physically distinctive groups such as Mexicans and Japanese during this period also aroused fear and hostility among those who disliked the newcomers.

During World War I thousands of impoverished black southerners were attracted to the cities of the North and the Midwest. This began a trend that would continue throughout the twentieth century whenever wars or economic booms made it possible for blacks to overcome discrimination and get factory jobs.

1921–1959: Immigration by Quota and Refugee Status

Agitation against the "new" immigrants and demands for "Oriental exclusion" led Congress to establish a quota system that drastically altered the flow of newcomers into the nation after 1920. At first the quotas were fixed at 3 percent of the number of people from each country who had been counted as U.S. residents in the 1910 census. But because that formula still allowed large numbers of "new" immigrants to enter the nation, in 1924 the law was changed so that quotas would be based on the national origins of the foreign-born population as of the 1890 census. This biased the quota system in favor of northern Europeans; in fact, Asians were explicitly excluded.

World War II and the cold war produced millions of homeless and stateless refugees, especially from Germany, Poland, and the Soviet Union. Almost half a million of these displaced persons arrived in the United States between 1948 and 1950; another 2 million came between

power—especially control of political institutions like legislatures, courts, the military, and the police. Thus, by the beginning of the twentieth century white Anglo-Saxon males (or WASPs, as they are sometimes called) had established dominance over U.S. society even though they no longer constituted a clear majority of the population.

Between 1820 and 1960 approximately 42 million people immigrated to the United States. Since 1960 an average of at least 300,000 immigrants have arrived each year, with the number increasing to more than 900,000 in the past few years. Since its earliest history the United States has had one of the most liberal immigration policies of any nation, a situation that is in part a consequence of the fact that many of its inhabitants can trace their origins to other countries. And except in the case of the Native Americans, who were expelled from their ancestral lands and forced to resettle on reservations, no area of the nation is seen as the ancestral homeland of a particular people. Unlike many nations, the United States is a multiethnic society where notions of "blood" (descent) do not coincide with claims to ancestral land. In the United States, therefore, there are no separatist movements that claim territory and threaten

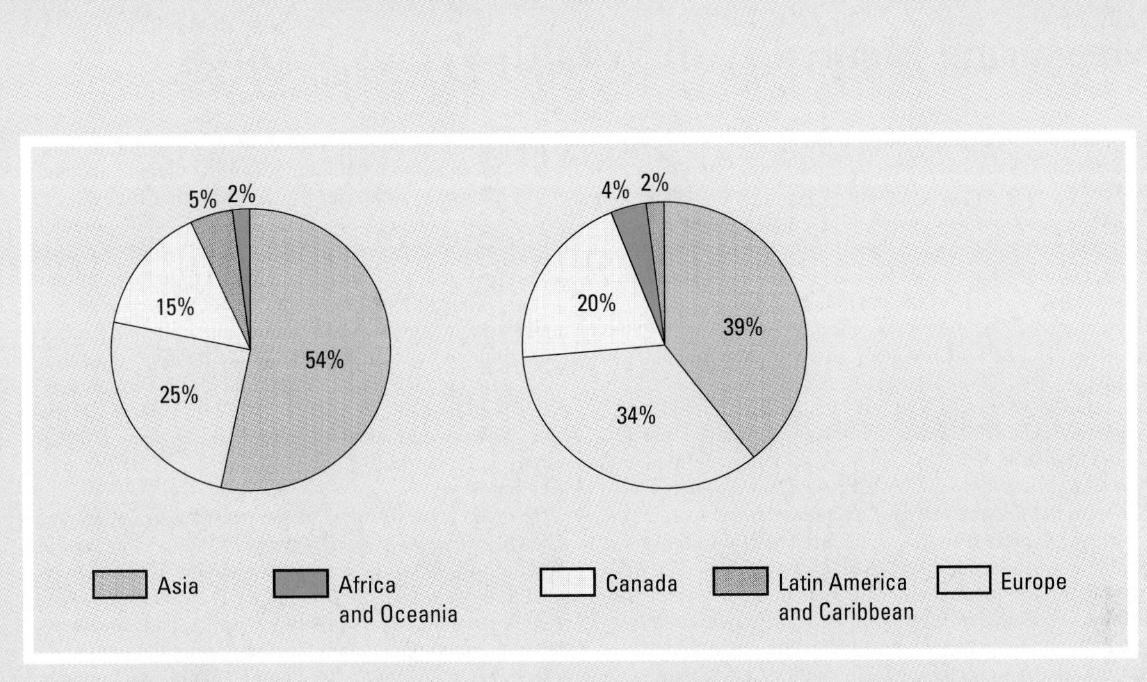

Source: Lee, 1998.

Immigration by Region and Selected Countries of Origin, 1951–1960 and 1981–1996

1950 and 1957. Their arrival swelled the populations of ethnic enclaves in America's industrial cities.

1960 to the Present: Worldwide Immigration

In the 1960s national quotas were replaced with a system based on preference for skilled workers and professionals, regardless of country of origin, as well as refugees and people with families already in the United States. Gradually Congress increased the number of immigrants who may enter the country legally. Today the United States is once again the foremost immigrant-receiving nation in the world. The largest streams of immigrants are from Mexico, other Central American and Caribbean nations, Asia (especially Korea, China, India, and Pakistan), and South America. Untold numbers of undocumented, or illegal, immigrants also arrive each year—especially from Mexico, Canada, and the Dominican Republic—but studies show that a high proportion of these newcomers eventually become legal residents or return to their country of origin (Lee, 1998; Portes & Rumbaut, 1996).

national unity. There are, however, severe inequalities that affect particular populations, especially Native Americans, African Americans, Latinos, and Appalachian whites. Many of these inequalities date back to major periods of settlement and ethnic group formation, which are summarized in Box 13.1.

Immigrant groups that are distinct because of racial features or culture (language, religion, dress, etc.) are often treated badly in the new society and hence develop the consciousness of being a minority group. Louis Wirth (1945), a pioneer in the study of racial and ethnic relations, defined a **minority group** as a set of "people who, because of their physical or cultural characteristics, are singled out from the others in the society in which they live for differential and unequal treatment, and who therefore regard themselves as objects of collective discrimination" (p. 347). The existence of a minority group in a society, Wirth explained, "implies the existence of a corresponding dominant group with a higher social status and greater privileges. Minority status carries with it exclusion from full participation in the life of the society" (p. 347).

In the United States the term *minority* often suggests "people of color," meaning African Americans, American

BOX 13.2 SOCIOLOGICAL METHODS

Measuring Inequality of Wealth Versus Income

The accompanying illustrations from Melvin Oliver and Thomas Shapiro's study, *Black Wealth/White Wealth* (1995), make an important point. If we look at African American households consisting of a married couple and their children and compare them to similar white households, the black family has somewhat less income (almost $5,000 a year) but far less wealth. On average, the black household's net worth is about $46,000 less, and its net financial assets are lower by an average of $27,000. In other words, income inequality is not nearly as severe as inequality of wealth. But how are these differences measured? And why bother making such measurements?

In Oliver and Shapiro's study, *net worth* is defined as "the straightforward value of all assets less any debts." The term *net financial assets* "excludes equity accrued in a home or vehicle from the calculation of a household's available resources" (Oliver & Shapiro, 1995, p. 58). *Income* is the stream of cash that comes into the household through jobs and earnings of all who participate in the economy in any way. But most income is spent quite quickly, as we all know too well, and it is much more difficult to divert income into more stable forms of wealth such as homes, savings accounts, mutual funds, and the like.

Most of the data on which Oliver and Shapiro base their analysis come from a single major survey of income and wealth among American households, the Survey of Income and Program Participation (SIPP). It contains data on income and, more to the point, extensive data on assets (checking accounts, savings accounts, equity in homes and other property, mutual funds, stocks, and much more) and debts (credit cards, medical bills, home mortgages, personal debts, and so on). Like all surveys, the SIPP tends to undersample populations that are difficult to reach, such as unemployed young males and the very rich. In consequence, the differences in income and wealth reported are quite conservative. The disparities would be even wider if the extremes of poverty and wealth were accurately represented. In any case, when one looks at middle-class segments of the survey, such as married-couple families, the errors are less important.

Why do these differences in wealth matter? Clearly, when a family can pass a home or its value to succeeding generations, that represents an advantage. When there is wealth in the household, the family is better able to withstand economic shocks or sudden reversals of fortune. When there is some wealth accumulating, the family is in a far better position to afford higher-quality educational opportunities for its young people. Recognizing these and many other advantages, Oliver and Shapiro argue that all forms of discrimination that prevent African Americans and others from accumulating wealth (for example, discrimination in mortgage assistance and other lending policies) are especially pernicious and must be corrected immediately.

Indians, Mexicans (many of whom have the darker coloring of Amerindian ancestry), and Asians. But the term can also be applied to people with lighter skin coloring. In Great Britain, for example, the Irish were a conquered people who were subjected to economic and social discrimination by the English. It is not surprising, therefore, that when the Irish began immigrating to the United States in the nineteenth century they were treated as an inferior minority by Americans of English ancestry. This attitude was especially prevalent in cities like Boston and New York, to which the Irish came in large numbers.

It should also be noted that the term *minority* does not always imply that the population is numerically inferior to the dominant group. There are counties in some states in the South, and entire cities in the North, in which African Americans constitute a numerical majority yet cannot be considered the dominant group because they lack the power, wealth, and prestige enjoyed by the white population. (See Box 13.2.)

Ethnic and Cultural Minority Groups Ethnic groups often form in cities or metropolitan regions for reasons of choice as well as necessity. In fact, very often the identity of a group is defined through the experience of coping with life in the new society as well as through the group's desire to hold on to its language, food, and other cultural ways (Handlin, 1992; Rodriguez, 1992).

The Costs of Being Black

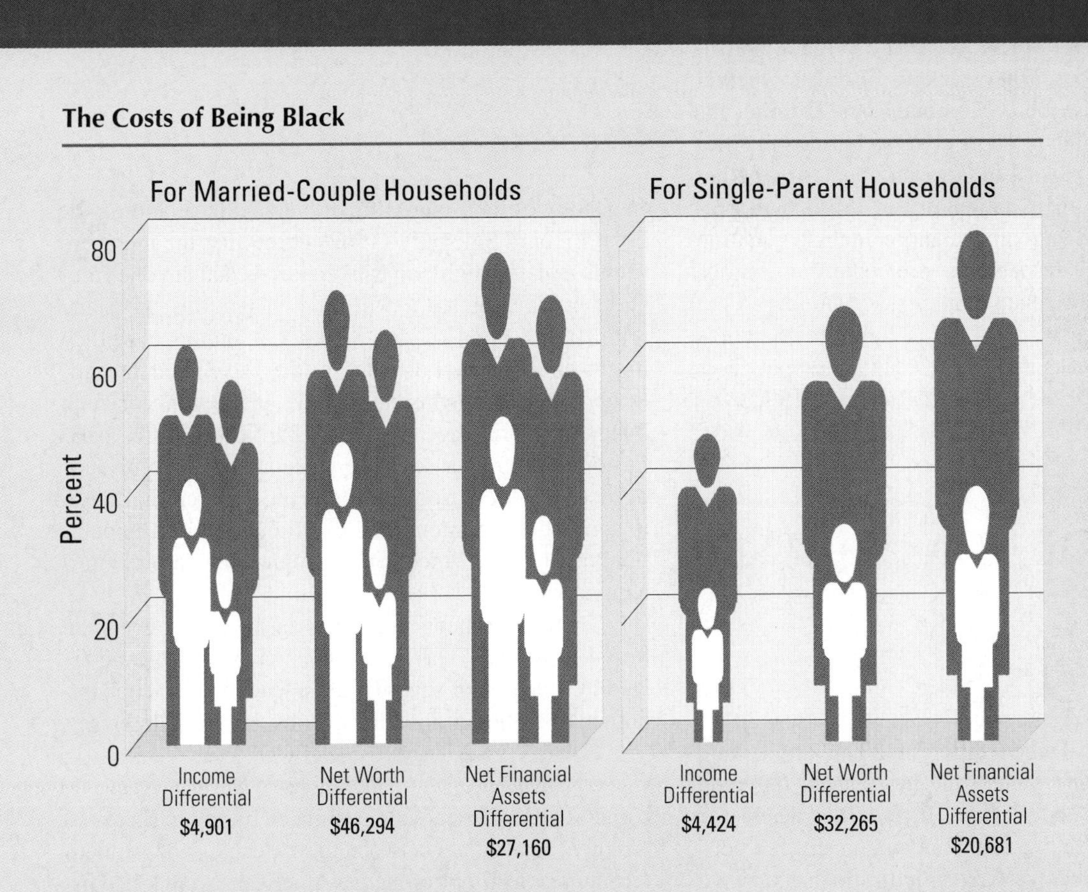

Note: Percent not explained by controlling for differences between similar white and black households.

Source: Oliver & Shapiro, 1995.

In one of the first and best studies of ethnic group formation, W. I. Thomas (1971/1921) described how in the early decades of the twentieth century Italian immigrants settled in small neighborhoods in Manhattan where they could live with people who had come from the same region and spoke the same dialect of Italian. At first these immigrants identified themselves as Calabrians or Sicilians or Piedmontese, the way they had before leaving their home regions. Thomas found that they identified with those home regions rather than with the nation of Italy. In New York, however, they could speak to each other and appreciate each other's food, and they had reason to join together to compete with other ethnic groups for jobs and homes. They were also called Italians by people outside their neighborhoods. Gradually, therefore, they began to think of themselves as Italians or Italian Americans, but this identity was forged out of their experience in the United States and did not mean the same thing as being an Italian in Italy.

Most ethnic populations have had similar experiences that caused them to develop a group consciousness and create organizations to represent their interests. Homosexuals and other populations that are defined in terms of their behavior and cultural traits may become organized into cultural minority groups, but they differ from true ethnic groups because they do not trace their ancestry to another society.

Today sociologists are interested in knowing whether a similar process of ethnic group formation is occurring in regions where there are large Spanish-speaking populations. They seek to discover whether Mexicans, Puerto Ricans, Colombians, and Dominicans are beginning to think of themselves as Latinos in cities where they are all present in large numbers, or whether their allegiance to their nationality of origin will prevent them from taking on the more inclusive Latino identity. Likewise, researchers are asking questions about whether a new Asian ethnicity will be forged out of the experience of various Asian immigrant groups, especially the Koreans and Chinese, in American cities (Moore & Pachon, 1985; Portes & Rumbaut, 1996).

WHEN WORLDS COLLIDE: PATTERNS OF INTERGROUP RELATIONS

Throughout history, when different racial and ethnic groups have met and mixed, the most usual outcome has been violence and warfare. In fact, the desire for peaceful and cooperative relations among diverse peoples has emerged relatively recently. In this section we explore a continuum of relations between dominant and minority groups that extends from complete intolerance to complete tolerance, as shown in the study chart below. At one extreme is extermination or genocide; at the other is assimilation.

Genocide

In a study of the Siane tribe of New Guinea, anthropologist Richard Salisbury (1962) found that the members of this isolated highland tribe believed that anyone from another tribe wanted to kill them. Therefore, the Siane felt that they must kill any member of another tribe they might encounter. (Fortunately, they excluded the anthropologist from this norm.) We often think of such behavior as primitive, savage, or barbarous. Yet barbarities on a far greater scale have been carried out by supposedly advanced societies. The most extreme of these is **genocide,** the intentional extermination of a population, defined as a "race" or a "people," by a more dominant population.

There have been numerous instances of genocide in recent history. Those incidents have been characterized by a degree of severity and a level of efficiency unknown to earlier civilizations. Consider the following:

- The Native American populations of North, Central, and South America were decimated by European explorers and settlers between the sixteenth and twentieth centuries. Millions of Native Americans were killed in one-sided wars, intentional starvation, forced marches, and executions. The population of Native Americans in

STUDY CHART A Continuum of Intergroup Relations

INTOLERANCE → TOLERANCE

Genocide	Expulsion	Slavery	Segregation	Assimilation
The intentional extermination of a population, defined as a "race" or a "people."	The forcible removal of a population from a territory claimed by another population.	Ownership of one population by another, which can buy and sell members of the enslaved population and controls every aspect of their lives.	Ecological and institutional separation of races or ethnic groups.	The process by which a minority group blends into the majority population and eventually disappears as a distinct people within the larger society.

Note: The placement of slavery to the right of expulsion is not meant to imply that slavery is less severe than expulsion. The reason for the placement is that slaves and slave masters are interdependent populations within a single society, whereas expulsion excludes the subordinated population from any form of membership in the society.

North America was reduced from more than 4 million in the eighteenth century to less than 600,000 in the early twentieth century (Thornton, 1987).

- When England, France, Germany, Portugal, and the Netherlands were engaged in fierce competition for colonial dominance of Africa during the nineteenth and early twentieth centuries, millions of native people were exterminated. The introduction of the Gatling machine gun made it possible for small numbers of troops to slaughter thousands of tribal warriors.

- Six million Jews, 400,000 gypsies, and about 2 million Russian civilians were systematically killed by the Nazis during World War II; thousands of Pakistanis and Indians were slaughtered after the partition of India in 1947; and thousands of Tamils living in Sri Lanka were exterminated by the Sinhalese in the 1980s. "Ethnic cleansing" in Africa and the former Yugoslavia, where entire villages have been wiped out, is a recent example of genocidal action.

Mass executions and other forms of genocide are almost always rationalized by the belief that the people who are being slaughtered are less than human and in fact are dangerous parasites. Thus the British and Dutch slaughtered members of African tribes like the Hottentots in the belief that they were a lower form of life, a nuisance species unfit even for enslavement. The Nazis rationalized the extermination of Jews and gypsies by the same twisted reasoning.

Expulsion

In many societies, extended conflicts between racial or ethnic groups have ended in **expulsion,** the forcible removal of one population from territory claimed by the other. Thus on the earliest map of almost every major American city there appears a double line drawn at the edges where the streets end. This is the Indian Boundary Line, and it symbolizes the expulsion of Native Americans from lands that were taken from them in order to create a city in which they would be strangers.

Expulsion was the usual fate of North America's native peoples, who also experienced massacres, deadly forced marches, and other genocidal attacks. These actions were rationalized, in part, by the nineteenth-century doctrine of "manifest destiny," which asserted the inevitability of the westward expansion of Europeans across the territories of the Southwest and West (Davis, 1995). As white settlement expanded westward, Native Americans were continually expelled from their tribal lands. After the 1848 gold rush and the rapid settlement of the West Coast, the pressure of white settlement pushed the Indian tribes into the high plains of the West and Southwest. Between 1865 and the 1890 massacre of Sioux Indians at Wounded Knee, South Dakota, the remaining free tribes in the West were forced to settle on reservations. In the process the Indians lost more than their ancestral lands. As the famous Sioux chief Black Elk put it, "A people's dream died" (quoted in Brown, 1970, p. 419).

The forced settlement of Native Americans on reservations is only one example of expulsion. Asian immigrants in the American West also suffered as a result of intermittent attempts at expulsion referred to as the Oriental exclusion movement. In an effort to prevent the large-scale importation of Chinese laborers into California and other states, Congress passed the Chinese Exclusion Act of 1882, which excluded Chinese laborers from entry into the United States for 10 years. But this legislation did little to relieve the hostility between whites and Asians. Riots directed against Chinese workers were common throughout the West during this period, and Chinese immigrants were actually expelled from a number of towns. In 1895, for example, a mob killed 28 Chinese immigrants in Rock Springs, Wyoming, and expelled the remaining Chinese population from the area (Lai, 1980).

The most severe example of expulsion directed against Asians in the United States occurred in 1942 after the Japanese attack on Pearl Harbor and the American entry into World War II. On orders from the U.S. government, more than 110,000 West Coast Japanese, 64 percent of whom were U.S. citizens, were ordered to leave their homes and their businesses and were transported to temporary assembly centers (Kitano, 1980). They were then assigned to detention camps in remote areas of California, Arizona, Idaho, Colorado, Utah, and Arkansas. When the U.S. Supreme Court declared unconstitutional the incarceration of an entire ethnic group without a hearing or formal charges (*Endo* v. *United States,* 1944), the Japanese were released, but by then many had lost their homes and all their possessions. In 1989 Congress finally voted to pay modest reparations to the families of Japanese Americans who had been imprisoned during the war.

Slavery

Somewhat farther along the continuum between genocide and assimilation is slavery. **Slavery** is the ownership of a population, defined by racial or ethnic or political criteria, by another population that not only can buy and sell members of the enslaved population but also has complete control over their lives. Slavery has been called "the peculiar institution" because, ironically, it has existed in some of the world's greatest civilizations. The socioeconomic systems of ancient Greece and Rome, for example, were based on the labor of slaves. And the great trading cities of late medieval Europe, such as Venice, Genoa, and Florence, developed plantation systems in their Mediterranean colonies that were based on slave labor. In fact, the foremost student of slave systems, Orlando Patterson, makes the following comment:

> There is nothing notably peculiar about the institution of slavery. It has existed from before the dawn of human history right down to the twentieth century, in the most primitive of human societies and in the most civilized. . . . Probably there is no group of people whose ancestors were not at one time slaves or slaveholders. (1982, p. vii)

Figure 13.1 indicates the magnitude of the transatlantic slave trade; the widths of the arrows represent the relative size of each portion of that terrible traffic in humanity. The arrows do not show, however, that for the Americas to acquire 11 million slaves who survived the voyage on the slave ships and the violence and diseases of the New World, approximately 24 million Africans had to be captured and enslaved (Fyfe, 1976; Patterson, 1982, 1991). It is evident from Figure 13.1 that the United States imported a proportionately small number of slaves. However, although somewhat less than 10 percent of all slaves were sold in the United States, by 1825 almost 30 percent of the black population in the Western Hemisphere was living in the United States (Patterson, 1982). This was due to the high rate of natural increase among the American slaves. In Brazil, by contrast, the proportion of slave imports was relatively high, but there was also a high mortality rate among the slaves owing to disease and frequent slave revolts.

At the time of the first U.S. census in 1790, there were 757,000 blacks counted in the overall population. By the outbreak of the Civil War, the number had increased to 4.4 million, of whom all but about 10 percent were slaves. In the southern states that fought in the war, one third of white families owned slaves, the average being nine "chattels" per owner (Farley & Allen, 1987). (The term *chattels* refers to living beings that are considered property, including slaves and, in some cultures, women.)

It can be inferred from the great increase in the U.S. slave population that slaveholders in the United States treated their slaves less badly than did slaveholders in other parts of the Western Hemisphere. This does not mean, however, that slaves in the Americas did not bitterly resent their condition and struggle against it. Slave revolts were extremely common in the Caribbean and elsewhere in the New World. Even when slaves could not rebel openly, they carried out many forms of resistance. Patterson points out that the slave has always striven, against all odds, "for some measure of regularity and predictability in his social life. . . . Because he was considered degraded, he was all the more infused with the yearning for dignity" (1982, p. 337). Patterson concludes that one of the chief ironies of slavery throughout history has been that "without [it] there

FIGURE 13.1 The Transatlantic Slave Trade

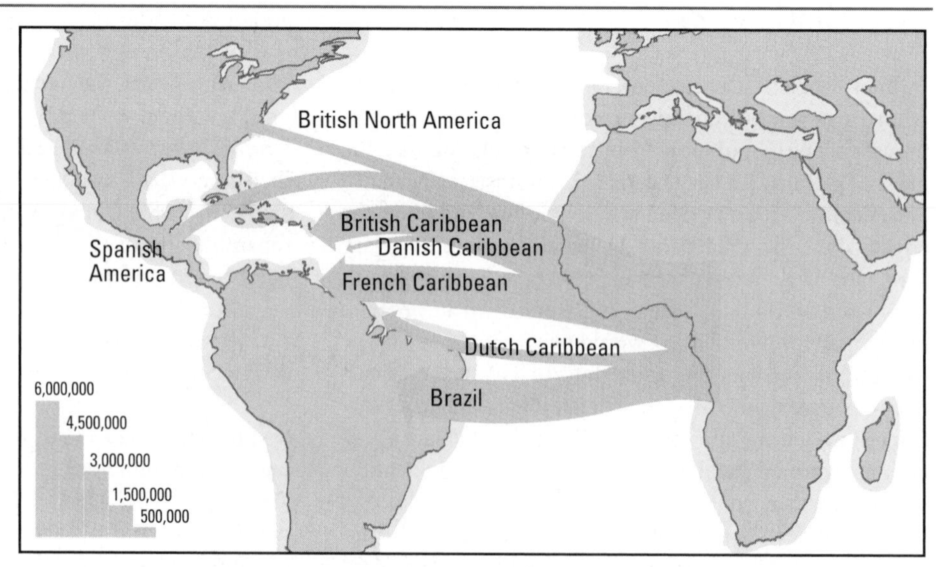

From the end of the sixteenth century to the early decades of the nineteenth, approximately 11 to 12 million Africans were imported to the New World (Patterson, 1982). The map shows that the largest numbers of slaves were sent to the European colonies in the Caribbean and to Brazil.

Source: Curtin, 1969.

This striking photo of fugitive slaves during the Civil War was taken at Cumberland, Virginia, on May 14, 1862. Note the look of proud determination on the people's faces, despite their obvious extreme poverty and exhaustion.

would have been no freedmen" (p. 342). In other words, the very idea of freedom developed in large part from the longing of slaves to be free.

Segregation

Although African American slaves gained their freedom during the Civil War and became citizens of the United States, this did not mean that they became fully integrated into American society. A long period of segregation followed. **Segregation** is the ecological and institutional separation of races or ethnic groups. The segregation of racially or ethnically distinct peoples within a society may be voluntary, resulting from the desire of a people to live separately and maintain its own culture and institutions. The Amish, the Hutterites, and the Hasidic Jews are examples of groups that voluntarily segregate themselves. But segregation is often involuntary, resulting from laws or other norms that force one people to be separate from others. Involuntary separation may be either **de jure**—created by laws that prohibit certain peoples from interacting with others or place limits on such interactions—or **de facto**—created by unwritten norms that result in segregation, just as if it were "in fact" legally required.

South Africa recently experienced a stunning reversal of its system of de jure segregation. Known as *apartheid,* South Africa's racial laws insisted that blacks, whites, and "coloreds" (people of mixed ancestry) remain separate. Intermarriage and integrated schools and communities were forbidden. In 1991, after years of struggle and violence, the white-dominated government agreed to repeal the apartheid laws. A constitutional convention—at which all segments of the population were represented—created a democratic regime in which all citizens' votes count equally. The African majority, led by Nelson Mandela and the African National Congress, assumed power in a relatively peaceful transition. Nevertheless, whites continue to hold most of the nation's wealth, while the majority of South African blacks are trapped in dire poverty and illiteracy, just as the former slaves were in the United States after the Civil War.

In the United States, legally sanctioned segregation no longer exists, but this has been true only in recent years. Before the civil rights movement of the 1960s, de jure segregation was common, especially in the southern states. The system that enforced segregation was supported by **Jim Crow** laws. This term refers to laws that enforced or condoned segregation, barred blacks from the polls, and the like. (Jim Crow was a nineteenth-century white minstrel who performed in blackface and thereby reinforced black stereotypes.) The system was in effect for about a hundred years, from just after the Civil War until the early 1970s. During that period the so-called color line was applied throughout the United States to limit the places where blacks could live, where they could work and what kinds of jobs they could hold, where they could go to school, and under what conditions they could vote.

At first the color line was an unwritten set of norms that barred or restricted black participation in many social institutions. By the turn of the century, however, segregation had become officially sanctioned through legislation and court rulings. This official segregation was rationalized by the "separate but equal" doctrine set forth by the Supreme Court in *Plessy* v. *Ferguson* (1896). Under this doctrine, separate facilities for people of different races were legal as long as they were of equal quality. In addition, de facto segregation and the existence of a "job ceiling" based on race served to keep blacks in subordinate jobs and segregated ghettos, as we saw in Chapter 12 in the case of Chicago's "black metropolis."

Only after years of struggle by opponents of de jure segregation did the U.S. Supreme Court finally decide, in the landmark case of *Brown* v. *Board of Education, Topeka* (1954), that "separate but equal" was inherently unequal. The *Brown* ruling put an end to legally sanctioned segregation of schools, hospitals, public accommodations, and the like. But it took frequent, often violent demonstrations and the mobilization of thousands of citizens in support of civil rights to achieve passage of the Civil Rights Act of 1964. That act mandated an end to segregation in private accommodations, made discrimination in the sale of housing illegal, and initiated a major attack on the job ceiling through the strategy known as affirmative action. Despite these judicial and legislative victories, however, de facto segregation remains a fact of life in the United States, especially in large cities (Massey & Denton, 1993).

Ecological studies of racial segregation show that residential segregation persists in almost all U.S. cities. Table 13.1 is based on a measure known as the *segregation index.* It shows "the minimum percentage of nonwhites who would have to change the block on which they live in order to produce an unsegregated distribution—one in which the percentage of nonwhites living on each block is the same throughout the city (0 on the index)" (Taeuber & Taeuber, 1965, p. 30). A value of 100 on the index would mean that all nonwhite people live on segregated blocks—that is, the city is 100 percent segregated. In 1970, for example, 81.2 percent of the nonwhite population in Boston would have had to move to another block to produce an unsegregated distribution of whites and nonwhites. By 1990 that proportion had declined to 68.2 percent, clearly an improvement.

TABLE 13.1 Trends in Black-White Segregation in 20 Metropolitan Areas With Largest Black Populations, 1970–1990

Metropolitan Area	1970	1980	1990
Northern Areas			
Boston	81.2	77.6	68.2
Chicago	91.9	87.8	85.8
Cleveland	90.8	87.5	85.1
Detroit	88.4	86.7	87.6
Kansas City	87.4	78.9	72.6
Los Angeles–Long Beach	91.0	81.1	73.1
Milwaukee	90.5	83.9	82.8
New York	81.0	82.0	82.2
Philadelphia	79.5	78.8	77.2
St. Louis	84.7	81.3	77.0
San Francisco–Oakland	80.1	71.7	66.8
Northern average	84.5	80.1	77.8
Southern Areas			
Atlanta	82.1	78.5	67.8
Baltimore	81.9	74.7	71.4
Birmingham	37.8	40.8	71.7
Dallas–Fort Worth	86.9	77.1	63.1
Houston	78.1	69.5	66.8
Memphis	75.9	71.6	69.3
New Orleans	73.1	68.3	50.3
Washington, DC	81.1	70.1	66.1
Southern average	75.5	68.3	66.5

Source: Adapted from Massey & Denton, 1993.

Despite such improvements in many metropolitan regions, Table 13.1 shows that most cities, especially in the North, continue to be highly segregated. Although segregation in some regions has fallen below the 70 percent level, high segregation indexes remain a fact of American life. Urban ecologists point out that most changes in segregation patterns are due to the movement of blacks into newer suburban communities and older white neighborhoods in the cities (Jaynes & Williams, 1989; Massey & Denton, 1993). (See the Mapping Social Change feature on pp. 414–415.)

Other problems of inequality follow from these persistent patterns of racial segregation. It is extremely difficult, for example, to have integrated schools, especially in the primary grades, when children live in segregated neighborhoods. And it is difficult to establish tolerance and understanding among racial and ethnic groups whose members have grown up in segregated neighborhoods and schools.

Assimilation

One of the factors that led to segregation was the fear of racial intermarriage, since the ideology of white supremacy held that intermarriage would weaken the white race. As late as 1950, 30 states had laws prohibiting such marriages, and even after racist sentiments began to diminish in the 1950s, 19 states (17 of them in the South) maintained such laws until the Supreme Court declared them unconstitutional in 1967 (Holt, 1980). Today interracial marriages account for about 2 percent of all marriages, but as we see in Figure 13.2, this represents a dramatic rise since 1960. And the popularity of Tiger Woods, Mariah Carey, and other celebrities from multiracial backgrounds suggests that the old norms against interracial families will continue to erode.

Intermarriage between distinct racial and ethnic groups is an important indicator of **assimilation,** the pattern of intergroup relations in which a minority group is forced or encouraged or voluntarily seeks to blend into the majority population and eventually disappears as a distinct people within the larger society. (We saw in Chapter 3 that assimilation refers to the process by which a culturally distinct group adopts the language, values, and norms of a larger society. Here we are using the term in a broader sense, to include social as well as cultural blending.) Needless to say, it makes a great deal of difference whether an ethnic or racial group has been the victim of forced assimilation or has been allowed to absorb the majority culture at its own pace. Many Latin American societies offer examples of peaceful, long-term assimilation of various racial and ethnic groups. For example, as a result of generations of intermarriage, Brazilians distinguish among many shades of skin color and other physical features rather than relying on crude black–white distinctions (Fernandes, 1968).

In the United States, assimilation has had a more troubled history. In an influential treatise on this subject, Milton Gordon (1964) identified three "ideological tendencies" that have affected the treatment of minority groups at various times. These ideologies specify how ethnic or racial groups should change (or resist change) as they seek acceptance in the institutions and culture of American society. They can be summarized as follows:

1. *Anglo-conformity*—the demand that culturally distinct groups give up their own cultures and adopt the norms and values of the dominant Anglo-Saxon culture
2. *The melting pot*—the theory that there would be a biological merging of ethnic and racial groups, resulting in a "new indigenous American type"
3. *Cultural pluralism*—the belief that culturally distinct groups can retain their communities and much of their culture even though they participate in the institutions of the larger society

Because these ideological tendencies have played an important role in intergroup relations in the United States, we will examine them in some detail.

Anglo-Conformity The demand for Anglo-conformity rests on the belief that the persistence of ethnic cultures, ethnic and racial communities, and foreign languages in

FIGURE 13.2 **Interracial Marriages, United States, 1960–1990**

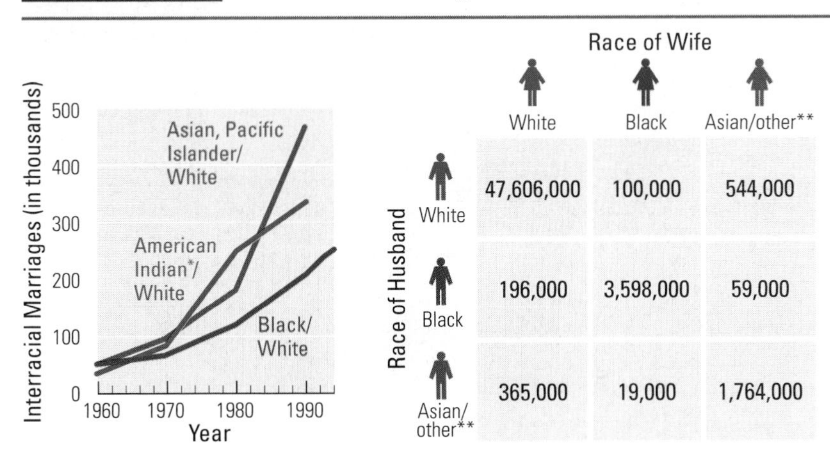

Race of Husband \ Race of Wife	White	Black	Asian/other**
White	47,606,000	100,000	544,000
Black	196,000	3,598,000	59,000
Asian/other**	365,000	19,000	1,764,000

* Includes Eskimo and Aleut for 1980–1992.
** Race includes Asian, American Indian, and Pacific Islanders; Hispanics may be of any race.

Source: Data from Census Bureau.

MAPPING SOCIAL CHANGE

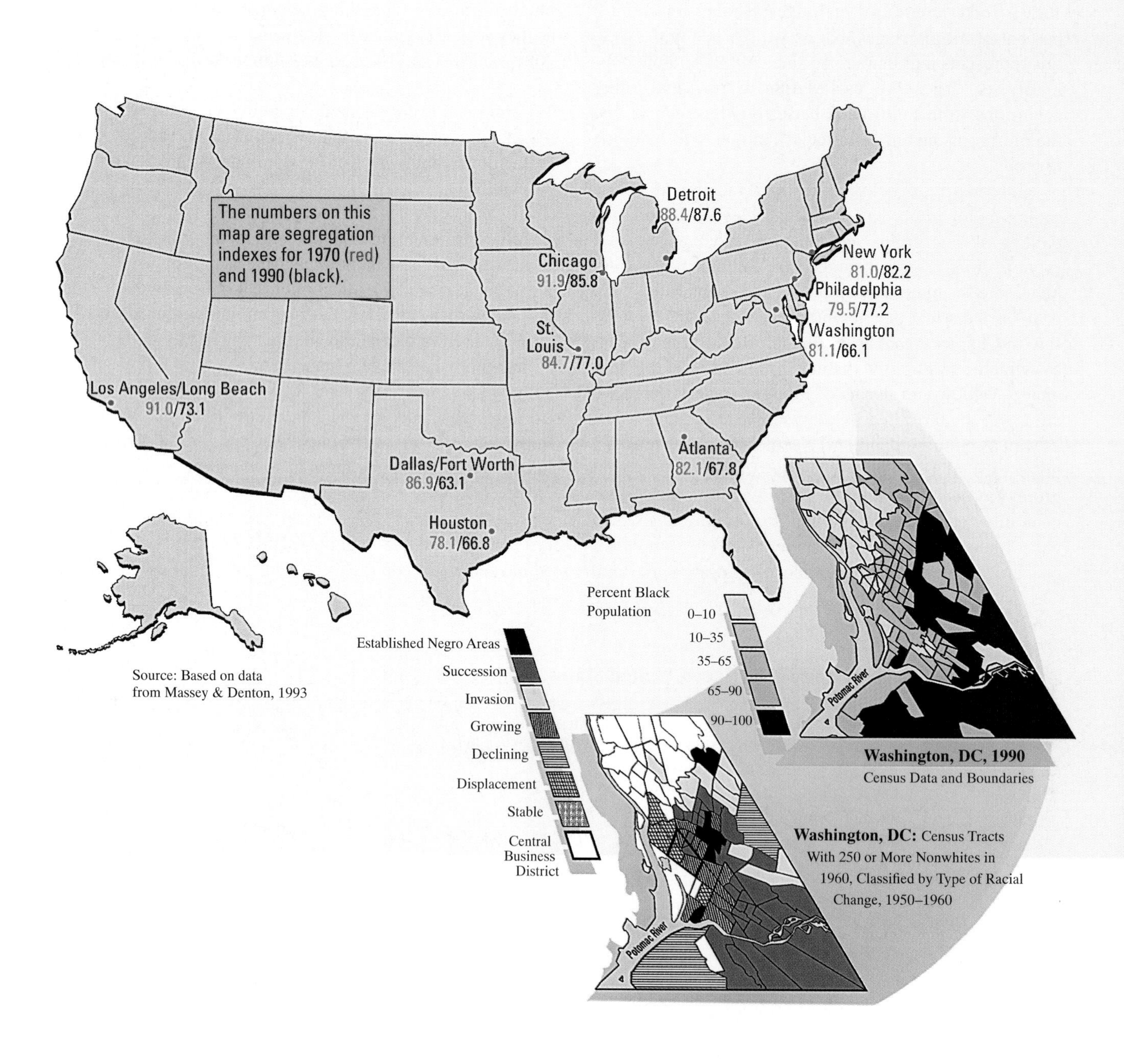

The numbers on this map are segregation indexes for 1970 (red) and 1990 (black).
Detroit
88.4/87.6
Chicago
91.9/85.8
New York
81.0/82.2
Philadelphia
79.5/77.2
Washington
81.1/66.1
St. Louis
84.7/77.0
Los Angeles/Long Beach
91.0/73.1
Atlanta
82.1/67.8
Dallas/Fort Worth
86.9/63.1
Houston
78.1/66.8
Percent Black Population
0–10
10–35
35–65
65–90
90–100
Potomac River
Washington, DC, 1990
Census Data and Boundaries
Established Negro Areas
Succession
Invasion
Growing
Declining
Displacement
Stable
Central Business District
Potomac River
Washington, DC: Census Tracts With 250 or More Nonwhites in 1960, Classified by Type of Racial Change, 1950–1960

Source: Based on data from Massey & Denton, 1993

The Persistence of Urban Residential Segregation

From the standpoint of U.S. race relations, Washington is of primary importance because of its world stature and visibility. Unlike other heavily segregated industrial cities, such as Detroit, Cleveland, Gary, or Birmingham, Washington never attracted a large African American working-class population. Its African American community is quite old, extending back through the period of slavery. The map of neighborhood racial change from demographers Alma and Conrad Taueber's classic book *Negroes in Cities* (1965) shows that between 1950 and 1960 there were areas along the Potomac River, such as the well-known community of Georgetown, where the African American population was being displaced by whites. The census tract analysis shows that in the same period there were large areas, especially in the eastern portion of the city, where blacks were replacing whites and racial segregation was increasing.

The African American population in Washington grew rapidly during the middle decades of the twentieth century, when the combination of federal antidiscrimination policies and the expansion of government agencies made it possible for African Americans to compete with whites for jobs in the federal government and in companies that serve the government's needs. In consequence, many of Washington's black neighborhoods are economically secure, but they are nonetheless segregated. The city also has some quite poor African American neighborhoods where high levels of addiction, violence, and homelessness are a persistent problem.

Other neighborhoods are undergoing a process of renewal and racial integration. Young white families and business owners, along with members of new immigrant groups, are mingling with their black neighbors. If these changes accelerate, coming census surveys may find that segregation has diminished. But as the 1990 census tract map shows, the separation of the races in residential neighborhoods is still a dominant social fact in the nation's capital.

an English-speaking society should be aggressively discouraged. The catchword of this ideology is *Americanization,* the idea that immigrants and their children must become "100 percent American" by losing all traces of their "foreign" accents, abandoning their ethnic cultures, and marrying nonethnic Americans. This demand is reflected in the movement to make English the official language of the United States, along with opposition to bilingual education. In some cases discrimination against members of certain ethnic groups is rationalized by the statement that they are not yet fully American.

The Melting Pot In an 1893 paper, a young historian named Frederick Jackson Turner challenged the notion that American culture and institutions had been formed by the nation's original Anglo-Saxon settlers. Turner's essay argued that the major influences on American culture were the experiences of the diverse array of people who met and mixed on the western frontier. Turner held that "in the crucible of the frontier the immigrants were Americanized, liberated, and fused into a mixed race" (1920/1893, pp. 22–23).

Turner's "crucible" became widely known as the "melting pot," a term taken from a play by Israel Zangwell about a Russian immigrant. David, the play's hero, makes the following declaration:

> America is God's crucible, the great Melting Pot where all the races of Europe are melting and re-forming! Here you stand, good folk . . . in your fifty groups, with your fifty languages and histories, and your fifty blood hatreds and rivalries. . . . Germans and Frenchmen, Irishmen and Englishmen, Jews and Russians—into the Crucible with you all! God is making the American. (1909, p. 37)

The melting-pot view of assimilation attracted many scholars, artists, and social commentators, but sociologists were not convinced that it was an accurate view of what actually happens to ethnic groups in American society. Their research showed that certain ethnic groups have not been fully assimilated, either through intermarriage or through integration into the nation's major institutions. Instead, there remain distinct patterns of **ethnic stratification;** that is, different groups are valued differently depending on how closely they conform to Anglo-Saxon standards of appearance, behavior, and values. People of Scandinavian or northern European descent, for example, are more readily accepted into the

The young black woman shown in this 1888 engraving from *Harper's Weekly* titled "Their Pride" is about to go away to a teacher-training college. All newcomers to urban America, be they blacks from the rural South or immigrants from European countries, have recognized that education is the best route to upward social mobility.

top levels of corporate management than people of Mediterranean descent, who in turn are more readily accepted than black, Hispanic, and Asian minorities.

In an early empirical study of assimilation, Ruby Jo Reeves Kennedy (1944) analyzed rates of ethnic intermarriage in New Haven, Connecticut. She found that intergroup marriages increased in frequency between 1870 and 1940 but that while there was a growing tendency for people to marry outside their *ethnic* group, there remained a very strong tendency for people not to marry outside their own *religious* group. From this finding, Kennedy developed the hypothesis of the "triple melting pot," the idea that assimilation occurs first among groups that share the same religion and later among groups with different religions. More recently, in *Beyond the Melting Pot* (1970), Nathan Glazer and Daniel Patrick Moynihan also concluded that ethnic and racial assimilation is far from inevitable. Ethnicity does not disappear as a result of assimilation; ethnic subcultures are continually being created and changed. Thus *Beyond the Melting Pot* pointed to the emergence and significance of cultural pluralism in American life.

Cultural Pluralism As new waves of immigrants from all over the world have streamed into the United States—Jews fleeing religious intolerance in Russia, Italians and Poles fleeing economic depression in their communities, Central and Latin Americans fleeing dictatorships and poverty, refugees from Asian countries torn by conflict—all have tended to join other members of their nationality group who settled in the United States in earlier decades. Although the influx of new immigrants has led to conflicts between older and newer residents of some communities, in many cases it has resulted in new growth in those communities—in new ethnic businesses such as restaurants and grocery stores, and new social institutions such as churches and social clubs. It has also reinforced pride in ethnic identity as expressed in language and other aspects of ethnic subcultures. This infusion of new energy into ethnic communities confirms Glazer and Moynihan's thesis that the emergence of cultural pluralism is a significant aspect of life in American society.

The recognition that ethnic groups maintain their own communities and subcultures even while some of their members are assimilated into the larger society gave support to the concept of cultural pluralism. A **pluralistic society** is one in which different ethnic and racial groups are able to maintain their own cultures and lifestyles even as they gain equality in the institutions of the larger society. Michael Waltzer's (1980) comparative research on pluralism in the United States and other societies has shown that although white ethnic groups like the Italians, Jews, and Scandinavians may have the option of maintaining their own subcultures and still be accepted in the larger society, blacks and other racial minorities frequently experience attacks on their subcultures (e.g., opposition to bilingual education or African studies courses) and, at the same time, are discriminated against in social institutions. Waltzer concludes that "racism is the great barrier to a fully developed pluralism" (p. 787). In his study of Latinos in the United States, Earl Shorris (1992) essentially agrees with Waltzer. He finds that despite intense pressure to assimilate, many people from Spanish-speaking nations continue to speak their native language and are using their growing influence in some parts of the country to promote bilingualism. Shorris also notes, however, that Latinos are embracing their own version of the American dream and are not advancing separatist claims.

The problems of pluralism are illustrated by the case of the French-speaking Quebecois minority in Canada. The Quebecois account for almost 28 percent of the Canadian population and comprise the majority in the province of Quebec. This leads to a situation in which a group with its own culture and language seeks protection against pressure to assimilate into the dominant culture, and yet demands equal access to the society's political, economic, and cultural institutions. The resulting tensions and hardships have led to the demand that Canada become a bilingual nation, with English and French given equal status. Although the Canadian government has taken firm steps to ensure that French is Quebec's first language, the memory of past discrimination has given rise to a social movement calling for the creation of an independent French-speaking nation. This movement is especially strong in small towns and rural areas whose entire population is French-speaking. This population strongly backed the separation referendum held in Quebec in 1995. However, the referendum was defeated by an extremely narrow margin, largely owing to voters in the more multicultural districts of Montreal, Quebec's largest and most diverse city.

As this example shows, a truly pluralistic society is very difficult to achieve. The various ethnic and racial subgroups within a society often feel a strong sense of cultural identity, which they wish to preserve. They also demand equal access to the society's institutions: access to better schools, opportunities to obtain jobs in every field, opportunities to hold important positions—in short, a fair share of the wealth and power available in the society. These desires sometimes conflict, with the result that some groups may be tempted to "go it alone"—that is, form their own cultural and political institutions (their own businesses, newspapers and other media, labor unions, etc.)—or else give up their ethnic identity in order to gain greater access to the society's major institutions.

CULTURE AND INTERGROUP RELATIONS

Why do ethnic stratification and inequality persist in societies that are becoming increasingly pluralistic, like those of the United States and Canada? Sociologists have proposed a variety of theories to explain racial and ethnic inequality. Before we explore them, however, we need to understand the cultural basis of ethnic diversity and intergroup hostility—that is, the underlying values and attitudes that shape people's consciousness of other groups and, hence, their behavior toward members of those groups. Chief among these are the tendency to view members of other groups in terms of stereotypes and to use those stereotypes to justify differential attitudes (prejudice) and behaviors (discrimination) toward such individuals.

Stereotypes

People often express the opinion that specific traits of members of certain groups are responsible for their disadvantaged situation. Thus in South Africa it was common for whites to assert that blacks were not ready for full citizenship because "they remain childlike and simple." In the United States, the fact that Hispanics are more likely to be found in low-paying jobs is explained by the assertion that "they don't want to learn English." And the fact that black unemployment rates are generally twice as high as white unemployment rates is explained by the statement that "they don't want to work; they like sports and music, but not hard work, especially in school." These explanations are **stereotypes,** inflexible images of a racial or cultural group that are held without regard to whether they are true.

Sociologist William Helmreich (1982) conducted a study of widely held stereotypes regarding America's major ethnic and racial groups. He found that "every single stereotype discussed turns out to have a reason, or reasons" (p. 242). Those reasons usually stem from earlier patterns of intergroup relations. For example, jokes about stupid Poles stem from a period in the nineteenth century when uneducated Polish peasants immigrated to the United States. The idea that blacks are good at music or sports also has some basis, not because blacks are naturally superior in those areas but because when blacks were barred from other avenues to upward mobility they were able to succeed in entertainment and sports; as a result, many young blacks have developed their musical and athletic talents more fully than whites have. But even though stereotypes usually have some basis, they never take account of all the facts about a group. As the famous social commentator Walter Lippmann once quipped, "All Indians walk in single file, at least the one I saw did."

Prejudice and Discrimination

The fact that many people hold stereotypical ideas about other groups may be an indication that they are ignorant or prejudiced, but it does not imply that they will actually discriminate against people whom they perceive as different. In a classic study of prejudice, social psychologist Richard LaPiere (1934) traveled throughout the United States with a Chinese couple, stopping at about 250 restaurants and hotels. Only one of the establishments refused them service. Six months later LaPiere wrote to each establishment and requested reservations for a Chinese couple. More than 90 percent of the managers responded that they had a policy of "nonacceptance of Orientals." This field experiment was replicated for blacks in 1952, with very similar results (Kutner, Wilkins, & Yarrow, 1952; Shibutani & Kwan, 1965).

The purpose of such experiments is to demonstrate the difference between prejudice and discrimination. **Prejudice** is an attitude that prejudges a person, either positively or negatively, on the basis of real or imagined characteristics (stereotypes) of a group to which that person belongs. **Discrimination,** on the other hand, refers to actual unfair treatment of people on the basis of their group membership.

The distinction between attitude and behavior is important. Prejudice is an attitude; discrimination is a behavior. As Robert Merton (1948) pointed out, there are people who are prejudiced and who discriminate against members of particular groups. There are also people who are not prejudiced but who discriminate because it is expected of them. With these distinctions in mind, Merton constructed the typology shown in Figure 13.3.

Merton's typology is valuable because it points to the variety of attitudes and behaviors that exist in multicultural and multiracial societies. However, it fails to account for situations in which certain groups are discriminated against regardless of the attitudes and behaviors of individuals. This form of discrimination is part of the "culture" of a social institution; it is practiced by people who are simply conforming to the norms of that institution; hence, it is known as institutional discrimination.

At its simplest, **institutional discrimination** is the systematic exclusion of people from equal access to and participation in a particular institution because of their race, religion, or ethnicity. But over time this intentional exclusion leads to another type of discrimination,

		PREJUDICE: Yes	PREJUDICE: No
DISCRIMINATION	Yes	TRUE BIGOT: does not believe in the American creed and acts accordingly	WEAK LIBERAL: not prejudiced, yet afraid to go against the bigoted crowd
DISCRIMINATION	No	CAUTIOUS BIGOT: does not believe in the American creed but is afraid to discriminate	STRONG LIBERAL: not prejudiced and refuses to discriminate

Source: Adapted from pp. 99–126 of *Discrimination and National Welfare,* edited by R. M. MacIver. Copyright © 1948 by the Institute for Religious and Social Studies. Reprinted by permission of Harper & Row Publishers, Inc.

FIGURE 13.3 **Merton's Typology of Prejudice and Discrimination**

which has been described as "the interaction of the various spheres of social life to maintain an overall pattern of oppression" (Blauner, 1972, p. 185). This form of institutional discrimination can be quite complex.

The conditions that led to the riot that broke out in South Central Los Angeles in 1992 after the acquittal of the police officers who beat a black motorist, Rodney King, conform very well to what sociologists have observed in similar communities where blacks or other minority groups are trapped in a self-perpetuating set of circumstances due largely to historical patterns of discrimination (Blauner, 1989; Tuch & Martin, 1997). Blocked educational opportunities result in low skill levels, which together with job discrimination limit the incomes of minority group members. Low income and residential discrimination force them to become concentrated in ghettos. The ghettos never receive adequate public services such as transportation, thus making the search for work even more difficult. In those neighborhoods, also, the schools do not stimulate achievement, thereby repeating the pattern in the next generation. The young often grow up bitter and angry and may form violent gangs that engage in vandalism and other activities that the outside world sees as antisocial. At the same time, the police patrol ghettos to the point of harassment, with the result that young blacks are more likely to be arrested—and to be denied jobs because of their arrest records. Recent research shows that dark-skinned African American males with a criminal record have a jobless rate of 54 percent, compared with 41.7 percent for light-skinned African American males and 25 percent for white males with similar records (Johnson & Farrell, 1995). When these conditions are combined with a precipitating event such as the savage beating of a black person and the acquittal of the police officers responsible for the beating, the results often take the form of violence like that which erupted in South Central Los Angeles.

All of the institutions involved—employers, local governments, schools, real estate agencies, agencies of social control—may claim that they apply consistent standards in making their decisions: They hire the most qualified applicants; they sell to the highest bidder; they apply the law evenhandedly; in short, they do not discriminate. Yet in adhering to its institutional norms each perpetuates a situation that was created by past discrimination.

Ethnic and Racial Nationalism

The conditions that contributed to the Los Angeles riot are an example of the pervasive and discouraging effects of institutional discrimination. In the face of such discrimination, racial and ethnic groups frequently organize their members into movements to oppose social inequality. Those social movements often appeal to ethnic or racial nationalism—that is, to the "we feeling" or sense of peoplehood shared by the members of a particular group.

Ethnic (or **racial**) **nationalism** is the belief that one's ethnic group constitutes a distinct people whose culture is and should be separate from that of the larger society. Feelings of nationalism among America's ethnic and racial groups have often been strongly affected by nationalist movements outside the United States. American Jews, for example, were deeply influenced by Zionism, a movement that arose in the late nineteenth century with the goal of creating a Jewish homeland in the holy land of Palestine (now Israel). Similarly, Irish Americans have been influenced by the struggle of the Catholic minority in Northern Ireland to gain independence from England. Often nationalist movements produce a new or renewed ideology of pluralism that replaces the goal of

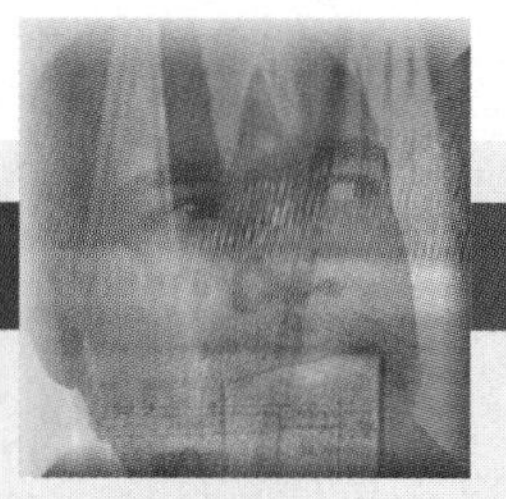

THE PERSONAL CHALLENGE

Social Change and Racism

Detective Mark Fuhrman's admitted racism in the trial of O. J. Simpson, the arrest of three white men in Texas who chained a black man to the back of their pickup truck and dragged him to his death, "ethnic cleansing" in Kosovo, religious and ethnic strife in India . . . where will it strike next? Will it ever end? Perhaps more to the point, what can we as individuals do in the face of the seemingly overwhelming racial and ethnic divisions in our society and the world? One answer is that we can try not to tolerate racism and bigotry and not fail to speak out when confronted by these forms of ignorance and hatred. This goes as much for the victims of these behaviors as it does for those who witness them.

An example of a person who refused to accept racism is Earl G. Graves, Jr., a senior executive at *Black Enterprise* magazine. Graves was mistakenly detained and frisked by detectives on a railroad platform in New York City's Grand Central Terminal. The detectives were seeking an alleged perpetrator who was reported to be African American, 5 feet 10 inches tall, with a moustache, and carrying a concealed weapon. Graves is an African American but is 6 feet 4 inches tall and has no moustache. Mortified at having been victimized in front of fellow commuters with whom he rides every day, Graves protested to railroad officials, who then published the accompanying apology in local newspapers.

Metro-North wants its riders and Mr. Graves's neighbors and fellow commuters to know that Mr. Graves was not the subject its detectives were looking for and that Metro-North detectives were wrong in making the decision to detain and pat down Mr. Graves.

Metro-North deeply regrets this most unfortunate and insensitive incident, and seeks by this notice to apologize publicly to Mr. Graves and his family.

Metro-North wants to assure its ridership that it is committed to taking the steps necessary to reduce the risk of future such incidents and to assure that all of its riders are treated with respect and sensitivity, without regard to their race, color, or national origin.

Donald Nelson

Donald N. Nelson
President

METRO-NORTH COMMUTER RAILROAD COMPANY

What can we as individuals do when we witness hostility directed against people just because they are different? We may find it difficult to confront racism or ethnic prejudice in people who are our peers and with whom we must interact on a daily basis. Often we look away or pretend to "go along," hiding our feelings of shame and confusion. Another strategy is to confront the problem head-on, as Graves did. It may take courage, but we can tell someone that their behavior is racist or prejudiced. Of course, this could result in the loss of a friend, but perhaps that is not the end of the world. We can befriend others who are not hampered by racism or ethnic bigotry.

It can also help to try to understand the origins of prejudice in people we know and love. Try to see how their attitudes have been affected by social changes such as new forms of competition with recently arrived ethnic or racial groups, or appeals to fear and prejudice by political leaders who stand to benefit from divisions among working people. You won't get very far by giving sociological lectures to those close to you, but by applying your sociological imagination you may come up with some strategies for changing their minds—or at least giving them an opportunity to see things a bit differently.

complete assimilation into the larger society. We can see this very clearly among African Americans.

In his famous study *An American Dilemma* (1944), the Swedish social scientist Gunnar Myrdal noted that African Americans were in fact "exaggerated Americans." Myrdal believed that blacks had assimilated American values and norms more than any other ethnic or racial group and that they had no distinctive subculture of their own. This theme was repeated in the influential work of E. Franklin Frazier (1957), the leading black sociologist of his generation. Frazier, like Myrdal, believed that African Americans had no distinctive culture. They had the same religions, language, and values as white Americans, and the folkways they had developed (e.g., musical

Racial violence and rioting invariably disrupt neighborhoods, often for long periods. During the 1919 Chicago race riots, many African Americans were evicted from homes in integrated working-class neighborhoods and forced back into the ghetto. During the 1992 Los Angeles riots, so many stores and businesses were burned that people like this mother and child felt severe disorientation long after the rioting had ended.

forms like jazz and the blues) had become part of mainstream American culture.

Events of the late 1950s and the 1960s drastically changed these views of black cultural assimilation. Along with the civil rights movement of the 1960s came a wave of nationalism. Black intellectuals and community leaders began calling not merely for integration but for cultural pluralism. Leaders like Stokely Carmichael, of the Student Nonviolent Coordinating Committee (SNCC), and Malcolm X, the charismatic Black Muslim leader, preached that "black is beautiful" and that African Americans should drop the term *Negro* in favor of *black.* In Carmichael's words, "They oppress us because we are black and we are going to use that blackness. . . . Don't be ashamed of your color" (quoted in Bracey, Meier, & Rudwick, 1970, p. 471). In recent years many black Americans have gone further and adopted the term *African American* in order to emphasize their common heritage and to deemphasize the racial character of their shared background.

People in the United States are not alone in struggling with issues of nationalism. Now that the cold war between the former Soviet Union and the United States has ended, ethnic and religious nationalism is the primary threat to peace throughout the world. Although North America is relatively free of the extreme nationalism that results in bloodshed, at various times in North American history ethnic or racial nationalism has strained the social fabric. Entire populations, such as African Americans, Puerto Ricans, or French Canadians, have yearned to become entirely separate from the larger culture and society.

At other times nationalism has been a creative social force, welding a racial or ethnic group into a more politically conscious community able to struggle more effectively for its rights. As the Native American sociologist Russell Thornton notes, there are some highly positive aspects of nationalism or, in the case of Native Americans, tribalism:

> Research shows that American Indians make considerable efforts to reaffirm tribalism in urban areas by living in Indian neighborhoods, by maintaining contacts with reservation areas and extended families, and by creating urban American Indian Centers. Frequent results are a new tribalism for urban American Indians and,

> sometimes, "bicultural" individuals, that is, American Indians who live successfully in Indian and non-Indian worlds. (1987, p. 239)

A key to Indian survival, Thornton concludes, is the ability to maintain an Indian identity while interacting with non-Indians in an urban environment. The same point is often made by Hispanic and African American sociologists. One difficulty in maintaining a bicultural existence is that it requires assertion of the minority group's values, its ways of expression, its history, and its demands for greater equality. Frequently the demand for greater equality is expressed in terms of the need for affirmative action.

Affirmative Action

What separates relatively peaceful multiethnic and multiracial nations from those in which racial, ethnic, or religious hatreds result in war and massacres? One important answer is the rule of law. When minority populations believe that their rights can be protected by government and laws, when they feel that their past grievances will be corrected in time, they will lend their support to the government and laws of their nation. However, in too many countries—even in the United States at times—minorities do not hold these beliefs. It is for this reason that the quest for racial and ethnic equality is a persistent issue in national affairs (Hacker, 1995).

Our society's foremost governmental and economic institutions are besieged with demands for *affirmative action*—that is, policies designed to correct persistent racial and ethnic inequalities in promotion, hiring, and access to other opportunities. Even more forceful in many parts of the country are the demands to do away with affirmative action policies. Conservatives bitterly oppose affirmative action (Glazer, 1975; Sowell, 1972), whereas liberals feel that such policies are necessary if our society is to undo the effects of past discrimination (Ezorsky, 1991; Horowitz, 1979).

At present the conservative view is in the ascendance. Congress has done away with preferential treatment for minority-owned firms in awarding small government contracts. In California steps have been taken to eliminate affirmative action policies in the state's universities and colleges. It is likely that efforts to eliminate affirmative action will continue and will be increasingly successful. But it is also true, as Martin Kilson points out, that affirmative action has been essential to the advancement of African Americans and, increasingly, Latinos in job markets that were formerly out of bounds to minorities. In 1970, for example, "barely 2 percent of the officers in the armed forces were black; twenty years later, thanks (partly) to affirmative action, 12 percent of officers were African Americans" (Kilson, 1995, p. 470).

We saw earlier that the Supreme Court struck down the doctrine that "separate but equal" facilities and institutions did not violate the constitutional rights of African Americans. It did so largely because social-scientific evidence showed that separate institutions are inherently unequal. Similar arguments have been advanced in affirmative action cases that have reached the Court. If a fire department in a city whose inhabitants are 30 percent black and Hispanic has no firefighters from those minority groups, it can be demonstrated that there is a pattern of discrimination that can be changed only if the department is required to hire a certain number of minority applicants—a quota—within a designated time. However, members of the majority may then feel that they are victims of "reverse discrimination," in which they are being penalized for the wrongs of earlier generations. Thus, difficult choices remain: Should employers mix or replace hiring decisions based on experience and merit with decisions based on race and ethnicity? In essence, the courts have said that they must.

Affirmative action remains an extremely controversial issue in American life. Table 13.2 shows that blacks and whites hold widely divergent views on government

Abby Abinanti, a Yurok Indian, has worked as a lawyer in private practice and public interest law. "In 1970 when I was graduating from college," she comments, "the Bureau of Indian Affairs announced a scholarship program. At that time nationally, only a handful of Indian people had become lawyers, and many of them were not 'Indian' identified in their work. The importance of lawyers to my community was emphasized to me by the older women. They understood our increasing need to look to the courts to protect our rights against encroachment . . . and to enforce promises given at an earlier time."

TABLE 13.2 Blacks' and Whites' Opinions on Racial Policy Issues (in percent)*

Should federal spending programs that assist blacks be increased, decreased, or kept about the same?

	Blacks	Whites
Increased	69.5	18.4
Kept the same	28.8	56.5
Decreased	1.7	25.2

Some people say that, because of past discrimination, blacks should be given preference in hiring and promotion. Others say that such preference in hiring and promotion of blacks is wrong because it gives blacks advantages they haven't earned. What about your opinion—are you for or against preferential hiring for blacks?

	Blacks	Whites
Strongly favor	53.4	6.2
Favor	10.1	7.9
Oppose	4.9	18.3
Strongly oppose	21.6	67.6

*N = 5,012

Source: Tuch, Sigleman, & Martin, 1997.

efforts to narrow the gaps in achievement by race. In fact, both whites and blacks tend to favor race-targeted policies that offer compensatory programs, such as job training and education, but the majority of whites and a significant minority of blacks oppose employment and educational programs that require preferential treatment in hiring and college admissions decisions (Tuch, Sigleman, & Martin, 1997). The divisions revealed by these attitude data confirm that Americans remain quite divided along racial lines over how much government should be doing to correct the cumulative impact of slavery and racial discrimination.

It should be noted that affirmative action applies to women as well as to racial and ethnic minorities (Epstein, 1995). The effects of institutional discrimination against women are explored in the next chapter.

THEORIES OF RACIAL AND ETHNIC INEQUALITY

For large numbers of people a dominant aspect of life in American society is racial or ethnic inequality and, often, hostility. How can we explain the persistence of these patterns of hostility and inequality? The first thought that comes to mind is that many people are prejudiced against anyone who is different from them in appearance or behavior. This may seem to explain phenomena like segregation and discrimination, but it fails to explain the variety of possible reactions to different groups. Social-psychological theories that focus on prejudice against members of out-groups find the origins of racism and ethnic inequality in individual psychological processes, but there are also more sociological theories that view prejudice as a symptom of other aspects of intergroup relations.

Social-Psychological Theories

The best-known social-psychological theories of ethnic and racial inequality are the frustration-aggression, projection, and authoritarian personality theories. All of these see the origins of prejudice in individual psychological orientations toward members of out-groups, but they differ in important ways.

Frustration-Aggression The frustration-aggression hypothesis, which is associated with the research of John Dollard, Neil Miller, and Leonard Doob (1939), holds that the origin of prejudice is a buildup of frustration. When that frustration cannot be vented on the real cause, the individual feels a "free-floating" hostility that may be taken out on a convenient target, or **scapegoat.** For example, in the case of workers in eastern Germany who have lost jobs in antiquated factories as a result of unification with western Germany, with its far more modern industries, a convenient scapegoat may be found in gypsies, Turks, or even Jews—all groups that have historically been accused of causing negative conditions for native-born Germans. To justify the hostility directed at the out-group, the prejudiced individual often grasps at additional reasons (usually stereotypes) for hating the "others," such as "Gypsies steal children"; "Jews are usurers"; or "Turks will work for almost nothing."

Projection The concept of projection is also used to explain hostility toward particular ethnic and racial groups. **Projection** is the process whereby we attribute to other people behaviors and feelings that we are unwilling to accept in ourselves. John Dollard (1937) and Margaret Halsey (1946) applied the concept of projection to white attitudes toward black sexuality. Observers had noted that southern whites frequently claimed that black males are characterized by an uncontrollable and even vicious sexuality. The theory of projection explains this claim as resulting from the white males' attraction to black females, an attraction that was forbidden by strong norms against interracial sexual contact. Thus the white male projected his own forbidden sexuality onto blacks and developed an attitude that excused his own sexual involvement with black women.

The Authoritarian Personality The frustration-aggression and projection explanations of prejudice are quite general in that anyone can develop pent-up frustrations that will engender hostility under certain conditions, and anyone can project undesirable traits onto others. A more specific explanation of prejudice is the theory of the authoritarian personality, which emerged from attempts to discover whether there is a particular type of person who is likely to display prejudice.

In 1950 a group of social scientists led by Theodor Adorno published an influential study titled *The Authoritarian Personality.* Their research had found consistent correlations between prejudice against Jews and other minorities and a set of traits that characterized what they called the authoritarian personality. Authoritarian individuals, they found, were punished frequently as children. Consequently, such individuals feel an intense anger that they fail to examine (Pettigrew, 1980). They submit completely to people in positions of authority, greatly fear self-analysis or introspection, and have a strong tendency to blame their troubles on people or groups whom they see as inferior to themselves. Unfortunately for Jews, blacks, and other minorities that have been subordinated in the past, the anger and hostility of the authoritarian personality is often directed against them.

Interactionist Theories

Not far removed from these social-psychological theories of intergroup hostility are theories derived from the interactionist perspective in sociology. But instead of locating the origins of intergroup conflict in individual psychological tendencies, interactionists tend to look at how hostility or sympathy toward other groups, or solidarity within a group, is produced through the norms of interaction and the definitions of the situation that evolve within and between groups. A few examples will serve to illustrate how the interactionist perspective is applied.

From his analyses of interaction in different groups, Georg Simmel concluded that groups often find it convenient to think of nonmembers or outsiders as somehow inferior to members of the group. But why does this familiar in-group–out-group distinction develop? Simmel explained it as arising out of the intensity of interactions within the group, which leads its members to feel that other groups are less important. Once they have identified another group as inferior, it is not a great leap to think of its members as enemies, especially because doing so increases their sense of solidarity (Coser, 1966; Simmel, 1904).

Conflict and hostility between racial or ethnic groups can be overcome by creating situations that require the groups to cooperate to achieve a common goal. Such situations occasionally occur—for example, when people from different ethnic and racial backgrounds compete in sports or when they work side by side in school and in industry. Unfortunately, the integration and friendship found in such settings are not sufficient to overcome the more deep-seated prejudices and fears of many racist individuals. As film director Spike Lee observes, racists place all black people in one of two categories: "entertainers and niggers." By this he means that people with racist sentiments tend to exclude famous entertainers and sports figures from their negative feelings about blacks as a group, thereby accentuating their failure to see all blacks as individual human beings.

Functionalist Theories

The difficulty of reducing racism and discrimination through interaction leads to the question of whether racial and ethnic inequalities persist because they function to the advantage of certain groups. One answer was provided by the functionalist theorist Talcott Parsons (1968), who wrote that "the primary historic origin of the modern color problem lies in the relation of Europeans to African slavery" (p. 366). Parsons was not denying that racism is produced through interactions. Rather, he was pointing out that the specific form taken by those interactions—oppression, subordination, domination of blacks by whites—is directly related to the perceived need of white colonialists and traders to use blacks for their own purposes. The whites could abduct, enslave, and sell Africans because their societies had developed technologies (e.g., oceangoing ships and navigational instruments) and institutions (e.g., markets and trading corporations) that made them immensely more powerful than the Africans.

From the functionalist perspective, inequalities among ethnic or racial groups exist because they have served important functions for particular societies. Thus in South Africa it was functional for the white government to insist on maintaining apartheid because to do otherwise would mean that whites would become a minority group in a black-dominated society. But as world opinion continued to condemn the white regime and blacks continued to build group solidarity and challenge that regime, it became less and less "functional" for the white government to insist on complete apartheid. Indeed, some have speculated that the nation's white rulers released Nelson Mandela and took steps to legalize the formerly banned African National Congress as a way of showing the world that they were willing to begin negotiations toward some form of shared power. Their move incurred the wrath of many South African whites, who feared the consequences of majority black rule under a new constitution. Similarly, the Israeli government's negotiations with the Palestine Liberation Organization enraged many Jewish settlers living in areas that would be placed under the control of Palestinians. This rage resulted in the shocking assassination of Israeli prime minister Yitzhak Rabin in November 1995 and continues to dominate politics in the Middle East.

In both Israel and South Africa it appears that the status quo of intergroup relations had become dysfunctional. Change had to come either through peaceful negotiations or through bloodshed. The processes whereby power is redistributed and new social relations develop are often best understood with the help of conflict theories.

Conflict Theories

Conflict theories do a better job than functionalist theories of explaining why groups with less power and privilege, like the South African blacks or the Palestinians in the Middle East, do not accept their place in the status quo. Conflict theories try to explain why these groups often mobilize to change existing intergroup relations (Blauner, 1989). Conflict changes societies, as we see clearly in the case of South Africa, where many years of conflict waged by the disenfranchised blacks led to the creation of a new constitution and majority black rule. But conflict can also destroy societies and nations, as we see all too clearly in the war-torn nations of the former Yugoslavia, where a long history of ethnic conflict has prevented the achievement of a stable, well-functioning society.

Conflict theories trace the origins of racial and ethnic inequality to the conflict between classes in capitalist societies. Marx believed, for example, that American wage earners were unlikely to become highly class conscious because ethnic and racial divisions continually set them against one another and the resulting strife could be manipulated by the capitalist class. Thus in American history we see many examples of black and Mexican workers being brought in as strikebreakers by the owners of mines and mills, especially during the 1920s and 1930s. Strikebreakers from different racial and ethnic groups often absorbed the wrath of workers, anger that might otherwise have been directed at the dominant class. To forge class loyalties despite the divisions created by racial and ethnic differences, Marx believed, it would be necessary for workers to see that they were being manipulated by such strategies.

i

Internal Colonialism Conflict perspectives on racial inequality also include the theory of **internal colonialism.** According to this theory, many minority groups, especially racial minorities, are essentially colonial peoples within the larger society. Four conditions mark this situation:

1. The "colonial" people did not enter the society voluntarily.
2. The culture of the "colonial" people has been destroyed or transformed into a version of the dominant culture that is considered inferior.
3. The "colonial" population is controlled by the dominant population.
4. Members of the "colonial" people are victims of racism; that is, they are seen as inferior in biological terms and are oppressed both socially and psychologically (Blauner, 1969; Davidson, 1973; Hechter, 1974). This is why African Americans, Native Americans, and Jewish Americans react with such anger to suggestions that they may be different from other populations on measures of intelligence or any other biological trait. They have seen such notions used to rationalize slavery or genocide and hence are not willing to let them pass as harmless speculation or "value-free" science.

Although these characteristics describe colonial peoples everywhere, the theory of internal colonialism asserts that they also apply to subordinated ethnic and racial groups in societies like England, the United States, Canada, and the nations of the former Soviet Union. Michael Hechter (1974) extended the theory to show that societies that have created colonial or "ghettoized" populations within their boundaries also develop a "cultural division of labor" in which the subordinated group is expected to perform types of work that are considered too demeaning to be done by members of the dominant population. The South African institution of *Baaskop* was (and in some parts of the nation still is) an example of this phenomenon. It is a set of norms specifying that lower-status, physically exhausting work is appropriate for blacks and higher-status work is appropriate for whites, and that whites should never accept "black" work nor allow themselves to be subordinate to blacks.

Ecological Theories

Are the segregated ghetto communities of black and Hispanic Americans a product of internal colonialism? The answer to this question depends on whether the residents of those communities are able to achieve upward mobility. According to ecological theories of intergroup relations, such mobility should occur naturally in the course of a group's adaptation to the culture and institutions of the larger society. The eventual outcome should

As a result of their unsuccessful resistance to the European invasion, Native Americans were segregated on reservations, many of which lacked adequate resources to permit them to share in the American Dream.

be the existence of racially and ethnically integrated communities.

Ecological theories explore the processes by which conflict between racial or ethnic groups develops and is resolved. Along these lines, Robert Park (1914) devised a cyclical model to describe intergroup relations in modern cities. That model consisted of the following stages:

1. *Invasion:* One or more distinct groups begin to move into the territory of an established population.
2. *Resistance:* The established group attempts to defend its territory and institutions against newcomers.
3. *Competition:* Unless the newcomers are driven out, the two populations begin to compete for space and for access to social institutions (housing, jobs, schooling, recreational facilities, etc.); this extends to competition for prestige in the community and power in local governmental institutions.
4. *Accommodation and Cooperation:* Eventually the two groups develop relatively stable patterns of interaction. For example, they arrive at understandings about segregated and shared territories (Suttles, 1967).
5. *Assimilation:* As accommodation and cooperation replace competition and conflict, the groups gradually merge, first in secondary groups and later through cultural assimilation and intermarriage. They become one people. A new group arrives, and the cycle begins again.

Of course, this is an abstract model. It represents what Park and other human ecologists believe are the likely stages in intergroup relations. The Korean American sociologist Ilsoo Kim (1981) found that Korean merchants in New York City met resistance both from white merchants and from black residents of ghetto communities when they purchased stores in those communities, but eventually they reached an accommodation with those groups. This finding tends to support Park's model, although the extent to which assimilation will occur remains an open question. On the other hand, critics point out that there is not always a steady progression from one of these stages to the next. Moreover, accommodation, cooperation, and assimilation do not occur in every case. The ecological model fails to explain why and how groups compete for power and under what conditions they eventually come to cooperate. Nevertheless, the model presents a general picture of the stages that culturally distinct groups often (but by no means always) go through over time.

A PIECE OF THE PIE

Up to this point we have explored the patterns of intergroup relations, the cultural basis of those patterns, and some of the theories that have been proposed to explain them. But we have not discussed how groups actually win, or fail to win, a fair share of a society's valued statuses and other rewards—that is, a piece of the social pie. In particular, the persistence of racial inequality in the United States requires more attention. This is a complex problem that is a source of continuing controversy.

How can we explain the fact that so many social problems in the United States today are associated with race? As William Julius Wilson (1984) has stated, "Urban crime, drug addiction, out-of-wedlock births, female-headed families, and welfare dependency have risen dramatically in the last several years and the rates reflect a sharply uneven distribution by race" (p. 75). Wilson and others doubt that racial prejudice and discrimination adequately account for the severity of these problems, for the period since the early 1970s has seen more antidiscrimination efforts than any other period in American history. The answers are to be found, Wilson argues, in how older patterns of racism and discrimination affect the present situation.

In an important study of this issue, sociologist Stanley Lieberson (1980) asked why the European immigrants who arrived in American cities in the late nineteenth and early twentieth centuries have fared so much better, on the whole, than blacks. Sociologist Marta Tienda asks similar questions about people of Latino ancestry (Tienda & Singer, 1995; Tienda & Wilson, 1992). Wilson, Lieberson, and Tienda all agree that the problem of lagging black mobility is a complex one. For African Americans, the situation can be summarized as follows:

1. African Americans have experienced far more prejudice and discrimination than any immigrant group, partly because they are more easily identified by their physical characteristics. As a result of the legacy of slavery, which labeled blacks as inferior, they have been excluded from full participation in American social institutions far longer than any other group.
2. Black families have higher rates of family breakup than white families. The problems of the black family are not part of the legacy of slavery, however. As Herbert Gutman (1976) has shown, slavery did not destroy black families to the extent that earlier scholars believed it did. Nor did the migration of blacks to the North during industrialization. A comprehensive review of the status of black Americans (Jaynes & Williams, 1989) notes that "there was no significant increase in male-absent households even after the massive migration to the urban North." Until the 1960s three quarters of black households with children under age 18 included both husband and wife. "The dramatic change came only later, and in 1986, 49 percent of black families with children under age 18 were headed by women" (Jaynes & Williams, 1989, p. 528). The report reasons that if black two-parent families remained the norm through slavery, the Great Depression, migration, urban disorganization, and ghettoization, then it appears unlikely that there is a single cause for the dramatic decline in two-parent families during the last two decades.
3. Structural changes in the American economy—first the shift from work on farms to work in factories and then the shift away from factory work to high-technology and service occupations—have continually placed blacks at a disadvantage. No sooner had they begun to establish themselves as workers in these economic sectors than they began to suffer the consequences of structural changes in addition to job discrimination.

Recent research by social scientists points to the decrease in manufacturing jobs and the increase in service-sector jobs as contributing to the sharp decline in the fortunes of black males. (Black females also suffer, but they are more readily recruited into clerical employment.) Throughout the second half of the twentieth century, the unemployment rate for black males was twice that for white males. This was as true in the 1990s as it was in the 1970s, but the situation is even worse for black men who do not have high school diplomas. Their unemployment rate is more than three times higher than that for white males (Oliver & Shapiro, 1990). A great deal of research shows that difficulty in securing adequate employment contributes to the problems young couples experience in forming lasting relationships (Danziger & Gottschalk, 1993; Garfinkel & McLanahan, 1986).

It is significant that among Hispanic groups in the United States, Puerto Ricans have experienced northward migration and problems of discrimination that are very similar to those experienced by African Americans in northern cities. Still a dependent territory of the United States, since World War II Puerto Rico has sent hundreds of thousands of working people to cities like New York, Philadelphia, and Chicago, which have been rapidly losing manufacturing jobs. The status of Puerto Ricans in the population is very similar to that of blacks. There is a growing Puerto Rican middle class and an even more rapidly growing number of Puerto Ricans among the poor. This situation is due largely to changes in the types of jobs available to unskilled blacks and Puerto Ricans (Sandefur & Tienda, 1988).

It is worth noting that social mobility is more available to blacks today than it was before the civil rights movement. Today there are millions of middle-class African Americans, and the earnings of black males who have completed college are about 85 to 90 percent of those of whites with comparable education. But although the black middle class is growing, the majority of black workers remain dependent on manual work in industry or lower-level service jobs—types of jobs that

Marta Tienda, a leading scholar of poverty among Latino Americans, chose this research area in part because she herself came from a Mexican American family of modest means.

are in decreasing supply and in which unemployment is common. Another important segment of black workers, male and female, are employed in the public sector. Drastic cuts in public budgets in the 1990s drove down their earnings as well. Chronic unemployment is associated with family breakup, alcohol and drug addiction, and depression. These social problems, in turn, severely hamper the ability of individuals to learn the attitudes and skills they need for entry into available jobs (Oliver & Shapiro, 1990; Wilson, 1987).

Nor is the life of middle-class blacks free from the continuing consequences of racism and prejudice. In interviews with a sample of middle-class black Americans in 16 cities, sociologist Joe R. Feagin found that "they reported hundreds of instances of blatant and subtle bias in restaurants, stores, housing, workplaces, and on the street" (1991, p. A44). And in recent years the number of crimes against minority individuals of all social classes has increased.

RESEARCH ON THE CUTTING EDGE: Migrating to Achieve Greater Equality

Reynolds Farley, one of the nation's most knowledgeable experts on the sociology of population change in the United States, has ransacked the U.S. censuses for data about Americans' efforts to achieve greater equality. His recent research shows that more and more Americans of African American and Latino descent are migrating out of the older industrial states to the more rapidly growing states of the South and the West. The data also show that whites, especially those in their retirement years, are continuing to move away from the Northeast and Midwest, but the flow of African Americans and Latinos is especially notable because it reverses the historic South-to-North migration pattern.

For 18 decades after the American Revolution, Farley explains, the population of the South grew far more slowly than the national average. The South relied on agriculture, whereas the most rapid growth of manufacturing was centered in the Northeast and the Midwest. By 1970 the South's share of the nation's total population had dropped to 30 percent. Now, Farley writes, "The drop has ended. The restructuring of employment and the growth of the service sector mean that many jobs are no longer bound to mines, deep water ports, or rail lines. And three southern metropolises—Houston, Miami, and Washington—are now major ports of entry for immigrants, so the South grows rapidly" (Farley, 1996, p. 281).

Since the era of rapid development just before and during World War II, the West, with California in the lead, has been growing extremely rapidly, but as Table 13.3 shows, the South leads all regions. (The figures in the table show each region's net gain or loss of migrants: the difference between the flow of migrants into the region and the flow out to other regions.) Figures 13.4 and 13.5 show the volume of migrant flows between regions. The differences between inflow and outflow on these and similar maps (for total migrants, migrants with college educations, and migrants over age 65) are indicated in Table 13.3. The right-hand column of the table is a rate of migration per 1,000 people, which allows us to compare these migration streams independently of the region's original population.

The changes shown in the maps and table have enormous consequences for racial and ethnic inequality. With the exception of migrants in their retirement years, most people on the move are seeking to relocate in areas of the nation where they hope to find better employment opportunities. Over the past 15 or 20 years, employers in the South have been most successful in attracting new plants and new businesses from all parts of the world. This success stimulates the migration of

TABLE 13.3 Interregional Exchange of Migrants, 1985–1990

Region	Net Gain or Loss of Interregional Migrants (in thousands)	Net Migrants per 1,000 Residents
Total Migrants		
Northeast	−1,116	−23
Midwest	−847	−15
South	+1,425	+18
West	+538	+11
Migrants Age 65 or Over		
Northeast	−217	−30
Midwest	−118	−15
South	+276	+26
West	+59	+10
Migrants With Complete College Educations		
Northeast	−108	−18
Midwest	−196	−34
South	+133	+17
West	+171	+30
Non-Hispanic Black Migrants		
Northeast	−153	−32
Midwest	−63	−12
South	+181	+13
West	+35	+14
Hispanic Migrants		
Northeast	−94	−31
Midwest	−9	−7
South	+61	+11
West	+42	+5

Source: Adapted from Farley, 1996.

FIGURE 13.4 Interregional Exchange of Hispanic Migrants, 1985–1990 (in thousands)

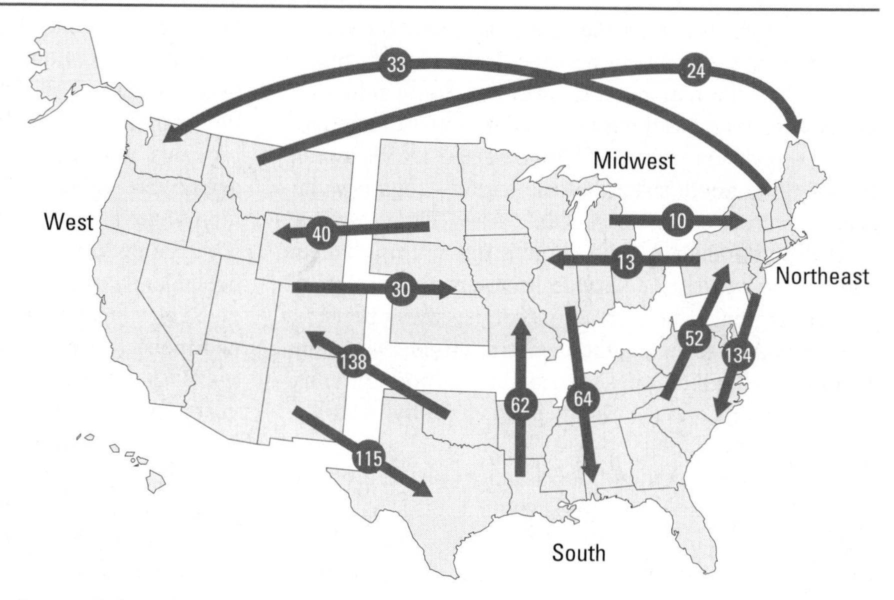

Source: Farley, 1996.

FIGURE 13.5 Interregional Exchange of Non-Hispanic Black Migrants, 1985–1990 (in thousands)

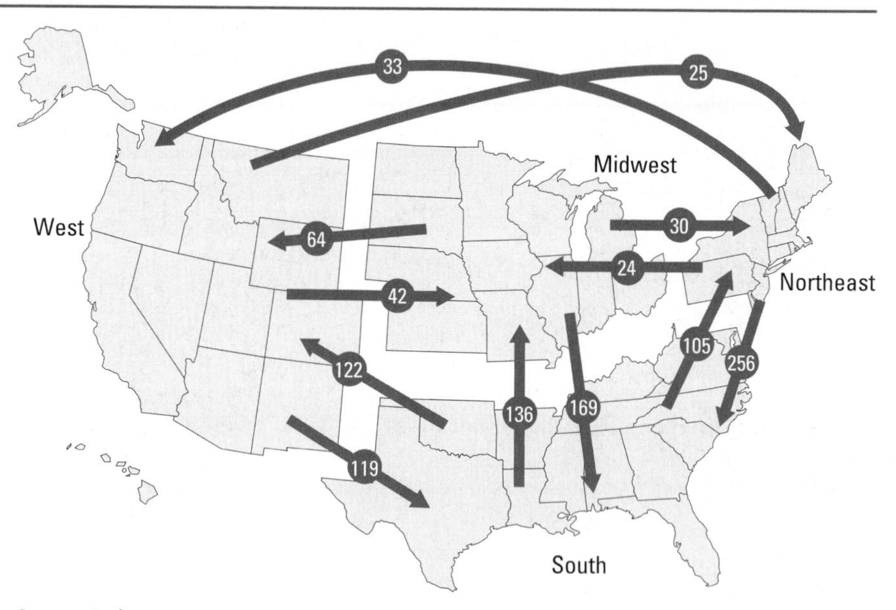

Source: Farley, 1996.

younger wage earners and educated job seekers from other regions. Their movement to growing metropolitan areas of the South, in turn, further stimulates the service, real estate, government, and health care sectors of those areas' economies.

African Americans are moving from the North to the South in record numbers, largely in order to benefit from the economic growth occurring in cities like Atlanta, Birmingham, and Houston, to name only the largest. Latinos are moving to the West in larger numbers but are also making their presence felt in southern states, especially Florida and Texas. The data in Table 13.3 show, however, that the Midwest and the Northeast are "exporting" college-educated migrants, many of whom are Latinos and African Americans, to the South and West. This migration pattern is worsening the situation of less educated, less skilled blacks and Latinos in the older manufacturing regions, where more secure manufacturing jobs and better service sector jobs are scarce. In short, the regional patterns of migration revealed by Farley's analysis are widening the gap between the haves and the have-nots. As more affluent and educated minority group members leave metropolitan areas, it is increasingly difficult for the "sending" areas to come up with the funds to continue investing in education, health care, and other services for their disproportionately poor inner-city populations. At the same time, many of the cities of the Midwest and Northeast, Chicago and New York in particular, continue to receive large inflows of black and Hispanic immigrants, especially from Latin America and Africa, who require education, health services, and employment training. As these immigrants gain experience and skills and improve their socioeconomic status, they or their children will likely join the net outflow of migrants to more rapidly growing metropolitan areas. The states of the Midwest and Northeast that "export" these migrants have a legitimate claim on the national government to compensate investments made in the migrants' "human capital" (education, health care, and so on). But along with losses in population go losses in political power, leaving the "sending" regions at a disadvantage in pressing their claims for help from Congress. On the other hand, the influx of a more racially and ethnically diverse population into the South is having important consequences for southern politics, a subject to which we will return in Chapter 20.

VISUAL SOCIOLOGY

Maya Lin: A Strong, Clear Vision

What measures can a society take to heal the wounds of civil strife and bloodshed? This question is being raised throughout the world as nations seek to diminish intergroup hatreds and cope with the corrosive memories of civil war and genocide. In the United States one of the strongest examples of successful efforts to address painful memories in a healing fashion is found in the commemorative sculpture of Maya Lin.

Commemoration, in the sense in which it is used to speak about Lin's work, refers to the effort a society makes to officially remember the major events that have marked its history. In the case of the Vietnam Veterans Memorial, the names of all U.S. military personnel who were killed in the war are carved into the shiny black granite of a long, gracefully proportioned wall. People from around the nation come to the wall to remember their loved ones or simply to remember the war and its many consequences for American society. As they stand in contemplation, their own images are reflected by the sheen of the polished stone. Few people leave this monument without being deeply affected by the experience.

Commemorative sculpture for major public sites is never the work of a single person. A committee usually selects from a number of possible approaches and artists. Thus in the case of the Vietnam Veterans Memorial much of the credit for the success of the monument must also go to the committee members who were able to recognize the power of Lin's vision. The same must be said for the committee representing the Southern Poverty Law Center, which commissioned Lin to design the Civil Rights Memorial in Montgomery, Alabama.

The civil rights movement was characterized by a great deal of heroism and martyrdom, as well as the kind of hatred and violence that became too familiar in photos like this famous one of anti–civil rights demonstrators pouring food and abuse onto civil rights workers seeking to integrate a lunch counter. A commemorative monument that focused on the heroism and the brutality, however, would not be as likely to result in emotional uplift as a quieter, more contemplative installation. On the other hand, a commemorative monument must also give clear recognition to the movement's heroes and martyrs. Lin seems to have solved this problem brilliantly.

Like the Vietnam Veterans Memorial, the Civil Rights Memorial draws the viewer into a personal meditation. Water from a deep pool on the monument's upper level "spills down in even sheets across the face of the wall, on which is a gold-leafed inscription of Martin Luther King's inspiring statement that the fight will continue (in the words of the Bible) 'until justice rolls down like waters and righteousness like a mighty stream.' An elliptical table, on which are engraved the landmark events in the civil rights era from 1955 to 1968, including the names of 40 people who died in the struggle, glistens with a film of water from above. People walk slowly around the monument reading the engraved text, and as they do so many feel compelled to run their fingers across the wet surface in an 'act of communion'" (Stein, 1994, p. 70).

SUMMARY

In biology, the term *race* refers to an inbreeding population that develops distinctive physical characteristics that are hereditary. However, the choice of which physical characteristics to consider in classifying people into races is arbitrary. And human groups have exchanged their genes through mating to such an extent that it is impossible to identify "pure" races.

The social concept of race has emerged from the interactions of various populations over long periods of human history. It varies from one society to another, depending on how the people of that society feel about the importance of certain physical differences among human beings. *Racism* is an ideology based on the belief that an observable, supposedly inherited trait is a mark of inferiority that justifies discriminatory treatment of people with that trait.

Ethnic groups are populations that have a sense of group identity based on a distinctive cultural pattern and shared ancestry. They usually have a sense of "peoplehood" that is maintained within a larger society. Ethnic and racial populations are often treated as *minority groups*—people who, because of their physical or cultural characteristics, are singled out from others in the society for differential and unequal treatment.

Intergroup relations can be placed on a continuum ranging from intolerance to tolerance, or from genocide through assimilation. *Genocide* is the intentional extermination of one population, defined as a "race" or a "people," by a more dominant population. It is almost always rationalized by the belief that the people who are being slaughtered are less than human.

Expulsion is the forcible removal of one population from territory claimed by another population. It has taken a variety of forms in American history, including the expulsion of Native Americans from their ancestral lands, the Oriental exclusion movement of the nineteenth century, and the detention of Japanese Americans during World War II.

Slavery is the ownership of a population, defined by racial or ethnic or political criteria, by another population that has complete control over the enslaved population. Slavery has been called "the peculiar institution" because it has existed in some of the world's greatest civilizations, including the United States.

Although African American slaves gained their freedom during the Civil War, a long period of segregation followed. *Segregation* is the ecological and institutional separation of races or ethnic groups. It may be either voluntary or involuntary. Involuntary segregation may be either *de jure* (created by laws that prohibit certain groups from interacting with others) or *de facto* (created by unwritten norms).

Assimilation is the pattern of intergroup relations in which a minority group is forced or encouraged or voluntarily seeks to blend into the majority population and eventually disappears as a distinct people within the larger society. In the United States, three different views of assimilation have prevailed since the early nineteenth century. They are Anglo-conformity, the demand that culturally distinct groups give up their own cultures and adopt the dominant Anglo-Saxon culture; the melting pot, the theory that there would be a social and biological merging of ethnic and racial groups; and cultural pluralism, the belief that culturally distinct groups can retain their communities and much of their culture and still be integrated into American society.

Stereotypes are inflexible images of a racial or cultural group that are held without regard to whether they are true. They are often associated with *prejudice,* an attitude that prejudges a person, either positively or negatively, on the basis of characteristics of a group to which that person belongs. *Discrimination* refers to actual behavior that treats people unfairly on the basis of their group membership; *institutional discrimination* is the systematic exclusion of people from equal participation in a particular social institution because of their race, religion, or ethnicity. Social movements whose purpose is to oppose institutional discrimination are often supported by *ethnic nationalism,* the belief that one's ethnic group constitutes a distinct people whose culture is and should be separate from that of the larger society. Policies designed to correct persistent racial and ethnic inequalities in promotion, hiring, and access to other opportunities are referred to as affirmative action.

Social-psychological theories of ethnic and racial inequality argue that a society's patterns of discrimination stem from individual psychological orientations toward members of out-groups. Interactionist explanations go beyond the individual level to see how hostility or sympathy toward other groups is produced by the norms of interaction that evolve within and between groups. The functionalist perspective generally seeks patterns of social integration that help maintain stability in a society. Conflict theories trace the origins of racial and ethnic inequality to the conflict between classes in capitalist societies. The conflict perspective includes the theory of *internal colonialism,* which holds that many minority groups are essentially colonial peoples within the larger society. Finally, ecological theories of race relations explore the processes by which conflict between racial or ethnic groups develops and is resolved.

The persistence of racial inequality in the United States is a source of continuing controversy. This complex problem is a result of a number of factors besides racial prejudice and discrimination. Other factors are high rates of family breakup and the effects of structural changes in the American economy. Although social mobility is more available to blacks today than it was before the civil rights movement, the majority of blacks are in insecure working-class jobs or unemployed.

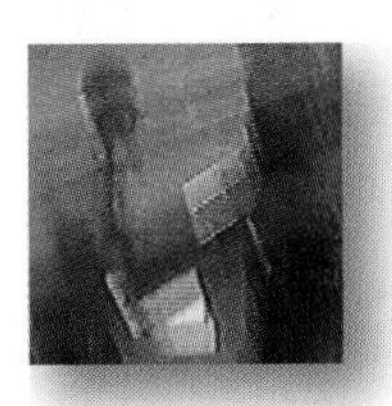

SOCIOLOGY VERSUS IDEOLOGY

Recent surveys show that Americans are severely divided in their opinions about how much the government should be doing to address inequalities in educational and employment opportunity based on race. African Americans tend to support affirmative action measures; whites tend to oppose them, or at least to oppose programs that seem to depend on quotas, which strike many white citizens as a form of "reverse discrimination." The plea that affirmative action is necessary to overcome a past characterized by severe racial inequity persuades some white people to favor affirmative action, but most continue to regard it as unfair.

What can sociological research bring to bear on the dispute that might lead to a calmer discourse between the races? Perhaps the best the sociologist can do is point to research that might lead both sides to reconsider the strength of their views.

To begin with, it is important to recognize that affirmative action in the military, large corporations, and government service is the major reason why there has been a narrowing of the income and opportunity gap between the races. The data show conclusively that without various forms of affirmative action that insist on racial integration of workplaces and institutions of higher education it would not be possible to overcome the effects of generations of Jim Crow norms.

Affirmative action, on the other hand, is being used extensively in American politics as a "wedge issue" to drive apart racially diverse constituencies like working-class Americans. Supporters of "race specific" affirmative action policies tend to lose the allegiance of voters whose votes they need to win elections, a situation that has doomed many such policies.

Given these rather contradictory considerations, supporters of affirmative action might be well advised to seek compromises and reformulations of policies that will result in the desired outcomes but will be perceived as fairer for all who might be affected. More funding for higher education scholarships and grants based on need, for example, will appeal to all racial groups but will offer proportionately greater help to low-income minority groups.

GLOSSARY

race: an inbreeding population that develops distinctive physical characteristics that are hereditary. (p. 398)

racism: an ideology based on the belief that an observable, supposedly inherited trait is a mark of inferiority that justifies discriminatory treatment of people with that trait. (p. 399)

ethnic group: a population that has a sense of group identity based on shared ancestry and distinctive cultural patterns. (p. 400)

minority group: a population that, because of its members' physical or cultural characteristics, is singled out from others in the society for differential and unequal treatment. (p. 403)

genocide: the intentional extermination of one population by a more dominant population. (p. 406)

expulsion: the forcible removal of one population from a territory claimed by another population. (p. 407)

slavery: the ownership of one racial, ethnic, or politically defined group by another group that has complete control over the enslaved group. (p. 407)

segregation: the ecological and institutional separation of races or ethnic groups. (p. 409)

de jure segregation: segregation created by formal legal sanctions that prohibit certain groups from interacting with others or place limits on such interactions. (p. 409)

de facto segregation: segregation created and maintained by unwritten norms. (p. 409)

Jim Crow: the system of formal and informal segregation that existed in the United States from the late 1860s to the early 1970s. (p. 410)

assimilation: a pattern of intergroup relations in which a minority group is absorbed into the majority population and eventually disappears as a distinct group. (p. 411)

ethnic stratification: the ranking of ethnic groups in a social hierarchy on the basis of each group's similarity to the dominant group. (p. 414)

pluralistic society: a society in which different ethnic and racial groups are able to maintain their own cultures and lifestyles while gaining equality in the institutions of the larger society. (p. 415)

stereotype: an inflexible image of the members of a particular group that is held without regard to whether it is true. (p. 416)

prejudice: an attitude that prejudges a person on the basis of a real or imagined characteristic of a group to which that person belongs. (p. 416)

discrimination: behavior that treats people unfairly on the basis of their group membership. (p. 416)

institutional discrimination: the systematic exclusion of people from equal participation in a particular institution because of their group membership. (p. 416)

ethnic (or racial) nationalism: the belief that one's own ethnic group constitutes a distinct people whose culture is and should be separate from that of the larger society. (p. 417)

scapegoat: a convenient target for hostility. (p. 421)

projection: the psychological process whereby we attribute to other people behaviors and attitudes that we are unwilling to accept in ourselves. (p. 421)

internal colonialism: a theory of racial and ethnic inequality that suggests that some minorities are essentially colonial peoples within the larger society. (p. 423)

WHERE TO FIND IT

BOOKS

Threatened People, Threatened Borders: World Migration and U.S. Policy (Michael S. Teitlebaum and Myron Weiner; Norton, 1995). A review of world patterns of population movement and U.S. policy responses to the growing demand for entry into North America, by a panel of renowned experts on migration and immigration.

When Work Disappears: The World of the New Urban Poor (William Julius Wilson; University of Chicago Press, 1996). An invaluable empirical study of one of the nation's largest and most visible urban ghettos. Wilson uses his research to make a strong case for "race-blind" social policies that will aid all low-income communities.

Black Wealth/White Wealth: A New Perspective on Racial Inequality (Melvin L. Oliver and Thomas M. Shapiro; Routledge, 1995). An outstanding study of racial inequality based on income versus wealth.

Drylongso: A Self Portrait of Black America (John Langston Gwaltney; Vintage, 1981). A detailed portrait of black American culture as expressed in interviews; a complex view of a stable and rich consciousness.

American Apartheid: Segregation and the Making of the Underclass (Douglas S. Massey and Nancy A. Denton; Harvard University Press, 1993). An up-to-date review of the continuing persistence of residential segregation in American cities.

Ethnic America: A History (Thomas Sowell; Basic Books, 1981). A comparative treatise on the success and contributions of the major nationality groups that have made the United States their home.

A Common Destiny: Blacks and American Society (Gerald David Jaynes and Robin M. Williams, Jr., eds.; National Academy Press, 1989). A volume covering most aspects of racial inequality and racial progress in the United States. Widely regarded as the most important recent summary of the status of black Americans, it was compiled by a distinguished group of social scientists convened by the National Academy of Sciences and including most of the nation's foremost students of American racial inequality.

Latinos: A Biography of the People (Earl Shorris; Norton, 1992). A beautifully written and thoughtful book describing Spanish-speaking peoples' search for cultural unity in the face of historical, socioeconomic, and racial diversity.

U.S. Race Relations in the 1980's and 1990's: Challenges and Alternatives (Gail S. Thomas, ed.; Hemisphere, 1990). Covers the recent rise in racism, racially motivated incidents, and race-hatred groups, as well as efforts to address issues of education, poverty, and other problems that disproportionately affect racial minorities. Argues that improvement in race relations will be difficult in the years ahead and will require a far greater national commitment than exists at present.

OTHER SOURCES

Harvard Encyclopedia of American Ethnic Groups. A valuable collection of essays about ethnic and racial relations in American society. Covers almost every imaginable group in the United States and offers summary essays on such sociologically relevant subjects as slavery, assimilation, ethnic accommodation, and conflict.

Asian Americans: Diverse and Growing (Sharon M. Lee; Population Reference Bureau, *Population Bulletin,* June 1998). A fine overview of the changing demographics of people of various Asian backgrounds in the United States.

INTERNET RESOURCES

Interracial Voice (www.webcom.com/~intvoice/). An electronic publication that promotes the establishment of a multiracial category on the year 2000 census.

The National Fair Housing Advocate (www.fairhousing.com/). Keeps track of legal cases concerning housing discrimination throughout the nation, links organizations that advocate fair housing practices, and provides updated information on federal guidelines, job openings, articles, news, and so forth.

International Sociological Organization (www.uem.es/). The Web site for the world organization of sociologists. Has links to many researchers who are dealing with issues of ethnicity and race in the developed and developing world.

European Research Centre on Migration and Ethnic Relations (ERCOMER) (www.ercomer.org). A university-based research institute in the Netherlands that promotes "peaceful coexistence, justice, and harmony in inter-ethnic relations." Offers access to research papers and important research findings on ethnic and racial relations.

The Movies, Race, and Ethnicity
(www.lib.berkeley.edu/MRC/EthnicImagesVid.htm). Provides videographies, bibliographies, and full-text articles of a sample of Hollywood films emphasizing issues faced by African Americans, Asians, Latinos, Jews, and Native Americans.

CHAPTER 14

INEQUALITIES OF GENDER

For most people the word *safari* conjures up images of wild animals—lions, elephants, and rhinos roaming the savannahs. I wanted to go and experience all that Kenya had to offer and yet there was something more that tugged at me. I felt a calling from deep within me to go to East Africa, to go and see the beginnings of all humankind.

The writer is sociologist Carol Chenault of Calhoun Community College in Decatur, Alabama. She and her husband, Ellis Chenault, traveled to Kenya to observe the way of life of the Maasai people. "I am confident," she says, "that the teaching of culture in college sociology classes for 15 years is a major source of my interest in the Maasai." For her and for the students with whom she now shares her experiences, this sociological voyage was an eye-opening one.

> What I observed was an extremely independent people whose entire lives revolve around the care and herding of their cattle. They truly believe that all of the cattle on earth are their gift from God (Engai). One Maasai warrior asked me if I had any cows. When I told him no, he said, "Good, because we would come get them if you did."

For Chenault, some of the most challenging aspects of life among the Maasai centered on the differences between the roles of men and women. "The Maasai culture has a very traditional division of labor between men and women," she writes.

> Women are nurturing, loving mothers who spend a great deal of their time hugging, holding, and kissing their children. However, care of the children is a responsibility that is also shared with the fathers. Early childhood has few responsibilities; time is spent learning the ways of the adults, the language skills and play. The girls gradually begin to learn the tasks of the women such as care of the animals, making clothing, food preparation, and doing beadwork. Girls are initiated into womanhood through circumcision; however, their attitude toward this is not one

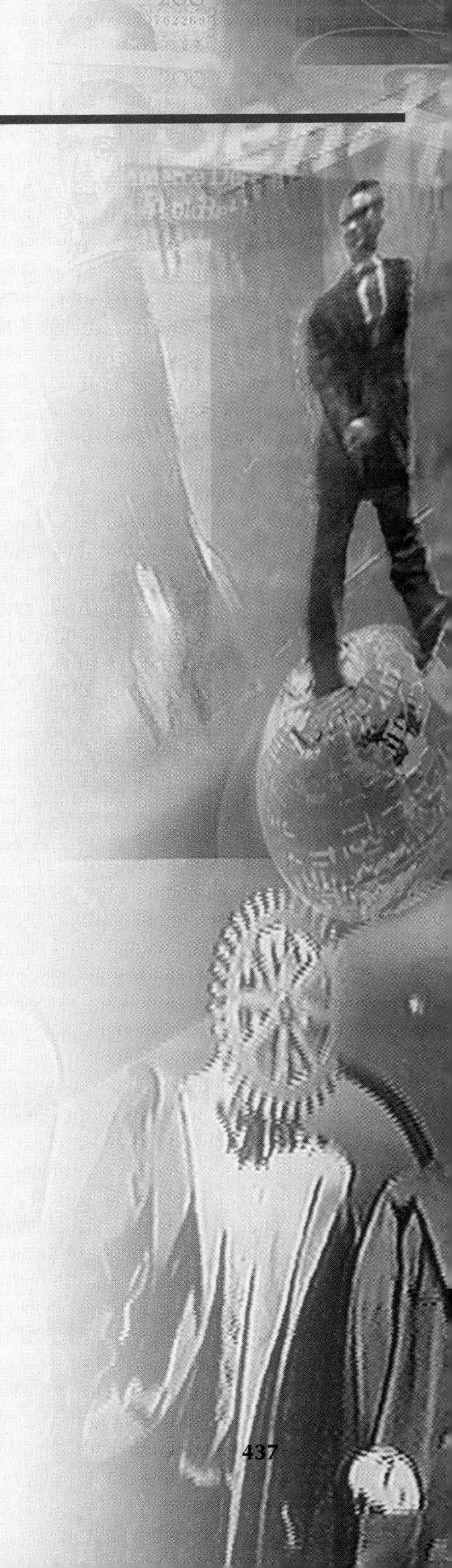

of fearful anticipation and bravery. The young women see this as a time when they have to give up their freedom for a strict married life. No longer will they be free to choose their lovers as they once did. Circumcision occurs in the mother's house and does not require the public demonstration of courage and bravery that is required of the males in their circumcision rituals. An elderly Maasai woman told me that it is unclean not to be circumcised.

Maasai girls are betrothed during infancy, or in some cases before they are born. A young couple may be approached and asked if their firstborn is a girl, then may she become the wife of their son. The wedding ceremony is a simple one in which the bride's head is shaved and then adorned with bands of beadwork. Her clothing at the wedding is very colorful in that it is all new and dyed with the familiar red ochre. Both the bride and groom are washed with milk at the time of the wedding. (Chenault, 1996, pp. 1–3)

GENDER AND INEQUALITY

Many women in the nations of Asia and Africa are confronting the cultural norms that keep them in subservient positions and prevent them from asserting their own choices about their futures. The Maasai are unusual in the contemporary world. They are pastoralists who are relatively isolated from the powerful social forces that are rapidly changing other tribes in Kenya and elsewhere in Africa. In consequence, the Maasai insist on the necessity of continuing to adhere to the norms of their own tribal culture. For the sociologist, the example of the Maasai is extremely useful because it challenges assumptions about gender and society.

As observant sociologists like Carol Chenault travel to other cultures, they invariably recognize that despite cultural differences, the differences in the life chances of men and women remain one of the major dimensions of inequality in human societies. This chapter therefore begins by presenting some of the key patterns of gender inequality in the United States and throughout the world. Then the second section examines the social and biological concepts of gender and sex that are essential to understanding the inequalities people experience as a consequence of their gender or sexual orientation. Gender stratification and gender roles are discussed in the third section.

The fourth section of the chapter will explore cultural norms and ideologies dealing with gender inequality, especially as they are manifested in industrial societies. The next two sections explore the women's movement and the situation of women in the workplace, where some of the most exciting and difficult changes in patterns of gender inequality are occurring today. The final section highlights current research being conducted at the grassroots level to identify and address the problems of women in the world today.

Patterns of Gender Inequality

Gender inequality stunts women's social and economic development throughout the world. It also impedes the development of entire societies, although it often takes different forms in developed and less developed nations. No wonder the subject has become a central theme of global social science. The United Nations International Conference on Population and Development, held in Cairo in 1994, concluded that "the empowerment of women and improvement of their status are important ends in themselves and are essential for the achievement of sustainable development" (United Nations, 1995). Some graphic examples of how gender inequality differs in the developed and less developed regions of the world are presented in Table 14.1 and Figure 14.1.

Table 14.1 shows that in the United States there is a continuing gap in income between men and women in the same occupations. We will see later in the chapter that even as increasing numbers of women join the labor force and compete for jobs and careers in ways that were almost unheard of just a few decades ago, men still tend to move into the higher-paying jobs in each occupation. Figure 14.1 shows quite dramatically that in the less developed regions women shoulder a far greater number of domestic burdens than men do. Much of the difference may seem to be due to the fact that women take primary responsibility for bearing and nurturing children. And one might argue that in a developed country like the United States, the differences in men's and women's incomes could be attributed to the fact that so many women take time out from work and

TABLE 14.1 Median Weekly Earnings of Full-Time Wage and Salary Workers, 1998

Occupation	Men	Women
Managerial and professional	$875	$632
Technical, sales, and administrative support	588	403
Service	372	282
Precision production	569	382
Operators, fabricators, and laborers	436	313
Farming, forestry, and fishing	302	257

Source: *Statistical Abstract,* 1998.

FIGURE 14.1 **A Woman's Day or a Man's Day?**

Which would you prefer if you could choose? The Swedish International Development Agency outlined a typical day for a man and a woman in a family that grows both cash crops and its own food supply, in its journal *Striking a Balance.* The family lives "somewhere in Africa."

The woman's day does not change if she is pregnant. There is little time to spare to visit a medical center for regular prenatal care.

A WOMAN'S DAY	A MAN'S DAY
rises first	
kindles the fire	
breast-feeds the baby	
fixes breakfast/eats	rises when breakfast is ready
washes and dresses childen	eats
walks 1 km to fetch water	
walks 1 km home	walks 1 km to field
gives the livestock food and water	
washes cooking utensils, etc.	works in the field
walks 1 km to fetch water	
walks 1 km home	
washes clothing	
breast-feeds the baby	
walks 1 km to field with food for husband	
walks 1 km back home	eats when wife arrives with food
walks 1 km to her field	
weeds fields	works in the field
breast-feeds the baby	
gathers firewood on the way home	
walks 1 km home	walks 1 km home
pounds maize	rests
walks 1 km to fetch water	
walks 1 km home	
kindles the fire	
prepares meal/eats	eats
breast-feeds the baby	walks to village to visit other men
puts house in order	goes to bed
goes to bed last	

Source: Sadik, 1995.

careers to bear and raise children. These familiar (and flawed) arguments raise further questions about biological versus social causes of gender inequality, which, in turn, may create confusion about the differences between sex and gender and the role of sexuality and sexual orientation in defining an individual's life chances. Before exploring the contours of gender inequality further, therefore, we turn to a discussion of the basic concepts of sex and gender.

SEX VERSUS GENDER

Masculine and *feminine* are among the most confusing words used both in everyday speech and in the social sciences. The confusion stems largely from the fact that our ideas of what is male and female in human individuals are based on overlapping influences from biology and culture. As in all such situations, it is difficult to sort out the relative contributions of each source of influence. But the underlying concepts are not hard to grasp: One's sex is primarily a biological quality, while one's gender is formed largely (but not entirely) by the cultural forces we experience through socialization from infancy on (Connell, 1995).

Sex refers to the biological differences between males and females, including the primary sex characteristics that are present at birth (i.e., the presence of specific male or female genitalia) and the secondary sex characteristics that develop later (facial and body hair, voice quality, and so on). Note that these biological qualities differ considerably among individuals, so that the differences between the sexes are not always as marked as the male–female distinction suggests. Moreover, some people are born as **hermaphrodites;** their primary sexual organs have features of both male and

BOX 14.1 USING THE SOCIOLOGICAL IMAGINATION

The Lenses of Gender

"Throughout the history of Western culture," Sandra Bem writes,

> three beliefs about women and men have prevailed: that they have fundamentally different psychological and sexual natures, that men are inherently the dominant or superior sex, and that both male-female difference and male dominance are natural. Until the mid-nineteenth century, this naturalness was typically conceived in religious terms, as part of God's grand creation. Since then, it has typically been conceived in scientific terms, as part of biology's—or evolution's—grand creation.
>
> Consequently, most Americans did not see any inconsistency between commitment to equality and denial of political rights to women until the appearance of the women's rights movement in the mid-nineteenth century. This first wave of feminist advocacy not only established women's basic political rights; it also made the inconsistency between ideology and the treatment of women widely visible for the first time in U.S. history.
>
> Beginning in the 1960s, the second major wave of feminist advocacy raised social consciousness still further by exposing—and naming—the "sexism" in all policies and practices that explicitly discriminate on the basis of sex. This second feminist challenge gradually enabled people to see that restricting the number of women in professional schools or paying women less than men for equal work was not a natural requirement of a woman's biological and historical role as wife and mother but an illegitimate form of discrimination based on outmoded cultural stereotypes. Even political reactionaries began to espouse the principle of equal work. (Bem, 1993, p.1)

Despite the major changes that have taken place over the past century and a half, Bem and other feminist scholars point out that we still cling to deeply ingrained assumptions about gender. As a result, we "systematically reproduce male power in generation after generation" (Bem, 1993, p. 2). Bem calls these often unquestioned assumptions "lenses of gender." Her purpose in identifying them and writing about how they shape our lives is to render them visible and thus subject to positive change.

The first of the three lenses of gender she identifies is *androcentrism,* or male-centeredness. This is the "historically crude perception that men are inherently superior to women," which leads to the even more insidious perception that male experiences are the natural standard or norm and that "female experience is a sex-specific deviation from that norm."

The second "lens" that Bem analyzes in her research is *gender polarization.* This is the crude idea that men and women are fundamentally different from each other, which leads to the pervasive use of that perceived difference "as an organizing principle for the social life of the

female organs, making it difficult to categorize the person as male or female. Hermaphrodism is uncommon, but it occurs in societies throughout the world.

Another ambiguous sexual category consists of the many thousands of **transsexuals,** people who feel very strongly that the sexual organs they were born with do not conform to their deep-seated sense of what their sex should be. Transsexuals sometimes undergo a course of endocrine hormonal treatments to change their secondary sex characteristics and may eventually have irreversible sex-change operations. Most such operations are performed to change the individual from a male to a female.

One of the most famous transsexuals in the contemporary world is the writer Jan Morris. As a male, James Morris reported on an expedition to Mount Everest and engaged in many activities that are associated with masculinity. But as Morris explains it, "I was born with the wrong body, being feminine by gender but male by sex, and I could achieve completeness only when the one was adjusted to the other" (quoted in Money & Tucker, 1975, p. 31). In making this distinction, Morris is calling attention to the difference between her biologically determined sex and her socially and emotionally influenced gender.

Gender refers to the culturally defined ways of acting as a male or a female that become part of an individual's personal sense of self. The vast majority of people develop a "gut-level" sense of themselves as male or female, boy or girl, early in life; however, as we will see shortly, some people's gender identities are more ambiguous. It is not entirely clear exactly how a person's gender identity is formed, but much of the research done to date suggests that the assignment of a gender at culture." Modes of dress, speech, social roles, "and even ways of expressing emotion and experiencing sexual desire" are subject to gender polarization.

The third lens of gender in our culture is that of *biological essentialism,* "which rationalizes and legitimizes both other lenses by treating them as the natural and inevitable consequences of the intrinsic biological natures of women and men." Of course, Bem does not deny that there are obvious biological differences between men and women, but she and other feminist scholars argue that "these facts have no fixed meaning independent of the way that a culture interprets and uses them, nor any social implications independent of their historical and contemporary context" (Bem, 1993, p. 2).

Bem realizes that eradication of these lenses of gender would require almost revolutionary changes. But the first step toward bringing about this revolution is to open your imagination to a future in which the distortions caused by existing gender lenses are absent.

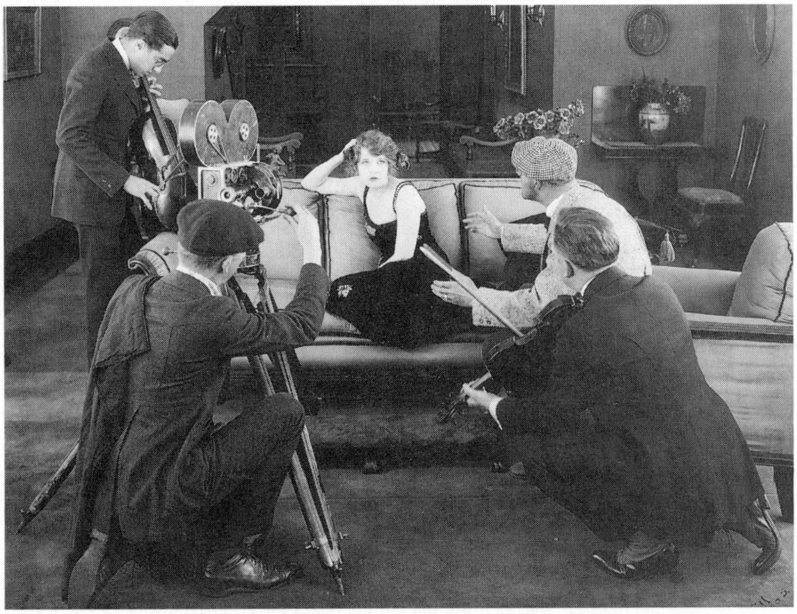

Here we literally see the lenses of gender at work. Note that all the production crew members are male. This 1919 photo of a film star on a studio set is clearly a study in gender polarization.

birth has a strong influence during the early years of life. In other words, children's feelings of being a boy or a girl are defined more by how they they are treated by their parents than by their actual biological sex characteristics (Stockard & Johnson, 1992).

Gender Socialization and Gender Separation

In Chapter 5 we introduced the concept of gender socialization and discussed its powerful influence on how men and women develop as members of society. We showed, for example, that the values of beauty and youth, especially as they are applied to women by the media, have a strong and often negative influence on how women feel about their bodies and about themselves (Wolf, 1991). Current research on gender inequality is focusing increasing attention on these issues, not only for women but for men as well. Social scientists who are critical of cultures that produce and perpetuate gender inequality point to the many ways in which these inequalities are justified by values, norms, and the workings of major social institutions. Sandra Bem, a feminist scholar and a prominent researcher on gender socialization, comments on these often unquestioned cultural processes in Box 14.1.

According to many comparative studies and international population reports, a central problem of gender socialization throughout the world is that the traditional separation of men and women in socializing agencies leads to the failure of men to recognize their responsibilities as sexual partners and fathers later in life. As a recent United Nations report states:

> It is clear that women cannot adequately protect their sexual and reproductive health in the context of power imbalances with their male partners. . . . Men must be involved because they share the responsibility of reproductive health, and both men and women do not understand their own and each other's bodies. (Sadik, 1995, p. 23)

In the United States, a major theme in social-scientific research is the way gender socialization, particularly in schools, tends to separate males and females into different social worlds with their own forms of activity and language. The theory that boys and girls are socialized to perform in more or less separate worlds receives a great deal of support in the sociological literature (Fine, 1987; Gilligan, 1982; Lever, 1978; Tannen, 1990). However, we will see shortly that the "separate worlds" thesis also has many limitations (Thorne, 1993).

A broad version of the gender separation theory as experienced in urban industrial societies is that children absorb ideas about their gender identities from their parents and the media. By the time they are preschoolers they have strong ideas about what types of dress are appropriate. Boys especially are mortified if they are caught wearing something they feel is inappropriate for their gender. But children's expressions of gender identity are also shaped through interactions in play groups throughout many years of schooling. From their separate, gender-segregated interactions with peers in school, it is argued, boys and girls develop different values, different ways of bonding with each other, different types of conflict and ways of resolving conflict. In many ways they develop separate cultures with their own norms, values, and languages. These differences are thought to persist through life and to exert a lasting influence on how the individual fares in the job market (Kanter, 1977).

In their insightful observation of preadolescent boys' interactions on Little League baseball teams, sociologist Gary Allen Fine and popular writer Stephen King were struck by the depth of male bonding that occurs as the boys undergo the pressure of competition and support each other through their successes and failures on the playing field (Fine, 1987; King, 1994). Fine notes that adult males often want their sons to join teams so that they can "teach their preadolescent boys to see the world from their perspective," which includes teaching what the men consider core values as these values emerge in sports competition. However, "whereas adults focus on hustle, teamwork, sportsmanship, and winning and losing"—as desirable moral behaviors developed through team sports—"the preadolescent transforms these concerns" (Fine, 1987, p. 78). The team members respect the adults' teachings, but they are more concerned with their self-presentation, which includes some of the adult values and some of those that are part of the preadolescent moral code—learning to be tough or fearful under the appropriate conditions; learning to control aggressions, fears, and tears; not being a "rat" (telling on others); and hustling in order to help the team win games. The boys engage in a great deal of sexual talk, much of which expresses their ambiguity about relationships with girls, in which their desire to have sexual experiences is offset by their fears of embarrassment and loss of male friends. The boys' banter with each other also expresses their fear of not being masculine enough. Although they have little or no experience with homosexuality, they use words like *faggot* or *gay* to tease each other or denigrate other boys.

Sociologists who study gender and socialization often point out that the strong same-sex bonds formed by boys in preadolescence and during their teen years may prove invaluable in the corporate world. In their research on life in American corporations, Rosabeth Moss Kanter and her associates find that sports analogies and language are commonplace. The male business leader typically thinks of those he works with as members of a team, and as a result of his early socialization experiences he is more comfortable with male team members than with females (Kanter & Stein, 1980).

Studies of girls' play groups note that although boys and girls are increasingly likely to play some games together, the separation of the genders in playgrounds and schoolyards remains a norm (Sutton-Smith & Rosenberg, 1961; Whiting & Edwards, 1988). These studies confirm that girls' games are likely to involve small groups, rather than large ones with more complex structures and hierarchies of status. They are interrupted not by arguments over rules or specific plays but by emotional issues, feelings of being slighted, or other hurts—which the girls repair in private conversations rather than in public protests to umpires or other adults. In addition, girls tend to engage in turn-taking games that emphasize cooperation, while boys more often engage in competitive team sports in which they become extremely concerned with establishing hierarchies of skill and success (e.g., batting averages, won–lost records) (Lever, 1976, 1978).

Of course, from their parents and playmates children learn many other social distinctions besides gender. Barrie Thorne, who observed children's peer interactions in two primary schools, makes this observation:

> In the lunchrooms and on the playgrounds of the two schools, African-American kids and kids whose main language was Spanish occasionally separated themselves into smaller, ethnically homogeneous groups. . . . I found that students generally separate first by gender and then, if at all, by race or ethnicity. (1993, p. 33)

Problems With the Gender Separation Theory

How fully are men and women socialized into separate social worlds with separate cultures? As mentioned earlier, there is a great deal of popular and social-scientific writing in support of this thesis. The immense success of Deborah Tannen's analysis of men's and women's spoken communication in her popular book *You Just Don't Understand: Women and Men in Conversation* suggests that men and women do feel separate and different in their everyday lives. But recent research on gender differences is quite critical of the idea of separate gender cultures (Thorne, 1993).

While there is an undeniable tendency for boys and girls to be socialized in gender-segregated peer groups, they are also raised together in families and often spend a great deal of time together. Boys and girls are increasingly participating in the same types of activities in their schools and communities. In some communities these changes are controversial, but in many others teachers, parents, and administrators seek to avoid the worst effects of gender separation (Meier, 1995). Girls' and women's team sports, for example, are undergoing rapid and steady growth. And as they review the literature on separate gender cultures, or conduct new research, students of gender often find that the previous emphasis on gender separateness was based on what they refer to as "the Big Man bias"—the tendency to observe only the activities of the most powerful and socially successful individuals. More isolated, less successful children are less likely to be subjects of research, yet their experiences are very important since they do not adhere strictly to the norms of gender separation (see Box 14.2). Finally, as Thorne (1993) argues, generalizations about "girls' culture" are based primarily on research with upper-class white girls, while the experiences of girls from other class, race, and ethnic backgrounds tend to be neglected.

Gender and Sexuality

Sexuality refers to the manner in which a person engages in the intimate behaviors connected with genital stimulation, orgasm, and procreation. Like most areas of human behavior, sexuality is profoundly influenced by cultural norms and social institutions like the family and the school, as well as by social structures like the class system of a society. Sociologists and historians believe that there have been major changes in sexuality as a result of changes in the economy, politics, and the family. Over the past 300 years, for example, the meaning and place of sexuality in American life have changed: from a family-centered system for ensuring reproduction and social stability during the colonial period; to a romantic and intimate sexuality in nineteenth-century marriage, with many underlying conflicts; to a commercialized sexuality in the modern period, when sexual relations are expected to provide personal identity and individual happiness apart from reproduction (D'Emilio & Freedman, 1988).

While these changes are important and controversial, they hardly begin to exhaust the range of sexual norms in cultures throughout the world. Polygyny (the practice of having more than one wife), which is illegal in North America and Europe, is condoned throughout much of the Islamic world. Marriage between adult men and preadolsecent girls is practiced in some Asian cultures but would be be considered a form of child abuse in the West. Women bare their breasts on the beaches of Europe but may be issued summonses for the same behavior on the beaches of North America. On American television screens women and men expose their bodies in a fashion that is horrifying to people in many parts of Asia. Catholic priests and nuns are expected to remain celibate, while Protestant ministers are encouraged to have families, as are Jewish rabbis. With all these contrasting cultural norms governing sexuality, can it be said that there are any universal sexual norms?

Universal cultural norms exerting social control over sexuality include the *incest taboo, marriage,* and *heterosexuality,* but even these norms include variations and differing degrees of sanction. As noted in Chapter 3, the incest taboo is known in every existing and historical society and serves to protect the integrity of the family. Incest norms prohibit sexual intimacy between brothers and sisters and between parents and children, and usually specify what other family relatives are excluded. However, the specific relatives involved vary in different cultures; for example, sex between first cousins is permissible in some societies and strictly taboo in others.

Every society also has marriage norms that specify the relationships within which sexual intimacy is condoned. Marriage norms protect the institution of the family, confer legitimacy on children, and specify parental rights and obligations. But while marriage norms are found in all cultures, there are vast differences in how they operate. In addition to variations having to do with polygyny, there are some important variations with regard to the strength of sanctions on adultery (sexual intimacy outside marriage). The majority of cultures prohibit adultery, but the sanctions vary widely. Former French President François Mitterrand, who died early in 1996, specified in his funeral instructions that both his wife and his mistress were to be present at his funeral. Although adultery is not openly condoned in France, it is far less strongly sanctioned than in the United States, where evidence that a presidential candidate may have

BOX 14.2 SOCIOLOGICAL METHODS

Problems in Research on Peer Socialization

Gary Alan Fine charted the friendship choices among the 11- and 12-year-old players on the Little League teams he observed for his study *With the Boys: Little League Baseball and Preadolsecent Culture.* A typical result is the sociogram depicted in the accompanying chart. Note that the arrows in the diagram may be either one-way or reciprocal, depending on whether the choices are one-way or mutual. Clearly, on this team there was a "ruling clique" consisting of Justin, Harry, and Whit, each of whom also chose each other and who were extremely likely to be chosen by the other boys, even when the three leaders did not reciprocate.

It should come as no surprise that these three "sociometric stars" were the best ballplayers on the team. They were also most admired for their toughness and masculinity. The most popular, Justin, was extremely fair with his teammates but was especially aggressive, prone to call opposing players "faggots" and to use racial epithets. Fine also found that Justin had the closest contact with girls, who also looked up to him as a leader, and was able to avoid the jibes of his male peers about this because of his sports ability and leadership. Fine notes that not all the teams had this particular clique structure and that emphasis on the activities of the elite boys on teams and in peer groups tends to neglect the experiences of the larger number of boys who are not the stars.

In her studies of what she calls "gender play," Barrie Thorne is quite critical of research that focuses on central figures like Justin because they tend to become models from which generalizations are made. "Assertions about gender differences in actual behavior refer, at best," she argues, "to *average* differences between girls and boys, or between groups of girls and groups of boys" (1993, p. 103). She cites studies of boys' and girls' play groups that conclude, from observations of the most dominant boys or most popular girls in the study, that boys are more likely to enjoy "rough and tumble" play. In fact, if one looks at data for the less dominant boys, it appears that the gender differences on this issue are not wide at all. While this bias is one that Fine tries to avoid, Thorne points out that it occurs in numerous other studies:

> The literature on "the boys' world" suffers from a "Big Man bias" akin to the skew found in anthropological research that equates male elites with men in general. In many observational studies of children in preschools and early elementary school, large, bonded groups of boys who are physically assertive, engage in "tough talk," and actively devalue girls anchor descriptions of "the boys' world" and themes of masculinity. Other kinds of boys may be mentioned, but not as the core of the gender story. (p. 98)

While "being tough" and "being nice" may be core values that distinguish boys' and girls' peer groups, Thorne can point to many instances in her field observations in which girls act tough and even beat up the boys. The more one builds racial, ethnic, and class diversity into the sample of children under observation, the more the gender behaviors overlap and the less rigid the gender borders appear in the data from the study.

had an extramarital affair often becomes a major issue in the campaign (Stockard & Johnson, 1992).

Heterosexuality refers to sexual orientation toward the opposite sex, in contrast to **homosexuality,** which refers to sexual orientation toward the same sex, and **bisexuality,** which refers to sexual orientation toward either sex. Heterosexuality is a norm in every society, but it too is subject to wide variations. Norms of heterosexuality function to ensure that there is genital sexual intercourse between men and women in the interest of population replacement and growth. The Shakers of colonial North America were an example of a society that attempted to curtail or eliminate heterosexuality in favor of celibacy in the interest of religious piety. Because Shaker society could continue only through recruitment of new members, it eventually declined and essentially disappeared.

While heterosexuality is practiced in some form in all societies, about one third of them actively ban homosexual practices, while many other societies tolerate them in some form (Murdock, 1983). Historically, the best-known examples of heterosexual societies that condoned some forms of homosexuality and bisexuality were ancient Greece, Confucian China, and Hawaii before it was colonialized by Europeans (Connell, 1995). In most Western cultures the norm of heterosexuality is quite strong, but older taboos against homosexualty and bisexuality are changing as well.

Homosexuality and Bisexuality The intensity of debate over homosexuality in the United States is an indication, on the one hand, of how controversial the behavior remains and, on the other, of the growing, if grudging, acceptance of alternatives to traditional norms

Sociogram of Friendship Choices in a Little League Team

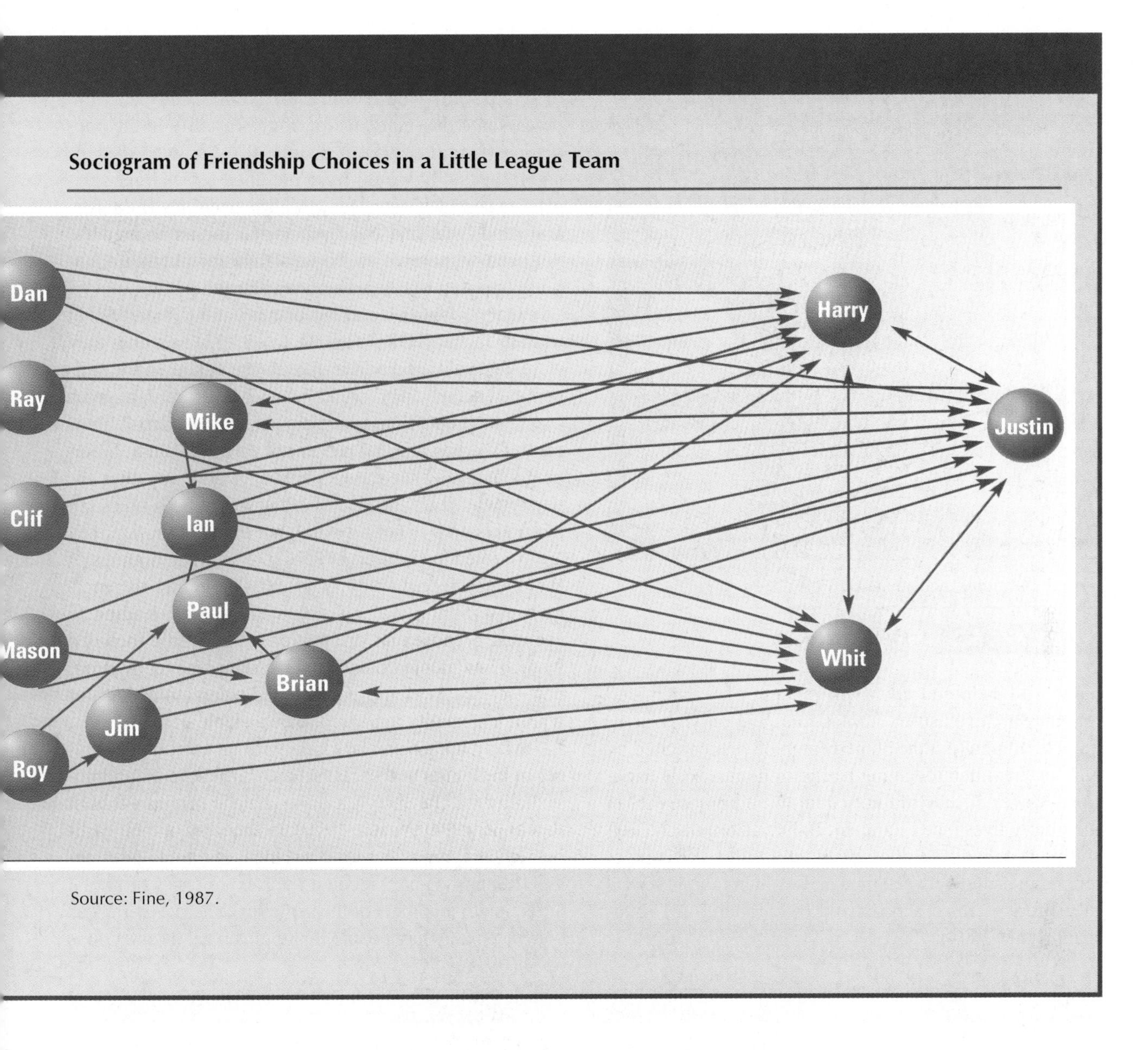

Source: Fine, 1987.

of heterosexuality. In the mid-twentieth century homosexuality was a taboo subject. Homosexual men and women remained secretive or "in the closet" about their true sexual feelings. Today there is at least one homosexual member of Congress and many other openly homosexual elected officials and public figures. At the same time, fear and loathing of homosexuals—gay men and lesbian women—is a major social problem that continues to cause immense human suffering.

Estimates of the incidence of homosexuality—that is, the proportion or number of women and men in a population who are sexually attracted to people of the same sex—raise a number of social-scientific and political problems. In his pathbreaking and courageous studies of male and female sexual behavior, Alfred Kinsey interviewed thousands of Americans about their sexual preferences and behavior. Kinsey described sexuality as a continuum extending from exclusive heterosexuality to equal attraction for the same and the opposite sex (bisexuality) to exclusive homosexuality. By his estimates, about 4 percent of men and 2 percent of women were exclusively homosexual, while far more, perhaps as many as one third of men and about 12 percent of women, said that they had had a homosexual experience leading to orgasm at least once in their lives (Kinsey, Pomeroy, & Martin, 1948, 1953). Subsequent reinterpretations of Kinsey's data led social scientists to estimate that 10 percent of the population was homosexual or had strong homosexual tendencies. This estimate became a commonly cited statistic in the politics of the gay rights movement. However, the most recent major study of sexuality in the United States, conducted by the National Opinion Research Center, finds this estimate to be too high. For instance, in the NORC sample about 4

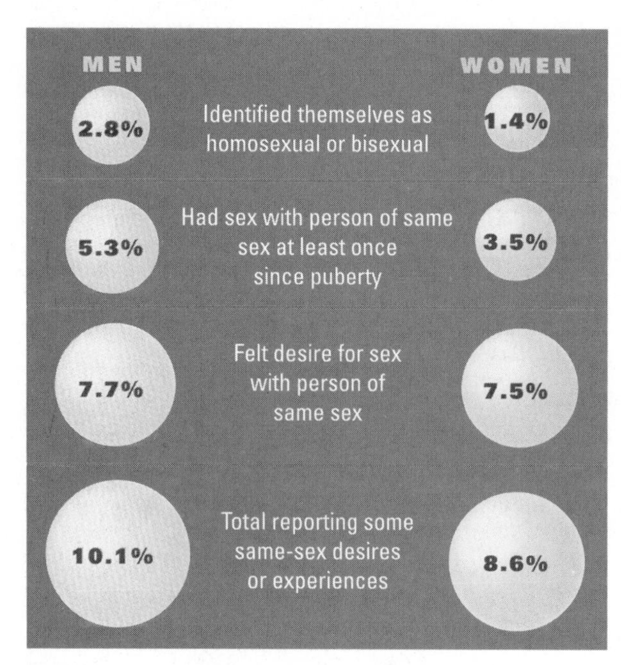

Source: Data from Laumann et al., 1994.

FIGURE 14.2 **Defining "Gay"**

percent of men and about 1.4 percent of women said that they are exclusively homosexual, but only about 10 percent of men (not the 30 percent to 35 percent cited by Kinsey) admitted to having had prior homosexual experiences even if they did not claim to be homosexual in their daily lives (see Figure 14.2). Data from the NORC study also indicate that there are wide differences among men and between men and women in how sexual pleasure is defined and derived (see Table 14.2) (Laumann et al., 1994).

Given the anxiety that many people feel about discussing their sexual orientation and behavior, however, there are some who feel that the NORC study may underestimate the incidence of homosexuality in the population (Gould, 1995). It should also be noted that the ambitious study would have included a far larger sample and somewhat greater scientific accuracy if the federal funds that had been scheduled for its budget had not been withdrawn by conservative members of Congress opposed to research on sexuality.

Only about 0.8 percent of men and 0.9 percent of women in the NORC sample (n=3,432) said that they were bisexual. But when they were asked more indirectly whether they were sexually attracted to both sexes, the proportions who said yes increased to 4.1 percent of women and 3.9 percent of men. Martin S. Weinberg observes that many people "would say that the person who feels the sexual attraction to both sexes but never acts on it is not bisexual, but in my definition they are" (quoted in Gabriel, 1995b, p. A12). The findings of Weinberg's recent study of bisexuality refute the common assumption that bisexuality is a stage leading to eventual homosexual orientation. His sample included women and men who had come to their bisexuality from both heterosexual and homosexual orientations and for whom bisexuality appeared to be a stable lifestyle.

Many homosexuals, along with an increasing number of biologists and social scientists, believe that homosexuality will be shown to have genetic origins (Tobach & Rosoff, 1994; Wilson, 1979). In accounts of their earliest sexual feelings homosexual men, and to a lesser degree lesbian women, recount their experiences of sexual attraction to members of the same sex. These experiences convince many gay people that their sexual orientation is

TABLE 14.2 **The Appeal of Various Sexual Practices (percentages of respondents)**

	Appealing to Men Ages 18–44				Appealing to Women Ages 18–44			
	Very	Somewhat	Not Really	Not at All	Very	Somewhat	Not Really	Not at All
Vaginal intercourse	83%	12%	1%	4%	78%	18%	1%	3%
Watching partner undress	50	43	3	4	30	51	11	9
Receiving oral sex	50	33	5	12	33	35	11	21
Giving oral sex	37	39	9	15	19	38	15	28
Active anal intercourse	5	9	13	73	—	—	—	—
Passive anal intercourse	3	8	15	75	1	4	9	87
Group sex	14	32	20	33	1	8	14	78
Same-sex partner	4	2	5	89	3	3	9	85
Sex with a stranger	5	29	25	42	1	9	11	80
Forcing someone to do something sexual	0	2	14	84	0	2	7	91
Being forced to do something sexual	0	3	13	84	0	2	6	92

Source: Laumann et al., 1994.

not merely a "lifestyle choice," as critics often claim. However, as yet there is insufficient evidence from genetic studies to resolve this issue (Kemper, 1990).

Homophobia Most of the young boys in Barrie Thorne's study of gender play used negative images of homosexuals well before they ever encountered a gay person. Fear of homosexuals and same-sex attraction is known as *homophobia.* Although it is quite common in the United States, its causes are not entirely clear. Most social-scientific (as opposed to religious or ideological) explanations of homophobia hinge on an analysis of the problems of masculinity and the male role in Western societies.

Identification with the male gender and the ability to take male roles, as these are defined for young boys and later for men, are two different aspects of what it means to "be a man." Every society has its notions of what distinguishes men from women, and seeks to teach boys and girls how to perform the roles assigned to their gender. In Western societies the male role is often depicted in movies, on television, and in advertising as distinct from that of the female and imbued with more strength, power, and rationality—a subject to which we return later in the chapter. Because as children they lack the power and strength they admire in images of masculinity, young boys typically become concerned about being seen as lacking in masculinity and the ability to take on male roles. Thus their vocabulary is rich with terms of abuse for anyone they believe to be lacking in masculinity: wimp, nerd, turkey, sissy, lily liver, yellowbelly, candy ass, ladyfinger, cream puff, mother's boy, dweeb, geek, and so on (Connell, 1995). As they grow older, a small but socially significant proportion of boys may become violently homophobic and go out of their way to abuse males whom they consider effeminate or "queer."

Girls may also develop homophobic attitudes early in life, often through identification with their male siblings, but homophobia among females often occurs during the teenage years. Female socialization stresses the ability to attract men, to appear feminine, and anxiety over these aspects of the female role may lead young women to reject other women whom they do not regard as feminine enough.

These early patterns can have lasting consequences. Throughout their lives the early habit of avoiding emotions for fear of seeming effeminate tends to make men in the United States and other Western societies wary of expressing their feelings and of admitting vulnerability: "Men are thus denied an important part of their . . . well-being when they cannot touch and cannot express their tender feelings for other men. Needless to say, gay men are damaged by the negative implications of homophobia" (Blumenfeld, 1992, p. 37). And of course homophobia causes untold suffering for gay men and women (Richmond-Abbott, 1992).

Fundamentalist religious beliefs are another important source of negative attitudes and strong disapproval of homosexuality, including vehement homophobia. For example, 77 percent of fundamentalist Baptists disagree or strongly disagree with the statement "Even if homosexuality is wrong, the civil rights of gays should be protected." They believe that homosexuality is immoral and inimical to the propagation of the species. In the United States it is common for prominent fundamentalist Christian leaders to state that AIDS is God's punishment for the homosexual lifestyle (Ammerman, 1990). Despite the vehemence of such religiously motivated attacks, gay men and women in smaller towns and communities with fundamentalist congregations are increasingly choosing to assert their right to live as they choose as long as they do not violate the rights of others (Miller, 1989).

Controversies over male and female roles, homosexuality, norms of sexual conduct, and the origins of sexual orientation all point to the extent to which these central areas of social life are subject to change and reactions to change. In the study of gender inequalities, the origins and patterns of change in the condition of women and the continuing inequalities between the sexes are a central focus, to which we now turn.

GENDER STRATIFICATION

After class and race, the most important dimensions of inequality in modern societies are gender and age. As stated earlier, gender refers to a set of culturally conditioned traits associated with maleness or femaleness. There are two sexes, male and female; these are biologically determined ascribed statuses. There are also two genders, masculine and feminine; these are socially constructed ways of being a man or a woman. **Gender roles** are the sets of behaviors considered appropriate for individuals of a particular gender. Controversies over whether women in the armed forces can serve in combat or whether men with children ought to be eligible for family leave from work are examples of issues arising out of the definition of gender roles.

All human societies are stratified by gender, meaning that males and females are channeled into specific statuses and roles. "Be a man"; "She's a real lady"—with these familiar expressions we let each other know that our behavior is or is not conforming to the role expectations associated with our particular gender. When women's roles are thought to require male direction, as is the case in many households and organizations, the unequal treatment of men and women is directly related to gender roles. The roles assigned to men and women are accorded differing amounts of income, power, or prestige, and these patterns of inequality contribute to the society's system of stratification.

Gender Roles: A Cultural Phenomenon

Until quite recently it was assumed that there were two spheres of life, one for women and the other for men. The chief agents of socialization—church, family, and school—worked effectively together to transmit and legitimize the notion that boys would grow up to be leaders in the world outside the home, while girls were expected to become wives and mothers whose involvement in the world outside the home would be more indirect. Out of this gender-based division of labor, which defined the activities that were appropriate for men and women, grew the notion of differences in men's and women's abilities and personalities. These differences were thought to be natural—an outgrowth of biological and psychological differences between males and females (Epstein, 1985). Behaviors that did not fit these patterns were viewed as deviant and in some cases as requiring severe punishment.

In the twentieth century, evidence from the social sciences has called into question the assumption that there are innate biological or psychological reasons for the different roles and temperaments of men and women. Margaret Mead's famous research in New Guinea directly challenged this assumption. Mead was one of the first social scientists to gather evidence to show that gender-specific behavior is learned rather than innate. In her study of gender roles in three tribes, Mead (1950) found that different tribes had different ways of defining male and female behavior. In one tribe, the Mundugumor, men and women were equally aggressive and warlike, traits that Westerners usually associate only with men. In a second tribe, the Tchambuli, the men spent their time gossiping about women and worrying about their hairdos, while the women shaved their heads and made rude jokes among themselves. In the third tribe, the Arapesh, both men and women behaved in sympathetic, cooperative ways and spent a great deal of time worrying about how the children were getting along, all behaviors that Westerners traditionally associate with women.

Mead's research has been criticized in that she may have been actively looking for gender-role patterns that differed from those that Westerners usually associate with males and females. Nevertheless, her study was highly significant, since it began a line of inquiry that established that gender roles are not innate. Nor are gender roles wholly determined by a society's relationship to its environment. Although hunting-and-gathering societies sent men out to hunt while women cared for the home, in early agrarian societies there was a less rigid division of labor. In early horticultural societies women had more power than they did in hunting-and-gathering societies or in later feudal societies. Women maintained the grain supply and knew the lore of cultivation, and they were priestesses who could communicate with the harvest and fertility gods (Adler, 1997; Balandier, 1971/1890).

The lesson of this cross-cultural research is that gender roles are heavily influenced by culture. Although the relationship of earlier societies to their natural environment often required that women tend the hearth and home while men went out to hunt for big game, women were also hunting for small game around the encampment and experimenting with new seeds and agricultural techniques. The division of labor by gender was never fixed; it could always be adapted to new conditions. In industrial societies, as the greater strength of males becomes less important as a result of advances in technology, it made less sense to maintain the earlier divisions of labor. Indeed, modern societies have demanded more involvement of women in a broader range of tasks. Women are now competing with men as military and police officers, engineers, scientists, judges, political leaders, and the like; they may be found in many roles that were assumed to be unsuitable for women less than half a century earlier.

Table 14.3 shows that women have increased their share of employment in many occupational groups that were formerly male "turf." The proportions of female executives and managers have increased dramatically, as have those of technicians and professionals. But older barriers and assumptions continue to stand in the way of equal access to male-dominated occupations such as precision production. Conversely, women continue to be disproportionately represented in "pink-collar" occupational sectors such as clerical and administrative support and domestic service (Bianchi & Spain, 1986).

Historical Patterns of Gender Stratification

In preindustrial societies gender often plays a greater role in social stratification than wealth and power. Fewer members of such societies are wealthy or powerful than is true in modern societies, but all are male or female. Cross-cultural research has provided some insight into the development of gender-based stratification in such societies. Among the main findings of that research are the following:

- Preindustrial societies are usually rigidly sex-segregated. Males and females tend to pass through

TABLE 14.3 Percentage of Women in Major Occupational Groups, 1970–1997

Major Occupational Group	1970	1997
Executive, administrative, and managerial	18.5%	48.9%
Professional specialty	44.3	53.3
Technicians and related support	34.4	51.9
Sales occupations	41.3	50.2
Administrative support, including clerical	73.2	78.8
Private household	96.3	95.4
Protective service	6.6	17.9
Other service	61.2	64.0
Precision production, craft, and repair	7.3	8.9
Machine operators, assemblers, and inspectors	39.7	37.7
Transportation and material-moving occupations	4.1	9.6
Handlers, equipment cleaners, helpers, and laborers	17.4	20.3
Total	38.0	46.2

Sources: Data from Bianchi & Spain, 1986; and *Statistical Abstract,* 1998.

life in cohesive peer groups. Even after marriage, women and men tend to spend more time with their same-sex peers than they do with their spouses.

- As societies increase in size and complexity, women usually become subordinated to men. There have been few if any societies in which women as a group controlled the distribution of wealth or the exercise of power (Harris, 1980; Leacock, 1978).

Larger, more complex societies also exhibit distinct patterns of gender-role stratification, but these are more likely to be part of a multidimensional system of stratification in which class, race, ethnicity, gender, and age are intertwined.

The origins of gender inequality in most modern societies can be traced to their feudal periods. Most industrial societies developed out of feudal societies, either as a result of revolutions (in Europe) or through changes brought about by colonialism (in North and South America, Asia, and parts of Africa). Although they are no longer as easily justified, many of the norms that specify separate spheres of activity for males and females, as well as the subordination of women to men, were carried over into modern societies.

In some feudal societies these norms were far more repressive than those found in modern societies; in others, the subordination of women was disguised as reverence or worship. In European feudal societies, for example, the norms of chivalry and courtly love seemed to elevate women to an exalted status, but in reality they reinforced practices that kept women in undervalued roles. They gave rise to a set of norms that specified that women do not initiate sexual activity, do not engage in warfare or politics but wait for men to resolve conflicts, and do not compete with men in any sphere of life beyond those reserved for women.

This medieval illustration depicts the norms of chivalry in action. Notice the women standing on the battlements, aloof from the fray. Their lot is to watch the combat and hope that their hero will not be skewered by his opponent.

Then and Now

From Babushka to "Cosmo Girl"

Who would have predicted that the international magazine publishing success story of the 1990s would be Russia's *Cosmo?* With the end of the Soviet empire, Russia and its former satellites have become one of the world's marketing hot spots. Under the stern control of Stalinist communist leaders, the Russian people longed for access to some of the consumer goods produced in the West. Women were especially oppressed under Soviet communism, despite that system's claims to have ended gender stratification. A few Soviet women were known to the world as graceful figure skaters or gymnasts, but the dominant image of the Soviet woman was as a toiler in the field, a sullen factory hand, or a swaddled street sweeper in a lumpy gray overcoat and babushka. More educated and accomplished women also spoke openly about how unromantic and cloddish they felt their men to be (Gray, 1990). But none of this explains why the Russian version of *Cosmopolitan* would be a runaway hit.

Cosmo's publisher, Anne Marie Van Gaal, had started the *Moscow Times,* a wildly successful English-language newspaper. Observing the success of *Cosmopolitan* in the United States, Van Gaal realized that a magazine containing explicit articles about how to be romantic and glamorous, with a generous element of the beauty ideal and much appeal to women's fantasy lives, could be successful with the rising Russian middle and upper classes, just as it had been first in the United States and then in many other nations with increasing numbers of educated young women. She was right: *Cosmo*'s sales now exceed half a million per issue.

The situation of Guinevere in the tales of King Arthur aptly illustrates this point (de Rougemont, 1983; Elias, 1978/1939). Courtly love applied only to women of the nobility. It specified that a woman must be chaste—either a virgin or entirely faithful to her husband at all times. Like Guinevere, she could be worshiped from afar by a noble knight (Lancelot), but she could not be touched or even spoken to without the consent of her male guardian (King Arthur). Figuratively she was set upon a pedestal. In reality, however, she was imprisoned—kept forever separate from the knightly suitor who worshiped her from a distance and engaged in chivalrous acts in her name. Adultery and the subordination of women were common in the lower orders of feudal society, but the women of the nobility were more likely to be constrained by the norms of chastity.

As feudal societies developed and changed, the norms of courtly love were extended to the new middle classes, and in greatly modified forms they still exist today. They can be seen, for example, in the traditions of courtship that persist in Spanish-speaking nations, in which a man serenades a woman (or hires someone to do it on his behalf) and the woman is never without a chaperone. It should be noted that courtly (or romantic) love eventually elevated the status of women from that of property to that of significant other, a change that had far-reaching implications for the position and influence of women in the family. At the same time, in many Western cultures norms derived from medieval notions of courtly love continue to justify the notion that there are "good women," whom one reveres and protects, and "bad women," who are available for sexual exploitation. This denial of normal female sexuality and the assertion of the male's sexual needs is associated with a host of psychological and social problems, including the conflict that some women feel about their sexuality and the inability of some men to relate to their wives as sexual partners.

GENDER INEQUALITY IN INDUSTRIAL SOCIETIES

In modern industrial societies age and gender interact to shape people's views of what role behavior is appropriate at any given time. Before puberty, boys and girls in the United States tend to associate in sex-segregated peer groups. Because they model their behavior on what they see in the home and on television, girls spend more of their time playing at domestic roles than boys do; boys meanwhile play at team sports more than girls do. These patterns are changing at different rates in different social classes, but they remain generally accepted norms of behavior. And they have important consequences: Women are more likely to be socialized into the "feminine" roles of mother, teacher, secretary, and so on, while men are more likely to be socialized into roles that are considered "masculine," such as those of corporate manager or military leader. It is expected that men will concern themselves with earning and investing while women occupy themselves with human relationships (Rossi, 1980; Witt, 1997).

Childhood socialization explains some of the inequalities and differences between the roles of men and women, but we also need to recognize the impact of social structures. In the United States, for example, it was assumed until fairly recently that boys and girls needed to be segregated in their games. Boys were thought to be much stronger and rougher than girls, and girls were thought to need protection from unfair competition with boys. This widespread belief translated into school rules that did not permit coeducational sports. Those rules, in turn, reinforced the more general belief that girls' roles needed to be segregated from those of boys. Such patterns have significant long-term effects. As sociologist Cynthia Epstein points out, human beings have an immense capacity "to be guided, manipulated, and coerced into assuming social roles, demonstrating behavior, and expressing thoughts that conform to socially accepted values" (1988, p. 240). Through such means, gender roles become so deeply ingrained in many people's consciousness that they feel threatened when women assert their similarities with men and demand equal opportunity and equal treatment in social institutions.

In their adult years men enjoy more wealth, prestige, and leisure than women do. Working women earn less than men do, and they are frequently channeled into the less prestigious strata of large organizations. Even as executives they are often shunted into middle-level positions in which they must do the bidding of men in more powerful positions. Similar patterns are found in all advanced industrial nations.

The Gender-Poverty Ratio

Earlier in the chapter (see Table 14.1) we showed that women earn less than men in the same occupations. Coupled with existing patterns of gender inequality, these income disparities mean that women are more likely to be living at or below the poverty line. The U.S. Census shows that about 16 percent of women and about 12 percent of men are officially classified as poor. "The ratio of women to men's poverty—the gender-poverty ratio—was 1.30 in 1991 [following the last major census], which means that women are 30 percent more likely to be poor than men" (Casper, McLanahan, & Garfinkel, 1994, p. 594). To compare this gender-poverty gap to that in other industrial nations, sociologists use an international study of income and poverty known as the Luxembourg Income Study. That study classified people as poor if they "live in a household whose disposable income (after taxes and government cash transfers) is less than 50 percent of the median disposable income for all households in that country" (Casper, McLanahan, & Garfinkel, 1994, p. 595). This definition (with adjustments for differing household size) is based on a relative measure rather than the absolute income standard used to determine the official "poverty line" in the United States. It allows investigators "to examine differences in poverty between men and women in different countries relative to the common standard of living in those countries" (p. 596).

Some findings of this international comparison are presented in Figures 14.3 and 14.4. While the figures show that the United States lags behind other industrial nations in reducing the gender-poverty gap, an analysis of the findings reveals that the causes of these differences are quite varied (Pressman, 1998). Disparities in employment, with women concentrated in lower-paying jobs, is a major factor in explaining the situation in the United States, but some nations do not have high rates of female labor force participation. In those nations—Italy, for example—female poverty is reduced by the tendency for women to marry men who can support them, and to remain married. In still other nations social policies play an important role. In "social democratic" nations like Sweden there is less gender integration in occupations than in the United States. Single women with young children are not obliged to work but are given generous subsidies through income support, child care, education, and other policies (Kammerman & Kahn, 1993). As a result, rates of female poverty are low.

In sum, comparative research indicates that gender inequality and female poverty rates are very much related to the specific workings of other social institutions—particularly those of the economy and government—and that they can be reduced when the society has the resources and will to do so.

Sexism

Gender stratification is reflected in attitudes that reinforce the subordinated status of women. The term **sexism** is used to refer to an ideology that justifies prejudice or discrimination based on sex. It results in the channeling of women into statuses considered appropriate for women and their exclusion from statuses considered appropriate for men. Sexist attitudes also tend to "objectify" women, meaning that they treat women as objects for adornment or sex rather than as

FIGURE 14.3 Poverty Rates of Women and Men (percentages)

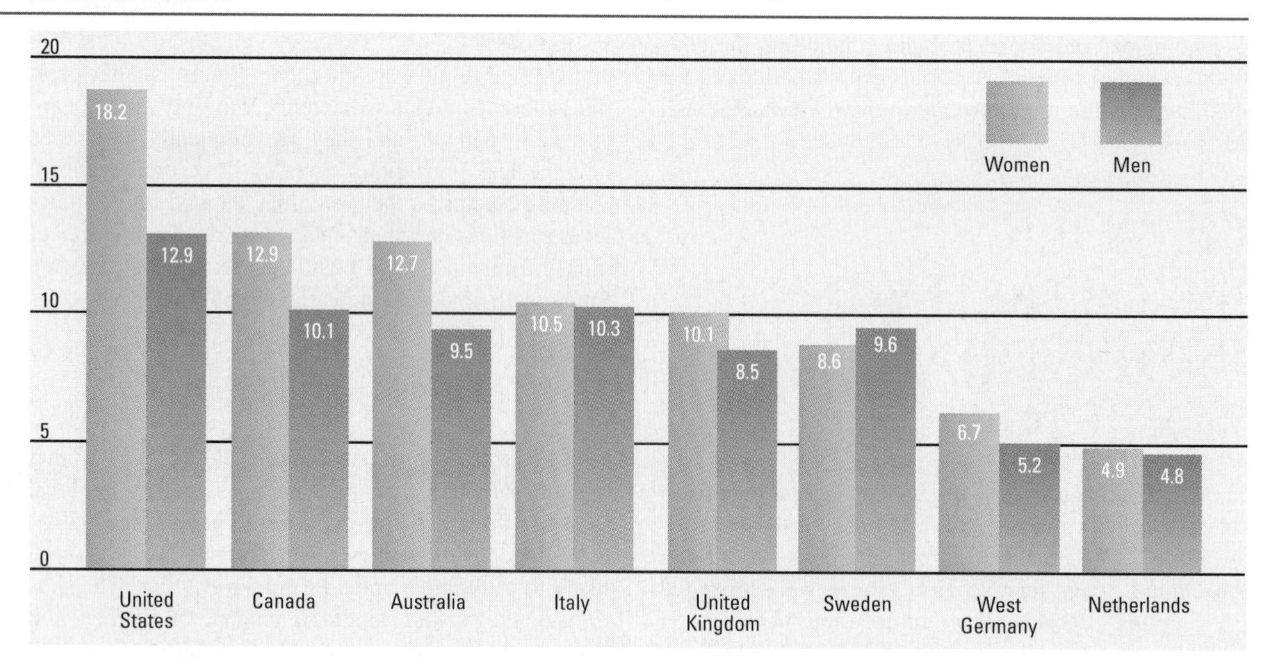

Source: Pressman, 1998.

individuals worthy of a full measure of respect and equal treatment in social institutions. This can be seen in the case of beautiful women. Such women receive special treatment from both men and women, but their beauty is a mixed blessing. The beautiful woman is often viewed as nothing more than an object for admiration. Being a woman is a master status (see Chapter 4) in that gender tends to outweigh the person's achieved statuses. This is even more painfully true for beautiful women. Someone like Marilyn Monroe is thought of only in terms of her beauty; the person beneath the surface is ignored.

The objectification of women can be seen in the beauty contest, which came into being in the United States in the summer of 1921 when the first Costume and Beauty Show was held at a bathing beach on the Potomac River. There the women wore tunic bathing suits and hats, but later that year a similar contest was held at Atlantic City, New Jersey, that eventually developed into the Miss America Pageant. In that contest women wore one-piece bathing suits that showed their calves and thighs and created a sensation in the tabloid newspapers. These contests and the publicity they generated made beauty a way for women to gain celebrity and wealth, but they also reflected the dominant male view that the most extraordinary women are those with the most stunning faces and the shapeliest figures (Allen, 1931). Women have been struggling against this view throughout the modern era, a struggle that has frequently been opposed not only by men but also by women who feel threatened by changes in their traditional statuses.

Sexism is also expressed in violence against women. Two million American women are severely beaten in their homes every year, and 20 percent of visits by women to hospital emergency rooms are caused by battering. (Thousands of men are battered by their wives each year as well, but they are far outnumbered by women victims.) In popular American culture,

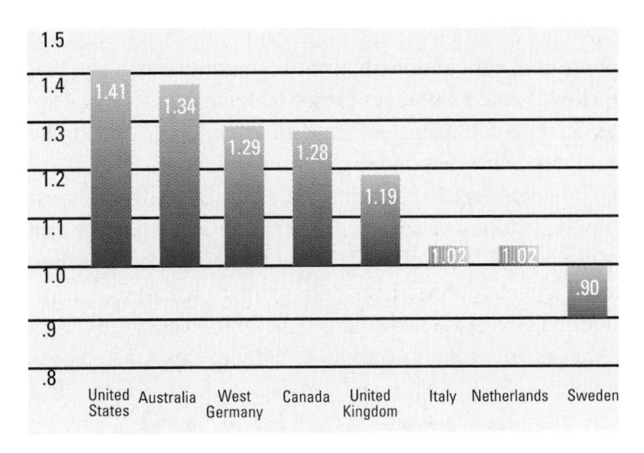

Source: Casper, McLanahan, & Garfinkel, 1994.

FIGURE 14.4 **Ratio of Women's to Men's Poverty Rates**

This photo from the 1950s shows the degree of explicit sexism in the airlines industry at the time. We have not found an equivalent photo for male flight crews. Although female airline employees still must conform to dress codes, they have formed unions and have won more stable careers and greater job security.

Men are often compelled by gender norms to behave in stereotypically "masculine" ways in public.

especially movies and television, violence against women is not condoned, but it is presented as a form of thrilling entertainment. Films like *Nightmare on Elm Street* and *Looking for Mr. Goodbar* suggest that the brightest and most independent women are most likely to be victimized. Examples like these hardly exhaust the overt types of sexism that exist in employment and other aspects of social life in the United States, but they are aspects of American culture that too often go unnoticed. Despite advances in women's rights and changes in women's access to careers, sexism remains commonplace in many areas of American life (Benokraitis & Feagin, 1986; Connell, 1995).

THE WOMEN'S MOVEMENT

In the 1950s, the women's movement was dormant. In the late nineteenth and early twentieth centuries women had organized social movements to gain full citizenship rights and greater control over reproduction (e.g., through family planning and birth control). In the United States they were known as suffragettes because they campaigned vigorously for woman suffrage—the right of women to vote. But in 1920, after the Nineteenth Amendment to the Constitution extended suffrage to women, the movement faded.

The victory of the suffragettes did not change the patterns of gender stratification, however. For example, in their famous studies of Middletown, Helen and Robert Lynd (1929, 1937) found that women who had worked their way into good factory jobs and professional occupations were severely set back during the Great Depression. Like blacks and members of other minority groups, women were the last to be hired and the first to be fired when times were bad. And although women were hired as factory workers in unprecedented numbers during World War II, they often faced discrimination and harassment and were usually "bumped off" their jobs by returning GIs after the war (Archibald, 1947). Nor did this situation change greatly when women began entering the professions and the corporate world in the 1950s. Women still tended to be limited to roles that were either subordinate to those of men or part of an entirely separate sphere of "women's work."

Early in the twentieth century, women in the suffrage movement were fighting for the right to vote. Their movement often met with organized resistance from men.

In her review of the origins of the modern women's movement, Jo Freeman (1973) wrote that even in the mid-1960s the resurgence of the women's movement "caught most social observers by surprise." After all, women were widely believed to have come a long way toward equality with men. Their average educational level was rising steadily and they were gaining access to the better professions and even to positions in the top ranks of corporate management. Freeman suggested that the movement developed out of an existing network of women's organizations that made many women aware of the inequality still prevailing in American society and encouraged them to organize to demand equal rights. Influential writing by women activists, such as Betty Friedan's *The Feminine Mystique* (1963), and studies of gender inequality by social scientists like Alice Rossi (1964) and Jessie Bernard (1964) played a major part in this growing awareness.

Meanwhile another, less formal network of women was developing. Women who were experiencing sexism within other social movements—the antiwar movement, the civil rights movement, the environmental protection movement, and the labor movement—began to form small "consciousness-raising" groups (Morgan, 1970). By questioning traditional assumptions and providing emotional and material support, these groups attempted to develop a sense of "sisterhood." Their members formed a loosely connected network of small groups that could mobilize resources and develop into a larger and better-organized movement.

The women's movement achieved significant victories during the 1960s and 1970s. In 1963 Congress passed "equal pay for equal work" legislation, and Title VII of the 1964 Civil Rights Act prohibited discrimination against women. In the 1970s the federal Equal Employment Opportunity Commission began to enforce laws barring gender-based discrimination in employment. Lawsuits filed under the Civil Rights Act and other federal laws forced many employers to pay more attention to women's demands for equality in pay and promotion and an end to sexual harassment on the job. Research shows that men very slowly began to accept roles that had previ-

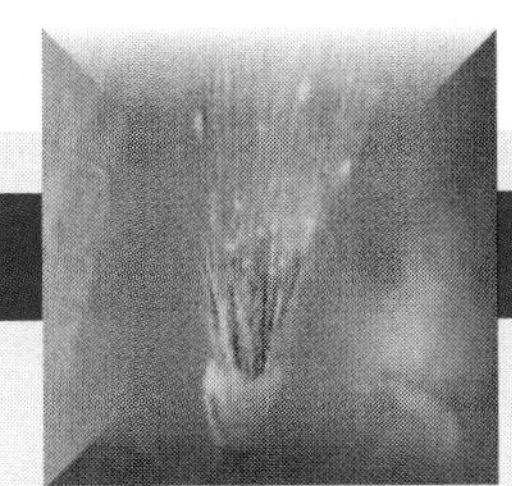

ART AND THE SOCIOLOGICAL EYE

Feminism and Art

Each day thousands of visitors crowd into the museums of major European and American cities. By far the majority of the artists whose works they come to view are white males. Is this because white males have so much more talent than women or people from minority backgrounds? Obviously the answer is no. It was not until the late nineteenth century that women began to have the opportunity to study art and display their talents—but once they did, they proved themselves just as capable of artistic expression as men. The same is true of African Americans and members of other minority groups, who were once excluded from access to careers in art.

After entering the museums, visitors often spend time looking at male representations of the nude female form, works that are usually explained (through the unacknowledged "lens" of androcentrism) as expressing a universal vision of beauty. But if people visit the city's more innovative museums and galleries, the works they view are just as likely to have been created by women as by men. And whether the artists are white women or women of color, their work is likely to express strong feminist themes. The art of Barbara Kruger, shown here, is one of many examples one might offer. Kruger's work appropriates the everyday techniques of newspaper headlines and photos to make strong but slightly ambiguous statements about the situation of people, especially women, confronted by institutional sexism, racism, or bureaucratic power. And like other feminist artists in the United States today, Kruger's work often challenges stereotypes about powerlessness, submission, sexuality, and other androcentric representations of women in society.

"Untitled (We have received orders not to move)," by Barbara Kruger, 1982. Photograph, 72″ × 48″. Courtesy Mary Boone Gallery, New York.

TABLE 14.4 Women's Earnings (per $1,000 received by men)

Occupation	1983	1995
More than 10% improvement		
Chefs and cooks	$711	$885
Realtors	683	794
Production inspectors	563	649
Waiters and waitresses	721	822
Public administrators	701	786
Computer analysts	773	860
Less than 10% improvement		
Journalists	$782	$855
Retail sales	636	693
Insurance adjustors	651	691
Financial managers	638	674
Education administrators	671	708
Janitors and cleaners	810	844
Engineers	828	862
Accountants	706	734
College faculty	773	781
Deterioration		
High school teachers	$886	$881
Health technicians	839	813
Electronic assemblers	857	808
Lawyers	890	818
Physicians	816	649

Source: Hacker, 1997.

ously been considered "women's work" (Rosen, 1987), and at work they were more likely to have women as supervisors than was true only a decade earlier. Throughout American society the attitudes of both men and women were becoming more favorable toward a more equal allocation of political and economic roles (NORC, 1998).

Despite these changes, gender equality is far from complete. Some survey data indicate that there is greater sharing of household tasks, with men doing more of what was once viewed as "women's work" around the home (Gallup, 1997). However, a closer analysis shows that it is premature to assume there have been far-reaching decreases in the sexual division of labor within the average American household. Studies that ask men and women whether the men are doing more housework, and what kinds of work they are actually doing, tend to come up with quite optimistic figures compared to those obtained in similar surveys a generation ago. But studies based on detailed time-use diaries find far less actual participation by men in housework chores than is reported in the opinion polls. This indicates that men and women may be embarrassed to admit that men are not taking on a fair share of responsibilities for housework and child care, and hence may inflate their estimates in response to pollsters' questions. Time diaries call for detailed information about how time is actually used, and in these diaries people tend to report their time use more accurately (Press & Townsley, 1998).

The Second Shift

Sociologist Arlie Hochschild (1989) coined the term *the second shift* to describe the extra time working women spend doing household chores after working at a job outside the home. This term emphasizes the expectation that women who work will also perform the bulk of domestic and child care duties. This is an example of the persistence of **patriarchy,** the dominance of men over women. Although women often attempt to influence their male partners to share household chores that were formerly considered "women's work," the men continue to resist despite slow changes—especially in two-career families in which both partners are highly educated. These topics are discussed further in Chapter 16.

Some sociologists have proposed that as women gain greater parity with men in occupations, the disparities between men and women's roles will diminish and gender roles in the household will become more symmetrical (Bernard, 1982; Willmott & Young, 1971). But as Hochschild (1989) and others report, this remains a speculative hypothesis despite the exceptions one might cite here and there. On the average, data on time budgets show that working women in the United States have at least 10 hours a week less leisure time than their husbands or partners because they shoulder far more of the domestic responsibilities (Press & Townsley, 1998). The same disparities do not exist in all industrial nations. France and Holland, for example, incorporate gender equality into national social policies. These nations provide excellent universal child care, family leave, and longer vacations than do lagging nations like the United States, and these policies ease the burdens of dual-earner families in coping with the demands of work and family life (Hacker, 1997).

Outside the home, patterns of gender inequality are also proving to be extremely persistent. Table 14.4 shows that between 1983 and 1995 there were occupations in which the gap between women's and men's earnings for comparable work narrowed, but in many others the decrease was less than 10 percent, and in some, such as high school teachers and health technicians, there was actually a widening of the pay gap. This persistent pattern of wage inequality, together with the tendency for women to be segregated in such occupations as secretary, bank teller, and elementary school teacher, has led activists in the women's movement to call for new policies designed to reduce inequality. Increasingly they have pressed for "comparable worth"—that is, for increases in the salaries paid in traditionally female occupations to levels comparable to those paid in similar, but traditionally male, occupations (Reid, 1998).

WOMEN AT WORK

In the industrial nations of North America, Europe, and Oceania, the increasing proportion of women who are in the labor force as paid employees is one of the most important aspects of social change in those societies. At midcentury, for example, 34 percent of U.S. women were in the labor force; but that proportion increased steadily. By the 1990s, about 70 percent of women considered themselves to be employees, either at work or looking for work (*Statistical Abstract,* 1998). Among women with children over age 20 the trends are similar. In 1960, just over 20 percent of these older women had jobs or were looking for jobs; by the late 1990s, this figure had more than doubled, to almost 47 percent (*Statistical Abstract,* 1998).

In fact, women assumed an important economic role well before industrialization created a sharp distinction between work in the paid labor force and work at home or in the fields. When the United States was an agrarian society, women planted and harvested crops, including extensive household gardens, and were also expected to take responsibility for housework and child rearing, both of which were highly labor intensive in the pre-electricity era. And because women have higher life expectancies than men, widows routinely ran farms or businesses after their husbands died, and many middle-class women gained additional income by taking in lodgers and selling handicrafts or other products of their labor. With the industrial revolution came growing demand for factory workers. Although male workers were often preferred, thousands of women and children also swelled the ranks of the new industrial working class. A recent study has shown that many of the working women in the early years of the industrial revolution were immigrants or poor single women from rural backgrounds. Large numbers of African American women were also working for wages in fields and factories and as domestic servants (Hacker, 1997).

World War II marked a significant turning point in women's labor force participation. Although returning servicemen "bumped" women (and minority men) from jobs in factories and offices, there were thousands of war widows and single women who had to continue earning wages whether they wanted to or not. As a result, although in the 1950s women's labor force participation was still far lower than it is today, in fact female employment was increasing at a far faster rate than male employment.

At the same time, the 1950s were a time of relative economic prosperity. The middle classes were expanding, and a suburban home became a central feature of the American dream. Popular culture, including the powerful new medium of television, emphasized the ideal norms of the middle-class nuclear family in which the mother was a homemaker and the father worked in the labor force. This "feminine mystique" asserted that

Today much research and policy making in developing nations emphasizes the problems of women and girls. In this photo, prostitutes hired by a Calcutta health clinic to mobilize against AIDS teach others to read in the clinic's courtyard.

FIGURE 14.5 Improvement in Women's Share in Selected Occupations

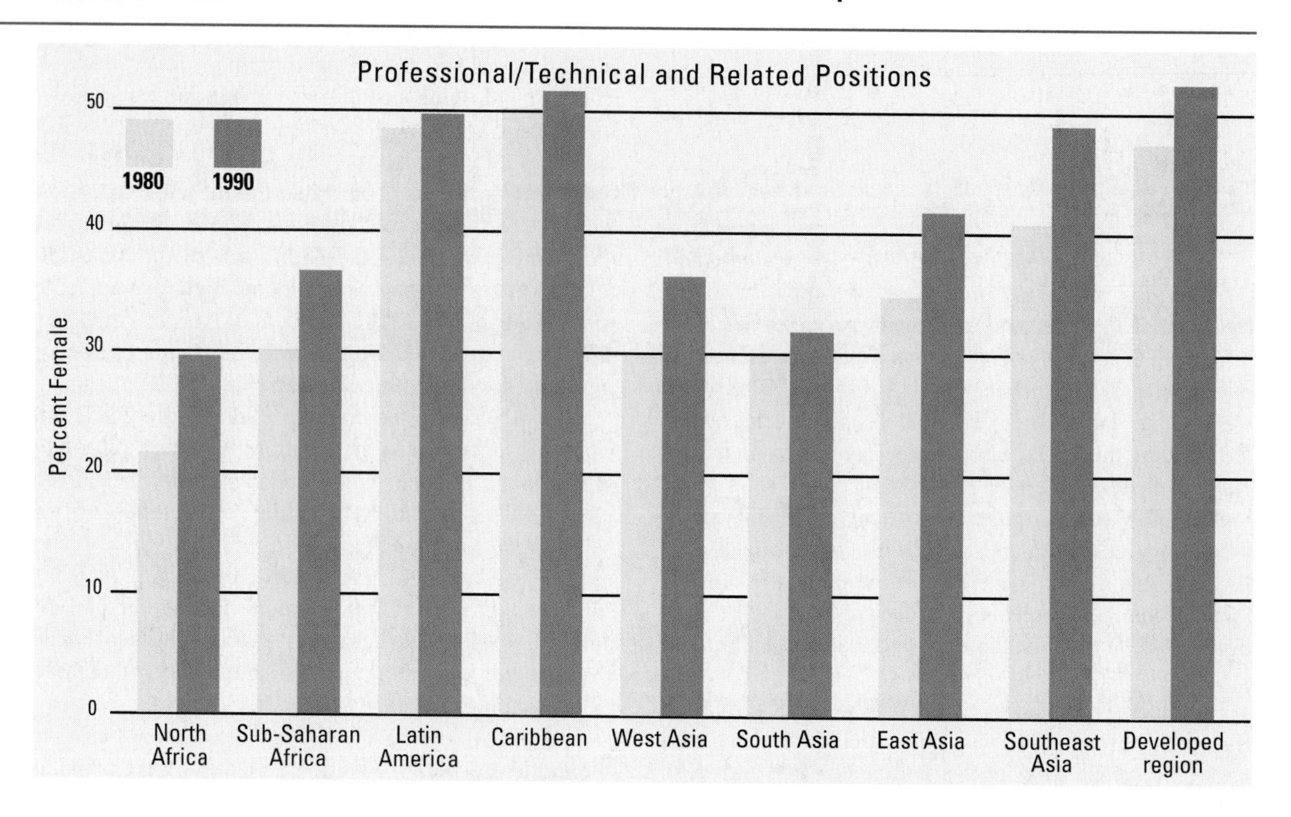

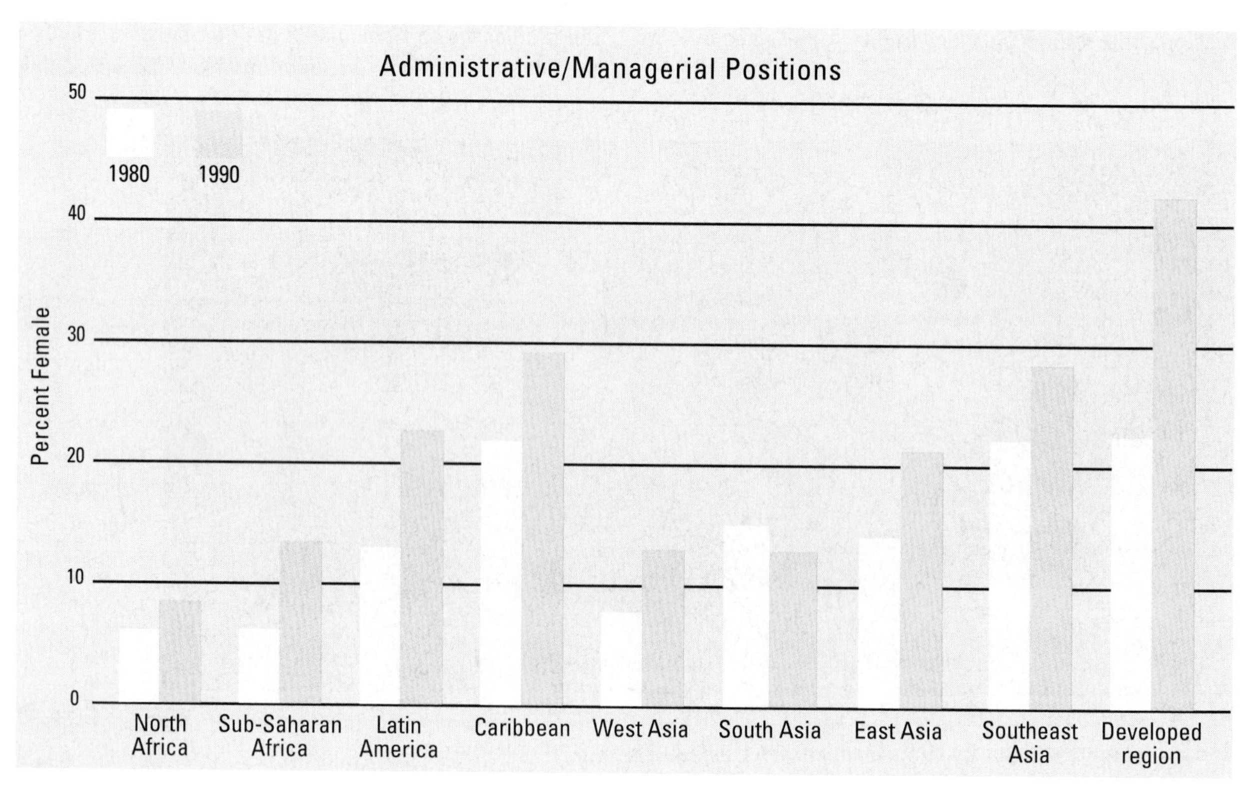

Source: Sadik, 1995.

women would find fulfillment as wives and mothers. In reality, thousands of women were entering the labor force in a trend that would continue throughout the second half of the century (Epstein, 1988).

Whether they worked outside the home because they wanted to or because necessity forced them to do so, women in the labor force encountered problems of gender segregation and discrimination. In a survey of 150,000 working women conducted during the mid-1980s by the National Commission on Working Women, the most frequently cited problems were low wages, differentials in fringe benefits, dead-end jobs with little opportunity for training or advancement, sexual harassment on the job, lack of child care or difficulty obtaining it, stress over multiple roles, and lack of leisure time (Richmond-Abbott, 1992).

An enduring problem for women in the labor force is segregation into what are known as "pink-collar ghettos." Secretarial and clerical work especially remain heavily gender segregated, but child care, nursing, and dental assistants also remain largely female occupations despite some increases in the proportions of males in these fields in recent decades. Largely as a result of antidiscrimination laws and women's efforts to break into occupations that previously were more or less closed to them, women have made considerable gains in some areas. Occupations like bus driver, psychologist, and others show significant gains in female employment, but clerical occupations remain a pink-collar ghetto: Clerical work employs one in five working women, a figure that has not changed since the 1950s (Roberts, 1995).

Throughout the world new occupations are emerging in professional and technical fields, and as women achieve greater educational parity with men the comparative statistics show that they are sharing these positions quite equally with men. Figures 14.5 and 14.6 show that this is the case in Latin America, the Caribbean, and the economically developed regions of the world, but that in much of Asia and Africa, where women still lag in education, there are wide disparities.

In administrative and managerial positions the situation is far less favorable for women. Competition for these desirable positions is much more intense and is often based not only on education but on access to social networks and political power. Since women are far less likely than men to have such access, their entry into the managerial ranks is stymied. An exception is found in the island nations of the Caribbean, where women have had access to education and training in business schools longer than in many other regions. Caribbean women often accumulate money and power through their own business activities, which enable them to become owners, managers, and administrators at a greater rate than is true in the more affluent regions of the world.

FIGURE 14.6 **Growth in Female Literacy Levels**

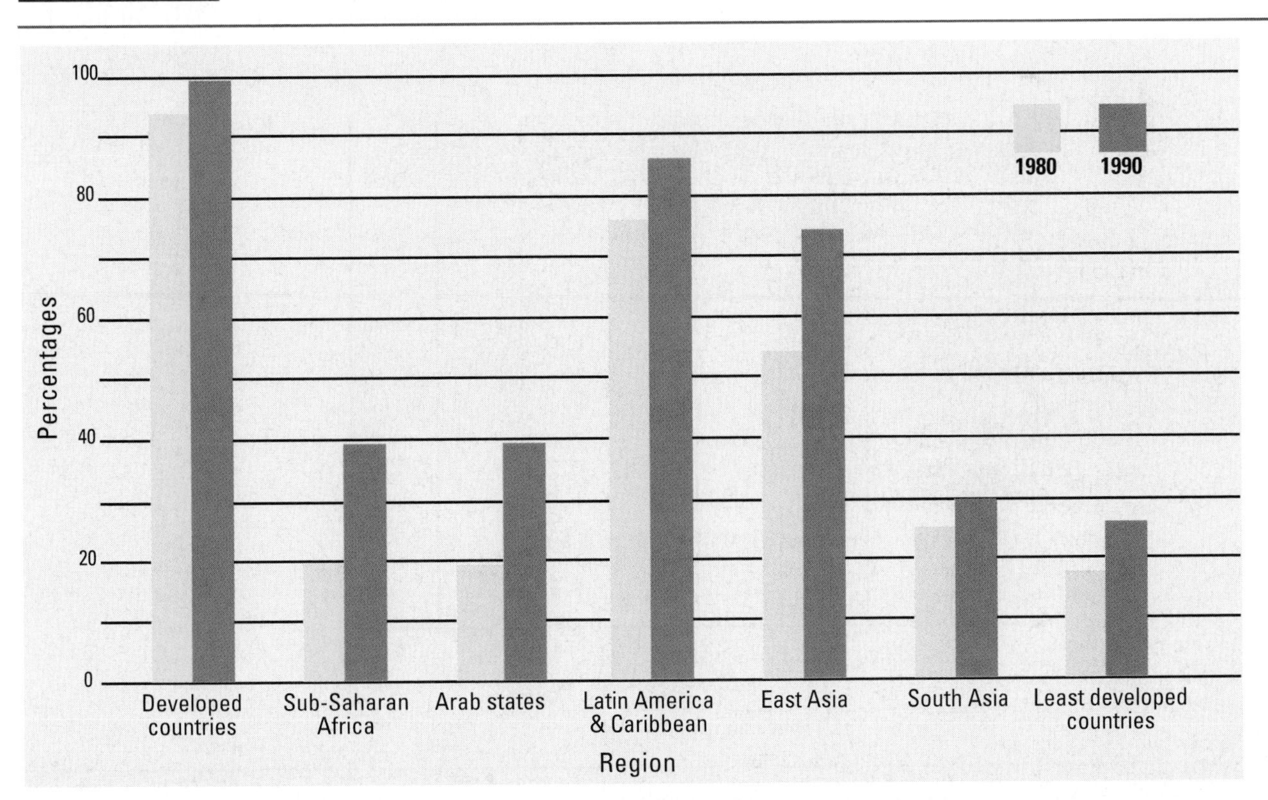

Source: Sadik, 1995.

RESEARCH ON THE CUTTING EDGE: Grassroots Warriors

The term *empowerment* is often used in discussions of women and development. The term refers to the process whereby relatively powerless people of any gender, race, ethnicity, or social class organize to assert their needs and overcome obstacles to their full participation in the institutions of their societies. The term can also refer to the processes whereby people gain the ability to assert their needs in more personal contexts—in their families, work groups, schools, or communities.

Empowerment of women is a major global issue because such a large proportion of the world's females are trapped in patriarchal societies. Indeed, the extent to which different nations move toward greater parity between male and female holders of elective office (as shown in Figure 14.7) is an international measure of relative progress in dealing with patriarchy. In the United States, which tends to lag behind other Western democracies in electing women to federal office, the empowerment of women is also a major sociological subject because so many women are victims of abuse, violence, poverty, gender discrimination, and other effects of patriarchy.

The empowerment of women is of particular concern to feminist sociologists, both male and female. Ever since the nineteenth-century movement to gain the vote for women, feminist social scientists and activists have been struggling to use theoretical reasoning, objective evidence, and persuasive argument to convince other citizens of the need for women to share political power at all levels of society, from the family to the community to the state and nation. An excellent recent example of feminist empirical sociology is Nancy Naples's research on the empowerment of women through community politics.

Naples (1998) uses a "biographical narrative approach" to explore the experience of women who became community activists in the late 1960s and who remained active in politics over the next 20 or more years. This method is a form of qualitative research in which a "panel" of representative subjects are reinterviewed periodically over a period long enough to trace their development through different stages of life. The

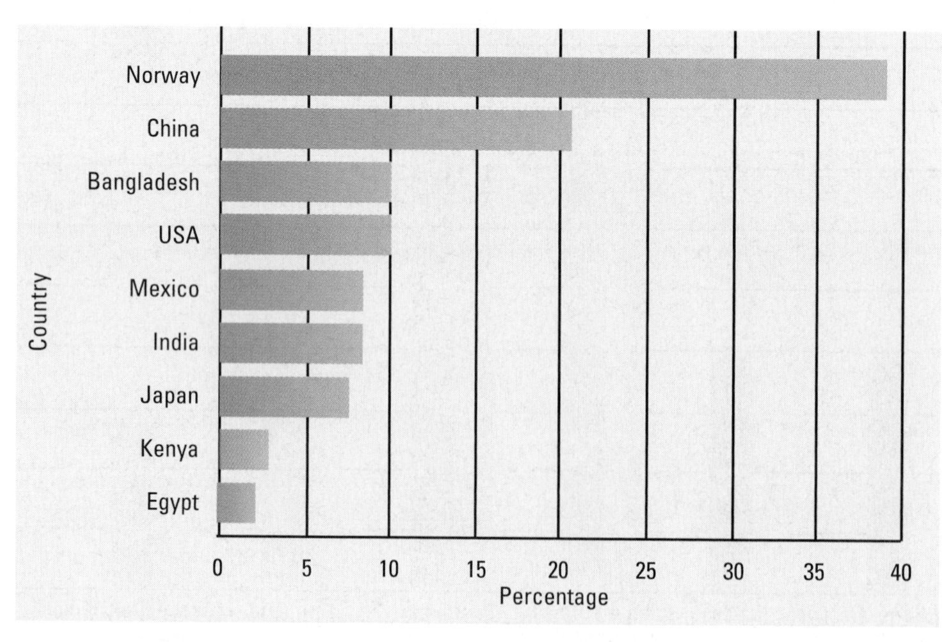

Source: United Nations, 1995.

FIGURE 14.7 **Women's Share of National Legislatures in Selected Countries**

sociologist encourages each subject—in this case mothers from low-income, predominantly minority communities in Philadelphia and Boston—to recount her biography, focusing on specific themes. In particular, Naples was interested in recording their stories (narratives) of how they began to become active in local movements to address poverty and racism in their communities. She also wanted to find out how they managed to juggle the often conflicting demands of motherhood, marriage, and jobs while maintaining their commitment to improving life in their communities. She selected the women from existing community action projects that had hired them to perform a variety of jobs. In 1995 she reinterviewed about a third of the sixty-plus women she had included in her panel more than 20 years earlier. She also interviewed the daughters of some of the women in order to "explore some unintended and rarely acknowledged intergenerational effects of mothers' community work" (1998, p. 8).

The women who told their stories to Naples were representative of the communities she visited. For the most part they were working-class women of African American, Puerto Rican, or white European descent. As mothers in low-income households—some with husbands, some without—they faced a variety of daunting challenges to their ability to be effective mothers and homemakers while also becoming active in the social movements that were developing in their communities. In the following passage Wilma North, one of the women in Naples's study, describes how she became involved in her community:

> Well, see the PTA was not my kind of thing. Back in those days the only thing PTAs did was have teas and sell candy. . . . [Well] that's the way I saw it anyway. There was too much to be done to be sitting around gossiping and having tea parties, and I never had time for that kind of stuff. I didn't then and I don't now. And there were things to be done like . . . [closing] a pinball parlor across the street from the school. Somebody needs to take some action. Or if you need crossing guards, or if there are dilapidated houses, or if you're not getting the kind of response from the police department that you feel the community needs to get. . . . And we did get it! (Naples, 1998, p.19)

North and other young mothers were recruited into what became known as community action projects. These were organizations funded by the federal government during the Johnson administration's War on Poverty. They used federal funding to provide training and jobs to people from low-income backgrounds, especially minorities and women, who would be paid to develop programs like day care, elder care, and preventive health services. North was trained in office management and administration, a career she has pursued ever since.

The community action projects were extremely controversial. Designed to give "maximum feasible participation" to residents of impoverished inner-city and rural communities, they often came into conflict with local political leaders. They were also favorite targets of critics who were opposed to government intervention in job creation and early education. But Naples's extensive interviews with women like North, whose lives were changed by their participation in the projects, yields a far more detailed account of the actual experience of the War on Poverty. Above all, Naples finds that the women almost always developed a sense of empowerment as mothers, with rights and responsibilities to help build better local institutions—day care centers, preschools, public schools, after-school centers, and the like—that would help them raise their children well despite economic deprivation in their communities and households.

VISUAL SOCIOLOGY

Gender Roles Among the Maasai

When sociologist Carol Chenault and her photographer husband, Ellis, spent time with the Maasai people in 1995, they sometimes felt as if they had wandered back in time to the biblical era. For, as Carol Chenault explains, the Maasai are true pastoralists:

> The people are Nilotic, migrating from the Nile regions of North Africa into the Great Rift Valley of Kenya and Tanzania. Myth has it that they lived in a deep crater where life was plagued with drought, famine, and discouragement. The only ray of hope for them was to follow the birds that kept bringing green grass and twigs. Scouts went up and over the steep escarpment and found green, fertile valleys with rivers and streams, and this is where they made their home. The language they speak is Maa; hence the term Maasai, meaning Maa-speaking people. The only transmission of the language is through the spoken word; their language is not written.

As we saw at the beginning of the chapter, gender role distinctions are very marked among the Maasai. But unlike Western industrial societies (where women are expected to spend more time on personal adornment), among the Maasai attention to beauty is a preoccupation of both males and females:

> Even the smallest of children are dressed with necklaces and anklets. Ear piercing is done during childhood, well before circumcision. A slit is cut in the earlobe and a green stick is placed in it. Over time this is replaced with successively larger [sticks] until the opening is large enough to be decorated with an earring and leather-beaded strands.

Chenault notes that despite their relative lack of acculturation to Western ways, she saw a Maasai man wearing a 35mm plastic film canister as an ear adornment, a vivid reminder of the spread of Western culture. The Maasai were extremely open and willing to share their culture with the visiting sociologist. "They are a proud people," Chenault concludes, "committed to passing their culture on to their future generations. I think this is the quality that I admire most about them."

These photos provide some clues about social status and gender roles among the Maasai. Notice that the elderly woman with close-cropped hair wears a limited number of personal adornments. Like most elderly women in traditional societies, she seems to feel no need to impress the viewer with her physical appearance. Not so for the woman with the distinctive headpiece and gorgeous array of necklaces. She is dressed in the fashion of the Maasai warrior herdsmen, but only the women wear these headpieces. In another photo the same woman, the wife of a prominent warrior, is shown with her children. The Maasai women bear children almost as long as they are physically capable of doing so.

The man with a shawl over his shoulders looking quizzically into the camera was one of a group who conducted their own "interview" with the sociologist. They thought it was terrible that the American sociologist was no longer having babies. The women also asked whether American women were circumcised. "When we told them we were not," said Chenault, "they clapped their hands down at their sides, indicating their disapproval."

The Maasai man playing the flute also makes sandals out of old tires. These are called "thousand milers" because the Maasai walk endlessly and need tough, cheap footwear. Making musical instruments and footgear are examples of tasks that are specifically "men's work" among the Maasai. The man wearing the distinctive feathered headdress and smiling at the sociologist is a professional guide. He has a high school education and moves easily between Westerners and Maasai. With one foot in the traditional world and another in the tourist industry, his role allows the Maasai people to remain culturally distinct while benefiting from their contact with interested visitors.

SUMMARY

Sex refers to the biological differences between males and females, including the primary sex characteristics that are present at birth and the secondary sex characteristics that develop later. Some people are born as *hermaphrodites;* their primary sexual organs have features of both male and female organs, making it difficult to categorize the person as male or female. Another ambiguous sexual category consists of *transsexuals,* who feel very strongly that the sexual organs they were born with do not conform to their deep-seated sense of what their sex should be.

Gender refers to the culturally defined ways of acting as a male or a female that become part of an individual's personal sense of self. Gender socialization in the family and in schools tends to separate males and females into different social worlds with their own forms of activity and language. However, boys and girls are increasingly participating in the same types of activities in their schools and communities, and many teachers, parents, and administrators seek to avoid the worst effects of gender separation.

Sexuality refers to the manner in which a person engages in the intimate behaviors connected with genital stimulation, orgasm, and procreation. It is profoundly influenced by cultural norms and social institutions like the family and the school, as well as by social structures like the class system of a society. Universal cultural norms exerting social control over sexuality include the incest taboo, marriage, and heterosexuality, but even these norms include variations and differing degrees of sanction.

Heterosexuality refers to sexual orientation toward the opposite sex, in contrast to *homosexuality,* which refers to sexual orientation toward the same sex, and *bisexuality,* which refers to sexual orientation toward either sex. Norms of heterosexuality function to ensure that there is genital sexual intercourse between men and women in the interest of population replacement and growth. Fear of homosexuals and same-sex attraction is known as homophobia.

All human societies are stratified by gender, meaning that males and females are channeled into specific statuses and roles. Until quite recently it was assumed that there were two separate spheres of life for men and women. Out of this gender-based division of labor grew the notion of differences in men's and women's abilities and personalities. These differences were thought to be based on biological and psychological differences between males and females. In the twentieth century, however, evidence from the social sciences established that *gender roles* are not innate but are strongly influenced by culture.

Preindustrial societies are usually rigidly sex-segregated. As societies increase in size and complexity, women usually become subordinated to men. The origins of gender inequality in most modern societies can be traced to their feudal periods. In Europe, for example, the norms of courtly love specified that women do not engage in warfare or politics or compete with men in any sphere of life beyond those reserved for women.

In modern industrial societies boys and girls are socialized into "masculine" and "feminine" roles. In their adult years men enjoy more wealth, prestige, and leisure than women do. Gender stratification in modern societies is reflected in attitudes that reinforce the subordinate status of women. *Sexism* refers to an ideology that justifies prejudice or discrimination based on sex.

The modern women's movement arose in the mid-1960s out of an already existing network of women's organizations, together with a less formal network of women in consciousness-raising groups. The movement won significant victories during the 1960s and 1970s and began to change the way men and women think about gender roles.

Women who work outside the home are also expected to perform the bulk of domestic and child care work. This "second shift" is an example of the persistence of *patriarchy,* the dominance of men over women. The most frequently cited problems of women in the workplace are low wages, differentials in fringe benefits, dead-end jobs, sexual harassment, lack of child care, stress over multiple roles, and lack of leisure time. Another problem is the segregation of women into the pink-collar ghettos of secretarial and clerical work.

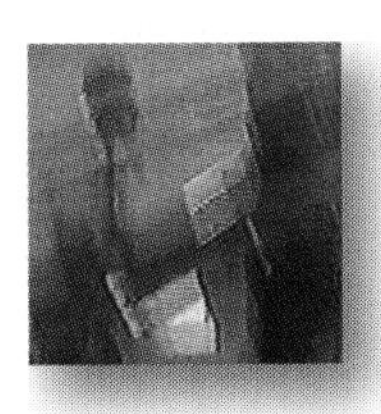

SOCIOLOGY VERSUS IDEOLOGY

Here is a stanza from a poem, "The Princess," by the English poet Alfred, Lord Tennyson:

> Men for the field and women for the hearth,
> Men for the sword and for the needle she,
> Man with the head and woman with the heart,
> Man to command and woman to obey,
> All else confusion.

This poem expresses the ideology of those who argue that changes in the status quo of male power and dominance upset the "natural order" and will lead to chaos. Feminist theory and research argues against this viewpoint and shows that justifications for androcentrism, whether found in religious practice, political institutions, or educational institutions, deprive women of the opportunity to realize their human potential.

A balanced sociological view of the issues must admit that there is some truth in the fears expressed in Tennyson's poem and similar arguments for the status quo. Any major social change entails confusion and disruption of the normal order of life for many people. But a sociologist who reviewed the available data, the way we have in this chapter, would ask how long this disruption would last. The sociologist would be forced to conclude that if women (who are, after all, half of any society's population) were to gain equal opportunities with men in education, politics, and the economy, the benefits for all members of society would far outweigh the temporary disruptions of male privilege and comfort.

GLOSSARY

sex: the biological differences between males and females, including the primary sex characteristics that are present at birth (i.e., the presence of specific male or female genitalia) and the secondary sex characteristics that develop later (facial and body hair, voice quality, etc.). (p. 439)

hermaphrodite: a person whose primary sexual organs have features of both male and female organs, making it difficult to categorize the individual as male or female. (p. 439)

transsexuals: people who feel very strongly that the sexual organs they were born with do not conform to their deep-seated sense of what their sex should be. (p. 440)

gender: the culturally defined ways of acting as a male or a female that become part of an individual's personal sense of self. (p. 440)

sexuality: the manner in which a person engages in the intimate behaviors connected with genital stimulation, orgasm, and procreation. (p.443)

heterosexuality: sexual orientation toward the opposite sex. (p. 444)

homosexuality: sexual orientation toward the same sex. (p. 444)

bisexuality: sexual orientation toward either sex. (p. 444)

gender role: a set of behaviors considered appropriate for an individual of a particular gender. (p. 447)

sexism: an ideology that justifies prejudice and discrimination based on sex. (p. 452)

patriarchy: the dominance of men over women. (p. 456)

WHERE TO FIND IT

BOOKS

Manhood in America: A Cultural History (Michael S. Kimmel; Free Press, 1996). A sociological analysis of changing norms of masculinity and ideas of manhood in North America.

Women and Children First: Environment, Poverty, and Sustainable Development (Filomena Cioma Steady; Schenkman, 1993). A recent review of international issues of gender, inequality, and development.

Women in Law, 2nd ed. (Cynthia Fuchs Epstein; University of Illinois Press, 1993). A fine empirical study of the structural and cultural barriers women face in the legal profession.

The Lenses of Gender (Sandra Bem; Yale University Press, 1993). A well-reasoned argument for the feminist perspective in the social sciences and for social policy changes. Based on extensive research by one of the nation's foremost social psychologists.

Gender Advertisements (Erving Goffman; Harvard University Press, 1979). A subtle treatment of the symbolism of gender and its use in everyday life, with excellent examples of visual material used as sociological data.

Gender, Family, and Economy: The Triple Overlap (Rae Lesser Blumberg, ed.; Sage, 1991). A collection of original essays based on recent research on gender stratification. Includes some excellent material on sharing of household chores by heterosexual couples, and valuable historical material on the stratification of minority women in the United States and elsewhere.

JOURNALS

Signs. A quarterly journal published by the University of Chicago Press that presents new research and theory in gender studies and feminist social science.

Gender & Society. The official publication of Sociologists for Women in Society.

INTERNET RESOURCES

National Organization for Women (www.now.org/). Features calls for action, press releases, and the NOW newspaper, all of which exemplify the activities of a mainstream feminist organization.

The Eagle Forum (www.eagleforum.org/). A conservative political organization that speaks out on issues of gender and inequality.

Women's Studies Program at the University of Maryland (www.inform.umd.edu:8080/EdRes/Topic/WomensStudies/). Offers reports on issues like women's status in higher education, women in the work force, sex discrimination, and sexual harassment. Also offers links to international organizations for comparative data.

CHAPTER 15

INEQUALITIES OF YOUTH AND AGE

Like many of the world's industrialized and urbanized nations, the United States has an aging population. Consider this simple fact: In 1960 the number of centenarians, people over the age of 100, in the United States was slightly over 3,200. Today it is over 60,000. Does this momentous change mean that the United States is becoming a society in which to be old is to be respected and appreciated? Most experts on aging would say no. Our culture, as judged by the messages we continually recieve on TV or in other media, values youth, vitality, strength, speed, and youthful beauty. No wonder the Irish poet William Butler Yeats wrote in his declining years that an aged person is:

> . . . but a paltry thing,
> a tattered cloak upon a stick,
> unless soul clap its hands and sing,
> and louder sing for every tatter
> in its mortal dress.

But how can an elderly person's soul sing in the face of declining health and the gnawing isolation that many elderly people experience? Some answers can be found in the efforts of people with strong sociological imaginations to change our perceptions of the elderly. One of the best examples of this is found in the photographic work of Imogen Cunningham.

Cunningham, one of the twentieth century's most influential photographers, spent most of her long life in San Francisco. In her later years she devoted much of her work to photographing people who were more than 90 years old but "still alive and kicking." Her photographs of elderly people, some of which appear in the Visual Sociology section of this chapter, are studies of dignity and courage in the face of advanced age and imminent death.

Ironically, despite the high value placed on youth, young people are another age class that is experiencing severe social stress. This is true in affluent societies like the United States as well as in many developing nations. The specific causes may differ widely from one region to another, but wherever mothers of children and young people are under stress due to poverty and violence, the children suffer as well. Of course, people in other age categories also experience inequalities based on age, but as we will see in this chapter, the young and the aged are particularly subject to age-based inequality.

SOURCES OF AGE STRATIFICATION

In many societies age determines a great deal about the opportunities open to a person and what kind of life that person leads. Only a little over half a century ago, for example, becoming old in the United States almost automatically meant becoming poor. Today children are the most impoverished and vulnerable population group in many nations, including the United States. This chapter examines how age stratification—the roles assigned to young people, adults, and the elderly—contributes to inequality. From a discussion of the life course and age structure of societies, the chapter proceeds to a closer examination of the situation of young people. The remainder of the chapter is devoted to the growing sociological subfield known as **gerontology,** the study of aging and the elderly in this and other societies.

The Life Course—Society's Age Structure

All societies divide the human life span into "seasons of life" (Bengston & Schaie, 1999; Neugarten, 1996). This is done through cultural norms that define periods of life, such as adulthood and old age, and channel people into **age grades**—sets of statuses and roles based on age. These systems of age grades create predictable social groupings and turning points that everyone in a society can easily recognize (Neugarten, 1996). Graduations, communions, weddings, retirements, and funerals are among the ceremonies that are used to mark these turning points.

Age strata are rough divisions of people into layers according to age-related social roles. We speak of infants, preschoolers, elementary school children, teenagers, young adults, and so on; these categories form a series of younger-to-older layers, or strata, in the population. People in different age strata command different amounts of scarce resources like wealth, power, and prestige (Moody, 1998). Numerous laws establish inequalities between youth and adults; they include laws governing the rights to vote, to purchase alcoholic beverages, to incur debt, and the like. In theory, a person who lacks the rights of adult citizenship will be protected by adults, who are responsible for providing him or her with adequate food, shelter, and education (and are presumed to have the resources to do so). In practice, however, hundreds of thousands of children and teenagers do not receive the care that is intended to offset their unequal status under the law.

Social scientists often refer to the **life course,** which may be defined as a "pathway along an age-differentiated, socially created sequence of transitions" (Hagestad & Neugarten, 1985, p. 36; see also Cain, 1964; Clausen, 1968; Elder, 1981). The cultural norms that specify the life course and its important transitions create what is thought of as the normal and predictable life cycle (Neugarten, 1996). We expect that we will go to school, find a job, get married, have children, and so on at certain times in our lives, and we consider it somewhat abnormal not to follow this pattern. Social scientists often refer to ceremonies that mark the transition from one phase of life to another as **rites of passage** (Van Gennep, 1960/1908). The confirmation, the bar mitzvah, the graduation, and the retirement party are examples of rites of passage in modern societies.

In the United States and other Western cultures, the life course is constructed from categories like childhood, adolescence, young adulthood, adulthood, mature adulthood, and old age. But our definitions of these categories lack the stability and uniformity of the age grades found in many traditional societies. For example, the French historian Philippe Ariès (1962) showed that in Western civilization the concept of childhood as a phase of life with distinct characteristics and needs did not develop until the late seventeenth century. Before that time children were treated as small adults. They were expected to perform chores and to conform to adult norms to the extent possible. When they reached puberty they were usually married, often to spouses to whom they had been promised in infancy. Today many studies of childhood argue that there is again a blurring of the cultural definitions of childhood and adulthood as children are exposed to adult themes in the media and many children are prosecuted for adult crimes (Applebome, 1998).

Norms regarding gender are closely linked to the life course established by a society. Thus Ariès's study of the emergence of childhood revealed that ideas about the appropriate forms of play and education for boys, and indeed the very concept of boyhood, developed at least a century before the concept of girlhood emerged. In eighteenth-century European societies boyhood was conceived of as a time when male children could play among themselves and receive education in the skills they would need as adults. Girls, in contrast, were treated as miniature women who were expected to work alongside their mothers and sisters.

Cohorts and Age Structures

When we think about age, we tend to think in terms of **age cohorts,** or people of about the same age who are passing through the life course together (Bosworth & Burtless, 1998). We measure our own successes and failures against the standards and experiences of our own cohorts—our schoolmates, our workmates, our senior circle—as we pass through life.

Demographers use the cohort concept in studying how populations change. If we divide populations into 5-year cohorts, grouped vertically from age 0 to 100+ and divided into male and female, we can form a *population pyramid,* a useful way of looking at the influence of age on a society. Figure 15.1 compares the age structure of the population of developing nations with that of the more developed ones, including the United States. The base of the population structure of the developing nations is wide because of high birthrates. But the structure narrows in each cohort because of high rates of mortality, due largely to lack of access to up-to-date medical care and lack of preventive health care. High mortality rates for people over age 40 result in the pyramidal form that is characteristic of the age structure of developing nations.

The developed nations have a more cylindrical population age structure, a consequence of far lower birthrates and, hence, smaller early-childhood and youth cohorts. For example, the 0–4 age cohort is smaller than the 5–9 cohort, reflecting the continuing decrease in birthrates in the industrialized and urbanized regions of the world. Note also that death rates are far lower in affluent countries than they are in the developing nations, so that in the absence of war or deadly epidemics the age cohorts pass through the age structure with little attrition until well into middle age, when increasing rates of mortality produce slightly smaller age cohorts.

In the developing nations, the age cohorts from birth to age 19 are termed the "critical" cohorts because the reproductive behavior of these cohorts will largely determine the future size of the world's population. There are about 2 billion people below the age of 20 living in less developed nations, of which more than 400

FIGURE 15.1 **Population of Less Developed and More Developed Nations, by Age and Sex**

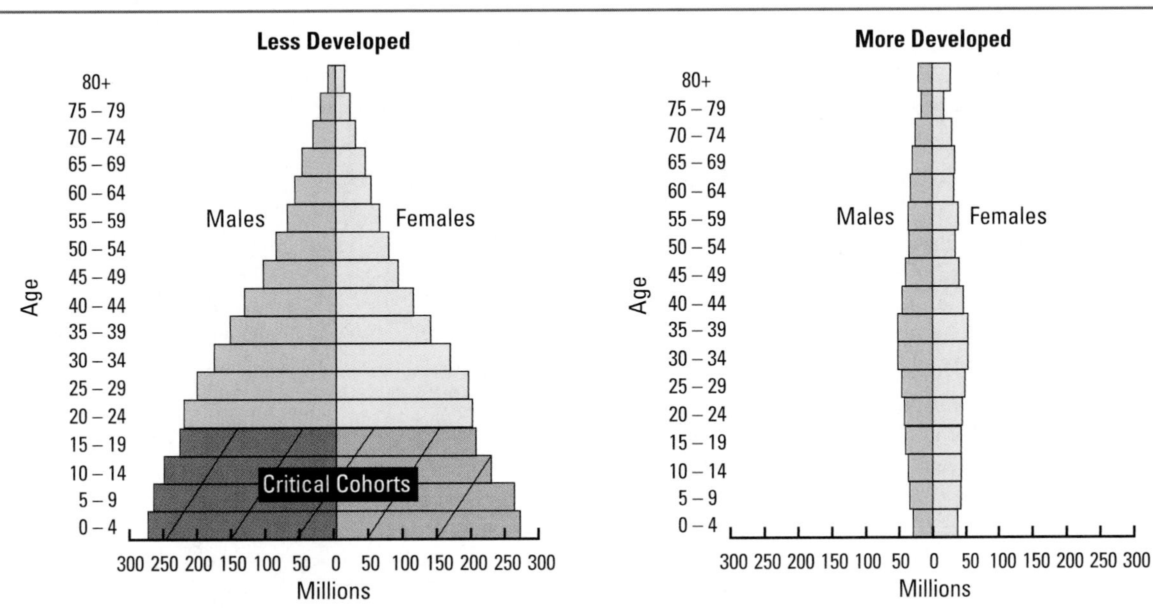

Source: Population Reference Bureau, 1998.

million are already in their early childbearing years (ages 15–19) (Population Reference Bureau, 1998). In some industrialized nations, about 1 percent of teenagers in these cohorts give birth, but in parts of Africa and Asia the rate can be as high as 24 percent. Overall, in Africa below the Sahara and above the nations of southern Africa, 18 percent of teenagers give birth each year. Were that rate to continue, the region's population would double in 23 years. In Latin America, where 8 percent of teenagers give birth each year, the population doubling time would be 42 years. Concerned about these statistics, representatives of more than 180 nations agreed in 1994 to rapidly increase investments in health care, family planning, and education for young women in the less developed regions of the world in the hope of stabilizing world population growth.

The Baby Boom When a population experiences marked fluctuations in fertility, there are bulges in its population pyramid that have important effects. Perhaps nowhere in the world has this phenomenon been more thoroughly studied than in the United States. The "baby boom" cohorts, which were produced by rapid increases in the birthrate from about 1945 through the early 1960s, have profoundly influenced American society and will continue to do so for the next three decades.

Throughout Europe and North America the baby boom generations did not have nearly as many children as their parents had. A mean family size of 2.1 children per couple is required for a population to remain constant over time (rapid growth requires a mean number of children closer to 3.0 per family). But since the 1970s the mean number of children per family in industrial societies has been about 1.85 (and much lower in nations like Japan and Germany); as a result, the baby boom has been followed by a relative shortage of children known as the "baby bust" (Keyfitz, 1986; Morgan, 1998). One way to visualize the impact of the baby boom and baby bust is to compare the population pyramids in Figure 15.2.

As these cohorts mature, they make new demands on the society's institutions. During the 1960s and 1970s, for example, when the baby boom cohorts passed through their college years, the nation's universities and colleges expanded; the slogan of the day was "Never trust anyone over 30." Now the baby boom cohorts are moving into the dominant age groups of the population. The Clintons, Tipper and Al Gore, the Bush brothers (George W. and Jeb), and many other well-known political leaders are in these cohorts. Their

FIGURE 15.2 Impact of the Baby Boom on the U.S. Population, 1950–2020

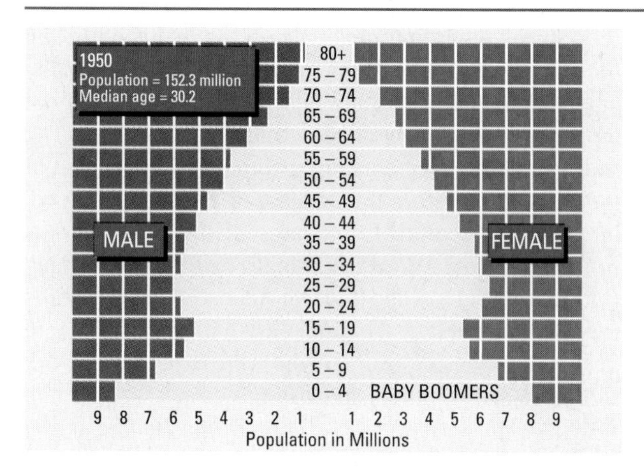

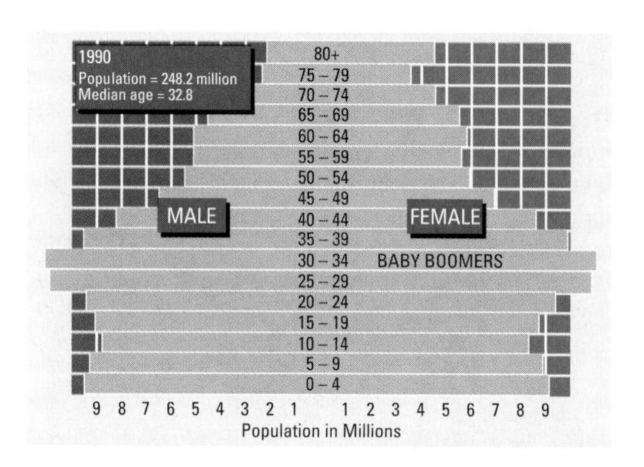

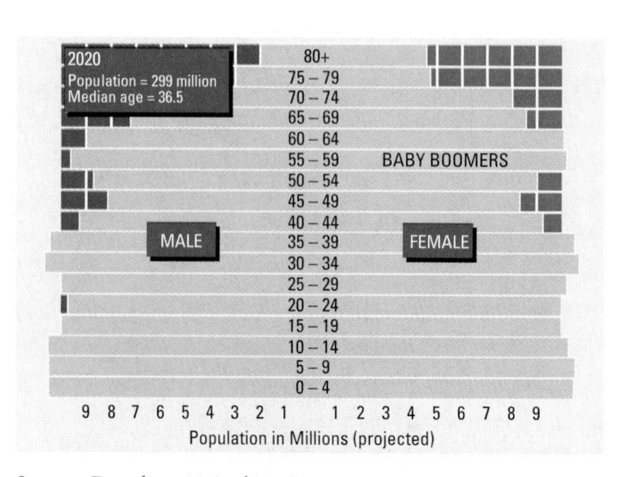

Source: Data from United Nations.

succession to national leadership during the 1990s represents a transfer of power from the generation that fought in World War II and is now in its seventies and eighties, to the generation that experienced the Vietnam War and the social movements of the 1960s. But sociologists note that the baby boom cohorts are quite diverse and cannot be easily categorized as liberal or conservative, religious or secular, and so on (Brauss, 1995).

The Baby Boom Echo Sometimes referred to as the "baby boomlet" or the "baby boom echo," the children of parents who were part of the original baby boom are themselves an important demographic phenomenon. Note in Figure 15.3 that in the late 1980s the number of births exceeded 4 million per year, a threshold reached over a longer period during the original baby boom, which extended from the 1940s to the 1960s. Like any "bulge" in a society's age structure, the wave of children born during the 1980s and early 1990s promises to have significant effects on social institutions like schools and businesses.

Figure 15.4 shows that a sizable proportion of the "second wave" or "boomlet" children are members of minority groups. This trend, combined with high rates

FIGURE 15.3 **Number of Births in the United States Each Year, 1920–1933 (in millions)**

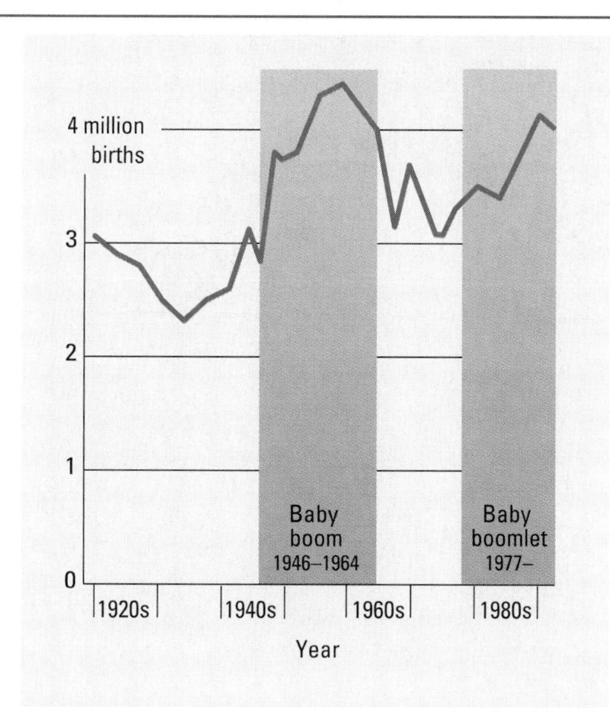

Source: Data from Census Bureau.

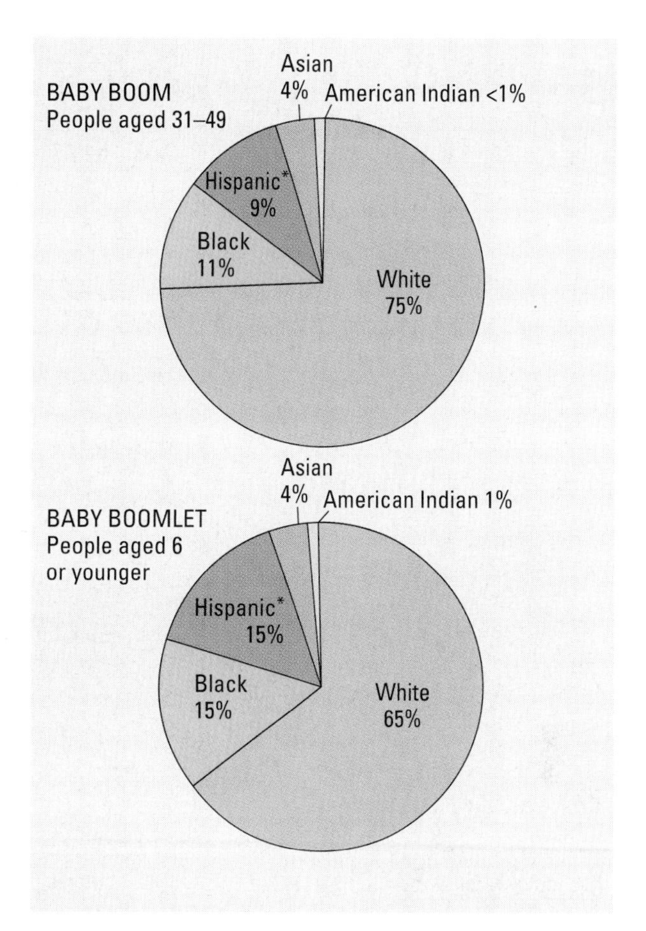

Note: Because of rounding, figures do not add to 100%.
*Hispanic people can be of any race.

Source: Data from Census Bureau.

FIGURE 15.4 **Racial and Ethnic Composition of the Baby Boom and Boomlet Generations**

of immigration, will increase the proportion of minority children moving through the nation's schools and colleges in coming decades. Between 1988 and 1993, for example, kindergarten enrollments rose by 7.2 percent and preschool enrollments by 14 percent, both trends that presage increasing enrollments in many school districts. This demographic effect, in turn, increases debate over the investment of public funds in schools versus other uses, such as medical care for the poor and elderly (Gabriel, 1995a).

Graying Populations Europe, North America, and Oceania have "graying" populations, in contrast to regions where the proportions of people age 65 and over remain relatively small. These differences are shown in

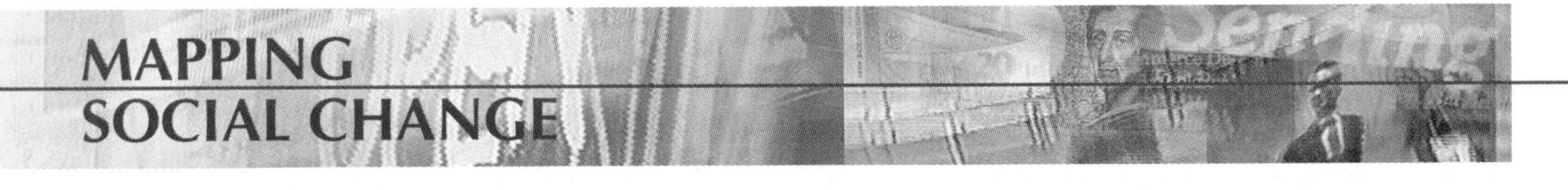
MAPPING
SOCIAL CHANGE

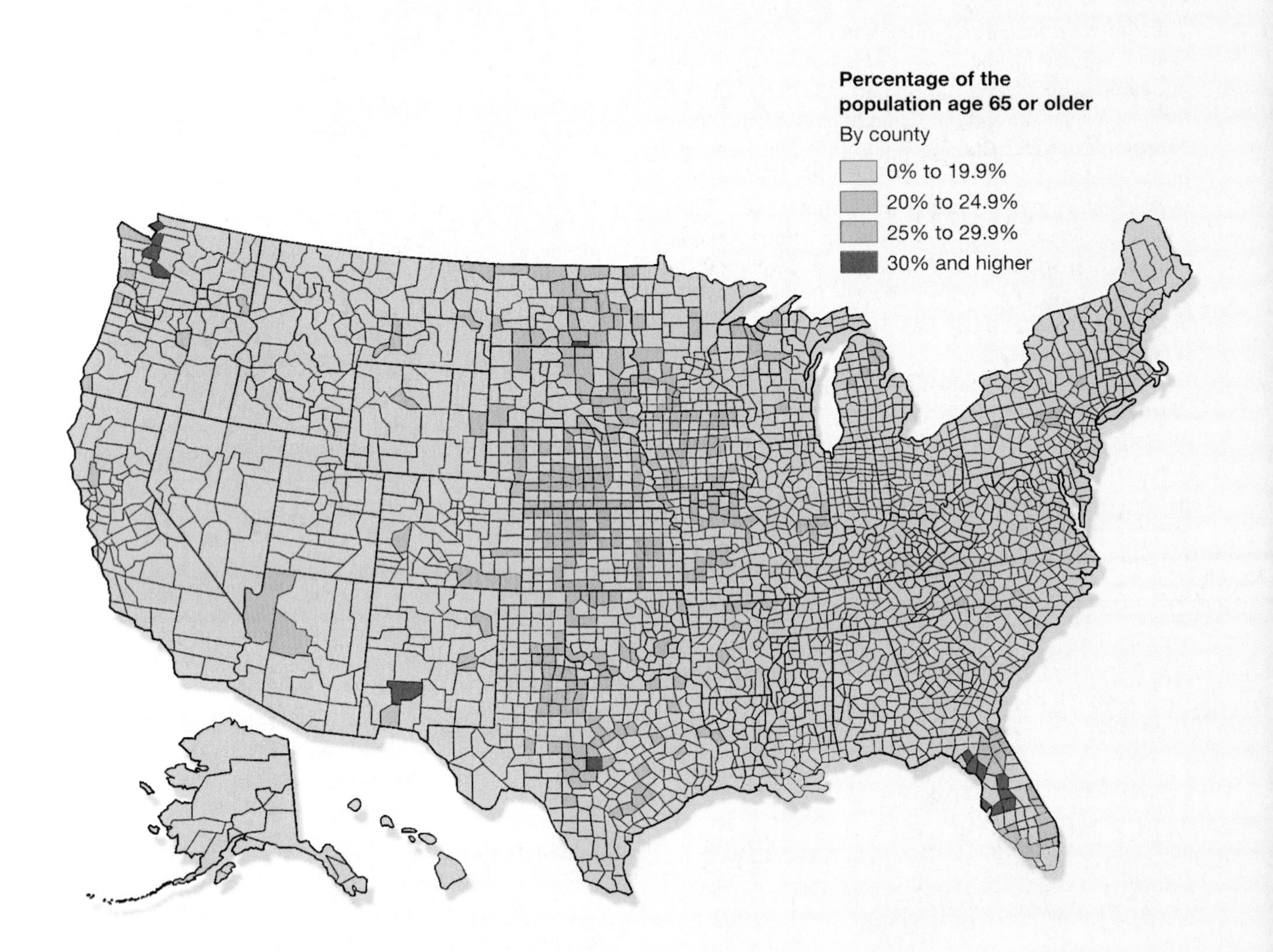
Percentage of the population age 65 or older
By county
0% to 19.9%
20% to 24.9%
25% to 29.9%
30% and higher

The Age Structure of Sun City, Arizona

The aging of the U.S. population is creating major migration streams and new patterns of urban and suburban growth. The accompanying map, which shows the percentage of the U.S. population age 65 or older by county, offers a detailed view of the ecology of aging. We are used to thinking of Florida, Arizona, and now Nevada as destinations for retired people, but why are there such high proportions of elderly people in the Great Plains states? In the Dakotas, Nebraska, and Minnesota, cold weather is a severe deterrent to in-migration by the elderly, so what is the reason for the projected growth of this population there? The answer has more to do with the movement of young people out of these states. As young people leave the Great Plains in search of opportunities in California and the rapidly growing states of the Southeast and Southwest, older people comprise a greater proportion of the remaining population.

Age
85+
80 – 84
75 – 79
70 – 74
65 – 69
60 – 64
55 – 59
50 – 54
45 – 49
40 – 44
35 – 39
30 – 34
25 – 29
20 – 24
15 – 19
10 – 14
5 – 9
0 – 4
Male
Female
16 12 8 4 0 0 4 8 12 16
Percent of Population

In many of the Plains states the elderly must be extremely vigorous in order to survive in the harsh environment and must be highly dedicated to their communities in order to maintain them in the absence of enough younger people to fill available positions, especially in nonpaid volunteer work.

The migration patterns of the elderly also create some unusual inversions of the normal age pyramid in places where they become the most numerous population segment. The age pyramid for Sun City, Arizona, a retirement community in the Sun Belt, is a case in point. In this retirement community, only 0.1 percent of the population is under age 15 and nearly 85 percent is over age 65. Note also that women outnumber men, as would be expected in view of their greater life expectancy.

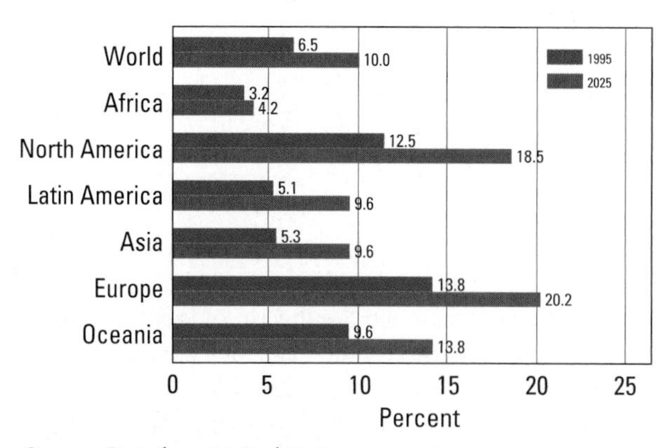

Source: Data from United Nations.

FIGURE 15.5 **Population Age 65 and Older, by World Regions, 1995 and 2025**

Figure 15.5. The dramatic increase in the proportion of elderly persons in the industrialized nations projected for the year 2025 is due largely to the aging of the baby boom cohorts. This increase, often referred to in the United States as the "graying of America," is already resulting in greater concern about the needs of the elderly and will augment the influence of the aged on American culture and social institutions.

Demographers and biologists disagree over estimates of how many very elderly people there will be in the U.S. population during the next century. Most Census Bureau demographers estimate that the population over age 85 will increase from about 3.3 million persons (mainly women) in 1990 to about 18.7 million in 2080. But demographers working with the U.S. Institute of Aging argue that there may be no biological limits on life expectancy. With advances in medicine and better prevention of disease, more people may live into their nineties. If that happens there could be as many as 70 million people over age 85 in the U.S. population late in the twenty-first century—a change that would have enormous effects on all aspects of social life (Kolata, 1992).

Life Expectancy

At age 65 and beyond, more than 50 percent of North American women are widowed, whereas only 13.6 percent of men have lost their wives. This is because the life expectancy of females is at least 7 years longer than that of males. (By **life expectancy** we mean the average number of years a member of a given population can expect to live beyond his or her present age.) Typically, men die before their wives do. And they tend to marry younger women, making widows the single largest category in the elderly population.

In the future, if male and female roles become more similar and women experience the stresses and risks that are thought to cause earlier death in men, the

Social movements among the elderly are increasing in strength as the older population grows in numbers. Older Americans have taken a very active interest in recent efforts to reform the nation's health care system.

gap between the life expectancies of the sexes may narrow. Research on sex differences in the causes of death indicates that women may be less vulnerable to death for genetic reasons, but it is extremely difficult to prove this scientifically. The research of Lois M. Verbrugge and colleagues (Verbrugge, Gruber-Baldini, & Fozard, 1996) on health trends among males and females indicates that the lifestyles of men and women in the United States have become similar; in particular, women's lives are more like men's. This change has unfortunate implications for women's health and longevity. For example, women are now smoking almost as much as men. Smoking among men declined from 50.2 percent in 1965 to 27.0 percent in 1995, while the rate for women decreased from 31.9 percent to 22.6 percent (*Statistical Abstract,* 1998). Among older smokers, men were twice as likely to quit as women. It is no surprise that lung cancer now almost equals breast cancer as a leading cause of death for women.

Age and Dependency People in the working adult cohorts—that is, those between the ages of 18 and 64 (although many continue to work well after age 64 and may start before age 18)—contribute disproportionately to the well-being of the young and the elderly. Of course, societies justify this pattern of dependency by recognizing that adults are merely doing in their turn what was done for them as children or will be done for them when they are among the frail elderly. Institutions of modern societies, such as public education and Social Security, ensure that a share of wealth passes to the dependent cohorts.

When there are very large numbers of children in a society, as is the case in the developing nations, or increasing numbers of elderly people, as is the case in the older nations of western Europe and North America, working adults may shoulder an increased burden. Figure 15.6 vividly illustrates the dependency problem. It charts the ratio of workers in the U.S. labor force to people receiving Social Security benefits. That ratio has been decreasing sharply with the aging of the U.S. population, so that now there are close to three workers for every retired person drawing on the Social Security fund. The graph also projects the ratio well into the twenty-first century in order to demonstrate that the dependency situation will continue to worsen as population aging accelerates. According to these projections, by the end of the twenty-first century's second decade more funds will be drawn out of the Social Security system than are being paid into it by active workers. In consequence, there is a lively debate among legislators about how to modify the system in the face of these changing demographics. Should the retirement age increase? Should Social Security taxes rise? Should the system be changed so that wealthier people receive lower benefits than less affluent citizens? These are all

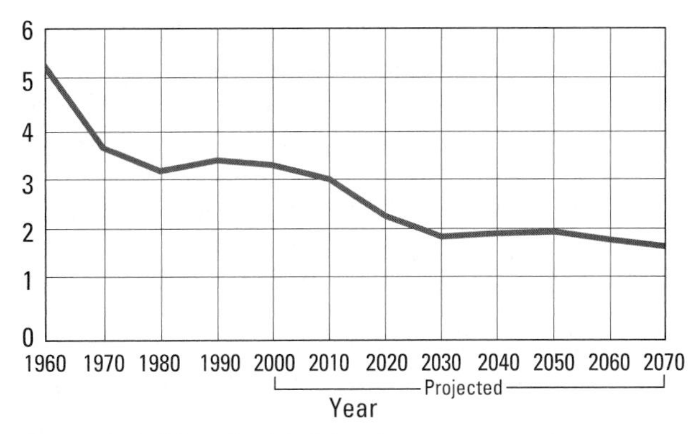

Source: Social Security Administration.

FIGURE 15.6 **Ratio of U.S. Workers to Social Security Beneficiaries**

extremely controversial issues that need to be addressed to ensure that the major gains in well-being made by elderly people since the 1930s are not eroded in coming decades.

AGE STRATIFICATION AND INEQUALITY

In urban industrial societies there are distinct patterns of stratification in which age defines the roles one plays and the rewards one can expect. We speak of the "age of majority"—the age at which a person crosses the legal boundary between childhood and adulthood. In fact, this age is not always clearly defined. A person can vote at age 18 but cannot legally consume alcohol until age 21. In addition to the age of majority, there are other distinct ages that mark the passage toward the full rights of adulthood. At age 18 one can join the armed forces without parental permission. At age 16 in many states one is no longer obliged by law to attend school, even though parents still share responsibility for the behavior of the school dropout. Teenagers can drive at age 16 in many U.S. counties, but their parents remain responsible, through insurance systems, for their actions and can be held liable for the consequences. We also make clear distinctions between children in the primary grades and teenagers in high school, and in most homes a child's passage through these age grades is accompanied by various privileges and responsibilities. So although the passage to full adult status may be somewhat vague, it is clear that to be young is to be less equal.

Later in life, as people become elderly, they may yield some of their autonomy to their grown children—either willingly through trusts and wills, or unwillingly as they are committed to nursing homes because their care has become too great a burden for their children. Thus it is at the early and late extremes of youth and age that the relationship between age and inequalities of power and material resources are most evident.

There are also age-based inequalities that may affect the life chances of nonelderly adults. As corporations downsize—that is, lay off employees in order to increase profit margins—higher-paid people in their fifties may find that their jobs are combined with other work and offered to younger, less well-paid replacements. These changes in the economy are producing major changes in the mature labor force. By the end of the 1990s some 66 percent of males between the ages of 55 and 64 were active workers, a decrease of about 20 percent since the mid-twentieth century. But the opposite was true for women. Among women in the same age group the proportion of active workers increased to 49 percent in 1998 from only 27 percent in 1950 (Haub & Pollard, 1998). More men are leaving full-time work in their fifties in order to pursue other interests, while more women are taking up the slack by finding full-time employment, often after working most of their lives at home as well as in the labor force.

Given the opportunity, elderly men and women can have a profound influence on the lives of children in their families and neighborhoods.

Race and Gender: The Double and Triple Binds of Aging

Inequalities of age are compounded by those of race and gender to produce particular forms of inequality among the aged in industrial societies, especially the United States. If being elderly is a disadvantage, being an elderly woman often places one at a double disadvantage. Being elderly, female, and black or Hispanic places one in triple jeopardy for being isolated, ill, and impoverished (Holden, 1996). Women of advanced age, especially those over age 85, are three times more likely than older men to be living in poverty. The situation is worse for elderly black and Hispanic women. Because large numbers of these women either were never married or experienced divorce (a subject to which we return in the next chapter), in their later years higher proportions live alone and have lower incomes than their white counterparts.

As we have seen, age influences life chances in important ways and helps produce the patterns of inequality that we often take for granted, especially among the young and the old. And age compounded by race and gender often entails further inequality. But if we study age stratification and inequality from a comparative perspective, we find some surprising differences. In smaller-scale, more traditional agrarian or pastoral societies, age stratification works in very different ways than in urban industrial societies. For example, among the Maasai, whom we met in the preceding chapter, there are fixed age grades for males and females that strictly define an individual's roles and opportunities throughout life. Children and adolescents have no individual status in Maasai society. It is not until their passage through circumcision and other rituals during puberty that they achieve status as autonomous individuals.

Initiation into adult status occurs for a group of Maasai of roughly similar age, who then constitute an age grade. There are four age grades in the Maasai system, each covering about 15 years and involving a mandatory set of activities, especially for males. Young initiates engage in military activity. Young elders are married adults with family and economic responsibilities. Elders have political decision-making power, and senior elders have religious ritual power. Each of these age grades has a name in the Maasai language, and passage into it is marked by rituals that convey differing degrees of authority and responsibility (Chenault, 1996). Other smaller-scale societies throughout the world organize age grades according to their own cultural norms, but most, like the Maasai, invest the different age grades with distinct degrees of power and control over the society's wealth.

Throughout the world the forces of social change unleashed by colonialism, industrialization, urbanization, and population growth have tended to disrupt the formal age grade systems of smaller, more isolated societies. Often the loss of their traditional ways of commemorating the passage from one age to another has been among the more disorienting aspects of social change for people in those societies. The passage into old age is a good example. In most small-scale tribal and peasant societies the elderly are revered as guardians of moral values, rituals, and religious beliefs. In urban industrial societies, by contrast, they are typically treated as dependents with little status and few rights. The situation of children is similar. In traditional societies children have well-defined roles. The young girls are typically helpers while the boys are expected to play games that teach them the skills they will need as adult males. But all this changes in urban industrial societies. In those societies the status of children is extremely fluid and subject to many conflicting ideas and rapid changes in norms (Ariès, 1962).

The Challenge of Youth

In *The Challenge of Youth* (1965), his seminal work on youth in Western societies, Erik Erikson observed that in societies with high levels of youth rebellion, deviance, and even suicide, the problem often lies in the failure of adults to provide young people with clear goals and paths toward a constructive future. Kenneth Keniston (1965), another expert on the problems of growing up in modern societies, adds that the challenges young people face are further complicated by such rapid and far-reaching changes that they are often bewildered about such questions as what careers to pursue, what standards to apply to personal relationships, how much to devote to oneself and how much to the community and the society. Many young people have parents or other adults in their lives who can help them solve these dilemmas, but many others do not. As we saw in Chapter 7, they may react to the problems and failures of those around them (e.g., alcoholism or divorce) by withdrawing into a world of anomic, alienated youth that explicitly rejects the values of the larger society, sometimes through gang violence.

In this regard, one of the most fascinating unanswered questions in the social sciences is why some children growing up under extremely adverse social conditions appear to be resilient, to be able to succeed in school and in other social settings, while so many others are set back and defeated by the same adverse conditions (Conger, Ge, & Elder, 1994; Garmezy, 1993; Williams & Kornblum, 1994). On a broader sociological level, however, Keniston and others note that changing norms of childhood and youth, on the one hand, and the decreasing well-being of young people in industrial nations, on the other, strongly influence the fortunes of youth everywhere.

Variations in the Value Placed on Children Social definitions of childhood—what we expect of children, how we value the child and treat the child—differ immensely throughout the industrial world as well as between modern and traditional societies. In her study of the changing meanings of childhood, sociologist Viviana Selizer shows that in the United States there occurred a "profound transformation in the economic and sentimental value of children—fourteen years of age or younger—between the 1870s and the 1930s" (1985, p. 3). During that period child labor laws ended a common practice, at least among poorer families, of requiring children to work for wages. Earlier in the nineteenth century the incomes of the middle classes had risen enough so that they could keep their young ones out of the labor force; by the 1930s this became the norm for all classes.

As a result of this transformation, Selizer points out, children became economically "worthless" but emotionally "priceless"—and this is true today as well. We think of our children as priceless beings to be nurtured, protected, socialized, and gradually introduced to the world of work and careers—not forced into that world abruptly after a very brief period of innocent childhood, as was the case in the early years of industrialization. Studies of children's responsibilities in the home find that people justify giving their children chores to do on the grounds that doing so builds character, discipline, responsibility, and so on, not because it contributes to the economic well-being of the family. In fact, in many middle-class homes the allowance system operates in lieu of any form of wages; the allowance is seen as a form of token money that the child has a right to receive but must also learn to spend responsibly.

But Selizer also shows that this transformation in the value placed on children was surrounded by a great deal of social conflict as the institutions of the larger society, especially courts and schools, were given the responsibility for enforcing the new norms about child protection. Debates over definitions of childhood and the proper roles of children continue today. For example, the home schooling movement has arisen among people who reject the values of the public schools and wish to maintain the absolute authority of parents over their children by educating them in the home.

More important than these conflicts are the class contradictions inherent in the value placed on childhood. Many families are not able to give their children the array of goods and opportunities that are available to children in more affluent families. The worst situations of material deprivation experienced by children

Then and Now

From Adulthood in Miniature to Childhood—and Back?

These photos offer versions of ideal behavior among children in the eighteenth century, the mid-twentieth century, and the present. They raise questions about the ever-shifting norms and expectations of childhood. In the eighteenth century, as depicted in the famous Gainsborough portrait known as the *Blue Boy,* the ideal of childhood was one of adulthood in miniature. By age 7 or 8, children "were dressed and treated as little adults. They dressed the same, did the same work, and entered into the sexual community of adults. . . . Even in the United States, the age of sexual consent was under 10 in half the states until the end of the 19th century" (Applebome, 1998, p. 3). The idea that children are different and have their own rights to education and protection under the law emerged in the early twentieth century. By mid-century the idea of healthy childhood was solidly reinforced in the mainstream culture, as the *Saturday Evening Post* illustration from the 1950s clearly demonstrates. Today, experts on childhood worry that cultural understandings about childhood have again become blurred. The public's fascination with another "miniature adult," JonBenet Ramsey (the winner of the 1995 Little Miss Colorado beauty pageant, who was murdered in 1996 at the age of 6), is only a small part of the story. Perhaps more ominous is the fact that by 1996 some 49 states had passed laws allowing the prosecution and sentencing of 14-year-olds as adults.

and young people, along with their parents, are often hidden from view in racial and ethnic ghettos and class-segregated communities "across the tracks" or in isolated trailer parks (Kozol, 1995). Overall trends in child welfare show that there is a widening gap between "priceless" children and children who bear a heavy burden of poverty and deprivation.

Indicators of Well-Being Among Youth There is growing consensus among social scientists on the need for a quantitative measure of well-being among youth. In recent years, as ideological debates over social problems like teenage fertility have become more rancorous, this need has become even more pressing (Zill, 1995). Toward this end, a group of eminent sociologists has been working

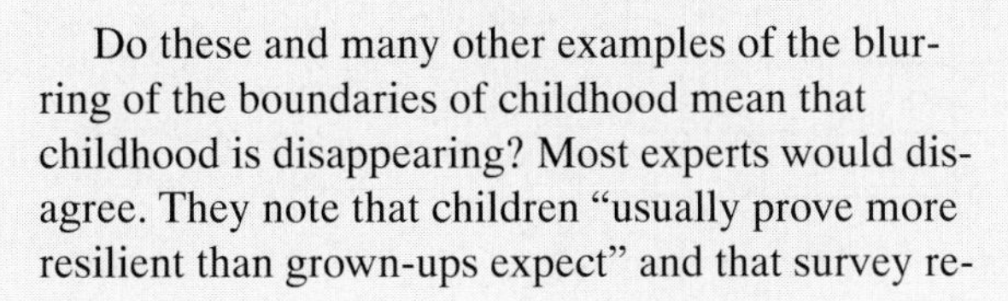

Do these and many other examples of the blurring of the boundaries of childhood mean that childhood is disappearing? Most experts would disagree. They note that children "usually prove more resilient than grown-ups expect" and that survey research shows "American teenagers to be worldly in ways previous generations were not, but sharing most of the values and sensibilities of earlier times" (Applebome, 1998, p. 3).

to create a set of data books titled *Kids Count.* These data books provide an invaluable and up-to-date array of indicators of trends in child well-being in the United States. They are available for use by local officials, scholars, and those who work with the youth of their communities.

Although in many communities the large majority of children are healthy and well cared for, the *Kids Count* data indicate that increasing numbers of young people are experiencing poverty and near-poverty conditions. Senator Daniel Patrick Moynihan of New York, himself a skilled social scientist, has noted that "a child in America is almost twice as likely to be poor as an adult. This is a condition that has never before existed in our history" (1988, p. 5).

TABLE 15.1 Indicators of the Well-Being of U.S. Children

Indicator	1985	1997
Percentage of low birth weight babies	6.8	7.3
Infant mortality rate (per 1,000 live births)	10.6	8.0
Child death rate, ages 1–14 (per 100,000 children)	33.8	29.0
Percentage of all births that are to single teens	7.5	9.0
Juvenile violent crime arrest rate, ages 10–17 (per 100,000 youths)	305	517
Percentage of teens not in school and not in labor force, ages 16–19	5.3	9.0
Teen violent death rate, ages 15–19 (per 100,000 teens)	62.8	69.0
Percentage of children in poverty	20.8	21.0
Percentage of children in single-parent families	22.7	26.0

Sources: *Focus,* Spring 1995; Annie E. Casey Foundation, 1998.

Table 15.1 presents some indicators of well-being among American youth. Measures of children in poverty, rates of violent death, infant mortality, teenagers not in school, and births to single teenagers are among the most important ways of looking at the proportion of children and young people whose well-being is questionable and whose daily lives are fraught with risk of further trouble. Perhaps the most controversial of these indicators are those dealing with births to teenagers. In the United States and other urban industrial societies there is a great deal of debate about increasing rates of teenage childbearing. But the facts show that, contrary to what many people believe, rates of births to teenagers decreased from 1960 through 1980, when they began increasing again but never reached the same levels as those prevailing at the end of the 1950s (see Figure 15.7). But by separating out births to teenagers aged 15–17 and 18–19, one sees that births to older teenagers have been decreasing since 1970, while births to younger teenagers increased during the 1960s and, although they are not nearly as high now, have not returned to earlier levels.

FIGURE 15.7 Births to Teenage Mothers in the United States, 1960–1992

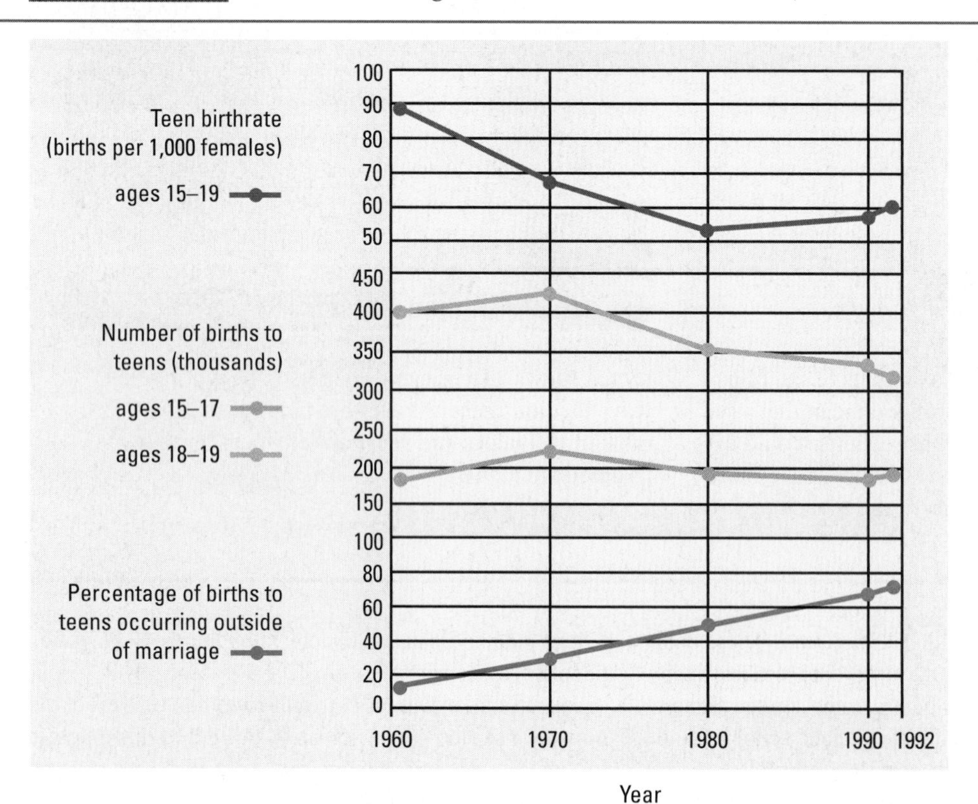

Source: Data from *Focus,* Spring 1995, p. 27.

The data on births outside marriage are most revealing. The most significant change is not so much the overall rate of fertility among teenage women as the rate of marriage among teenagers, which has declined drastically. This fact is at the center of debates over "family values." While there is no question that children have a better chance of avoiding poverty if there are two parents present in the household, it is also true that separation and divorce are among the leading causes of poverty for young mothers and children. In consequence, the effort to influence and oblige young fathers, whether married or not, to take responsibility for their children is an important social policy trend in American communities and throughout the world (United Nations, 1995).

It should be noted that some of the indicators presented in Table 15.1, such as the rate of violent death among teenagers, have actually improved in the last year or two. Others, however, such as the percentage of low birth weight babies, have worsened. These indicators are sensitive to changes in society (declining use of crack-cocaine, efforts to reduce truancy, dropout prevention programs, etc.). The low birth weight indicator is especially sensitive to changes in the economic situation of low-income households.

Throughout the world the infant mortality rate is considered the single most important indicator of the relative well-being of a nation's people (see Chapter 21). It measures the extent and efficacy of a nation's investment in the health of its people. The infant mortality rate in the United States—8.0 per 1,000 live births—ranks among the world's best, but it is higher than those in Japan (3.8), the Netherlands (5.7), or Australia (5.3), while it is much lower than that in Russia (17.0) and far lower than those in relatively impoverished nations like India (72.0) (Population Reference Bureau, 1998). Thus the United States, one of the world's most affluent nations, actually ranks on the lower end of the scale of infant mortality among the developed nations with which it is often compared. Much of the difference is due to the large number of underweight babies born in the United States.

The infant mortality rate for African Americans in the United States (14.2) is more than twice the rate for whites (6.0) (*Statistical Abstract,* 1998). Much of this disparity is due to the high percentage of babies with low birth weights among members of minority groups, especially those in poverty. Birth weights below 5.5 pounds are a cause of infant death and are also associated with higher rates of illness in later years as well as a higher incidence of neurological and developmental handicaps and poor academic performance (Partin & Palloni, 1995). Figure 15.8 shows that the incidence of low birth weight babies among African Americans declined from more than 14 percent in 1970 to a little over

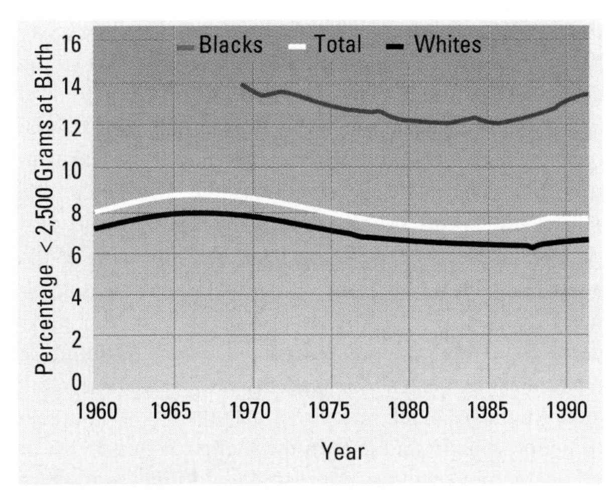

Source: Partin & Palloni, 1995.

FIGURE 15.8 **Percentage of Low Birth Weight Babies, by Race, 1960–1990**

12 percent in the mid-1980s, but began climbing again after 1984.

In explaining these trends, sociologists Melissa Partin and Alberto Palloni (1995) point out that while poverty is always associated with low birth weight, a woman's access to prenatal care in her community combined with her decision not to have an abortion are important factors. As more poor minority women choose not to have abortions and find it more difficult to obtain adequate care during pregnancy, the risk of low birth weight increases. Neither side in the abortion debate wishes to see more abortions among poor women. The central implication of these findings is that the rate of low birth weight babies would be reduced by improved access to health care for young mothers and more education about the need to decrease behaviors such as smoking during pregnancy.

Age and Inequality

At the turn of the twentieth century the largest segment of the U.S. population living in poverty or near-poverty conditions was the elderly (Preston, 1984). This was also true throughout much of Europe. Until the development of modern systems of employee pensions, Social Security, and basic health insurance (both private insurance systems and public systems like Medicare), people over age 60 were highly likely to be poor. Although poor and working-class elderly people without substantial wealth or savings might receive assistance from adult children and other family members, even those relatively fortunate individuals often felt deep resentment about having worked all their lives only to end up powerless and dependent (Atchley, 1991).

The Social Security Act of 1935 established federal old-age pensions based on employee and employer contributions, as well as unemployment insurance and other antipoverty programs. The results of these programs, and the Medicare system instituted in 1965, have been dramatic. Rates of poverty among the elderly, which were close to 40 percent in the early decades of the twentieth century, declined to about 28 percent in the early 1960s and then to about 14 percent after Medicare went into effect. At present 10.8 percent of people age 65 and over in the United States live at or below the poverty line, compared to 13.7 percent of the population as a whole (*Statistical Abstract,* 1998).

Among elderly members of minority groups, however, the situation is not nearly so positive. Over 25 percent of African Americans and 24 percent of Hispanics over age 65 are living in poverty (*Statistical Abstract,* 1998). These differences are often due to the fact that elderly people who previously worked at lower-paying jobs, which contributed little or nothing to private pension plans, may be forced to live almost exclusively on their federal Social Security payments, which by themselves are not enough to raise them above the official poverty line. And elderly blacks and Hispanics are especially likely to have had jobs that did not provide pensions.

As more people join the ranks of the elderly, concern about their economic situation becomes an ever more powerful theme in American life. Many Americans in their early sixties are subject to mandatory retirement. Once they are forced into retirement, they are limited in what they can earn without experiencing cuts in their Social Security payments and decreases in Medicare support. Matilda White Riley, a leading authority on the sociology of aging, asserts that there is a growing gap between the number of skilled and energetic elderly people in the population and the availability of "meaningful opportunities in work, family, and leisure" (Riley, Kahn, & Foner, 1994). We return to this theme shortly.

Age and Disability As life expectancies have increased throughout the world, so have the numbers of people with major disabilities that seriously impair their ability to function effectively in their daily lives. In the United States the average life expectancy for people age 60 is 19 years for women and 15 years for men. On the average, women who live that many years after their sixtieth birthday can expect to experience major disabilities for 5 of those years (4 years for men). In Germany, another affluent industrial nation, life expectancies for women and men over age 60 are 22 years and 18 years, respectively, and the chances of experiencing major disabilities are the same as those in the United States. In Egypt, a developing nation, the comparable figures are

i

Robert Frost, shown here in his advanced years, was far more than a poet of rural America. Among other things, he advocated the need for people to remain passionate participants in the world around them throughout their lives.

13 years of life expectancy for women after age 60 and 12 years for men, but because of less advanced health care the average Egyptian woman can expect to experience 6 years of disability (4 years for men) (Haber & Dowd, 1994). People who live longer do not automatically suffer major disabilities. As they age and become more frail, however, the experience of living with disability inevitably affects higher proportions of women because of their greater life expectancy. (See Box 15.1.)

The problems of health and disability among the elderly are compounded by inequalities of socioeconomic status. The more affluent and educated people are as they enter their sixties, the healthier they tend to be. In their last years, however, the effects of socioeconomic status diminish (House et al., 1994). People who have worked for decades tending machines in factories, often in polluted and stressful conditions, are more likely to have health problems and to have difficulty obtaining regular medical care than people who have worked in more physically and emotionally favorable conditions. Moreover, data on aging and health show that people from poor and working-class backgrounds are more likely to engage in **psychosocial risk behaviors** like smoking and heavy drinking (90 or more drinks a month). They also tend to experience a higher than average number of **lifetime negative experiences** (e.g., the death of a child or spouse, divorce, physical assault), which cause long-term stress. Even when risk behaviors and sources of chronic stress are considered, however, quantitative studies show a strong positive relationship between higher socioeconomic status and better health in elderly people (House et al., 1994).

Loss of Social Functions With Age As people age, they experience more medical problems and disabilities—but does this mean that they must inevitably withdraw from social life? Can anyone deny the biological facts of aging and death? Isn't it only natural for the elderly to perform less important roles as their physical capacities diminish? The answer, of course, is that it is not obvious at all. The increasing participation of elderly widows in securities investment clubs throughout the nation is a good example of how elderly people may perform socially important roles even as their physical strength diminishes. It is not uncommon for members of these clubs to impress far younger stockbrokers with their skill as players in the volatile securities market.

Sexual behavior among the elderly is another area in which popular notions about the influence of biology on age roles have been disproved. In the late nineteenth century and well into the twentieth it was widely believed that people lose their sexual desire and potency after middle age. In fact, however, a number of studies have shown that the image of the elderly as lacking sexual desire and the ability to enjoy sex is an ageist stereotype. For example, in a pioneering study, Eric Pfeiffer, Adrian Verwoerdt, and Glenn Davis (1972) gathered data from several samples of elderly people—including individuals as old as 94—that showed conclusively that although sexual interest and activity tended to decline with age, sex remained an important aspect of the subjects' lives. The researchers also found, however, that elderly men are more interested in sex than women of the same age. The explanations for this difference are cultural rather than biological. Elderly men are in short supply. They tend to be married to women with whom they have had a long-standing relationship that includes an active sex life. If they are not married, they are in such great demand that they have less difficulty than women of the same age in finding a sexually compatible partner (Greeley, Michael, & Smith, 1990; Kornblum & Julian, 1998).

Aging and Structural Lag As noted earlier, social scientists who study the situation of the elderly increasingly recognize that longer life spans need to be accompanied by new concepts of social roles in more advanced years. The well-known gerontologists Matilda White Riley and John W. Riley, Jr., believe there is a growing mismatch between the strengths and capacities of older people and their roles in society. As people live longer, they often find themselves living alone with few constructive roles that demand their time and attention. Neglect of the elderly further reduces their mental and physical strength. The Rileys believe that "increasing numbers of competent older people and diminishing role opportunities cannot long coexist" (1989, p. 28). In their view, there is an urgent need for many small-scale programs that create work and volunteer opportunities in which older people can use their skills and feel needed (Riley, Kahn, & Foner, 1994).

Shifts in age statuses can have dysfunctional effects for society as well. A research project of the Carnegie Foundation attempted to assess the impact of an aging society (i.e., one in which the median age is increasing) on social roles. For example, whose responsibility is it to pay for long-term care of the elderly? A family may have to pay $40,000 a year or more to provide nursing home care for a disabled elderly mother or father. Should the sons and daughters take on this responsibility? Most of us would say yes, if we are able to do so. The Carnegie study found, however, that an increasing proportion of people say, in effect, "No sir, that's a public responsibility, let them go on Medicaid." But in that case we would force the elderly to sell all their assets in order to qualify for government aid, since Medicaid is available only to people who have exhausted their own funds before applying to the government for assistance (Pifer & Bronte, 1986).

The question of responsibility for caring for the frail elderly is a controversial policy issue. The conservative

BOX 15.1 USING THE SOCIOLOGICAL IMAGINATION

i

Staying Involved to Stay Healthy

Recent social-scientific evidence confirms what many elderly people already know: To age well, it helps to pursue an active life. The data also disprove the worrisome idea that as increasing numbers of frail and disabled elderly people are kept alive through medical science, there will be an accumulation of older people living in pain or with crippling disabilities.

The accompanying graphs are taken from a major study conducted by the National Long Term Care Surveys. They show that if 1982 disability rates had held steady through 1995, there would have been an increase of almost 40 percent in the number of disabled people over age 85. Instead, because of better lifestyles and improved health care, the increase in the proportion of disabled elderly persons was only 10 percent, while for those age 85 and over it was slightly less than 30 percent.

The graphs also show that the total number of people in all the age categories, and especially in the eldest one, grew far more rapidly than did the proportion of disabled elderly people (Kolata, 1996).

The new data also show that education plays a highly significant role in explaining the lower rates of disability and disabling diseases among the elderly. Demographer Samuel Preston has calculated that "in 1980 more than 40 percent of Americans from 85 to 89 had fewer than 8 years of schooling. But by the year 2015, that figure will have dropped to 10 percent or to 15 percent" (quoted in Kolata, 1996, p. C3). Researchers agree that the effects of education on health in old age demand further study. It seems certain, however, that education helps people understand how to take better care of themselves, may lead to higher incomes (and thus to better health care), and may allow people to remain mentally active throughout life (Brody, 1996).

Gerontological research at Pennsylvania State University indicates that self-education as well as formal education is important in maintaining cognitive activity. People who read, travel, take courses, and otherwise continue to develop their minds are able to maintain intellectual functioning longer than people who become "couch potatoes." Intellectual activity leads, in turn, to greater attention to personal health, social activities, and moderate physical exertion. Current research shows that social involvement, mental stimulation, and the presence of social support are important to a healthy and relatively vigorous advanced old age. Social support—the presence of family and friends in one's daily life—is highly correlated with physical and mental health among the elderly.

These studies might seem to suggest that if an elderly person becomes ill and disabled or loses his or her mental abilities, it is a result of poor health habits and lack of education. However, a study of Alzheimer's disease among Catholic nuns shows that the propensity to develop this mentally debilitating and fatal disease is independent of education and that signs of the disease may appear very early in some individuals. Among people who do not have genetic predispositions toward certain diseases and did not experience highly stressful work and family lives when they were younger, continued mental development and social activity may be the keys to a robust old age (Brody, 1996).

position is that family members or close relatives should be required to pay for such care until funds are no longer available. The liberal position is that individuals should not have to be dependent on possibly reluctant family members; moreover, the high costs of long-term care for the elderly are an especially heavy burden for households with low incomes. The elderly themselves, who form a major segment of the U.S. voting population, are increasingly concerned about threats to their economic and social security, especially as represented by efforts to reduce the extent of health coverage under the Medicare system. They also are increasingly sensitive to any slights or discriminatory behavior based on their age.

Ageism The attitude known as **ageism** is similar to sexism (see Chapter 14). The term refers to an ideology that justifies prejudice or discrimination based on age. Ageism limits people's lives in many ways, both subtle and direct. It may label the young as incapable of learning. It labels the elderly as mentally incapable or asexual or too frail to get around. But people of all ages increasingly reject these notions. In their everyday lives in families and communities, for example, older people continually struggle against the debilitating effects of ageism. "Just because I need help crossing the street doesn't mean I don't know where I'm going," an elderly woman said to community researcher Jennie Keith (1982, p. 198).

Gerontologist Robert Butler observes that "ageism allows the younger generation to see older people as different from themselves; thus they subtly cease to identify with their elders as human beings" (1989, p. 139). Butler, a physician and social scientist, has found that as the proportion of older people in a society increases (as is occurring in the United States and Europe), the prevalence of ageism also increases. The younger generations, he notes, tend to fear that the older, increasingly frail and dependent generations will deprive them of opportunities for advancement. This fear is expressed in demands for

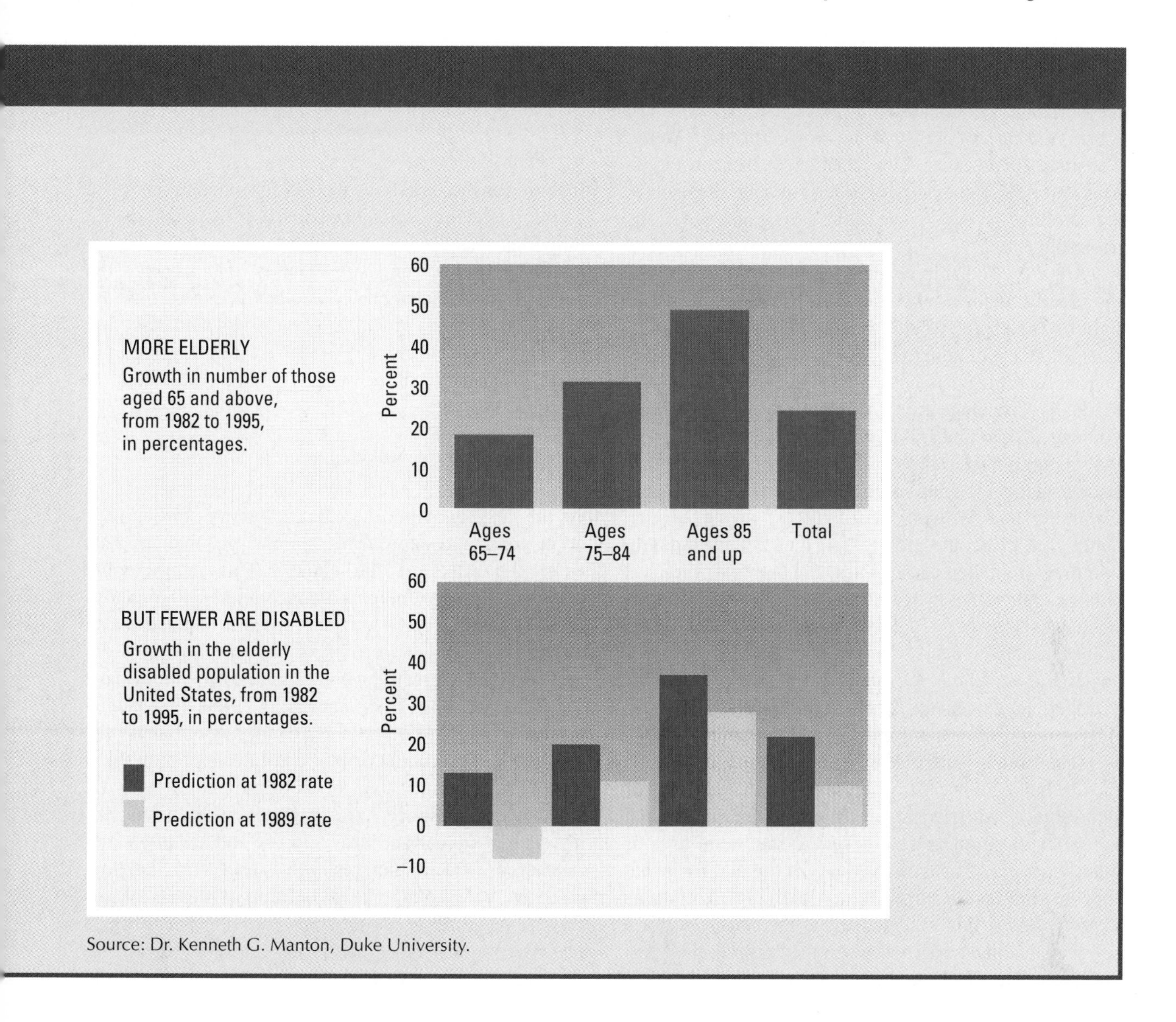

Source: Dr. Kenneth G. Manton, Duke University.

reduced spending on Medicare and other programs that assist the elderly, as well as in the belief that the elderly are affluent and do not need social supports.

Stereotyping of the Elderly A form of ageism that can be corrected through greater sociological knowledge about aging is the tendency to view all old people as frail, dependent, incapable, and asexual. But as this proportion of the population increases, older people will be found doing many things that are viewed as inappropriate for their age. A woman in her eighties recently set the age record for solo sailing across the Atlantic. Just before reaching the age of 100, George Burns dedicated a book "to his past five doctors."

Just as other groups in the population reject labels that deprive them of their individuality and self-esteem, elderly people reject ageist labels that relegate them to dependent, unequal status. There will of course be large numbers of dependent older people, but there will also be large numbers of vigorous people of advanced age who need and demand encouragement and support in maintaining their vitality and in continuing to live life to its fullest. Younger people who can apply their sociological imagination to these issues will be highly valued (Kornblum & Julian, 1998).

SOCIAL MOVEMENTS AMONG THE ELDERLY

Although they have been less far-reaching than the women's movement, social movements among the elderly—led by organizations like the Gray Panthers and the American Association of Retired Persons—have had a significant impact on American society. And as the population continues to age, we can expect to see more evidence of the growing power of the elderly. Consider, for instance, how changes in the consciousness and political activity of elderly people themselves are altering the way sociologists formulate questions about old age.

Until the social movements of the 1960s prompted the elderly to form movements to oppose ageism and fight for their rights as citizens, the most popular social-scientific view of aging was *disengagement theory.* Numerous empirical studies had shown that old people gradually disengage from involvement in the lives of younger people and from economic and political roles that require responsibility and leadership. In a well-known study of aging people in Kansas City, Elaine Cumming and William Henry (1971) presented evidence that as people grow older, they often gradually withdraw from their earlier roles, and that this process is a mutual one rather than a result of rejection or discrimination by younger people. From a functionalist viewpoint, disengagement is a positive process both for society as a whole (because it opens up roles for younger people) and for the elderly themselves (because it frees them from stressful roles in their waning years).

The trouble with disengagement theory is that, on the one hand, it appears to excuse policy makers' lack of interest in the elderly and, on the other hand, it is only a partial explanation of what occurs in the social lives of elderly people. An alternative view of the elderly is that they need to be reengaged in new activities. Known as *activity theory,* this view states that the elderly suffer a sense of loneliness and loss when they give up their former roles. They need activities that will serve as outlets for their creativity and energy (Palmore, 1981).

Theories of aging are not simply abstract ideas that are taught in schools and universities. The disengagement and activity theories lead to different approaches that often impose definitions of appropriate behavior on people who do not wish to conform to those definitions. Today gerontologists and elderly activists tend to reject both theories. They see older people demanding opportunities to lead their lives in a variety of ways based on individual habits and preferences developed earlier in life. Elderly people themselves express doubt that activity alone results in successful adjustment to aging or happiness in old age. For example, in her study of a French retirement community, Jennie Keith (1982) made this observation:

> The residents . . . seem to offer support to the gerontologists who have tried to mediate the extreme positions, disengagement vs. activity, by introducing the idea of styles of aging. . . . Some people are happy when they are very active, others are happy when they are relatively inactive. From this point of view, life-long patterns of social participation explain the kinds and levels of activity that are satisfying to different individuals. (p. 59)

In sum, for the elderly as well as for women, there is a growing tendency among social scientists to emphasize individual needs and capabilities. The social movements for gender and age equality also advance the needs of individuals, but in a collective manner, by asserting the needs of entire populations and rejecting preconceived notions of what is best for all women, all youth, all men, or all the elderly.

The Hospice Movement

As the large baby boom generation moves through the adult years and approaches middle age, their parents enter the ranks of the elderly. This demographic change has led to increased concern about the quality of life of the elderly and about death and the dying process. The questions of how best to prepare to die and whether one can have a "good death" might have seemed strange to Americans of an earlier time, when so many people died as a result of sudden, acute illnesses like pneumonia or tragic accidents in mines and mills. Today almost two thirds of the people who die each year experience long struggles with chronic illnesses like cancer and heart disease and are often "sequestered" from the rest of society in hospital intensive-care units. The proportion of deaths occurring in hospitals, nursing homes, and other institutions has risen rapidly since the 1930s and now exceeds 80 percent of deaths. Sociologists who study death and dying believe that this is a major reason for poll results showing that a majority of Americans support the right of patients to receive a lethal overdose from their doctors (Horgan, 1997). It is also a reason why so many health care experts and activists are seeking alternatives to hospital care for dying loved ones.

People who are dying often experience a period between life and death known as the living–dying interval (Kübler-Ross, 1989). For many dying people and their loved ones, this is a time of extreme importance. During this time there may be reconciliations, clarifications, expressions of love and understanding, and, under the best circumstances, a sense of closure and repose for the dying person. But most people die in hospitals. Nurses, doctors, and orderlies may intrude on these emotional scenes to minister to the patient. The hospital is dedicated to prolonging life, even in a dying patient. As the Rileys express it: "The hospital is geared for treatment and cure, it functions according to standards of efficiency and bureaucratic rules, its environment is sterile and unwelcoming to those who would visit patients *in extremis*" (1989, p. 27).

In recent decades, first in Great Britain and later in the United States, a social movement has emerged that is known as the hospice movement. The term *hospice* refers to a place, a set of services, or both; increasingly it refers to a service that can be brought into the dying person's home. The purpose of the hospice is to offer palliative care, which makes the patient as comfortable as possible so that all concerned can use the living–dying interval effectively and humanely. In a hospice the patient and his or her surrounding social group are viewed as the relevant social unit. Medical personnel and other professionals, such as social workers, are available to help everyone involved.

FIGURE 15.9 **Number of Patients in U.S. Hospices, 1985–1995**

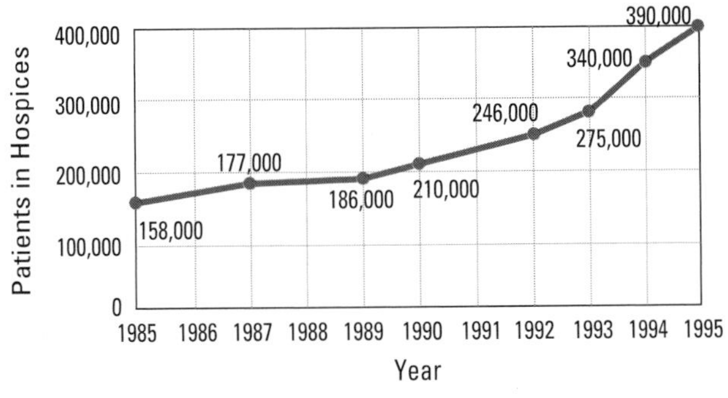

Source: Horgan, 1997.

Although it is still true that many people die in hospitals, hospice care is available in most parts of the United States and is an increasingly popular alternative to hospital death (see Figure 15.9). This change is a result of modifications in the insurance laws, which now allow Medicare reimbursements for home hospice care. Since 1983, when these rule changes were made in response to pressure from the hospice movement, the number of people who choose to die at home has more than doubled, and it continues to increase despite the additional sacrifices that home hospice care requires from loved ones (Belkin, 1992; Horgan, 1997).

Dying is never an easy experience, and the hospice movement cannot make it so. Nevertheless, the hospice movement is providing opportunities for new definitions of the last social role in the life course—the dying role—as well as the roles of those who will remain behind to continue their own life course.

RESEARCH ON THE CUTTING EDGE: Identities in the Making

Just because a population is aging does not imply that its young people are no longer an essential human resource. In fact, as we have seen in this chapter, the younger generations will inherit extra burdens and responsibilities in aging societies. In that context their values, skills, and energy will be more important than ever. But are today's young people ready to meet the challenges of rapidly changing societies? This question is being asked by a number of social scientists, notably James Youniss and Miranda Yates in their theoretical and empirical work on youth and community service.

Youniss and Yates (1997) have investigated Erik Erikson's thesis that societies that "fail to harness the services of youth in their historical aims" leave young people "to seek their own combinations, in small groups occupied with serious games, good-natured foolishness, cruel prankishness, and delinquent warfare. In no other stage of the life cycle, then, are the promise of finding oneself and the threat of losing oneself so closely allied" (Erikson, 1965, p. 11). On the basis of an extensive review of literature about youth and society Youniss and Yates have developed Erikson's insight further, as they explain in this passage:

> Peer culture does not suddenly loom up at adolescence pitted against adult norms. Children have coexisted in the two domains during their development, with ideas, rules, and moral stances continuously crossing between them. By adolescence, peer and adult spheres are clearly delineated, but also begin to be integrated. Youth need to learn how to translate rules and ethics from one sphere to the other in a two-way process. Hence, children's social knowledge is not an internalized version of adults' ideas, but is constructed by interacting with adults and by reformulating ideas in interactions with peers, and vice versa. (1997, p.28)

This view implies that societies need to create opportunities for young people to interact with adults as they jointly seek to address society's problems and, in this fashion, to realize their society's "historical aims."

The Research Background

Youniss and Yates examined all the available research on participation in youth groups like Girl and Boy Scouts, 4-H Clubs, Future Farmers of America, YMCA and YWCA, and many more. Social-scientific evaluations of young people's experiences offer evidence that they encourage involvement in the community and inhibit antisocial activities. Research on young people's participation in more focused efforts to deal with social problems, however, show even deeper and more lasting influences over the individual's life. For example, a major follow-up study of men and women who joined the civil rights movement during the Freedom Summer of 1964 and went to Mississippi to volunteer in Freedom Schools and help in voter registration found that more than 20 years later "the volunteer alumni continued to devote their energies to work that changes society in contrast to the no-shows [volunteers who decided not to go to the South], who took up more conventional occupations" (McAdam, 1988, quoted in Youniss & Yates, 1997, p. 33).

Critics of such studies argue that today's young people are not as motivated by social causes as they were a generation ago. Today's teenagers and young adults, it is claimed, are more likely to be motivated by the desire for personal mobility, excitement and adventure, and trendy pleasures. "The recent surge in calls for service," they add, "also has roots in the perception that contemporary youth seem politically apathetic and morally rudderless" (Youniss & Yates, 1997, p.11). Maryland now requires high school graduates to perform 72 hours of community service; Vermont and Pennsylvania have been considering similar laws.

But the authors' review of historical studies and surveys over the past half-century shows that participation by young people in a variety of political causes or volunteer activities has not decreased. This leads them to argue against laws that require community service. Young people have many different needs, just as adults and elderly people do. Some need to work for pay after school, even though this may not always be the most desirable course of action. Those who can volunteer are often urged to work for charitable causes or in "caring institutions" like nursing homes and hospitals. These are worthy activities, but the authors agree with sociologist Harry Boyte, who distinguishes between involvement based on "care" and involvement that actually promotes "participation" with adults in a socially active "problem solving politic" (Boyte, 1991). Volunteerism that is mandated and that results in a ritual form of care-giving may be no more effective in changing values and building future activism than no service at all.

The Empirical Study

To explore their thesis, Youniss and Yates found a program in a Catholic high school whose student body consisted largely of African Americans from middle-class backgrounds. In their senior year the students took a required course in "social justice." They read sociological studies and critical analyses of social institutions, studied examples of Christian activism, and worked one day a week at a local soup kitchen that served meals to 300 to 400 people each day. Most of those who came for meals were "black men in their 30s and 40s. Some had served in the military. Many looked physically ill. Drug abuse was a pervasive problem, and there were difficulties on several occasions during the year in trying to keep the dining room free of drugs and alcohol" (Youniss & Yates, 1997, p. 44). The students often met adult volunteers with whom they conversed, but the adults who came to the kitchen on court-mandated "community service" sentences were usually far more isolated and reticent than the students. Back in the classroom the students held lively discussions about their experiences and about the sociology of social class, poverty, and homelessness.

As part of their required work for the semester, the students were required to write four short essays about their experiences in the soup kitchen and with the homeless people and the adults who worked among them. The researchers collected and analyzed these essays. Through them they were able to see how the students' experiences were (and sometimes were not) actually changing their feelings about themselves and about social issues in small but gratifying ways.

By no means did the social justice course produce uniform political thinking or uniformly high levels of social commitment among the students. The researchers revisited the students during and after their college years and asked them to again write essays or letters about their ideas of service. Some retained the idealism they had gained in the religion class. Others had shed their idealism, and some had become somewhat cynical about social activism. The most important finding from the researchers' viewpoint, however, is that all the students agreed that the experience had "jarred them out of a passive acceptance of society as it is and started them on a path of critical thinking about society and what it might be" (Youniss & Yates, 1997, p.126). Here is how two of the most committed students summed up the experience:

I am currently working (volunteer) to help the homeless find jobs and homes. My lifelong aspiration is to be able to start a program where all homeless can be helped in some way. My expectation is that if this program can catch on and a bill—like the health bill—is passed, we can eliminate this plague.

Right now I have the world in my hand. I am three months from a Harvard degree. I will be attending either Harvard or Yale law school in fall. I am almost guaranteed more money, success, and material goods than any black kid from northeast DC could ever imagine. However, I plan to dedicate myself to creating the same opportunities for other kids in the inner city that I had. Social Justice class did this. The impact of that class was immeasurable. (quoted in Youniss & Yates, 1997, p. 126)

VISUAL SOCIOLOGY

After 90

Imogen Cunningham's life spanned almost a century, and she spent it making loving photographic portraits of people in the cities where she lived. Born in Scranton, Pennsylvania, at the end of the nineteenth century, she learned photography in its early days, when camera equipment was extremely cumbersome and unpredictable. With the encouragement of her artistic and thoughtful father, shown here in a photo she took when he was 90, Cunningham built her own darkroom and learned every aspect of the fledgling art of photographic portraiture.

In her adult years Cunningham lived and worked mainly in Seattle and San Francisco, but her career gave her opportunities to take photographs of people all over the world. During the period before World War II she was one of the most celebrated photographers working for the innovative magazine Vanity Fair. The subjects of her portraits range from world-famous artists and scientists to unknown but singular characters such as the tattooed woman who spent her life in the carnival.

At the end of her life Cunningham became interested in a sociological question: What ideas and activities occupy the minds of very old people? The photos shown here are a representative sample from a collection of stirring portraits she made of people who were, as she put it, "alive and kicking." She photographed many people

She is a distinguished radiobiologist who asked me to photograph her. I wasn't taking on commissions anymore, but I did it because she didn't care if she looked old and she didn't hate her face.

My father at 90.

She said to me, "When you come here nobody knows where you are."

This woman had been at the carnival all her life, but I found her in the hospital. It looks like lace, doesn't it?

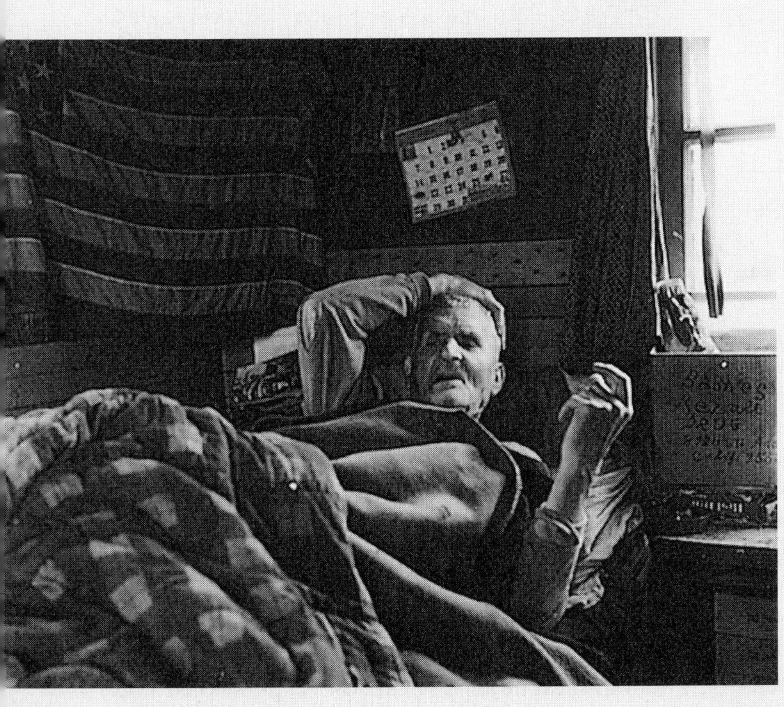

Old Norton lives by himself in his shack.

in their nineties, like Old Norton in his shack and the nursing home resident feeling forgotten and forlorn, for whom the main challenge of life was merely to live through each day. She also photographed people who, like Cunningham herself, had the good fortune to have a passion for work and activity and felt little concern about their physical deterioration because they were absorbed in their creative efforts.

Cunningham died at the age of 93, just as the volume containing these photographs was about to be published. Although she missed the chance to celebrate the appearance of her book about the very old, she herself lived long enough to have the pleasure (denied to many artists) of seeing her work honored during her lifetime. Even from the few photos presented here we can readily see that her work captures the mystery of the human spirit and demonstrates that extreme age does not extinguish the love of life.

SUMMARY

In many societies age determines a great deal about the opportunities open to a person and what kind of life that person leads. The study of aging and the elderly is termed *gerontology.*

All societies channel people into *age grades,* or sets of statuses and roles based on age. The transitions among these age grades create a *life course* and are often marked by ceremonies known as *rites of passage.*

Age cohorts are people of about the same age who are passing through life's stages together. The baby boom cohorts, which were produced by rapid increases in the birthrate from about 1945 through the early 1960s, have profoundly influenced American society. A sizable proportion of the children of the baby boom generation, the "baby boom echo," are members of minority groups.

By *life expectancy* we mean the average number of years a member of a given population can expect to live beyond his or her present age. As life expectancy in a population increases, the proportion of the population that is dependent on the adult cohorts also increases.

In urban industrial societies there are distinct patterns of stratification in which age defines the roles one plays and the rewards one can expect. The forces of social change unleashed by colonialism, industrialization, urbanization, and population growth have tended to disrupt the formal age grade systems of smaller, more isolated societies.

Social definitions of childhood differ immensely throughout the industrial world as well as between modern and traditional societies. As a result of increasing incomes and the passage of child labor laws, children became economically "worthless" but emotionally "priceless." However, there is a growing gap between "priceless" children and children who bear a heavy burden of poverty and deprivation.

A century ago, the largest segment of the U.S. population living in poverty or near-poverty conditions was the elderly. As a result of programs such as Social Security and Medicare, rates of poverty among the elderly have decreased dramatically. However, the situation is not nearly so positive for elderly members of minority groups.

As people age, they experience more medical problems and disabilities, but this does not mean that they must inevitably withdraw from social life. Social scientists who study the situation of the elderly point out that longer life spans need to be accompanied by new concepts of social roles in more advanced years.

Ageism is an ideology that justifies prejudice or discrimination based on age. As the proportion of older people in a society increases, as is occurring in the United States and Europe, the prevalence of ageism also increases.

As the population as a whole has aged, the impact of the elderly on American society has increased. This is changing the way sociologists view old age. Before the 1970s the most popular social-scientific view of aging was disengagement theory, the belief that as people grow older they gradually disengage from their earlier roles. An alternative view of the elderly, known as activity theory, states that the elderly need activities that will serve as outlets for their creativity and energy. Today gerontologists tend to reject both of these theories, seeing older people demanding opportunities to lead their lives in a variety of ways based on individual habits and preferences.

The growing proportion of elderly people in the population has led to increased concern about the quality of life of the elderly and about death and the dying process. One outcome of this concern is the hospice movement, which attempts to provide dying people and their loved ones with a comfortable, dignified alternative to hospital death.

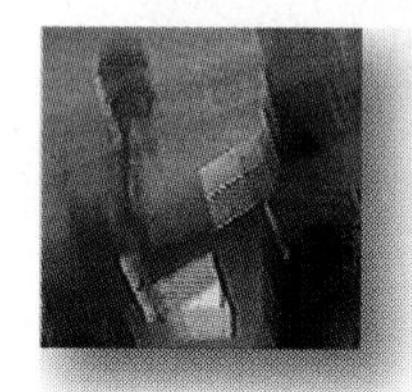

SOCIOLOGY VERSUS IDEOLOGY

You will frequently hear people begin sentences this way: "Old people are always . . . " or "Young people can't seem to understand that . . ." They may be liberals who are lamenting what they see as the failure of young people to take a more active part in social causes. They may be conservatives who resent paying into the Social Security fund so that others who are less frugal or less productive can be assured of a modest level of living in their declining years. From a sociological perspective, however, you will no doubt be quick to take issue with any statements that begin with such broad generalizations.

As we have shown in this chapter, it is impossible to make blanket assertions about old people or young people (or any major population segment, for that matter). What makes sociology so fascinating—and frustrating to those who tend to think in "sound bites"—is the enormous diversity of human populations. We learn that there is no single theory that will explain the needs of all young people or all elderly people. Sociologists can make a difference by conducting empirical research that reveals the categories of young or elderly people who are in greatest need as a result of economic and social inequalities.

GLOSSARY

gerontology: the study of aging and older people. (p. 470)

age grade: a set of statuses and roles based on age. (p. 470)

life course: a pathway along an age-differentiated, socially created sequence of transitions. (p. 470)

rite of passage: a ceremony marking the transition to a new stage of a culturally defined life course. (p. 470)

age cohort: a set of people of about the same age who are passing through the life course together. (p. 471)

life expectancy: the average number of years a member of a given population can expect to live. (p. 476)

psychosocial risk behaviors: behaviors that are detrimental to health, such as smoking and heavy drinking. (p. 485)

lifetime negative experiences: experiences that cause long-term stress, such as the death of a child or spouse. (p. 485)

ageism: an ideology that justifies prejudice and discrimination based on age. (p. 486)

WHERE TO FIND IT

BOOKS

Community Service and Social Responsibility in Youth (James Youniss and Miranda Yates; University of Chicago Press, 1997). A case study of the effects of student activism matched with a course in social justice on identity formation in young people, with excellent bibliographic background.

The Sociology of Death (David Clard, ed.; Blackwell, 1994). A fine collection of theoretical and empirical studies of death and dying in Western societies, including research on AIDS deaths, war memorials, cremation or burial, the British hospice movement, and much more.

Age and Structural Lag (Matilda White Riley, Robert L. Kahn, and Anne Foner; Wiley, 1994). A comprehensive collection of articles covering many recent research initiatives. Dr. Riley is one of the world's foremost authorities on the demography of aging and the comparative situation of elderly people throughout the world.

Pricing the Priceless Child: The Changing Social Value of Children (Viviana Selizer; Free Press, 1985). An extremely innovative sociological and historical analysis of changing norms of childhood in the United States and other Western nations. Develops a thesis about the changing value placed on childhood as the economy and society change.

The Coming of Age (Simone de Beauvoir; Warner, 1973). A classic statement by one of the twentieth century's most influential thinkers about the combined effects of gender and age on human development and society.

Number Our Days (Barbara Myerhoff; Simon & Schuster, 1978). A haunting study of how elderly people struggle to survive and die with dignity. The setting is a senior citizen center in Venice Beach, California, and the people are elderly Jewish immigrants and refugees from Europe.

OTHER SOURCES

Kids Count Data Book. A timely and scientifically accurate source of comparative demographic data on the situation of children in the United States and elsewhere.

Age in America: Trends and Projections. A chartbook that analyzes demographic and socioeconomic trends affecting the United States and its elderly population.

Handbook of Aging and the Social Sciences. A valuable source of interdisciplinary research and policy studies dealing with the condition of the aged in the United States and other Western societies.

INTERNET RESOURCES

Socionet (www.socio.com). The Web site for the journal *Sociometrics.* It is devoted to the study of social groups and networks and how they are related to social issues of all kinds (for example, adolescence, health, disability).

PART 4

SOCIAL INSTITUTIONS

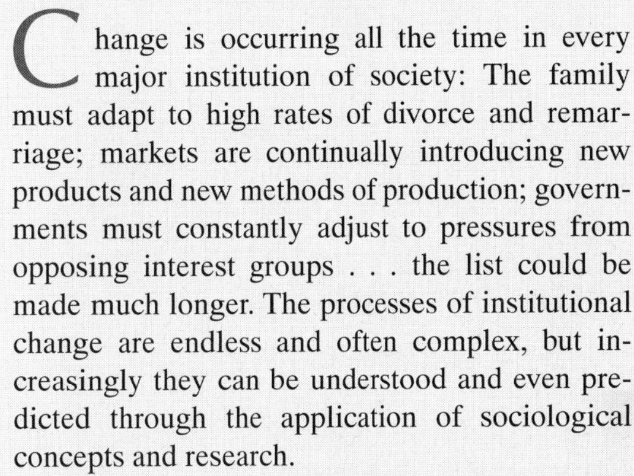

Change is occurring all the time in every major institution of society: The family must adapt to high rates of divorce and remarriage; markets are continually introducing new products and new methods of production; governments must constantly adjust to pressures from opposing interest groups . . . the list could be made much longer. The processes of institutional change are endless and often complex, but increasingly they can be understood and even predicted through the application of sociological concepts and research.

In this set of chapters the concepts and perspectives discussed earlier in the book will be applied to an analysis of several major social institutions. Chapter 16 explores the changing nature of the family, and especially research on the ways in which family statuses and roles are changing and the social forces that seem to account for these changes. The other chapters in Part 4 explore other institutions—proceeding from an analysis of how each is typically organized to a discussion of how it is changing. Chapters 17 and 18 discuss three major cultural institutions—religion, education, and the media—whose role is to transmit and reinforce a society's most cherished values. Chapter 19 deals with the roles and statuses that generate jobs and income. Chapter 20 shows how political institutions determine "who gets what, when, and how." It also deals with the grave difficulties facing nations attempting to institute the rule of law. Finally, in Chapter 21 the institutions of science and technology are analyzed, with special attention to their role in health care and environmental issues.

CHAPTER 16

THE FAMILY

The two photos shown here of "the birth of a family" convey much of the joy of family intimacy: the young couple just before she gives birth and the new father thrilled by his first experience with their child as the mother lays the baby on his chest. In the first photo we see a couple about to become a family. They will quickly take on a host of new roles with new demands and gratifications that will challenge their love in ways that they can hardly anticipate. In the second photo our eyes linger on the baby's tiny, perfect fingers and toes and the rapt gaze of the new mother. Is there the slightest hint in the man's somewhat rigid fingers that the crying newborn's vulnerability may tinge his joy with feelings of awkwardness? We cannot know this for sure, but the early days and months of parenthood are filled with many emotions—and love, awkwardness, joy, anxiety, and fatigue are prominent among them.

If we examine these photos from a historical perspective, we will surmise that this is a contemporary couple. There are few clues to this in the scenes themselves. Her flannel nightgown could be from any decade of this century or the last. The plain gold band on his left hand is a timeless symbol of matrimony. What reveals them as people of our time is the photographer's choice of them as subjects. No doubt couples in earlier decades behaved in exactly similar ways, but they were not often encouraged to display their joy and intimacy in public. In

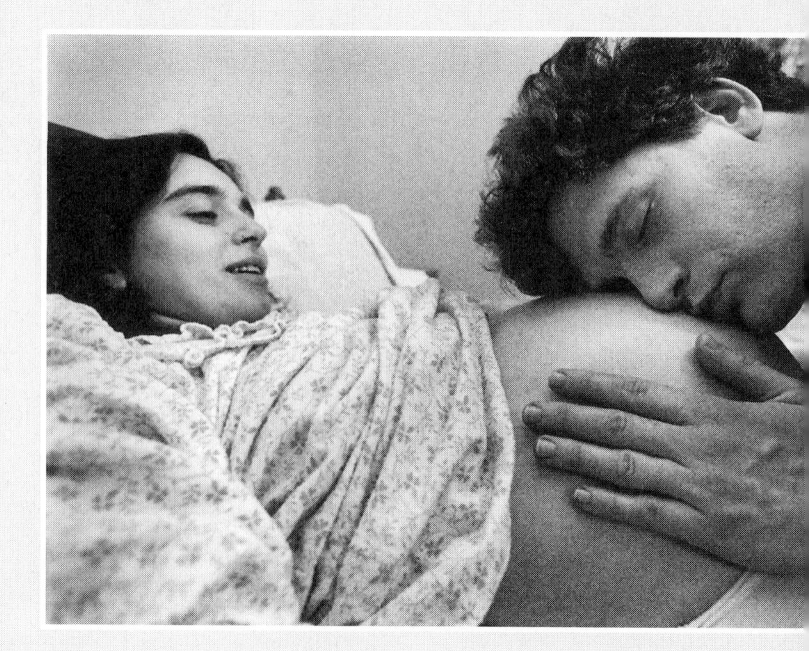

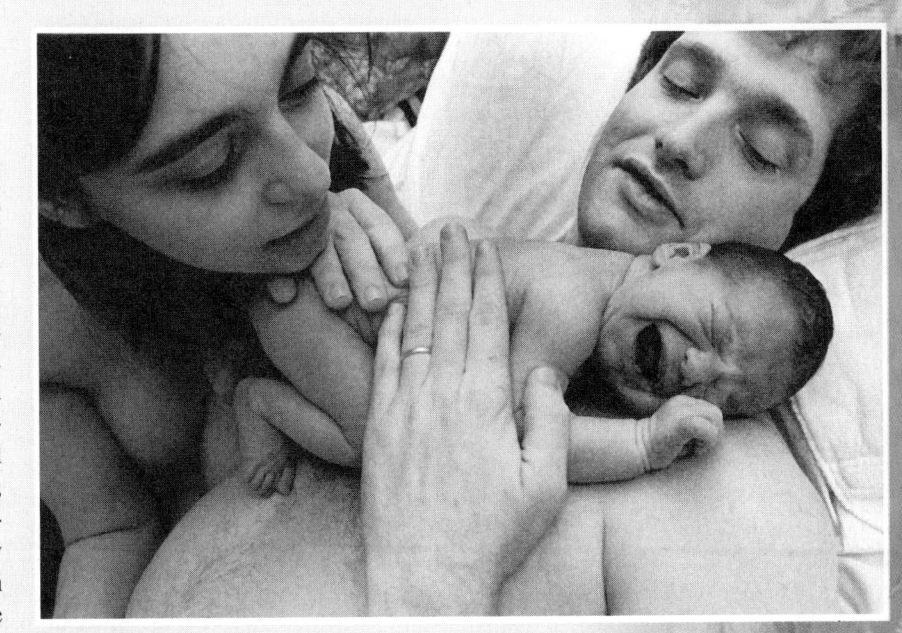

the early twentieth century it was more common for photographers to capture the gender-specific aspects of childbirth. The woman's joy at receiving the baby, often with adoring grandmothers or nurses in the background, expressed the heavily gender-specific quality of the childbirth scene. The man was portrayed in a stereotypical fashion, pacing in the hospital corridor, perhaps with one cigarette in hand and another still burning in a nearby ashtray, symbols of his ineffectual nervousness. Other photos of early family moments would focus on the father's first glimpse of the baby through the windows of the hospital nursery. Contemporary approaches to childbirth encourage more direct involvement of the father in the birthing process. They advocate more immediate contact between the new parents and the infant. For thousands of couples, going through the experience of childbirth together provides some of life's most thrilling memories.

As we gaze at these most personal glimpses into family life, we may also be reminded of the mystery of human love and attraction. We come back to these mysteries in the Visual Sociology section of the chapter, which captures people in a variety of intimate embraces, all of which convey the power of love and intimacy. Between this intimate chapter opening and the return to intimacy at the chapter's end, however, we will be dealing with many of the challenges to love and intimacy that arise in today's world. The risk in taking this sociological journey into the institutions of family and marriage is that we may forget to what extent love—between a couple, or among siblings, or between parents and children—is a force that can overcome many of the shocks and strains of modern life. On the other hand, we will learn that there are ways in which society makes love of various kinds possible and even inevitable.

THE NATURE OF FAMILIES

We continually hear warnings of the "death of the family" in modern societies. As early as 1934, sociologists like William Fielding Ogburn were writing about the decline of the family. "Prior to modern times," according to Ogburn, "the power and prestige of the family was due to . . . functions it performed. . . . The dilemma of the modern family is caused by the loss of many of these functions in recent times" (Ogburn & Nimkoff, 1934, p. 139). More recent commentators have pointed to the so-called sexual revolution as a threat to this basic social institution. Others insist that even in the face of trends toward single parenthood, cohabitation without marriage, divorce and remarriage, and smaller family size, the family is "here to stay" (Goode, 1971; National Commission on America's Urban Families, 1993).

In this chapter we will see that there are so many variations in family form even within a single society that it has become increasingly difficult to speak of "the family" as a single set of statuses, roles, and norms. Nevertheless, it remains true that in all known societies almost everyone is socialized within a network of family rights and obligations known as *family role relations* (Goode, 1964). In the first section of this chapter, therefore, we examine the family as an institution and show how family role relations have changed in the twentieth century. The second section focuses on how the basic sociological perspectives explain changes in family roles and functions and in the interactions between parents and children. In the third section we look at the dynamics of family formation—especially mate selection, marriage, divorce, and remarriage.

The Family as an Institution

There is a big difference between a particular family and the family as a social institution. Your family, for example, has a particular structure of statuses (mother, father, sister, brother, grandmother, etc.), depending on who and where its members are. Your immediate family is also part of an extended kinship system (aunts, uncles, cousins, grandparents, etc.) that also varies according to how many branches you and your immediate family actually keep track of. The family as a social institution, however, comprises a set of statuses, roles, norms, and values devoted to achieving important social goals. Those goals include the social control of reproduction, the socialization of new generations, and the "social placement" of children in the institutions of the larger society (colleges, business firms, and so forth).

These are not the only goals that the family is expected to fulfill. In many societies, especially tribal or peasant societies, the family performs almost all the functions necessary to meet the basic needs of its members. Those needs are listed in Figure 16.1, which compares less differentiated gemeinschaft or village-type societies with more differentiated gesellschaft societies like that of the United States. It shows that most of the functions that were traditionally performed by the family are now performed either partly or wholly by other social institutions that are specially adapted to performing those functions.

Figure 16.1 only begins to illustrate the institutional complexity of modern societies. Not only are there many more separate institutions in modern societies, but basic social functions are often divided among several institutions. Thus the institutions that meet protective and social-control needs include governments at all levels, the military, the judiciary, and the police. Replacement needs are met by the family through mating and reproduction. In addition to the family, religious institutions, education, and other cultural institutions meet the need to socialize new generations.

The family's ability to meet the goals society expects it to meet is often complicated by rapid social change. For example, many families in industrial communities throughout the United States were quite good at preparing their children for blue-collar manufacturing jobs, but those jobs have been eliminated owing to automation and the globalization of the economy. For many families, therefore, it is no longer clear how they can equip their offspring to compete in the new job market.

SOCIAL FUNCTION	Less Differentiated (gemeinschaft)	More Differentiated (gesellschaft)
	TYPE OF SOCIETY	
Communication among members: first through language alone and then through specialized institutions devoted to communication	Family, kin networks	Mass media
Production of goods and services: from the basic items required for survival to items designed to satisfy new, more diverse needs created by increasing affluence	Family groups	Economic institutions
Distribution of goods and services: within societies at first, and later, with the rise of trade and markets, between societies as well	Extended family, local markets	Markets, transportation institutions
Protection and defense: including protection from the elements and predators and extending to defense against human enemies	Family, village, tribe	Armed forces, police, insurance agencies, health care institutions
Replacement of members: both the biological replacement of the deceased and the socialization of newcomers to the society	Family	Family, schools, religious institutions
Control of members: to ensure that the society's institutions continue to function and that conflict is reduced or eliminated	Family	Family, religious institutions, government in all forms

Source: Adapted from Lenski & Lenski, 1982.

FIGURE 16.1 **The Family: Institutional Differentiation**

When there are major changes in other institutions in a society, such as economic institutions, families must adapt to those changes. Similarly, when the family changes, other institutions will be affected. As an example, consider the impact of the "sexual revolution" that began in the 1920s. During that decade, often called the Roaring Twenties, young adults began to upset the moral code that had regulated the behavior of couples for generations, at least among people who considered themselves "respectable." According to this traditional set of norms, women were the guardians of morality. Men were more likely to give in to sexual desire, "but girls of respectable families were supposed to have no such temptations" (Allen, 1931, p. 74). After World War I, however, "respectable" young women began to reject the **double standard,** by which they had to adhere to a different and more restrictive moral code from that applied to men. They began smoking cigarettes, wearing short skirts, and drinking bootleg gin in automobiles or at "petting parties." These changes in the behavior of couples produced far-reaching changes in other institutions. Women demanded more education and more opportunities to earn income. This in turn led to demands for new institutions such as coeducational colleges and integrated workplaces, as well as greater access by women to existing institutions such as medicine, law, and science. These changes in women's expectations and behavior, along with the increasing educational and occupational mobility of men after the world wars, contributed immensely to changes in traditional family roles—especially the weakening of the role of the father as the central and most powerful authority figure in the family. The consequences of these fundamental changes in family roles are still felt today.

Changes in the institutions in one area of social life can place tremendous pressure on those in other sectors. This is especially true in the case of the family, as will be evident throughout this chapter. But before we examine the ways in which families have changed in this century, it is necessary to define some terms that are frequently used in discussing the family.

Defining the Family

The family is a central institution in all human societies, although it may take many different forms (Cherlin, 1996). A **family** is a group of people related by blood, marriage, or adoption. Blood relations are often called *consanguineous attachments* (from the Latin *sanguis,* meaning "blood"). Relations between adult persons living together according to the norms of marriage or other intimate relationships are called *conjugal relations.* The role relations among people who consider themselves to be related in these ways are termed **kinship.**

The familiar kinship terms—*father, mother, brother, sister, grandfather, grandmother, uncle, aunt, niece, nephew, cousin*—refer to specific sets of role relations that may vary greatly from one culture to another. In many African societies, for example, "mother's brother" is someone to whom the male child becomes closer than he does to his father and from whom he receives more day-to-day socialization than he may from his father. It must be noted that biological or "blood" ties are not

These photos show a conventional nuclear family, with a father, mother, and two children, and an extended family, with three generations of family members, including aunts, uncles, and cousins. Although these may once have been the most common forms of traditional and modern family structures, we have seen that the number of single-parent families, and of families that share custody of the children after divorce, is increasing.

necessarily stronger than ties of adoption. Adopted children are usually loved with the same intensity as children raised by their biological parents. And many family units in the United States and other societies include "fictive kin"—people who are so close to members of the family that they are considered kin despite the absence of blood ties (Stack, 1974). Finally, neither blood ties nor marriage nor adoption adequately describes the increasingly common relationship between unmarried people who consider themselves a couple or a family.

The smallest units of family structure are usually called **nuclear families.** This term is usually used to refer to a wife and husband and their children, if any. Nowadays one frequently hears the phrase "the traditional nuclear family" used to refer to a married mother and father and their children living together. But as we will see throughout this chapter, there is no longer a "typical" nuclear family structure. Increasingly, therefore, sociologists use the term *nuclear family* to refer to two or more people related by consanguineous or conjugal ties or by adoption who share a household; it does not require that both husband and wife be present in the household or that there be any specific set of role relations among the members of the household (Benokraitis, 1996).

The nuclear family in which a person is born and socialized is termed the **family of orientation.** The nuclear family a person forms through marriage or cohabitation is known as the **family of procreation.** The relationship between the two is shown in Figure 16.2.

Kinship terms are often confusing because families, especially large ones, can be rather complex social structures. It may help to devote a little time to working through Figure 16.2. "Ego" is the person who is taken as the point of reference. You can readily see that Ego has both a family of orientation and a family of procreation. So does Ego's spouse. Ego's parents become the in-laws of the spouse, and the spouse's parents are Ego's in-laws. But like the vast majority of people, Ego also has an **extended family** that includes all the nuclear families of Ego's blood relatives—that is, all of Ego's uncles, aunts, cousins, and grandparents. Ego's spouse also has an extended family, which

FIGURE 16.2 **Nuclear and Extended Family Relationships**

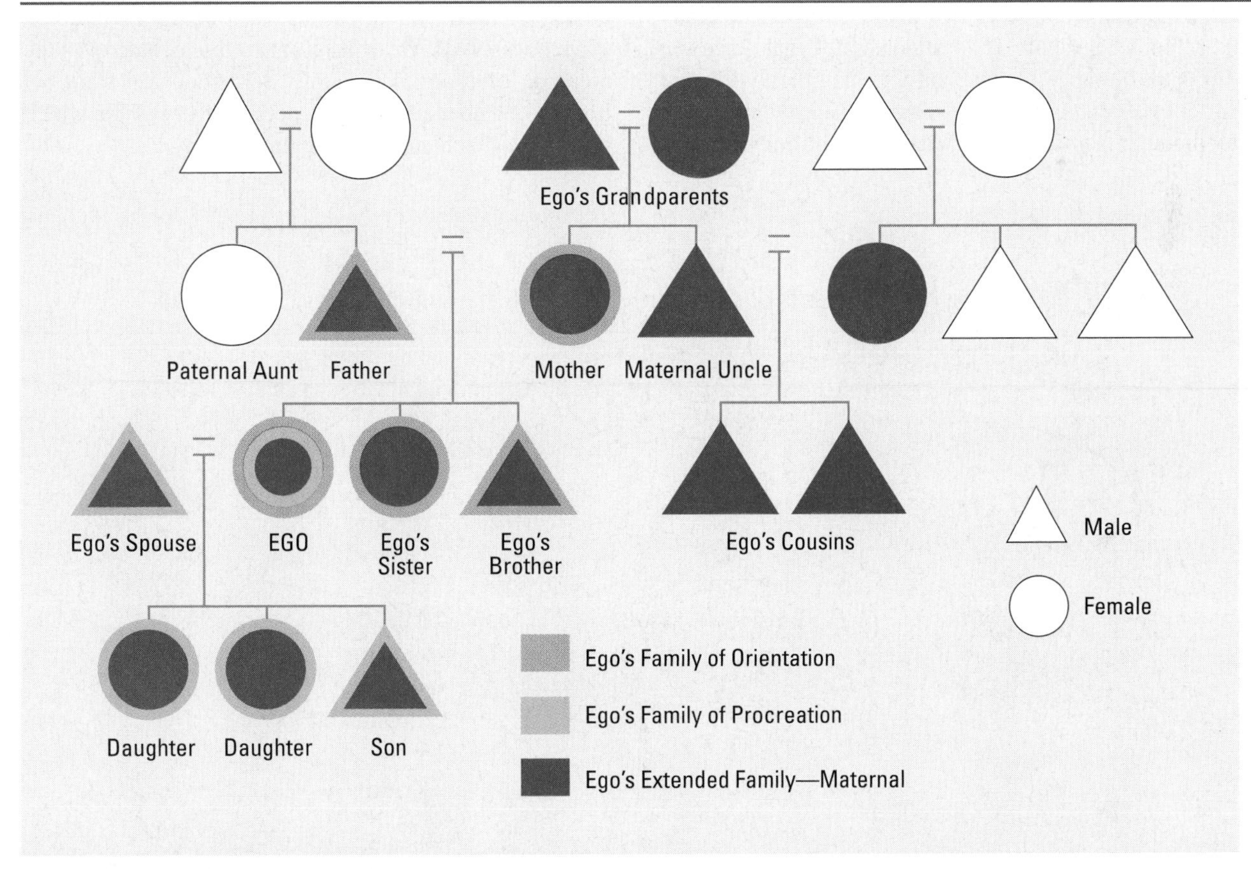

This diagram shows the family relationships of a hypothetical individual, Ego. Ego was born and socialized in what sociologists call the family of orientation. Ego formed a family of procreation through marriage or cohabitation.

is not indicated in the figure, nor is it defined as part of Ego's extended family. But relationships with the spouse's extended family are likely to occupy plenty of Ego's time. Indeed, as the figure shows, the marriage bond brings together far more than two individuals. Most married couples have extensive networks of kin to which they must relate in many, ways throughout life.

Variations in Family Structure

For the past several years sociologists have been demonstrating that the traditional household consisting of two parents and their children is no longer the typical American family. This is not a value judgment that sociologists applaud or deplore, but a matter of demographic fact. In the 1940s there were about eight times as many households headed by a married couple as there were households headed by unrelated individuals. By 1980 there were only about two and a half times as many. While the majority of children in the United States still live with a married mother and father who are the child's biological parents, that arrangement is becoming less common. This fact is illustrated in Figure 16.3, which shows that between 1970 and 1990 the proportion of children in two-parent families declined by 12 percent. Figure 16.4 demonstrates that since the 1960s the number of husband-wife families with one earner or no earner has declined quite sharply, while the numbers of families with dual earners and families headed by a woman with no spouse present have risen. This historical chart makes the point that two-parent families continue to be the statistical norm but that dual-earner families (in which both parents are in the labor force) began to outnumber gender-differentiated families (father in the labor force, mother in the home) in 1980 and now outnumber single-earner families by at least 7 million families.

This trend points to a growing diversity of family types, including far more people living alone and far more women raising children alone. There are also more households composed of unrelated single people who live together not just because they are friends but because only by doing so can they afford to live away from their families of orientation. Thus, when they discuss family norms and roles, sociologists must be careful not to represent the traditional nuclear family, or even the married couple, as typical. As an institution, the contemporary family comprises a far greater array of household types than ever before.

Sociologists are divided on the meaning and implications of these facts. For some, the trend toward greater diversity of family forms means that the family as an institution is adapting to change (Skolnick, 1996; Stacey, 1990). For others, the decline of the family composed of two married parents living with their biological children poses a serious threat to the well-being of children (Popenoe, 1994). We will return to this debate over the meaning of changes in family form and function, but first we consider additional aspects of the ways in which families have changed during this century.

FIGURE 16.3 **Living Arrangements of Children Under Age 18, United States, 1970–1990**

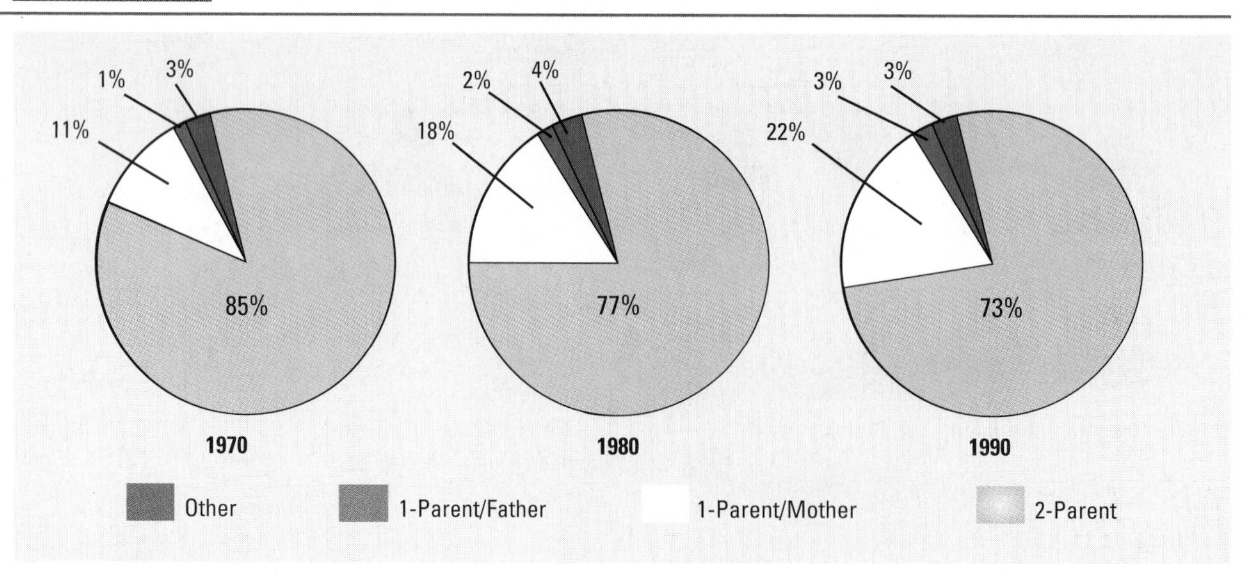

Source: Based on data from Census Bureau.

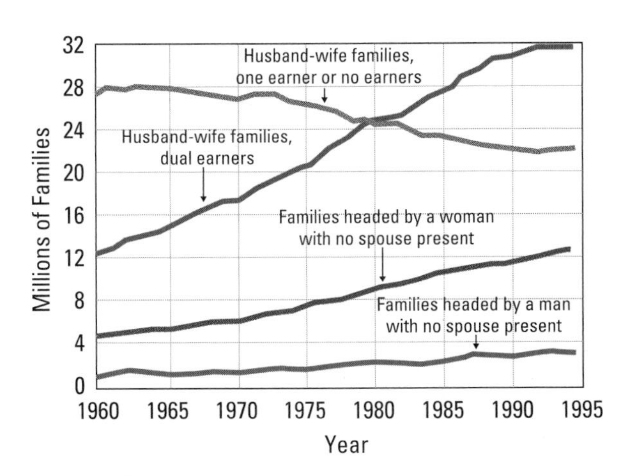

Source: Farley, 1996.

FIGURE 16.4 **Number of Families, by Type, United States, 1960–1994**

Parenting, Stepparenting, and Social Change

It has never been easy to be a parent. Keeping a family intact and raising children who are confident and capable and will become loving, capable parents themselves is a domestic miracle. For better or (sometimes) worse, most people learn to be parents by following the examples set by their own parents when they were children. Much as we may swear that when our turn comes we will improve on the way our parents performed their roles, we often find ourselves acting toward our children much as our parents acted toward us. In fact, many parents seek help from family counselors because they are shocked to find themselves doing and saying the same things to their children that their parents did to them (Scarf, 1995).

Many other adults and children will experience major reorganizations of their families as couples divorce and remarry. When stepfamilies are formed, family members must adapt to new role relations as stepchildren and stepparents, often while maintaining many of their obligations in the original family. Sharon L. Hanna, an expert on stepfamilies, notes that there are more than 20.6 million stepparents in the United States and estimates that in the early years of the new century almost 50 percent of all Americans will be part of a stepfamily (Hanna, 1994).

"Stepfamilies," sociologists Frank Furstenberg and Andrew Cherlin point out, "are a curious example of an organizational merger; they join two family cultures into a single household" (1991, p. 83). While this joining of family cultures also occurs in first marriages, in those marriages the couple usually has time to work out differences before children are born. In the case of stepfamilies, the task of working out a mutually acceptable concept of the family may be more stressful than the partners anticipated.

Sociologists point out that positive relationships in any family do not just happen but are worked on consciously through much learning and discussion. These efforts are particularly necessary in stepfamilies (Hanna, 1994). Some studies of stepfamilies show that new stepparents often conceive of themselves as "healers," people who can use the force of love to form strong bonds of attachment between two sets of children. Very often, however, the children cling to the fantasy that their original parents will get back together, and the stepparents' efforts may meet with frustration. Stepparents who manage to reorganize the family successfully tend to be diplomatic and sensitive and to avoid competing with biological parents. They gain the support of their spouses and wait for opportunities to gain the acceptance of their stepchildren (Papernow, 1988).

However, resolving the complexity and ambiguity of roles in reorganized families is no simple matter. Many factors come into play when a new family is formed. For example, Furstenberg and Cherlin point out that legal norms "reinforce the second-class status of stepparents" as nonbiological partners (1991, p. 80). A biological father is held responsible for child support even if he has never lived with the child, while a stepfather incurs no financial responsibility for a child even if he has lived with that child for many years. "The only legal recourse for stepparents who want to claim rights and responsibilities is to adopt their stepchildren, an act that seems to symbolize a deeper and more permanent tie, more like a biological relationship" (Furstenberg & Cherlin, 1991, p. 80).

Members of all kinds of families experience particular stress in times of economic hardship or when other social changes influence the family. Stress and change in the larger society may cause parents to feel unable to perform their roles as providers of material and emotional support. In the early decades of the twentieth century, for example, the combination of frequent severe recessions, high rates of industrial accidents, and lack of adequate health care had a significant impact on families. Fathers were often forced to leave their families in search of work; infection often claimed the lives of mothers during childbirth. Demographers estimate that at the beginning of the century, when divorce rates were still extremely low, almost 25 percent of all children lost at least one parent through death before reaching the age of 15 (Uhlenberg, 1980).

By midcentury, improvements in health care had reduced this proportion to 5 percent, but divorce rates had increased to 11 percent, and another 6 percent of children were born to unmarried parents. Thus even in the 1950s—the most stable decade for families and children—demographers estimate that 22 percent of children were being raised in families that had only one parent as a result of death, separation, divorce, or single parenthood

(Bumpass, Raley, & Sweet, 1995). Even this relative stability lasted only a decade. By the end of the 1950s divorce rates had begun their precipitous rise, and so had rates of births to unmarried parents. By 1995 almost 30 percent of births in the United States were to unmarried mothers (National Center for Health Statistics, 1995).

These profound changes in family composition have many causes and many consequences. As families age, children move through the school years, adolescence, and young adulthood while parents pass through adulthood to their senior years. Throughout this period all families seek to accomplish certain basic socialization goals. As the diversity of families increases, the ways in which they attempt to reach these goals differs, but the goals themselves do not change nearly as much.

The Family Life Cycle

Sociologist Paul Glick, an innovator in the field of family demography and ecology, developed the concept of the *family life cycle,* the idea that families pass through a sequence of five stages:

1. Family formation: first marriage
2. Start of childbearing: birth of first child
3. End of childbearing: birth of last child
4. "Empty nest": marriage of last child
5. "Family dissolution": death of one spouse (Glick & Parke, 1965)

These are typical stages in the life cycle of conventional families. Although there are other stages that could be identified within each of these—such as retirement (the years between leaving the workforce and the death of one spouse) or the "baby" stage (during which the couple is rearing preschool-age children)—the ones that Glick listed are most useful for comparative purposes.

As the "typical" family structure becomes ever more difficult to identify owing to changes in family norms, the stages of family development also vary. In fact, the stages of the family life cycle have become increasingly useful as indicators of change rather than as stages that all or most families can be expected to experience. The Census Bureau estimates, for example, that there are at least 4 million heterosexual couples in the United States who are cohabiting. This number

FIGURE 16.5 **Estimated Median Age at First Marriage, United States, 1890–1990**

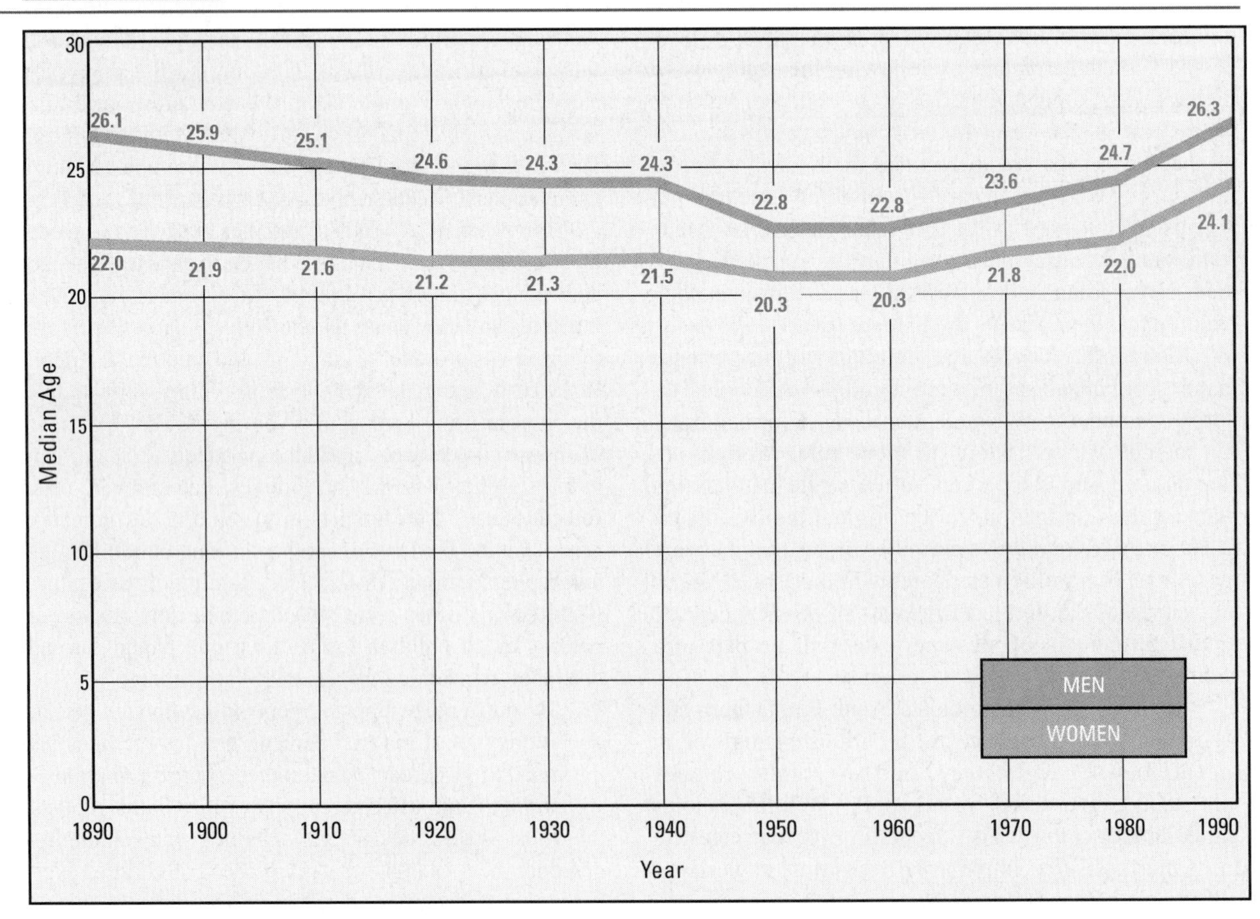

Source: Data from Census Bureau.

represents an eightfold increase since 1970, when an estimated 500,000 cohabiting couples were counted (*Statistical Abstract,* 1998). (Note that the census figures include only heterosexual couples and therefore vastly underestimate the total number of cohabiting couples.) Although the majority of cohabiting couples will eventually marry, about half will break up or continue to live together without marrying (Farley, 1996).

When sociologists look at the median age at which people experience the various stages of the family life cycle, significant trends emerge. Consider age at first marriage. In 1890 the median age of Americans at first marriage was 26.1 years for men and 22 years for women; in the 1950s and 1960s it reached the historic lows of 22.8 and 20.3. Then it began rising again slowly, until by 1990 it was 26.3 for men and 24.1 for women (see Figure 16.5). Age at first marriage tends to be somewhat higher for males than for females and is associated with educational attainment; more highly educated people are more likely to delay marriage. There is also a tendency for members of minority groups to marry later owing to the proportionately lower income levels of these populations. People with lower incomes tend to delay marriage or not to marry because they lack the material means to sustain a marital relationship (Farley, 1996).

As it passes through the family life cycle, every family experiences changes in its system of role relations. In analyzing these changes, social scientists often modify Glick's stages in order to focus more sharply on interactions within the family. An example is the set of developmental stages shown in Figure 16.6. There are major emotional challenges at each of these stages. For example, families with adolescent children must adapt to the children's growing independence. This may involve going through a stage of negotiation over such issues as money, cars, and dating. But researchers who study family life note that parents are often confused

FIGURE 16.6 **Stages of the Family Life Cycle**

STAGE	EMOTIONAL PROCESS	REQUIRED CHANGES IN FAMILY STATUS
1. Between families: The unattached young adult	Accepting parent–offspring separation	a. Differentiation of self in relation to family of origin b. Development of intimate peer relationships c. Establishment of self in work
2. The joining of families through marriage: The newly married couple	Commitment to new system	a. Formation of marital system b. Realignment of relationships with extended families and friends to include spouse
3. The family with young children	Accepting new generation of members into the system	a. Adjusting marital system to make space for children b. Taking on parenting roles c. Realignment of relationships with extended family to include parenting and grandparenting roles
4. The family with adolescents	Increasing flexibility of family boundaries to include children's independence	a. Shifting of parent–child relationships to permit adolescents to move in and out of system b. Refocus on midlife marital and career issues c. Beginning shift toward concerns for older generation
5. Launching children and moving on	Accepting a multitude of exits from and entries into the family system	a. Renegotiation of marital system as a dyad b. Development of adult-to-adult relationships between grown children and their parents c. Realignment of relationships to include in-laws and grandchildren d. Dealing with disabilities and death of parents (grandparents)
6. The family in later life	Accepting the shifting of generational roles	a. Maintaining own and/or couple functioning and interests in face of physiological decline; exploration of new familial and social role options b. Support for a more central role for middle generation c. Making room in the system for the wisdom and experience of the elderly; supporting the older generation without overfunctioning for them d. Dealing with loss of spouse, siblings, and other peers, and preparation for own death; experiencing life review and integration

Source: Adapted from McGoldrick & Carter, 1982.

about how to interact with their adolescent children. They may assume that it is normal for adolescents to leave the family circle and to become enmeshed in their own peer groups, which often get into trouble. However, as Carol Gilligan (1987) has pointed out, adolescents also want the continuing guidance and involvement of adults.

Social changes that create hardships for young people—especially declines in less skilled manufacturing jobs, increases in low-wage service-sector jobs, and shortages of low-rent housing—have a major impact on the transition from dependence to independence for many young adults in the United States and elsewhere in the world. The Population Reference Bureau reports that "compared with the 1970s, more young adults (ages 18 to 24) are living at home with their parents. . . . Over half (54 percent) of all 18–24 year olds lived with their parents in 1991" (Ahlburg & De Vita, 1992, p. 9). Sociologist Paul Attewell, who is conducting research on what he terms "incompletely launched young adults," observes that continuing difficulties in finding work will oblige many young adults to live with their family of orientation, a situation that often frustrates family members' expectations about "normal" movement through the life course (cited in Lehman-Haupt, 1999).

In later stages of family life the parents must be willing to watch their grown children take on the challenges of family formation while they themselves worry about maintaining their marital roles or caring for their own parents. The latter issue is taking on increasing importance in aging societies like the United States. And since women are still expected to be more nurturing and emotionally caring than men, it often falls to women to worry about the question of "where can Mom live?" (Hochschild, 1989). According to Elaine Brody, a leading researcher on this issue, "It's going to be primarily women for a long time. Women can go to work as much as they want, but they still see nurturing as their job." Moreover, "with many more very old people, and fewer children per family, almost every woman is going to have to take care of an aging parent or parent-in-law" (quoted in Lewin, 1989, p. A1).

As if these stages were not stressful enough, consider the complications resulting from divorce, remarriage, and the combining of children of different marriages in a new family. It is increasingly common, for example, for teenagers and young adults to have parents who are dating and marrying, and for children to have four parental figures in their lives instead of two. These and many other changes in the family pose profound problems for sociologists who seek to develop theoretical perspectives on the family as a central institution in contemporary societies, as we will see in the next section.

PERSPECTIVES ON THE FAMILY

High divorce rates do not indicate that the family is about to disappear as an institution in modern societies. Even though many marriages end in divorce and increasing numbers of young adults postpone marriage or decide not to marry at all, the large majority do marry, and the majority of those who divorce will eventually remarry. Far from disappearing, the family is adapting to new social values and to changes in other institutions, especially economic ones. But one need only listen to an hour or two of talk radio or read a newspaper to find fierce debates about whether it is morally justified for single people to have children or for members of same-sex unions to be legally married or for unmarried adults to live together outside of wedlock, along with many other issues involving "family values." Most sociologists leave the moral debates to people with ideological commitments to one or another concept of the ideal family. Social scientists devote more thought to theoretical issues and to conducting research designed to obtain factual information about the changing family in the United States and elsewhere in the world. In this section, therefore, we review the basic sociological perspectives on family roles and relationships, as well as research that applies those perspectives.

The Interactionist Perspective

Interactions within the family cover a wide range of emotions and may take very different forms in different families. Families laugh and play together, work together, argue and bicker, and so on. All of these aspects of family interaction are important, but frequently it is the arguing and bickering that drives family members apart. Studies of family interaction therefore often focus on the sources of tension and conflict within the family.

Problems of family interaction can stem from a variety of sources. Often problems arise in connection with critical life stages or events, such as the loss of a job or the time when adolescent children begin to assert their independence in ways that threaten established family roles and arrangements. And conflicts often occur because of the particular ways in which the family's experiences are shaped by larger social structures. In the armed forces, for example, families often experience severe stress because of frequent moves from one base to another (Shaw, 1979). In many cases such moves draw family members closer together, but in other cases family interactions are marked by tension because children resent their inability to maintain stable friendships.

The context within which family life occurs can affect family interactions in other ways as well. At the lower levels of a society's stratification system, for example, money (or the lack of it) is often a source of conflict between parents or between parents and children. But the rich are by no means immune to problems of family interaction. Because they do not have to be concerned about the need to earn a living as adults and because their parents can satisfy any desires they may have, the children of the very rich often develop a sadness that resembles anomie (Wixen, 1979). Their lack of clear goals, which sometimes expresses itself in a compulsion to make extravagant purchases, may give rise to conflict between them and their parents.

In order to study these situations, either to help a family resolve its problems or to understand the nature of family conflict more thoroughly, the social scientist must interact with and observe the family, either at home, in a lab, or in a therapy setting (Minuchin, 1974; Satir, 1972; Whitchurch & Constantine, 1993). For instance, consider the case described by two family therapists, Augustus Napier and Carl Whitaker (1980), in which the parents are deeply troubled by the behavior of their 17-year-old daughter, the oldest of their three children. The father is a prominent attorney, the mother a college-educated woman who has devoted herself to homemaking. The parents' definition of the situation that has brought them into therapy is that they are worried about their daughter. She disobeys, is delinquent, and is depressed enough to make them fear that she might commit suicide. The mother and daughter fight bitterly, but usually the mother retreats. The father often attempts to defend the daughter, but eventually he sides with the mother.

After a few sessions Napier and Whitaker begin to challenge the parents' assumption that it is the daughter who is "the problem." They guess that her depression, anger, and delinquent behavior are symptoms of larger problems in the family. After some gentle probing, they identify a triangle of family conflict in which "two parents are emotionally estranged from each other and in their terrible aloneness they overinvolve their children in their emotional distress" (p. 83). The children blame themselves for their parents' problems and develop a low sense of self-worth. The therapists help the couple face the issues that divide them and that motivate their children to offer themselves as scapegoats or intermediaries who divert attention from the parents' basic problems.

Families like the one just described need to resolve a contradiction inherent in the institution of the family: the need to maintain the *individuality* of each member while providing love and support for him or her within a set of *interdependent* relationships. Many families never succeed in developing ways of encouraging each member to realize his or her full potential within the context of family life. And research shows that the core problem is usually the failure of the adult couple, even in intact families, to understand and develop their own relationship. Such a couple may become what John F. Cuber and Peggy B. Haroff (1980) term "conflict-habituated" or "devitalized." Couples of the first type have evolved ways of expressing their hostility toward each other through elaborate patterns of conflict that persist over many years. In contrast, the devitalized or "empty-shell" marriage may have begun with love and shared interests, but the partners have not grown as a couple and have drifted apart emotionally. Each has the habit of being with the other, a habit that may be strongly supported by the norms of a particular ethnic or religious community. Neither partner is satisfied by the relationship, but neither feels that he or she can do anything to change the situation. Thus the conflicts that might have produced change are reduced to indifference.

Today social scientists who study family interaction must deal with family structures that are more complex than the traditional nuclear family. Divorce and remarriage create many situations in which children have numerous sets of parental figures—parents and stepparents, grandparents and surrogate grandparents, and so on. These changes in family form result in new patterns of family interaction. For example, in a study of 2,000 children conducted over a 5-year period, sociologist Frank Furstenberg found that 52 percent of children raised by their mothers do not see their fathers at all—partly because some fathers are absent by choice but also because opportunities for visits decrease as parents remarry or move away (cited in Aspaklaria, 1985). This produces situations in which parents strive to maintain long-distance relationships with their children and occasionally have brief, intense visits with them. Although this specific type of relationship may not be the most desirable, it appears that if parents and children can express love and affection even when the parents are divorced, the children's ability to feel good about themselves and to love others in their turn may not be impaired.

The Conflict Perspective

The conflict perspective on families and family interactions assumes that "social conflict is a basic element of human social life." Conflict exists "within all types of social interaction, and at all levels of social organization. This is as true of the family as it is of any other type of social entity" (Farrington & Chertok, 1993,

p. 368). From this perspective one can observe actual family interactions (the micro level) and ask why conflict occurs, what the issues are, and how they are resolved. At a more macro level, one can ask how conditions of inequality and class conflict influence actual families or the laws and policies governing the family as a central institution in society.

"Family Values" and Class Conflict As mentioned earlier, citizens of the United States are engaged in a major public debate over issues related to the family, including homosexual marriage, freedom to obtain abortions, enforcement of child support and child abuse laws, and a host of other controversies that have a direct bearing on the family as an institution in society. The federal budget battles of the past decade also represent an arena of conflict over family policies, for they centered on the social safety net of welfare agencies, payments to families with dependent children, and many other spending categories that transfer money from more affluent families and individuals to those with fewer resources and, often, great need.

There are approximately 7.7 million families in the United States living at or below the official poverty line, and 10.5 million living at or below 125 percent of the poverty line. Figure 16.7 reveals that poverty affects a far greater proportion of black and Hispanic families than white families, even though the majority of poor families are white. Box 16.1 on pages 512 and 513 presents additional information on family poverty. (Also refer again to Box 12.2, which discusses how poverty in families is measured.)

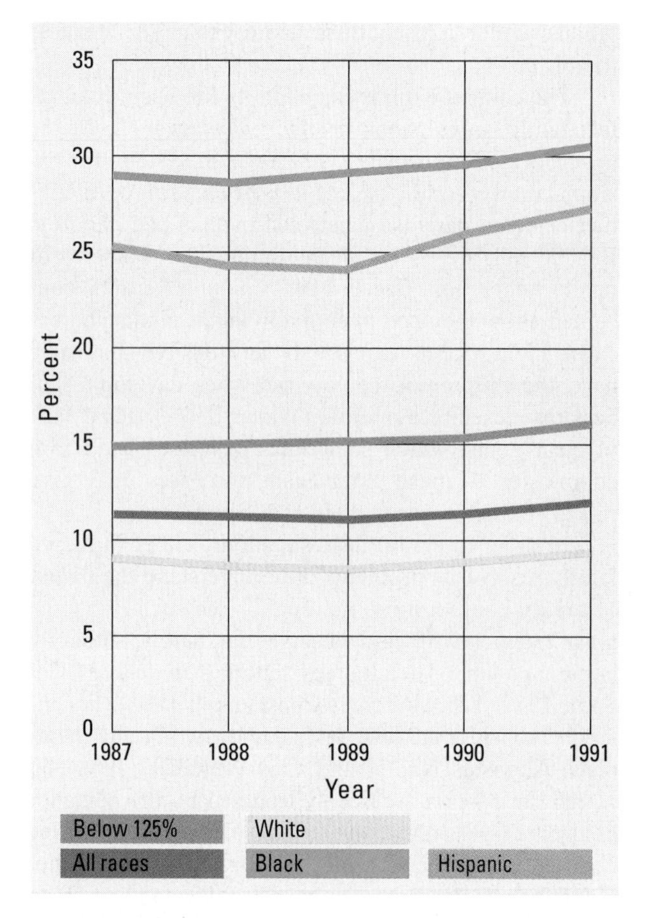

Source: Data from *Statistical Abstract,* 1993.

FIGURE 16.7 **Percent of Families in Poverty, United States, 1987–1991 (including value of all benefits)**

Sociologists who study poverty argue that most families are poor for the obvious reason that they do not earn or receive enough income. But critics of liberal social-welfare policies argue that self-destructive behaviors of family members themselves, such as dropping out of school, drug use, and out-of-wedlock childbearing, are causes of poverty. Social-welfare policies that are too generous or take the form of entitlements that do not have to be earned are said to encourage behaviors that cause poverty.

Social scientists who study class conflict and poverty policies, notably Herbert Gans (1995) and Christopher Jencks (1993), point out that class conflict helps explain the views of different population groups on family poverty. In the early decades of the twentieth century, they note, a large proportion of impoverished families consisted of elderly couples who did not have Social Security or Medicaid, two programs that have drastically reduced poverty among elderly Americans. Yet at the time that these policies were being debated, many members of the upper classes opposed them, viewing them as undemocratic transfers of wealth to less deserving members of society. The situation today is similar, as members of the upper classes oppose adequate funding for day care and other policies that can help move adult family members from poverty and dependence on public support to employment at decent jobs.

Functionalist Views of the Family

i

From a functionalist perspective, the family evolves in both form and function in response to changes in the larger social environment. As societies undergo such major changes as industrialization and urbanization, the family must adapt to the effects of those changes. Functionalist theorists like Talcott Parsons and William Goode have called attention to the loss of family functions that occurs as other social institutions like schools, corporations, and social-welfare agencies perform functions that were previously reserved for the family.

We have discussed numerous examples of this trend in earlier chapters—the tendency of families to have fewer children as the demand for agricultural labor decreased, the changing composition of households as people were required to seek work away from their families of orientation, the increasing number of dual-income households, and so on. The functionalist explanation of these changes is that as the division of labor becomes more complex and as new, more specialized institutions arise, the family too must become a more specialized institution. Thus modern families no longer perform certain functions that used to be within their domain, but they do play an increasingly vital part in early-childhood socialization, in the emotional lives of their members, and in preparing older children for adult roles in the economic institutions of industrial societies (Cherlin, 1996; Parsons & Bales, 1955).

A good example of empirical studies of how family functioning adapts to social change is research on the effects of child care on infant development. Figure 16.8 shows that the percentage of women aged 18 to 44 who are in the labor force after having a child in the last year rose from slightly more than 30 percent in 1976 to over 50 percent in 1990; it crept higher each year during the 1990s. The figure also shows that there is a wide array of child care arrangements. For very young children, day care is far less frequent than care by other relatives or care by nonrelatives outside an institution. But by the time children of working parents reach the age of 5, day care centers account for fully 75 percent of child care arrangements. An often cited study (Belsky, 1988; see also Belsky & Steinberg, 1991) offered some evidence that infants who are given over to a nonparent early in life may suffer developmental problems. This finding has been widely challenged. More recent research using more sophisticated methods shows that with quality care there are no discernible disadvantages for children of working mothers. Of course, the larger issue is how to provide quality care while parents are away working (Eckholm, 1992).

FIGURE 16.8 **Working Mothers and Child Care Arrangements**

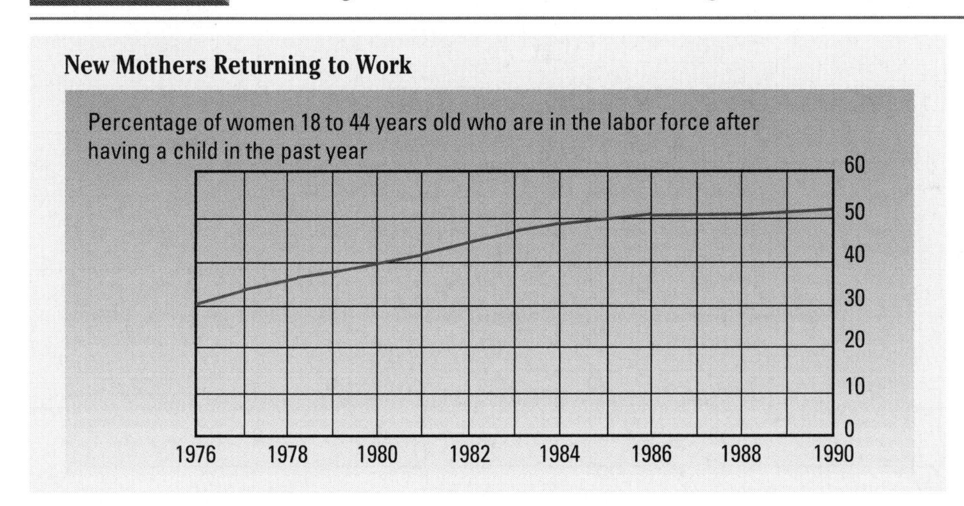

Child Care Arrangements

Percentage of children in each age group*

	less than 1	1 year old	2 years old	3 years old	4 years old	5 years old
Children in relative care	24%	24%	19%	21%	18%	15%
Children in nonrelative care	17%	19%	20%	19%	15%	17%
In center-based program	7%	11%	19%	41%	65%	75%

*Columns do not add up to 100 because some children participated in more than one type of day care.

Sources: Data from Census Bureau (top); U.S. Department of Education, National Center for Education Statistics (bottom).

BOX 16.1 SOCIOLOGICAL METHODS

Interpreting Statistics on Family Poverty

The number and characteristics of families in poverty in the United States are subjects of controversy. How one defines family poverty has a major bearing on the number of families identified as poor. The way one interprets statistics on families living in poverty influences the way one explains the causes and consequences of family poverty. Social scientists and policy makers can consult the same statistics and produce widely differing explanations. The accompanying charts illustrate these points.

Chart A shows that in 1959 there were just over 8.3 million families in poverty, meaning that total family income was below the official poverty line (based on

Chart A **Families Below Poverty Level, United States, 1959–1991**

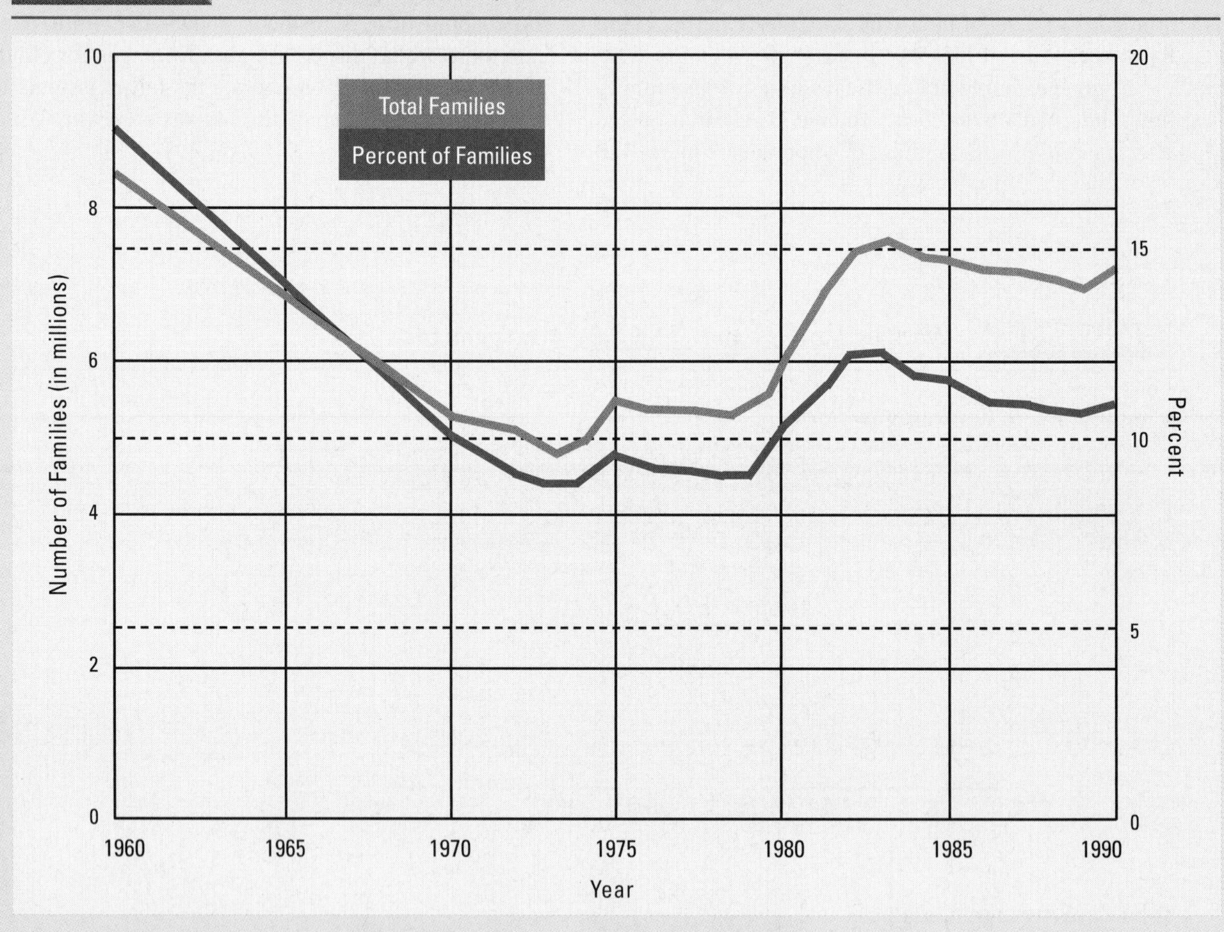

Source: Data from *Statistical Abstract,* 1993.

Researchers who approach the issue of child care from a functionalist perspective ask how societies cope with the dilemma faced by families in which both parents work outside the home. In an influential study, Kammerman and Kahn (1981) found that advanced industrial societies are struggling with the question: "Can adults manage productive roles in the labor force at the same time as they fulfill productive roles within the family—at home?" (p. 2). Many European nations answered this question by developing policies to assist working parents, including the development of high-quality child and infant care services (e.g., universal day care and school meal programs), paid parental leave, and family allowances that ease the financial burden on families with several children. In the United States, however, there is a great deal of debate about number of family members and income, but not counting the value of other benefits such as food stamps). That number accounted for 18.5 percent of all families. By 1973 the figure had declined to 4.8 million, or 8.8 percent of all families—a historic low. Family poverty began to rise again in the early 1980s; by 1986, 10.9 percent of families fell below the official poverty line.

At the time, some policy makers and social scientists argued that these trends were misleading because they did not take into account the value of food stamps, Medicaid, school lunches, and other benefits received by families. These sources of "income" were factored into the calculations, but the new figures did not differ significantly from the earlier ones. Even when the value of federal and state benefits was factored in, the results showed that 10.7 percent of families were poor.

A report titled *Families First,* issued by a task force convened by the Bush administration, included Chart B, which shows that poverty among single-mother families rose precipitously during the 1960s and 1970s, leveled off during the 1980s, and began rising again at the end of that decade. Taken by itself, this chart could make a casual observer think that poverty in families has risen dramatically since the 1960s and that the increase must be a result of single parenthood. The implication is that families are poor because there is only one breadwinner present. An alternative interpretation is that poor people have trouble forming families or keeping them together. When they become poor because of unemployment, their number increases.

Economic conditions can produce drastic changes in the number of families with incomes below the official poverty line. From 1989 to 1991, for example, there was a severe recession in the U.S. economy. Hundreds of thousands of women and men lost their jobs. During this period 254,000 black families, 231,000 Hispanic families, and 551,000 white families joined the millions already in poverty. If we consider all families earning or receiving benefits that fall below 125 percent of the poverty level (meaning another 25 percent of income above the official poverty line), we see that slightly more than 10 million U.S. families (30 percent) are poor by this standard.

It is also true that when adolescents have children out of wedlock, these families are extremely likely to have incomes that fall below the official poverty line. Thus in 1991, of the 7,712,000 families in poverty, 937,000 (35.5 percent) were headed by a householder who was between 15 and 24 years old. Not all of these families were composed of a single adolescent or young adult mother and her children, but at least three quarters were of that type.

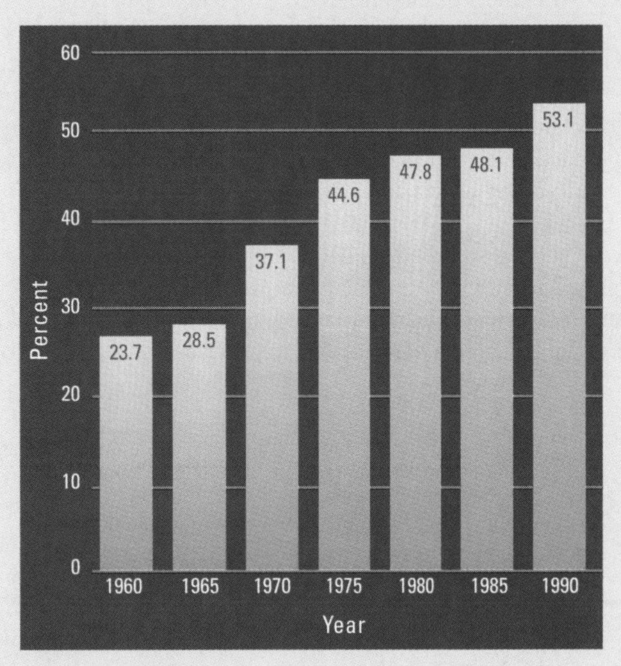

Source: Data from Census Bureau.

Chart B **Single-Mother Families as a Percent of All Poor Families, United States, 1960–1990**

One can conclude from these statistics that the number of families in poverty is very large in absolute terms but remains a small proportion of all families. Poverty in families increases as a result of changes in patterns of employment and unemployment, but it has also been increasing as a consequence of more out-of-wedlock births and a higher rate of family breakup.

whether such policies are appropriate functions of government or whether they lead to higher taxes, forcing more mothers into the labor force (Pittman, 1993).

In related studies, sociologist Arlie Hochschild (1989, 1997) asked how adult family members adjust to what she calls the "second shift." Only a few decades ago, housework and child care were viewed as primarily the mother's duty even if she worked outside the home. Today, many working parents cooperate and share domestic roles and responsibilities. But it is often difficult to achieve such cooperation. Hochschild found that "most egalitarian women—those with strong feelings about sharing—did one of two things. They married men who planned to share at home or they actively tried to change their husband's understanding of his role at home" (1989, p. 193). Many of the women

Hochschild interviewed were like Adrian Sherman, who "took the risky step of telling her husband 'It's share the second shift or it's divorce.' She staged a 'sharing showdown' and won" (p. 193). Other women do not go so far as to stage a showdown. They may make incremental efforts to change their husbands' attitudes. Still others discuss the issue with their husbands, but the process is not an easy one. More than half of the working women Hochschild interviewed had tried in one way or another to change domestic roles. As Hochschild notes, "If women lived in a culture that presumed active fatherhood, they wouldn't need to devise personal strategies to bring it about" (p. 193).

In sum, research from all the major sociological perspectives supports the view that the family is a resilient institution; it adapts to changing economic conditions and changing values. As Napier observes:

> Every family is a miniature society, a social order with its own rules, structure, leadership, language, style of living. . . . The hidden rules, the subtle nuances of language, the private rituals and dances that define every family as a unique microculture may not be easy for an outsider to perceive at first glance, but they are there. (Napier & Whitaker, 1980, p. 78)

At the macro level, the strength of the family as an institution does not mean the divorce rate will decrease rapidly or families that experience severe stress due to unemployment, ill health, and the like will have an easier time remaining intact. Whatever the perspective from which they study the family, sociologists recognize that problems like family violence and family breakup are pervasive and require ever more effective intervention techniques to protect vulnerable family members and strengthen the family as an institution. We turn, therefore, to examples from contemporary research that show how the basic sociological perspectives on the family can help us understand and intervene more successfully in different situations of family violence.

Family Violence: Applying the Perspectives

Family researchers Jeffrey L. Edleson and Ngoh Tiong Tan (1993) describe cases of family violence that can be analyzed in terms of each of the main sociological perspectives on the family. Their research compares Chinese and North American families in which the wives have experienced incidents of abuse by their husbands.

Spouse Abuse in a Chinese Family Mrs. Lee is referred to a social worker after seeking aid at a police station when she was struck in the face by her husband. In Chinese and other Asian cultures, the wife is expected to leave her family of orientation to take up residence with her husband and his extended family. Mrs. Lee's husband's family is dominated by her mother-in-law, a powerful widow who treats Mrs. Lee as the person with the lowest status in the family. Mrs. Lee wishes her mother-in-law would be less autocratic and complains about her dictatorial behavior to her own parents and to her husband. Her husband, on the other hand, believes his wife should conform to the ancient norms of Chinese family life and obey his mother. He and his mother consider Mrs. Lee a "traitor" because she has appealed to authorities outside the family.

From a functionalist perspective the police and social workers are representatives of institutions in the larger society that seek to help families avoid violence. But they cannot simply impose a new form of power on this complex extended family. Instead, they apply an interactionist perspective to the conflict as they understand it. The social worker convenes a meeting of the family to which Mrs. Lee's older brother is also invited. Together the group works out a detailed understanding of the roles of all the family members. The social worker explains that Mrs. Lee is not seeking to challenge her mother-in-law. In fact, however, the social worker has demonstrated to the extended family that Mrs. Lee has a support network and resources outside the family—a new notion for a traditional Chinese family. No doubt there will be further conflict, but power relations within the family have been subtly altered as a result of this intervention.

A North American Family Copes With Violence Susan was a regular victim of her husband Bob's violent anger. The beatings began on their honeymoon and ended 20 years later when he struck her with a baseball bat, almost killing her. During those years their children suffered emotionally from the violent interactions they witnessed and the unhappiness their mother experienced. But Susan loved Bob and struggled to make the marriage work. Now the couple are in court because after the last assault Bob was arrested. He faces a jail sentence, and Susan has abandoned hope of making the marriage work. She understands that she must make a new life for herself. She is seeking opportunities to do so but knows that she also needs protection from possible future assaults.

Unlike Mr. and Mrs. Lee, neither Bob nor Susan lives near an extended-family network. Many Americans do not live in close contact with relatives; on the average, "extended families do not figure as prominently in the lives of Caucasian Americans as they do in other subcultures and in many Asian societies" (Edleson & Tan, 1993, p. 384). Bob and Susan had left their extended families in search of economic opportunity. Susan now has nowhere to turn for help except the police and social-welfare agencies. This social safety net will act to protect her so that she and her children can live in peace. From a functionalist perspective, Susan will also have to rely on these agencies to enforce Bob's responsibilities toward his children. The functionalist sociologist argues, therefore, that it is of vital importance that the agencies function effectively in this and similar cases.

Susan needs the help of outside agencies to gain relief from the deadly conflict that has destroyed her family. Yet these very agencies came into being as a result of conflict in the larger society. They emerged after decades of protests and demands by women and social activists for ways to protect innocent family members from their abusers. But as we learn almost daily from reports of incidents of spouse and child abuse, it is never easy to ensure that the agencies society has created will intervene effectively. Even if they have the necessary resources, there is no guarantee that they will always act wisely to balance the conflicting values of individual rights and preservation of the family.

Domestic abuse is the most frequently reported issue in American policing (Defina & Wetherbee, 1997). These cases only touch the surface of an immense social problem, but they offer some insight into how sociologists contribute to understanding these extremely complex instances of violence at the micro level. Contemporary law enforcement officers are trained in the basic sociology and psychology of family violence. They understand that the batterer and the domestic victim are often caught in a cycle of repeated hostility and violence. Too often the violent spouse, usually the husband or male partner, expresses contrition after the violent act and is forgiven, and so the cycle resumes. The authorities and the neighbors believe the household is a troubled one but do not feel that they can do very much about it.

In many communities, however, new policies are insisting on more stringent approaches to domestic violence. The norms and sanctions are changing to prevent fatalities and trauma to children and spouses. Gradually the focus is shifting from merely "maintaining the peace" to arresting offenders, protecting victims, and referring battered women to shelters and other community resources available to help victims of domestic violence. Better understanding of the dynamics of violent family interactions, more forceful demands by women and community residents for protection from batterers, and more effective techniques for creating protective organizations are all applications of the basic perspectives of sociology.

Of course, all the training in the world does not explain why so many loving relationships become violent. On the other hand, with so much stress and such great economic, social, and emotional demands placed on husbands and wives, an even better question might be why, despite high rates of divorce and family breakup, so many unions are successful and so many individuals crave love and a stable relationship more than almost any other cherished value. We turn, therefore, to a sociological consideration of romance and marriage.

DYNAMICS OF MATE SELECTION AND MARRIAGE

We may think of mate selection and marriage as matters that affect only the partners themselves, but in reality the concerns of parents and other family members are never very far from either person's consciousness. And as we will see shortly, the values of each partner's extended family often have a significant impact on the mate-selection process.

Marriage as Exchange

People in Western cultures like to think that interpersonal attraction and love are the primary factors in explaining why a couple forms a "serious" relationship and eventually marries. But while attraction and love are clearly important factors in many marriages, social scientists point out that in all cultures the process of mate selection is carried out according to basic rules of bargaining and exchange (Daniel, 1996). Sociologists and economists who study mate selection and marriage from this exchange perspective ask who controls the marriage contract, what values each family is attempting to maximize in the contract, and how the exchange process is shaped by the society's stratification system.

Among the upper classes of China and Japan before the twentieth century, marriage transactions were controlled by the male elders of the community—with the older women often making the real decisions behind the scenes. In many societies in the Middle East, Asia, and preindustrial Europe, the man's family negotiated a "bride price" with the woman's family. This price usually consisted of valuable goods like jewelry and clothes, but in some cultures it took the form of land and cattle. Throughout much of Hindu India, in contrast, an upper-class bride's family paid a "groom price" to the man's family.

Although such norms appear to be weakening throughout the world, arranged marriages remain the customary pattern of mate selection in many societies. The following account describes factors that are often considered in arranging a marriage in modern India:

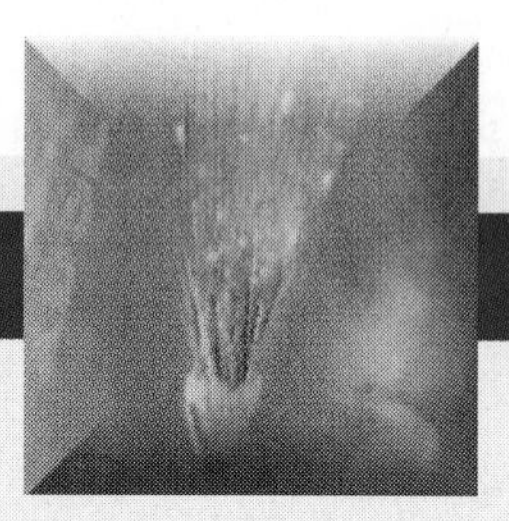

ART AND THE SOCIOLOGICAL EYE

Symbols of Marriage

The famous wedding portrait painted by Jan van Eyck in 1434 can be interpreted on a number of levels. On the surface, it is simply a painting of a bride and groom, people of obvious wealth and social standing, in their bedroom. But at a deeper level of meaning, the painting is a study in the symbols associated with marriage in Europe during the fifteenth century. The wedding bed was understood as a symbol of marital fidelity, the dog as a symbol of trust; and the couple have removed their shoes to show that they stand on the holy ground of matrimony. These and other symbols in the painting could not be understood, however, without some study of the way people of that time thought about marriage and its relationship to religious belief.

A contemporary artist would use other symbols and perhaps try to convey other emotions in depicting a newly married couple, just as the photographer captures a few very telling symbols in the portraits of the young couple and their new baby at the beginning of this chapter. But the van Eyck painting is considered a great work of art because of the timeless and universal quality of the married couple's expression as they gaze out on an approving society.

The Marriage of Giovanni Arndini and Giovanna Cenani, by Jan van Eyck, 1434. Panel, 33″ × 22½″.

> Every Sunday one can peruse the wedding ads in the classifieds. Many people still arrange an alliance in the traditional manner—through family and friend connections. Caste is becoming less important a factor in the selection of a spouse. Replacing caste are income and type of job. The educational level of the bride-to-be is also a consideration, an asset always worth mentioning in the ad. A faculty member at a college for girls has estimated that 80–90 percent of the students there will enter an arranged marriage upon receiving the B.A. Perhaps 10 percent will continue their studies. In this way, educating a daughter is parental investment toward securing an attractive, prosperous groom. (Smith, 1989)

In all of these transactions the families base their bargaining on considerations of family prestige within the community, the wealth of the two families and their ability to afford or command a given price, the beauty of the bride and the attractiveness of the groom, and so on. Different cultures may evaluate these qualities differently, but in each case the parties involved think of the coming marriage as an exchange between the two families (Goode, 1964). But do not get the idea that only selfish motives are involved in such marriages. Both families are also committing themselves to a long-standing relationship because they are exchanging their most precious products, their beloved young people. Naturally, they want the best for their children (as this is defined in their culture), and they also want a climate of mutual respect and cooperation in their future interfamily relationships.

Changing Norms of Mate Selection

Endogamy/Exogamy All cultures have norms that specify whether a person brought up in that culture may marry within or outside the cultural group. Marriage within the group is termed **endogamy;** marriage outside the group and its culture is termed **exogamy.** In the United States, ethnic and religious groups normally put pressure on their members to remain endogamous—that is, to choose mates from their own group. These rules tend to be especially strong for women. Among Orthodox Jews, for example, an infant is considered to have been born into the religion only if the mother is Jewish; children of mixed marriages in which the mother was not born a Jew are not considered Jewish. The conflict between Orthodox and Reform Jews over the status of children born to non-Jewish mothers—even when the mothers have converted to Judaism and the children have been raised as Jews from birth—is an example of conflict over endogamy/exogamy norms. Many African tribes have developed norms of exogamy that encourage young men to find brides in specific villages outside the village of their birth. As we saw in Box 4.1, such marriage systems tend to promote strong bonds of kinship among villages and serve to strengthen the social cohesion of the tribe while breaking down the animosity that sometimes arises between villages within a tribe.

Homogamy Another norm of mate selection is **homogamy,** the tendency to marry a person from a similar social background. The parents of a young woman from a wealthy family, for example, attempt to increase the chances that she will associate with young men of the same or higher social-class standing. She is encouraged to date boys from "good" families. After graduating from high school she will be sent to an elite college or university, where the pool of eligible men is likely to include many who share her social-class background. She may surprise her parents, however, and fall in love with someone whose social-class, religious, or ethnic background is considerably different from hers. But when this happens she will invariably have based her choice on other values that are considered important in the dating and marriage market—values like outstanding talent, good looks, popularity, or sense of humor. She will argue that these values outweigh social class, especially if it seems apparent that the young man will gain upward mobility through his career. Often the couple will marry and not worry about his lower social-class background. On the other hand, "the untalented, homely, poor man may aspire to a bride with highly desirable qualities, but he cannot offer enough to induce either her or her family to choose him, for they can find a groom with more highly valued qualities" (Goode, 1964, p. 33).

Homogamy in mate selection generally serves to maintain the separateness of religious groups. Because the Census Bureau does not collect systematic data on religious preferences, it is extremely difficult to obtain accurate data on religious intermarriage. Yet sociologists and religious leaders agree that although parents continue to encourage their children to marry within their religion, there is a trend away from religious homogamy, particularly among Protestants and Catholics (Scanzoni, 1995). This trend may affect rates of divorce and separation. Recent research shows that families who maintain an active religious life tend to have lower divorce rates than those who do not participate in religious worship together (National Commission on America's Urban Families, 1993). Similarly, a study by Howard Weinberg (1994) found that shared religion has the strongest effect on the likelihood that couples will be able to overcome a period of separation and achieve a successful reconciliation.

Interracial Marriage The norm of homogamy also applies to interracial marriage. Before 1967, when the U.S. Supreme Court struck them down as unconstitutional, many states had laws prohibiting such marriages. After that decision, marriages between blacks and whites began to increase rapidly. Figure 16.9 shows that although they remain less than 12 percent of all new marriages involving at least one African American, there has been a steep increase in black-white marriages since 1970, and this increase is continuing (Besharov & Sullivan, 1996). Intermarriage of Hispanic and non-Hispanic whites also increased during the same period, from about half a million marriages in 1960 to almost 1.3 million in 1994 (*Statistical Abstract,* 1995). Both of these trends are signs that more people are choosing to marry across racial and ethnic lines, but since the vast majority of marriages are within racial categories, it is clear that the norm of racial homogamy remains strong in the United States.

Romantic Love

Although exchange criteria and homogamy continue to play significant roles in mate selection, for the past century romantic attraction and love have been growing in importance in North American and other Western cultures. Indeed, the conflict between romantic love and the parental requirement of homogamy is one of the great themes of Western literature and drama. "Let me not to the marriage of true minds/Admit impediments," wrote Shakespeare in a famous sonnet. The "star-crossed lovers" from feuding families in *Romeo and Juliet*—like the lovers who have forsaken family fortunes to be together in innumerable stories and plays written since Shakespeare's time—attest to the strong value we place on romantic love as an aspect of intimate relationships between men and women (Cancian, 1994).

FIGURE 16.9 **Percentage of All New Marriages Involving at Least One Black Person**

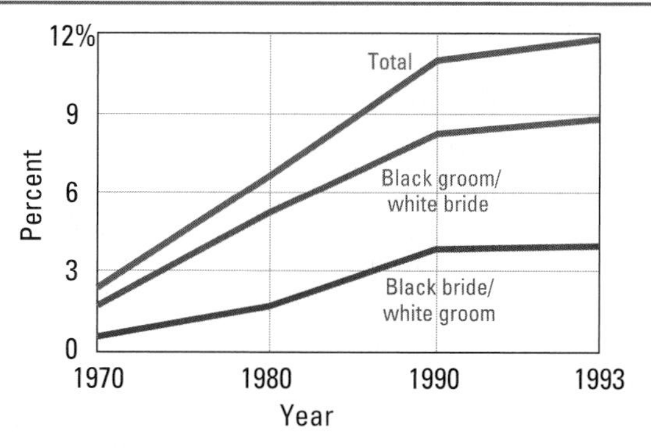

Source: Besharov & Sullivan, 1996.

In his classic study of worldwide marriage and family patterns, William J. Goode (1964) found that compared with the mate-selection systems of other cultures, that of the United States "has given love greater prominence. Here, as in all Western societies to some lesser degree, the child is socialized to fall in love" (p. 39). Yet although it may be taken for granted now that people will form couples on the basis of romantic attachment, major changes in the structure of Western societies had to occur before love as we know it could become such an important value in our lives. In particular, changes in economic institutions as a result of industrialization required workers with more education and greater maturity. These changes, in turn, lengthened the period of socialization, especially in educational institutions. This made it possible for single men and women to remain unattached long enough to gain the emotional maturity they needed if they were to experience love and make more independent decisions in selecting their mates.

However familiar it may seem to us, love remains a mysterious aspect of human relationships. We do not know very much—from a verifiable, scientific standpoint—about this complex emotional state. We do not know fully what it means to "fall in love" or what couples can do to make their love last. But two theories that have stimulated considerable research on this subject are Winch's theory of *complementarity* and Blau's theory of *emotional reciprocity.*

Complementary Needs and Mutual Attraction Robert F. Winch's (1958) theory of complementary needs, based on work by the psychologist Henry A. Murray, holds that people who fall in love tend to be alike in social characteristics such as family prestige, education, and income but different in their psychological needs. Thus, according to Winch, an outgoing person often falls in love with a quiet, shy person. The one gains an appreciative audience, the other an entertaining spokesperson. A person who needs direction is attracted to one who needs to exercise authority; one who is nurturant is attracted to one who needs nurturance; and so on.

Winch and others have found evidence to support this theory, but there are some problems with this research. It is difficult to measure personal needs and the extent to which they are satisfied. Moreover, people also show a variety of patterns in their choices of mates. Some people, in fact, seem to be attracted to each other because of their similarities in looks and behavior rather than because of their differences.

Attraction and Emotional Reciprocity Peter Blau's (1964) theory of emotional reciprocity as a source of love is based on his general theory that relationships usually flourish when people feel satisfied with the exchanges between them. When people feel that they are loved, they are more likely to give love in return. When they feel that they love too much or are not loved enough, they will eventually come to feel exploited or trapped and will seek to end the relationship. In research on 231 dating couples, Zick Rubin (1980; see also Peplau, Hill, & Rubin, 1993) found that among those who felt this equality of love, 77 percent were still together 2 years later, but only 45 percent of the unequally involved couples were still seeing each other. As Blau explained it, "Only when two lovers' affection for and commitment to one another expand at roughly the same pace do they mutually tend to reinforce their love" (quoted in Rubin, 1980, p. 284).

Blau's exchange approach confirms some popular notions about love—particularly the ideas that we can love someone who loves us and that inequalities in love can lead to separation. And yet we still know little about the complexities of this emotion and how it translates into the formation of the most basic of all social groups: the married couple. This is ironic, because books, movies, and popular songs probably pay more attention to love than to just about any other subject.

Francesca M. Cancian (1987), a leading researcher on love in a changing society, notes that when men and women place a high value on individuality the price may be "a weakening of close relationships." As people spend time enhancing their own lives they may become more self-centered and feel less responsible for providing love and nurturance. Cancian's research on loving relations among couples of all kinds has convinced her that people are increasingly seeking a form of love that "combines enduring love with self-development." She regards this kind of relationship as based on interdependencies in which each member of the couple attempts to assist the other in realizing his or her individual potential and at the same time seeks to strengthen the bond between them. But she finds that in order to foster a loving interdependent relationship couples often need to sacrifice a certain amount of independence and career advancement—not a simple matter in a culture that places a high value on individual achievement.

The nature of loving relationships is likely to be a subject of highly creative sociological research in coming years. Since the capacity to love and be loved depends on individual characteristics as well as on social conditions, much of the research in this area will be done by social psychologists. An example is presented in Box 16.2.

Marriage and Divorce

More than any other ritual signifying a major change in status, a wedding is a joyous occasion. Two people are legally and symbolically joined before their kin and friends. It is expected that their honeymoon will be pleasant and that they will live happily ever after. But about 20 percent of first marriages end in annulment or divorce within 3 years (Cherlin, 1996). Of course, divorce can occur at any time in the family life cycle, but the early years of family formation are the most difficult for the couple because each partner experiences new stresses that arise from the need to adjust to a complex set of new relationships. As Monica McGoldrick and Elizabeth A. Carter (1982) point out, "Marriage requires that a couple renegotiate numerous personal issues that they have previously defined for themselves or that were defined by their parents, from when to sleep, have sex, or fight, to how to celebrate holidays and where and how to live, work, and spend vacations" (p. 178). For people who were married before, these negotiations can involve former spouses and shared children, resulting in added stress for the new couple.

In the United States and other Western societies, the rate of divorce rose sharply after World War II, accelerated even more dramatically during the 1960s and 1970s, and has decreased only slightly since then. These statistics often lead sociologists to proclaim that there is an "epidemic" of divorce in the United States. But demographer Donald Bogue has concluded that "the divorce epidemic is not being created by today's younger generation. It has been created by today's population aged thirty or more, who married in the 1960s and before" (1985, p. 190). This generation was noted for its search for self-realization, often at the expense of intimate family relationships. It is not yet clear whether subsequent generations, who appear to be somewhat more pragmatic, will continue this trend. If they do, we can expect high divorce rates to continue.

Most states in the United States now have some form of no-fault divorce that reduces the stigma of divorce by making moral issues like infidelity less relevant than issues of child custody and division of property. While the growing acceptance of divorce helps account for why divorce rates are so much higher in the United States than in other nations with highly educated populations, Americans today also place a higher value on successful marriage than their parents may have. This means that they often divorce in the expectation of forming another, more satisfying and mutually sustaining relationship (Ahlburg & De Vita, 1992).

Trial Marriage In the 1980s it was widely believed that the practice of "trial marriage," or cohabitation before marriage, would result in greater marital stability: Couples who lived together before marriage would gain greater mutual understanding and a realistic view of marital commitment, and this would result in a lower divorce rate among such couples after they actually married. However, by the 1990s it had become evident

BOX 16.2 USING THE SOCIOLOGICAL IMAGINATION

Love of a Lifetime?

It takes a lot of imagination to sustain love. Insight into both oneself and one's mate is vital, but so is a sociological imagination. It is not hard to know how social conditions affect a person's feelings in the long and short term, but it is surprising how few people develop the capacity to see the world the way those close to them see it. When caught up in the early passion of love this may not be a problem. But as a relationship develops, people need as much imagination and understanding as possible. An awareness of the various dimensions of intimate love may help (Sternberg, 1999).

Social psychologists who study love in married and unmarried couples are finding new evidence that confirms what many people have long known intuitively: It is in fact quite difficult for couples to maintain the level of passion they experienced in the early stages of their marriage. "People don't know what they are in for when they fall in love," asserts Yale's Robert Sternberg. "The divorce rate is so high, not because people make foolish choices, but because they are drawn together for reasons that matter less as time goes on" (quoted in Goleman, 1985b, p. C1).

Sternberg believes that love has three components: intimacy, passion, and commitment. Intimacy is a shared sense that the couple can reveal their innermost feelings to each other even as those feelings change. Passion is largely a matter of physical attraction and sexuality. Commitment is a shared sense that each member of the couple is permanently devoted to the other. Research shows that the fullest love demands all three of these qualities but that over a long relationship passion is the first to fade; intimacy develops slowly and steadily as a result of shared experiences and values; and commitment develops more gradually still (see the accompanying graphs).

In the early stages of a love relationship, the couple may become so caught up in their passion that they do not help each other develop as autonomous individuals. This can produce serious problems in later stages of the relationship. At the same time, commitment alone cannot substitute for the other two qualities of love. "You have to work constantly at rejuvenating a relationship," Sternberg explains. "You can't just count on its being OK, or it will tend toward a hollow commitment, devoid of passion and intimacy. People need to put the kind of energy into it that they put into their children or career" (quoted in Goleman, 1985b, p. C5).

Studies by Sternberg and many others find that children who have been deprived of parental love, especially by the parent of the opposite sex, often have trouble developing commitment and sharing intimacy (Sternberg, 1999). Often they avoid feeling vulnerable and dependent on another person by avoiding strong emotional ties altogether.

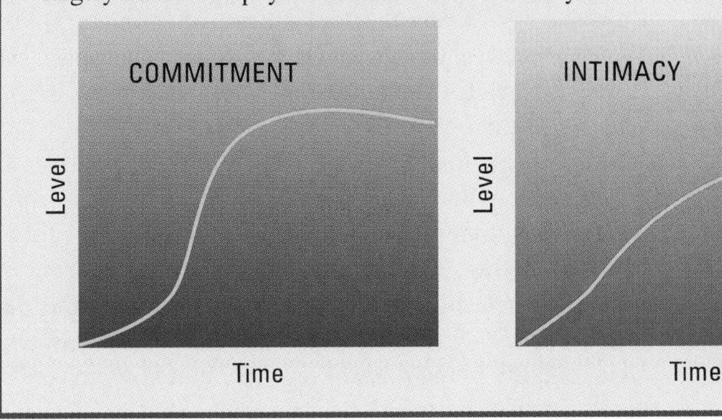

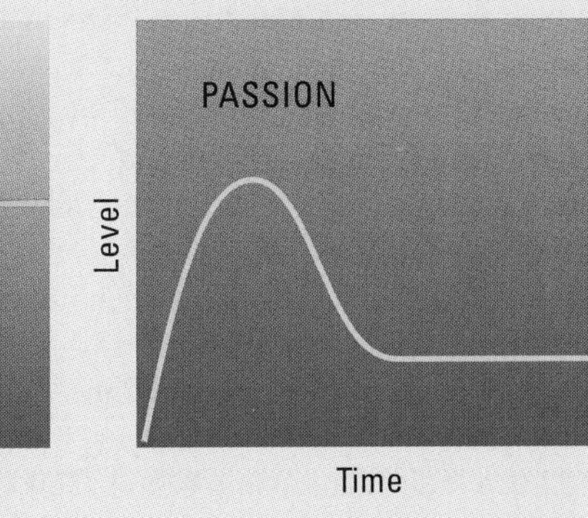

that these expectations were unfounded; in fact, the divorce rate among couples who had lived together before marriage was actually higher than the rate for couples who had not done so. Within 10 years of the wedding, 38 percent of those who had lived together before marriage had divorced, compared to 27 percent of those who had married without cohabiting beforehand.

On the basis of an analysis of data from a federal government survey of more than 13,000 individuals, Larry Bumpass and James Sweet (1989; see also Bumpass & Raley, 1995; Bumpass, Raley, & Sweet, 1995) concluded that couples who cohabit before marriage are generally more willing to accept divorce as a solution to marital problems. They also found that such couples are less likely to be subject to family pressure to continue a marriage that is unhappy or unsatisfactory. In addition, cohabitation has become a predictable part of the family life cycle, not only before marriage but in the interval between divorce and remarriage. Once again, these changes in families mean that family life has become ever more variable and uncertain for children. Thus while some advocate a return to traditional family norms and values, others urge greater tolerance for a variety of family types and accommodation to their needs. This conflict is most evident in discussions of same-sex relationships.

Gay and Lesbian Relationships

In 1995 the Walt Disney Company, long viewed as a defender of traditional family values, shocked many conservative Americans by announcing that it would provide spousal benefits (health insurance, pension rights, etc.) for its employees who were part of a same-sex relationship or marriage. In taking this action, Disney was simply acknowledging that norms are changing and it is becoming less acceptable for companies to impose their definitions of the family on their employees.

Margaret L. Andersen, a noted expert on issues of sex and gender, makes this observation:

> Sociological and popular understanding of gay and lesbian relationships has been greatly distorted by the false presumption that only heterosexual relationships are normal ways of expressing sexual intimacy and love. We live in a culture that tends to categorize people into polar opposites: men and women, Black and white, gay and straight. (1993, p. 57)

However, an unknown but significant number of people do not conform to conventional heterosexual behaviors. They may have felt a preference for intimacy with members of the same sex very early in life without knowing how to identify their feelings. Often they went through long periods of confusion and self-doubt before learning to understand and accept their homosexuality.

Our culture makes homosexual love into a form of deviance. Homosexuals, in consequence, are victims of homophobia of all descriptions (from individual acts of insensitivity or aggression to antigay legislation), which places additional burdens on homosexual relationships (Meyer, 1990; Sullivan, 1996). Because homosexual couples are not welcome or do not feel comfortable in many social situations, they often feel a need to keep their relationship secret. This deprives homosexual men and women of the social approval that heterosexual couples enjoy—and take for granted—in their peer and work relations.

In large measure owing to the stigma attached to homosexual relationships, there are no accurate census counts of homosexual families, nor is there much research about life in gay and lesbian households. The research that has been done, however, shows that fewer children live in male homosexual families than in female homosexual ones, mainly because women tend to win custody of their children in divorce cases. Research on gay fathers shows that compared with heterosexual fathers they tend to be better at setting limits on their children's behavior, are somewhat more nurturing, and place more emphasis on verbal communication with their children—features of family life that are often the province of the mother in a heterosexual family (Bigner & Bozett, 1989).

Sources of Marital Instability

i

Because it generally causes a great deal of emotional stress and pain for everyone involved, to say nothing of its financial costs, divorce is a subject of intense study by social scientists. Space does not permit a full treatment of this subject here, but we can discuss some of the major variables associated with divorce.

Age at marriage is one of the leading factors in divorce. It seems that it is best not to marry too young or to wait too long before marrying. Women who marry while still in their teens are twice as likely to divorce as women who tie the knot in their twenties. But those who marry in their thirties are half again as likely to divorce as those who marry in their twenties (Ruggles, 1997). How can we explain these differences?

In a study of age and marital instability, Alan Booth and John N. Edwards (1985) examined data obtained from a national sample of more than 1,700 couples. Among the younger couples, a pattern that the authors identified as inadequate role performance—especially not being attuned to the other partner's sexual needs and not being comfortable with the role of husband or wife—seemed to explain much of the instability in their relationships. Among those who married at an older age, another pattern appeared. These couples were more likely to engage in bitter disputes over the division of labor within the family—that is, over the definition of gender roles. Women who marry at older ages, for example, are more likely to demand that both partners share equally in the housework.

It appears that until about the middle of the twentieth century, when women were more likely to assume the role of homemaker, marriages had a higher probability of lasting. The demand by women for sharing of household roles—what sociologists often call the demand for a "symmetrical family" in which partners share equally in the roles of provider and homemaker—creates new opportunities for both men and women, but it also places more stress on the family in its early stages. Symmetrical families are not necessarily more dissatisfied with their marriages than nonsymmetrical families, nor do they necessarily encounter more marital problems. It appears, however, that the greater ability of both partners to be economically independent may allow them to consider divorce sooner than would a couple in which the wife is not in the labor force (Bittman, 1993).

Even when a couple is doing well economically, researchers find that the husband's role in the couple's problem-solving efforts is a key predictor of whether or not the couple will divorce. Couples in which the husband "stonewalls," withdraws from arguments, and fails to make an effort to work things out are more likely to divorce (Brody, 1992).

A number of other factors have also been found to be correlated with marital instability and divorce. They include the following:

- The couple met "on the rebound" (i.e., after one or both partners had recently experienced a great loss or hurt), and the new relationship may be flawed as a result.
- One of the partners wants to live far away from his or her family of orientation, suggesting that feelings of hostility toward family members may complicate the relationship.
- The spouses' family backgrounds are markedly different in terms of race, religion, education, or social class, which may result in differences in values that can cause conflict between the spouses.
- The couple is dependent on one of the extended families for income, shelter, or emotional support.
- The couple married after an acquaintanceship of less than 6 months or after an engagement of more than 3 years.
- Marital patterns in either spouse's extended family were unstable.
- The wife became pregnant before or within the first year of marriage (McGoldrick & Carter, 1982; Bumpass, Raley, & Sweet, 1995).

These correlates of marital instability do not, of course, necessarily mean that marriages that occur under these conditions are doomed to break up. They merely suggest that on the basis of statistical probabilities such marriages have a higher chance of ending in divorce.

The Impact of Divorce

At midcentury it was still considered almost impossible for a divorced man to run for the presidency. Ronald Reagan was the first President who had previously been divorced. Senators Robert Dole and Phil Gramm, candidates in the 1996 presidential election, had each been divorced. Divorce is much more prevalent today. Nevertheless, it remains a significant event in people's lives—as significant as marriage itself. But unlike a wedding, a divorce is not a happy event. Although some divorces turn out well for both partners, the majority do not.

Research on divorce has shown that many of the most disruptive consequences are due to its economic impact. Women suffer an average decline of about 30 percent in their income in the year following separation, while men experience a 15 percent increase. In fact, the majority of women who apply for various forms of public support do so because they have recently experienced a drastic decline in income due to divorce, separation, or abandonment. In addition, almost 40 percent of divorced mothers (and the children in their custody) move within the first year after divorce, and another 20 percent move after a year, a rate far higher than that for married couples (Cherlin, 1996; Furstenberg & Cherlin, 1991). And as if the breakup of their families is not stressful enough, many children also experience the loss of friends and familiar neighborhoods.

Beyond the material effects of divorce, there are the longer-term effects on family roles and the feelings of family members. In an important longitudinal study, Wallerstein and Blakeslee (1989) tracked 60 families with a total of 131 children for a period of 10 to 15 years after divorce. Both parents and children were interviewed at regular intervals. The data from those interviews show that the turmoil and stress of divorce may continue for a year or more. Many divorced adults continue to feel angry, humiliated, and rejected as much as 18 months later, and the children of divorced parents tend to exhibit a variety of psychological problems. Moreover, both men and women have a diminished capacity for parenting after divorce. They spend less time with their children, provide less discipline, and are less sensitive to their needs. Even a decade after the divorce, the parents may be chronically disorganized and unable to meet the challenges of parenting. Instead, they come to depend on their children to help them cope with the demands of their own lives, thereby producing an "overburdened child"—one who, in addition to handling the normal stresses of childhood, also must help his or her parent ward off depression (Wallerstein & Blakeslee, 1989).

There is a silver lining to the dark cloud of divorce, however. Since so many adults who are now marrying for the first time come from families that have experienced divorce, they are likely to take more time in selecting their mates in an effort to make sure that their choice is best for both partners; in addition, they try to become economically secure before marrying, thereby eliminating a major source of stress in a new marriage (Cherlin, 1996). Thus the recent modest downturn in the divorce rate may be expected to continue in the future. In any case, the data on the effects of divorce on adults and children suggest that societies need to do more to ease the stress experienced by young families—for example, by providing more day care facilities, establishing more flexible work schedules, and offering opportunities for family leave so that they can care for family members in times of need.

Research on the problems of African American families adds emphasis to the relationship between poverty and problems of family dissolution. Research on black families also confirms the basic need for family-oriented social policies.

RESEARCH ON THE CUTTING EDGE: The Black Family

The headlines announced a stunning new research finding: "BIRTH RATE FALLS TO A 40-YEAR LOW AMONG UNWED BLACK WOMEN: EDUCATION AND CONTRACEPTION ARE SEEN AS CRUCIAL" (Holmes, 1998). The news came from a research project headed by demographer Stephanie J. Ventura of the National Center for Health Statistics. Her analysis of recent fertility data and health statistics shows that "there were 74.4 births per 1,000 unmarried black women in 1996—the last year for which complete data were available. That rate is significantly below the peak of 90.7 per 1,000 reached in 1989." This decline appears to be part of a decrease in fertility among all black women—teenagers and adults, married and unmarried. As Figure 16.10 shows, the birthrate among unmarried white women, both adults and teenagers, is far lower, although it was rising through much of the 1990s.

In seeking explanations for this significant shift in births to unmarried black women, Ventura and others point to increases in education programs in African American communities, which encourage teenagers and younger adults to stay in school and to use contraception. Donna E. Shalala, secretary of health and human services, also notes that increased condom use is associated with greater awareness of the risk of AIDS. "We really did scare people," she comments, and of course AIDS is a matter worthy of fear and great precaution.

Encouraging news about declines in births to unmarried teenagers and young adults is only part of the larger and far more complex picture that describes the strengths and problems of African American families. Poverty rates are higher for black and other minority families than for white families, even though in absolute terms there are many more poor white families. Adding to the confusion is the controversy over "family values." In the case of the black family, the central issue is whether high rates of poverty, crime, and school failure in low-income, segregated black communities are a cause or a consequence of the large number of black single-parent families.

This issue is not a new one. More than half a century ago the eminent African American sociologist E. Franklin Frazier argued in *The Negro Family in the United States* (1939) that the violence of the Middle Passage, in which Africans were captured, shipped in chains, and sold into slavery, had wiped out their previous family ties. Frazier believed that as a result of that experience, the only enduring family form among blacks was the bond between mother and child. He urged black Americans to emulate the white middle-class family. In 1965 his views were endorsed by the famous Moynihan Report, which attributed the higher rates of social problems found in poor black communities to inadequate family functioning and the "tangle of pathology" that results from fatherlessness.

Critics of this theory argue that it neglects the effects of both past and present discrimination, which have been particularly severe for black men in America. The extreme difficulty experienced by black men in gaining the economic security necessary to sustain a family, it is argued, becomes a formidable obstacle to marriage (Bumpass, Raley, & Sweet, 1995). Focusing on the problems of the black family suggests that there is something wrong with poor black people rather than with the larger society (Miller, 1993).

Demographer Valerie Kincade Oppenheimer (1994) notes that the number of men between the ages of 20 and 35 who are not in the labor force has been rising since the 1950s and has accelerated since 1970. This trend is particularly evident among black men, especially

FIGURE 16.10 Birthrates, by Race and Marital Status, United States, 1980–1996

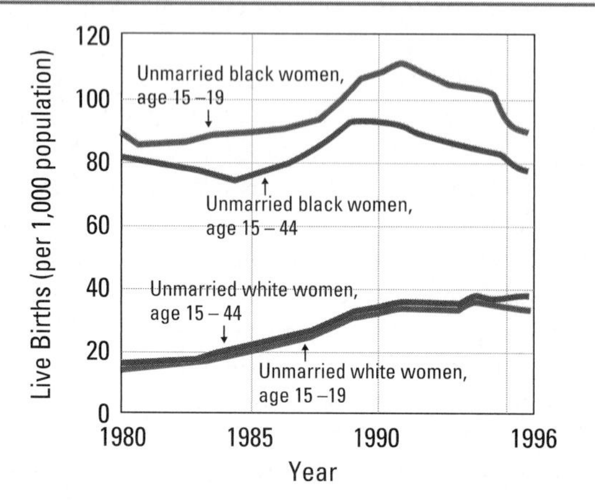

Source: Data from National Center for Health Statistics.

those with no college education. In 1960 approximately 26 percent of this group were not in school or in the labor force; by 1990 that figure had risen to more than 50 percent. (The comparable figures for white men are 11 percent and 23 percent, respectively.) One consequence of this trend has been a marked increase in delay of marriage and in rates of nonmarriage. Oppenheimer observes that among black women "there seems little doubt that, in addition to greater delays in marriage, the proportion who will never marry has been rising significantly and is historically unprecedented. It is unclear whether this represents a rejection of marriage or is indicative of the difficulties blacks have been experiencing in making a satisfactory marriage" (p. 306).

Most sociologists agree that the rapid rise in female-headed African American families is due primarily to the difficulties faced by young black men with limited education when seeking jobs. Harriet P. McAdoo, a specialist in the study of disadvantaged women and children, makes the following point:

> The tradition used to be that if a woman got pregnant, they were expected to get married, if not forced to get married, to supply income and a name for the child. But if the man has no job, there's no impetus for them to get married. Her parents would have three people to support instead of just two. The boy would become an additional burden on the family. Parents feel: "Why should I force her to marry him? He'll be a drain." (quoted in Cummings, 1983, p. 56)

Other sociologists believe that the problem is multidimensional. As family sociologist Bonnie Thornton Dill points out, "Poverty, unemployment, welfare programs that penalize families: these things explain a major portion of the variance between blacks and whites in the number of female-headed households, not all of it." She notes that changes in the nature of urban poverty that increasingly isolate poor people from quality education and employment opportunities also deprive them of the opportunity to marry. "Teen-age pregnancy among blacks has not increased so much," Dill concludes, "but the rate of marriage among teen-agers has dropped and the number of children born to married couples has dropped. These things are related to perceptions of opportunity" (quoted in Suro, 1992, p. A12).

Even though the data show that the phenomena of teenage pregnancy, female-headed families, and children being raised in poverty are related to social class rather than to race, the fact that more blacks than whites are found in the lower income levels of society means that these problems are proportionally more serious for blacks. In the words of sociologist Joyce Ladner (1973), "Life in the black community has been conditioned by poverty, discrimination, and institutional subordination" (p. 425). Under these conditions, she concludes, the black family has shown surprising resilience. Although within a given black family there may be households headed by single women, the extended family often provides substantial support for its less fortunate members.

VISUAL SOCIOLOGY

Intimate Relationships

At the beginning of the chapter we saw a young couple become a family of three as they shared the intimate experience of childbirth. In closing the chapter we return to some emotional photos of intimate relationships. All the photos come from a book titled *Life and Love: A Book of Embraces.* They were created by professional photographers trying to capture moments of love and its expression in a diverse social world.

The prom night, the wedding scene, the gay lovers expressing their deep devotion and sadness, the rapturous young couple lost in their intimacy—

Prom night
Dan Habib, 1994

Vietnamese wedding banquet
Donna Ferrato, 1994

Alicia and Buddy
Jeanne Strongin, 1990

all convey something of the joy of human love. But each of the photos may also make one wonder about the particular story behind it. Will the relationship of the lovers at their prom last through the next phases of life as each partner continues to grow and meet new people? Will Alicia and Buddy learn to adapt to the shocks and stresses with which life confronts love at every turn? If these questions sound as if they could be from a soap opera, there is a good reason for that. After all, love stories are featured on daytime television because, of all subjects, it is closest to our hearts and our desires and our doubts. "Love," the poet Emily Dickinson wrote, "is the thing with feathers." So often, when it seems to be within our grasp, it takes wing. But once we know it at all, we seek it forever.

Person with AIDS being comforted by a friend
Frank Fournier, 1985

SUMMARY

In all known societies almost everyone is socialized within a network of family rights and obligations that are known as family role relations. In simple societies the family performs a large number of other functions as well, but in modern societies most of the functions that were traditionally performed by the family are performed partly or entirely by other social institutions.

A *family* is a group of people related by blood, marriage, or adoption, and the role relations among family members are known as *kinship* relations. The smallest unit of family structure is the *nuclear family,* consisting of two or more people related by consanguineous ties or by adoption who share a household. The nuclear family in which a person is born and socialized is the *family of orientation,* and the nuclear family a person forms through marriage or cohabitation is the *family of procreation.* An *extended family* includes an individual's nuclear family plus all the nuclear families of his or her blood relatives.

The traditional household consisting of two parents and their children is no longer the typical American family. Since the 1940s there has been a dramatic increase in female-headed single-parent families and in nonfamily households, as well as in the numbers of women and men living alone and in the numbers of unmarried same-sex couples. These changes are often complicated by divorce, remarriage, and the combining of children of different marriages in stepfamilies.

The typical stages of the family life cycle are family formation, start of childbearing, end of childbearing, empty nest, and family dissolution (i.e., the death of one spouse). As it passes through this cycle, every family experiences changes in its system of role relations.

The structural context within which family life occurs can affect family interactions in a variety of ways. Problems may arise in connection with the demands placed on the family by institutions of the larger society, or as a result of its position in the society's stratification system. A basic contradiction inherent in the institution of the family is the need to maintain the individuality of each member while providing love and support for him or her within a set of interdependent relationships.

From a conflict perspective, changes in the family as an institution cannot occur without conflict both within the family and between the family and other institutions. Such conflict is illustrated by public debates over family policies and "family values."

Functionalist theorists have called attention to the loss of family functions that occurs as other social institutions assume functions that were previously reserved for the family. At the same time, they note that modern families play a vital part in early-childhood socialization, in the emotional lives of their members, and in preparing older children for adult roles.

In all cultures the process of mate selection is carried out according to basic rules of bargaining and exchange. In many societies the customary pattern of mate selection is the arranged marriage, in which the families of the bride and groom negotiate the marriage contract. All cultures also have norms that specify whether a person brought up in that culture may marry within or outside the cultural group. Marriage within the group is termed *endogamy;* marriage outside the group is termed *exogamy.* In societies in which marriages are based on attraction and love, individuals tend to marry people similar to themselves in social background, a tendency that is referred to as *homogamy.* Homogamy generally serves to maintain the separateness of religious and racial groups.

Compared with the mate selection systems of other cultures, that of the United States gives love greater prominence. Yet from a scientific standpoint little is known about this complex emotional state. It appears that people who fall in love tend to be alike in social characteristics but different in their psychological needs; however, this is not always the case. There is also considerable evidence that love relationships are more lasting when the partners' affection for each other is roughly equal.

In the United States and other Western societies, the rate of divorce has risen sharply since World War II. In the 1980s it was widely believed that the practice of cohabitation before marriage would result in greater marital stability, but in fact the divorce rate among couples who lived together before marriage is higher than the rate for couples who have not done so.

Age at marriage has been found to be one of the leading factors in divorce. Marriages that take place when the woman is in her teens or in her thirties are much more likely to end in divorce than marriages that take place when the woman is in her twenties. Among other factors that have been found to be correlated with divorce are marked differences in the family backgrounds of the spouses, dependence on either spouse's extended family, patterns of marital instability in either spouse's extended family, and early pregnancy. Studies of the impact of divorce have found that the turmoil and distress of divorce may continue for a year or more. Both men and women have a diminished capacity for parenting after divorce and may come to depend on their children to help them cope with the demands of their own lives.

SOCIOLOGY VERSUS IDEOLOGY

Why are sociologists so "politically correct"? To be politically correct is to bend over backwards not to offend people who are different or groups that are claiming their rights in society. People who dislike the thought of gay marriage, for example, criticize sociolgists for discussing the number of gay marriages, or research on the subject, or the feelings of gay couples. These critics often claim that such "neutral" sociological discussions actually give tacit approval to a behavior that potentially weakens the institution of the family. One hears similar arguments about sociological analysis of interracial marriage or teenage fertility.

If sociologists are politically correct from this perspective, it is partly because they encounter people in all their diversity. They are not sheltered from people of different races, religions, sexual preferences, and so forth. This does not mean that they cannot have their own moral values. Indeed, their values often guide their research (but not to the point of bias in the methods and analysis they employ). But their role is to present empirical facts and trends as best they can.

GLOSSARY

double standard: the belief that women must adhere to a different and more restrictive moral code than that applied to men. (p. 501)

family: a group of people related by blood, marriage, or adoption. (p. 502)

kinship: the role relations among people who consider themselves to be related by blood, marriage, or adoption. (p. 502)

nuclear family: two or more people related by blood, marriage, or adoption who share a household. (p. 503)

family of orientation: the nuclear family in which a person is born and raised. (p. 503)

family of procreation: the nuclear family a person forms through marriage or cohabitation. (p. 503)

extended family: an individual's nuclear family plus the nuclear families of his or her blood relatives. (p. 503)

endogamy: a norm specifying that a person brought up in a particular culture may marry within the cultural group. (p. 518)

exogamy: a norm specifying that a person brought up in a particular culture may marry outside the cultural group. (p. 518)

homogamy: the tendency to marry a person who is similar to oneself in social background. (p. 518)

WHERE TO FIND IT

BOOKS

Time for Life (John P. Robinson amd Geoffrey C. Godbey; Pensylvania State University Press, 1997). An excellent source of comparative data on how much time men and women spend on various activities in their families and households.

Divided Families: What Happens to Children When Parents Part, rev. ed. (Frank F. Furstenberg, Jr.; Harvard University Press, 1994). An assessment of the impact of divorce on the emotional, educational, and economic lives of children who experience the divorce of their parents.

The Reconstruction of Family Policy (Elaine A. Anderson and Richard C. Hula, eds.; Greenwood, 1991). A series of original essays on the problems of developing family policies, with special emphasis on conflicting interests of parents, children, and society. Includes good material on latchkey children and after-school care, welfare policies, and problems of the elderly in families.

Public and Private Families: An Introduction (Andrew J. Cherlin; McGraw-Hill, 1996). A balanced and comprehensive overview of family sociology.

The Family Crucible (Augustus Y. Napier and Carl Whitaker; Bantam, 1980). An introduction to the dynamics of family interaction and family therapy.

Stepfamilies: Who Benefits? Who Does Not? (Alan Booth and Judy Dunn, eds; Erlbaum, 1994). Covers essential topics on emerging family forms and socialization in stepfamilies.

JOURNALS

Journal of Marriage and the Family. Published quarterly by the National Council on Family Relations, Minneapolis, Minnesota. The best journal on family research in the United States.

Sex Roles: A Journal of Research. A monthly journal published by Plenum Press.

OTHER SOURCES

Current Population Reports. A series of special reports on household composition, marriage and divorce, and labor force participation; available in the census section of your library or from the nearest office of the U.S. Department of Labor or U.S. Publications Office. These reports present detailed demographic and economic data and include comparisons with earlier years.

Vital Statistics of the United States, Vol. 3: Marriage and Divorce. An annual compilation that presents a complete count of marriages and divorces in the preceding year. Data are broken down by age, race, previous marital status, and other characteristics.

INTERNET RESOURCES

American Sociological Association, Section on the Family (www.asanet.org/family.htm). Contains references to sociologists doing original research on the family.

Administration for Children and Families of the U.S. Government (www.rcp.dhhs.gov). A rich mine of official information about family stability and change; suggests many other useful sites to browse for research on the family in the United States and elsewhere.

Anthropology Department at the University of Manitoba (www.umanitoba.ca/anthropology/kintitle.html). Offers "Kinship and Social Organization," an online interactive tutorial that is engaging and informative and will reinforce or greatly expand your knowledge of family systems.

U.S. Census Bureau (www.census.gov/population/socdemo/). Offers an array of statistical information, including information about the estimated median age at first marriage, the number of interracial couples, the number of young adults living at home, and language preferences within homes.

CHAPTER 17

RELIGION

Sister Gertrude Morgan is considered one of the most distinctive religious artists in the United States. Born in 1900 to a devoutly Christian African American family in Lafayette, Alabama, Morgan experienced a number of religious visions that shaped her life and led her, at the age of 23, to begin drawing and painting. In an unschooled but inspiring and highly personal style, she portrays the religious life of her society in the form of a spiritual "autobiography." Her work and that of the other artists featured in the Visual Sociology section at the end of this chapter are outstanding examples of how religious feelings and thoughts find expression in artistic works by people who did not necessarily have artistic training and did not necessarily think of themselves as artists.

Morgan, for example, regarded her art as a form of private communication between herself and her God. The images and the words she wrote on them came pouring out of her in response to the experience of religious inspiration. But she never exhibited her work or had any intention of selling it. In the 1970s art lovers in New Orleans, her adopted hometown, discovered her work and began featuring it in exhibitions. Much against her will, Morgan became something of an artistic celebrity. This acclaim disturbed her; she felt that it distracted her from her devotion and dedication to good works. As a result, she stopped producing art, believing that she had been so ordered by God.

After her death in 1980, Morgan's art became even more highly acclaimed. In 1982 it was prominently featured in a historic exhibition titled "Black Folk Art in America" at Washington's Corcoran Gallery. Its simple beauty and evidence of saintly devotion continue to inspire. Morgan's story is not dissimilar to that of many other folk artists whose religious vision seems unquenchable.

As we experience other societies and cultures, it is quite remarkable how much of our time is spent looking at and thinking about the expression of spiritual feelings in their art, literature, and architecture. Anyone traveling through the capitals of Europe or Asia or any other continent will find that the greatest examples of a culture's achievements are often inspired by religious feelings and institutions. Michelangelo's breathtaking sculptures and frescos in the Vatican's Sistine Chapel, the great mosques of Istanbul, the prayer wheels on Tibetan mountain crags, the processions of carved saints in a Mexican Indian village festival, and the whimsically spiritual work of American religious folk artists like Sister Gertrude Morgan are only a few of the thousands of examples of expressions of faith that form the heart of so many of the world's cultures. But this observation should not surprise us. Religion is one of the most powerful and yet most mysterious forces in human life. Religions exert an immense influence over our understanding of right and wrong, good and evil. They also embody humankind's most universal longings to make sense of life and death. Faced with powerful feelings of awe and mystery, people throughout the world often seek to express those feelings through art. Morgan and the other religious artists featured in this chapter find themselves compelled to express their spiritual feelings in their art. While their work may not hold us in the same awe as that of Michelangelo, it is inspiring evidence of the timeless power of the spiritual in human life.

Lord I don't want to be buried in the storm by Sister Gertrude Morgan, 1970. Acrylic and ink on cardboard, 19½ cm × 9½ cm.

RELIGION IN SOCIETY

Religion, one of the oldest human institutions, is also among the most changeable and complex. On the one hand, religion expresses our deepest yearnings for spiritual enlightenment and understanding; on the other, conflicts over religious beliefs and practices have given rise to persecution, wars, and much human suffering, as can be seen, for example, in the violence between Muslim and Serbian Orthodox Christians in war-torn Bosnia and Kosovo. Compared to that tragic conflict, the dispute among Southern Baptists in the United States is mild but nonetheless important. Little wonder, therefore, that the founders of sociology—including Émile Durkheim, Karl Marx, and Max Weber—all wrote extensively about the power of religion and the great changes religion has undergone as societies have evolved.

Defining Religion

Religion is not easy to define. One could begin with a definition that has a concept of God as its core, but many religions do not have a clear concept of God. One could define religion in terms of the emotions of spirituality, oneness with nature, awe, mystery, and many other feelings, but that would not be a very helpful definition because emotions are extremely difficult to capture in words. Taking another tack, one might think in terms of organized religion—churches, congregations, ministers, rabbis, and so on—but clearly the organizational aspect of religion is just one of its many dimensions. It is frustrating to have to work so hard to define something that seems so commonplace, yet without a good working definition of religion it is impossible to compare different religions or refer to particular aspects of religion.

We can approach a working definition of religion by saying that **religion** is any set of coherent answers to the dilemmas of human existence that makes the world meaningful. From this point of view, religion is how human beings express their feelings about such ultimate concerns as sickness or death or the meaning of human life. Almost all religions involve their adherents in a system of beliefs and practices that express devotion to the supernatural and foster deep feelings of spirituality. In this sense, we say that religion functions to meet the spiritual needs of individuals.

But religion has also been defined in terms of its social function: It is a system of beliefs and rituals that serves to bind people together through shared worship, thereby creating a social group. **Rituals** are formal patterns of activity that express symbolically a set of shared meanings; in the case of religious rituals such as baptism or communion, the shared meanings are sacred. The term **sacred** refers to phenomena that are regarded as extraordinary, transcendent, and outside the everyday course of events—that is, supernatural. That which is sacred may be represented by a wide variety of symbols, such as a god or set of gods; a holy person such as the Buddha; various revered writings like the Bible, the Torah, or the Koran; holy objects such as the cross or the star of David; holy cities like Jerusalem or Mecca; and much else. The term **profane** refers to all phenomena that are not sacred (Kurtz, 1995).

Religion as an Institution

Religion is a major social institution because it carries out important social functions and encompasses a great variety of organizations (e.g., churches, congregations, charities), each with its own statuses and roles (e.g., ministers, priests, rabbis, parishioners, fund-raisers) and specific sets of norms and values (e.g., the Ten Commandments, the Golden Rule, the Koranic rules). As an institution, religion carries out the function of helping people express their feelings of spirituality and faith. Religion is often said to be a cultural institution because it guides a society's mental life, especially its ideas about morality, goodness, and evil. Of course, religion is not alone in performing these functions, but it remains a powerful source of moral precepts (Gilbert, 1997).

Throughout most of human history, until the past century or two, religion dominated all cultural life, as it still does in many societies. This photo shows Muslim worshippers praying outdoors in Mecca, Saudi Arabia, the most sacred site of the Islamic faith.

Religion also serves to confer legitimacy on a society's norms and values. Families seek the "blessing of holy matrimony" in wedding ceremonies. Baptisms, bar mitzvahs, confirmations, and other religious ceremonies mark the passage of children through their developmental stages and are occasions for statements about proper behavior and good conduct. Swearing on a Bible is common in courtrooms and on other occasions when norms of truth and fairness are being enforced. The political institutions of society also frequently look to religion for legitimation. The monarchs of eighteenth-century Europe invoked the will of God in their activities and sometimes claimed to rule by "divine right." The Pledge of Allegiance includes the phrase "under God" to reinforce the feeling that the destiny and unity of the nation and the values of liberty and justice are human efforts to carry out an even higher purpose.

Our society insists on separating the influence of religion from the laws of government. There is always controversy, however, over how much legitimacy religions can confer on political behavior (Swift, 1998). Fundamentalist Muslims may deny women rights that would be viewed as routine in our own culture (such as the right to drive a car); in the United States, Christian fundamentalists may claim that God does not favor political candidates who support abortion or homosexual rights. But these claims that religion legitimates particular norms and that God does not condone their transgression by political leaders usually clash with more secular interpretations of the society's norms. Thus debates over the role of religion in the political process can become extremely controversial.

The Power of Faith

Until comparatively recent times religion dominated the cultural life of human societies. Activities that are now performed by other cultural institutions, particularly education, art, and the media, used to be the province of religious leaders and organizations. In hunting-and-gathering bands and in many tribal societies, the holy person, or shaman, was also the teacher and communicator of the society's beliefs and values. In early agrarian societies the priesthood was a powerful force; only the priests were literate and, hence, able to interpret and preserve the society's sacred texts, which represented the culture's most strongly held values and norms. For example, in ancient Egypt, where the pharaoh was worshiped as a god, his organization of regional and local priests controlled the entire society.

Today religion continues to play an important part in the lives of people throughout the world even though the influence of organized religions is diminishing in many societies. In the United States, for example, the

Gallup Poll routinely asks Americans whether they believe in God. In 1944, 96 percent of the population said they were believers; in the 1990s the proportion of believers remains over 90 percent (National Opinion Research Center, 1998). The strength of religious attitudes and the influence of some religions can also be seen in the conflict over abortion, which plays such a prominent role in American politics, and in the controversies generated by Christian fundamentalists who believe in the literal interpretation of the Bible and may therefore deny the validity of evolutionary theory. Outside the United States, the Islamic world is torn by religious strife between liberals and fundamentalists, and Northern Ireland and Lebanon remain deeply divided owing largely to conflict between Protestants and Catholics or between Christians and Muslims.

At the same time that religion is a source of division and conflict, however, it can also be a force for healing social problems and moving masses of people toward greater insight into their common humanity. This occurs at the micro level of interaction—for example, in groups like Alcoholics Anonymous, in which spirituality is an essential part of the recovery program. At the macro level, the power of faith can be seen in impoverished rural and urban communities throughout Latin America. In those communities Catholic church leaders, parish priests, and lay parishioners have embraced the ideals of a "social gospel" that seeks the liberation of believers from poverty and oppression. These "base Christian communities," as they are often called, have become a powerful force in the movement for social justice and other far-reaching changes in their societies (Tabb, 1986).

Secularization and Its Limits Despite the great power of spirituality, since medieval times the traditional dominance of religion in many spheres of life has been greatly reduced. The process by which this has occurred is termed **secularization.** This process, according to Robert A. Nisbet (1970), "results in . . . respect for values of utility rather than of sacredness alone, control of the environment rather than passive submission to it, and, in some ways most importantly, concern with man's present welfare on this earth rather than his supposed immortal relation to the gods" (p. 388). Secularization usually accompanies the increasing differentiation of

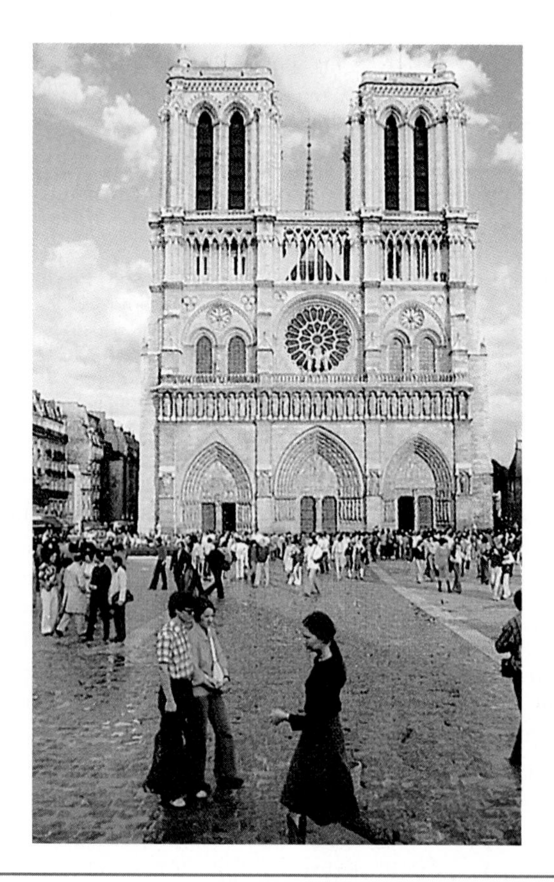

These two pictures of cathedrals provide a graphic illustration of secularization. In the medieval town, the cathedral dominated the landscape; it was the highest, and symbolically the most important, building. In the modern city, even the great cathedral is dwarfed by the greater scale and importance of the commercial buildings that surround it.

cultural institutions—that is, the separation of other institutions from religion. In Europe during the Middle Ages, for example, there were no schools separate from the church. The state, too, was thought of as encompassed by the church or at least as legitimated by the official state religion and church organization. Laws and courts were guided by religious doctrine, and clerical law could often be as important as civil law—indeed, to be tried as a heretic often meant torture and death. Churches engaged in large-scale economic activity, owned much land and property, and often mounted their own armies.

The Renaissance, the Enlightenment, and the revolutions of the eighteenth and nineteenth centuries all speeded the process of differentiation in which schools, science, laws, courts, and other institutions gained independence from religious control. However, this process has not occurred at the same rate throughout the world. For example, the removal of education from the control of religious institutions has occurred more slowly in some societies than in others. In Eastern Europe all education was controlled by the state until very recently. In most countries of Western Europe and the Americas there are religious schools, but these are separate from and overshadowed by the state-run educational system (Kurtz, 1995).

The emergence of cultural institutions like the secular public school and the weakening of the influence of religion on government do not result in complete secularization. People who are free to determine their own religious beliefs and practices may attend church less often or not at all, but total secularization does not occur (Finke & Stark, 1992). Moreover, in almost every society that has experienced secularization one can find examples of religious revival. Indeed, modern communication technologies, especially television, have contributed immensely to the revival of interest in religion, as witnessed by the popularity of "televangelists" like Oral Roberts and Pat Robertson (Iannaccone, Stark, & Finke, 1998).

VARIETIES OF RELIGIOUS BELIEF

Religious sentiments and behavior persist even in highly secularized societies like the United States. As sociologist Robert Wuthnow (1988) points out,

> The assumption that religion in modern societies would gradually diminish in importance or else become less capable of influencing public life was once widely accepted. That assumption has now become a matter of dispute. . . . Modern religion is resilient and yet subject to cultural influences; it does not merely survive or decline, but adapts to its environment in complex ways. (p. 474)

Many of the adaptations that religions make to social change become evident when one examines the varieties of religious belief in the world today, and especially when one studies the major world religions.

Major World Religions

In a 1913 essay Max Weber commented that "by 'world religions' we understand the five religions or religiously determined systems of life-regulation which have known how to gather multitudes of confessors around them" (1958/1913, p. 267). Among these Weber included, in addition to Christianity, "the Confucian, Hinduist, Buddhist, and Islamic religious ethics." He added that despite its small population of adherents, Judaism should also be considered a world religion because of its influence on Christianity and Islam as well as on Western ethics and values even outside the religious sphere of life.

i

In discussing religion, sociologists often refer to the "Islamic world" of the Middle East, the "Roman Catholic world" of Latin America and southern Europe, the "Hindu world" of the Indian subcontinent, and the "Buddhist world" of the Far East. The United States, northern Europe, and Australia are among the societies in which Protestantism is strongest. There are also, of course, the nations of Eastern Europe and the former Soviet Union, where until recently communism as a civil religion was the only legitimate belief system (although millions of people resisted the state's efforts to eradicate traditional religious faiths) (Robertson, 1985).

Figure 17.1 shows the distribution of the world's major religions. Note that in many parts of the world, particularly in Asia and Africa, large numbers of people practice other religions. Many are indigenous peoples whose religious practices may be influenced by the major world religions but continue to be based on beliefs and ways of worship that are unique to that culture. For example, in her sociological portrait of Mama Lola, a Haitian priestess, Karen McCarthy Brown notes that many Haitians practice voodoo, which is derived from African religious beliefs. Some of Mama Lola's followers may worship in Catholic churches yet continue to practice the traditional rituals of voodoo as well (Brown, 1991, 1992).

Classification of Religious Beliefs

The religions practiced throughout the world today vary from belief in magic and supernatural spirits to complicated ideas of God and saints, as well as secular

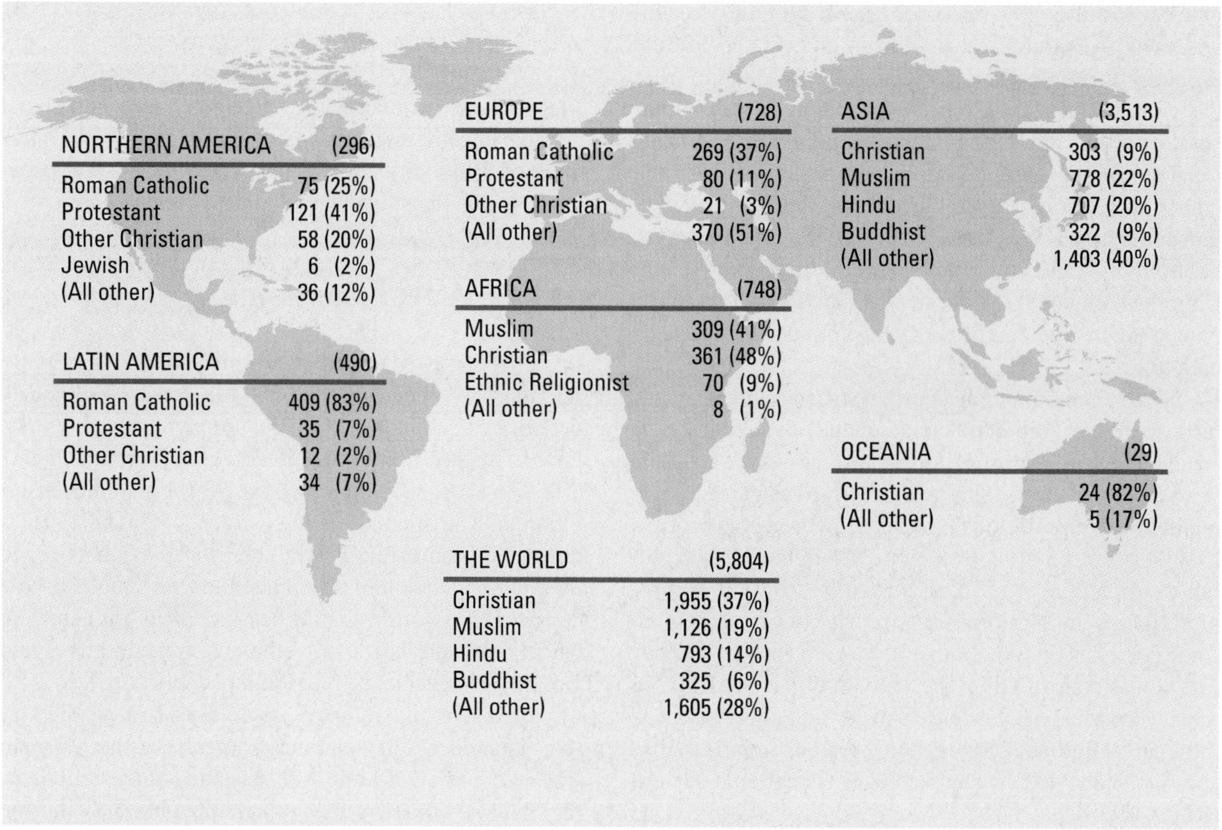

NORTHERN AMERICA	(296)
Roman Catholic	75 (25%)
Protestant	121 (41%)
Other Christian	58 (20%)
Jewish	6 (2%)
(All other)	36 (12%)

LATIN AMERICA	(490)
Roman Catholic	409 (83%)
Protestant	35 (7%)
Other Christian	12 (2%)
(All other)	34 (7%)

EUROPE	(728)
Roman Catholic	269 (37%)
Protestant	80 (11%)
Other Christian	21 (3%)
(All other)	370 (51%)

AFRICA	(748)
Muslim	309 (41%)
Christian	361 (48%)
Ethnic Religionist	70 (9%)
(All other)	8 (1%)

ASIA	(3,513)
Christian	303 (9%)
Muslim	778 (22%)
Hindu	707 (20%)
Buddhist	322 (9%)
(All other)	1,403 (40%)

OCEANIA	(29)
Christian	24 (82%)
(All other)	5 (17%)

THE WORLD	(5,804)
Christian	1,955 (37%)
Muslim	1,126 (19%)
Hindu	793 (14%)
Buddhist	325 (6%)
(All other)	1,605 (28%)

Note: Percentages may not add up to 100 because of rounding.

Source: Data from *Britannica Book of the Year,* 1997.

FIGURE 17.1 Estimated Religious Membership, by Continent (in millions)

religions in which there is faith but not God (Bowker, 1997). With such a wide range of religious beliefs and practices to consider, it is useful to classify them in a systematic way. One often-used system classifies religions according to their central belief. In this scheme the multiplicity of religious forms is reduced to a more manageable list consisting of five major types: simple supernaturalism, animism, theism, abstract ideals, and civil religion. (See the study chart shown on the next page.) In this section we describe each type briefly. Be warned, though, that not all religions fit neatly into these basic categories.

Simple Supernaturalism In less complex and rather isolated societies, people may believe in a great force or spirit, but they may not have a well-defined concept of God or a set of rituals involving God. Studies by anthropologists have found that some isolated peoples—for example, South Pacific island cultures and Eskimo tribes—believe strongly in the power of a supernatural force but do not attempt to embody that force in a visualized conception of God. In this form of religion, called **simple supernaturalism,** there is no discontinuity between the world of the senses and the supernatural; all natural phenomena are part of a single force. Consider these remarks by an Inuit Eskimo:

> When I was small I knew a man who came from the polar bears. He had a low voice and was big. That man knew when he was a cub and his bear mother was bringing him to the land from the ocean. He remembered it. (quoted in Steltzer, 1982, p. 111)

Animism More common among hunting-and-gathering societies is a form of religion termed **animism,** in which all forms of life and all aspects of the earth are inhabited by gods or supernatural powers. Most of the indigenous peoples of the Western Hemisphere were animists, and so were many of the tribal peoples of Africa before the European conquests. Europeans almost invariably branded American Indians "heathens and barbarians" because, among other things, the Indians believed that "people journeyed into supernatural realms and returned, animals conversed with each other

i

STUDY CHART Forms of Religion

FORM	DESCRIPTION	EXAMPLE
Simple Supernaturalism	A form of religion in which there is no discontinuity between the world of the senses and the supernatural; all natural phenomena are part of a single force	Some Inuit Eskimo cultures
Animism	A form of religion in which all forms of life and all aspects of the earth are inhabited by gods or supernatural powers	Native American culture; some African tribal cultures
Theism	A form of religion in which gods are conceived of as separate from humans and from other living things on the earth, although the gods are in some way responsible for the creation of humans and for their fate	
Polytheism	A form of theism in which there are numerous gods, all of whom occupy themselves with some aspect of the universe and of human life	The pantheon of gods of the ancient Greeks and Romans
Monotheism	A form of theism that is centered on belief in a single all-powerful God who determines human fate and can be addressed through prayer	Christianity, Islam, Judaism
Abstract Ideals	A form of religion that is centered on an abstract ideal of spirituality and human behavior	Buddhism, Confucianism
Civil Religion	A collection of beliefs, and rituals for communicating those beliefs, that exists outside religious institutions	Marxism-Leninism; some versions of humanism

and humans, and the spirits of rocks and trees had to be placated" (Jennings, 1975, p. 48). The same can be said of European attitudes toward African religions. Determined to subjugate nature and make the earth yield more wealth for new populations, the Europeans could not appreciate the meanings of animism for people who lived more closely in touch with nature.

Yet if one takes some time to read about the perceptions of animistic religions, it becomes clear that they contain much wisdom for our beleaguered planet. In one beautifully written account, an Oglala Sioux medicine man, Black Elk, speaks "the story of all life that is holy and is good to tell, and of us two-leggeds sharing in it with the four-leggeds and the wings of the air and all green things; for these are children of one mother and their father is one Spirit" (Neihardt, 1959/1932, p. 1). Black Elk's prayer continues:

> Grandfather, Great Spirit, lean close to the earth that you may hear the voice I send. You towards where the sun goes down, behold me; thunder Beings, behold me! You where the White Giant lives in power, behold me! You where the sun shines continually, whence come the day-break star and the day, behold me! You in the depths of the heavens, an eagle of power, behold! And you, Mother Earth, the only Mother, you who have shown mercy to your children! (p. 5)

Traces of animism can also be seen in the religious beliefs of the ancient Egyptians, Greeks, and Romans. The Greeks, for example, spoke of naiads inhabiting rivers and springs, and of dryads inhabiting forests. These varieties of nymphs were believed to be part of the natural environment in which they dwelled, but sometimes they took on semihuman qualities. They thus bridged the gap between a quasi-animistic religion and the more familiar theistic systems that evolved in Greece and Rome.

Theism Religions whose central belief is **theism** usually conceive of gods as separate from humans and from other living things on the earth—although these gods are in some way responsible for the creation of humans and for their fate. Many ancient religions were **polytheistic,** meaning that they included numerous gods, all of whom occupied themselves with some aspect of the universe and of human life. In the religion of the ancient Greeks, warfare was the concern of Ares; music, healing, and prophecy were the domain of Apollo; his sister Artemis was concerned with hunting;

Among the Quechua-speaking Indian peoples of the Andes, Christianity and older religious practices often coexist. Shown here is a ritual in which guinea pigs, believed to possess magical powers, are sacrificed for the well-being of the community.

Poseidon was the god of seafaring; Athena was the goddess of handicrafts and intellectual pursuits; and so on. A similar division of concerns and attributes could be found among the gods of the Romans and, later, among the gods of the Celtic tribes of Gaul and Britain.

The ancient Hebrews were among the first of the world's peoples to evolve a **monotheistic** religion—one centered on belief in a single all-powerful God who determines human fate and can be addressed through prayer. This belief is expressed in the central creed of the Jews: "Hear O Israel, the Lord our God, the Lord is One." Jewish monotheism, based on the central idea of a covenant between God and the Jewish people (as represented in the written laws of the Ten Commandments, for example), helped stimulate the codification of religious law and ritual, so that the Jews became known as "the people of the book." As they traveled and settled throughout the Middle East, the Jews were able to take their religion with them and hold on to the purity of their beliefs and practices (Johnson, 1987; Kurtz, 1995).

Christianity and Islam are also monotheistic religions. The Roman Catholic version of Christianity envisions God as embodied in a Holy Trinity consisting of God the Father, Christ the Son, and the Holy Spirit of God, which has the ability to inspire the human spirit. The fundamental beliefs of Islam are similar in many respects to those of Judaism and Christianity. Islam is a monotheistic religion centering on the worship of one God, Allah, according to the teachings of the Koran as given by Allah to Mohammed, the great prophet of the Muslim faith. In his early preachings, Mohammed appears to have believed that the followers of Jesus and the believers in Judaism would recognize him as God's messenger and realize that Allah was the same as the God they worshiped. The fundamental aim of Islam is to serve God as he demands to be served in the Koran.

Another basically monotheistic religion, Hinduism, is difficult to categorize. On the one hand, it incorporates the strong idea of an all-powerful God who is everywhere yet is "unsearchable"; on the other hand, it conceives of a God who can be represented variously as the Creator (Brahma), the Preserver (Vishnu), and the Destroyer (Shiva). Each of these personifications takes a number of forms in Hindu ritual and art. Of all the great world religions, Hinduism teaches most forcefully that all religions are roughly equal "paths to the same summit."

Abstract Ideals In China, Japan, and other societies of the Far East, religions predominate that are centered not on devotion to a god or gods but on an abstract ideal of spirituality and human behavior. The central belief of Buddhism, perhaps the most important of these religions, is embodied in these thoughts of Siddhārtha Gautama, the Buddha:

> Life is a Journey
> Death is a return to the Earth
> The universe is like an inn
> The passing years are like dust

Like all the world's major religions, Buddhism has many branches. The ideal that unifies them all, however, is the teaching that worship is not a matter of prayer to God but a quest for the experience of godliness within oneself through meditation and awareness.

Another important religion based on abstract ideals is Confucianism, which is derived from the teachings of the philosopher Confucius (551–479 B.C.). The sayings of Confucius are still revered throughout much of the Far East, especially among the Chinese, although the formal study of Confucius's thought has been banned since the communist revolution of the late 1940s and early 1950s. The central belief of Confucianism is that one must learn and practice the wisdom of the ancients. "He that is really good," Confucius taught, "can never be unhappy. He that is really wise can never be perplexed. He that is really brave can never be afraid" (quoted in McNeill, 1963, p. 231).

In Confucianism the central goal of the individual is to become a good ruler or a good and loyal follower and thus to carry out the *tao* of his or her position. *Tao* is an untranslatable word that refers to the practice of virtues that make a person excellent at his or her discipline. As is evident even in this brief description, Confucianism is a set of ideals and sayings that tend toward conservatism and acceptance of the status quo, although the wise ruler should be able to improve society for those in lesser positions. Little wonder that under communism this ancient and highly popular set of moral principles and teachings was banned in favor of what sociologists call a civil religion.

Civil Religion In the last few decades some social scientists, notably Robert Bellah (1970), have expanded the definition of religion to include so-called **civil religions.** These are collections of beliefs, and rituals for communicating those beliefs, that exist outside religious institutions (Swift, 1998). Often, as in the former Soviet communist societies, they are attached to the institutions of the state. Marxism-Leninism can be thought of as a civil religion, symbolized by the reverence once paid to Lenin's tomb. Central to communism as a civil religion is the idea that private property is evil while property held in common by all members of the society (be it the work group, the community, or the entire nation) is good. The struggle against private property results in the creation of the socialist personality, which values all human lives and devalues excessive emphasis on individual success, especially success measured by the accumulation of property. Although the communist regimes of the Soviet Union and Eastern Europe have fallen, there are millions of people in those nations and in China who were socialized to believe in these principles.

In the United States, certain aspects of patriotic feeling are sometimes said to amount to a civil religion: Reverence for the flag, the Constitution, the Declaration of Independence, and other symbols of America is cited as an example. Thus most major public events, be they commencements, political rallies, or Super Bowl games, begin with civil-religious rituals like the singing of the national anthem or the recitation of the Pledge of Allegiance, in which a nonsectarian God is invoked to protect the nation's unity ("one nation under God").

Although there is no doubt that Lenin's image and the American flag may be viewed as sacred in some contexts, neither communism nor American patriotism can compete with the major world religions in the power of their central ideals and their spirituality. In consequence, sociologists tend to concentrate on religions in the traditional sense—that is, on the enactment of rituals that represent the place of sacred beliefs in human life. In the remainder of this chapter we discuss the structure of religious institutions and the processes by which new ones arise.

RELIGION AND SOCIAL CHANGE

Now that the nations behind the former "Iron Curtain" are enjoying new freedoms, the role of religion in bringing about social change is an important aspect of life in those societies. Indeed, in many parts of the world, religion is one of the primary forces opposing or supporting change. However, it is not always a simple matter to predict whether religion will encourage change or hinder it. In Israel, for example, highly orthodox Jews, though they account for a small minority of the electorate, often hold the votes necessary to keep the ruling party in power. The orthodox political parties favor continued Jewish settlement on the West Bank of the Jordan River and oppose the creation of a Palestinian state near Israel's borders.

i

In the United States, the Catholic church plays an active role in seeking social change. The church strongly opposes women's right to obtain abortions legally. Instead, it supports a return to the traditional view of abortion as a crime, which is based on the belief that humans must submit to the will of God and not use their technological skills to achieve power over life and death. In this instance, therefore, although the church is promoting social change, the change represents a return to an earlier moral standard. Throughout much of Latin America, in contrast, the Catholic church is fighting in support of the masses of urban and rural poor who seek social justice and equitable economic development. In Brazil, for example, the typical Catholic priest or nun favors the political left, which seeks a more egalitarian distribution of wealth and income and is highly critical of the rich.

We could add many other examples of the role played by religion and religious organizations in social change. But in attempting to generalize about the relationship between religion and social change it is useful to return briefly to classic sociological theories. The pioneering European sociologists, particularly Karl Marx and Max Weber, noted the prominent role of religion in social change. But they wondered whether the influence of religious faith is a determining force in social change, or whether religious sentiments and the activities of religious organizations are an outgrowth of changes in more basic economic and political institutions.

Marx, as we have seen, believed that economic institutions are fundamental to all societies and that they are the source of social change. In his view, religion and other cultural institutions are shaped by economic and political institutions; they are a "superstructure" that simply reflects the values of those institutions—of markets, firms, the government, the military, and so on. The function of cultural institutions, especially religion, is to instill in the masses the values of the dominant class. In this sense they can be said to shape the consciousness of a people, but they do so in such a way as to justify existing patterns of economic exploitation and the existing class structure. Religion, in Marx's words, is "the opium of the people" because it eases suffering through prayer and ritual and deludes the masses into accepting their situation as divinely ordained rather than organizing to change the social system.

For Weber, on the other hand, religion can be the cause of major social change rather than the outcome or reflection of changes in other institutions. Weber set forth this thesis in one of his most famous works, *The Protestant Ethic and the Spirit of Capitalism* (1974/1904). Noting that the rise of Protestantism in Europe had coincided with the emergence of capitalism, Weber hypothesized that the Protestant Reformation had brought about a significant change in cultural values and that this was responsible for the more successful development of capitalist economic systems in Protestant regions. As Weber explained it, Protestantism instilled in its followers certain values that were conducive to business enterprise, resulting in the accumulation of wealth. Because the early Protestants believed that wealth was not supposed to be spent on luxuries or "the pleasures of the flesh," the only alternative was to invest it in new or existing business enterprises—in other words, to contribute to the rapid economic growth characteristic of capitalist systems (Kurtz, 1995). This view was reinforced by the belief—also part of the Protestant ethic—that a person who worked hard was likely to be among those predestined for salvation.

Some have questioned the validity of Weber's thesis regarding Protestantism and early capitalism, but there can be no doubt that religious institutions are capable of assuming a major role in shaping modern societies. Throughout the Islamic world there are currents of orthodoxy and reform that threaten to cause both civil and international wars. And as noted earlier, in many Latin American communities the Catholic church leads the movement for large-scale social change. So Marx was wrong in his claim that religion functions largely to maintain the existing values of more basic social institutions. On the contrary, religion can often lead to new ways of organizing societies—to new political and economic institutions as well as whole new lifestyles.

The Marxian view of religion is still relevant to those who are critical of the influence of religious institutions. These critics assert that religion, along with other cultural institutions such as education, serves to reaffirm and perpetuate inequalities of wealth, prestige, and power. When the poor are encouraged to pray for a better life, for example, they are further oppressed by a religion that prevents them from realizing that they need to marshal their own power to challenge the status quo. Thus there is still some question as to whether (and how much) religious institutions change society in any fundamental way.

To make informed judgments about this and related issues, we need to have a better understanding of the nature of religious organizations and how they function in modern societies. In the next section, therefore, we describe the main types of religious organizations. We then focus on trends in religious belief and practice in the United States.

STRUCTURE AND CHANGE IN MODERN RELIGIONS

Religion today is a highly structured institution, with numerous statuses and roles within a variety of organizations as well as many kinds of smaller, less bureaucratic groups. This was not always so. The religions of tribal peoples were not highly institutionalized; that is, there were no separate organizations like churches or interfaith councils or youth fellowships. It is true that the occupational status of holy person or priest might exist. Thus most Native American peoples had spiritual leaders who specialized in the rituals and symbols through which the members of the tribe could address the Great Spirit and the sacred spirits of their ancestors. But even in societies that had spiritual leaders or priests, religious practice was intertwined with tribal and family life. There was no concept of the church as a separate institution specializing in religious rituals.

Religion as a fully differentiated institution developed in agrarian societies, and it was in such societies that formal religious organizations first appeared. As we saw in Chapter 4, agrarian societies produce enough surplus food to support a class of priests and other specialists in religious rituals. In those less complex societies religion was incorporated into village and family life; it had not yet become differentiated into a recognized, separate institution with its own statuses and roles (Warner, 1998). Over time, however, the development of religious institutions resulted in a wide variety of organizations devoted to religious practice. Today those organizations include the church, the sect, and the denomination.

Forms of Religious Organization

Churches and Sects A **church** may be defined as a religious organization that has strong ties to the larger society. Often in its history it has enjoyed the loyalty of most of the society's members; indeed, it may have been linked with the state itself (Weber, 1963/1922). An example is the Church of England, or Anglican church. A **sect,** by contrast, is an organization that rejects the religious beliefs or practices of established churches. Whereas the church distributes the benefits of religious participation to anyone who enters the sanctuary and stays to follow the service, the sect limits the benefits of membership (i.e., salvation, fellowship, common prayer) to those who qualify on narrower grounds of membership and belief (McGuire, 1987; Weber, 1963/1922).

Sects require strong commitment on the part of their members and usually are formed when a small group of church members splits off to form a rival organization. The sect may not completely reject the beliefs and rituals of the church from which it arose, but it changes them enough to be considered a separate organization. Most "storefront churches" are actually sects that have developed their own particular interpretations of religious ritual.

An important difference between churches and sects is that churches draw their adherents from a large social environment—that is, from a large pool of possible members—whereas the size of the population from which a sect draws its members tends to be small (Iannaccone, Stark, & Finke, 1998). Also, churches make relatively limited demands on their members while sects make heavy claims on their members' time, money, and emotional commitment. As Robert Wuthnow (1988) writes, "Churches attempt to regulate or fulfill a few of the activities or needs of large numbers of people; sects attempt to regulate or fulfill many of the activities or needs of small numbers of people" (p. 495).

Denominations A third type of religious organization is the **denomination.** Unlike a sect, a denomination is on good terms with the religious institution from which it developed, but it must compete with other denominations for members. An example of a denomination is the United Methodist Church, a Protestant denomination that must compete for members with other Protestant denominations such as Presbyterians, Episcopalians, and Baptists. Denominations sometimes evolve from sects. This occurs when the sect is successful in recruiting new members and grows in both size and organizational complexity. Sociologists who study religion have found that the bureaucratic growth of a sect and its increasing influence over nonreligious matters in the community is associated with a decline in the fervor with which the sect pursues its spiritual ideals and a decrease in its efforts to remain faithful to the claims that originally made it so different. This aspect of religious change is an active area of research (Wuthnow, 1994).

Cults Still another type of religious body, the **cult,** differs in significant ways from the organizations just described. Cults are usually entirely new religions whose members hold beliefs and engage in rituals that differ from those of existing religions. Some cults have developed out of existing religions. This occurred in the case of early Christianity, which began as a cult of Jews who believed that Jesus of Nazareth was the Messiah and who practiced rituals that were often quite different from those of Judaism. Cults may also be developed by people who were not previously involved in a church or sect, such as individuals who become active in pagan cults like those based on ancient forms of witchcraft (Barrett, 1996). Most major religions began as seemingly insignificant cults, but new cults are formed every day throughout the world, and very few of them last long enough to become recognized religions. The Mormons, who began as a cult in 1830 with six members, grew into a large religious organization with more than 60,000 members by 1850 and today is one of the fastest

This photograph of a religious festival in the province of Chiapas in Mexico clearly shows the social organization of the village church. The male officials of the church (visible in the background) march with banners and candles while the women bear the saint through the village.

growing Christian churches in the world (Finke & Stark, 1992).

In a study of trends in church, sect, and cult membership, sociologists Roger Finke and Rodney Stark find that, contrary to common assumptions, the rate at which cults are formed is no higher today than it was in earlier historical periods. Moreover, compared to some of the earlier cults, contemporary cults often have less success in gaining members:

> No cult movement of the sixties or early seventies seems ever to have attracted more than a few thousand American members, and most of even the well-publicized groups counted their true membership in the hundreds, not the thousands. For example, after more than thirty years of missionizing, it is estimated that there have never been more than 5,000 followers of the Unification Church (the Moonies) in the United States, and some of them are from abroad. (Finke & Stark, 1992, p. 241; see also Melton, 1989)

In a similar example, Barry Kosmin and his colleagues (1991) estimate, on the basis of the largest sample survey of religious beliefs ever conducted in the United States ($n=113{,}000$), that actual membership in the Church of Scientology very likely does not exceed 45,000, despite the cult's claims that its domestic membership is more than 450,000. On the basis of findings like these, Finke and Stark conclude that cults have always interested some relatively small proportion of U.S. citizens and still do. Despite sensational media accounts of the activities of cults like the Branch Davidians, the rate of cult formation in the United States has not increased appreciably in recent decades, nor are cults more successful in gaining members than they were before.

One reason cults and sects do not always grow in numbers or influence is that established religions often absorb them through the process known as *co-optation.* This term is not limited to religious institutions; we encountered it in Chapter 8 in connection with social movements. It refers to any process whereby an organization deals with potentially threatening individuals or groups by incorporating them into its own organizational structure (Selznick, 1966). For example, over the centuries the Roman Catholic church has been particularly successful at co-opting regional Catholic sects by including their leaders in the panoply of lesser saints, thereby allowing people to worship a holy person of their own culture while remaining true to the world church.

Religious Interaction and Change

Sects and cults are a major source of change in religious organizations. People who are not satisfied with more established churches and denominations, or are otherwise alienated from society, often form or join a cult or sect (Barrett, 1996). One of the most convincing explanations of the emergence of sects was suggested by H. Richard Niebuhr (1929), borrowing from Max Weber's (1922) pioneering analysis of churches and sects. According to Weber, churches tend to justify the presence of inequality and stratification because they must appeal to people of all classes. Sects, on the other hand, may be led by charismatic individuals who appeal to people who have felt the sting of inequality. Niebuhr agreed with Weber that class conflict was a primary cause of sect formation. But he observed that as a sect becomes more successful and better organized, it becomes more like a church and begins to justify existing systems of stratification. This creates the conditions in which new sects may emerge.

Another motivation for the formation of sects or cults is dissatisfaction with the interactions that occur in more established organizations. In church rituals, for example, prayer is often led by a priest or other religious professional and is relatively restrained, whereas in sects and cults communication between God and the individual is more direct and typically allows the individual to express deep emotions. The different styles of interaction in different types of religious organizations can be illustrated by the contrast between the hierarchy of statuses and roles that characterizes the Catholic church (with its pope, cardinals, bishops, priests, and other well-defined statuses) and the seemingly greater equality and looser structure of a cult like Krishna Consciousness or the Unification Church.

Often people who are attracted to cults are influenced by a charismatic leader who inspires them to new and very personal achievements, such as ecstatic experiences, a sense of salvation, or a release from physical or psychological suffering. Some become cult members simply because they are lonely; others are born to cult members and are socialized into the cult.

Some but by no means all cults are extremely authoritarian and punitive. Their leaders may demand that members cut themselves off entirely from family and friends and sacrifice everything for the sake of the cult. The leaders may also insist that they themselves are above the moral teachings to which their followers must adhere. Under these conditions of isolation and submission to a dominant authority, cult members may be driven to incredible extremes of behavior—even mass suicide, as occurred in the case of the followers of Heaven's Gate in 1997. However, not all cults are so dangerous or so easily condemned, and there is an ongoing conflict between norms that protect the right of individuals to belong to cults and efforts to protect people from the harm that can occur when cult leaders place themselves above morality and the law.

Sociological Insights Into Waco and Wounded Knee The confrontation that took place in Waco, Texas, on February 28, 1993, between the Branch Davidians and agents of the Bureau of Alcohol, Tobacco, and Firearms (BATF) revealed the need for greater understanding of alternative religious groups (Foster et al., 1998). Among sociologists who study religion and social change, there is growing consensus that the term *cult* can be dangerous when it is applied as a pejorative label to all religious groups that are outside the mainstream of religious organizations.

Nancy Ammerman, one of the foremost authorities on Protestant religious groups in the United States, notes that "One of the primary interpretive lenses through which the general public views groups such as the Branch Davidians is the lens supplied by 'cult awareness' groups and 'exit counselors.' Namely, most people think members of a group like the Branch Davidians must have been 'brainwashed' into joining" (Foster et al., 1998, p.27). Analyses of the Waco tragedy now place greater emphasis on the deep misunderstandings

Ghost dance.

Big Foot, leader of the Sioux, lies frozen on the battlefield of Wounded Knee, South Dakota. Photograph ca. 1890.

that existed between the government agencies, primarily the FBI and the BATF, that laid siege to the Branch Davidian community, and the members of the community—especially its leader, David Koresh. These misunderstandings appear to have been deepened by hysterical press and television coverage during the episodes of violence and siege (Bates, 1999).

Sociologist Joel Martin observes that the Branch Davidians were not newcomers but an established alternative religious group that had lived in the Waco area since the mid-1930s. The group did believe in an unconventional variant of Seventh-Day Adventist millenarianism and that Armageddon and the end of time were close at hand. They also had some unorthodox and disturbing sexual practices, but these charges, and the charge that they were stockpiling weapons, could have been addressed by quietly apprehending the group's leader for questioning rather than engaging in the large-scale operation that resulted in an armed standoff and the eventual death of the entire group in a blaze of fire and bullets.

Many sociologists point to the similarities between the Waco disaster and the tragedy at Wounded Knee, South Dakota, in 1890. In both cases (and many others like them), a new religion attracts men and women with visions of a new heaven and earth. As they seek to live in accord with these visions, they withdraw into their own community, altering their behavior in ways that may seem strange and even threatening to others. People outside the community begin to spread rumors about them. Government authorities get nervous and plan to move against the group. Journalists congregate, anticipating a big story. When a confrontation takes place, gunfire erupts and a few of the authorities are killed. In the end, the millenarian community is destroyed. After the tragedy, officials, scholars, historians, and journalists interpret what happened. Many conclude that the community's members had been caught up in a "messiah craze." Thus the dead are blamed for their own deaths.

Such was the case with the Lakota Indian Ghost Dancers who were killed at Wounded Knee in 1890. In the Ghost Dance religion the Indians began to believe they might be invulnerable to white attackers. In a sense, they courted death at the hands of white men. Many believed that if the Ghost Dance did not protect them, they would at least achieve speedy transport to a happier existence than the one they led among the conquering white people (Hackett, 1995). The situation of the Branch Davidians killed outside Waco a century later was similar. The issue is not whether either group was right or wrong, but how people in a position to bring understanding to these tragic situations can help prevent similar events from occurring over and over again.

TRENDS IN RELIGION IN THE UNITED STATES

In studying religion in the United States, sociologists are unable to use census data because the Census Bureau does not collect information on religious affiliation. However, some statistics are available from smaller sample surveys such as those conducted by the National Opinion Research Center (NORC). The available data indicate that more than half of Americans over age 18 identify themselves as Protestant, about one quarter are Roman Catholic, and about 2 percent identify themselves as Jewish. All other religions combined account for about 6 percent of the population. The remaining 8 percent say that they have no religious affiliation; they constitute the third-largest category of religious preference in the United States (*Statistical Abstract*, 1998). (See Figure 17.2.)

The religious preferences of Americans have been changing slowly but steadily over the last two generations. The Protestant majority has gradually declined, owing partly to upward mobility and the tendency of more affluent and educated people to be less active in religious institutions. Meanwhile the proportion of the population who identify themselves as Catholics has been growing, primarily as a result of the large numbers of Hispanic immigrants who have entered the United States in recent years. Also, since the mid-1960s there has been an increase in the percentage of people who express no religious preference. Thus it is possible that as the twenty-first century begins, less than half of all Americans will be Protestant (Iannaccone, Stark, & Finke, 1998; National Opinion Research Center, 1998).

FIGURE 17.2 **Percentage Distribution of Religious Preference in the United States (persons 18 years of age and older)**

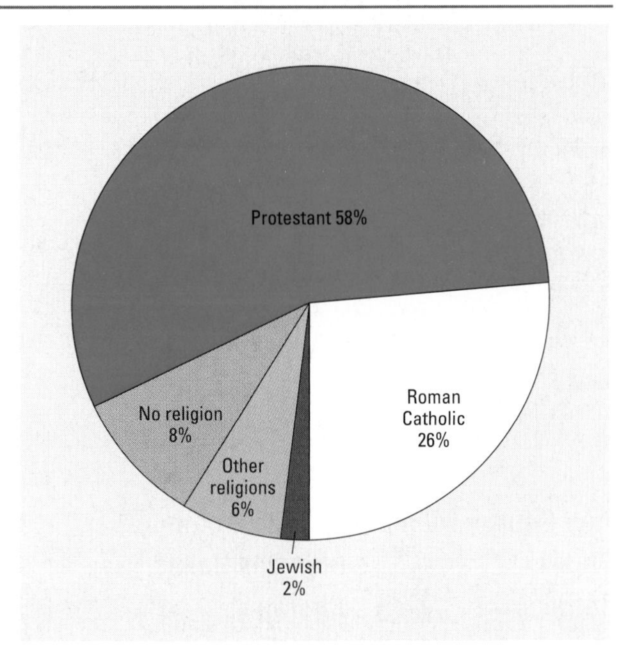

Source: Data from *Statistical Abstract,* 1998.

Membership in a religious organization is quite different from identification with a religious faith, and this is reflected in statistics on Christian church membership in the United States. Because data on membership are obtained from the organizations themselves, there are significant differences in who is included. (Some churches, for example, count all baptized infants, whereas others count only people above a certain age who are enrolled as members.) The Mapping Social Change feature on pages 548 and 549 illustrates the geographic distribution of reported church membership in the United States. As can readily be seen, different regions of the country are quite different with respect to the church membership of the majority of their residents. Protestants outnumber members of all other religions in the South, for example, and in the West and Southwest the Roman Catholic church is dominant. There are also significant differences in church membership by size of place: Protestants are most likely to live in rural places, small cities, and suburbs of smaller cities, while native-born Catholics are found in medium-sized cities and suburbs of larger cities. Much of the rapid growth of Catholic congregations in the United States is accounted for by immigrants to major cities like Miami, Los Angeles, and New York (Niebuhr, 1995). Jews, who originally migrated to the nation's largest cities, are most numerous in major metropolitan regions like New York, Los Angeles, Chicago, and Miami (National Opinion Research Center, 1998).

Religion has always had an especially important place in the lives of African Americans, and a major proportion of the black population is deeply religious. Many of the functions of religion as a central cultural institution—meeting the spiritual needs of individuals, binding together the members of social groups, and sometimes engendering social change—are illustrated by the history of the black church in America, summarized in Box 17.1 on page 550.

Unofficial Religion

Religion in the United States is increasingly voluntaristic. People are less likely to practice a religion simply because their parents did so. They are more likely than ever to join a religious group that appeals to their desire

MAPPING SOCIAL CHANGE

Christian Religious Affiliation in the United States

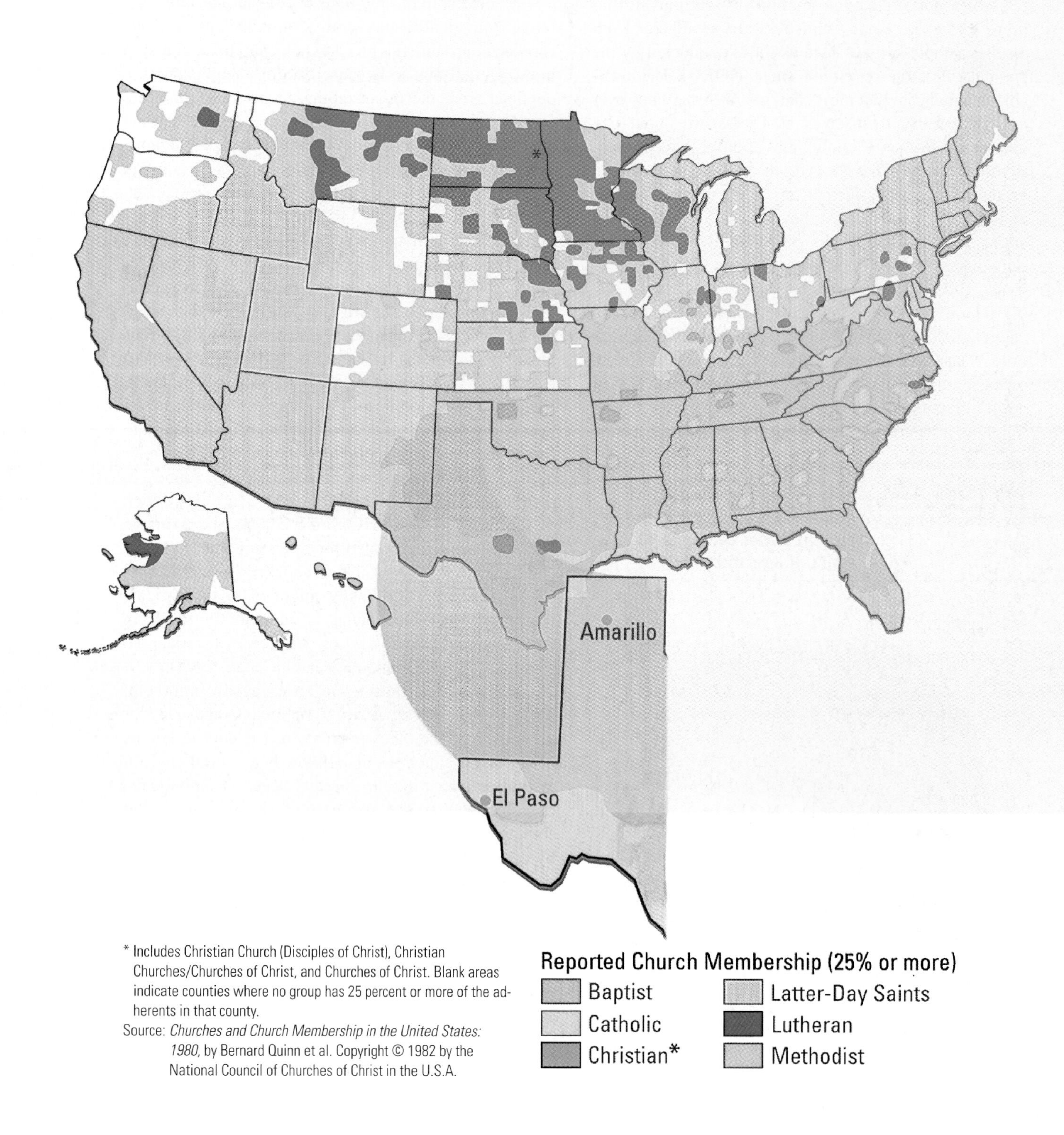

* Includes Christian Church (Disciples of Christ), Christian Churches/Churches of Christ, and Churches of Christ. Blank areas indicate counties where no group has 25 percent or more of the adherents in that county.

Source: *Churches and Church Membership in the United States: 1980*, by Bernard Quinn et al. Copyright © 1982 by the National Council of Churches of Christ in the U.S.A.

Religious Pluralism in the "Bible Belt"

The wide swath of states in the South and Southwest where Baptist church congregations predominate is often known as the "Bible Belt." This term refers to the strong religious convictions of many of the region's residents and to the powerful influence of Baptist churches in thousands of small communities. On the accompanying map, the Bible Belt stands out quite clearly. But maps showing the distribution of religious beliefs often mask the changes occurring in regions that appear more homogeneous than they are. Migration, urbanization, and differences in the growth of specific religious groups can lead to more religious pluralism (or less, depending on the region) even though a map shows a static distribution.

When we look closely at communities in west Texas, at the western end of the area of heavy Baptist church membership we find higher levels of Catholic church membership. This is due in large part to the growing presence of Latino residents, especially people of Mexican and Chicano backgrounds. Wherever there are changing patterns of church membership (represented here by photos of Baptist and Catholic congregations in this region), there is also likely to be more interfaith communication and more frequent discussion of the need for religious tolerance.

Similar evidence of pluralism can be found in other regions of the United States. In the heavily Catholic Northeast, for example, the map of Christian congregations does not indicate non-Christian, particularly Jewish, congregations, which are concentrated in the larger cities of this region. Catholic-Jewish interfaith dialogues are strongest in these urban centers. So while ecological maps like this one are useful, they only begin to suggest the important sociological facts that will be revealed by further research.

BOX 17.1 USING THE SOCIOLOGICAL IMAGINATION

The Black Church in America

The respected black sociologist E. Franklin Frazier was a firm believer in the principles of the Chicago school of ecological research and an astute student of race relations. He specialized in two of the most basic social institutions—the family and the church—as they affected and were affected by the situation of black Americans.

Frazier's (1966) study of the history of the black church in America emphasized the role of religion in the social organization of blacks ever since they were imported from Africa to the New World to be sold as slaves. Few aspects of the slaves' African culture were able to survive the experience of abduction and slavery, but one of those that did was dancing, which Frazier refers to as the most elemental form of religious expression. Although the slaves were encouraged to dance, the religious beliefs expressed in their dances were not permitted to flourish in America.

With the coming of Baptist and Methodist missionaries, the slaves found a form of religion in which they could express their most intense feelings. The religious services in which they participated served to bind them together into close-knit groups, replacing the bonds of kinship and tribal membership they had lost when they were sold into slavery. However, the freed slaves resented the subordinate status of blacks in the white-dominated churches. They left those churches and established all-black churches.

Out of these black churches, Frazier noted, other forms of black organizational activity emerged. The churches began to play an important role in the organization of the black community in such areas as economic cooperation and the building of educational institutions. The church also became the center of political life for black Americans, the arena in which they learned how to compete for power and position.

The large-scale urbanization of blacks that began in World War I transformed the black church. In particular, the church became more secular, placing less emphasis on salvation and turning its attention to the serious problems of blacks in the here and now. Changes in the class structure of the black community have also been reflected in the black church, with many middle-class blacks shifting to the Presbyterian, Episcopalian, and Congregational churches. A countertrend may be seen in the emergence of "storefront churches" (i.e., sects) that satisfy the spiritual needs of those who prefer a more intimate and expressive form of worship. And some blacks have joined so-called holiness cults that seek to return to an earlier form of expressive Christianity.

At the same time, blacks have become more fully integrated into American society. This has naturally affected the organization of black communities and the black church. Today the church is no longer the dominant institution for blacks that it once was (Patterson, 1999). Nevertheless, as Frazier pointed out, the black church has left its imprint on most aspects of African American life. And it is clear that the black church (including the Black Muslims and other Islamic sects) provides much of the leadership and many of the rank-and-file supporters of movements for political and economic justice among minority groups in the United States.

Internationally acclaimed singer Aretha Franklin attributes her success to the vitality of the black church in the United States, which sustains an immense market for gospel songs and records.

for membership in a community of like-minded peers and that upholds their particular moral standards (Swift, 1998). People in the United States and Canada are also more likely to engage in what sociologists term *unofficial religion*—a set of beliefs and practices that are "not accepted, recognized, or controlled by official religious groups" (McGuire, 1987, p. 89). Sometimes called folk or popular religion because it is practiced by ordinary people rather than by religious professionals in formal organizations, unofficial religion takes many forms. It is engaged in by people who purchase religious books and magazines, follow religious programs on television, make religious pilgrimages, or practice astrology, faith healing, transcendental meditation, occult arts, and the like—and who may belong to organized churches at the same time. Contrary to older sociological theories that viewed these practices as holdovers from rural folk cultures, research has shown that urban societies continually produce their own versions of popular religion (Fischer, 1987).

Effects of Immigration

Largely as a consequence of recent waves of immigration from Asia and the Near East, membership in Islamic and Hindu religious groups is growing rapidly. By the mid-1990s it was estimated that there were between 4.5 and 5 million Muslims in the United Sates and Canada, and more than 1,000 Islamic centers with about half a million Muslims attending mosque services as official members. The largest numbers of Muslims are found in Chicago, Washington, New York, Toronto, and southern California. The largest ethnic groups among North American Muslims are African American and Indo-Pakistani, with Arab Americans ranked third. The oldest continuously functioning mosque in the United States is located in Cedar Rapids, Iowa, and is known as the Mother Mosque of America.

Most Muslims in the United States belong to the Sunni sect, but membership in the more fundamentalist and militant Shiite sect is also increasing. Muslims, particularly those of the first immigrant generation, tend to hold views on moral issues quite similar to those of conservative Jews and Christians, and to oppose abortion, pornography, and extramarital sex. First-generation Muslim immigrants, like other first-generation immigrants, tend to be concerned about whether their children will marry within the faith and continue to share their Islamic beliefs and values (Swift, 1998).

Immigration from Pakistan, India, and nations of Africa and the Caribbean that were formerly part of the British Empire has caused a rapid increase in the number of Hindus in North America in the past decade or so. An estimated 400,000 Hindus now reside in the United States, most of them concentrated in Southern California, New York, and Washington, D.C. Most Hindus worship at home, but Hindu temples have recently been established in several major U.S. cities. A far smaller proportion of practicing Hindus in the United States are native-born Americans who are attracted to the religion's values of inner peace and enlightenment and have studied with Indian gurus like Bagwan Shree Rajneesh (Swift, 1998).

Another trend, while not strictly a result of immigration, nonetheless reflects the influence of other cultures on contemporary religious practices. Sociologists studying religion in the United States note the small but significant increase in rates of participation in pagan religions, witchcraft, or Wicca. Wicca is particularly attractive to feminists who reject what they perceive as the patriarchal tendencies of Judaism and Christianity. Pagan holy days and rituals are associated with, but not limited to, the solstices and equinoxes. Pagans believe that women participate in the divine nature of the Goddess, the source of human life and of all nature. Many pagans are reluctant to reveal their religious beliefs and practices because of the history of Christian suppression of paganism (Adler, 1997).

Religiosity

Neither church membership figures nor self-reports of religious identification are accurate indicators of the aspect of religious behavior known as **religiosity.** This term refers to the depth of a person's religious feelings and how those feelings are translated into religious behavior. Responses to questions about whether one believes in God and how strongly, whether one believes in a life after death, whether one's religious beliefs provide guidance in making major decisions, and the frequency of one's church or temple attendance all can be used to measure religiosity.

When sociologists study religiosity as opposed to church membership, some important results emerge. While a common stereotype of Americans describes them as materialistic, individualistic, pleasure seeking (hedonistic), and secularized, data on their religious faith and behavior do not confirm the stereotype. Compared with citizens of other urban industrial nations, Americans are more religious. In fact, the most recent World Values Survey of the University of Michigan reports that

> Fully 44 percent of Americans attend church once a week, excluding funerals and christenings, compared with 27 percent of Britons, 21 percent of the French, 4 percent of Swedes, and 3 percent of Japanese. Not only do they go to church, but 53 percent of Americans say religion is very important in their lives, compared with just 16 percent of the British, 14 percent of the French, and 13 percent of Germans. (Reese, 1998)

Of course, these figures mask major variations among Americans in their commitment to religious practice, and in fact sociologists are engaged in a lively debate about the meaning of the survey data. Closer examinations of church attendance statistics have found that a sizable proportion of Americans who say they attend church or synagogue regularly actually exaggerate their estimates. Analysis of their responses suggests that the figures for American church attendance are inflated, in some instances by 50 percent or more (Hadaway & Marler, 1998). But other sociologists note that successful congregations have a core group of regular attenders who tend to be most active in all the congregation's activities. Those on the margins of this core group tend to be those who also overestimate their attendance (Ploch & Hastings, 1995; Reese, 1998).

There is far less disagreement over what the poll data reveal about other aspects of religiosity, particularly about beliefs as opposed to behavior. About 90 percent of Americans believe in the existence of God, and 86 percent express certainty in a life after death. Adults in the Pacific states, which tend to have the lowest rates of church membership, do not differ very much on these measures from those in other regions. The lower rates of church membership on the West Coast are correlated with the greater spatial mobility and lower median age of the populations of those states. People who move frequently tend to sever their attachments not only to churches but to all organizations in the community. Yet people who do not belong to organized churches can nevertheless hold deeply cherished religious beliefs.

Data on church attendance do not support the notion that the United States is becoming an increasingly secular society. In fact, as can be seen in Table 17.1, for some major religions church attendance has been increasing since 1950 (Hout & Greeley, 1987; Ploch & Hastings, 1995). There are no clear patterns in rates of church attendance between 1939 and 1990. Although Catholics have the highest weekly attendance rates and Jews the lowest, those rates can change considerably from one year to the next, for reasons that are not fully understood. Nevertheless, it is clear that the data do not bear out the secularization hypothesis. It should also be noted that increases in immigration during the 1980s and 1990s are significantly increasing the diversity of religious practice in the United States. As noted earlier, Muslims, Hindus, Buddhists, and others are well represented among the new immigrants.

Religious Pluralism

Differences in the distribution of religious belief are important because they show how religion helps account for cultural differences among populations within a society. Thus we use the expression "Bible Belt" to refer to the most strongly Protestant areas of the nation—the South and Midwest. The large Catholic populations in some cities make them quite different in cultural terms from the Protestant-dominated suburbs. And the large Jewish populations in New York and Miami are reflected in the cultural life of those cities—in joint Hanukkah/Christmas celebrations in offices and schools, for example, and the incorporation of Jewish holidays into the school calendar.

The United States is an example of a society in which the state protects religious pluralism. The First Amendment to the Constitution guarantees freedom of religion, which in turn allows different religions to exist and to compete for adherents if they wish to do so. Religious pluralism is also protected in most European nations today, although before the Protestant Reformation of the sixteenth century many European states made Roman Catholicism the official state religion and attempted to force all citizens to convert to Catholicism. And in the early 1990s the religious diversity and tolerance of the former Yugoslavia dissolved into a genocidal religious and nationalist war.

In societies in which religious pluralism is protected, one usually can observe the continual formation

TABLE 17.1 Percent of Adults Attending Services Last Week, by Religion and Year, United States, 1939–1990

	Religion				
Year	Protestant	Catholic	Jewish	Other	None
1939	40	64	12	—	1
1969	39	64	8	18	0
1979	40	53	19	40	5
1990	47	50	30*	31*	6*

*Small sample size accounts for large fluctuations in these percentages.

Sources: Hout & Greeley, 1987; National Opinion Research Center, 1998.

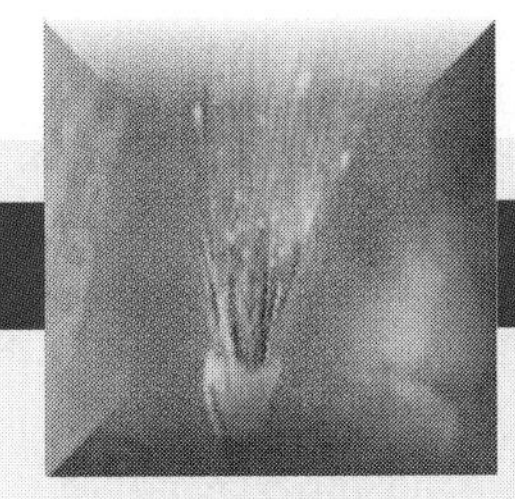

THE PERSONAL CHALLENGE

Social Change and Religious Pluralism

As societies become more culturally diverse as a result of global migrations, there is an increase in the occurrence of behaviors that seem to violate widely held norms and values. Debates and conflicts over such behaviors become more and more frequent. In France, where the Islamic population has been greatly augmented by recent immigration, there have been numerous conflicts over Muslim women wearing veils and head scarves to school or work. In California, Sikh students have been banned from wearing *kirpans,* sheathed blades worn next to the skin that are considered sacred. The state of Oregon refused to pay unemployment benefits to a Native American who was fired for using peyote, a drug used in tribal religious rituals. A Florida law banning the ritual sacrifice of chickens by practitioners of Santeria, a religion with African roots, was overturned by the Supreme Court.

A recent case in St. Paul, Minnesota, shows what can happen when religious practices run counter to local customs or laws. A veiled Muslim woman shopping in a downtown mall was detained by police for concealing her identity "by means of a robe, mask or other disguise." When she refused to take off the veil, she was taken to a room where the police forced her to remove it. The incident outraged Muslims throughout the nation; to them, it was a violation of religious freedom. The woman later commented, "Why is it that people can walk around half naked and nobody will say anything? I choose to cover up and people have a problem with it" (quoted in Pesce, 1994).

Incidents like this are not uncommon in communities that are growing more diverse. The wave of arson directed against African American churches in many regions of the United States is another indication that cultural diversity is difficult for some people to tolerate. The burning of black churches is a strong symbolic act against norms of religious and cultural tolerance and diversity. The recent attacks are reminiscent of similar incidents during the early days of the civil rights movement. In this regard, Ralph Reed, former director of the Christian Coalition, asserts the need to "work to make racial conciliation a major priority" (quoted in Herbert, 1996, p. A27).

As immigrant groups move from major metropolitan centers to smaller cities and towns, they are likely to encounter incidents of hostility. If the number of such incidents continues to increase, citizens will be forced to choose among several alternatives. Should they try to discourage immigrants and the diversity they bring? Should they try to help one another understand and accommodate people who are different? Should they insist that the newcomers conform to their norms, or should they welcome new customs and beliefs? These are not abstract issues. What choices will you make in response to the pluralist challenge?

Like Muslims, practioners of Santeria, an offshoot of Yoruba African religion, are subject to prejudice. Shown here is a Santeria religious offering.

of new religious organizations. New religious groups tend to arise either out of schisms within existing organizations or as a result of the teaching of charismatic leaders who attract people to new religious movements. Many new religious movements are also stimulated by severe inequalities and the widening gap between rich and poor. Whether it originates in schisms within existing religious organizations, as a result of the emergence of charismatic religious leaders, or in response to severe inequalities, religious pluralism presents many personal questions and challenges. Two examples of these challenges are found in movements for social change within the Catholic church and the rise of fundamentalism in the United States and elsewhere.

Social Change and the Catholic Church It is a striking fact that throughout the world one finds deeply devout individuals involved in movements for social change. Religious workers were killed in El Salvador because they advocated greater democracy and equality. In Poland before the fall of the Soviet empire, the Catholic church was a leader in the movement to democratize that nation. American bishops speak out against poverty and inequality in the United States. The civil rights movement has tended to be led by clergymen and bases many of its appeals on biblical principles of justice and equity. Yet on the other hand we could cite many examples of religious intolerance and an apparently high correlation between devotion to religious faith and resistance to social change.

Support for and resistance to social change are relative ideas—relative, that is, to the position of their advocates. If one supports a woman's right to choose whether to have an abortion, the social change in question is tolerance for abortion rights (which does not necessarily mean that one favors abortion). If one is opposed to abortion rights, the change one seeks is a return to a situation in which abortion is illegal. As noted earlier in the

The pope, shown here mingling with a crowd of devout followers, often takes a conservative position on social issues like abortion and women's roles. In many parts of Latin America, however, the Catholic church is an important force for social change, and its leaders play an important part in the struggle against severe economic inequality.

chapter, by opposing abortion rights the Catholic church stands for social change in a conservative direction.

In religious life the same institution can stand for conservative social change in one place and for more radical change in another. For example, the conservative Episcopalian (Anglican) church in Northern Ireland favors the continuation of British colonial rule there, while the Episcopalian church in the United States supports liberal causes such as abortion rights and environmental action. The relationship between the Catholic church and social change is even more complex. In the United States the church is accused by some of its own leaders of having too conservative a view of social change. In Brazil, by contrast, the church is accused of being too liberal. How can a single institution be perceived so differently in two societies?

Part of the answer is that a church that is so large and embraces so many separate organizations can never be free from internal politics. Many Catholic theologians agree with Father Richard P. McBrien of Notre Dame University that "since his election in 1978, Pope John Paul II has been determinedly appointing a certain type of cleric to important archdioceses and dioceses all around the world" (1990, p. A17). According to McBrien and others, these bishops and cardinals tend to be "uncritically loyal to the Pope" and "rigidly authoritarian and solitary" in their leadership style. They are unlikely to take their parishioners' views into account when they differ from those of the pope and his curial associates. For their part, high church officials argue that their function is to keep the church on the right track in defending the sanctity of life while seeking constructive social change where possible. Hence, on many social issues—abortion rights, the rights of homosexuals, treatment of AIDS patients, ordination of women, economic reform—the leadership of the Catholic church (though not always its rank-and-file adherents) tends to act as a conservative force in the United States and other industrial nations.

The opposite is true in Brazil and other Latin American nations, where the Catholic church is identified with movements for radical social reforms. In Brazil, theologian Leonardo Boff and Bishop Dom Helder Camara have developed and taught the idea that spiritual salvation is not possible without action to liberate the oppressed. Religion itself could help transform the miserable conditions of the peasantry and the Indians and the impoverished working people (Schubeck, 1995). This view has come to be known as "liberation theology."

As liberation theology took hold during the 1960s and 1970s, church leaders began to develop what they called "the preferential option for the poor." Recognizing that the so-called miracle of Brazilian economic growth had not helped most of the nation's poor people, they proposed to correct this situation even at the expense of the spiritual needs of the affluent. Church leaders and lay activists organized thousands of small groups known as Communidades Eclesial de Base, or CEBs. These 30- to 40-person groups were the building blocks of a powerful social movement to bring both spiritual comfort and economic uplift to the nation's poor masses. Never a very popular strategy in Rome, the CEBs seem nevertheless to have had a real impact in moving Brazil's elite classes toward the realization that if they do not do more to distribute the benefits of economic development they might be swept out of power in coming elections.

At this writing it is not clear what the future holds either for liberation theology or for the liberal–conservative dispute among Catholics in the United States. During his visit to Mexico and the United States in 1999, the Pope made the rising gap between rich and poor throughout the world a major issue in his addresses. But within the church hierarchy there are ongoing efforts to ban the teaching of liberation theology, as well as criticism of the preferential option for the poor (*National Catholic Reporter,* 1997). It is clear, however, that like all the major religions of the world, Catholicism will continue to face the challenges of social change. Among the most severe of those challenges is the rise of fundamentalist religious movements, which pose major threats to religious pluralism in Islamic nations, among Indian Hindus, among Jews in Israel, and in many congregations throughout the United States.

Fundamentalism Religious *fundamentalists* are believers (and their leaders) who are devoted to the strict observance of ritual and doctrine. They hold deep convictions about right and wrong in matters of faith and lifestyle, with little tolerance for differences in belief and practice, and they are fiercely opposed to astrology, magic, unorthodox conceptions of religion, and any form of civil religion. In all the world religions there are divisions between liberal and conservative approaches to the religious norms that guide daily life. Among fundamentalists, however, there is a belief in the

> basic, intrinsic, essential, inerrant truth about humanity and deity; that this essential truth is fundamentally opposed by forces of evil which must be vigorously fought; that this truth must be followed today according to the fundamental, unchangeable practices of the past; and that those who believe and follow these fundamental teachings have a special relationship with the deity. (Hunsberger, 1995, p. 9)

Clearly, these beliefs, passionately held as they are, present enormous challenges for societies in which there is a good deal of religious diversity. (The consequences of Islamic fundamentalism, particularly for the lives of women, are presented in Box 17.2.) In the United States, fundamentalist believers may be found in every major religion, but fundamentalism is especially strong among some Protestant churches and sects.

BOX 17.2 GLOBAL CHANGE AND U.S. SOCIETY

Women in Islam

More than 1 billion people adhere to the Islamic faith. Most of them live in the Arab states of the Middle East and North Africa, but Muslims may be found in growing numbers in every corner of the globe, including the United States.

Relations between the Islamic world and Western societies are colored by a great deal of misinformation and bias. An example from the social sciences is the theory that social changes do not affect the Islamic world in the same way that they do, or did, affect the Western nations. In particular, it is argued that because of the subordinate status of women in the Islamic religion, the movements for gender equality, limiting of family size, and greater equality in politics that have brought about such dramatic change in the West will not occur in the Islamic world (Caldwell & Caldwell, 1988). This hypothesis attributes the slow pace of demographic change in many Islamic nations to orthodox Islam's traditional values, such as polygyny, which permits men to father many children with as many as four wives. The tradition of female submission in orthodox Islam is also said to account for the persistence of gender inequalities in those nations. Islamic sociologists point out, however, that there are significant differences among the various Islamic societies and that some have long-standing norms supporting birth control and education for women.

When we examine empirical data about the Islamic nations, many ironies become apparent. There are nations where traditional Islamic norms are quite powerful and women cannot hope for educational or occupational equality with men, and there are others where these differences are diminishing rapidly. For example, in Saudi Arabia patriarchal norms are in effect and the subordination of women is strictly enforced. Women are not allowed to drive, yet that nation is the world's foremost producer of petroleum. As shown in the accompanying table, in Saudi Arabia only 65 percent of girls of primary school age are enrolled in school, among the lowest rates in the Arab world. Extremely poor, peripheral Islamic nations like Somalia (with only 13 percent female primary school attendance) are even less advanced, but in these nations it is usually the challenge of day-to-day survival rather than religious orthodoxy that prevents children from attending school. In Iraq, on the other hand, orthodoxy is less powerful and women's average level of education is among the highest in the Islamic world.

Many other differences among the Arab nations in the heart of the Islamic world show up clearly in the table. (Note that Iran is not included in this list because, although it is an Islamic nation, its culture is Persian, not Arab.) There are major differences among these nations in basic social and demographic indicators. All have high fertility rates, but in more urbanized Arab nations like Kuwait, where education for women is more easily available and polygyny is waning, rates of fertility and child mortality are closer to those of other more developed nations. In more agrarian societies like Egypt, which also tend to have larger populations, fertility remains extremely high, as does child mortality. Sociologists from these societies find, however, that Islamic teachings themselves do not account for these disparities so much as economic conditions such as poverty and concentration of the population in rural villages and farms (Obermeyer, 1992).

Demographic and Socioeconomic Indicators in Arab Countries

	Population (millions)	Gross National Product per Capita (US$, 1989)	Percent Urban	Percent of Age Group Enrolled in Primary School*X		Total Fertility Rate 1985–1990	Infant Mortality Rate 1985–1990	Life Expectancy 1985–1990 (years)	
				Male	Female			Male	Female
	(1)	(2)	(3)	(4)	(5)	(6)	(7)	(8)	(9)
Algeria	24.9	2,230	51.7	105	87	5.4	74	63	65
Bahrain	0.5	6,340	82.9	111	110	4.1	16	69	73
Egypt	52.4	640	46.7	100	79	4.5	65	58	60
Iraq	18.9	—	71.3	105	91	6.4	69	63	65
Jordan	4.0	1,640	68.0	98	99	6.2	44	64	68
Kuwait	2.0	16,150	95.6	95	92	3.9	18	71	75
Lebanon	2.7	—	83.7	105	95	3.8	48	63	67
Libya	4.5	5,310	70.2	128	119	6.9	82	59	63
Mauritania	2.0	500	46.8	61	42	6.5	127	44	48
Morocco	25.1	880	48.0	85	56	4.8	82	59	63
Oman	1.5	5,220	10.6	103	92	7.2	40	62	66
Qatar	0.4	15,500	89.4	121	121	5.6	31	67	72
Saudi Arabia	14.1	6,020	77.3	78	65	7.2	71	62	65
Somalia	7.5	170	36.4	26	13	6.6	132	43	47
Sudan	25.2	480	22.0	59	41	6.4	108	49	51
Syria	12.5	980	50.4	115	104	6.8	48	63	67
Tunisia	8.1	1,260	54.3	126	107	4.1	52	65	66
United Arab Emirates	1.6	18,430	77.8	98	100	4.8	26	69	73
Yemen, YAR (North)	9.2	640	25.0	141	40	8.0	120	50	50
Yemen, PDRY (South)	2.5	430	43.3	96	35	6.7	120	49	52

*Enrollment ratios are expressed as a percent of the total population of primary school age. For countries with universal education, the gross enrollment ratio may exceed 100 percent because some pupils are above or below the official primary school age.

Source: Obermeyer, 1992.

Historically, the influence of religious fundamentalism on the culture and politics of American life reached its height in 1919, when the Eighteenth Amendment to the U.S. Constitution prohibited the manufacture, sale, or transportation (and hence the consumption) of "intoxicating liquors," regardless of whether one supported the fundamentalist view of alcohol consumption. The repeal of Prohibition in 1933 represented a defeat for religious fundamentalism in American culture and politics.

Another famous episode involving fundamentalism was the so-called "monkey trial" of 1925, in which John T. Scopes, a biology teacher in Tennessee, was charged with teaching the then forbidden theory of evolution, which many religious fundamentalists oppose because it contradicts the account of creation presented in the Bible. The prominent lecturer and religious fundamentalist (and former presidential candidate) William Jennings Bryan prosecuted the controversial case, which received worldwide attention. Despite a strong defense by Scopes's lawyer, Clarence Darrow, Scopes was convicted and fined $100. (Although courts later reversed the conviction on a technicality, it would be more than 30 years before the Tennessee legislature repealed the law that prohibited teaching evolution theory in state-supported schools.) Despite the conviction, the Scopes trial is seen as a turning point for the intellectual opponents of fundamentalism. Although fundamentalism continued to pose a challenge to mainline Baptist, Methodist, and Presbyterian denominations, after the Scopes trial it lost influence—until recently—as a religious ideology (Simpson, 1983; Swift, 1998).

Sociologists of religion tend to view the rise of fundamentalist religious movements as part of a larger effort by people in modern societies to make moral and spiritual sense of their lives. In the Islamic world, fundamentalism is also associated with the failure of elites to deliver on their promises of well-being for the masses of people in their societies (Hussein, 1994). In more affluent nations like the United States, however, the appeal of fundamentalism seems to be more often based on rejection of moral relativism, racial integration, and cultural pluralism, and on the desire to find moral absolutes that can guide people through complex and ambiguous social environments.

This fact is reflected in the remarks of Robert Bellah, a well-known commentator on American culture. For several centuries, he writes, Americans "have been embarked on a great effort to increase our freedom, wealth, and power. For over a hundred years, a large part of the American people, the middle class, has imagined that the virtual meaning of life lies in the acquisition of ever-increasing status, income, and authority, from which genuine freedom is supposed to come" (Bellah et al., 1985, p. 284). Yet many Americans seem uneasy about their lives despite their material comfort. They seem to yearn for spiritual values without necessarily wanting to return to traditional religious practices. They adhere to the values of individualism, but at the same time they long for the stronger sense of community and commitment that one finds in religious congregations. Bellah predicts that Americans will continue to seek self-actualization as individuals, but that increasingly they will express their desire for community attachments and higher values, either in traditional religions or in civil-religious practice. Research on the challenges faced by Southern Baptists, however, suggests that the divisions between more secular and more fundamentalist believers will continue to strain the institutions of pluralistic societies like the United States.

RESEARCH ON THE CUTTING EDGE: Social Change and the Southern Baptists

Sociologist Nancy Tatom Ammerman views issues of pluralism and social change as central to understanding the split between fundamentalist and moderate Southern Baptists. She writes:

> The disruption against which fundamentalists struggle is often labeled "modernity." Whatever else that label may mean, the "modern" world is one in which change is a fact of life, in which people of multiple cultures live side by side, and in which religious rules have been largely relegated to a private sphere of influence. (1990, p. 150)

To assess differences among Baptists in the way they view social change and diversity, Ammerman surveyed more than 900 Baptist leaders in congregations throughout the United States. Her survey sample included men who were pastors and church deacons and women who were presidents of their local Baptist women's organizations. After examining their responses to key questions about their beliefs, Ammerman developed categories (or "theological parties") with labels corresponding to the terms respondents often used to describe themselves (e.g., *fundamentalist, conservative, moderate*). She also developed a scale for measuring individuals' degree of opposition to modernity. The questions on which the scale is based and the method of scoring responses are discussed in Box 17.3. Table 17.2 shows that "among those who disapprove of pluralism, fundamentalists outnumber moderates nearly ten to one" (Ammerman, 1990, p. 151). But the data in the table and in Box 17.3 also show that Baptists classified as "conservative," whose views fall somewhere between those of the fundamentalists and moderates, are the most numerous category and are quite divided in their views of pluralism.

When she examined how respondents felt about modernity in relation to such variables as education, occupation, and rural versus urban residence, Ammerman found some surprising results. Although people often think of individuals who hold strong fundamentalist beliefs and reject modernity as uneducated people living in remote farming hamlets, and of individuals who approve of modernity as educated middle-class urban dwellers, Ammerman's results challenge this conventional wisdom:

> Those among Southern Baptists who were the most skeptical of change, choice, and diversity were blue collar workers, those with middle incomes, people who had moved from farm to city, and those who had been to college, but did not have a degree. And combinations were important: people who had moved from farm to city *and* had some college or people with middle income *and* blue collar households, for instance. These were people who knew exactly what the modern world was all about, and they were less enthusiastic about embracing it than were any other Southern Baptists. And the less enthusiasm they had for modern attitudes, the more likely they were to adopt fundamentalist beliefs and identity. (1990, p. 155)

Eventually, Ammerman believes, these differences in outlook may lead to an irremediable schism in which moderates may form a new Baptist denomination. She is not convinced that religious fundamentalism poses dire threats. The rise of Christian fundamentalism does, in her view, challenge widespread beliefs about the inevitable progression of modern societies toward ever greater "differentiation, individualism, and rationalism as 'the way things are'" (1997, p. 202). Modern frameworks for understanding the social world, she argues, emphasize the "individualized 'meaning system' that would be carved out of differentiation and pluralism." But we live in a world

> where organizational boundaries are more fluid, where mergers and out-sourcing and flextime and telecommuting are as common as the time clock and the stockholder corporation. We live in a world where rationality and the scientific method are valued, but also critiqued, where multiple sources of wisdom are finding a voice. We live in a world where people are both more rooted in particularistic ethnic and religious communities that refused to melt and more

TABLE 17.2 Percent in Different Theological Parties, by Responses to Pluralism

Theological Party	Strong Approval of Pluralism	Moderate Approval of Pluralism	Disapproval of Pluralism	Total
Self-Identified Moderate	21%	5%	3%	9%
Moderate Conservative	16	8	3	9
Conservative	38	55	43	48
Fundamentalist Conservative	18	24	27	23
Self-Identified Fundamentalist	8	9	23	11
Total	101%	101%	99%	100%
(Number of Cases)	(256)	(500)	(187)	(944)

Note: Difference statistically significant at $p < .001$. Some percentages do not total 100 due to rounding.

Source: Ammerman, 1990.

BOX 17.3 SOCIOLOGICAL METHODS

Scales and Composite Scores

Sociologists who are analyzing questionnaire responses often create *scales* in order to arrive at a composite score for a particular variable. Nancy Tatom Ammerman created scales to measure the strength of a person's fundamentalist beliefs and opposition to modernity. (See the accompanying tables.) A scale is a set of statements that fit together conceptually in that they deal with aspects or examples of the same attitude or behavior. For each statement, the respondent may be asked to choose among responses ranging, for instance, from 1 ("Strongly disagree") to 5 ("Strongly agree"), with "Unsure" in the middle. The statements in a particular scale (e.g., fundamentalist beliefs) are scattered throughout the questionnaire to avoid *response set,* the tendency to answer all such items similarly without giving them much thought; response set is a frequent problem when similar items are grouped together. To ensure that the statements in a scale actually measure the behavior or attitude the researcher is studying, great care must be taken in their selection and wording.

Once the questionnaires have been completed and the data entered into a computer, the researcher can arrive at a simple composite score by adding up the scores on all the items in a scale. Then the researcher can either use each respondent's composite score as a continuous variable or establish cutoff points that will create categories for analysis, as Ammerman did. In her measurement of opposition to modernity, for example (see table), Ammerman decided that those who agreed or strongly agreed with all the items (i.e., whose scores ranged from 5 to 9) could be classified as "strongly agreeing with pluralism." Those who agreed with most of the items but not all of them (i.e., scored from 10 to 14) were classified as showing "moderate approval." Those who scored 15 or above—in other words, those who disagreed as often as they agreed—were classified as "disapproving."

Fundamentalist Beliefs

ITEMS	SCORING
The Scriptures are the inerrant Word of God, accurate in every detail.	Strongly agree = 5
God recorded in the Bible everything He wants us to know.	Strongly agree = 5
The Genesis creation stories are there more to tell us about God's involvement than to give us a precise "how and when."	Strongly disagree = 5
The Bible clearly teaches a premillennial view of history and the future.	Strongly agree = 5
It is important that Christians avoid worldly practices such as drinking and dancing.	Strongly agree = 5

DISTRIBUTION

RANGE	PERCENT
5–10	5
11–15	14
16–20	47
21–25	34

Opposition to Modernity

ITEMS	SCORING
I like living in a community with lots of different kinds of people.	Very untrue = 5
Public schools are needed to teach children to get along with lots of different kinds of people.	Strongly disagree = 5
I sometimes learn about God from friends in other faiths.	Very untrue = 5
One of the most important things children can learn is how to deal creatively with change.	Strongly disagree = 5
Children today need to be exposed to a variety of educational and cultural offerings so they can make informed choices.	Strongly disagree = 5

DISTRIBUTION

RANGE	PERCENT
5–9	29
10–14	52
15–19	17
20–25	2

> aware of the larger world and the choices that have brought them to their current practices. (1997, p. 202)

A clear implication of Ammerman's brilliant empirical research, and the conclusions she draws from her findings, is that fundamentalist communities are the result of unquestioned religious socialization of some people and the rational choices of others. These congregations may present severe challenges to religious pluralism, but in the long run it is in pluralistic societies that they will eventually find a secure, though limited, place.

VISUAL SOCIOLOGY

Expulsion, by Edgar Tolson.

Religious Folk Art

The religious art of Sister Gertrude Morgan is only one of many possible examples of how religious inspiration can motivate artistic expression in people without formal training in the visual arts. Morgan's work is sought after by art collectors today, as is the art of Grandma Moses, one of the first unschooled American artists to become famous. But Morgan's art was always spiritually motivated, and all her works express the religious feelings that welled up within her throughout her life as an evangelical preacher.

The Barefoot Prophetesses, by Sister Gertrude Morgan, New Orleans, Louisiana, c. 1971. Watercolor on paper, 11½″ × 16″.

Crucifixion, by Elijah Pierce. Carved and painted wood with glitter; mounted on wood panel, 47½″ × 30½″.

Edgar Tolson (1904–1984) is another highly acclaimed folk artist whose work was often inspired by religious feelings. Tolson was born in Lee City, Kentucky, and worked for most of his life in the small town of Campton, Kentucky. A specialist in wood carvings, which he often referred to as his "dolls," Tolson was a whittler who became something of a regular at craft fairs and folk festivals. His work expresses the traditional fundamentalist beliefs of the Appalachian people. Although Tolson often carved secular pieces, his strongest inspiration came from the Book of Genesis and from stories told by country preachers for generations.

Elijah Pierce (1892–1984) was born in Baldwyn, Mississippi, and worked for much of his life in Columbus, Ohio. An African American from extremely humble origins, Pierce was a renowned storyteller and folk artist who plied his craft between the haircuts he administered to clients in the barbershop where he worked. His carved and painted wood scene of the crucifixion (now in the Columbus Museum of Art) depicts the story of Christ as well as the temptations of evil. It is a powerful example of the religious impulse moving through the soul of an unschooled but gifted artist.

These are but a few of the many well-known religious folk artists. Somewhere in your area there may be undiscovered folk artists creating similar works, and it is likely that even if their work is largely secular, some of it expresses their deepest spiritual feelings.

SUMMARY

Religion is among the oldest and most changeable and complex of human social institutions. It has been defined as any set of coherent answers to the dilemmas of human existence that makes the world meaningful. It has also been defined as a system of beliefs and rituals that serves to bind people together into a social group. *Rituals* are formal patterns of activity that express a set of shared meanings; in the case of religious rituals, the shared meanings are *sacred,* pertaining to phenomena that are regarded as extraordinary, transcendent, and outside the everyday course of events.

Until comparatively recent times religion dominated the cultural life of human societies. Since medieval times, however, the traditional dominance of religion over other institutions has been reduced by a process termed *secularization.* This process is never complete; religion continues to play an important role in the contemporary world.

In simpler and rather isolated societies, people may believe in a great force or spirit, but they do not have a well-defined concept of God or a set of rituals involving God. This form of religion is called *simple supernaturalism.* More common among hunting-and-gathering societies is *animism,* in which all forms of life and all aspects of the earth are inhabited by gods or supernatural powers. *Theism,* in contrast, comprises belief systems that conceive of a god or gods as separate from humans and from other living things on the earth. Many ancient religions were *polytheistic,* meaning that they included numerous gods. The ancient Hebrews were among the first of the world's peoples to evolve a *monotheistic* religion, one centered on belief in a single all-powerful God.

In China, Japan, and other societies of the Far East, religions predominate that are centered not on devotion to a god or gods but on an abstract ideal of spirituality and human behavior. In addition, some social scientists have expanded the definition of religion to include *civil religions,* or collections of beliefs and rituals that exist outside religious institutions.

A major controversy in the study of religious institutions has to do with the role they play in social change. Karl Marx believed that cultural institutions like religion are shaped by economic and political institutions, and they function to instill in the masses the values of the dominant class. Max Weber, on the other hand, argued that religion can cause major social change by instilling certain values in the members of a society, in turn causing changes in other institutions.

Religion today is a highly structured institution, with numerous statuses and roles within a variety of organizations as well as many kinds of smaller, less bureaucratic groups. The three main types of religious organizations are the church, the sect, and the denomination. A *church* is a religious organization that has strong ties to the larger society and has at one time or another enjoyed the loyalty of most of the society's members. A *sect* rejects the religious beliefs or practices of an established church and usually is formed when a group of church members splits off to form a rival organization. A third type of religious organization is the *denomination,* which is on good terms with the religious institution from which it developed but must compete with other denominations for members.

A *cult* is an entirely new religion. Along with sects, cults are a major source of change in religious organizations. People who are not satisfied with more established churches and denominations may form or join a cult or sect. New religious movements arise when a "religious innovator" attracts a number of followers. This is particularly likely when traditional religions fail to meet the needs of their members or when a society is undergoing rapid secularization.

Membership in a religious organization is quite different from identification with a religious faith, and this is reflected in trends in religion in the United States. Among those trends are a growing tendency to practice unofficial or "folk" religion and an emphasis on religiosity as opposed to church membership. *Religiosity* refers to the depth of a person's religious feelings and how those feelings are translated into religious behavior. Studies of religiosity find high percentages of Americans believing in the existence of God and in a life after death.

In societies characterized by religious pluralism, one usually can observe the continual formation of new religious organizations.

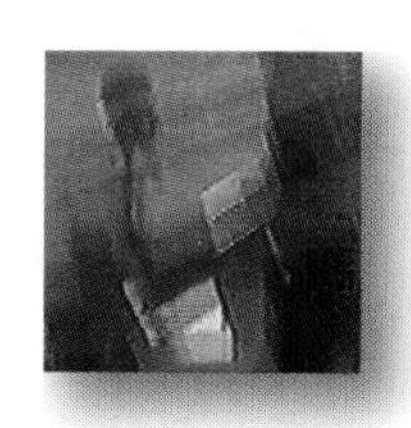

SOCIOLOGY VERSUS IDEOLOGY

At a certain point on the continuum of religious belief, sociology comes in for some harsh criticism. Sociology and the other social sciences, along with all science and much philosophy, are considered branches of "secular humanism." After all, sociology is a product of the enlightenment and its attack on orthodoxy. It contributes to the worldview of people who tend to doubt authority and question received wisdom of all kinds. From the point of view of most religious fundamentalists, as well as many religious conservatives, sociology is a product of a materialistic, individualistic, overly rational society. Thus even sociologists of religion like Nancy Ammerman, whose work seeks to help others understand the thoughts and feelings of Christian fundamentalists, eventually find themselves cast out of the fold of true believers because of their nonscriptural interpretation of facts.

The intellectual and spiritual disapproval that sociologists encounter, however, should not distract them from empirical research and writing. I hope you will find evidence in this chapter that sociologists seek to appreciate and learn more about the religious feelings and behavior of people in different cultures. If they are not religious themselves (although many sociologists of religion are in fact quite religious), they are at least dedicated to helping more secular people understand what religious people experience and, therefore, what more secular people may be missing in life.

GLOSSARY

religion: any set of coherent answers to the dilemmas of human existence that makes the world meaningful; a system of beliefs and rituals that serves to bind people together into a social group. (p. 534)

ritual: a formal pattern of activity that expresses symbolically a set of shared meanings. (p. 534)

sacred: a term used to describe phenomena that are regarded as extraordinary, transcendent, and outside the everyday course of events. (p. 534)

profane: a term used to describe phenomena that are not considered sacred. (p. 534)

secularization: a process in which the dominance of religion over other institutions is reduced. (p. 536)

simple supernaturalism: a form of religion in which people may believe in a great force or spirit but do not have a well-defined concept of God or a set of rituals involving God. (p. 538)

animism: a form of religion in which all forms of life and all aspects of the earth are inhabited by gods or supernatural powers. (p. 538)

theism: a belief system that conceives of a god or gods as separate from humans and from other living things on the earth. (p. 539)

polytheistic: a term used to describe a theistic belief system that includes numerous gods. (p. 539)

monotheistic: a term used to describe a theistic belief system centered on belief in a single all-powerful God. (p. 540)

civil religion: a collection of beliefs and rituals that exist outside religious institutions. (p. 541)

church: a religious organization that has strong ties to the larger society. (p. 543)

sect: a religious organization that rejects the beliefs and practices of existing churches; usually formed when a group leaves the church to form a rival organization. (p. 543)

denomination: a religious organization that is on good terms with the institution from which it developed but must compete with other denominations for members. (p. 543)

cult: a new religion. (p. 543)

religiosity: the depth of a person's religious feelings. (p. 551)

WHERE TO FIND IT

BOOKS

One God: Peoples of the Book (Edith S. Engel and Henry W. Engel, eds.; Pilgrim Press, 1990). An introduction to the major monotheistic religions with a message of peace, commonality, and openness.

"The Sociology of Religion" (Robert Wuthnow; in Neil Smelser, ed., *Handbook of Sociology;* Sage, 1988). An extremely useful review of current trends in the sociology of religion by one of the field's leading scholars. Includes an excellent bibliography of classic and recent sociological research on religion.

Base Communities and Social Change in Brazil (W. E. Hewitt; University of Nebraska Press, 1991). A fine case study of the influential Catholic ecclesiastical base communities movement in Latin America, based primarily on research in São Paulo, Brazil. A good example of empirical research on religion and social change at the community level.

Gods in the Global Village (Lester Kurtz; Pine Forge Press, 1995). A good overview of religion and trends in religious organization from a global perspective.

Religion and the American Experience (Donald C. Swift; Sharpe, 1998). A sociologically informed history of religious movements and church organization in the United States from 1765 to 1997.

Religion: The Social Context, 3rd ed. (Meredith B. McGuire; Wadsworth, 1992). A comprehensive text on the sociology of religion with excellent material on religion and social change.

JOURNALS

Journal for the Scientific Study of Religion. Available in most college and university libraries; publishes recent research on religious practices, religiosity, and changes in religious institutions.

OTHER SOURCES

The Encyclopedia of American Religions, 4th ed. (Gale Research). Contains useful descriptions of religions in America; covers beliefs, organization, distribution in the population, and other aspects.

INTERNET RESOURCES

Boston College Center for International Higher Education (www.bc.edu). Provides links to research centers and international news and library information dealing with the Catholic religion throughout the world.

The Anti-Defamation League (www.adl.org). Monitors hate activities in the United States and throughout the world; includes addresses of other useful research sites.

Ontario Consultants on Religious Tolerance (www.religioustolerance.org). Explores many religions and states; does not promote or denounce a specific belief. The site does list religious beliefs and news topics among more than 20 denominations.

Andrew Greeley: Author, Priest, Sociologist (www.agreeley.com). Offers insights into Andrew Greeley's recent writings. The site includes previews of his recently published works and articles addressing recent issues in Catholicism and other religious groups.

CHAPTER 18

EDUCATION AND COMMUNICATIONS MEDIA

RAISE THE STANDARDS. CREATE A LEVEL PLAYING FIELD. END SOCIAL PROMOTION. USE VOUCHERS FOR SCHOOL CHOICE. These are only a few of the most popular slogans people use in debates over education. There is no end to the controversies over what should and should not be accomplished in the nation's schools and colleges. After all, education is one of the few institutions on which we can pin our hopes for equality of opportunity. From preschool to college, educational institutions are part of the fabric of our lives. They are central institutions in our communities. But for too many children and adolescents they fail to live up to expectations. Faced with ever greater demands for achievement, school systems that are experiencing high levels of failure and poor performance are under immense pressure. And often these same school systems receive disproportionately lower shares of public support per student. Fortunately, it is in cities with overburdened schools that one often finds educational pioneers like Dr. James Comer.

The activism, theories, and writing of the Yale University psychiatrist and educational philosopher guide current school reform efforts in urban America. School districts in Prince George's County (Maryland), Chicago, Miami, New Haven, Dallas, and many other areas are undertaking major efforts at educational reform. Nowhere are their goals quite as ambitious as those of the Prince George's County school district. The Comer-inspired reform efforts involve schoolchildren in collaborative planning with their parents, teachers, and principals. "What I've tried to do here . . . is to shape the system, the school, so that it becomes the advocate and support for the kid, and a believer in the kid in the same way that my parents were" (quoted in Schorr, 1988, p. 232).

Comer traces his educational innovations to his experiences as an African American child in East Chicago, Indiana: "My three friends with whom I started elementary school—one died at an early age from alcoholism; one spent most of his life in jail; and one has been in and out of mental institutions all of his life. I was the only one to survive whole." Comer's father, a steelworker, and his mother, a domestic, played an active role in the education of their sons: "[They] came to school if there was a problem, and knew how to make sure that people were sensitive and concerned about us." Their concern and their active involvement were essential to Comer's success in school at every level through medical training. His realization that too many lives and minds were being lost through school failure was a major factor in his decision to devote his career to research on schools and child development.

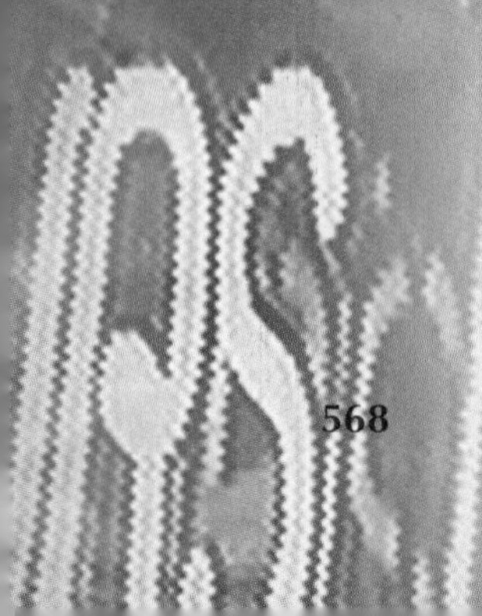

Comer's research and experience convinced him that children from neighborhoods that are undergoing severe strain — owing to poverty, racism, unemployment, violence, drug addiction, and related social problems — enter school with major deficits. They are often "underdeveloped in their social, emotional, linguistic, and cognitive growth." As a result, they often withdraw, act up, and do not learn. Of course, not all such students have these problems. Those who do, however, are labeled "slow learners" or "behavior problems" and consequently fall even further behind their classmates. Comer's solution appears very simple on the surface: Change the "climate of demoralized schools by paying much more attention to child development and to basic management of the school" (quoted in Schorr, 1988, p. 233).

Of course, the process of changing school climates and management is much more difficult than this simple formula suggests. The Comer model of institutional change calls for the formation of a School Planning and Management Team in each school; the team is directed by the principal and has from 12 to 14 members, including teachers, aides, and parents. A second team is made up of helping professionals — the school psychologist, social workers, special education teachers, and counselors; it acts as an advisory group. The purpose of the teams is to ally the parents with the school so that, in Comer's words, "you reduce the dissonance between home and school and you give the kid a long-term supporter for education at home" (quoted in Schorr, 1988, p. 233; see also Comer, 1997; O'Neil, 1997).

EDUCATION FOR A CHANGING WORLD

Efforts to change the goals and methods of education are not new. One of the most famous trials in history took place in ancient Greece when Socrates was accused of corrupting the morals of Athenian youth with his innovative ideas and educational methods. Today, the fact that educational reforms like those proposed by James Comer are receiving so much hopeful attention is evidence of the importance of education in modern societies. There is great concern with the need to improve education in order to train new generations of workers. We will see at many points in this chapter that there is a widespread desire to reform schools without significantly increasing school funding. However, although critics of the public schools abound, there is little consensus about how to achieve better student performance (Torres & Mitchell, 1998).

Education may be defined as the process by which a society transmits knowledge, values, norms, and ideologies and in so doing prepares young people for adult roles and adults for new roles; in other words, it transmits the society's culture to the next generation. Education thus is a form of socialization that is carried out by institutions outside the family, such as schools, colleges, preschools, and adult education centers. Each of these is an educational institution because it encompasses a set of statuses and roles designed to carry out specific educational functions—it is devoted to transmitting a specific body of knowledge, values, and norms of behavior. A particular school or college may be referred to as an *institution* in everyday language, but in sociological terms it is an *organization* that exemplifies an educational institution. Thus, El Centro College in Dallas is an organization that exemplifies the institution of higher education. The high school you attended was also an organization, but its curriculum and norms of conduct were those of the institution known as secondary education.

Educational institutions have a huge effect on communities in the United States and other modern societies. Upwardly mobile couples often base their choice of a place to live on the quality of the public schools in the neighborhood. Every neighborhood has at least an elementary school, and every large city has one or more high schools and at least one community college or four-year college. Cities usually also have a variety of school administrations—public and private—and some owe their existence, growth, and development to the presence of a college or university (Ballantine, 1997).

The Nature of Schools

Educational institutions affect not only the surroundings but also the daily lives of millions of Americans: children and their parents, college and university students, teachers and professors. Hence, education is a major focus of social-scientific research. To the sociologist, the most common educational institution, the school, is a specialized structure with a special function: preparing children for active participation in adult activities. Schools are sometimes compared with total institutions (see Chapter 5), in which a large group of involuntary "clients" is serviced by a smaller group of staff members. The staffs of such institutions tend to emphasize the maintenance of order and control, and this often leads to the development of elaborate sets of rules and monitoring systems. This comparison cannot be taken too literally (schools are not prisons, although some of the "inmates" may think of them as such), but the typical school does tend to be characterized by a clearly defined authority system and set of rules. In fact, sociologists often cite schools as examples of bureaucratic organizations (Ballantine, 1997; Mulkey, 1993).

A more interactionist viewpoint sees the school as a set of behaviors; that is, the central feature of the school is not its bureaucratic structure but the kinds of interactions and patterns of socialization that occur there. From this perspective, we need to examine all the roles involved and see how they interact—how they mesh or fail to mesh. For example, homework is assigned, but who besides the student is involved in completing the assignment? Are the parents involved? In

This photo of Tibetan refugee children in a classroom in northen India illustrates the similarity of classroom organization in many parts of the world.

what ways—as helpers or merely as disciplinarians? Do the parents ever ask what the teacher did with a homework assignment? The interactionist perspective thus insists on examining all the factors involved in learning, often in an effort to determine how they may be strengthened or challenged so as to create more effective learning environments (Torres & Mitchell, 1998).

Conflict and the Reproduction of Inequality Conflict theorists, by contrast, view education in modern societies as serving to justify and maintain the status quo (Aronowitz & Giroux, 1985; Castells et al., 1998). Two contemporary studies of the school experiences of students from lower social classes offer strong support for "reproduction theory," the idea that the educational experiences of children of a given social class serve to keep them in that class once their schooling is completed. Paul Willis's research in an industrial community in England draws on the experience of a group of boys from low-income homes who thought of themselves as "the lads." They delighted in making fun of higher achievers in their school, whom they called "the earholes." In many different ways, however, the lads indicated that they believed their teachers were pushing them into low-prestige futures. Willis interpreted the hostility and alienation of these students as a way of resisting the social forces they encountered in school and in their communities that were channeling them into working-class careers (Willis, 1983).

In a study that confirms some aspects of Willis's research and challenges others, Jay MacLeod (1995) spent a great deal of time observing and interacting with two groups of students from working-class or poor backgrounds in an American high school in a poor community. MacLeod's white subjects, who called themselves the "Hallway Hangers," were very much like "the lads." In school they were angry and alienated. They believed that there was no point in raising their educational aspirations above those of their parents and older siblings. Their school experiences did nothing to change their resigned outlook. In contrast, a black peer group, "the Brothers," had taken on the higher aspirations of an African American version of the American dream of hard work and eventual success. Although objectively they were not doing much better in school than the Hallway Hangers, in MacLeod's view their greater optimism and resolve calls into question the apparent inevitability of social-class reproduction in the public schools. While MacLeod does not deny the existence of educational forces that tend to reproduce inequalities from one generation to another, he believes that there is always the possibility of "agency"—that is, determined efforts by individuals and groups to challenge the status quo of inequality, even if this challenge often produces very mixed results (MacLeod, 1995).

As we will see shortly, this critical perspective challenges the more popular view that education is the main route to social mobility and that it can offset inequalities in family background (Bell, 1973). When sociologists analyze the impact of educational institutions on society, they generally conclude that the benefits of education are unequally distributed and tend to reproduce the existing stratification system (Fullan, 1993; Jencks et al., 1972).

Who Goes to School?

The idea that all children should be educated is a product of the American and French revolutions of the late eighteenth century. In the European monarchies the suggestion that the children of peasants and workers should be educated would have been considered laughable. In those societies children went to work with adults at an early age, and adolescence was not recognized as a distinct stage of development. Formal schooling, generally reserved for the children of the elite, typically lasted 3 or 4 years, after which the young person entered a profession.

Even after the creation of republics in France and the United States and the beginning of a movement for universal education, the development of a comprehensive system of schools took many generations. In the early history of the United States, the children of slaves, Native Americans, the poor, and many immigrant groups, as well as almost all female children, were excluded from educational institutions. The norm of segregated education for racial minorities persisted into the twentieth century and was not overturned until 1954 in the Supreme Court's famous ruling in *Brown* v. *Board of Education of Topeka.* Even after that decision it took years of civil rights activism to ensure that African Americans could attend public schools with whites. Thus, although the idea of universal education in a democracy arose early, it took many generations of conflict and struggle to transform it into a strong social norm (Cremin, 1980; Krueger, 1998).

The idea of mass education based on the model created in the United States and other Western nations has spread throughout the world. Mass education differs from elite education, which is designed to prepare a small number of privileged individuals (generally sons of upper-class families) to run the institutions of society (the military, the clergy, the law, etc.). Mass education focuses instead on the socialization of all young people for membership in the society. It is seen as a way for young people to become citizens of a modern nation-state. Mass education establishes an increasingly standardized curriculum and tries to link mastery of that curriculum with personal and national development (Benavot et al., 1991; Meyer, Ramirez, & Soysal, 1992).

Figure 18.1 indicates that by 1980 all the nations of the world had adopted the basic model of a mass educational system (although many had not extended educational opportunities to the majority of young citizens). Note the sharp upsurge in the 1950s. This was the decade when large numbers of colonial nations in Africa and Asia became independent. Commitment to a mass educational system became a hallmark of modernity for these and other new nations. It was also a time when agencies like the World Bank, UNESCO (United Nations Educational, Social, and Cultural Organization), and the U.S. Agency for International Development began to actively encourage and provide financial support for mass educational institutions.

An extension of mass education in the United States was the expansion of public institutions of post-secondary education (and the parallel expansion of private colleges and universities, often with federal aid and research funding). This expansion was fueled by the post–World War II baby boom and the massive increase in the college-age population in the 1960s. During those years there was a parallel boom in employment for elementary school teachers and then for high school teachers and college professors. But after the boom came the bust: The birthrate fell sharply beginning in the late 1950s, and by the mid-1970s college enrollments also began to decline. These changes had dramatic effects on primary schools, but at the college level they were partially offset by an unprecedented countertrend: the immense increase in the number of older students seeking higher education. Today large numbers of adults are returning to college. Unlike the typical student of earlier years, they are in the labor force, are married and living with their spouses, are going to school part-time, and are seeking skills and knowledge to enhance their careers (Mulkey, 1993; Torres & Mitchell, 1998).

The return of so many adults to educational institutions has led many social scientists to describe a future in which education will be a lifelong process in which people of all ages will move in and out of educational institutions. Nevertheless, the most rapidly growing area of education is preschool programs, an important trend that is discussed later in the chapter.

Schools and Adolescent Society

A key feature of education is the fact that schools structure the lives of children and adolescents. This is particularly true at the high school level. A famous study of

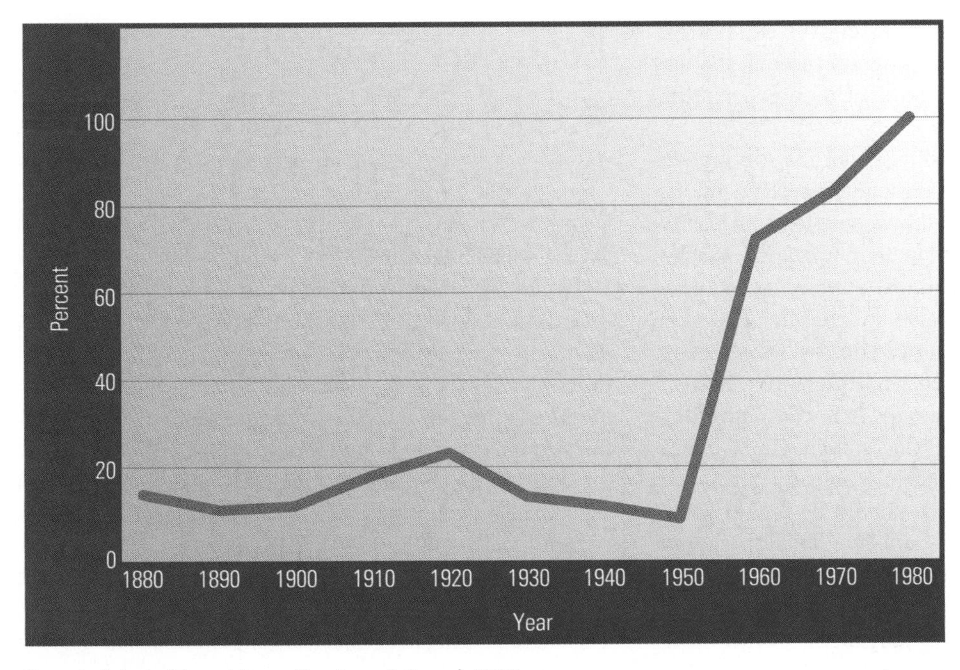

Source: Adapted from Meyer, Ramirez, & Soysal, 1992.

FIGURE 18.1 **Percentage of Nations Developing Systems of Mass Education, by Decade**

the effect of schools on adolescents and youth, *The Adolescent Society,* was published by James Coleman in 1961. Its main point was that schools help create a social world for adolescents that is separate from adult society. According to Coleman, this is an almost inevitable result of the growing complexity of industrial societies, in which, as we saw earlier, functions that were formerly performed by the family are increasingly shifted to other institutions, especially educational institutions. Yet schools cannot provide the same kind of support and individual attention that the family can. As a result, according to Coleman, the student is pushed into associations primarily with others of the same age. These small groups of age-mates come to "constitute a small society, one that has most of its important interaction within itself, and maintains only a few threads of connection with the outside adult society" (1961, p. 3).

Coleman's research was not unique. Other social scientists have analyzed what has come to be known as the "youth culture" in terms of changes in the structure of American society. Briefly, what has happened is that the rising level of expectations regarding educational attainment has placed more and more demands on the student. Often this pressure is not matched with clear means of meeting the new demands. This mismatch between ends and means can lead to the emergence of deviant individuals and groups, as we saw in our discussion of Merton's typology of deviance in Chapter 7. Often the resulting frustration leads teenage peer groups to become extremely ambivalent and even cynical about adult expectations. They develop fierce loyalty and conformity to the peer group, strictly observing group norms and not tolerating any deviance from those norms, which themselves may deviate from those of their teachers and parents.

The risk for society in the development of isolated adolescent cultures is that teenagers will fail to learn how they can become involved in shaping their own society. Great cultural gaps between youth and their parents are often a signal that the older generation has not offered young people enough opportunities to work with adults for constructive social change (Gilligan, cited in Norman, 1997; Moscos, 1988).

Education and Citizenship

Another important aspect of education in the United States is the relationship between education and citizenship. Throughout its history this nation has emphasized public education as a means of transmitting democratic values, creating equality of opportunity, and preparing new generations of citizens to function in society. In addition, the schools have been expected to help shape society itself. During the 1950s, for example, efforts to combat racial segregation focused on the schools. Later, when the Soviet Union launched the first orbiting satellite, American schools and colleges came under intense pressure and were offered many incentives to improve their science and mathematics programs so that the nation would not fall behind the Soviet Union in scientific and technological capabilities.

Education is often viewed as a tool for solving social problems, especially social inequality. The schools, it is thought, can transform young people from vastly different backgrounds into competent, upwardly mobile adults. Yet these goals seem almost impossible to attain (Cahill, 1992). In recent years, in fact, public education has been at the center of numerous controversies arising from the gap between the ideal and the reality. Part of the problem is that different groups in society have different expectations. Some feel that students need better preparation for careers in a technologically advanced society; others believe children should be taught basic job-related skills; still others believe education should not only prepare children to compete in society but also help them maintain their cultural identity (and, in the case of Hispanic children, their language). On the other hand, policy makers concerned with education emphasize the need to increase the level of student achievement and to involve parents in their children's education (Wilson, 1993).

Some reformers and critics have called attention to the need to link formal schooling with programs designed to address social problems. Sociologist Charles Moscos, for example, is a leader in the movement to expand programs like the Peace Corps, VISTA, and Outward Bound into a system of voluntary national service. National service, as Moscos defines it, would entail "the full-time undertaking of public duties by young people—whether as citizen soldiers or civilian servers—who are paid subsistence wages" and serve for at least a year (1988, p. 1). In return for this period of service, the volunteers would receive assistance in paying for college or other educational expenses.

Advocates of national service and school-to-work programs believe that education does not have to be confined to formal schooling. In devising strategies to provide opportunities for young people to serve their society, they emphasize the educational value of citizenship experiences gained outside the classroom. Early in his administration President Bill Clinton, a believer in national service, implemented a modest program known as Americorps. Americorps is a volunteer community service program targeted at young people who normally would not apply to existing volunteer programs like the Peace Corps. However, social programs like Americorps and Teach America, which recruits recent college graduates to teach in disadvantaged communities, are continually criticized by proponents of fiscal conservatism and privatization of public schools—even when participants in the programs are unpaid.

ATTAINMENT, ACHIEVEMENT, AND EQUALITY

While some educational reformers focus on the need to expand learning opportunities through nonschool service experiences, by far the majority of scholars and administrators seek improvements in educational institutions themselves. Their efforts often focus on *educational attainment,* or the number of years of schooling that students receive, and *educational achievement,* or the amount of learning that actually takes place. Both aspects of education are closely linked to economic inequality and social mobility (see Chapter 12).

Educational Attainment

In any discussion of education as a major social institution, the concept of **educational attainment** (number of years of school completed) holds a central place. Educational attainment is correlated with income, occupation, prestige, attitudes and opinions, and much else. It is essential, therefore, for social scientists to understand the impact of recent trends in school enrollment on the educational attainment of the population as a whole and of various subgroups of the population.

Table 18.1 shows the median number of years of school completed by the population as a whole in the decades since 1940, and Table 18.2 presents data on educational attainment by race, ethnicity, and sex. Both tables apply to the population aged 25 and older. It is immediately clear that the average American today has much more education than the average American of the early 1940s. It is also clear that whites are more likely to complete high school than blacks or Hispanics, and considerably more likely to attend college. The data

TABLE 18.1 Median Years of School Completed

Year	Median Years of School Completed
1990	12.7
1980	12.5
1970	12.2
1960	10.6
1950	9.3
1940	8.6

Source: Data from Census Bureau.

TABLE 18.2 Educational Attainment, by Race, Ethnicity, and Sex: 1947–1997

	Percent High School Graduate						Percent College Graduate					
	White		Black		Hispanic		White		Black		Hispanic	
Year	M	F	M	F	M	F	M	F	M	F	M	F
1997	83%	83%	74%	76%	55%	55%	27%	22%	13%	14%	11%	10%
1991	80	80	67	67	51	51	25	19	11	12	10	9
1987	77	77	63	64	52	50	25	17	11	10	10	8
1980	71	70	51	51	46	44	22	14	8	8	10	6
1970	57	58	32	35	NA	NA	15	9	5	4	NA	NA
1962	47	50	23	26	NA	NA	12	7	4	4	NA	NA
1947	33	37	13	15	NA	NA	7	5	2	3	NA	NA

NA=Data not available.

Sources: Census Bureau, 1988, 1992; *Statistical Abstract,* 1998.

indicate the increasingly high value placed on education, especially college education. People over age 75, who were born in the early decades of the twentieth century, have, on the average, considerably lower levels of educational attainment than their children and grandchildren (Farley, 1996; *Statistical Abstract,* 1998).

Barriers to Educational Attainment

Tracking and Inequality The rise of mass education gave the middle and lower classes greater opportunities for upward social mobility through educational attainment. But as early as the 1920s many schools began to use "tracking" systems in which higher-achieving students were placed in accelerated classes while others were shunted into vocational and other types of less challenging classes. Today tracking remains a major problem in public schools. Parents of "gifted" children seek educational challenges for their sons and daughters and do not want them to be held back by slower learners. However, tracking systems can make average students feel less valued, and there is a danger that gifted but alienated students will be labeled as nonachievers.

Figure 18.2 is based on a national study of more than 14,000 eighth grade students in public schools. It shows quite clearly that white and Asian students are far more likely than black and Hispanic students to be tracked into high-ability groups, and that students from high socioeconomic backgrounds are also far more likely to be tracked into such groups. (Ability grouping is synonymous with tracking.) Some social scientists argue that these differences are a result of ability differences among racial and class groups; others assert that they are a consequence of race and class bias (Wolfe, 1998). Since there are far more Hispanic and black students in the lower socioeconomic classes, there is little question that social factors outweigh biological ones in explaining these results. In any case, tracking separates children and is increasingly viewed as leading to educational inequalities. A current trend in educational practice, therefore, is to institute detracking programs and to provide highly gifted students with additional challenges through cooperative education (in which they have opportunities to teach others) and through after-school programs (Betts, 1998; Mansnerus, 1992).

Dropping Out The educational attainment of various subgroups of the population is virtually identical up to the age of 13 (Farley, 1996). The picture begins to change in the early secondary school years, however. Until quite recently the dropout rate for black and Hispanic students in high school was almost twice the rate for white students. In 1970, for instance, 10.8 percent of white students dropped out without finishing high school, while among black students the proportion was 22.2 percent. By 1996, the situation had improved remarkably as a result of strenuous efforts by students, parents, and teachers across the United States. In that year 9.2 percent of white students and 11.0 percent of black students dropped out (*Statistical Abstract,* 1998). Table 18.3 on page 577 shows that high school graduation rates (the opposite of dropout rates) vary among the states, with the highest in Alaska at 92.1 percent and the lowest in Kentucky at 75.4 percent.

The main reason for dropping out of school is poor academic performance, but there are other reasons as well. Students often drop out because of the demands of work and family roles; many are married, or unmarried and pregnant, and/or working at regular jobs. Whatever

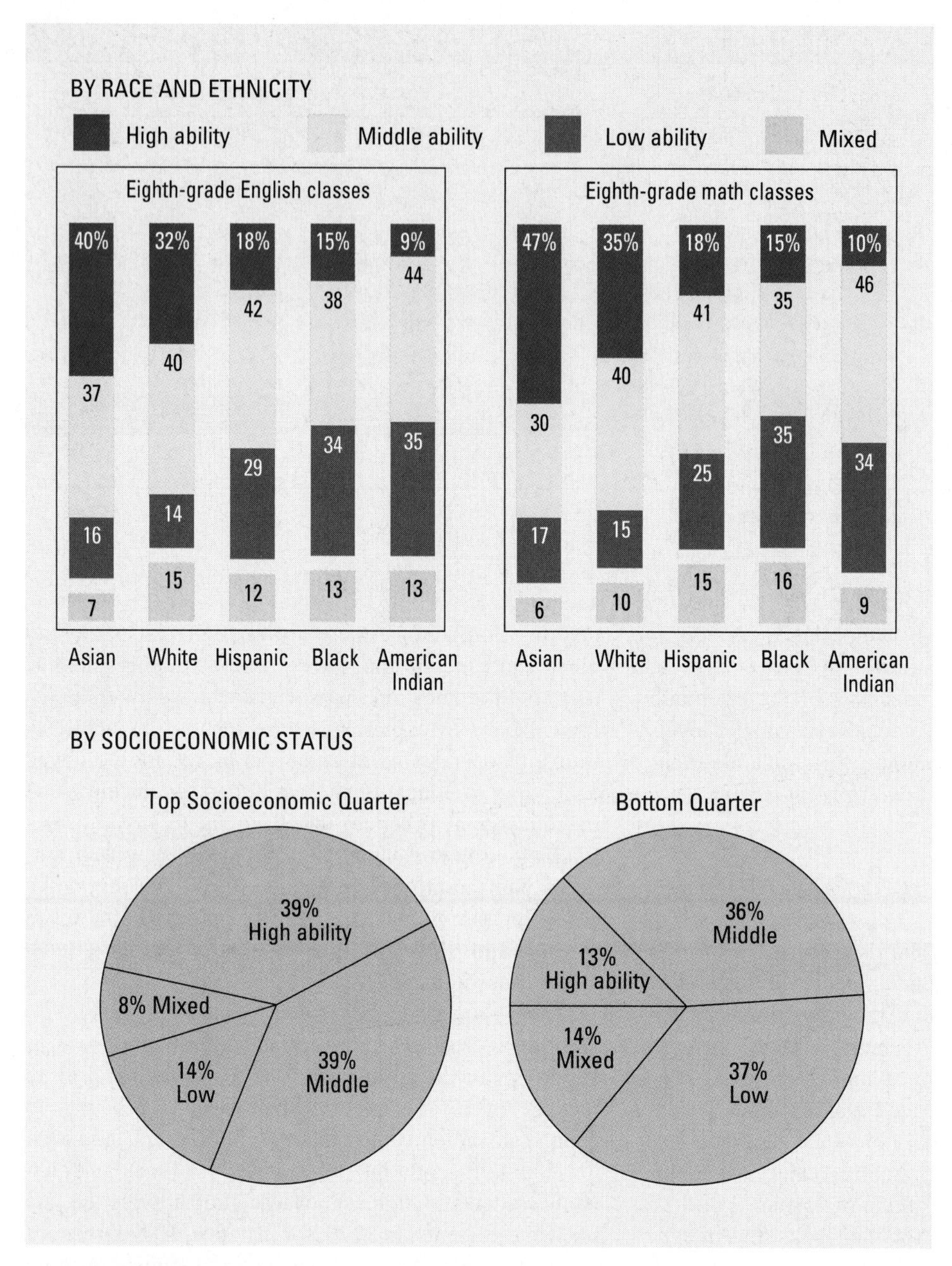

Source: Adapted from Mansnerus, 1992.

FIGURE 18.2 **Ability Grouping ("Tracking") in U.S. Public Schools**

the reason, the effects of dropping out can be serious. Dropouts have less chance of joining the labor force than high school graduates; whatever jobs they find tend to be low-paying ones. From 1974 to 1992 the income advantage from completing college increased for all individuals, regardless of gender or minority status. But black–white differentials are found at each level of attainment. White college graduates, for example, earned at least 20 percent more than black college graduates of the same age (National Center for Educational Statistics, 1995).

Degree Inflation The trend toward increasingly higher levels of educational attainment has had an unexpected effect known as "degree inflation" (Pedersen, 1997). Employers have always paid attention to the educational credentials of potential employees, but today they require much more education than in the past. For example, in the early decades of the twentieth century a person could get a teaching job with just a high school diploma; now a bachelor's or master's degree is usually required. The same is true of social work. And secretaries, who formerly could get by without a high school

TABLE 18.3 High School Graduation Rates, by State (persons age 25 and over)

State	Percent
Alabama	77.6
Alaska	92.1
Arizona	82.6
Arkansas	76.9
California	80.7
Colorado	87.6
Connecticut	84.0
Delaware	84.4
District of Columbia	80.3
Florida	81.4
Georgia	78.8
Hawaii	83.7
Idaho	85.7
Illinois	84.4
Indiana	81.9
Iowa	86.7
Kansas	88.1
Kentucky	75.4
Louisiana	75.7
Maine	85.8
Maryland	84.7
Massachusetts	85.9
Michigan	86.0
Minnesota	87.9
Mississippi	77.5
Missouri	80.1
Montana	88.6
Nebraska	86.0
Nevada	85.4
New Hampshire	85.1
New Jersey	84.8
New Mexico	78.0
New York	80.0
North Carolina	78.4
North Dakota	82.6
Ohio	86.2
Oklahoma	85.2
Oregon	84.7
Pennsylvania	82.4
Rhode Island	77.5
South Carolina	77.3
South Dakota	85.6
Tennessee	76.1
Texas	78.5
Utah	89.5
Vermont	84.4
Virginia	81.3
Washington	88.8
West Virginia	77.3
Wisconsin	87.1
Wyoming	91.3

Source: Adapted from *Statistical Abstract,* 1998.

diploma, now are often required to have at least some college education or, in some cases, a college degree.

Degree inflation is discouraging to some students and prevents them from continuing their education. It also adds to the expense of education (both directly and indirectly in terms of lost income), and hence prevents less advantaged students from undertaking advanced studies. Degree inflation also increases the amount of time that must be devoted to formal education. It therefore raises questions about the meaning of educational achievement—that is, the value of the time spent attaining educational credentials.

Educational Achievement

Differences in levels of educational attainment are viewed as a sign that public education is not meeting the expectations of society in terms of the quantity of education provided to citizens. There is also controversy over the quality of education, or educational achievement as reflected in scores on standardized tests like the Scholastic Aptitude Test (SAT). **Educational achievement** refers to how much the student actually learns, measured by mastery of reading, writing, and mathematical skills. It is widely believed that the average level of educational achievement has declined drastically in the past two decades. In a historic 1983 report titled *A Nation at Risk,* the National Commission on Excellence in Education pointed to the decline in the average test scores of high school students since the mid-1960s and stated that the schools have failed to maintain high educational standards. In recent years standardized scores in math have risen somewhat, but verbal scores (reading and vocabulary) have not (Jencks & Phillips, 1999). Many observers attribute the decline in test scores to a variety of social conditions such as too much television viewing and changing values related to family life.

Critics of excessive reliance on standard tests might agree with prominent educational reformer Ted Sizer, who points out that "none of the major tests used in American elementary and secondary education correlates well with long-term success or failure. SAT scores, for example, suggest likely grades in the freshman year at college; they do not predict much thereafter" (1995, p. 58). For better or worse, however, the SAT and other standard tests continue to be used as primary measures of individual and school performance.

Another problem related to educational achievement is the high rate of "functional incompetency" among Americans. A large number of people are unable to read, write, keep a family budget, and the like. Although more and more people are obtaining a college education, many others are being left behind—particularly members of the lower social classes, people for whom English is a second language, and people with learning disabilities. As increasing amounts of education are required for better jobs, this cleavage between educational haves and have-nots becomes an ever more dangerous trend. (See Box 18.1.)

Recent research on school achievement offers even more disturbing evidence of deficiencies in the educational achievement of U.S. students, beginning at an early age. Tests of achievement by thousands of schoolchildren in comparable U.S. and Asian cities—Minneapolis, Chicago, Sendai (Japan), Beijing (China), and Taipei (Taiwan)—found that the mathematics scores of American first-graders were lower than those of Asian first-graders. In some American schools first-graders' scores were similar to those of Asian students, but by the fifth grade all the American students had fallen far behind. In computation, for example, only 2.2 percent of children in Beijing scored as low as the mean score for U.S. fifth-graders. On a test consisting of word problems, only 10 percent of fifth-graders in Beijing scored as low as the average U.S. student. Gaps also appeared in reading skills, although vast differences in written languages make it somewhat difficult to compare achievement in those skills (H. W. Stevenson, 1992, 1998).

Americans are likely to suggest that these differences occur because citizens of Asian nations value education more than Americans do and that in Asian schools children are placed under great stress, learn by rote and drill, and spend long hours in school. The researchers found that only the first of these explanations is correct. It is true that Asian parents value education more than American parents do, but it is not true that Asian schoolchildren experience more stress, do more rote learning, and spend more time at their desks. In fact, the opposite is true: Asian children experience less stress and anxiety than American children, enjoy school more, and have far more opportunities for recreation and breaks from classroom work despite their longer school day. The frequency of recreation breaks appears to help them pay more attention to their classroom work than is typical of children in U.S. schools.

Close observation of teaching methods revealed that Asian teachers tend to use work that the students do at their desks (seatwork) as an opportunity to pay attention to individual students. Teachers in U.S. schools are less likely to circulate through the class and comment on individual students' efforts.

Finally, Figure 18.3 on page 582 shows important differences in the way parents perceive schooling in Asia and the United States. Mothers in the United States tend to be far more positive than Asian mothers about their children's schoolwork at all grade levels. They are also more likely to believe that the school is doing an excellent job in educating their children. American parents tend to believe far more in the effects of innate ability and less in the value of effort in school than Asian parents do. This difference is especially important because it seems to lead U.S. parents to more readily accept their children's performance in school and to demand less of them. Parents' belief in innate ability also appears to make them far less critical of the schools than they might otherwise be (H. W. Stevenson, 1998).

This research has many practical implications for social change in the schools. It is apparent, for example, that simply lengthening the school day in the United States would not have the desired results. Indeed, it might well have negative results without far greater provision for time when children can play and socialize with their peers as they do in Asian schools.

Education and Social Mobility

Studies by educational sociologists have consistently found a high correlation between social class and educational attainment and achievement. As we see in Table 18.4 on page 583, over the past 20 years there has been a marked erosion of the income of non–college graduates.[1] Only college graduates have experienced significant increases in earnings. This fact indicates that social stratification in the United States is increasing, rather than declining as a result of greater equality of opportunity.

Changes in the nature of work and global competition at all skill levels place ever greater pressure on educational institutions to produce better results. Nevertheless, there is evidence that the promise of equal education for all remains far from being fulfilled. Educational institutions have been subjected to considerable criticism by observers who believe that they hinder, rather than enhance, social mobility (Fullan, 1993).

As noted earlier, critical sociologists often argue that the schools actually reproduce existing patterns of stratification and inequality rather than promote social mobility. The point here is that working-class and poor children do not automatically inherit their lower-class position through the failure of education to reach them or through their own failure to accept the value of education. The influence of specific family and school environments has been demonstrated numerous times in empirical research on how inequality is reinforced by school experiences, especially in schools in low-income neighborhoods (MacLeod, 1995; Torres & Mitchell, 1998).

Inequality in the Classroom The studies by Paul Willis and Jay MacLeod described earlier are based on close observation of students in inner-city communities

(continued on page 582)

[1] The table also reveals a significant gender gap in earnings. Inequalities of gender are discussed in Chapter 14.

BOX 18.1 USING THE SOCIOLOGICAL IMAGINATION

Poverty and Illiteracy

There is a strong correlation between illiteracy and poverty throughout the world (see map on next page). The poor nations of the Sahel, such as Mali, Chad, Niger, Ethiopia, and the Sudan, suffer from some of the highest rates of illiteracy in the world. High rates of illiteracy are also evident in much of South Asia, especially on the Indian subcontinent. On the other hand, the correlation is not absolute: Illiteracy rates are quite low in poor nations like Mexico and Cuba, where the ideology of social development places strong emphasis on educating the mass of citizens to the fullest extent possible.

Sociological theories of modernization stress the need for populations to become literate so that their members will be better informed voters, more highly skilled workers, more careful parents, and generally better able to realize their human potential. Reductions in the level of ignorance yield improvements in every aspect of a nation's social and civic life.

This line of argument is clearly supported by the correlation between poverty and illiteracy in India (see the accompanying chart). Overall, the chart shows that as illiteracy decreases, the amount of money a household is able to spend each month increases. The chart also shows that illiteracy and the poverty associated with it are much more prevalent among women than among men, with rural women showing the highest rates of illiteracy and poverty (about 90 percent illiterate in the high-poverty category). That men are far more likely than women to become literate in India is a reflection of the immense gap in prestige between the sexes in that society; men are considered far more worthy of education than women.

Research on the effects of literacy on vital measures of social change such as reduced fertility clearly show the importance of educating women. Data from research conducted in Thailand and other Southeast Asian nations suggest that until women gain access to at least minimal educational opportunities, fertility rates in those nations will remain high. This research confirms the hypothesis that "demographic change is unlikely if the movement towards mass schooling is confined largely to males" (London, 1992, p. 306).

These relationships among gender, illiteracy, and social indicators like poverty and fertility offer a warning that investments in literacy alone are necessary but not sufficient to accelerate social change in a population. Investments in literacy must be accompanied by strategies to reach the most impoverished and discriminated-against segments of the population (such as rural women in many societies). Such strategies, in turn, are difficult to develop in a society in which the powerful may fear their effects. In nations where poor rural women are offered more education, for example, the women often begin to demand greater equality. It takes farsighted leadership to actively promote such strategies.

Another consequence of increasing literacy and education in poor societies is that some of the "best and brightest" are attracted by opportunities in richer societies. This phenomenon is known as the "brain drain" and is a significant problem for developing nations (Alam, 1988; Cortes, 1980). Relatively well-educated people, often with professional training, are an important source of migration from Third World nations to the wealthier nations of North America and Western Europe. Indeed, immigration policies in North America actively encourage immigration by well-trained individuals and establish quotas that discourage less educated persons from seeking entry.

In order to address the problems created by the brain drain phenomenon, the poorer nations have developed policies requiring that people trained at public expense (e.g., through scholarships and training programs) must serve in their own nation until it has recouped its investment in their education. Such policies are not popular with educated people in those nations, especially if they are aware of more attractive opportunities outside their own society. But if the full benefits of literacy and education are to be applied to national development, such policies are necessary.

(continued)

Poverty and Illiteracy in India

Percentage of Illiterates Age 15 and Over
Female/rural
Female/urban
Male/rural
Male/urban
100
80
60
40
20
0
40–50
70–85
125–150
250–300
Household per Capita Expenditure
(rupees a month)

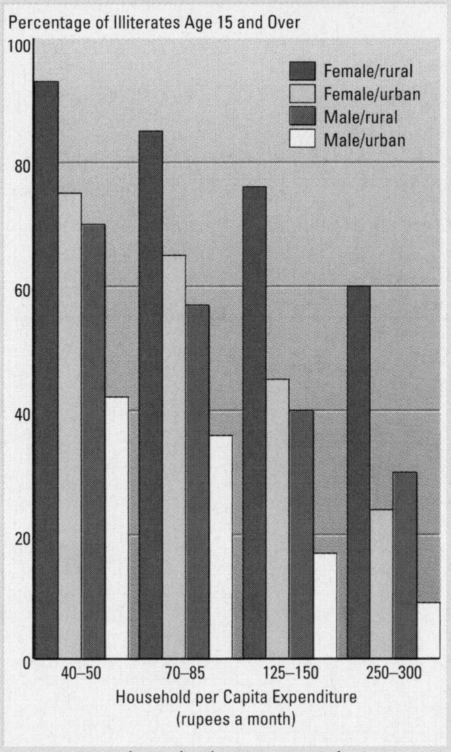

Source: Data from third quinquennial survey on employment and unemployment, Ministry of Planning, New Delhi, India.

BOX 18.1 continued

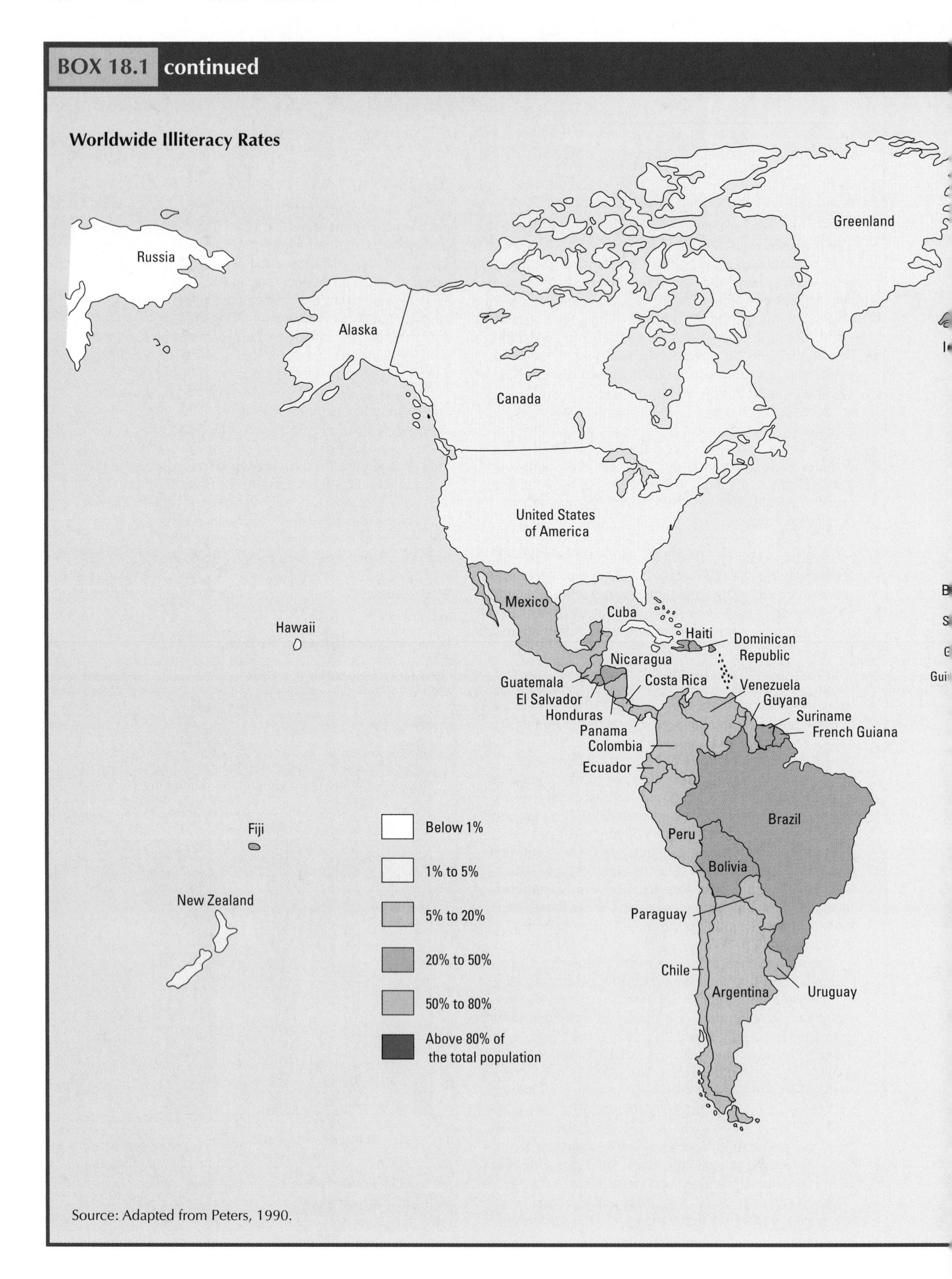

Source: Adapted from Peters, 1990.

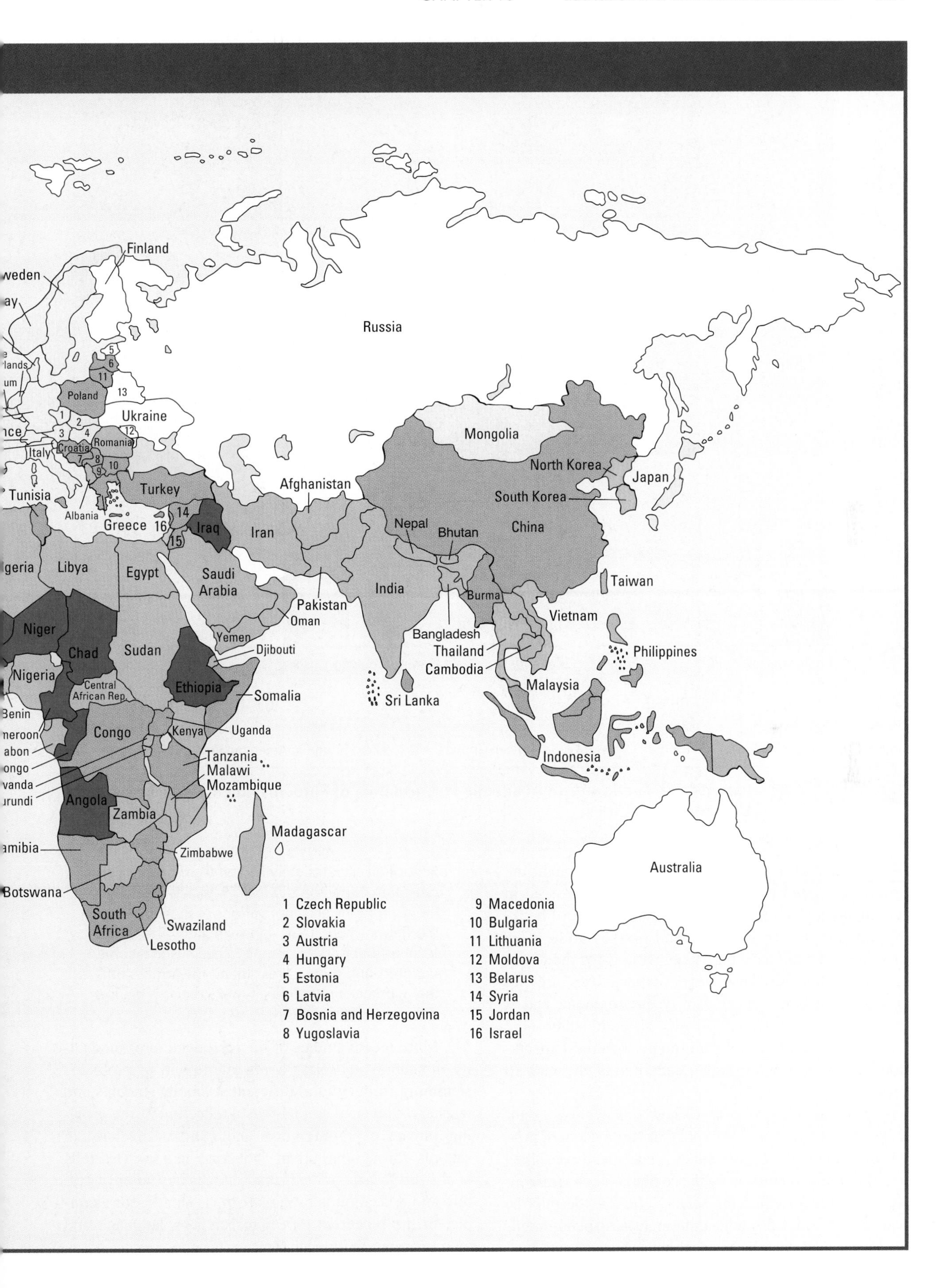
Finland
weden
ay
lands
um
Poland
13
Ukraine
nce
Romania
Italy
Croatia
Tunisia
Albania
Greece
Turkey
14
16
15
Iraq
Iran
Afghanistan
Russia
Mongolia
North Korea
Japan
South Korea
China
Nepal
Bhutan
Taiwan
geria
Libya
Egypt
Saudi
Arabia
Pakistan
Oman
India
Burma
Vietnam
Bangladesh
Thailand
Cambodia
Philippines
Niger
Chad
Sudan
Yemen
Djibouti
Nigeria
Central
African Rep.
Ethiopia
Somalia
Sri Lanka
Malaysia
Benin
neroon
abon
ongo
vanda
urundi
Congo
Kenya
Uganda
Tanzania
Malawi
Mozambique
Indonesia
Angola
Zambia
Madagascar
amibia
Zimbabwe
Australia
Botswana
South
Africa
Swaziland
Lesotho
1 Czech Republic
2 Slovakia
3 Austria
4 Hungary
5 Estonia
6 Latvia
7 Bosnia and Herzegovina
8 Yugoslavia
9 Macedonia
10 Bulgaria
11 Lithuania
12 Moldova
13 Belarus
14 Syria
15 Jordan
16 Israel

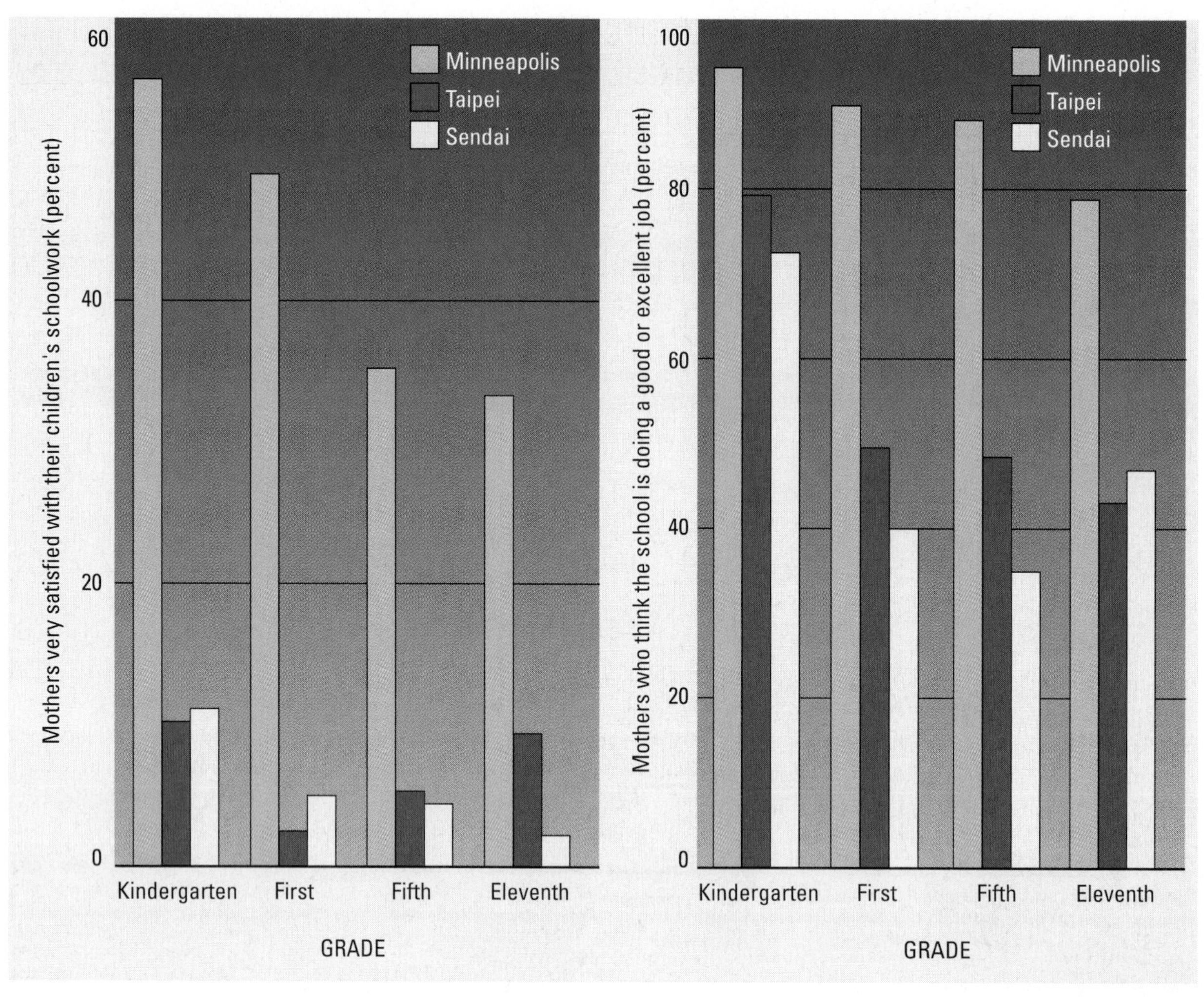

Source: From "Learning From Asian Schools," by H. W. Stevenson.

FIGURE 18.3 **Parents' Satisfaction With Academic Performance of Students and Schools in Selected U.S. and Asian Cities**

in England and the United States. They document the processes whereby class inequalities are reinforced in school as children from working-class and poor backgrounds are shown in subtle and not-so-subtle ways that they need not have high expectations for their educational future. If these lower expectations are not countered at home in the family environment, and if the students are not exceptionally gifted in school, the resulting cycle of school failure and lowered expectations for mobility through education is difficult to break.

In a classic early study of how this cycle is established, Ray C. Rist (1973) found that "the system of public education in the United States is specifically designed to aid in the perpetuation of the social and economic inequalities found within the society" (p. 2)—despite the widespread belief that education increases social mobility. He reached the following conclusion:

> Schooling has basically served to instill the values of an expanding industrial society and to fit the aspirations and motivations of individuals to the labor market at approximately the same level as that of their parents. Thus it is that some children find themselves slotted toward becoming workers and others toward becoming the managers of those workers. (1973, p. 2)

More recent studies of the reproduction of inequality in schools and classrooms often point to problems stemming from cultural differences among students and teachers. This new emphasis is a reflection of the growing impact of immigration and cultural diversity in schools throughout urban industrial nations (Castells et al., 1998). One of the primary issues in studies of culture and education is bilingualism, as shown, for example, in the recent empirical research on language and culture among Latino students.

TABLE 18.4 Average Earnings, by Level of Education, 1979 and 1995

	1979	1995
Without a High School Education		
Males	$29,723	$23,338
Females	17,093	16,319
High School Graduates		
Males	37,508	32,708
Females	21,473	21,961
College Graduates		
Males	55,751	61,717
Females	30,915	37,924

Note: Table reflects average annual earnings in 1995 dollars.

Sources: National Center for Education Statistics; Bureau of Labor Statistics; Current Population Survey, Census Bureau.

Language and Inequality A study by Lourdes Diaz Soto (1997) traces how the recent retreat from bilingual education has affected children from homes in which Spanish is the primary language. Although they were always controversial, programs that offered instruction in the student's native language as well as in English are being cut back in U.S. schools, largely in response to demands for "English only" instruction and what is often known as "total immersion" in a single language environment. Diaz Soto gathered data from participant observation and interviews of teachers, students, and parents in a small industrial city in Pennsylvania where a bilingual program had been replaced with an English-only curriculum. Her research documents the frustration and disappointment felt by Spanish-speaking students and their parents as opportunities to enhance the students' ability to speak, read, and write Spanish are withheld. Most painful is the message that only English is permitted in the schools. Diaz Soto's research shows that while this message may be motivated by the desire to have all students master one language (English), it has the negative effect of communicating to Latino students and their parents that their culture is devalued and that their voices will be silenced.

The Spanish-speaking parents argued that there was a need for earlier opportunities for students to learn a second language in elementary school so that children whose home language was not English could see that their language and culture were valued and would be taught in school. Diaz Soto, like many other educational researchers, notes the waste of human resources as children's abilities in languages other than English are lost. She points out that elsewhere in the world "second- and multiple-language learning is common: Danish schools' educational system includes compulsory second-language learning at age eleven; Swedish schools initiate second-language learning in the lower grades; France is initiating language learning for children under five; three-year-olds in northern parts of Germany are experimenting with second-language learning" (1997, p. 94). But in the United States, despite the growing emphasis on global relations and global social change, language education is extremely weak and "bilingual education continues to be suspect." Unfortunately, students who feel that their language and culture are devalued in school, be they Hispanic, Native American, Asian American, or others, often withdraw from participation and fail to develop their full potential.

Important as language and other cultural issues may be in explaining school achievement, economic class may be even more significant. It remains true that students from poor families are likely to attend schools with far fewer resources, less favorable student–teacher ratios, and more staff turnover than the schools attended by middle- and upper-class students (Torres & Mitchell, 1998). These class differences are also striking at the level of higher education.

Inequality in Higher Education Inequality in higher education is primarily a matter of access—that is, ability to pay. Ability to pay is unequally distributed among various groups in society, and as a result students from poor, working-class, and lower-middle-class families, as well as members of racial minority groups, are most likely to rely on public colleges and universities. Social-scientific evidence indicates that such inequalities are alleviated by federal aid to students from families with low or moderate income. It is likely that without Pell grants, work-study funds, and student loan programs the proportion of students from such families would be far lower than it currently is. Even with such aid, the proportion of low-income students in colleges and universities is lower than that of students from high-income families (Fuller & Elmore, 1996).

Sandra Baum (1987) examined the educational careers of 2,000 students who were high school seniors in 1980. She found that college attendance rates were higher for students from high-income families than for students from low-income families (60 percent versus 46 percent). However, she found that achievement in school seemed to account for an even larger difference: Students from low-income families who scored high on achievement tests had far higher rates of college attendance than low-scoring students from affluent homes. Thus achievement in the primary school grades seems to be at least as important as family income in explaining a student's success in higher education—but remember that in the lower grades students from more affluent homes tend to achieve more than students from less advantaged homes.

Baum and other researchers have also found that low-income students tend to be concentrated in two-year colleges and that students in such colleges are much more likely to drop out before completing a degree than students who enter four-year colleges. On the basis of findings like these, most educational researchers agree that without college assistance for needy students, differences in educational attainment and achievement would be much greater. And since public institutions of higher education remain the primary route to college degrees for low-income students, increased support for such institutions also tends to diminish inequality in access to higher education (Torres & Mitchell, 1998).

Education for Equality

i

The question of whether and in what ways education leads to social mobility remains open. Some social scientists view education as an investment like any other: The amount invested is reflected in the future payoff. This is known as *human capital theory.* In this view, differences in payoffs (jobs and social position) are justified by differences in investment (hard work in school and investment in a college education). However, critics of the educational system point out that the resources required to make such "investments" are not equally available to all members of society.

The question of whether the society as a whole, or only its more affluent members, will invest in its "human capital" through such means as preschool programs and student loans is a major public policy issue. This is especially true as technological advances coupled with degree inflation increase the demand for educated people (Castells et al., 1998; Reich, 1992a). Empirical evidence of the effectiveness of such investments may be seen in the results of numerous evaluations of Head Start and other early-education and prekindergarten educational programs (Kagan & Neuman, 1998). One of these evaluations, conducted by the High/Scope Educational Research Foundation and completed in the early 1980s, thoroughly demonstrated the lasting positive effects of a high-quality preschool program.

The subject of the study was a preschool program at the Perry Elementary School, located in a low-income black neighborhood of Ypsilanti, Michigan, a small industrial city on the outskirts of Detroit. The researchers randomly selected 123 three-year-old boys and girls and assigned them either to the preschool group or to another group that would not go to the preschool program. The latter children, like the majority of children at that time, would not be enrolled in school for another 2 years.

About twice a year for the next 20 years the researchers traced the experiences of the experimental (preschool) and control (no-preschool) groups. The data in Table 18.5 show that there were important differences between the two groups as a consequence of the preschool experience. Members of the preschool group achieved more in school, found better jobs, had fewer arrests, and had fewer illegitimate children; in short, they experienced more success and fewer problems than members of the control group. These findings confirm what other, less well-known studies had shown. For example, pioneering studies by Benjamin Bloom (1976) and by Piaget and Inhelder (1969) had indicated that as much as 50 percent of variance in intellectual development takes place before the age of 4.

The most significant contribution of studies like the High/Scope research is their documentation of the actual improvements that a good preschool program can produce in disadvantaged children. The High/Scope

TABLE 18.5 Major Findings at Age 19 in the Perry Preschool Study

Category	Number Responding*	Preschool Group	No-Preschool Group
Employed	121	59%	32%
High school graduation (or its equivalent)	121	67%	49%
College or vocational training	121	38%	21%
Ever detained or arrested	121	31%	51%
Females only: teen pregnancies, per 100**	49	64	117
Functional competence (APL Survey: possible score 40)	109	24.6	21.8
% of years in special education	112	16%	28%

*Total *N*=123.
**Includes all pregnancies.

Source: Data from Berrueta-Clement et al., 1984.

study was also able to translate its findings into dollar figures. A year in the Perry program cost about $1,350 per child. The researchers showed that the "total social benefit" from the program (measured by higher tax revenues, lower welfare payments, and lower crime costs) was equivalent to $6,866 per person.

This research became a classic in the social-scientific study of education because it demonstrated the undeniable benefits of preschool education to society (Kagan & Neuman, 1998). It also appeared during a period when funding for public preschool programs was in danger of being drastically reduced. The findings of the High/Scope study gave congressional advocates of preschool education the evidence they needed to justify continued federal funding. In many instances, however, social scientists know what needs to be done to improve educational achievement and increase equality of educational opportunity but are frustrated by the resistance to change that is inherent in large-scale bureaucratic systems such as the public school systems of many states and municipalities.

THE STRUCTURE OF EDUCATIONAL INSTITUTIONS

A significant barrier to educational reform is the bureaucratic nature of school systems. We noted earlier that sociologists view the school as a specialized structure with a special socializing function and that it is also a good example of a bureaucratic organization. As any student knows, there is a clearly defined status hierarchy in most schools. At the top of the hierarchy in primary and secondary schools is the principal, followed by the assistant principal and/or administrative assistants, the counselors, the teachers, and the students. Although the principal holds the highest position in the system, his or her influence on students usually is indirect. The teacher, on the other hand, is in daily command of the classroom and therefore has the greatest impact on the students. In this section we discuss several aspects of the structure of educational institutions and attempts to change those institutions.

Schools as Bureaucracies

As the size and complexity of the American educational system have increased, so has the tendency of educational institutions to become bureaucratized (Torres & Mitchell, 1998). The one-room schoolhouse is a thing of the past; today's schools have large administrative staffs and numerous specialists such as guidance counselors and special education teachers. Teachers themselves specialize in particular subject areas or grade levels. Schools are also characterized by a hierarchy of authority. The number of levels in the hierarchy varies, depending on the nature of the school system. In large cities, for example, there may be as many as seven levels between the superintendent and school personnel, making it difficult for the superintendent to control the way policies are carried out. Similarly, in any given school it may be difficult for the principal to determine what actually happens in the classroom.

The reforms sought by educational leaders like James Comer are being implemented in schools throughout the United States. Large cities like Chicago and New York are instituting more opportunities for parents to get involved in school issues and seeking to reduce the size of schools so that teachers, students, and administrators feel that they are part of a "learning community" in which they all know one another and are concerned with one another's well-being—conditions that may not be possible in larger schools where bureaucratic rules and regulations often seem to take precedence over personal relations (Comer, 1997; Meier, 1995).

In many school districts, however, the pace of reform is slowed by bitter disagreements about how to fund the schools in an era of reduced public spending. There are also controversies over issues such as sex education, tolerance for homosexual students, distribution of condoms to prevent AIDS and teenage pregnancy, bilingual education, and school prayer. These can divide communities and prevent structural changes like the move toward smaller schools and more local responsibility for educational decisions (McFadden, 1998; Rose, 1995).

The Classroom In most modern school systems the primary school student is in the charge of one teacher, who instructs in almost all academic subjects. But as the student advances through the educational structure, the primary school model (which evolved from the one-room school with a single teacher) is replaced by a "departmental" structure in which the student encounters a number of specialized teachers. The latter structure is derived largely from that of the nineteenth-century English boarding school. But these two basic structures are frequently modified by alternative approaches like the "open" primary school classroom, in which students are grouped according to their level of achievement in certain basic skills and work in these skill groups at their own pace. The various groups in the open classroom are given small group or individual instruction by one or more teachers rather than being expected to progress at the same pace in every subject.

Open classrooms have not been found to produce consistent improvements in student performance, but they have improved the school attendance rates of students from working-class and minority backgrounds.

Students in open classrooms tend to express greater satisfaction with school and more commitment to classwork. The less stratified authority structure of the open classroom and the greater amount of cooperation that occurs in such settings may help students enjoy school more and, in the long run, cause them to have a more positive attitude toward learning.

The Teacher's Role Given the undeniable importance of classroom experience, sociologists have done a considerable amount of research on what goes on in the classroom. Often they start from the premise that, along with the influence of peers, students' experiences in the classroom are of central importance to their later development. It is rare, however, that social scientists have the opportunity to trace the influence of teachers over students' lifetimes. A classic exception to this is a study (Pedersen & Faucher, 1978) that examined the impact of a single first grade teacher on her students' subsequent adult status (see Box 18.2). The surprising results of this study have important implications. It is evident that good teachers can make a big difference in children's lives, a fact that gives increased urgency to the need to improve the quality of primary school teaching. The reforms carried out by educational leaders like James Comer suggest that when good teaching is combined with high levels of parental involvement the results can be even more dramatic.

Because the role of the teacher is to change the learner in some way, the teacher–student relationship is an important part of education. Sociologists have pointed out that this relationship is asymmetrical or unbalanced, with the teacher being in a position of authority and the student having little choice but to passively absorb the information provided by the teacher. In other words, in conventional classrooms there is little opportunity for the student to become actively involved in the learning process. On the other hand, students often develop strategies for undercutting the teacher's authority: mentally withdrawing, interrupting, and the like (Darling-Hammond, 1997; Rose, 1995). Much current research assumes that students and teachers influence each other instead of assuming that the influence is always in a single direction.

Changing the System

The bureaucratic organization of school systems makes them highly resistant to change. The basic classroom unit has remained essentially unchanged since the days of the one-room school, as has the structure in which one teacher is in charge of a roomful of students. But despite the difficulties, there are endless demands for school reform and changes in the way students and teachers are evaluated. A full treatment of controversies over educational policy is offered in most courses on the sociology of education. Here we discuss three current controversies that have a direct bearing on whether schools can deliver on their promise of equality of opportunity. They concern class size, school choice, and desegregation.

Class Size Although the basic organization of the classroom has not changed much over time, there is wide variation in the ratio of students to teachers. Higher-quality education is associated with lower student–teacher ratios—that is, fewer students per teacher. It goes without saying that the more students there are in the classroom, the more difficult it is for the teacher to spend time with individual students. One of the reasons parents who can afford to do so often send their children to private schools is so their children will receive more attention and direct instruction from their teachers.

In public schools, classrooms in lower-income communities tend to have the highest student–teacher ratios. Classrooms in suburban schools in affluent communities have among the lowest student–teacher ratios in public education. This is one reason that so many families seek to move to those communities (Ballantine, 1997). Extensive studies in Tennessee and in various urban school districts provide quantitative support for the hypothesis that student performance is indirectly correlated with class size: The lower the number of students per teacher, the higher average school performance will be in terms of grades and performance on standardized tests (Gursky, 1998). These results are extremely convincing. Some critics believe that more research is required in order to take into consideration the effects of teacher experience, availability of new teaching technologies, and other factors. Nevertheless, Congress and the Clinton administration have been trying to work out a compromise on proposals to create federal funding to increase the number of teachers in the United States by 100,000 in order to achieve the goal of 18 students on average per teacher in the first three grades. In many disadvantaged schools there are as many as 28 students per teacher in these critically important lower grades.

School Choice One of the most controversial issues confronting educational institutions today is that of school choice. In fact, "school choice" involves several types of choices. As educational sociologist Carol H. Weiss notes:

> Choice plans include magnet schools: public schools with special emphases and/or facilities that draw students from across a district, such as technology schools or music and art schools. Choice plans include charter schools, authorized by a number of state legislatures to be free of most local school district regulations; students apply for admission. (quoted in Fuller & Elmore, 1996, p. vii)

BOX 18.2 SOCIOLOGICAL METHODS

Tracing the Influence of a Remarkable Teacher

In an unusual study of teacher–student interaction, Eigil Pedersen and Terese Annette Faucher (1978) demonstrated the persisting value of an outstanding primary school teacher. The study is unusual because the social scientists were actually looking for a general pattern of low teacher expectations. Instead, they found an exception that demonstrated the powerful effects an exceptional teacher can have on students. In scientific research, an unexpected result that leads to a new insight is known as *serendipity.*

Pedersen and Faucher began their study of IQ and achievement patterns among disadvantaged children at a school that was marked by high rates of failure. A high percentage of the school's graduates failed in their first year of high school and dropped out. The researchers attempted to explain these failure rates in terms of the concept of the "self-fulfilling prophecy," the idea that if teachers expect students to do poorly, the students are likely to perform accordingly. But if the negative prophecy seems to work in many instances, can we find evidence that a *positive* self-fulfilling prophecy—the belief that students can perform well—will also work? This is what makes Pedersen and Faucher's study so interesting.

As they examined the IQ scores of pupils in the school's first grade classes, the researchers found a clear association between changes in a pupil's IQ and the pupil's family background, first grade teacher, and self-concept. Further examination of the school's records revealed a startling fact: The IQs of pupils in one particular teacher's first grade class were significantly more likely to increase in subsequent years than the IQs of pupils in first grade classes taught by other teachers. And pupils who had been members of that class were more than twice as likely to achieve high status as adults as pupils who had been members of other first grade classes (see the accompanying table).

What was so special about Miss A, as the outstanding teacher was labeled? First, she was still remembered by her students when they were interviewed 25 years after they had been in her class. More than three quarters of those students rated her as very good or excellent as a teacher. "It did not matter what background or abilities the beginning pupil had; there was no way that the pupil was not going to read by the end of grade one." Miss A left her pupils with a "profound impression of the importance of schooling, and how one should stick to it" and "gave extra hours to the children who were slow learners." In nonacademic matters, too, Miss A was unusual:

> When children forgot their lunches, she would give them some of her own, and she invariably stayed after hours to help children. Not only did her pupils remember her, but she apparently could remember each former pupil by name even after an interval of 20 years. She adjusted to new math and reading methods, but her success was summarized by a former colleague this way: "How did she teach? With a lot of love!" (pp. 19–20)

In summing up their findings, Pedersen and Faucher stated that their data "suggest that an effective first-grade teacher can influence social mobility" (p. 29). Their findings differ from those of Coleman (1976) and Jencks and colleagues (1972), who believe that there is little correlation between school experiences and adult status. Pedersen and Faucher agree that further research on the relationship between teacher effects and adult status is needed. However, "In the meantime, teachers . . . should not accept too readily the frequent assertion that their efforts make no long-term difference to the future success of their pupils" (p. 30). As debates continue over how much difference good teaching can make and as governments and school systems wonder whether they should do more to reward good teaching, the results of this study are worthy of careful review.

Adult Status, by First Grade Teacher

	First Grade Teacher			
Adult Status	**Miss A**	**Miss B**	**Miss C**	**Others**
High	64%	31%	10%	39%
Medium	36	38	45	22
Low	0	31	45	39
Total	100%	100%	100%	100%
(*N*)	(14)	(16)	(11)	(18)
Mean adult status	7.0	4.8	4.3	4.6

(*N*)=Number of students who could be located and interviewed 25 years later.

Note: "Adult status" was determined from interviews that included questions on occupational status and work history, highest grade completed, rent paid and number of rooms, and related indicators of social position.

Source: Eigil Pedersen and Terese Annette Faucher, "A New Perspective on the Effects of First-Grade Teachers on Children's Subsequent Adult Status," *Harvard Educational Review,* 1978, 48:1, 1–31.

Among conservative school reformers, school choice often means plans to allow students to choose private schools as well as public ones by giving parents vouchers to pay for tuition and leaving them free to decide where and how they will use the vouchers. These "free market" school choice plans have been popular among conservatives since the 1950s, when they were first developed by the economist Milton Friedman.

At this writing there is a great deal of ongoing research designed to evaluate one or another of the school choice initiatives. Note, however, that the range of choice is so wide that no single study can demonstrate the overall effect of school choice on all the desirable outcomes of education (knowledge, mobility, equality, morality, citizenship, and so forth). And there are so many actual school choice behaviors—including moving to better school districts, home schooling of children, vouchers or education grants (such as the Pell grants in higher education), and the choice among parochial, private, and public schools—that researchers are hard pressed to respond to the challenges presented by these changes in educational institutions (Fuller & Ellmore, 1996). But most liberal critics of school choice plans are afraid that while greater choice is attractive to parents in poor school districts as well as to those in more affluent ones, the overall consequence of increasing choices among different types of schools will be to increase the segregation of students by class and race. And as such segregation increases, so will rates of failure in schools that have been abandoned by better students and more aware parents (Torres & Mitchell, 1998). Indeed, from this critical perspective, increasing racial segregation in public and private education can already be viewed as an outcome of school choice.

The Retreat From School Desegregation Perhaps the most significant change in educational institutions in the past quarter century was desegregation. Efforts to desegregate public schools began in 1954 with the Supreme Court's ruling in *Brown* v. *Board of Education of Topeka,* when the Court held that segregation had a negative effect on black students even if their school facilities were "separate but equal" to those of white students. In the first few years after this landmark decision, most of the states in which schools had been legally segregated instituted desegregation programs. However, in some states, particularly in the Deep South, desegregation orders were resisted, and by 1967 only 26 percent of the nation's school districts were desegregated (Orfield & Eaton, 1996).

Further progress toward desegregated schools was made after the passage of the Civil Rights Act of 1964 and the Elementary and Secondary Education Act of 1965; by 1973, de jure (legally sanctioned) segregation had been almost entirely eliminated. However, de facto (in-practice) segregation remained common in many parts of the nation. This pattern was created by two factors: school districts' traditional policy of requiring students to attend schools in their own neighborhoods, and "white flight," the tendency of whites to move to the suburbs in order to take advantage of suburban housing opportunities and, in some cases, to avoid sending their children to central city schools with large percentages of black students. On the basis of an extensive study of the impact of desegregation programs, Coleman (1976) concluded that "policies of school desegregation which focus wholly on within-district segregation . . . are increasing, rather than reducing or reversing, the tendency for our large metropolitan areas to consist of black central cities and white suburbs" (p. 12).

Coleman's findings had a significant influence on the policy debates that took place throughout the 1970s and 1980s over what could be done to increase racial equality in public education, especially the issue of whether schoolchildren should be bused across city lines. Educational policy makers are now seeking ways of promoting racial integration, not just by means of busing plans but also through changes in school curriculums and instructional methods.

When education critic Jonathan Kozol visited public schools in central cities and suburban communities, he was shocked by the stark evidence of continuing racial and class segregation he found:

> What startled me most was the remarkable degree of racial segregation that persisted almost everywhere. . . . Moreover, in most cities, influential people that I met showed little inclination to address this matter and were sometimes even puzzled when I brought it up. Many people seemed to view the segregation issue as "a past injustice" that had been sufficiently addressed. . . . I was given the distinct impression that my inquiries about this matter were not welcome. (1991, pp. 2–3)

The children Kozol interviewed understood quite well the way class and racial segregation affected their own educational opportunities. Samantha, an eighth-grader in the impoverished schools of East St. Louis, Illinois, described to Kozol the great differences in resources between the schools she attended and those in the nearby white suburbs—lack of adequate textbooks, no computers, fewer after-school programs, and the like. Her mother had tried to get her into one of the better schools but had been told that she had to stay "in her jurisdiction." "Is it a matter of race?" Kozol asked Samantha. He notes that in answering him the girl chose her words slowly and with great care: "Well, the two things, race and money, go so close together what's the difference? I live here, they live there, and they don't want me in their school" (pp. 30–31). Clearly, the process of racial integration still has a long way to go. Although school choice plans may increase racial segregation in schools, residential segregation has a much greater impact on school segregation. As sociologist John Yinger points out, "segregation in housing obviously leads to segregation in schools" (1995, p. 142). In his view, the failure to ensure freedom of residential choice for African Americans and other people of color remains a major obstacle to equality of education through desegregation.

THE COMMUNICATIONS MEDIA

If education is the process by which a society transmits knowledge, values, norms, and ideologies, as discussed in the first sections of this chapter, it is easy to see how education and information go hand in hand. In the remainder of the chapter, therefore, we turn to another major institution of modern societies: the communications media, or *mass media.* The decade of the 1990s began with a stirring lesson on the importance of communications media in the modern world. Throughout the world people were riveted to their television sets as they learned about the overthrow of a tyrant in Romania. For several weeks they had watched as one after another of the communist states of Eastern Europe faltered and toppled—first Poland, then Hungary, then Czechoslovakia, then East Germany. But the situation in Romania was far more violent—and much more public. In that nation, in fact, television played a dominant role in the revolution that put an end to communist rule.

Shortly before the uprising the Romanian dictator, Nicolae Ceausescu, had ordered security forces to massacre citizens who had participated in a demonstration in the provincial capital of Timisoara. The shootings were ordered over a closed-circuit television network. Following the massacre Ceausescu called for a "spontaneous" demonstration by his supporters in the main square of the capital, Bucharest; the demonstration would be televised and broadcast throughout the nation. However, by giving this order he in effect organized the demonstration that would lead to his own downfall.

Students at the University of Bucharest were on their first day of winter vacation when they heard the news of the Timisoara massacre. Many went to the main square, where the university is located. There they encountered the official demonstration in support of the dictator. When Ceausescu tried to defend the killings he had ordered, the students hissed and booed. The dictator reacted with a stunned look and paused in his speech. After a few minutes the television broadcast was stopped. For the first time, viewers throughout the nation became aware not only that a massacre had taken place but also that the dictator who had ordered it was vulnerable himself. In the following days demonstrations against Ceausescu evolved into a full-scale revolution. The revolutionary forces occupied the national television studios, and people around the world watched the downfall of a notorious tyrant on their television screens.

The media institutions in the United States are arguably the most active and powerful of any in the world, both because of their diversity and revenue and because they reach a larger proportion of homes than their counterparts anywhere else. Thus events like the impeachment of President Clinton, the trial of O. J. Simpson, and the bombing of the federal office building in Oklahoma City take on national significance in part because people throughout the nation can share in the immediate experience of the situation. It is not surprising, then, that television and other media play an immense role in U.S. politics. To take just one example, in November 1998 a significant number of voters, especially members of minority groups, came to the polls vowing to cast a ballot for or against the impeachment of President Clinton after having viewed the congressional debates on television.

We could give many other examples of the overwhelming power of television in American life, but the media play a significant role in many other societies as well. In many parts of the world the media are strictly controlled and are viewed as institutions for maintaining stability and legitimacy. In others the media are more independent and often promote social change.

Media Institutions in Modern Societies

Communications media are institutions that specialize in communicating information, images, and values about ourselves, our communities, and our society. Typical media institutions in modern societies are the print media (newspapers and magazines), movies, radio, and television, as well as the Internet. In the United States the messages communicated by the media can be political or nonpolitical, religious or secular, educational or purely entertaining, but in every case they use symbols to tell us something about ourselves and our environment.

Many social scientists have been deeply impressed by the media's ability to incorporate people into a society's national life and bring about changes in their traditional values. The media are run by professional communicators, people who are skilled in producing and transmitting news and other communications. These skills can be used to enhance the ability of national leaders to influence and persuade the masses. For this reason, the media are always under pressure to communicate the information and values that people in power want to have communicated.

The influence of the media is a source of continual controversy as different groups strive for greater control over media communications. In many societies the media are subject to strict censorship. The very idea of news and entertainment institutions that are free from censorship by political or religious institutions is a relatively recent development. But even in societies like the United States, which pride themselves on laws that protect the freedom of the press and other media, we can find numerous examples of conflict between norms that establish media freedom and norms that are designed to control the media. During the Persian Gulf War of 1991, military officials formed a pool of reporters who were taken through preapproved sites. Media coverage of the war was tightly controlled. On the domestic front, a reporter was sentenced to prison for refusing to divulge his sources of information regarding an organized crime figure. His appeal, which was not considered by the U.S. Supreme Court, claimed that the law guaranteed his right to maintain the confidentiality of his informants. We could add many other examples. The point is that the freedom of the press and the other media and their existence as separate institutions can never be taken for granted.

In the remainder of this section we first look at the issue of media differentiation: What is the basic structure of media institutions as we know them, and how has that structure developed in relation to other institutions, especially government and the economy? We then examine some typical patterns of media consumption in the United States and at how the media influence individual behavior, followed by a discussion of television and violence. Finally, we turn to the question of how powerful the media are and whether there are any limits on their power. We will see evidence of the immense influence of the media in American political life, together with other evidence suggesting that the media's ability to determine people's behavior is far from absolute.

Differentiation of Media Institutions

In describing the development of media institutions, a good place to begin is Robert Park's (1967/1925b) essay on the development of the newspaper, which was one of the first communication institutions to become differentiated from other institutions. Pointing out that "the power of the press may be roughly measured by the number of people who read it" (p. 80), Park emphasized the role of urbanization in increasing the size of the reading public: "Reading, which was a luxury in the country, has become a necessity in the city" (p. 81). The growth of newspapers thus is closely linked to the growth of cities.

The earliest newspapers, Park observed, were newsletters whose purpose was to keep readers informed of the latest gossip, whether of the court or of the village. As the number of people to be gossiped about increased, coverage began to be limited to the most prominent individuals among them. When the paper could no longer cover every incident that would interest its readers, it had to choose the ones that were in some way romantic or symbolic and treat them as human interest stories. Thus the newspaper became "an impersonal account of manners and life" (p. 84).

As time went by, different kinds of newspapers developed. There were politically oriented papers whose purpose was to report on the doings of government. Such papers tended to come under the control of a particular party, and eventually some rebelled. The rebels—the so-called independent press—included *The New York Times,* which attacked and eventually overthrew the political machine run by Boss Tweed. Over time there also developed a variety of newspapers known as the *yellow press.* The goal of such papers, according to Park, was "to write the news in such a way that it would appeal to the fundamental passions. The formula was: love and romance for the women; sport and politics for the men" (p. 95). Newspapers of this type attracted many more readers than the party papers or the independent press, thereby extending the "newspaper habit" to the mass public.

The outcome of this process of differentiation was what is now known as "the press," a cultural institution that is also a business. Newspaper publishers are interested in making a profit by offering news and features that will interest the greatest possible number of readers. However, the press continually struggles to remain independent from other institutions. The freedom of the press is never fully guaranteed. To cite just one example, the outspoken columnist Sidney Schanberg was forced to resign from his position at *The New York Times* after some of his statements provoked an angry outcry from landlords and real estate financiers. (The story of Schanberg's partnership with a Cambodian journalist is told in the film *The Year of Living Dangerously.*)

Today the dominant communications institutions in the United States are the electronic media: broadcasting corporations like CBS, NBC, ABC, and Turner Broadcasting, which own television and radio stations of their own and have affiliated local stations throughout the nation. Their growth has been based primarily on the effectiveness of these media as vehicles for advertising, which is their main source of revenue. Can they be depended on to broadcast news that is free from the influence of the advertisers that support them and, by extension, the owners and managers of large corporations? This too is an area of much social-scientific research. In what is still one of the best studies of this subject, Herbert J. Gans (1979) conducted observations and interviews at many of the nation's major newspapers, news magazines, and television networks. He reached the following conclusion:

> Since national news is produced commercially, one might imagine that story selectors are under constant pressure to choose news which will attract the most profitable audience. In practice, however, they are not. In the news media I studied, as in most others, editorial and business departments operate independently of each other. Business departments would like to influence editorial decisions in order to increase audience size and attract advertisers, but they can only make proposals. (p. 214)

More recent research on the media and its coverage of controversial social issues would seem to dispute Gans's conclusion. Todd Gitlin (1992) found evidence that the commercial interests of advertisers do influence what stories get told in the media. A notable example is the inverse correlation between the amount of money a magazine receives for tobacco advertisements and the likelihood that it will publish information about the dire consequences of smoking—the more tobacco ad revenue a magazine receives, the less often such stories appear in it. An even more sensational example is the scandal that erupted in 1995 over the decision by *60 Minutes,* one of television's most respected "magazine" news shows, not to broadcast an interview with a former Philip Morris executive because of possible legal consequences (the executive had signed an agreement not to discuss his knowledge of the industry with the media even after retiring). Thus the issue of how powerful and objective the media are remains an open field of research.

Patterns of Media Consumption

"Everyone agrees," wrote Harold Wilensky (1964) in an influential paper, "that abundance everywhere brings a rise in mass communications, through radio, television, and press; the development of mass education and the concomitant spread of literacy; and finally, mass entertainment on a grand scale" (p. 173). But he and other sociologists also believe that the media have produced a mass culture in which an increasing number of people have similar cultural tastes and political values, which are shaped by the media. At the middle and lower levels of the stratification system, according to this view, people form an increasingly homogeneous mass with similar thoughts and feelings. However, those thoughts and feelings are not anchored in spiritual and moral values; they are fluid and can be easily manipulated by those who control media organizations.

But what are the facts? As usual, the first responsibility of the sociologist when faced with claims like these is to try to support or refute them with empirical evidence. This effort begins with measures of how much time people actually spend as consumers of different kinds of media communications. For example, survey data show that most Americans have their television sets turned on for as much as 4 hours per day. To understand what such numbers signify, it is necessary to compare television viewing with the use of other media and to see how patterns of media consumption change over time. Data from National Opinion Research Center (NORC) polls are presented in Table 18.6. They indicate that there has been a decrease in

TABLE 18.6 Media Consumption by American Adults, 1975–1996

The Press	**1975**	**1985**	**1990**	**1996**
Read every day	66.9%	53.6%	52.5%	48.3%
Once a week	8.5	22.2	22.1	23.0
Less than once a week	5.4	6.6	10.7	10.0
Never	4.3	5.8	4.9	5.0
Radio (hrs./day)	**1975**	**1985**	**1990**	**1996**
None	1.0%	0.0%	0.7%	NA
1–4 hours	77.0	75.5	74.7	NA
5 or more hours	22.0	24.1	25.3	NA
Television (hrs./day)	**1975**	**1985**	**1990**	**1996**
None	3.4%	4.5%	3.1%	4.0%
1–4 hours	78.1	79.0	80.6	84.0
5 or more hours	18.3	16.7	16.2	13.0

Note: Total number of respondents in the sample varies between 1,380 and 1,520. Percentages may not add up to 100 percent owing to small number of "Don't know" answers and because of rounding.

Source: Data from National Opinion Research Center (NORC).

Then and Now

Pigskin Mania

Football emerged in more or less its present form in the 1890s as a popular sport among wealthy young men at elite eastern universities like Yale and Harvard. In North America, football has become not only a sport but a generator of enormous revenues and a cultural form that many people almost feel they could not live without. The growth in popularity of the sport was intimately associated with the powerful influence of the mass media (see Box 18.3), which can broadcast games to ever-larger audiences. But the growth of media institutions, in turn, is dependent on the revenues made possible by the huge audiences attracted to major games. Thus there is a symbiotic relationship between television and the educational institutions that can afford to support football teams (Kriegel, 1993; Lawton, 1984).

More than 50 million fans attend one or more football games in a given year, and no one attending a college or university in the United States needs to be told that football games are central events in the life of many campuses. In fact, for many men and an increasing number of women football has come to resemble a civil religion. There are fans whose lives become subordinated to the rituals of preparation for the Saturday game. Legendary coaches like Alabama's Bear Bryant or Penn State's Joe Paterno assume a status that is well below that of a god but far above that of an ordinary mortal (C. R. Wilson, 1987). The rivalries that emerge between regional powerhouse teams and are played out each year become material for local legends. Many will admit to having prayed for a successful third-down conversion.

The metaphors inherent in football—success, hard work, strength, speed, organization, strategy—all make the sport a study in cultural ritual that mirrors the values of the larger society in which it has flourished (Brokaw, 1994). These cultural qualities of the game also make it extremely popular to television viewers and enable broadcasters to make large payments to the universities; these, in turn, help support other, less profitable sports. The popularity of football also enhances the university's prestige and its ability to compete for students and faculty.

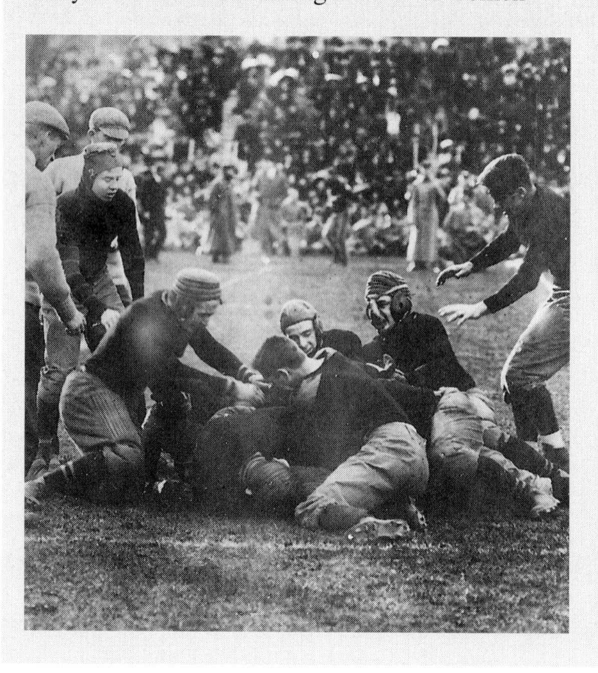

newspaper reading, yet at the same time they show that for large numbers of American adults the daily paper remains an important source of news, information, and entertainment. The data also show that contrary to what many people think, television did not make radio obsolete. In fact, radio competes well with television for people's attention, especially at work and in cars, where television generally is not available. Still, television ranks far above all other media in surveys of how people use their leisure time.

The data in Table 18.6 also suggest that there have been small but potentially significant decreases in television watching. Remember that in large populations even small percentage changes can mean large changes in absolute numbers. In fact, the major networks and communications corporations are worried about this trend, for it could mean huge losses of advertising revenue if it continues. Their concern is heightened by large increases in the amount of time people spend in front of their home computers. Although it represents just a fraction of the average time spent watching television, the average amount of time spent at the computer screen and on the Internet is expected to increase from 1 hour per month in 1990 to 39 hours in 2001, and it is likely that this trend will continue to accelerate. In sum, the data do not lend much support to the mass culture theory. They do show that people consume a great deal of media communication, but their tastes are diverse, and they continue to read as well as to watch and listen. As a result, they are not really homogeneous in their tastes, and the decrease in television viewing time suggests that they are not unduly influenced by media communications.

Television and Violence

Does watching a lot of television lead people to commit violent acts? Few questions about the media have generated as much research and debate as this one (Reiss & Roth, 1993). A 1995 CBS/New York Times opinion poll indicates that parents believe the portrayal of violence in the popular media, especially television, is a cause of violence among adolescents. Some of the key results of this survey are presented in Figure 18.4. The television shows most often mentioned as forbidden to children were MTV's *Beavis and Butt-Head* and Fox's *Mighty Morphin Power Rangers.* However, it has proved far more difficult to establish an actual behavioral link between media consumption and violence than the results of opinion polls would suggest. Indeed, this question is a good test of the media's power to influence behavior, as opposed to attitudes and opinions.

Each hour of prime-time television programming presents an average of five acts of violence. Estimates made in the early 1990s calculated that the average child

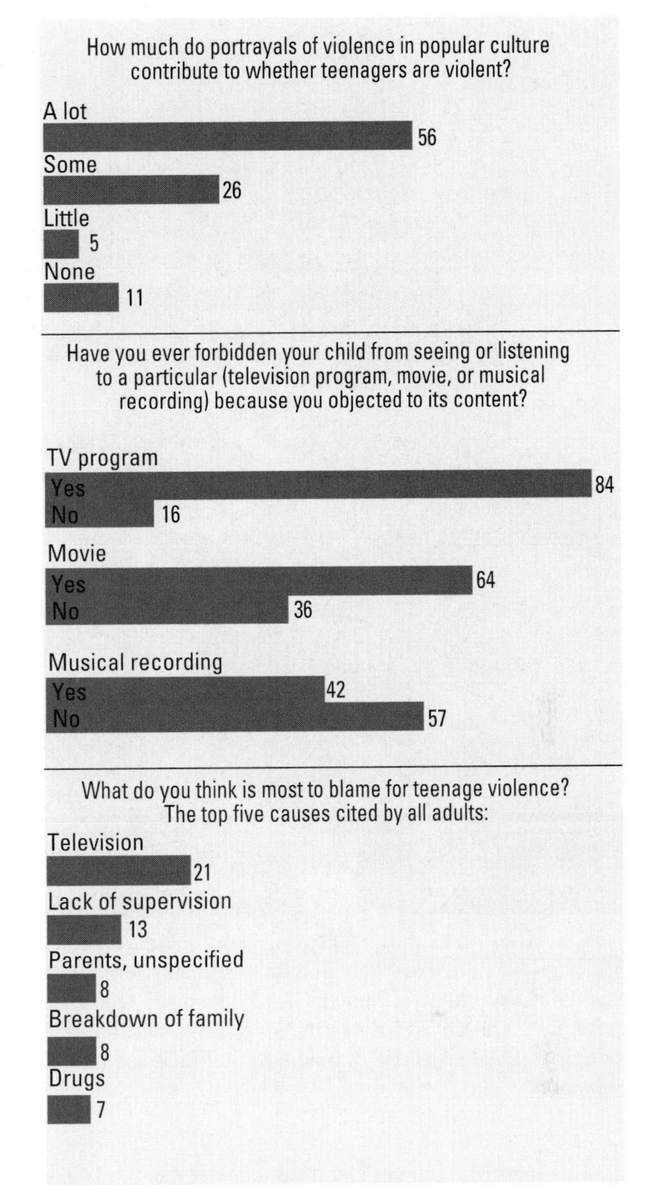

Source: Kolbert, 1995.

FIGURE 18.4 **Attitudes About the Effects of Televised Violence on Behavior (percent of U.S. adults)**

entering middle school or junior high school had already watched at least 8,000 murders and more than 100,000 acts of violence on television (Hamburg, 1992). With the increased availability of cable television and video games, even more violent programming is available to children (Bok, 1998). Various studies have examined the connection between televised violence and violent behavior, especially in children and teenagers. Many of these studies found a causal relationship between the viewing of violence on television and later aggressive behavior, although some earlier research argued

Although they are extremely popular, especially among young male television viewers, the Mighty Morphin Power Rangers have been criticized by opponents of violent television programming. The critics believe that exposing children to violence, even when it is highly stylized, may create patterns of imitation that increase the level of violence in children's behavior.

that such a relationship cannot be demonstrated (e.g., Kaplan & Singer, 1976; Milavsky, 1977). By 1993, however, there was far greater consensus among sociological and psychological researchers that violence does have both short- and long-term effects. In 1993 the American Psychological Association issued a report that concluded that "there is absolutely no doubt that higher levels of viewing violence on television are correlated with increased acceptance of aggressive attitudes and increased aggressive behavior" (American Psychological Association, 1993, quoted in Bok, 1998, p. 59).

These conclusions are based on analyses of many individual studies of television watching and aggression using a technique known as *meta-analysis.* This method involves the statistical analysis of numerous studies in an attempt to discover what common patterns they may reveal. Such analyses indicate that very frequent exposure to aggression on television leads to aggression in children and adults. The explanations for this finding are still being debated. The most common explanation is that television violence produces a form of "social learning"—that is, televised violence provides models showing the viewer how to act violently and also provides an approved social context for this learning: television watching in the home (Cole, 1995; Paik, 1991).

Increased aggression is not the only effect of a heavy diet of TV violence. George Gerbner, a pioneer in monitoring the frequency of violent incidents on TV and measuring its effects, finds what he describes as a heightened sense of insecurity and vulnerability in children exposed to large amounts of TV violence. Such children suffer from the "mean world syndrome," a sense that the world is mean and gloomy; they feel at high risk of victimization from violence and believe that their neighborhoods are unsafe (Bok, 1998; Signorelli, Gerbner, & Morgan, 1995).

It is obvious that not every viewer of televised violence is tempted to act out that violence. It is equally clear that there is some not-yet-determined relationship between televised violence and aggressive behavior in some individuals. The relationship appears to depend at least in part on the viewer's emotional condition. Contending that the research findings are inconclusive, the major television networks have resisted efforts to regulate the content of their programming, although the first hour of prime time, from 8:00 to 9:00 P.M., has been labeled the "family hour" and is limited to programs deemed appropriate for family viewing.

Media Power and Its Limits

A familiar expression in modern societies is "information is power." And because the media control such a large and diverse flow of information, they have immense power. Questions about the power of the media become especially urgent when one imagines what could happen if control of the media fell into the hands of groups that oppose democratic institutions. In such a case, could the persuasive power of the media be used to destroy individual and political freedom?

George Orwell addressed this issue in his famous book *1984,* which portrays a society in which everyone is constantly watched on two-way television. Not only the actions but also the thoughts of each member of the society are monitored in this way by a powerful central government. Originally published in 1949, *1984* presents a terrifying vision of the potential power of the media when used by a dictator to control the thoughts and behavior of the population. But it is not merely an imaginative fantasy. It is based on the efforts of totalitarian regimes, especially those of Hitler and Stalin, to control the media and use them to control the masses.

In contemporary societies where totalitarian regimes try to control the minds of citizens (e.g., Romania before the revolution of 1989–1990) or in which oppressive regimes attempt to suppress democratic movements (e.g., China), total control over the media is a major goal of the state. In the United States, as noted earlier, media institutions are protected against undue political influence and are governed by professional norms regarding the gathering and broadcasting of information. There is also an increasingly wide range of channels to choose among, which limits somewhat the power of major broadcasters (Samuelson, 1998). However, even in the United States and other democratic societies in which freedom of the press is guaranteed, there are still many problems related to access to the media and their power to attract large audiences. For example, as we will see in Chapter 20, the media exert a powerful influence over the conduct of American politics.

On the other hand, in a democratic society television and other media can be a two-edged sword, conferring power on those in the spotlight but also subjecting them to sometimes embarrassing public scrutiny. Richard Nixon, Bill Clinton, and O. J. Simpson are only a few of the many public figures who have risen, and fallen, and perhaps risen again, in front of a national television audience. The public has a stake, therefore, in determining whether the media are adequately and evenhandedly investigating the actions of the powerful and the famous.

Technological Limits When media institutions are well differentiated from political and other institutions, it is actually quite difficult for powerful individuals or groups to manipulate mass audiences. This becomes even more true as changing technologies give people more opportunities to choose the types of messages they receive via the media. Cable television offers the potential for much greater diversity in program content: Viewers can watch everything from public affairs to pornography. Videocassette equipment also makes possible a wider range of choices. So although it is entirely likely that the size of the television audiences for special events like the Super Bowl will continue to increase (see Box 18.3), the audience of media consumers is becoming ever more diverse and fickle and, hence, ever more difficult to reach as a mass audience. Even in societies like China, in which the media are agencies of the state and may broadcast only material that has been approved by political leaders, new technologies like VCRs and the Internet promise to make it more difficult to control the flow of information.

Social Limits Another limit on the power of the media is the nature of communication itself. Researchers have not found a direct link between persuasive messages and actual behavior. People do not change their cultural values and norms just because the media tell them to. Instead, investigators have identified a **two-step flow of communication:** The messages communicated by the media are evaluated by certain respected individuals, who in turn influence the attitudes and behavior of others (Katz, 1966). Such individuals function as **opinion leaders** in matters ranging from voting to the purchase of paper towels, which is one reason advertisements so often portray a knowledgeable person praising a product to eager, if ignorant, listeners. The role of influential individuals in the process is a reminder that communication via the media has not

BOX 18.3 USING THE SOCIOLOGICAL IMAGINATION

The National Hookup

It happens every January. After a long season of grueling competition, the two surviving professional football teams face each other in the Super Bowl. The day is now called Super Bowl Sunday. More than 100 million people from coast to coast will watch the game on television. Each minute of advertising time will cost almost $2.5 million, and America's most successful corporations and their advertising agencies will bid against one another for the time. Truly we have in the Super Bowl an occasion when mass interest in a national sports event creates an audience that crosses all lines of class, race, ethnicity, and religion.

But this level of national attention to one game—and the profits it generates for the team owners, the players, and many others—could not have been imagined before the technology of television made it possible to broadcast events simultaneously across the continent. Television gave football the national hookup it needed to become a billion-dollar industry. At the same time, football and other professional sports created the media's largest audiences and influenced the scheduling of television time.

In this example two social institutions are intimately connected. Sports is a social institution that is structured in different ways throughout the society. In communities, the most common sports organizations are Little Leagues and school teams. At some large colleges, sports are structured in ways that will generate revenues that can be used to support the college's entire athletic program. The bigger and more competitive a college's football program is, the more it depends on television revenues as well as ticket sales, and the more heavily recruited its players are. Finally, the biggest, fastest, and most talented players are drafted into the business of professional football, which is intimately linked with the nation's most pervasive media institution, television.

Sociologists who conduct research on sports and the media point out that in many societies sports are not conducted as a business that depends on television and advertising. For example, in her study of soccer (the most popular team sport in the world) Janet Lever (1983) found that players may have professional careers but that their teams are not organized as businesses. In Brazil, as in many other nations, sports are engaged in by social clubs in cities and towns. The clubs pay their players, and the clubs with the largest dues-paying memberships get the best players. Volunteer directors are elected by councilmen, who in turn are elected by all the shareholders of a club. To qualify as a club director, a man must be a shareholder; making large donations to the club also helps.

Soccer in Brazil thus may not be as profitable or as "national" as football in America, but Lever has shown that it also creates a "national hookup." In fact, soccer stars like the famed Pélé are national heroes, as can be seen in the following description:

> [Returning from the Mexican World Cup] the team flew to Rio de Janeiro, where they were greeted by close to 2 million ecstatic people waiting in the rain. The secretary of tourism coordinated a special Carnival parade of samba schools. . . . The major clubs from the major cities had contributed players to the winning team. Animosities were suspended for a time of alliance. In a land where communication is problematic, the team's contribution to national unity was of great value. (Lever, 1983, pp. 68–69)

The essential similarity between sports in nations like Brazil and those in which professional sports are a major business is that at the national level, where audiences for championship games are drawn from homes throughout the society, it is television that has enhanced the appeal of sports and given it such a central place in people's identification with their city and their nation.

replaced oral, interpersonal communication, nor is it likely to (Bok, 1998).

Another example of social limits on television's power to shape individual behavior comes from research on people's actual behavior while the television set is on. Comparative research in societies throughout the world suggests that television is an extremely powerful medium for reaching masses of people and involving them in shows that raise important issues of morality—serials and story shows are popular wherever they appear. However, the fact that the television set is turned on for several hours each day does not necessarily mean that people are actually absorbed in watching the shows. Many people have become used to television as an accompaniment to their daily activities, but they do not necessarily always watch it or allow themselves to be influenced by it (Signorelli, Gerbner, & Morgan, 1995). On the other hand, the media convey a great deal of information about our cultural environment and cannot help but affect how we perceive that environment. The extent to which they shape our perceptions is a subject of ongoing research and debate.

Studies that report on the number of hours Americans watch television can be misleading because people are often involved in other activities while the television is on and are not actually watching it. It is also true, however, that the constant presence of the television may distract attention from family activities or schoolwork.

RESEARCH ON THE CUTTING EDGE: The Media and Agenda Setting

The media may not always be successful in telling people what to think or how to act, but they are extremely successful in telling us what to think about. This viewpoint is the basis of an influential line of social-scientific research on the media. It emphasizes the power of the media to shape public discourse. The discussions, debates, news shows, reports, in-depth analyses, and other approaches to current events can be seen as proceeding according to an unwritten and always changing agenda that places priorities on different subjects at different times.

The *agenda-setting process* is an ongoing competition to gain the attention of media professionals, the public, and policy makers (Dearing & Rogers, 1996). Research on agenda setting explores how social problems become public issues, who the supporters (proponents) of a given issue are, and how successful they are at getting media attention for their cause or issue. From this perspective, getting on the media's agenda is a "zero sum game," meaning a competition in which one issue can get more attention only if others get less, because time in the public spotlight is a finite resource. Unlike studies of media institutions, research on agenda setting focuses on the interactions among various "gatekeepers" within media institutions, such as press agents and publicists. Media gatekeepers are people at newspapers or radio and television stations who are in a position to actually influence whether or not an issue is featured. Figure 18.5 presents a schematic view of influences on the media's agenda, on the public agenda (public opinion), and on the policy agenda of legislatures.

Following is an example of the agenda-setting process at work:

> An issue proponent might be a newsperson covering a famine in an African nation who shoots a spectacular 3½-minute news story in a refugee camp that is broadcast on U.S. evening television news. Because of the investment of time, effort, and firsthand experience, the reporter

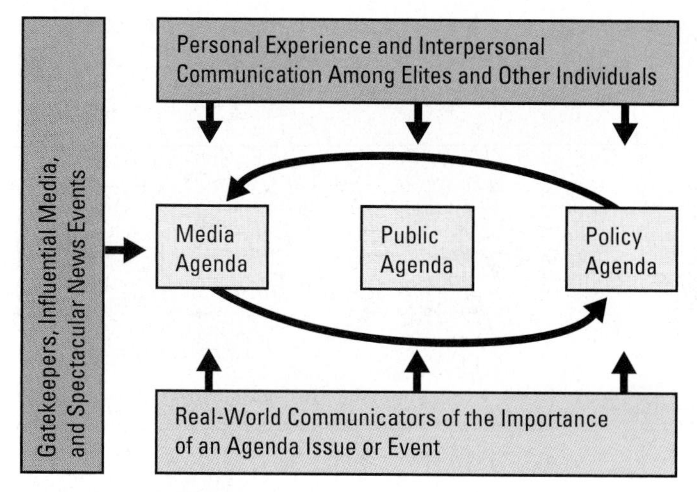

Source: Dearing & Rogers, 1996.

FIGURE 18.5 **Components of the Agenda-Setting Process**

becomes a proponent of the famine as an important issue worthy of news attention and publication. Attention to an issue, whether by media personnel, members of the public, or policymakers, represents power by some individuals or organizations to influence the decision process. The reporter covering the famine may have been influenced to shoot the story from a certain perspective because of discussions with a foreign government official who was frustrated with his or her country's lack of response to the famine. The visual power of the video footage, in turn, may influence an editor's decision about the relative importance of the famine news story in relation to other possible news stories. The news, when broadcast, influences millions of people in a variety of ways. Thousands of television viewers call an 800 telephone number to donate money and food. Some viewers work to change U.S. foreign policy about disaster relief to the African nation. A Senate staff member crafts legislation in the name of her boss. Hundreds of newspaper editors and other media gatekeepers decide that the famine deserves prominent news coverage. Several newspaper readers write letters to the editor to protest U.S. government food aid in the face of poverty in America. Thus, the famine becomes a two-sided issue. Within a few weeks, the very real but little-known famine problem is transformed into the "famine issue" and climbs to the top of the media agenda in the United States. The reporter gets a promotion. (Dearing & Rogers, 1996, pp. 3–4)

The sociologists note that the famine may continue to attract attention, or it may not. The outcome depends on competition from other issues and their proponents, and on the ability of the proponents of famine relief to keep the famine in the news. Any item on the public's or the media's issue agenda can remain salient only for a relatively brief period before it is displaced by other issues. (*Salience* refers to the perception that an issue is at the forefront of public and media concern.) As an example, consider the rise and fall of the so-called "War on Drugs" during the late 1980s and early 1990s (see Figure 18.6). In this example, media salience is measured by the number of drug war stories published by the *Washington Post.* Salience on the public agenda is measured by the percentage of respondents who said that drugs are the most important problem facing the United States today. The issue became highly salient in both the media and the public mind with the death of basketball star Len Bias in 1986, Nancy Reagan's "Just Say No" campaign, and growing awareness of the ravages of crack-cocaine in the middle and late 1980s. Eventually the issue waned in salience as a result of competition from the Persian Gulf War and other pressing issues.

Agenda-setting research is a sociological framework that helps us understand the complexity of public

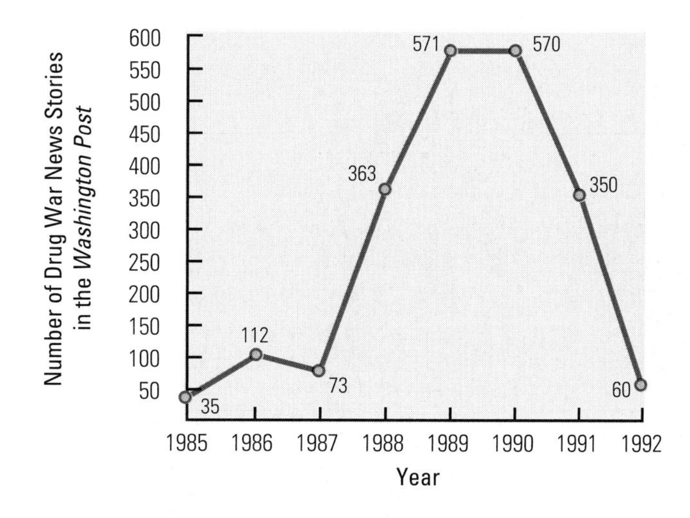

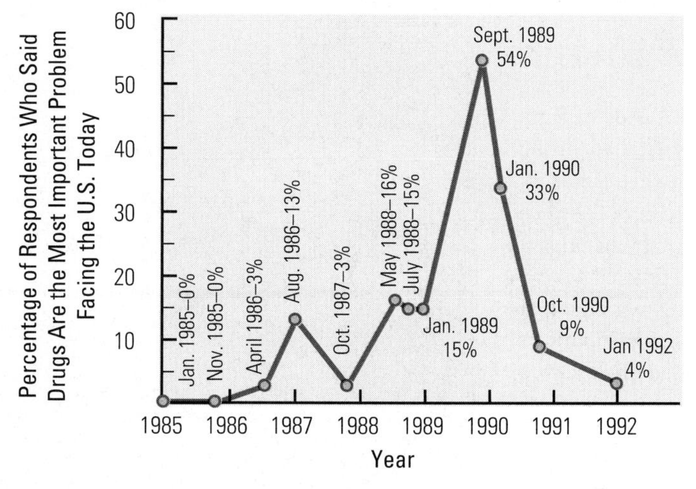

Source: Dearing & Rogers, 1996.

FIGURE 18.6 **The Drug Issue on the U.S. Media Agenda (above) and on the Public Agenda (below)**

concern about social issues and the media's coverage of those issues. It helps us see how naive it is to lament that the media do not give more attention to "feel-good" news stories and positive rather than negative events. After all, there are many issue proponents working with all their talent and creativity to get their issues on the various agendas, and unless we understand this agenda-setting process, it is almost impossible to make sense of how social issues rise and fall in the media and in the public mind.

VISUAL SOCIOLOGY

Implementing Comer's Vision of School Development

Photos of children, parents, and teachers working in school settings tend to evoke scenes from our own childhoods. But such photos rarely capture the quality of the educational experiences that are occurring within the photo's frame. However, these photos are taken from the highly influential Web site of the Yale School Development Program, which is headed by James Comer. These scenes are from schools throughout the United States that have been adopting and modifying Comer's vision and mission of school development. As stated, this vision and mission seem like something all can share without much debate. In practice, however, implementing the Comer process is more difficult. It requires restructuring the ways in which the school district, the schools, principals, teachers, parents, and students work together. The model hinges on efforts to have decisions made at the level where they are to be put into practice, rather than at higher levels. This means more authority for parents, teachers, and students, but that authority is accompanied by greater responsibility for thinking through the issues.

A central concept of this form of school restructuring is that each school should establish its own academic and social goals through collaborative relationships based on trust and respect. Each school forms three adult teams: a planning and management team for developing and implementing a school improvement plan; a student and staff support team, which deals with psychological and social welfare issues; and a parent team that is encouraged to become actively involved in school activities at all levels of responsibility.

The Web site offers a link to Northwestern University's extensive evaluation of the Comer schools model

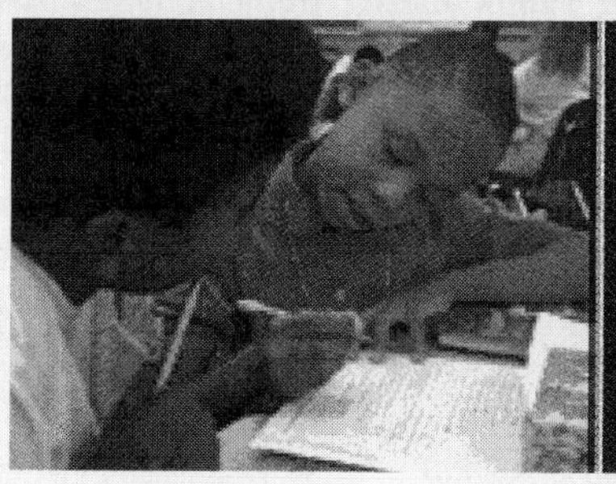

as implemented in Chicago. The evaluation was directed by Thomas Cook, one of the nation's most prominent evaluation research specialists. In Chicago, 10 schools that restructured ("Comer schools") and 9 schools that did not restructure ("control schools") were evaluated. Cook's research team found that "the Comer schools are now improving at faster rates in both math and reading than control s-chools. The Comer schools were among the lowest achieving in the city at the project's inception 4 years ago but have now caught up to and surpassed performance in the control schools" (Tremmel, 1999, p. 1).

SUMMARY

Education is the process by which a society transmits knowledge, values, norms, and ideologies and in so doing prepares young people for adult roles and adults for new roles. It is accomplished by specific institutions outside the family, especially schools and colleges.

Schools are often cited as examples of bureaucratic organizations, since they tend to be characterized by a clearly defined authority system and set of rules. A more interactionist viewpoint sees the school as a distinctive set of interactions and patterns of socialization. Conflict theorists view schools as institutions whose purpose is to maintain social-class divisions and reproduce the society's existing stratification system.

The post–World War II baby boom caused a bulge in elementary school enrollments beginning in 1952 and an expansion of the college-age population in the 1960s. These trends were reversed in the 1960s and 1970s, but at the same time more and more adults have sought additional education.

A key feature of education is the fact that schools structure the lives of children and adolescents. They help create a social world for adolescents that is separate from adult society. Another important aspect of education in the United States is the relationship between education and citizenship. Education is also viewed as a tool for solving social problems, especially social inequality.

Educational attainment refers to the number of years of school a person has completed. It is correlated with income, occupation, prestige, and attitudes and opinions. The average American today has much more education than the average American of the early 1940s. One effect of higher levels of educational attainment is "degree inflation," in which employers require more education of potential employees.

Educational achievement refers to how much the student actually learns, measured by mastery of reading, writing, and mathematical skills. The steady decline in mean SAT scores in the 1970s and 1980s was viewed as a sign that average levels of educational achievement have declined. Cross-cultural research has shown that these deficiencies are apparent from an early age. American parents are more likely than Asian parents to be satisfied with their children's schoolwork and to believe that the schools are doing a good job; they therefore are less likely to be critical of the schools or to demand more of their children.

Educational institutions have been criticized by observers who believe that they hinder, rather than enhance, social mobility. Higher levels of educational attainment provide the credentials required for better jobs, and students who are able to obtain those credentials usually come from the middle and upper classes. A factor that has been shown to affect students' school careers is teacher expectations regarding students, which are affected by the teacher's knowledge of the student's family background.

Inequality in higher education is primarily a matter of access—that is, ability to pay. Students from poor, working-class, and lower-middle-class families, as well as members of racial minority groups, are most likely to rely on public colleges and universities. Most educational researchers agree that without college assistance for needy students, differences in educational attainment and achievement would be much greater.

The American educational system is highly bureaucratized, a fact that acts as a major barrier to educational change. Nevertheless, when they have been required to change—as in the case of desegregation—schools have often proved to be very adaptable.

Communications media are institutions that specialize in communicating information, images, and values about a society and its members. The media are run by professional communicators who are skilled in producing and transmitting news and other communications. In some societies the media are subject to strict censorship, and even in societies in which the freedom of the press and other media is protected, that freedom is often threatened by attempts to control the nature of the information communicated by the media.

The differentiation of the communications media from other institutions is illustrated by the development of newspapers. Over time "the press" became an independent cultural institution that is also a business whose goal is to make a profit. The same can be said of the electronic media, the dominant communications media today.

Some sociologists have criticized the media for producing a "mass culture" in which an increasing number of people have similar cultural traits and values and can be easily manipulated by those who control media institutions. However, data on media consumption do not support the mass culture theory. They show that listeners and viewers are not homogeneous in their tastes and are not unduly influenced by the media. Other research has shown that there is a complex relationship between the viewing of violent acts on television and subsequent aggressive behavior in some subgroups of the population.

Because the media in modern societies control such a large and diverse flow of information, they have immense power. In the United States, they exert a powerful influence over the conduct of politics. Political candidates without large budgets to spend on media advertising cannot compete effectively in election campaigns, and presidents and other powerful leaders must be able to perform well on television. On the other hand, changing technologies have given media consumers a wider range of choices, thereby limiting the power of the media to influence mass audiences.

SOCIOLOGY VERSUS IDEOLOGY

Why are schools in the United States and other democratic nations a subject of so much ideological controversy and debate? Why do so many experts claim that the schools are failing when most people feel that their children's schools are doing a rather decent job? The answers to both questions are closely related and draw on many sociological insights about educational and media institutions.

A person with strong ideological preferences—conservative or liberal, religious or secular—is likely to blame the schools for being too permissive or lax in discipline. So many hopes for equality of opportunity and social mobility and the preparation of new generations for citizenship are pinned on the schools that it is no wonder they are targets of ideologically based criticism. But what does sociology have to contribute to these debates?

As always, the sociologist can help cut through the ideological debates by providing well-designed research and fair-minded empirical evidence. Are the schools actually failing to live up to our expectations? The research we have explored says that they are, and that this is especially true for specific groups of students in particular communities. Will the students achieve more if the nation sets higher standards for school achievement? Sociological research on class size and other features of the learning environment strongly suggests that raising standards can boost achievement only so much without additional resources to entice well-qualified teachers to embattled school districts. These answers may not be convincing to everyone, and sociologists often find themselves in the situation of the messenger who was executed for bringing bad news. But all sciences face the problem of presenting facts that do not satisfy everyone.

GLOSSARY

education: the process by which a society transmits knowledge, values, norms, and ideologies and in so doing prepares young people for adult roles and adults for new roles. (p. 570)

educational attainment: the number of years of school an individual has completed. (p. 574)

educational achievement: how much the student actually learns, measured by mastery of reading, writing, and mathematical skills. (p. 577)

communications media: institutions that specialize in communicating information, images, and values about ourselves, our communities, and our society. (p. 589)

two-step flow of communication: the process in which messages communicated by the media are evaluated by certain respected individuals, who in turn influence the attitudes and behavior of others. (p. 595)

opinion leader: an individual who consistently influences the attitudes and behavior of others. (p. 595)

WHERE TO FIND IT

BOOKS

The Power of Their Ideas: Lessons for America From a Small School in Harlem (Deborah Meier; Beacon, 1995). The portrait of a model school that has played a seminal role in the current school reform movement, by one of the most innovative educators in the world today.

Savage Inequalities: Children in America's Schools (Jonathan Kozol; Crown, 1991). A chilling view of the persistent inequalities of race and social class that bar children and teachers in the nation's urban public schools from achieving educational success.

Who Chooses? Who Loses? Culture, Institutions, and the Unequal Effects of School Choice (Bruce Fuller and Richard F. Elmore; Teachers College Press, 1996). An extremely useful overview of school choice alternatives and evaluations of their consequences for students, families, and communities.

The Sociology of Education, 4th ed. (Jeanne H. Ballantine; Prentice Hall, 1997). A comprehensive text in the sociology of education, with many additional bibliographic sources.

Mayhem: Violence as Public Entertainment (Sissela Bok; Perseus, 1998). An impassioned but well-documented study of media violence and its effects on children and society.

Deciding What's News (Herbert Gans; Pantheon, 1979). The best sociological study of how the news we see on television is processed by media institutions.

JOURNALS

Social Problems. A sociological journal devoted to research and policy analysis related to major social problems. Publishes good articles on family research.

Sociology of Education. A quarterly journal published by the American Sociological Association. Presents recent research on education and human social development.

Variety. The weekly magazine of the entertainment business. Offers fascinating insights into the power of media institutions.

INTERNET RESOURCES

The Carnegie Foundation (www.carnegie.org/). A major sponsor of strategic research on education and educational policy in the United States. A good place to start searching for research studies.

Education Resources Information Center (ERIC) (www.aspensys.com/eric/index.html). The world's largest database for research and information about schools and education.

U.S. Department of Education (www.ed.gov). Provides information about federal education programs and initiatives, with links to other major education sites.

United Nations Educational, Scientific, and Cultural Organization (UNESCO) (www.unesco.org/). The UN's major education agency. Sponsors conferences and research programs on issues of global education, literacy, community development, professional development, and much more.

National Center on Educational Statistics (www.ed.gov/NCES/pubs/ce/index.html). Offers a wealth of data on school performance, especially in the United States.

CHAPTER 19

ECONOMIC INSTITUTIONS

E.L. Winthrop, an American businessman from a major city north of the Rio Grande, was on a vacation trip to Mexico. While there he traveled to a remote and rather quaint village somewhere in the province of Oaxaca. While wandering through the dusty streets of that little pueblo, which had neither electricity nor running water, he came upon an Indian man squatting on the earthen-floored porch of his palm hut. The Indian was weaving baskets.

These were not ordinary baskets. Winthrop had never seen such colors and such fine detail as the Indian was weaving into the little containers, using only natural fibers and dyes from the nearby jungle. Winthrop immediately decided that he must buy one no matter what the price.

When Winthrop learned that the Indian was selling his wares for only 50 centavos, perhaps a tenth of what he had expected to pay, Winthrop could not help thinking that here was a fine opportunity to start a thriving business. The baskets would make ideal containers for gift candy; he could sell them at home for three or four times the price. And even counting shipping costs and other expenses, both Winthrop and the Indian would surely make out very well.

"I've got big business for you, my friend," he announced to the surprised Indian. "Do you think you can make me one thousand of those little baskets?"

Winthrop imagined that the Indian would enlist the help of his family and kin in the project, and at the prices he could pay they all stood to gain tremendously.

That might be true enough, the Indian admitted, but who then would look after the corn and beans and goats? Yes, the Indian could see that he and his kin would have plenty of money to buy these things as well as many other items that they did not yet possess, but, he explained, "Of the corn that others may or may not grow, I cannot be sure to feast upon."

Not easily discouraged, Winthrop proceeded to show the Indian how wealthy he and his entire village could become even if they continued to sell the baskets for only 50 centavos. The Indian seemed interested, which spurred Winthrop on to more detailed descriptions of the basket assembly line they could begin together in the dusty little town. But after it was all explained, the Indian was still unmoved. The thousands of pesos that could be earned in this way were more than he could reckon with, since he had always used the few pesos from his basket selling only to supplement the modest needs of his family. There was another reason, too:

> Besides, señor, there's still another thing which perhaps you don't know. You see, my good lordy and caballero,

I've to make these canastitas my own way and with my song in them and with bits of my soul woven into them. If I were to make them in great numbers there would no longer be my soul in each, or my songs. Each would look like the other with no difference whatever and such a thing would slowly eat up my heart. Each has to be another song which I hear in the morning when the sun rises and when the birds begin to chirp and the butterflies come and sit down on my baskets so that I may see a new beauty, because, you see, the butterflies like my baskets and the pretty colors on them, that's why they come and sit down, and I can make my canastitas after them. And now, señor jefecito, if you will kindly excuse me, I have wasted much time already, although it was a pleasure and a great honor to hear the talk of such a distinguished caballero like you. But I'm afraid I've to attend to my work now, for day after tomorrow is market day in town and I got to take my baskets there. Thank you, señor, for your visit. Adios. (Traven, 1966, p. 72)

SOCIOLOGY AND ECONOMICS

This simple story captures the essence of what Max Weber meant when he observed that as societies grow in size, complexity, and rationality there is less enchantment in human life. The Indian could never imagine himself as a captain of industry, the manager of a workforce of Indians who would buy their corn in a store with a portion of their wages. But neither could Winthrop appreciate the full meaning of the Indian's refusal and the metaphor of his song. The Indian weaver feared that mass production would rob his baskets of their poetry and their maker of his soul. The businessman was used to dealing with markets and with the logic of profit and loss. He believed that he would be creating a "win-win" situation. He would profit, and he would be helping the Indian earn a better living for himself and his family. After all, the Indian was engaged in selling his baskets in the market. We will see in the Visual Sociology section at the end of this chapter that in the modern world there is a constant struggle to come out ahead in markets yet at the same time to maintain one's sense of creativity—one's soul. Nor is this the only dilemma forced upon us by modern economic institutions.

In this chapter we turn to an analysis of the major economic institutions of modern industrial societies. As we do so, it will be well to keep in mind examples from nonindustrial societies like that of the basket weaver in order to appreciate the enormous impact of economic and social change in the past two centuries. Such examples will also show that there is not just one path toward economic modernization, nor only one blueprint for how the economic institutions of a modern society should work.

In any society economic institutions specialize in the production of goods and services. How we survive in modern societies depends on whether we have jobs and income, on the nature of markets, on the public policies that involve governments in economic affairs, on worker–management relations, and so on. Thus it is little wonder that sociologists have dedicated much research to these questions. But if that is true, what distinguishes sociology from economics? In our review of the sociology of economic institutions we begin by distinguishing between these two social-scientific fields.

The well-known economist Paul Samuelson (1998) defines *economics* as the study of how people and social groups choose to employ scarce resources to produce various commodities and distribute them for consumption among various persons and groups in society. Sociologists are also concerned with how individuals and societies make choices involving scarce resources like time, talent, and money. But sociologists do not assume that scarcity or supply and demand are the only reasons for making such choices (Elster, 1997). Sociologists are deeply concerned with showing how the norms of different cultures affect economic choices. For example, we cannot assume that people with equal amounts of money or talent are free to choose the same values or activities. We often find that women who might have chosen to pursue careers in business management have been channeled into such occupations as teaching or social work. Until quite recently, American culture defined the latter occupations as appropriate for women, whereas managerial roles were viewed as appropriate for men. Women therefore did not have the same choices open to them even when they had equal amounts of scarce resources like talent.

It would not be accurate to stress only the differences between sociology and economics, however. Increasingly there are sociologists (and economists) who use the central theory of economics to study all situations in which people choose to allocate scarce resources to satisfy competing ends (Coleman, 1986; Hechter & Kanazawa, 1996). By "central theory" we mean the central proposition of economics: *People will attempt to maximize their pleasure or profit in any situation and will also try to minimize their loss or pain.* Applying this idea to nonbusiness situations, the Nobel Prize–winning economist and sociologist Gary S. Becker has conducted research on racial discrimination, marriage, and household choices. In this research he applies an economic "rational choice" approach to the study of the behavior of people in groups and organizations that do not produce goods or services. Becker points out that even the choice of a mate or the decision to have a child can be looked upon as an economic decision because people who make these decisions intuitively calculate the "price" of their actions in terms of opportunities gained or lost (Becker, 1997). Having a child means giving up other scarce resources, such as free time or income that could be used in other ways. And we saw in Chapter 16 how intuitive calculations enter into the choice of a marriage partner.

Following the lead of rational-choice theorists like Becker, sociologist Michael Hechter (1987) has studied the larger social consequences of the individual propensity to maximize pleasure and avoid pain. He concludes, among other things, that the larger the number of individuals in an organization, the more the organization will be required to invest in inspectors, managers, security personnel, and the like in order to control its members' propensity to "do their own thing." The same point was made over a century ago by Karl Marx: "All combined labour on a large scale requires . . . a directing authority. . . . A single violin player is his own conductor, an orchestra requires a separate one" (1962/1867, pp. 330–331).

Although it is tempting to apply principles of rational choice or exchange theory to noneconomic institutions like the family and the church, in this chapter we focus specifically on behavior in economic institutions such as businesses and markets and on the social implications of changes in these institutions. We will see that sociologists are continually comparing the ways in which such institutions are organized in different societies. The comparative study of how these differences relate to other features of a society's culture and social structure are very much a part of the sociological perspective on economic institutions.

Sociologists who study economic institutions are also interested in how markets for goods and services change as people learn to use new technologies and to divide up their labor in increasingly specialized ways (Drucker, 1992). They also look at how competition for profits may lead to illegal activities and cause governments to try to regulate economic institutions. In addition, sociologists attempt to show how the labor market is organized and how professions develop. Thus markets and the division of labor, the interactions between government and economic institutions, and the nature of jobs and professions are the major subjects of sociological research on economic institutions (Wallerstein, 1998). In the rest of this chapter we discuss each of these subjects in turn.

MARKETS AND THE DIVISION OF LABOR

A hallmark of an industrial society is the production of commodities and services to be exchanged in markets. In London, Hong Kong, New York, and other major cities throughout the world there are markets for gold, national currencies, stocks and bonds, and commodities ranging from pork bellies to concentrated frozen orange juice. However, not all societies have economies dominated by industry and markets. Subsistence economies, where people live in small villages like that of the Mexican basket weaver, do not have highly developed markets. In such economies people seek to produce enough food and other materials to enable them to meet their own needs, raise their children, and maintain their cultural traditions. The basic unit of production is the family, rather than the business firm as is the case in a capitalist economy. But there are few, if any, purely subsistence economies in the world today. Most subsistence economies must engage in some trade in international markets to obtain goods such as medicine or tools. These markets are dominated by the larger firms and entrepreneurs of the global capitalist economy, as we saw in the case of the basket weaver and Mr. Winthrop.

Socialist economies like Cuba or China also buy and sell products in international markets. Within their societies, however, they usually try to limit the influence of markets. Prices for essential goods, especially foodstuffs, are not entirely subject to the laws of supply and demand; often they are prevented by law from rising too high or falling too low. Many types of goods and services, including firms themselves, cannot be sold to private buyers because they are said to belong to all the people of the society. We will return to the conceptual differences among these types of economies later in the chapter. For now, remember that few contemporary societies conform to one or another of the basic economic types. Most nations encompass capitalist markets, subsistence economies, and elements of socialism as well. Nevertheless, the market is an increasingly dominant economic institution throughout the world.

The Nature of Markets

Markets are economic institutions that regulate exchange behavior. In a market, different values or prices are established for particular goods and services, values

BOX 19.1 USING THE SOCIOLOGICAL IMAGINATION

Markets in a Subsistence Economy

In precolonial times the Tiv, a small West African tribal society in what is now Nigeria, were a good example of a *subsistence economy,* one in which producers try to meet the needs of their immediate and extended families and do not produce goods or services for export beyond the family, village, or tribe. Among the Tiv the land was divided up into garden plots on which each family grew its own food. When the harvest was completed, the remaining vegetation was slashed and burned. The ashes served to fertilize the soil, but sometimes a plot was allowed to lie fallow for a few seasons. This meant that there was always a lot of discussion about which plots were to be used by which families in any given season. Never was a garden plot bought or sold—the idea that land could be bought and sold was not part of Tiv culture. The land belonged to the tribe as a whole. Individual families were granted the right to use the land, but there was no such thing as private property in the form of land that could be bought and sold.

Nor were the products of the land thought to be the property of the grower. Food was distributed according to need, and a person who accepted food was expected to return the gift at some future time. Only a few goods, mainly items of dress worn on ceremonial occasions, were traded for currency—in this case, short brass bars. The Tiv also accepted brass bars from foreign traders in exchange for carvings or woven cloths. But they would have laughed at the idea of trading the bars for food. They would have shared their food with strangers without any expectation of payment.

The conquest of Tivland by the British in the early twentieth century brought immense changes to this nonmarket society. The British colonial administrators were shocked when they observed Tiv weddings, in which it appeared to them that young women were being traded for ceremonial goods. In reality, these exchanges were not market transactions; the items that were given in exchange for a bride were a symbol of respect for the bride's family.

The British also noted that the Tiv seemed always to be at war, either with neighboring tribes or among themselves. To them this was a further sign that they needed to "civilize" the Tiv; in fact, however, the warfare was an outcome of the colonial system of taxation. Each tribal chief was required to pay a certain sum, either in British currency or in a commodity such as pigs. Because the Tiv had no currency and had to struggle to produce the required commodities, they were forced to cultivate more land. Their conflicts with neighboring clans or tribes were caused by the need to acquire more land. Thus the wars that the British observed were an almost immediate result of their own policies.

When the British succeeded in ending the warfare that had broken out among the Tiv, the only remaining source of cash to pay the taxes was wage labor on the colonial plantations. The Tiv therefore sent their sons and daughters to work for the British, and on their own land they began to grow crops that they could sell. Before long they were part of the British market economy, selling their crops and labor for cash with which they could buy manufactured goods. Eventually, like many other small tribal societies, they were engulfed by the larger market system of the more powerful European nations (Bohannon & Bohannon, 1953, 1962).

that vary according to changing levels of supply and demand and are usually expressed in terms of a common measure of exchange, or currency. A market is not the same thing as a marketplace. As an economic institution, a market governs exchanges of particular goods and services throughout a society. This is what we mean when we speak of the "housing market," for example. A marketplace, on the other hand, is an actual location where buyers and sellers make exchanges. Buyers and sellers of jewelry, for instance, like to be able to gather in a single place to examine the goods to be exchanged. The same is true for many other goods, such as clothing and automobiles.

Market transactions are governed by agreements or contracts in which a seller agrees to supply a particular item and a buyer agrees to pay for it. Exchanges based on contracts are a significant factor in the development of modern societies. As the social theorist Talcott Parsons pointed out, the use of contracts makes impersonal relations possible. Contracts neutralize the relevance of the other roles of the participants, such as kinship and other personal relationships, that govern exchanges in nonmarket situations. In contractual relations, for example, the fact that people are friends or kin does not, in principle, change the terms of their agreement and the need to repay debts (Parsons, 1991).

Among hunting-and-gathering peoples and in relatively isolated agrarian societies before the twentieth century, markets in the modern sense of the term did not exist. Think of the Mexican Indian basket weaver who could not imagine buying corn and other foods. He was used to growing foodstuffs himself and exchanging them with members of his extended family. The idea of buying them with currency was foreign to him. In social-scientific terms, a society cannot be said to have a fully developed market economy if many of the

In this idealized depiction of British colonial rule in Africa, white administrators confer in the foreground while happy natives work in the background. In reality, the "blessings" of colonialism were extremely mixed. As we see in the case of the Tiv, the experience of becoming part of the world market system made it difficult to maintain a small-scale society and culture.

commodities it produces are not exchanged for a common currency at prices determined by supply and demand.

The spread of markets into nonmarket societies has been accelerated by political conquest and colonialism as well as by the desire among tribal and peasant peoples to obtain the goods produced by industrial societies. To illustrate this point and to show what happens as a smaller-scale society becomes integrated into world markets, consider the case of the Tiv, a tribal society living in what is now Nigeria. The story of the Tiv is told in Box 19.1.

Markets and the World Economic System

In the late fifteenth and early sixteenth centuries, according to sociologist Immanuel Wallerstein (1998), a "European world-economy" came into existence. The new economy was a kind of social system that the world had not known before. It was based on economic relationships, not on political empires; in fact, it encompassed empires, city-states, and the emerging nation-states.

Great empires had been a feature of the world scene for at least 5,000 years before the dawn of the modern era. But the empires of China, India, Africa, the Mediterranean, and the Middle East were primarily political rather than economic systems. Wallerstein argues that because the great empires dominated vast areas inhabited by peoples that lacked military and political power, they were able to establish a flow of economic resources from the outlying regions to the imperial centers. The means used were taxation, tribute (payments for protection by the imperial army), and trade policies in which the outlying societies were

BOX 19.2 GLOBAL CHANGE AND U.S. SOCIETY

Two Faces of U.S. Corporations Overseas

The two news stories shown here, which appeared in *The New York Times,* graphically illustrate two contrasting and extremely common situations created by the globalization of business activity. As U.S. corporations move their manufacturing operations overseas in order to benefit from cheaper labor and capture new markets, their presence is greeted with a combination of enthusiasm and criticism, which also affects how Americans are regarded by people in other societies. This is true not only of American firms but of those from all the nations involved in the expansion of manufacturing and trade throughout the less developed world. In the case of American firms rushing to do business in Vietnam, however, there is a special irony. The billboard shown in the photo is located outside the main airport of Ho Chi Minh City (formerly Saigon)—but far from being seen as a sign of renewed American imperialism, it is a welcome indication that the United States is pouring vital capital into a former enemy nation.

Although social critics in the developing nations fear the growing influence of global capital, represented by major U.S. and other firms, most leaders welcome the flow of dollars regardless of political

Editorial and Op-Ed pages, 12-13
Education Advertising
Careers in Education and
Health Care Employment

The New York Times
Sunday, December 17, 1995

Week in Review
Section 4

Yankee Come Back

American Money Makes the Whole World Sing

By ALLEN R. MYERSON

ONCE again, American guns, carried by American soldiers, are heading into a distant fray, their reception in Bosnia at best uncertain. Memories of Somalia and, still, Vietnam, haunt every military venture overseas.

American butter, spread by American corporations investing abroad at a record pace, has meanwhile never been so widely welcomed. Although United States battleships and B-52's have in many places overstayed their welcome, Boeings filled with business executives and Commerce Department aides can land almost anywhere, even at Noi Bai Airport in Hanoi.

The new spirit has spread so thoroughly and rapidly, but with so little comment, that it's easy to forget how recently the United States was seen as an economic Godzilla. Nobody talks about Coca-Colonization anymore — "Coke welcomes you to Hanoi," say a pair of signs in the shape of giant bottles bracketing the airport highway — and seldom are Yankees bearing briefcases told to go home.

More Jobs at Home

As most economists agree, direct foreign investment, as distinct from portfolio investment in stocks and bonds, generally raises demand for American goods and services and creates jobs and wealth back home. Although some foreign spending has shifted jobs abroad, most goes to penetrate overseas markets.

In the first nine months of this year, American direct investment abroad was running at a record $74 billion annual rate, more than double the 1992 pace. By now, the world's thirst for American investment is taken for granted, so much so that exceptions seem downright quaint.

When noisy Indian nationalists recently halted construction of a $3 billion American power plant and health inspectors shut New Delhi's first Kentucky Fried Chicken, nobody in other countries applauded. Even many Indians snickered. "The Tale of Two Flies," they called the Kentucky Fried Chicken case, noting that by local standards, a restaurant with so few deserved to be celebrated, not punished.

The rest of the world is competing for American capital, to finance factories, public works, telecommunications, technology transfers and jobs. This nation's capitalists are eager to oblige, and not just in countries recuperating from Communism or rambunctious self-reliance.

The Vietnamese, among others, are happy to have former imperialists' investment, which is visible on the road to the Ho Chi Minh City airport.

forced to produce certain goods for the imperial merchants.

But this system—exemplified most clearly in the case of the Roman empire—required a huge military and civil bureaucracy, which absorbed much of the imperial profit. Local rebellions and wars continually increased the expense of maintaining imperial rule. Political empires thus can be viewed as a primitive means of economic domination. "It is the social achievement of the modern world," Wallerstein (1974) comments, "to have invented the technology that makes it possible to increase the flow of the surplus from the lower strata to the upper strata, from the periphery to the center, from the majority to the minority" (p. 16) without the need for military conquest.

What technologies made the new world system possible? They were not limited to tools of trade, such as the compass or the oceangoing sailing vessel, or to tools of domination like the Gatling machine gun. They also included organizational techniques for bringing land, labor, and local currencies into the larger market economy: ways of enclosing and dividing up land in order to charge rent for its use; financial and accounting systems that led to the creation of new economic institutions like banks; and many others. We discuss the role of science and technology in social change more thoroughly in Chapter 21, but it is important to note here that the term **technology** refers not only to tools but also to the procedures and forms of social organization that increase human productive capacity (Faulkner, 1997; Polanyi, 1944).

Although the European colonial powers (and the United States) often used political and military force to bring isolated societies into their markets, in the twentieth century they allowed their former colonies to gain independence yet still maintained economic control over them. This occurred because the economies of the colonial societies had become dependent on the

Business Day

The New York Times

In Peru, a Fight for Fresh Air

U.S.-Owned Smelter Makes Residents Ill and Angry

Photographs by Vera Lentz for The New York Times

The Southern Peru Copper's smelter in Ilo, Peru, spews 2,000 tons of sulfur dioxide into the air daily, casting a heavy fog that forces motorists to use headlights during the day. The company denies the plant has caused health problems.

By CALVIN SIMS

ILO, Peru — Every night before she goes to bed, Anna Gilapa gets down on her knees and stuffs cloth in the cracks around her doors and windows to protect her family from the acrid smoke released by the nearby smelter of the Southern Peru Copper Corporation.

Mrs. Gilapa fears for the health of her 1-year-old son, Brian, who suffers from severe respiratory problems and often coughs up thick mucous. And for that, she and other residents of this bleak coastal town 580 miles south of Lima blame the largest local company, Southern Peru Copper, which environmentalists say spews 2,000 tons of sulfur dioxide into the air each day, or 10 to 15 times the limit for similar plants operating in the United States.

At times, the smoke from the smelter is so thick that it hovers over the city like a heavy fog, forcing motorists to turn on their headlights during the day and sending residents to hospitals and clinics coughing, wheezing and vomiting. On those days, children are told to

Ana Gilapa stuffs cloth in the cracks of her doors and windows to prevent the smelter's acrid smoke from entering her home. Her 1-year-old son, Brian, suffers from severe respiratory problems. She blames the smelter.

ideology. In the 1990s direct investment by U.S. corporations to fund economic activities in other nations amounted to more than $70 billion, almost double the amount invested in the 1980s. Throughout the world there is fierce competition "for American capital, to finance factories, public works, telecommunications, technology transfers and jobs" (Myerson, 1995, p. 1). But the article about U.S.-owned smelting plants in Peru exemplifies the darker side of the rapid industrial growth that occurs when global firms establish plants in developing nations.

Southern Peru Copper's smelting plant in the city of Ilo emits 2,000 tons of noxious sulfur dioxide into the air each day. The pollution created by these emissions regularly forces drivers to use headlights in the middle of the afternoon. Families whose members work in the plant suffer from extremely high rates of respiratory illness. Brown sludge from the plant pollutes the nearby ocean shores. Ilo's mayor, Ernesto Herrera Becerra, says that emissions from the smelter have destroyed a large portion of the agricultural and fishing resources in the area. He adds: "This situation would not be allowed to exist in the United States" (quoted in Sims, 1995, p. D1). But plant officials note that their activities do not violate any local environmental or labor laws. For their part, the mayor and others point out that Peru had to have extremely liberal laws governing pollution, as well as lax enforcement of safety standards, in order to attract global firms in the first place.

technologies and markets controlled by the Western powers. Today former colonies are struggling to develop independent economic systems, but their ability to compete effectively in world markets is limited by the increasing power of **multinational corporations,** or *multinationals.* These are economic enterprises that have headquarters in one country and conduct business activities in one or more other countries (Barnet, 1994). (See Box 19.2.)

Multinationals are not a new phenomenon. Trading firms like the Hudson's Bay Company and the Dutch East India Company were chartered by major colonial powers and granted monopolies over the right to trade with native populations for furs, spices, metals, gems, and other valued commodities. Thus exploitation of the resources of colonial territories has been directed by multinational corporations for over two centuries. Modern multinationals do not generally have monopolies granted by the state, yet these powerful firms, based primarily in the United States, Europe, and Japan, are transforming the world economy by "exporting" manufacturing jobs from nations in which workers earn high wages to nations in which they earn far less. This process, which is particularly evident in the shoe, garment, electronics, textile, and automobile industries, has accelerated the growth of industrial working classes in the former colonies while greatly reducing the number of industrial jobs in the developed nations.

Changes in worldwide production patterns challenge Wallerstein's thesis that there is a world economic order dominated by the former colonial powers. In particular, industrial nations like the United States and Japan are increasingly losing manufacturing jobs while newly industrializing nations like Brazil and Indonesia are gaining them. But as Box 19.3 shows, the economic survival of millions of rural workers continues to be heavily influenced by changing markets in the highly developed nations.

BOX 19.3 GLOBAL CHANGE AND U.S. SOCIETY

Addictive Substances and World Markets

In 1983, when sociologist Edmundo Morales returned to his birthplace in northern Peru to conduct research on land reform, he found that the peasants of the Andes villages had become dependent on the worldwide traffic in cocaine. Their efforts to gain cooperative ownership of the land they farmed had faded. Instead, most of the able-bodied men in the villages were busy packing for the long trip down the eastern face of the Andes. Most of them would work for months in coca-processing laboratories hidden in the jungle. Some would earn cash income by selling food, especially pork and beef, to the managers of the cocaine factories. The demand for cocaine in the United States and other societies had changed their lives.

Morales was dismayed to see so much urgently needed food being shipped from the impoverished highlands to the jungles. He felt that the enormous growth of the cocaine market had led his people down a false path. They were gaining cash income, perhaps, but in the bargain they were becoming increasingly addicted to cocaine and were neglecting the need for more basic economic progress in their own villages (Morales, 1986, 1989).

Morales was able to use his knowledge of the culture and his training as a sociologist to identify the connections between the plight of the Andean peasants and the world market for addictive substances. He knew that cocaine was simply another in a long line of substances for which huge markets had emerged in the richer nations, with drastic results for the less developed world. Foremost among those substances was sugar. In the thirteenth century, sugar was a luxury available only to royalty. But as its use as a source of energy and as a basis for alcoholic beverages—especially rum—became more widespread, the demand for sugar production became one of the main causes of European expansion into the tropical regions of the world. Moreover, sugar production required that huge amounts of land be devoted to growing sugarcane, and many cane cutters were needed to harvest the crop. Thus along with the expansion of sugar production went slavery, first in parts of North Africa and the Azores and later in the Caribbean islands and Latin America (Bennett, 1954; Mintz, 1985). And sugar was only the beginning. The cultivation of wheat, tea, and coffee; the extraction of iron, tin, gold, silver, and copper; and the widespread cutting of primeval forests were all spurred by the emergence of mass markets in the developed nations of the West. Andean coca for cocaine and Indochinese poppies for heroin were latecomers in a series of crops that had changed the earth and all its peoples.

Some scholars argue that economic development in the world's commercial capitals actually produced *underdevelopment*—dependency and poverty and overpopulation—in huge parts of Africa, Asia, and Latin America. But Morales questions the notion of "development of underdevelopment." He believes that despite their tendency to become addicted to cocaine, the villagers of northern Peru are becoming more sophisticated in their market dealings and their understanding of politics. When they turn their attention once again to land reform, as eventually they must, their experiences in recent years will have made them less culturally isolated and more politically effective.

ECONOMICS AND THE STATE

The earliest social scientists were deeply interested in understanding the full importance of modern economic institutions. In fact, since the eighteenth century almost all attempts to understand large-scale social change have dealt with the question of how economic institutions operate. But the age-old effort to understand economic institutions has not been merely an academic exercise. The fate of societies throughout the world has been and continues to be strongly influenced by theories about how economic institutions operate, or fail to do so, in nation-states with different types of political institutions. The major economic ideologies thus are also political ideologies, and economics is often called *political economics* (Heilbroner, 1995). In this section we review three economic ideologies—mercantilism, capitalism, and socialism—and some variations of them. (See the study chart below.)

Political-Economic Ideologies

i

Mercantilism The economic philosophy known as **mercantilism,** which was prevalent in the sixteenth and seventeenth centuries, held that a nation's wealth could be measured by the amount of gold or other precious metals possessed by the royal court. The best economic system, therefore, was one that increased the nation's exports and thereby increased the court's holdings of gold.

The mercantilist theory had important consequences for economic institutions. For example, the guilds, or associations of tradespeople, that had arisen in medieval times were protected by the monarch, to whom they paid tribute. The guilds controlled their members and determined what they produced. In this way they were able to produce goods cheaply, and those goods were better able to compete in world markets, thereby increasing exports and bringing in more wealth for the court. But the workers were not free to seek the best jobs and wages available. Rather, the guilds required them to work at assigned tasks for assigned wages, and the guildmasters fixed the price of work (wages), entry into jobs, and all working conditions. Land in mercantilist systems was not subject to market norms either. As in feudalism, land was thought of not as a commodity that can be bought and sold—that is, a commodity subject to the market forces of supply and demand—but as a hereditary right derived from feudal grants.

Laissez-Faire Capitalism The ideology of **laissez-faire capitalism** attacked the mercantilist view that the wealth of nations could be measured in gold and that

STUDY CHART Political-Economic Ideologies

IDEOLOGY	DESCRIPTION	EXAMPLE
Mercantilism	An economic philosophy based on the belief that the wealth of a nation can be measured by its holdings of gold or other precious metals and that the state should control trade.	European nations under feudalism
Laissez-Faire Capitalism	An economic philosophy based on the belief that the wealth of a nation can be measured by its capacity to produce goods and services and that these can be maximized by free trade.	England at the time of the industrial revolution
Socialism	An economic philosophy based on the concept of public ownership of property and sharing of profits, together with the belief that economic decisions should be controlled by the workers.	The Soviet Union before 1989; China and Cuba today
Democratic Socialism	An economic philosophy based on the belief that private property may exist at the same time that large corporations are owned by the state and run for the benefit of all citizens.	Holland and the Scandinavian nations
Welfare Capitalism	An economic philosophy in which markets determine what goods will be produced and how, but the government regulates economic competition.	The United States

the state should dominate trade and production in order to amass more wealth. The laissez-faire economists believed that a society's real wealth could be measured only by its capacity to produce goods and services—that is, by its resources of land, labor, and machinery. And those resources, including land itself, could best be regulated by free trade in world markets (Halevy, 1955; Smith, 1910/1776).

The ideology of laissez-faire capitalism also sought to free workers from the restrictions that had been imposed by the feudal system and maintained under mercantilism. Thus it is no coincidence that the first statement of modern economic principles, Adam Smith's *The Wealth of Nations,* was published in 1776. Revolution was in the air—not only political revolution but the industrial revolution as well. And some of the most revolutionary ideas came from the pens of people like Adam Smith, Jeremy Bentham, and John Stuart Mill, who understood the potential for social change contained in the new capitalist institutions of private property and free markets.

Private property (as opposed to communal ownership such as existed among the Tiv) is not merely the possession of objects but a set of rights and obligations that specify what their owner can and cannot do with them. The laissez-faire economists believed that the owners of property should be free to do almost anything they liked with their property in order to gain profit. Indeed, the quest for profit would provide the best incentive to produce new and cheaper products. This, in turn, required free markets in which producers would compete to provide better products at lower prices.

These economic institutions are familiar to us today, but the founders of laissez-faire capitalism had to struggle to win acceptance for them. No wonder they thought of themselves as radicals. In fact, their economic and political beliefs were so opposed to the rule of monarchs and to feudal institutions like guilds that they could readily be seen as revolutionary. They believed that the state should leave economic institutions alone (which is what *laissez-faire* means). In their view, there is a natural economic order, a system of private property and competitive enterprise that functions best when individuals are free to pursue their own interests through free trade and unregulated production. As Smith put it, "Every individual . . . intends only his own gain, and he is in this, as in many other cases, led by an invisible hand to promote an end which was no part of his intention" (quoted in Halevy, 1955, p. 90). One of the most famous images in the history of the social sciences, Smith's "invisible hand" refers to the pervasive influence of the forces of supply and demand operating in markets for goods and services. Laissez-faire economists believe that in the long run these forces will improve the lives of people everywhere.

Socialism As an economic and political philosophy, **socialism** began as an attack on the concepts of private property and personal profit. These aspects of capitalism, socialists believed, should be replaced by public ownership of property and sharing of profits. As we have noted in earlier chapters, this attack on capitalism was motivated largely by horror at the atrocious living conditions caused by the industrial revolution. The early socialists thought of economics as the "dismal science" because it seemed to excuse a system in which a few people were made rich at the expense of the masses of workers. They detested the laissez-faire economists' defense of low wages and wondered how workers could benefit from the industrial revolution instead of becoming "wage slaves." They proposed the creation of smaller-scale, more self-sufficient communities that would produce modern goods but would do so within a cooperative framework.

Karl Marx viewed these ideas as utopian dreams. He taught that the socialist state must be controlled by the working class, led by their own trade unions and political parties, which would do away with markets, wage labor, land rent, and private ownership of the means of production. These aspects of capitalism would be replaced by socialist economic institutions in which the workers themselves would determine what should be produced and how it should be distributed.

Marx never completed his blueprint of how an actual socialist society might function. That chore was left to the political and intellectual leaders of the communist revolutions—Lenin, Leon Trotsky, Rosa Luxemburg—and, finally, to authoritarian leaders like Joseph Stalin, Mao Ze-dong, and Fidel Castro. They believed that all markets and all private industry must be eliminated and replaced by state-controlled economic planning, collective farms, and worker control over industrial decision making. Unless capitalist economic institutions were completely rooted out, they believed, small-scale production and market dealings would give rise to a new bourgeois class.

In the socialist system as it evolved under the communist regimes of the Soviet Union and China, centralized planning agencies and the single legal party, the Communists, would have the authority to set goals and organize the activities of the worker collectives, or soviets. Party members and state planners would also devise wage plans that would balance the need to reward skilled workers against the need to prevent the huge income inequalities found in capitalist societies. In Soviet-style societies, markets were not permitted to regulate demand and supply; this vital economic function was supposedly performed by government agencies. Societies that are managed in this fashion are said to have *command economies.* The state commands economic institutions to supply a specific amount (a quota) of each product and to sell it at a particular price. (Note

These children worked in a Rhode Island textile mill in the 1890s. Children were preferred as workers in this industry because it was believed that their nimble fingers helped maintain the pace of production.

that not all command economies are dominated by communist parties. Germany under the Nazis and Italy under the fascists were also command economies.)

With the collapse of the Soviet political empire, the economies of Russia and the other nations of the former Soviet Union (as well as its satellites in Eastern Europe) are attempting a transition to capitalist economic institutions. It has become evident that command economies provided these nations with overdeveloped industrial infrastructures and limited capability to compete in world markets for goods and services. Indeed, it has been said that the command economic system of the former Soviet Union gave it "the most impressive nineteenth century industrial infrastructure in the world, 75 years too late" (Ryan, 1992, p. 22).

The Soviet system was notorious for the inefficiency of its economic planning and industrial production. Under the command system, factory managers must continually hoard supplies, or raid supplies destined for similar factories in other regions, in order to meet their production quotas. Or they must trade favors with other factory managers to obtain supplies, a system that creates a hidden level of exchange based on bribes and favoritism. And because there are no free markets, goods that are desired by the public often are not available simply because the planners have not ordered them. In fact, in the Soviet-style economies there were always clandestine markets that operated outside the control of the authorities and supplied goods and services to those with the means to pay for them. These underground markets thereby increased inequality in those societies and generated greater public disillusionment with command economies and one-party communist rule.

The massive social upheavals of 1989 and 1990 brought an end to the communist command economic system, but they also initiated a period of great economic and social uncertainty in the former Soviet-dominated nations. At present, efforts to privatize their economies and use market forces to regulate the supply of goods and services, including labor, have begun to produce economic growth in some of those nations. Rapid changes have also produced new forms of insecurity and corruption that have caused many people there to wish for the relative stability and security of the communist past (Ash, 1999; Woo, 1994). These problems are made even worse by the collapse of the political system in some nations of the former Soviet empire, which has enabled organized criminal gangs known as the *mafiya* (not related to the Mafia) to flourish. Without the rule of law established by a legitimate system of government, it is extremely difficult to modernize an economy and prevent severe abuses of power through violence and terror. (The role of legitimacy is discussed further in Chapter 20.) In recent elections in Russia, Slovakia, and the former East Germany, many people voted for former Communist party officials who wish to slow down the conversion to a capitalist economy (Tagliabue, 1998).

Democratic Socialism A far less radical version of socialism than that attempted in the Soviet bloc is known as **democratic socialism.** This economic philosophy is practiced in the Scandinavian nations, especially Sweden, Denmark, and Norway, as well as in Holland and to a lesser extent in Germany, France, and Italy. It holds that the institution of private property must continue to exist because people want it to, and that competitive markets are needed because they are efficient ways of regulating production and distribution. But large corporations should be owned by the nation or, if they are in private hands, required to be run for the benefit of all citizens, not just for the benefit of their stockholders. In addition, economic decisions should be made democratically.

Democratic socialists look to societies like Sweden and Holland for examples of their economic philosophy in practice (Harrington, 1973). In Sweden, for instance, workers can invest their pension benefits in their firm and thereby gain a controlling interest in it. This process is intended to result in socialist ownership of major economic organizations.

In the United States there is a long history of conflicted relations between workers and owners of capital. Cooperative systems in which authority and even ownership are shared are developing, but more slowly than in the social democracies of Europe. Social scientists who study the U.S. economy note that in recent years there has been a reaction among owners of capital against the principles of democratic socialism, especially those that stress cooperation between labor and management and the right of workers to organize unions and engage in collective bargaining (Freeman, 1993). Employers have also shown diminished commitment to welfare capitalism, the political-economic model that until recently has had a great deal of influence in the noncommunist modern economies (Harrison, 1994).

Welfare Capitalism Emerging to some extent as a response to the challenge posed by the Russian Revolution (which called attention to many of the excesses of uncontrolled or laissez-faire capitalism), **welfare capitalism** represented a new way of looking at relationships between governmental and economic institutions. Welfare capitalism affirms the role of markets in determining what goods and services will be produced and how, but it also affirms the role of government in regulating economic competition (e.g., by attempting to prevent the control of markets by one or a few firms).

Welfare capitalism also stresses the role that governments have always played in building the roads, bridges, canals, ports, and other facilities that make trade and industry possible. Expanding on this role, the theory of welfare capitalism asserts that the state should also invest in the society's human resources—that is, in the education of new generations and the provision of a minimum level of health care. Welfare capitalism also guarantees the right of workers to form unions in order to reach collective agreements with the owners and managers of firms regarding wages and working conditions. It creates social-welfare institutions like Social Security and unemployment insurance. And in order to stimulate production and build confidence in times of economic depression, welfare capitalism asserts that governments must borrow funds to finance large-scale public works projects like the construction of the American interstate highway system during the 1950s.

The theory of welfare capitalism is associated with the writings of John Maynard Keynes, Joan Robinson, John Kenneth Galbraith, and James Tobin, all of whom contributed to the revision of laissez-faire economic theory. Welfare capitalism dominated American economic policy from World War II until the 1970s, when a succession of economic crises—inflation, energy shortages, and unemployment—turned the thoughts of many Americans once again toward laissez-faire capitalism. Today, however, the increasingly evident gap between the haves and the have-nots in American society has given rise to renewed interest in enhancing the institutions of the welfare state—unemployment insurance, health insurance, low-income housing, public education, and others. At the same time, fear of governmental control over people's lives, and its costs in the form of higher taxes, has generated a revolt against the "welfare state" in many segments of the population in the United States and other industrial nations. This trend has resulted in conservative electoral victories and unprecedented reductions in social spending. What is not clear is whether the public's desire for relative security from economic stress will outweigh the desire for lower taxes and reductions in the scope of government bureaucracies.

The United States: A Postindustrial Society?

Although many people in the United States see the failure of socialism in the Soviet bloc as a victory for capitalism, sociologists and poll takers have called attention to a growing sense of unease about the U.S. economy. As demographer Valerie Kincade Oppenheimer observes, "Since about 1970, the average real earnings position of young men has deteriorated considerably. This has been true not only for high school dropouts but also for high school and college graduates" (1994, p. 331). (Figure 19.1 shows the value of continuing one's education through college and beyond.) The declining incomes of U.S. workers over the past quarter century have contributed to the growth in dual-earner households.

The disparity between the incomes of people with professional and technologically sophisticated jobs and those of people with less intellectually demanding jobs is likely to increase as economic growth centers on computers, telecommunications, and biological technologies. Jobs that require lower levels of education and

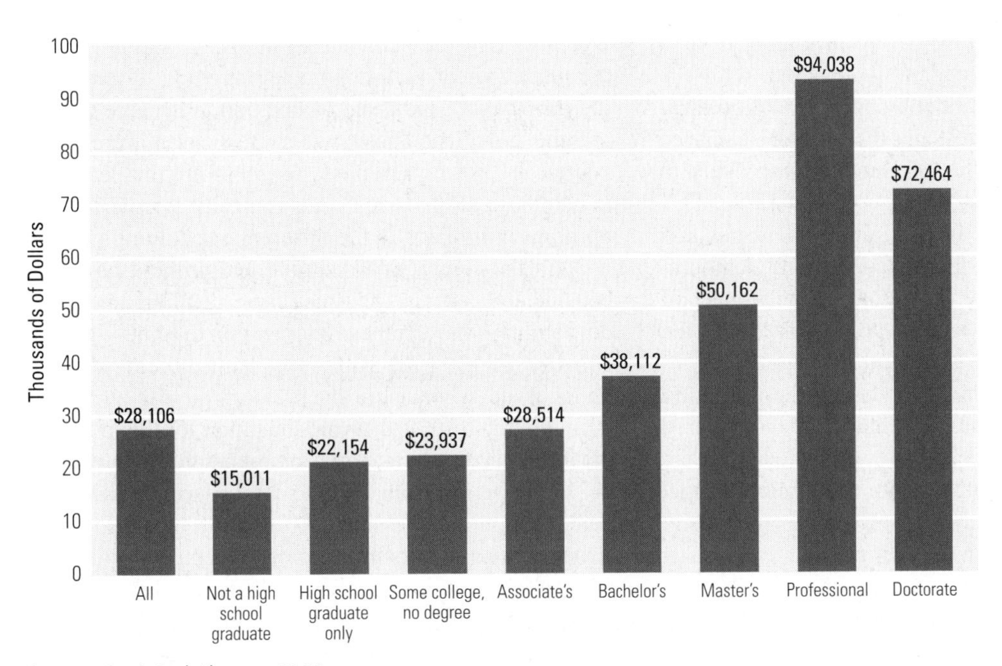

Source: *Statistical Abstract,* 1998.

FIGURE 19.1 **Mean Earnings, by Highest Degree Earned, 1997**

training may be more onerous, but the number of people who can perform them is relatively large. The basic economic laws of supply and demand therefore drive up the wages of highly educated workers while those of less-educated workers are driven downward.

In an influential study, *The Coming of Post Industrial Society,* sociologist Daniel Bell (1973) reviewed changes in American economic institutions indicating that we are undergoing a transition from an industrial to a postindustrial society. A postindustrial society "emphasizes the centrality of theoretical knowledge as the axis around which new technology, economic growth, and the stratification of society will be organized" (p. 116). As societies undergo the transition to postindustrial economic institutions, an "intellectual technology" based on information arises alongside machine technology. Industrial production does not disappear, but it becomes less important. New industries devoted to providing knowledge and information become the primary sources of economic growth (Block, 1990; Sabel, 1995).

Changes in the distribution of jobs in the goods-producing and service-producing sectors of the American economy clearly show the growing importance of services, especially those employing professional and technical workers. For Bell, this change is inevitable. Industries decline, but new ones appear. A century ago, for example, more than 70 percent of Americans were employed in agriculture and related occupations. Now this figure is below 3 percent. Similar transformations are taking place in manufacturing today, and they can be made less painful if the society is willing to invest in education and the retraining of workers in declining industries like mining, steel production, or manufacturing for military purposes (Harrison, 1994).

Other social scientists view these changes less optimistically. Some note that declines in industrial production have pushed thousands of skilled workers into lower-paying jobs in the service sector—for example, in the fast-food industry or as janitors or security guards—which helps explain the decline in average wages (Aronowitz, 1996; Fallows, 1996). Others are concerned that the gains made by blacks and women in access to better-paying industrial jobs since World War II are being erased. William Julius Wilson (1998), for example, has observed that access to control over the means of production is increasingly based on educational criteria and thus threatens to solidify the subordinate position of the black lower class.

European sociologists have criticized the theory of postindustrial society on even broader grounds, saying that it tends to promote belief in the inevitability of an increasingly impoverished working class with even more reason to demand a more equal distribution of wealth (Touraine, 1998). In the United States, the theory of postindustrial society has been criticized on the ground that even though manufacturing accounts for a declining share of the labor force, the society still depends on manufactured goods of all kinds. Critics argue that the United States cannot relinquish its manufacturing base simply because a theory says it is outmoded (Wilson, 1998).

A Changing Social Contract? Conversion from a wartime to a peacetime economy, together with the massive changes brought about by economic globalization, technological advances, and the growth of service industries, is increasing the pressure on employers to keep their costs down in order to be competitive. This results, as we have seen, in declining real wages and widening economic inequalities. In the social democracies of Europe, especially Scandinavia, the impact of these changes is eased by social-welfare policies. In the United States, the welfare state and its "safety net" of economic and social policies (Medicare, the minimum wage, unemployment insurance, etc.) are considered important by those who need to be protected from the harshest consequences of sudden economic change. Others, however, attack them as wasteful and costly government intervention in the economy.

The 1994 congressional elections brought in a conservative leadership committed to a "contract with America" that promised to balance the federal budget, reduce the size of government, and cut taxes. These and related measures are consistent with laissez-faire principles that argue against government intervention in the economy and insist that individual choices about how to spend income and wealth should be protected to the extent possible. But these are controversial issues. As can be seen in Figure 19.2, there is a deep division between those who believe government should reduce income inequality and those who oppose taxation or other policies that reduce the wage and income gap. (Also see Box 19.4.)

Note that the percentages in Figure 19.2 represent the responses of people who felt strongly, one way or the other, about the issue of taxation to achieve greater income equality. Those whose responses were more neutral are not represented. Yet these are the people in the "middle of the road" whose shifting allegiances and opinions will make all the difference in coming years.

As the figure shows quite clearly, there has been considerable change in Americans' opinions about the desirability of government intervention to achieve greater equality. In 1994 the public was rather evenly divided on the issue, whereas for the previous quarter century or more the majority had favored taxation for this purpose. Note, however, how volatile opinions on this issue can be. In the period from 1988 to 1991, during a rather severe economic recession, 32 percent of respondents strongly favored taxation to reduce inequality; that proportion fell to only 24 percent in the relatively better economic climate of 1994. It appears that the conservative attack on the welfare state has persuaded large numbers of Americans to oppose taxation for the purpose of reducing inequality, but it remains to be seen how stable this trend will be as the gap between rich and poor widens.

In his analysis of postindustrial society, Daniel Bell (1973) assumed that the "social contract" in the United States and other advanced nations would be maintained. In return for hard work, saving, and efforts to improve their skills, citizens could expect modest increases in their living standards, adequate health care, protection

FIGURE 19.2 **Trends in Opinion on the Question "Should Government Reduce Income Inequality?" (United States, 1972–1994)**

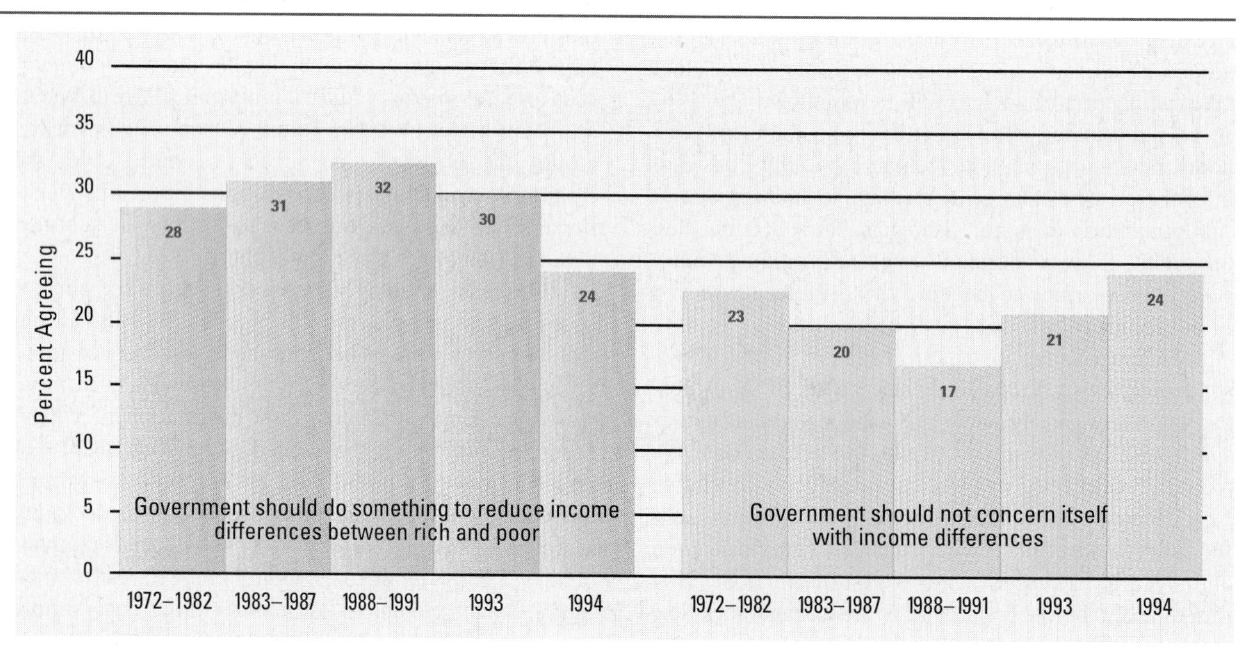

Source: National Opinion Research Center, 1994.

BOX 19.4 SOCIOLOGICAL METHODS

How the NORC Asks Americans About Policies to Reduce Inequality

Much of the data about how people in the United States view social issues comes from the General Social Survey (GSS) conducted by the National Opinion Research Center (NORC) at the University of Chicago. One of the most prestigious opinion surveys in the nation, the GSS is a random sample survey that has been conducted annually since 1972. It therefore allows social scientists to track changes in opinion over time.

The question about government intervention to reduce income inequality is an example of a question for which the GSS provides longitudinal data. Note that Figure 19.2 presents data from the two ends of the scale, but not from the middle, in order to compare changes in the distribution of liberal and conservative opinions on this critical issue. The actual data, shown here in a table from the *NORC Codebook,* include other categories of responses along with hundreds of nonapplicable responses from people who were not asked this particular question in a given year.

Note that this question emphasizes actions that "the government in Washington ought" to take. In a sense, therefore, it combines two issues: whether income inequality is a problem that should be addressed, and what policies should be used to address it—that is, whether government ought to raise taxes on the wealthy and/or provide assistance to people with limited incomes. No doubt many people respond to this question on the conservative side of the scale not because they feel comfortable with the existing income distribution but because they do not want government to have the dominant role in dealing with this or any other major social issue. To identify the independent effects of these different attitudes, the investigator would ask additional questions about income inequality and would cross-classify the responses to those questions with the responses to the question about government intervention.

*Some people think that the government in Washington ought to reduce the income differences between the rich and the poor, perhaps by raising the taxes of wealthy families or by giving income assistance to the poor. Others think that the government should not concern itself with reducing this income difference between the rich and the poor.**

Here is a card with a scale from 1 to 7. Think of a score of 1 as meaning that the government ought to reduce the income difference between rich and poor, and a score of 7 as meaning that the government should not concern itself with reducing income differences. What score between 1 and 7 comes closest to the way you feel?

(CIRCLE ONE)

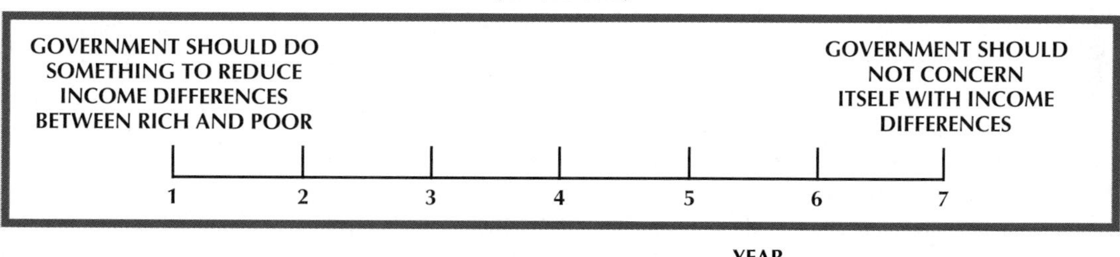

RESPONSE**	YEAR 1972–1982	1983–1987	1988–1991	1993	1994
Government should 1	392	1,237	764	184	289
....2	218	605	460	122	183
....3	368	977	709	201	322
....4	447	1,135	793	187	420
....5	261	719	440	130	292
....6	163	431	278	81	167
Goverment should not 7	321	784	402	130	303
Don't know.............................8	51	98	91	18	27
No answer..............................9	8	22	11	4	8
Not applicable ....BK	11,397	1,534	1,959	549	981

*This is the exact wording of the question as presented to respondents.

**The denominators used in calculating the percentages presented in Figure 19.2 are the totals of the responses in which an opinion was expressed (items 1–7). The figures in the "Don't know," "No answer," and "Not applicable" categories are not included.

Source: Adapted from 1994 *NORC Codebook.*

from crime, decent public education for their children, and protection of their rights. However, the Reagan administration of the 1980s and the conservative Congress of the 1990s have challenged the assumption that government would provide these benefits if they were not provided by the economy itself. This alteration in the basic social contract that has guided social policy for the past half century will have far-reaching consequences (Blau, 1993; Davey, 1995).

WORKERS, MANAGERS, AND CORPORATIONS

During the 1990s the United States enjoyed its longest period of sustained economic growth in the twentieth century. Unemployment dropped to lows not seen since the post–World War II boom years. Worker productivity increased steadily. In the financial markets there were record increases in the prices of securities and mutual funds, and since a large proportion of U.S. employees have their pension funds in these securities, there was a widespread sense of well-being during most of the decade. This positive outlook was not shared by workers in the lower ranks of labor, who continued to experience erosion of their spending ability. However, in relation to the major European and Asian economies, where economic crisis and chronic unemployment were serious problems, the U.S. economy assumed a position of almost undisputed dominance.

In the midst of this overall prosperity, sociologists note that there remain unresolved social problems. Feminist sociologists point to the persistence of discrimination against women in the workplace. Others warn of worker alienation, while still others focus on the increasing power and dominance of corporations over the lives of individuals (Drennan, Tobier, & Lewis, 1996; Odaka, 1998; Touraine, 1998).

Gender and Workplace Diversity

One of the most outstanding results of affirmative action efforts over the past generation has been the entry of women into jobs and professions that were formerly closed to them. While affirmative action to correct racial discrimination and inequality has drawn fire from critics of quotas and admissions policies in education, conflicts arising in the workplace have also been fierce on occasion. In 1998, for example, Texaco Inc. agreed to settle a gender discrimination claim by a group of its female employees. The women were working at all levels of the company but were paid less than their male counterparts for the same work. While not admitting its guilt, Texaco agreed to pay $3.1 million in back pay to the 186 women in the lawsuit brought on their behalf by the U.S. Department of Labor. This settlement sent a clear signal to other major domestic and global corporations that the push for gender equality in workplaces continues (Stevenson, 1999). In fact, in the same year the Supreme Court ruled against the Virginia Military Institute's exclusion of women and in so doing made it clear that affirmative action to achieve gender equality is justified on economic grounds (Skaggs, 1998).

Feminist sociologists note that as important as these victories may be, they should not convey a sense of uniform progress toward greater gender equality and tolerance of diversity in the workplace (Bradley, 1998; Epstein, 1995). Major corporations like Texaco often adopt far more conciliatory positions on these issues than do smaller corporations, which are less likely to be in the public spotlight. For example, the Walt Disney Company recognized the growing diversity of its labor force by conceding the right of cohabiting adults, no matter what their gender, to receive the same health and other employee benefits as heterosexual married couples. But for every company like Disney there are scores of others in which the forms of inequality that produce differences in pay for women and members of minority groups persist.

Globalization of the activities of major corporations has an immense impact on gender relations outside the United States. On the Mexican border, for example, there are hundreds of so-called *maquilladora* industries. These are factories that do production work of all kinds, often on a contract basis, for larger corporations. Women and men employed in these factories often experience conditions and hours of work that would be illegal in the United States. But in Mexico workers are often thankful to have the work and feel that they must endure the hardships. As *maquilladora* workers begin to protest and seek to join labor unions, they find themselves enmeshed in class conflict and labor struggles that may be entirely new to them. The experiences of earning their own wages and, especially, taking on leadership roles at work often motivate female workers to challenge traditional male-dominated family and community relations as well. Thus experiences in the workplace produce a movement for more egalitarian gender relations in other institutions. Nor is this trend limited to Mexico. Sociologists who study economic globalization and gender relations are finding many instances of these changes throughout the world (Cowie, 1998; Nelson, 1998).

Increases in the demand for part-time workers, the trend toward contracting for services by freelance workers (who may be highly skilled professionals), decreases in the willingness of employers to pay health care benefits, and efforts to prevent employees from joining labor unions are only a few of the major trends in the economies of industrialized nations that are causing workplace conflict and employee alienation. As we see next, alienation is not a new problem, but it takes on new aspects as the nature of work and economic institutions changes.

Employee Mentoring Versus Worker Alienation

Many sociologists use the concept of alienation to explain the gap between workers and managers. Karl Marx first applied the term to the situation of workers under early capitalism. The worker in a factory performed only a fraction of the work that went into a product and therefore could feel little sense of ownership of the final product. And because the worker typically could not control the work process very much, he or she came to feel like a mere cog in a giant machine, a feeling that produced a sense of not being able to control one's own actions on the job (Fromm, 1961; Marx, 1961/1844). Today the term **alienation** is often used to describe the feeling of being powerless to control one's own destiny (Burawoy, 1996; Hunter, 1998). At work, people may feel alienated because their labor is divided up into activities that are meaningless to them. For example, in a classic study of autoworkers, oil workers, and others in different factory settings, Robert Blauner (1964) found that workers who perform highly repetitive tasks on an assembly line are more alienated than workers in groups whose tasks involve teamwork.

In a study of organizational techniques that have been used successfully in Japan, Ezra Vogel (1979) pointed out that Japanese firms tend to avoid worker alienation by strengthening the role of the small group:

> The essential building block of a company is not a man with a particular role assignment and his secretary and assistants. The essential building block of the organization is the section. . . . The lowly section, within its sphere, does not await executive orders but takes the initiative. . . . For this system to work effectively leading section personnel need to know and to identify with company purposes to a higher degree than persons in an American firm. They achieve this through long experience and years of discussion with others at all levels. (pp. 143–145)

The irony of this finding and others like it is that even before World War II American and European industrial sociologists had demonstrated the benefits of team approaches to work (McCord, 1991). The famous studies by Elton Mayo, discussed in Chapter 2, were designed to find out how workers at any level of an organization could be made to feel less alienation and a greater sense of ownership of their jobs. This research demonstrated the importance of primary-group relationships at work. But the function of the manager is to motivate employees to work more efficiently and at higher levels, a goal that is not easily achieved (Hickins, 1998).

Large companies are often extremely concerned about worker alienation. Because they invest heavily in personnel training, losing employees to other businesses or having their productivity decrease because of feelings of discontent can result in lower profits as well as poor morale. Employee mentoring programs are a popular approach to preventing alienation. Mentoring usually matches an experienced employee with a newcomer; it is also used to groom employees for advancement in the organization.

In essence, a mentor is a personal teacher and helper. The experienced employee shows the newcomer how to do the work more effectively, how to avoid "burnout" (the sense that one has no more energy to devote to job), how to deal with the quirks and foibles of certain supervisors, and much more. The mentoring approach is widely used in business, education, health care, and government—and of course some programs are more effective than others.

Sociologists or people with sociological training are often asked to evaluate mentoring programs. For example, in one large metropolitan hospital evaluators studied more than 600 professional, administrative, and clerical employees who were involved in a mentoring program. They found a strong negative correlation between time spent in mentoring and measures of employee alienation. This means that the more mentoring an individual received, the less alienation he or she felt. The evaluation also suggested ways to measure the effectiveness of mentoring aside from merely looking at whether the employee had a mentor or how much time the mentor and the employee spent together. Overall, organizations that institute mentoring programs and attempt to improve them through ongoing monitoring and evaluation are extremely pleased with the results (Koberg, Chapell, & Ringer, 1994).

Individuals and Corporations

One possible source of worker alienation is the increasing power of the corporation relative to the individual. In the United States and other industrialized societies, the corporation has gained ever greater dominance over other economic institutions. Although they may have been established by entrepreneurs like Andrew Carnegie or John D. Rockefeller, who took great risks to build personal empires, modern corporations are bureaucratic organizations whose executive leadership may, at least in principle, be replaced. Sociologists point out that the idea of the *corporate actor* was essential to the development of the modern corporation (Zafirovski & Levine, 1997). The pioneering economic sociologist James S. Colemen noted that "as economic enterprise outgrew the family, it became useful for several persons to join together to carry out the enterprise. They were able to do this only if they were protected [as individuals] from the liabilities that the joint enterprise might incur" (1982, p. 8). This was accomplished by the creation of a new class of fictional "persons"—corporations—that could incur debts and liabilities on their own account. The liability of any individual member of the corporation was limited to the amount that he or she had invested in it.

The emergence of the concept of the corporate actor had far-reaching effects. By the beginning of the twentieth century, the importance of corporate actors (most of which are bureaucratic organizations) had begun to outweigh the importance of individuals in most of the institutional sectors of society. These trends are illustrated in Figures 19.3 and 19.4, which show the dramatic rise in the number of corporations over the past half century and the changing balance in the participation of individuals versus corporate actors in cases heard in the New York State Court of Appeals. "What these changes suggest," Coleman concluded, "is a structural change in society over the past hundred years in which corporate actors play an increasing role and natural persons play a decreasing role" (1982, p. 15).

A further effect of the dominance of corporate actors is that the individual is at a disadvantage in dealings with these entities. Today, corporate actors generally have access to much more information, derived from surveys, advertising, credit ratings, and the like (Faulkner, 1997). In conflicts that reach the courts, although the corporate actor and the individual are equal parties in the eyes of the law, in fact the corporate actor usually has the upper hand as it is backed up by numerous specialists, whereas the individual, even with the best legal aid, usually can draw on only limited resources.

Social movements like the labor movement and the consumer and environmental protection movements

FIGURE 19.3 **Growth in Number of Corporations in the United States, 1916–1968**

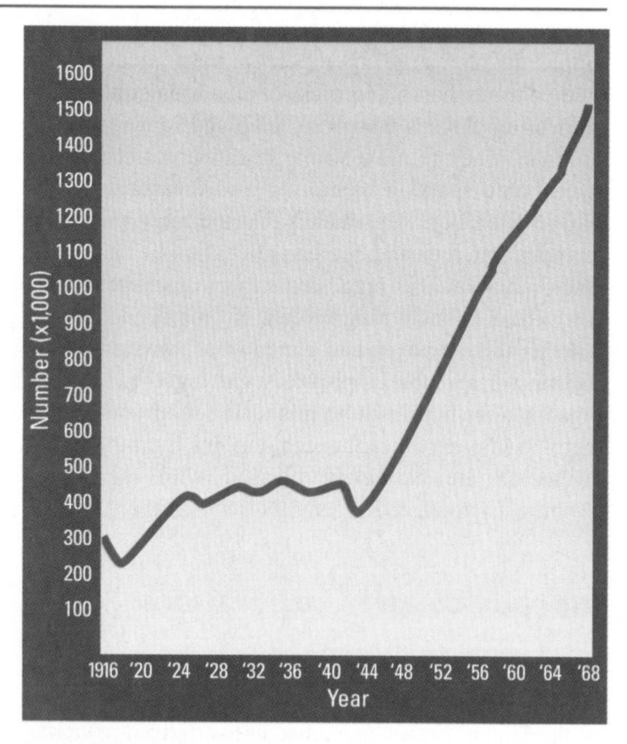

Source: "Distribution of Economic Resources in the United States," an unpublished paper prepared for *Actions of Individuals and Groups as Indicators of Power Distribution,* a project sponsored by the National Science Foundation. National Opinion Research Center, Chicago, 1974.

FIGURE 19.4 **Participation of Persons and Corporate Actors in Court Cases: New York State Court of Appeals, 1853–1973**

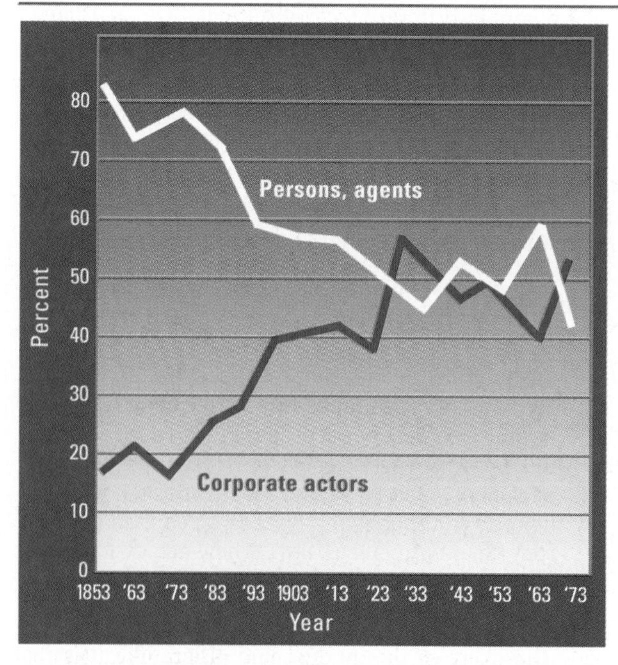

Source: "A Study of the Realtive Participation of Persons and Corporate Actors in Court Cases," an unpublished paper prepared for *Actions of Individuals and Groups as Indicators of Power Distribution,* a project sponsored by the National Science Foundation. National Opinion Research Center, Chicago, 1974.

attempt to reduce the asymmetry or imbalance in these relationships (for example, they seek to regulate corporations by making them reveal information). Countermovements try to restore the rights of corporate actors (as in the movement to deregulate business and industry). Such movements are one consequence of the imbalance between the individual and the corporate actor. Another consequence may be an increase in the number of deviant acts directed against corporations of all kinds. People may cheat or steal from corporate actors—for example, they may take out their frustrations by destroying public telephones, scrawling graffiti on mass transit facilities, or illegally tapping into cable television boxes—without feeling nearly as guilty as they might if their actions were directed against individuals. Sociological perspectives on work and workers can offer some leads to further progress in dealing with such problems in the future.

Sociological Perspectives on the Workplace

Industrial sociology is concerned with the social organization of work and the types of interactions that occur in the workplace. Like sociologists in other fields, industrial sociologists use the functionalist, conflict, and interactionist perspectives in their research. The results of that research have been used to support various approaches to labor–management relations.

Human Relations in Industry: A Functionalist Perspective The *human relations* perspective on management is associated with the research of Elton Mayo and his colleagues at Western Electric's Hawthorne plant and in aircraft and metal production plants. Their goal was to use experimental methods and observation of workers and managers on the job in an attempt to understand how the factory's formal organization and goals are affected by patterns of informal organization within the workplace. This was in sharp contrast with earlier approaches to labor–management relations, especially the *scientific management* approach developed early in the twentieth century by Frederick W. Taylor.

Scientific management, or Taylorism as it is often called, was one of the earliest attempts to apply objective standards to management practices. After rising through the ranks from laborer to chief engineer in a large steel company, Taylor turned his attention to the study of worker productivity. He noticed that workers tend to adhere to informal norms that require them to limit their output. For example, he found that the rate of production in a machine shop was only about one third of what might normally be expected. Taylor decided to use the authority of management to speed up the workers. He fired stubborn men, hired "green hands" who did not know the norms of the experienced workers, and experimented with ways of breaking down the labor of each worker into its components. Every job, he claimed, could be scientifically studied to determine how it could be performed most efficiently. Such "time and motion studies," combined with piecework payment systems that induced workers to produce more because they were paid for each unit produced above a set number, became the hallmarks of scientific management (Miller & Form, 1964). Taylor's principles were quickly incorporated into the managerial practices of American businesses, but they were often resisted by workers and their unions (Braverman, 1974).

Mayo's experiments were intended to determine what conditions would foster the highest rates of worker productivity. His observations convinced him that increased productivity could be obtained by emphasizing teamwork among workers and managers, rather than through pay incentives or changes in such variables as lighting, temperature, and rest periods (Bendix, 1974). We saw in Chapter 2 that Mayo's research showed that it was the attention given to the workers that mattered, not the various experimental conditions. But Mayo drew another, more important conclusion from the Hawthorne experiments:

> The major experimental change was introduced when those in charge sought to hold the situation humanly steady . . . by getting the cooperation of the workers. What actually happened was that six individuals became a team and the team gave itself wholeheartedly and spontaneously to cooperating in the experiment. (Mayo, 1945, pp. 72–73)

Although Mayo's research focused on the interactions between workers and managers, the human relations approach that grew out of that research can be said to represent the functionalist perspective on the workplace because it stresses the function of managerial efforts in increasing worker productivity. The functionalist perspective is also illustrated by William F. Whyte's classic study of the restaurant industry, which we encountered in Chapter 4. In a small restaurant, the organization's structure is simple: "There is little division of labor. The owner and employees serve together as cooks, countermen, and dishwashers." But when the restaurant expands, a number of supervisory and production occupations are added to its role structure. According to Whyte, this magnifies old problems and gives rise to new ones:

> In a large and busy restaurant a waitress may take orders from fifty to one hundred customers a day (and perhaps several times at each meal) in addition to the orders (much less frequent)

> she receives from her supervisor. When we add to this the problem of adjusting to service pantry workers, bartenders, and perhaps checkers, we can readily see the possibilities of emotional tension—and, in our study, we did see a number of girls break down and cry under the strain. (1949, p. 304)

Whyte discovered that tension and stress could be reduced, and customers served more happily, if the restaurant was organized in such a way that lower-status employees were not required to give orders directly to higher-status ones. For example, waiters and waitresses should not give orders directly to cooks but should place their orders with a pantry worker or use a system of written orders. By changing the organization of statuses and roles in the restaurant (and, thus, the way they functioned as a social system) it would be possible to increase work satisfaction and output.

Conflict at Work The human relations approach just discussed seeks to improve cooperation between workers and managers in order to achieve the organization's goals. Industrial sociologists who take a conflict perspective feel that this approach automatically condones the goals of managers and fails to consider more basic causes of worker–management conflict such as class conflict (Touraine, 1998). This results in continual fine-tuning of organizational structures rather than in more thorough reforms.

In an influential study of a midwestern metal products factory, Michael Burawoy (1980) found that even after changes were made on the basis of the human relations approach, the workers continued to limit their output. The workers called this "making out" and saw it as a way of maximizing two conflicting values: their pay and their enjoyment of social relations at work. But instead of wondering why the workers refused to produce more, Burawoy asked why they worked as hard as they did. He concluded that the workers were actually playing into the hands of the managers and owners because neither they nor the unions ever questioned management's authority to control basic production decisions. The norms of the shop-floor culture, of which "making out" is an example, caused the workers to feel that they were resisting the managers' control. As a result, they did not feel a need to engage in more direct challenges to the capitalist system itself (Burawoy, 1996).

A fundamental problem for conflict sociologists is that class conflict must be shown to exist; it cannot be assumed to exist simply because Marxian theory defines workers and business owners as opposing classes. In fact, there is considerable evidence that despite their misfortunes workers do not view their interests as automatically opposed to those of owners and managers. As workers and supervisory personnel lost their jobs in the severe recession of the early 1990s, they often expressed discontent with the nation's leadership and with company managers. Generally, however, the anguish of the unemployed tends to be turned inward, taking the form of depression rather than militant class consciousness. Class consciousness may be popularized in the lyrics of a Bruce Springsteen song—"They're closing down the textile mill across the railroad track/Foreman says these jobs are going boys/and they ain't coming back to your hometown"—but it has not produced much overt class conflict or new social movements (Bensman & Lynch, 1987).

In addition to class conflict between workers and management, there are problems of inequality in the workplace. As mentioned earlier, during the decade of the 1990s the U.S. economy experienced unprecedented growth and low inflation, just about the best possible conditions for business and workers (Nasar, 1999). Jobs were being created in record numbers. Civilian employment stood at 133.1 million, up from 108.4 million in 1992. And all this was occurring despite severe economic crises in Asia and parts of Latin America. This powerful surge in the U.S. economy vastly aided the absorption of immigrants into the labor force without increasing intergroup tensions due to competition for scarce jobs. On the other hand, as we see in Figure 19.5, unemployment rates among black and Hispanic workers remained twice or almost twice that of whites. And the unemployment rate among teenagers, usually a far higher figure, remained over 14 percent in 1999.

The data on causes of unemployment included in Figure 19.5 are based on the federal government's Current Population Survey, which interviews a representative sample of workers and job seekers as well as people without work. Note that of the 6.1 million people out of work, 2.7 million were people who had lost a job. However, even more people were unemployed because they had voluntarily left a job, were returning to the labor market after an absence, or were seeking their first job. That total, 3.4 million, far exceeds the number of people who have been forced out of work. A situation in which workers are more likely to leave jobs voluntarily than to be laid off creates what is known as a "tight labor market," in which many employers are struggling to maintain adequate labor forces.

But if workers are beginning to be in scarce supply, why is inflation so low? A tight labor market should push wages up as employers compete for available employees. Sociologists believe that the explanation may be found in changes in economic institutions. They point to the number of people classified in the "hidden unemployment" categories. As shown in Figure 19.5, these categories include 3.4 million people who are working part-time but would rather find full-time work

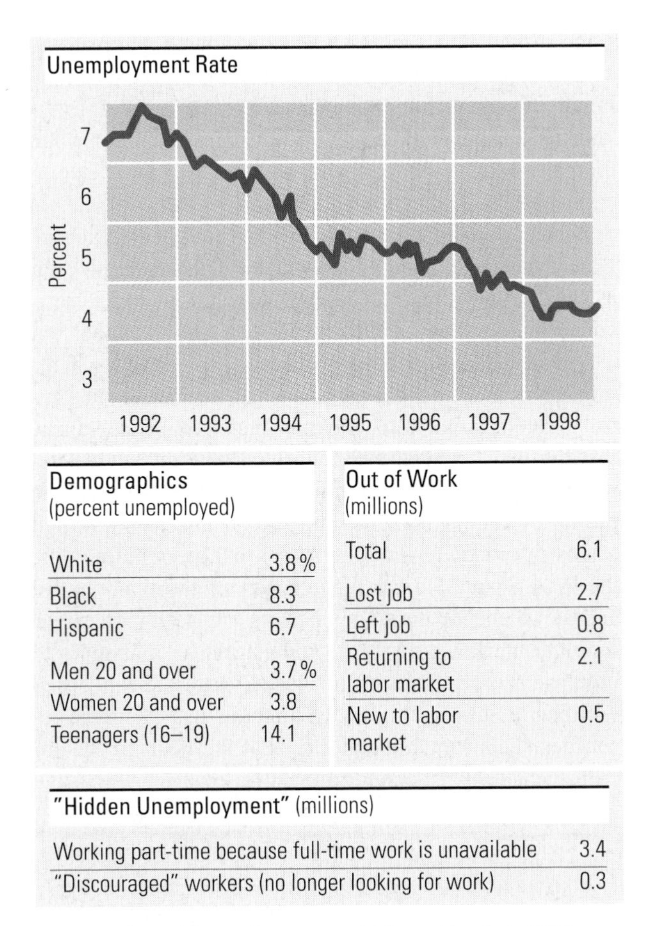

Unemployment Rate
7
6
5
4
3
Percent
1992 1993 1994 1995 1996 1997 1998
Demographics
(percent unemployed)
White 3.8%
Black 8.3
Hispanic 6.7
Men 20 and over 3.7%
Women 20 and over 3.8
Teenagers (16–19) 14.1
Out of Work
(millions)
Total 6.1
Lost job 2.7
Left job 0.8
Returning to labor market 2.1
New to labor market 0.5
"Hidden Unemployment" (millions)
Working part-time because full-time work is unavailable 3.4
"Discouraged" workers (no longer looking for work) 0.3

Source: Adapted from *The New York Times,* March 6, 1999.

FIGURE 19.5 Conditions in the U.S. Labor Market, 1999

(which would also be more likely to pay full health and other benefits) and about 300,000 "discouraged workers" who looked for work during the past 12 months but have given up their search for some reason. Increases in part-time work in the booming service sectors of the economy help keep wage inflation down because part-time workers often do not receive employee benefits and pay raises.

These employment figures present a picture of a nation with extremely high rates of employment but with underlying problems of employment inequity (for example, black–white differences in unemployment) that often become far worse when the economy enters a downward phase of the business cycle. Another trend that worries both economists and sociologists is the rise in rates of consumer debt and bankruptcies in the United States, as shown in Figure 19.6. About two thirds of domestic production in the United States is purchased by U.S. consumers, a fact that helps explain the intense pressure on those consumers to go to the mall or to order merchandise from mail-order catalogs. But credit card and other forms of consumer debt, which fuel the economy in very good times, can become a major problem if the economy weakens and bankruptcies rise. This is especially true for people employed in part-time service work or in manufacturing, where workers are most vulnerable to plant closings and layoffs.

Professions: An Interactionist Perspective Factory workers and other low-status employees are not the only subjects of research by sociologists who study work and employment. Because they are a growing segment of modern economic institutions, professionals have also been studied extensively. A **profession** is an occupation with a body of recognized knowledge and a developed intellectual technique. Its knowledge is transmitted by a formal educational process, and entry into the profession is regulated by testing procedures. A code of professional ethics governs each profession and regulates relations with colleagues, clients, and the public. Most professions are licensed by the state (Cant & Sharma, 1998; Montagna, 1977). This definition could be applied to a variety of occupations. For example, doctors and attorneys have long been considered professionals, but what about nurses and stockbrokers? Sociologists have addressed this question in their research on the "professionalization" of occupations—that is, on the way different occupations attempt to achieve the status of professions.

According to Everett C. Hughes (1959), a profession is a set of role relationships between "experts" and "clients," in which the professional is an expert who

FIGURE 19.6 Consumer Debt and Personal Bankruptcies, 1991–1996

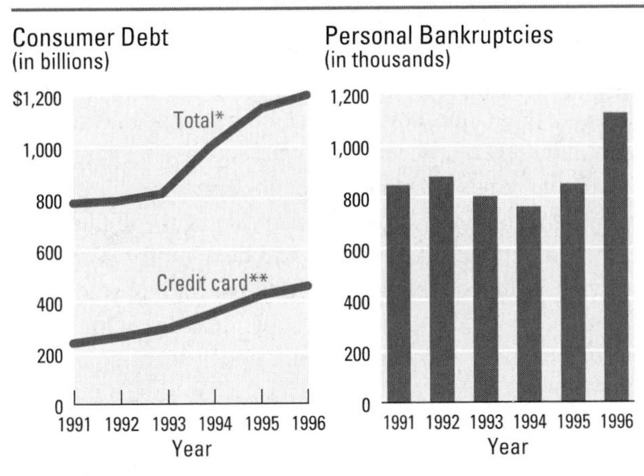

*Excludes home mortgages.

**Includes overdraft protection.

Source: Federal Reserve Board, Administrative Office of U.S. Courts.

offers knowledge and judgments to clients. Often the professional must assume the burden of the client's "guilty knowledge": The lawyer must keep secret the client's transgressions, and the physician must hide any knowledge of the sexual behavior or drug use of famous and not-so-famous patients. Professions thus "rest on some bargain about receiving, guarding, and giving out communications" (Hughes, 1959, p. 449; see also Abbott, 1993).

Interactionist sociologists pay special attention to the role relations that develop in various professions. They are particularly interested in the processes of *professional socialization*—learning the profession's formal and informal norms. For example, in a well-known study of professional socialization in medical schools, Howard Becker and his colleagues (1961) found that in the later years of their training, medical students' attention is devoted more and more to learning the informal norms of medical practice. The formal science they learn in class is supplemented by the practical knowledge of the working doctor, and the student's skill at interacting with higher-status colleagues becomes extremely important to achieving the status of a professional physician. In a highly acclaimed study on the same subject, Charles Bosk (1979) showed that resident surgeons in a teaching hospital are forgiven if they make mistakes that the older surgeons believe to be "normal" aspects of the learning process, but not when their mistakes are repeated or are thought to be due to carelessness. The point of these and similar studies is that role performance is learned through interaction. Role expectations vary from one profession to another, as do the ways in which these expectations are experienced by members of the profession (Abbott, 1988).

Our discussion of professions can serve as a reminder of a basic characteristic of economic institutions, one that we noted early in the chapter: Economic institutions are concerned with allocating scarce resources to satisfy competing ends. Thus, just as the guilds of feudal societies controlled access to skilled trades in medieval times, professional associations like the American Medical Association attempt to control access to the professions. Efforts by nation-states to regulate markets and trade, and efforts by managers to increase the productivity of workers, are all concerned with achieving the most advantageous use of scarce resources. This fundamental fact is at the heart of the sociological study of economic institutions.

RESEARCH ON THE CUTTING EDGE: How People Find Jobs

How do people find their jobs? Do they rely on the want ads? Family contacts? College placement services? From a strictly economic viewpoint the question is answered by relying on market mechanisms. Employers seek workers at given wages for particular kinds of work. Potential employees seek employers who need their skills and will pay the highest rate for their services. The supply and demand for particular services will explain changes in the prevailing wages for those services. But how do people actually learn about available jobs, and how do they obtain them once they know about them? Market models of job-seeking behavior often assume that there is a perfect flow of information. In fact, however, sociological research uncovers some surprising gaps in the flow of information and reveals that both employers and employees actually behave in ways that are not the most rational from an economic viewpoint.

Mark Granovetter (1974, 1995), one of the nation's foremost economic sociologists, is largely responsible for discovering what is now known as "the strength of weak ties." This discovery turns out to be a key to explaining how a large proportion of people actually go about finding their jobs. By "weak ties" Granovetter refers to people who are part of our social networks even if we are not always aware that they are. They are acquaintances or friends of acquaintances. It turns out that these weak ties, as opposed to stronger ties of kinship, can be essential in the success of job searches. Here is how Granovetter formulated his research problem and collected the data for his pioneering research.

While a graduate student in sociology at Harvard, Granovetter developed an interest in chains of contact and flows of information through them. His mentor, the economic sociologist Harrison White (1970), had made a number of important discoveries about how information in markets flows through chains of acquaintance, and Granovetter wanted to extend this research to labor market and job-seeking behavior. He decided to interview a sample of prospective employees in order to determine whether they also relied on chains of acquaintance within social networks to help them find jobs. To reduce the possible variance among different occupations, and because he could not afford to interview more than about 280 respondents, Granovetter decided to limit his sample to professional, technical, and managerial workers. He conducted the interviews in the Boston suburb of Newton, obtaining the names of eligible male respondents through a commercial directory service that lists residence, occupation, and related information for each entry. (In subsequent research he broadened his study to include women.) In collecting the data, he combined in-person interviews with career histories and basic demographic information—age, education, race or ethnicity, place of birth, and so on.

The first surprises in Granovetter's research came as he looked at the data about which job-finding methods worked for which types of job seekers. Examples of these findings from his original survey are presented in Tables 19.1 and 19.2. In no case do formal job search methods—that is, the use of impersonal intermediaries (newspaper ad, placement service, employment agency, professional association, and the like)—account for the majority of job placements. Direct application, in which the applicant shows up at the employer's office without an appointment, and personal contacts through social networks are more important than formal searches, especially at the middle levels of the salary scale. Table 19.1 shows that 154 out of 275 (56 percent) of respondents found their jobs through personal contacts, clearly the most effective of the three methods of job seeking. Table 19.2 shows that personal contacts are particularly important when the position being filled is newly created. Indeed, the lengthy interviews Granovetter conducted with his respondents revealed a pattern in which a personal contact mentioned a job that another contact in a certain company had mentioned as just being created—thus providing the job seeker with "insider information" in seeking the position.

Many readers may feel that these findings are not actually surprising. Don't we often hear it said that "it's not what you know, but who you know" that explains how people succeed in getting jobs? Indeed, this is a time-honored cliché—but Granovetter explored its meaning more fully. After all, "who you know" should include family members and close kin. Yet family and close kin contacts and referrals accounted for a minority of the personal referrals that paid off in terms of

TABLE 19.1 Level of Income of Respondent in Present Job, by Job-Finding Method Used

	Method Used				
Income	Formal Means	Personal Contacts	Direct Applications	Other	Total
Less than $10,000	28.0%	22.7%	50.0%	5.3%	27.6%
$10,000–14,999	42.0%	31.8%	30.8%	26.3%	33.1%
$15,000–24,999	24.0%	31.2%	15.4%	52.6%	28.4%
$25,000 or more	6.0%	14.3%	3.8%	15.8%	10.9%
N	50	154	52	19	275

Source: Adapted from Granovetter, 1995.

TABLE 19.2 Origin of Job, by Job-Finding Method of Respondent

	Method Used				
Origin of Job	Formal Means	Personal Contacts	Direct Application	Other	Total
Direct replacement	47.1%	40.5%	58.0%	38.9%	44.9%
Added on	31.4%	15.7%	18.0%	27.8%	19.9%
Newly created	21.6%	43.8%	24.0%	33.3%	35.3%
N	51	153	50	18	272

Note: Percentages may not add up to 100 due to rounding.

Source: Adapted from Granovetter, 1995.

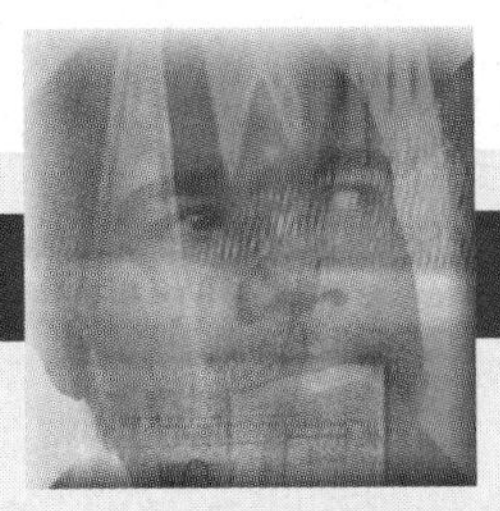

THE PERSONAL CHALLENGE

Social Change and the Choice of an Employer

As you think about how economic conditions will affect your future in the job market, there are a number of lessons to take notice of. First, and perhaps most obvious, is the realization that it is impossible to overestimate the importance of the investments you and your family make in your education and training. The longer a person stays away from higher education and postgraduate training in favor of "temporary" work situations, the more difficult it is to find the time and financial resources to return. People who drop out of high school or college are at a double disadvantage. Not only will they earn significantly less than people who complete their education, but they face a greater likelihood of experiencing economic insecurity.

Another, less obvious, lesson is that because of changes in the U.S. economy in recent years you need to think carefully in choosing an employer. If you join a company with more than 100 employees, you will improve the odds of receiving adequate pension and medical benefits (see the accompanying charts). If you choose to work for a government agency or one that relies on public funds, you will probably receive better benefits, but you also risk the possibility that budget reductions will lead to job insecurity. Finally, whatever your opinion of unions, you will be better off in an industry that recognizes unions, such as public utilities or transportation manufacturing, than in one that does not, such as textiles or wholesale and retail trade. The average amount spent on employees by employers in the United States is over $14,000 per year; in textiles it is about $8,000, while in public utilities it is more than $20,000. The heavily unionized transportation manufacturing sector (including automobile manufacturing) pays more than $3,500 per year per employee in medical benefits, while the far less unionized wholesale and retail trade sector pays about $2,300 (Uchitelle, 1995).

In coming years, as employees realize that vital benefits are higher in some industries than in others, there will likely be a resurgence of labor unrest and efforts to form unions (Hunter, 1998). If you choose an industry with relative labor peace, in which union benefits are the rule rather than the exception, the data indicate that you will also be gaining a great deal of economic security.

Percentage of Companies Offering Various Benefits, by Size of Company

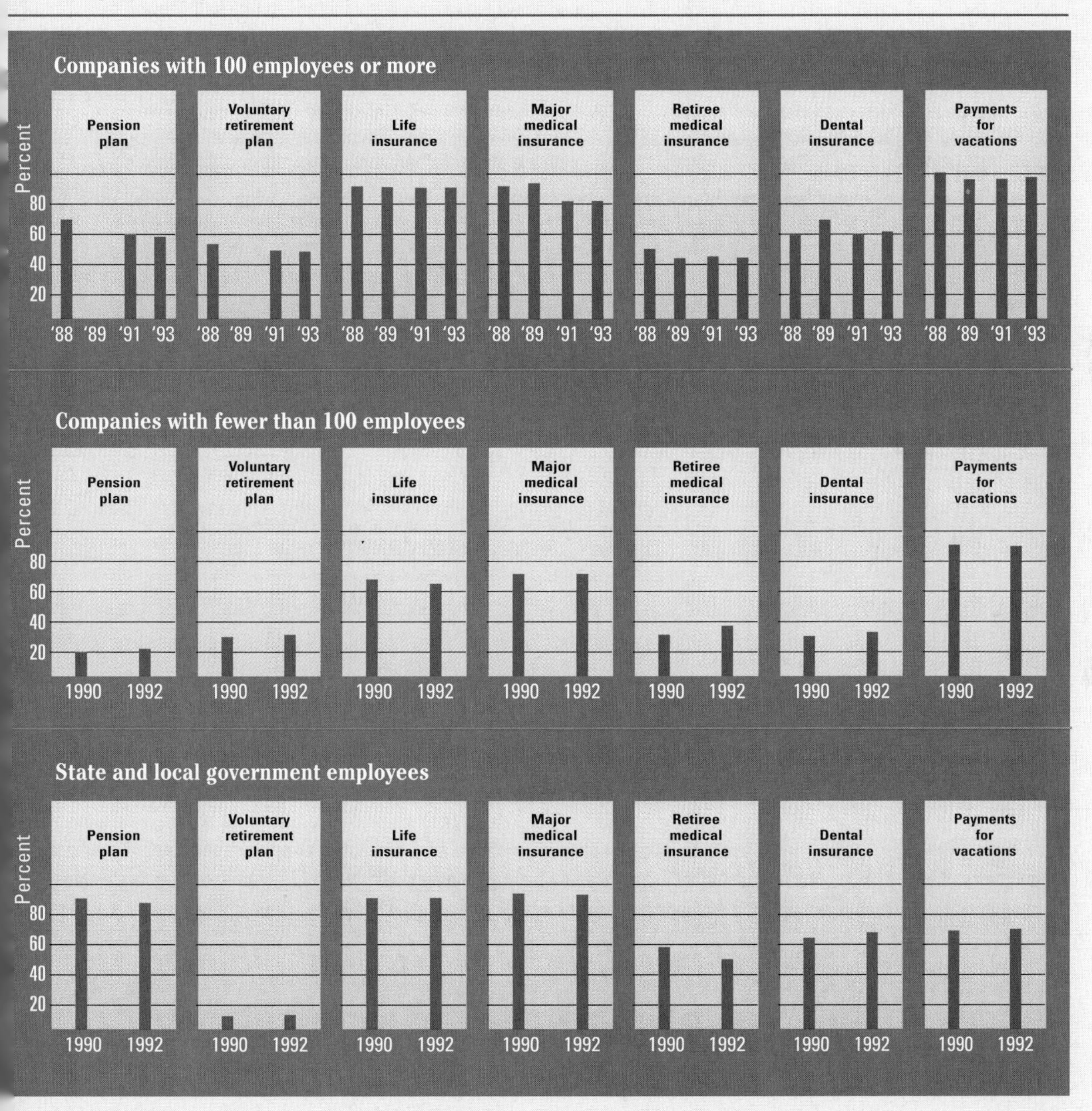

Sources: U.S. Chamber of Commerce employee benefits surveys (average employer contribution); Employee Benefits Research Institute Databook on Employee Benefits, 3rd edition (benefits by size of company).

obtaining a job. Far more important were links between friends and acquaintances. The following three cases from Granovetter's original study illustrate how weak ties actually work:

> George C. was working as a technician for an electrical firm, with a salary of about $8,000 and little apparent chance for advancement. While courting his future wife, he met her downstairs neighbor, the manager of a candy shop, a concession leased from a national chain. After they were married, Mr. C. continued to see him when visiting his mother-in-law. The neighbor finally talked him into entering a trainee program for the chain, and arranged an interview for him. Within three years, Mr. C. was earning nearly $30,000 in this business.

> Herman D. was the owner of a fruit and vegetable store, which he sold (at age 45) because of ill health. He took a vacation; meanwhile his brother, a business executive, attended a meeting where a colleague mentioned that he was looking for someone to do inventory management. Mr. D. had done similar work before buying his store, and his brother therefore suggested him. He was hired several days later.

> Gerald F. was a salesman for a wholesale liquor distributor. A friend who was a doctor asked him if he would be interested in managing a nursing home, and if so, to put together a resume. One of the references Mr. F. used for the resume was his wife's cousin, owner of a fashionable antique shop. When the nursing home job didn't

> come through, the wife's cousin, now aware that Mr. F. was considering changing jobs, offered him a job as business manager of his shop, which he accepted. (Granovetter, 1974, p. 49)

These are not the family ties that one often associates with "who you know." In fact, another surprising finding of Granovetter's research is that on the average it takes about two or more contacts in the chain of network associations to get the job applicant to appear as a likely candidate for the prospective employer, so it is not the direct contact but the contact through network chains that makes the difference in the successful job search.

One implication of this research is that people who are on the margins of society because of their race, gender, or social class are less likely to have the network resources to draw upon when looking for a job. Granovetter used this finding to argue in favor of affirmative action in hiring. African Americans and other groups that have experienced systematic exclusion from particular types of jobs and careers have plenty of personal contacts and social networks, Granovetter explains. But because they are underrepresented in the existing structures of employment, these contacts do not produce chains of referral into these careers. Once a core of formerly excluded people are included, largely through intentional efforts like affirmative action, the normal network system of weak ties will work for them as it does for others.

VISUAL SOCIOLOGY

Crafts Traditions in Modern Economies

As these photos show, crafts traditions are flourishing. The Indian basket maker whom we met at the beginning of the chapter is not alone. In fact, in industrial nations like England and the United States many people are engrossed in using their creative powers to make traditional and modern craft objects of all kinds. Millions of people throughout the industrial world are facing the same problem as the Indian basket weaver: how to market their goods while maintaining their pride in their skills and their love of the process.

Contrary to the dire predictions of many social critics and early sociologists, such as Karl Marx and Max Weber, who lived more than a century ago during the earlier phases of capitalism, the enchantment people find in creating arts and crafts has not withered in the face of mass production and mass marketing. On the contrary, the easy availability of mass-produced goods seems to have made people even more eager to buy unique objects made by people who are carrying out enduring cultural traditions. And global forms of marketing like the World Wide Web are proving to be extremely powerful institutions for disseminating information about craftspeople and their products.

While craftspeople will always face an uphill struggle to make a living from their work, particularly in

William Morris

traditional societies, we see proof that the struggle can pay off. Early in the era of industrial capitalism, William Morris, a man with keen sociological insights and a passion for arts and crafts, started what became known as the Arts and Crafts Movement. Morris, a contemporary of Marx, was deeply concerned that industrial expansion and urbanization would wipe out craft traditions throughout the world. If he were alive today, he would still be concerned about the loss of traditional skills and techniques for producing baskets and other objects of everyday life, but he would be extremely gratified to see the burgeoning activity in this area of life, as we can by simply browsing the Web.

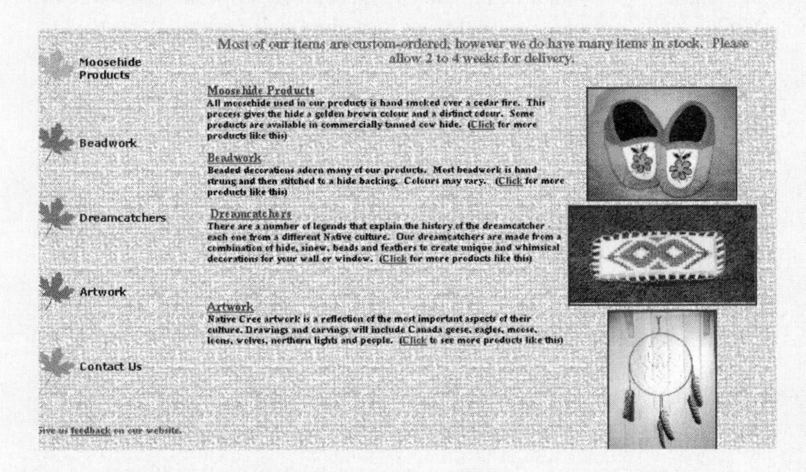

SUMMARY

Economics is the study of how individuals and societies choose to employ scarce resources to produce various commodities and distribute them among various groups in the society. Sociologists are also concerned with how individuals and societies make such choices, but much of their research focuses on how cultural norms affect those choices. The main subjects of sociological research in this area are markets and the division of labor, the interactions between government and economic institutions, and the nature of jobs and professions.

A hallmark of an industrial society is the production of commodities and services to be exchanged in markets. A *market* is an economic institution that regulates exchange behavior. In a market, different values are established for particular goods and services; those values are usually expressed in terms of a common measure of exchange, or currency. Market transactions are governed by agreements, or contracts, in which a seller agrees to supply a particular item and a buyer agrees to pay for it. The spread of markets throughout the world began in the late fifteenth century as a result of the development of new *technologies* that facilitated trade. Today world markets are dominated by *multinational corporations,* economic enterprises that have headquarters in one country and conduct business activities in one or more other countries.

The fate of societies throughout the world has been strongly influenced by the economic ideologies of mercantilism, capitalism, and socialism. *Mercantilism* held that the wealth of a nation could be measured by its holdings of gold, so the best economic system was one that increased the nation's exports and thereby increased its holdings of gold. *Laissez-faire capitalism,* on the other hand, argued that a society's wealth could be measured only by its capacity to produce goods and services—that is, its resources of land, labor, and machinery. The major institutions of capitalism are private property and free markets.

Socialism arose out of the belief that private property and personal profit should be replaced by public ownership of property and sharing of profits. According to Marx, the socialist state would be controlled by the workers, who would determine what should be produced and how it should be distributed. Soviet-style socialist societies were characterized until recently by command economies, in which the state commands economic institutions to supply a specific amount of each product and to sell it at a particular price. *Democratic socialism* holds that private property must continue to exist but that large corporations should be owned by the nation or, if they are in private hands, required to be run for the benefit of all citizens.

In *welfare capitalism,* markets determine what goods and services will be produced and how, but the government regulates economic competition. Welfare capitalism also asserts that the state should invest in the society's human resources through policies promoting education, health care, and social welfare.

In the United States, the transition from an economy based on the production of goods to one based on the provision of services has resulted in the displacement of thousands of skilled workers into lower-paying jobs in the service sector. These trends appear to signal a change in the basic social contract in which individuals who work hard can expect increases in their living standards.

Many sociologists use the concept of *alienation* to explain this gap. This term refers to the feeling of being powerless to control one's own destiny. At work, people may feel alienated because their labor is divided up into activities that are meaningless to them. Mentoring programs have been found to raise employee morale and prevent burnout.

In the United States and other industrialized societies, the corporation has gained ever-greater dominance over other economic institutions. Corporations are "fictional persons" that can incur debts and liabilities on their own account. During the last hundred years or so, the importance of corporations has come to outweigh that of individuals in most of the institutional sectors of society, and the individual is often at a disadvantage in dealings with corporations.

The scientific management approach to labor–management relations attempted to increase productivity by determining how each job could be performed most efficiently and by using piecework payment systems to induce workers to produce more. Efforts to determine what conditions would result in the highest rates of worker productivity led to the recognition that cooperation between workers and managers is an important ingredient in worker satisfaction and output. This gave rise to the human relations approach to management, which seeks to improve cooperation between workers and managers.

Conflict theorists feel that the human relations approach fails to consider the basic causes of worker–management conflict. They study how social class and status both at work and outside the workplace influence relations between workers and managers. Interactionist theorists have devoted considerable study to professionalization, or the way in which occupations attempt to gain the status of *professions,* and to the processes of professional socialization (i.e., learning the formal and informal norms of the profession).

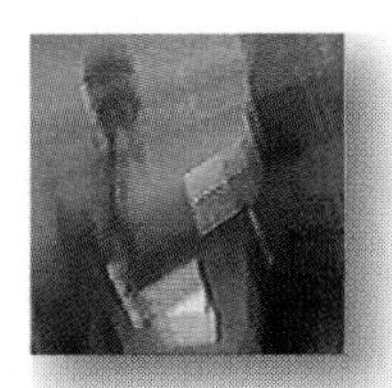

SOCIOLOGY VERSUS IDEOLOGY

Sociologists tend to have more liberal views than economists. Whereas economists are likely to favor the interests of businesspeople, sociologists more often identify with those who experience the consequences of persistent inequality and exploitation. This makes them sometimes look like "bleeding hearts" in the eyes of "tough-minded" economists. Of course, there are sociologists, especially those who apply rational-choice models to social behavior, who appear, at least on the surface, to be as conservative as laissez-faire economists. In fact, both economists and sociologists differ among themselves in their political ideologies.

In evaluating the writing and research of both economists and sociologists, a critical reader needs to be able to identify instances in which they allow their values and assumptions to lead their thinking, and also to be able to know when they are basing their judgments on hard facts. Values and judgments are important. So are hard facts. Each is appropriate in its proper context. Hard facts about social phenomena, however, are open to various interpretations, which often lead to new researchable questions and deeper understandings of complex social phenomena like employment and unemployment. For newcomers to these fields this is often frustrating. We all want clarity and truth. We also all struggle with our own biases and ideologies. If one always asks what assumptions and ideological values are guiding the social scientists' choice of subjects and their interpretations, it will be easier to sort out the ideology from the facts.

GLOSSARY

market: an economic institution that regulates exchange behavior through the establishment of different values for particular goods and services. (p. 609)

technology: tools, procedures, and forms of social organization that increase human productive capacity. (p. 612)

multinational corporation: an economic enterprise that has headquarters in one country and conducts business activities in one or more other countries. (p. 613)

mercantilism: an economic philosophy based on the belief that the wealth of a nation can be measured by its holdings of gold or other precious metals and that the state should control trade. (p. 615)

laissez-faire capitalism: an economic philosophy based on the belief that the wealth of a nation can be measured by its capacity to produce goods and services (i.e., its resources of land, labor, and machinery) and that these can be maximized by free trade. (p. 615)

socialism: an economic philosophy based on the concept of public ownership of property and sharing of profits, together with the belief that economic decisions should be controlled by the workers. (p. 616)

democratic socialism: an economic philosophy based on the belief that private property may exist at the same time that large corporations are owned by the state and run for the benefit of all citizens. (p. 617)

welfare capitalism: an economic philosophy in which markets determine what goods will be produced and how, but the government regulates economic competition. (p. 618)

alienation: the feeling of being powerless to control one's own destiny; a worker's feeling of powerlessness owing to inability to control the work process. (p. 623)

profession: an occupation with a body of knowledge and a developed intellectual technique that are transmitted by a formal educational process and testing procedures. (p. 627)

WHERE TO FIND IT

BOOKS

Capitalism, Socialism, and Democracy (Joseph A. Schumpeter; Harper, 1950). A classic description of capitalist and socialist economic institutions and the political consequences of the ways in which they operate.

Lean and Mean: The Changing Landscape of Corporate Power in the Age of Flexibility (Bennett Harrison; Basic Books, 1994). An analysis of the origins of corporate downsizing in a global economy and its consequences for stratification.

The Winner-Take-All Society (Robert H. Frank and Philip J. Cook; Free Press, 1995). An influential analysis of the U.S. economy that shows that as more Americans compete for fewer and bigger prizes, income inequality and economic waste are the consequences.

Out to Work (Alice Kessler Harris; Oxford University Press, 1982). A history of wage-earning women in the United States, with much insight into the impact of changing economic institutions on women and their careers.

Capitalism and Freedom (Milton Friedman; University of Chicago Press, 1962). A lucid statement of the laissez-faire perspective on economic and governmental institutions.

Getting a Job: A Study of Contexts and Careers, 2nd ed. (Mark Granovetter; University of Chicago Press, 1995). A model of well-reasoned empirical research in the field of economic sociology.

Post Industrial Possibilities: A Critique of Economic Discourse (Fred Block; University of California Press, 1990). A sociologist skilled in economic and sociological analysis illuminates the quandaries and possibilities of postindustrial society.

JOURNALS

Monthly Labor Review. Published by the U.S. Department of Labor. An excellent source of data on employment and unemployment, plant closings and openings, and other economic trends. Also contains analytical articles that interpret major economic trends and changes in the labor market.

The Labor Relations Reporter. A valuable tool for anyone interested in research on labor–management issues. Consult bound volumes for weekly news and trend analysis and longer articles on important economic events, which are cited in the weekly summaries. Usually available in law, business, or other specialized libraries.

OTHER SOURCES

Current Population Reports: Economic Characteristics of Households in the United States. A quarterly report published by the Bureau of the Census. Presents monthly averages of household income and participation in cash and noncash transfer programs.

Handbook of Economic Statistics. Published annually by the Central Intelligence Agency. Compares economic statistics for communist, OECD, and selected other countries.

Handbook of Labor Statistics. Published annually by the Bureau of Labor Statistics. A compilation of statistical data on labor conditions and labor force characteristics in the United States and selected foreign countries.

Employment and Earnings. Published by the Bureau of Labor Statistics. A monthly report that presents current statistics on U.S. employment, unemployment, hours worked, and earnings.

INTERNET RESOURCES

International Monetary Fund (www.cgs.edu/acit/help/imf.html). Provides access to a great deal of comparative data on how the world's nations are developing—or not developing—along a variety of important economic and social dimensions.

The World Bank (www.worldbank.org/) and the **Organization for Economic Cooperation and Development (OECD)** (www.oecd.org/). Two sources of extensive data on social indicators such as health and education in different nations and regions.

WWW Virtual Library on Economics (http://netec.wusti.edu/WebEc.html). Offers innumerable resources on the many applications of economic and political-economic research throughout the world.

African Studies Department at the University of Pennsylvania (www.sas.upenn.edu/African_Studies/AS.html). A valuable site for those interested in the economics and politics of a rapidly changing continent.

CHAPTER 20

POLITICS AND POLITICAL INSTITUTIONS

Alma Adamkus never believed that her husband would adjust to retirement, and she was right. After almost three decades as a high-level bureaucrat in the U.S. Environmental Protection Agency's midwestern division, Valdas Adamkus claimed that he wanted to live out his golden years golfing and puttering about the couple's home in the Chicago suburb of Hinsdale. Alma didn't buy it for a second. "He's the kind of person," she says, "who can't just sit home and do nothing."

This was a major understatement. Adamkus, 71, is now president of Lithuania, his country of origin to which he dreamed of someday returning.

During numerous visits to the former Soviet state, his friends began convincing him that he not only should return but should think seriously of taking a lead in rebuilding the nation's political institutions after years of Soviet rule. In Lithuania he was regarded as a national hero. An anti-Stalinist freedom fighter, he had been forced to flee to the United States, where he became a citizen and a successful government official.

Alma and Valdas Adamkus

Lithuania's 3.8 million people had suffered a half century of invasion, communism, and political corruption. The modest but sturdy Adamkus appeared to offer stability and good sense. "He ran as someone who could quiet the country's political conflicts," says Jurate Novagrockiene, a professor of international relations at Vilnius University in the Lithuanian capital. "So far it has worked out."

President Adamkus himself admits to feeling a bit overwhelmed. "I'm still too full of enthusiasm to have a real sense of the difficulties ahead," says Adamkus, who hopes to restore the nation's social and economic fabric so that Lithuania may stand proudly among the other members of the European Union. Since his inauguration on February 26, 1998 (one day after giving up his U.S. citizenship), Adamkus has applied his austere personal style to his unexpected presidency. In the garage of the fourteenth-century palace where he will work, he parks the family car, a practical Volvo. Lithuania's first lady shops on foot, and the couple prefer to stay home at night so that his guards and staff can be at home with their families.

Adamkus escaped Lithuania after World War II and the Soviet occupation by clinging to the bottom of a tanker train heading into the West, where he and his family were reunited in a refugee camp. He emigrated to America in 1949, arriving with one suitcase, five dollars, and his family. They settled in Chicago because of its strong Lithuanian community. Adamkus worked in a factory while studying at night in a public college to earn a degree in civil engineering. He became active in the local Lithuanian independence movement. Later, as his career with the EPA developed, he took every chance to return to his homeland to implement environmental projects and to address his compatriots over the Voice of America radio station.

"I knew the day would come," he says, "when my country would be free."

One of the new president's first acts was to seek legislation to abolish the nation's death penalty. And in the face of the multitude of problems confronting his tiny nation, Adamkus finds cause for optimism in Lithuania's youth. He is impressed at how they have retained their language and culture despite years of Soviet occupation. "These people," he says proudly, "will never give up" (O'Neill, 1998).

Lithuania is fortunate to have had a peaceful election and to have a democratically elected leader. Many other nations of the former Soviet empire are struggling to develop their democratic institutions in the face of severe ethnic hostilities and growing corruption. Elections are a vital aspect of any democratic political regime. But elections alone, as we will see in the Visual Sociology section at the end of this chapter, do not ensure democracy and the rule of law.

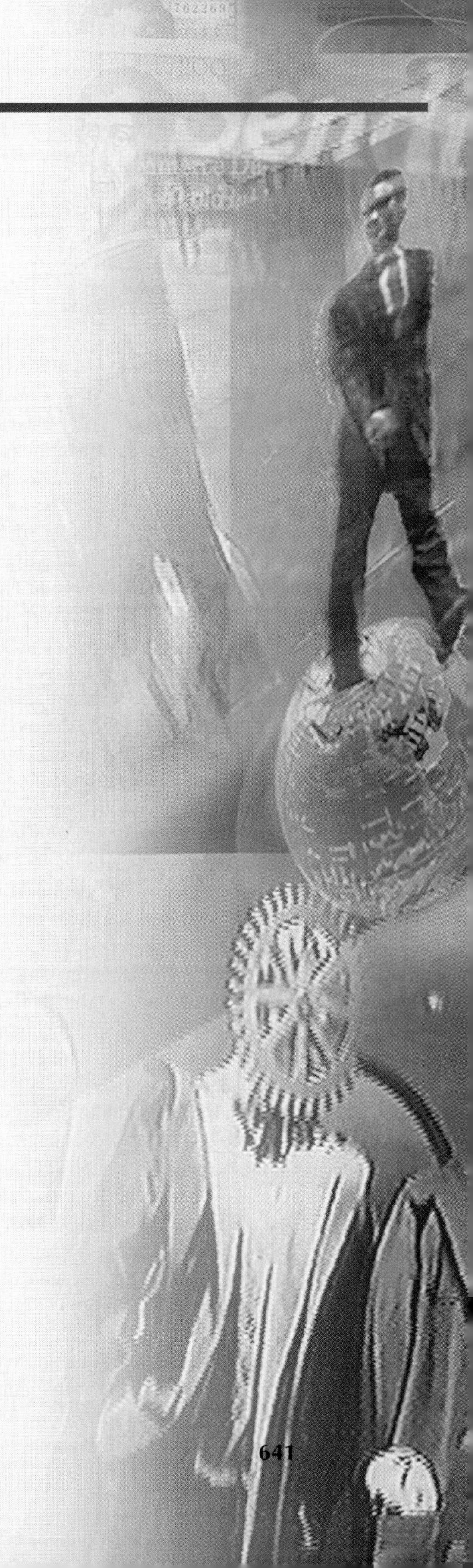

THE NATURE OF POLITICS AND POLITICAL INSTITUTIONS

"Politics," Max Weber liked to tell his students, "is the slow boring of hard boards." By this he meant that political change is almost never achieved easily. Creating new political institutions or changing old ones, even just changing the leadership of existing institutions, usually requires years of effort. There will be endless meetings, ideological debates, fund-raising, negotiations, and campaigning. At times, however, especially during times of revolutionary social change, the pace of political change is fast and furious.

We are living through a period of severe crisis and dramatic change in world politics. It is not at all clear that the democratic nation-state, the rule of law, and guarantees of the rights of minorities can be established in many parts of the world (Mestrovic, 1991). In the view of many political sociologists, the twentieth century is ending with the world in a state of political unrest and nationalistic conflict much like that at the end of the nineteenth century (Abu-Lugod, 1996). This uncertainty in world and national politics makes the comparative study of political institutions a major growth area in sociology.

The fall of the Berlin Wall in 1989 symbolized the beginning of a new era in world politics. The cold war between the former Soviet Union and the Western capitalist democracies had dominated the politics of many parts of the world for more than 40 years. Although many nations, including the United States, still feel the effects of the arms race that accompanied the competition between the superpowers, with the end of the cold war has come an end to the nuclear "balance of terror" and the suppression of political expression within the former Soviet empire. Nationality groups in the former Soviet republics are asserting their desire for independence or, at least, for protection of their rights within a multiethnic state.

The fall of authoritarian communist parties means that there is often no central authority to restrain populations who yearn to form their own independent nation or who covet the lands of their neighbors. Despite the success of Lithuania, the situation is not encouraging. The political upheavals in the former Yugoslavia, Africa, and India, the struggle to transform South Africa into a truly democratic state, the potential for disunion in Canada, and the rise of militias and other violently antigovernment hate groups in the United States—all these changes and others promise to make the comparative study of politics a central area of social-scientific research for years to come.

We can gain much insight into what the nationality groups in Eastern Europe, the black majority in South Africa, and others who are trying to bring about political change are experiencing by looking more carefully at politics and political institutions and the place they occupy in national life. This section of the chapter reviews some of the essential concepts of politics, with special emphasis on power, authority, and legitimacy. The second section explores the evolving nature of the nation-state, while the third section describes the political institutions that are typical of modern nations and compares the political regimes of states and territories throughout the world. The fourth section explores the major theories that have been proposed to explain why political institutions operate as they do and how they change or resist change. In the final section we look at the role of the military in the political life of contemporary nations.

Politics, Power, and Authority

In any society, Harold Lasswell (1936) argued, politics determines "who gets what, when, and how." Different societies develop their own political institutions, but everywhere the basis of politics is competition for power. **Power** is the ability to control the behavior of others, even against their will. To be powerful is to be able to have your way even when others resist (Mills, 1959). The criminal's power may come through a gun and the threat of injury—but many people wield power over others without any threat of violence, merely through the agreement of the governed. We call such power *authority*. **Authority** is institutionalized power—power whose exercise is governed by the norms and statuses of organizations. Those norms and statuses specify who can have authority, how much authority is attached to different statuses, and the conditions under which that authority can be exercised.

Political institutions are sets of norms and statuses that specialize in the exercise of power and authority. The complex set of political institutions—judicial, executive, and legislative—that operate throughout a society form the **state** (see Figure 20.1). This chapter will concentrate primarily on these explicitly political institutions, but keep in mind that in modern societies many institutions that are active in politics are not part of the state. Labor unions, for example, are economic institutions because they represent their members in bargaining with business owners and managers, but they play an active role in politics when they support particular political candidates or when they lobby for government-funded benefits for workers and their families.

When we look at the way in which conflicts over scarce resources like wealth, power, and prestige occur, we are looking at politics. From this perspective there can be a politics of the family, in which its members vie for attention, respect, use of the family car, money, and other resources. There can be a politics of the school, in which teachers compete for benefits such as smaller class size or better students. And of course the politics of government at all levels of society determines how the society's resources are allocated among various groups or classes. This view of politics, which looks at competition for power and at conflict over the use of power in a variety of settings, was reflected in the works of Karl Marx and Max Weber, two pioneering sociologists who devoted much of their work to the analysis of politics and change in political institutions. According to Marx, the way power is distributed in a society's institutions is a feature of its system of stratification. People and groups that have power in economic institutions, for example, are most likely to have power in political institutions as well. However, much research in political sociology is devoted to the analysis of competition for power in political parties and electoral campaigns (Lipset, 1998). A central problem of political sociology is whether the electoral process can sustain people's belief in legitimate authority in a society despite inequalities in power and wealth (Kloby, 1997).

Legitimacy and Authority

A basic dilemma that every political institution must solve is how to exercise legitimate authority—that is, how to govern with the consent and goodwill of the governed. "You can't sit on bayonets," goes an old political expression. This is a way of saying that although a state can exercise its power through the use of police force or coercion, eventually this will not be sufficient to govern a society. When the Soviet Union abandoned the use of force to control its satellite states in Eastern Europe, the governments of those nations (Romania,

FIGURE 20.1 **Political and Military Institutions of the United States**

Institutional Sector	Major Institutions	Examples of Key Organizations
The State	Executive	The President and his advisers The Cabinet
	Legislative	The U.S. Senate The U.S. House of Representatives
	Judicial	The United States Supreme Court Federal courts State courts
	Military	The armed forces The National Guard
	National security	The FBI The CIA Local police forces
Civil Politics	Political parties	The Democratic party The Republican party
	Interest groups	National Organization for Women National Rifle Association

Czechoslovakia, East Germany) were soon toppled by popular revolts. The communist parties in those nations had governed for too long without the consent of the governed—that is, without legitimacy.

But why do people consent to be governed without the use of force? For political sociologists since Weber's time, the answer to this question begins with the concept of legitimacy. As defined by political sociologist Seymour Martin Lipset (1981), **legitimacy** is the capacity of a society "to engender and maintain the belief that the existing political institutions are the most appropriate for the society" (p. 64). In other words, legitimacy results from citizens' belief in the norms that specify how power is to be exercised in their society. Even if they disagree with some aspects of their political institutions or dislike their current leaders, they still hold to an underlying belief in their political system.

Traditional, Charismatic, and Legal Authority Weber (1947) recognized that shared beliefs about legitimate authority may differ widely in different groups or societies. These differences result in three basic types of authority—traditional, charismatic, and legal—that have important effects on the way societies cope with social change. (See the study chart below.)

In tribal and feudal societies, Weber observed, **traditional authority** prevails. The rulers usually attain authority through heredity, by succeeding to power within a ruling family or clan. Traditional authority is legitimated by the people's idea of the sacred, and traditional leaders are thought to derive their authority from God. The absolute monarchs of Europe, for example, were believed to rule by "divine right."

Traditional leaders usually embrace all the traditional values of the people they rule. This gives immense scope to their authority but makes it difficult for them and their systems of government to adapt to social changes that challenge those values. Thus the absolute monarchs of Europe had difficulty adapting their methods of governing to the growing power of new social classes like the proletariat and the bourgeoisie. The traditional chiefs of African societies today find it difficult to retain their authority as people in those societies gain more education and wealth and abandon their traditional village and agrarian way of life.

Charismatic authority, Weber's second type, finds its legitimation in people's belief that their leader has God-given powers to lead them in new directions. A charismatic leader, such as Jesus or Joan of Arc, comes to power not through hereditary succession but through a personal calling, often claimed to be inspired by supernatural powers. The charismatic leader usually develops a new social movement that challenges older traditions. Charismatic leaders may appear in all areas of life, "as prophets in religion, demagogues in politics, and heroes in battle. Charismatic authority generally functions as a revolutionary force, rejecting the traditional values and rebelling against the established order" (Blau & Scott, 1962, pp. 30–31).

Charismatic authority tends to bring about important social changes but is usually unstable because it depends on the influence of a single leader. Moreover, as noted in Chapter 8, charisma tends to become institutionalized as the charismatic leader's followers attempt to build the values and norms of their movement into an organization with a structure of statuses and roles and its own traditions. This process occurred in the early history of Christianity as the disciples built organizations that embodied the spirit and teaching of Christ.

Weber's third type of authority, **legal authority,** is legitimated by people's belief in the supremacy of the law. This type of authority "assumes the existence of a

STUDY CHART Weber's Three Types of Authority

TYPE	DEFINITION	EXAMPLE
Traditional Authority	Authority that is hereditary and is legitimated by traditional values, particularly people's idea of the sacred.	Tribal chiefs; absolute monarchs
Charismatic Authority	Authority that comes to an individual through a personal calling, often claimed to be inspired by supernatural powers, and is legitimated by people's belief that the leader does indeed have God-given powers.	Joan of Arc; Mahatma Gandhi
Legal Authority	Authority that is legitimated by people's belief in the supremacy of the law; obedience is owed not to a person but to a set of impersonal principles.	Presidents; prime ministers

formally established body of social norms designed to organize conduct for the rational pursuit of specified goals. In such a system obedience is owed not to a person—whether a traditional chief or a charismatic leader—but to a set of impersonal principles" (Blau & Scott, 1962, p. 31). The constitutions of governments establish legal forms of authority. All formal organizations, not only modern nation-states but also factories, schools, military regiments, and so on, have legal forms of authority.

Ideally, legal forms of authority adhere to the principle of "government of laws, not of individuals." We may be loyal to a leader, but that leader, in turn, must adhere to the laws or regulations that establish the rights and obligations of the leader's office. In practice, individual leaders often abuse their office by using their authority for private gain, as we have seen in cases of government corruption ranging from Watergate to the savings and loan scandal. But legal systems and the rule of law cannot long maintain their legitimacy if official wrongdoing is not punished (Breiner, 1996).

Of Weber's three basic types, legal systems of authority are the most adaptable to social change. In legal systems, when the governed and their leaders face a crisis, they can modify the laws to adapt to new conditions. This does not always occur easily or without political conflict, however, especially when momentous changes are under way, as described in Box 20.1.

Political Culture Because the stability of political institutions depends so directly on the beliefs of citizens, we can readily see that political institutions are supported by cultural norms, values, and symbols like the Statue of Liberty. These are commonly referred to as the society's *political culture.* When we look at other societies, it is clear that their political cultures differ quite markedly from ours. Americans justify a political system based on competitive elections by invoking the values of citizen participation in politics and equality of political opportunity. Chinese Communist party leaders justify their single-party political system by downplaying the value of citizen participation in politics and asserting the need for firm leaders and a centralized state that will hold the society together and reduce inequalities. Since there is widespread skepticism about the ability of most people to participate in the political process, political sociologists believe that it will be difficult to promote democracy and the rule of law in many parts of the world (McCrone, 1996; Moore, 1968).

Nationalism Perhaps the strongest and most dangerous political force in the world today is nationalism. **Nationalism** is the belief of a people that they have the right and the duty to constitute themselves as a nation-state. Religion, language, and a history of immigration or oppression are among the shared experiences that can cause a people to feel that they ought to have their own state. The situation of the French-speaking population of Quebec is a good example. Their language sets the Quebecois apart from English-speaking Canadians. So does their history of discrimination in English-dominated workplaces. In 1995 the people of Quebec narrowly defeated a separatist referendum, but it is still possible that nationalist sentiments in Quebec may result in the formation of a new nation on the North American continent.

In the case of Quebec, where political institutions and the rule of law are well established, such a change would not necessarily be traumatic. But elsewhere in the world the rise of ethnic hostility and nationalism threaten to produce chaos. As former secretary of state Warren Christopher noted in a statement before the House Foreign Relations Committee, "If we don't find some way that the different ethnic groups can live together in a country, how many countries will we have? We'll have 5,000 countries rather than the one hundred plus we have now" (quoted in Blinder & Crossette, 1993, p. A1). While the numbers may be exaggerated, few political sociologists doubt that new nations will be created as a consequence of nationalist conflicts in coming years.

Crises of Legitimacy Societies occasionally undergo political upheavals like those that are occurring at this writing in Nigeria, Liberia, the former Yugoslavia, and the former Soviet Union. When these periods of unrest and instability involve enough of the citizens to such a degree that they challenge the legitimacy of the nation's political institutions, sociologists call the situation a *crisis of legitimacy.* For example, the Declaration of Independence was written during a crisis in which the American colonists challenged the legitimacy of British rule. In modern South Africa, where the black majority now rules with the cooperation and participation of the white minority, it still remains to be seen whether political legitimacy can be maintained in the face of extremes of wealth and poverty (Kloby, 1997).

In sum, as we scan the globe today, it is evident that the world's political ecology includes many states that are extremely stable and many others in which political instability threatens to develop into a crisis of legitimacy.

THE NATION-STATE IN CRISIS

The most important political territory in world affairs is the **nation-state,** the largest territory within which a society's political institutions can operate without having

BOX 20.1 USING THE SOCIOLOGICAL IMAGINATION

Lincoln at Gettysburg

The documents that establish the legal authority of the government of the United States, particularly the Declaration of Independence and the Constitution, are often said to be "living documents." By this we mean that their meaning is not forever fixed, either in our laws or in our minds. We constantly interpret and reinterpret their meaning in light of our changing experience as a people and as a nation. What did the founders of the American nation mean by such ringing words as "All Men are created equal" or the rights to "Life, Liberty, and the Pursuit of Happiness"? Do these lofty phrases imply, for example, that we must strive to undo the consequences of past wrongs, such as slavery and colonial domination, through contemporary policies like affirmative action? To what extent are we responsible for trying to achieve "liberty and justice for all"?

In every era of American history there have been controversies over these questions. But at no time did debate over justice and law so threaten the nation's future as during the Civil War. That war was one of the bloodiest and most fratricidal in history. At the Battle of Gettysburg (July 1–3, 1863) alone there were more than 50,000 casualties. This battle was the turning point of the war, but the bloodshed and destruction continued for almost 2 more years. When President Abraham Lincoln agreed to dedicate the military cemetery at Gettysburg, therefore, he did so in the conviction that his role as President required him to interpret the meaning of the soldiers' sacrifice and of the war itself.

Political historian Gary Wills (1992) notes that in his Gettysburg Address (November 19, 1863) Lincoln did not condemn the Constitution as a document that legitimated slavery. Instead, through his interpretive words he "altered the document from within, by appeal from its letter to the spirit" (p. 38). Lincoln told the vast throng gathered to mourn the war dead that the nation was "conceived in Liberty, and dedicated to the proposition that all men are created equal." This opening line drew applause from the mourners, as did other lines in the brief but moving speech. But the almost revolutionary implications of the speech did not please all listeners then, nor do they please all Americans now. *The Chicago Times*'s review of the Gettysburg Address reminded readers that the dead officers and soldiers had given their lives to uphold the U.S. Constitution and the Union it created. "How dare [Lincoln], then standing on their graves, misstate the cause for which they died, and libel the statesmen who founded the government? They were men possessing too much self respect to declare that negroes were their equals, or were entitled to equal privileges" (from Wills, 1992, p. 39).

Time has vindicated Lincoln's broad interpretation of the Declaration and the Constitution. "For most people now," Wills concludes, "the Declaration means what Lincoln told us it means, as a way of correcting the Constitution itself without overthrowing it" (1992, p. 147). Although there are many people who argue for a stricter interpretation of the Constitution, one that does not necessarily grant equality to all and insists on the rights of states to set their own policies on questions of civil rights, the revolution in thought that Lincoln created with the Gettysburg Address holds sway over the American spirit. In Wills's words:

> The proponents of states' rights may have arguments, but they have lost their force, in courts as well as in the popular mind. By accepting the Gettysburg Address, its concept of a single people dedicated to a proposition, we have been changed. Because of it, we live in a different America. (p. 147)

The Lincoln Memorial is a shrine not only to a great president but also to the essential values of American political culture. It evokes feelings of reverence for the ideas of national unity and the rule of law.

to face challenges to their sovereignty. Throughout the world, however, we see nation-states in crisis. In Africa, there has been persistent civil strife in many nations, particularly Somalia, Sierra Leone, Liberia, and Angola. In India, the world's most populous democracy, strife between Hindus and Muslims continues to threaten peace and political stability. In Indonesia, economic crisis and rampant corruption led to the fall of the Suharto regime. The nations of Central America are threatened by dictatorship and civil strife. At this writing, the stability of

Russia's national government and its ability to effect a peaceful leadership transition is very much in question. Nor is the United States immune from terrorism directed against the legitimacy of the nation—as we saw in the case of the Oklahoma City bombing and in the threats of terrorism by armed militia groups that challenge the state's legitimacy, detest the United Nations and other agencies of world government, and claim that the U.S. federal government's protection of abortion clinics or the rights of homosexuals and other minorities justifies armed resistance (Horowitz, 1999).

Fortunately, one can also point to improvements in the stability of some nations that have faced severe threats to their existence and to social peace within their borders. Northern Ireland, Chile, and Nigeria are examples of nations that have experienced marked improvements in national legitimacy in the recent past. On balance, however, the world enters a new millennium facing serious challenges to the stability and viability of many of its nations.

We noted in Chapter 4 that nation-states claim a legitimate monopoly over the use of force within their borders. By this we mean that within a nation-state's borders, only organizations designated by the state (i.e., the government) may use force. But if the citizens believe that their government is legitimate, why is force necessary at all? The answer is that even when citizens believe their government to be legitimate, force (or, more often, the threat of force) may be needed to maintain order and ensure compliance with the law. For example, citizens may agree in principle that some of their income should be taken by the state through taxes to cover its expenses, but some individuals or groups (such as the Freemen in the United States) may attempt to avoid paying taxes.

In many parts of the world bribery and corruption among public officials are commonplace. When elites do not observe the rule of law, it is difficult to convince less powerful and less wealthy citizens to do so. Conflicts arising from this issue take the form of scandals such as the incidents leading to the impeachment of President Clinton or the collusion between politicians and corrupt businesspeople in Russia. The point is that those charged with enforcing the laws should be beyond reproach. Otherwise the legitimacy of government and the laws is diminished in the minds of citizens—and in their behavior.

People who feel that the state uses its power in an illegitimate manner are not allowed to form private armies to resist the state. They may challenge the legitimacy of a particular law in the courts, where rational arguments and interpretations of law replace violent conflict. Nevertheless, all the institutions of the state, including the courts, are ultimately based on the state's monopoly over the use of force (Breiner, 1996). In societies in which the state is weak or in which its legitimacy is denied by particular groups, it is common to find bandits, guerrilla armies, or terrorist groups.

Citizenship and the Rule of Law

The term **citizenship** refers to the status of membership in a nation-state. Like all statuses, citizenship is associated with a specific set of rights and obligations (Transaction,

These new citizens at a swearing-in ceremony have had to demonstrate knowledge of basic English and awareness of the rights and responsibilities of United States citizens.

Then and Now

Television Reshapes the Campaign Trail

In 1948 Harry Truman campaigned across the United States, making speeches from a special car at the back of a train. This was one of the last "whistle stop" campaigns in American presidential politics. The older style of campaign, in which the candidate attempted to meet as many voters as possible, was accompanied by a great deal of "back room" politics—the candidate met with local political leaders, who bargained for favors in return for bringing out their supporters on election day. These dealings, of course, were not visible to the people who gathered at the "whistle stops."

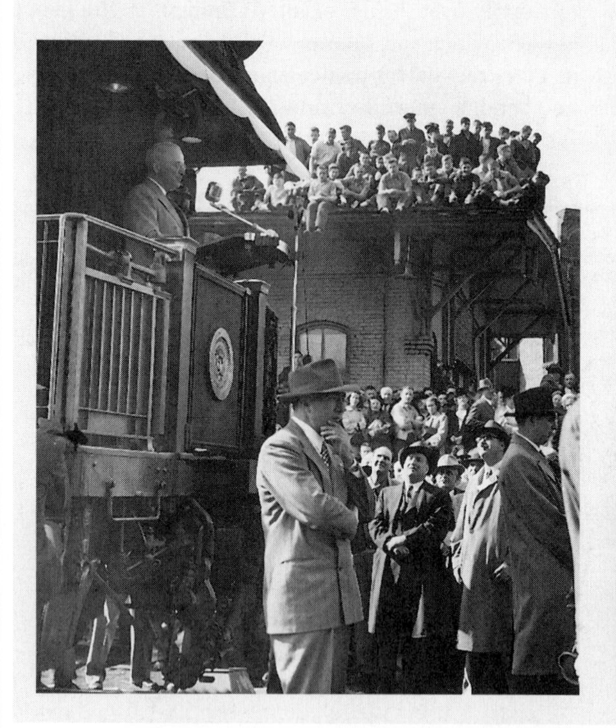

Today the power of television brings the candidate and the campaign message into the homes of millions of voters, diminishing the importance of town-by-town campaigning. But television demands camera skills that many politicians do not necessarily possess. Television has also vastly increased the importance of fund-raising to pay the high costs of television advertising. The influence of local political party leaders has declined, but "back room" politics and the influence of wealthy and powerful contributors have yet to disappear.

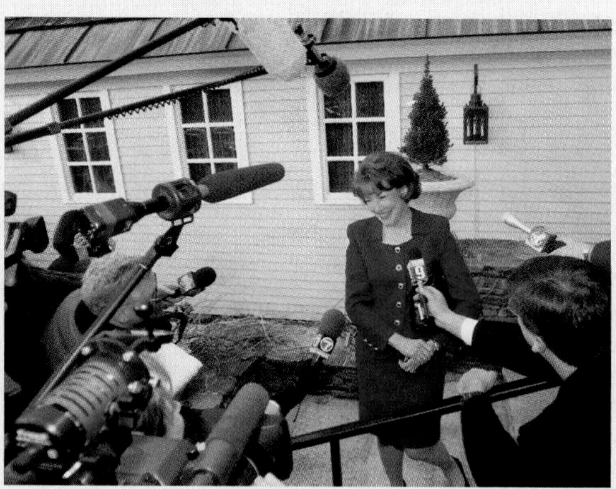

1997). As political sociologist Reinhard Bendix put it, "A core element of nation-building is the codification of the rights and duties of all adults who are classified as citizens" (1969, p. 89). In other words, the roles of citizens in the society's political life must be made clear to all.

In feudal societies most of the members of the society did not participate in political life. The needs of various groups in the population were represented, if at all, by the edicts of powerful landholders, generals, and clergymen. In modern nation-states as a result of major social movements this form of representation has been replaced by a form based on citizen participation, in which representatives elected by the citizens are entitled to vote on important public issues. But as societies adopted the principle of citizen participation, conflicts arose over the question of who would be included among a society's citizens. This is illustrated by the violent struggle over citizenship and political participation for blacks in South Africa, led by Nelson Mandela and the African National Congress, that finally resulted in a new constitution and majority rule.

Rights of Citizenship The rights of citizenship include much more than the right to vote. T. H. Marshall (1964) defined those rights as follows:

- *Civil rights* such as "liberty of person, freedom of speech, thought and faith, the right to own property and to conclude valid contracts, and the right to justice."
- *Political rights* such as the right to vote and the right of access to public office.
- *Social rights* ranging from "the right to a modicum of economic welfare and security to the right to share fully in the social heritage and to live the life of a civilized being according to the standards prevailing in the society" (pp. 71–72).

This list does not necessarily imply that these rights actually exist in a given society or that they are shared equally by all of the society's citizens. In the United States, for example, before 1920 women were considered citizens but were denied the right to vote. No society in which these rights are denied to citizens can be said to have established the rule of law within its boundaries.

Participation in Local Politics The question of who is entitled to full participation in politics is a key issue at the local level as well as at the national level. At the national level, participation is vital to a group's position in society, as can be seen in the struggles by women and blacks to win the right to vote. But in the cities and towns and communities in which daily life is lived, the same guarantee of full participation plays a role in social mobility.

W. E. B. Du Bois's study, *The Philadelphia Negro,* which we discussed in Chapter 1, includes this poignant petition by black community leaders to the city's mayor:

> We are here to state to your excellency that the colored citizens of Philadelphia are penetrated with feelings of inexpressible grief at the manner in which they have thus far been overlooked and ignored by the Republican party in this city. . . . We are therefore here, sir, to earnestly beseech of you as a faithful Republican and our worthy chief executive, to use your potent influence as well as the good offices of your municipal government, if not inconsistent with the public weal, to procure for the colored people of this city a share at least, of the public work and the recognition which they now ask for and feel to be justly due to them. (1967/1899, p. 374)

At the turn of the twentieth century Philadelphia, like most American cities, was dominated by leaders who were among the town's richest and most powerful citizens. Immigrant workers and the poor were effectively excluded from political participation. The Great Depression drastically altered this pattern. Fearing that the misery created by the Depression would lead to unrest among the lower classes that could result in social disorder and even revolution, upper- and middle-class Americans began to look more favorably on the call for the creation of new social-welfare institutions. Such institutions would ease the plight of the poor and the unemployed without fundamentally changing capitalist economic institutions. At the same time, the poor and the working class began to exercise the right to vote in increasing numbers.

In the century and a half since the Civil War, the extension of full rights of citizenship and political participation has helped the United States avoid crises of legitimacy and civil strife. As we will see in the next section, however, by no means do all nations experience the same political developments, nor are they necessarily striving to extend the rights of participation in democratic political institutions to all citizens. Nor have all nations been as fortunate as the United States and others that have managed to solve their political crises without civil strife.

POLITICAL INSTITUTIONS: A GLOBAL VIEW

i

The central problem of contemporary politics, according to Seymour Martin Lipset, is this: "How can a society incorporate continuous conflict among its members and social groups and yet maintain social cohesion and legitimacy of state authority?" (1979, p. 108). This question has taken on particular urgency since World War II, when the colonial empires of Europe crumbled in the face of nationalistic movements and hundreds of new nations were created. In comparing the political institutions of the world's nations, it is helpful to begin by outlining the nature of the political institutions one expects to find in any modern state and noting the problems associated with their development.

A Rational Administration Modern nations are governed by elected and appointed officials whose authority is defined by laws. To the extent possible, they are expected to use their authority for the good of all citizens rather than for their own benefit or for that of particular groups. They are forbidden to use their authority in illegitimate ways—that is, in ways that violate the rights of the citizens.

Modern political institutions usually specify some form of separation of powers. The authors of the U.S. Constitution, for example, were careful to create a system of checks and balances among the nation's legislative, executive, and judicial institutions so that abuses of authority by one could be remedied by the others. Thus when agents of the Nixon campaign organization were caught breaking into the Democratic party headquarters during the 1972 presidential election campaign and the President and some of his advisers were found to have covered up their role in this and other illegal activities, the impeachment proceedings initiated by the legislative and judicial branches of the government led to the President's resignation.

In many other nations abuses of state authority are far more common. The power of rulers is unchecked by other political institutions, and although citizens may question the legitimacy of their rule, they are powerless to prevent the rulers' use of coercion in violation of their rights. Such states are often ruled by **demagogues,** or leaders who use personal charisma and political symbols to manipulate public opinion. Demagogues appeal to the fears of citizens and essentially trick them into giving up their rights of political participation. Hitler's ability to sway the suffering German masses made him the outstanding example of demagoguery in the twentieth century.

A Party System Organizations of people who join together in order to gain legitimate control of state authority—that is, of the government—are **political parties.** Parties may be based on ideologies, or they may simply represent competing groups with the same basic values. Many American political sociologists assert that nations must make certain that other political parties are able to compete with the ruling party (Janowitz, 1968; Lipset, 1995). Failure to protect the existence of an opposition party or parties leads to **oligarchy**—rule by a few people who seek to stay in office indefinitely rather than for limited terms of office.

Parties that seek legitimate power and accept the rule of other legitimate parties form a "loyal opposition" that monitors the actions of the ruling party, prevents official corruption, and sustains the hopes of people whose needs are not adequately met by the ruling party. Revolutionary political parties, it should be noted, do not view the state as legitimate. They therefore do not agree to seek authority through legitimate procedures like elections. For this reason they tend to be banned by most governments. On the other hand, many modern nations, including the United States, have banned or repressed nonrevolutionary communist parties, largely because of their opposition to private property and their sympathy for the communist-dominated regime of the former Soviet Union.

Institutions for Maintaining Order We have seen that states control the use of force within their borders. They also seek to protect their territories against attacks by other states. For these purposes most states maintain police forces and armies. But sometimes the state's leaders have difficulty controlling these institutions, and quite often military factions seize power in what is called a *coup d'état* (or simply *coup*).

A coup usually results in the establishment of an oligarchy in which the state is ruled by a small elite that includes powerful members of the military. The Latin American *junta* is a special type of oligarchy in which military generals rule, usually with the consent of the most powerful members of the nation's nonmilitary elite. Their rule is commonly opposed by members of the intellectual professions, especially journalists, professors, writers, and artists. The dissent expressed by these individuals often leads to censorship and further repression of nonmilitary political institutions.

Regimes that accept no limits to their power and seek to exert their rule at all levels of society, including the neighborhood and the family, are known as **totalitarian regimes;** Nazi Germany and the Soviet Union under Stalin are examples. Such regimes cannot exercise total power without the cooperation of the military. It would be incorrect, however, to attribute the existence of oligarchies and totalitarian regimes to the power of the military alone. On many occasions military leaders have led coups that deposed the state's existing rulers and then turned over state power to nonmilitary institutions.

Nevertheless, in modern societies military institutions have become extremely powerful. Even before the end of World War II Harold Lasswell (1941) warned that nations might be moving toward a system of "garrison states—a world in which the specialists in violence are the most powerful group in society" (p. 457). This theme was echoed by Dwight D. Eisenhower, one of the nation's celebrated soldier-Presidents. As he was leaving office, Eisenhower warned that the worldwide arms buildup, together with the increasing sophistication of modern weapons, was producing an "industrial-military complex." By this he meant that the military and suppliers of military equipment were gaining undue influence over other institutions of the state. (We discuss military institutions more fully in the final section of the chapter.)

Democratic Political Systems

In contrast to oligarchies and totalitarian regimes, democratic societies offer all of their citizens the right to participate in public decision making. Broadly defined, **democracy** means rule by the nation's citizens (the Greek *demos,* from which the word *democracy* is derived, means "people"). In practice, democratic political institutions can take many different forms as long as the following conditions are met:

1. The political culture legitimizes the democratic system and its institutions.
2. One set of political leaders holds office.
3. One or more sets of leaders who do not hold office act as a legitimate opposition (Lipset, 1981, 1995).

The two most familiar forms of democratic political rule are the British and American systems. In the British *parliamentary* system, elections are held in which the party that wins a majority of seats in the legislature "forms a government," meaning that the leader of the party becomes the head of the government and appoints other party members to major offices. Once formed, the government generally serves for a specified length of time. If no party gains a majority of the legislative seats, a coalition may be formed in which smaller parties with only a few seats can bargain for positions in the government. Such a system encourages the formation of smaller parties.

In the American *representative* system, political parties attempt to win elections at the local, state, and national levels of government. In this system the President is elected directly, and the party whose candidate is elected President need not have a majority of the seats in the legislature. Success therefore depends on the election of candidates to national office. In the United States two political parties, the Republicans and the Democrats, have developed the resources and support necessary to achieve this. It is difficult for smaller parties to gain power at the national level because voters do not believe they have much chance of electing national leaders, and even if they do win a few legislative seats they will not be asked to form coalitions, as they would in a parliamentary system.

One-Party Systems Suppose that a nation has only one political party. Does this mean that it is undemocratic? Not in principle. A one-party state or region can be democratic as long as the citizens are free to form other parties, as long as the party's leaders can be replaced through democratic processes, and as long as the leaders do not violate the rights of citizens. But how likely are these conditions to be met if there are no opposing parties? The evidence suggests that the odds are against democracy in one-party states and regions, as we see in communist one-party states and many noncommunist one-party dictatorships (McCrone, 1996).

A Typology of Regimes

Social scientists use analyses of political institutions to develop typologies of political regimes. One such typology, which is widely used, was developed by J. Denis and Ian Derbyshire (1996). They classify national regimes into these categories:

1. *Liberal democracies.* These regimes are marked by multiparty elections, competitive parties, separation of powers, and guarantees of the rights of minorities and individuals. Examples are the United States, Canada, Great Britain, France, Brazil, and Japan.
2. *Emergent democracies.* These have constitutions that specify all or most of the institutions and processes of liberal democracies but are marked by problems in fully establishing democratic processes due to one-party dominance, insurgencies, corruption, and so forth. They are often viewed as "liberal democracies on trial." Examples are Chile, Ivory Coast, Mali, Haiti, Morocco, Tunisia, and the Philippines.
3. *Communist regimes.* These are run by a "revolutionary dictatorship" and a single communist party that in principle is serving the interests of the working class. There are very limited guarantees of individual or minority rights. Political command over economic institutions is widespread but subject to market experimentation. China, North Korea, and Cuba are examples of existing communist regimes.
4. *Nationalistic socialist regimes.* These are similar to communist regimes, with a single socialist party. However, they are more inclined to promote the interests of one national group over others and to allow private commerce. There is little or no protection of individual or minority rights. Examples are Iraq, Libya, Tanzania, and Syria.
5. *Authoritarian nationalist regimes.* The extreme nationalism of these regimes leads to intolerance and exclusion of other races and creeds, often in the most brutal or genocidal fashion as in Nazi Germany or contemporary Kosovo or Rwanda.
6. *Military regimes.* These are ruled by a military elite or junta, usually with extremely limited protection of citizens' rights and no free elections. Current examples are Nigeria, Sierra Leone, Sudan, and Myanmar, although many emergent democracies and some liberal democracies are plagued by problems of civilian control over the military.

TABLE 20.1 World Political Regimes, 1989 and 1995

	Number of States by Regime Type							
Year	Lib-Dem	Em-Dem	Communist	Nat-Soc	Auth-Nat	Military	Islam-Nat	Absolutist
1989	50	33	16	21	16	16	0	12
1995	73	72	5	8	13	7	2	12
Change	+23	+39	–11	–13	-3	-9	+2	—

Key

Lib-Dem Liberal Democratic
Em-Dem Emergent Democratic
Nat-Soc Nationalistic Socialist
Auth-Nat Authoritarian Nationalist
Islam-Nat Islamic Nationalist

Source: Denis & Derbyshire, 1996.

7. *Islamic nationalist regimes.* These are ruled by nationalistic political regimes devoted to fundamentalist Islam. Afghanistan and Iran are the two existing examples of this type of regime.
8. *Absolutist regimes.* These are usually ruled by an absolute monarch who passes power to successors through a hereditary line. Constitutional forms of government, popular assemblies, judiciary rules that counter the executive power, and political parties are banned. Sultanates, emirates, and traditional monarchies such as Jordan, Saudi Arabia, and Swaziland are examples of this type of regime.

The large majority of the world's people live in liberal or emergent democracies, although China, the world's most populous nation with more than 1 billion people, is ruled by a communist regime. Table 20.1 shows that there has been a trend toward more democratic regimes over the last few years, but this trend cannot be viewed as an inevitable progression. There are far too many obstacles to true democracy, even in the most stable democratic regimes, to allow us to take the future of democracy for granted. Indeed, we will see in the next section that persistent problems of poverty and inequalities in the distribution of power are among the most serious obstacles to the emergence of democratic institutions.

PERSPECTIVES ON POLITICAL INSTITUTIONS

Our description of democratic political institutions leaves open the question we posed earlier: How can a society accommodate competition or conflict among its members and social groups and yet maintain social cohesion and legitimacy of state authority (Lipset, 1998)? One answer is that a society can resolve conflicts through democratic processes. But this leads us to ask what conditions allow democratic institutions to form and, once formed, what ensures that they actually function to reduce the inequalities that engender conflict. The broadest test of democratic institutions is not whether they are embodied in formal organizations like legislatures and courts but whether they are able to address the problems of inequality and injustice in a society.

There are at least three schools of sociological thought regarding these questions. The first, derived from the functionalist perspective, asserts that democratic political institutions can develop and operate only when certain "structural prerequisites," such as a large middle class, exist in a society. The second school of thought, often referred to as the **power elite model,** is based on the conflict perspective. It is highly critical of the functionalist view, supporting its criticism with evidence of the ways in which so-called democratic political institutions actually operate to favor the affluent. A third position, known as the **pluralist model,** asserts that the existence of ruling elites does not mean that a society is undemocratic, as long as there are divisions within the elite and new groups are able to seek power and bargain for policies that favor their interests.

Structural Prerequisites of Democracy and the Rule of Law

Seymour Martin Lipset (1981, 1994) argues that democratic political institutions are relatively rare because if they are to exist and function well, the society must have attained a high level of economic and cultural development. To prove his theory, Lipset surveyed data on elections, civil rights, freedom of the press, and party systems in 48 nation-states, and subsequently in many other countries (Lipset, 1994). He found that the presence of these institutions was correlated with a nation's level of economic development, its degree of urbanization, the literacy of its citizens, and the degree to which its culture values equality and tolerates dissent. Table 20.2 and Box 20.2 present Lipset's findings in more detail.

TABLE 20.2 A Comparison of European and Latin American Democracies and Dictatorships, by Wealth, Industrialization, Education, and Urbanization

A. Indices of Wealth

	Per Capita Income in $	Thousands of Persons per Doctor	Persons per Motor Vehicle	Telephones per 1,000 Persons	Radios per 1,000 Persons	Newspaper Copies per 1,000 Persons
European and English-speaking stable democracies	685	0.86	17	205	350	341
European and English-speaking unstable democracies and dictatorships	305	1.4	143	58	160	167
Latin American democracies and unstable dictatorships	171	2.1	99	25	85	102
Latin American stable dictatorships	119	4.4	274	10	43	43

B. Indices of Industrialization

	Percentage of Males in Agriculture	Per Capita Energy Consumed
European stable democracies	21	3.6
European dictatorships	41	1.4
Latin American democracies	52	0.6
Latin American stable dictatorships	67	0.25

C. Indices of Education

	Percentage Literate	Primary Education Enrollment per 1,000 Persons	Post-Primary Enrollment per 1,000 Persons	Higher Education Enrollment per 1,000 Persons
European stable democracies	96	134	44	4.2
European dictatorships	85	121	22	3.5
Latin American democracies	74	101	13	2.0
Latin American dictatorships	46	72	8	1.3

D. Indices of Urbanization

	Percent in Cities Over 20,000	Percent in Cities Over 100,000	Percent in Metropolitan Areas
European stable democracies	43	28	38
European dictatorships	24	16	23
Latin American democracies	28	22	26
Latin American stable dictatorships	17	12	15

Source: From *Political Man: The Social Bases of Politics,* by Seymour Martin Lipset (Baltimore: Johns Hopkins University Press, 1981, expanded and updated edition). Reprinted by permission of the author.

BOX 20.2 SOCIOLOGICAL METHODS

Diagramming Social Change

The accompanying chart is an example of how a diagram can effectively communicate ideas about the directions and influences of social change. This diagram presents the social structures from which democratic political institutions arise (listed on the left) and some possible further consequences of the emergence of those political institutions (on the right).

The arrows in the diagram indicate either conditions that contribute to democracy or possible consequences of the emergence of democratic institutions. As in most social-scientific analysis, the author, political scientist Seymour Martin Lipset, is careful not to make direct assertions about causality. He does not state, for example, that an open class system (measured by upward and downward social mobility) inevitably leads to the emergence of democratic regimes. Instead, social mobility and the other aspects of social structure listed on the left are "necessary but not sufficient" preconditions for the emergence of democracy. That is, they must exist to some degree in a society, but they are not sufficient to "cause" democracy. For that to happen, other conditions must also be present, such as the leadership of people who can sacrifice their desire for personal power in the interest of creating democratic political institutions. George Washington, for example, is revered as a pioneer of democracy, because when he had the chance to become a monarch he chose to relinquish personal power and lead the newly independent colonies toward the creation of a constitutional monarchy.

Note that the arrows on the right side of the diagram suggest that democratic rule is associated with the emergence of social phenomena that run counter to democracy. Bureaucracy, for example, is a common consequence of democratic regimes. Laws passed by democratic processes tend to spawn administrations designed to deliver services, provide protection, enforce regulations, and much more. These bureaucracies, in turn, may stifle democracy by replacing the will of the majority with rules, regulations, and procedures. Note, however, that these excesses can be corrected by the actions of democratic institutions (e.g., further legislation). The arrows merely indicate that democratic rule has certain counter-democratic consequences.

In an expansion of the comparative research on which this diagram is based, Lipset makes the following comment:

> Not long ago, the overwhelming majority of the members of the United Nations had authoritarian systems. As of the end of 1993, over half, 107 out of 186 countries, have competitive elections and various guarantees of political and individual rights—that is, more than twice the number two decades earlier in 1970. (1994, p. 1)

But Lipset, a veteran of many decades of close observation of democratic rule in nations throughout the world, is quite cautious about predicting a trend toward the emergence of democratic regimes. Too many nations, like Nigeria and Russia, have turned toward more authoritarian rule after promising attempts to develop democratic political institutions. Above all, Lipset warns:

> [If nations] can take the high road to economic development they can keep their political houses in order. The opposite is true as well. Governments that defy the elementary laws of supply and demand will fail to develop and will not institutionalize genuinely democratic systems. (1994, p. 1)

In this cross-cultural research and in his research on democracy in the United States, Lipset attempted to show that the growth of a large middle class is essential to democracy. The middle-class population tends to be highly literate and, hence, able to make decisions about complex political and social issues. Moreover, middle-class citizens feel that they have a stake in their society and its political institutions. Accordingly, they often support policies that would reduce the class and status cleavages—the distinctions between the haves and the have-nots—that produce social conflict.

Sociologists who study voting behavior tend to support Lipset's thesis that the stability of democratic institutions rests on structural features that diminish conflict in a society. But their research has revealed something less than complete stability. Morris Janowitz (1978) made this observation:

> Since 1952, there has been an increase in the magnitude of shifts in voting patterns from one national election to the next. . . . Increasingly important segments of the electorate are prepared to change their preference for president and also to engage in ticket splitting [voting for candidates of different parties]. (p. 102)

These shifts in voting behavior may represent new alignments of voters that could lead to major changes in the nation's public policies. The important point, however,

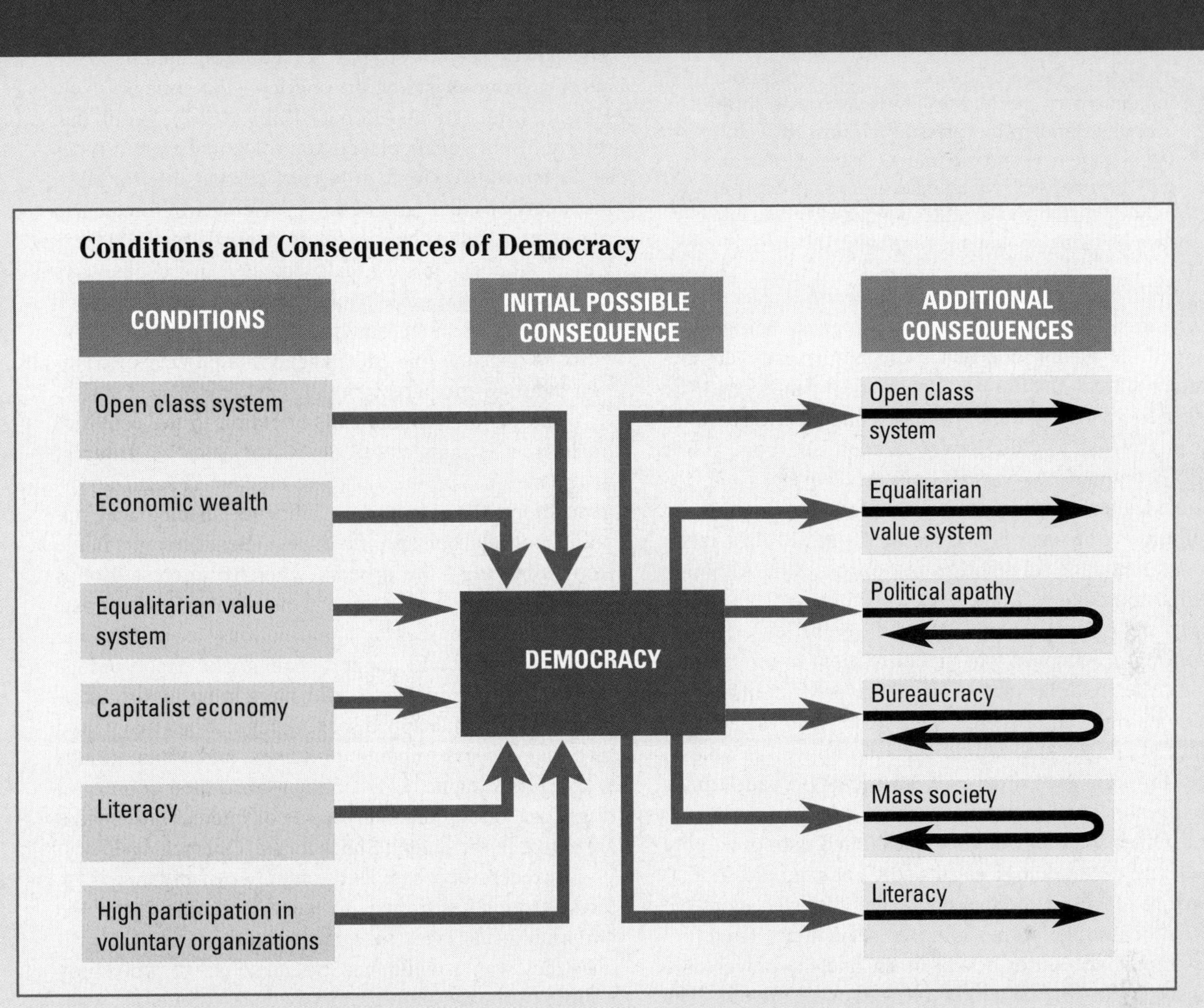

Source: From *Political Man: The Social Bases of Politics,* by Seymour Martin Lipset (Baltimore: Johns Hopkins University Press, 1981, expanded and updated edition). Reprinted by permission of the author.

is that these realignments, while they influence which parties and political leaders gain or lose power, do not affect the process of democratic competition itself.

The Power Elite Model

However important elections are to the functioning of democratic institutions, there are strong arguments against the idea that they significantly affect the way a society is governed. Some social scientists find, for example, that political decisions are controlled by an elite of rich and powerful individuals. This "power elite" tolerates the formal organizations and procedures of democracy (elections, legislatures, courts, etc.) because it essentially owns them and can make sure that they act in its interests no matter what the outcome of elections may be. C. Wright Mills, the chief proponent of this point of view, has described the power elite as follows:

> The power elite is composed of men whose positions enable them to transcend the ordinary environments of ordinary men and women. They are in positions to make decisions having major consequences. . . . They are in command of the major hierarchies and organizations of modern society. They rule the big corporations. They run the machinery of the state. . . . They direct the

> military establishment. They occupy the strategic command posts of the social structure. . . .
>
> The power elite are not solitary rulers. . . . Immediately below the elite are the professional politicians of the middle levels of power, in the Congress and in the pressure groups, as well as among the new and old upper classes of town and city and regions. Mingling with them in curious ways . . . are those professional celebrities who live by being continually displayed. (1956, p. 4)

When it was first published, Mills's *The Power Elite* created a stir among sociologists. Mills challenged the assumption that societies that have democratic political institutions are in fact democratic. He asserted instead that party politics and elections are little more than rituals. Power is exercised by a ruling elite of immensely powerful military, business, and political leaders that can put its members into positions of authority whenever it wishes to do so. Mills's claim was significant for another reason as well: Although Mills appreciated Marx's views on the role of class conflict in social change, he did not believe that the working class could win power without joining forces with the middle class. He therefore attempted to demonstrate the existence of a ruling elite to an educated public, which would then, he hoped, be able to see through the rituals of political life and make changes through legitimate means.

Other sociologists have questioned the power elite thesis on methodological grounds. For example, Talcott Parsons (1960) noted that the power elite was supposed to act behind the scenes rather than publicly. Therefore its actions could not be observed, and the power elite thesis could not be either proved or disproved. The power elite thesis therefore was not scientifically sound, according to Parsons.

Numerous adherents of the power elite thesis have attempted to show that a ruling elite does indeed exist and that its activities can be observed. One of the best known of these researchers is Floyd Hunter (1953), whose classic studies of Atlanta's "community power structure" attempted to show that no more than 40 powerful men were considered to have the ability to make decisions on important issues facing the city and its people. Most of these men were conservative, cost-conscious business leaders. The Hunter study stimulated many attempts to find similar power structures in other cities and in the nation as a whole (Domhoff, 1978, 1983). But although the term *power structure* has found its way into the language of politics, these studies have been strongly criticized for basing their conclusions on what people say about who has power rather than on observations of what people with power actually do (Walton, 1970).

The Pluralist Model

The power elite thesis has not gone unchallenged. In another famous study of politics—this one in New Haven, Connecticut—Robert Dahl (1961) found that different individuals played key roles in different types of decisions. No single group was responsible for all of the decisions that might affect the city's future. No power elite ruled the city, Dahl argued. Instead, there were a number of elites that interacted in various ways on decisions that affected them. In situations in which the interests of numerous groups were involved, a plurality of decision makers engaged in a process of coalition building and bargaining.

The pluralist model calls attention to the activities of interest groups at all levels of society. **Interest groups** are not political parties; they are specialized organizations that attempt to influence elected and appointed officials on specific issues. These attempts range from **lobbying**—the process whereby interest groups seek to persuade legislators to vote in their favor on particular bills—to making contributions to parties and candidates who will support their goals. Trade unions seeking legislation that would limit imports, organizations for the handicapped seeking regulations that would give them access to public buildings, and ethnic groups seeking to influence U.S. policy toward their country of origin are among the many kinds of interest groups that are active in the United States today (Oberschall, 1996).

In recent decades, as the number of organized interest groups has grown, so has the complexity of the bargaining that takes place between them and the officials they want to influence. Social scientists who study politics from the pluralist perspective frequently wonder whether the activities of these groups threaten the ability of elected officials to govern effectively. Thus a study of a federal economic development project in Oakland, California, found that the personnel in the agency responsible for such projects had to learn to deal with supporting and opposing interest group leaders and elected officials throughout the life of the project. It took 6 years of constant negotiation on many unanticipated issues—affirmative action in employment, environmental concerns, design specifications—before a firm could win a contract to build an airplane hangar (Pressman & Wildavsky, 1984).

Politics and Social Interaction

From the interactionist perspective, political institutions, like all others, are "socially constructed" in the course of human interaction. They do not simply come into being as structures of norms, statuses, and roles and then continue to function without change. Instead, they are continually being shaped and reshaped through interaction. This view closely parallels the popular notion that democratic political institutions must be continually challenged if they are to live up to their ideals. Otherwise they will become oligarchies that rule for the benefit of small cliques rather than for the mass of citizens (Oberschall, 1996).

The question of how social interaction affects a society's political institutions was a central concern of the leading philosophers of ancient Greece. In his *Politics,* for example, Aristotle "held that humans were made for life in society just as bees were made for life in the hive" (Bernard, 1973, p. 30). Because humans are "political animals," Aristotle believed, their constant discussions of political issues and their ability to form coalitions allow them to arrive at a consensus regarding what a good society is and how it should be governed. Unfortunately, he concluded, existing political systems were flawed. The divisions and gaps in interaction created by wealth prevented consensus and gave rise to conflict. Moreover, people in occupations like farming did not have enough time to examine all the sides of an issue and work toward agreement with their fellow citizens. In a good society, therefore, all the citizens must be free to devote their full attention to political affairs. (In practice, Greek democracy required an economy based on the labor of slaves.)

Nearly 2,000 years later the Italian political adviser Niccolò Machiavelli (1469–1527) again took up the relationship between human interaction and political institutions. Machiavelli believed that political institutions had to be based on the recognition that all human beings are capable of evil as well as good. In his *Discourses* he made this observation:

> All those who have written upon civil institutions demonstrate . . . that whosoever desires to found a state and give it laws, must start by assuming that all men are bad and ever ready to display their vicious nature, whenever they may find an occasion for it. (1950/1513, vol. 1, p. 3)

Some people, he admitted, are merciful, faithful, humane, and sincere, but even though there are virtuous people in every society, political leaders must anticipate the worst possible behavior. In his most famous work, *The Prince,* Machiavelli suggested that the wise ruler or prince would be a master of astuteness—the ability to "read" the intentions of allies and opponents in the tiniest of gestures and reactions—and would also have mastered the skills of diplomacy. He used the following comparison to make this point:

> A prince . . . must imitate the fox and the lion, for the lion cannot protect himself from the traps, and the fox cannot defend himself from wolves. One must therefore be a fox to recognize traps, and a lion to frighten wolves. Those that wish to be only lions do not understand this. (1950/1513, vol. 1, p. 66)

In the five centuries since they were originally formulated, these ideas have had a powerful influence on political thinkers. The authors of the U.S. Constitution, for example, attempted to avoid situations in which "foxes" or "lions" could take advantage of the weaknesses of others. They anticipated the more self-serving aspects of political interaction rather than simply assuming that a new society would bring about cooperation among citizens: "If men were angels, no government would be necessary" (*The Federalist,* No. 51). It is for this reason, as we noted earlier in the chapter, that they planned a government in which each of the major branches would be able to check any abuse of power by the other branches.

Political Interaction in the "Democratic Experiment" In the 1830s Alexis de Tocqueville, a young French aristocrat, visited the world's first experiment in national democracy. Although the United States was by no means a true democracy, since it condoned slavery, indentured servants, and war against Native Americans, European social thinkers took a keen interest in the model of citizen rule through democratic institutions that was being established in the new nation. Tocqueville was stunned by the "tumult" of American political interactions, which "must be seen in order to be understood." In America, he wrote, political activity was in evidence everywhere:

> A confused clamor is heard on every side; and a thousand simultaneous voices demand the immediate satisfaction of their social wants. Everything is in motion around you; here, the people of one quarter of a town are met to decide upon the building of a church; there, the election of a representative is going on; a little further, the delegates of a district are traveling in a hurry to the town in order to consult upon some local improvements; or, in another place, the labourers of a village quit their plows to deliberate upon the project of a road or a public school. (1980/1835, p. 78)

Tocqueville expressed surprise at the fact that although they were denied the vote, in the United States "even the women frequently attend public meetings, and listen to political harangues as recreation after their household labours" (p. 79). In fact, the culture of America was so suffused with political issues that he believed Americans could not converse except to discuss issues. The typical American, he observed, "speaks to you as if he were addressing a meeting" (p. 79).

BOX 20.3 USING THE SOCIOLOGICAL IMAGINATION

Politics and the English Language

George Orwell, the author of *Animal Farm* and *1984*, addressed the abuse of language for political purposes in his essay "Politics and the English Language":

> Political speech and writing are largely the defense of the indefensible. Things like the continuance of British rule in India, the Russian purges and deportations, the dropping of the atom bombs on Japan, can indeed be defended, but only by arguments which are too brutal for most people to face, and which do not square with the professed aims of political parties. Thus political language has to consist largely of euphemism, question-begging, and sheer cloudy vagueness. Defenseless villages are bombarded from the air, the inhabitants driven out into the countryside, the cattle machine-gunned, the huts set on fire with incendiary bullets: this is called *pacification*. Millions of peasants are robbed of their farms and sent trudging along the roads with no more than they can carry: this is called *transfer of population* or *rectification of frontiers*. People are imprisoned for years without trial, or shot in the back of the neck or sent to die of scurvy in Arctic lumber camps: this is called *elimination of unreliable elements*. Such phraseology is needed if one wants to name things without calling up mental pictures of them. Consider for instance some comfortable English professor defending Russian totalitarianism. He cannot say outright, "I believe in killing off your opponents when you can get good results by doing so." Probably, therefore, he will say something like this:
>
> While freely conceding that the Soviet regime exhibits certain features which the humanitarian may be inclined to deplore, we must, I think, agree that a certain curtailment of the right to political opposition is an unavoidable concomitant of transitional periods, and that the rigours which the Russian people have been called upon to undergo have been amply justified in the sphere of concrete achievement. (1950, p. 136)

If Orwell sounds highly pessimistic about the conduct of politics, remember that he was writing in a time of even greater political cruelty and chaos than our own. In his lifetime he had seen nations with the most advanced constitutions commit the most brutal acts of war and repression. The point of his essay is that to be politically objective, to seek the true meanings of political acts and the consequences of political beliefs, we must "start at the verbal end":

> The great enemy of clear language is insincerity. When there is a gap between one's real and one's declared aims, one turns as it were instinctively to long words and exhausted idioms, like a cuttlefish squirting out ink. . . . If you simplify your English . . . when you make a stupid remark its stupidity will be obvious, even to yourself. Political language . . . is designed to make lies sound truthful and murder respectable, and to give an appearance of solidity to pure wind. One cannot change this all in a moment, but one can at least change one's own habits, and from time to time one can even, if one jeers loudly enough, send some worn-out and useless phrase . . . into the dustbin where it belongs. (p. 140)

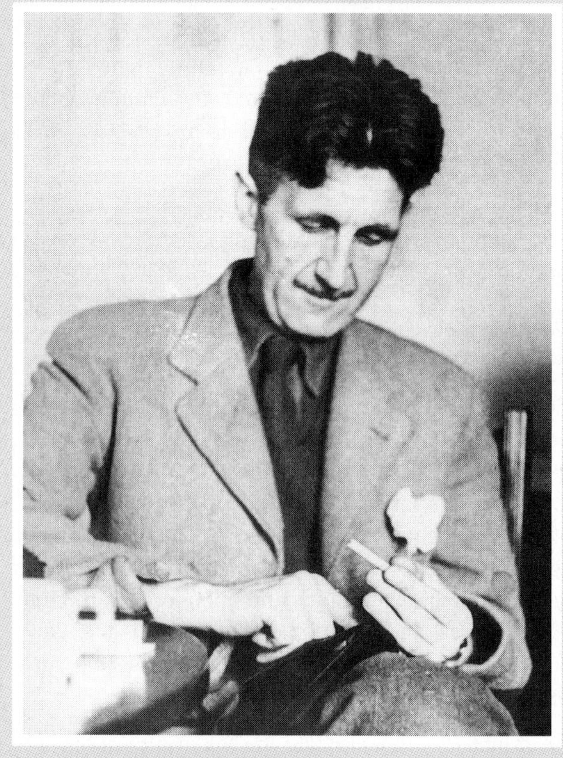

George Orwell

Tocqueville was concerned that democratic societies might encourage the rise of demagogues who could manipulate the masses. But his observations persuaded him that the immense number of competing groups in the American political system would prevent this from happening. And although he worried about the possibility of political conflict tearing the young nation apart, he became convinced that Americans' love of liberty and belief in the legitimacy of their political institutions would carry them through any crisis that might arise.

Political Communication Today For those out of power as well as those in power, political communication and the ability to gain the attention of the mass media are major concerns. Sociologist Todd Gitlin (1993) reminds us how important this aspect of politics was to the student radicals who challenged U.S. policies in Vietnam in the 1960s and 1970s. They were only one among many social movements of that period that discovered the great power of television in a society with mass publics (see Chapter 8). Through skillful use of television coverage of staged confrontations, sit-ins, and demonstrations, it was possible for a relatively small group of activists to mobilize larger numbers of supporters. The labor movement had used similar tactics in earlier decades, but by the 1960s it had become easier to communicate to mass audiences and sway public opinion—at least in a society in which the press, including the television news, is relatively free from governmental interference.

In a world in which the techniques of communication are increasingly sophisticated, it becomes ever more important to analyze political communications and to read between the lines of political rhetoric. Often political jargon masks deeds that leaders would rather not admit to. George Orwell, a master of political commentary, made this point in his essay "Politics and the English Language," part of which is presented in Box 20.3.

MILITARY INSTITUTIONS

The problem of civilian control of the military remains the most important issue in many of the world's nation-states. At this writing, for example, Nigeria, the largest emergent democratic regime in Africa, is experiencing a peaceful transition from military absolutism to constitutional democracy. After years of harsh military dictatorship, extreme political and economic corruption, and suppression of the media, universities, and the courts by a military "strongman," two major parties are competing peacefully for electoral support. The press is thriving again, and exiled intellectuals and artists are returning (*Dare,* 1999). Nigeria, however, is not an isolated example of the problem of social control over the military. Throughout the world the problem of how to maintain the allegiance of the military often looms as the largest threat to democracy and to the legitimacy of governments. These situations raise a major question in political sociology: How can states control their military institutions? Answers are to be found in knowledge about the nature of the modern military as a social institution as well as in the political culture of different nations.

A key question about military institutions in all societies, and especially in democracies, is how the military can be made to submit to the guidance of civilian political institutions. A democratic society cannot remain democratic for very long if the military usurps the authority granted to it by the civilian institutions of the nation-state. This is a common occurrence—taking the form of coups led by military "strongmen"—in nations in which the military is not adequately controlled by other institutions.

The problem of civilian control of the military was recognized by Tocqueville, who noted that in aristocratic nations the military could be controlled by the government because the military leaders were aristocrats and therefore were part of the court. "The officer is noble, the soldier is a serf," he observed. "The one is naturally called upon to command, the other to obey" (1980/1835, p. 128). Tocqueville formed the hypothesis that greater equality would make a nation less likely to go to war, since its citizens would have more to lose in wars and more power to prevent them. Yet he also feared the ambitions of military leaders in democratic nations. "In democratic armies," he wrote, "all the soldiers may become officers, which . . . immeasurably extends the bounds of military ambition" (p. 129).

Morris Janowitz (1960), a highly innovative political sociologist, responded to Tocqueville's concerns in his analysis of the history of military institutions in democratic nations. He argued that the military has often played a crucial role in establishing and protecting democratic institutions. In the Greek city-states and the Roman republic, as well as the democracies that emerged from the revolutions of the late eighteenth century, citizens were obligated to serve in the armed forces. The military institution could function only through the enlistment of "citizen soldiers." Because the citizen soldier was committed to democratic institutions and was serving because such service was a requirement of citizenship, the military would be unlikely to take over the functions of democratic political institutions. Soldiers would presumably place the values of their society above the demands of the military.

When Janowitz reviewed the impact of new methods of warfare on the military, he concluded that because of the increasing sophistication of military technology, the military in many modern nations had become staffed by "professional soldiers" for whom service in the armed forces is a career. In addition, the shift to a volunteer military, rather than one that depends on citizen soldiers, may threaten the control of the military by civilian institutions (Janowitz, 1960).

The Economic Role of the Military

Even in the United States, where the threat of a coup is not considered great, the military has so much influence on the economy that social control of the military becomes difficult. The function of the armed forces is, of course, the defense of the nation-state. However, in the United States the military also serves an important economic function as a producer of jobs and revenue. The economies of states like Alaska, Connecticut, Maryland, California, and Virginia became highly dependent on the wealth gained from defense contracts during the decades of the cold war and the arms race. In addition, the private companies that vie for military contracts became a major source of employment, often providing some of the most highly paid and secure jobs in their regions. Throughout the United States the influence of the military "pork barrel"—federal spending on defense contracts and the resulting employment—often makes it difficult for legislators to support cuts in military budgets and programs.

The growth of military production in different parts of the United States also adds to the potential for armed conflict elsewhere in the world. Faced with declining domestic sales, U.S. arms makers seek to supply advanced weapons to buyers in other nations. Indeed, weapons account for a significant portion of U.S. exports; the United States is the world's leading exporter of conventional weapons (Boese, 1997).

Military Socialization

Social control of the military is partially explained by the professional socialization of military personnel. This point was vividly illustrated by Tom Wolfe in *The Right Stuff* (1980). "There were many pilots in their thirties," Wolfe wrote, "who to the consternation of their wives, children, mothers, fathers, and employers, volunteered to go active in the reserves and fly in combat in the Korean War." This was in vivid contrast to the attitude of the foot soldiers in that war, whose morale during some periods of the prolonged conflict "was so bad it actually reached the point where officers were prodding men forward with gun barrels and bayonets" (p. 32).

This contrast between the "fighter jocks," socialized to want to prove they had "the right stuff" and thus win promotion to higher ranks, and the far more cautious draftees and recruits on the ground, raises a series of questions: What motivates a person to face death in war? Is it desire for glory and advancement? Is it fear of punishment? Is it the pressure of collective action? From an interactionist perspective, this contrast results from the ways in which people are socialized in military institutions, coupled with their definitions of the situation. The fighter jocks define the situation as one in which they must act as they do—it is the job for which they have volunteered. The draftees do not define the situation in the same way; they are more interested in their own survival and in adhering to group norms like "Never volunteer."

In his classic study of socialization in a military academy, Sanford Dornbusch (1955) showed that traditional military socialization processes are designed to develop a high level of motivation and commitment to the institution. Those processes include the following:

- *Suppression of previous statuses.* Through haircuts, uniforms, and the like, the recruit is deprived of visible clues to his or her previous social status.

- *Learning of new norms and rules.* At the official level, the recruit is taught obedience to the rules of the military; through informal socialization, he or she is taught the culture of the military institution.

- *Development of solidarity.* Both informal socialization and harsh discipline build solidarity and lasting friendships among recruits; they learn to depend on one another.

- *The bureaucratic spirit.* The recruit is taught unquestioning acceptance of tradition and custom; orders are taken and given from morning to night.

The controversy that greeted President Bill Clinton's order that the military cease discriminating against homosexuals is an indication of the sensitivity that surrounds issues of military socialization. Gay men and women have been serving in the armed forces for decades but have had to deny their sexual preferences in military exams and screening procedures. Objections to openly admitting homosexuals into the military centered on the possibility that gay sexuality would be a divisive influence and a detriment to fighting morale. Gay activists in and outside the military answered that there is no reason why the norms of military conduct would not apply to openly gay soldiers just as they always have to gay and nongay military personnel.

Despite his personal opposition to the President's order, General Colin L. Powell, then chairman of the Joint Chiefs of Staff, stated that "we know that there will be changes and . . . you and the Commander in Chief [the President] can always count on us for faithful support and execution of your decisions" (quoted in Schmitt, 1993, p. A1). General Powell was referring not only to the issue of acceptance of homosexuals in the military but also to the issue of cuts in military budgets. His statement is indicative of the control the constitutional rule of law provides over the powerful institution of the military.

Social Change and Military Institutions

Like all major social institutions, the military has undergone many changes, some of which can be seen in the transformation of American military institutions in the past two centuries. In the era of the citizen soldier, people expected to be called into service only in emergency situations and to be trained by a small cadre of professional officers. Today, in contrast, the permanent armed forces have more than 1.5 million uniformed members, and the annual military budget amounts to over $260 billion (*Statistical Abstract,* 1998).

Today the requirements of highly technological forms of warfare and the need to maintain a complex set of military organizations throughout the world have changed our concept of a military career. The military as a modern social institution requires professionally trained officers who commit themselves to spending much of their working lives in the military. Thus, as the military has evolved, so has its need to recruit and train career officers in specialized military academies.

Over time, the institutions that train professional soldiers have had to adapt to certain changes. For example, as the larger society has accepted the demand for gender equality and women have gained access to occupations that were previously reserved for men, the military has had to redefine its historical perception that women soldiers are better equipped for desk duty than for service in combat. Although to date few women have fought in ground warfare, women in the military are now being trained for combat roles. The idea of women engaging in ground combat remains controversial among military traditionalists—primarily males—but the increasing automation of air and ground combat makes it more reasonable for women to assume the same military roles as men, as some did in the Persian Gulf War (West, 1998).

At least in the United States, the military tends to be rather responsive to civilian demands. Yet cuts in military budgets as a result of the end of the cold war offer the U.S. military one of its greatest challenges since the end of World War II: how to remain an effective fighting force, adapted to new forms of conflict (e.g., antiterrorism, small wars, and peacekeeping interventions) while it is scaled down in an era of increasing demand for reductions in military spending.

RESEARCH ON THE CUTTING EDGE: Is the United States Becoming a Nation of Solitary Bowlers?

In "Bowling Alone," a famous and controversial article, political scientist Robert Putnam (1995) presented evidence that Americans have become far less a "nation of joiners" than they were when Tocqueville visited the United States. In fact, they are becoming less inclined to join social groups, to become informed about current events, to trust other Americans, and perhaps more important, to vote in elections. Putnam's phrase "bowling alone" is a metaphor or catch phrase for a range of phenomena that point to growing social isolation and what he and other social scientists term a "decline in social capital."

Social capital refers to all forms of engagement in civic life in communities, congregations, political parties, and voluntary associations. This engagement can be thought of as "capital" because it supplies the much-needed energy and individual activity that make civic life possible. As we saw earlier in the chapter, the legitimacy of democratic political institutions depends on citizen participation. If voters are alienated from political institutions and do not believe that government acts in their interests, the consequences can be devastating. Political alienation results in apathy and the feeling that no political action matters (Putnam, 1995). It results in resentment of government and a weakening of the rule of law. Catastrophic violence like the Oklahoma City bombing may be a symptom of a growing crisis of political legitimacy. As Putnam points out:

> Consider the well-known decline in turnout in national elections over the last three decades. From a relative high point in the early 1960s, voter turnout had by 1990 declined by nearly a quarter; tens of millions of Americans had forsaken their parents' habitual readiness to engage in the simplest act of citizenship. Broadly similar trends also characterize participation in state and local elections.
>
> It is not just the voting booth that has been increasingly deserted by Americans. A series of identical questions posed by the Roper Organization to national samples ten times each year over the last two decades reveals that since 1973 the number of Americans who report that "in the past year" they have "attended a public meeting on town or school affairs" has fallen by more than a third (from 22 percent in 1973 to 13 percent in 1993). Similar (or even greater) relative declines are evident in responses to questions about attending a political rally or speech, serving on a committee of some local organization, and working for a political party. By almost every measure, Americans' direct engagement in politics and government has fallen steadily and sharply over the last generation, despite the fact that average levels of education—the best individual-level predictor of political participation—have risen sharply throughout this period. Every year over the last decade or two, millions more have withdrawn from the affairs of their communities. (1995, p.4)

Putnam and others who study changing patterns of civic engagement trace these declines to major generational shifts over the course of the twentieth century. The generations born earlier in the century fought the world wars and experienced devastating economic depression and the rebuilding of the economy through concerted corporate and civic action. The boom in participation in higher education after World War II, spurred by the GI Bill and growing demand for a skilled labor force, also contributed enormously to a sense of civic involvement, which translated into participation in groups of all kinds. Now these generations are being replaced by people born after the mid-1940s, the baby boomers and those in the so-called "Generation X," who account for about 60 percent of the nation's adult population (Putnam, 1996). These cohorts appear far less interested in civic engagement than their parents and grandparents were, but the causes of this change are not well established.

Putnam attributes much of the change to the influence of television, the most powerful of the new media that became available in the second half of the twentieth century. He admits that available data do not fully prove this hypothesis but maintains that they are extremely suggestive. Two pieces of evidence for this claim are revealed in Figures 20.2 and 20.3, which relate group membership to education and media consumption. Group membership is measured in terms of the number of groups people say they belong to (including church congregations, labor unions, sports teams, and so on). Adding the education variable helps control for the effects of years of schooling on media consumption. The two graphs deal with newspaper readership and hours of television viewing. They offer a striking visual comparison of the effects of different media on civic engagement at each level of education. Clearly, the effects of the two media diverge. At each level of education, the more people read newspapers, the more group

FIGURE 20.2 Group Membership, by Newspaper Readership and Education

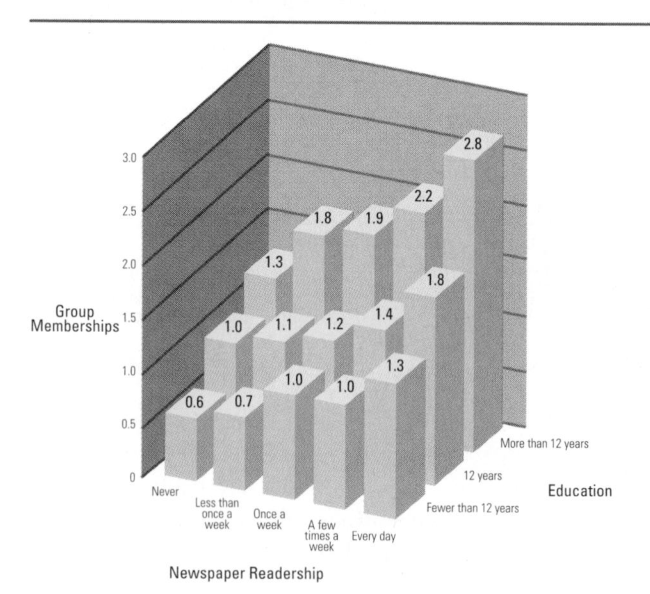

Source: Data from NORC.

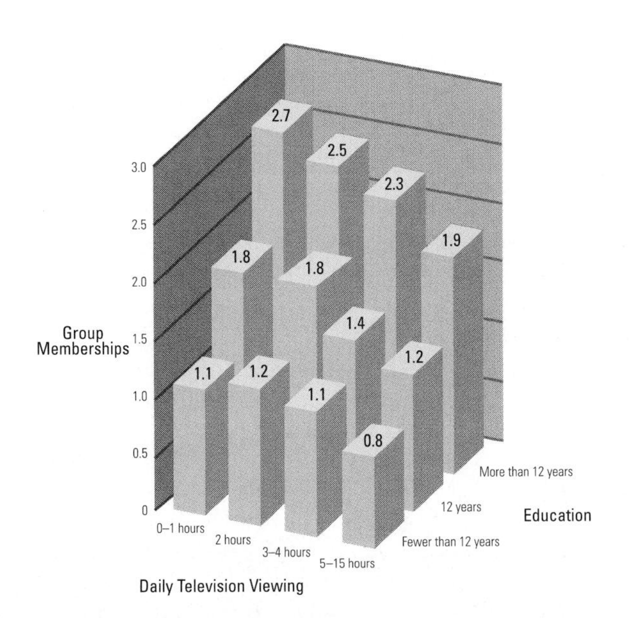

Source: Data from NORC.

FIGURE 20.3 **Group Membership, by Television Viewing and Education**

membership they report. But the more television they watch, the less group affiliation they report. These effects do not change with age; in fact, younger people are more likely to watch television rather than read newspapers and less likely to report membership in social groups than people in older generations. Note also that group membership and political participation are highly correlated—in other words, the same effects are found for voting.

These are disturbing trends. No social scientist can predict whether they will continue or how and why they might improve. Putnam himself believes that more Americans will be "bowling alone" in the future. He offers this comment:

> The most whimsical yet discomfiting bit of evidence of social disengagement in contemporary America that I have discovered is this: more Americans are bowling today than ever before, but bowling in organized leagues has plummeted in the last decade or so. Between 1980 and 1993 the total number of bowlers in America increased by 10 percent, while league bowling decreased by 40 percent. (Lest this be thought a wholly trivial example, I should note that nearly 80 million Americans went bowling at least once during 1993, nearly a third more than voted in the 1994 congressional elections and roughly the same number as claim to attend church regularly. Even after the 1980s' plunge in league bowling, nearly 3 percent of American adults regularly bowl in leagues.)
>
> The rise of solo bowling threatens the livelihood of bowling-lane proprietors because those who bowl as members of leagues consume three times as much beer and pizza as solo bowlers, and the money in bowling is in the beer and pizza, not the balls and shoes. The broader social significance, however, lies in the social interaction and even occasionally civic conversations over beer and pizza that solo bowlers forgo. Whether or not bowling beats balloting in the eyes of most Americans, bowling teams illustrate yet another vanishing form of social capital. (1995, p. 10)

VISUAL SOCIOLOGY

Elections

Few political institutions are more important than free, contested elections. But how difficult it is to protect this vital institution! We need only think of the Philippines, where for more than 20 years, until the victory of Corazón Aquino in 1986, the regime led by Ferdinand Marcos was able to manipulate elections in order to control state power. Nor is the right of citizens to participate in elections ever fully guaranteed; rather, it must be protected by laws and by the vigilance of an educated citizenry.

Images of our own political campaigns and elections serve to remind us that the right to vote has had to be won by certain groups in the course of American history. Here we see women and blacks voting for the first time after long struggles to win this fundamental right of citizenship.

The existence of elections is not in itself a measure of democracy. There must also be a contest, and it must make a difference who wins. In China, national elections are considered to be important events, and most people vote. But these elections are usually mere rituals in which citizens are expected to vote for a single slate of candidates and thus affirm their support of the government. Access to state authority is not changed by the election itself.

Wherever there are single-party states, elections are usually either uncontested or rigged. Latin American states are often characterized by such elections, in which citizens are eager to vote yet have little hope that the outcome will change their lives. Even so, the existence of elections holds out the promise that a time may come when an election will be free from coercion and the outcome will make a difference in how the state is governed. But for this to happen there must be other institutions too, such as an independent legal system that protects the rights of citizens who support opposition candidates. And the elections themselves must be protected so that ballots are secret and are counted fairly. Although these conditions are extremely difficult to establish, we have seen time and again that given an opportunity, people will press for a free, competitive electoral system.

SUMMARY

Politics determines "who gets what, when, and how." The basis of politics is competition for *power*, or the ability to control the behavior of others, even against their will. *Authority* is institutionalized power, or power whose exercise is governed by the norms and statuses of organizations. Sets of norms and statuses that specialize in the exercise of power and authority are *political institutions*, and the set of political institutions that operate in a particular society forms the *state*.

Although a state can exercise its power through the use of force, eventually this will not be enough to govern a society. When people consent to be governed without the use of force, the state is said to be legitimate. *Legitimacy* is a society's ability to engender and maintain the belief that the existing political institutions are the most appropriate for that society. It is the basis of a society's political culture—the cultural norms, values, and symbols that support and justify its political institutions. According to Max Weber, the political cultures of different societies give rise to three different types of authority: *traditional, charismatic,* and *legal.* Perhaps the strongest and most dangerous political force in the world today is *nationalism*, the belief of a people that they have the right and the duty to constitute themselves as a nation-state.

The most important political territory in world affairs is the *nation-state*, the largest territory within which a society's political institutions can operate without having to face challenges to their sovereignty. Throughout the world there are nation-states in crisis. Even when citizens believe that their government is legitimate, force may be needed to maintain order and ensure compliance with the law.

Citizenship is the status of membership in a nation-state. The rights of citizenship include civil rights (e.g., freedom of speech, thought, and faith), political rights (e.g., the right to vote), and social rights (e.g., the right to a certain level of economic welfare and security). The question of who is entitled to full participation in politics is a key issue at the local level as well as at the national level.

Modern nations are governed by elected and appointed officials whose authority is defined by laws. In order to prevent abuses of authority, modern political institutions usually specify some form of separation of powers in which abuses by one institution can be remedied by others. *Political parties* are organizations of people who join together in order to gain legitimate control of state authority. Parties that accept the rule of other legitimate parties form a "loyal opposition" that monitors the actions of the ruling party and prevents the emergence of *oligarchy*, or rule by a few people who stay in office indefinitely. Regimes that accept no limits to their power and seek to exert their rule at all levels of society are known as *totalitarian regimes.*

Democracy means rule by the nation's citizens: Citizens have the right to participate in public decision making, and those who govern do so with the explicit consent of the governed. In the British parliamentary system, elections are held in which the party that wins a majority of the seats in the legislature "forms a government": The leader of the party becomes the head of government and appoints other party members to major offices. In the American representative system, the party whose candidate is elected President need not have a majority of seats in the legislature.

National regimes can be classified into several categories: liberal democracies, emergent democracies, communist regimes, nationalistic socialist regimes, authoritarian nationalist regimes, military regimes, Islamic nationalist regimes, and absolutist regimes. The majority of the world's people live in liberal or emergent democracies.

Functionalist theorists assert that certain "structural prerequisites" must exist in a society for democratic political institutions to develop and operate. Among these are high levels of economic development, urbanization, and literacy, as well as a culture that tolerates dissent. The *power elite model* holds that the presence of democratic institutions does not mean that a society is democratic; political decisions are actually controlled by an elite of rich and powerful individuals. This view is challenged by the *pluralist model*, which holds that political decisions are influenced by a variety of *interest groups* through a process of coalition building and bargaining.

Many political thinkers have been concerned with the relationship between social interaction and political institutions. Among the most influential was Machiavelli, who argued that political institutions must be based on the recognition that human beings are capable of evil as well as good. This recognition played a major part in the planning of the government of the United States, in which each branch of the government is able to check abuses of power by the other branches.

In modern nation-states political communication is a major concern for those out of power as well as those in power. Analyzing political communications and reading between the lines of political rhetoric is very important. This is particularly true in societies characterized by mass publics and sophisticated communication techniques.

A major question in political sociology is how states can control their military institutions. A society cannot remain democratic very long if the military usurps the authority granted to it by civilian institutions. One factor that contributes to this problem is the fact that the military is staffed by professional soldiers for whom service in the armed forces is a career. Another factor is the immense influence of the military on the economy. On the other hand, military socialization instills norms and values that may contribute to social control of the military.

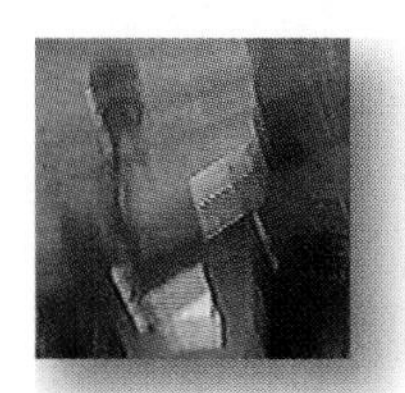

SOCIOLOGY VERSUS IDEOLOGY

At the end of the 1950s the American sociologist Daniel Bell wrote an influential book titled *The End of Ideology.* His thesis was that the world's democratic nations were entering a new era in which the vast gulf between the ideologies of the left and the right would cease to explain social change. The revolutionary period was largely over, and communism and fascism were largely obsolete in the face of the growing influence of the modern welfare state. Bell's was a brilliant analysis, but it was a premature forecast, to say the least. Ideologies that command people's passions still threaten world peace, just as they endanger democratic institutions in the United States.

Here we are not talking about conservatives and liberals who agree to compete within the legitimate institutions of government and take their chances on persuading the electorate. Sociological research indicates that the most extreme ideologues are those for whom the ends—such as denying women the right to choose abortion or homosexuals the right to engage in their chosen lifestyle—justify violence or other illegitimate means. From this ends–means reasoning, terrorism often emerges, and the state often responds with counterviolence that may not appear very different from terrorism. The sociologist, therefore, understands that conservative and liberal ideologies, whether they are applied to religion, economics, or social welfare, are not nearly as much of a problem as are ideologues and demagogues. These extremists pose a serious threat to democratic institutions because in the name of ideological or moral purity they are willing to sacrifice the rights and even the lives of innocent people.

GLOSSARY

power: the ability to control the behavior of others, even against their will. (p. 642)

authority: power whose exercise is governed by the norms and statuses of organizations. (p. 643)

political institution: a set of norms and statuses pertaining to the exercise of power and authority. (p. 643)

state: the set of political institutions operating in a particular society. (p. 643)

legitimacy: the ability of a society to engender and maintain the belief that the existing political institutions are the most appropriate for that society. (p. 644)

traditional authority: authority that is hereditary and is legitimated by traditional values, particularly people's idea of the sacred. (p. 644)

charismatic authority: authority that comes to an individual through a personal calling, often claimed to be inspired by supernatural powers, and is legitimated by people's belief that the leader does indeed have God-given powers. (p. 644)

legal authority: authority that is legitimated by people's belief in the supremacy of the law; obedience is owed not to a person but to a set of impersonal principles. (p. 644)

nationalism: the belief of a people that they have the right and the duty to constitute themselves as a nation-state. (p. 645)

nation-state: the largest territory within which a society's political institutions can operate without having to face challenges to their sovereignty. (p. 645)

citizenship: the status of membership in a nation-state. (p. 647)

demagogue: a leader who uses personal charisma and political symbols to manipulate public opinion. (p. 650)

political party: an organization of people who join together to gain legitimate control of state authority. (p. 650)

oligarchy: rule by a few people who stay in office indefinitely rather than for limited terms. (p. 650)

totalitarian regime: a regime that accepts no limits to its power and seeks to exert its rule at all levels of society. (p. 650)

democracy: a political system in which all citizens have the right to participate in public decision making. (p. 651)

power elite model: a theory stating that political decisions are controlled by an elite of rich and powerful individuals even in societies with democratic political institutions. (p. 652)

pluralist model: a theory stating that no single group controls political decisions; instead, a plurality of interest groups influence those decisions through a process of coalition building and bargaining. (p. 652)

interest group: an organization that attempts to influence elected and appointed officials regarding a specific issue or set of issues. (p. 656)

lobbying: the process whereby interest groups seek to persuade legislators to vote in their favor on particular bills. (p. 656)

WHERE TO FIND IT

BOOKS

"Politics as a Vocation," in ***From Max Weber*** (Hans Gerth and C. W. Mills, eds.; Oxford University Press, 1946). A fascinating sociological statement about what it takes to live the life of a professional politician and what kinds of people and professions are best suited to that life.

Politics: Who Gets What, When and How (Harold Lasswell; McGraw-Hill, 1936). A classic treatment of the meaning of politics and political life. Contains some exceptional insights into political interaction and the personalities of politicians.

Inequality, Power and Development: The Task of Political Sociology (Jerry Kloby; Humanities Press International, 1995). An up-to-date analysis of political sociology from a global perspective, with emphasis on the problems of poverty and inequalities of power.

Foundations of Political Sociology (Transaction, 1997). A comprehensive text on political sociology.

Political Man (Seymour Martin Lipset; Johns Hopkins University Press, 1981). A revision of a classic of political sociology. Traces the rise of democratic political institutions to other changes in a nation's stratification system and economic institutions.

Continental Divide: The Values and Institutions of the United States and Canada (Seymour Martin Lipset; Canadian-American Committee, 1989). Lipset is the dean of American political sociologists. In this empirical study comparing the United States and Canada, he shows why Canada has had more success in adapting its political institutions to address social problems.

Shooting an Elephant and Other Essays (George Orwell; Harcourt Brace Jovanovich, 1950). A collection of Orwell's essays; includes "Politics and the English Language" and his writing on colonialism and class politics.

JOURNALS

Political Science Quarterly. A leading journal in the field of political science. Publishes good articles on voting behavior and change in political institutions.

Public Opinion Quarterly. A journal of public opinion studies. Often presents recent research on how Americans and other publics view recent political and economic events.

OTHER SOURCES

World Almanac and Book of Facts. Published annually by World Almanac Books, Mahwah, NJ.

America Votes. Published biennially by the Elections Research Center, Washington, DC. Presents data on both federal and state elections.

Statistics of the Presidential and Congressional Election. Official statistics on federal elections, collected by the clerk of the House and published annually.

INTERNET RESOURCES

The New York Times (www.nytimes.com/). An unparalleled source for political events and trends in the world. To receive the full range of available materials, you will have to register at this address.

Stanford University, Hoover Institution on War, Revolution, and Peace (www.hoover.stanford.edu/). Analyzes social, political, and economic change in the world.

The White House (www.whitehouse.gov/). A comprehensive site that features an Interactive Citizens' Handbook providing information about the U.S. federal government. The site also offers a list of commonly requested federal services and ways to communicate with the President and Vice President via electronic mail.

United States Federal Judiciary (www.uscourts.gov/). The function of the courts' official Web site is to provide a clearinghouse for information from and about the judicial branch of the U.S. government. The site offers guidelines for understanding the federal court system and publishes press releases about recent decisions.

CHAPTER 21

SCIENCE, TECHNOLOGY, AND MEDICINE

Some of its opponents call it the "billion dollar shaft." Its supporters hope that it will bring rapid economic growth to a remote region of Nevada. The project is the federal government's Yucca Mountain nuclear waste repository, which is under construction and scheduled to receive enormous quantities of radioactive waste products from the nation's nuclear power plants by 2010. As the nation waits for its opening, nuclear waste is being stored in a smaller facility in New Mexico. Controversies rage over the safety of both sites.

Sociologist Kai Erikson has conducted field research in the small communities that dot the region of Yucca Mountain. In his view, the threat of nuclear contamination, however remote, that troubles people who live there is a "new species of trouble" (see "Research on the Cutting Edge: Coping With Disasters in Urban Communities" in Chapter 9). Although they are continually told that the spent radioactive fuel will not cause them any harm, the people of the region are concerned about the presence of dangerous, cancer-producing materials in the area. "We live in a precarious world," Erikson writes, "and those people who must make their way through it without the capacity to forget those perils from time to time are doomed to a good deal of anxiety" (1998, p. 154). What once was a semidesert region of ranchers and miners will experience the growth spurt associated with industrialization. The daily arrival of huge transports of potentially deadly wastes will become a part of everyday life. All this would be less troubling if there was consensus among scientists that the disposal facility is safe.

When the storage facility is completed, there will be a labyrinth of more than 100 miles of underground tunnels. Into them will go about 12,000 massive containers of spent fuel from nuclear power reactors and about 4,400 smaller containers of highly radioactive waste from former nuclear weapons production plants, (Carter & Pigford, 1998). The scientific studies and environmental assessments that have been done to determine the safety of the disposal facility could fill a small library. But there are still debates among the scientists. Some point to evidence of seismic activity under the mountain. Others believe that there is a distinct possibility of radioactive material seeping into the region's groundwater over many centuries. Although these concerns are countered by assurances from other scientists and engineers that the odds of their occurring are extremely low, those experts cannot promise absolute security.

Many people who wonder about the safety of any nuclear waste disposal facility, no matter how deep in the ground, will cite Murphy's Law, which says that anything that can go wrong will, eventually, go wrong. Scientists explain this "law" in terms of probabilities of bad events occurring and by pointing out all the turbulence that exists in nature, which is often unpredictable and makes all forms of regularity and stability suspect (Matthews, 1997). None of this uncertainty makes people in the region sleep any easier. Nor does it help people who live on the coasts and fear the ultimate effects of global warming, or who live near nuclear plants and fear earthquakes, or who live near any of a hundred types of chemical or other manufacturing facilities that produce hazardous wastes. An extremely problematic aspect of our technologically sophisticated consumer society is found in the waste and pollution problems it produces. At best, scientists can tell us that our chances of not being poisoned by it are extremely small—but never zero.

THE NATURE OF SCIENCE AND TECHNOLOGY

Our optimism about science and technology is frustrated not only by problems of nuclear waste disposal but by many other consequences of technological change. Depletion of the ozone layer in the atmosphere, air and water pollution, controversies over access to AIDS drugs, global warming, unemployment due to technological change—these and many other problems result from our dependence on modern technologies and the scientific research that develops them. But if technologies are a source of negative change and global problems, they are also the basis of much of our material well-being. For better and, too often, for worse, science and technology are among the most powerful forces for social change in our world. In this chapter, therefore, we explore the ways in which science and technologies develop and are controlled and how they can be used to meet the challenges of our future existence on the earth.

Science is a major institutional sector of modern societies: A hallmark of the modern social order is the conduct of scientific research in universities and other research organizations. And as we saw in Chapter 19, the control of information, especially scientific and technical information, is a source of prestige and power in postindustrial societies. The study of science and technology is therefore an increasingly important sociological specialty.

Science is, essentially, knowledge. (The Latin word *scientia,* from which *science* is derived, means "knowledge.") But science is a particular kind of knowledge, knowledge that has been obtained through the scientific method (see Chapter 2)—that is, as a result of the process of developing and testing hypotheses. Even so, science encompasses an extremely diverse array of subjects—"from subatomic reactions to mental processes; from mathematical laws of thermodynamics to the economics of race relations; from the births and deaths of stars to the migration of birds; from the study of ultramicroscopic viruses to that of extragalactic nebulas; from the rise and dissolution of cultures and crystals to the rise and dissolution of atoms and universes" (*Encyclopaedia Britannica,* 1967, Vol. 20, p. 7). These diverse topics are grouped together under the label "Science" because they all involve systematic and repeated observations.

Science is customarily divided into "pure" and "applied" branches. *Pure science* is scientific investigation that is devoted exclusively to the pursuit of knowledge for its own sake, with no immediate concern for using that knowledge to solve practical problems. For example, for more than 30 years Barbara McClintock did genetics research on corn plants, which had no immediate application to agricultural technology but eventually was recognized as having applications in molecular biology. *Applied science,* in contrast, is the application of known scientific principles to a practical problem; the outcome in many cases is new technologies. And those technologies can have a tremendous impact on the quality of human life.

We saw in Chapter 3 that **technology** is an aspect of culture. It can be defined as the use of tools and knowledge to manipulate the physical environment in order to achieve desired practical goals. Although modern technologies are often based on scientific discoveries, technology is much older than science; it has its roots in the tool-making and fire-building skills of the earliest human groups. People have always needed to find better ways of doing things or ways of making their lives more comfortable. Technological innovations have met those needs for thousands of years. In fact, technological change proceeded without the benefit of scientific knowledge for the bulk of human history. Science as we know it today is a relatively recent development. Although its origins may be traced to the mathematicians and philosophers of the ancient world, its emergence as a separate sphere of knowledge, with its own norms and values, dates from the late sixteenth century. From that time on, science tested and experimented on nature and progressed far beyond what nature spontaneously had ever revealed (Hess, 1997).

Accelerated by the discoveries of modern science, technology rapidly expanded the human capacity to live

BOX 21.1 USING THE SOCIOLOGICAL IMAGINATION

The Gaia Hypothesis

The late Lewis Thomas, a writer and research physician and former chancellor of Memorial Sloan-Kettering Cancer Center, provided these speculations about the relationship between human beings and the earth:

> The notion that life on Earth resembles, in detail, the sort of coherent, connected life we attribute to an organism is now something more than a notion. Thanks in large part to the studies begun in the 1970s by James Lovelock, Lynn Margulis, and their associates, we now know that planetary life, the "biosphere," regulates itself.
>
> It maintains in precision the salinity and acid-base balance of its oceans, holds constant over millions of years the exactly equilibrated components of its atmosphere with the levels of oxygen and carbon dioxide at just the optimal levels for respiration and photosynthesis. It lives off the sun, taking in the energy it requires for its life and reflecting away the rest into the unfillable sink of space. This is the "Gaia hypothesis," the new idea that the Earth itself is alive.
>
> The one biological function the Earth does not yet perform to qualify for the formal definition of an organism is reproduction. But wait around, and keep an eye on it. In real life, this may turn out to be what it started to do twenty years ago [on the occasion of the first moon walk].
>
> Given enough time, and given our long survival as working parts of the great creature (a chancy assumption for the moment), the Earth may be entering the first stages of replication, scattering seeds of itself, perhaps in the form of microorganisms similar to those dominating the planet's own first life for the first two billion years of its Precambrian period. Atmospheres similar to ours may emerge over planets or moons now uninhabitable somewhere in our solar system, and then, with enough time and luck, out in the galaxy, even beyond.
>
> So the first moon walk brought forth two new possibilities of viewing ourselves and our home. First, the Gaia idea of a living Earth, not at all the mystical notion that it would have seemed a few years back (but still carrying the same idea of "oneness" that has long preoccupied the mystics among us), is now becoming the most practical, down-to-earth thought ever thought. And second is the idea of the Earth reproducing itself, and the possible role we might be playing, consciously or unconsciously, in the huge process.
>
> Finally, as something to think about, there is the strangest of all paradoxes: the notion that an organism so immense and complex, with so many interconnected and communicating central nervous systems at work, from crickets and fireflies to philosophers, should be itself mindless. I cannot believe it. (1989, p. 25)

in and exploit different habitats. Indeed, in the past two centuries it has changed the face of the earth (Merton, 1996). Airports, skyscrapers, housing developments, oil refineries, superhighways, the Panama Canal, the rocket launching pad at Cape Canaveral are all products of technology that have transformed the surroundings in which human beings live and work. It is this interrelationship between science and technology that is responsible for the breathtaking speed of technological change in our century. (Box 21.1 introduces the Gaia hypothesis, the idea that humans and their cultures and technologies are also part of the living earth and must act accordingly.)

The benefits of technological change are not always distributed equally. Quite often, in fact, technologies allow people in wealthier nations to produce substances that formerly could be obtained only from the poor nations. For example, sugar beets grown in Europe and the United States are replacing the sugarcane produced in tropical nations as a source of sugar. At the same time, new technologies designed to increase crop yields—the fertilizers and pesticides that created the so-called "green revolution" in parts of Asia—appear to be adding excessive amounts of pollution to the environment, thereby creating huge costs for future generations. These problems of technology and industrial side effects have led many social scientists and environmental activists to call for **sustainable development,** or development that does not damage the environment or create new environmental burdens (World Commission on Environment and Development, cited in Keyfitz, 1991). Sustainable development leads to economic and

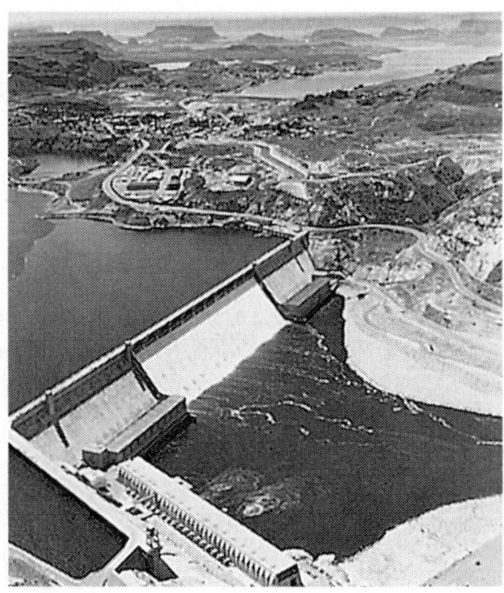

These two photos, one of an ancient irrigation system in Iran (c. A.D. 900) and one of the Grand Coulee Dam on the Colorado River, show the impact of human technology on the earth's surface. Today there are few parts of the globe that do not bear the traces and scars of human technology.

social growth without compromising the ability of future generations to meet their needs.

Because technology has had such an enormous impact on modern life, we devote a large portion of this chapter to the ways in which new technologies arise, how they affect social life, and how they lead to social change. As an especially revealing example of how technological change can shape the values and institutions of a society, we will examine the evolution and social impact of medical technology. But first we need to explore the nature of scientific institutions, since they generate the knowledge on which much technological change is based.

SCIENTIFIC INSTITUTIONS: THE SOCIOLOGICAL VIEW

The Sociology of Science

Sociologists who study scientific institutions generally follow one of two basic approaches, described as *interactionist* and *institutional* (Ben-David, 1984). Those who use the interactionist approach observe how scientists interact among themselves—for example, how they divide and coordinate work in laboratories and how they approach scientific problems. Those who use the institutional approach study the role of the scientist in different countries, the structure of scientific organizations, and the culture of scientific institutions (i.e., the norms and values of science). This distinction is not total, however; there is a great deal of overlap between the two approaches.

The Interactionist Approach Studies using the interactionist approach have focused on the scientific community—that is, on "the network of communication and social relationships between scientists working in given fields or in all fields" (Ben-David, 1984, p. 3). The questions asked by sociologists who take this approach pertain to how scientists go about the daily work of research and why scientific "revolutions" sometimes occur. In a well-known study titled *The Structure of Scientific Revolutions,* Thomas Kuhn (1962) explored the nature of the scientific community. He found that the rules of the scientific method are not adequate to describe what scientists do. Rather than spending their time testing and refuting existing hypotheses in order to establish new, more valid ones, they often take it for granted that existing theories are valid and use them in their efforts to solve specific problems. In other words, the researcher uses existing theories and methods as a **paradigm,** or model, to guide future research.

This view of the scientific community (or, rather, communities of specialized researchers) implies that science is insulated from the rest of society. Scientists are guided by the tradition of research in their field, which is passed along from one generation of scientists to the next. The problems they choose to solve are determined by that tradition, as are the methods they use in trying to solve them. This process continues, according to Kuhn, until the paradigm is no longer useful—

that is, until enough members of the scientific community believe a particular set of observations can no longer be explained by existing theories and procedures. Then the community becomes more open to outside influences. Its members explore a variety of ideas not directly related to the dominant paradigm in their field, ideas that in some cases lead to a scientific revolution. When this occurs, the old paradigm is set aside in favor of a new one that will henceforth guide the work of the members of a particular scientific community. Some of the most famous scientific revolutions have occurred in physics—for example, in the shift from Newtonian physics to Einstein's theory of relativity and quantum theory and most recently in the discovery of the existence of subatomic particles.

The Institutional Approach The institutional approach to the study of science does not contradict the interactionist view. Instead, it asks why science develops differently in different societies, and with what consequences. In this approach, certain conditions encourage the development of scientific institutions. Those conditions include the recognition of empirical research as a legitimate way of gaining new knowledge. In addition, science must be independent from other fields, such as theology or philosophy. Under such conditions separate institutions devoted to scientific research, such as graduate schools and institutes of technology, can develop. This has been the case in the United States, where the introduction of graduate training in the sciences, together with research related to professional training (as occurs, for example, in medical schools) led to the establishment of fully equipped research institutes at major colleges and universities. One effect of the presence of such institutes is a large proportion of Americans among the winners of Nobel Prizes in the sciences (see Table 21.1).

The institutionalization of science in the United States has had dramatic and far-reaching effects: "In agriculture, education, sociology, and eventually in nuclear research the universities pioneered research on a scale that far exceeded the needs of training students and was, from the very outset, an operation distinct from teaching" (Ben-David, 1984, p. 146). Attempts have been made to establish similar research organizations in Europe, but these were hampered by the rigid structure of European universities. European universities have a closed system of professorships that cannot accommodate scientists who want to conduct research. A number of specialized research institutes have been established outside the universities, but none is as extensive or as influential as the research institutes associated with American universities.

In the 1950s, during the cold war, there was widespread recognition that the United States needed to improve its scientific research capacity. As a result, government and universities began cooperating even more closely than they had before. As Daniel Bell (1973) pointed out in his seminal study of technology and postindustrial society, the rush to invest in scientific research had some important consequences:

> [It produced] the expansion of the universities as research institutions, the creation of large scientific laboratories at universities supported by government (the jet propulsion lab at Cal Tech, the Argonne atomic lab at the University of Chicago, MITRE and Lincoln Lab at MIT, the Riverside electronics lab at Columbia, and the like), [and] the growth of "consortiums" such as the Brookhaven lab on Long Island managed by a half-dozen universities. After these have come the large government health-research centers such as those at the National Institutes of Health, the major National

TABLE 21.1 Nationality of Nobel Prize Winners in the Sciences, 1901–1996

	1901–1915	1916–1930	1931–1945	1946–1960	1961–1975	1976–1990	1991–1995	1996
United States	3	3	14	38	41	63	18	5
United Kingdom	7	8	11	14	20	9	—	1
Germany	15	12	11	4	8	7	3	—
France	10	3	2	—	5	2	2	—
Russia	2	—	—	4	3	1	—	—
Japan	—	—	—	1	2	1	—	—
Other countries	15	15	11	13	13	15	4	2

Note: This table shows changes in the quality of science in various nations, as measured by an international standard—the Nobel Prize. Wealthier nations that direct more funds to scientific institutions (such as the United States) account for a large proportion of Nobel Prize winners. Although the size of a nation's population is also important, some wealthy nations with large populations (such as Japan) are not significantly increasing their share of Nobel Prizes, largely because they are not investing in the scientific institutions that conduct basic research.

Source: Data from *Statistical Abstract,* 1998.

> Science Foundation–supported laboratories, the creation of a vast number of nonprofit research "think tanks" such as Rand, the Institute of Defense Analysis, the Aerospace Corporation, and so on. (p. 248)

One consequence of this unplanned growth of scientific organizations in universities, government, the military, and the private sector, Bell concludes, is that it became impossible to create a single set of policies for the support of science. The various organizations must compete for resources and are vulnerable to changing national needs as well as new demands for scientific knowledge by business and industry. This adds to the complexity and competitiveness of scientific institutions.

These aspects of scientific institutions may also explain their continuing differentiation in the United States and other Western nations. For example, the departmental structure of American colleges and universities has been shown to encourage the growth of new disciplines (Merton, 1996). Interdisciplinary programs and new fields of study can be sponsored by existing departments until they can compete for support as independent disciplines. An example is the development of statistics as a separate field of study. Originally a branch of mathematics, statistics has been studied and taught by mathematicians and physicists in Europe and Great Britain since the seventeenth century. In the United States, however, departments of biology, education, psychology, economics, and other fields (e.g., demography) developed specialties in statistics. Eventually separate departments of statistics were established.

In sum, since its origins in the independent and often secretive experiments of philosophers and clerics, scientific research has become institutionalized in complex organizations. Yet we will see at many points in this chapter that the process of differentiation in which science becomes an institution separate from others is never complete. The work of scientists must be paid for, and the more their research is "pure" (in that it has no apparent uses that generate profits), the more it must be supported by other institutions like government or industry. This dependence of science on other institutions continually subjects scientists to pressure to make their work relevant to the needs of business or the military. Conflict between scientists and their sponsors thus has been a feature of science since its origins. To understand that conflict more fully, we will take a more detailed look at the norms of scientific institutions.

The Norms of Science

We saw in Chapter 3 that every social institution develops norms that specify how its special functions are to be carried out. This is readily illustrated by the institutions of science. The function of those institutions is to extend knowledge by means of a specific set of procedures (i.e., the scientific method). The norms of science are derived from that function (Merton, 1996).

Universalism One of the basic norms of scientific institutions is *universalism,* which holds that the truth of scientific knowledge must be determined by the impersonal criteria of the scientific method, not by criteria related to race, nationality, religion, social class, or political ideology. This would seem to be self-evident until one remembers that international rivalries have been part of the history of science since the Renaissance. And consider the case of the Russian geneticist Trofim D. Lysenko, who, on the basis of some extremely unscientific research on plant genetics during the Stalinist era, claimed that acquired characteristics of plants could be inherited by the next generation (Rossianov, 1993). This claim seemed to offer hope for improvement of the Soviet Union's faltering agricultural production. It also fit well with Soviet ideology, which held that better human beings could be created through adherence to the ideals of the revolution. To Stalin and his advisers, science seemed to have proved the value of the Soviet culture and social system. Lysenko was granted a virtual dictatorship over biological research in the Soviet Union, and hundreds of geneticists lost their jobs (Sakharov, 1990). Lysenko was deposed during the Khrushchev era, but the damage done to Soviet agriculture and biological research in the name of ideology lasted many years longer.

Common Ownership Another norm of science is *common ownership of scientific findings.* Those findings are a result of collaboration and hence are not the property of any individual, although in some cases they may bear the name of the person who first published them, as in "Darwin's theory of evolution" or "Einstein's theory of relativity." One outcome of this norm is frequent conflicts over scientific priority—that is, over who was the first to discover or publish a particular item of scientific knowledge. Thus there may be disagreements over who discovered the differential calculus—Newton or Leibniz—but there are no limitations on the use of that calculus. A further consequence of the norm of common ownership is the norm of *publication*—the requirement of full and open communication of scientific findings in journals accessible to all. Secrecy is out of place in science.

However, because scientific research is so often conducted in the interests of national defense or under the sponsorship of private firms that hope to profit from applications of the findings, the norms of common ownership and publication are often suspended. Such situations have led to innumerable conflicts in scientific

BOX 21.2 USING THE SOCIOLOGICAL IMAGINATION

Galileo and the Inquisition

The first person to use a telescope to study the skies was Galileo Galilei, an Italian mathematician who lived from 1564 to 1642. His observations convinced him that the earth revolved around the sun. Up to that time it had been taken for granted that the earth was the center of the universe, and this belief was strongly entrenched in the doctrines of the Catholic church. Galileo's views were so radical that he was tried by the Inquisition, ordered to deny what he knew to be the truth, and forced to spend the last 8 years of his life under house arrest.

Galileo's fate illustrates a principle we have mentioned at numerous points in this book: As societies become more complex, the process termed *differentiation* removes various functions from existing institutions and creates new institutions to perform them. Galileo was tried by the Inquisition because in his time science had not yet become differentiated from philosophy and religion. A scientist must on no account discover anything that contradicted the doctrines of the church.

In his play *Galileo,* Bertolt Brecht paints a vivid picture of the constraints placed on Galileo by the situation of science in his day: An assistant has delivered a gift from the Court of Naples—a model of the sky according to the wise men of ancient Greece—and has asked him to explain it. "You see the fixed ball in the middle?" says Galileo. "That's the earth. For two thousand years man has chosen to believe that the sun and all the host of stars revolve about him. Well. The Pope, the cardinals, the princes, the scholars, captains, merchants, housewives, have pictured themselves squatting in the middle of an affair like that." Galileo goes on to predict that before long people "will be learning that the earth rolls round the sun, and that their mothers, the captains, the scholars, the princes, and the Pope are rolling with it."

The assistant is not convinced, but he admits that he has mentioned Galileo's ideas to his mother, Galileo's housekeeper. The housekeeper says to Galileo, "Last night my son tried to tell me that the earth goes round the sun. You'll soon have him saying that two times two is five." Later Galileo says to the assistant, "Andrea, I wouldn't talk about our ideas outside." "Why not?" asks Andrea. "Certain of the authorities won't like it," replies Galileo. His statement is confirmed by a friend: "How can people in power leave a man at large who tells the truth, even if it be the truth about the distant stars?"

Today scientists are studying subatomic particles called quarks. They have proposed that dinosaurs had feathers rather than scales, and they have suggested that the universe began with a big bang and that stars eventually become black holes. They have discovered the process by which the continents were formed and the structure of human genes. In none of these cases have the findings been challenged by "the authorities," religious or otherwise. Rather, they have been judged by the standards of scientific investigation, one of the functions of the institution we call science.

But the process of differentiation is never complete. In recent years the ancient tension between science and religion has taken a new turn: The scientific theory of evolution has been challenged by fundamentalist religious groups because it contradicts statements made in the Bible. These groups have put pressure on publishers to delete discussions of evolution from school and college textbooks, or at least to mention "creation science" as well as evolution. Although their efforts have had only limited success, they have not been ignored.

circles. An outstanding example is the case of J. Robert Oppenheimer, one of the leaders in the development of the atomic bomb during World War II. Although Oppenheimer's sympathy for certain radical causes was well known, he was given a full security clearance both during and after the war, when he continued his pioneering research on the applications of nuclear physics. But when he publicly stated his support of international sharing of findings in nuclear physics—and opposed the development of a thermonuclear or hydrogen bomb—his opponents brought up the old charge that he was a subversive and could not be trusted with scientific secrets. In 1953 President Dwight D. Eisenhower ruled that Oppenheimer was to be denied access to secret scientific information, which meant that he would also be denied access to the laboratories where the most important research in nuclear physics was being conducted (Lakoff, 1970; Parshall, 1998).

Disinterestedness A further norm of scientific institutions is *disinterestedness.* The scientist does not allow the desire for personal gain to influence the reporting and evaluation of results; fraud and irresponsible claims are outlawed. In fact, more than most other activities, scientific research is subject to the scrutiny of others. This is part of the nature of that research, which involves the search for results that can be verified; in other words, science is, in a sense, self-policing. The norm of disinterestedness does not imply that scientists cannot hope to profit from their findings, and there are many instances in which scientists have held lucrative patents for their discoveries. But it does imply that related norms of scientific research, such as unbiased observation and thoroughness in reporting findings, must take precedence over any selfish motives.

Sociologists are concerned that the pressure on scientists to make discoveries that will earn large sums of money will damage the credibility of scientific institutions. The trend toward partnerships between private corporations and scientific institutions may also put pressure on scientists to violate scientific norms. The Monsanto Chemical Corporation, for example, gave $23.5 million to Washington University for biological research, and a West German pharmaceutical company invested $40 million in research at Harvard Medical School. Dorothy Nelkin (1996), one of the nation's leading experts on scientific policy, warns that when businesses form alliances with universities to conduct research in such areas as biological technologies, the scientists involved must keep their findings secret until patents have been applied for. Nelkin concludes that such secrecy violates the norm that science should be shared.

Similarly, when chemists at universities in Utah and Great Britain announced in 1989 that their collaboration had achieved "cold fusion," the production of energy through the fusion of hydrogen atoms at low temperatures (a reaction that scientists believe requires extremely high temperatures), it appeared that a new era of low-cost energy might be on the horizon. But other scientists were highly skeptical. The so-called discoveries had not been published in scientific journals prior to their announcement to the press—a violation of the norm of publication. The cold fusion apparatus was also kept secret for a time so that the scientists and their universities could apply for patents and enlist the support of major corporations for further research. These actions made scientists suspicious. Before too long the results were found to be incorrect, an outcome perhaps of too much hope for profit and not enough careful measurement (Frazier, 1998).

The norms just described are well-established aspects of modern science, which is increasingly differentiated from other institutions, particularly religion and the state. But science was not always viewed as a legitimate institution or a respectable occupation. In fact, early in its history science was often regarded as a dangerous activity with the potential to threaten the existing social order. This can be seen quite clearly in the repression of Galileo by the Inquisition, described in Box 21.2 on page 677.

TECHNOLOGY IN MODERN SOCIETIES

We noted earlier that a significant aspect of modern science is its contribution to the rapid pace of technological change. The technologies produced by scientific research are applied to all aspects of human life and hence are a major force in shaping and changing other institutions in addition to scientific institutions themselves. An example is the impact of technological change on the institutions of mass communication. As we saw in Chapter 18, the advent of radio and then television dramatically changed the ways in which social and cultural values are transmitted to various groups in society. And now it is possible for people in regions of the world that lack up-to-date telephone systems to "leapfrog" over technological problems by purchasing cellular phones, which do not require a vast telecommunications infrastructure. We will note other examples of the impact of technology in the following pages. But first it is necessary to enlarge our understanding of what is meant by the term *technology.*

Dimensions of Technology

In Chapter 19 we pointed out that there is more to technology than tools and skills, that ways of organizing work are also part of technology. Technology has three dimensions, which may be summarized as follows:

1. Technological tools, instruments, machines, gadgets, which are used in accomplishing a variety of tasks. These material objects are best referred to as *apparatus,* the physical devices of technical performance
2. The body of technical skills, procedures, routines—all *activities* or behaviors that employ a purposive, step-by-step, rational method of doing things
3. The *organizational networks* associated with activities and apparatus (Winner, 1977, pp. 11–12)

The last of these dimensions may be clarified by an example. Organizational networks are sets of statuses and roles. All technologies establish or modify such networks. Thus the automobile owner is part of a network that includes dealers, mechanics, parts suppliers, insurers, licensing agents, and junkyard owners. Our great-great-grandparents were probably part of a network of horse dealers, harness makers, buggy suppliers, and blacksmiths, a network that has been largely eliminated by the advent of motorized vehicles.

Technological change can occur in any or all of the dimensions just listed. The most far-reaching changes involve all three, especially the third. For example, the industrial revolution completely changed the organization of economic institutions and also had significant effects on other institutions, such as the family. Likewise, the internal combustion engine, which made possible the development of the automobile, has completely transformed the ecology of North America (Flink, 1988). Since 1969 the total number of vehicles in the United States has grown six times faster than the total population (Wald, 1997). On the other hand, some technological changes are limited to modifications in the apparatus or technical skills needed for a particular task (the surgical stapler is an example) and do not affect large numbers of people or have major social impacts.

Not only do technological changes affect various groups and institutions within a society, and sometimes transform a society, but technology itself is affected by the social conditions prevailing at any given time. The acceptance of a particular technological innovation may depend on prior changes in other aspects of a society. Thus television might not have had as great an impact if it had been invented in the nineteenth century, when working people had far less leisure time than they do today. Other innovations have failed to gain acceptance because they appeared too soon. An example is the Sony Corporation's unsuccessful attempt to introduce tape recorders in Japan in 1950. Japanese consumers did not perceive a need or use for them, and they went unsold. Much the same thing is happening today in the case of home computers. Once the thrill of computer games wears off, the computer often stands idle because the average household has no other uses for it. In consequence, the home computer industry is hoping that by selling computers equipped with modems, which enable users to browse the World Wide Web and interact in electronic communities on the Internet, they will maintain consumer interest in the new information technologies (Rheingold, 1993; Winner, 1997).

Technological Dualism

It should be noted that the effects of new technologies are not always positive. The phrase *technological dualism* is sometimes used to refer to the fact that technological changes often have both positive and negative effects. The introduction of diesel locomotives, for example, greatly increased the efficiency of railroad operations, but it also led to the decline and eventual abandonment of railroad towns whose economies were based on the servicing of steam locomotives (Cottrell, 1951). Another example is the automation of industrial production. Automation has greatly improved manufacturing processes in many industries. It has increased the safety of certain production tasks and led to improved product quality in many cases. But it has also replaced thousands of manual workers with machines, and significant numbers of those workers find themselves unemployed and lacking the skills required by the high-tech occupations of postindustrial society (see Box 21.3).

Some observers go so far as to say that technology is a danger to the modern world. They feel that it has become an autonomous force, that it is out of control. This is a recurrent theme in movies and science fiction—HAL, the computer that takes over the ship in *2001: A Space Odyssey,* is a good example. But it is also claimed that technology is increasingly independent from human control in the real world. Events like the accident at the Three Mile Island nuclear power plant in 1979; the toxic gas leak that killed more than 2,000 people in Bhopal, India, in 1984; the disaster at the Chernobyl nuclear power plant in the Soviet Union in 1986; and the year 2000 computer failures seem to indicate that human beings cannot control the technologies they have created.

Sociologists who have studied this issue point out that the problem is not one of humans being dominated by machines but, rather, one of depending on technology to meet a wide and growing range of human needs. The Bhopal facility produced a pesticide that made possible larger harvests of much-needed grain; computers with the year 2000 problem ran billions of important programs before becoming obsolete. The result of our dependence on the benefits of complex technologies is an increasingly complex set of organizations and procedures for putting those technologies to work. This

BOX 21.3 GLOBAL CHANGE AND U.S. SOCIETY

Worker Responses to High Technology

The rapid growth of employment in industrial regions like California's "Silicon Valley," Route 128 outside Boston, and the "Silicon Prairie" in the Dallas–Fort Worth area, as well as in hundreds of industrial parks throughout the United States, is viewed as one of the benefits of investment in high-technology industries. The term *high technology* is associated with computers, advanced electronics, genetic engineering, and other frontiers of technological change, but it is rarely well defined. And without a clear definition of this term it is almost impossible to assess its impact on society.

As the term is used by academics, policy makers, and journalists, high technology refers to at least one of the following features of technology:

- An extensive degree of technological sophistication embodied in a product
- A rapid rate of employment growth associated with an innovative product
- A large research and development effort associated with production (Markson & Bloch, 1985)

One implication of this definition is that it includes job-creating processes like research and development as well as technologies like computers, which also have created new growth in employment. Yet the employment-producing features of high technology can be problematic. Many high technologies, such as robotics and computer-aided design, are intended to reduce employment by substituting the work of machines guided by computers for human production of all kinds. Early machine technologies tended to replace human labor power, but high technology tends to reduce the need for human brain power. Employment in occupations like drafting and industrial drawing in engineering and architecture, for example, is threatened by the accelerating use of computer design and graphics programs (Shaiken, 1985).

During the industrial revolution in England, workers sometimes rioted and wrecked the machinery that was replacing their labor. Such violence against machines was termed *Luddism* after an English stocking maker, Ned Ludd, who smashed his employer's knitting frame with a sledgehammer and thereby touched off the famous Luddite riots of 1811 and 1812 (Hobsbawm, 1975; Rudé, 1964). Outbursts of collective behavior directed against automation have been relatively rare, however. Instead, workers have generally accepted their fate and looked for other sources of employment.

Nevertheless, the effects of high technology are so threatening to so many workers that unions have taken a number of initiatives designed to alleviate those effects. An example occurred at Britain's Lucas Aerospace, a maker of landing gear and electronic equipment, in the mid-1970s. Automation had reduced the Lucas workforce from 18,000 to 13,000. At that point the employees and their union leaders conducted an ambitious study that showed that highly skilled workers were losing their jobs at the same time that many of the society's needs remained unmet. The study suggested hundreds of ways in which the company could employ its skilled workforce to meet these needs and still earn profits. Although the plan was rejected, it has served as a model for other worker strategies.

This spectacular display of nature's power is a reminder of the precarious existence of humans on the earth. Thunderstorms, tornadoes, and hurricanes destroy billions of dollars' worth of property and disrupt untold numbers of lives every year.

requires more human effort and skill, and the chances of error and breakdown are greater. The point is not that technology is out of control but that often there is a lag between the introduction of new technologies and the development of adequate controls over the application of those technologies (Winner, 1996).

One of the main issues that concerns social scientists in this regard is why such lags occur. Consider the problem known as wind shear. Numerous major plane crashes have been caused by sudden downdrafts associated with severe thunderstorms on the approaches to airport runways. These downdrafts literally slam large aircraft to the ground. Industry and government officials have known about the wind shear problem since at least 1975, when an Eastern Airlines jet crashed at New York's Kennedy Airport under such conditions. But not until a similar crash occurred in New Orleans in 1984 were airports ordered to install instruments that could predict such conditions. Why the delay? Some sociologists have argued that the

Here is a vivid example of the risks and unintended effects of technology. This bizarre accident, which occurred in France early in the twentieth century, proved to some skeptics that humans could not control the power of their technology. As technology advances, the risks generally increase in proportion to the benefits.

pressure on corporations to show a profit causes them to neglect spending on safety measures. Others, taking a more functionalist view, tend to explain such lags in terms of the time required for organizations to recognize the causes of the problem and develop new statuses and roles to cope with them (Perrow, 1984, 1996).

Technology and Social Change

Among catalysts of social change, technological innovation ranks extremely high, along with major social forces such as population growth, war, epidemics, and rising expectations for better lives. Technological change can also bring about large-scale social change quite rapidly, often well within an individual life span. One need only think of the impact of television. This powerful communications medium, introduced in the United States on a mass basis in the 1950s, has significantly altered the way people receive information about their society and the world, how they behave as consumers, how they engage in politics, and how they use their leisure time.

In Chapter 10 we discussed William Fielding Ogburn's theory of **cultural lag,** the classic sociological treatment of how societies adjust to technological change and the problems that can result. One example of cultural lag can be seen in the case of cable and satellite television. The actual technological change occurred rapidly, with the result that many new broadcasting channels became available. But the social development of new material for those channels requires further investment, creativity, and organization. Thus there is a lag between the immediate effects of technological change and the social adjustments necessary to make effective use of the new technology.

A problem with the cultural lag theory is that it often fails to account for the effects of social power. For example, AIDS activists and public health officials would like to see condoms advertised on television. They argue that people who are already suffering from AIDS, as well as people who might be exposed to AIDS and other sexually transmitted diseases, need as much information as possible about safe sex. Condom companies want to advertise on television, as is common in other nations, especially in Europe. However, television stations generally do not accept condom ads because they are worried about the reaction of individuals and organizations that oppose public discussion of sex and birth control. The power to block advertising of condoms on television is an example of a cultural lag that will not necessarily be overcome, because it stems from a deep and unresolved cultural conflict in American society. Other cultural lags resulting from technological change may be more amenable to intentional efforts to address them, as can be seen in the case of air pollution.

The lags described by Ogburn can be at least partially reduced by the process of **technology assessment,** or efforts to anticipate the consequences of particular technologies for individuals and for society as a whole. For instance, the massive plan to reduce air pollution in the Los Angeles basin requires careful assessment. According to the National Environmental Policy Act and related state laws, any major action by a public agency that affects the environment must be assessed for its impact on the environment and on the citizens involved. Laws that require technology assessment—especially those that require corporations to abide by the findings of such assessments—tend to increase the power of citizens in communities affected by technological change. They are therefore a source both of conflict and of movements for social reform. In the Los Angeles case, a number of small, inadequately funded environmental organizations have succeeded in forcing the California Environmental Policy Administration to fund the plan to reduce air pollution (Ayres, 1995; Weisman, 1989).

Theories that view technological innovations as a source of social change must also recognize that technological changes do not occur at an even pace. Some analysts, such as economist N. D. Kondratieff, believe that technological innovation follows a cyclical pattern. They have shown that the growth of particular industries produces a "long boom," a period of economic expansion and prosperity that lasts about 25 years and is followed by a period of decline and depression of about the same duration or slightly longer (Schilling, 1991). This pattern is illustrated in Table 21.2, which shows the cycles of industrial development in Britain since the late 1700s.

It may be that the new technologies of computers and automation will begin another long boom or wave of economic growth in the next decade, as many people in advanced industrial societies hope. But it is clear that in the late 1970s the previous long boom, stimulated in part by the availability of cheap energy, was over. Some sociologists and economists believe that a new long boom is occurring based on computers, telecommunications, and biotechnologies, and perhaps on new

TABLE 21.2 Long Booms and Associated Industries

Dates	Industry	
1790–1815	Cotton	Mechanization of spinning
1848–1873	Textiles	Mechanization of spinning and weaving
	Engineering	Production by machine of textile machinery, steam engines, and locomotives
1896–1921	Engineering	Batch production; semiautomatic machinery; marine engineering; automobiles
	Electrical	
	Chemical	Rise of science-based industries
	Steel	Bulk production
1945–1974	Automobiles	
	Mechanical and electrical consumer durables	Assembly-line mass production
	Petrochemicals	Continuous-flow process production
Present	Computers	Information and telecommunications technologies

Source: Adapted from CSE Microelectronics Group, 1980. Reprinted with permission from CSE Books, London.

breakthroughs in energy technologies, such as electric automobiles (Korotz, 1998).

The Quest for Energy

Throughout human history a central aspect of technological change has been the quest for new sources of energy to meet the needs of growing populations. That quest has given rise to a succession of energy technologies, each more sophisticated than the last. Animal power gave way to steam-driven machinery, which in turn was replaced by the internal combustion engine. Reliance on oil and its derivatives, especially gasoline, encouraged the growth of powerful energy corporations, which often lobby government agencies for assistance in developing new technologies like nuclear energy. And today the technologically advanced nations are attempting to control the fusion reaction, in which hydrogen atoms are fused into helium, thereby producing an enormous release of energy. The implications of this energy technology, if it can be achieved, are staggering. Fusion promises to bring about a major revolution in human existence. It could make possible the colonization and exploration of space, the rapid development of the less developed nations, the elimination of energy technologies based on oil and coal (which pollute the environment), and much else. But the effort to develop fusion power is also indicative of a fundamental crisis in modern life: the dwindling supply of finite energy resources.

The problem of oil depletion is only the most recent in a series of energy crises that began with the depletion of the supply of game animals by hunting in Paleolithic times. The shortage of meat created conditions that spurred the development of agriculture. Later, in the waning years of the Roman empire, a shortage of labor power to grind flour encouraged the use of water power. The industrial revolution had its origins in the depletion of the supply of wood during the Renaissance. Coal was plentiful, and experiments with its use as an energy source led to the development of new techniques for producing energy—and new machinery and processes for manufacturing goods. Today, as supplies of oil and coal diminish, the search for new energy sources continues (Lovins, 1991, 1996).

It would seem from what we have said so far that the quest for energy is a positive force that results in new, sometimes revolutionary technologies that greatly improve the quality of human life. Many people believe that societies can meet their growing energy needs by continually investing in more sophisticated technologies. This approach has led to the development of huge nuclear power plants to replace oil-fueled generators, and it is widely hoped that investment in fusion, an even more complex technology, will eliminate the dangers posed by nuclear power.

This view is subject to considerable criticism, however. Amory Lovins (1977, 1998), for example, distinguishes between "hard" and "soft" energy paths. The former "relies on rapid expansion of centralized high technologies to increase supplies of energy, especially in the form of electricity." The latter "combines a prompt and serious commitment to efficient use of energy [and] rapid development of renewable energy sources" (p. 25). Present and proposed energy policies favor the hard path, which involves intensive use of available coal, oil, and natural gas plus heavy investment in nuclear power. These are *capital-intensive* technologies because they rely heavily on sophisticated equipment (capital) rather than labor power.

Soft energy technologies depend on renewable sources like sun and wind and tend to be *labor-intensive* in that larger numbers of people are needed to produce a given amount of energy. They are more diverse than hard energy technologies and are more directly matched to energy needs. (Solar energy, for example, can be used to heat water without first being converted into electricity.) But the major difference between the two paths, according to Lovins, is that whereas the soft path depends on "pluralistic consumer choice in deploying a myriad of small devices and refinements, the hard path depends on difficult, large-scale projects [e.g., nuclear power plants and fusion reactors] requiring a major social commitment under centralized management" (1977, p. 54). Such projects are characterized by a "remote and . . . uncontrollable technology run by a faraway, bureaucratized, technical elite who have probably never heard of you" (p. 55).

Whether or not one accepts Lovins's thesis, the use of nuclear power to generate electricity remains a major social and political issue (Perrow, 1996). Underlying the conflict over the safety of nuclear power plants is the issue of control. In the ancient world those who controlled the irrigation systems were the ruling elite; in the United States the "robber barons" of the late nineteenth century often gained both wealth and political influence from their control of oil and coal supplies. Thus much of the opposition to nuclear power plants stems from the recognition that control over energy supplies is a key source of economic and political power.

IMPACTS OF TECHNOLOGY

Technological Systems

The propensity of people in highly industrialized societies to seek technological solutions to problems like pollution or hazardous waste disposal is often referred to as the "technological fix." The belief that new technologies can solve existing environmental and other problems can itself have adverse consequences. This is especially true in relation to technologies like nuclear energy, which require complex social systems to operate them. As Charles Perrow (1984) has written, "Human-made catastrophes appear to have increased with industrialization as we built devices that could crash, sink, burn, or explode" (p. 11). Perrow also points out that the increasing complexity of modern technology has led to a new kind of catastrophe: the failure of whole systems (i.e., activities and organizational networks as well as apparatus)—for example, the Three Mile Island accident or the *Challenger* disaster.

The enormous risks associated with complex technologies have led many observers to call for more thorough assessment of the potential impact of new technologies before they are put into operation. According to Perrow, it is important to study technological systems in their entirety rather than focusing on individual components of those systems. For example, in the case of Three Mile Island, the accident was not a simple matter of a faulty valve but the consequence of a combination of factors—an overworked maintenance staff, equipment failures, ineffective safety precautions, inadequate training, and the unwillingness of scientists and bureaucrats to admit that they might be mistaken. Similar conditions led to the explosion of the *Challenger* space shuttle in 1986. Once again we are reminded that technology consists not just of apparatus that can malfunction but also of knowledge and skills that may be deficient and of organizational networks that occasionally break down.

Environmental Stress

New technologies often cause new forms of pollution and environmental stress. *Pollution* may be defined as the addition to the environment of agents that are potentially damaging to the welfare of humans or other organisms (Ehrlich & Ehrlich, 1997). *Environmental stress* is a more general term that refers to the effects of society on the natural environment. Pollution is the most common form of environmental stress, but it is not the only one.

One example of environmental stress resulting from technology is the surprising finding that winter fish kills in Wisconsin lakes were caused by snowmobiles. Heavy snowmobile use on a lake compacts the snow, thereby reducing the amount of sunlight filtering through the ice and thus interfering with photosynthesis by aquatic plants. As the plant life dies, its decomposition further reduces the amount of oxygen in the water. The fish then die of asphyxiation.

The fish–plant–oxygen relationship is a natural ecological system. The snowmobile is a technological innovation whose potential uses are highly varied. The production, marketing, and use of snowmobiles are elements of a social system. It is this social system that is responsible for the environmental stress resulting from snowmobile use. The land available for snowmobiling is increasingly scarce in an urban society like the United States. Frozen lakes near urban centers thus seem ideal for this purpose, but snowmobiles cause environmental stress in the form of fish kills and thereby create the need for new social controls over the uses of this technology.

Often the need for such controls does not become apparent until a great deal of damage has been done. Nor is it ever entirely clear that new social controls or new technologies can solve the problem at hand. For example, as described at the beginning of this chapter, in a recent study of comparative environmental risks, sociologist Kai T. Erikson (1994b) examined the vast area of Nevada where the U.S. government carried out its nuclear bomb testing and where it now plans to bury large quantities of high-level nuclear wastes. These wastes can remain extremely dangerous for thousands of years. Erikson points out that scientists can assess the geological risks of storing nuclear waste deep underground but have much less experience in assessing the social risks. As seen in the Visual Sociology feature at the end of the chapter, symbols for danger that are familiar to us could easily be as undecipherable to our descendants in 10,000 years as the writing on many ancient tombs is to us today.

Studies of the impact and social control of technologies are an active frontier of sociological research. The Environmental Sociology section of the American Sociological Association routinely publishes research reports that assess the polluting and environmentally stressful impacts of technology. Many such studies have shown that the people who bear the heaviest burden of pollution are most often those who are least able to escape its effects. The poor, minorities, and workers and their families in industrial regions are exposed to higher levels of air, water, and solid-waste pollution than more affluent people are (Erikson, 1994a). This is not a surprising finding. People whose livelihoods depend on polluting industries generally learn to tolerate and even ignore the pollution associated with those industries. In fact, when environmental activists protest against the polluting effects of mines and smelters, they often find that their most vocal opponents are those who are most negatively affected by the pollution. In the past 20 years, however, there has been a significant change in attitudes, especially on the part of trade union leaders in polluting industries; such leaders are more likely to press for pollution controls than they were in the past.

The place of technology in modern societies is a subject of continuing controversy. Key issues include

Then and Now

The Impact of Major Disasters

In 1993 more than 500 hundred miles of the Mississippi and Ohio Rivers reached flood levels after weeks of drenching rains. The floods were the worst in decades, and certainly the worst in the contemporary era, in which modern systems of levees and dikes have been constructed in the great river valley. Billions of dollars' worth of property was destroyed. Thousands of people lost their homes. Many more thousands were forced to dig out their homes, which had been polluted by oil-soaked mud. For months a five-state area in the Midwest was a federal disaster area. Even now the rebuilding continues, and fierce controversies rage about how much of the built environment in the flood plain can be restored and how much should be declared ineligible for flood insurance, which would preclude rebuilding. So thousands of families in the agricultural heartland of the United States continue to face agonizing uncertainties and severe economic, social, and psychological hardships caused by the flooding. And yet in the national press this is considered an old story, and one almost never reads about it outside the region. Other disasters have crowded it out.

In the summer of 1998 the worst hurricane ever to hit Central America devastated much of coastal Honduras, Nicaragua, El Salvador, and parts of Guatemala. The storm, known as Hurricane Mitch, struck a region of extreme poverty. Many thousands of people lost their lives. Almost 300,000 were made homeless. The storm unleashed a desperate new wave of migration to the U.S. border that continues at this writing (Zarembo, 1999). It will take years to rebuild the economies of Honduras and Nicaragua. For years people will suffer the consequences of a storm that commanded the world's attention for a few weeks and resulted in news articles for a few months. The point here is that public attention is short but the consequences of natural disasters are long, and much of the poverty that one witnesses in the United States and elsewhere in the world is brought on by disruptions in people's efforts to build their local economies. Technologies of weather prediction, flood control, and water containment cannot keep up with population growth and the many attractions of living near rivers and coasts, no matter how dangerous that environment may become.

not only pollution and other impacts of technology but also the need to control the development and uses of technological innovations so that they benefit all sectors of society. The complex interactions between technology and other aspects of the social order can be seen in the case of medical technology.

THE CASE OF MEDICAL TECHNOLOGY

Throughout most of human history, limitations on food production, together with lack of medical knowledge, have placed limits on the size of populations. Dreadful diseases like the bubonic plague have actually reduced populations. In England the plague, known as the Black Death, was responsible for a drastic drop in the population in 1348 and for the lack of population growth in the seventeenth century (see Figure 21.1). In 1625 alone, 35,417 residents of London died of the plague. Smallpox and dysentery have had similar, though less dramatic, effects (Davis, 1992; Wrigley, 1969).

Until relatively recently physicians were powerless either to check the progress of disease or to prolong life. In fact, they often did more harm than good—their remedies were more harmful than the illnesses they were intended to cure. As Lewis Thomas (1979) has stated:

> Bleeding, purging, cupping, the administration of infusions of every known plant, solutions of every known metal, every conceivable diet including total fasting, most of these based on the weirdest imaginings about the cause of the disease, concocted out of nothing but thin air—this was the heritage of medicine up until a little over a century ago. (p. 133)

Thomas's point is that before the nineteenth century, when scientists finally began to understand the nature of disease, physicians based their treatments on folklore and superstition. In fact, with few exceptions the practice of healing, like many other aspects of science, was closely linked to religion. In ancient Greece people who suffered from chronic illnesses and physical impairments would journey to the temple of Asclepius, the god of healing, in search of a cure. In medieval times pilgrims flocked to the cathedral at Lourdes in France (as many still do today) in the belief that they would thereby be cured of blindness, paralysis, or leprosy. Not until Louis Pasteur, Robert Koch, and other researchers developed the germ theory of

FIGURE 21.1 **Long-Term Population Trends in England and Wales, 1000–1800**

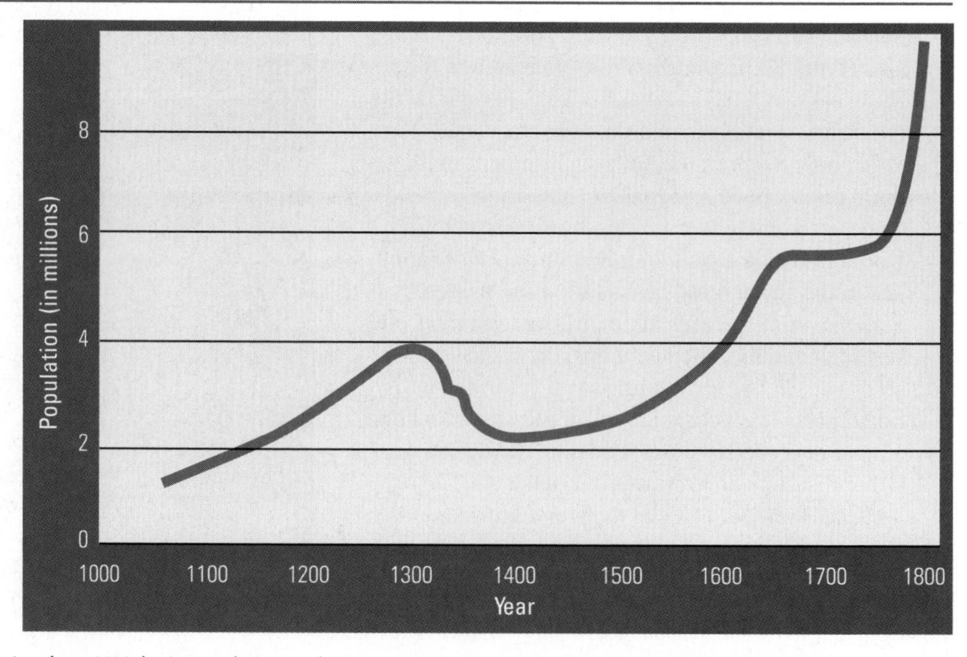

Source: Reprinted with permission from *Wrigley's Population and History,* 1969. Copyright © McGraw-Hill Book Company.

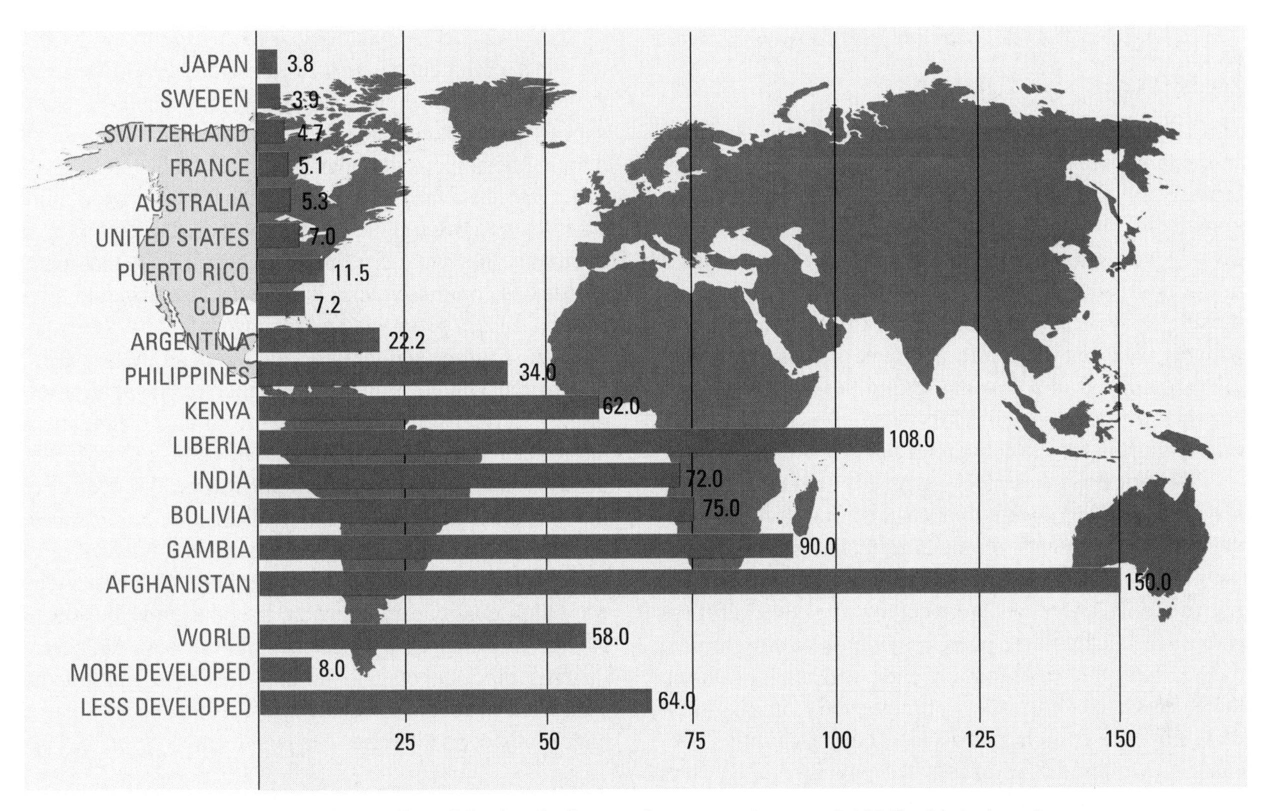

Note: The infant mortality rate is the number of deaths of infants under 1 year of age per 1,000 live births in a given year.

Source: Data from *Population Reference Bureau,* 1998.

FIGURE 21.2 **Infant Mortality Rates for Selected Countries**

disease did medicine become fully differentiated from religion. Their discoveries, together with progress in internal medicine, pathology, the use of anesthesia, and surgical techniques, led to the twentieth-century concept of medicine as a scientific discipline (Cockerham, 1998).

During the nineteenth century scientific research resulted in the discovery of the causes of many diseases, but at first this progress led physicians to do less for their patients rather than more: They began to allow the body's natural healing processes to work and ceased to engage in damaging procedures like bloodletting. At the same time, they made major strides toward improving public health practices. They learned about hygiene, sterilization, and other basic principles of public health, especially the need to separate drinking water from waste water. These innovations, which occurred before the development of more sophisticated drugs and medical technologies, contributed to a demographic revolution that is still under way in some parts of the world. Suddenly rates of infant mortality decreased dramatically, births began to outnumber deaths, and life expectancy increased. As we saw in Chapter 9, this change resulted not from the highly sophisticated techniques of modern medicine but largely from the application of simple sanitation techniques and sterilization procedures (Rosner, 1995). In fact, these simple technologies have had such a marked effect on infant survival that the rate of infant mortality in a society is often used as a quick measure of its social and economic development (see Figure 21.2).

In sum, as medical science progressed toward greater understanding of the nature of disease and its prevention, new public health and maternal care practices contributed to rapid population growth. In the second half of the nineteenth century, such discoveries as antiseptics and anesthesia made possible other life-prolonging medical treatments. In analyzing the effects of these technologies, sociologists ask how people in different social classes gain access to them and how they can be more equitably distributed among the members of a society. The ways in which medical technologies have been institutionalized in hospitals and the medical profession are a central focus of sociological research on these questions.

The Hospital: From Poorhouse to Healing Institution

In the twentieth century the nature of medicine changed dramatically as scientific investigation expanded our knowledge of the causes and cures of disease. That knowledge led to the development of a vast array of technologies for the prevention and cure of many known illnesses, as well as the long-term care of terminally ill patients. Because the more complex of these technologies are applied in a hospital setting, it is worthwhile to consider the development of the hospital as the major social institution for the delivery of health care.

Historically, hospitals evolved through several stages, beginning as religious centers and eventually developing into centers of medical technology (Cockerham, 1998). The first hospitals were associated with the rise of Christianity; they were community centers for the care of the sick and the poor, providing not only limited medical care but also food, shelter, and prayer. During the Renaissance, hospitals were removed from the jurisdiction of the church and became public facilities. Because they offered food and shelter to the poor regardless of their health, they soon became crowded with invalids, the aged, orphans, and the mentally ill. The third phase in the development of hospitals began in the seventeenth century, when physicians gained influence over the care of patients in hospitals. Gradually the nonmedical tasks of hospitals disappeared and the hospital took on its present role as an institution for medical care and research.

The modern hospital began to emerge at the end of the nineteenth century as a result of the development of the science of medicine. Especially important were advances in bacteriology and increased knowledge of human physiology, along with the use of ether as an anesthetic. Because the new medical technologies were more complex and often more expensive than earlier forms of treatment, they were centralized in hospitals so that many physicians could use them. Physicians also began to refer patients of all social classes to hospitals, and those patients paid for the services provided to them there.

In the United States the number of hospitals grew rapidly in the twentieth century—from a few hundred at the turn of the century to more than 6,500 in the 1990s. Today hospitals play an important role in the control of medical practice and access to medical care. For example, doctors who want to practice in a particular hospital must be accepted by the hospital's medical board. Patients who want high-quality care in private hospitals must be able to pay the fees charged by those hospitals or have the necessary insurance coverage. And hospitals have a monopoly on advanced medical technologies, a fact that has had a major impact on the American health care system.

Hypertrophy in Health Care?

Today the technologies available for the diagnosis and treatment of serious illnesses are often described as "miracles of modern medicine." The MRI scanner, for example, allows hospital technicians to observe a

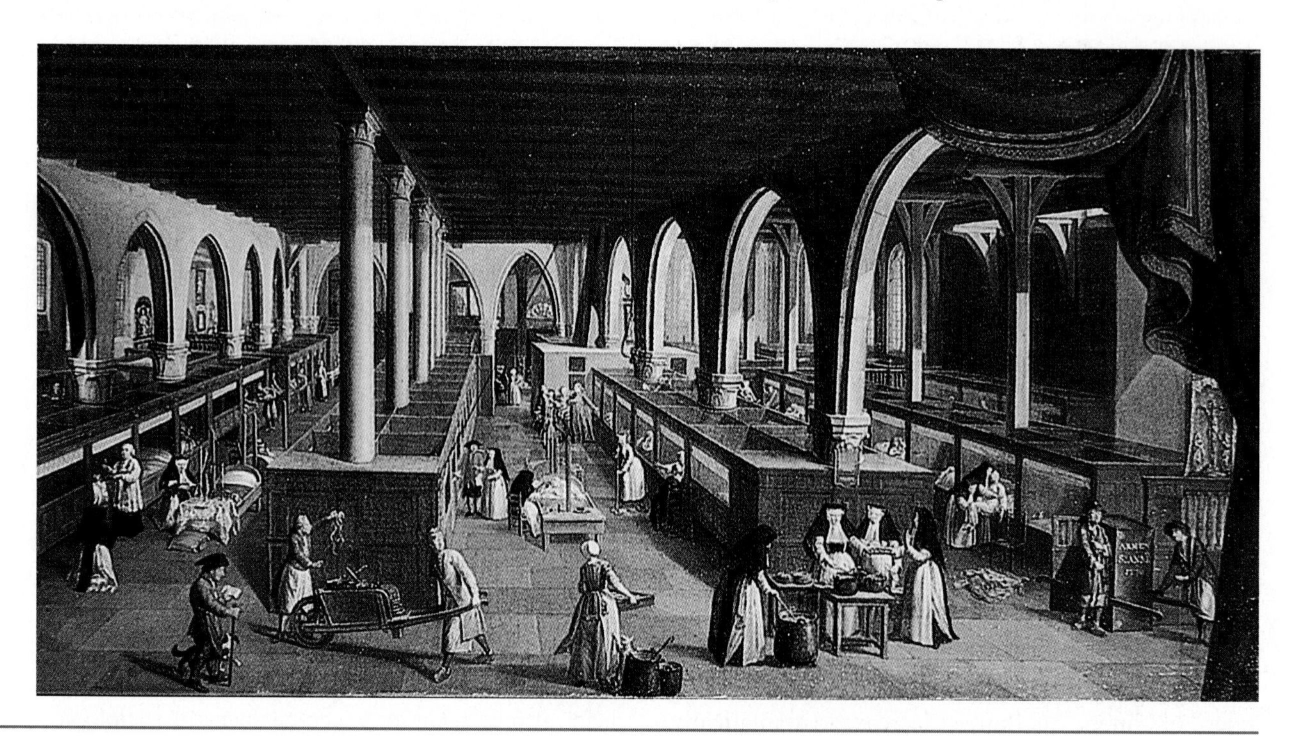

This painting by Jan Beerblock (1739–1806) depicts the sick wards in Sint-Janshospital in Bruges, Belgium, a bleak forerunner of modern hospital organization.

patient's internal organs without the use of X rays; renal dialysis is used to prevent patients from dying of kidney failure; open-heart surgery is practically a routine operation, with the patient kept alive during the process by an external heart-lung machine; other surgical procedures involve the use of laser beams and fiber optics to perform delicate operations. All of these technologies require the use of extremely expensive equipment and highly trained personnel.

The development of increasingly sophisticated and costly medical technologies, together with the practice of requiring patients (or their insurance companies) to pay for hospital services, has led to a crisis in American medical care. The high cost of medical care has become a major public issue, as has the fact that some groups in the population are unable to obtain adequate care. (Over 30 percent of Americans have inadequate medical insurance, or none at all.) Some critics claim that the American health care system is suffering from *hypertrophy*, by which they mean that it has expanded to a size and complexity at which it has become dysfunctional. In their view, excessive emphasis on technological progress has created a situation in which the needs of the patient are subordinated to those of the providers of health care.

According to Paul Starr (1982), the problems of the American health care system stem from the way in which medical institutions evolved. As medical knowledge increased and technological advances were made, physicians developed narrow specialties and hospitals invested in specialized equipment. The physicians referred their patients to the hospitals for sophisticated medical testing and treatment. At the same time, the institution of health insurance emerged in response to demands for a more equitable distribution of health care. Insurance companies or the government began to pay for the services provided by hospitals. Physicians and hospitals became highly interdependent, so much so that they began to "assert their long-run collective interests over their short-run individual interests" (Starr, 1982, p. 230). Their collective interests involve continued investment in complex technologies, with the result that medical care is becoming more and more expensive. The high cost of medical care makes it more difficult for the poor, the elderly, and other groups to afford high-quality care and heroic life-preserving measures. In the closing years of the twentieth century, efforts to make health care less costly while preserving the insurance systems that provide medical coverage for the majority of Americans are proving to be one of the most controversial and rancorous issues in U.S. political life (Starr, 1995).

Medical Sociology

Starr's 1982 study of the evolution of health care institutions had a major impact on the health care reform movement of the 1990s. It is an excellent example of *medical sociology.* This relatively new field of study has emerged in response to the development of medicine as a major institution of modern societies. Many sociologists are employed by health care institutions, and some medical schools have established faculty positions for sociologists. These trends are further evidence of the increasing role of sociology in assessing the effects of technological change on other aspects of society.

In recent years medical sociologists have faced a new and serious challenge: helping society cope with the ethical issues that arise as it becomes increasingly possible to prolong human life by artificial means. (Some of these issues are discussed in the Research on the Cutting Edge section that follows.) Procedures such as heart transplants are extremely expensive and cannot possibly be made available to all patients who need them. Are they to be limited to those who can pay for them? If not, how should the patients who will benefit from such procedures be chosen? Medical sociologists are frequently asked to conduct research that will affect decisions of this nature—a form of technology assessment.

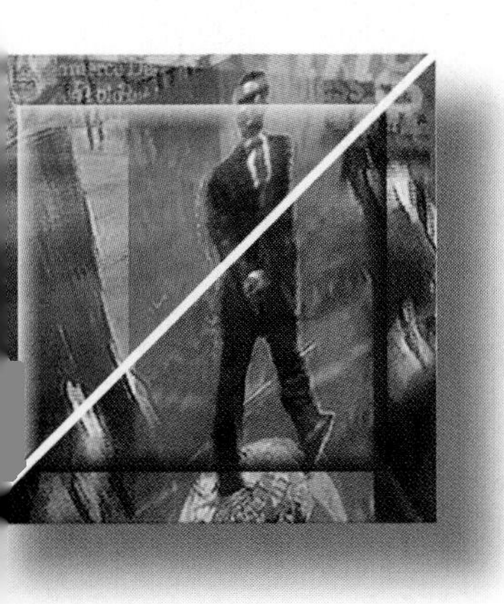

RESEARCH ON THE CUTTING EDGE: Patients' Rights and Medical Practice

Another issue that has arisen as a result of advances in medical technology—as well as the sensational actions of Dr. Jack Kevorkian (see Chapter 2)—is what exactly constitutes death. Should a person whose brain has stopped functioning be considered dead even if other life functions can be maintained by artificial means? Do patients have a "right to die"—that is, to request that life-sustaining equipment be disconnected if the brain has ceased to function? These are extremely difficult questions, and health care institutions, whose function is to strive to maintain life whatever the cost, are ill equipped to deal with them. Critics charge that medical technologies simply prolong the dying process, but courts and legislatures have been reluctant to accept the notion of a right to die—a good example of what Ogburn meant by the phrase *cultural lag.*

The courts have recognized, however, that patients do have the rights to decide what medical treatment they will receive and when to terminate treatment. This has resulted in the development of what are called "living wills," in which a patient signs a statement expressing the wish not to be resuscitated in the event of incurable medical conditions in which only artificial life support systems would maintain life. Patients often wish not to cause themselves or their families needless suffering when death is the inevitable outcome. "Heroic" medical interventions like CPR or other efforts to prevent death may interfere with what is, after all, the natural outcome of a severe illness. In consequence, the living will and instructions not to resuscitate are becoming more common in hospitals in the United States and elsewhere in the world.

Aside from the ethical questions that may arise in situations in which doctors and family members must decide how inevitable death may be and how soon it may occur, a number of important sociological issues emerge as well. These are well illustrated in a recent national study sponsored by the Robert Wood Johnson Foundation, which examined doctors' actual behavior in situations in which patients have signed living wills.

Under the leadership of William Knaus of the University of Virginia, investigators followed doctors and nurses in five major teaching hospitals as they treated 4,301 desperately ill patients. Among their findings:

- While a third of the patients had asked not to be revived with cardiopulmonary resuscitation, half the time "Do Not Resuscitate" was never written on their charts.

- Nearly 40 percent of patients spent at least 10 days in intensive care, kept alive only by breathing machines.
- Half the patients able to communicate in their last three days of life reported that they were in severe pain.

Joanne Lynn of George Washington University, a co-director of the study, observed that "We don't decide to let patients die in peace until almost the last moment. This is hard on patients, their families, and the health professionals who care for them."

The investigators were convinced that poor communication is the primary cause of disregard of the patient's wishes and instructions. They designed a follow-up study of another 4,800 dying patients to test this hypothesis. Half the patients were treated in conventional fashion. The other half were assisted by a specially trained nurse who "consulted with patients and their families, forced them to confront the realities of dying, and kept doctors informed about their patients' conditions and wishes." The investigators were stunned to find that the nurses' intervention had no significant effect on the outcome. "The tools that experts thought would work didn't," Lynn comments. "Physicians are taught to save lives, that death is failure," she says. "Patients and families have come to expect miracles in every case. No one wants to give up too soon."

Even if doctors, nurses, and often the patients' families are unwilling to end heroic efforts to prolong life, sociologists observe forces within medical institutions that are putting pressure on doctors to honor patients' right to die in peace (Larson, 1996). About 30 percent of all Medicare payments go to patients in their last year of life. This suggests that medical insurance companies and health maintenance organizations will increasingly place pressure on doctors to question what they see as a moral obligation to use life-saving technologies even when there is no chance of success. "We have to recognize," Lynn concludes, "that there are alternatives to ending life hooked up to high-tech machines" (quoted in Jaroff, 1996, p. 76).

VISUAL SOCIOLOGY

12001 A.D. Are You Listening? Can You Understand?

As sociologist Kai Erikson learned in his study of new forms of risk and danger, the disposal of dangerous radioactive wastes poses a staggering variety of technical and social dilemmas. From a sociological standpoint, however, few are as challenging as that of how to communicate with people in a future civilization that might have nothing in common with our own.

Radioactive waste products remain dangerous for thousands of years. Even if these wastes are buried in caverns deep below the desert, it will be necessary to post warnings in case, for whatever reason, people penetrate the barriers and come near the dangerous materials. Conventional symbols or signs, such as "DANGER, RADIOACTIVE WASTES—DO NOT ENTER" might suffice if we could assume that the intruders read English. But can we make that assumption?

Archaeologists can often interpret the purpose of ancient monuments like the Sphinx, although they may not be entirely sure of their interpretations. For example, they cannot always decipher hieroglyphics like those shown here, found in ancient Assyria. How to communicate in a universal fashion with people who do not know our language and culture is the subject of a recent study by the Sandia National Laboratories. The photos and diagrams shown here illustrate some of

the possible warnings that could be posted near the materials buried at Yucca Mountain, Nevada, the most likely site for the deposition of nuclear wastes in the United States.

The Sandia researchers doubt that the well-known symbol for radiation will stand the test of time. They

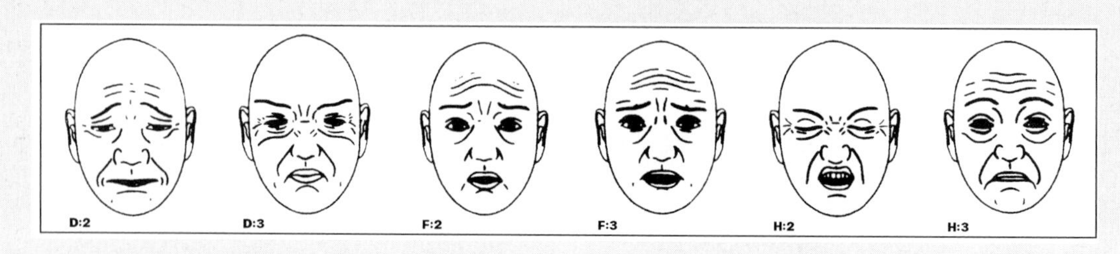

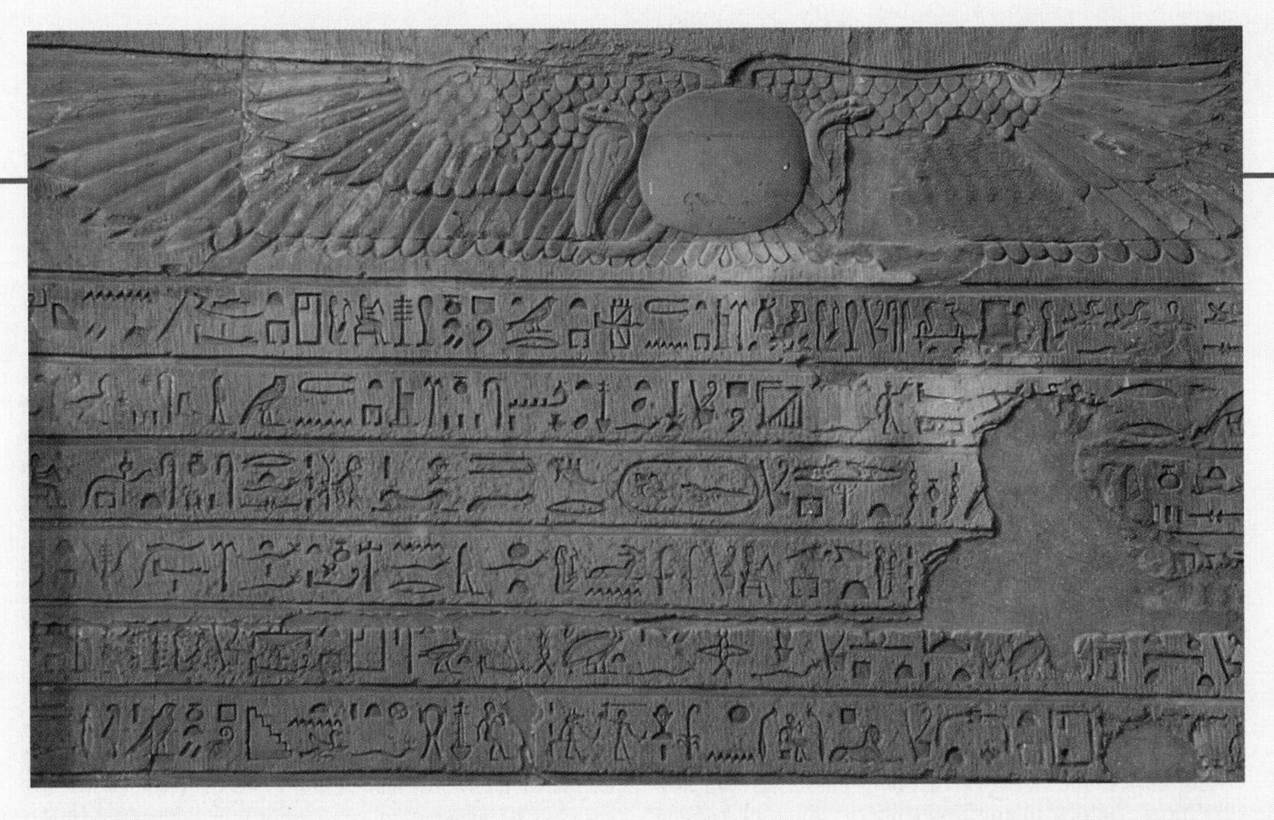

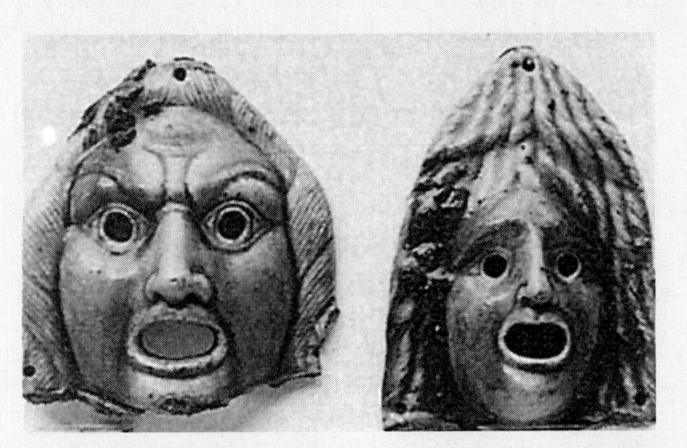

believe that if we are to communicate with people from a future civilization, we must find more universal symbols. For example, some ancient monuments used masks that might evoke dread, like those pictured here from Medinet el Fayyum, Egypt. Perhaps warnings using such universal symbols of fright would work. Another proposed solution involves using dangerous shapes such as the "landscape of thorns" to warn away the curious without relying on language. Still another noncultural, non-linguistic means of communication is the "black hole" design that evokes fear.

SUMMARY

A hallmark of the modern social order is the conduct of scientific research in universities and other research organizations. *Science* is knowledge obtained as a result of the process of developing and testing hypotheses. Pure science is scientific investigation devoted to the pursuit of knowledge for its own sake. Applied science is the application of known scientific principles to a practical problem.

Technology involves the use of tools and knowledge to manipulate the physical environment to achieve desired practical goals. Technology is much older than science, but the discoveries of modern science are creating new technologies at a rapid rate and have greatly expanded the human capacity to live in and exploit different habitats. Negative effects of technological change have led to calls for *sustainable development,* or development that does not damage the environment or create new environmental burdens.

The interactionist approach to the study of scientific institutions focuses on the scientific community or communities, the network of communication and social relationships among scientists. A well-known study based on this perspective found that in many cases scientific researchers do not test and refute existing theories but, rather, assume that they are valid and use them as a *paradigm* for future research. This view of the scientific community implies that science is insulated from the rest of society and that the problems scientists choose to solve are determined by the tradition of research in their field.

The institutional approach to the study of science asks why science develops differently in different societies. In this view, certain conditions encourage the development of scientific institutions. They include recognition of empirical research as a legitimate way of acquiring new knowledge. In addition, science must be independent from other fields. However, scientific institutions are never entirely separate from other institutions because scientific research is often supported by government or industry, a situation that puts pressure on scientists to meet the practical needs of those institutions.

One of the most basic norms of scientific institutions is universalism, which holds that the truth of scientific knowledge must be determined by the impersonal criteria of the scientific method. Another norm of science is the common ownership of scientific findings (although such findings sometimes bear the name of the person who first published them). A third norm of scientific institutions is disinterestedness, meaning that scientists do not allow the desire for personal gain to influence the reporting and evaluation of results.

The technologies produced by scientific research are a major force in shaping and changing other institutions. The basic dimensions of technology are physical devices or apparatus, the activities associated with their use, and the organizational networks within which those activities are carried out. The fact that technological changes often have both positive and negative effects is sometimes referred to as technological dualism.

Sociologists recognize that social institutions are often slow to adapt to changing technologies. This recognition forms the basis of the theory known as *cultural lag.* The time required for social institutions to adapt to technological change can be reduced by the process of *technology assessment,* or efforts to anticipate the consequences of particular technologies for individuals and for society as a whole.

The increasing complexity of modern technology has led to a new kind of catastrophe: the failure of whole systems. The enormous risks associated with complex technologies have led many observers to call for a more thorough assessment of the potential impact of new technologies before they are put into operation. This is especially important from the standpoint of protection of the environment.

The complex interactions between technology and other aspects of the social order are illustrated by the case of medical technology. Until relatively recently, physicians were powerless either to check the progress of disease or to prolong life. Scientific research led to the discovery of the causes of many diseases during the nineteenth century, but in the twentieth century a vast array of technologies was developed both for the prevention and cure of illnesses and for the long-term care of terminally ill patients.

The technologies used in the diagnosis and treatment of serious illnesses require extremely expensive equipment and highly trained personnel. This has caused health care to become very expensive, and as a result some groups in the population are unable to obtain adequate care. Some critics claim that extreme emphasis on technological progress has created a situation in which the needs of the patient are subordinated to those of the providers of health care. In recent years medical sociologists have been faced with the challenge of helping society cope with the ethical issues that arise as it becomes increasingly possible to prolong human life by artificial means.

SOCIOLOGY VERSUS IDEOLOGY

Is global warming a genuine threat or a sensational scare drummed up by environmentalists? Is protection of the Amazon rain forest something that people in other regions of the world should actually try to do something about? Is the security of food supplies endangered by new technologies of genetic engineering? One could endlessly pose questions like these and not exhaust the supply of contentious issues. How does a sociologically informed person think about them in order to arrive at a considered opinion?

One suggestion is to ask yourself whether the person or group making a claim like those just mentioned has a selfish interest in taking that position. For example, a lobbying group advocating policies that favor a particular form of energy, such as petroleum, is immediately suspect. But are scientists, who are supposedly disinterested, automatically more believable than such groups? Hardly. Their claims also must stand up to tests. They may have selfish interests in that they seek research grants. They may be working for advocacy groups with an ideological interest in a particular issue. This means that one needs to read their communications carefully. See what the evidence is and how it has been gathered. Use your sociological imagination to become a hard-nosed consumer of scientific communications.

Often those with direct economic interests in an issue will have reason to slant scientific findings. Almost as often, however, the scientific findings themselves are ambiguous. Frequently we just cannot be sure. In such cases it is helpful to place an imaginary bet. Suppose, for example, we bet that global warming is not a threat, and continue to increase carbon dioxide emissions into the atmosphere. Are the consequences of losing that bet worse than the consequences of taking action to reduce emissions? If we lose the bet that global warming is real, we may face worse disasters than Hurricane Mitch or the Midwest floods in our children's lifetimes. If you read articles about this issue carefully, you will see that the community of scientists, with only a few exceptions, has reached a consensus that the world needs to bet on the probability that the threat is real. You will also recognize that this consensus is not ideologically based.

GLOSSARY

science: knowledge obtained as a result of the process of developing and testing hypotheses. (p. 672)

technology: the use of tools and knowledge to manipulate the physical environment in order to achieve desired practical goals. (p. 672)

sustainable development: development that does not damage the environment or create new environmental burdens. (p. 673)

paradigm: a model comprising existing theories and methods; a general way of seeing the world that dictates what kind of scientific work should be done and what kinds of theory are acceptable. (p. 674)

cultural lag: the time required for social institutions to adapt to a major technological change. (p. 681)

technology assessment: efforts to anticipate the consequences of particular technologies for individuals and for society as a whole. (p. 682)

WHERE TO FIND IT

BOOKS

So Human an Animal (Rene Dubos; Transaction, 1998). A classic statement of human ecology and the environment, updated by medical sociologists.

Energy, Society and Environment (David Elliott; Routledge, 1997). An up-to-date overview of energy systems and their social and environmental impacts.

The Social Transformation of American Medicine (Paul Starr; Basic Books, 1982). A sociological history that shows why American medical institutions have become increasingly cumbersome and costly.

The Jobless Future: Sci Tech and the Dogma of Work (Stanley Aronowitz and William DiFazio; University of Minnesota Press, 1994). A critical reading of the changes brought about by technology and automation and their impact on the prospects for work in the twenty-first century.

On Social Structure and Science (Robert K. Merton; University of Chicago Press, 1996). A collection of original essays and research by one of the founders of the sociology of science. Extremely useful for concepts and theoretical frameworks.

The Healing Experience: Readings on the Social Context of Health Care (William Kornblum and Carolyn D. Smith, eds.; Prentice Hall, 1994). An anthology that explores medical and health-related problems from the perspectives of well-known health care providers and medical sociology researchers.

Nature's Economy (Donald Worster; Cambridge University Press, 1994). Perhaps the best available history of American ecological ideas and their influence on social thought and action.

JOURNALS

Scientific American. A well-known monthly magazine that usually includes at least one important article dealing with the impact of new technologies on society.

Bulletin of the Atomic Scientists. A renowned and respected journal, founded by a coalition of physicists and social scientists, that focuses on the control of nuclear energy and weapons.

Journal of Health and Social Behavior. A quarterly journal that presents sociological research on problems of human health and illness. Also features articles on change in social institutions and organizations as a consequence of new technologies.

OTHER SOURCES

Resources, Environment and Population: Present Knowledge and Future Options (Kingsley Davis and Mikhail S. Bernstam, eds.; Population Council, 1991). A special edition of *Population and Development Review* containing invaluable reviews of issues like sustainable development, energy development, world population forecasts, pollution, and climate.

INTERNET RESOURCES

National Institutes of Health (www.nih.gov). Health information, press releases about scientific discoveries, and a link to scientific resources are all found at this site. The site also has links to major support services, including CancerNet, AIDS information, and clinical alerts.

UNAIDS: The Joint United Nations Programme on HIV/AIDS (www.unaids.org/). A good place to begin searching for health resources on a global basis. The AIDS page provides many links to other health resources.

Women's Health Interactive (www.womens-health.com/). Offers interactive sessions on a wide variety of critical health issues affecting women throughout the world.

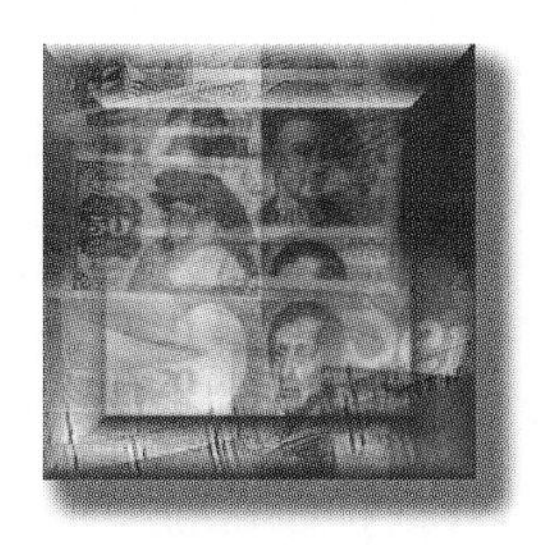

APPENDIX: CAREERS IN SOCIOLOGY

What could be more stimulating than to spend a lifetime learning and teaching about human society? But can one really make a living as a sociologist? And what use is a sociology major if one does not go on to do graduate work in sociology?

In attempting to answer these questions, it is helpful to distinguish between careers in sociology itself and applications of sociology in other fields. It may also help to know that throughout the world many government and business leaders are sociologists. In Brazil, for example, the nation's president has a doctorate in sociology.

In North America most sociologists with graduate degrees teach and conduct research in colleges and universities. This has been true since the early decades of the twentieth century, but there is a growing trend toward the application of sociology in many fields outside academia. More than 25 percent of sociologists now work for nonacademic employers (Lyson & Squires, 1993). But sociological training does not have to lead to a sociological career. It can be valuable preparation for careers in a wide variety of other professions, including law, business, education, architecture, medicine, social work, politics, and public administration. Whatever the specific occupation, the emphasis will be on understanding human behavior and relationships in different kinds of social settings.

For most careers in sociology, a graduate degree is necessary—a Ph.D. in the case of academic sociology, and an M.A. or Ph.D. for most jobs in applied fields. However, a B.A. in sociology can often be of value in occupations outside of sociology. In addition, at any educational level a sociologist is likely to specialize in a particular area, such as the family, education, health, environmental issues, sex roles, the military, or law enforcement.

The majority of sociologists earn more than $50,000 a year. As in many professions, men tend to earn more than women. The American Sociological Association continues to promote equality of opportunity for women and minorities; a survey of sociologists found that about 70 percent of the women thought their opportunities for promotion and advancement were either very good or pretty good (Lyson & Squires, 1993). The association has also established a minority scholarship program and continually seeks to improve the representation of minority groups in sociology.

ACADEMIC CAREERS

Colleges and universities may place differing amounts of emphasis on teaching and on conducting research. At major universities, professors of sociology are expected to devote a significant portion of their time to original research and writing for publication, while at smaller colleges there is more emphasis on teaching. This is not to say that sociologists in smaller institutions do not do research, just that more importance is attached to their teaching.

A professor of sociology at a large university must have a Ph.D. He or she may teach four or five courses a year, including advanced seminars for graduate students. But research and writing are likely to play a greater role in his or her career development. Research projects may be financed by grants from federal or other research agencies, and graduate students may be hired as research assistants using funds from the grant. The professor then can use the project as a means of teaching at the graduate level, while concentrating on publishing scholarly articles and, sometimes, books. The latter are an important element in the university's decision regarding tenure, or guaranteed employment, which is generally made during the individual's fifth or sixth year on the faculty.

A sociology instructor in a community college may have an M.A., although many of these institutions now require a Ph.D. He or she typically does a great deal of teaching—perhaps five courses a semester. Although this teaching load becomes less demanding as courses are repeated, the instructor must continually seek out up-to-date materials and be alert to innovations in teaching techniques. In addition to preparing for courses, he or she is expected to serve on academic and research committees, meet with individual students, and keep abreast of new developments in sociology and related fields.

Another important consideration is that sociologists at colleges and universities not only teach sociology to graduate and undergraduate sociology majors, but also teach students in other disciplines such as law, education, medicine, social work, and nursing. Almost eight out of ten practicing sociologists are teaching in universities, colleges, and high schools.

CAREERS IN APPLIED SOCIOLOGY

Opportunities for sociologists in government, business, and not-for-profit organizations are extremely varied and can be categorized in a number of ways. One approach (Rossi & Whyte, 1983) is to divide applied sociology into three main branches: applied social research, social engineering, and clinical sociology. *Applied social research* is similar in many ways to the basic research conducted at universities, except that it is conducted in order to obtain empirically based knowledge of applied issues. It may take the form of descriptive studies such as victimization surveys, analytical research such as studies of voter preferences, or evaluation research—that is, systematic attempts to estimate the potential effects of a proposed social program or a new approach to management in a business firm. *Social engineering* goes a step further and attempts to use sociological knowledge to design social policies or institutions with a specific purpose. An example is the federally funded preschool program known as Head Start. *Clinical sociology* refers to the use of sociological knowledge in providing assistance to individuals and organizations. A clinical sociologist might, for example, work with a large corporation to find ways of improving employee morale.

To get an idea of the variety of job opportunities in applied sociology, consider the following examples (American Sociological Association, 1995):

Research Director in a Health Center Sociologists employed by a state-supported health science center teach future physicians, nurses, and other health care professionals about the sociological aspects of health care. Among other things, they provide data about the population groups served by the center and the distribution of disease and health care needs in those groups.

Staff Researcher in a Federal Agency Many sociologists are employed in various positions by federal agencies. An example is a staff member of a program that administers grants to researchers at universities throughout the country. This involves discussing research plans with applicants and monitoring grants that have been awarded.

Member of the Planning Staff in a State Department of Transportation Opportunities for sociologists in state agencies are also varied. An example is performing long-range forecasting for the transportation department, which entails projecting population shifts in major urban and suburban areas. The sociologist must keep informed about relevant research and prepare frequent reports and analyses based on recent studies.

Staff Administrator in a Public Assistance Agency In local government, a sociologist might serve as program coordinator for a city social-service department. This job would include processing reports and legal forms as well as direct contact with clients of the agency—particularly individuals who are poor, disabled, aged, or members of minority groups. In addition, it would involve helping clients solve personal problems by referring them to other public and private agencies that provide assistance of various kinds.

Staff Member of a Research Institute A sociologist employed by a private research institute might help carry out studies of specific problems at the request of government agencies, businesses, or political organizations. This could involve developing new research projects (including writing grant proposals) as well as supervising existing ones.

CAREERS IN OTHER FIELDS

People with sociological training often work in other occupations that use their training, even if they are not specifically referred to as sociologists. For example, a personnel manager in a small manufacturing firm would be considered a business executive rather than a sociologist. But training in industrial sociology and social psychology can be immensely valuable in carrying out the responsibilities of a personnel manager—for instance, in setting strategies and programs for hiring, training, supervising, and promoting employees. Many businesses and other organizations have come to recognize that sociological training is worthwhile for administrators and executives.

Kenneth Donow, a senior adviser to the Public Service Satellite Consortium in Washington, D.C., is an example of a sociologist who has made a successful transition from an academic career to a career in business. Although his training was geared toward a university position, Donow's interest in communications and information policy steered him toward the business world. He was hired by a consulting firm to analyze the worldwide semiconductor industry. It soon became clear to him that business analysis involves the study of formal organizations and social networks, something for which sociologists are ideally suited. He was able to convince other business clients to incorporate sociological approaches into their operations and notes that today "sociological practice has become fully integrated into the largest U.S. corporations" (1990, p. 6). Donow believes that people with sociological training will be increasingly sought by American businesses during the next few decades.

There are many other areas in which sociological training is viewed as appropriate preparation for employment. A publication of the American Sociological Association (1995) lists the following positions among those that can be obtained with an undergraduate degree in sociology: interviewer, research assistant, recreation worker, group worker, teacher (if certified), administrative assistant, probation and parole worker, career counselor, community planner, editorial assistant, social worker (not certified), and statistical assistant. A graduate degree can also be useful outside of sociology. For example, if you seek a career in a particular industry or field—such as transportation—participating in original research in that field as part of a graduate degree program can provide a means of entering the field with the help of individuals who notice and respect your work.

EMPLOYMENT PROSPECTS

The directors of the American Sociological Association believe that "the future appears bright for sociology. The next century may be the most exciting and critical period in the field's history." The Association anticipates a surge in the hiring of younger sociologists as professors, research personnel, and administrators with sociological backgrounds who were part of the baby boom generation retire from the work force. Moreover, "internationalization of both higher education and the profession of sociology will also lead to new opportunities inside academia and in applied settings" (American Sociological Association, 1995, p. 22).

In the 1980s a study of the specific types of jobs held by recipients of M.A. and Ph.D. degrees in sociology found that close to half the M.A.s and one third of the Ph.D.s were employed by government agencies. About one third of the Ph.D.s were employed by not-for-profit organizations, as were about one fourth of the M.A.s. About a fifth of each group were employed in the private sector (Huber, 1984). These patterns hold true today as well, but with some significant changes. More sociologists are working in the private sector—especially in advertising, telecommunications, computers and software development, and personnel—and somewhat fewer are working in the public sector.

Not-for-profit foundations and companies continue to look to people with sociological training. In this era of corporate downsizing and reductions in funding for public institutions, people with diverse training in sociological methods and theories often find that they have the flexibility required in today's employment market. The American Sociological Association notes that "solid training in sociology at the undergraduate or graduate levels forms a foundation for flexible career development. The better your training and the more skills you have acquired, the better you will be able to take advantage of new opportunities" (1995, p. 22).

PREPARING FOR A CAREER IN SOCIOLOGY

A bachelor's degree in sociology provides excellent preparation for either graduate study or employment. However, few employers look specifically for sociology B.A.s in preference to other liberal arts degrees. To enhance the value of a B.A. in sociology relative to degrees in other fields, it is advisable to emphasize courses in research methods and statistics.

A graduate degree is necessary to become a practicing sociologist. It is required for any career in academic sociology and for most careers in applied sociology. An M.A. or M.S. is sufficient for teaching in high schools or two-year colleges, or for employment in public agencies and private firms. A Ph.D. is usually necessary for teaching and research in universities and for high-level positions in research institutes, private industry, and government agencies.

The Ph.D. generally requires four or five years of study, often including the completion of an M.A. In many graduate schools, work is available as part of the learning experience; the graduate student may be employed as a teaching assistant (TA) or research assistant (RA). A TA typically grades exams, meets with students who need extra help, and leads discussion sections. As noted earlier, RAs are generally hired by faculty members who have received grants for specific research projects.

Most graduate degree programs begin with courses on basic theoretical issues, research methods, and statistics. After a year of such courses the student may take an examination and perhaps be awarded the Master's degree. Subsequently the emphasis shifts from course work to actual research. In this phase the student is exposed to a wide variety of research methods, including computer skills, in-depth interviewing, participant observation, and use of census materials, questionnaires, and surveys. This is followed by a set of Ph.D. examinations, which may be written or oral, and then by the writing of a dissertation. The dissertation must be an original piece of scholarship. Usually it entails submitting a formal proposal to a committee of faculty advisers, who also preside over the student's oral defense of the dissertation when it is completed.

More than 130 universities in the United States offer a Ph.D. in sociology, and most of them also offer an M.A. or M.S. More than 150 other schools offer only an M.A. Some graduate programs in sociology are designed specifically to prepare students for nonacademic careers in business or government. They may, for example, substitute internships in agency offices for the traditional TA or RA experience.

Applicants to graduate sociology programs do not have to have an undergraduate major in sociology, but the requirements of graduate departments differ considerably in such areas as foreign language skills and statistics courses. Many departments require applicants to take the Graduate Record Examination (GRE) and have scores available by the beginning of the calendar year. This is particularly important for financial aid decisions. Letters of recommendation can greatly influence the decisions of graduate admissions committees.

The reputation of the institution from which an individual receives a graduate degree can have a significant effect on career opportunities. In addition, graduate departments differ in the quality of their programs in various specialties, such as criminology, demography, the study of the family, and so on. The prospective degree candidate should pay careful attention to the department's reputation in specific areas as well as the reputation of the college or university itself. And, of course, it is a good idea to apply to several graduate programs.

To aid students in choosing a graduate program in sociology, the American Sociological Association issues an annual publication titled *Guide to Graduate Departments of Sociology.* This guide contains information about more than 200 degree programs in the United States and Canada—including degrees awarded, names and specialties of faculty members, special programs, tuition and fees, availability of fellowships and assistantships, deadlines for applications, and how to obtain further information and application forms. The guide is available in college libraries or can be ordered from the American Sociological Association, 1722 N Street N.W., Washington, D.C. 20036.

REFERENCES

American Sociological Association. (1995). *Careers in sociology.* Washington, DC: American Sociological Association.

Donow, K. R. (1990, April). On the transition from the academy to a career in business. *Footnotes* (American Sociological Association), p. 6.

Huber, B. J. (1984). *Career possibilities for sociology graduates.* Washington, DC: American Sociological Association.

Lyson, T. A., & Squires, G. D. (1993, February). The "lost generation" of sociologists. *Footnotes* (American Sociological Association), pp. 4–5.

Rossi, P. H., & Whyte, W. F. (1983). The applied side of sociology. In H. E. Freeman, R. R. Dynes, P. H. Rossi, and W. F. Whyte (Eds.), *Applied sociology.* San Francisco: Jossey-Bass.

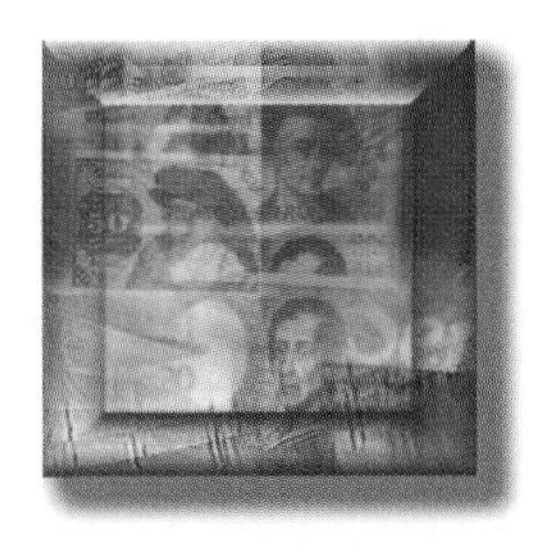

GLOSSARY

accommodation: the process by which a smaller, less powerful society is able to preserve the major features of its culture even after prolonged contact with a larger, stronger culture.

acculturation: the process by which the members of a civilization incorporate norms and values from other cultures into their own.

achieved status: a position or rank that is earned through the efforts of the individual.

age cohort: a set of people of about the same age who are passing through the life course together.

age grade: a set of statuses and roles based on age.

ageism: an ideology that justifies prejudice and discrimination based on age.

agencies of socialization: the groups of people, along with the interactions that occur within those groups, that influence a person's social development.

agents of socialization: individuals who socialize others.

alienation: the feeling of being powerless to control one's own destiny; a worker's feeling of powerlessness owing to inability to control the work process.

animism: a form of religion in which all forms of life and all aspects of the earth are inhabited by gods or supernatural powers.

anomie: a state of normlessness.

anticipatory socialization: socialization that prepares an individual for a role that he or she is likely to assume later in life.

ascribed status: a position or rank that is assigned to an individual at birth and cannot be changed.

assimilation: a pattern of intergroup relations in which a minority group is absorbed into the majority population and eventually disappears as a distinct group; the process by which culturally distinct groups in a larger civilization adopt the norms, values, and language of the host civilization and are able to gain equal statuses in its groups and institutions.

authority: power that is considered legitimate both by those who exercise it and by those who are affected by it, and whose exercise is governed by the norms and statuses of organizations.

behaviorism: a theory stating that all behavior is learned and that this learning occurs through the process known as conditioning.

bisexuality: sexual orientation toward either sex.

bureaucracy: a formal organization characterized by a clearly defined hierarchy with a commitment to rules, efficiency, and impersonality.

capitalism: a system for organizing the production of goods and services that is based on markets, private property, and the business firm or company.

caste: a social stratum into which people are born and in which they remain for life.

charisma: a special quality or "gift" that motivates people to follow a particular leader.

charismatic authority: authority that comes to an individual through a personal calling, often claimed to be inspired by supernatural powers, and is legitimated by people's belief that the leader does indeed have God-given powers.

church: a religious organization that has strong ties to the larger society.

citizenship: the status of membership in a nation-state.

civil religion: a collection of beliefs and rituals that exist outside religious institutions.

civil society: the sphere of nongovernmental, nonbusiness social activity carried out by voluntary associations, congregations, and the like.

civilization: a cultural complex formed by the identical major cultural features of a number of societies.

class consciousness: a group's shared subjective awareness of its objective situation as a class.

class: a social stratum that is defined primarily by economic criteria such as occupation, income, and wealth.

closed question: a question that requires the respondent to choose among a predetermined set of answers.

closed society: a society in which social mobility does not exist.

closed stratification system: a stratification system in which there are rigid boundaries between social strata.

cohort: all persons born in a specified time span.

collective behavior: nonroutine behavior that is engaged in by large numbers of people responding to a common stimulus.

communications media: institutions that specialize in communicating information, images, and values about ourselves, our communities, and our society.

community: a set of primary and secondary groups in which the individual carries out important life functions.

conditioning: the shaping of behavior through reward and punishment.

confidentiality: the promise that the information provided to a researcher by a respondent will not appear in any way that can be traced to that respondent.

conflict perspective: a sociological perspective that emphasizes the role of conflict and power in society.

control group: in an experiment, the subjects who do not experience a change in the independent variable.

controlled experiment: an experimental situation in which the researcher manipulates an independent variable in order to observe and measure changes in a dependent variable.

core state: a technologically advanced nation that has a dominant position in the world economy.

correlation: a specific relationship between two variables.

counterculture: a subculture that challenges the accepted norms and values of the larger society and establishes an alternative lifestyle.

crime: an act or omission of an act that is prohibited by law.

crowd: a large number of people who are gathered together in close proximity to one another.

crude birthrate: the number of births occurring during a year in a given population, divided by the midyear population.

crude death rate: the number of deaths occurring during a year in a given population, divided by the midyear population.

cult: a new religion.

cultural evolution: the process by which successful cultural adaptations are passed down from one generation to the next.

cultural lag: the time required for social institutions to adapt to a major technological change.

cultural relativity: the recognition that all cultures develop their own ways of dealing with the specific demands of their environments.

culture: all the modes of thought, behavior, and production that are handed down from one generation to the next by means of communicative interaction rather than by genetic transmission.

de facto segregation: segregation created and maintained by unwritten norms.

de jure segregation: segregation created by formal legal sanctions that prohibit certain groups from interacting with others or place limits on such interactions.

deference: the respect and esteem shown to an individual.

demagogue: a leader who uses personal charisma and political symbols to manipulate public opinion.

demeanor: the ways in which individuals present themselves to others through body language, dress, speech, and manners.

democracy: a political system in which all citizens have the right to participate in public decision making.

democratic socialism: an economic philosophy based on the belief that private property may exist at the same time that large corporations are owned by the state and run for the benefit of all citizens.

demographic transition: a set of major changes in birth and death rates that has occurred most completely in urban industrial nations in the past 200 years.

denomination: a religious organization that is on good terms with the institution from which it developed but must compete with other denominations for members.

dependent variable: the variable that a hypothesis seeks to explain.

developing nation: a nation that is undergoing a set of transformations whose effect is to increase the productivity of its people, their health, their literacy, and their ability to participate in political decision making.

deviance: behavior that violates the norms of a particular society.

differential association: a theory that explains deviance as a learned behavior determined by the extent of a person's association with individuals who engage in such behavior.

differentiation: the processes whereby sets of social activities performed by one social institution are divided among different institutions.

discrimination: behavior that treats people unfairly on the basis of their group membership.

double standard: the belief that women must adhere to a different and more restrictive moral code than that applied to men.

downward mobility: movement by an individual or group to a lower social stratum.

dramaturgical approach: an approach to research on interaction in groups that is based on the recognition that much social interaction depends on the desire to impress those who may be watching.

dyad: a group consisting of two people.

education: the process by which a society transmits knowledge, values, norms, and ideologies and in so doing prepares young people for adult roles and adults for new roles.

educational achievement: how much the student actually learns, measured by mastery of reading, writing, and mathematical skills.

educational attainment: the number of years of school an individual has completed.

ego: according to Freud, the part of the human personality that is the individual's conception of himself or herself in relation to others.

endogamy: a norm specifying that a person brought up in a particular culture may marry within the cultural group.

endogenous force: pressure for social change that builds within a society.

equality of opportunity: equal opportunity to achieve desired levels of material well-being and prestige.

equality of result: equality in the actual outcomes of people's attempts to improve their material well-being and prestige.

ethnic (or racial) nationalism: the belief that one's own ethnic group constitutes a distinct people whose culture is and should be separate from that of the larger society.

ethnic group: a population that has a sense of group identity based on shared ancestry and distinctive cultural patterns.

ethnic stratification: the ranking of ethnic groups in a social hierarchy on the basis of each group's similarity to the dominant group.

ethnocentrism: the tendency to judge other cultures as inferior to one's own.

ethnomethodology: the study of the underlying rules of behavior that guide group interaction.

exogamy: a norm specifying that a person brought up in a particular culture may marry outside the cultural group.

exogenous force: pressure for social change that is exerted from outside a society.

experimental group: in an experiment, the subjects who are exposed to a change in the independent variable.

expulsion: the forcible removal of one population from a territory claimed by another population.

extended family: an individual's nuclear family plus the nuclear families of his or her blood relatives.

family of orientation: the nuclear family in which a person is born and raised.

family of procreation: the nuclear family a person forms through marriage or cohabitation.

family: a group of people related by blood, marriage, or adoption.

feral child: a child reared outside human society.

field experiment: an experimental situation in which the researcher observes and studies subjects in their natural setting.

folkways: weakly sanctioned norms.

formal organization: a group that has an explicit, often written, set of norms, statuses, and roles that specify each member's relationships to the others and the conditions under which those relationships hold.

frequency distribution: a classification of data that describes how many observations fall within each category of a variable.

functionalism: a sociological perspective that focuses on the ways in which a complex pattern of social structures and arrangements contributes to social order.

gemeinschaft: a term used to refer to the close, personal relationships of small groups and communities.

gender role: a set of behaviors considered appropriate for an individual of a particular gender.

gender socialization: the ways in which we learn our gender identity and develop according to cultural norms of masculinity and femininity.

gender: the culturally defined ways of acting as a male or a female that become part of an individual's personal sense of self.

generalized other: a person's internalized conception of the expectations and attitudes held by society.

genocide: the intentional extermination of one population by a more dominant population.

gerontology: the study of aging and older people.

gesellschaft: a term used to refer to the well-organized but impersonal relationships among the members of modern societies.

ghetto: a section of a city that is segregated either racially or culturally.

group: any collection of people who interact on the basis of shared expectations regarding one another's behavior.

Hawthorne effect: the unintended effect that results from the attention given to subjects in an experimental situation.

hegemony: undue power or influence.

hermaphrodite: a person whose primary sexual organs have features of both male and female organs, making it difficult to categorize the individual as male or female.

heterosexuality: sexual orientation toward the opposite sex.

homogamy: the tendency to marry a person who is similar to oneself in social background.

homosexuality: sexual orientation toward the same sex.

horticultural society: a society whose primary means of subsistence is raising crops, which it plants and cultivates, often developing an extensive system for watering the crops.

human ecology: a sociological perspective that emphasizes the relationships among social order, social disorganization, and the distribution of populations in space and time.

hypothesis: a statement that specifies a relationship between two or more variables that can be tested through empirical observation.

id: according to Freud, the part of the human personality from which all innate drives arise.

ideas: ways of thinking that organize human consciousness.

identification: the social process whereby an individual chooses role models and attempts to imitate their behavior.

ideologies: systems of values and norms that the members of a society are expected to believe in and act on without question.

impression management: the strategies one uses to "set a stage" for one's own purposes.

in-group: a social group to which an individual has a feeling of allegiance; usually, but not always, a primary group.

independent variable: a variable that the researcher believes causes a change in another variable (i.e., the dependent variable).

informal organization: a group whose norms and statuses are generally agreed upon but are not set down in writing.

informed consent: the right of respondents to be informed of the purpose for which the information they supply will be used and to judge the degree of personal risk involved in answering questions, even when an assurance of confidentiality has been given.

institution: a more or less stable structure of statuses and roles devoted to meeting the basic needs of people in a society.

institutional discrimination: the systematic exclusion of people from equal participation in a particular institution because of their group membership.

interactionism: a sociological perspective that views social order and social change as resulting from all the repeated interactions among individuals and groups.

interest group: an organization that attempts to influence elected and appointed officials regarding a specific issue or set of issues.

intergenerational mobility: a change in the social class of family members from one generation to the next.

internal colonialism: a theory of racial and ethnic inequality that suggests that some minorities are essentially colonial peoples within the larger society.

intragenerational mobility: a change in the social class of an individual within his or her own lifetime.

Jim Crow: the system of formal and informal segregation that existed in the United States from the late 1860s to the early 1970s.

kinship: the role relations among people who consider themselves to be related by blood, marriage, or adoption.

labeling: a theory that explains deviance as a societal reaction that brands or labels as deviant people who engage in certain behaviors.

laissez-faire capitalism: an economic philosophy based on the belief that the wealth of a nation can be measured by its capacity to produce goods and services (i.e., its resources of land, labor, and machinery) and that these can be maximized by free trade.

laws: norms that are written by specialists, collected in codes or manuals of behavior, and interpreted and applied by other specialists.

legal authority: authority that is legitimated by people's belief in the supremacy of the law; obedience is owed not to a person but to a set of impersonal principles.

legitimacy: the ability of a society to engender and maintain the belief that the existing political institutions are the most appropriate for that society.

life chances: the opportunities an individual will have or be denied throughout life as a result of his or her social-class position.

life course: a pathway along an age-differentiated, socially created sequence of transitions.

life expectancy: the average number of years a member of a given population can expect to live.

lifetime negative experiences: experiences that cause long-term stress, such as the death of a child or spouse.

linguistic-relativity hypothesis: the belief that language determines the possibilities for thought and action in any given culture.

lobbying: the process whereby interest groups seek to persuade legislators to vote in their favor on particular bills.

macro-level sociology: an approach to the study of society that focuses on the major structures and institutions of society.

market: an economic institution that regulates exchange behavior through the establishment of different values for particular goods and services.

mass public: a large population of potential spectators or participants who engage in collective behavior.

mass: a large number of people who are all oriented toward a set of shared symbols or social objects.

master status: a status that takes precedence over all of an individual's other statuses.

material culture: patterns of possessing and using the products of culture.

megalopolis: a complex of cities distributed along a major axis of traffic and communication, with a total population exceeding 25 million.

mercantilism: an economic philosophy based on the belief that the wealth of a nation can be measured by its holdings of gold or other precious metals and that the state should control trade.

metropolitan area: a central city surrounded by a number of smaller cities and suburbs that are closely related to it both socially and economically.

micro-level sociology: an approach to the study of society that focuses on patterns of social interaction at the individual level.

middle-level sociology: an approach to the study of society that focuses on relationships between social structures and the individual.

minority group: a population that, because of its members' physical or cultural characteristics, is singled out from others in the society for differential and unequal treatment.

modernization: a term used to describe the changes experienced by societies and individuals as a result of industrialization, urbanization, and the development of nation-states.

modernization: a term used to describe the changes that societies and individuals experience as a result of industrialization, urbanization, and the development of nation-states.

monotheistic: a term used to describe a theistic belief system centered on belief in a single all-powerful God.

mores: strongly sanctioned norms.

multinational corporation: an economic enterprise that has headquarters in one country and conducts business activities in one or more other countries.

nation-state: the largest territory within which a society's political institutions can operate without having to face challenges to their sovereignty.

nationalism: the belief of a people that they have the right and the duty to constitute themselves as a nation-state.

natural selection: the relative success of organisms with specific genetic mutations in reproducing new generations with the new trait.

nonterritorial community: a network of relationships formed around shared goals.

normative order: the array of norms that permit a society to achieve relatively peaceful social control.

norms: specific rules of behavior.

nuclear family: two or more people related by blood, marriage, or adoption who share a household.

objective class: in Marxian theory, a social class that has a visible, specific relationship to the means of production.

occupational prestige: the honor or prestige attributed to specific occupations by adults in a society.

oligarchy: rule by a few people who stay in office indefinitely rather than for limited terms.

open question: a question that does not require the respondent to choose from a predetermined set of answers; instead, the respondent may answer in his or her own words.

open society: a society in which social mobility is possible for everyone.

open stratification system: a stratification system in which the boundaries between social strata are easily crossed.

opinion leader: an individual who consistently influences the attitudes and behavior of others.

out-group: any social group to which an individual does not have a feeling of allegiance; may be in competition or conflict with the in-group.

paradigm: a model comprising existing theories and methods; a general way of seeing the world that dictates what kind of scientific work should be done and what kinds of theory are acceptable.

participant observation: a form of observation in which the researcher participates to some degree in the lives of the people being observed.

pastoral society: a society whose primary means of subsistence is herding animals and moving with them over a wide expanse of grazing land.

patriarchy: the dominance of men over women.

peer group: an interacting group of people of about the same age that has a significant influence on the norms and values of its members.

percent analysis: a mathematical operation that transforms an absolute number into a proportion as a part of 100.

peripheral area: a region that supplies basic resources and labor power to more advanced states.

plea bargaining: a process in which a person charged with a crime agrees to plead guilty to a lesser charge.

pluralist model: a theory stating that no single group controls political decisions; instead, a plurality of interest groups influence those decisions through a process of coalition building and bargaining.

pluralistic society: a society in which different ethnic and racial groups are able to maintain their own cultures and lifestyles while gaining equality in the institutions of the larger society.

political institution: a set of norms and statuses pertaining to the exercise of power and authority.

political party: an organization of people who join together to gain legitimate control of state authority.

political revolution: a set of changes in the political structures and leadership of a society.

polytheistic: a term used to describe a theistic belief system that includes numerous gods.

power elite model: a theory stating that political decisions are controlled by an elite of rich and powerful individuals even in societies with democratic political institutions.

power: the ability to control the behavior of others, even against their will.

prejudice: an attitude that prejudges a person on the basis of a real or imagined characteristic of a group to which that person belongs.

primary deviance: an act that results in the labeling of the offender as deviant.

primary group: a social group characterized by intimate, face-to-face associations.

privacy: the right of a respondent to define when and on what terms his or her actions may be revealed to the general public.

profane: a term used to describe phenomena that are not considered sacred.

profession: an occupation with a body of knowledge and a developed intellectual technique that are transmitted by a formal educational process and testing procedures.

projection: the psychological process whereby we attribute to other people behaviors and attitudes that we are unwilling to accept in ourselves.

psychosocial risk behaviors: behaviors that are detrimental to health, such as smoking and heavy drinking.

public opinion: the values and attitudes held by mass publics.

race: an inbreeding population that develops distinctive physical characteristics that are hereditary.

racism: an ideology based on the belief that an observable, supposedly inherited trait is a mark of inferiority that justifies discriminatory treatment of people with that trait.

rate of reproductive change: the difference between the crude birthrate and the crude death rate for a given population.

recidivism: the probability that a person who has served a jail term will commit additional crimes and be jailed again.

reference group: a group that an individual uses as a frame of reference for self-evaluation and attitude formation.

relative deprivation: deprivation as determined by comparison with others rather than by some objective measure.

religion: any set of coherent answers to the dilemmas of human existence that makes the world meaningful; a system of beliefs and rituals that serves to bind people together into a social group.

religiosity: the depth of a person's religious feelings.

resocialization: intense, deliberate socialization designed to change major beliefs and behaviors.

rite of passage: a ceremony marking the transition to a new stage of a culturally defined life course.

ritual: a formal pattern of activity that expresses symbolically a set of shared meanings.

role conflict: conflict that occurs when in order to perform one role well a person must violate the expectations associated with another role.

role expectations: a society's expectations about how a role should be performed, together with the individual's perceptions of what is required in performing that role.

role strain: conflict that occurs when the expectations associated with a single role are contradictory.

role taking: trying to look at social situations from the standpoint of another person from whom one seeks a response.

role: the way a society defines how an individual is to behave in a particular status.

sacred: a term used to describe phenomena that are regarded as extraordinary, transcendent, and outside the everyday course of events.

sample survey: a survey administered to a selection of respondents drawn from a specific population.

sample: a set of respondents selected from a specific population.

sanctions: rewards and punishments for abiding by or violating norms.

scapegoat: a convenient target for hostility.

science: knowledge obtained as a result of the process of developing and testing hypotheses.

scientific method: the process by which theories and explanations are constructed through repeated observation and careful description.

secondary deviance: behavior that is engaged in as a reaction to the experience of being labeled as deviant.

secondary group: a social group whose members have a shared goal or purpose but are not bound together by strong emotional ties.

sect: a religious organization that rejects the beliefs and practices of existing churches; usually formed when a group leaves the church to form a rival organization.

secularization: a process in which the dominance of religion over other institutions is reduced.

segregation: the ecological and institutional separation of races or ethnic groups.

semiperipheral area: a state or region in which industry and financial institutions are developed to some extent but that remains dependent on capital and technology provided by other states.

sex: the biological differences between males and females, including the primary sex characteristics that are present at birth (i.e., the presence of specific male or female genitalia) and the secondary sex characteristics that develop later (facial and body hair, voice quality, etc.).

sexism: an ideology that justifies prejudice and discrimination based on sex.

sexuality: the manner in which a person engages in the intimate behaviors connected with genital stimulation, orgasm, and procreation.

significant other: any person who is important to an individual.

simple supernaturalism: a form of religion in which people may believe in a great force or spirit but do not have a well-defined concept of God or a set of rituals involving God.

slavery: the ownership of one racial, ethnic, or politically defined group by another group that has complete control over the enslaved group.

social category: a collection of individuals who are grouped together because they share a trait deemed by the observer to be socially relevant.

social change: variations over time in the ecological ordering of populations and communities, in patterns of roles and social interactions, in the structure and functioning of institutions, and in the cultures of societies.

social conditions: the realities of the life we create together as social beings.

social control: the set of rules and understandings that control the behavior of individuals and groups in a particular culture.

social Darwinism: the notion that people who are more successful at adapting to the environment in which they find themselves are more likely to survive and to have children who will also be successful.

social group: a set of two or more individuals who share a sense of common identity and belonging and who interact on a regular basis.

social mobility: movement by an individual or group from one social stratum to another.

social movement: organized collective behavior aimed at changing or reforming social institutions or the social order itself.

social revolution: a complete transformation of the social order, including the institutions of government and the system of stratification.

social stratification: a society's system for ranking people hierarchically according to such attributes as wealth, power, and prestige; the process whereby the members of a society are sorted into different statuses.

social structure: the recurring patterns of behavior that people create through their interactions, their exchange of information, and their relationships.

socialism: an economic philosophy based on the concept of public ownership of property and sharing of profits, together with the belief that economic decisions should be controlled by the workers.

socialization: the processes whereby we learn to behave according to the norms of our culture.

society: a population that is organized in a cooperative manner to carry out the major functions of life.

sociobiology: the hypothesis that all human behavior is determined by genetic factors.

socioeconomic status (SES): a broad social-class ranking based on occupational status, family prestige, educational attainment, and earned income.

sociological imagination: according to C. Wright Mills, the ability to see how social conditions affect our lives.

sociology: the scientific study of human societies and human behavior in the groups that make up a society.

spatial mobility: movement of an individual or group from one location or community to another.

state: the set of political institutions operating in a particular society.

status group: a category of people within a social class, defined by how much honor or prestige they receive from the society in general.

status symbols: material objects or behaviors that indicate social status or prestige.

status: a socially defined position in a group.

stereotype: an inflexible image of the members of a particular group that is held without regard to whether it is true.

stigma: an attribute or quality of an individual that is deeply discrediting.

structural mobility: movement of an individual or group from one social stratum to another caused by the elimination of an entire class as a result of changes in the means of existence.

subculture: a group of people who hold many of the values and norms of the larger culture but also hold certain beliefs, values, or norms that set them apart from that culture.

subjective class: in Marxian theory, the way members of a given social class perceive their situation as a class.

superego: according to Freud, the part of the human personality that internalizes the moral codes of adults.

sustainable development: development that does not damage the environment or create new environmental burdens.

technologies: the products and the norms for using them that are found in a given culture.

technology assessment: efforts to anticipate the consequences of particular technologies for individuals and for society as a whole.

technology: tools, procedures, and forms of social organization that increase human productive capacity; the use of tools and knowledge to manipulate the physical environment in order to achieve desired practical goals.

territorial community: a population that functions within a particular geographic area.

theism: a belief system that conceives of a god or gods as separate from humans and from other living things on the earth.

theoretical perspective: a set of interrelated theories that offer explanations for important aspects of social behavior.

theory: a set of interrelated concepts that seeks to explain the causes of an observable phenomenon.

total institution: a setting in which people undergoing resocialization are isolated from the larger society under the control of a specialized staff.

totalitarian regime: a regime that accepts no limits to its power and seeks to exert its rule at all levels of society.

traditional authority: authority that is hereditary and is legitimated by traditional values, particularly people's idea of the sacred.

transsexuals: people who feel very strongly that the sexual organs they were born with do not conform to their deep-seated sense of what their sex should be.

triad: a group consisting of three people.

two-step flow of communication: the process in which messages communicated by the media are evaluated by certain respected individuals, who in turn influence the attitudes and behavior of others.

unobtrusive measures: observational techniques that measure behavior but intrude as little as possible into actual social settings.

upward mobility: movement by an individual or group to a higher social stratum.

urbanization: a process in which an increasing proportion of a total population becomes concentrated in urban settlements.

values: the ideas that support or justify norms.

variable: a characteristic of an individual, group, or society that can vary from one case to another.

voluntary association: a formal organization whose members pursue shared interests and arrive at decisions through some sort of democratic process.

welfare capitalism: an economic philosophy in which markets determine what goods will be produced and how, but the government regulates economic competition.

REFERENCES

Abbott, A. D. (1988). *The system of professions: An essay on the division of expert labor.* Chicago: University of Chicago Press.

Abbott, A. D. (1993). The sociology of work and occupations. *Annual Review of Sociology, 19,* 187–209.

Abu-Lugod, J. (1996, January). Personal communication.

Adler, M. (1997). *Drawing down the moon: Witches, druids, goddess-worshippers, and other pagans in America today.* New York: Viking.

Adorno, T. (1950). *The authoritarian personality.* New York: Harper.

Aganbegyan, A. (1989). *Perestroika 1989.* New York: Scribner.

Ahlburg, D. A., & De Vita, C. J. (1992, August). New realities of the American family. *Population Bulletin* (Population Reference Bureau).

Alam, M. (1988). *Studies on brain drain.* Monticello, IL: Vance Bibliographies.

Albrecht, D. G. (1997). The changing structure of U.S. agriculture. *Rural Sociology, 62,* 474–490.

Aldous, J. (1982). From dual-earner to dual-career families and back again. In J. Aldous (Ed.), *Two paychecks: Life in dual-career families.* Newbury Park, CA: Sage.

Alland, A. (1973). *Human diversity.* Garden City, NY: Doubleday.

Allen, F. L. (1931). *Only yesterday: An informal history of the 1920s.* New York: Harper.

Allen, G. (1997). Abolishing parole saves lives and property. *Corrections Today, 59,* 22.

Altman, D. (1987). *AIDS in the mind of America.* Garden City, NY: Doubleday.

American Assembly (1990). The American Assembly on population growth and the global environment. *Population and Development Review, 16,* 384–386.

Ammerman, N. T. (1990). *Baptist battles: Social change and religious conflict in the Southern Baptist Convention.* New Brunswick, NJ: Rutgers University Press.

Ammerman, N. T. (1997). Organized religion in a voluntaristic society. *Sociology of Religion, 58,* 203–216.

Andersen, M. L. (1993). *Thinking about women: Sociological perspectives on sex and gender* (3rd ed.). New York: Macmillan.

Anderton, D. L. (1997). *The population of the United States* (3rd ed.). New York: Free Press.

Annie E. Casey Foundation. Annual. *Kids count: A pocket guide on America's youth.* Washington, DC: Population Reference Bureau.

Applebome, P. (1998, May 10). No room for children in a world of little adults. *New York Times,* sec. 4, pp. 1, 3.

Archibald, K. (1947). *Wartime shipyard.* Berkeley: University of California Press.

Ariès, P. (1962). *Centuries of childhood.* New York: Vintage.

Aron, R. (1955). *The century of total war.* Boston: Beacon.

Aronowitz, S. (1996). Losing and winning in a conservative age. *Social Policy, 27,* 10–21.

Aronowitz, S. (1998). *From the ashes of the old.* Boston: Houghton Mifflin.

Aronowitz, S., & DiFazio, W. (1994). *The jobless future: Sci-tech and the dogma of work.* Minneapolis: University of Minnesota Press.

Aronowitz, S., & Giroux, H. A. (1985). *Education under siege.* South Hadley, MA: Bergin and Garvey.

Asch, S. E. (1966). Effects of group pressure upon the modification and distortion of judgments. In H. Proshansky & B. Seidenberg (Eds.), *Basic studies in social psychology.* Fort Worth: Holt, Rinehart and Winston.

Ash, T. G. (1999, February 15). Helena's kitchen. *New Yorker,* pp. 32–39.

Ashmawi, M. S. (1998). *Against Islamic extremism: The writings of Muhammad Sa'id al-'Ashmawy.* Gainesville: University Press of Florida.

Aspaklaria, S. (1985, September 12). A divorced father, a child, and a summer visit together. *New York Times,* p. C1.

Atchley, R. C. (1991). *Social forces and aging: An introduction to social gerontology.* Belmont, CA: Wadsworth.

Autman, S. (1998, March 26). Experts blame youth violence on the culture, access to guns. *St. Louis Post-Dispatch,* p. 8.

Ayala, V. (1996). *Falling through the cracks: AIDS and the urban poor.* Putnam Valley, NY: Social Change Press.

Ayres, B. D., Jr. (1995, November 3). California smog cloud is cleaning up its act. *New York Times,* p. A14.

Bacci, M. (1997). *A concise history of world population* (2nd ed.). Malden, MA: Blackwell.

Backer, B. (1982). *Death and dying: Individuals and institutions.* New York: Wiley.

Bakke, E. W. (1933). *The unemployed man.* London: Nisbet.

Balandier, A. (1971/1890). *The delight makers.* Orlando: Harcourt Brace Jovanovich.

Balbus, I. (1978). Commodity form and legal form: An essay on the relative autonomy of the law. In C. E. Reasons & R. M. Rich (Eds.), *Sociology of law: A conflict perspective.* Toronto: Butterworths.

Balch, E. G. (1910). *Our Slavic fellow citizens.* New York: Charities Publication Committee.

Baldassare, M. (1986). *Trouble in paradise: The suburban transformation in America.* New York: Columbia University Press.

Bales, R. F., & Slater, P. E. (1955). Role differentiation in small decision-making groups. In T. Parsons & R. F. Bales (Eds.), *Family, socialization, and interaction process.* New York: Free Press.

Ballantine, J. H. (1997). *The sociology of education: A systematic analysis* (4th ed.). Upper Saddle River, NJ: Prentice Hall.

Barnet, R. J. (1994, December 19). Lords of the global economy: Stateless corporations. *The Nation,* pp. 754–758.

Barrett, D. V. (1996). *Sects, "cults," and alternative religions: A world survey and sourcebook.* London: Blandford.

Bashi, V. (1991). Mentoring of at-risk students. *Focus* (University of Wisconsin, Institute for Research on Poverty), *13,* 26–32.

Bassuk, E. L. (1984, July). The homelessness problem. *Scientific American,* 40–45.

Bates, A. (1999, February 15). The siege at Waco: Deadly inferno. *Booklist,* pp. 1050–1051.

Batson, C. D., Batson, J. G., & Todd, R. M. (1995). Empathy and the collective good: Caring for one or the others in a social dilemma. *Journal of Personality and Social Psychology, 68,* 619–631.

Battaglia, L., & Zecchin, F. (1989). *Chroniques siciliennes.* Paris: Centre Nationale de la Photographie.

Baum, S. (1987). *Financial aid to low-income college students: Its history and prospects.* (Institute for Research on Poverty Discussion Paper No. 846-87). Madison: University of Wisconsin.

Baumgartner, M. P. (1988). *The moral order of a suburb.* New York: Oxford University Press.

Bayer, R. (1988). *Private acts, social consequences: AIDS and the politics of public health.* New York: Free Press.

Bayer, R. (1991). The great drug policy debate. *Millbank Quarterly, 69,* 341–364.

Bearak, B. (1997, April 28). Eyes on glory. *New York Times,* p. A1.

Becker, G. S. (1997, December 29). Why every married couple should sign a contract. *Business Week,* p. 30.

Becker, H. S. (1961). *Boys in white: Student culture in medical school.* Chicago: University of Chicago Press.

Becker, H. S. (1963). *The outsiders: Studies in the sociology of deviance.* New York: Free Press.

Beebe, S. A. (1997). *Communication in small groups: Principles and practices* (5th ed.). New York: Longman.

Beer, F. A. (1981). *Peace against war: The ecology of international violence.* San Francisco: Freeman.

Belkin, L. (1992, March 2). Choosing to die at home: Dignity has its burdens. *New York Times,* pp. A1, B4.

Bell, D. (1962). Crime as an American way of life. In D. Bell, *The end of ideology.* New York: Free Press.

Bell, D. (1973). *The coming of post industrial society: A venture in social forecasting.* New York: Basic Books.

Bellah, R. N. (1970). *Beyond belief.* New York: Harper.

Bellah, R. N., Madison, R., Sullivan, W. M., Swidler, A., & Tipton, S. M. (1985). *Habits of the heart: Individualism and commitment in American life.* Berkeley: University of California Press.

Belsky, J. (1988). The "effects" of infant day care reconsidered. *Early Childhood Research Quarterly, 3,* 235–272.

Belsky, J., & Steinberg, L. D. (1991). *Infancy, childhood & adolescence: Development in context.* New York: McGraw-Hill.

Bem, S. L. (1993). *The lenses of gender: Transforming the debate on sexual inequality.* New Haven, CT: Yale University Press.

Ben-David, J. (1984). *The scientist's role in society: A comparative study.* Chicago: University of Chicago Press.

Benavot, A., et al. (1991). Knowledge for the masses: World models and national curricula, 1920–1986. *American Sociological Review, 56,* 85–91.

Bendix, R. (1969). *Nation-building and citizenship.* Garden City, NY: Doubleday.

Bendix, R. (1974). *Work and authority in industry.* Berkeley: University of California Press.

Bendix, R., & Lipset, S. M. (1966). *Class, status, and power: Social stratification in comparative perspective* (2nd ed.). New York: Free Press.

Bengston, V., & Schaie, K. W. (Eds.) (1999). *Handbook of theories of aging.* New York: Springer.

Bennett, M. K. (1954). *The world's food.* New York: Harper.

Benokraitis, N. V. (1996). *Marriages and families: Changes, choices, and constraints* (2nd ed.). Upper Saddle River, NJ: Prentice Hall.

Benokraitis, N. V., & Feagin, J. R. (1986). *Modern sexism: Blatant, subtle, and covert discrimination.* Englewood Cliffs, NJ: Prentice Hall.

Bensman, D., & Lynch, R. (1987). *Rusted dreams: Hard times in a steel community.* New York: McGraw-Hill.

Bentham, J. (1789). *An introduction to the principles of morals and legislation.* London: T. Payne.

Berger, B. (1961). The myth of suburbia. *Journal of Social Issues, 17,* 38–48.

Berger, B. (1968). *Working class suburb: A study of auto workers in suburbia.* Berkeley: University of California Press.

Bergmann, B. R., & Hartmann, H. (1995, May 1). A program to help working parents. *The Nation,* pp. 592ff.

Berlin, I. (1998, May 14). My intellectual path. *New York Review of Books,* pp. 53–60.

Berman, S. (1997). *Children's social consciousness and the development of social responsibility.* Albany: State University of New York Press.

Bernard, J. (1964). *Academic women.* University Park: Pennsylvania State University Press.

Bernard, J. (1973). *The sociology of community.* Glenview, IL: Scott, Foresman.

Bernard, J. (1982). *The future of marriage* (2nd ed.). New Haven, CT: Yale University Press.

Berrueta-Clement, J. R., Schweinhart, L. J., Barnett, W. S., Epstein, A. S., & Weikart, D. P. (1984). *Changed lives: The effects of the Perry preschool program on youths through age 19.* Ypsilanti, MI: High/Scope Press.

Berry, B. (1978). Latent structure of urban systems: Research methods and findings. In L. S. Bourne & J. W. Simmons (Eds.), *Systems of cities: Readings on structure, growth, and policy.* New York: Oxford University Press.

Besharov, D. J., & Sullivan, T. S. (1996, Fall). Welfare reform and marriage. *Public Interest,* pp. 81–95.

Betts, J. R. (1998, March). The two-legged stool: The neglected role of educational standards in improving America's public schools. *Federal Reserve Bank of New York Economic Policy Review,* pp. 97–117.

Bianchi, S. M., & Spain, D. (1986). *American women in transition.* New York: Russell Sage.

Bierstedt, R. (1963). *The social order.* New York: McGraw-Hill.

Bigner, J. J., & Bozett, F. W. (1989). Parenting by gay fathers. *Marriage and Family Review, 14,* 155–175.

Birdwhistell, R. (1970). *Kinesics and context: Essays on body motion and communication.* Philadelphia: University of Pennsylvania Press.

Bittman, M. (1993). Australians' changing use of time, 1974–1987. *Social Indicators Research, 30,* 91–109.

Black, D. (Ed.) (1984). *Toward a general theory of social control.* Orlando: Academic Press.

Blau, J. R. (1993). *Social contracts and economic markets.* New York: Plenum.

Blau, P. (1964). *Exchange and power in social life.* New York: Wiley.

Blau, P., & Duncan, O. D. (1967). *The American occupational structure.* New York: Wiley.

Blau, P. M., & Scott, R. W. (1962). *Formal organizations: A comparative approach.* San Francisco: Chandler.

Blauner, R. (1964). *Alienation and freedom.* Chicago: University of Chicago Press.

Blauner, R. (1969). Internal colonialism and ghetto revolt. *Social Problems, 16,* 393–408.

Blauner, R. (1972). *Racial oppression in America.* New York: Harper.

Blauner, R. (Ed.) (1989). *Black lives, white lives: Three decades of race relations in the United States.* Berkeley: University of California Press.

Blinder, D., & Crossette, B. (1993, February 7). As ethnic wars multiply, U.S. strives for a policy. *New York Times,* p. A1.

Bloch, M. (1964). *Feudal society* (Vol. 1). Chicago: University of Chicago Press.

Block, F. (1990). *Post industrial possibilities: A critique of economic discourse.* Berkeley: University of California Press.

Bloom, B. S. (1976). *Human characteristics and school learning.* New York: McGraw-Hill.

Bluestone, B. (1992). Negotiating the future: A labor perspective on American business. New York: Basic Books.

Blumenfeld, W. J. (Ed.) (1992). *Homophobia.* Boston: Beacon.

Blumer, H. (1969a). Elementary collective groupings. In A. M. Lee (Ed.), *Principles of sociology* (3rd ed.). New York: Barnes & Noble.

Blumer, H. (1969b). *Symbolic interactionism.* Englewood Cliffs, NJ: Prentice Hall.

Blumer, H. (1978). Elementary collective behavior. In L. E. Genevie (Ed.), *Collective behavior and social movements.* Itasca, IL: Peacock.

Bobrick, B. (1997). *Angel in the whirlwind: The triumph of the American Revolution.* New York: Simon & Schuster.

Boese, W. (1998). U.S. retains top spot in latest UN conventional arms register. *Arms Control Today, 28,* 27.

Boggs, V. W., & Meyersohn, R. (1988). The profile of a Bronx salsero: Salsa's still alive! *Journal of Popular Music and Society, 11,* 7–14.

Bogue, D. J. (1969). *Principles of demography.* New York: Wiley.

Bogue, D. J. (1985). *The population of the United States: Historical trends and future projections.* New York: Free Press.

Bohannon, L., & Bohannon, P. (1953). *The Tiv of central Nigeria.* London: International African Institute.

Bohannon, L., & Bohannon, P. (1962). *Markets in Africa.* Evanston, IL: Northwestern University Press.

Bok, S. (1998). *Mayhem: Violence as public entertainment.* Reading, MA: Perseus.

Bonner, R. (1994, August 25). Hutu and Tutsi mill the rice and set an example. *New York Times,* p. A1.

Bonnett, A. W. (1981). *Institutional adaptation of West Indian immigrants to America.* Lanham, MD: University Press of America.

Booth, A., & Edwards, J. N. (1985, February). Age at marriage and marital instability. *Journal of Marriage and the Family,* pp. 67–75.

Bornstein, D. (1996). *The price of a dream: The story of the Grameen Bank and the idea that is helping the poor to change their lives.* New York: Simon & Schuster.

Bose, C. E., & Rossi, P. H. (1983). Gender and jobs. *American Sociological Review, 48,* 316–330.

Bosk, C. (1979). *Forgive and remember: Managing medical failure.* Chicago: University of Chicago Press.

Bosworth, B., & Burtless, G. (Eds.) (1998). Aging societies: The global dimension. Washington, DC: Brookings Institution.

Bott, E. (1977). Urban families: Conjugal roles and social networks. In S. Leinhardt (Ed.), *Social networks: A developing paradigm.* Orlando: Academic Press.

Bottomore, T. (Ed.) (1973). *Karl Marx.* Englewood Cliffs, NJ: Prentice Hall.

Bowker, J. W. (1997). *World religions.* New York: DK Publishers.

Boyte, H. C. (1991). Community service and service education. *Phi Delta Kappan, 72,* 765–767.

Bracey, J. H., Meier, A., & Rudwick, E. (1970). *Black nationalism in America.* Indianapolis: Bobbs-Merrill.

Bradley, H. (1998). A new gender(ed) order? Researching and rethinking women's work. *Sociology, 32,* 869–874.

Braidwood, R. S. (1967). *Prehistoric man* (7th ed.). Glenview, IL: Morrow.

Brake, M. (1980). *The sociology of youth cultures and youth subcultures.* London: Routledge and Kegan Paul.

Brandon, K. (1998, May 24). Accused youth had good home, caring folks. *Chicago Tribune,* p. 1.

Braudel, F. (1976/1949). *The Mediterranean and the Mediterranean world in the age of Philip II.* New York: Harper.

Braudel, F. (1981). *The structures of everyday life: Vol. 1. Civilizations and capitalism: 15th–18th century.* New York: Harper.

Braudel, F. (1984). *The perspective of the world: Vol. 3. Civilizations and capitalism: 15th–18th century.* New York: Harper.

Braudy, L. (1997). *The frenzy of renown: Fame and its history.* New York: Oxford University Press.

Brauss, P. (1995). The baby boom at mid-decade. *American Demographics, 17,* 40–46.

Braverman, H. (1974). *Labor and monopoly capital.* New York: Monthly Review Press.

Breese, G. (1966). *Urbanization in newly developing countries.* Englewood Cliffs, NJ: Prentice Hall.

Breiner, P. (1996). Max Weber and democratic politics. Ithaca, NY: Cornell University Press.

Brimelow, P. (1995). *Alien nation: Common sense about America's immigration disaster.* New York: Random House.

Brody, D. (1960). *Steelworkers in America: The non-union era.* Cambridge, MA: Harvard University Press.

Brody, J. (1992, August 11). To predict divorce, ask 125 questions. *New York Times,* pp. C1, C9.

Brody, J. E. (1996, February 28). Good habits outweigh genes as key to a healthy old age. *New York Times,* p. C9.

Brokaw, T. (1994, January 2). Learning lessons for life in football. *New York Times,* sec. 8, p. 11.

Bronfenbrenner, U. (1981). Children and families. *Society, 18,* 38–41.

Bronfenbrenner, U., & Ceci, S. J. (1994). Nature-nurture reconceptualized in developmental perspective: A bioecological model. *Psychological Review, 101,* 568–586.

Brooke, J. (1997, April 28). Former cult member warns of continuing cult loyalties. *New York Times,* p. A16.

Brown, C. (1966). *Manchild in the Promised Land.* New York: Macmillan.

Brown, D. (1970). *Bury my heart at Wounded Knee.* New York: Washington Square Press.

Brown, K. M. (1991). *Mama Lola: A Vodou priestess in Brooklyn.* Berkeley: University of California Press.

Brown, K. M. (1992, April 15). Writing about "the other." *Chronicle of Higher Education,* p. A56.

Brown, L. R., Flavin, C., & French, H. F. (1997). *State of the world 1998: A Worldwatch Institute report on progress toward a sustainable society.* New York: Norton.

Brown, R. (1965). *Social psychology.* New York: Free Press.

Browne, M. W. (1987, September 1). Relics of Carthage show brutality amid the good life. *New York Times,* pp. C1, C10.

Brundtland, G. H. (1989, September). How to secure our common future. *Scientific American,* p. 190.

Bulmer, M. (1986). *The Chicago school of sociology.* Chicago: University of Chicago Press.

Bumpass, L. L., & Raley, R. K. (1995). Redefining single-parent families: Cohabitation and changing family reality. *Demography, 32,* 97–110.

Bumpass, L. L., Raley, R. K., & Sweet, J. A. (1995). The changing character of stepfamilies: Implications of cohabitation and nonmarital childbearing. *Demography, 32,* 425–437.

Bumpass, L. L., & Sweet, J. A. (1989). Children's experience in single-parent families: Implications of cohabitation and marital transitions. *Family Planning Perspectives, 21,* 256–260.

Burawoy, M. (1980). *Manufacturing consent.* Chicago: University of Chicago Press.

Burawoy, M. (1996). A classic of its time. *Contemporary Sociology, 25,* 296–300.

Burgess, E. W. (1925). The growth of the city: An introduction to a research project. In R. E. Park & E. W. Burgess (Eds.), *The city.* Chicago: University of Chicago Press.

Burstein, R. (1991). Legal mobilization as a social movement tactic: The struggle for equal employment opportunity. *American Journal of Sociology, 96,* 1201–1225.

Burtless, G. (1998, Summer–Fall). Can the labor market absorb three million welfare recipients? *Focus* (Institute for Research on Poverty), pp. 1–6.

Buruma, I. (1984). *Behind the mask: On sexual demons, sacred mothers, transvestites, gangsters, drifters and other Japanese cultural heroes.* New York: Pantheon.

Bury, J. B. (1932). *The idea of progress: An inquiry into its origin and growth.* New York: Dover.

Butler, R. (1989). Dispelling ageism: The cross-cutting intervention. *Annals of the American Academy of Political and Social Science, 503,* 138–148.

Butterfield, F. (1992a, July 19). Are American jails becoming shelters from the storm? *New York Times,* p. E4.

Butterfield, F. (1992b, January 13). Studies find a family link to criminality. *New York Times,* pp. A1, A16.

Butterfield, F. (1998a, August 9). Prison population growing although crime rate drops. *New York Times,* p. 18.

Butterfield, F. (1998b, March 29). Reason for dramatic drop in crime puzzles the experts. *New York Times,* p. 16.

Cabral, E. (1998, Spring–Summer). Taking microlending to the next level. *Ford Foundation Report,* p. 6.

Cahill, S. E. (1992). The sociology of childhood at and in an uncertain age. *Contemporary Sociology, 21,* 669–672.

Cain, L. D. (1964). Life course and social structure. In R. F. Faris (Ed.), *Handbook of modern sociology.* Chicago: Rand McNally.

Caldwell, J., & Caldwell, P. (1988). Women's position and child mortality and morbidity in LDCs. In *Conference on women's position and demographic change in the course of development.* Liège: IUSSP.

Califano, J. A. (1998, February 21). A punishment-only prison policy. *America,* pp. 3–4.

Cameron, W. B. (1966). *Modern social movements: A sociological outline.* New York: Random House.

Campbell, B. (1984). *Wigan Pier revisited: Poverty and politics in the eighties.* London: Virago.

Cancian, F. (1987). *Love in America: Gender and self development.* New York: Cambridge University Press.

Cancian, F. M. (1994). *Romantic longings: Love in America, 1830–1980.*

Cant, S., & Sharma, U. (1998). Reflexivity, ethnography and the professions (complementary medicine): Watching you watching me watching you (and writing about both of us). *Sociological Review, 46,* 244–264.

Cantril, H., with Gaudet, H., & Herzog, J. (1982/1940). *The invasion from Mars.* Princeton, NJ: Princeton University Press.

Caplan, A. L. (1978). *The sociobiology debate: Readings on ethical and scientific issues.* New York: Harper.

Caplow, T. (1969). *Two against one: Coalition in triads.* Englewood Cliffs, NJ: Prentice Hall.

Caplow, T., Bahr, H. M., Chadwick, B. A., Hill, R., & Williamson, M. H. (1983). *Middletown families: Fifty years of change and continuity.* New York: Bantam.

Caplow, T., Bahr, H. M., Modell, J., & Chadwick, B. A. (1991). *Recent social trends in the United States 1960–1990.* Montreal: McGill-Queen's University Press.

Carnoy, M., Castells, M., & Benner, C. (1997). Labour markets and employment practices in the age of flexibility: A case study of Silicon Valley. *International Labour Review, 136,* 27–48.

Carter, L. J. (1975). *The Florida experience.* Baltimore: Johns Hopkins University Press.

Carter, L. J., & Pigford, T. H. (1998). Getting Yucca Mountain right. *Bulletin of the Atomic Scientists, 54,* 56–62.

Casper, L. M., McLanahan, S. S., & Garfinkel, I. (1994). The gender-poverty gap: What can we learn from other countries? *American Sociological Review, 59,* 594–605.

Cassidy, J. (1995, October 16). Who killed the middle class? *New Yorker,* pp. 113–124.

Castells, M., Flecha, R., Freire, P., Giroux, H. A., Macedo, D., & Willis, P. (1998). *Critical education in the new information age.* Totowa, NJ: Rowman and Littlefield.

Cecco, J. P., & Parker, D. A. (1995). The biology of homosexuality: Sexual orientation or sexual preference? *Journal of Homosexuality, 28,* 1–27.

Chambliss, D. F. (1996). *Beyond caring: Hospitals, nurses, and the social organization of ethics.* Chicago: University of Chicago Press.

Chambliss, W. J. (1973, December). The Saints and the Roughnecks. *Society,* pp. 23–31.

Chaudhary, V. (1998, March 13). Bullying of gays "rife in schools." *Guardian,* p. 8.

Chenault, C. (1996). *Maasai.* Unpublished manuscript, Calhoun Community College, Decatur, AL.

Cherlin, A. J. (1981). *Marriage, divorce, remarriage.* Cambridge, MA: Harvard University Press.

Cherlin, A. J. (1996). *Public and private families: An introduction.* New York: McGraw-Hill.

Chernin, K. (1981). *The obsession: Reflections on the tyranny of slenderness.* New York: Harper.

Chernow, R. (1998). *Titan: The life of J. D. Rockefeller Sr.* New York: Random House.

Chirot, D. (1986). *Social change in the modern era.* San Diego: Harcourt Brace Jovanovich.

Chirot, D. (1994a). *How societies change.* Thousand Oaks, CA: Pine Forge.

Chirot, D. (1994b). *Modern tyrants: The power and prevalence of evil in our age.* New York: Free Press.

Chodorow, N. (1978). *The reproduction of mothering: Psychoanalysis and the sociology of gender.* Berkeley: University of California Press.

Choldin, H. M. (1997). How sampling will help defeat the undercount. *Society, 34,* 26–30

Chomsky, N. (1965). *Aspects of the theory of syntax.* Cambridge, MA: MIT Press.

Chwast, J. (1965). Value conflicts in law enforcement. *Crime and Delinquency, 2,* 151–161.

Clairborne, W. (1998, July 24). Farmers face harvest of despair. *Washington Post,* p. 4.

Clark, W. C. (1989, September). Managing planet Earth. *Scientific American,* pp. 46–54.

Clausen, J. A. (1968). *Socialization and society.* Boston: Little, Brown.

Clay, G. (1980). *Close-up: How to read the American city.* Chicago: University of Chicago Press.

Clay, G. (1994). *Real places: An unconventional guide to America's landscape.* Chicago: University of Chicago Press.

Cloward, R., & Ohlin, L. (1960). *Delinquency and opportunity: A theory of delinquent gangs.* New York: Free Press.

Cochran, M., Larner, M., Riley, D., Gunnarson, L., & Henderson, C. R., Jr. (1990). *Extending families: The social networks of parents and their children.* Cambridge, England: Cambridge University Press.

Cockerham, W. C. (1998). *Medical sociology* (7th ed.). Upper Saddle River, NJ: Prentice Hall.

Cohen, J. E. (1998, October 8). How many people can the earth support? *New York Review of Books,* pp. 29–31.

Cohen, N. (1961). *The pursuit of the millennium.* New York: Harper.

Cole, J. (1995). *The UCLA television monitoring project.* Los Angeles: UCK+LA Center for Communications Policy.

Coleman, J. S. (1961). *The adolescent society.* New York: Free Press.

Coleman, J. S. (1964). Research chronicle: The adolescent society. In P. E. Hammond (Ed.), *Sociologists at work.* Garden City, NY: Doubleday.

Coleman, J. S. (1976). Liberty and equality in school desegregation. *School Policy, 6,* 9–13.

Coleman, J. S. (1982). *The asymmetric society.* Syracuse, NY: Syracuse University Press.

Coleman, J. S. (1986, May). Social theory, social research, and a theory of action. *American Journal of Sociology, 91,* 1309–1335.

Collins, E. G. C., & Scott, P. (Eds.) (1978). Everyone who makes it has a mentor. *Harvard Business Review, 56,* 89–101.

Comer, J. P. (1997). *Waiting for a Miracle: Why schools can't solve our problems and how we can.* New York: Dutton.

Comte, A. (1971/1854). The positive philosophy. In M. Truzzi (Ed.), *Sociology: The classic statements.* New York: Random House.

Conger, D., Ge, X., & Elder, G. H., Jr. (1994). Economic stress, coercive family process, and developmental problems of adolescence. *Child Development, 65,* 541–561.

Connell, R. W. (1995). *Masculinities.* Berkeley: University of California Press.

Cookson, P. W. (1997). New kid on the block? A closer look at America's private schools. *Brookings Review, 15,* 22–25.

Cookson, P. W., & Persell, C. H. (1985). *Preparing for power: America's elite boarding schools.* New York: Basic Books.

Cooley, C. H. (1909). *Social organization: A study of the large mind.* New York: Scribner.

Cooley, C. H. (1956/1902). Human nature and the social order. New York: Free Press.
Coon, C. (1962). *The origin of races.* New York: Knopf.
Corsaro, W. (1997). The sociology of childhood. Thousand Oaks, CA: Pine Forge.
Cortes, C. E. (Ed.) (1980). *The Latin American brain drain.* New York: Arno.
Coser, L. A. (1966). *The functions of social conflict.* New York: Free Press.
Cottrell, W. F. (1951). Death by dieselization: A case study in the reaction to technological change. *American Sociological Review, 16,* 358–365.
Cowie, J. (1998). The terror of the machine: Technology, work, gender and ecology on the U.S.-Mexico border. *Labor History, 39,* 502.
Cramer, E., & Boyd, J. (1995). The tenure track and the parent track: A road guide. *Wilson Library Bulletin, 69,* 41–42.
Cranz, G. (1982). *The politics of park design: A history of urban parks in America.* Cambridge, MA: MIT Press.
Creel, M. W. (1988). *"A peculiar people": Slave religion and community-culture among the Gullahs.* New York: New York University Press.
Cremin, L. A. (1980). *American education: The national experience 1783–1876.* New York: Harper.
Cressey, D. R. (1971/1953). *Other people's money: A study in the social psychology of embezzlement.* Belmont, CA: Wadsworth.
Cronon, W. (1983). *Changes in the land: Indians, colonists, and the ecology of New England.* New York: Hill and Wang.
Crowder, M. (1966). *A short history of Nigeria* (rev. ed.). New York: Praeger.
Cuber, J. F., & Haroff, P. B. (1980). Five types of marriage. In A. Skolnick & J. Skolnick (Eds.), *Family in transition.* Boston: Little, Brown.
Cumming, E., & Henry, W. (1971). *Growing old: The process of disengagement.* New York: Basic Books.
Cummings, J. (1983, November 20). Breakup of black family imperils gains of decades. *New York Times,* p. 56.
Curtin, P. D. (1969). *The Atlantic slave trade.* Madison: University of Wisconsin Press.
Curtiss, S. (1977). *Genie: A psycholinguistic study of a modern-day "wild child."* New York: Academic Press.

Dahl, R. (1961). *Who governs?* New Haven, CT: Yale University Press.
Dahrendorf, R. (1990). *The modern social conflict: An essay on the politics of liberty.* Berkeley: University of California Press.
Dahrendorf, R. (1997). *After 1989: Morale, revolution, and civil society.* New York: St. Martin's Press.
Daniel, K. (1996). The marriage premium. In Tommasi, M., & Tommasi, K. (Eds.), *The new economics of human behavior.* Cambridge, England: Cambridge University Press.
Dankert, C. E., Mann, F. C., & Northrup, H. R. (Eds.) (1965). *Hours of work.* New York: Harper.
Danziger, S., & Gottschalk, P. (Eds.) (1993). *Uneven tides: Rising inequality in America.* New York: Russell Sage.
DARE. (1999, March 8). Letter from Lagos. *The Nation,* p. 8.
Darling-Hammond, L. (1997, November). What matters most: 21st-century teaching. *Education Digest,* pp. 4–10.
Das, S., & Srinivasan, K. (1997). Duration of firms in an infant industry: The case of Indian computer hardware. *Journal of Development Economics, 53,* 157–167.
Davey, J. D. (1995). *The new social contract: America's journey from welfare state to police state.* Westport, CT: Praeger.
Davidson, D. (1973). The furious passage of the black graduate student. In J. Ladner (Ed.), *The death of white sociology.* New York: Vintage.
Davis, K. (1939). Illegitimacy and the social structures. *American Journal of Sociology, 45,* 215–233.
Davis, K. (1947). Final note on a case of extreme isolation. *American Journal of Sociology, 52,* 432–437.
Davis, K. (1955). The origin and growth of urbanization in the world. *American Journal of Sociology, 60,* 429–437.
Davis, K. (1968). The urbanization of the human population. In S. F. Fava (Ed.), *Urbanism in world perspective: A reader.* New York: Crowell.
Davis, K. (1991). Population and resources: Fact and interpretation. In K. Davis and M. S. Bernstam (Eds.), *Resources, environment, population: Present knowledge, future options.* New York: Oxford University Press.
Davis, K., & Moore, W. E. (1945). Some principles of stratification. *American Sociological Review, 10,* 242–249.
Davis, K. C. (1995, September 3). Ethnic cleansing didn't start in Bosnia. *New York Times,* sec. 4, pp. 1, 6.
Davis, L. (1992). *Natural disasters: The black plague.* New York: Facts on File.
Davis, M. (1998). *Ecology of fear: Los Angeles and the imagination of disaster.* New York: Metropolitan Books.
Davis, N. J. (1975). *Deviance: Perspectives and issues in the field.* Dubuque, IA: Brown.
Dearing, J. W., & Rogers, E. M. (1996). *Agenda setting.* Thousand Oaks, CA: Sage.
Deaux, K., & Wrightsman, L. S. (1988). *Social psychology* (5th ed.). Pacific Grove, CA: Brooks/Cole.
Defina, M. P., & Wetherbee, L. (1997, October). Advocacy and law enforcement: Partners against domestic violence. *FBI Law Enforcement Bulletin,* pp. 22–26.
Deierlein, K. (1994). Ideology and holistic alternatives. In W. Kornblum & C. D. Smith (Eds.), *The healing experience: Readings on the social context of health care.* Englewood Cliffs, NJ: Prentice Hall.
D'Emilio, J., & Freedman, E. B. (1988). *Intimate matters: A history of sexuality in America.* New York: Harper.
Denis, J., & Derbyshire, I. (1996). *Political systems of the world.* New York: St. Martin's Press.
Denno, D. W. (1990). *Biology and violence: From birth to adulthood.* New York: Cambridge University Press.
de Rougemont, D. (1983). *Love in the western world.* Princeton, NJ: Princeton University Press.
de Tocqueville, A. (1955/1856). *The old regime and the French revolution* (S. Gilbert, Trans.). Garden City, NY: Doubleday.
de Tocqueville, A. (1956/1840). *Democracy in America.* New York: Vintage.
de Tocqueville, A. (1980/1835). *On democracy, revolution, and society* (J. Stone & S. Mennell, Trans.). Chicago: University of Chicago Press.

Dewey, R. (1948). Charles Horton Cooley: Pioneer in psychosociology. In H. F. Barnes (Ed.), *An introduction to the history of sociology.* Chicago: University of Chicago Press.

De Witt, K. (1994, August 15). Wave of suburban growth is being fed by minorities. *New York Times,* pp. A1, B6.

Diaz Soto, L. (1997). *Language, culture, and power: Bilingual families and the struggle for quality education.* Albany: State University of New York Press.

Djilas, M. (1982). *The new class: An analysis of the communist system.* Orlando: Harcourt Brace Jovanovich.

Dobzhansky, T. (1962). *Mankind evolving.* New Haven, CT: Yale University Press.

Dodgson, R. A. (1987). *The European past: Social evolution and spatial order.* London: Macmillan.

Dollard, J. (1937). *Caste and class in a southern town.* New Haven, CT: Yale University Press.

Dollard, J., Miller, N., & Doob, L. (1939). *Frustration and aggression.* New Haven, CT: Yale University Press.

Domhoff, G. W. (1978). *The powers that be.* New York: Random House.

Domhoff, G. W. (1983). *Who rules America now?* New York: Simon & Schuster.

Donow, K. R. (1990, April). On the transition from the academy to a career in business. *Footnotes* (American Sociological Association), p. 6.

Dornbusch, S. (1955). The military as an assimilating institution. *Social Forces, 33,* 316–321.

Dowd, M. (1985, November 17). Youth, art, hype: A different bohemia. *New York Times Magazine,* pp. 26ff.

Drake, S. C., & Cayton, H. (1970/1945). *Black metropolis: Vol. 1. A study of Negro life in a northern city* (rev. ed.). Orlando: Harcourt Brace Jovanovich.

Dreifus, C. (1998, April 14). She talks to apes, and, according to her, they talk back. *New York Times,* p. F4.

Drennan, M. P., Tobier, E., & Lewis, J. (1996). The interruption of income convergence and income growth in large cities in the 1980s. *Urban Studies, 33,* 63–83.

Drucker, P. F. (1992). The new society of organizations. *Harvard Business Review, 70,* 95–105.

Du Bois, W. E. B. (1967/1899). *The Philadelphia Negro: A social study.* New York: Schocken.

Duncan, G. J., & Brooks-Gunn, J. (Eds.) (1997). *Consequences of growing up poor.* New York: Russell Sage.

Duneier, M. (1992). *Slim's table: Race, respectability, and masculinity.* Chicago: University of Chicago Press.

Durkheim, E. (1964/1893). *The division of labor in society* (2nd ed.). New York: Free Press.

Durkheim, E. (1951/1897). *Suicide, a study in sociology.* New York: Free Press.

Dyer, J. (1996, November–December). Ground zero. *Utne Reader,* pp. 80ff.

Eckholm, E. (1985, June 25). Kanzi the chimp: A life in science. *New York Times,* pp. C1, C3.

Eckholm, E. (1992, October 6). Learning if infants are hurt when mothers go to work. *New York Times,* pp. A1, A21.

Edleson, J. I., & Tan, N. T. (1993). Conflict and family violence: The tale of two families. In P. G. Boss, W. J. Doherty, R. LaRossa, W. R. Schumm, & S. K. Steinmetz (Eds.), *Sourcebook of family theories and methods: A contextual approach.* New York: Plenum.

Egan, T. (1995, September 3). Many seek security in private communities. *New York Times,* pp. A1, A22.

Ehrlich, P. R., & Ehrlich, A. H. (1997). Erlichs' fables. *Technology Review, 100,* 38–48.

Eibl-Eibesfeldt, I. (1989). *Human ethology.* Hawthorne, NY: Aldine.

Eisenberg, S. (1998). *We'll call if we need you: Experiences of women working in construction.* Ithaca, NY: Cornell University Press.

Eisley, L. (1961). *Darwin's century.* Garden City, NY: Doubleday Anchor.

Eisley, L. (1970). *The invisible pyramid.* New York: Scribner.

Elder, G. H. (1981). History and the life course. In D. Berteaux (Ed.), *Biography and society: The life history approach to the social sciences.* Newbury Park, CA: Sage.

Eldridge, C. C. (1978). *Victorian imperialism.* Atlantic Highlands, NJ: Humanities Press.

Elias, N. (1978/1939). The civilizing process. In N. Elias (Ed.), *The development of manners.* New York: Urizen.

Elkin, F., & Handel, G. (1989). *The child and society: The process of socialization* (5th ed.). New York: Random House.

Elkind, D. (1970). *Children and adolescents: Interpretative essays on Jean Piaget.* New York: Oxford University Press.

Ellul, J. (1964). *The technological society.* New York: Vintage.

Ellwood, D. (1988). *Poor support: Poverty in the American family.* New York: Basic Books.

Elshtain, J. B. (1997, May 5). Heaven can wait. *New Republic,* p. 23.

Elster, J. (1997). Accounting for tastes. *University of Chicago Law Review, 64,* 749–764.

Epstein, C. (1995, Fall). Affirmative action. *Dissent,* pp. 463–465.

Epstein, C. F. (1985). Ideal roles and real roles. *Research in Social Stratification and Mobility, 4,* 29–51.

Epstein, C. F. (1988). *Deceptive distinctions: Sex, gender, and the social order.* New Haven, CT: Yale University Press.

Epstein, C. F. (1995, Fall). Affirmative action. *Dissent,* pp. 463–466.

Erikson, E. (1963). *Childhood and society.* New York: Norton.

Erikson, E. (Ed.) (1965). *The challenge of youth.* Garden City, NY: Doubleday Anchor.

Erikson, K. T. (1962). Notes on the sociology of deviance. *Social Problems, 9,* 307–314.

Erikson, K. T. (1966). *Wayward puritans: A study in the sociology of deviance.* New York: Wiley.

Erikson, K. T. (1976). *Everything in its path.* New York: Simon & Schuster.

Erikson, K. T. (1994a). *A new species of trouble.* New York: Norton.

Erikson, K. T. (1994b, March 6). Out of sight, out of our minds. *New York Times Magazine,* pp. 34ff.

Erikson, K. T. (1998). Trauma at Buffalo Creek. *Society, 35,* 153–162.

Etzkowitz, H., & Glassman, R. (1991). *The renaissance of sociological theory: Classical and contemporary.* Itaska, IL: Peacock.

Ezorsky, G. (1991). *Affirmative action.* Ithaca, NY: Cornell University Press.

Faison, S. (1997, August 17). Chinese happily break the "one child" rule. *New York Times,* pp. 1, 10.

Fallows, D. (1985). *A mother's work.* Boston: Houghton Mifflin.

Fallows, J. (1996, May). They only look dead: Why progressives will dominate the next political era. *Washington Monthly,* pp. 48–50.

Farley, R. (1996). *The new American reality.* New York: Russell Sage.

Farley, R., & Allen, W. R. (1987). *The color line and the quality of life in America.* New York: Russell Sage.

Farnsworth, C. H. (1994, August 15). With her songs, Eskimo bares her people's pain. *New York Times,* p. A4.

Farrington, K., & Chertok, E. (1993). Social conflict theories of the family. In P. G. Boss, W. J. Doherty, R. LaRossa, W. R. Schumm, & S. K. Steinmetz (Eds.), *Sourcebook of family theories and methods: A contextual approach.* New York: Plenum.

Faulkner, R. R. (1997). Karl Polanyi meets the masters of the universe. *Contemporary Sociology, 26,* 688–692.

Fava, S. (1956). Suburbanism as a way of life. *American Sociological Review, 21,* 34–38.

Fava, S. (1985). Residential preferences in the suburban era: A new look? *Sociological Focus, 18,* 109–117.

Feachem, R. G. A., et al. (Eds.) (1993). *The health of adults in the developing world.* New York: Oxford University Press.

Feagin, J. R. (1991, November 27). Blacks still face the malevolent reality of white racism. *Chronicle of Higher Education,* p. A44.

Felson, M. (1998). *Crime and everyday life.* Thousand Oaks, CA: Pine Forge.

Fernandes, F. (1968). The weight of the past. In J. H. Franklin (Ed.), *Color and race.* Boston: Beacon.

Fine, G. A. (1987). *With the boys: Little League baseball and preadolescent culture.* Chicago: University of Chicago Press.

Finke, R., & Stark, R. (1992). *The churching of America 1776–1992.* New Brunswick, NJ: Rutgers University Press.

Fischer, C. S. (1976, 1987). *The urban experience.* Orlando: Harcourt Brace Jovanovich.

Fischer, C. S. (1982). *To dwell among friends: Personal networks in town and city.* Chicago: University of Chicago Press.

Fishman, R. (1987). *Bourgeois utopias: The rise and fall of suburbia.* New York: Basic Books.

Fiske, M., & Chiriboga, D. A. (1990). *Change and continuity in adult life.* San Francisco: Jossey-Bass.

Flink, J. J. (1975). *The car culture.* Cambridge, MA: MIT Press.

Flink, J. J. (1988). *The automobile age.* Cambridge, MA: MIT Press.

Flores, J. (1988, Fall). Rappin', writin', and breakin': Black and Puerto Rican street culture in New York. In *In search of New York* [Special issue]. *Dissent,* pp. 580–585.

Fontana, B. (1993). *Hegemony and power.* Minneapolis: University of Minnesota Press.

Foster, J. (1990). *Villains: Crime and community in the inner city.* London: Routledge.

Foster, L., Martin, J. W., Chidester, D., & Ammerman, N. T. (1998). Forum: Interpreting Waco. *Religion and American Culture, 8,* 1–30.

Foucault, M. (1973). *Madness and civilization: A history of insanity in the Age of Reason.* New York: Vintage.

Foucault, M. (1984). *The Foucault reader.* New York: Pantheon.

Frank, A. G. (1966). The development of underdevelopment. *Monthly Review, 18,* 3–17.

Frank, R. H. (1988). *Passions within reason: The strategic role of the emotions.* New York: Norton.

Frazier, E. F. (1957). *The Negro in the United States* (rev. ed.). New York: Macmillan.

Frazier, E. F. (1966). *The Negro church in America.* New York: Schocken.

Frazier, K. (1998, May–June). Cold fusion saga ends at its University of Utah birthplace. *Skeptical Inquirer,* pp. 8–9.

Fredman, L. (1995). Bullied to death in Japan. *World Press Review, 42,* 25.

Freeman, J. (1973). The origins of the women's liberation movement. In J. Huber (Ed.), *Changing women in a changing society.* Chicago: University of Chicago Press.

Freeman, R. (1994). *Working under different rules.* New York: Russell Sage.

Freeman, R. B. (1993). How much has de-unionization contributed to the rise in male earnings inequality? In S. Danziger & P. Gottschalk (Eds.), *Uneven tides: Rising inequality in America.* New York: Russell Sage.

Fried, M. (1963). Grieving for a lost home. In L. J. Duhl (Ed.), *The urban condition.* New York: Basic Books.

Friedan, B. (1963). *The feminine mystique.* New York: Dell.

Friedman, M. (1962). *Capitalism and freedom.* Chicago: University of Chicago Press.

Fromm, E. (1961). *Marx's concept of man.* New York: Ungar.

Fullan, M. (1993). *Change forces: Probing the depths of educational reform.* New York: Falmer.

Fuller, B., & Elmore, R. F. (1996). *Who chooses? Who loses? Culture, institutions, and the unequal effects of school choice.* New York: Teachers College Press.

Furedi, F. (1997). *Population and development: A critical introduction.* New York: St. Martin's Press.

Furstenberg, F. (1976). *Unplanned parenthood: The social consequences of unplanned childbearing.* New York: Free Press.

Furstenberg, F. F., & Cherlin, A. J. (1991). *Divided families: What happens to children when parents part.* Cambridge, MA: Harvard University Press.

Fyfe, C. (1976). The dynamics of African dispersal. In M. L. Kilson & R. I. Rothberg (Eds.), *The African diaspora.* Cambridge, MA: Harvard University Press.

Gabriel, T. (1995a, February 12). A generation's heritage: After the boom, a boomlet. *New York Times,* pp. 1, 34.

Gabriel, T. (1995b, June 12). A new generation seems ready to give bisexuality a place in the spectrum, *New York Times,* p. A12.

Gaines, D. (1998). Teenage wasteland (2nd ed.). Chicago: University of Chicago Press.

Galbraith, J. K. (1995, September 19). Blame history, not the liberals. *New York Times,* p. A21.

Gallup Organization. (1997, March). *The Gallup Poll Monthly,* pp. 8–13.

Galtung, J. (1985). War. In A. Kuper & J. Kuper (Eds.), *The social science encyclopedia.* London: Routledge and Kegan Paul.

Gamson, J. (1994). *Claims to fame: Celebrity in America.* Berkeley: University of California Press.

Gamson, W. A. (1992). *Talking politics.* Cambridge, England: Cambridge University Press.

Gans, H. (1962, 1984). *The urban villagers.* New York: Free Press.

Gans, H. (1967, 1976). *The Levittowners: Way of life and politics in a new suburban community.* New York: Pantheon.

Gans, H. (1979). *Deciding what's news.* New York: Pantheon.

Gans, H. (1985). The uses of poverty: The poor pay all. In W. Feigelman (Ed.), *Sociology: Full circle* (4th ed.). Fort Worth: Holt, Rinehart and Winston.

Gans, H. (1995). *The war against the poor.* New York: Basic Books.

Garcia, A. (1996). Moral reasoning in interactional context: Strategic uses of care and justice arguments in mediation hearings. *Sociological Inquiry, 66,* 197–214.

Gardner, H. (1983). *Frames of mind: The theory of multiple intelligences.* New York: Basic Books.

Garfinkel, H. (1967). *Studies in ethnomethodology.* Englewood Cliffs, NJ: Prentice Hall.

Garfinkel, I., & McLanahan, S. S. (1986). *Single mothers and their children: A new American dilemma.* Washington, DC: Urban Institute.

Garmezy, N. (1993). Children in poverty: Resilience despite risk. *Psychiatry, 56,* 127–136.

Garner, R., & Tenuto, J. (1997). *Social movement theory and research: An annotated bibliographical guide.* Lanham, MD: Scarecrow Press.

Garreau, J. (1991). *Edge city: Life on the new frontier.* Garden City, NY: Doubleday Anchor.

Garrett, M. (1997). The effects of infant child care on infant-mother attachment security: Results of the NICHD Study of Early Child Care. *Child Development, 68,* 860–879.

Geertz, C. (1973). The growth of culture and the evolution of mind. In C. Geertz, *The interpretation of culture.* New York: Basic Books.

Genevie, L. E. (Ed.) (1978). *Collective behavior and social movements.* Itasca, IL: Peacock.

Geoghegan, T. (1991). *Which side are you on?* New York: Farrar Straus Giroux.

Gerbner, G. (1990). *Violence profile.* Philadelphia: Annenberg School of Communications.

Gerth, H., & Mills, C. W. (1958). *From Max Weber: Essays in sociology.* New York: Oxford University Press.

Geschwender, J. A. (1977). *Class, race and worker insurgency.* New York: Cambridge University Press.

Gibbs, J. P. (1994). *A theory about control.* Boulder, CO: Westview.

Giddens, A. (1984). *The constitution of society.* Berkeley: University of California Press.

Gilbert, J. B. (1997). *Redeeming culture: Religion in an age of science.* Chicago: University of Chicago Press.

Gilder, G. (1982). *Wealth and poverty.* New York: Basic Books.

Gilligan, C. (1982). *In a different voice: Psychological theory and women's development.* Cambridge, MA: Harvard University Press.

Gilligan, C. (1987). Adolescent development reconsidered. In C. E. Irwin, Jr. (Ed.), *New directions for child development: No. 37. Adolescent social behavior and health.* San Francisco: Jossey-Bass.

Gilligan, C. (1988). Adolescent development revisited. In Gilligan, C., Ward, J. V., Taylor, J. M., & Bardige, B. (Eds.), *Mapping the moral domain.* Cambridge, MA: Harvard University Press.

Gilligan, C., Ward, J. V., Taylor, J. M., & Bardige, B. (Eds.) (1998). *Mapping the moral domain.* Cambridge, MA: Harvard University Press.

Gist, N. P. (1968). Urbanism in India. In S. F. Fava (Ed.), *Urbanism in world perspective: A reader.* New York: Crowell.

Gitlin, T. (1992, March–April). Following the money: Where does your advertising dollar go? *The Democratic Left,* pp. 3–5.

Gitlin, T. (1993). *The sixties.* New York: Bantam.

Glausiusz, J. (1997, August). The ecology of language. *Discover,* p. 30.

Glazer, N. (1975). *Affirmative discrimination: Ethnic inequality and public policy.* New York: Basic Books.

Glazer, N., & Moynihan, D. P. (1970). *Beyond the melting pot: The Negroes, Puerto Ricans, Jews, Italians, and Irish of New York City* (2nd rev. ed.). Cambridge, MA: MIT Press.

Glick, P. C., & Parke, R., Jr. (1965). New approaches in studying the life cycle of the family. *Demography, 2,* 187–202.

Glueck, S., & Glueck, E. T. (1950). *Unraveling juvenile delinquency.* New York: Commonwealth Fund.

Goffman, E. (1958). Deference and demeanor. *American Anthropologist, 58,* 488–489.

Goffman, E. (1959). *The presentation of self in everyday life.* Garden City, NY: Doubleday.

Goffman, E. (1961). *Asylums.* Garden City, NY: Doubleday.

Goffman, E. (1963). *Stigma: Notes on the management of spoiled identity.* Englewood Cliffs, NJ: Prentice Hall.

Goffman, E. (1965). *Interaction ritual: Essays on face-to-face behavior.* Garden City, NY: Doubleday.

Goffman, E. (1972). Territories of the self. In E. Goffman (Ed.), *Relations in public.* New York: Harper.

Goldhamer, H., & Shils, E. (1939). Types of power and status. *American Journal of Sociology, 45,* 171–182.

Goleman, D. (1985a, March 19). Dislike of own body found common among women. *New York Times,* pp. C1, C5.

Goleman, D. (1985b, September 10). Patterns of love charted in studies. *New York Times,* pp. C1, C5.

Gonzalez-Mena, J. (1998). *The child in the family and the community.* Upper Saddle River, NJ: Merrill.

Goodall, J. V. L. (1968). A preliminary report on expressive movements and communications in Gombe Stream chimpanzees. In P. Jay (Ed.), *Primates: Studies in adaptation and variability.* Fort Worth: Holt, Rinehart and Winston.

Goodall, J. (1994). Digging up the roots: Relationships with the Gombe chimpanzees. *Orion, 13,* 21–21.

Goode, E. (1992). *Collective behavior.* Fort Worth: Harcourt Brace.

Goode, E. (1994). *Deviant behavior* (4th ed.). Englewood Cliffs, NJ: Prentice Hall.

Goode, W. J. (1964). *The family.* Englewood Cliffs, NJ: Prentice Hall.

Goode, W. J. (1971, November). World revolution and family patterns. *Journal of Marriage and the Family,* pp. 624–635.

Goodman-Draper, J. (1995). Health care's forgotten majority: Nurses and their frayed white collars. Westport, CT: Auburn House.

Gordon, M. (1964). *Assimilation in American life.* New York: Oxford University Press.

Goslin, D. A. (1965). *The school in contemporary society.* Glenview, IL: Scott, Foresman.

Gottmann, J. (1978). Megalopolitan systems around the world. In L. S. Bourne & J. W. Simmons (Eds.), *Systems of cities: Readings on structure, growth, and policy.* New York: Oxford University Press.

Gould, S. J. (1981). *The mismeasure of man.* New York: Norton.

Gould, S. J. (1995, February). Ghosts of bell curves past. *Natural History,* pp. 12–19.

Gould, S. J. (1996). *Full house: The spread of excellence from Plato to Darwin.* New York: Harmony Books.

Gramlich, E. M., Kasten, R., & Sammartino, F. (1993). Growing inequality in the 1980s: The role of federal taxes and cash transfers. In S. Danziger & P. Gottschalk (Eds.), *Uneven tides: Rising inequality in America.* New York: Russell Sage.

Gramling, R., & Krogman, N. (1997). Communities, policy and chronic technological disasters. *Current Sociology, 45,* 41–57.

Gramsci, A. (1971). *Selections from prison notebooks.* London: Routledge and Kegan Paul.

Gramsci, A. (1992/1965). *Prison notebooks.* New York: Columbia University Press.

Gramsci, A. (1995). *Further selections from the prison notebooks.* Minneapolis: University of Minnesota Press.

Granovetter, M. (1974, 1995). *Getting a job: A study of contacts and careers.* Chicago: University of Chicago Press.

Gray, F. D. (1990). *Soviet women: Walking the tightrope.* Garden City, NY: Doubleday.

Greeley, A., Michael, R. T., & Smith, T. (1990). Americans and their sexual partners. *Society, 27,* 36–42.

Grodin, D., & Lindlof, T. R. (Eds.) (1996). Constructing the self in a mediated world. Thousand Oaks, CA: Sage.

Gross, J. (1992, March 29). Collapse of inner-city families creates America's new orphans. *New York Times,* pp. 1, 20.

Gugler, J. (Ed.) (1997). *Cities in the developing world: Issues, theory, and policy.* New York: Oxford University Press.

Gursky, D. (1998). Class size does matter. *Education Digest, 64,* 15–19.

Gusfield, J. (1966). *Symbolic crusade: Status politics and the American temperance movement.* Urbana: University of Illinois Press.

Gusfield, J. (1981). *The culture of public problems: Drinking, driving and the symbolic order.* Chicago: University of Chicago Press.

Gusfield, J. (1996). *Contested meanings: The construction of alcohol problems.* Madison: University of Wisconsin Press.

Gutman, H. (1976). *The black family in slavery.* New York: Pantheon.

Haber, L. D., & Dowd, J. E. (1994, January). *A human development agenda for disability: Statistical considerations* (Working Paper for the Statistical Division of the United Nations Secretariat). New York: United Nations.

Hacker, A. (1995, Fall). Affirmative action. *Dissent,* pp. 465–467.

Hacker, A. (1997). *Money: Who has how much and why.* New York: Simon & Schuster.

Hackett, D. G. (Ed.) (1995). *Religion and American culture: A reader.* New York: Routledge.

Hadaway, C. K., & Marler, P. L. (1998, May 6). Did you really go to church this week? Behind the poll data. *The Christian Century,* pp. 472–476.

Hagestad, G. O., & Neugarten, B. L. (1985). Age and the life course. In R. H. Binstock & E. Shanas (Eds.), *Handbook of aging and the social sciences* (2nd ed.). New York: Van Nostrand Reinhold.

Halevy, E. (1955). *The growth of philosophic radicalism* (M. Morris, Trans.). Boston: Beacon.

Hall, J. R. (1987). *Gone from the promised land: Jonestown in American cultural history.* New Brunswick, NJ: Transaction.

Halle, D. (1984). *America's working man: Work, home and politics among blue-collar property owners.* Chicago: University of Chicago Press.

Halsey, M. (1946). *Color blind.* New York: Simon & Schuster.

Hamburg, D. (1992). *Today's children.* New York: New York Times Books.

Hamilton, C. V. (1969). The politics of race relations. In C. U. Daley (Ed.), *The minority report.* New York: Pantheon.

Handlin, O. (1992). The newcomers. In P. S. Rotherberg (Ed.), *Race, class and gender in the United States.* New York: St. Martin's Press.

Hanna, S. L. (1994). *Person to person: Positive relationships don't just happen.* Englewood Cliffs, NJ: Prentice Hall.

Harbison, F. H. (1973). *Human resources as the wealth of nations.* New York: Oxford University Press.

Hare, A. P. (1992). *Groups, teams, and social interaction.* Westport, CT: Praeger.

Hare, A. P., Blumberg, H. H., Davis, M. F., & Kent, V. (1994). *Small group research: A handbook.* Norwood, NJ: Ablex.

Harlow, H. F. (1986). *From learning to love: The selected papers of H. F. Harlow.* New York: Praeger.

Harlow, H. F., & Harlow, M. K. (1962, November). Social deprivation in monkeys. *Scientific American,* pp. 137–147.

Harper, C. L. (1993). *Exploring social change* (2nd ed.). Englewood Cliffs, NJ: Prentice Hall.

Harper, D. A. (1982). *Good company.* Chicago: University of Chicago Press.

Harrington, M. (1973). *Socialism.* New York: Bantam.

Harrington, M. (1987). *The next lift: The history of a future.* New York: Holt.

Harris, M. (1980). *Culture, people, nature: An introduction to general anthropology.* New York: Harper.

Harris, M. (1983). *Cultural anthropology.* New York: Harper.

Harrison, B. (1994). When government gets it right. *Technology Review, 97,* 66.

Harrison, B. B., Tilly, C., and Bluestone, B. (1986, March–April). Wage inequality takes a great U-turn. *Challenge,* pp. 26–32.

Haub, C. (1997). World population rises to 5.840 billion in 1997. *Population Today,* 25.

Haub, C., & Pollard, K. (Eds.) (1998, April). Speaking graphically. *Population Today,* p. 6.

Hauser, P. M. (1957). The changing population pattern of the modern city. In P. K. Hatt & A. J. Reiss (Eds.), *Cities and society.* New York: Free Press.

Hawkins, G. (1976). *The prison: Policy and practice.* Chicago: University of Chicago Press.

Hechter, M. (1974). *Internal colonialism.* Berkeley: University of California Press.

Hechter, M. (1987). *Principles of group solidarity.* Berkeley: University of California Press.

Hechter, M., & Kanazawa, S. (1996, March 29). Sociological rational choice theory. In Elster, J. (Ed.), Doing our level best (rational-choice theory—a symposium). *Times Literary Supplement,* pp. 12–14.

Hechter, M., & Kanazawa, S. (1997). Sociological rational choice theory. *Annual Review of Sociology, 23,* 191–214.

Heilbroner, R. L. (1995). Putting economics in its place. *Social Research, 62,* 883–898.

Heitzeg, N. (1996). *Deviance: Rulemakers and rulebreakers.* Minneapolis: West.

Helmreich, W. B. (1982). *The things they say behind your back.* Garden City, NY: Doubleday.

Henry, D. O. (1989). *From foraging to agriculture.* Philadelphia: University of Pennsylvania Press.

Henslin, J., & Briggs, M. (1971). Dramaturgical desexualization: The sociology of the vaginal examination. In J. Henslin (Ed.), *Studies in the sociology of sex.* New York: Appleton-Century-Crofts.

Herbert, B. (1996, June 21). Burning their bridges. *New York Times,* p. A27.

Herlihy, D. (1995). Biology and history: The triumph of monogamy. *Journal of Interdisciplinary History, 25,* 571–583.

Herrnstein, R., & Murray, C. (1994). *The bell curve.* New York: Free Press.

Hess, D. (1997). *Science studies: An advanced introduction.* New York: New York University Press.

Hewes, G. W. (1954). A conspectus of the world's cultures in 1500 A.D. *University of Colorado Studies,* no. 4, pp. 1–22.

Heyman, S. J., & Earle, A. (1997, Summer). Working conditions faced by poor families and the case of children. *Focus* (Institute for Research on Poverty), pp. 56–60.

Hickins, M. (1998). Reconcilable differences. *Management Review, 87,* 54–59.

Hill, J. (1978). Apes and language. *Annual Review of Anthropology, 7,* 89–112.

Hinton, W. (1966). *Fanshen: A documentary of revolution in a Chinese village.* New York: Vintage.

Hirschi, T., & Gottfredson, M. R. (1980). *Understanding crime: Current theory and research.* Newbury Park, CA: Sage.

Hirschi, T., & Gottfredson, M. R. (Eds.) (1994). *The generality of deviance.* New Brunswick, NJ: Transaction.

Hobsbawm, E. (1975). *The age of capital.* New York: Scribner.

Hochschild, A. (1989). *The second shift: Working parents and the revolution at home.* New York: Viking.

Hochschild, A. (1997). *The time bind: When work becomes home and home becomes work.* New York: Henry Holt.

Hodge, R. W., & Treiman, D. J. (1968). Class identification in the United States. *American Journal of Sociology, 73,* 535–547.

Hoecker-Drysdale, S. (1992). *Harriet Martineau: First woman sociologist.* New York: Berg Publishers.

Hoffman, L. M. (1989). *The politics of knowledge: Activist movements in medicine and planning.* Albany: State University of New York Press.

Holden, C. (1996). New populations of old people. *Science, 273,* 46–48.

Holcombe, S. (1995). *Managing to empower: The Grameen Bank's experience of poverty alleviation.* London: Zed Books.

Hollingshead, A. (1949). *Elmtown's youth.* New York: Wiley.

Holmes, S. A. (1998, July 1). Birth rate falls to 40-year low among unwed black women. *New York Times,* pp. A1, A16.

Holt, T. C. (1980). Afro-Americans. In *Harvard encyclopedia of American ethnic groups.* Cambridge, MA: Belknap.

Holy, L. (1985). Groups. In A. Kuper & J. Kuper (Eds.), *The social science encyclopedia.* London: Routledge and Kegan Paul.

Homans, G. (1950). *The human group.* Orlando: Harcourt Brace Jovanovich.

Homans, G. (1951). The Western Electric researches. In S. D. Hoslett (Ed.), *Human factors in management.* New York: Harper.

Homans, G. (1961). *Social behavior: Its elementary forms.* Orlando: Harcourt Brace Jovanovich.

Hopkins, T. K., & Wallerstein, I. (Eds.) (1996). *The age of transition: Trajectory of the world-system 1945–2025.* London: Zed Books.

Horgan, J. (1997, May). Seeking a better way to die. *Scientific American,* pp. 100–105.

Horowitz, M. (1979). The jurisprudence of Brown and the dilemmas of liberalism. *Harvard Civil Rights–Civil Liberties Review, 14,* 599–610.

Horowitz, R. (1985). *Honor and the American dream: Culture and identity in a Chicago neighborhood.* New Brunswick, NJ: Rutgers University Press.

Horowitz, T. (1999, March 15). Run, Rudolph, run: How the fugitive becomes a folk hero. *New Yorker,* pp. 46–52.

House, J. S., Lepkowski, J. M., Kinny, A. M., Mero, R. P., Kessler, A., & Herzog, R. (1994). The social stratification of aging and health. *Journal of Health and Social Behavior, 35,* 213–234.

Hout, M., & Greeley, A. M. (1987). The center doesn't hold: Church attendance in the United States, 1940–1984. *American Sociological Review, 52,* 325–345.

Howe, I. (1983). *1984 revisited: Totalitarianism in our century.* New York: Harper.

Huang, C. (1994). *Immigration and the underclass.* Unpublished doctoral dissertation, City University of New York Graduate School.

Hubbard, J. (1991). *Shooting back: A photographic view of life by homeless children.* San Francisco: Chronicle Books.

Hughes, E. (1945). The dilemmas and contradictions of status. *American Journal of Sociology, 50,* 353–359.

Hughes, E. (1958). *Men and their work.* New York: Free Press.

Hughes, E. (1959). The study of occupations. In R. K. Merton, L. Broom, & L. S. Cottrell, Jr. (Eds.), *Sociology today: Problems and prospects.* New York: Basic Books.

Hummin, D. (1990). *Commonplaces: Community ideology and identity in American cities.* Albany: State University of New York Press.

Humphreys, L. (1970, 1975). *Tearoom trade: Impersonal sex in public places.* Hawthorne, NY: Aldine.

Hunsberger, B. (1995). Religion and prejudice: The role of religious fundamentalism, quest, and right-wing authoritarianism. *Journal of Social Issues, 51,* 113–130.

Hunt, M. (1985). *Profiles of social research: The scientific study of human interactions.* New York: Russell Sage.

Hunter, F. (1953). *Community power structure: A study of decision makers.* Chapel Hill: University of North Carolina Press.

Hunter, M. (1998, September 7). Labor's resurgence. *Maclean's,* p. 37.

Hussein, M. (1994, December). Behind the veil of fundamentalism. *UNESCO Courier,* pp. 25–28.

Hutchinson, R. (Ed.) (1997). *New directions in urban sociology.* Greenwich, CT: JAI Press.

Iannaccone, L., Stark, R., & Finke, R. (1998). Rationality and the "religious mind." *Economic Inquiry, 36,* 373–390.

Ianni, F. A. J. (1998). New Mafia: Black, Hispanic and Italian styles. *Society, 35,* 115–129.

Iglesia, R. (1990). *Cronistas e historiadores de la conquista de Mexico.* Mexico City: Biblioteca de la Ciudad de Mexico.

Jackman, M. R., & Jackman, R. W. (1983). *Class awareness in the United States.* Berkeley: University of California Press.

Jackson, B. (1972). *In the life: Versions of the criminal experience.* New York: NAL.

Jackson, K. (1985). *The crab grass frontier: The suburbanization of the United States.* New York: Oxford University Press.

Jacobson, J. L. (1992). Improving women's reproductive health. In L. R. Brown (Ed.), *State of the world.* New York: Norton.

Jahoda, M. (1982). *Employment and unemployment: A social-psychological analysis.* London: Cambridge University Press.

Jahoda, M., Lazarsfeld, P., & Zeisel, H. (1971). *Marienthal: A study of an unemployed community.* Hawthorne, NY: Aldine.

Janowitz, M. (1960). *The professional soldier: A social and political portrait.* New York: Free Press.

Janowitz, M. (1968). Political sociology. In D. Sills (Ed.), *The international encyclopedia of the social sciences.* New York: Free Press.

Janowitz, M. (1978). *The last half century: Societal change and politics in America.* Chicago: University of Chicago Press.

Janowitz, M., & Shils, E. A. (1948). The cohesion and disintegration of the Wehrmacht in World War II. *Public Opinion Quarterly, 12,* 280–315.

Jargowsky, P. A. (1997). Poverty and place: Ghettos, barrios, and the American city. New York: Russell Sage.

Jaroff, L. (1995, December 4). Knowing when to stop: Doctors go to heroic lengths to keep terminally ill patients alive—often against their wishes. *Time,* p. 76.

Jaynes, G. D., & Williams, R. M., Jr. (1989). *A common destiny: Blacks and American society.* Washington, DC: National Academy Press.

Jencks, C. (1993). *Rethinking social policy: Race, poverty and the underclass.* Cambridge, MA: Harvard University Press.

Jencks, C. (1994). *The homeless.* Cambridge, MA: Harvard University Press.

Jencks, C., & Phillips, M. (1999, January). The black-white test score gap: How to reduce it. *Current,* pp. 9–16.

Jencks, C., Smith, M., Acland, H., Bane, M. J., Cohen, D., Gintis, H., Heyns, B., & Michelson, S. (1972). *Inequality: A reassessment of the effect of family and schooling in America.* New York: Basic Books.

Jennings, F. (1975). *The invasion of America: Indians, colonialism and the cant of conquest.* New York: Norton.

Johnson, D. K. (1988). Adolescents' solutions to dilemmas in fables: Two moral orientations—two problem solving strategies. In C. Gilligan, J. V. Ward, J. M. Taylor, & B. Bardige (Eds.), *Mapping the moral domain.* Cambridge, MA: Harvard University Press.

Johnson, J. H., Jr., & Farrell, W. C., Jr. (1995, July 7). Race still matters. *Chronicle of Higher Education,* p. A48.

Johnson, P. (1987). *A history of the Jews.* New York: Harper.

Joravsky, D. (1971). *The Lysenko affair.* Cambridge, MA: Harvard University Press.

Kagan, S. L., & Neuman, M. J. (1998). Lessons from three decades of transition research. *Elementary School Journal, 98,* 365–380.

Kagay, M. R., & Elder, J. (1992, August 9). Numbers are no problem for pollsters. Words are. *New York Times,* sec. 4, p. 5.

Kahl, J. A. (1965). *The American class structure.* Fort Worth: Holt, Rinehart and Winston.

Kahneman, D., Knetsch, J., & Thaler, R. (1986). Fairness and the assumptions of economics. *Journal of Business, 59,* S285–S300.

Kammerman, S. B. (1995). *Starting right: How America neglects its youngest children and what we can do about it.* New York: Oxford University Press.

Kammerman, S., & Kahn, A. (1981). *Child care, family benefits, and working parents: A study in comparative policy.* New York: Columbia University Press.

Kammerman, S., & Kahn, A. J. (1993, Fall). What Europe does for single parent families. *The Public Interest,* pp. 70–86.

Kanter, R. M. (1977). *Men and women of the corporation.* New York: Basic Books.

Kanter, R. M., & Stein, B. A. (Eds.) (1980). *Life in organizations: Workplaces as people experience them.* New York: Basic Books.

Kaplan, D. E. (1998, April 13). Yakuza Inc.: U.S. investors snapping up bad loans from Japanese banks will collide with organized crime. *U.S. News & World Report,* pp. 40–41.

Kaplan, R. M., & Singer, R. D. (1976). Television violence and viewer aggression: A reexamination of the evidence. *Journal of Social Issues, 32,* 35–70.

Kasarda, J. D. (1988). *Metropolis era.* Newbury Park, CA: Sage.

Kasarda, J. D. (1989). Urban industrial transition and the underclass. *Annals of the American Academy of Political and Social Science, 501,* 26–47.

Katz, E. (1957). The two step flow of communication: An up-to-date report on an hypothesis. *Public Opinion Quarterly, 21,* 61–78.

Katz, E. (1966). Communication research and the image of society: Convergence of two traditions. In A. G. Smith (Ed.), *Communication and culture: Readings in the codes of human interaction.* Fort Worth: Holt, Rinehart and Winston.

Keith, J. (1982). *Old people, new lives: Community creation in a retirement residence* (2nd ed.). Chicago: University of Chicago Press.

Kelley, T. (1998, February 3). Charting a course to ethical profits. *New York Times,* sec. 3, p. 1.

Kemper, T. D. (1990). *Social structure and testosterone: Explorations of the socio-bio-social chain.* New Brunswick, NJ: Rutgers University Press.

Keniston, K. (1965). Social change and youth in America. In E. H. Erikson (Ed.), *The challenge of youth.* Garden City, NY: Doubleday Anchor.

Keniston, K. (1977). *All our children.* Orlando: Harcourt Brace Jovanovich.

Kennedy, R. J. R. (1944). Single or triple melting pot? *American Journal of Sociology, 49,* 331–339.

Kessler, S. J., & McKenna, W. (1978). *Gender: An ethnomethodological approach.* Chicago: University of Chicago Press.

Keyfitz, N. (1986). The population that does not reproduce itself. In K. Davis, M. S. Bernstam, & R. Ricardo-Campbell (Eds.), *Below replacement fertility in industrial societies: Causes, consequences, policies. Population and Development Review, 12* (Suppl.), 139–145.

Keyfitz, N. (1991). Toward a theory of population-development interaction. In K. Davis & M. S. Bernstam (Eds.), *Resources, environment and population: Present knowledge and future options* (special edition of *Population and Development Review*). New York: Population Council.

Kidder, T. (1985). *House.* Boston: Houghton Mifflin.

Killian, L. (1952). Group membership in disaster. *American Journal of Sociology, 57,* 309–314.

Kilson, M. (1995, Fall). Affirmative action. *Dissent,* pp. 469–470.

Kim, I. (1981). *New urban immigrants: The Korean community in New York.* Princeton, NJ: Princeton University Press.

Kimmel, M. S. (1990). *Revolution: A sociological interpretation.* Philadelphia: Temple University Press.

Kincaid, D., & Portes, A. (1994). *Comparative National Development.* Chapel Hill: University of North Carolina Press.

King, S. (1994, May). Diamonds are forever. *Life,* p. 26.

Kinsey, A. C., Pomeroy, W. B., & Martin, C. E. (1948). *Sexual behavior in the human male.* Philadelphia: Saunders.

Kinsey, A. C., Pomeroy, W. B., & Martin, C. E. (1953). *Sexual behavior in the human female.* Philadelphia: Saunders.

Kitano, H. H. I. (1980). *Race relations.* Englewood Cliffs, NJ: Prentice Hall.

Klandermans, B. (1997). *The social psychology of protest.* Cambridge, MA: Blackwell.

Klass, P. (1988, October 30). Wells, Welles and the Martians. *New York Times Book Review,* pp. 1, 48–49.

Kleinberg, O. (1935). *Race differences.* New York: Harper.

Kloby, J. (1997). *Inequality, power and development: The task of political sociology.* Atlantic Highlands, NJ: Humanities Press.

Koberg, C. S., Chappell, D., & Ringer, R. C. (1997). Correlates and consequences of protégé mentoring in a large hospital. *Annual Review of Sociology, 23,* 191–215.

Kohlberg, L., & Gilligan, C. (1971). The adolescent as a philosopher: The discovery of the self in a post-conventional world. *Daedalus, 100,* 1051–1086.

Kolata, G. (1992, November 11). New views on life span alter forecasts on elderly. *New York Times,* p. A1.

Kolata, G. (1996, February 27). New era of robust elderly belies the fears of scientists. *New York Times,* pp. A1, C3.

Kolbert, E. (1995, August 20). Americans despair of popular culture. *New York Times,* sec. 2, pp. 1, 23.

Koler, S. R., & Freeman, B. J. (1994). Analysis of environmental deprivation: Cognitive and social development in Romanian orphans. *Journal of Child Psychology and Psychiatry and Allied Disciplines, 35,* 769–781.

Kornblum, W. (1974). *Blue collar community.* Chicago: University of Chicago Press.

Kornblum, W. (1995). Times Square as a field research site. In R. P. McNamara, *Sex, scams, and street life: The sociology of New York City's Times Square.* Westport, CT: Greenwood.

Kornblum, W., & Julian, J. (1998). *Social problems* (9th ed.). Englewood Cliffs, NJ: Prentice Hall.

Kornblum, W., & Lawler, K. (1999). *Handbook of park user research.* Washington, DC: Urban Institute and Lila Wallace Readers Digest Foundation.

Kornhauser, W. (1952). The Negro union official: A study of sponsorship and control. *American Journal of Sociology, 57,* 443–452.

Korotz, G. (1998, January 12). What's moving today's economy? *Business Week,* p. 32.

Kosmin, B. (1991). *Research report: The national survey of religious identification.* New York: Center for Jewish Studies, City University of New York Graduate Center.

Kozol, J. (1995). *Amazing grace: The lives of children and the conscience of a nation.* New York: Crown.

Kremer, P. (1998, February 4). Sociologists rediscover the links between suicide and economic crisis. *Le Monde,* p. 8.

Kriegel, L. (1993). From the catbird seat: Football, baseball, and language. *Sewanee Review, 101,* 213–225.

Kroc, R. (1977). *Grinding it out: The making of McDonald's.* Chicago: Henry Regnery.

Kronus, C. L. (1977). Mobilizing voluntary associations into a social movement: The case of environmental quality. *Sociological Quarterly, 18,* 267–283.

Krueger, A. B. (1998, March). Reassessing the view that American schools are broken. *Federal Reserve Bank of New York Economic Policy Review,* pp. 29–44.

Krugman, P. W. (1992). *The sum of diminished expectations: U.S. economic policy in the 1990s.* Cambridge, MA: MIT Press.

Krugman, P. W. (1995, August 21). The wealth gap is real and it's growing. *New York Times,* p. A15.

Kuhn, T. (1962). *The structure of scientific revolutions.* Chicago: University of Chicago Press.

Kunen, J. S. (1995, July 10). Teaching prisoners a lesson. *New Yorker,* pp. 34–39.

Kurtz, L. (1995). *Gods in the global village.* Thousand Oaks, CA: Pine Forge.

Kutner, B., Wilkins, C., & Yarrow, P. R. (1952). Verbal attitudes and overt behavior involving racial prejudice. *Journal of Abnormal and Social Psychology, 47,* 649–652.

Kwamena-Poh, M., Tosh, J., Waller, R., & Tidy, M. (1982). *African history in maps.* Burnt Hill, England: Longman.

Kwong, P. (1998). *Forbidden workers: Illegal Chinese workers and American labor.* New York: New Press.

Ladner, J. (Ed.) (1973). *The death of white sociology.* New York: Vintage.

Lai, H. M. (1980). The Chinese. In S. Thornstrom (Ed.), *Harvard encyclopedia of ethnic groups.* Cambridge, MA: Harvard University Press.

Lakoff, S. (1970, Autumn). Science and conscience. *International Journal,* pp. 754–765.

Land, K. C. (1989). Review of *Predicting Recidivism Using Survival Models,* by Peter Schmidt and Ann Dryden Witte. *Contemporary Sociology, 18,* 245–246.

LaPiere, R. (1934). Attitudes vs. actions. *Social Forces, 13,* 230–237.

Larson, E. (1996, January 22). The soul of an HMO. *Time,* pp. 44–53.

Lash, S. (1992). *Modernity and identity.* Cambridge, MA: Blackwell.

Laslett, P. (1972). Introduction. In P. Laslett & R. Wall (Eds.), *Household and family in past time.* Cambridge, England: Cambridge University Press.

Laslett, P. (1983). *The world we have lost* (3rd ed.). London: Methuen.

Lasswell, H. (1936). *Politics: Who gets what, when and how.* New York: McGraw-Hill.

Lasswell, H. D. (1941). The garrison state. *American Journal of Sociology, 46,* 455–468.

Latané, B., & Darley, J. (1970). *The unresponsive bystander: Why doesn't he help?* New York: Meredith.

Laumann, E. O., Gagnon, J. H., Michaels, R. T., & Michaels, S. (1994). *The social organization of sexuality: Sexual practices in the United States.* Chicago: University of Chicago Press.

Lawton, J. (1984). *The all American war game.* Oxford, England: Blackwell.

Lazreg, M. (1994). *The eloquence of silence: Algerian women in question.* New York: Routledge.

Leacock, E. (1978). Women's status in egalitarian society: Implications for social evolution. *Current Anthropology,* 19, 247–275.

LeBon, G. (1947/1896). *The crowd.* London: Ernest Bonn.

Lee, S. M. (1998). Asian Americans: Diverse and growing. *Population Bulletin,* p. 53.

Lehman-Haupt, R. (1999, January 24). Rooming with a guy named Mom. *New York Times,* p. 9.

LeMasters, E. E. (1975). *Blue collar aristocrats.* Madison: University of Wisconsin Press.

Lenski, G., & Lenski, J. (1982). *Human societies* (4th ed.). New York: McGraw-Hill.

Lerman, H. (1996). *Pigeonholing women's misery.* New York: Basic Books.

Lever, J. (1976). Sex differences in the games children play. *Social Problems, 23,* 478–487.

Lever, J. (1978). Sex differences in the complexity of children's play and games. *American Sociological Review, 43,* 471–483.

Lever, J. (1983). *Soccer madness.* Chicago: University of Chicago Press.

Levi, P. (1989). *The drowned and the saved.* New York: Vintage International.

Levi-Strauss, C. (1995). *Saudades do Brasil: A photographic memoir.* Seattle: University of Washington Press.

Levy, S. (1997, April 7). Blaming the Web. *Newsweek,* pp. 46–47.

Lewin, T. (1989, November 14). Aging parents: Women's burden grows. *New York Times,* pp. A1, B12.

Lewis, M., & Fiering, C. (1982). Some American families at dinner. In L. M. Laosa & I. E. Sigel (Eds.), *Families as learning environments for children.* New York: Plenum.

Lewontin, R. C. (1982). *Human diversity.* New York: Scientific American Library.

Lewontin, R. C. (1995, April 20). The social organization of sexuality. *New York Review of Books,* pp. 24–29.

Lieberson, S. (1980). *A piece of the pie: Blacks and white immigrants since 1880.* Berkeley: University of California Press.

Liebow, E. (1967). *Tally's corner: A study of Negro streetcorner men.* Boston: Little, Brown.

Lieske, J. (1993). Regional subcultures of the United States. *Journal of Politics, 55,* 888–913.

Lin, M. (1995, November 1). Interview by Terri Gross on *Fresh Air,* National Public Radio.
Linton, R. (1936). *The study of man.* New York: Appleton.
Lipset, S. M. (1979). *The first new nation.* New York: Norton.
Lipset, S. M. (1981). *Political man.* Baltimore: Johns Hopkins University Press.
Lipset, S. M. (1994). The social prerequisites of democracy revisited. *American Sociological Review, 59,* 1–22.
Lipset, S. M. (1995, December). America today: Malaise and resiliency. *Current,* pp. 3–11.
Lipset, S. M. (1998). One nation after all. *Wilson Quarterly, 22,* 100–102.
Lofland, J. F. (1981). Collective behavior: The elementary forms. In N. Rosenberg & R. H. Turner (Eds.), *Social psychology: Sociological perspectives.* New York: Basic Books.
Lofland, L. (1998). *The public realm: Exploring the city's quintessential social territory.* Hawthorne, NY: Aldine.
Lombroso, C. (1911). *Crime: Its cause and remedies.* Boston: Little, Brown.
London, B. (1992). School-enrollment rates and trends, gender, and fertility: A cross-national analysis. *Sociology of Education, 65,* 305–318.
Lovins, A. (1977). *Soft energy paths: Toward a durable peace.* Cambridge, MA: Ballinger.
Lovins, A. (1991). Energy, people and industrialization. In K. Davis & M. S. Bernstam (Eds.), *Resources, environment and population: Present knowledge and future options.* New York: Population Council.
Lovins, A. (1996, September). The next energy crisis? *Popular Science,* pp. 89–91.
Lovins, A. (1998, October 10). On the rebound. *New Scientist,* p. 521.
Lowey, I. S. (1990). World urbanization in perspective. In K. Davis & M. S. Bernstein (Eds.), Resources, environment, and population. *Population and Development Review, 15* (Suppl.), 148–178.
Lowry, E. H. (1993). *Freedom and community: The ethics of interdependence.* Albany: State University of New York Press.
Luce, I. (1964). *Letters from the Peace Corps.* Washington, DC: Robert B. Luce.
Lukes, S., & Scull, A. (Eds.) (1983). *Durkheim and the law.* New York: St. Martin's Press.
Lurie, A. (1981). *The language of clothes.* New York: Random House.
Lye, K. (1995). *The complete atlas of the world.* Austin: Raintree Steck-Vaughn.
Lynd, R. S., & Lynd, H. M. (1929). *Middletown: A study in American culture.* Orlando: Harcourt Brace Jovanovich.
Lynd, R. S., & Lynd, H. M. (1937). *Middletown in transition: A study in cultural conflicts.* Orlando: Harcourt Brace Jovanovich.
Lyson, T. A., & Squires, G. D. (1993, February). The "lost generation" of sociologists. *Footnotes* (American Sociological Association), pp. 4–5.

Machiavelli, N. (1950/1513). *The prince* and *The discourses.* New York: Modern Library.
Machlis, G., & Tichnell, D. L. (1985). *State of the world's parks: An international assessment.* Boulder, CO: Westview.
MacLeod, J. (1987, 1995). *Ain't no makin' it: Leveled aspirations in a low-income neighborhood.* Boulder, CO: Westview.
MacWhinney, B. (1998). Models of the emergence of language. *Annual Review of Psychology, 49,* 199–227.
Majundar, R. C. (Ed.) (1951). *The history and culture of the Indian people.* London: Allen and Unwin.
Makela, K. (1996). *Alcoholics Anonymous as a mutual-help movement: A study in eight societies.* Madison: University of Wisconsin Press.
Malinowski, B. (1927). *Sex and repression in savage society.* London: Harcourt Brace.
Malson, L. (1972). *Wolf children and the problem of human nature* (E. Fawcett, P. Aryton, & J. White, Trans.). New York: Monthly Review Press.
Malthus, T. (1927–1928/1798). *An essay on population.* New York: Dutton.
Mannheim, K. (1941). *Man and society in an age of reconstruction.* Orlando: Harcourt Brace Jovanovich.
Mansnerus, L. (1992, November 1). Should tracking be derailed? *Education Life, New York Times,* pp. 14–16.
Mark, N. (1998). Beyond individual differences: Social differentiation from first principles. *American Sociological Review, 63,* 309–330.
Marks, J. (1994, December). Black, white, other. *Natural History,* pp. 32–35.
Marks, P. M. (1998). *In a barren land: American Indian dispossession and survival.* New York: Morrow.
Markson, A. R., & Bloch, R. (1985). Defensive cities: Military spending, high technology, and human settlements. In M. Castells (Ed.), *High technology, space, and society.* Newbury Park, CA: Sage.
Marshall, T. H. (1964). *Class, citizenship and social development.* Garden City, NY: Doubleday.
Martinson, R. (1972, April 29). Planning for public safety. *New Republic,* pp. 21–23.
Marwell, G., & Ames, R. E. (1985). Experiments on the provision of public goods II: Provision points, stakes, experience, and the free-rider problem. *American Journal of Sociology, 90,* 926–937.
Marx, K. (1961/1844). *Economic and philosophical manuscripts of 1844.* Moscow: Foreign Languages Publishing House.
Marx, K. (1962/1867). *Capital: A critique of political economy.* Moscow: Foreign Languages Publishing House.
Marx, K. (1963/1869). *The eighteenth Brumaire of Louis Bonaparte.* New York: International Publishers.
Marx, K., & Engels, F. (1955). Wage labor and capital. In *Selected works in two volumes.* Moscow: Foreign Languages Publishing House.
Marx, K., & Engels, F. (1969/1848). *The communist manifesto.* New York: Penguin.
Mascie-Taylor, C. G. N. (1990). *Biosocial aspects of social class.* New York: Oxford University Press.
Massey, D. S., & Denton, N. A. (1993). *American apartheid: Segregation and the making of the underclass.* Cambridge, MA: Harvard University Press.

Matthews, R. A. J. (1997). The science of Murphy's law. *Scientific American, 276,* 88–92.

Mauss, M. (1966/1925). *The gift.* New York: Free Press.

Mawson, A. R. (1989). Review of *Homicide,* by Martin Daly and Margo Wilson. *Contemporary Sociology, 18,* 238–240.

Maxwell, M. (1991). *The sociobiological imagination.* Albany: State University of New York Press.

Mayfield, L. (1984). *Teenage pregnancy.* Doctoral dissertation, City University of New York.

Mayo, E. (1945). *The social problems of an industrial civilization.* Boston: Harvard University, Graduate School of Business Administration.

McAdam, D. (1988). *Freedom summer.* New York: Oxford University Press.

McAdam, D. (1992). Gender as a mediator of the activist experience: The case of Freedom Summer. *American Journal of Sociology, 97,* 1211–1240.

McAdam, D., McCarthy, J. D., & Zald, N. (1988). Social movements. In N. J. Smelser (Ed.), *The handbook of sociology.* Newbury Park, CA: Sage.

McAdam, D., McCarthy, J. D., & Zald, M N. (Eds.) (1996). *Comparative perspectives on social movements: Political opportunities, mobilizing structures, and cultural framings.* New York: Cambridge University Press.

McAdam, D., & Snow, D. S. (1997). *Social movements: Readings on their emergence, mobilization, and dynamics.* Los Angeles: Roxbury.

McAndrew, M. (1985). Women's magazines in the Soviet Union. In B. Holland (Ed.), *Soviet sisterhood.* Bloomington: Indiana University Press.

McBrien, R. P. (1990, March 12). A papal attack on Vatican II. *New York Times,* p. A17.

McCall, W. (1998, May 28). Biological makeup cited on question of why kids can go on killing sprees. *Chicago Tribune,* p. 2.

McCord, W. (1991). *The dawn of the Pacific century: Implications for three worlds of development.* New Brunswick, NJ: Transaction.

McCrone, D. (1996). *The sociology of nationalism.* London: Routledge.

McFadden, R. B. (1998). Review of *Waiting for a miracle,* by James P. Comer. *Black Issues in Higher Education, 15,* 30.

McFalls, J. H. (1998). Population: An introduction. *Population,* p. 53.

McGoldrick, M., & Carter, E. A. (1982). *The family life cycle in normal family processes.* London: Guilford.

McGuire, M. B. (1987). *Religion: The social context* (2nd ed.). Belmont, CA: Wadsworth.

McIntyre, M. (1995). Altruism, collective action, and rationality: The case of LeChambon. *Polity, 27,* 537–557.

McKendrick, N., Brewer, J., & Plump, J. H. (1982). *The birth of a consumer society.* Bloomington: Indiana University Press.

McKenzie, E. (1994). *Privatopia: Homeowner associations and the rise of residential private government.* New Haven, CT: Yale University Press.

McKinlay, J. B., & McKinlay, S. M. (1977). The questionable contribution of medical measures to the decline of mortality in the United States in the twentieth century. *Health and Society, 53,* 405.

McNamara, R. P. (1994). *Crime displacement: The other side of prevention.* East Rockway, NY: Cummings and Hathaway.

McNamara, R. P. (1995). *Sex, scams, and street life: The sociology of New York City's Times Square.* Westport, CT: Greenwood.

McNeill, W. (1963). *The rise of the West: A history of the human community.* Chicago: University of Chicago Press.

McNeill, W. H. (1982). *The pursuit of power: Technology, armed force, and society since A.D. 1000.* Chicago: University of Chicago Press.

Mead, G. H. (1971/1934). *Mind, self and society.* Chicago: University of Chicago Press.

Mead, M. (1950). *Sex and temperament in three primitive societies.* New York: NAL Mentor.

Mead, M. (1971). Comment. In J. Tanner & B. Inhelder (Eds.), *Discussions on child development.* New York: International Universities Press.

Meier, D. (1995). *The power of their ideas: Lessons from a small school in Harlem.* Boston: Beacon.

Melton, J. G. (1989). *The encyclopedia of American religions* (3rd ed.). Detroit: Gale Research.

Merton, R. K. (1938). Social structure and anomie. *American Sociological Review, 3,* 672–682.

Merton, R. K. (1948). Discrimination and the American creed. In R. M. MacIver (Ed.), *Discrimination and national welfare.* New York: Institute for Religious and Social Studies; dist. by Harper.

Merton, R. K. (1968). *Social theory and social structure* (3rd ed.). New York: Free Press.

Merton, R. K. (1996). *On social structure and science.* Chicago: University of Chicago Press.

Merton, R. K. (1998). Personal communication.

Merton, R. K., & Kitt, A. (1950). Contributions to the theory of reference group behavior. In R. K. Merton & P. Lazarsfeld (Eds.), *Continuities in social research.* New York: Free Press.

Mestrovic, S. G. (1991, September 25). Point of view: Why East Europe's upheavals caught social scientists off guard. *Chronicle of Higher Education,* p. A56.

Meyer, J. (1990). Guess who's coming to dinner this time? A study of gay intimate relationships and the support for those relationships. *Marriage and Family Review, 14,* 59–82.

Meyer, J. W., Ramirez, F. O., & Soysal, Y. N. (1992). World expansion of mass education, 1870–1980. *Sociology of Education, 65,* 128–149.

Milavsky, W. R. (1977, August 19). *TV and aggressive behavior of elementary school boys.* Paper delivered at the meeting of the American Psychological Association, San Francisco.

Milgram, S. (1970). The experience of living in cities. *Science, 167,* 1461–1468.

Milgram, S. (1974). *Obedience to authority: An experimental view.* New York: Harper.

Miller, A. T. (1993). African American families. In M. B. Katz (Ed.), *The "underclass" debate: View from history.* Princeton, NJ: Princeton University Press.

Miller, D. C., & Form, W. H. (1964). *Industrial sociology* (2nd ed.). New York: Harper.

Miller, N. (1989). *In search of gay America: Women and men in a time of change.* New York: Atlantic Monthly Press.

Miller, W. B. (1958). Lower class culture as a generating milieu of gang delinquency. *Journal of Social Issues, 14,* 5–19.

Millman, M. (1976). *The unkindest cut: Life in the backrooms of medicine.* New York: Morrow.

Mills, C. W. (1951). *White collar.* New York: Oxford University Press.

Mills, C. W. (1956). *The power elite.* New York: Oxford University Press.

Mills, C. W. (1959). *The sociological imagination.* New York: Oxford University Press.

Mintz, S. (1985). *Sweetness and power: The place of sugar in modern history.* New York: Viking.

Minuchin, S. (1974). *Families and family therapy.* Cambridge, MA: Harvard University Press.

Mitchell, J. D. (1998). Before the next doubling. *World Watch, 11,* 20–27.

Mitra, A. (1997). Diasporic Web sites: Ingroup and outgroup discourse. *Critical Studies in Mass Communication, 14,* 158–181.

Modell, J., & Hareven, T. K. (1973). Urbanization and the malleable household: An examination of boarding and lodging in American families. *Journal of Marriage and the Family, 35,* 466–479.

Mollenkopf, J. (1985). *The contested city.* Princeton, NJ: Princeton University Press.

Money, J., & Tucker, P. (1975). *Sexual signatures.* Boston: Little, Brown.

Monnier, M. (1993). *Mapping it out: Expository cartography for the humanities and social sciences.* Chicago: University of Chicago Press.

Montagna, P. D. (1977). *Occupations and society.* New York: Wiley.

Moody, H. R. (1998). *Aging: Contexts and controversies* (2nd ed.). Thousand Oaks, CA: Pine Forge.

Moore, B. (1968). *The social origins of dictatorship and democracy: Lord and peasant in the making of the modern world.* Boston: Beacon.

Moore, H. E. (1958). *Tornados over Texas.* Austin: University of Texas Press.

Moore, J. W. (1991). *Going down to the barrio: Homeboys and homegirls in change.* Philadelphia: Temple University Press.

Moore, J., & Pachon, H. (1985). *Hispanics in the United States.* Englewood Cliffs, NJ: Prentice Hall.

Morales, E. (1986). Coca and the cocaine economy and social change in the Andes of Peru. *Economic Development and Cultural Change, 35,* 143–161.

Morales, E. (1989). *Cocaine: White gold rush in Peru.* Tucson: University of Arizona Press.

Morgan, D. L. (1998). Introduction: The aging of the baby boom. *Generations, 22,* 5–10.

Morgan, R. (Ed.) (1970). *Sisterhood is powerful: An anthology of writings from the women's liberation movement.* New York: Random House.

Morris, W. F., & Foxx, J. J. (1987). *Living Maya.* New York: Abrams.

Moscos, C. A. (1988). *Call to civic service.* New York: Free Press.

Mosley, W. H., & Cowley, P. (1991, December). *The challenge of world health.* Washington, DC: Population Reference Bureau.

Moynihan, D. P. (1988, Spring). Our poorest citizens. *Focus,* (Institute for Research on Poverty), p. 5.

Moynihan, P. (1965). *The Negro family.* Internal report of the Office of Policy Planning and Research, U.S. Department of Labor.

Mulkey, L. M. (1993). *Sociology of education: Theoretical and empirical investigations.* Fort Worth: Harcourt Brace.

Murdock, G. (1949). *Social structure.* New York: Macmillan.

Murdock, G. (1983). *Outline of world cultures* (6th ed.). New Haven, CT: Human Relations Area Files.

Murray, C. (1984). *Losing ground.* New York: Basic Books.

Murray, C. A. (1988). *In pursuit: Of happiness and good government.* New York: Simon & Schuster.

Myerson, A. R. (1995, December 17). American money makes the whole world sing. *New York Times,* sec. 4, pp. 1, 14.

Myrdal, G. (1944). *An American dilemma.* New York: Harper.

Myrdal, J. (1965). *Report from a Chinese village.* New York: Random House.

Nagel, T. (1994, May 12). Freud's permanent revolution. *New York Review of Books,* pp. 34–39.

Naipaul, V. S. (1998). Beyond belief: Islamic excursions among the converted peoples. New York: Random House.

Naisbitt, J., & Aburdene, P. (1990). *Megatrends 2000.* New York: Morrow.

Napier, A. Y., & Whitaker, C. (1980). *The family crucible.* New York: Bantam.

Naples, N. (1998). *Grassroots warriors: Activist mothering, community work, and the War on Poverty.* London: Routledge.

Nasar, S. (1999, March 6). Economic growth still at fast pace as hiring surges. *New York Times,* p. 10.

National Catholic Reporter (1997, April 11). Low-intensity war against liberation theology. P. 28.

National Center for Educational Statistics (1995). *The educational progress of black students.* Washington, DC: U.S. Department of Education.

National Center for Health Statistics (1995). *Vital statistics of the United States: Vol. 1. Natality.*

National Commission on America's Urban Families (1993). *Families first.* Washington, DC: U. S. Government Printing Office.

National Opinion Research Center (NORC). Annual. *General social survey, cumulative codebook.* Chicago: University of Chicago Press.

Neihardt, J. G. (1959/1932). *Black Elk speaks: Being the life story of a holy man of the Oglala Sioux.* New York: Washington Square Press.

Nelkin, D. (1996). Conjuring science. *Nature, 384,* 423.

Nelson, B. (1998, November). Motivating workers worldwide. *Workforce.*

Neugarten, B. L. (1996). *The meanings of age.* Chicago: University of Chicago Press.

Newcomb, T. (1958). Attitude development as a function of reference groups: The Bennington study. In E. Maccoby, T. M. Newcomb, & E. L. Hartley (Eds.), *Readings in social psychology* (3rd ed.). Fort Worth: Holt, Rinehart and Winston.

Newman, K. S. (1988). *Falling from grace.* New York: Free Press.

Niebuhr, G. (1995, October 3). With every wave of newcomers, a church more diverse. *New York Times,* p. B6.

Niebuhr, H. R. (1929). *The social sources of denominationalism.* New York: Meridian.

Ning, P. (1995). *Red in tooth and claw: Twenty-six years in communist Chinese prisons.* New York: Grove.

Nisbet, R. A. (1969). *Social change and history.* New York: Oxford University Press.

Nisbet, R. A. (1970). *The social bond.* New York: Knopf.

Norman, M. (1997, November 9). From Carol Gilligan's chair. *New York Times Magazine,* p. 50.

Nussbaum, M., & Sen, A. (Eds.) (1993). *The quality of life.* New York: Oxford University Press.

Nyden, P. (1984). *Steelworkers rank and file.* New York: Praeger.

Obermeyer, C. M. (1992). Islam, women, and politics: The demography of Arab countries. *Population and Development Review, 18,* 33–60.

Oberschall, A. (1996). *Social movements.* New Brunswick, NJ: Transaction.

Odaka, K. (1998). Portraits of the Japanese workplace: Labor movements, workers, and managers. *Journal of Comparative Economics, 26,* 825–829.

Office of Technology Assessment, U.S. Congress (1986). *Technology, public policy, and the changing structure of agriculture: Vol. 2. Background papers: Part D. Rural communities.* Washington, DC.

Ofosu-Amaah, V. (1998). Declines in fertility levels evident in Africa, notes UN Population Fund. *UN Chronicle, 35,* 20–21.

Ogburn, W. F. (1942). Inventions, population and history. In American Council of Learned Societies, *Studies in the history of culture.* Freeport, NY: Books for Libraries Press.

Ogburn, W. F. (1957). Cultural lag as theory. *Sociology and Social Research, 41,* 167–174.

Ogburn, W. F., & Nimkoff, M. F. (1934). *The family.* Boston: Houghton Mifflin.

O'Hare, W. (1996). A new look at poverty in America. *Population Bulletin,* vol. 51, no. 2.

Oldenburg, R. (1997). *The great good place.* Marlowe.

Oliver, M. L., & Shapiro, T. M. (1990). Wealth of a nation: A reassessment of asset inequality in America shows at least one third of households are asset-poor. *American Journal of Economics and Sociology, 49,* 129–150.

Oliver, M., & Shapiro, T. (1995). *Black wealth/white wealth: A new perspective on racial inequality.* New York: Routledge.

Olojede, D. (1995, June 5). Chaos lurks in Nigeria. *Newsday,* pp. A6, A22.

Olson, M. (1965). *The logic of collective action.* Cambridge, MA: Harvard University Press.

O'Neil, J. (1997). Building schools as communities: A conversation with James Comer. *Educational Leadership, 54,* 6–11.

O'Neill, A-M. (1998, June 29). Prodigal president. *People Weekly,* pp. 129ff.

Opie, I., & Opie, P. (1969). *Children's games in street and playground.* Oxford, England: Oxford University Press.

Oppenheimer, V. K. (1994). Women's rising employment and the future of the family in industrial societies. *Population and Development Review, 20,* 293–342.

Orfield, G., & Eaton, S. E. (1996). *Dismantling desegregation: The quiet reversal of Brown v. Board of Education.* New York: New Press.

Ornish, D. (1998). *Love and survival: The scientific basis for the healing power of intimacy.* New York: HarperCollins.

Orshansky, M. (1965, January). Counting the poor: Another look at the poverty profiles. *Social Security Bulletin,* pp. 3–26.

Orwell, G. (1949). *1984.* New York: New American Library.

Orwell, G. (1950). *Shooting an elephant and other essays.* New York: Harcourt Brace Jovanovich.

Paik, H. J. (1991). *The effects of television violence on aggressive behavior: A meta-analysis.* Doctoral dissertation, Syracuse University.

Palmore, E. (1981). *Social patterns in normal aging.* Durham, NC: Duke University Press.

Papernow, P. (1988). Stepparent role development: From outsider to intimate. In W. R. Beer (Ed.), *Relative strangers.* Totowa, NJ: Rowman and Littlefield.

Park, R. E. (1914). Racial assimilation in secondary groups. *Publications of the American Sociological Society, 8,* 66–72.

Park, R. E. (1967/1925a). The city: Suggestions for the investigation of human behavior in the urban environment. In R. E. Park & E. W. Burgess (Eds.), *The city.* Chicago: University of Chicago Press.

Park, R. E. (1967/1925b). The natural history of the newspaper. In R. E. Park & E. W. Burgess (Eds.), *The city.* Chicago: University of Chicago Press.

Park, R. E. (1967/1926). The urban community as a spatial pattern and a moral order. In R. H. Turner (Ed.), *Robert E. Park on social control and collective behavior.* Chicago: University of Chicago Press.

Park, R. E., & Burgess, E. W. (1921). *Introduction to the science of sociology* (rev. ed.). Chicago: University of Chicago Press.

Park, R. E., & Burgess, E. W. (Eds.) (1967/1925). *The city.* Chicago: University of Chicago Press.

Parsons, T. (1937). *The structure of social action.* New York: McGraw-Hill.

Parshall, G. (1998, August 17–24). Brotherhood of the bomb. *U.S. News & World Report,* pp. 64–68.

Parsons, T. (1940). An analytic approach to the theory of social stratification. *American Journal of Sociology, 45,* 841–862.

Parsons, T. (1951). *The social system.* New York: Free Press.

Parsons, T. (1960). *Structure and process in modern societies.* New York: Free Press.

Parsons, T. (1966). *Societies: Evolutionary and comparative.* Englewood Cliffs, NJ: Prentice Hall.

Parsons, T. (1968). The problem of polarization along the axis of color. In J. H. Franklin (Ed.), *Color and race.* Boston: Beacon.

Parsons, T. (1991). The integration of economic and sociological theory. *Sociological Inquiry, 61,* 10–60.

Parsons, T., & Bales, R. F. (1955). *Family, socialization, and interaction process.* New York: Free Press.

Partin, M., & Palloni, A. (1995, Spring). Accounting for the recent increases in low birth weight among African Americans. *Focus* (Institute for Research on Poverty), pp. 33–38.

Patterson, O. (1982). *Slavery and social death.* Cambridge, MA: Harvard University Press.

Patterson, O. (1991). *Freedom: Vol. 1. Freedom in the making of Western culture.* New York: Basic Books.

Patterson, O. (1999). *Rituals of blood: Consequences of slavery in two American centuries.* Washington, DC: Civitas/Counterpoint.

Pavlov, I. (1927). *Conditioned reflexes: An investigation of the physiological activity of the cerebral cortex* (G. V. Anrep, Trans. & Ed.). London: Oxford University Press.

Pear, R. (1993, July 8). Big health gap, tied to income, is found in U.S. *New York Times,* pp. A1, B10.

Pedersen, D. (1997, March 3). When an A is average. *Newsweek,* p. 64.

Pedersen, E., & Faucher, T. A., with Eaton, W. W. (1978). A new perspective on the effects of first-grade teachers on children's subsequent adult status. *Harvard Educational Review, 48,* 1–31.

Peek, C. W., Zsembik, B. A., & Coward, R. T. (1997). The changing caregiving networks of older adults. *Research on Aging, 19,* 333–361.

Peplau, L. A., Hill, C. T., & Rubin, Z. (1993). Sex role attitudes in dating and marriage: A 15-year follow-up of the Boston couples study. *Journal of Social Issues, 49,* 3–25.

Perin, C. (1977). *Everything in its place: Social order and land use in America.* Princeton, NJ: Princeton University Press.

Perrow, C. (1984). *Normal accidents: Living with high-risk technologies.* New York: Basic Books.

Perrow, C. (1996). Prosaic organizational failure. *American Behavioral Scientist, 39,* 1040–1056.

Perutz, M. F. (1992). The fifth freedom. *New York Review of Books, 39,* 5–7.

Pesce, C. (1994, October 5). In Minn., a crime of cover-up. *USA Today,* p. A3.

Peters, A. (1990). *Peters atlas of the world.* New York: HarperCollins.

Pettigrew, T. F. (1980). Prejudice. In *Harvard encyclopedia of American ethnic groups.* Cambridge, MA: Belknap.

Pfeiffer, E., Verwoerdt, A., & Davis, G. (1972). Sexual behavior in middle life. *American Journal of Psychiatry, 128,* 1262–1267.

Phillips, T. (1998). The end of the rope, part three: The capital punishment revival. *Contemporary Review, 272,* 181–186.

Piaget, J., & Inhelder, B. (1969). *The psychology of the child.* New York: Basic Books.

Pifer, A., & Bronte, L. (Eds.) (1986, Winter). The aging society. *Daedalus.*

Pillari, V. (1998). *Human behavior in the social environment: Families, groups, organizations, and communities.* Pacific Grove, CA: Brooks/Cole.

Pittman, J. F. (1993). Functionalism may be down, but it surely is not out. In P. G. Boss, W. J. Doherty, R. LaRossa, W. R. Schumm, & S. K. Steinmetz (Eds.), *Sourcebook of family theories and methods: A contextual approach.* New York: Plenum.

Piven, F. F. (1997). *The breaking of the American social contract.* New York: New Press.

Platt, A. M. (1977). *The child savers* (2nd ed.). Chicago: University of Chicago Press.

Ploch, D. R., & Hastings, D. W. (1995). Some church; some don't. *Journal for the Scientific Study of Religion, 34,* 507–516.

Polansky, N. A., Chalmers, M. A., Buttenweiser, E., & Williams, D. P. (1981). *Damaged parents, an anatomy of child neglect.* Chicago: University of Chicago Press.

Polanyi, K. (1944). *The great transformation.* Boston: Beacon.

Polsby, N. (1980). *Community power and political theory* (2nd ed.). New Haven, CT: Yale University Press.

Popenoe, D. (1994). Family decline and scholarly optimism. *Family Affairs, 6,* 9–10.

Population Reference Bureau. (1998). Parents whose children need day care often face daily crisis. *Population Today, 26,* 1–2.

Population Reference Bureau. (1998). *1998 world population data sheet.* Washington, DC: U.S. Government Printing Office.

Portes, A., & Rumbaut, R. G. (1996). *Immigrant America: A portrait.* Berkeley: University of California Press.

Portes, A., & Stepick, A. (1985). Unwelcome immigrants. *American Sociological Review, 50,* 493–514.

Press, J. E., & Townsley, E. (1998). Wives' and husbands' housework reporting: Gender, class, and social desirabilty. *Gender and Society, 12,* 188–219.

Pressman, J. L., & Wildavsky, A. (1984). *Implementation* (3rd ed.). Berkeley: University of California Press.

Pressman, S. (1998). The gender poverty gap in developed countries: Causes and cures. *Social Science Journal, 35,* 275–287.

Preston, J. (1998, March 6). Mexico's overtures to the Zapatistas bring tensions in Chiapas to a new boiling point. *New York Times,* p. A8.

Preston, S. H. (1984). Children and the elderly: Divergent paths for America's dependents. *Demography, 21,* 435–457.

Preston, S. H., Elo, I. T., & Foster, A. (1998). Reconstructing the size of the African American population by age and sex, 1930–1990. *Demography, 35,* 1–21.

Putnam, R. (1995). Bowling alone: America's declining social capital. *Journal of Democracy, 6,* 65–78.

Putnam, R. D. (1996, Winter). The strange disappearance of civic America. *American Prospect,* pp. 34–49.

Pyle, R. E. (1996). *Persistence and change in the Protestant establishment.* Westport, CT: Praeger.

Quadagno, J. S. (1979). Paradigms in evolutionary theory—Sociobiological models of natural selection. *American Sociological Review, 44,* 100–109.

Quinney, R. (1978). The ideology of law: Notes for a radical alternative to legal oppression. In C. E. Reasons & R. M. Rich (Eds.), *Sociology of law: A conflict perspective.* Toronto: Butterworths.

Quinney, R. (1980). *Class, state, and crime.* White Plains, NY: Longman.

Radelet, M. L., & Bedau, H. A. (1992). *In spite of innocence: Erroneous convictions in capital cases.* Boston: Northeastern University Press.

Randall, V. (1998). *Political change and underdevelopment: A critical introduction to Third World politics* (2nd ed.). Durham, NC: Duke University Press.

Rathje, W. L. (1993). Less fat? Aw, baloney. Garbage Project studies indicate misreporting of fatty meat consumption. *Garbage, 5,* 22–23.

Reckless, W. C. (1933). *Vice in Chicago.* Chicago: University of Chicago Press.

Reddy, M. A. (Ed.) (1994). *Statistical abstract of the world.* Detroit: Gale Research.

Redfield, R. (1947). The folk society. *American Journal of Sociology, 52,* 293–308.

Reese, S. (1998, August). Religious spirit. *American Demographics,* p. 62.

Reich, R. B. (1992, Winter). Training a skilled work force. *Dissent,* pp. 42–46.

Reid, L. L. (1998). Devaluing women and minorities: The effects of race/ethnic and sex composition of occupations on wage levels. *Work and Occupations, 25,* 511–537.

Reid, S. T. (1993). *Criminal justice.* New York: Macmillan.

Reiss, A. J., Jr., & Roth, J. A. (Eds.) (1993). *Understanding and preventing violence.* Washington, DC: National Academy Press.

Reissman, F. (1984). *The self-help revolution.* New York: Human Sciences Press.

Rejwan, N. (1998). *Arabs face the modern world: Religious, cultural, and political responses to the West.* Gainesville: University of Florida Press.

Renner, M. (1989). Enhancing global security. In L. R. Brown et al., *State of the world 1989: A Worldwatch Institute report on progress toward a sustainable society.* New York: Norton.

Rheingold, H. (1993). *The virtual community: Homesteading on the electronic frontier.* New York: Harper.

Richards, E. (1989). *The knife and gun club: Scenes from an emergency room.* New York: Atlantic Monthly Press.

Richmond-Abbott, M. (1992). *Masculine and feminine: Gender roles over the life cycle* (2nd ed.). New York: McGraw-Hill.

Riesman, D. (1957). The suburban dislocation. *Annals of the American Academy of Political and Social Science, 314,* 123–146.

Riis, J. A. (1890). *How the other half lives: Studies among the tenements of New York.* New York: Scribner.

Riley, M. W., Foner, A., & Waring, J. (1988). The sociology of age. In N. E. Smelser (Ed.), *The handbook of sociology.* Newbury Park, CA: Sage.

Riley, M. W., Kahn, R. L., & Foner, A. (1994). *Age and structural lag.* New York: Wiley.

Riley, M. W., & Riley, J. W., Jr. (1989). The lives of older people and changing social roles. *Annals of the American Academy of Political and Social Science, 503,* 14–28.

Rist, R. C. (1973). *The urban school: A factory for failure.* Cambridge, MA: MIT Press.

Robbins, W. (1985, February 10). Despair wrenches farmers' lives as debts mount and land is lost. *New York Times,* pp. 1, 30.

Roberts, S. (1995, April 27). Women's work: What's new, what isn't. *New York Times,* p. B6.

Robertson, R. (1985). The sacred and the world system. In P. Hammond (Ed.), *The sacred in a secular age.* Berkeley: University of California Press.

Robinson, J. (1988). *The rhythm of everyday life.* Boulder, CO: Westview.

Robinson, J. P., & Godbey, G. (1997). *Time for life: The surprising ways Americans use their time.* University Park: Pennsylvania State University Press.

Robinson, P. (1994). *Freud and his critics.* Berkeley: University of California Press.

Rochon, T. R. (1998). *Culture moves: Ideas, activism, and changing values.* Princeton, NJ: Princeton University Press.

Rock, P. (1985). Symbolic interactionism. In A. Kuper & J. Kuper (Eds.), *The social science encyclopedia.* London: Routledge and Kegan Paul.

Rodriguez, C. E. (1992). The Puerto Rican community in the South Bronx: Contradictory views from within and without. In P. S. Rotherberg (Ed.), *Race, class and gender in the United States.* New York: St. Martin's Press.

Rose, M. (1995). *Possible lives: The promise of public education.* Boston: Houghton Mifflin.

Rosen, E. I. (1987). *Bitter choices: Blue-collar women in and out of work.* Chicago: University of Chicago Press.

Rosenhan, D. L. (1973). On being sane in insane places. *Science, 179,* 250–258.

Rosner, D. (1995). *Hives of sickness.* New Brunswick, NJ: Rutgers University Press.

Rosow, I. (1965). Forms and functions of adult socialization. *Social Forces, 44,* 38–55.

Ross, H. L. (1963). *Perspectives on the social order.* New York: McGraw-Hill.

Rossi, A. (1964). Equality between the sexes: An immodest proposal. In R. J. Lifton (Ed.), *The woman in America.* Boston: Beacon.

Rossi, A. (1977). The biosocial basis of parenting. *Daedalus, 106,* 1–31.

Rossi, A. (1980). Aging and parenthood in the middle years. In P. B. Baltes & O. G. Brim, Jr. (Eds.), *Life-span development and behavior* (Vol. 3). Orlando: Academic Press.

Rossianov, K. D. (1993). Editing nature: Joseph Stalin and his "new" Soviet biology. *Isis, 84,* 728–745.

Roszak, T. (1969). *The making of a counterculture.* Garden City, NY: Doubleday.

Rothenberg, P. S. (Ed.) (1992). *Race, class and gender in the United States* (2nd ed.). New York: St. Martin's Press.

Rothenberg, R. (1990, October 5). Surveys proliferate, but answers dwindle. *New York Times,* pp. A1, D4.

Rubin, Z. (1980). The love research. In A. Skolnick & J. H. Skolnick (Eds.), *Family in transition.* Boston: Little, Brown.

Rubington, E., & Weinberg, M. S. (1996). *Deviance: The interactionist perspective* (6th ed.). Boston: Allyn & Bacon.

Rudé, G. (1964). *The crowd in history.* New York: Wiley.

Ruggles, P. (1990). *Drawing the line: Alternative poverty measures and their implications for public policy.* Washington, DC: Urban Institute.

Ruggles, P. (1992). Measuring poverty. *Focus* (Institute for Research on Poverty), *14,* 2.

Ruggles, S. (1997). The rise of divorce and separation in the United States, 1880–1990. *Demography, 34,* 455–467.

Ruskin, C. (1988). *The quilt: Stories from the NAMES Project.* New York: Pocket Books.

Rutstein, S. O. (1991). *Levels, trends and differentials in infant and child mortality in the less developed countries.* Paper presented at the seminar on Child Survival Interventions: Effectiveness and Efficiency, at the Johns Hopkins University School of Hygiene and Public Health, Baltimore, June 20–22.

Ryan, A. (1992). Twenty-first century limited. *New York Review of Books, 39,* 20–24.

Rybczynski, W. (1987). *Home: A short history of an idea.* New York: Penguin.

Rymer, R. (1992a, April 13). A silent childhood. *New Yorker,* pp. 41–53.

Rymer, R. (1992b, April 20). A silent childhood, part II. *New Yorker,* pp. 43–47.

Sabel, C. F. (1995). Bootstrapping reform: Rebuilding firms, the welfare state, and unions. *Politics & Society, 23,* 5–49.

Sadik, N. (1995). Decisions for development: Women, empowerment and reproductive health. In *The state of the world population 1995.* New York: United Nations Population Fund.

Safire, W. (1997, December 7). Sampling is not enumerating. *New York Times,* sec. 4, p. 17.

Sagarin, E. (1975). *Deviants and deviance: A study of disvalued people and behavior.* New York: Praeger.

Sahagun, L., & Stammer, L. B. (1998, February 20). Entire Promise Keepers' staff to be laid off. *Los Angeles Times.*

Sahlins, M. D. (1960, September). The origin of society. *Scientific American,* pp. 76–87.

Sahlins, M. D. (1976). *The use and abuse of biology: An anthropological critique of sociobiology.* Ann Arbor: University of Michigan Press.

Sakharov, A. D. (1990, May 14). The poisonous legacy of Trofim Lysenko. *Time,* p. 61.

Salgado, S. (1990). *An uncertain grace.* New York: Aperture.

Salgado, S. (1993). *Workers.* New York: Aperture.

Salisbury, R. F. (1962). *From stone to steel.* Parkville, Australia: Melbourne University Press.

Samuelson, P. A. (1998). How Foundations came to be ("Foundations of Economic Analysis"). *Journal of Economic Literature, 36,* 1375–1387.

Samuelson, R. J. (1998, July 13). Down with the media elite!? *Newsweek,* p. 47.

Sanchez-Jankowski, M. S. (1991). *Islands in the street: Gangs and American urban society.* Berkeley: University of California Press.

Sandefur, G. D., & Tienda, M. (Eds.) (1988). *Divided opportunities: Minorities, poverty, and social policy.* New York: Plenum.

Sanderson, W., & Scherbov, S. (1997). Doubling of world population unlikely. *Nature, 387,* 803–805.

Sapir, D. (1997, October). Women in the front line. *The Unesco Courier,* pp. 27–29.

Sapolsky, R. (1998, April). How the other half heals: Links between health and socioeconomic status. *Discover,* pp. 46–72.

Sassen, S. (1991). *Global city.* Princeton, NJ: Princeton University Press.

Sassen, S. (1994). *Cities in a world economy.* Thousand Oaks, CA: Pine Forge.

Sassen, S. (1998). *Globalization and its discontents: Essays on the new mobility of people and money.* New York: Free Press.

Satir, V. (1972). *Peoplemaking.* Palo Alto, CA: Science and Behavior Books.

Savage-Rumbaugh, E. S., & Shanker, S. G. (1998). *Ape language and the human mind.* New York: Oxford University Press.

Scanzoni, J. (1995). *Contemporary families and relationships: Reinventing responsibility.* New York: McGraw-Hill.

Scarf, M. (1995). *Intimate worlds: Life inside the family.* New York: Random House.

Schaller, G. B. (1964). *The year of the gorilla.* Chicago: University of Chicago Press.

Schilling, A. G. (1991, March 18). This could be the eve of a false recovery. *Forbes,* pp. 74–75.

Schmitt, E. (1993, February 6). The top soldier is torn between 2 loyalties. *New York Times,* p. A1.

Schooler, C., & Miller, J. (1985). Work for the household: Its nature and consequences for husbands and wives. *American Journal of Sociology, 90,* 97–124.

Schor, J. B. (1992). *The overworked American: The unexpected decline of leisure.* New York: Basic Books.

Schorr, L. B. (1988). *Within our reach: Breaking the cycle of disadvantage.* Garden City, NY: Doubleday.

Schubeck, T. L. (1995). Ethics and liberation theology. *Theological Studies, 56,* 107–123.

Schuman, H., & Scott, J. (1989). Generations and collective memories. *American Sociological Review, 54,* 359–381.

Schumpeter, J. A. (1950). *Capitalism, socialism, and democracy* (3rd ed.). New York: Harper.

Schur, E. M. (1971). *Labeling deviant behavior: Its sociological implications.* New York: Harper.

Schur, E. M. (1984). *Labeling women deviant: Gender, stigma, and social control.* New York: Random House.

Schwartz, B. (1976). Images of suburbia: Some revisionist commentary and conclusions. In B. Schwartz (Ed.), *The changing face of the suburbs.* Chicago: University of Chicago Press.

Scott, T. G. (1998). *The United States of suburbia.* Amherst, NY: Prometheus.

Scull, A. T. (1988). Deviance and social control. In N. J. Smelser (Ed.), *The handbook of sociology.* Newbury Park, CA: Sage.

Seligson, M. A., & Passe-Smith, J. T. (1993). *Development and underdevelopment: The political economy of inequality.* Boulder, CO: Renner.

Selizer, V. (1985). *Pricing the priceless child: The changing social value of children.* New York: Free Press.

Selznick, P. (1952). *The organizational weapon: A study of Bolshevik strategy and tactics.* New York: McGraw-Hill.

Selznick, P. (1966). *TVA and the grass roots: A study in the sociology of formal organization.* New York: Harper.

Sen, A. (1992). *Inequality reexamined.* New York: Russell Sage.

Sen, A. (1993, May). The economics of life and death. *Scientific American,* pp. 40–47.

Sen, A. (1997). Maximization and the act of choice. *Econometrica, 65,* 745–779.

Sennett, R., & Cobb, J. (1972). *The hidden injuries of class.* New York: Random House.

Shaiken, H. (1985). *Work transformed: Automation and labor in the computer age.* Fort Worth: Holt, Rinehart and Winston.

Shaw, C. R. (1929). *Delinquency areas: A study of the geographic distribution of school truants, juvenile delinquents, and adult offenders in Chicago.* Chicago: University of Chicago Press.

Shaw, J. A. (1979). The child in the military community. In J. D. Call, J. D. Noshpitz, R. L. Cohen, & I. N. Berlin (Eds.), *Basic handbook of child psychiatry.* New York: Basic Books.

Shibutani, T., & Kwan, K. M. (1965). *Ethnic stratification.* New York: Macmillan.

Shils, E. (1970). Tradition, ecology and institution in the history of sociology. *Daedalus, 99,* 760–825.

Shils, E. (1985). Sociology. In A. Kuper & J. Kuper (Eds.), *The social science encyclopedia.* London: Routledge and Kegan Paul.

Shorris, E. (1992). *Latinos: A biography of the people.* New York: Norton.

Shotland, R. L. (1985). When bystanders just stand by. *Psychology Today, 19,* 50–55.

Siegel, A. (1992, March). Felice Schwartz tries to set the record straight. *Working Woman,* p. 17.

Signorelli, N., Gerbner, G., & Morgan, M. (1995). Violence on television: The cultural indicators project. *Journal of Broadcasting & Electronic Media, 39,* 278–283.

Silberman, C. (1980). *Criminal violence, criminal justice.* New York: Random House.

Silk, L. (1989, May 12). Rich and poor: The gap widens. *New York Times,* p. D2.

Simmel, G. (1904). The sociology of conflict. *American Journal of Sociology, 9,* 490ff.

Simmons, J. L. (1985). The nature of deviant subcultures. In E. Rubington & M. S. Weinberg (Eds.), *Deviance: The interactionist perspective.* New York: Macmillan.

Simpson, C. E., & Yinger, J. M. (1953). *Racial and cultural minorities: An analysis of prejudice and discrimination.* New York: Harper.

Simpson, J. H. (1983). Moral issues and status politics. In R. C. Liebman & R. Wuthnow (Eds.), *The new Christian right: Mobilization and legitimation.* Hawthorne, NY: Aldine.

Sims, C. (1995, December 12). In Peru, a fight for fresh air. *New York Times,* pp. D1, D4.

Siwolop, S. (1995, August 27). Have I got a business opportunity for you. *New York Times,* pp. D1, D8.

Sizer, T. (1995, January 8). What's wrong with standard tests. *Education Life, New York Times,* p. 58.

Sjoberg, G. (1968). The preindustrial city. In S. Fava (Ed.), *Urbanism in world perspective: A reader.* New York: Crowell.

Skaggs, J. M. (1998). Justifying gender-based affirmative action under *United States v. Virginia's* "exceedingly persuasive justification" standard. *California Law Review, 86,* 1169–1210.

Skinner, B. F. (1976). *Walden two.* New York: Macmillan.

Skocpol, T. (1979). *States and social revolutions.* New York: Cambridge University Press.

Smelser, N. J. (1962). *Theory of collective behavior.* New York: Free Press.

Smelser, N. J. (1966). The modernization of social relations. In M. Weiner (Ed.), *Modernization.* New York: Basic Books.

Smigel, E. O. (1964). *The Wall Street lawyer: Professional organizational man.* New York: Free Press.

Smiley, J. (1992). *A thousand acres.* New York: Ballantine.

Smith, A. (1910/1776). *The wealth of nations.* London: University Paperbacks.

Smith, C. D. (1994). *The absentee American: Repatriates' perspectives on America and its place in the contemporary world.* Putnam Valley, NY: Aletheia.

Smith, C. D. (Ed.) (1996). *Strangers at home: Essays on the effects of living overseas and coming "home" to a strange land.* Putnam Valley, NY: Aletheia.

Smith, D. (1989). *Promises* (Letter from Madurai). Oberlin, OH: Oberlin Shansi Memorial Association.

Smith, R. A. (1968). Los Angeles, prototype of supercity. In S. F. Fava (Ed.), *Urbanism in world perspective: A reader.* New York: Crowell.

Snipp, C. M. (1991). *American Indians: The first of this land.* New York: Russell Sage.

Snow, D. A., & Benford, R. D. (1992). Master frames and cycles of protest. In A. D. Morris & C. M. Mueller (Eds.), *Frontiers in social movement theory.* New Haven, CT: Yale University Press.

Sorenson, A. B., Weinert, F. E., & Sherrod, L. R. (Eds.) (1986). *Human development and the life course: Multidisciplinary perspectives.* Hillsdale, NJ: Erlbaum.

Sorokin, P. (1937). *Social and cultural dynamics: Vol. 3. Fluctuation of social relationships, war, and revolution.* New York: American Book.

Sowell, T. (1972). *Black education: Myths and tragedies.* New York: McKay.

Spencer, H. (1874). *The study of sociology.* New York: Appleton.

Spengler, O. (1965/1918). *The decline of the West.* New York: Modern Library.

Spitz, R. A. (1945). Hospitalism: An inquiry into the genesis of psychiatric conditions in early childhood. In A. Freud (Ed.), *The psychoanalytic study of the child.* New York: International Universities Press.

Stacey, J. (1990). *Brave new families: Stories of domestic upheaval in late twentieth century America.* New York: Basic Books.

Stack, C. (1974). *All our kin.* New York: Harper.

Staggenborg, S. (1998). *Gender, family, and social movements.* Thousand Oaks, CA: Pine Forge.

Starr, P. (1982). *The social transformation of American medicine.* New York: Basic Books.

Starr, P. (1995, Winter). What happened to healthcare reform? *American Prospect,* pp. 20–31.

Statistical abstract of the United States. Annual. Washington, DC: U.S. Bureau of the Census.

Steen, G. R. (1996). *DNA and destiny: Nature and nurture in human behavior.* New York: Plenum.

Stein, J. (1994, December). Space and place. *Art in America,* pp. 66–71.

Steltzer, U. (1982). *Inuit: The north in transition.* Chicago: University of Chicago Press.

Sternberg, R. J. (1999). *Love is a story: A new theory of relationships.* New York: Oxford University Press.

Stevens, W. K. (1988, December 20). Life in the Stone Age: New findings point to complex societies. *New York Times,* pp. C1, C15.

Stevenson, H. W. (1992, December). Learning from Asian schools. *Scientific American,* pp. 70–76.

Stevenson, H. W. (1998, March). A study of three cultures: Germany, Japan, and the United States. *Phi Delta Kappan,* pp. 524–530.

Stevenson, R. W. (1992, July). U.S. judge orders $7.5 million award to whistleblower. *New York Times,* pp. 1, 12.

Stevenson, R. W. (1999 January 6). Texaco is said to set payment over sex bias. *New York Times,* p. C1.

Steward, J. H. (1955). *The theory of culture change: The methodology of multilinear evolution.* Urbana: University of Illinois Press.

Stewart, G. R. (1968). *Names in the land.* Boston: Houghton Mifflin.

Stimson, C. (1980). Women and the American city. *Signs, 5* (Suppl.).

Stockard, J., & Johnson, M. M. (1992). *Sex and gender in society.* Englewood Cliffs, NJ: Prentice Hall.

Stolberg, C. G. (1997, November 23). Gay culture weighs sense and sexuality. *New York Times,* sec. 4, pp. 1, 6.

Stouffer, S. A., Suchman, E. A., DeVinney, L. C., Star, S. A., & Williams, R. A., Jr. (1949). *Studies in social psychology in World War II: Vol. 1. The American soldier: Adjustment during army life.* Princeton, NJ: Princeton University Press.

Strange, M. (1988). *Family farming: A new economic vision.* Lincoln: University of Nebraska Press.

Sudman, S., & Bradburn, N. (1982). *Asking questions: A practical guide to questionnaire design.* San Francisco: Jossey-Bass.

Sudnow, D. (1967). *Passing on: The social organization of dying.* Englewood Cliffs, NJ: Prentice Hall.

Sullivan, K. (1998, November–December). Defining democracy down. *American Prospect,* pp. 91–96.

Sumner, W. G. (1940/1907). *Folkways.* Boston: Ginn.

Sumner, W. G. (1963/1911). *Social Darwinism: Selected essays.* Englewood Cliffs, NJ: Prentice Hall.

Suro, R. (1992, May 26). For women, varied reasons for single motherhood. *New York Times,* p. A12.

Sutherland, E. H. (1940). White collar criminality. *American Sociological Review, 5,* 1–12.

Suttles, G. (1967). *The social order of the slum.* Chicago: University of Chicago Press.

Suttles, G. (1972). *The social construction of communities.* Chicago: University of Chicago Press.

Sutton-Smith, B., & Rosenberg, B. G. (1961). Sixty years of historical change in the game preferences of American children. *Journal of American folklore, 74,* 17–46.

Swedberg, R. (1994). Explanations in economic sociology. New York: Russell Sage.

Swift, D. C. (1998). *Religion and the American experience:* A social and cultural history. Armonk, NY: Sharpe.

Szelenyi, I. (1983). *Urban inequalities under state socialism.* New York: Oxford University Press.

Tabb, W. (1986). *Churches in struggle: Liberation theologies and social change.* New York: Monthly Review Press.

Taeuber, K. E., & Taeuber, A. F. (1965). *Negroes in cities.* Hawthorne, NY: Aldine.

Tagliabue, J. (1998, August 19). Communists, remodeled, keep trying in Germany. *New York Times,* p. A5.

Talbot, M. (1998, May 24). Attachment theory: The ultimate experiment. *New York Times Magazine,* pp. 24–30.

Tannen, D. (1990). *You just don't understand: Women and men in conversation.* New York: Morrow.

Tarrow, S. G. (1994). *Power in movement: Social movements, collective action, and politics.* Cambridge, England: Cambridge University Press.

Tawney, R. A. (1966/1932). *Land and labor in China.* Boston: Beacon.

Taylor, J. M., Gilligan, C., & Sullivan, A. M. (1995). *Between voice and silence: Women and girls, race and relationship.* Cambridge, MA: Harvard University Press.

Teitelbaum, M., & Weiner, M. (1995). *Threatened peoples, threatened borders: World migration and U.S. policy.* New York: Norton.

Temkin, L. S. (1993). *Inequality.* New York: Oxford University Press.

Tewksbury, R., & Taylor, J. M. (1996). The consequences of eliminating Pell Grant eligibility for students in post-secondary correctional education programs. *Federal Probation, 60,* 60–63.

Thio, A. (1998). *Deviant behavior* (5th ed.). New York: Longman.

Thomas, L. (1979). *The medusa and the snail.* New York: Bantam.

Thomas, L. (1989, July 15). Beyond the moon's horizon—Our home. *New York Times,* p. 25.

Thomas, W. I. (1971/1921). *Old world traits transplanted.* Montclair, NJ: Patterson.

Thomas, W. L. (1956). *Man's role in changing the face of the earth.* Chicago: University of Chicago Press.

Thorne, B. (1993). *Gender play: Boys and girls in school.* New Brunswick, NJ: Rutgers University Press.

Thornton, R. (1987). *American Indian holocaust and survival: A population history since 1492.* Norman: University of Oklahoma Press.

Tienda, M., & Singer, A. (1995). Wage mobility of undocumented workers in the United States. *International Migration Review, 29,* 112–138.

Tienda, M., & Wilson, F. D. (1992). Migration and the earnings of Hispanic men. *American Sociological Review, 57,* 661–678.

Tilly, C. (1993). *European revolutions, 1492–1992.* Oxford, England: Blackwell.

Tobach, E., & Rosoff, B. (Eds.) (1994). *Challenging racism and sexism: Alternatives to genetic explanations.* New York: Feminist Press at the City University of New York.

Toffler, A. (1970). *Future shock.* New York: Bantam.

Tönnies, F. (1957/1887). *Community and society* (C. P. Loomis, Trans. & Ed.). East Lansing: Michigan State University Press.

Torres, C. A., & Mitchell, T. R. (Eds.) (1998). *Sociology of education: Emerging perspectives.* Albany: State University of New York Press.

Touraine, A. (1998). Social transformations of the twentieth century. *International Social Science Journal, 50,* 165–172.

Toynbee, A. J. (1972). *A study of history.* New York: Oxford University Press.

Transaction Publishers (1997). *Foundations of political sociology.* New Brunswick, NJ: Transaction.

Traub, J. (1984, September 9). A village in India: Reluctant progress. *New York Times Magazine,* pp. 106ff.

Traven, B. (1966). Assembly line. In *The night visitor, and other stories.* New York: Hill and Wang.

Tremmel, P. (1999, January 20). Comer school reform project improves both academics and behavior. *Northwestern News,* pp. 1–3.

Trent, J. W., Jr. (1995, July–August). Suffering fools. *The Sciences,* pp. 18–22.

Trillin, C. (1998, February 2). New Orleans unmasked. *New Yorker,* pp. 38–43.

Truzzi, M. (1971). *Sociology: The classic statements.* New York: McGraw-Hill.

Tuch, S. A., & Martin, J. K. (1997). *Racial attitudes in the 1990s: Continuity and change.* Westport, CT: Praeger.

Tuch, S. A., Sigelman, L., & Martin, J. K. (1997). Fifty years after Myrdal: Blacks' racial policy attitudes in the 1990s. In Tuch, S. A., & Martin, J. K., *Racial attitudes in the 1990s: Continuity and change.* Westport, CT: Praeger.

Tucker, C. (1998, November 29). Working poor deserve voice. *Denver Post,* p. G-02.

Tumin, M. M. (1967). *Social stratification: The forms and functions of inequality.* Englewood Cliffs, NJ: Prentice Hall.

Turk, A. T. (1978). Law as a weapon in social conflict. In C. E. Reasons & R. M. Rich (Eds.), *Sociology of law: A conflict perspective.* Toronto: Butterworths.

Turner, F. J. (1920/1893). *The frontier in American history.* Fort Worth: Holt, Rinehart and Winston.

Turner, R. H. (1974). The theme of contemporary social movements. In R. E. L. Faris (Ed.), *Handbook of modern sociology.* Chicago: Rand McNally.

Uchitelle, L. (1995, July 16). For employee benefits, it pays to wear the union label. *New York Times,* sec. F, p. 10.

Uhlenberg, P. (1980). Death and the family. *Journal of Family History, 5,* 313–320.

UNDP (United Nations Development Programme) (1997). *Human Development Report.* New York: Oxford University Press.

United Nations (1995, March). *Adoption of the declaration and programme of action of the World Summit for Social Development.* New York: United Nations.

United Nations (1995). *Gender, population & development: The role of the United Nations Population Fund.* New York: United Nations, UNFPA.

United Nations (1995, March). *Summary of the programme of action of the International Conference on Population and Development.* New York: United Nations, Department of Public Information.

U.S. Bureau of the Census (1990). *Census of population: 1990.* Washington, DC: U.S. Bureau of the Census.

Van Gennep, A. (1960/1908). *The rites of passage.* Chicago: University of Chicago Press.

Vanneman, R., & Cannon, L. W. (1987). *The American perception of class.* Philadelphia: Temple University Press.

Vass, W. K. (1979). *The Bantu speaking heritage of the United States.* Los Angeles: Center for Afro-American Studies, University of California.

Verbrugge, L. M., Gruber-Baldini, A. L., & Fozard, J. L. (1996). Age differences and age changes in activities: Baltimore longitudinal study of aging. *Gerontologist, 36,* 342–350.

Vilas, C. M. (1993). The hour of civil society. *NACLA Report on the Americas, 27,* 38–43.

Vining, D. R., Jr. (1985, April). The growth of core regions in the Third World. *Scientific American,* pp. 42–49.

Vogel, E. (1979). *Japan as number one: Lessons for America.* Cambridge, MA: Harvard University Press.

Wald, M. L. (1997, September 21). Number of cars is growing faster than human population. *New York Times,* p. 35.

Wallerstein, I. (1974). *The modern world system: Capitalist agriculture and the origins of the European world-economy in the sixteenth century.* Orlando: Academic Press.

Wallerstein, I. (1999). The heritage of sociology, the promise of social science. Presidential address, XIVth World Congress of Sociology, Montreal, July 26, 1998. *Current Sociology, 47,* 1–43.

Wallerstein, J., & Blakeslee, S. (1989). *Second chances: Men, women and children a decade after divorce.* New York: Ticknor & Fields.

Walsh, E. J., & Warland, R. H. (1983). Social movement involvement in the wake of a nuclear accident: Activists and free riders in the TMI area. *American Sociological Review, 48,* 764–780.

Walton, J. (1970). A systematic survey of community power research. In M. Aiken & P. Mott (Eds.), *The structure of community power.* New York: Random House.

Walton, J. (1992). *Western times and water wars: State, culture, and rebellion in California.* Berkeley: University of California Press.

Waltzer, M. (1980). Pluralism. In *Harvard encyclopedia of American ethnic groups.* Cambridge, MA: Belknap.

Warhus, M. (1997). *Another America: Native American maps and the history of our land.* New York: St. Martin's Press.

Warner, R. S. (1998). Work in progress toward a new paradigm for the sociological study of religion in the United States. *American Journal of Sociology, 98,* 1044–1093.

Warner, W. L., Meeker, M., & Calls, K. (1949). *Social class in America: A manual of procedure for the measurement of social status.* Chicago: Science Research Associates.

Watson, J. B. (1930). *Behaviorism.* New York: Norton.

Weaver, M. A. (1994, September 12). A fugitive from justice. *New Yorker,* pp. 46–60.

Webb, E., Campbell, D. T., Schwarz, R. D., & Sechrest, L. (1966). *Unobtrusive measures: Nonreactive research in the social sciences.* Chicago: Rand McNally.

Weber, M. (1922). *Gesammelte aufsatze zur Religionssoziologie.* Tubingen, Germany: Mohr.

Weber, M. (1947). *The theory of social and economic organization* (A. M. Henderson & T. Parsons, Trans.). New York: Free Press.

Weber, M. (1949). *The methodology of the social sciences* (E. A. Shils & H. A. Finch, Eds. & Trans.). New York: Free Press.

Weber, M. (1958/1913). World religion. In H. Gerth & C. W. Mills (Trans. & Eds.), *From Max Weber: Essays in sociology.* New York: Oxford University Press.

Weber, M. (1958/1922). Economy and society. In H. Gerth & C. W. Mills (Trans. & Eds.), *From Max Weber: Essays in sociology.* New York: Oxford University Press.

Weber, M. (1962/1921). *The city.* New York: Collier.

Weber, M. (1963/1922). *The sociology of religion* (E. Fischoff, Trans.). Boston: Beacon.

Weber, M. (1968). The concept of citizenship. In S. N. Eisenstadt (Ed.), *Max Weber on charisma and institution building.* Chicago: University of Chicago Press.

Weber, M. (1974/1904). *The Protestant ethic and the spirit of capitalism* (T. Parsons, Trans.). New York: Scribner.

Weinberg, H. (1994). Marital reconciliation in the United States: Which couples are successful? *Journal of Marriage and the Family, 56,* 80–88.

Weisman, A. (1989, July 30). L.A. fights for breath. *New York Times Magazine,* pp. 14ff.

Wekerle, G. (Ed.) (1980). *New space for women.* Boulder, CO: Westview.

Wellins, S. (1990). *Children's use of television in England and the United States.* Doctoral dissertation, City University of New York.

Wells, M. J. (1996). *Strawberry fields: Politics, class, and work in California agriculture.* Ithaca, NY: Cornell University Press.

West, C. (1992, February 8). Learning to talk of race. *New York Times Magazine,* pp. 24–25.

West, W. (1998, March 9). The U.S. armed forces: Ready to fight—nicely? *Insight on the News,* p. 48.

Westin, A. (1967). *Privacy and freedom.* New York: Atheneum.

Wheeler, S. R. (1998, April 10). Future showing promise: Donors revitalize religious group. *Denver Post,* pp. A1, A5.

Whitchurch, G. G., & Constantine, L. L. (1993). Systems theory. In P. G. Boss, W. J. Doherty, R. LaRossa, W. R. Schumm, & S. K. Steinmetz (Eds.), *Sourcebook of family theories and methods: A contextual approach.* New York: Plenum.

White, H. (1970). *Chains of opportunity.* Cambridge, MA: Harvard University Press.

White, M. J. (1987). *American neighborhoods and residential differentiation.* New York: Russell Sage.

Whiting, B. B., & Edwards, C. P. (1988). *Children of different worlds.* Cambridge, MA: Harvard University Press.

Whorf, B. L. (1961). The relation of habitual thought and behavior to language. In J. B. Carroll (Ed.), *Language, thought, and reality: Selected writings of Benjamin Lee Whorf.* Cambridge, MA: MIT Press.

Whyte, W. F. (1943). *Street corner society.* Chicago: University of Chicago Press.

Whyte, W. F. (1949). The social structure of the restaurant. *American Journal of Sociology, 54,* 302–310.

Whyte, W. F. (1984). *Learning from the field.* Newbury Park, CA: Sage.

Wilensky, H. (1964). Mass society and mass culture: Interdependence or independence? *American Sociological Review, 29,* 173–193.

Wilford, J. N. (1997, November 18). New clues show where people made the great leap to agriculture. *New York Times,* pp. F1, F6.

Wilkinson, A. (1989, July 24). Sugarcane. *New Yorker,* pp. 56ff.

Williams, T. (1989). *The cocaine kids.* Reading, MA: Addison-Wesley.

Williams, T. (1992). *Crack house.* Reading, MA: Addison-Wesley.

Williams, T., & Kornblum, W. (1994). *The uptown kids: Struggle and hope in the projects.* New York: Putnam.

Willis, P. (1983). Cultural production and theories of reproduction. In L. Barton & S. Walker (Eds.), *Race, class and education.* London: Croom-Helm.

Willmott, P., & Young, M. (1971). *Family and class in a London suburb.* London: New American Library.

Wills, G. (1992). *Lincoln at Gettysburg: The words that remade America.* New York: Simon & Schuster.

Wilson, B. L. (1993). *Mandating academic excellence: High school responses to state curriculum reform.* New York: Teachers College Press.

Wilson, C. R. (1987). The death of Bear Bryant: Myth and ritual in the modern South. *South Atlantic Quarterly, 86,* 282–295.

Wilson, E. O. (1975). *Sociobiology.* Cambridge, MA: Belknap.

Wilson, E. O. (1979). *On human nature.* New York: Bantam.

Wilson, E. O. (1998). *Consilience: The unity of knowledge.* New York: Knopf.

Wilson, J. Q. (1977). *Thinking about crime.* New York: Vintage.

Wilson, W. J. (1984). The urban underclass. In L. W. Dunbar (Ed.), *The minority report.* New York: Pantheon.

Wilson, W. J. (1987). *The truly disadvantaged: The inner city, the underclass, and public policy.* Chicago: University of Chicago Press.

Wilson, W. J. (1996). When work disappears: The world of the new urban poor. Chicago: University of Chicago Press.

Wilson, W. J. (1998). Inner-city dislocations. *Society, 35,* 270–278.

Winch, R. F. (1958). *Mate selection.* New York: Harper.

Winner, L. (1977). *Autonomous technology: Technics-out-of-control as a theme in political thought.* Cambridge, MA: MIT Press.

Winner, L. (1996, February–March). Know-nothing technology policy. *Technology Review,* p. 55.

Winner, L. (1997, August–September). The neverhood of Internet commerce. *Technology Review,* p. 31.

Wirth, L. (1945). The problem of minority groups. In R. Linton (Ed.), *The science of man in the world crisis.* New York: Columbia University Press.

Wirth, L. (1968/1938). Urbanism as a way of life. In S. F. Fava (Ed.), *Urbanism in world perspective: A reader.* New York: Crowell.

Witt, S. D. (1997). Parental influence on children's socialization to gender roles. *Adolescence, 32,* 253–360.

Wittfogel, K. (1957). *Oriental despotism: A comparative study of total power.* New Haven, CT: Yale University Press.

Wixen, B. N. (1979). Children of the rich. In J. D. Call, J. D. Noshpitz, R. L. Cohen, & I. N. Berlin (Eds.), *Basic handbook of child psychiatry.* New York: Basic Books.

Wolf, A. (1998). *One nation after all.* New York: Viking.

Wolf, E. R. (1984a, November 4). The perspective of the world. *New York Times Book Review,* pp. 13–14.

Wolf, E. R. (1984b, November 4). Unifying the vision. *New York Times Book Review,* p. 11.

Wolf, N. (1991). *The beauty myth.* New York: Morrow.

Wolfe, A. (1998, October 25). The black-white test score gap. *New York Times Book Review,* p. 15.

Wolfe, T. (1980). *The right stuff.* New York: Bantam.

Wolff, E. N. (1995). *Top heavy: A study of the increasing inequality of wealth in America.* New York: Twentieth Century Fund.

Wolfgang, M. E., & Riedel, M. (1973). Race, judicial discretion, and the death penalty. *Annals of the American Academy of Political and Social Science, 407,* 119–133.

Woo, W. T. (1994). The art of reforming centrally planned economies: Comparing China, Poland, and Russia. *Journal of Comparative Economics, 18,* 276–308.

World Bank (1997). *World development indicators.* Washington, DC: World Bank.

Wren, D. J. (1997). Adolescent females' "voice" changes can signal difficulties for teachers and administrators. *Adolescence, 32,* 463–470.

Wright, E. O. (1979). Class structure and economic determination. Orlando: Academic Press.

Wright, E. O. (1989). *The debate on classes.* New York: Verso.

Wright, E. O. (1997). *Class counts: Comparative studies in class analysis.* New York: Cambridge University Press.

Wright, E. O., Costello, C., Hachen, D., & Sprague, J. (1982). The American class structure. *American Sociological Review, 47,* 702–726.

Wrigley, E. A. (1969). *Population and history.* New York: McGraw-Hill.

Wrigley, J. (1995). *Other people's children.* New York: Basic Books.

Wrong, D. H. (1961). The oversocialized conception of man in modern sociology. *American Sociological Review, 24,* 772–782.

Wu, H., & Wakeman, C. (1995). *Bitter winds: A memoir of my years in China's gulag.* New York: Wiley.

Wuthnow, R. (1988). Sociology of religion. In N. J. Smelser (Ed.), *The handbook of sociology.* Newbury Park, CA: Sage.

Wuthnow, R. (1994). *Sharing the journey: Support groups and America's new quest for community.* New York: Free Press.

Wuthnow, R. (1994). *God and Mammon in America.* New York: Free Press.

Yinger, J. (1995). *Closed doors, opportunities lost: The continuing costs of housing segregation.* New York: Russell Sage.

Youniss, J., & Yates, M. (1997). *Community service and social responsibility in youth.* Chicago: University of Chicago Press.

Zafirovski, M., & Levine, B. B. (1997). Economic sociology reformulated: The interface between economics and sociology. *American Journal of Economics and Sociology, 56,* 109–138.

Zangwell, I. (1909). *The melting pot.* New York: Macmillan.

Zarembo, A. (1999, March 15). A hurricane's orphans. *Newsweek,* p. 47.

Zeisel, H. (1982). *The limits of law enforcement.* Chicago: University of Chicago Press.

Zill, N. (1995, Spring). Back to the future: Improving child indicators by remembering their origins. *Focus* (Institute for Research on Poverty), pp. 17–24.

Zolberg, A. (1981). International migrations in political perspective. In M. M. Kritz, C. B. Keely, & S. M. Tomasi (Eds.), *Global trends in migration: Theory and research on international population movements.* New York: Center for Migration Studies.

Zonana, H. (1997). The civil commitment of sex offenders. *Science, 278,* 1248–1249.

Zurcher, L. A., & Snow, D. A. (1981). Collective behavior and social movements. In M. Rosenberg & R. Turner (Eds.), *Social psychology: Sociological perspectives.* New York: Basic Books.

Zweigenhaft, R., & Domhoff, G. W. (1998). Diversity in the power elite. New Haven, CT: Yale University Press.

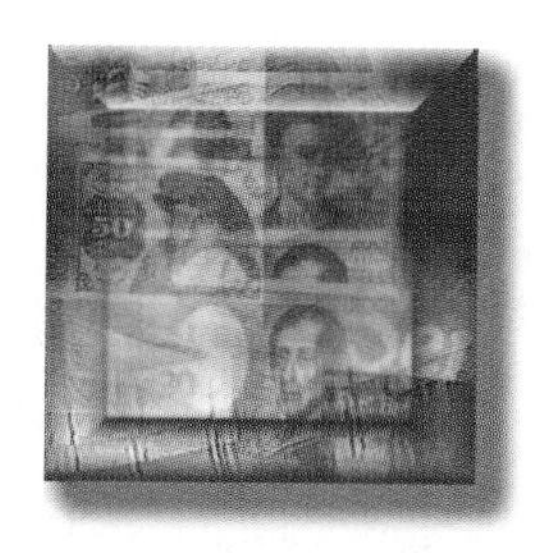

CREDITS AND ACKNOWLEDGMENTS

Chapter 1 p. xxxii: (top) © Crandall/The Image Works, (middle) © Mary Kate Denny/PhotoEdit, (bottom) ©Jeffrey Jay Foxx; p. 1: © Eugene Richards; pp. 2–3: © Jeff Greenberg/The Image Works; p. 6: © M. Dwyer/Stock Boston; p. 9: (left) © Gina Doggett, Paris, (top right) courtesy of Manjula Giri, (bottom right) Harriet Martineau by Richard Evans, by courtesy of the National Portrait Gallery, London; p. 10: Culver Pictures Inc.; p. 11: (left) Corbis/Bettmann, (right) Culver Pictures Inc.; p. 14: UPI/Corbis-Bettmann; p. 17: Reuters/Corbis-Bettmann; p. 20: © Marleen Daniels/Gamma Liaison Network; p. 22: (top) © Crandall/The Image Works, (bottom) © Zbigniew Bzzdak/The Image Works; p. 23: (top) © Alain Evrard/Photo Researchers, Inc., (middle) © Nour/Gamma Liaison, (bottom) AP/Wide World Photos.

Chapter 2 p. 29: © Mary Kate Denny/Photo Edit; p. 32: (left) UPI/Corbis-Bettmann, (right) AP/Wide World Photos; p. 33: © Blake Discher/Sygma; p. 37: From Chroniques Siciliennes, #27; p. 38: © Douglas Mazonowicz/Monkmeyer; p. 39: Egyptian Museum, Cairo; pp. 42, 52–53:AP/Wide World Photos.

Chapter 3 p. 59: © Dan Groshong/Sygma; p. 60: © Chenault Photography Studio, Decatur, AL; p. 63: (left) © Photofest, (right) © Matthew Neal McVay/Stock Boston; p. 65: © Gaye Hilsenrath/The Picture Cube; p. 66: Coram Foundation, London/Bridgeman Art Library, London; p. 73: © Richard B. Levine; p. 79: personal property of Margaret Washington; pp. 84–85: all photographs © Jeffrey Jay Foxx.

Chapter 4 p. 91: © Eugene Richards; p. 96: © Photopia; p. 100: © Gerhard Hinterleitner/Gamma Liaison Network; p. 103: (top) copyright Nicolas Sapieha/Art Resource, NY, (bottom) © John Neubauer; p. 108: © Rick Maiman/Sygma; p. 109: © Paul De Maria/Sygma; p. 113: AP/Wide World Photo; pp. 116–117: © Eugene Richards.

Chapter 5 p. 122: (top) © Ken Fisher/Tony Stone Images, (middle) photograph © Margaret Morton, author/photographer *The Tunnel* (Yale University Press, 1995), (bottom) © Matt Herron 1987; p. 123: © Robert Brenner/PhotoEdit; p. 124: © Ken Fisher/Tony Stone Images; p. 125: AP/Wide World Photos; p. 130: (left) AP/Wide World Photo, (right) © David Pokress/NEWSDAY; p. 131: S. Curtiss, *Genie: A Psycholinguistic Study of a Modern Day "Wild Child"* (Academic Press, 1977), used by permission; p. 138: The New York Public Library, Astor, Lenox & Tilden Foundations; p. 143: © Cathlyn Melloan/Tony Stone Images; p. 148: © David Young-Wolff/PhotoEdit; pp. 150–151: © Chris Heflin/Shooting Back, © Charlene Williams/Shooting Back, © Shawn Nixon/Shooting Back, © Jim Hubbard/Shooting Back, © Calvin Stewart/Shooting Back.

Chapter 6 pp. 156–157: © Frances M. Roberts; p. 159: © Gregg Hadel/Tony Stone Images; p. 165: (top) Corbis-Bettmann, (bottom) AP/Wide World Photo; p. 182: © Don Pollard; p. 184: (top left) © Michael Newman/PhotoEdit, (all others) © Ken Heyman; p. 185: all photographs © Ken Heyman.

Chapter 7 p. 191: photograph © Margaret Morton, author/photographer *The Tunnel* (Yale University Press, 1995); p. 193: Essex Institute, Salem, Massachusetts; p. 195: © Brooksfilms Limited; p. 198: New York City Department of Health; p. 199: © Rafiqur Rahman/Reuters/Corbis-Bettmann; p. 200: (left) Stock Montage, Inc., (right) Corbis/Bettmann; p. 201: (left) © Patrick Ward/Stock Boston, (right) © A. Ramey/PhotoEdit; p. 213: © Alon Reininger/Contact/Woodfin Camp & Associates; p. 215: (top) The Granger Collection, New York, (bottom) AP/Wide World Photo; pp. 220–221: photographs © Margaret Morton, author/photographer *The Tunnel* (Yale University Press, 1995).

Chapter 8 p. 227: Courtesy of William Kornblum; p. 229: (top) Chicago Historical Society, (bottom) © Bob Tur/Los Angeles New Service/Robert E. Clark/Los Angeles News Service; p. 238: (top) Popperfoto/Archive Photos, (bottom) Culver Pictures Inc.; p. 241: © David Young-Wolff/PhotoEdit; p. 242: Museo del Prado, Madrid; p. 243: Cliches des Musees Nationaux; p. 246: © Jeff Greenberg/Photo Researchers, Inc.; p. 249: (left) © Philippe Ledru/Sygma, (right) © Kevin Winter/DMI; pp. 250–251: © Matt Herron 1987.

Chapter 9 p. 257: © Robert Brenner/PhotoEdit; p. 263: © Alon Reininger/Woodfin Camp & Associates; p. 267: © Peter Menzel/Stock Boston; p. 268: (left) Corbis/Bettmann, (right) © Kirk Condyles/Impact Visuals; p. 274: © R. Richards/Gamma Liaison; p. 276: (top left) © J. B. Diederich/Contact Press Images, (top right) © Freda Leinwand, (bottom) © Dennis Stock/Magnum; p. 280: © Jean Wentworth/The Picture Cube; p. 284: © Kirk Condyles/Impact Visuals; p. 286: (top)courtesy of the National Park Service, Frederick Law Olmsted National Historic Site, (bottom) © 1992 Bruce Davidson/Magnum Photos, Inc.; p. 287: (top) © 1999 Bruce Davidson/Magnum Photos, Inc., (bottom) © 1995 Bruce Davidson/Magnum Photos, Inc.

Chapter 10 p. 292: © Betty Press/Woodfin Camp & Associates; p. 298: © Michael Grecco/Stock Boston; p. 299:

Corbis-Bettmann; p. 301: (left) Corbis-Bettmann, (right) © Traver/Gamma Liaison; p. 304: Corbis-Bettmann; p. 307: © Brian Willer/NYT Pictures; p. 308: Musee d'Orsay, Paris/Cliche des Musees Nationaux; p. 318: (top) Independent Picture Service, (bottom left) courtesy of United Nations/DPI, (bottom right) photographs by Eliot Elisofon, 1969, Slide no. TIVC2, Eliot Elisofon Photographic Archives, National Museum of African Art; p. 319: (top) photograph by Eliot Elisofon, 1972, Slide no. HEBR 2, Eliot Elisofon Photographic Archives, National Museum of African Art, (bottom) © F. Willett.

Chapter 11 p. 324: (top) © Bernard Pierre Wolff/Magnum Photos, Inc., (middle) © Michael Jacobson-Hardy, (bottom) Chenault Studios/Decatur, AL; p. 325: (top) AP/Wide World Photo, (bottom) © Steve Skjold/PhotoEdit; p. 326: © Miguel Rio Branco/Magnum Photos, Inc.; p. 336: © PhotoEdit; p. 338: © Hulton Deutsch Collection; p. 339: Corbis-Bettmann; p. 342: © Carolyn D. Smith; p. 344: (left) AP/Wide World Photo, (right) © David Kampfner/Gamma Liaison Network; p. 352: (top) © Bernard Pierre Wolff/Magnum Photos, Inc., (bottom) © Sebastio Salgado/Contact Press Images; p. 353: all photographs © Sebastio Salgado/ Contact Press Images.

Chapter 12 p. 358: © Jose More; p. 361: © Jeff Greenberg/PhotoEdit; p. 362: The Art Institute of Chicago; p. 363: Philadelphia Museum of Art: given by Wright S. Ludington; p. 374: (top) L'Illustration ca. 1850/Corbis-Bettmann, (bottom) © A. Ramey/Woodfin Camp & Associates; p. 375: © Barbara Norfleet; p. 378: © Elliot Erwitt/Magnum; p. 387: © Nathan Benn/Woodfin Camp & Associates; pp. 390–391: all photographs © Michael Jacobson-Hardy.

Chapter 13 p. 396: AP/Wide World Photos; p. 397: © Steve Skjold/PhotoEdit; p. 400: AP/Wide World Photos; p. 401: © Maggie Steber; p. 409: reproduced from the Collections of the Library of Congress, LC-B8171-383; p. 412: © Folio Inc.; p. 414: New York Public Library Picture Collection/New York Public Library, Astor, Lenox & Tilden Foundations; p. 419: (left) Chicago Historical Society, (right) © C. Simonpietri/Sygma; p. 422: © Nita Winter; p. 424: © Dan Dry/Dan Dry & Associates; p. 426: © Fujihira/Monkmeyer; p. 430: AP/Wide World Photos; p. 431: (top) AP/Wide World Photos, (bottom) © William Johnson/Stock Boston.

Chapter 14 pp. 436–437: © Chenault Studios, Decatur, AL; p. 441: The Kobal Collection; p. 449: Bibliotheque Royale Albert I, Brussels; p. 450: (left) © P. Lisenkin/Sovfoto/Eastfoto, (right) © Hearst Magazines International; p. 453: (left) © Catherine Ursillo/Photo Researchers, (right) Continental Airlines; p. 454: Stock Montage, Inc.; p. 455: "Untitled, 1982 (we have received orders not to move)" by Barbara Kruger, courtesy Mary Boone Gallery, New York; p. 457: Celia Dugger/NYT Pictures; pp. 452–453: all photographs © Chenault Studios/Decatur, AL.

Chapter 15 p. 469: The Imogen Cunningham Trust; p. 475: Kevin Moloney/NYT Pictures; p. 476: © Farrell Grehan/Photo Researchers; p. 478: © Hanson Carroll/Archive Photos; p. 480: The Huntington Library, Art Collections, and Botanical Gardens, San Marino, California/SuperStock; p. 481: (left) The Saturday Evening Post © The Curtis Publishing Company, (right) © Dave Sartin/Liaison; p. 484: © Terry Ashe/Gamma Liaison; pp. 492–493: all photographs The Imogen Cunningham Trust.

Chapter 16 p. 496: (top) © A. Ramey/Woodfin Camp & Associates, (middle) © Michael Newman/PhotoEdit, (bottom) © Jeanne Strongin, 1990; p. 497: collection of the Museum of American Folk Art, New York, gift of Mr. And Mrs. Edwin C. Braman; pp. 498–499: © Eugene Richards; p. 502: (top) © Paul Conklin/Monkmeyer, (bottom) © Stephanie Maze/Woodfin Camp & Associates; p. 517: The National Gallery, London; p. 526: (top) © Dan Habib/Impact Visuals, (bottom) © Donna Ferrato, 1994; p. 527: (top) © Jeanne Strongin, 1990, (bottom) © 1994 Frank Fournier/Contact Press Images.

Chapter 17 p. 533: Collection of Sandra S. Jaffe/Preservation Hall; p. 535: © Leo de Wys; p. 536: (left) © David Moore, Black Star, (right) © Porterfield-Chickering/Photo Researchers; p. 540: © Edmundo Morales; p. 544: courtesy of Susan Kornblum; p. 545: © A. Keler/Sygma; p. 546: copyright Smithsonian Institution; p. 549: © Bob Daemmrich/ Stock Boston; p. 550: © Bob Daemmrich/The Image Works; p. 551: © Ross-Marino/Sygma; p. 554: © John Berry/Gamma Liaison; p. 556: Corbis-Bettmann; p. 562: (top) Columbus Museum of Art, Ohio: museum purchase, (bottom) Collection of Ellsworth and Ann Taylor courtesy of Hackley Gallery, Winchester, Kentucky; p. 563: Collectin of the Museum of American Folk Art, New York, gift of Mr. And Mrs. Edwin C. Braman.

Chapter 18 p. 569: courtesy Office of Public Affairs/Yale University; p. 571: © George Holton/Photo Researchers; p. 592: (left) Brown Brothers, (right) © Mark Lewis/Liaison International; p. 594: © Saban Entertainment, Inc.; p. 597: © Chuck Fishman/Woodfin Camp & Associates; p. 600: (top) © Michael Marsland/Office of Public Affairs/Yale University, (bottom) courtesy Office of Public Affairs/Yale University; p. 601: all photographs courtesy Office of Public Affairs/Yale University.

Chapter 19 p. 607: © Susan Spano; p. 611: The Granger Collection, New York; pp. 612–613: copyright © 1995 by The New York Times Co., reprinted by permission; p. 614: © Edmundo Morales; p. 617: Library of Congress; pp. 634–635: © Billy E. Barnes/PhotoEdit, Brown Brothers, Canadian Native Cree Web Page courtesy of Susan Dueckman, © Gary A. Conner/PhotoEdit, © Michael Newman/PhotoEdit.

Chapter 20 p. 640: AP/Wide World Photos; p. 651: © Jon Feingersh/The Stock Market; p. 646: © Andrew Holbrooke/ Gamma Liaison; p. 647: (left) Brown Brothers, (right) AP/Wide World Photos; p. 658: Popperfoto/Archive Photos; p. 661: © Jim Anderson/Woodfin Camp & Associates; p. 664: (top) Culver Pictures, Inc., (bottom) New York Public Library Picture Collection/New York Public Library, Astor, Lenox & Tilden Foundations; p. 665: (top) © Marc & Evelyne Bernheim/

Woodfin Camp & Associates, (middle) © Eastfoto, (bottom) AP/Wide World Photos.

Chapter 21 p. 671: © A. Ramey/Woodfin Camp & Associates; p. 674: (left) The Oriental Institute/University of Chicago, (right) Water and Power Resources Service, U.S. Department of the Interior; p. 677: New York Public Library Picture Collection/New York Public Library, Astor, Lenox & Tilden Foundations; p. 680: © Warren Faidley/Weatherstock; p. 681: © H. Roger-Viollet; p. 685: (top) © Eric Sander/Gamma Liaison, (bottom) © Tom Prettyman/PhotoEdit; pp. 692–693: © C. J. Pickerell/FPG International; © Joel Simon/Tony Stone Images; © C. H. Hortsjo, in I. Eibl-Eibesfeldt, *Human Ethology,* reprinted by permission of Aldine de Gruyter and I. Eibl-Eibesfeldt; © concept by Michael Brill, illustration by Safdar Abidi; Metropolitan Museum of Art, New York/Carnarvon Collection, Edward S. Harkness Gift, 1926; courtesy Nuclear Regulatory Commission; © concept by Michael Brill, illustration by Safdar Abidi.

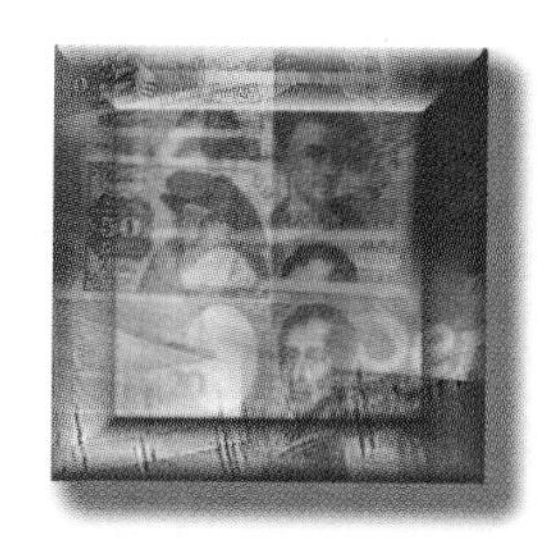

NAME INDEX

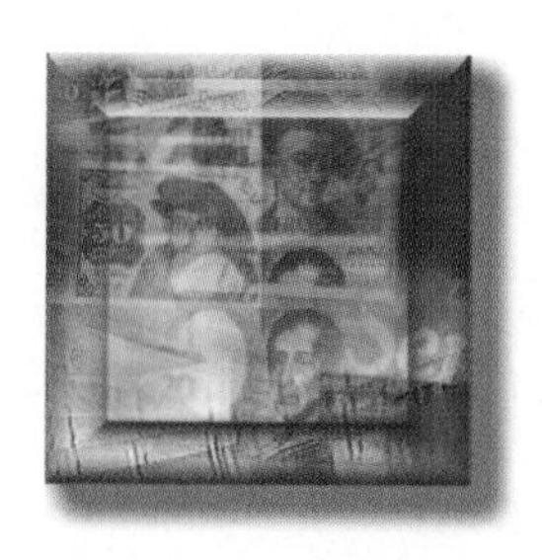

SUBJECT INDEX